exploring physical
geography

About the Cover

This photograph by Michael Collier shows farms atop a hill in West Virginia. Morning mist hanging down in the valleys outlines the network of streams, like Grove Creek and Fish Creek, that collect and drain into the Ohio River near Moundsville. The underlying rocks are sedimentary and in places contain seams of coal that were formed from ancient swamps that covered the area during the Pennsylvanian Period. This area is part of the Appalachian Plateau Province, which was uplifted to its current elevation of about 1,200 feet above sea level by tectonic compression that created the Appalachian Mountains. In addition to coal mining, agriculture has always been important to this region. Hardwood forests have been removed along the ridges to create fields for crops and cattle.

Michael Collier received his B.S. in geology at Northern Arizona University, M.S. in structural geology at Stanford, and M.D. from the University of Arizona. He rowed boats commercially in Grand Canyon in the late 1970s and early 1980s. He now lives in Flagstaff, Arizona, where he practices family medicine. Collier has published books about the geology of Grand Canyon National Park, Death Valley, Denali National Park, and Capitol Reef National Park. He has done books on the Colorado River basin, glaciers of Alaska, and climate change in Alaska. He recently completed a three-book series on American mountains, rivers, and coastlines, designed around his spectacular photographs taken from the air. As a special-projects writer with the USGS, he wrote books about the San Andreas fault, climate change, and downstream effects of dams, with each book featuring his many photographs. Collier has produced an iPad app about seeing landscapes from the air. He received the USGS Shoemaker Communication Award in 1997, the National Park Service Director's Award in 2000, and the American Geological Institute's Public Contribution to Geosciences Award in 2005.

STEPHEN J. REYNOLDS
Arizona State University

ROBERT V. ROHLI
Louisiana State University

JULIA K. JOHNSON
Arizona State University

PETER R. WAYLEN
University of Florida

MARK A. FRANCEK
Central Michigan University

exploring physical geography

CYNTHIA C. SHAW
Lead Illustrator, Art Director

EXPLORING PHYSICAL GEOGRAPHY, SECOND EDITION

Published by McGraw-Hill Education, 2 Penn Plaza, New York, NY 10121. Copyright © 2018 by McGraw-Hill Education. All rights reserved. Printed in the United States of America. Previous edition © 2015. No part of this publication may be reproduced or distributed in any form or by any means, or stored in a database or retrieval system, without the prior written consent of McGraw-Hill Education, including, but not limited to, in any network or other electronic storage or transmission, or broadcast for distance learning.

Some ancillaries, including electronic and print components, may not be available to customers outside the United States.

This book is printed on acid-free paper.

2 3 4 5 6 7 QVS 21 20 19 18 17

ISBN 978-1-259-54243-5
MHID 1-259-54243-2

Chief Product Officer, SVP Products & Markets: *G. Scott Virkler*
Vice President, General Manager, Products & Markets: *Marty Lange*
Vice President, Content Design & Delivery: *Betsy Whalen*
Managing Director: *Thomas Timp*
Senior Director, Content Design & Delivery: *Linda Avenarius*
Brand Manager: *Michael Ivanov, Ph.D.*
Director, Product Development: *Rose Koos*
Director of Digital Content: *Philip Janowicz, Ph.D.*
Product Developer: *Jodi Rhomberg*
Market Development Manager: *Tamara Hodge*
Marketing Manager: *Noah Evans*
Digital Product Analyst: *Patrick Diller*
Digital Product Developer: *Joan Weber*
Program Manager: *Lora Neyens*
Content Project Managers: *Laura Bies, Tammy Juran & Sandy Schnee*
Buyer: *Sandy Ludovissy*
Design: *Matt Backhaus*
Content Licensing Specialists: *Lori Hancock & Melisa Seegmiller*
Cover Image: *Michael Collier*
Compositor: *SPi Global*
Printer: *Quad/Graphics*

All credits appearing on page or at the end of the book are considered to be an extension of the copyright page.

Library of Congress Cataloging-in-Publication Data

Names: Reynolds, Stephen J., author.
Title: Exploring physical geography / Stephen J. Reynolds, Arizona State
    University, Robert V. Rohli, Louisiana State Univresity, Julia K. Johnson,
    Arizona State University, Peter R. Waylen, University of Florida, Mark A.
    Francek, Central Michigan University.
Description: Second edition. | New York, NY : McGraw-Hill, [2017]
Identifiers: LCCN 2016034542 | ISBN 9781259542435 (alk. paper)
Subjects: LCSH: Physical geography—Textbooks.
Classification: LCC GB54.5 .R49 2017 | DDC 910/.02—dc23 LC record available at
https://lccn.loc.gov/2016034542

The Internet addresses listed in the text were accurate at the time of publication. The inclusion of a website does not indicate an endorsement by the authors or McGraw-Hill Education, and McGraw-Hill Education does not guarantee the accuracy of the information presented at these sites.

BRIEF CONTENTS

CONTENTS

CHAPTER 4:
ATMOSPHERIC MOISTURE   108

CHAPTER 5:
WEATHER SYSTEMS
AND SEVERE WEATHER   142

CHAPTER 6:
ATMOSPHERE-OCEAN-CRYOSPHERE INTERACTIONS   180

CHAPTER 7:
CLIMATES AROUND THE WORLD   212

CHAPTER 11: VOLCANOES, DEFORMATION, AND EARTHQUAKES 344

CHAPTER 12: WEATHERING AND MASS WASTING 384

CHAPTER 18:
BIOMES   564

PREFACE

TELLING THE STORY . . .

WE WROTE *EXPLORING PHYSICAL GEOGRAPHY* so that students could learn from the book on their own, freeing up instructors to teach the class in any way they want. I (Steve Reynolds) first identified the need for this type of book while I was a National Association of Geoscience Teachers' (NAGT) distinguished speaker. As part of my NAGT activities, I traveled around the country conducting workshops on how to infuse active learning and scientific inquiry into introductory college science courses, including those with upwards of 200 students. In the first part of the workshop, I asked the faculty participants to list the main goals of an introductory science course, especially for non-majors. At every school I visited, the main goals were similar to those listed below:

- to engage students in the process of scientific inquiry so that they learn what science is and how it is conducted,
- to teach students how to observe and interpret landscapes and other aspects of their physical environment,
- to enable students to learn and apply important concepts of science,
- to help students understand the relevance of science to their lives, and
- to enable students to use their new knowledge, skills, and ways of thinking to become more informed citizens.

I then asked faculty members to rank these goals and estimate how much time they spent on each goal in class. At this point, many instructors recognized that their activities in class were not consistent with their own goals. Most instructors were spending nearly all of class time teaching content. Although this was one of their main goals, it commonly was not their top goal.

Next, I asked instructors to think about why their activities were not consistent with their goals. Inevitably, the answer was that most instructors spend nearly all of class time covering content because (1) textbooks include so much material that students have difficulty distinguishing what is important from what is not, (2) instructors needed to lecture so that students would know what is important, and (3) many students have difficulty learning independently from the textbook.

In most cases, textbooks drive the curriculum, so my coauthors and I decided that we should write a textbook that (1) contains only important material, (2) indicates clearly to the student what is important and what they need to know, and (3) is designed and written in such a way that students can learn from the book on their own. This type of book would give instructors freedom to teach in a way that is more consistent with their goals, including using local examples to illustrate geographic concepts and their relevance. Instructors would also be able to spend more class time teaching students to observe and interpret landscapes, atmospheric phenomena, and ecosystems, and to participate in the process of scientific inquiry, which represents the top goal for many instructors.

COGNITIVE AND SCIENCE-EDUCATION RESEARCH

To design a book that supports instructor goals, we delved into cognitive and science-education research, especially research on how our brains process different types of information, what obstacles limit student learning from textbooks, and how students use visuals versus text while studying. We also conducted our own research on how students interact with textbooks, what students see when they observe photographs showing landscape features, and how they interpret different types of scientific illustrations, including maps, cross sections, and block diagrams that illustrate the evolution of environments. *Exploring Physical Geography* is the result of our literature search and of our own science-education research. As you examine *Exploring Physical Geography*, you will notice that it is stylistically different from most other textbooks, which will likely elicit a few questions.

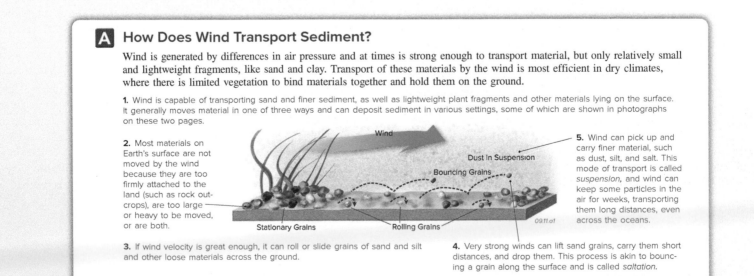

A How Does Wind Transport Sediment?

Wind is generated by differences in air pressure and at times is strong enough to transport material, but only relatively small and lightweight fragments, like sand and clay. Transport of these materials by the wind is most efficient in dry climates, where there is limited vegetation to bind materials together and hold them on the ground.

1. Wind is capable of transporting sand and finer sediment, as well as lightweight plant fragments and other materials lying on the surface. It generally moves material in one of three ways and can deposit sediment in various settings, some of which are shown in photographs on these two pages.

2. Most materials on Earth's surface are not moved by the wind because they are too firmly attached to the land (such as rock outcrops), are too large or heavy to be moved, or are both.

5. Wind can pick up and carry finer material, such as dust, silt, and salt. This mode of transport is called *suspension*, and wind can keep some particles in the air for weeks, transporting them long distances, even across the oceans.

3. If wind velocity is great enough, it can roll or slide grains of sand and silt and other loose materials across the ground.

4. Very strong winds can lift sand grains, carry them short distances, and drop them. This process is akin to bouncing a grain along the surface and is called *saltation*.

Wind — Dust in Suspension — Bouncing Grains — Stationary Grains — Rolling Grains

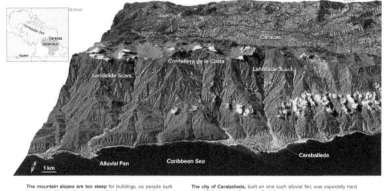

CHAPTER

12 Weathering and Mass Wasting

THE BREAKDOWN OF SURFACE MATERIALS—weathering—produces soils and can lead to unstable slopes. Such slope instability is called *mass wasting*, which is the movement of material downslope in response to gravity. Mass wasting can be slow and barely perceptible, or it can be catastrophic, involving thick, dangerous slurries of mud and debris. It is a type of erosion that strips material off a landscape and transports that material away. What physical and chemical weathering processes loosen material from solid rocks and lead to mass wasting? What factors determine if a slope is stable, and how do slopes fail? In this chapter, we explore weathering and mass wasting, which help sculpt natural landscapes.

The Cordillera de la Costa is a steep 2 km-high mountain range that runs along the coast of Venezuela, separating the capital city of Caracas from the sea. This image, looking south, has topography overlain with a satellite image taken in 2000. The white areas are clouds and the purple areas are cities. The Caribbean Sea is in the foreground. The map below shows the location of Venezuela on the northern coast of South America.

In December 1999, torrential rains in the mountains caused landslides and mobilized soil and other loose material as debris flows and flash floods that buried parts of the coastal cities. Some light-colored landslide scars are visible on the hillsides in this image.

How does soil and other loose material form on hillslopes? What factors determine whether a slope is stable or is prone to landslides and other types of downhill movement?

Huge boulders smashed through the lower two floors of this building in Caraballeda and ripped away part of the right side (▼). The mud and water that transported these boulders are no longer present, but the boulders remain as a testament to the strength of the event.

The mountain slopes are too steep for buildings, so people built the coastal cities on the less steep fan-shaped areas at the foot of each valley. These flatter areas are alluvial fans composed of mountain-derived sediment that has been transported down the canyons and deposited along the mountain front.

What are some potential hazards of living next to steep mountain slopes, especially in a city built on an active alluvial fan?

The city of Caraballeda, built on one such alluvial fan, was especially hard hit in 1999 by debris flows and flash floods that tore a swath of destruction through the town. Landslides, debris flows, and flooding killed more than 19,000 people and caused up to $30 billion in damage in the region. The damage is visible as the light-colored strip through the center of town.

How can loss of life and destruction of property by debris flows and landslides be avoided or at least minimized?

This aerial photograph (◄) of Caraballeda, looking south up the canyon, shows the damage in the center of the city caused by the debris flows and flash floods. Many houses were completely demolished by the fast-moving, boulder-rich mud.

1999 Venezuelan Disaster

A *debris flow* is a slurry of water and debris, including mud, sand, gravel, pebbles, boulders, vegetation, and even cars and small structures. Debris flows can move at speeds up to 80 km/hr (50 mph), but most are slower. In December 1999, two storms dumped as much as 1.1 m (42 in.) of rain on the coastal mountains of Venezuela. The rain loosened soil on the steep hillsides, causing many landslides and debris flows that coalesced in the steep canyons and raced downhill toward the cities built on the alluvial fans.

In Caraballeda, the debris flows carried boulders up to 10 m (33 ft) in diameter and weighing 300 to 400 tons each. The debris flows and flash floods raced across the city, flattening cars and smashing houses, buildings, and bridges. They left behind a jumble of boulders and other debris along the path of destruction through the city.

After the event, USGS geoscientists went into the area to investigate what had happened and why. They documented the types of material that were carried by the debris flows, mapped the extent of the flows, and measured boulders (▼) to investigate processes that occurred during the event. When the scientists examined what lay beneath the foundations of destroyed houses, they discovered that much of the city had been built on older debris flows. These deposits should have provided a warning of what was to come.

Weathering and Mass Wasting 385

384

Exploring Physical Geography promotes inquiry and science as an active process. It encourages student curiosity and aims to activate existing student knowledge by posing the title of every two-page spread and every subsection as a question. In addition, questions are dispersed throughout the book. Integrated into the book are opportunities for students to observe patterns, features, and examples before the underlying concepts are explained. That is, we employ a *learning-cycle approach* where student exploration precedes the introduction of geographic terms and the application of knowledge to a new situation. For example, chapter 12 on slope stability, pictured above, begins with a three-dimensional image of northern Venezuela and asks readers to observe where people are living in this area and what natural processes might have formed these sites.

Wherever possible, we introduce terms after students have an opportunity to observe the feature or concept that is being named. This approach is consistent with several educational philosophies, including a learning cycle and just-in-time teaching. Research on learning cycles shows that students are more likely to retain a term if they already have a mental image of the thing being named (Lawson, 2003). For example, this book presents students with maps showing the spatial distribution of earthquakes, volcanoes, and mountain ranges and asks them to observe the patterns and think about what might be causing the patterns. Only then does the textbook introduce the concept of tectonic plates.

Also, the figure-based approach in this book allows terms to be introduced in their context rather than as a definition that is detached from a visual representation of the term. We introduce new terms in italics rather than in boldface, because boldfaced terms on a textbook page cause students to immediately focus mostly on the terms, rather than build an understanding of the concepts. The book includes a glossary for those students who wish to look up the definition of a term to refresh their memory. To expand comprehension of the definition, each entry in the glossary references the pages where the term is defined in the context of a figure.

WHY ARE THE PAGES DOMINATED BY ILLUSTRATIONS?

Physical geography is a visual science. Geography textbooks contain a variety of photographs, maps, cross sections, block diagrams, and other types of illustrations. These diagrams help portray the spatial distribution and geometry of features in the landscape, atmosphere, oceans, and biosphere in ways words cannot. In geography, a picture really is worth a thousand words.

Exploring Physical Geography contains a wealth of figures to take advantage of the visual and spatial nature of geography and the efficiency of figures in conveying geographic concepts. This book contains few large blocks of text—most text is in smaller blocks that are specifically linked to illustrations. Examples of our integrated figure-text approach are shown throughout the book. In this approach, each short block of text is one or more complete sentences that succinctly describe a geographic feature, geographic process, or both of these. Most of these text blocks are connected to their illustrations with leader lines so that readers know exactly which feature or part of the diagram is being referenced in the text block. A reader does not have to search for the part of the figure that corresponds to a text passage, as occurs when a student reads a traditional textbook with large blocks of text referencing a figure that may appear on a different page. The short blocks are numbered if they should be read in a specific order.

This approach is especially well suited to covering geographic topics, because it allows the text to have a precise linkage to the geographic location of the aspect being described. A text block discussing the Intertropical Convergence Zone in Costa Rica can have a leader that specifically points to the location of this feature. A cross section of atmospheric circulation can be accompanied by short text blocks that describe each part of the system and that are linked by leaders directly to specific locations on the figure. This allows the reader to concentrate on the concepts being presented, not deciding what part of the figure is being discussed.

The approach in *Exploring Physical Geography* is consistent with the findings of cognitive scientists, who conclude that our minds have two different processing systems, one for processing pictorial information (images) and one for processing verbal information (speech and written words). This view of cognition is illustrated in the figure below. Cognitive scientists also speak about two types of memory: *working memory* involves holding and processing information in short-term memory, and *long-term memory* stores information until we need it (Baddeley, 2007). Both the verbal and pictorial processing systems have a limited amount of working memory, and our minds have to use much of our mental processing space to reconcile the two types of information in working memory. For information that has both pictorial and verbal components, as most geographic information does, the amount of knowledge we retain depends on reconciling these two types of information, on transferring information from working memory to long-term memory, and on linking the new information with our existing mental framework. For this reason, this book integrates text and figures, as in the example shown here.

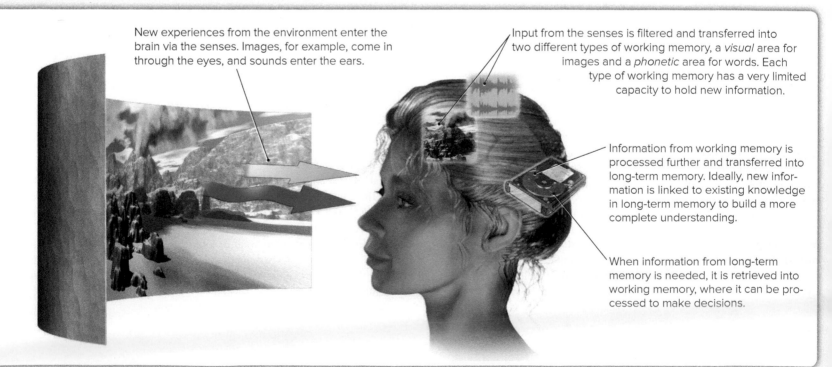

New experiences from the environment enter the brain via the senses. Images, for example, come in through the eyes, and sounds enter the ears.

Input from the senses is filtered and transferred into two different types of working memory, a *visual* area for images and a *phonetic* area for words. Each type of working memory has a very limited capacity to hold new information.

Information from working memory is processed further and transferred into long-term memory. Ideally, new information is linked to existing knowledge in long-term memory to build a more complete understanding.

When information from long-term memory is needed, it is retrieved into working memory, where it can be processed to make decisions.

WHY ARE THERE SO MANY FIGURES?

This textbook contains more than 2,600 figures, which is two to three times the number in most introductory geography textbooks. One reason for this is that the book is designed to provide a concrete example of each process, environment, or landscape feature being illustrated. Research shows that many college students require concrete examples before they can begin to build abstract concepts (Lawson, 1980). Also, many students have limited travel experience, so photographs and other figures allow them to observe places, environments, and processes they have not been able to observe firsthand. The numerous photographs, from geographically diverse places, help bring the sense of place into the student's reading. The inclusion of an illustration for each text block reinforces the notion that the point being discussed is important. In many cases, as in the example on this page, conceptualized figures are integrated with photographs and text so that students can build a more coherent view of the environment or process.

Exploring Physical Geography focuses on the most important geographic concepts and makes a deliberate attempt to eliminate text that is not essential for student learning of these concepts. Inclusion of information that is not essential tends to distract and confuse students rather than illuminate the concept; thus, you will see fewer words. Cognitive and science-education research has identified a redundancy effect, where information that restates and expands upon a more succinct description actually results in a decrease in student learning (Mayer, 2001). Specifically, students learn less if a long figure caption restates information contained elsewhere on the page, such as in a long block of text that is detached from the figure. We avoid the redundancy effect by including only text that is integrated with the figure.

The style of illustrations in *Exploring Physical Geography* was designed to be more inviting to today's visually oriented students who are used to photo-realistic, computer-rendered images in movies, videos, and computer games. For this reason, many of the figures were created by world-class scientific illustrators and artists who have worked on award-winning textbooks, on Hollywood movies, on television shows, for *National Geographic,* and in the computer-graphics and gaming industry. In most cases, the figures incorporate real data, such as satellite images, weather and climatological data, and aerial photographs. Our own research shows that many students do not understand cross sections and other subsurface diagrams, so nearly every cross section in this book has a three-dimensional aspect, and many maps are presented in a perspective view with topography. Research findings by us and other researchers (Roth and Bowen, 1999) indicate that including people and human-related items on photographs and figures attracts undue attention, thereby distracting students from the features being illustrated. As a result, our photographs have nondistracting indicators of scale, like dull coins and plain marking pens. Figures and photographs do not include people or human-related items unless we are trying to (1) illustrate how geographers study geographic processes and features, (2) illustrate the relevance of the processes on humans, or (3) help students connect and relate to the human dimension of the issue.

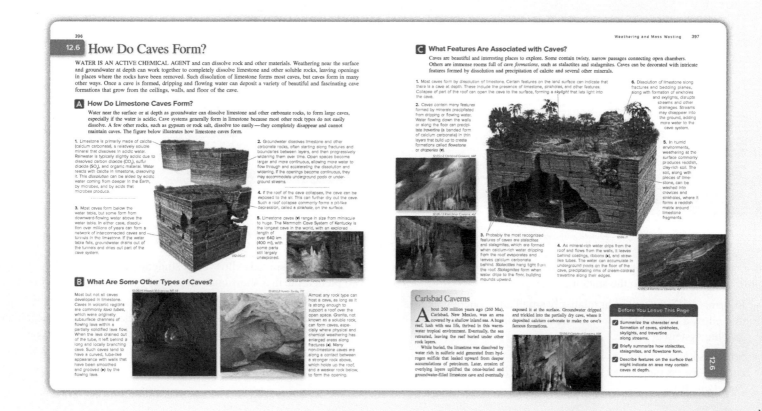

WHY DOES THE BOOK CONSIST OF TWO-PAGE SPREADS?

This book consists of two-page spreads, most of which are further sub-divided into sections. Research has shown that because of our limited amount of working memory, much new information is lost if it is not incorporated into long-term memory. Many students keep reading and highlighting their way through a textbook without stopping to integrate the new information into their mental framework. New information simply displaces existing information in working memory before it is learned and retained. This concept of cognitive load (Sweller, 1994) has profound implications for student learning during lectures and while reading textbooks. Two-page spreads and sections help prevent cognitive overload by providing natural breaks that allow students to stop and consolidate the new information before moving on.

Each spread has a unique number, such as 6.10 for the tenth topical two-page spread in chapter 6. These numbers help instructors and students keep track of where they are and what is being covered. Each two-page spread, except for those that begin and end a chapter, contains a *Before You Leave This Page* checklist that indicates what is important and what is expected of students before they move on. This list contains learning objectives for the spread and provides a clear way for the instructor to indicate to the student what is important. The items on these lists are compiled into a master *What-to-Know List* provided to the instructor, who then deletes or adds entries to suit the instructor's learning goals and distributes the list to students before the students begin reading the book. In this way, the *What-to-Know List* guides the students' studying.

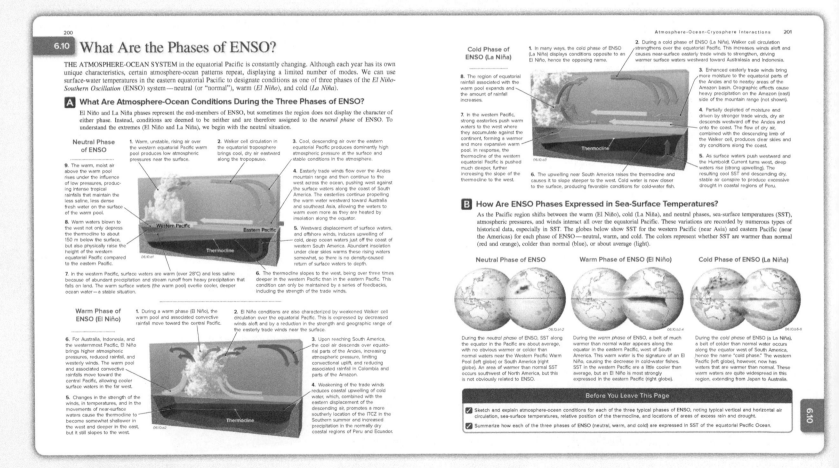

Two-page spreads and integrated *Before You Leave This Page* lists offer the following advantages to the student:

- Information is presented in relatively small and coherent chunks that allow a student to focus on one important aspect or geographic system at a time.
- Students know when they are done with this particular topic and can self-assess their understanding with the *Before You Leave This Page* list.

- Two-page spreads allow busy students to read or study a complete topic in a short interval of study time, such as the breaks between classes.
- All test questions and assessment materials are tightly articulated with the *Before You Leave This Page* lists so that exams and quizzes cover precisely the same material that was assigned to students via the *What-to-Know* list.

The two-page spread approach also has advantages for the instructor. Before writing this book, the authors wrote most of the items for the *Before You Leave This Page* lists. We then used this list to decide what figures were needed, what topics would be discussed, and in what order. In other words, *the textbook was written from the learning objectives*. The *Before You Leave This Page* lists provide a straightforward way for an instructor to tell students what information is important. Because we provide the instructor with a master *What-to-Know* list, an instructor can selectively assign or eliminate content by providing students with an edited *What-to-Know* list. Alternatively, an instructor can give students a list of assigned two-page spreads or sections within two-page spreads. In this way, the instructor can identify content for which students are responsible, even if the material is not covered in class. Two-page spreads provide the instructor with unparalleled flexibility in deciding what to assign and what not to cover. It allows this book to be easily used for one-semester and two-semester courses.

CONCEPT SKETCHES

Most items on the *Before You Leave This Page* list are by design suitable for student construction of concept sketches. Concept sketches are sketches that are annotated with complete sentences that identify geographic features, describe how the features form, characterize the main geographic processes, and summarize histories of landscapes (Johnson and Reynolds, 2005). An example of a concept sketch is shown to the right.

Concept sketches are an excellent way to actively engage students in class and to assess their understanding of geographic features, processes, and history. Concept sketches are well suited to the visual nature of geography, especially cross sections, maps, and block diagrams. Geographers are natural sketchers using field notebooks, blackboards, publications, and even napkins, because sketches are an important way to record observations and thoughts, organize knowledge, and try to visualize the evolution of landscapes, circulation in the atmosphere and oceans, motion and precipitation along weather fronts, layers within soils, and biogeochemical cycles. Our research data show that a student who can draw, label, and explain a concept sketch generally has a good understanding of that concept.

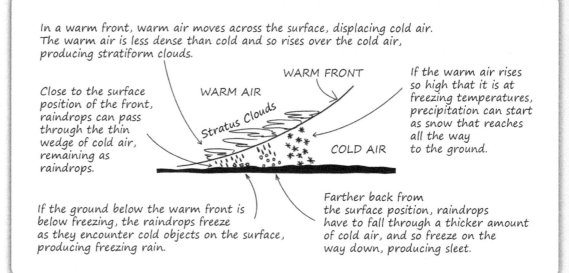

In a warm front, warm air moves across the surface, displacing cold air. The warm air is less dense than cold and so rises over the cold air, producing stratiform clouds.

Close to the surface position of the front, raindrops can pass through the thin wedge of cold air, remaining as raindrops.

If the warm air rises so high that it is at freezing temperatures, precipitation can start as snow that reaches all the way to the ground.

If the ground below the warm front is below freezing, the raindrops freeze as they encounter cold objects on the surface, producing freezing rain.

Farther back from the surface position, raindrops have to fall through a thicker amount of cold air, and so freeze on the way down, producing sleet.

WARM FRONT
WARM AIR
Stratus Clouds
COLD AIR

REFERENCES CITED

Baddeley, A. D. 2007. *Working memory, thought, and action*. Oxford: Oxford University Press, 400 p.

Johnson, J. K., and Reynolds, S. J. 2005. Concept sketches—Using student- and instructor-generated annotated sketches for learning, teaching, and assessment in geology courses. *Journal of Geoscience Education*, v. 53, pp. 85–95.

Lawson, A. E. 1980. Relationships among level of intellectual development, cognitive styles, and grades in a college biology course. *Science Education*, v. 64, pp. 95–102.

Lawson, A. 2003. *The neurological basis of learning, development & discovery: Implications for science & mathematics instruction*. Dordrecht, The Netherlands: Kluwer Academic Publishers, 283 p.

Mayer, R. E. 2001. *Multimedia learning*. Cambridge: Cambridge University Press, 210 p.

Roth, W. M., and Bowen, G. M. 1999. Complexities of graphical representations during lectures: A phenomenological approach. *Learning and Instruction*, v. 9, pp. 235–255.

Sweller, J. 1994. Cognitive Load Theory, learning difficulty, and instructional design. *Learning and Instruction*, v. 4, pp. 295–312.

HOW IS THIS BOOK ORGANIZED?

Two-page spreads are organized into 18 chapters that are arranged into five major groups: (1) introduction to Earth, geography, and energy and matter; (2) atmospheric motion, weather, climate, and water resources; (3) introduction to landscapes, earth materials, sediment transport, plate tectonics, and tectonic processes (e.g., volcanoes and earthquakes); (4) processes, such as stream flow and glaciation, that sculpt and modify landscapes; and (5) soils, biogeography, and biogeochemical cycles. The first chapter provides an overview of geography, including the scientific approach to geography, how we determine and represent location, the tools and techniques used by geographers, and an introduction to *natural systems*—a unifying theme interwoven throughout the rest of the book. Chapter 2 covers energy and matter in the Earth system, providing a foundation for all that follows in the book.

The second group of chapters begins with an introduction to atmospheric motion (chapter 3), another theme revisited throughout the book. It features separate two-page spreads on circulation in the tropics, high latitudes, and mid-latitudes, allowing students to concentrate on one part of the system at a time, leading to a synthesis of lower-level and upper-level winds. Chapter 3 also covers air pressure, the Coriolis effect, and seasonal and regional winds. This leads naturally into chapter 4, which is a thorough introduction to atmospheric moisture and the consequences of rising and sinking air, including clouds and precipitation. Chapter 5 follows with a visual, map-oriented discussion of weather, including cyclones, tornadoes, and other severe weather. The next chapter (chapter 6), unusual for an introductory geography textbook, is devoted entirely to interactions between the atmosphere, oceans, and cryosphere. It features sections on ocean currents, sea-surface temperatures, ocean salinity, and a thorough treatment of ENSO and other atmosphere-ocean oscillations. This leads into a chapter on climate (chapter 7), which includes controls on climate and a climate classification, featuring a two-page spread on each of the main climate types, illustrated with a rich blend of figures and photographs. These spreads are built around globes that portray a few related climate types, enabling students to concentrate on their spatial distribution and control, rather than trying to extract patterns from a map depicting all the climate types (which the chapter also has). The climate chapter also has a data-oriented presentation of climate change. This second part of the book concludes with chapter 8, which presents the hydrologic cycle and water resources, emphasizing the interaction between surface water and groundwater.

The third part of the book focuses on landscapes and tectonics. It begins with chapter 9, a visually oriented introduction to understanding landscapes, starting with familiar landscapes as an introduction to rocks and minerals. The chapter has a separate two-page spread for each family of rocks and how to recognize each type in the landscape. It presents a brief introduction to weathering, erosion, and transport, aspects that are covered in more detail in later chapters on geomorphology. Wind transport, erosion, and landforms are integrated into chapter 9, rather than being a separate, sparse-content chapter that forcibly brings in non-wind topics, as is done in other textbooks. It also covers relative and numeric dating and how we study the ages of landscapes. It is followed by chapter 10 on plate tectonics and regional features. Chapter 10 begins with having students observe large-scale features on land and on the seafloor, as well as patterns of earthquakes and volcanoes, as a lead-in to tectonic plates. Integrated into the chapter are two-page spreads on continental drift, paleomagnetism, continental and oceanic hot spots, evolution of the modern oceans and continents, the origin of high elevations, and the relationship between internal and external processes. The last chapter in this third part (chapter 11) presents the processes, landforms, and hazards associated with volcanoes, deformation, and earthquakes. It also explores the origin of local mountains and basins, another topic unique to this textbook.

The fourth group of chapters concerns the broad field of geomorphology—the form and evolution of landscapes. It begins with chapter 12, a more in-depth treatment of weathering, mass wasting, and slope stability. This chapter also has two-page spreads on caves and karst topography. Chapter 13 is about streams and flooding, presenting a clear introduction to drainage networks, stream processes, different types of streams and their associated landforms and sediment, and how streams change over time. It ends with sections on floods, calculating stream discharges, some examples of devastating local and regional floods, and the many ways in which streams affect people. Chapter 14 covers glaciers and glacial movement, landforms, and deposits. It also discusses the causes of glaciation and the possible consequences of melting of ice sheets and glaciers. Chapter 15 covers the related topic of coasts and changing sea levels. It introduces the processes, landforms, and hazards of coastlines. It also covers the consequences of changing sea level on landforms and humans.

The fifth and final group of chapters focuses on the biosphere and begins with chapter 16, which explores the properties, processes, and importance of soil. This chapter covers soil characterization and classification, including globes showing the spatial distribution of each main type of soil. It ends with a discussion of soil erosion and how soil impacts the way we use land. Chapter 17 provides a visual introduction to ecosystems and biogeochemical cycles. It addresses interactions between organisms and resources within ecosystems, population growth and decline, biodiversity, productivity, and ecosystem disturbance. The last part of chapter 17 covers the carbon, nitrogen, phosphorus, and sulfur cycles, the role of oxygen in aquatic ecosystems, and invasive species. The final chapter in the book, chapter 18, is a synthesis chapter on biomes. It discusses factors that influence biomes and then contains a two-page spread on each major biome, with maps, globes, photographs, and other types of figures to convey where and why each biome exists. It includes a section on sustainability and ends with a synthesis that portrays biomes in the context of many topics presented in the book, including energy balances, atmospheric moisture and circulation, climate types, and soils.

TWO-PAGE SPREADS

Most of the book consists of *two-page spreads*, each of which is about one or more closely related topics. Each chapter has four main types of two-page spreads: opening, topical, connections, and investigation.

Opening Two-Page Spread

Opening spreads introduce the chapter, engaging the student by highlighting some interesting and relevant aspects and posing questions to activate prior knowledge and curiosity.

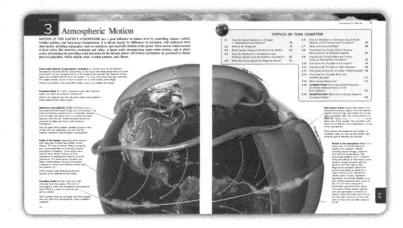

Topical Two-Page Spread

Topical spreads comprise most of the book. They convey the geographic content, help organize knowledge, describe and illustrate processes, and provide a spatial context. The first topical spread in a chapter usually includes some aspects that are familiar to most students, as a bridge or scaffold into the rest of the chapter. Each chapter has at least one two-page spread illustrating how geography impacts society and commonly another two-page spread that specifically describes how geographers study typical problems.

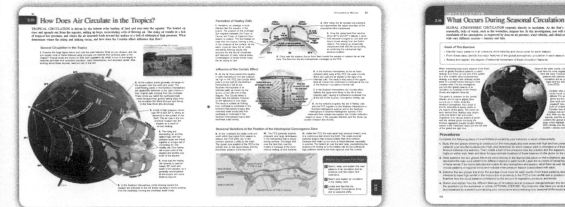

Connections Two-Page Spread

The next-to-last two-page spread in each chapter is a *Connections spread* designed to help students connect and integrate the various concepts from the chapter and to show how these concepts can be applied to an actual location. *Connections* are about real places that illustrate the geographic concepts and features covered in the chapter, often explicitly illustrating how we investigate a geographic problem and how geographic problems have relevance to society.

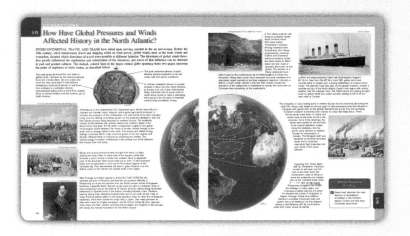

Investigation Two-Page Spread

Each chapter ends with an *Investigation* spread that is an exercise in which students apply the knowledge, skills, and approaches learned in the chapter. These exercises mostly involve virtual places that students explore and investigate to make observations and interpretations and to answer a series of geographic questions. Investigations are modeled after the types of problems geographers investigate, and they use the same kinds of data and illustrations encountered in the chapter. The Investigation includes a list of goals for the exercises and step-by-step instructions, including calculations and methods for constructing maps, graphs, and other figures. These investigations can be completed by students in class, as worksheet-based homework, or as online activities.

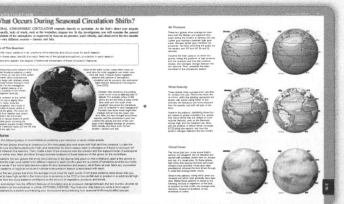

NEW IN THE SECOND EDITION

The second edition of *Exploring Physical Geography* represents a significant revision. The style, approach, and sequence of chapters are unchanged, but every chapter received new photographs, revised figures, major to minor editing of text blocks and, in some cases, minor reorganization. We revised text blocks to improve clarity and conciseness and to present recent discoveries and events. Most chapters contain the same number of two-page spreads, but the content on some spreads was extensively revised. Nearly all changes were made in response to comments by reviewers, students, and instructors who are using the materials. The most important revisions are listed below.

- This edition features completely different fonts from the first edition. The new fonts were chosen partly to improve the readability on portable electronic devices, while retaining fidelity to a quality printed book. This font replacement resulted in countless small changes in the layout of individual text blocks on every two-page spread. In addition to replacing all of the fonts within the text, all figure labels were replaced with the new font, a process that required opening, editing, and commonly resizing every illustration. In addition, all labels were incorporated into the actual artwork, rather than overlaying text on the artwork using the page-layout program, as was done for many figures in the first edition. This involved adding labels to hundreds of illustrations, but has the benefit of having every label as an integral part of its associated art file, a useful feature for constructing PowerPoint files.

- This edition contains more than 100 new photographs, with a deliberate intention to represent a wider geographic diversity, providing students with local examples from their region. For the geomorphology chapters, many photographs from the first edition were reprocessed from the original to improve clarity and provide more detail.

- This edition contains 120 new and heavily revised illustrations. Figures from the first edition were replaced with new versions to update information so that it is more recent, to improve student understanding of certain complex topics, and for improved appearance. Investigations in several chapters were completely revised.

CHAPTER 1: There are six new photographs and minor revisions to several illustrations. As with other chapters in the book, there are numerous minor edits to the text.

CHAPTER 2: For this chapter we replaced three photographs, including one of the Earth from the Moon. In addition, we produced new versions of 14 figures, including new globes for the investigation. A number of these revisions reflect the decision to refer to the Sun using Sun angle instead of Zenith angle, which some students found confusing.

CHAPTER 3: This chapter has one new photograph, and we revised or created ten figures, mostly concerning flow between the upper and lower parts of the troposphere. We also added labels showing the locations of important high- and low-pressure zones.

CHAPTER 4: One photograph was replaced in this chapter, and we revised 21 figures, mostly for font changes. Heavily revised figures include new maps and rendered globes of specific humidity and precipitation, upper-level convergence, and extreme precipitation. The Investigation for this chapter was completely redone, now having students observe and explain global patterns of humidity, water vapor, and land cover.

CHAPTER 5: For this chapter, we replaced four photos and revised more than a dozen figures (30 if you count all the font replacements). Revised figures include revisions to several maps, figures related to the formation of lightning, and figures depicting upper-level divergence. In addition, we replaced a world map of tornado frequency with more visually appealing and less distorting globes of these data. The Connections spread received a change in layout, and the Investigation was heavily revised with six new maps and new procedures.

CHAPTER 6: Chapter 6 contains 24 revised figures, mostly for fonts, but including a number of new versions for important figures. These include new renders of globes for the opening two-page spread, globes about the relationship of winds to ocean currents, and globes about ENSO. There are new figures about the Walker cell, El Niño SST anomalies, and other topics. We deleted a figure and associated text on salinity in Section 6.7.

CHAPTER 7: In this chapter we replaced just one new photograph, but we revised 54 figures, mostly for fonts. There are new versions of globes in the opening two-page spread and in the introduction to the climate types, as well as newly rendered globes of every type of climate. There are new maps of the urban heat islands and new graphs depicting the data for climate change, revising the graphs to incorporate the most recent data. There are new figures for sea-level change, the decrease of Arctic ice, and the frequency of storms over time. We also revised the maps in the Investigation.

CHAPTER 8: We deleted one photograph from this chapter, and incorporated six new photographs. New photographs are from Central Texas, Maryland, the Black Hills, and the state of Washington. Most of the figure revisions were font replacements and adding labels to globes, but larger revisions included figures about global water budgets and newly rendered water-balance globes. There is also a new map of the Ogallala aquifer and water-level changes in the aquifer. We changed some values in the table for the Investigation.

CHAPTER 9: This chapter has 19 new photographs, including new multi-specimen photographs of common minerals. There are also new close-ups of rocks and new photographs of landscapes. We also revised the geologic timescale to reflect new dates. We made minor revisions to three figures.

CHAPTER 10: Chapter 10 has one new photograph, but a number of photographs were reprocessed from the original. As with all other chapters, all figures with labels were revised, including 16 in this chapter, and more extensive revisions occurred in figures regarding paleomagnetism.

CHAPTER 11: For this chapter we deleted one photograph and two satellite images, and replaced five photographs. The layout and text were revised for the opening two-page spread and for floods associated with volcanic eruptions. Legends were added to the global earthquake maps, and there are new versions of two illustrations of the Alaskan earthquake. A number of other figures were revised, mostly as a result of replacing fonts.

CHAPTER 12: This chapter has seven new photographs, and a number of existing photographs were reprocessed for clarity. The new photographs expand the geographic diversity, including photographs from Central Texas and Florida. There is a new map showing the location of Venezuela and a new map of the global distribution of karst. There are a number of minor revisions to other figures.

CHAPTER 13: There are nine new photographs for this chapter, including ones from Alaska, Central Texas, and the Potomac River. Twenty-nine illustrations in the chapter were significantly revised or had fonts replaced. For example, all the hydrographs were revised, as were the figures for the Upper Mississippi flood.

CHAPTER 14: Chapter 14 has 13 new photographs from Alaska, Wyoming, Colorado, and other states. Text was revised in conjunction with the new photographs.

CHAPTER 15: For this chapter there are 11 new photographs, mostly from Florida and coastal Alabama. Figure revisions for this chapter consisted mostly of font replacements, of which there were many.

CHAPTER 16: We replaced four photographs for this chapter. The position of Sections 16.5 and 16.6 were swapped, putting the discussion of climate and soil ahead of the discussion of terrain, parent material, and time.

CHAPTER 17: Chapter 17 has five new photographs, with accompanying word changes. Significant changes were made to more than a dozen figures, including the opening 3D perspective. There are new figures in the sections regarding biogeochemical cycles for nitrogen, sulfur, and oxygen.

CHAPTER 18: We replaced 17 photographs in this chapter, adding photographs from the Everglades, Mississippi, Alaska, and the southern Rocky Mountains. There are also new photographs of coral reefs. For figure revisions, there is a new version of maps of tropical rain-forest deforestation, coral-reef distribution, and the Panama connection.

McGraw-Hill Connect®
Learn Without Limits

Connect is a teaching and learning platform that is proven to deliver better results for students and instructors.

Connect empowers students by continually adapting to deliver precisely what they need, when they need it, and how they need it, so your class time is more engaging and effective.

73% of instructors who use **Connect** require it; instructor satisfaction **increases** by 28% when **Connect** is required.

Analytics

Connect Insight®

Connect Insight is Connect's new one-of-a-kind visual analytics dashboard—now available for both instructors and students—that provides at-a-glance information regarding student performance, which is immediately actionable. By presenting assignment, assessment, and topical performance results together with a time metric that is easily visible for aggregate or individual results, Connect Insight gives the user the ability to take a just-in-time approach to teaching and learning, which was never before available. Connect Insight presents data that empowers students and helps instructors improve class performance in a way that is efficient and effective.

Mobile

Connect's new, intuitive mobile interface gives students and instructors flexible and convenient, anytime–anywhere access to all components of the Connect platform.

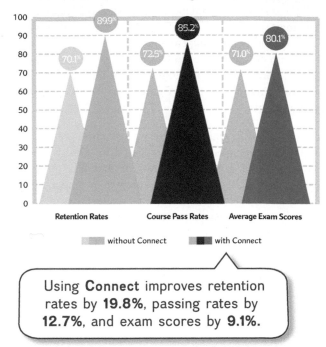

Connect's Impact on Retention Rates, Pass Rates, and Average Exam Scores

Retention Rates: 70.1% without Connect, 89.9% with Connect
Course Pass Rates: 72.5% without Connect, 85.2% with Connect
Average Exam Scores: 71.0% without Connect, 80.1% with Connect

without Connect | with Connect

Using **Connect** improves retention rates by **19.8%**, passing rates by **12.7%**, and exam scores by **9.1%**.

Impact on Final Course Grade Distribution

without Connect	Grade	with Connect
22.9%	A	31.0%
27.4%	B	34.3%
22.9%	C	18.7%
11.5%	D	6.1%
15.4%	F	9.9%

Students can view their results for any **Connect** course.

Adaptive

THE **ADAPTIVE** **READING EXPERIENCE** DESIGNED TO TRANSFORM THE WAY STUDENTS READ

More students earn **A's** and **B's** when they use McGraw-Hill Education **Adaptive** products.

SmartBook®

Proven to help students improve grades and study more efficiently, SmartBook contains the same content within the print book, but actively tailors that content to the needs of the individual. SmartBook's adaptive technology provides precise, personalized instruction on what the student should do next, guiding the student to master and remember key concepts, targeting gaps in knowledge and offering customized feedback, and driving the student toward comprehension and retention of the subject matter. Available on tablets, SmartBook puts learning at the student's fingertips— anywhere, anytime.

Over **8 billion questions** have been answered, making McGraw-Hill Education products more intelligent, reliable, and precise.

www.mheducation.com

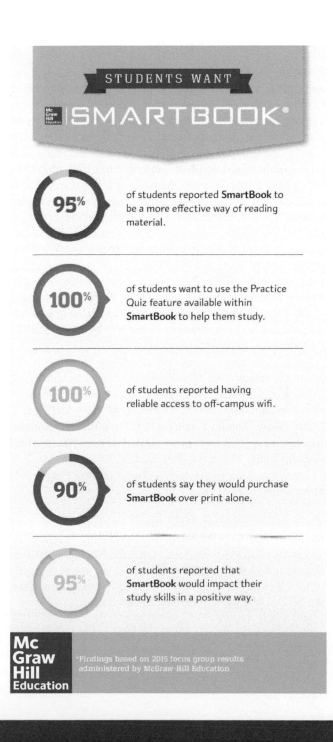

STUDENTS WANT

SMARTBOOK®

95% of students reported **SmartBook** to be a more effective way of reading material.

100% of students want to use the Practice Quiz feature available within **SmartBook** to help them study.

100% of students reported having reliable access to off-campus wifi.

90% of students say they would purchase **SmartBook** over print alone.

95% of students reported that **SmartBook** would impact their study skills in a positive way.

McGraw Hill Education

*Findings based on 2015 focus group results administered by McGraw-Hill Education

ACKNOWLEDGMENTS

Writing a totally new type of introductory geography textbook would not be possible without the suggestions and encouragement we received from instructors who reviewed various drafts of this book and its artwork. We are especially grateful to people who contributed entire days either reviewing the book or attending symposia to openly discuss the vision, challenges, and refinements of this kind of new approach. Our colleagues Paul Morin and Mike Kelly contributed materials in various chapters, for which we continue to be grateful.

This book contains over 2,600 figures, several times more than a typical introductory geography textbook. This massive art program required great effort and artistic abilities from the illustrators and artists who turned our vision and sketches into what truly are pieces of art. We are especially appreciative of Cindy Shaw, who was lead illustrator, art director, and a steady hand that helped guide a diverse group of authors. For many figures, she extracted data from NOAA and NASA websites and then converted the data into exquisite maps and other illustrations. Cindy also fine-tuned the authors' layouts, standardized illustrations, and prepared the final figures for printing. Chuck Carter produced many spectacular pieces of art, including virtual places featured in the chapter-ending Investigations. Susie Gillatt contributed many of her wonderful photographs of places, plants, and creatures from around the world, photographs that helped us tell the story in a visual way. She also color corrected and retouched most of the photographs in the book. We also used visually unique artwork by Daniel Miller, David Fierstein, and Susie Gillatt. Suzanne Rohli performed magic with GIS files, did the initial work on the glossary, and helped in many other ways. We were ably assisted in data compilation and other tasks by geography students Emma Harrison, Abeer Hamden, Peng Jia, and Javier Vázquez, and by Courtney Merjil. Terra Chroma, Inc., of Tucson, Arizona, supported many aspects in the development of this book, including funding parts of the extensive art program and maintenance of the ExploringPhysicalGeography.com website.

Many people went out of their way to provide us with photographs, illustrations, and advice. These helpful people included Susie Gillatt, Vladimir Romanovsky, Paul McDaniel, Lawrence McGhee, Charles Love, Cindy Shaw, Sandra Londono, Lynda Williams, Ramón Arrowsmith, John Delaney, Nancy Penrose, Dan Trimble, Bixler McClure, Michael Forster, Vince Matthews, Ron Blakey, Doug Bartlett, Ed DeWitt, Phil Christensen, Scott Johnson, Peg Owens, Emma Harrison, Skye Rodgers, Steve Semken, and David Walsh.

We used a number of data sources to create many illustrations. Reto Stöckli of the Department of Environmental Sciences at ETH Zürich and NASA Goddard produced the Blue Marble and Blue Marble Next Generation global satellite composites. We are very appreciative of the NOAA Reanalysis Site, which we used extensively, and for other sites of the USDA, NASA, USGS, NRCS, and NPS.

We have treasured our interactions with the wonderful Iowans at McGraw-Hill Higher Education, who enthusiastically supported our vision, needs, and progress. We especially thank our current and previous publishers and brand managers Michelle Vogler, Michael Ivanov, Ryan Blankenship, and Marge Kemp for their continued encouragement and excellent support. Jodi Rhomberg and Laura Bies skillfully and cheerfully guided the development of the book during the entire publication process, making it all happen. Lori Hancock helped immensely with our ever-changing photographic needs. We also appreciate the support, cooperation, guidance, and enthusiasm of Thomas Timp, Marty Lange, Kurt Strand, Noah Evans, Matt Garcia, Lisa Nicks, David Hash, Traci Andre, Tammy Ben, and many others at McGraw-Hill who worked hard to make this book a reality. Kevin Campbell provided thorough copy editing, and reviewed the glossary and index. Angie Sigwarth and Rose Kramer provided excellent proofreading that caught small gremlins before they escaped. Our wonderful colleague Gina Szablewski expertly directed the development of LearnSmart materials and provided general encouragement.

Finally, a project like this is truly life consuming, especially when the author team is doing the writing, illustrating, photography, near-final page layout, media development, and development of assessments, teaching ancillaries, and the instructor's website. We are extremely appreciative of the support, patience, and friendship we received from family members, friends, colleagues, and students who shared our sacrifices and successes during the creation of this new vision of a textbook. Steve Reynolds thanks the ever-cheerful, supportive, and talented Susie Gillatt; John and Kay Reynolds; and our mostly helpful book-writing companions, Widget, Jasper, and Ziggy. Julia Johnson thanks Annabelle Louise and Hazel Johnson, and the rest of her family for enthusiastic support and encouragement. Steve and Julia appreciate the support of their wonderful colleagues at ASU and elsewhere.

Robert Rohli is grateful to his wife Suzanne, a geographer herself, for her patient and unflagging assistance with so many aspects of this project. In addition, their son, Eric, and daughter, Kristen, also contributed in various direct and indirect ways. Their support and enthusiasm, and the encouragement of so many other family members and friends, particularly Bob's and Suzanne's parents, was an important motivator. Rohli also feels deep appreciation for so many dedicated mentors who stimulated his interest in physical geography while he was a student. These outstanding educators include John Arnfield, David Clawson, Carville Earle, Keith Henderson, Jay Hobgood, Merrill Johnson, Ricky Nuesslein, Kris Preston, John Rayner, Jeff Rogers, Rose Sauder, and many others. And finally, Rohli thanks the many students over the years whose interest in the world around them makes his job fun.

Peter Waylen thanks his wife, Marilyn, for her continued unstinting support and encouragement in this and all his other academic endeavors. He would also like to acknowledge geographers who have been very influential in guiding his satisfying and rewarding career, the late John

Thornes, Ming-ko Woo, and César Caviedes. He also thanks Germán Poveda for the stream of stimulating new ideas, including Daisy World with a hydrologic cycle. Peter thanks his coauthors, especially Steve Reynolds and Julia Johnson, for providing the opportunity to participate in this novel and exciting project.

Mark Francek wishes to thank his wife, Suezell, who from the onset, said, "You can do this!" despite his initial doubts about being able to find the time to complete this project. His five kids and two grandkids have also been supportive, making him smile and helping him not to take his work too seriously. Mark's academic mentors over the years, including Ray Lougeay, Lisle Mitchell, Barbara Borowiecki, and Mick Day, have instilled in him a love of field work and physical geography. He also thanks the hundreds of students he taught over the years. Their eagerness to learn has always pushed him to explore new academic horizons. Finally, Mark appreciates working with all his coauthors. He marvels at their patience, kindness, and academic pedigrees.

Cindy Shaw, lead illustrator, is grateful to John Shaw and Ryan Swain, who were of enormous help with final art-file preparation. She particularly appreciates the support of her ever-patient husband, Karl Pitts, who during the project adapted to her long working hours and a steady diet of take-out food. As a scientist, he was always interested and happy to bounce ideas around and clarify any questions. Finally, Cindy thanks all the authors for being a pleasure to work with.

All the authors are very grateful for the thousands of students who have worked with us on projects, infused our classrooms with energy and enthusiasm, and provided excellent constructive feedback about what works and what doesn't work. We wrote this book to help instructors, including us, make students' time in our classes even more interesting, exciting, and informative. Thank you all!

REVIEWERS

Special thanks and appreciation go out to all reviewers. This book was improved by many beneficial suggestions, new ideas, and invaluable advice provided by these reviewers. We appreciate all the time they devoted to reviewing manuscript chapters, attending focus groups, surveying students, and promoting this text to their colleagues.

We would like to thank the following individuals who wrote and/or reviewed learning goal–oriented content for *LearnSmart*.

Florida Atlantic University, Jessica Miles
Northern Arizona University, Sylvester Allred
Roane State Community College, Arthur C. Lee
State University of New York at Cortland, Noelle J. Relles
University of North Carolina at Chapel Hill, Trent McDowell
University of Wisconsin—Milwaukee, Gina Seegers Szablewski

University of Wisconsin—Milwaukee, Tristan J. Kloss
Elise Uphoff

Special thanks and appreciation go out to all reviewers, focus group and Symposium participants. This first edition (through several stages of manuscript development) has enjoyed many beneficial suggestions, new ideas, and invaluable advice provided by these individuals. We appreciate all the time they devoted to reviewing manuscript chapters, attending focus groups, reviewing art samples, and promoting this text to their colleagues.

PHYSICAL GEOGRAPHY REVIEWERS

Antelope Valley College, Michael W. Pesses
Arizona State University, Bohumil Svoma
Austin Peay State University, Robert A. Sirk
Ball State University, David A. Call
California State University–Los Angeles, Steve LaDochy
California State University–Sacramento, Tomas Krabacker
College of Southern Idaho, Shawn Willsey
College of Southern Nevada, Barry Perlmutter
Eastern Washington University, Richard Orndorff
Eastern Washington University, Jennifer Thomson
Florida State University, Holly M. Widen
Florida State University, Victor Mesev
Frostburg State University, Phillip P. Allen
Frostburg State University, Tracy L. Edwards
George Mason University, Patricia Boudinot
Las Positas College, Thomas Orf
Lehman College, CUNY, Stefan Becker
Long Island University, Margaret F. Boorstein
Mesa Community College, Steve Bass
Mesa Community College, Clemenc Ligocki
Metro State, Kenneth Engelbrecht
Metro State, Jon Van de Grift
Minnesota State University, Forrest D. Wilkerson
Monroe Community College, SUNY, Jonathon Little
Moorpark College, Michael T. Walegur
Normandale Community College, Dave Berner
Northern Illinois University, David Goldblum
Oklahoma State University, Jianjun Ge
Oregon State University, Roy Haggerty
Pasadena City College, James R. Powers
Rhodes College, David Shankman
Samford University, Jennifer Rahn
San Francisco State University, Barbara A. Holzman
South Dakota State University, Trisha Jackson
South Dakota State University, Jim Peterson
Southern Illinois University–Edwardsville, Michael J. Grossman
Southern Utah University, Paul R. Larson
State University of New York at New Paltz, Ronald G. Knapp
Texas State University–San Marcos, David R. Butler
The University of Memphis, Hsiang-te Kung
Towson University, Kent Barnes
United States Military Academy, Peter Siska
University of Calgary, Lawrence Nkemdirim

University of Cincinnati, Teri Jacobs
University of Colorado–Boulder, Jake Haugland
University of Colorado–Colorado Springs, Steve Jennings
University of Georgia, Andrew Grundstein
University of Missouri, C. Mark Cowell
University of Nevada–Reno, Franco Biondi
University of North Carolina–Charlotte, William Garcia
University of Oklahoma, Scott Greene
University of Saskatchewan, Dirk de Boer
University of Southern Mississippi, David Harms Holt
University of Tennessee, Derek J. Martin
University of Tennessee–Knoxville, Julie Y. McKnight
University of Wisconsin–Eau Claire, Christina M. Hupy
University of Wisconsin–Eau Claire, Joseph P. Hupy

University of Wisconsin–Eau Claire, Garry Leonard Running
University of North Dakota, Paul Todhunter
Weber State University, Eric C. Ewert

PHYSICAL GEOGRAPHY FOCUS GROUP AND SYMPOSIUM PARTICIPANTS

Ball State University, Petra Zimmermann
Blinn College, Rhonda Reagan
California State University–Los Angeles, Steve LaDochy
Georgia State University, Leslie Edwards

Indiana Purdue University–Indianapolis (IUPUI), Andrew Baker
Kansas State University, Doug Goodin
Mesa Community College, Steven Bass
Minnesota State University, Ginger L. Schmid
Northern Illinois University, Lesley Rigg
Northern Illinois University, Mike Konen
South Dakota State University, Bruce V. Millett
Texas A&M University, Steven Quiring
University of Alabama, Amanda Epsy-Brown
University of Colorado–Boulder, Peter Blanken
University of North Carolina–Greensboro, Michael Lewis
University of Oklahoma, Scott Greene
University of Wisconsin–Oshkosh, Stefan Becker

STEPHEN J. REYNOLDS

Stephen J. Reynolds received an undergraduate degree from the University of Texas at El Paso, and M.S. and Ph.D. degrees in geosciences from the University of Arizona. He then spent ten years directing the geologic framework and mapping program of the Arizona Geological Survey, completing a new *Geologic Map of Arizona*. Steve currently is a professor in the School of Earth and Space Exploration at Arizona State University, where he has taught various courses about regional geology, earth resources, evolution of landscapes, field studies, and teaching methods. He was president of the Arizona Geological Society and has authored or edited nearly 200 maps, articles, and reports on the evolution of Western North America. He also coauthored several widely used textbooks, including the award-winning *Exploring Geology* and *Exploring Earth Science*. His current science research focuses on regional geology, geomorphology, and resources of the Southwest. He has done science-education research on student learning in college science courses, especially the role of visualization. He was the first geoscientist with his own eye-tracking laboratory, where he and his students have researched student learning, including the role of textbooks and other educational materials. Steve is known for innovative teaching methods, has received numerous teaching awards, and has an award-winning website. As a National Association of Geoscience Teachers (NAGT) distinguished speaker, he traveled across the country presenting talks and workshops on how to infuse active learning and inquiry into large introductory geology classes. He is commonly an invited speaker to national workshops and symposia on active learning, visualization, and teaching.

ROBERT V. ROHLI

Robert Rohli received a B.A. in geography from the University of New Orleans, an M.S. degree in atmospheric sciences from The Ohio State University, and a Ph.D. in geography from Louisiana State University (LSU). He currently serves as professor of geography at LSU, coordinator of the Louisiana Geographic Education Alliance, and faculty director of the LSU Residential Colleges Program. Previously, he was assistant professor of geography at Kent State University (KSU) and regional climatologist at the Southern Regional Climate Center. His teaching and research interests are in physical geography, particularly synoptic and applied meteorology/climatology, atmospheric circulation variability, and hydroclimatology. He has taught Physical Geography, Climatology, Meteorology, World Climates, Methods in Synoptic Climatology, Applied Meteorology, Analysis of Spatial Data, Water Resources Geography, and others. Major themes in his teaching include the systems approach to physical geography, collaboration among students from different disciplines in producing group research projects, and development of applied problem-solving skills. He has been an active supporter of undergraduate education initiatives, including living-learning communities, LSU's Communication across the Curriculum program, improved teaching assessment methods, and outreach activities—especially those that promote geography. He has published more than 45 refereed research articles, mostly on topics related to synoptic or applied climatology, and over 20 loosely refereed manuscripts, encyclopedia articles, proceedings papers, and technical reports. He has also coauthored *Climatology*, a widely used textbook, and *Louisiana Weather and Climate*.

JULIA K. JOHNSON

Julia K. Johnson is currently a full-time faculty member in the School of Earth and Space Exploration at Arizona State University. Her M.S. and Ph.D. research involved structural geology and geoscience education research. She teaches introductory geoscience to more than 1,500 students per year, both online and in person, and supervises the associated in-person and online labs. She also coordinates the introductory geoscience teaching efforts of the School of Earth and Space Exploration, helping other instructors incorporate active learning and inquiry into large lecture classes. Julia coordinated an innovative project focused on redesigning introductory geology classes so that they incorporated more online content and asynchronous learning. This project was very successful in improving student performance, mostly due to the widespread implementation of concept sketches and partly due to Julia's approach of decoupling multiple-choice questions and concept-sketch questions during exams and other assessments. Julia is recognized as one of the best science teachers at ASU and has received student-nominated teaching awards and very high teaching evaluations in spite of her challenging classes. Her efforts have dramatically increased enrollments. She coauthored the widely used *Exploring Geology, Exploring Earth Science,* and publications on geology and science-education research, including an article in the *Journal of Geoscience Education* on concept sketches. She is the lead author of *Observing and Interpreting Geology*, an innovative laboratory manual in which all learning is built around a virtual world. She also developed a number of websites used by students around the world, including the *Visualizing Topography* and *Biosphere 3D* websites.

PETER R. WAYLEN

Peter Waylen is Professor of Geography and associate dean in the college of Liberal Arts and Sciences at the University of Florida. He holds a B.Sc. in geography from the London School of Economics, England, and a Ph.D. from McMaster University, Canada. He has also served as assistant professor at the University of Saskatchewan, visiting associate professor at the University of Waterloo, Canada, Hartley Visiting Research Fellow at the University of Southampton, England, and visiting scholar in the Department of Engineering Hydrology, University College Galway, Ireland. His teaching and research interests are in the fields of hydrology and climatology, particularly the temporal and spatial variability of risks of such hazards as floods, droughts, freezes, and heat waves, and the way in which these vary in the long run, driven by global-scale phenomena like ENSO. He has worked throughout Anglo- and Latin America, and several parts of Africa. He teaches Introductory Physical Geography, Principles of Geographic Hydrology, and Models in Hydrology, and is a former University of Florida Teacher of the Year. His research is principally interdisciplinary and collaborative with colleagues and students. It has been supported variously by the Natural Research Council of Canada, NSF, NOAA, NASA, and the Inter-American Institute for Global Change Research. Results appear in over 100 geography, hydrology, and climatology refereed outlets and book chapters.

MARK A. FRANCEK

Mark Francek is a geography professor at Central Michigan University (CMU). He earned his doctorate in geography from the University of Wisconsin-Milwaukee, his master's in geography from the University of South Carolina, and his bachelor's degree in geography and psychology from the State University College at Geneseo, New York. He has teaching and research interests in earth science education, physical geography, and soil science. Mark has pedaled three times across America and teaches biking geography field classes in and around the Great Lakes region and the Appalachian Mountains. He has authored and coauthored more than 30 scholarly papers, funded in part by the NSF and the State of Michigan, and has presented his research at national and state conferences. At CMU, Francek has served as acting director of the Environmental Studies Program, director of the Science and Technology Residential College, and now interim chair of the Geography Department. He has received state and national teaching awards, including the CMU Teaching Excellence Award, the Carnegie Foundation for the Advancement of Teaching Michigan Professor of the Year, the Presidents Council State Universities of Michigan Distinguished Professor of the Year, the National Council for Geographic Education Distinguished Teaching Award, and Michigan Science Teachers Association College Teacher of the Year. His "Earth Science Sites of the Week" Listserv, which highlights the best earth-science websites and animations, reaches thousands of K–16 educators from around the world.

Illustrators and Artists

CYNTHIA SHAW

Cynthia Shaw holds a B.A. in zoology from the University of Hawaii–Manoa as well as a master's in education from Washington State University, where she researched the use of guided illustration as a teaching and learning tool in the science classroom. Now focusing on earth science, mapping, and coral reef ecology, she writes and illustrates for textbooks and museums, and develops ancillary educational materials through her business, Aurelia Press. Her kids' novel, *Grouper Moon*, is used in many U.S. and Caribbean science classrooms and is making a real impact on shaping kids' attitudes toward fisheries conservation. Currently landlocked in Richland, Washington, Cynthia escapes whenever possible to travel, hike, and dive the reefs to field-sketch and do reference photography for her projects.

CHUCK CARTER

Chuck Carter has worked in the artistic end of the science and entertainment industries for more than 30 years. He helped create the popular computer game *Myst* in 1992. Chuck worked on more than two dozen video games as an artist, art director, computer graphics supervisor, and group manager. He has a decades-long relationship with *National Geographic* as an illustrator and helping launch National Geographic Online. Carter worked as a digital matte painter for science fiction shows like *Babylon 5, Crusade,* and *Mortal Kombat*, as well as art and animation for motion rides like Disney's Mission to Mars and Paramount's Star Trek: the Experience. His illustration clients include *Wired* magazine, *Scientific American*, and numerous book publishers. He is co-founder of Eagre Games Inc.

SUSIE GILLATT

Susie Gillatt grew up in Tucson, Arizona, where she received a bachelor of arts degree from the University of Arizona. She has worked as a photographer and in different capacities in the field of video production. She is president of Terra Chroma, Inc., a multimedia studio. Initially specializing in the production of educational videos, she now focuses on scientific illustration and photo preparation for academic books and journals. Many of the photographs in this book were contributed by Susie from her travels to experience different landscapes, ecosystems, and cultures around the world. For her own art, she especially enjoys combining photography with digital painting and exploring the world of natural patterns. Her award-winning art has been displayed in galleries in Arizona, Colorado, and Texas.

exploring physical
geography

The Nature of Physical Geography

THE EARTH HAS A WEALTH of intriguing features, from dramatic mountains to intricate coastlines and deep ocean trenches, from lush, beautiful valleys to huge areas of sparsely vegetated sand dunes. Above the surface is an active, ever-changing atmosphere with clouds, storms, and variable winds. Occupying all these environments is life. In this chapter and book, we examine the main concepts of physical geography, along with the tools and methods that physical geographers use to study the landscapes, oceans, climate, weather, and ecology of Earth.

The large globe spanning these two pages is a computer-generated representation of Earth, using data collected by several satellites. On land, brown colors depict areas of rock, sand, and soil, whereas green areas have a more dense covering of trees, bushes, grasses, and other vegetation. Oceans and lakes are colored blue, with greenish blue showing places where the water is shallow or where it contains mud derived from the land. Superimposed on Earth's surface are light-colored clouds observed by a different satellite, one designed to observe weather systems.

What are all the things you can observe from this portrait of our planet? What questions arise from your observations?

Most questions that arise from observing this globe are within the domain of *physical geography*. Physical geography deals with the landforms and processes on Earth's surface, the character and processes in oceans and other bodies of water, atmospheric processes that cause weather and climate, and how these various aspects affect life, and much more.

01.00.a2 Santorini, Greece

Natural hazards, including volcanic eruptions and earthquakes, are a major concern in many parts of the world. In the Greek Island of Santorini (◄), people live on the remains of a large volcano that was mostly destroyed in a huge eruption 3,600 years ago, an eruption that probably gave rise to the story of Atlantis.

What occurs during a volcanic eruption? Do all volcanoes erupt in the same way, and how can we recognize a volcano in the landscape?

01.00.a3 Morocco

The Sahara Desert, on the opposite side of the Mediterranean Sea from Greece, has a very different climate. Here is a very dry environment, resulting in huge areas covered by sand dunes (▲) with sparse vegetation.

What do the features of the landscape— the landforms—tell us about the surface processes that are forming and affecting the scenery? What causes different regions to have different climates, some that are hot and dry, and others that are cold and wet? Is the climate of the Sahara somehow related to the relative lack of clouds over this area, as shown on the globe?

TOPICS IN THIS CHAPTER

01.00.a4 Tibet

Water is the most important resource on the planet, and Earth's temperatures allow water to occur in three states of matter—solid, liquid, and vapor. Examine this photograph (◄) and identify all the ways in which water is expressed on the surface and in the atmosphere. Is some water likely present but not visible? Geographers are concerned with where resources are, what causes a resource to be where it is, and how to reconcile the inevitable economic, environmental, and cultural trade-offs involved in using a resource.

How does water occur in the atmosphere, how is its presence expressed, and what is its role in severe weather? How does water occur and move on Earth's surface, and what landforms result from running water?

01.00.a5 Indonesia

Oceans cover about 70% of Earth's surface. Ocean temperatures, currents, and salinity all play a major role in global weather, climate, and the livability of places, even for those far from the coast. The oceans and nearby lands (▲) represent important habitats for plants and animals, which can be greatly impacted by human activities.

How do satellites help us measure the temperature, salinity, and motion of the oceans, and how do changes in any of these factors affect plants and animals that live in or near the sea?

01.00.a1

The Ancient and Modern Discipline of Geography

Geographers seek to understand the Earth. They do this by formulating important and testable questions about the Earth, employing principles from both the natural and social sciences. Geographers use these principles to portray features of the Earth using maps and technologically intensive tools and techniques that are distinctly geographical. Geographers synthesize the diverse information revealed by these tools to investigate the interface between the natural and human environments. The study of the spatial distribution of natural features and processes occurring near Earth's surface, especially as they affect and are affected by humans, is physical geography.

The ancient discipline of geography is especially relevant in our modern world, partly because of the increasing recognition that many problems confronting society involve complex interactions between natural and human dimensions. Such problems include the spatial distribution and depletion of natural resources; contamination of air, water, and soils; susceptibility of areas to landslides, flooding, and other natural disasters; formation of and damage caused by hurricanes, tornadoes, and other severe weather; the current and future challenges of global environmental change; and the environmental implications of globalization. The topics and questions introduced on these pages provide a small sample of the aspects investigated by physical geographers and are discussed more fully in the rest of the book. We hope you enjoy the journey learning about the fascinating planet we call home.

1.1 What Is Physical Geography?

PHYSICAL GEOGRAPHY IS THE STUDY of spatial distributions of phenomena across the landscape, processes that created and changed those distributions, and implications for those distributions on people. Geography is both a natural and a social science. Geographers think broadly, emphasizing interconnections and complex issues, solving complicated problems such as resource management, environmental impact assessment, spread of disease, and urban planning. Although many such occupations do not have the title of *geographer*, they require a geographic perspective. Let's have a closer look at what the geographic perspective entails.

A What Approach Do Geographers Use to Investigate Important Issues?

Geographers approach problems from different perspectives than other natural and social scientists. Specifically, geographers think *spatially*, meaning they emphasize the setting, such as location, in addressing problems, and *holistically*, integrating ideas from a wide variety of the natural and social sciences. In many ways, it is not *what* is studied that makes it geography, but instead *how* it is studied. The decision of whether to drill for oil in Alaska's Arctic National Wildlife Refuge (ANWR) is a complicated issue that can be best understood using the geographic approach.

1. This figure (▶) shows a three-dimensional perspective of the central part of ANWR, looking south with the ice-covered Arctic Ocean in the foreground. ANWR is well known for its abundant caribou and other Arctic animals. Before reading on, examine this scene and think about all the information you would need if you wanted to understand how drilling for oil and gas might impact the caribou.

2. To understand this issue, you might ask a series of questions. Where do the caribou live? Since they migrate seasonally, where are they at different times of the year? What do they eat, where are these foods most abundant, and what factors control these abundances? Where is water available, and how much rain and snow do different parts of the region receive? Is the precipitation consistent from year to year? When is the mating season, and where do the mothers raise their young?

01.01.a1

3. You could also ask questions about the subsurface oil reserves. Where is the oil located, and what types of facilities will be required to extract and transport the oil? How much land will be disturbed by such activities, and how will this affect the caribou?

4. The issues of ANWR nicely illustrate why we would use a geographic approach. Most of the questions we asked here have a *spatial component*, as indicated by the word "where," and could be best answered with some type of map. The questions also have an explicit or implicit societal component, such as how development could affect the traditional way of life of the native people of the region.

01.01.a2

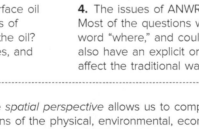

Prudhoe Bay Oil Field ○

ANWR

500 km N

01.01.a3 ANWR, AK

5. The *spatial perspective* allows us to compare the locations of the physical, environmental, economic, political, and cultural attributes of the issue. ANWR (◀) is the large area outlined in orange. Its size is deceptive since Alaska is huge (by far the largest state in the U.S.). For comparison, ANWR is only slightly smaller than the state of South Carolina.

6. Directly to the west of ANWR is the Prudhoe Bay oil field, the largest oil field in North America. Not all of ANWR is likely to contain oil and natural gas, and an assessment of the oil resources by the U.S. Geological Survey (USGS) identified the most favorable area as being near the coast. To consider the question about oil drilling, we would want to know where this favorable area is, how much land will be disturbed by drilling and associated activities, when these disturbances will occur, and how these compare with the location of caribou at different times of the year, especially where they feed, mate, and deliver their young.

7. The *holistic perspective* allows us to examine the interplay between the environment and the aesthetic, economic, political, and cultural attributes of the problem. Most of ANWR is a beautiful wilderness area (▲), as well as being home to caribou, native people, and various plants and animals.

B How Does Geography Influence Our Lives?

Observe this photograph, which shows a number of different features, including clouds, snowy mountains, slopes, and a grassy field with horses and cows (the small, dark spots). For each feature you recognize, think about what is there, what its distribution is, and what processes might be occurring. Then, think about how these factors influence the life of the animals and how they would influence you if this were your home.

01.01.b1 Henry Mtns., UT

1. The snow-covered mountains, partially covered with clouds, indicate the presence of water, an essential ingredient for life. The mountains have a major influence on water in this scene. Melted snow flows downhill toward the lowlands, to the horses and cows. The elevation and shape of the land influence the spatial distribution and type of precipitation (rain, snow, and hail) and the pattern of streams that develop to drain water off the land.

2. The horses and cows roam on a flat, grassy pasture, avoiding slopes that are steep or barren of vegetation. The steepness of slopes reflects the strength of the rocks and soils, and the flat pasture resulted from loose sand and other materials that were laid down during flooding along a desert stream. The distribution of vegetation is controlled by steepness of slopes, types of soils and other material, water content of the soil, air temperatures, and many other factors, all of which are part of physical geography. The combined effect of such factors in turn affect, and are affected by, the human settlements in the area to make every place, including this one, distinctive and unique.

01.01.b2

01.01.b3

01.01.b4

5. This image shows the shape of the land across the region, including the mountains (the pasture is on the left part of the map). Colors indicate the average amount of precipitation, with green showing the highest amounts. The mountains, on average, receive the most rain and snow.

3. A better view of the spatial distribution of the green pasture is provided by this aerial photograph (a photograph taken from the air, like from a plane or drone). This view of the pasture and adjacent areas reveals the shape of the pasture, and we could measure its length, width, and area. Such measurements would help us decide how many horses and cows the land could support.

4. Geographers calculate various measures of the landscape, like the steepness of slopes, and then overlay this information on the original map or image. In the figure above, red shading shows the steepest slopes, along and below the pinkish cliff. Yellow and green indicate less steep slopes, and relatively flat areas are unshaded. Such a map would help us decide which areas could be new pastures.

Before You Leave This Page

✔ Describe the geographic approach.

✔ List some examples of information used by physical geographers and how these types of information could influence our lives.

1.1

1.2 How Do We Investigate Geographic Questions?

PHYSICAL GEOGRAPHERS STUDY DIVERSE PROBLEMS, ranging from weather systems and climate change to ocean currents and landscape evolution. The types of data required to investigate each of these problems are equally diverse, but most geographers try to approach the problem in a similar, objective way, guided by spatial information and relying on various geographic tools. Geography utilizes approaches from the natural and social sciences, blending them together in a geographic approach. Like other scientists, geographers pose questions about natural phenomena and their implications, propose a possible explanation (hypothesis) that can be tested, make predictions from this hypothesis, and collect data needed to critically evaluate whether the hypothesis passes the tests.

A How Do Geographers Approach Problems?

Geographers ask questions like the following:

- Where is it?
- Why is it where it is?
- How did it get where it is?
- Why does it matter where it is?
- How does "where it is" influence where other things are and why they are there?

The conceptual basis of these questions lies in the notion that the *location* of something affects, and is a product of, other features or processes in both the natural and human environment, and of interactions between the natural and human environments. Natural and human phenomena are constantly changing and constantly impacting other features in new ways, influencing aspects like site selection and risk of natural hazards. To address such complex issues, we use a variety of tools and methods, such as maps, computer-simulation models, aerial photographs, satellite imagery, statistical methods, and historical records. The figure to the right illustrates some aspects to consider.

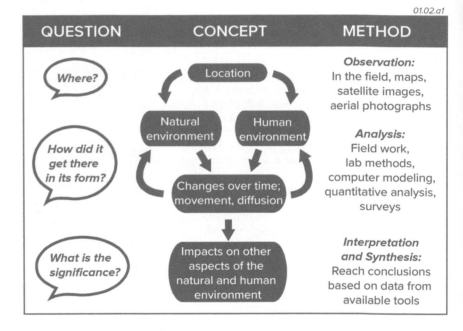

01.02.a1

B What Is the Difference Between Qualitative and Quantitative Data?

Geographers approach problems in many ways, asking questions about Earth processes and collecting data that help answer these questions. Some questions can be answered with qualitative data, but others require quantitative data, which are numeric and are typically visualized and analyzed using data tables, calculations, equations, and graphs.

01.02.b1 Augustine Island, AK

01.02.b2 Augustine Island, AK

01.02.b3 Augustine Island, AK

When Augustine volcano in Alaska erupts, we can make various types of observations and measurements. Some observations are *qualitative*, like descriptions, and others are measurements that are *quantitative*. Both types of data are essential for documenting natural phenomena.

Qualitative data include descriptive words, labels, sketches, or other images. We can describe this picture of Augustine volcano with phrases like "contains large, angular fragments," "releases steam," or "the slopes seem steep and unstable." Such phrases can convey important information about the site.

Quantitative data involve numbers that represent measurements. Most result from scientific instruments, such as this thermal camera that records temperatures on the volcano, or with measuring devices like a compass. We could also collect quantitative measurements about gases released into the air.

C How Do We Test Alternative Explanations?

Science proceeds as scientists explore the unknown—making observations and then systematically investigating questions that arise from observations that are puzzling or unexpected. Often, we try to develop several possible explanations and then devise ways to test each one. The normal steps in this *scientific method* are illustrated below, using an investigation of groundwater contaminated by gasoline.

Steps in the Investigation

Observations

1. Someone makes the *observation* that groundwater from a local well near an old buried gasoline tank contains gasoline. The first step in any investigation is to make observations, recognize a problem, and state the problem clearly and succinctly. Stating the problem as simply as possible simplifies it into a more manageable form and helps focus our thinking on its most important aspects.

Questions Derived from Observations

2. The observation leads to a *question*—Did the gasoline in the groundwater come from a leak in the buried tank? Questions may be about what is happening currently, what happened in the past, or, in this case, who or what caused a problem.

Proposed Explanations and Predictions from Each Explanation

3. Scientists often propose several explanations, referred to as *hypotheses*, vetted by initial evidence, to explain what they observe. A hypothesis is a causal explanation that can be tested, either by conducting additional investigations or by examining data that already exist.

----

4. One explanation is that the buried tank is the source of contamination.

5. Another explanation is that the buried tank is not the source of the contamination. Instead, the source is somewhere else, and contamination flowed into the area.

6. We develop *predictions* for each explanation. A prediction for the explanation in number 4 might be that the tank has some kind of leak and should be surrounded by gasoline. Also, if the explanation in number 4 is true, the type of gasoline in the tank should be the same as in the groundwater. Next, we plan some way to *test* the predictions, such as by inspecting the tank or analyzing the gasoline in the tank and groundwater.

Results of Investigation

7. To study this problem, an early step is to compile all the necessary data. This might include maps showing the location of water wells, the direction of groundwater flow, and locations of gas stations and other possible sources of gasoline. In our case, investigation discovered no holes in the tank or any gasoline in the soil around the tank. Records show that the tank held leaded gasoline, but gasoline in the groundwater is unleaded. We compare the results of any investigation with the predictions to determine which possible explanation is most consistent with the new data.

Conclusions

8. Data collected during the investigation support the conclusion that the buried tank is not the source of contamination. Any explanation that is inconsistent with data is probably incorrect, so we pursue other explanations. In this example, a nearby underground pipeline may be the source of the gasoline. We can devise ways to evaluate this new hypothesis by investigating the pipeline. We also can revisit the previously rejected hypothesis if we discover a new way in which it might explain the data.

01.02.c1

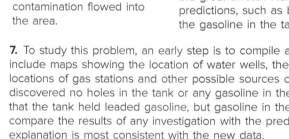

9. The goal is to collect data, assemble information, and draw conclusions without letting our personal bias interfere with carrying out good science. We want to reach the explanation that best explains all the data. Few things are ever "proved" in science, some can be "disproved," but generally we are left to weigh the pros and cons of several still-viable explanations. We choose the one that, based on the data, is most likely to be correct.

Before You Leave This Page

✓ Summarize some of the aspects commonly considered using a geographic approach.

✓ Explain the difference between qualitative and quantitative data, providing examples.

✓ Explain the logical scientific steps taken to critically evaluate a possible explanation.

1.3 How Do Natural Systems Operate?

EARTH HAS A NUMBER OF SYSTEMS in which matter and energy are moved or transformed. These involve processes of the solid Earth, water in all its forms, the structure and motion of the atmosphere, and how these three domains (Earth, water, and air) influence life. Such systems are *dynamic*, responding to any changes in conditions, whether those changes arise internally *within* the system or are imposed externally from *outside* the system.

A What Are the Four Spheres of Earth?

Earth consists of four overlapping spheres—the atmosphere, biosphere, hydrosphere, and lithosphere—each of which interacts with the other three spheres. The atmosphere is mostly gas, but includes liquids (e.g., water drops) and solids (e.g., ice and dust). The hydrosphere represents Earth's water, and the lithosphere is the solid Earth. The biosphere includes all the places where there is life—in the atmosphere, on and beneath the land, and on and within the oceans.

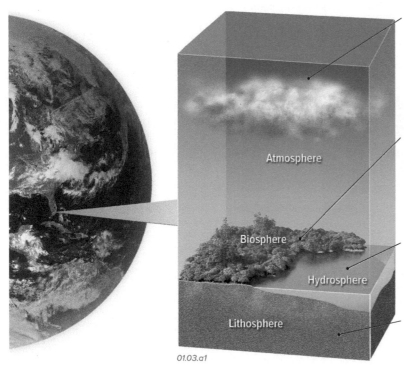

01.03.a1

1. The *atmosphere* is a mix of mostly nitrogen and oxygen gas that surrounds Earth's surface, gradually diminishing in concentration out to a distance of approximately 100 kilometers, the approximate edge of outer space. In addition to gas, the atmosphere includes clouds, precipitation, and particles such as dust and volcanic ash. The atmosphere is approximately 78% nitrogen, 21% oxygen, less than 1% argon, and smaller amounts of carbon dioxide and other gases. It has a variable amount of water vapor, averaging less than 4%.

2. The *biosphere* includes all types of life, including humans, and all of the places it can exist on, above, and below Earth's surface. In addition to the abundant life on Earth's surface, the biosphere extends about 10 kilometers up into the atmosphere, to the bottom of the deepest oceans, and downward into the cracks and tiny spaces in the subsurface. In addition to visible plants and animals, Earth has a large population of diverse microorganisms.

3. The *hydrosphere* is water in oceans, glaciers, lakes, streams, wetlands, groundwater, moisture in soil, and clouds. Over 96% of water on Earth is salt water in the oceans, and most fresh water is in ice caps, glaciers, and groundwater, not in lakes and rivers.

4. The *lithosphere* refers generally to the solid upper part of the Earth, including Earth's crust. Water, air, and life extend down into the lithosphere, so the boundary between the solid Earth and other spheres is not distinct, and the four spheres overlap.

B What Are Open and Closed Systems?

Many aspects of Earth can be thought of as a system—a collection of matter, energy, and processes that are somehow related and interconnected. For example, an air-conditioning system consists of some mechanical apparatus to cool the air, ducts to carry the cool air from one place to another, a fan to move the air, and a power source. There are two main types of systems: *open systems* and *closed systems*.

01.03.b1 Noxubee NWR, MS

01.03.b2

1. An *open system* allows matter and energy to move into and out of the system. A tree (◄) is an open system, taking in water and soil-derived nutrients, extracting carbon dioxide from the air to make the carbon-rich wood and leaves, sometimes shedding those leaves during the winter as shown here, and expelling oxygen as a by-product of photosynthesis, fueled by externally derived energy from the Sun.

2. A *closed system* does not exchange matter, or perhaps even energy, with its surroundings. The Earth as a whole (►) is fundamentally a closed system with regard to matter, except for the escape of some light gases into space and the arrival of occasional meteorites. It is an open system for energy, which is gained via sunlight and can be lost to space.

C How Do Earth Systems Operate?

Systems consist of matter and energy, and they respond to internally or externally caused changes in matter and energy, as a tree responds to a decrease in rain (matter) or colder temperatures during the winter (energy). Systems can respond to such changes in various ways, either reinforcing the change or counteracting the change.

System Inputs and Responses

1. One of Earth's critical systems involves the interactions between ice, surface water, and atmospheric water. This complex system, greatly simplified here (▶), remains one of the main challenges for computer models attempting to analyze the causes and possible consequences of climate change.

2. Liquid water on the surface *evaporates* (represented by the upward-directed blue arrows), becoming water vapor in the atmosphere. If there is enough water vapor, small airborne droplets of water accumulate, forming the low-level clouds illustrated here.

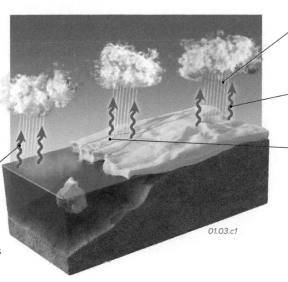

01.03.c1

3. Under the right conditions, the water freezes, becoming snowflakes or hail, which can fall to the ground. Over the centuries, if snow accumulates faster than it melts, the snow becomes thick and compressed into ice, as in *glaciers,* which are huge, flowing fields of ice.

4. The water molecules in snow and ice can return directly to the atmosphere via several processes.

5. If temperatures are warm enough, snow and ice can melt, releasing liquid water that can accumulate in streams and flow into the ocean or other bodies of surface water. Alternatively, the meltwater can evaporate back into the atmosphere. Melting also occurs when icebergs break off from the glacier.

6. The movement of matter and energy carried in the various forms of water is an example of a *dynamic system* — a system in which matter, energy, or both, are constantly changing their position, amounts, or form.

Feedbacks

7. The system can respond to changes in various ways, which can either reinforce the effect, causing the overall changes to be amplified (increased in effect), or partially or completely counteract the effect, causing changes to be dampened (decreased in effect). Such reinforcements or inhibitors are called *feedbacks.*

8. In our example, sunlight shines on the ice and water. The ice is relatively smooth and light-colored, reflecting much of the Sun's energy upward, into the atmosphere or into space. In contrast, the water is darker and absorbs more of the Sun's energy, which warms the water.

9. If the amount of solar energy reaching the surface, or trapped near the surface, increases, for whatever reason, this may cause more melting of the ice. As the front of the ice melts back, it exposes more dark water, which absorbs more heat and causes even more warming of the region. In this way, an initial change (warming) triggers a response that causes even more of that change (more warming). Such a reinforcing result is called a *positive feedback.*

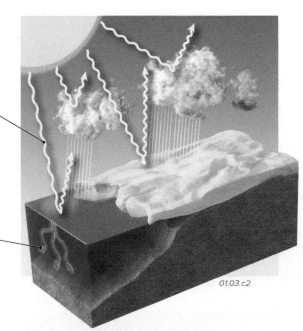

01.03.c2

10. The warming of the water results in more evaporation, moving water from the surface to the atmosphere, which in turn may result in more clouds. Low-level clouds are highly reflective, so as cloud cover increases they intercept more sunlight, leading to less warming. This type of response does not reinforce the change but instead dampens it and diminishes its overall effect. This dampening and resultant counteraction is called a *negative feedback.*

11. As this overly simplified example illustrates, a change in a system can be reinforced by positive feedbacks or stifled by negative ones. Both types of feedbacks are likely and often occur at the same time, each nudging the system toward opposite behaviors (e.g., overall warming or overall cooling). Feedbacks can leave the system largely unchanged, or the combined impact of positive and negative feedbacks can lead to a stable but gradually changing state, a condition called *dynamic equilibrium.*

Before You Leave This Page

- ✓ Describe Earth's four spheres.
- ✓ Explain what is meant by open and closed systems.
- ✓ Sketch and explain examples of positive and negative feedbacks.

1.3

1.4 What Are Some Important Earth Cycles?

MATTER AND ENERGY MOVE within and between each of the four spheres. A fundamental principle of all natural sciences is that energy and matter can be neither created nor destroyed, but only transferred from one form to another—the *First Law of Thermodynamics*. A second principle is that energy and matter tend to become dispersed into a more uniform spatial distribution—the *Second Law of Thermodynamics*. As a result, matter and energy are stored, moved, dispersed, and concentrated as part of natural *cycles*, in which material and energy move back and forth among various sites within the four spheres.

A What Is Cycled and Moved in the Atmosphere?

Atmospheric processes involve the redistribution of *energy* and *matter* from one part of the atmosphere to another. Moving air masses have *momentum*, which can be transferred from one object to another.

Energy
01.04.a1

Storage and transfer of energy are the drivers of Earth's climate and weather. Energy can be moved from one part of the atmosphere to another, such as by air currents associated with storms (▲). Also, energy is released or extracted from the local environment when water changes from one state of matter to another, such as from a liquid to a gas.

Matter
01.04.a2

Water in all of its forms, along with other matter, moves globally, tending to disperse, but other factors prevent an even spatial distribution. As a result, some regions are more humid and cloudy than others, as shown in this satellite image (▲) of water vapor (blue is more, brown is less). Also, water cycles between vapor, liquid, and solid states.

Momentum
01.04.a3

Moving air masses have mass and velocity, so they have *momentum*, which is defined as mass times velocity. A dense, fast-moving dust storm has more momentum than a gentle breeze in dust-free air. Winds near the surface are slowed by interactions with trees, hills, buildings, etc. Winds aloft are faster (▲) and can transfer their momentum downward.

B How Are Matter and Energy Moved in the Hydrosphere?

Many processes in the Earth occur as part of a cycle, a term that describes the movement of matter and energy between different sites in Earth's surface, subsurface, and atmosphere. The most important of these is the *hydrologic cycle*, which involves local-to-global-scale storage and circulation of water and associated energy near Earth's surface.

1. Water vapor in the atmosphere can form drops, which can remain suspended in the air as clouds or can fall to the ground as *precipitation*. Once on the surface, water can coat rocks and soils, and it causes these solids to decompose and erode.

6. Water evaporates from the oceans, surface waters on land, and from soils and plants, returning to the atmosphere and completing the hydrologic cycle. Phase changes between a solid, liquid, and gas involve energy transformations.

2. When winter snows don't melt completely, as is common at higher elevations and at polar latitudes, ice accumulates in glaciers. Glaciers transport sediment and carve the underlying landscape.

3. Moving water on the surface can flow downhill in *streams*, encountering obstacles, like solid rock and loose debris. The flowing water and the material it carries breaks these obstacles apart and picks up and transports the pieces. Flowing water is the most important agent for sculpting Earth.

4. The uppermost part of oceans is constantly in motion, partly due to friction (and transfer of momentum) between winds in the atmosphere and the surface of the oceans. Winds in the oceans cause waves that erode and shape shorelines.

5. Water can sink into the ground and travel through cracks and other empty spaces in rocks and soils, becoming *groundwater*. Groundwater can react chemically with rocks through which it flows, dissolving or depositing material. It typically flows toward lower areas, where it may emerge back on Earth's surface as springs.

01.04.b1

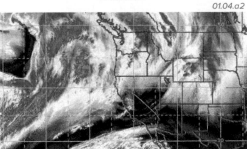
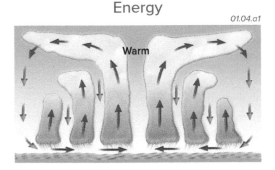

Precipitation · Streams · Evaporation · Ocean Currents · Groundwater Flow

C How Does the Rock Cycle Affect Materials of the Lithosphere?

Matter and energy are also stored and moved on Earth's surface and subsurface. The *rock cycle*, summarized below and discussed in more detail later, describes the movement of matter and energy on and below Earth's surface at timescales from seconds to billions of years, involving such processes as erosion, burial, melting, and uplift of mountains.

1. *Weathering:* Rock is broken apart or altered by chemical reactions when exposed to sunlight, rain, wind, plants, and animals. This weathering creates loose pieces of rock called *sediment.*

2. *Erosion and Transport:* Sediment is then stripped away by *erosion,* and moved (transported) by gravity, glaciers, flowing water, or wind.

3. *Deposition:* After transport, the sediment is laid down, or deposited, at any point along the way, such as beside the stream, or when it reaches a lake or the sea.

01.04.c1

4. *Burial and Formation of Rock:* Sediment is eventually buried, compacted by the weight of overlying sediment, and perhaps cemented together by chemicals in the water to form a harder rock.

8. *Uplift:* Deep rocks may be uplifted back to the surface where they are again exposed to weathering.

7. *Solidification:* As magma cools, either at depth or after being erupted onto the surface, it begins to crystallize and solidify into rock.

6. *Melting:* A rock exposed to high temperatures may melt and become molten, forming magma.

5. *Deformation and Related Processes:* Rock can be subjected to strong forces that squeeze, bend, and break the rock, causing it to deform. The rock might also be heated and deformed so much that it is changed into a new kind of rock.

D What Cycles and Processes Are Important in the Biosphere?

The biosphere includes life and all of the places it exists. It overlaps the atmosphere, hydrosphere, and lithosphere, extending from more than 10 kilometers above the surface to more than 10 kilometers beneath sea level, both on the seafloor and within Earth's subsurface. Life interacts with the other three spheres, forming a number of important cycles, several of which are described later.

1. Plants exchange gases with the atmosphere. They extract carbon dioxide (CO_2) from the atmosphere and use the carbon for their leaves, stems, roots, spines, and other leafy or woody parts. Plants also release oxygen, a key ingredient in life. Life on Earth is currently the main source of the oxygen and CO_2 in the atmosphere.

2. The Sun is the ultimate source of energy for photosynthesis, as well as movement of matter and energy in the atmosphere and most movement of material on Earth's surface.

3. Life interacts with the hydrologic cycle. Plants take in water from the rocks and soil, which may have arrived from lakes and other bodies of surface water, or directly from the atmosphere, as during precipitation. Plants then release some water back into the environment.

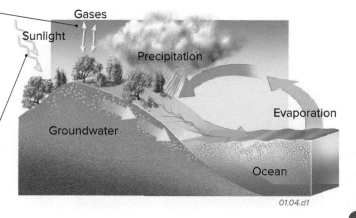

01.04.d1

4. Life also interacts with aspects of the rock cycle. Plants help break down materials on Earth's surface, such as when a plant root pushes open fractures in rocks and soil. Plants help stabilize soils, inhibiting erosion, by slowing down the flow of water, allowing it to remain in contact with rocks and soil longer. This increases the rate at which weathering breaks down materials. Humans have altered the surface of Earth by removing vegetation that would compete with crops, villages, and cities.

5. Water, nutrients, and other materials are stored and cycled through the biosphere, at local to global scales. We use the term *biogeochemical cycle* to indicate that plants, animals, and bacteria, in addition to chemical and physical processes, are involved in the cycling of a chemical substance through different parts of the environment. For example, the movement of carbon between the biosphere, atmosphere, hydrosphere, and lithosphere is a biogeochemical cycle known as the *carbon cycle.*

Before You Leave This Page

☑ Describe some examples of transfer of energy, matter, and momentum in the atmosphere.

☑ Sketch and describe the hydrologic cycle, the rock cycle, and how the biosphere exchanges matter with the other spheres.

1.5 How Do Earth's Four Spheres Interact?

ENERGY AND MATTER MOVE between the land, water, atmosphere, and biosphere—between the four spheres. There are various expressions of these interactions, many of which we can observe in our daily lives. In addition to natural interactions, human activities, such as the clearing of forests, can affect interactions between the spheres. Changes in one component of one sphere can cause impacts that affect components of other spheres.

A What Are Some Examples of Energy and Matter Exchanges Between Two Spheres?

The four spheres interact in complex and sometimes unanticipated ways. As you read each example below, think of other interactions—observable in your typical outdoor activities—that occur between each pair of spheres.

Atmosphere-Hydrosphere

01.05.a1 Indonesia

The Sun's energy evaporates water from the ocean and other parts of the hydrosphere, moving the water molecules into the atmosphere. The water vapor can remain in the atmosphere or can condense into tiny drops that form most clouds. Under certain conditions, the water returns to the surface as precipitation.

Atmosphere-Lithosphere

01.05.a2 Mount Veniaminof, AK

Active volcanoes emit gases into the atmosphere, and major eruptions release huge quantities of steam, sulfur dioxide, carbon dioxide, and volcanic ash. In contrast, weathering of rocks removes gas and moisture from the atmosphere. Precipitation accumulates on the land, where it can form standing water, groundwater, or erosion-causing runoff.

Atmosphere-Biosphere

01.05.a3 Indonesia

Plants and animals utilize precipitation from the atmosphere, and some plants can extract moisture directly out of the air without precipitation. Broad-scale circulation patterns in the atmosphere are a principal factor in determining an area's climate, and the climate directly controls the types of plants and animals that inhabit a region.

Hydrosphere-Lithosphere

01.05.a4 Glacier Parkway, Alberta, Canada

Channels within a stream generally bend back and forth as the water flows downhill. The water is faster and more energetic in some parts of the stream than in others, and so erodes into the streambed and riverbank. In less energetic sections, sediment will be deposited on the bed, like the gravel in this photograph. Earth's surface can be uplifted or dropped down, as during an earthquake, and the resulting changes can influence the balance of erosion and deposition.

Hydrosphere-Biosphere

01.05.a5 Raja Ampat, Indonesia

Oceans contain a diversity of life, from whales to algae, and everything in between. Coral reefs represent an especially life-rich environment, formed when living organisms extract materials dissolved in or carried by seawater to produce the hard parts of corals, shells, and sponges. At greater ocean depths, where waters are colder, shells and similar biological materials dissolve, transferring material back to the seawater.

Lithosphere-Biosphere

01.05.a6 Near Badlands National Park, SD

The clearest interaction between the lithosphere and biosphere is the relationship between plants and soils. The type of soil helps determine the types of plants that can grow, and in turn depends on the types of starting materials (rocks and sediment), the geographic setting of the site (e.g., slope versus flat land), climate, and other factors. Plants remove nutrients from the soil but return material back to the soil through roots and annual leaf fall, or plant death and decay.

B   To What Extent Do Humans Influence Interactions Between the Spheres?

Anyone who has flown in an airplane or spent some time using Google Earth® appreciates the amazing amount of human influence on the landscape. The intent of development is almost always to improve the human condition, but the complex chain reaction of impacts that cascade through the system can cause unintended and often harmful impacts elsewhere in one or more of the four spheres, as illustrated in the examples below. Some consequences of human impacts are not felt immediately but only appear much later, after the activity has continued for many years.

01.05.b1 Western Alabama

01.05.b2 Grand Coulee Dam, WA

01.05.b3 Phoenix, AZ

Humans clear forests, a critical part of the biosphere, to provide lumber and grow food. In addition to the loss of habitat for plants and animals, deforestation reduces the amount of CO_2 that can be extracted out of the atmosphere and stored in the carbon-rich trunks, branches, and leaves of plants. Removing plant cover also causes increased runoff, which enhances soil erosion and leads to the additional loss of plant cover—an unintended consequence and a positive feedback.

Over 80,000 dams exist in the U.S., providing water supplies, generating electricity, protecting towns from flooding, and providing recreational opportunities. Dams also alter the local water balance by interrupting the normal seasonal variations in flows of water and by capturing silt, sand, wood, and other materials that would normally go downstream. Construction and filling of the reservoir disrupts ecosystems, displaces people, and threatens or destroys plant and animal communities.

Local warming of the atmosphere occurs near cities because of normal urban activities (lighting, heating, etc.) and because many urban materials, like dark asphalt, capture and store more heat than natural open space. Heat is also released from car exhausts and industrial smokestacks. Non-natural drainage systems cause rapid accumulation and channeling of water. Development infringes on natural plant and animal communities, disturbs or covers soil, and alters erosion rates.

Geography in Our Modern World

The science of geography has ancient origins, arising from the need of early civilizations for maps for navigation, planning, and other purposes. The examples above illustrate current societal issues that should also be considered from a spatial perspective—the percentage decrease in the area of rain forests, the outline of areas that will be flooded by construction of a dam, or the patterns of population growth and the resulting changes in local temperatures. Understanding location and spatial distributions is as important in our modern world as it was in ancient times, because these factors are crucial in understanding the environment or identifying possible sites for any human activity. For example, what spatial factors should be considered when planning a new subdivision or a new business? Geographic factors can be the difference between success and failure, and an understanding of natural and human environments around the world is as important as ever. The map shown here depicts the average precipitation in the lower 48 states, with purple and blue designating the highest average annual precipitation, red and orange indicating the lowest precipitation, and green and yellow designating intermediate precipitation. How would you describe some of the main patterns? Where are the highest and the lowest precipitation amounts? Why are these wet or dry areas located where they are? What are some implications of the spatial distribution of high-precipitation versus low-precipitation regions? How might these variations in precipitation influence agriculture or the water supplies of the growing desert cities of the U.S. Southwest? Geographers address these and many other types of questions as part of their work.

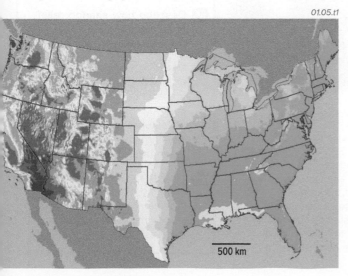
01.05.t1
500 km

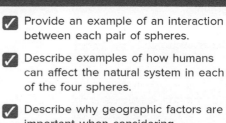

Before You Leave This Page

✓ Provide an example of an interaction between each pair of spheres.

✓ Describe examples of how humans can affect the natural system in each of the four spheres.

✓ Describe why geographic factors are important when considering environmental issues or when evaluating potential sites for a new agricultural area or business.

1.5

1.6 How Do We Depict Earth's Surface?

EARTH'S SURFACE DISPLAYS various features, including mountains, hillslopes, and river valleys. We commonly represent such features on the land surface of an area with a *topographic map* or *shaded-relief map*, each of which is useful for certain purposes. Some maps allow us to visualize the landscape and navigate across the land, whereas others permit the quantitative measurement of areas, directions, and steepness of slopes.

A How Do Maps Help Us Study Earth's Surface?

Maps are the primary way we portray the land surface. Some maps depict the shape and elevation of the land surface, whereas others, like a soil map, represent the materials on that surface. Views of SP Crater in northern Arizona provide a particularly clear example of the relationship between the land surface and different types of maps.

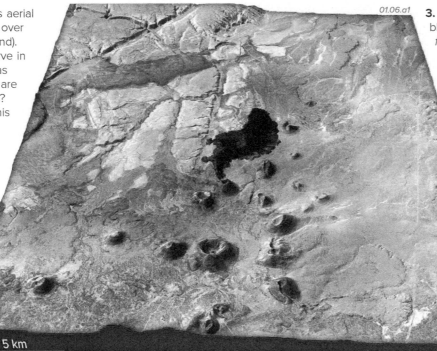

01.06.a1

1. This perspective view has aerial photography superimposed over topography (shape of the land). What features do you observe in the topography? Which areas are high in elevation? What are the most distinctive features? Take a minute to observe this scene before reading on.

2. The area has distinct, cone-shaped hills surrounded by broad, less steep areas. The hills are small volcanoes, which formed when fragments of molten rock were ejected into the air and settled around a volcanic vent.

3. In the center of the area is a nearly black feature, which is a solidified *lava flow* formed when fluid magma erupted onto the surface in the last 5,000 years. The volcano at the southern end of the lava flow is named *SP Crater*, and is well known to many physical geographers.

4. Examine other features in the scene. Note the light-gray areas in the upper left parts of the image, and the linear features formed by fractures that cut across the gray rocks. Different materials are forming different types of landscapes. This entire area has a relatively dry climate, with few trees to obscure the landform features.

5 km

01.06.a2 SP Crater, northern AZ

01.06.a3 SP Crater, northern AZ

5. This aerial photograph (◄) shows SP Crater and the dark lava flow that erupted from the base of the volcano. Note that the slopes of the volcano are much steeper than those of the surrounding land.

6. This photograph (▲), taken from the large crater south of SP Crater, shows the crater (on the left) and several other volcanoes. The view is toward the north. Try to match some of the features in this photograph with those shown in the larger perspective view. In starting such activities, it is always best to begin with a known feature (in this case SP Crater).

Before You Leave These Pages

✓ Describe how shaded relief and topographic maps depict the surface.

✓ Describe what contours on a topographic map represent, how the shapes of contours reflect the shapes of features, and how contour spacing indicates the steepness of a slope.

✓ Sketch and describe the meaning of the terms *elevation*, *depth*, *relief*, and *slope*; include how we express steepness of slope.

7. A *shaded-relief map* (▼) emphasizes the shape of the land by simulating light and dark shading on the hills and valleys. The individual hills on this map are volcanoes. The area is cut by straight and curving stream valleys that appear as gouges in the landscape. Simulated light comes from the upper left corner of the image.

8. A *topographic map* (▼) shows the elevation of the land surface with a series of lines called *contours*. Each contour line follows a specific elevation on the surface. Standard shaded-relief maps and topographic maps depict the shape of the land surface but give no specific and direct information about what lies beneath.

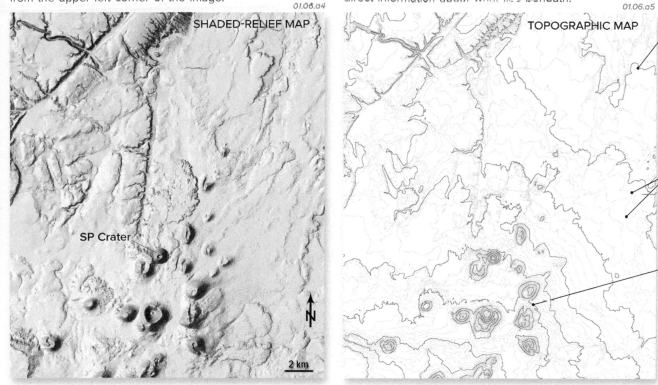

SHADED-RELIEF MAP

01.06.a4

SP Crater

2 km

TOPOGRAPHIC MAP

01.06.a5

9. Most topographic maps show every fifth contour with a darker line, to emphasize the broader patterns and to allow easier following of lines across the map. These dark lines are called *index contours*.

10. Adjacent contour lines are widely spaced where the land surface is fairly flat (has a gentle slope).

11. Contour lines are more closely spaced where the land surface is steep, such as on the slopes of the volcano. Note how the shapes of the contours reflect the shapes of the different volcanoes.

B | How Do We Refer to Differences in Topography?

Earth's surface is not flat and featureless, but instead has high and low parts. Topography is steep in some areas but nearly flat in others. We use common terms to refer to the height of the land and the steepness of slopes.

Elevation, Depth, and Relief

1. The height of a feature above sea level is its *elevation*. Scientists describe elevation in *meters* (m) or *kilometers* (km) above sea level, but some maps and most signs in the U.S. list elevation in feet (ft).

2. Beneath water, we talk about *depth*, generally expressing it as depth below sea level. We use *meters* for shallow depths and *kilometers* for greater ones.

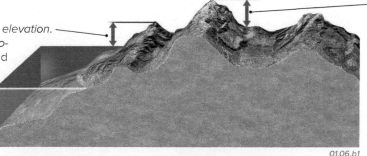

01.06.b1

3. We also refer to the height of a feature above an adjacent valley. The difference in elevation of one feature relative to another is *topographic relief.* Like elevation, we measure relief in meters or feet; we refer to rugged areas as having *high relief* and to flatter areas as having *low relief.*

Slope and Gradient

4. One way to represent the topography of an area, especially the steepness of the land surface, is to envision an imaginary slice through a terrain, like this one through SP Crater (▶). The dark line shows the change in elevation across the land surface, and is a *topographic profile.* Cliffs and slopes that drop sharply in elevation are *steep* slopes, whereas topography that is less steep is referred to as being *gentle*, as in a gentle slope.

W   E
26°

01.06.b2 *SP Crater, northern AZ*

5. We describe steepness of a slope in *degrees* from horizontal. The eastern slope of SP Crater has a 26-degree slope (26° slope). We also talk about *gradient*—a 26° slope drops 480 meters over a distance of one kilometer, typically expressed as 480 m/1,000 m or simply as 0.48.

1.6

1.7 What Do Latitude and Longitude Indicate?

IMAGINE TRYING TO DESCRIBE the location of an "X" on a featureless sphere. What system would you devise to convey the location? If the sphere did not have any markings or seams, we would need to first establish a frame of reference—a place on the sphere from which to reference the location of the X. For these reasons, we have devised systems of imaginary gridlines on the Earth. These are referenced as angles from the known points within or on the Earth. The most commonly used imaginary gridlines are *latitude* and *longitude*, which are displayed on many maps and are provided by the location capabilities of many cellular phones.

A How Do We Represent Locations on a Globe?

If you were trying to convey the location of the X on the sphere, or the location of a city on our nearly spherical planet, a good place to begin visualizing the problem is to establish a framework of imaginary gridlines. Another important aspect is to consider how lines and planes interact with a sphere.

Parallels

1. We could draw lines that circle the globe, each staying the same distance from the North or South Pole. The lines are parallel to one another and remain the same distance apart, and so are called *parallels*. In addition, these lines are parallel to imaginary cuts through the Earth, perpendicular to Earth's spin axis (which goes through the North and South Poles). The parallel that is halfway between the North and South Pole is the *equator*.

01.07.a1

2. If we traveled along one of these lines (i.e., along a parallel), we would stay at the same distance from the pole as we encircled the planet. In other words, our position in a north-south framework would not change.

Meridians

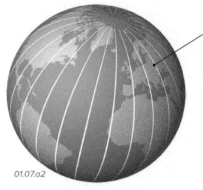
01.07.a2

3. Lines that encircle the globe from North Pole to South Pole are called *meridians*. Meridians do not stay the same distance apart and are not parallel. Instead, meridians are widest at the equator and converge toward each pole. A meridian would be the path you would travel if you took the most direct route from the North Pole to the South Pole, or from south to north.

4. The term *meridian* comes from a Latin term for midday because the Sun is along a meridian (i.e., is due south or north) at approximately noon. The terms A.M. (for before noon) and P.M. (for after noon) are also derived from this Latin term (e.g., post meridiem).

Great Circles

5. The intersection of a plane and a sphere is a curved, circular line that encircles the sphere. If the plane is constrained to pass through the center of the sphere, we call the resulting intersection a *great circle*. A great circle also represents the shortest distance between two points on a sphere and so is the path airlines travel over long distances.

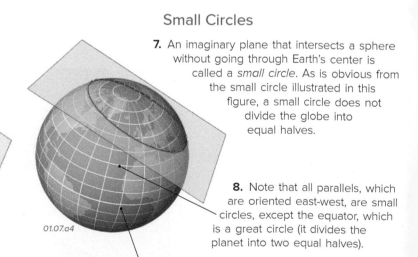
01.07.a3

6. A great circle divides the sphere into two equal halves. The equator is a great circle, separating the Earth into two hemispheres—the *Northern Hemisphere* north of the equator and the *Southern Hemisphere* south of the equator. A north-south oriented great circle is used to separate the *Western Hemisphere*, which includes North and South America, from the *Eastern Hemisphere*, which includes Europe, Asia, Africa, and Australia. Antarctica, over the South Pole, and the Arctic Ocean, over the North Pole, each straddle the great circle between the Eastern Hemisphere and Western Hemisphere.

Small Circles

7. An imaginary plane that intersects a sphere without going through Earth's center is called a *small circle*. As is obvious from the small circle illustrated in this figure, a small circle does not divide the globe into equal halves.

01.07.a4

8. Note that all parallels, which are oriented east-west, are small circles, except the equator, which is a great circle (it divides the planet into two equal halves).

9. In contrast, each north-south meridian, when paired with its counterpart on the other side of the globe, forms a great circle. Any such pair of meridians divides the globe into two equal halves. When viewed together, parallels and meridians divide the planet into a grid of somewhat rectangular regions. Such regions encompass greater area near the equator than near the poles, due to the convergence of meridians toward the poles.

B What Are Latitude and Longitude?

If you were a pilot flying from New York City to Moscow, Russia, how would you know which way to go? Our imaginary grid of parallels and meridians provides a precise way to indicate locations using latitude and longitude, which are expressed in degrees. Fractions of a degree are expressed as decimal degrees (e.g., 9.73°) or as minutes and seconds, where there are 60 minutes (indicated by ') in a degree and 60 seconds (") in a minute (e.g., 9° 43' 48").

1. This map (▶) illustrates the nature of the problem. If we want to navigate from New York to Moscow, we can tell from this map that we need to go a long way to the east and some amount to the north. These directions, although accurate, would not be good enough to guide us to Moscow. We need to specify the locations of each place more precisely and then figure out the shortest flight path. Fortunately, we can find on the Internet that the location of New York City, as given by latitude and longitude, is 40.7142° N, 74.0064° W. The location of Moscow is 55.7517° N, 37.6178° E. Now if we only knew what these numbers signify!

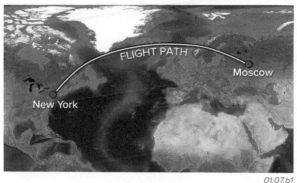

01.07.b1

2. Before you go any further, note the yellow line on this map, which represents the shortest route between New York and Moscow. It does not look like the shortest route on this flat, two-dimensional map, but it is indeed the shortest route on the three-dimensional globe. What type of path do you think this flight route follows? It is a great circle.

Latitude

3. The *latitude* of a location indicates its position north or south of the equator. Lines of latitude are *parallels* that encircle the globe east-west.

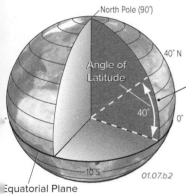

01.07.b2

4. The angle created by drawing lines connecting the position of an object on the Earth's surface to the center of the Earth, and then to the equator, defines the number of degrees of *latitude* of the object's position. In the Northern Hemisphere, latitude is expressed as degrees north. In the Southern Hemisphere, latitude is expressed as degrees south or as negative degrees.

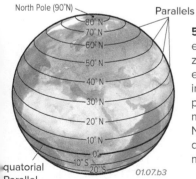

01.07.b3

5. Parallels of latitude run east-west around the Earth. The zero line of latitude is the equator, with the values increasing to 90 at the north and south poles. There are ten million meters from the equator to the North and South Poles, so one degree of latitude is approximately 111 km (69 miles).

Longitude

7. The *longitude* of a location indicates its east-west position. Lines of longitude are *meridians* that encircle the globe north-south.

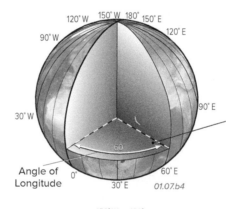

01.07.b4

8. As a starting point, a zero-degree meridian is defined as the north-south line that passes through Greenwich, U.K.—this is called the *Prime Meridian*. The angle created by the object's position, the center of the Earth, and the Prime Meridian defines that object's *longitude*, given as degrees east or west of the Prime Meridian. Meridians west of the Prime Meridian often are expressed as negative degrees.

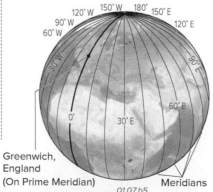

01.07.b5

Greenwich, England (On Prime Meridian)   Meridians

9. Meridians of longitude run north-south. They are widest at the equator (where a degree of longitude is also about 111 km) and converge at higher latitudes until they meet at the poles. Starting at the zero meridian through Greenwich, values increase toward 180° E and 180° W as they approach the *International Date Line*, an imaginary line that runs through the middle of Pacific Ocean (not shown; on the opposite side of the globe).

6. In addition to the equator, there are a few lines of latitude that are especially important. These include the *Tropic of Cancer* and *Tropic of Capricorn*, which are 23.5° north and south of the equator, respectively. Also important are the *Arctic Circle* and *Antarctic Circle*, which are 66.5° north and south of the equator (23.5° away from the corresponding pole). As discussed later, the 23.5° angle is how much the Earth's axis is tilted with respect to the Sun.

01.07.b6

Before You Leave This Page

☑ Sketch and explain what is meant by a parallel, meridian, great circle, and small circle.

☑ Sketch and explain the meaning of latitude and longitude, indicating where the zero value and maximum value are for each measurement.

1.8 What Are Some Other Coordinate Systems?

WE USE OTHER SYSTEMS besides latitude and longitude to describe location. These include the Universal Transverse Mercator (UTM) system, the State Plane Coordinate System (SPCS), and the Public Land Survey System (PLSS). Each is very useful for certain applications, and some are used to specify the location of real-estate properties appearing on legal documents associated with purchasing a house. Therefore, they are relevant to most citizens, even those who are not geographers.

A How Do We Use the UTM System?

Maps can show large regions, even the entire world. The main considerations for displaying large regions arise mostly from the fact that we live on a three-dimensional world (a sphere) and flat maps are two-dimensional. One solution to this challenge is the *Universal Transverse Mercator* (UTM) system, a method of identifying locations across the nonpolar part of the Earth. UTM is the most useful method of location for people who frequently hike or camp, or for people who work outdoors in nonurban settings.

1. The UTM system slices the nonpolar region into 60 north-south zones, each 6° of longitude wide. The slices are numbered from 1 to 60, with numbers increasing eastward from the International Date Line. A slice comprises two *UTM zones,* one in the Northern Hemisphere and another in the Southern Hemisphere. For example, most of Florida is in UTM zone 17 N, whereas the southern tip of South America is mostly in UTM Zone 19 S. What is the UTM zone for the place where you grew up or go to school?

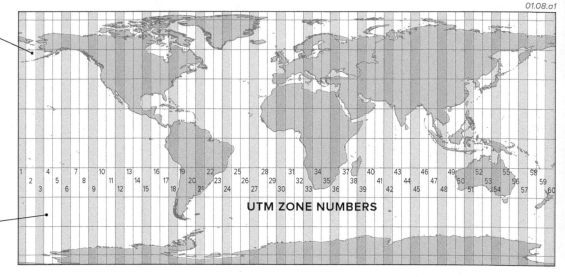

01.08.a1

UTM ZONE NUMBERS

2. The slices are further subdivided into grid zones, each 20° of latitude long, as shown by the rectangles on this map. The purpose of UTM zones is to ensure that location is portrayed accurately in the middle of each division, as distortion increases toward the edges. Due to large distortions that occur in the UTM system near the poles, UTMs are typically only used between 80° N and 80° S latitudes (we generally do not use UTM within 10° of the poles).

3. For a location within a grid zone, we specify coordinates as *eastings* and *northings*. Eastings are a measure of the number of meters east or west of the central meridian for that zone. Northings are a measure of the position north or south of the equator. The map below shows the aerial photograph of the horse and cow pasture shown earlier in this chapter, but this time with a UTM grid labeled with eastings (along the bottom of the map) and northings (along the left side of the map).

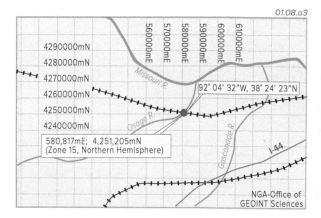

4290000mN
4280000mN
4270000mN
4260000mN
4250000mN
4240000mN

560000mE
570000mE
580000mE
590000mE
600000mE
610000mE

Missouri R.

92° 04' 32"W, 38° 24' 23"N

Osage R.

Gasconade R.

I-44

580,817mE; 4,251,205mN
(Zone 15, Northern Hemisphere)

01.08.a3

NGA-Office of GEOINT Sciences

4. The advantage of the UTM system is that it is a "square" grid system measured in meters rather than degrees, so it is convenient for measuring direction and distance. Note how useful this grid and UTM system would be if you were riding around trying to record the location of each horse in the pasture. Two horses (not visible here) are grazing at an easting of 495250 and a northing of 4214100; can you determine about where these horses are? Are they in the green pasture?

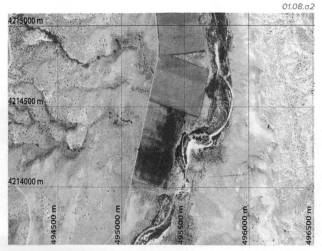

01.08.a2

4215000 m
4214500 m
4214000 m

494500 m
495000 m
495500 m
496000 m
496500 m

5. We can specify locations using several systems, and convert from one location system to another. The map above shows the position of a site expressed in both latitude-longitude (commonly called "lat-lon") and UTM coordinates. There are Internet sites that allow easy conversion from lat-lon to UTM and vice versa. To go from UTM to lat-lon, you have to specify the UTM zone, which can be determined using the large map near the top of this page.

B How Do We Describe Locations Using the State Plane Coordinate System?

The State Plane Coordinate System (SPCS) is a third system for mapping, used only in the U.S. SPCS ignores the distortion caused by the curvature of the Earth by treating the surface as a plane, so it should only be used for smaller areas like states or parts of states. As a result, the system can use X-Y coordinates to represent positions, simplifying land surveys and calculations of distances and areas. Another advantage is that the projection was chosen based on the geographic orientation of the state or section of the state, to minimize distortion for that area.

In the SPCS, most states are subdivided into two or more zones called *state plane zones*; some states are a single zone. Alaska has 10 zones and Hawaii has five zones. The boundaries of the zones generally are east-west or north-south, but are not straight, following local county boundaries (trying to keep a county within a single zone).

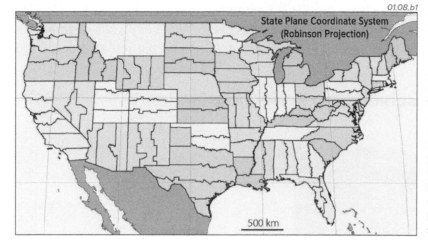

01.08.b1

State Plane Coordinate System
(Robinson Projection)

500 km

States that are elongated east-west, such as Tennessee, use different map approaches to generate the state plane coordinates than states like Illinois that are elongated north-south. The goal is to customize the drawing of the map so as to minimize the distortion that is always present when trying to show features of a spherical Earth on a flat piece of paper. So local U.S. maps, such as for flood zones, roads, or property delineation, are likely to use the SPCS. If you buy a house in the U.S., the legal documents will likely use SPCS to specify the location of the property, perhaps accompanied by a survey in UTM.

C How Do We Describe Location Using the Public Land Survey System?

The Public Land Survey System (PLSS) is another system used in the U.S. for describing the location of lands and for subdividing larger land parcels into smaller ones. When you hear someone refer to a "section of land" or a "quarter-section," they are talking about PLSS. The PLSS is also called the township-range system.

1. The Public Land Survey System was designed specifically for public lands, such as those administered by the U.S. Department of the Interior, and as a result is most widely used in states where there are federal lands (▼). It is not used in many eastern states, where there is little land that is not privately owned, and in Texas, which has much state-owned land. These two regions are shown in yellow.

States Included in the PLSS

01.08.c1

Included

2. PLSS is based around some initial point. From this point, a *Principal Meridian* extends both north and south and a *Base Line* extends both east and west. Beginning at the Principal Meridian, the land is subdivided into six-mile-wide, north-south strips of land called *ranges*. Beginning at the Base Line, the land is subdivided into six-mile-wide, east-west strips of land called *townships*.

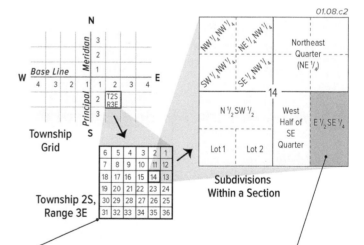

01.08.c2

Township Grid

Township 2S, Range 3E

Subdivisions Within a Section

3. Each square of the township-range grid is six miles in an east-west direction and six miles in a north-south direction, so it is 36 mi² in area. Each grid square is further subdivided into 36 sections that are each one square mile in area. Township and range lines and section boundaries are included on many topographic maps.

4. Each one-square-mile section can be further divided into quarters, eighths, and even smaller subdivisions. The rectangle in the southeastern corner of Section 14 would be described as being in the eastern half of the southeast quarter of Section 14, Township 2 South and Range 3 East. This is abbreviated: E1/2 SE1/4, S. 14, T2S, R3E.

Before You Leave This Page

✓ Describe the UTM system, how positions are expressed, and some of its advantages.

✓ Describe the State Plane Coordinate System and its main advantages.

✓ Describe the Public Land Survey System and how areas of land are subdivided.

1.8

1.9

How Do Map Projections Influence the Portrayal of Spatial Data?

EARTH IS NOT FLAT, so a flat map cannot portray all locations accurately. An ideal map would preserve directions, distances, shapes, and areas, but it is not possible to preserve all four of these accurately. Instead, either the *shape* of features on a map, such as country outlines, is preserved or the *area* of features is preserved, but never both at the same time. Many map projections depict both shape and area somewhat inaccurately, as a trade-off, so that neither will be shown more inaccurately. *Cartographers* (map makers) have developed different ways of projecting our three-dimensional world onto a flat map, and each approach is called a *map projection*. The particular type of projection is chosen based on the intended use of the map.

A What Is the Rationale Behind Map Projections?

1. A *map projection* is a mathematical algorithm used to represent places on a three-dimensional spherical Earth on a flat map. Imagine shining a light through a partially transparent globe and observing the image projected on the back wall (▶). This is what a map projection does, but in a quantitative way. While many projections exist, the best projection for a given map will introduce the least distortion for the key areas being shown. Whenever a map is made, some distortion is introduced by the projection. It is impossible to avoid distorting either shapes or areas, or doing some distortions of each.

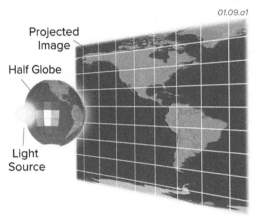

01.09.a1

2. Some map projections attempt to preserve shapes, and are called *conformal*. If shapes are preserved (▶), directions may be preserved but areas are distorted and scale will vary across the map. These imperfections get worse for maps that show larger areas.

3. In *equal-area* projections, areas are preserved but shape is distorted (▶). Compass directions cannot be shown correctly, so such a map should not be used for navigation. If the proper projection is selected for a given application, distortions are minimized for the aspect (e.g., shape) that is most important and for the region of most interest.

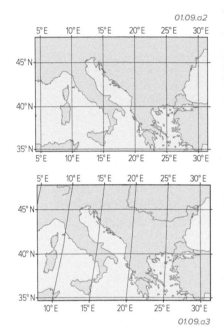

01.09.a2

01.09.a3

B What Are the Major Types of Projections and What Advantages Does Each Offer?

Sinusoidal

1. Perhaps the easiest projection to visualize conceptually is to imagine peeling an orange and slicing it in a few strategic places to allow it to be flattened without buckling (▼). *Sinusoidal* projections work on this same premise. If the map can be interrupted so that areas of lesser significance for a given application are not shown, then less distortion exists in the areas that are shown. Straight, parallel lines remain so, and have their correct length. Meridians become progressively longer toward the edges of each lobe of the map. While areas are preserved, shape distortion increases near the edges of each lobe.

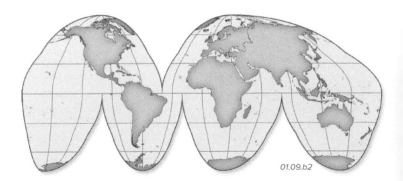

01.09.b2

01.09.b1

2. The shape distortion problem in such projections can be mitigated by increasing the number of central meridians around which accuracy is preserved (▲). However, this comes at the expense of having more areas of interrupted coverage. Notice how the central meridians are straightest and appear at right angles to the parallels at the equator. These are the areas that are depicted most accurately for this type of projection. The most common type of map using this projection strategy is called a *Goode projection*.

Cylindrical

3. In *cylindrical projections*, the globe is transformed to a flat page by projecting a globe outward onto a cylinder. The projection starts at a line, called the *standard line*, where the globe touches the cylinder, usually at the equator. These types of map projections have no distortion at the standard line (equator), but distortion becomes worse with increasing distance from the standard line. The resulting maps portray parallels of latitude as straight lines with the same length as the equator (that is, distorted in length) and depicts meridians also as straight lines intersecting the parallels at right angles.

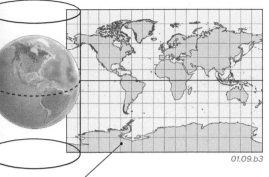

01.09.b3

4. Cylindrical projections (◄) depict compass directions as straight lines, so they are excellent for navigation. However, because the meridians are depicted (falsely) as being parallel to each other, east-west exaggeration of distances is severe, particularly in high latitudes. To allow these maps to be conformal (preserve shapes), north-south distances are stretched to match the east-west exaggeration. This makes high-latitude areas greatly exaggerated in size, but they retain shapes.

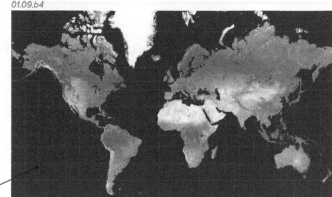

01.09.b4

5. High-latitude distortion increases to such an extent that the poles cannot be shown.

6. The most familiar type of cylindrical projection is the Mercator map, which became an important tool in the Age of Exploration. The part of the map at the right, which is a *Mercator projection*, portrays Greenland as being larger than the conterminous U.S. Is this true?

7. Some maps blend aspects of a cylindrical map with other types of projections. The *Robinson projection* (►) is a commonly used projection of a world map, especially in textbooks. It does not fully preserve areas or shapes, instead representing a compromise between conformal and equal-area projections. The meridians curve gently, and the parallels are straight lines horizontally across the map. A feature it shares with cylindrical maps is severe distortion near the poles.

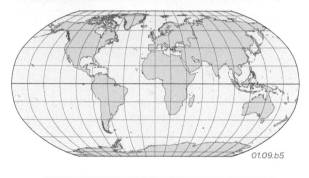

01.09.b5

Conical

8. *Conical projections* involve conceptualizing a cone over the globe, usually with the apex of the cone vertically above the pole. No distortion occurs along the arc where the globe touches the cone—the standard line, usually a parallel of latitude. If the cone slices through the globe and intersects the surface along two arcs (usually parallels of latitude), the projection is called *polyconical*. In either case, distortion increases with distance away from these arcs.

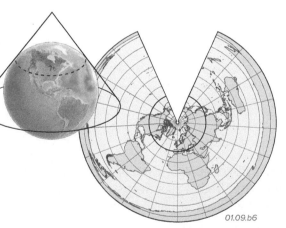

01.09.b6

9. In conical projections, parallels are concentric circles and meridians are lines radiating from the center of curvature of the parallels. This family of projections is neither conformal nor equal area.

10. Conical projections can only show areas within a single, complete hemisphere (since regions that curve underneath the globe cannot be projected). This type of map works best when the area mapped is small in latitudinal extent. Notice how poorly this projection performs for showing a large area (◄).

Planar

11. In *planar projections*, the plane onto which the map is projected touches the globe in a single point, which becomes the center of the map. Distortion increases away from this point, and any straight line from this point is a line of true direction. Again, only one full hemisphere can be shown on such a map.

01.09.b7

12. The pole is a focus in a type of planar projection called a *polar stereographic projection*. Scale becomes exaggerated toward the equator, but all lines connecting the shortest distance between two points on the sphere (great circles) are shown as straight lines. Planar projections are useful, therefore, for air navigation.

Before You Leave This Page

☑ Describe what a map projection is, and how different types of map projections are created.

☑ Summarize the principles that should be taken into account when selecting the proper map projection.

☑ Explain the advantages and disadvantages of sinusoidal, cylindrical, conical, and planar projections.

1.9

How Do We Use Maps and Photographs?

MAPS ARE AMONG OUR MOST IMPORTANT TOOLS for depicting and analyzing spatial information, whether we are interested in environmental issues or election results. Cartographers generate different kinds of maps that are designed to show Earth's landscape features, its weather and climate, and the distribution of plants, animals, or many other types of variables. Some cover small areas of Earth's surface, whereas others cover entire continents.

A How Much Area Do Maps Portray?

If we are hiking across the landscape, we want a detailed map that shows the location of every hill and valley. If we are interested in global climate change, we may want a map showing average temperatures for the entire planet. We use the general term *scale* to describe how much area the map shows. More specifically, scale is the ratio of the distance on a map to the actual distance (in the same units) on Earth.

Large-Scale Maps

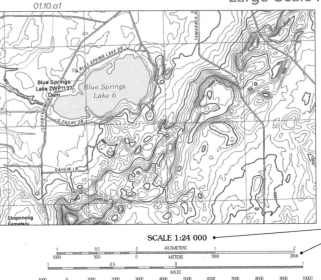

01.10.a1

SCALE 1:24 000

1. This topographic map (◄) shows hills and lakes that formed as glacial features in Kettle Moraine State Forest in central Wisconsin. We can convey the scale of the map in three ways. First, we can report the scale with *words*—on the original version of this map (reduced here to fit on the page), one cm on the map equals 24,000 cm on the surface. Second, we can report this same information as a *ratio* of a distance on the map to the actual distance on the ground, which is called the map's *representative fraction;* for the original version of this map the representative fraction was 1:24,000, as reported on the map. Third, most maps include some type of visual *bar scale*.

2. The original scale of this map, 1:24,000, is the typical scale used in the U.S. for topographic maps, with one inch equaling 2,000 feet. A map like this, which shows a local area, has a large representative fraction. It would require a relatively large map to show a large area—it is called a *large-scale map*.

Small-Scale Maps

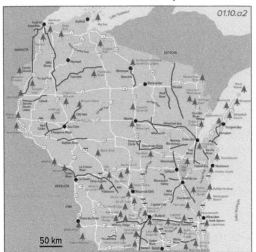

01.10.a2

50 km

3. The map above shows state parks, forests, and recreation areas in Wisconsin. This type of regional map portrays a relatively large area with a relatively small map—such a map has a small representative fraction and is a *small-scale map*.

B How Are Maps Made?

Originally, topographic maps were produced by sending a team of surveyors out in the field and having them map the area, drawing lines on paper maps, and taking notes. Today, such maps can be produced directly from laser and radar measurements from orbiting spacecraft or from pairs of photographs taken from slightly different perspectives.

1. Aerial photographs are typically taken from a plane or satellite as it flies across the terrain. The onboard, downward-pointing camera takes photographs at specific intervals in such a way that there is some overlap between the area captured by two successive photographs. The perspective of the camera is slightly different between the two photographs in the same way that our two eyes simultaneously have a slightly different perspective of the same scene. Test this concept by looking at your surroundings, closing one eye at a time, and noticing how objects shift slightly in position relative to one another. The apparent shift is related to their difference in distance from us.

01.10.b1 *Arabia Terra, Mars*

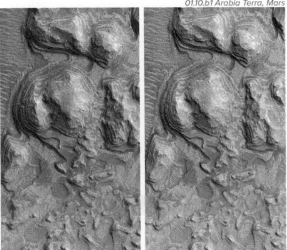

2. Two aerial photographs that have overlapping coverage and slightly different perspectives (◄) are called a *stereo pair*, usually generated by two images taken seconds apart, while the plane or satellite is moving. In other cases, both images are taken at the same time but by two different cameras placed a specific distance apart. When the two photographs of a stereo pair are placed at a proper distance side by side, a tool called a *stereoscope* enables us to see the scene in 3D, with the hills appearing to stand in relief above valleys. Such stereo pairs can be used to make a topographic map.

C How Can Maps Be Used for Reporting Information?

Sometimes we make new maps in the field, such as by using surveying equipment to make a topographic map that depicts the shape of topography. In most cases, we use existing maps, like the ones shown previously, and mark on the map the location of things we observe, such as the locations of glacial features or certain types of trees. In either case, this type of map actually produces new knowledge and is therefore a form of *primary data*.

The procedure is to visit the field site with an appropriately detailed map or aerial image, representing a *base map* upon which observations can be plotted. The base map can be a large-scale topographic map or a detailed aerial photograph, like the one shown to the right. Observations and other information are plotted directly on the base map or on a partially transparent overlay. Alternatively, locations can be determined with a handheld GPS device where the coordinates are saved and later mapped using a computer-based mapping program.

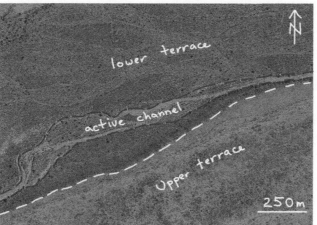

01.10.c1 Queen Creek, AZ

This aerial photograph shows different materials on the surface of several levels (elevations) along a desert river channel. The gray part, bounded by the dashed red lines, represents the active channel and related areas that are flooded during most years. The lower terrace is slightly higher in elevation above the channel, and is flooded less frequently. The upper terrace is high enough to avoid any flooding. This map was produced by walking through the field area and drawing on the aerial photograph the boundaries between different areas. This map would be useful for determining flooding potential and other types of land-use planning.

D How Can Maps Be Used to Analyze and Interpret the Environment?

Preexisting maps become the basis for various interpretations. For example, the annotated aerial photograph above could be used to plan the locations of a subdivision, especially deciding where *not* to build. A preexisting map that is used for providing the input for answering some other question is known as a *secondary data source*.

01.10.d1

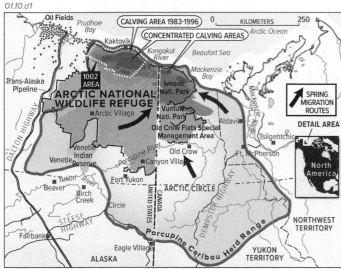

1. Many maps (◄) contain a combination of data and interpretations. Examine this map of the Arctic National Wildlife Refuge (ANWR), and identify aspects that are data versus those aspects that represent some type of interpretation. Then continue reading below.

2. *Data*—The locations of features on this map would be considered data. These include the outline of the coastline, the boundary between Alaska and Canada, the locations of rivers and roads, and the outline of ANWR.

3. *Interpretations*—Other aspects of the map are interpretations, which commonly represent an expert's opinion of a situation. On this map of ANWR, interpretations include the migration routes of caribou (the large black arrows) and the locations where caribou give birth to their calves (calving areas, in green). This map, consisting of data and interpretations, would be considered a secondary data source. It might be used to determine which areas are permissible for drilling.

01.10.d2

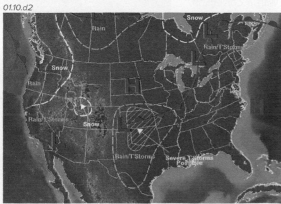

4. Surface weather maps (◄) likewise are a combination of data and interpretations. The edge of the continent and the outlines of states are clearly data. Weather maps also show analysis of the location of areas of relatively high atmospheric pressure (H), low atmospheric pressure (L), and weather fronts—lines interpreted to mark the boundary between air of very different temperatures and humidity. The triangles and semicircles point in the direction of air movement, another interpretation. Dashed lines outline areas of rain or snow. Such maps help predict today's and tomorrow's weather.

Before You Leave This Page

☑ Explain the difference between large-scale and small-scale maps and when you would use each.

☑ Describe how we can use maps to record new information

☑ Describe how we can use maps to analyze and interpret existing information about the environment.

1.10

1.11 How Do We Use Global Positioning Systems and Remote Sensing?

THE GLOBAL POSITIONING SYSTEM (GPS) and remote sensing have greatly increased the accuracy of geographic field studies and given geographers new methods for performing geographic analyses. GPS helps geographers define spatial relationships among Earth's surface features, and a wide variety of remote-sensing techniques help geographers define regional patterns and monitor changing environmental conditions.

A What Is GPS?

GPS is familiar as a navigation system in our cars, cellular telephones, or handheld devices used for location and guidance. GPS provides the accurate position on Earth's surface including latitude, longitude, elevation, and even how fast we are traveling. This information comes from a series of satellites orbiting Earth that send radio signals to ground-based receivers, like the ones on our dashboards, or in our phones or handheld GPS.

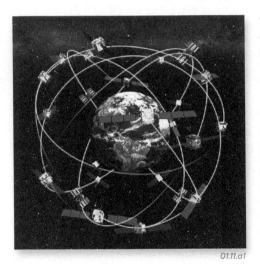

01.11.a1

The U.S. government launches, controls, and monitors a constellation of 24 satellites orbiting in six different planes around Earth (◄). Several generations of satellites currently operate in the GPS constellation (►), with newer generations being deployed to improve accuracy and reliability.

The time required for a radio signal from a satellite to reach a receiver on Earth is related to its distance to the receiver. A GPS receiver "knows" where each satellite is located in space at the instant when the GPS unit receives the signal. Calculating the distances from four or more satellites allows the GPS unit to calculate its own position, commonly with a precision and accuracy of several meters (for a handheld GPS unit). Higher precision can be achieved by occupying a single site for a long interval of time and then averaging the measurements.

01.11.a2

B How Do We Use GPS to Study Geographic Features?

GPS is used in a variety of applications from tracking wildlife migration or package delivery, to improving ocean and air travel. Even farmers use GPS to harvest crops and improve yield. Geographers use GPS for a variety of activities, including monitoring changes in the environment, collecting more accurate field data when surveying or mapping, and making decisions about how to best prevent or address natural disasters. Geographers employ two types of GPS devices, the familiar handheld GPS and the Differential GPS (DGPS).

01.11.b1

1. A handheld GPS device (◄) is a navigation tool for finding a location. These instruments operate on the same principles as all other GPS devices in that they receive radio signals from orbiting satellites that contain information about the position and distance of the satellite. GPS works best outside and with a clear view of the sky, but it can operate with reduced accuracy in settings where parts of the sky, and therefore view of the satellites, is partially blocked.

2. Geographers use handheld GPS mostly for field work, including mapping the locations of landscape features, determining locations of water and soil samples, and inventorying populations of plants and animals.

3. *Differential GPS* (DGPS) is the same as GPS but with a correction signal added to improve the precision and accuracy. Accuracy is enhanced because the correction signal performs an independent check of each GPS satellite's signal. DGPS can provide accuracy of less than several meters.

4. Geographers use DGPS when precision is important, such as in surveys (►) of changes in the land surface over timescales of decades or to gauge the erosion effects of a recent hurricane on a shoreline and its communities.

01.11.b2

C What Is Remote Sensing?

The term *remote sensing* refers to techniques used to collect data or images from a distance, including the processing of such data, and the construction of maps using these techniques. Remote sensing can be carried out using a helicopter, airplane, drone, satellite, balloon, ship, or other vehicles, or it can be performed with instruments fixed on the land surface. The instrument-carrying vehicle or site is called the *platform*, and the instrument that collects the images and other data is the *sensor*. There are two general types of remote sensing systems: *passive systems* and *active systems*.

Passive Remote Sensing

01.11.c1 New Orleans, LA after Hurricane Katrina

In passive remote sensing, the sensor points at the area of interest and records whatever light, heat, or other energy is naturally coming from that region. Aerial photography and most satellite images, like the one to the left, are recorded by passive sensors. The sensors are tuned to collect specific types and wavelengths of energy, such as infrared, visible, and ultraviolet energy. Most sensors collect an array of similar frequencies.

Active Remote Sensing

Radar Waves

01.11.c2

In active remote sensing, an energy source, usually on the same platform as the sensor, directs a beam of energy downward or sideways toward the area of interest. Such energy can include radar, as shown here, micro-waves, laser light, or other types of energy. The sensor then measures how much of this energy returns to the platform and whether this energy has been modified by its interaction with the surface or atmosphere.

D What Types of Remote Sensing Are Used By Geographers?

Geographers use a variety of remote-sensing techniques, measuring various types of energy, to study Earth's atmosphere, hydrosphere, lithosphere, and biosphere. Geographers also document and investigate patterns in land use, vegetation cover, erosion rates, extent of pollution, ocean temperatures, and atmospheric water content and circulation.

Visible and Near Infrared (IR)

01.11.d1 Washington, DC

1. Aerial photo-graphs typically record visible light reflected off an area, but some photo-graphs and many satellite images also record adjacent bands of infrared energy (near-IR). On near-IR images, vegetation commonly is depicted with a reddish tone, as in this image of Washington, D.C.

Thermal Infrared (IR)

01.11.d2 Providence, RI

2. Objects also emit energy, either from the internal heat of an object or from heat initially gained from the Sun. This image shows thermal-IR derived temperatures of Providence, Rhode Island, with lighter colors showing hotter areas in the city.

Microwave

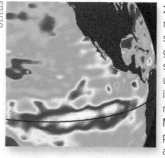

01.11.d3

3. Images from microwave-sensing satellites and ground-based stations provide us with weather images in nightly newscasts. Microwaves can penetrate clouds and haze, providing a clear view of the ground at all times. They also can measure the height of the sea surface, as shown here.

Multispectral

01.11.d4 Nili Patera, Mars

4. Some multipur-pose satellites collect data at multiple wave-lengths of energy and therefore have the name *multispectral*. Multispectral data are used for studying natural hazards, inventorying plant communities, tracking forest fires, and observing landscapes on other planets, as shown here.

Radar, Sonar, and Lidar

01.11.d5 Mount St. Helens, WA

5. Radar, sonar, and a newer technique called lidar all involve emitting waves of a certain wavelength and then measuring how much is reflected back to the sensor and the time required for the various beams to return. These data allow us to map the surface, like volcanic features.

Before You Leave This Page

☑ Explain what GPS is, how it works, and how we can use it to investigate geographic problems.

☑ Summarize what remote sensing is, explaining the difference between passive and active systems.

☑ Describe five types of remote-sensing data and one example of how each type is used.

1.12 How Do We Use GIS to Explore Spatial Issues?

MAPS ARE USED FOR REPORTING OBSERVATIONS and making interpretations from previously collected observations, and they can also be analyzed to create new maps. Maps created from aerial photographs, satellite imagery, and field observations can be stored in computer databases called *geographic information systems (GIS)*, where a variety of information can then be combined quickly and efficiently to examine relationships among the different features. For example, we might be interested in comparing the distribution of plant communities with the type of soil, average precipitation, and percentage of nights when it freezes. Modern geographic analysis using GIS is data-rich and computer-intensive, providing the clearest way to explore many types of spatial relationships.

A What Are Geographic Information Systems?

1. The first step required to analyze geographic problems is to decide what key data sets are needed to best understand the issues. Then, maps, images, and data are imported into a GIS database. Maps, aerial photographs, and satellite images need to be matched to standard geographic coordinates, such as UTM. The process of taking an image or map and matching it to standard coordinates is called *georeferencing* or *rectification*.

2. Once data are in the GIS, we can overlay different maps to compare different data sets. We commonly call each map a GIS *layer* because in the GIS computer we can arrange the different maps (layers) one on top of another, as shown in the figure to the right. Examining multiple kinds of data in this way would be useful, for example, in selecting a site for some type of facility.

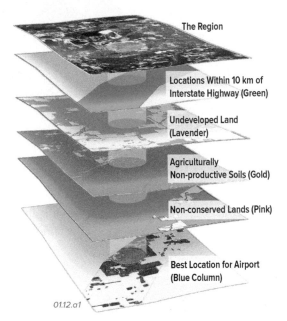

The Region

Locations Within 10 km of Interstate Highway (Green)

Undeveloped Land (Lavender)

Agriculturally Non-productive Soils (Gold)

Non-conserved Lands (Pink)

Best Location for Airport (Blue Column)

01.12.a1

3. Suppose city planners want to find the best location for a new airport. They first need to determine what already exists in the region so they examine a satellite image or aerial photograph (the top layer in the GIS layers shown here). Priority is given to sites that are within 10 km of an interstate highway (second layer from top). The site must contain a large area of undeveloped land (third layer), does not have soil that could be productively farmed (fourth layer), is not in a conservation area (fifth layer), and other considerations, such as who owns the property and whether the site has unstable slopes and soils (not shown).

4. Without a GIS, several individual maps would have to be compared by hand to determine the best possible sites. A GIS can store each of these maps digitally and allow the user to identify any locations that meet the specified criteria. The digital format also allows for easy updating, and inquiries can be run on the potential sites to glean further information, such as current land costs.

5. Another popular use of GIS is for route optimization. If you have ever used Google Maps® or a navigation system on your phone or in your car, you've made use of this feature of GIS. In the example below, GIS is used to show a route that minimizes the distance that a shipping company must travel to deliver packages to locations marked with an "X" on a given day. Similar inquiries can be used to optimize routes for garbage pickup, hurricane evacuation, or school bus stops.

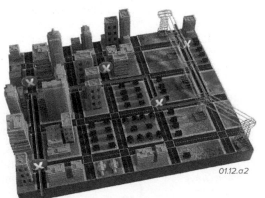

01.12.a2

6. The site selection and routing applications of GIS can be used together. For example, a power company may want to minimize the distance that power lines must be run, while still keeping the lines as far as possible from heavily populated and low-lying areas. Also, the power lines ideally are hidden from view as much as possible by local topography and forests.

7. GIS can be used in combination with remote sensing to classify areas on images. This can be very useful in determining land use and land cover for environmental applications. In its simplest form, areas could be classified as being urban, forested, agricultural, undeveloped, or water. More elaborate classifications could be used to identify types of wetlands and forests (▼) or other important aspects.

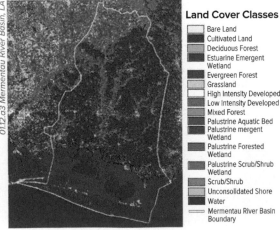

01.12.a3 Mermentau River Basin, LA

Land Cover Classes

- Bare Land
- Cultivated Land
- Deciduous Forest
- Estuarine Emergent Wetland
- Evergreen Forest
- Grassland
- High Intensity Developed
- Low Intensity Developed
- Mixed Forest
- Palustrine Aquatic Bed
- Palustrine mergent Wetland
- Palustrine Forested Wetland
- Palustrine Scrub/Shrub Wetland
- Scrub/Shrub
- Unconsolidated Shore
- Water
- Mermentau River Basin Boundary

8. Another key advantage of GIS is that it allows direct comparison of changes in an area over time, whether those changes are in land use, vegetation cover, population density, or some other variable. By comparing earlier or later images of the same area, rates of urban sprawl, deforestation, or wetland loss could be documented, allowing us to consider possible remedies in an informed manner.

B   What Kinds of Calculations Are Possible with GIS?

Spatial analysis is a cornerstone of geography, and GIS can evaluate the spatial distributions within the data, highlight correspondences among different variables, and automate the identification of properties of spatial distributions.

If the locations of observed features can be considered as points, such as stations that reported precipitation in the past 48 hours or sites in a stream where dissolved oxygen is low, physical geographers often want to know whether those features (1) occur in clusters, (2) are spaced approximately evenly apart (i.e., regular distribution), or (3) are distributed randomly across the landscape. *Point-pattern analysis* can answer this question objectively, and GIS can complete the geostatistical calculations automatically. Analysis proceeds on the principle of laying an imaginary grid over the study area and determining whether the object of interest is distributed significantly less evenly (i.e., clustered) or more evenly (i.e., regular) across the grid cells than would be expected in a random distribution.

Types of Spatial Distributions

1. Geographers are interested in the *distributions* of objects, whether of plants or animals or the locations of gas stations. There are a number of different types of distributions, three of which are shown here (▶). The name of each type conveys the overall character of the distribution, but there are usually exceptions (called *outliers*) to the general patterns.

Clustered   Regular   Random

01.12.b1

4. Objects can have *random* distribution. Some trees occur together, but not consistently. Point-pattern analysis indicates that the trees are neither clustered nor distributed regularly. The random arrangement fails to provide an obvious explanation for tree location.

2. Objects can have a *clustered* distribution, with objects tending to occur together rather than being dispersed widely apart. Point-pattern analysis would conclude that the trees on the left side of the diagram are clustered. We may conclude that individual trees prefer proximity to a feature present in only part of the area, such as next to ponds, streams, and other sources of water. In many cases, the explanation for a clustered distribution is not so obvious as in this case.

3. Objects can also have a more *regular* distribution, more or less evenly spaced apart and not spatially clustered. Point-pattern analysis would suggest a regular distribution of trees. Perhaps individual trees, when away from lakes and streams, are more likely to survive if spread out as far as possible from other trees, increasing access to precipitation. Perhaps they are planted by humans, as in fruit orchards, or as windbreaks.

Calculations on Lines and Polygons

5. In addition to analyzing the distributions of points, GIS can calculate distances and angles between points, as when a navigation system calculates the distance to a destination. GIS can also be used to quantify the extent of areas, represented as polygons, such as the park shown here. GIS can calculate areas enclosed within polygons, identify the center of an irregularly shaped polygon, or calculate the perimeter of any area, like this park.

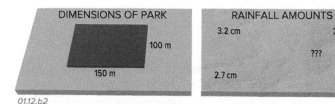

DIMENSIONS OF PARK
100 m
150 m
01.12.b2

RAINFALL AMOUNTS
3.2 cm   2.0 cm
        ???
2.7 cm   1.6 cm

Spatial Interpolation

6. There are statistical methods of estimating a data value at any location if we have values for nearby locations. Such analysis, commonly called *spatial interpolation*, is done by comparing the location to known values of other points at different distances. For example, the amounts of rainfall, in centimeters, during a storm were measured at four weather stations, but we need to estimate the amount received in a nearby location that lacks a weather station. There are various statistical strategies, with names like *kriging* and *inverse-distance methods*. A common approach is to weigh the known values at other points by the distance from known points to the unknown point.

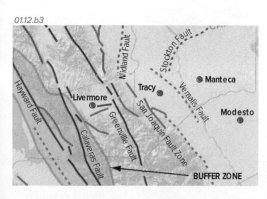

01.12.b3

Hayward Fault
Midland Fault
Stockton Fault
Calaveras Fault
Greenville Fault
San Joaquin Fault Zone
Vernalis Fault
Livermore
Tracy
Manteca
Modesto
BUFFER ZONE

Identification of Buffer Zones

7. We may not want two things in close proximity, like a school next to earthquake faults that are most active (◀). To solve such a problem, we need to compare the locations of schools with faults capable of causing earthquakes. GIS can contain all of these types of information, and can outline a *buffer zone* near such faults, within which school construction would be prohibited. Or a buffer zone may be drawn around a hurricane to indicate the zone of hurricane-force winds.

These buffered areas can then be overlain on other maps, such as one of population density, to identify the number of people whose homes are affected by the hurricane.

Before You Leave This Page

☑ Explain what GIS is, including the concept of map overlay.

☑ Give several examples of the types of spatial distributions tested in point-pattern analysis.

☑ Explain the types of functional calculations that GIS can do on spatial data.

1.12

1.13 What Is the Role of Time in Geography?

WE LIVE ON A GLOBE THAT ROTATES, causing locations on the surface to pass from day to night and back again. Not everyone witnesses sunrise at the same time, because the Sun rises at different times in different locations. Some ideas from geography, especially the concept of longitude, help us understand these differences and describe time so that society can operate in a more orderly manner. Most of us think of time as the hours, minutes, and seconds on a clock, but much longer units of time are used when considering Earth's long history.

A How Do We Define Time Globally?

Some units of time, like a year or length of day, arise from natural progressions of the Earth as it orbits around the Sun in a year and completes a full daily rotation in 24 hours. Other measurements of time are locally based, so in the 1800s the world had to agree on an international system for defining time, based on the *Prime Meridian* and the *International Date Line*.

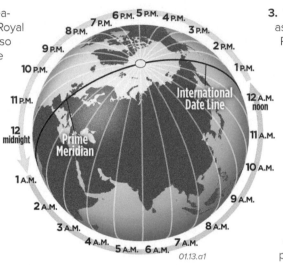

01.13.a1

1. *Prime Meridian* is defined as the 0° longitude measurement on the Earth, passing through the British Royal Observatory in Greenwich, U.K. This location was also chosen as the reference point for world time, a time called *Greenwich Mean Time* (GMT). *Coordinated Universal Time* (UTC) is based on atomic clocks and is the world standard. Time anywhere in the world is referenced relative to time at Greenwich.

2. The globe shown here has meridians spaced equally apart, so that there are 24 zones centered on the lines, one for each of the 24 hours in a day. If you could instantaneously travel from one meridian to the next, there would be a one-hour time difference. If political and other considerations did not intervene, the distribution of time zones could precisely follow lines of longitude, each 15° apart.

3. The *International Date Line* (IDL) is defined as the 180° measurement of longitude in the Pacific Ocean—the meridian on the exact opposite side of the Earth from Greenwich. Segments of the IDL are shifted east and west to accommodate the needs of some Pacific nations, so that travel and trade among those islands is easier.

4. If you cross the International Date Line, you cross into a different calendar day. Traveling westward across the IDL puts you one day ahead in the calendar relative to immediately east of the IDL; this is described as "losing a day." Moving eastward across the line you move to the previous date, so we say you "gain a day."

B How Are Time Zones Defined?

The world is divided into 24 time zones, based loosely on longitude. This map color-codes these 24 time zones, most of which have irregular boundaries because they follow natural or political boundaries or try to keep some population center in a single zone. The boundaries between the four time zones covering the contiguous U.S. are mostly drawn along state or county boundaries or natural features.

For most latitudes of the Earth, the Sun shines on all 360° of longitude sometime during the course of a 24-hour day, covering 15° of longitude of new territory each hour. The Earth is divided into 24 time zones, each about 15° of longitude wide. Areas within a time zone adopt the same time, and there is a one-hour jump from one time zone to the next. If you are in one time zone, the time zone to the west is one hour earlier, and the time zone to the east is one hour later.

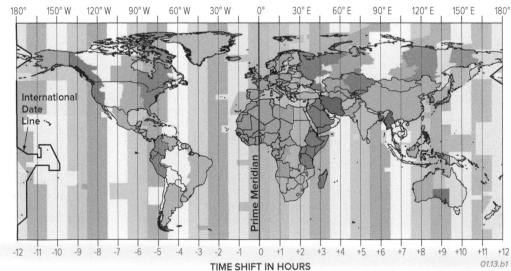

TIME SHIFT IN HOURS
01.13.b1

In most of the U.S. and Canada, clocks during the summer are set to one hour later for much of the year. This *Daylight Savings Time* (DST) provides daylight for an extra hour during the evening and one less hour of daylight in the morning. Some areas, like Saskatchewan and most of Arizona, do not observe DST, remaining instead on "standard time."

C How Do We Refer to Rates of Events and Processes?

Many aspects of physical geography involve the rates of processes, such as how fast a hurricane is moving toward a coastline or how fast water in a river is flowing. We calculate the rates of such processes in a similar way to how we calculate the speed of a car or a runner. For scientific work, units are metric, so we talk about millimeters per year, kilometers per hour, or similar units of distance and time.

Calculating Rates

1. A runner (▶) provides a good reminder of how to calculate rates. A rate is how much something changed divided by the time required for the change to occur. Generally, we are referring to how much distance something travels in a given amount of time.

01.13.c1

2. If this runner sprinted 40 meters in 5 seconds, the runner's average speed is calculated as follows:

$$distance/time = 40\ m/\ 5\ s = 8\ m/s$$

3. What is the runner's average speed if she runs 400 meters in 80 seconds? Go ahead, try it. We don't need to provide you the answer to this one.

Relatively Rapid Earth Processes

4. Some Earth processes are relatively rapid, occurring within seconds, minutes, or days. Relatively rapid natural processes include the velocity of the atmospheric jet stream (100s km/hr), speeds of winds inside a severe storm (100s km/hr), motion of the ground during earthquakes (5 km/s), movement of an earthquake-generated wave (tsunami) across the open ocean (100s km/hr), and the catastrophic advance of an explosive volcanic eruption (100s km/hr).

5. Hurricane Sandy, a huge and incredibly destructive storm that occurred in October 2012, is an example of a rapid natural process. Sandy originated as a tropical storm but migrated up the East Coast of the U.S. until it turned inland and struck New Jersey. While in the tropics, Sandy had winds estimated at 185 km/hr (115 mi/hr), but it had weakened considerably by the time the storm came ashore. The storm killed nearly 300 people along its path and caused damages of over $70 billion, mostly in New Jersey and New York.

May 21, 2009

November 5, 2012   USGS

01.13.c2–3 Seaside Heights, NJ

6. The two photographs above show Seaside Heights, N.J., before and after the storm. The yellow and red arrows point to the same houses in both photographs. Although the photographs were taken several years apart, nearly all the damage occurred within a 24-hour period. In this short time, the shape of the coastline was extensively rearranged, houses were destroyed, and the entire neighborhood was covered in a layer of beach sand washed in by the waves.

Relatively Slow Earth Processes

7. Other Earth processes are very slow, occurring over decades, centuries, or millions of years. Natural processes that are fairly slow include movement of groundwater (m/day), motion of continents (cm/yr), and uplift and erosion of the land surface (as fast as mm/yr, but typically much slower). Although these processes are relatively slow, the Earth's history is long (4.55 billion years), so there is abundant time for slow processes to have big results, such as uplift of a high mountain range.

8. Observe this photograph taken along a canyon wall (▶) and ask yourself how long each feature took to form. You do not need to arrive at any answers. The tan, brown, and yellowish rocks are all volcanic rocks, formed from molten rock and volcanic ash erupted from an ancient volcano.

01.13.c4 Superstition Mtns., AZ

9. Several questions about the rates of processes come to mind. Each layer in the volcanic rocks may represent a single pulse of eruption and could have accumulated rapidly, in minutes or hours. How long did it take to form all the layers? The landscape currently is being eroded, and this process has been occurring in this area for millions of years. How long will it take for the large brown blocks to fall or slide off the lower cliff? Some of these questions are about the present (how fast is erosion occurring), some are about the past (eruptions), and others are about the future (the blocks). The easiest questions to answer are usually about the present.

Before You Leave This Page

☑ Explain what GMT is and where the starting point is.

☑ Describe why we have time zones, and what influences time-zone boundaries and width.

☑ Explain how we calculate rates, giving some examples of relatively fast and slow natural processes.

1.13

1.14 How Did Geographers Help in the 2010 Gulf of Mexico Oil-Spill Cleanup?

ON APRIL 20, 2010, an explosion on the Deepwater Horizon oil rig in the northern Gulf of Mexico killed 11 workers, injured 17 others, and initiated the most disastrous oil spill in U.S. history. For the next 86 days, oil gushed into the Gulf. This oil spill is an example of a complex problem with interconnected environmental, economic, and social implications. The geographic approach is ideally suited to solving such problems. The oil spill provides a way to connect the various approaches and tools discussed in this chapter and apply these tools to an important and real-world example.

A Where Was the Oil from the Spill?

The question above seems straightforward, but a major controversy emerged during the summer of 2010 when scientists couldn't identify where the oil from the oil spill had traveled. The spatial perspective of geographers was indispensable in resolving this problem, and their work was in high demand.

01.14.a1 Gulf of Mexico, near LA

◄ Satellite imagery was one of the most effective tools in monitoring the location and movement of oil. This image shows the Gulf waters off Louisiana, with an oil slick visible in the center of the image. The satellite imagery was essential in locating and tracking the oil slick and documenting how the oil interacted with ocean currents and was dispersed in the Gulf.

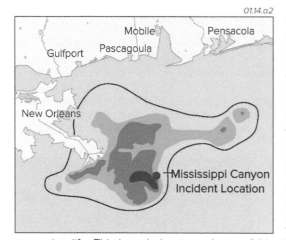

01.14.a2

Geographers converted satellite images into visualizations (◄) and incorporated these images into GIS databases, where the location and movement of oil could easily be conveyed to cleanup crews and to those concerned with the oil's impact on marine life. This knowledge was also useful in determining the impact of potential storm systems, including tropical storms.

B How Did the Oil Travel?

Geographers wanted to know how the oil traveled from the spill site to the coastal area to better understand what to expect if the spill continued and to also explore the impact of ocean currents, waves, and wind on dispersal patterns.

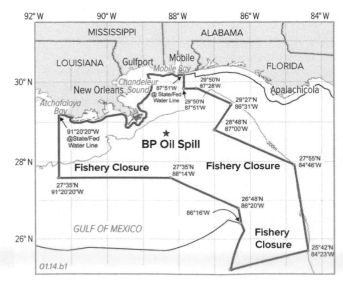

01.14.b1

◄ Physical geographers worked with other natural scientists to understand oceanic and atmospheric circulation patterns that influenced the speed and direction of oil migration, which guided decisions about which areas of the Gulf should be closed to fishing.

An understanding of the circulation patterns within and near the Gulf (►), including changes due to hurricanes and other strong storms, improved the ability to forecast the movement of the oil. In this map, red colors indicate faster currents.

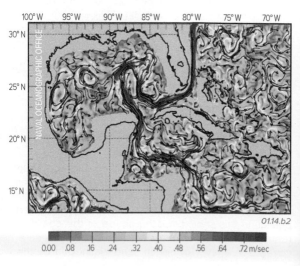

01.14.b2

0.00 .08 .16 .24 .32 .40 .48 .56 .64 .72 m/sec

C What Is the Effect of the Oil on the Earth-Ocean-Atmosphere-Human System?

The geographer's holistic perspective was needed to understand the "big picture" of impacts of the spill on the various "spheres," including the human realm. Examine the hypothetical place portrayed below, and think about all the aspects that could affect the distribution and dispersal of an oil spill, or be affected by the oil. Think about how the oil could relate to interactions among the natural spheres and to social systems of people who might live and work here. After you have thought about these issues, read the text around the figure, which describes a few of the impacts.

1. Oil reduces evaporation from the surface of the ocean and also reduces sunlight penetration into the Gulf. This could affect amounts and patterns of precipitation and runoff toward the sea.

2. Winds could bring volatilized oil (oil dispersed as vapors in the air) onto the land, depending on the directions of the wind. If the vapors are highly concentrated, such fumes would be unpleasant and unhealthy.

3. Winds and waves can lift oil onto the beach and coastal wetlands, where it can be harmful to the many species of birds, fish, and other animals that dwell there. A GIS database of animal densities could identify the most vulnerable, high-impact sites.

4. Sunlight can solidify oil, causing it to float as a "tarball" or sink to the seafloor. Some 25% of the oil in the 2010 spill is unaccounted for, presumably because it solidified, sank, and became part of the seafloor. Some oil was decomposed by microorganisms. The long-term impact is unknown, but likely harmful.

5. Oil harms many plants, fish, and other animals that ingest it directly. When these organisms are eaten, the predator ingests the oil, too. Fortunately, most oil is concentrated in parts of the animal that are not usually eaten by humans (shells, organs, etc.). Oil also allows types of bacteria that feed on it to proliferate at the expense of other forms of life, altering the food chain, and releasing gases to the atmosphere.

6. Ocean currents can disperse the oil and move it out of the immediate area, so we would want to understand the directions and rates of ocean currents interacting with the oil spill. Such maps would help us anticipate problems downcurrent.

01.14.c1

Applying the Geographic Perspective to Disaster Mitigation

Following any disaster, controversies arise, people try to understand what went wrong, and policies are considered to prevent future disasters. Decisions should only be made after understanding the connection between the physical and human systems involved. Over 40 years ago, geographer Werner Terjung described an ideal level of geographic problem solving, which he termed "physical-human-process-response systems," of the kind shown by the figure here. He meant that we need to understand the process involved in both natural and social systems, including our reaction to those processes, which can change our behaviors toward influencing future processes within the natural system, within the social system, and across the two systems.

Was the spill the result of inappropriate regulations, unenforced regulations, or poor decision making? Should dispersants have been used on the oil, or should it have only been collected with skimmers? Should a moratorium on future oil drilling in the Gulf have been implemented? How would such a moratorium affect our future energy supply and the economy of the region? What would be the impact on such a decision locally, regionally, nationally, and internationally?

Terjung also noted that those processes and the responses to them occur at a range of scales that themselves interact and impact the processes occurring at other scales, from millimeters to the global scale. Terjung also noted that no single study had yet accomplished this level of detail. However, he correctly observed that the day would come when we would need to consider problems with impacts in diverse topical and spatial scales, like an oil spill near a sensitive ecology and a densely populated coastline.

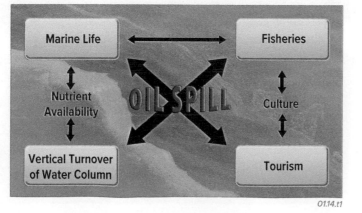

01.14.t1

Before You Leave This Page

☑ Summarize how geographers helped study the 2010 Gulf oil spill and how oil interacted with the environment.

What Might Happen If This Location Is Deforested?

YOU HAVE BEEN EMPLOYED by a county planning commission. You are asked to assess any possible impacts of logging (removing trees) of a mountainside in the area under your jurisdiction. To address the problem, you rely on your broad perspective and skills in the use of maps, satellite-image interpretation, physical geographic principles, and the scientific method. Then you will communicate your findings to policy makers and politicians on the commission.

Goals of This Exercise:

- Develop a list of observations about the area, accompanied by ideas of possible impacts of deforestation.
- Produce a list of questions you have about how different features and other aspects might be impacted by deforestation and how these might affect logging activities.
- Develop a list of the types of information you would need to answer your most important questions.

This three-dimensional perspective shows an area with a variety of terrains and climate. Make some observations about the landscape and record your observations on a sheet of paper or on the worksheet. As you study this landscape, record any questions you have about the impact of deforestation on the atmosphere, hydrosphere, lithosphere, and biosphere. The questions are what are important here, not the answers. You will be able to answer most or all of your questions by the end of the course. Include in your questions any you have about the potential impacts of deforestation on humans. Listed around the figure are some of the features and other aspects you should consider, but there are others not listed. Think broadly and be creative. We have listed a few questions to help you begin.

Patterns of atmospheric circulation, including prevailing wind direction and precipitation patterns:

Streams, lakes, and glaciers:

Slopes: How might the steepness of slopes interact with deforestation? Are steeper slopes more difficult or easier to log? What happens to soil if we remove vegetation on a steep slope? How about on a more gentle slope? Does erosion occur faster or slower on steep slopes?

Local climate: How cold is this place normally? Does most precipitation occur as snow or as rain, and how does this vary with the time of year? Is the precipitation spread out evenly over the year, or does it mostly occur during some seasons? If the winds are blowing from the water to the land, how do the mountains affect this moist air?

Farmland:

Types of Rocks, Sediment, and Soil:

Procedures

1. Examine the scene and make observations about the landscape. In your notes, write down what you consider to be the most important observations about the area.

2. Think about how different features or aspects could be affected by deforestation, or how that feature or aspect could affect the proposed logging. Write down the most important questions that arise from your considerations. For example, what are the implications if the area has a relatively wet climate compared to a drier climate? How would removal of the trees affect erosion along the stream? Consider any other factors that might be important with regard to the issue but might not be obvious in the scene. For example, does the economy of this region depend on logging?

3. List any types of information you would like to know about this location to answer your questions or further clarify the problem. When considering how logging could affect erosion of the slopes, for example, we would want to know whether the plan was to log steep slopes, gentle slopes, or both.

Water supplies from the streams, lake, and groundwater:

Impact on how sunlight reaches the bare ground:

Nearby cities and towns:

Coastal environment, which features a current flowing parallel to the coast:

Birds and other animals on land; fish in the lake, streams, and ocean; creatures on the seafloor:

Optional Activities

Your instructor may have you complete one or more of the additional activities listed below.

4. List all the layers you would you want to have in a GIS database to study the issue and any remote-sensing techniques that would be needed to collect data.

5. For two aspects that you consider during this investigation, generate a research question that addresses an important point capable of being answered using evidence in the form of data and observations. List the types of data you would need to answer these two questions and how you might collect them (field studies, satellites, etc.).

6. For one of your two research questions, develop a well-conceived hypothesis—a statement, not a question, that explains what you suspect the answer to your question will be. Think about the various tools that a physical geographer uses. Which tools would you use to test your specific hypothesis?

01.15.a1

1.15

2 Energy and Matter in the Atmosphere

ALMOST ALL NATURAL SYSTEMS on Earth derive their energy from the Sun, but not all areas receive the same amount of sunlight. Instead, the amount of energy reaching Earth's surface varies from region to region, from season to season, and from hour to hour. The interaction of energy with the Earth's atmosphere and surface determines the climate, weather, and habitability of an area.

02.00.a2 Namibia, Africa

Sunlight warms the land and oceans, which in turn warm our atmosphere (◄), making some regions, such as this spectacular desert in Namibia, warmer than others.

What type of energy is in sunlight, and does all of the Sun's energy make it to Earth's surface?

The Sun rises and sets each day, except in some polar places where the Sun shines 24 hours a day during the summer (▶). In other places, on the opposite pole of the planet, there is total darkness during the same 24 hours.

What causes variations in the number of daylight hours, both from place to place and from season to season?

02.00.a3 Philippines

Extratropical regions have seasons, changing from the warm days of summer to the cold, snowy times of winter (▶).

What causes the change from season to season, and do all areas experience summer at the same time?

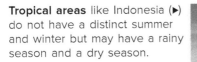

02.00.a4 Sapporo, Japan

Arctic Ocean

Asia

Pacific Ocean

Australia

Antarctica

Tropical areas like Indonesia (▶) do not have a distinct summer and winter but may have a rainy season and a dry season.

Why do some regions experience summer and winter but others do not?

02.00.a5 Banda Island, Indonesia

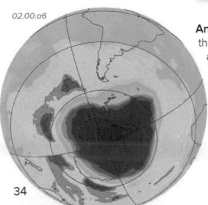

02.00.a6

Antarctica, during its winter, has a dramatic thinning of the overlying ozone layer in the atmosphere, with the area affected shown here in purple (◄).

What is ozone, what causes this thinning, and why is there so much global concern about this phenomenon when ozone makes up less than 0.001% of all the gas in the atmosphere?

TOPICS IN THIS CHAPTER

Ancient people, such as the builders of Stonehenge (▶) 4,000 years ago, used changes in the position of the Sun over time to schedule important activities, such as the planting of crops.

What causes the seasons, and what indicates the end of one season and the start of another?

02.00.a7 Stonehenge, Wiltshire, England

02.00.a8 West-Central IL

Greenland Arctic Ocean

Asia

Atlantic
Ocean

Africa

South
America

Indian
Ocean

Atlantic
Ocean

Antarctica

02.00.a1

Clouds, as in the thunderstorm above (▲), consist of small drops of water and ice crystals. The water to make the drops and crystals evaporated from the surface using energy from the Sun.

How much energy is needed to cause evaporation, and where does that energy ultimately go?

02.00.a9 Earth and Moon, Lunar Reconnaissance Orbiter

A spacecraft orbiting the Moon combined an image it took of Earth with those taken of the lunar surface to produce this spectacular portrait (▲) of our planet.

Where does light coming from Earth originate, and does this light indicate that Earth is emitting energy?

2.0

2.1 What Is the Atmosphere?

A RELATIVELY THIN LAYER OF GAS—the *atmosphere*—surrounds Earth's surface. The atmosphere shields us from harmful high-energy rays from space, is the source of our weather and climate, and contains the oxygen, water vapor, and other gases on which all life depends. What is the character and composition of the atmosphere, and how does it interact with light coming from the Sun?

A What Is the Character of the Atmosphere?

Examine this view of Earth, taken from a spacecraft orbiting high above the Earth. As viewed from space, Earth is dominated by three things: the blue oceans and seas, the multicolored land, and clouds. If you look closely at the very edge of the planet, you can observe a thin blue fringe that is the atmosphere. From this perspective, the atmosphere appears to be an incredibly thin

02.01.a1 Arabian Peninsula

02.01.a2 San Luis Valley, CO

As viewed from the ground, the atmosphere mostly appears as a blue sky with variable amounts of clouds, which are commonly nearly white

layer that envelops our planet, separating us from the dark vastness of space. Clouds, which are so conspicuous in any image of Earth taken from space, mostly circulate within the lower atmosphere, bringing rain and snow. The winds that move the clouds are also within the lower atmosphere.

or some shade of gray. During sunset and sunrise, the sky can glow reddish or orange. The colors of the sky and clouds are due to the way sunlight interacts with matter in the atmosphere.

B What Is the Structure of the Atmosphere?

The atmosphere extends from the surface of Earth upward for more than 100 km, with some characteristics of the atmosphere going out to thousands of kilometers. The atmosphere is not homogeneous in any of its attributes, but instead has different layers that vary in temperature, air pressure, and the amount and composition of gases. Each layer has the term "sphere" as part of its name, referring to the way each layer successively wraps around the Earth with a roughly spherical shape. Examine the figure below and then read the text from the bottom left, starting with the lowest and most familiar part of the atmosphere.

4. The top layer is the *thermosphere*, derived from the Greek word for heat because this layer, surprisingly, can become very hot (more than 1,500°C) as gas particles intercept the Sun's energy. It is the altitude where the spectacular *auroras* (i.e., "Northern Lights") originate from interactions of solar energy and energetic gas molecules.

3. Above the stratosphere is the *mesosphere*, where "meso" is Greek for "middle," as this layer is in the middle of the atmosphere. The mesosphere starts at 50 km, the top of the stratosphere, and goes up to more than 80 km (~50 miles) in altitude. The upper part of the mesosphere is very cold (−85°C, −120°F), and is considered by many scientists to be the coldest place within the Earth system. It is within the mesosphere that most small meteors burn up, producing the effect called "shooting stars." Radio waves from Earth bounce off this layer and the overlying layer, allowing us to hear radio stations from far away.

2. The next layer up is the *stratosphere*, beginning at an altitude of about 10 km above sea level, at about the elevation of Earth's highest peaks. The name is derived from a Latin term for spreading out, referring to its layered (not mixed) character. Temperatures are also stratified, varying from cooler lower altitudes to warmer upper ones. The lowest part of the stratosphere is an altitude at which many commercial jets fly because the air offers less resistance to motion, allowing appreciable fuel savings.

1. The lowest layer is the one with which we surface-dwellers interact. It contains the air we breathe, clouds, wind, rain, and other aspects of weather. This layer is the *troposphere*, with the name "tropo" being derived from a Greek word for turning or mixing, in reference to the swirling motion of clouds, wind, and other manifestations of weather.

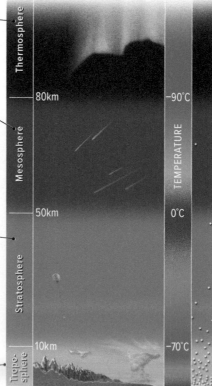
02.01.b1

5. On the right side of this figure are scattered bright dots that represent gas molecules in the atmosphere. The molecules are infinitely smaller and more abundant than shown here. Note that the molecules are concentrated lower in the atmosphere and become much more sparse upward. Over 70% of the mass of the atmosphere is in the lowest 10 km, that is, within the troposphere. The mesosphere and thermosphere contain only a few tenths of a percent of the atmosphere's mass.

C What Is the Composition of the Atmosphere?

The atmosphere is not completely homogeneous in its vertical, horizontal, or temporal composition, but chemists and atmospheric scientists have estimated its average composition, as represented in the graph and table below.

1. The atmosphere is held in place by the balance between gravity (which is directed downward and keeps the gases close to the surface) and a buoyancy force (which is directed upward and exists because material tends to flow toward the vacuum of outer space). Near the surface, the greater weight of the overlying atmosphere results in more molecules being tightly packed close to the surface of the Earth and a rapid thinning of the number of molecules with distance upward in the atmosphere, away from the surface. This effect of gravity also influences the composition of the atmosphere, with a higher proportion of heavier gases, like oxygen, low in the atmosphere (in the troposphere) and a higher proportion of lighter gases, such as hydrogen, higher up in the atmosphere.

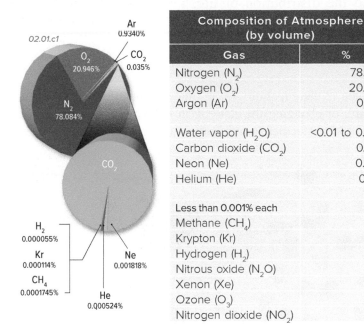

02.01.c1

Composition of Atmosphere (by volume)	
Gas	**%**
Nitrogen (N$_2$)	78.084
Oxygen (O$_2$)	20.946
Argon (Ar)	0.934
Water vapor (H$_2$O)	<0.01 to 0.400
Carbon dioxide (CO$_2$)	0.039
Neon (Ne)	0.002
Helium (He)	0.001

Less than 0.001% each
Methane (CH$_4$)
Krypton (Kr)
Hydrogen (H$_2$)
Nitrous oxide (N$_2$O)
Xenon (Xe)
Ozone (O$_3$)
Nitrogen dioxide (NO$_2$)

2. As shown by this diagram and table, the two dominant gases are nitrogen (78%) and oxygen (21%), followed by argon. Several other gases, such as a variable amount of water vapor, along with carbon dioxide (CO$_2$), methane (CH$_4$), nitrous oxide (N$_2$O), and ozone (O$_3$) play significant roles in global climate through their interaction with energy emitted by the Sun and re-emitted by the Earth.

3. The atmosphere also contains various types of solids and liquids called *aerosols*, such as dust, industrial pollutants, and tiny drops of liquid from volcanic eruptions. Aerosols play an important role in the energy balance of the Earth and aid in the formation of clouds and precipitation.

D How Does the Atmosphere Interact with Energy from the Sun?

Gas in the atmosphere interacts with visible light and other energy radiated by the Sun, as well as energy reflected from and radiated from Earth's surface. An understanding of the possible types of interactions helps explain what we see every day, such as colors, as well as the underlying causes of weather and climate. Here we discuss four types of interactions: *transmission, reflection, absorption,* and *scattering*. The discussion will emphasize light, but the principles are applicable to other forms of radiant energy, such as ultraviolet energy.

Transmission—An object can be entirely or mostly transparent to light and other forms of radiant energy, in the way that a clear glass sphere permits most light to pass through. Allowing such energy to pass through is called *transmission*. The atmosphere is largely transparent to visible light, allowing it to pass through and illuminate Earth's surface. For some types of radiant energy the atmosphere is only partially transparent, because certain molecules interact with some of the incoming radiation. Energy from the Sun is efficiently transmitted through space, which is nearly a vacuum, lacking many molecules that could interact with the energy.

Absorption—Objects retain some of the energy that strikes them, and this process of retention is called *absorption*. Objects have varying degrees of absorption. This dark, dull-textured sphere is highly absorbent, retaining most of the light that strikes it, accounting for the sphere's dark color (not much light coming back off the surface). Energy that is absorbed by an object can be released back in another form (like heat from the dark ball). The giving out of radiant energy is called *emission*.

02.01.d1

Reflection—Instead of passing through an object, some light can bounce off the object, the process of *reflection*, as illustrated by light reflecting off a polished metal sphere. Not all light and energy bounce off real objects, so reflectivity can be thought of as a continuum from objects being perfectly reflective to having no reflectivity. For visible light, objects that are white or made of shiny, polished metal are highly reflective, whereas dark, rough objects have low reflectivities. In the summer Sun, a white car, which reflects much of the Sun's heat, is cooler than a dark car that is not so reflective. If an object, like a leaf, preferentially reflects green light, we see the object as green.

Scattering—Most objects reflect some light, but in a way that disperses the energy in various directions, *scattering* the light. The shiny but rough sphere shown here has a highly reflective surface, but the roughness causes light to be scattered in various directions. Such scattering occurs when light and other energy strike the land, and it is also caused by certain types of aerosols and gases, like water vapor, in the atmosphere. Gases in the atmosphere preferentially scatter blue light, which spreads out in all directions through the atmosphere, causing the dominantly blue color of the sky.

Before You Leave This Page

☑ Describe what the atmosphere is, including its average composition.

☑ Sketch and describe the layers in the atmosphere.

☑ Describe four ways that matter interacts with light and other radiant energy.

2.1

2.2 What Is Energy and How Is It Transmitted?

THE TRANSMISSION OF ENERGY, and the interactions between energy and matter, define the character of our planet and control weather, climate, and the distribution of life, including humans. Here, we examine the fundamentals of energy, including what it is, where it comes from, and how it is moved from one place to another.

A What Is Energy?

All matter contains *energy*, which is the capability of an object to do work, such as pushing or pulling adjacent objects, changing an object's temperature, or changing the state of an object, as from a liquid to a gas. How is such energy expressed at the scale of atoms and molecules (combinations of atoms)?

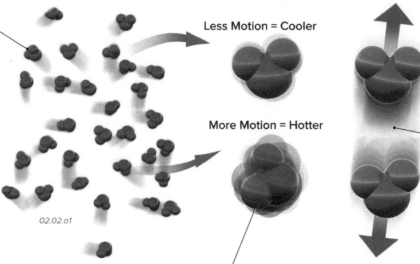

02.02.a1

Less Motion = Cooler

More Motion = Hotter

1. Energy is expressed at an atomic level by *motions* of atoms, molecules, and their constituent parts. These motions include changes in position and in-place vibrations and rotations. The more motion the atoms and molecules display, the more energy the system contains. Atoms and molecules in a gas can move at over 200 m/s, or about 500 mi/hr. These fast-moving objects collide with adjacent ones, causing atmospheric pressure in a similar fashion to the way in which air molecules hold out the walls of an inflated balloon.

2. The temperature of an object is a measure of the average energy level (motions) of its molecules and atoms. The molecules of all objects on Earth, regardless of temperature, are moving, some more than others. This type of energy, due to motions of objects, is called *kinetic energy*.

3. Energy can also be tied up *within* the atomic or molecular structure of matter. When we heat a liquid, we impart energy to the constituent molecules, causing some of the molecules to move apart, escaping as vapor. But these molecules also carry stored energy, which can be released when the molecules recombine into a liquid. This type of energy is called *potential energy*, because it is not being expressed directly, but could potentially be released. Where potential energy is related to a change in the *state of matter*, such as from a liquid to a gas, as in the example presented here, it is called *latent energy* (latent means hidden).

B How Does Energy Relate to the State of Matter?

In our everyday world, matter exists in three different forms or *states*—solid, liquid, and gas. The energy content of the material determines which of these states of matter dominates at any time and place.

02.02.b1

1. When matter is in a *solid state*, the constituent atoms and molecules are bound together, such as in the structured internal architecture of a crystal. The energy levels (motions) of the atoms and molecules are low enough that the solid can withstand the vibrations and other motions without coming apart. As a solid is heated up by an external energy source, like the flame, the motions become more intense.

2. If enough energy is added to a solid, the motions begin to break the bonds that hold the solid together. The material begins to melt, turning into a *liquid*, which is a collection of mobile atoms and molecules that more or less stay together but are not held into the rigid form of a solid. For example, adding energy increases the vibrations of water molecules in an ice crystal until the bonds holding the ice crystal together begin to disintegrate and the ice melts.

3. If even more energy is added, such as by placing the liquid water on a hot burner, the added energy causes even more energetic motions of water molecules in the liquid, allowing more and more molecules to break free of the liquid and enter the air as a *gas*—water vapor. As a result, atoms and molecules in their gaseous state are more energetic than in their solid or liquid equivalents. Not only are the molecules in a gas vibrating and rotating intensely, molecules can now move freely through the air at high speeds—having a large amount of kinetic energy.

C   What Occurs During the Processes of Warming and Cooling?

When something changes temperature, we say that it is warming or cooling, but what is happening during a change in temperature? For warming or cooling to occur, energy must be transferred from one object to another, such as from a flame to the air around it.

When an object increases in temperature, it is *warming*. Warming occurs where an object *gains energy* from the surroundings, resulting in the increase in temperature. You gain energy when you feel the warming benefit of a campfire (▶). There is no need to touch the flames to feel the warming effect. Energy is transmitted from the fire to the surroundings. When you heat up water in a pan on the stove, energy is transferred from the burner to the pan and then from the pan to the water. If you add enough energy, the water molecules become so warm and energetic that they start to escape the pan during the process of boiling.

02.02.c1

02.02.c2

Cooling is the loss of energy to the surroundings. When you are out in cold weather, you lose energy to the cold air, causing you to feel cold. A similar process occurs when you add warm water to ice cubes in a glass dish (◀). Initially, the water is warmer than the ice, but once the water and ice are in contact, molecular motion in the water is transmitted to the cooler ice. This transfer of energy adds more energy to the ice, eventually melting it. As the surrounding water loses energy to the ice, it cools, resulting in cold water. Cooling indicates a loss of energy, whereas warming indicates a gain of energy.

D   What Are the Four Types of Energy Transfer?

Heat is thermal energy transferred from higher temperature to lower temperature objects. Heat transfer, also called *heat flux* or *heat flow*, results when two adjacent masses have different temperatures. The four mechanisms of heat transfer are *conduction, radiation, convection*, and *advection*. An understanding of the various mechanisms of energy transfer is crucial in understanding weather, climate, evolution of landscapes, and many of our daily activities.

Conduction — A water-filled pan on a burner gets hot as thermal energy is transferred by direct contact between the burner and pan, and the pan and water. Heat transfer by direct contact is *conduction*, which involves transferring thermal energy from the warmer object (with more energy) to the cooler one (with less energy). Energy can only be transferred from the more energetic one to the less energetic ones. Molecules are most densely packed in solids and least densely packed in gases, so conduction is a more important mode of transfer in solids, whereas gases conduct energy much less efficiently.

02.02.d1

Advection — Moving a pan full of hot water away from the stove also transfers heat from one place to another. Energy transfer by the *horizontal movement* of a material, such as moving the pan sideways off the burner, is called *advection*. Fog along many coastlines results from warm air flowing sideways and mixing with cooler air over the coast. This horizontal transfer of energy via the moving air is advection.

Radiation — A hot burner on a stove can warm your hands a short distance away. Such warming occurs because heat from the burner radiates through the air, a process called *radiation* or *radiant heat transfer*. Radiation is energy transmission by means of electrical and magnetic fields. All objects constantly emit various kinds and quantities of this particular type of electrical and magnetic radiation, which is known as *electromagnetic radiation*. This form of energy is even capable of passing through a vacuum.

Convection — Energy is conducted through the base of the pot and into the lowest layer of water contacting the pot. These molecules move faster, requiring more room to do so, which causes the volume occupied by the warmer water to increase and its *density* (mass per unit volume) to decrease. Being less dense than the overlying cooler water, the warm water rises. When the rising water reaches the surface, it cools and flows back down the sides of the pot. This type of vertical heat transfer by flow of a gas, a liquid, or a weak solid is *convection*. If the material flows around a circular path, as in the pan, we use the term *convection cell*.

Before You Leave This Page

✓ Describe what constitutes the energy in an object.

✓ Describe how energy relates to states of matter.

✓ Describe the four major mechanisms of energy transfer and provide an example of each.

2.2

2.3 What Are Heat and Temperature?

THE TERMS HEAT AND TEMPERATURE are used every day, but what do they actually mean? *Temperature* is a measure of the object's internal kinetic energy—the energy contained within molecules that are moving, and *heat* is thermal energy transferred from one object to another. Moving molecules drive many processes in the Earth-ocean-atmosphere system, such as evaporation, precipitation, and erosion.

A What Is Sensible Heat?

1. The term heat is used in two ways. Scientists use *heat* to refer to the transfer of thermal energy from a warmer object to a cooler one or to the energy that is transferred in this way. The amount of heat is specified in a unit called a *Joule*, a measure of work or energy. Two common examples illustrate heat nicely.

2. What happens when you hold a cup of hot tea? Your hand feels heat coming from the cup. You are feeling the transfer of thermal energy from the cup to your hand. For this to happen, water molecules were heated and made to move. Once in the cup, the moving molecules collided with the inside of the cup, warming it. That heat is transferred through the cup and against your hand via conduction—from the burner to the bottom of the kettle, to the water, to the cup, and finally to your hand. Conduction and convection both help distribute heat within the kettle and the cup.

02.03.a1

3. What do you think happens when you hold a cold glass of ice water? In this case, the molecules in your hand are more energetic than the ones in the cold drink, so heat is transferred from your hand to the glass. Your hand feels cold because you are losing thermal energy to the cold glass.

4. This type of heat, which changes the temperature of two objects through exchange, is called *sensible heat* because we can sense it. But do we sense the actual temperature of an object or just the heat gain or loss? Try this experiment: Find a metal object and a wooden or plastic object in the same place. Place your hand on each and observe what you feel. Go do it, and then come back and continue reading. There, you no doubt sensed that the metal felt colder than the wood or plastic, but both have been in the room for a while and so are exactly the same temperature. This experiment shows that we sense heat gain and loss more than the actual temperature. In your experiment, metal conducted heat away from your hand faster, and so felt colder, but it wasn't.

B What Is Temperature and How Do We Measure It?

Temperature is a quantitative measure of the average kinetic energy of molecules in an object—in other words, the hotness or coldness. Measurement of temperature of an object, whether a solid, liquid, or gas, involves the transfer of sensible heat from the object to some type of measuring device, usually a thermometer. Official temperatures are measured in a variety of ways, depending on the accuracy that is required and the location where temperature is to be measured.

02.03.b1

02.03.b2 Redoubt volcano, AK

02.03.b3

02.03.b4

The mercury-in-glass thermometer is the most familiar tool to measure temperature. Mercury is a convenient element for this task because many of its physical properties remain consistent over the range of temperatures experienced on Earth. As the mercury's temperature increases, it expands and fills more of the tube. When it cools, the mercury contracts and withdraws down the tube.

Infrared thermometers calculate the temperature of a solid or liquid surface by pointing the sensor at the surface and measuring a range of wavelengths of energy emitted by that surface. Equations then relate wavelength to energy and energy to temperature. Infrared thermometry is a form of remote sensing, convenient when the surface is too far away, or too dangerous, to be measured directly.

When we need to measure sudden and slight temperature changes very precisely, we use special thermometers in which differences in energy content cause a thermoelectric response that can be wired to a computerized data recorder. The temperature can be calculated using specific equations that derive the amount of heat as a function of the amount of electrical current and the resistance of the electrical circuit.

Temperatures in the atmosphere are usually measured by weather balloons. These include instrument packages called *radiosondes* that measure a range of variables at various heights as the balloon ascends. Temperature is measured using thermoelectric principles. Wind, humidity, and other variables are also measured and relayed to the ground via radio signal.

C How Does the Fahrenheit Scale Relate to the Celsius and Kelvin Scales?

Most Americans are familiar with the *Fahrenheit temperature scale*, in which 32° represents the freezing point of water and 70° is a comfortable temperature. Nearly all nations except the U.S. use the *Celsius scale*, and scientists use the Celsius scale or a related scale called the *Kelvin scale*. We typically compare the scales with reference to the temperatures at which water freezes or boils, called the *freezing point* and *boiling point*, respectively.

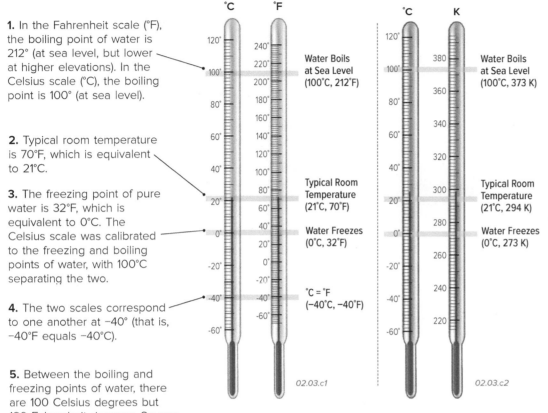

1. In the Fahrenheit scale (°F), the boiling point of water is 212° (at sea level, but lower at higher elevations). In the Celsius scale (°C), the boiling point is 100° (at sea level).

2. Typical room temperature is 70°F, which is equivalent to 21°C.

3. The freezing point of pure water is 32°F, which is equivalent to 0°C. The Celsius scale was calibrated to the freezing and boiling points of water, with 100°C separating the two.

4. The two scales correspond to one another at −40° (that is, −40°F equals −40°C).

Water Boils at Sea Level (100°C, 212°F)

Typical Room Temperature (21°C, 70°F)

Water Freezes (0°C, 32°F)

°C = °F (−40°C, −40°F)

02.03.c1

Water Boils at Sea Level (100°C, 373 K)

Typical Room Temperature (21°C, 294 K)

Water Freezes (0°C, 273 K)

02.03.c2

5. Between the boiling and freezing points of water, there are 100 Celsius degrees but 180 Fahrenheit degrees. So one Fahrenheit degree is only 100/180 (or 5/9) of a Celsius degree. This fact, along with the different "starting point" (i.e., the freezing point of water) forms the basis of converting between Fahrenheit and Celsius. To convert from Fahrenheit to Celsius, we must first subtract 32 degrees from the Fahrenheit temperature to allow for the fact that the starting point is offset by 32 degrees in the two systems. Then we must multiply by 5/9 to allow for the differences in the value of a degree on each scale. The equations for converting back and forth are as follows:

$$C = 5/9 \times (F - 32)$$

$$F = (C \times 9/5) + 32$$

6. Both the Celsius and the Fahrenheit scales are "arbitrary" in the sense that zero degrees doesn't mean that there is a lack of internal energy. Likewise, a doubling of the Fahrenheit temperature does not mean that there is twice as much internal energy. In scientific calculations, we need a temperature scale that allows us to relate changes in internal energy to the absolute amount of heat gained or lost by a system.

7. The Kelvin temperature scale (K) was devised as an "absolute" temperature scale to remedy these problems. In the Kelvin system, 0 K corresponds to the temperature at which no internal energy exists and all molecular motion theoretically ceases. This temperature is known as *absolute zero*, and is −273°C or −460°F. Doubling the internal energy of molecules would double their motion and be associated with a doubling of the Kelvin temperature.

8. In the Kelvin system, water freezes at 273 K and boils at 373 K. Converting from Celsius to Kelvin temperature is easy:

$$K = C + 273$$

or

$$C = K - 273$$

Conversions between Fahrenheit and Kelvin can be made by converting first to Celsius and then to Kelvin. Note that we do not use a degree symbol with the Kelvin scale.

D How Many Stations Report Temperature?

This globe (▶) shows part of the worldwide distribution of weather stations that have temperature data sets for at least many decades. The distribution is uneven, with most stations being on continents. Most stations are concentrated in densely populated areas, especially in the lowlands of more developed regions, such as the eastern U.S. Other regions, such as the center of South America (the Amazon rain forest), have very few stations to represent rather large areas. In recent decades, remote-sensing techniques have allowed truly global temperature coverage, but these data sets are not available as far back in time.

02.03.d1

Before You Leave This Page

✓ Explain what temperature is.

✓ Explain how various instruments to measure temperature work.

✓ Describe the strengths and weaknesses of the various temperature scales.

✓ Convert back and forth between the three temperature scales.

✓ Characterize the distribution of temperature-monitoring stations.

2.4 What Is Latent Heat?

WATER OCCURS IN ALL THREE PHYSICAL STATES—solid, liquid, and gas—at temperatures common on Earth. Although the chemical structure of water remains unchanged from state to state, the three states, also called *phases*, are differentiated by the physical spacing and connections of the water molecules. Considerable quantities of energy, contained as *latent heat*, are involved in these changes of state, and act as moderators of global climate.

A What Are the Forms of Latent Heat?

The chemical substance *water* consists of two hydrogen atoms bonded to one oxygen atom, with a chemical formula of H_2O. The change in state between any two of these phases requires an addition of energy or involves release of energy, depending on which direction the change is occurring (e.g., liquid to solid versus solid to liquid). In the figure below, a blue-to-red arrow indicates that a change (e.g., melting) requires energy to proceed, whereas a red-to-blue arrow indicates that energy is released.

1. When ice is placed in warmer surroundings, like an ice cube on a kitchen counter, energy from its environment flows into the ice, increasing the internal motions of water molecules in the solid, crystalline structure. At first the ice heats up (increases in temperature), but once it reaches a certain temperature (its melting point), it begins to melt. *Melting* requires energy to be added.

2. During melting, energy input into the system is stored (absorbed) in water molecules of the liquid—as *latent heat*. The latent heat associated with melting is called the *latent heat of fusion*. If enough energy is removed for the liquid water to be converted back into ice during *freezing*, the latent heat stored in the water molecules is released back into the surroundings as heat (also called *latent heat of fusion*). For freezing to continue, this released energy must be dispersed into the surroundings. The warm flow of air coming from the back or bottom of a household freezer is the heat being dispersed from the cooling system.

3. Conversion from a liquid phase to a vapor phase, *evaporation*, also requires an input of energy from the surroundings, as when you boil water in a pan. During evaporation, energy added to the liquid breaks the bonds holding the water molecules together, allowing molecules to escape as a gas. The liberated gas molecules are moving fast and so have increased kinetic energy and carry energy stored as *latent heat*. When the gas molecules are cooled and recombine into a liquid through the process of *condensation*, the latent heat is released back into the surroundings. The latent heat associated with evaporation and condensation is the *latent heat of vaporization*.

4. Water can also go directly from the solid state to a vapor, the process of *sublimation*. Sublimation requires energy from the surroundings and stores latent heat in the gas molecules —the *latent heat of sublimation*. The reverse process, converting water vapor directly into ice, is called *deposition* and is the main way snowflakes form. Deposition releases the latent heat back into the environment, but when it is snowing it is cold enough that we would not easily notice any addition of latent heat to the cold air.

Melting Freezing LATENT HEAT OF FUSION Evaporation Condensation LATENT HEAT OF VAPORIZATION Sublimation Deposition LATENT HEAT OF SUBLIMATION 02.04.a1

B What Happens to Temperature During Melting and Boiling?

1. An interesting thing happens to temperature when we melt ice and evaporate water, as shown by the graph to the right. This graph plots the energy input into the system (measured in a unit of energy called a kilojoule) versus the resulting temperature if we start with a kilogram of ice. The process begins in the lower left corner, with ice at −20°C, well below its melting temperature (0°C).

2. The initial input of energy into the system causes ice to increase in temperature, represented by the first inclined, red part of the line. The increase in energy is expressed as sensible heat.

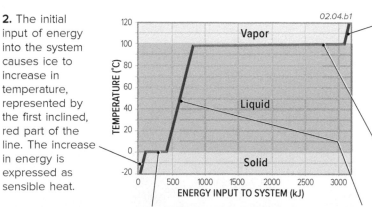

02.04.b1

6. Once all the water has become vapor, further heating causes the steam to increase temperature (inclined, red line). In this entire process, more energy was used to change states (brown horizontal lines) than was used to increase temperature (the inclined, red lines). Of the total amount of energy used, less than 20% went to change temperature and more than 80% was used to change state!

5. As the water starts to boil, it does not increase in temperature (long, horizontal brown line). All increase in energy is used to convert liquid into the vapor phase and is stored as latent heat.

3. When the temperature reaches 0°C, the melting point, the ice starts to melt. The temperature does not change during melting (as shown in the short, horizontal brown line). Instead, all the increase in energy is going into breaking the bonds and is stored as latent heat.

4. Once all the ice is melted, the increase in energy again causes an increase in temperature (sensible heat), as shown by the second inclined red line. The temperature of the water increases until it reaches the boiling point (100°C).

C What Does Latent Heat Do to the Surroundings?

Latent energy added to the water molecules or released by the water molecules allows the phase change to proceed, but it also impacts the temperature of the surrounding environment. More than five times the energy is involved in evaporating or condensing water (the latent heat of vaporization) than in raising the temperature of the same mass of water from the freezing point to the boiling point. The large quantities of latent heat have a huge role in many aspects of our world, including changes in atmospheric temperature.

Water: Vapor, Liquid, and Ice

This graph shows conditions under which each phase occurs for water. For reference, a pressure of 1.0 bar is the average atmospheric pressure at sea level, and 21°C is a typical room temperature.

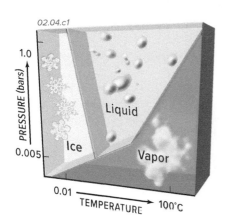
02.04.c1

Ice occurs at low temperatures, whereas liquid and water vapor are favored by higher temperatures. Higher pressure acts to hold the water molecules within the liquid rather than allowing them to escape into the air. If we cool vapor, we get liquid if at higher pressures or ice if at lower ones, causing the formation of clouds or precipitation, both of which can consist of liquid (drops of water) or ice.

Phases, such as vapor, that require more energy to form are called *high-energy states*, whereas less energetic ones are *low-energy states*.

Cooling and Heating the Air

When water in the atmosphere changes state, it releases or takes in thermal energy, heating or cooling the surrounding air. This diagram illustrates the change in air temperature during phase changes. Red arrows indicate that phase change in this direction causes the surrounding air to heat up.

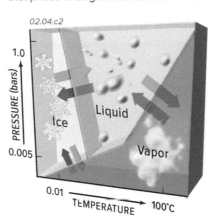
02.04.c2

Blue arrows indicate that the surrounding air must provide heat to the phase change, and so the air cools.

Energy must be released into the surroundings for water to go from a higher energy state to a lower energy state. Heat is released (red arrows) when water vapor forms droplets or ice crystals, or when liquid water freezes—this warms the surrounding air.

Heat is taken in from the surroundings (blue arrows) when ice melts or water evaporates, or ice sublimates. Any of these processes cool the air.

D What Are Examples of Latent Heat in the Environment?

02.04.d1 Hurricane Dolores

1. In 2015 Hurricane Dolores formed in the Eastern Pacific and came onshore in Southern California, where it caused heavy rainfall, flooding, and landslides. Hurricanes release a huge amount of latent heat as water vapor condensed into water drops in the clouds and as rainfall. In a single hurricane, the amount of latent heat released during condensation is equivalent to much more than all the electrical energy consumed by the world since electricity was first harnessed for human use.

3. A cold beverage in a can, bottle, or glass warms as a result of conduction from its warmer surroundings, but it warms even more quickly because condensation (as expressed by growing water drops) releases latent heat on the outside of the container.

02.04.d3

02.04.d2 Florida

2. Severe freezes are a serious threat to the citrus industry in the southern and southwestern U.S. Sustained subfreezing temperatures may ruin the crop and even destroy the tree. To prevent extensive losses, grove owners spray the trees with water, which freezes on the crop. At first sight the weight of the accumulated ice only seems to add to the damage by breaking off fruit and limbs, but each kilogram of water that freezes on the trees releases enough energy to the fruit and plant to prevent cold temperatures from destructively freezing water within the cells of the plant.

Before You Leave This Page

☑ Explain the relationship between latent heat and changes in the state of water.

☑ Compare the quantities of energy involved in the latent heats of fusion and vaporization, compared to changes in temperatures.

☑ Sketch and describe how latent heat affects the environment, providing some examples.

2.5 What Is Electromagnetic Radiation?

ELECTROMAGNETIC RADIATION is one of the fundamental entities of nature. It dominates our daily interactions with the world, determining the color of objects, the character of the air we breathe, and the physical characteristics of the water we drink. Electromagnetic radiation is essential to the operation of weather and our climate system and to all life forms on the Earth's surface, including us.

A What Are Some Common Examples of Electromagnetic Radiation?

Electromagnetic radiation (EMR) is all around us, although we may not be aware of all of its manifestations. Its most obvious expression is as light, which is the only kind of EMR that we can observe with our eyes. EMR is also what causes sunburns, is the heat we feel from a heat lamp, and is expressed in all sorts of wireless communications, including TV, radio, and WiFi.

Visible light is one type of EMR. Visible light is itself composed of different types of EMR, which we can observe by using a prism (▶) to spread out visible light into its rainbow colors. On the following pages, we explain what causes different colors.

02.05.a1

02.05.a2

There are many types of EMR that are not visible to us. These include radio waves (like WiFi), X-rays, and microwave ovens (◀).

B What Is the Character of Electromagnetic Radiation?

Electromagnetic radiation consists of energy radiated from charged particles and manifested as interacting electrical and magnetic fields. For most applications, it is useful to think of electromagnetic radiation as a series of waves that, if unconfined, radiate out in all directions. When observed in certain ways, such as at microscopic distances and exceptionally short timescales, EMR appears to be composed of particles. Here, we will treat EMR as waves.

1. To visualize the rather abstract concept of EMR, think of it as a series of waves (▶) of electrical and magnetic energy that are moving from one place to another, in this case, from left to right. Like any waves, some parts are higher and some are lower. The direction in which a wave is moving is the *direction of propagation*. The wave shown here is propagating from left to right. To envision how such a wave moves, think about what happens when you shake the end of a rope or string. The rope curves into a series of waves, which move from your hand outward to the end of the rope.

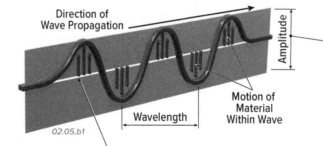

Direction of Wave Propagation

Amplitude

Motion of Material Within Wave

Wavelength

02.05.b1

2. In this type of wave, the motion of any part of the wave is mostly up and down, parallel to the small arrows on this figure—the rope stays in your hand, so it is not moving away from you, although the waves are. Motion within the wave (up and down) is perpendicular to the direction of propagation of the wave (left to right).

3. In describing such waves, whether in a rope or EMR, the term *amplitude* refers to the height of the wave, from trough to crest. The greater the difference in height between the top and bottom of the wave, the greater the amplitude. Most surfers like large-amplitude waves.

4. We use *wavelength* to describe the distance between two adjacent crests (tops) or two adjacent troughs (bottoms). The longer the distance between two adjacent crests or troughs, the greater the wavelength. If we know the speed of the wave, we can describe the *frequency*—the number of waves per second passing a point.

5. Electromagnetic waves travel through electric and magnetic fields, which can be envisioned as two mutually perpendicular planes. In this figure (▶), the electric field, containing the electrical component of the wave, is vertical (dark gray), whereas the magnetic field is horizontal (colored blue with the orange horizontal arrows).

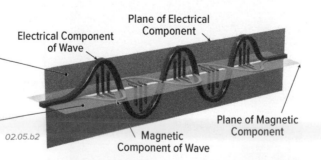

Plane of Electrical Component

Electrical Component of Wave

Plane of Magnetic Component

Magnetic Component of Wave

02.05.b2

6. The electrical and magnetic waves have the same wavelength, and the crests of the two types of waves are the same distance along the direction of wave propagation (i.e., the crests line up). Motion in both types of waves is perpendicular to the direction the wave propagates, as described in the previous figure. Electromagnetic waves get their name from these linked electrical and magnetic waves that move in unison.

C How Is Electromagnetic Energy Generated and Transmitted?

Electromagnetic radiation is generated by changes associated with charged particles—atoms and molecules produced by vibrations within the particle, by changes in the energy level of electrons, and by fusion of particles, as occurs in the Sun. The Sun is the dominant source of energy for Earth, emitting EMR at a variety of wavelengths.

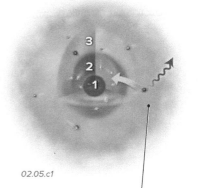

02.05.c1

1. Atoms have a tiny central core—the *nucleus*—that is so much smaller than the entire atom that it cannot be shown here. The red and yellow spheres are electrons. Groups of electrons travel around the nucleus at different distances, called *electron shells*. Each shell has a different level of energy, increasing away from the nucleus. If an electron in an outer, higher energy shell drops into an inner, lower energy shell, it must emit the extra energy as an EMR wave, which radiates out in all directions (shown here as one direction for simplification).

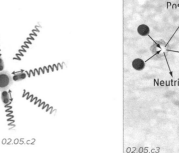

02.05.c2

2. Electrons within atoms and the bonds (shown here as springs) within molecules vibrate back and forth, changing positions slightly. This motion emits EMR, and the faster such motions occur, the higher the frequency of the EMR that is emitted.

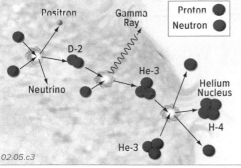

02.05.c3

3. The Sun is huge and extremely hot. It is nearly all composed of hydrogen and helium, the two lightest chemical elements. Through a series of steps, protons and neutrons join together to produce larger particles, such as helium—the process of *fusion*, which releases huge amounts of EMR that radiate outward.

4. Earth is approximately 150 million km from the Sun, and the nearly complete vacuum of space separating the two objects eliminates the possibility of conduction, convection, and advection, all of which require a medium for the transfer to occur (there is no medium in a vacuum). The Sun's immense mass means that very little material can escape the Sun's gravitational pull. Therefore, almost all energy comes to Earth via radiation, which can pass through a vacuum, as illustrated by the fact that light from a flashlight (▶) can easily pass through sealed flasks, one containing air and other containing a vacuum.

02.05.c4

D How Much Energy Does an Object Emit?

Scientists have discovered a number of important quantitative relationships, some of which are considered laws of nature, including those that govern the production and transfer of energy. One important relationship is the *Stefan-Boltzmann Law*, which relates an object's temperature to the amount of EMR it emits.

1. The Stefan-Boltzmann Law is represented graphically by this figure, which plots the *amount of emitted energy* as a function of temperature in Kelvin (K), for a specific type of object that absorbs all the energy that strikes it. We are simplifying this discussion a bit.

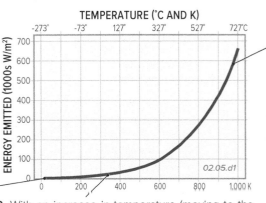

02.05.d1

2. The graph starts at a temperature of zero Kelvin (−273°C), and at this temperature the amount of emitted energy is zero. All molecular motion ceases at zero Kelvin—which is why it is called *absolute zero*—so at this temperature no radiation would be emitted. No motion means no EMR.

3. With an increase in temperature (moving to the right on the graph), the amount of energy emitted increases, slowly at first. What this indicates is that as an object increases in temperature, the object emits more energy (in the form of EMR). In other words, hot objects emit more radiation than do cold ones. Since temperature represents the amount of motion associated with the atoms and molecules within a material, the graph means that the amount of EMR increases as the amount of atomic-scale motion increases.

4. At higher temperatures, the amount of energy emitted increases dramatically as a function of temperature, much faster than it did at lower temperatures. Compare how much the amount of emitted energy increased from 200 to 400 K with how much it increases from 800 to 1,000 K. The shape of the curve indicates that hot objects emit much larger quantities of energy than cooler ones.

5. Using the equation from which this graph was derived, and factoring in some other aspects not discussed here, we can use an object's surface temperature to predict how much energy it should emit. The Earth's average temperature is 283 K, whereas the Sun's average surface temperature is 6,000 K (way off the graph)—we would predict the Sun emits much more energy than the Earth. From the calculations, the Sun should emit more than 200,000 times more energy than Earth!

Before You Leave This Page

☑ Explain what EMR is and how it is generated.

☑ Explain how the surface temperature of an object is related to the quantity of EMR it emits.

2.6 What Controls Wavelengths of Radiation?

ELECTROMAGNETIC RADIATION (EMR) takes many different forms, all of which have their own distinct properties, determined by their wavelengths. Variations in wavelength explain the existence of different colors and warming of the Earth due to climate change. How are differences in the wavelength of electromagnetic energy expressed in our world and the rest of the universe?

A   What Range of Wavelengths Does an Object Emit?

Not all molecules in an object vibrate or move at the same speed, so an object emits energy with some variation in the wavelengths and amounts of energy it emits. Some molecules vibrate or move faster, emitting EMR with shorter wavelengths and higher energies. Other molecules vibrate or move more slowly, resulting in EMR with longer wavelengths and lower energies. Temperature is a measure of the *average energy content* of all molecules in an object.

Molecules in a cube-shaped container (▶) are color coded by *energy level*; bluish green is lowest, brown is highest, and other colors are in between. Molecules possessing the average energy level (i.e., temperature of the object) are in green. Observe the relative proportions of the different energy levels (that is, the different amounts of each color of molecule). Which are most abundant and which are least abundant?

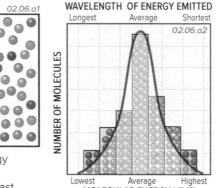

02.06.a1

WAVELENGTH OF ENERGY EMITTED

Longest   Average   Shortest

02.06.a2

NUMBER OF MOLECULES

Lowest   Average   Highest
MOLECULAR ENERGY LEVEL

This graph (◀) shows the number of molecules for each class of energy in the square object we just examined. The height of each column represents the number of molecules emitting energy at that wavelength.

The peak of the graph represents the wavelengths and energy levels that are most common. Columns to the left of the peak have lower energy and longer wavelengths than the peak, whereas those to the right have higher energy and shorter wavelengths. Taking into account all the molecules, there will be an average energy level emitted by this material at this temperature (near the peak).

B   What Type of Energy Is Emitted from an Object?

The motion of molecules within an object is related to the object's overall temperature. There is an important relationship between temperature and the wavelength of energy emitted by that object.

1. Examine the figure below, which shows three blocks of the same material but at different temperatures, with the blue block being cool, the brown block being warmer, and the red block being the hottest. Coming off each block are arrows depicting the amount of EMR being emitted (represented by the number of arrows) and the dominant wavelengths (size of waves on arrows). Observe how the amounts and wavelengths relate to temperature.

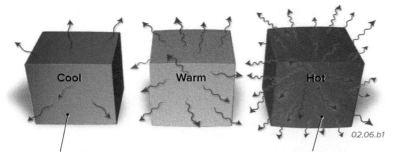

Cool   Warm   Hot

02.06.b1

2. If an object is relatively cool, the motion of its constituent atoms and molecules will be relatively slow. This results in less EMR emitted and energy with a longer wavelength.

3. As the temperature of an object increases, it not only emits more EMR (as shown by more arrows for the warm and hot blocks in the figure), but the wavelength of the EMR becomes shorter at higher temperatures. In a hot object, the atoms and molecules are moving relatively fast, and they change energy states more often, producing EMR with a shorter wavelength. If you shake a rope or string rapidly, you make shorter waves than if you shake it slowly. Go try it!

4. Whether an object is cool or hot, it emits a range of wavelengths of EMR, as described in the section above. However, scientists discovered a numeric relationship relating the dominant wavelength of EMR to temperature, a relationship known as *Wien's Law.*

5. The red curve on this graph (▶) represents Wien's Law, plotting the dominant wavelength of EMR emitted as a function of temperature. The curve is highest on the left, indicating that an object with a low temperature emits EMR with very long wavelengths.

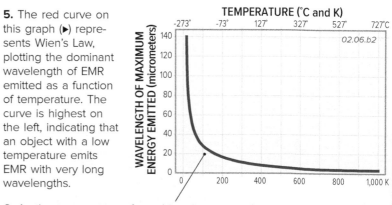

TEMPERATURE (°C and K)

-273°   -73°   127°   327°   527°   727°C

02.06.b2

WAVELENGTH OF MAXIMUM ENERGY EMITTED (micrometers)

0   200   400   600   800   1,000 K

6. As the temperature of an object increases, the wavelength decreases very rapidly at first and then more slowly toward higher temperatures. Wien's Law indicates that cold objects emit much longer wavelength energy than hot objects. Using this graph or the equation associated with Wien's Law, we can use the temperature of an object to then predict the dominant wavelength of EMR it emits. For the Sun's very high temperatures (off the curve to the right), we predict energy wavelengths of 0.5 μm (micrometers). The Earth's much cooler temperature yields longer wavelengths of about 10 μm.

C   What Is the Electromagnetic Spectrum?

Objects in the universe emit a diversity of wavelengths of electromagnetic radiation. EMR varies from relatively long-wavelength, low-energy radio waves to short-wavelength, high-energy X-rays and gamma rays. The different wavelengths of electromagnetic radiation are arranged in what is called the *electromagnetic spectrum*, shown below. Inspect this figure and read the associated text. In this figure, the longer wavelengths (measured in meters) are at the top. Accordingly, shorter frequencies, which indicate the number of waves per second (measured in Hertz), are at the top.

02.06.c2

1. The longest wavelengths of EMR are *radio waves*. These include the waves that carry signals from typical FM and AM radio stations, and also include even longer radio waves called VLF (for very low frequency), with wavelengths of up to more than 100 km. VLF waves can penetrate significant depths of water and so are used for communication with submarines.

2. *Microwaves* are another form of EMR, with shorter wavelengths, higher frequencies, and higher energies than radio waves. Microwave ovens use a specific frequency of microwave that energetically excites (heats) water molecules.

3. Next on the scale is *infrared energy* (IR). Although we cannot see IR, our skin is sensitive to it, so we often think of infrared radiation as heat. Infrared energy is incredibly important on Earth, playing a key role in keeping our planet a hospitable temperature. There are several types of IR, including *thermal-IR*, which is close to microwaves in wavelength, and *near-IR*, which is near to visible light, the next entry on the spectrum.

4. *Visible light* occupies a relatively narrow part of the spectrum. It varies from red colors at long wavelengths to violet colors at short ones. Orange, yellow, green, and blue are in between red and violet.

5. Next to, and with shorter wavelengths (higher frequencies) and more energy than visible light, is *ultraviolet light* (UV). Ultraviolet is more energetic than visible light and is known to cause skin cancers and possible genetic mutations.

6. The shortest wavelength (highest frequency and therefore most energetic) waves are *X-rays* and *gamma rays*. These are potentially harmful. The energy in X-rays is used in medical technology as it will pass through soft tissue, but not bone.

(Hz) — INCREASING FREQUENCY
(m) — INCREASING WAVELENGTH

1
10^2
10^4
10^6
10^8
10^{10}
10^{12}
10^{14}
10^{16}
10^{18}
10^{20}
10^{22}
10^{24}

10^8
10^6
10^4
10^2
10^0
10^{-2}
10^{-4}
10^{-6}
10^{-8}
10^{-10}
10^{-12}
10^{-14}
10^{-16}

Long Radio Waves
AM Radio
FM Radio
Micro-waves
Infrared
Ultraviolet
X-Rays
Gamma Rays

02.06.c1

Visible Spectrum

INCREASING WAVELENGTH IN MICROMETERS (µm)

0.7
0.6
0.5
0.4

7. The wavelengths of visible light are less than a millionth of a meter (10^{-6} m), which is a measure called a *micrometer* or *micron* for short and depicted as µm. The different colors of visible light are different wavelengths of energy, ranging between 0.4 µm (violet) and 0.7 µm (red). The component wavelengths of visible light can be observed (▲) when the light is split by an optical prism or in a rainbow.

8. The human eye is sensitive to radiation of the wavelengths between violet and red, and this is why this portion of the EMR spectrum is known as "visible light." The Sun's wavelengths are concentrated at 0.5 µm, where we could predict them based on Wien's Law. This wavelength is in the middle of visible light, coinciding with blue light. It is not a coincidence that the human eye developed to detect EMR of the wavelengths that carry most energy from the Sun.

9. The Sun produces huge quantities of energy per square meter of its surface, as indicated by the yellow area on the graph to the right. It also produces most of its energy near 0.5 µm, centered on the wavelength of visible light; these wavelengths are called *shortwave radiation*. The Sun emits a very low proportion of its energy below 0.1 µm or above 1.0 µm in wavelength.

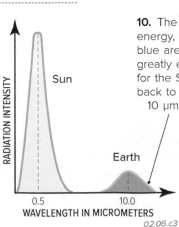

RADIATION INTENSITY
Sun
Earth
0.5   10.0
WAVELENGTH IN MICROMETERS
02.06.c3

10. The Earth emits low quantities of energy, as indicated by the relatively small blue area on the graph, which is here greatly exaggerated relative to the curve for the Sun. Most of that energy is emitted back to space at wavelengths of around 10 µm, called *longwave radiation*. As we will discover, the fact that Earth receives its energy at short wavelengths, and returns it at wavelengths 20 times longer, is of fundamental importance to maintaining temperatures on Earth favorable for life.

Before You Leave This Page

☑ Describe why an object emits a range of EMR wavelengths.

☑ Describe the relationship among molecular motions, wavelengths of emitted EMR, and temperature.

☑ Summarize the electromagnetic spectrum, indicating the relative order of different types of EMR on the spectrum.

2.6

2.7 What Causes Changes in Insolation?

THE ENERGY TRANSMITTED from the Sun to Earth, called incoming solar radiation, or *insolation*, has varied only slightly during the short time for which we have accurate measurements from satellites. How much energy do we receive from the Sun, and why does it vary at all?

A How Much Energy Is Transferred from the Sun to the Earth?

1. To help us envision how much energy Earth receives, imagine a thin circular disc (▶) that is the same diameter as Earth, making it just large enough to intercept all energy that would fall on Earth's more complex, spherical shape.

2. According to both measurements and calculations, a relatively consistent amount of energy is radiated on this disc every second. This relatively consistent amount of energy—equivalent to 1,366 watts per square meter (W/m²)—is called the *solar constant*. This equates to only about one two-billionth of the power emitted from the Sun, since the Sun's energy goes out in all directions, not just toward Earth. Each planet would have a different solar constant due to differences in their distance from the Sun, with planets farther from the Sun receiving less energy.

02.07.a1

4. During the course of a day, this amount of energy is spread out over the entire surface area of the planet and its atmosphere. A sphere has four times more surface area than a circle of the same radius, so the energy must be spread out across four times more area for the sphere

3. However, each place on Earth's surface does not get 1,366 W/m² of power from the Sun all the time. Any place experiencing night receives zero, whereas equatorial locations experiencing clear skies during the summer may get 600–700 W/m² at noon. Something happens to the rest of the insolation between the time it reaches the atmosphere and the time it reaches the surface.

compared to the disc. Dividing 1,366 W/m² by four yields 341.5 W/m². Geographers and others actually use 345 W/m² (slightly more than ¼ of 1,366) for the average insolation to the top of the atmosphere because of bending of light rays after the Sun goes down.

B How Does the Amount of Insolation Change During a Year?

The amount of insolation varies during the year because the Earth is closer to the Sun at certain times of the year, but not the times you would expect—Earth is closer to the Sun during the Northern Hemisphere winter.

1. Although the average Earth-Sun distance is 150 million km, Earth's orbit is slightly elliptical rather than circular. The orbit is so nearly circular that it is virtually impossible to notice the elliptical shape in an accurately drawn view (▶). However, Earth is slightly closer to the Sun at some times of year than others, with the Earth-Sun distance varying about 3% during the course of the year.

2. The date of closest approach is called the *perihelion*, where "peri" is a Greek word for "around" and "helio" refers to the Sun. Perihelion currently occurs in early January, but the exact date varies a bit each year, and moves forward by one day each 65 years. Over several hundred thousand years it may change season completely, causing changes in climate.

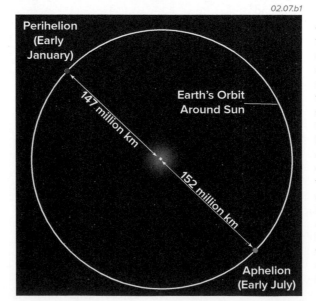

02.07.b1

Perihelion (Early January)

Earth's Orbit Around Sun

147 million km

152 million km

Aphelion (Early July)

4. This variation in Earth-Sun distance, from perihelion to aphelion and back again, constantly changes the average amount of insolation Earth's atmosphere receives. When the Earth is closest to the Sun (during the Northern Hemisphere winter) it intercepts more of the insolation. Changes in the amount of insolation striking the Earth on any given day can be calculated knowing the solar constant and the Earth-Sun distance on that day. From these calculations, the amount of energy is greatest at perihelion, decreases for half the year as Earth approaches aphelion, and then increases the next half of the year on the way to perihelion. The total amount of variation in energy related to Sun-Earth distance is about 7%.

3. The date of farthest approach is the *aphelion*, where "apo" comes from a Greek word meaning "from." The aphelion is, perversely for Northern Hemisphere dwellers, early in July. In other words, we are farthest from the Sun during our summer—the seasons cannot be explained by our changing distance from the Sun. As with perihelion, the actual date of aphelion varies slightly from year to year. For now and the immediate future, it occurs in early July.

C Does the Sun's Output of Energy Change Over Years, Decades, and Centuries?

Like most natural systems, the Sun's activity varies, sometimes increasing and other times decreasing. The Sun exhibits cycles that repeat over the course of a decade or so, and it has longer fluctuations that occur over multiple decades. These cycles influence the Sun's total output of energy, called the *total solar irradiance* (or simply TSI).

1. This top pair of images (▶) shows the Sun at two different times, approximately 9 years apart. In the left image, the Sun has a number of dark spots, termed *sunspots*. These are places that are slightly cooler than the rest of the Sun, and so show up darker. In the image to the right, the Sun lacks any sunspots. The number of sunspots varies from year to year.

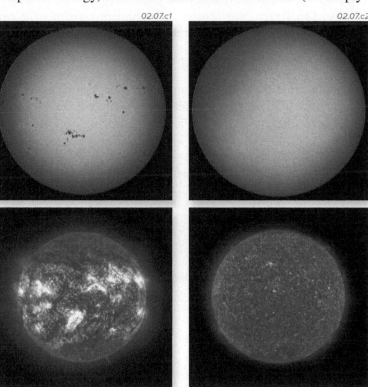

02.07.c1   02.07.c2

02.07.c5

2. This second pair of images shows the Sun at the same two times as in the top images, but displaying emissions at ultraviolet wavelengths, rather than visible light (▶). They indicate that the Sun is much more active and emitting more overall energy when there are more sunspots. So more sunspots means more energy output from the Sun, even though the sunspots themselves are "cooler" areas. Data collected by satellites since 1978 show small (< 1%) monthly changes in the output of solar energy and a correlation between sunspot activity and TSI.

02.07.c3   02.07.c4

3. Another expression of the Sun's activity are bursts of intense energy and matter directed into space, termed *solar flares* (▲). The bright regions producing solar flares in the image above are so large they could contain the Earth many times over. Solar flares pose a hazard to people and equipment in space and can affect wireless communications on Earth.

4. Scientists have been observing and counting the number of sunspots over time, calculating the average number present at any time, a value called the *sunspot number*. This graph (▶) plots monthly averages of sunspot number since 1950. From the data, there is an approximately 11-year cycle of sunspot activity, where the number of sunspots starts out at some minimum number and then increases to a maximum number and then decreases again. This full cycle from minimum to maximum and back to minimum is called a *solar cycle* (or sunspot cycle). Solar cycles are numbered sequentially from the 1700s (the first accurate sunspot data) to the present—we are currently in solar cycle 24.

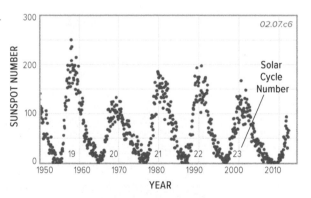

02.07.c6

5. As you can observe from this graph, the maximum number of sunspots (the heights of the "peaks" on this graph) varies from cycle to cycle. The higher the sunspot number, the more active the Sun. The duration of a cycle can also be more than or less than 11 years. The length of a cycle can help solar scientists predict the maximum sunspot number, and therefore solar activity, for the next cycle.

6. Sunspots have been counted for several centuries, revealing some important patterns (▶). This plot shows sunspot observations from the 1600s. The data are much more accurate for recent times, compared to the early observations, but the broad patterns shown are accepted by most scientists. Red points are early, less accurate measurements, blue data are more recent, and the black line is a smoothed average.

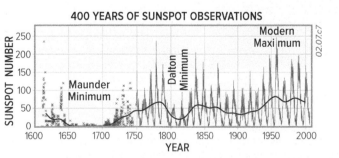

400 YEARS OF SUNSPOT OBSERVATIONS

02.07.c7

7. There have been several times, such as the *Maunder Minimum*, when there were few or no sunspots for many years in a row. The Maunder Minimum occurred during one of the coldest periods in the last 1,000 years, a time called the *Little Ice Age*. In contrast, our modern warming has occurred during a time of very high solar activity, as recorded by numerous sunspots.

Before You Leave This Page

✓ Explain the solar constant.

✓ Sketch, label, and explain Earth's perihelion and aphelion and their effect on insolation on Earth.

✓ Explain how sunspot activity varies over time.

2.7

2.8 Why Does Insolation Vary from Place to Place?

DESPITE A FAIRLY CONSISTENT supply of energy from the Sun, considerable differences in the quantity of insolation are experienced between the poles and equator, and also over several timescales—most noticeably, changes between seasons and between night and day. On Earth, variations in insolation are mostly related to latitude.

A What Controls the Insolation Reaching the Atmosphere and Earth's Surface?

The amount of sunlight varies from time to time and from place to place. Variations in temperature from season to season and from warm, tropical regions to cold, polar regions reflect variations in the amount of insolation. A main factor controlling the maximum possible amount of insolation is a site's position on the Earth, specifically the latitude.

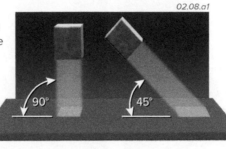
02.08.a1

1. When thinking about insolation, an important aspect to consider is the orientation of energy with respect to the surface being irradiated. In the figure to the right, a flat surface is illuminated by two identical lights, each with an energy equal to the amount of solar energy striking the top of the atmosphere (345 W/m²). The left light is shining directly down on the surface, so the surface receives the same amount of energy per square meter as the light emits (345 W/m²).

2. The second light is angled 45° relative to the surface, so when its light strikes the surface the light is spread out over a larger area. As a result, the amount of energy per area is less, and the area would be less brightly lit. For an angle of 45°, the energy is spread out over an area 1.41 times longer than for the left light, so the resulting energy is less, at 244 W/m². For the Sun, the angle between the incoming light and the surface is called the *Sun angle* and is 90° for the left light and 45° for the right one.

3. The geometric relationship shown in the previous figure applies to the entire Earth because of its spherical shape. The figure to the right shows how equal rays of solar energy strike different parts of the Earth, varying in Sun angle from a position along the equator to ones nearer the poles. Each orange bar is of equal width and represents insolation equal to the solar constant.

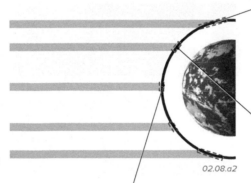
02.08.a2

6. In contrast to the equator, insolation arriving above polar regions strikes Earth's atmosphere at a very low Sun angle, much lower than the angled light in the figure above. As a result, the energy represented by the ray (the solar constant) spreads over a relatively large surface area (the dashed red box). Insolation striking that part of the atmosphere, as measured in energy amount per area, is considerably dispersed, diminished to perhaps only 20% of the solar constant.

5. In the mid-latitude regions north or south of the equator, insolation strikes the outside of Earth's atmosphere at a slightly oblique Sun angle and is distributed over a wider area than at the equator. Thus the amount of energy is distributed over a larger area compared to the equator, and the energy per area is less. The area shown (on the outside of the atmosphere) receives about 60% of the solar constant, more if the area is closer to the equator or less if it is farther away—that is, the amount of insolation varies as a function of latitude.

4. Some insolation strikes the outside of Earth's atmosphere perpendicularly, like a light pointed directly toward a surface. This part of the atmosphere is receiving the maximum insolation possible, or an amount of energy equivalent to the solar constant, at the "top" of the atmosphere, 1,366 W/m².

7. In addition to these geometric effects, the atmosphere further influences the amount of insolation that reaches Earth's surface, depending on the angle of the incoming energy. As shown in this figure (▶), energy that comes directly down at the surface, like a vertical light, passes through the atmosphere in the shortest distance possible. Some energy is scattered or absorbed by the atmosphere.

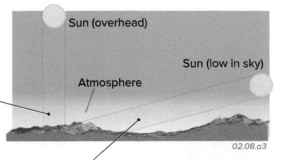
Sun (overhead)
Sun (low in sky)
Atmosphere
02.08.a3

8. Energy coming at a lower Sun angle must pass through more of the atmosphere to reach the surface. This increases the amount of atmospheric *attenuation* and causes less energy to reach the surface. This is a further reason why early morning and late afternoon sunlight does not feel as intense as that at noon, and why the winter Sun, which is low in the sky (low Sun angle), does not feel as intense as the summer Sun, which is higher in the sky.

9. When viewing the entire Earth (▶), latitude strongly influences the amount of atmospheric attenuation. Near the equator at noon, the Sun's energy is coming at a high Sun angle and experiences relatively less attenuation.

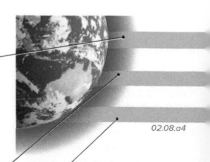
02.08.a4

10. In the mid-latitudes, the energy is more oblique to the Earth, so it must pass through more of the atmosphere. This results in more atmospheric attenuation, resulting in less energy reaching Earth's surface.

11. In polar and high-latitude regions, the Sun angle is very low, and the Sun's energy must pass through even more of the atmosphere. The amount of atmospheric attenuation is greater, so less energy passes through to the surface.

B How Does Sun Angle Affect Insolation?

1. The angle made by two lines, one drawn from the observer to the Sun and the other drawn horizontally, is the Sun angle. When the shadow of an object is directly underneath the object, as in the photograph below (▼), the Sun angle is 90°.

2. This figure (▶) shows the Sun at four different positions, corresponding to four different Sun angles (at noon). The 10 thin orange rays of sunlight represent equal amounts of energy (the solar constant). The brackets on the ground indicate the width over which that same amount of sunlight strikes the surface.

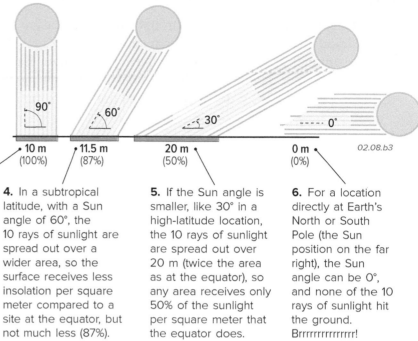

02.08.b3

10 m (100%)   11.5 m (87%)   20 m (50%)   0 m (0%)

Sun Angle

02.08.b1

02.08.b2

3. In this example, if the Sun angle is 90°, as happens near the equator and in the tropics, the 10 rays of sunlight intersect the Earth's surface over a width of 10 meters. The light is perpendicular to the surface, so the area receives the maximum amount (100%) of energy possible.

4. In a subtropical latitude, with a Sun angle of 60°, the 10 rays of sunlight are spread out over a wider area, so the surface receives less insolation per square meter compared to a site at the equator, but not much less (87%).

5. If the Sun angle is smaller, like 30° in a high-latitude location, the 10 rays of sunlight are spread out over 20 m (twice the area as at the equator), so any area receives only 50% of the sunlight per square meter that the equator does.

6. For a location directly at Earth's North or South Pole (the Sun position on the far right), the Sun angle can be 0°, and none of the 10 rays of sunlight hit the ground. Brrrrrrrrrrrrrr!

C How Does Insolation Vary by Time and Location on Earth's Surface?

The amount of sunlight that reaches the surface varies greatly with latitude, due to the reinforcing effects of Sun angle and atmospheric attenuation. These variations in insolation result in huge ranges in temperature between the warm, tropical regions of the world and the cold, polar regions.

02.08.c1 Raja Ampat, Indonesia

02.08.c2 Antarctica

02.08.c3 Hermosa, CO

The least variation in insolation through the year occurs in equatorial regions, which do not experience summer versus winter. The length of daylight in equatorial regions is always around 12 hours, and Sun angles (at noon) are always large. The Sun is directly overhead in March and September and nearly so the rest of the year, so temperatures remain warm throughout most of the year.

Polar regions experience the maximum variation in insolation. During the summer this region has 24 hours of sunlight, allowing fully lit outdoor excursions (▲), but during the winter there is no sunlight. Even during the summer, small Sun angles prevent insolation from increasing too much. The combination of months of darkness and low insolation causes polar areas to have low average temperatures.

Areas between the tropical and polar regions, such as those in the mid-latitudes, experience an intermediate amount of variability in insolation and so typically have distinct seasons. Mid-latitudes have some sunlight and darkness every day, with more daylight hours and higher Sun angles during the relatively warm summer. The winter has fewer daylight hours and smaller Sun angles, so it is colder.

Before You Leave This Page

✓ Sketch, label, and explain the relationship between latitude and the amount of insolation, including atmospheric effects.

✓ Sketch, label, and explain Sun angle and how it influences the amount of insolation an area receives.

✓ Summarize how Sun angle and variations in length of day are expressed in tropical, polar, and mid-latitude locations.

2.8

2.9 Why Do We Have Seasons?

MOST LOCATIONS progress through different seasons, from warmer summers to cooler winters and back again. The progression from season to season accompanies changes in the position of the Sun, such as from higher in the sky during the summer to lower in the sky during the winter. Except at the equator, the durations of daylight and darkness also vary, with longer days during the summer and longer nights during the winter. What causes most parts of Earth to have different seasons? Why do tropical areas not have summer or winter?

A What Causes the Seasons?

The annual "march of the seasons" indicates variations in the amount of sunlight received at different latitudes during the course of a year. The main cause of these variations is that Earth's axis of rotation, about which our planet spins once during a 24-hour day, is tilted relative to the plane in which we orbit the Sun, the *orbital plane*.

1. Earth's axis of rotation is tilted relative to our planet's orbital plane, as shown by the figure below, which is a sideways perspective of Earth's nearly circular orbit. The axis remains fixed in orientation with respect to the orbital plane and the stars throughout the year—the tilt does not change during a year. As Earth orbits the Sun, the Northern Hemisphere faces in the direction of the Sun during some times of the year, while the Southern Hemisphere faces the Sun at other times. Observe the relationship between the orientation of the rotation axis and the direction to the Sun.

7. The seasons therefore are controlled by the relationship between the tilt of Earth's rotation axis and the orbit of Earth around the Sun. Each season is bounded by a solstice and an equinox, which in turn represent days of either a maximum or minimum amount of apparent tilt relative to the Sun.

2. During December, the North Pole and Northern Hemisphere face away from the Sun. The Northern Hemisphere therefore receives less direct insolation and so experiences the cooler temperatures of winter. Rotation about the axis brings most parts of the hemisphere into the sunlight for part of the day.

September 21

23.5°

December 21

(COMPONENTS NOT TO SCALE)

23.5°

June 21

23.5°

23.5°

02.09.a1

March 21

6. In late June, the North Pole and Northern Hemisphere more directly face the Sun and so start to experience summer. The *June Solstice* (Summer Solstice in the Northern Hemisphere) marks the date when the North Pole's rotation axis is pointed most directly toward the Sun. At the same time, the South Pole and Southern Hemisphere face away from the Sun and start to experience winter.

3. In contrast, the South Pole and Southern Hemisphere more directly face the Sun and receive more direct insolation. The Southern Hemisphere therefore has its summer (at the same time the Northern Hemisphere has its winter). The *December Solstice* (Winter Solstice in the Northern Hemisphere, Summer Solstice in the Southern Hemisphere) is the date when the North Pole (marking Earth's rotation axis) is pointed farthest away from the Sun.

4. As Earth orbits around the Sun, the rotation axis no longer points directly away from the Sun. In March (and later again in September), Earth's axis is pointing sideways relative to the Sun—neither the North Pole nor the South Pole is inclined toward the Sun. In this position, neither hemisphere receives more insolation than the other, and so both experience the more moderate temperatures of spring and fall. The *March Equinox* and *September Equinox* are times when the axis is exactly sideways and the durations of daylight and darkness are equal (equinox is Latin for "equal night").

5. Between March and June, Earth's continued orbit causes the North Pole and Northern Hemisphere to increasingly face the Sun and to warm up as summer approaches. During this time, the South Pole and Southern Hemisphere begin to face away from the Sun, cooling down on the way to winter.

8. In thinking about the seasons, it is important to remember that Earth's orbit is nearly circular, as shown in this view looking straight down on Earth's orbital plane (▶). In this figure, as in the one above, the sizes of the Earth and Sun are greatly exaggerated relative to the size of the orbit, and the Earth is shown much larger than it actually is relative to the Sun.

9. The nearly circular geometry of Earth's orbit reinforces the fact that seasons are not caused in any way by differences in distance from the Sun. In fact, Earth is slightly closer to the Sun during the northern winter (perihelion) and slightly farther away during the northern summer (aphelion), opposite to what would be required to explain the seasons.

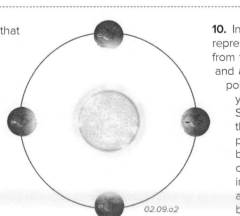

02.09.a2

10. In each globe (◀), Earth's rotation axis is represented by the small purple rod protruding from the North Pole (pointing toward the viewer and a little toward the left). Note that the axis points in the same direction throughout the year—it points toward Polaris, the North Star (not shown). Over thousands of years, the axis varies in orientation, not always pointing at Polaris. The amount of tilt varies by a degree either way (22.5° to 24.5°) over 40,000 years, but we are most interested here in the current tilt of 23.5°, an angle that will reappear throughout this book.

B What Factors Determine the Temporal and Spatial Variations in Insolations?

Earth's rotational axis is tilted relative to the plane in which we orbit the Sun (the orbital plane). The tilt of the axis is currently 23.5° from vertical to the orbital plane, an angle that is reflected in important geographic features on our planet, including how we define the Tropic of Cancer, Tropic of Capricorn, Arctic Circle, and Antarctic Circle.

The Tropics

1. These two globes show the Earth on the solstices—days when one of the poles is most exposed to the Sun and the other pole most faces away. During the December Solstice, the noon Sun is no longer directly overhead at the equator but instead is 23.5° to the south, directly over the *Tropic of Capricorn*. On this day, the equator has a Sun angle of 66.5° (90°–23.5°).

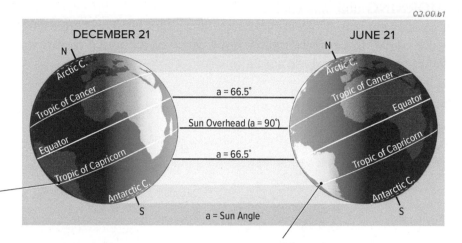

02.09.b1

4. The rest of the year, the overhead Sun migrates between the Tropics of Cancer and Capricorn, delivering relatively intense and constant insolation between these latitudes, causing a warm region called the *tropics*. In contrast, mid-latitude and polar regions receive a highly variable supply of energy and exhibit more variability during the year—that is, they exhibit more *seasonality*.

2. During the December Solstice, the Southern Hemisphere receives the maximum amount of insolation it will receive all year, marking the start of the southern summer. Also on this day, the Northern Hemisphere receives the least amount of insolation for the year, marking the beginning of the northern winter.

3. During the June Solstice, Earth is on the opposite side of the Sun, and the North Pole faces toward the Sun by its maximum amount. The overhead Sun, and the position of maximum insolation, have migrated to a latitude 23.5° north of the equator, the *Tropic of Cancer*. On the day of the June Solstice, the Northern Hemisphere receives the most insolation of the year, shifting into summer, while the Southern Hemisphere receives the least amount and moves into winter.

The Polar Circles

5. These two globes show the same two days as those above, the solstices. During the December Solstice, the northward limit of sunlight is the *Arctic Circle*, located at a latitude of 66.5° N, or 23.5° south of the North Pole. Along the Arctic Circle, the Sun will barely appear on the horizon at noon time and then disappear for a very long night. All places north of that latitude receive no insolation on this day—they have 24 hours of darkness.

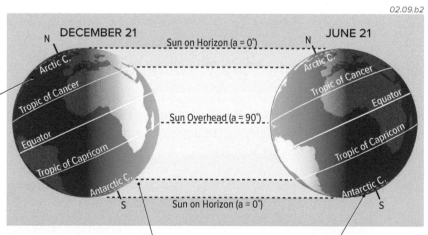

02.09.b2

8. At this time, the area within the Arctic Circle (66.5° N) is fully illuminated. Note that the same angles keep reoccurring—23.5° and 66.5° (which is 90° minus 23.5°). These values, for the two tropics (Cancer and Capricorn) and for the two polar circles (Arctic and Antarctic), are the same as the tilt angle of the rotation axis. The tilt of Earth's axis controls the locations of the tropics and polar circles.

6. Note that the entire *Antarctic Circle*, in the Southern Hemisphere, is illuminated, but with very oblique sunlight (low Sun angles). As the Earth completes its daily rotation, the South Pole and nearby areas remain in the sunlight all 24 hours—these places have 24 hours without darkness.

7. During the June Solstice, locations in the Southern Hemisphere now face away from the Sun. The southern limit of sunlight is the Antarctic Circle, at latitude 66.5° S. Here, the Sun will appear on the horizon for a brief moment at noon.

City	Lati-tude	Equi-nox	Dec. Sol.	June Sol.
Saskatoon, Canada	52.0° N	38.0°	14.5°	61.5°
Jacksonville, FL	30.5° N	59.5°	36.0°	83.0°
San José, Costa Rica	10.0° N	80.0°	56.5°	76.5°
Quito, Ecuador	0.0°	90.0°	66.5°	66.5°

9. Using the 23.5° angle and an area's latitude, we can calculate the Sun angle for any place on Earth. Examine this table (◄), which lists Sun angles for the equinoxes and both solstices for several cities. Note that the Sun angle varies by 47° from solstice to solstice—47° is two times 23.5°.

Before You Leave This Page

☑ Sketch and explain how the position of the Earth with respect to the Sun at various times of the year explains the seasons.

☑ Explain the solstices and equinoxes in terms of Earth-Sun position, seasons, and locations of tropics and polar circles.

2.9

2.10 What Controls When and Where the Sun Rises and Sets?

THE SUN RISES EACH MORNING and sets each evening, but at slightly different times from day to day. Also, the Sun does not rise or set in exactly the same direction every day, although the changes from day to day are so gradual as to be unnoticeable. Over the course of several months, however, we can notice significant changes in where and when sunrise and sunset occur. What accounts for these variations?

A Why Does the Sun Rise and Set?

1. At any moment in time, half of the Earth's surface area is sunlit, experiencing day, and half is in the darkness of night. The dashed line encircling the world separates the lighted and dark halves and is called the *circle of illumination*. When viewed straight on, as in these figures, the circle of illumination appears as a line, but it has a curved shape from any other perspective. The colored areas on these globes are time zones.

2. In the left globe, North America and western South America are on the side of the Earth hidden from the Sun, and so it is night. At the same time, eastern South America and Africa are on the side facing the Sun and so are in daylight. With a rotating Earth, this view is only an instantaneous snapshot of which areas are in sunlight and which are in darkness.

TIME 1

TIME 2
(4 Hours Later)

Direction
of Earth's
Rotation

Circles of
Illumination

02.10.a1

4. The *circle of illumination*, the boundary between day and night, moves westward across the surface as the planet rotates eastward. As this occurs, the Sun has not changed position—the Earth has simply rotated. It finishes a complete rotation in 24 hours.

3. In the right globe, the Earth has rotated an additional four hours—the globe rotates to the right when viewed in this perspective, or counterclockwise when viewed from above the North Pole. At this later time, South America and eastern North America have rotated into the sunlight (it is morning), but the west coast of the U.S. is still in the last hour of night.

B Why Does the Length of Daylight Vary Through the Year?

In high latitude parts of the world, there are significant differences between the length of daylight from season to season. In such regions, days are noticeably shorter during the winter than during the summer. In accordance with this, nights are longer during the winter and shorter during the summer. In contrast, at the equator, there are always 12 hours of day and 12 hours of night, irrespective of the time of year.

1. These globes show the circle of illumination at three different times of year: the December Solstice, an equinox, and the June Solstice. To help us visualize the circle of illumination, the three larger globes are depicted as if being observed by the small figure next to the corresponding small globes. The axial tilt remains fixed in orientation as Earth orbits the Sun, but here it is portrayed from different perspectives.

2. During an equinox (the large, center globe), the tilt axis is oriented neither toward nor away from the Sun. So, the pattern of light and dark is symmetrical with respect to the equator and other lines of reference. It takes the same amount of time for every location to rotate in and out of sunlight. At equinox, every location on Earth has 12 hours of sunlight and darkness.

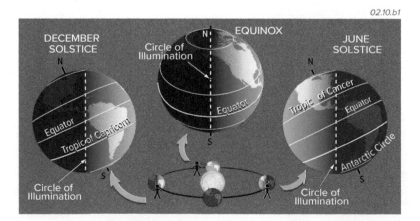

02.10.b1

DECEMBER
SOLSTICE

EQUINOX

Circle of
Illumination

JUNE
SOLSTICE

Tropic of Cancer

Equator

Tropic of Capricorn

Equator

Antarctic Circle

Circle of
Illumination

Circle of
Illumination

3. At other times of the year, Earth's axis appears tilted toward or away from the Sun, and so the circle of illumination is not symmetrical relative to lines of reference, such as the tropics. In the left globe, representing the December Solstice, any line of latitude in the Southern Hemisphere, such as the Tropic of Capricorn, is more in sunlight than in darkness. Therefore, days are longer and nights are shorter. The opposite is true for any latitude in the Northern Hemisphere, which is more in the dark than in the light, causing nights to be longer than days.

5. The variation in the lengths of day versus night from season to season increases with increasing latitude, either north or south of the equator. Such seasonal changes in day length are absent at the equator and greatest at the poles.

4. During the June Solstice and adjacent months, the opposite is true—more of the Northern Hemisphere is in sunlight than is in darkness. As a result, days are longer than nights in the Northern Hemisphere during this time (the northern summer). In contrast, more of the Southern Hemisphere is in darkness, so days are shorter and nights are longer during this time (the southern winter).

C Why Do Arctic Areas Sometimes Have 24 Hours of Sunlight or Darkness?

The most extreme variations in the lengths of daytime and nighttime occur in the highest latitudes, including the Arctic region around the North Pole and the Antarctic region around the South Pole. North of the Arctic Circle and south of the Antarctic Circle, summer days can have more than 24 hours of straight daylight. During the winter, it can remain dark for all 24 hours, night after night. Either condition can last for months. How is this so?

02.10.c1 Arctic Circle

02.10.c2

This image combines different photographs to show the path of the Sun during several hours at a location north of the Arctic Circle. The Sun remains low in the sky and dips toward the horizon at midnight, but never actually sets—this location has 24 hours of sunlight during the middle of summer. The low Sun angle means that insolation striking the land is spread out and so is relatively weak, and it has a long and attenuated path through the air.

The figure above shows sunlight on the north polar region during the December Solstice. All the area within the Arctic Circle is in darkness (it is on the side opposite to the Sun). As the Earth rotates about its axis, the entire area within the circle remains out of the sunlight—24 hours of darkness. In days following the solstice, sunlight begins to creep into the Arctic Circle, so less of the Arctic has 24-hour nights.

The opposite situation occurs during the June Solstice, shown here. Note that the entire Arctic Circle faces the Sun and will remain in sunlight as the Earth rotates about its axis. On days before and after the solstice (when the North Pole less directly faces the Sun), areas just inside the Arctic Circle would not be in constant sunlight, and so would have slightly less than 24 hours of sunlight and would have minutes to hours of nighttime.

D What Controls the Time and Direction of Sunrise and Sunset?

Except at the equator, the times of sunrise and sunset shift slightly from day to day, but typically by less than a few minutes every day. From month to month, however, we notice significant differences in the times of sunrise and sunset, and therefore in the duration of day and night. The changes in time are accompanied by gradual changes in the direction from which the Sun rises and the direction in which it sets.

1. This figure depicts where and when the Sun rises and sets at a location at 45° N latitude, which is halfway between the equator and the North Pole. Observe this figure and note how the locations of sunrise and sunset change by date.

2. The December and June dates, marking the two solstices, show the two extremes. The locations and times of sunrise and sunset fall between these two extremes for all other dates. Either equinox is halfway between the solstices. Each day, the Sun's path defines a circle, which lies on a plane that is inclined at an angle that is equal to the site's latitude (45° in this case).

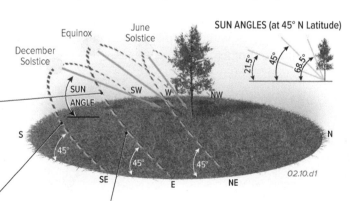
02.10.d1

3. At the December Solstice, the Northern Hemisphere faces away from the Sun the maximum amount, and so the Sun is as low in the sky as it ever gets. The Sun rises and sets as far south as any day of the year, rising in the southeast and setting in the southwest. This is the shortest day and longest night of the year.

4. At either equinox, neither pole faces the Sun, and so there are 12 hours of daylight and 12 hours of darkness at every location on Earth. Everywhere on Earth, the Sun rises due east of the site and sets due west on an equinox. If you wanted to determine directions without a compass, and it happened to be the date of an equinox, you could precisely determine an east-west direction by drawing a line from the direction of sunrise to the direction of sunset, as was done by ancient cultures. Also, you could determine your latitude by measuring the Sun angle at noon on this day—your latitude is equal to 90° minus the Sun angle.

5. At the June Solstice, the Northern Hemisphere faces the Sun, and so the Sun is the highest in the sky it will be all year. The Sun rises and sets as far north as any day of the year, rising in the northeast and setting in the northwest. This is the longest day and the shortest night of the year.

Before You Leave This Page

☑ Sketch, label, and explain why the length of day and night vary during the year, and which dates have the longest and shortest daylight hours.

☑ Sketch, label, and explain why a polar region can have 24 hours of daylight or 24 hours of darkness.

☑ Sketch, label, and explain where the Sun rises and sets, and has the highest and lowest Sun angles, for different times of the year.

2.10

2.11 How Does Insolation Interact with the Atmosphere?

INSOLATION REACHES THE EARTH but has to pass through the atmosphere before it reaches us. The atmosphere does not transmit all of the Sun's energy; some wavelengths of energy are partially or completely blocked by atmospheric components, such as gas molecules. The interactions between insolation and the atmosphere explain many aspects of our world, like blue skies, red sunsets, and even the existence of life.

A What Are the Principal Components of the Atmosphere?

The atmosphere is composed of gas molecules, especially nitrogen and oxygen, and small solid particles, including dust, and drops of water and other liquids. These particles and drops are together called *aerosols*, to convey that they are suspended in the atmosphere, floating with the moving air. Most gases and aerosols are produced by natural processes, but some are introduced into the atmosphere by activities of humans.

02.11.a1 Iceland

02.11.a2 Mount Etna, Italy

02.11.a3

Gases — The atmosphere is nearly all nitrogen (N_2) and oxygen (O_2), with lesser amounts of argon (Ar), and water vapor (H_2O), which occurs within the steam shown above. It contains trace amounts of other molecules, such as carbon dioxide (CO_2), methane (CH_4), nitrous oxide (N_2O), and sulfur dioxide (SO_2).

Solid Particles — If tiny enough, solid particles can be suspended in the atmosphere. Such particles include volcanic ash (▲), and wind-blown dust, salt, and pollen. They also include soot and smoke from natural and human-caused fires. Some small particles are produced by chemical reactions in the air.

Drops of Liquid — The atmosphere contains drops of water, with much smaller amounts of other liquids. Most water drops are tiny enough to remain suspended in the air as in clouds (▲). If small drops combine or otherwise grow, they may become too heavy to remain suspended, falling as rain.

B How Do Atmospheric Components Affect Insolation?

Insolation, like all types of electromagnetic radiation, can be affected by material through which it passes. As solar energy attempts to pass through the atmosphere, it interacts in various ways with the different atmospheric components. The types of interaction that occur depend on the size and physical nature of the component (e.g., solid versus a gas), the wavelength of the energy (blue versus red light, for example), and other factors.

02.11.b2

Reflection and Absorption

Reflection — Some atmospheric components can *reflect* incoming insolation, such as by this snow-flake. Reflected energy can be returned directly into space or can interact with other atmospheric components and remain in the atmosphere.

Snowflake

Absorption — An atmospheric component, such as this gas molecule, can instead *absorb* the energy, converting the incoming electromagnetic energy into kinetic energy expressed as motions of the molecule.

Gas Molecule
02.11.b1

Scattering

Scattering — Insolation can be *scattered* by atmospheric compo-nents (▶), which send the energy off in various directions. Some processes of scattering affect shorter wavelengths of EMR more than longer ones; for example blue light is scattered more than red light. Other processes affect all wavelengths of solar energy equally.

Sky Color — As insolation enters the atmosphere, blue and violet light are preferentially scattered by gases, and this scattered light causes us to see the sky as blue. The remaining light that passes through gives the Sun a yellowish white color. When sunlight passes through the atmosphere at a low angle, as during sunrise and sunset, most colors, except orange and red, have been scattered out. Scattering of the remaining orange and red light produces the familiar orange and red glow.

C How Do Different Layers of the Atmosphere Interact with Insolation?

If the atmosphere consisted of gases that did not interact with insolation, we would expect that its temperature would decline with distance from Earth. However, atmospheric temperatures exhibit some surprising changes with altitude, as a direct result of the interactions between insolation and atmospheric components, specifically gases.

1. Observations from high-flying planes, balloons, and rockets reveal that the atmosphere is divided vertically into four layers distinguished by their thermal properties. These layers are, from bottom to top, the *troposphere*, *stratosphere*, *mesosphere*, and *thermosphere*. In the troposphere and mesosphere, temperatures decline with increasing altitude, as expected, displaying a *normal temperature gradient*. Temperatures in the stratosphere and thermosphere, however, actually increase with altitude—a *temperature inversion* (reverse gradient). Why?

2. The unexpected change in temperature gradient in some layers is due to the absorption of insolation of some wavelengths of the electromagnetic (EM) spectrum, but not others that arrive at the outside of the atmosphere. Most energy is in visible-light wavelengths, with significant amounts of adjacent wavelengths of ultraviolet (UV) and infrared (IR) energy. Some wavelengths of energy interact with certain atmospheric molecules, transferring energy as they do so. Different interactions occur in different layers, accounting for the differences in temperature gradient, and therefore the distinctions among the different layers. The four atmospheric layers are separated by distinct breaks called *pauses*.

3. The very shortest and most energetic of EM radiation from the Sun are X-rays and gamma rays, with wavelengths approaching the size of gas molecules. These incoming wavelengths of EMR are effectively intercepted by the few molecules of nitrogen (N_2) and oxygen (O_2) in the uppermost parts of the atmosphere—the *thermosphere*. The greatest number of interceptions (and transfer of energy) occurs at the first opportunity for the gamma and X-rays to encounter these gases, at the outermost parts of the thermosphere. This causes the outer thermosphere to warm up, but so few molecules exist at such levels that you would freeze to death instantly, even at temperatures approaching 1,200°C. A progressively smaller proportion of these energetic rays penetrates lower in the thermosphere, so fewer energy exchanges occur, and temperatures decline downward across the thermosphere.

4. The next layer down, the *mesosphere*, possesses no particular properties to intercept wavelengths of insolation, so it displays a normal temperature gradient (temperature decreases upward). The boundary at the top of the mesosphere is the *mesopause*.

5. The *stratosphere* has relatively high concentrations of the trace gas *ozone* (O_3), which effectively absorbs UV wavelengths. The same principle prevails as in the thermosphere, with the greatest amount of absorption occurring near the top of the stratosphere, which is therefore relatively warm. Progressively less absorption of incoming UV occurs downward, resulting in a decrease in temperatures downward—a temperature inversion. The top of the stratosphere is the *stratopause*.

6. Conditions in the troposphere can be very complex with the presence of various sublayers of air and the influence of clouds, but temperatures usually decrease upward (a normal temperature gradient). The top of the troposphere is the *tropopause*.

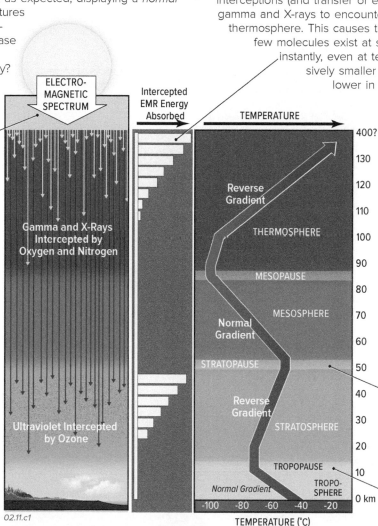

02.11.c1

7. The greatest amount of the Sun's emitted energy is at wavelengths of visible light, along with UV and IR. (▼) Interactions between solar energy and various components of the atmosphere (such as N_2) cause some wavelengths of energy to be intercepted through absorption, scattering, or reflection in the atmosphere. Therefore, the spectrum of EM energy that reaches the surface is different in detail from that which enters the top of the atmosphere, as depicted by the curve showing the amount of insolation at sea level on this graph. Certain wavelengths of EMR are absorbed by certain components, such as water (H_2O), oxygen (O_2), and carbon dioxide (CO_2), as expressed by dips in the curve for insolation received at sea level.

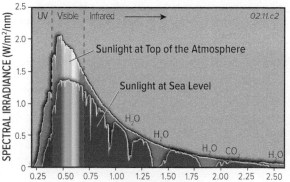

02.11.c2

Before You Leave This Page

✓ Summarize the principal components of the atmosphere.

✓ Sketch and explain reflection, absorption, and scattering, and how they explain the color of the sky.

✓ Sketch, label, and explain the four layers in the atmosphere, how they interact with insolation, and how these interactions explain the different temperature gradients.

2.11

What Is Ozone and Why Is It So Important?

OZONE IS AN ESSENTIAL GAS in the atmosphere, shielding life on the surface from deadly doses of ultraviolet radiation from the Sun. In the past several decades, there has been major concern about the loss of ozone in our atmosphere, particularly a seasonal decrease in ozone above Antarctica. How can a gas that generally constitutes less than one molecule in every 10 million in the air be so important, and why does the Antarctic region experience severe ozone loss?

A   What Is Ozone and Where Does It Occur?

Ozone is a molecule composed of three oxygen atoms bonded together (O_3), instead of the much more common arrangement of two oxygen atoms in a molecule of oxygen gas (O_2). More than 90% of ozone occurs in the stratosphere, but ozone also occurs in lesser amounts in the troposphere and mesosphere.

This graph (▶) plots the concentration of ozone in the atmosphere as a function of altitude above Earth's surface. Note that concentrations are very low overall, measured in parts per million (ppm—the number of ozone molecules for every one million molecules of atmosphere). The maximum concentration is in the middle of the stratosphere, a zone called the *ozone layer*, but even here the ozone concentration is only 6 to 8 ppm. Ozone concentrations decrease rapidly into the overlying mesosphere.

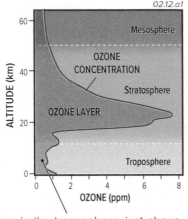

02.12.a1

02.12.a2 Los Angeles, CA

Ozone is a component of *smog* (◀), a type of air pollution common in larger cities, especially under certain atmospheric conditions. The term smog was derived from air pollution that looks like smoke and fog.

Ozone is also produced naturally due to ionization of oxygen gas molecules in the air near lightning. In such settings, we can sense ozone's distinctive smell, and this is the origin of the term *ozone*, after the Greek verb for "to smell."

02.12.a3

There is a slight increase in concentrations low in the troposphere, just above the surface. Whereas ozone in the ozone layer (stratospheric ozone) is beneficial to life, ozone near the surface (ground-level ozone) is a harmful air pollutant, produced by sunlight striking hydrocarbons, such as car exhaust, in the air.

B   How Is Ozone Produced and Destroyed in the Stratosphere?

1. Electromagnetic radiation with a wavelength of visible light or shorter (e.g., ultraviolet) has the ability to affect the chemical bonds in a compound, a process called *photodissociation*. Such processes occur throughout the atmosphere, but particularly in the stratosphere, where (1) much of the EMR arriving from the Sun has not yet been absorbed and (2) the concentration of molecules becomes high enough to allow photodissociation and subsequent formation of ozone. This figure shows how photodissociation forms and destroys ozone.

2. Oxygen in the atmosphere is mostly oxygen gas (O_2), composed of two oxygen atoms bonded together. These molecules absorb EMR from the Sun, including several types of ultraviolet (UV) energy. They are strongly affected by an energetic type of UV called UV-C, which has relatively short wavelengths. UV-C can break apart the two oxygen atoms, liberating a free (unbonded) oxygen atom (O).

3. The freed oxygen atom can quickly bond with another free oxygen atom, forming a new molecule of oxygen gas (O_2).

4. Alternatively, a freed oxygen atom can combine with an existing oxygen gas molecule to form a molecule of ozone (O_3). This process is the way ozone is produced in the atmosphere.

5. Ozone molecules are capable of absorbing a longer wavelength form of ultraviolet called UV-B, in addition to UV-C. UV-B has lower energy than UV-C, but more of it reaches Earth's surface. UV-B is the wavelength of ultraviolet radiation that causes sunburn and contributes to skin cancer. It also produces vitamin D in our bodies when it interacts with our skin.

6. UV-C and UV-B can photodissociate an ozone molecule, leaving a molecule of oxygen gas and a free oxygen atom. These can recombine with other atoms and molecules to form another molecule of ozone, so the process of formation and destruction of ozone is a cycle, with ozone and oxygen gas continuously being broken apart, combined, and broken apart again. If the rates of formation and destruction of ozone molecules are equal, the ozone concentration will remain constant. Variations in atmospheric conditions, however, change the relative rates, causing concentrations to change over time and from place to place.

UV-C

UV-C & UV-B

Ozone

02.12.b1

C What Is the Distribution of Ozone in the Atmosphere?

1. These two globes show the amount of ozone in the atmosphere at two different months, as measured from a satellite. In general, ozone concentrations are relatively low (green) over the equator, and are higher (yellows and oranges) in middle and high latitudes. The difference between the two globes shows that ozone concentrations change with the season, responding to changes in the patterns of insolation. Ozone amounts are expressed as Dobson units (DU).

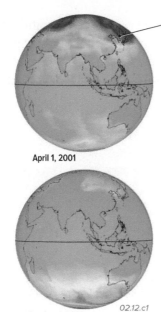

April 1, 2001

October 1, 2001

02.12.c1

OZONE (DU)

100   500

2. In April, the highest concentrations are in higher latitude regions of the Northern Hemisphere.

3. In October, the highest values are in high latitudes of the Southern Hemisphere, but there is a huge area with very low concentrations centered over the South Pole (at the bottom of the globe). This low exists because total darkness during the Antarctic winter lasts for several months, during which time there is no insolation to form ozone, but ozone continues to be chemically destroyed (see below).

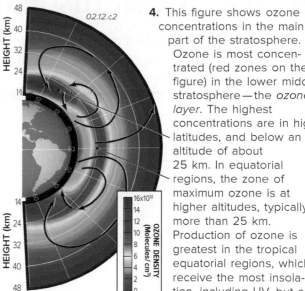

02.12.c2

HEIGHT (km)
48
40
32
24
14

HEIGHT (km)
14
24
32
40
48

OZONE DENSITY (Molecules/ cm³)
16x10¹²
14
12
10
8
6
4
2
0

4. This figure shows ozone concentrations in the main part of the stratosphere. Ozone is most concentrated (red zones on the figure) in the lower middle stratosphere—the *ozone layer*. The highest concentrations are in high latitudes, and below an altitude of about 25 km. In equatorial regions, the zone of maximum ozone is at higher altitudes, typically more than 25 km. Production of ozone is greatest in the tropical equatorial regions, which receive the most insolation, including UV, but a very slow-moving circulation pattern in the stratosphere moves the ozone upward and laterally toward the poles, accounting for the higher concentrations there.

D What Is Causing Depletion of Ozone and Formation of the "Ozone Hole"?

Ozone is critical to life because it shields us from dangerous UV-C and UV-B radiation. In the past half century, we became aware that our protective ozone shield was being depleted by human activities, especially the production of chlorofluorocarbons (CFC), chemicals released from aerosol cans, air conditioning units, refrigerators, and polystyrene. As a result, production of CFCs was limited by international agreement, the Montreal Protocol, adopted in the late 1980s.

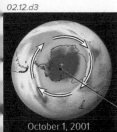

Sept. 16, 2011

02.12.d1

02.12.d3

October 1, 2001

April 1, 2001

1. This image shows ozone concentrations in the Southern Hemisphere as measured by a NASA satellite, with the smallest amounts in purple. The image, taken in September, shows 10.6 million square miles of the Antarctic region with severe ozone depletion. This depletion is described as "thinning of the ozone layer."

2. Most ozone is destroyed by natural processes, but humans have introduced chemicals that accelerate such losses. Chlorofluorocarbons (CFCs) contain *halogens*, elements like chlorine and bromine that easily bond with another element or molecule. Halogens can break apart ozone by attracting one of the oxygen atoms away, forming a new molecule, such as one with chlorine and oxygen (ClO).

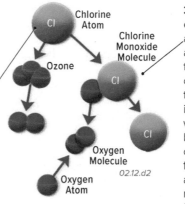

Chlorine Atom

Chlorine Monoxide Molecule

Ozone

Cl

Cl

Oxygen Molecule

Oxygen Atom

02.12.d2

3. The ClO molecule is short lived, and the chlorine atom breaks away to combine with other O atoms, typically outcompeting O_2 for bonding with O, thereby breaking apart ozone or keeping it from forming. On average, a single Cl may be responsible for the destruction of up to 10,000 ozone molecules. More than 70% of halogens in the atmosphere were introduced by humans.

4. Halogens are regarded as a major factor in the depletion of the ozone layer in the Southern Hemisphere, shown in the top globe. The bottom globe shows the Northern Hemisphere at an equivalent time of year (spring in both places). Why is there thinning of the ozone layer over Antarctica but less so over the North Pole?

5. The contrasting geography of the two polar regions plays a major role. Antarctica, in the center of the top globe, is a continent surrounded by ocean. This arrangement produces a zone of rapid circumpolar winds, which effectively exclude the import of ozone from nonpolar areas.

6. In contrast, the Arctic is an ocean surrounded by irregularly shaped continents (Eurasia and North America). The alternating pattern of oceans and continents induces far greater north-south movement and complex wind patterns, which encourage the exchange and mixing of gases, including ozone, into and out of the northern polar atmosphere.

Before You Leave This Page

✓ Explain what ozone is, where it is, and how it protects us.

✓ Describe the natural and human-related processes that contribute to ozone formation and destruction.

2.12

2.13 How Much Insolation Reaches the Surface?

NOT ALL INSOLATION reaches Earth's surface. Much of it is intercepted (absorbed, scattered, or reflected) by the atmosphere. Measurements and models allow us to account for the destination of insolation globally. Approximately 69% of the energy arriving at the top of the atmosphere remains in the Earth's system, of which 20% is stored (absorbed) in the atmosphere and 49% is absorbed by, and heats, the Earth's surface. The rest (31%) is lost back into space. This accounting of insolation is termed the *global shortwave-radiation budget*.

A How Is Insolation Intercepted in the Atmosphere?

02.13.a1

The atmosphere absorbs radiation of various wave-lengths, including ultraviolet (UV) light, an energetic shortwave EMR emitted by the Sun. This graph depicts the absorption of different wavelengths of UV (in Dobson units), along with the concentration of ozone (the red curve). UV is absorbed chiefly by oxygen gas, ozone, and nitrogen gas. As shown in this graph, the shortest-wavelength, highest-energy UV (UV-C) is absorbed first, followed by longer-wave-length, lower-energy UV-B. Nearly all UV-A, the longest-wavelength, least-energetic UV, is transmitted to the surface. When a wavelength of EMR passes through a substance, such as the atmosphere, without being intercepted, we say that the substance is "transparent to the energy." The atmosphere is nearly transparent to UV-A.

02.13.a2

Clouds can absorb, scatter, and reflect insolation, bouncing it back higher into the atmosphere and even into space. Clouds in the lower part of the troposphere, *low-level clouds*, can be highly reflective, like the ones shown here. The dark underside of the clouds indicates that little light is transmitted.

02.13.a3

Clouds higher in the atmosphere, such as these wispy *cirrus clouds*, are typically less reflective, but still reflect about 50% of insolation striking them. These *high-level clouds* occur in the middle to upper troposphere and are typically composed of ice crystals, unlike low-level clouds, which contain mostly water drops.

B What Happens to Insolation That Reaches Earth's Surface?

Earth's surface is diverse in its topography, rocks, soil, bodies of water, types of vegetation, and amount of human development. Each type of surface material has certain colors and other characteristics, so some surfaces reflect relatively little insolation and absorb much heat, while others reflect most insolation and absorb less heat.

1. The Apollo astronauts saw the Earth rise above the Moon's horizon (▼). The human eye is only sensitive to EMR in the visible (0.4–0.7 µm) wavelengths, and the Earth emits most of its energy at wave-lengths 20 times longer than this (infrared). What the astronauts saw was visible light originating from the Sun and reflecting from the Earth. All objects reflect some insolation, and the percentage of insola-tion that is reflected by an object is termed its *albedo*. Albedo varies from 0% for a theoretical black, rough object that reflects no light to 100% for a perfectly white, smooth one; no common objects have these end-member values. The average albedo for the entire Earth-ocean-atmosphere system is 31%.

2. Different land surfaces have different albedos. When humans use the land surface, they generally change the albedo compared to its natural state, modifying the energy balance, with local and perhaps regional implications. Examine the albedo values for different types of land cover and think how changes in the land cover would affect the energy balance.

02.13.b1

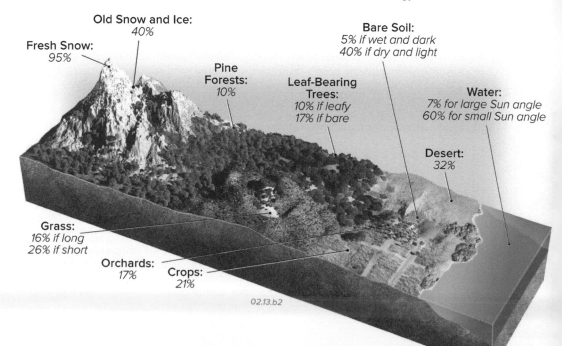

Fresh Snow: 95%

Old Snow and Ice: 40%

Pine Forests: 10%

Bare Soil: 5% if wet and dark 40% if dry and light

Leaf-Bearing Trees: 10% if leafy 17% if bare

Water: 7% for large Sun angle 60% for small Sun angle

Desert: 32%

Grass: 16% if long 26% if short

Orchards: 17%

Crops: 21%

02.13.b2

C What Percentage of Insolation Goes Where?

We can measure and model how much insolation is intercepted in the air versus how much reaches the surface. Some insolation that reaches the surface is reflected upward into the atmosphere, and some of this reflected energy goes all the way back into space. Most energy is absorbed by the water, land, and vegetation. The figure below shows the global shortwave-radiation budget—how much insolation goes where, a key component of the global energy balance.

1. We begin by considering the total amount of insolation arriving at the top of the atmosphere as being represented as 100%. Then, we can examine what percentages of this total amount end up where. The left side of this figure depicts reflection and scattering of insolation, whereas the right side represents absorption.

2. *Scattering* by various air components (gas molecules, dust, etc.) returns 7% of total insolation back to space. Such scattering causes blue skies, red sunsets, and red sunrises, and downward-scattered radiation is what illuminates the surface in shady areas and during overcast days. Scattering decreases the amount of insolation that reaches the surface.

3. The amount of energy *reflected by clouds* is variable and controlled by many complex factors. Different types of clouds have different albedos, and some lower-level clouds are obscured below other, higher clouds. The location of a cloud is also important—tropical clouds receive more direct overhead sunlight and so have more available insolation to reflect, whereas low clouds over the poles receive little direct light, or no light at all during winter. Also, the amount of the world covered by clouds varies, depending on the season and other aspects of weather. Considering all the factors, clouds on average reflect 20% of the planet's total insolation back to space.

4. Chemical constituents of the atmosphere, such as O_2 and N_2 gas, absorb EMR of various wavelengths, intercepting 17% of the total insolation. Events on Earth's surface may cause considerable changes in the number of aerosols in the troposphere. Volcanic eruptions emit gases, volcanic ash, and other aerosols into the atmosphere, and these effects persist for several years. These additions increase the amount of absorption in the atmosphere.

5. Clouds absorb only a small percentage of insolation, about 3% of total insolation. The complex interaction between reflection, scattering, and absorption by various layers of clouds makes their impact on climate difficult to quantify—the net effect of clouds is one of the most difficult aspects to represent in global climate models. Some types of clouds tend to hold in heat, warming the surface, whereas others have an overall cooling effect.

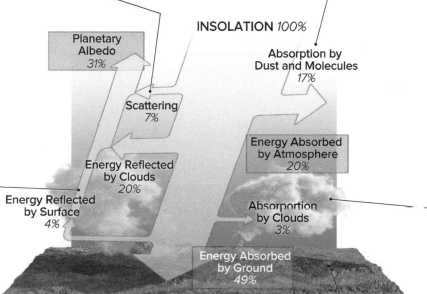

INSOLATION *100%*

Planetary Albedo *31%*

Scattering *7%*

Absorption by Dust and Molecules *17%*

Energy Reflected by Clouds *20%*

Energy Absorbed by Atmosphere *20%*

Energy Reflected by Surface *4%*

Absorption by Clouds *3%*

Energy Absorbed by Ground *49%*

02.13.c1

7. So of the total insolation, 20% is converted to sensible heat or latent heat after absorption in the atmosphere, causing an increase in temperature or a change in state (mostly evaporation of water), respectively. A total of 49% of the insolation is retained by the land, as what is called *ground heat*. The remaining 31% is reflected back into space, accounting for the remainder of the 100% of insolation (20% + 49% + 31% = 100%).

6. Earth's surface albedo is fairly low but varies widely. Snow is highly reflective (high albedo), but many rocks are dark and rough (low albedo). Water generally has a lower albedo than land, and more than 70% of the Earth is covered with water. Due to reflection, scattering, and absorption in the atmosphere, only 53% of insolation reaches Earth's surface. But 4% of this energy is reflected back to space once it hits the surface, leaving 49% to be absorbed.

Changing the Global Radiation Budget

In June 1991 Mount Pinatubo in the Philippines erupted (◄), throwing volcanic ash and gases up to 34 km into the troposphere and stratosphere. The ash circled the Earth, reducing insolation by about 1 watt per square meter (W/m²), and cooling global temperatures 0.5 C°. The image to the right shows, in red, the dust circulating around the tropical regions of the world the following December (more than a half year later). The eruption was also accompanied by a decrease in global vegetation, as measured by satellite, suggesting extra absorption of insolation by volcanic ash. The effects lasted for about two years until the ash either fell out of the atmosphere or was washed out by precipitation.

2 Luzon Island, Philippines

02.13.c3

Before You Leave This Page

✓ Summarize how the atmosphere intercepts insolation.

✓ Describe albedo, give five examples, and summarize the importance for Earth's energy balance.

✓ Sketch, label, and explain the interactions of insolation with the atmosphere and surface, noting the percentages of insolation lost and stored in various ways.

2.13

2.14 What Happens to Insolation That Reaches the Surface?

APPROXIMATELY HALF OF INSOLATION is transmitted to Earth's surface, and this energy is variably reflected, absorbed, and re-emitted. Earth absorbs energy of short wavelengths, including insolation, but re-emits it at longer wavelengths. Certain *greenhouse gases* in the Earth's atmosphere interact with this outgoing long-wavelength radiation, complicating the return of this energy to space and helping keep our planet at a hospitable temperature.

A How Does Insolation Interact with the Surface?

Insolation that reaches Earth's surface is at short wavelengths, centered on the visible spectrum. It also includes ultraviolet (UV) and infrared (IR) wavelengths adjacent to the visible spectrum, commonly called *near-UV* and *near-IR*, respectively. Such electromagnetic radiation with wavelengths less than 4 μm is called *shortwave radiation*. By contrast, Earth emits energy at longer wavelengths, or *longwave radiation*. Shortwave radiation that reaches Earth's surface can be converted into other forms of energy.

Shortwave Radiation Converted to Sensible Heat

1. Some energy that strikes Earth's surface is *absorbed* by molecules, increasing their temperature and sensible heat. The heat stored in the land and water is called *ground heat*. The amount of heating of the surface, and the flux of ground heat, depends on all the factors that influence the distribution of insolation, such as latitude, season, length of day, and cloud cover. It is also influenced by whether insolation strikes land or water, by moisture and mineral content of the surface, and other factors.

2. As Earth's land and water transfer sensible heat to the adjacent atmosphere, the air warms. Heating of Earth's land and water therefore warms the adjacent atmosphere—actually more than direct insolation does. The land and (especially) water surfaces heat up and cool down slowly. It takes both a while to exchange heat with the air, so the warmest surface temperatures of the day typically occur hours after noon. For the same reason, the coldest temperatures typically occur just before sunrise.

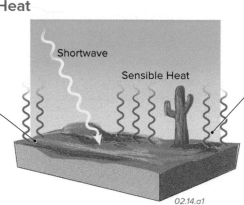

02.14.a1

Shortwave Radiation Converted to Latent Heat

3. Energy striking Earth's surface can also be converted to latent heat, such as when ice melts. As the shortwave energy of the insolation converts ice to liquid, molecules in the liquid begin to carry the associated latent heat of fusion. The water can flow away from the site of melting, or it can remain and refreeze, releasing the latent heat to the local environment.

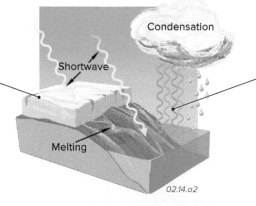

4. Shortwave energy from insolation can also cause water to evaporate, forming water vapor. Molecules in the vapor carry the *latent heat of vaporization* into the atmosphere. When the vapor condenses into water drops, such as in clouds or as precipitation, the molecules give back this latent heat to the surrounding atmosphere, warming it (sensible heat). Through this process, shortwave radiation is converted first to latent heat and then to sensible heat. Similar processes include conversion of ice to vapor (sublimation) and vapor to ice (deposition).

02.14.a2

Shortwave Radiation Absorbed and Re-emitted as Longwave Radiation

5. Shortwave radiation that strikes the surface can be absorbed by materials it encounters, such as by rocks, soils, wood, and water, increasing the motions of their constituent molecules. These materials emit some of this energy and are at temperatures appropriate for the energy emissions to be at long wavelengths (according to Wien's Law). In this way, shortwave radiation from the Sun is absorbed by the surface and then radiated as longwave radiation.

6. The emitted longwave radiation can return to space or be intercepted by clouds, gas molecules, or solid aerosols. Some of the energy absorbed by these atmospheric components is then re-emitted as longwave radiation in all directions—into space, into other parts of the atmosphere, or back toward Earth's surface. Longwave energy that is emitted downward is commonly called *counter-radiation*, and helps keep the Earth at temperatures suitable for life.

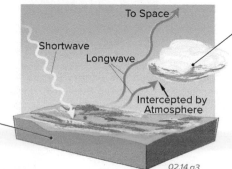

02.14.a3

B How Do Greenhouse Gases Interact with Electromagnetic Energy?

The atmosphere contains certain components, most notably water molecules, that absorb some wavelengths of *outgoing longwave radiation* (OLR). This energy is then emitted via longwave radiation to the surrounding atmosphere or even back down to Earth, keeping the planet warmer than it would be otherwise. Many people use the term *greenhouse effect* to refer to this warming influence, such as the way glass allows in sunlight but traps heat to keep a greenhouse warm. Atmospheric components that exhibit this behavior are called *greenhouse gases*.

1. The figure below shows the relative amount of absorption (vertical axis of each graph) of EMR at various wavelengths (horizontal axis), by individual greenhouse gases. It also shows the amount of radiation that is scattered by gases and particles (second graph from bottom). Peaks on each curve show where the gas absorbs a certain wavelength of energy. A value of 1 indicates very high absorption or scattering at that wavelength. The top graph shows the composite effect for all gases in the atmosphere.

2. Characteristic wavelengths of incoming visible light and adjacent wavelengths of UV are barely impacted by atmospheric gases. As a result, insolation at these wavelengths is mostly transmitted through the atmosphere, illuminating and heating Earth's surface.

3. Insolation at shorter wavelengths, such as UV, is nearly all absorbed by atmospheric gases, especially by oxygen gas, ozone, and scattering by particles. These atmospheric gases and particles shield us from this dangerous radiation.

4. The two shaded curves on the bottom of the diagram show the wavelengths of energy emitted by the Sun (yellow) versus those emitted by the Earth (blue). The majority of insolation entering the top of Earth's atmosphere is in the form of visible light (0.4–0.7 μm), whereas Earth emits its radiation back to space in the form of longwave radiation, mostly as *thermal infrared*, at wavelengths between about 8 and 20 μm. There is a clear separation between the wavelengths of *incoming shortwave radiation* and those of *outgoing longwave radiation*. The curves for individual gases show that each greenhouse gas intercepts some amounts of certain wavelengths of outgoing longwave radiation.

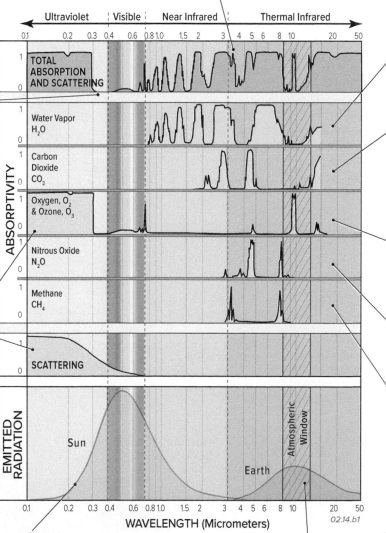

5. There is a band of wavelengths at 8–13 μm that pass through the atmosphere without much loss from scattering and other effects. This is expressed as the large trough in the top curve, which shows the total effect of absorption and scattering for the entire atmosphere. The trough represents a part of the thermal IR spectrum where most energy is transmitted through the atmosphere—this region of the longwave spectrum is called the *atmospheric window* because it allows most longwave radiation out through the atmosphere.

6. Water vapor (H_2O) is an abundant, and the most important, greenhouse gas. It absorbs a wide range of wavelengths of OLR, including most wavelengths greater than 15 μm. The quantity of atmospheric water vapor varies from place to place and from time to time. It is typically higher over the oceans than over land. Water vapor and clouds help the surface stay warm by emitting extra counter-radiation downward, moderating changes in temperature between day and night.

7. Carbon dioxide (CO_2) is an atmospheric gas present in trace amounts, measured in parts per million. Quantities of CO_2 in the atmosphere have increased as the world warmed since the last glacial advance. Some CO_2 released into the atmosphere is from natural sources, such as volcanic eruptions and the natural decay of vegetation, and some is from human activities.

8. Oxygen and ozone intercept only some wavelengths of OLR. The previously noted ability of these gases to absorb *incoming* UV radiation effectively can be seen clearly by high peaks on the far left side of this diagram.

9. Nitrous oxide (N_2O) is a by-product of many industrial chemical processes, particularly production of fertilizers. It absorbs some energy at thermal-IR wavelengths.

10. Methane (CH_4), sometimes known as "marsh gas," occurs naturally as the result of the decay of organic material. Methane can also be released from oil and gas production or when frozen ground thaws or the seafloor warms up. It absorbs some thermal-IR energy.

Before You Leave This Page

☑ Sketch, label, and explain what can happen to incoming shortwave radiation that reaches Earth's surface.

☑ Sketch, label, and explain the relationship between insolation, outgoing longwave radiation, and greenhouse gases.

2.15 How Does Earth Maintain an Energy Balance?

SIXTY-NINE PERCENT OF INSOLATION received at the outside of the Earth's atmosphere is available for sensible, ground, and latent heating. Ultimately all of this energy must be returned to space as longwave radiation in order to attain a balance between incoming and outgoing radiation. A greater loss to space would cool the global system, and a smaller loss would increase global temperatures. Just as there is a shortwave radiation budget, there is a budget of *global outgoing longwave radiation*. By interacting with outgoing longwave radiation (OLR), greenhouse gases help maintain Earth's hospitable temperature. Earth would be a very different planet if the greenhouse effect operated in a different manner or if the amount of greenhouse gases were different.

Sensible and Latent Heat Flux from Earth's Surface

1. There are various ways that Earth's surface and atmosphere transfer longwave energy. This page examines losses via the transfer of sensible and latent heat, and the facing page deals with losses through emission of longwave radiation. Read in a counterclockwise order, around the outside of both pages.

2. Of the 100 units of insolation (shortwave) that enter the top of Earth's atmosphere, 31 units are reflected or scattered by the atmosphere directly to space. That leaves 69 units within the surface and atmosphere.

3. Gas molecules, clouds, and various particles absorb 20 units, or 20% of insolation. For now, remember that this energy must be somehow released back into space.

4. The other 49 units of insolation are transmitted through the atmosphere. These units are then absorbed by Earth's surface, including land, oceans, and other bodies of water. These 49 units must somehow also escape the surface, or else the surface would keep heating up indefinitely as the Sun continued to transmit shortwave energy.

5. If 49 units of insolation reach Earth's surface, this is more than twice the amount (20 units) that is absorbed in the atmosphere. An implication of this is that the Sun heats Earth's surface more than it does the atmosphere, and in turn, the surface heats the atmosphere (by 29 units). Warming of the atmosphere from below in this way is one reason why air temperatures generally decrease upward with increasing altitude.

6. As the atmosphere is heated from below, the warmer air near the surface starts to rise upward, inducing convection in the troposphere. Approximately 7 units of energy are transmitted, mostly by convection, to the adjacent air as sensible heat. This flow of energy is called the *sensible heat flux*.

7. Most of the Earth is covered by ocean, and many land areas include lakes, wetlands, and heavily vegetated regions, so much of the energy reaching the surface goes to *latent heat flux* (melting ice, evaporating water, and transpiration from plants). Melting of ice only transfers energy between different parts of the surface (ice sheet to sea, for example), so it does not directly impact the atmospheric energy budget—but evaporation does. As the warm air rises convectively, it carries aloft the recently evaporated water vapor into the ever cooler air at higher altitudes. Eventually the moist air cools sufficiently to condense into water drops and form clouds, which then release the latent heat into the atmosphere. Almost half of all the energy reaching the surface of Earth (23 of 49 units) is returned to the atmosphere in this way.

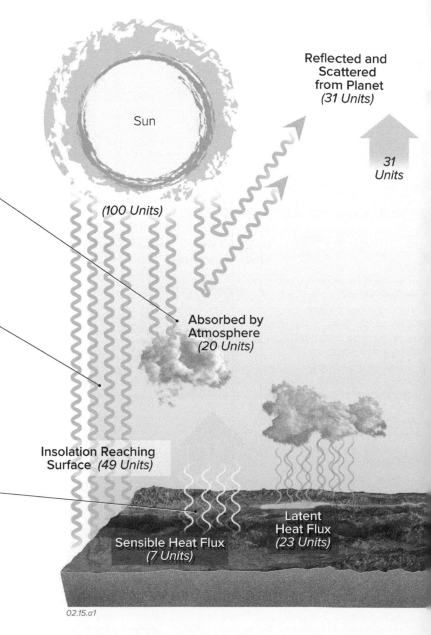

Reflected and Scattered from Planet (31 Units)

Sun

31 Units

(100 Units)

Absorbed by Atmosphere (20 Units)

Insolation Reaching Surface (49 Units)

Sensible Heat Flux (7 Units)

Latent Heat Flux (23 Units)

02.15.a1

8. The combined contributions of sensible and latent heat carry about 30 of the 49 units of the shortwave radiation stored at the surface into the atmosphere. Note that the numbers on these two pages add up to slightly more than 100 units (100%) because values have been rounded to whole numbers. When carried out with more precise numbers, it all adds up to 100 units.

Longwave Energy Flux From Surface and Atmosphere

13. We started with 100 units of insolation, so these 69 units plus the 31 units of shortwave insolation reflected (the planetary albedo) provide a perfect balance of input and output of energy to and from the Earth's land-ocean-atmosphere system. Keep in mind that these values are average annual values for the globe. Any individual place is unlikely to experience such a balance. Circulation of the atmosphere and oceans transfers energy from places that have an excess relative to the global average to those areas that have a relative deficit.

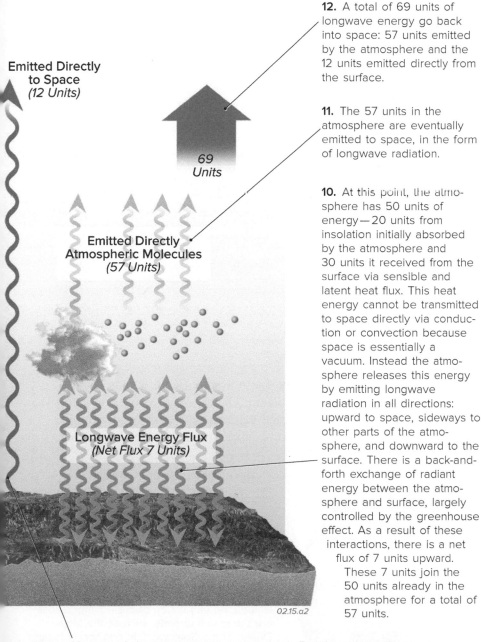

Emitted Directly to Space
(12 Units)

69 Units

Emitted Directly Atmospheric Molecules
(57 Units)

Longwave Energy Flux
(Net Flux 7 Units)

02.15.a2

12. A total of 69 units of longwave energy go back into space: 57 units emitted by the atmosphere and the 12 units emitted directly from the surface.

11. The 57 units in the atmosphere are eventually emitted to space, in the form of longwave radiation.

10. At this point, the atmosphere has 50 units of energy—20 units from insolation initially absorbed by the atmosphere and 30 units it received from the surface via sensible and latent heat flux. This heat energy cannot be transmitted to space directly via conduction or convection because space is essentially a vacuum. Instead the atmosphere releases this energy by emitting longwave radiation in all directions: upward to space, sideways to other parts of the atmosphere, and downward to the surface. There is a back-and-forth exchange of radiant energy between the atmosphere and surface, largely controlled by the greenhouse effect. As a result of these interactions, there is a net flux of 7 units upward. These 7 units join the 50 units already in the atmosphere for a total of 57 units.

9. Of the 49 units of shortwave radiation that reach Earth's surface, 12 units are emitted directly from the surface (water and land) to space as longwave radiation, without significant interactions with the atmosphere. This is possible because of the *atmospheric window* that allows certain IR wavelengths to radiate upward through the atmosphere with only minimal losses to absorption, reflection, and scattering.

A World without the Greenhouse Effect

Imagine two Earths, identical in all other ways (solar constant and planetary albedo) except the presence of a greenhouse effect. In the figure shown here, the lower globe has greenhouse gases in its atmosphere, whereas the upper globe does not.

From surface and satellite observations, average global surface temperatures on Earth currently are about +15°C (+59°F). We can estimate the average surface temperature of the imaginary planet using the Stefan-Boltzmann Law, relating the energy emission and temperature. Using this law, the surface temperature of the imaginary world is predicted to be −18°C (−9°F), significantly colder than the currently observed temperature of Earth.

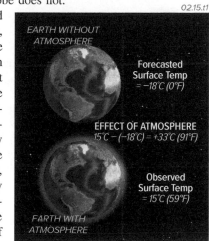

02.15.t1

EARTH WITHOUT ATMOSPHERE

Forecasted Surface Temp = −18°C (0°F)

EFFECT OF ATMOSPHERE 15°C − (−18°C) = +33°C (91°F)

Observed Surface Temp = 15°C (59°F)

EARTH WITH ATMOSPHERE

Some of the difference between these two estimates must be caused by greenhouse gases, which are keeping Earth's current temperatures warmer, thereby allowing water to exist in a liquid state, a key factor in supporting life as we know it. Calculating the actual contribution (in degrees of warming) of greenhouse gases is too complex to pursue here, because such calculations involve many other factors.

From these rough calculations, we can see why there is concern over the impacts of increased concentrations of greenhouse gases due to human-related emissions of carbon dioxide (CO_2), methane (CH_4), and nitrous oxide (N_2O). We address the changing concentrations of greenhouse gases and the broader topic of *climate change* in a later chapter on climate.

Before You Leave This Page

✓ Sketch, label, and explain the nature of the Earth's overall energy balance, including the longwave radiation balance.

✓ Explain the linkage between the longwave radiation balance and the "greenhouse effect" on Earth.

2.16 How Do Insolation and Outgoing Radiation Vary Spatially?

FLOWS OF ENERGY into and out of Earth's system vary spatially, depending on latitude, whether it is inland or over the ocean, cloud cover, and many other factors. The pattern of insolation also changes over several timescales, from daily rotation of the planet to the longer changes in season, causing spatial and temporal imbalances—zones of surplus energy and zones with an energy deficit, relative to the planetary average. These energy imbalances provide the driving force for global weather and climate.

A How Does Insolation Vary Spatially?

Insolation striking the top of the atmosphere, when averaged over a year, shows a smooth gradient, from higher amounts over the equator to much lower amounts at the poles. A more complex pattern emerges if we examine how much insolation actually reaches the surface during the course of a year, as shown in the figure below.

1. At the broadest scale, latitude controls insolation. The highest amounts are in low latitudes, and the lowest amounts are near the poles. Purples on this globe are as low as 120 W/m², whereas orange and red are more than 250 W/m².

2. Although insolation is strong in the tropics it is reduced somewhat due to absorption, reflection, and scattering by the abundant clouds that characterize these same regions.

3. The highest values are in subtropical deserts.

4. Universally low values mark the Antarctic and Arctic, which have days to months of total darkness. Because of the low Sun angles the lengthy summer days do not make up for the dark winters.

5. The mid-latitudes are between 30° and 60° latitude, between the lines shown on this globe. At these latitudes, there is a relatively steep gradient, with the amount of annual insolation decreasing toward the pole.

6. There is a marked contrast between the gradual variation in insolation patterns over oceans and the irregular patchwork of variations over continents. The more irregular character over land is because of its variable land covers and elevations, which in turn influence the albedo, cloud cover, and other factors.

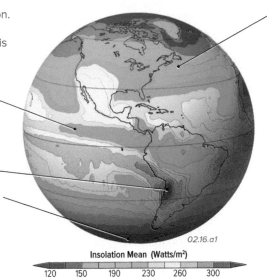

02.16.a1

Insolation Mean (Watts/m²)
120 150 190 230 260 300

B How Does Outgoing Longwave Radiation Vary Spatially?

Using satellites, we can measure the amount of longwave radiation that leaves the Earth system (land, oceans, and atmosphere). The figure below shows the total amount of outgoing longwave radiation (OLR), averaged over an entire year. The picture for any single day is more complex, with more irregular weather-related patterns. The amount of outgoing longwave radiation from a region greatly reflects the amount of incoming shortwave insolation that strikes the surface, so is lowest near the poles.

1. The quantity of energy emitted from an object depends on its surface temperature. Latitude strongly controls annual insolation, so it also controls larger patterns of both temperature and OLR.

2. The highest amounts of OLR are in the subtropics, which receive more insolation than regions farther from the equator. Some of this "extra" energy is directly returned as OLR from the surface and from the atmosphere.

3. The pattern of high OLR in the subtropics is complicated over land, such as a zone of lower OLR along the western side of South America. This is due to a combination of a high mountain range (the Andes) and a cold ocean current (the Humboldt Current) along the west coast, both of which reduce the temperature. Reduced temperatures affect OLR.

4. The Arctic and Antarctic differ from one another, although it is difficult to see from this perspective. The oceanic Arctic emits more OLR, appearing warmer, than the continental Antarctic. A similarly low OLR is associated with land in Greenland, compared to the adjacent ocean, which emits more OLR because it is warmer.

5. Conversely, some of the highest continental values of OLR are between 20° and 30° latitude, on either side of the equator. These are the global deserts, such as the Mojave Desert of the Southwest. In addition to having high temperatures, the absence of water vapor allows OLR to escape through the atmosphere.

6. The patterns in this globe and the one above change with the season, as the latitude of maximum insolation shifts north and south.

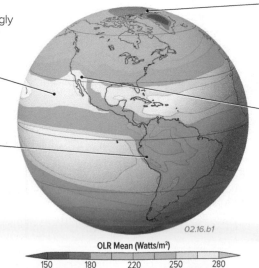

02.16.b1

OLR Mean (Watts/m²)
150 180 220 250 280

C How Does the Balance Between Incoming and Outgoing Radiation Vary Spatially?

Tropical regions receive much more insolation than do mid-latitude or polar regions, and this results in an unequal distribution of energy as a function of latitude—tropical regions have more Sun-induced energy than the poles. Energy flows from regions of relative surplus toward regions with relative deficits.

Radiative Balances

1. This graph shows the average amount of insolation at the top of the atmosphere versus the amount of outgoing energy (OLR) emitted, as a function of latitude, from the equator on the left to the North Pole on the right. A graph constructed from the equator to the South Pole would be similar.

2. Like a bank account, the difference between incoming and outgoing radiation is a location's *radiation balance*. On average, regions near the equator, tropics, and subtropics receive more energy from insolation than they emit in OLR—they have a positive balance or a *radiation surplus*. The region of surplus extends as far north (and south) as about 35° latitude, well past the Tropics of Cancer and Capricorn. The amount of insolation and OLR barely change across the tropics (leftmost part of the graph).

3. Poleward of the tropics, insolation declines rapidly, while the amount of outgoing radiation decreases more gradually.

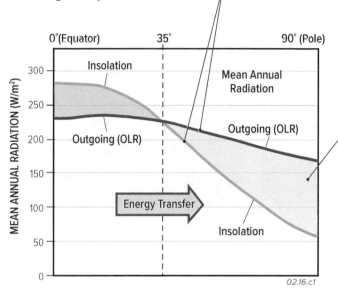

02.16.c1

4. The two curves cross around 35° latitude, represented by the vertical dashed line. At this location, the amount of outgoing radiation (OLR) equals the amount of incoming radiation, on average—at this latitude, there is a radiation balance of zero.

5. Poleward of 35°, regions emit more OLR than the shortwave radiation that they receive from insolation—they have a *radiation deficit*. Insolation continues declining rapidly at mid-latitudes, reaching very low values in the polar regions. The energy balance becomes increasingly negative toward the poles.

6. Energy is transferred from areas of surplus to those of deficit, toward the poles.

Global-Scale Patterns of Radiative Surplus and Deficit

7. This globe shows the mean annual radiation balance based on satellite observations. Orange, red, and dark pink represent zones of surplus, whereas yellow, green, and blue are regions of deficit. The white dotted lines show 35° N and 35° S. What patterns do you notice?

8. As expected, the regions with the highest energy surplus, depicted in red and orange, are concentrated along the equator, extending out across the tropics on both sides of the equator.

9. The patterns on land are locally more complex than the patterns in the ocean, suggesting the complicating influence of variations in elevation and the resulting complex patterns of clouds, snow, wind directions, etc.

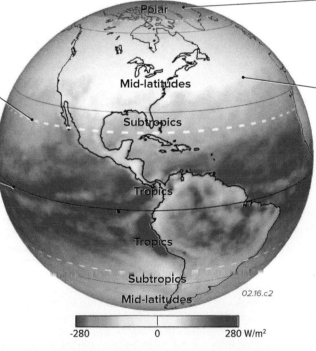

02.16.c2

-280   0   280 W/m²

10. The regions with the most severe energy deficits are near the poles. These regions lose much more energy to space than they gain from low-angle sunlight during the summer, and they receive no insolation during the winter.

11. The mid-latitudes, between the tropics and the polar circles, exhibit a slight radiative deficit, represented here by a pale pink color. They gain energy transferred poleward from the tropics.

12. More than half of the planet's surface area has an energy surplus, and less than half has an energy deficit. The total amount of surplus energy equals the amount of energy deficit, and the planet is in an overall energy balance. As we will explore later, the transfer of energy from areas of radiative surplus to those of deficit drive much of our wind patterns and weather.

Before You Leave This Page

✓ Sketch and explain the global patterns of insolation and outgoing longwave radiation.

✓ Sketch and explain global patterns of surplus and deficit energy, using a graph or map, and explain how these are expressed in global-scale patterns in the regional distribution of areas of energy excess and deficits.

2.16

2.17 Why Do Temperatures Vary Between Oceans and Continents?

WATER EXHIBITS VERY DIFFERENT thermal properties from those displayed by the rocks and soil. These differences in thermal properties cause oceans and land to warm and cool at different rates, leading to significant temperature variations between oceans and land. Such differences help explain major patterns of global temperature and climate.

A How Do Water and Earth Materials Respond to the Same Changes in Energy?

Heat Capacity

1. In evaluating how water, rocks, and other materials heat up or cool down, an important consideration is how much energy is needed to heat up an object, and how much heat that object can retain. A physical attribute called *heat capacity* expresses how much heat is required to change a volume's temperature by one Kelvin.

HEAT CAPACITY

Less 02.17.a1 *More*

2. The heat capacity of an object is determined by the kind of material in the object, such as rock versus water, and by the size of the object. The larger block above has a greater heat capacity than the smaller block, as long as both are composed of similar materials.

SPECIFIC HEAT

Land   Water

Less 02.17.a2 *More*

Specific Heat

3. To compare the inherent thermal responses of different materials, irrespective of how much of the material is present, we use a property called *specific heat capacity*, or simply *specific heat*. Specific heat is the amount of energy needed to increase a kilogram mass of a substance by 1 K (or 1 C°).

4. The specific heat of water is four times that of most rocks and materials. This means that it takes four times more energy to heat water than it takes to heat an equivalent mass of rock.

Thermal Responses of Water Versus Other Earth Materials

5. Due to their differences in specific heat, oceans (water) heat up during the day differently than land. If the same amount of insolation strikes water and land, the land will increase in temperature 4 degrees (K or C°) for every degree the water increases. As a result, land warms up much faster than the ocean, under the same environmental conditions, but some of this difference is offset by increased losses of heat from the land to the air.

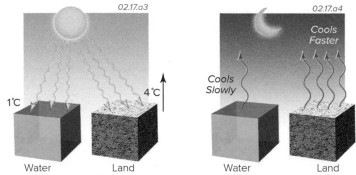

02.17.a3 / 02.17.a4

Water   Land   Cools Slowly   Cools Faster   Water   Land

1°C   4°C

6. At night, the water and land both lose energy to the cool night air. For the same energy loss, land cools more than does water. Also, the land became hotter during the day, and hot objects radiate more energy than cool ones, so the land loses energy faster than the water. As a result, land cools off much faster than water at night. Air heats more quickly over land and more slowly over water.

Depth of Heating, Cooling, and Mixing

7. Another factor in how water and land respond to changes in insolation and air temperatures is how deeply heat is able to enter each material. Nearly all water allows at least some transmission, so shortwave radiation can penetrate to depths of tens of meters or more. In contrast, rocks and soil are largely opaque, so insolation is confined to the surface, and heat must move downward into the land by conduction, which it does by only a meter or two during the day.

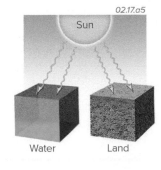

02.17.a5   Sun

Water   Land

8. Some materials, such as water, are relatively mobile, which allows them to flow and mix. Under calm ocean conditions (left column), limited mixing causes surface waters warmed by the Sun to remain near the surface, so there is a strong temperature contrast with depth. Surface winds induce waves (center column), resulting in turbulence, which carries warm waters downward, mixing them with cooler waters at a depth. Salt water (right column) is more dense than fresh water, so any waters that are saltier than normal, such as from partial evaporation, can sink, causing mixing of the water column. Mixing allows heat to be carried deeper into the water column (much faster than heat is conducted) and brings up cooler water that gives off energy to the atmosphere more slowly (because it is cool). As a result, mixed water heats up more slowly than does land, which experiences almost no vertical mixing.

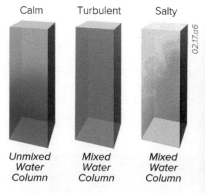

Calm   Turbulent   Salty

02.17.a6

Unmixed Water Column   *Mixed Water Column*   *Mixed Water Column*

Latent Heat

9. Water has another unique capacity relative to land—it can store abundant energy as latent heat. Insolation that strikes water is transformed into one of three different fluxes of energy. Some energy goes into heating the water (ground heat), some heats the air (sensible heat in the atmosphere), but a large amount goes into latent heat produced by evaporation. In contrast, insolation striking land goes mostly into ground heat and into sensible heat in the atmosphere. Land contains some water, but lesser amounts of its insolation go into latent heat.

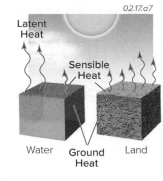

02.17.a7

Latent Heat

Sensible Heat

Water   Ground   Land
         Heat

Distribution of Continents and Oceans

10. Another factor that influences the global energy budget, and the balance of energy that falls on land versus the oceans, is the difference between the Northern and Southern Hemispheres. As can be observed on any map or globe, the Northern Hemisphere has the majority of the planet's landmasses, whereas the Southern Hemisphere is dominated by oceans. As a result, an equal amount of insolation striking both hemispheres will result in more latent heat being generated in the Southern Hemisphere than in the Northern Hemisphere. In December, when the Southern Hemisphere more directly faces the Sun, more insolation will fall on water than during June.

02.17.a8

B How Do Temperatures Reveal Thermal Differences Between Ocean and Land?

1. These various factors, from specific heat to latent heat, cause land and water to respond very differently to insolation and to the change from day to night. Land, with its relatively small specific heat, limited mixing, and limited amount of latent heating, warms up more quickly than water and reaches higher temperatures. At night, land's higher daytime temperatures cause it to lose heat more rapidly than does water. This keeps the night warm for a while, but eventually the cool night air dominates.

2. Water, with its large specific heat, partial transparency, and ability to mix vertically, heats up more slowly and does not reach as high a temperature. Also, much insolation is converted into latent heat that is transferred to the atmosphere via evaporation and condensation, so this energy is not available to heat the body of water. Large bodies of water therefore experience smaller temperature variations and more moderate temperatures overall, compared to land. Land areas adjacent to the water can partly experience the moderate temperatures caused by the unique thermal properties of water.

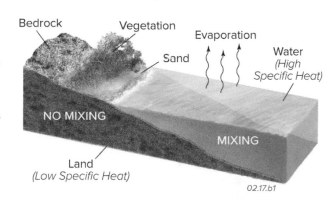

Bedrock   Vegetation
                    Evaporation
              Sand   Water
                           (High
                           Specific Heat)
NO MIXING
                    MIXING
Land
(Low Specific Heat)
02.17.b1

Mean Annual Temperature

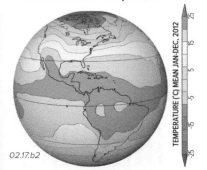

02.17.b2

TEMPERATURE (C) MEAN JAN-DEC, 2012

3. These three globes show the yearly average temperature, average January temperature, and average July temperature for a typical year. Observe the temperatures shown on each globe and use concepts presented in this chapter to try to explain the main patterns. Aspects to consider include variations in insolation due to latitude and clouds, land-sea contrasts, and the distribution of continents. These globes show data for the different times of year, so they express seasonal variations. In each globe, red and orange are hotter, blue and purple are colder, and yellow and green are intermediate in temperature. The overall average temperatures (◄) range from less than −50°C in Antarctica to locally greater than +20°C in some tropical and subtropical regions.

Average January Temperature

02.17.b3

TEMPERATURE (C) JAN 4, 2013

4. Average January temperatures (◄) go from less than −50°C near the North Pole to more than +25°C in the tropics.

5. Average July temperatures (►) are higher in the Northern Hemisphere than in the Southern Hemisphere.

Average July Temperature

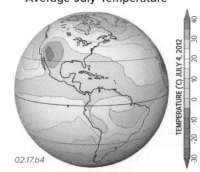

02.17.b4

TEMPERATURE (C) JULY 4, 2012

Before You Leave This Page

☑ Sketch, label, and explain all the factors that cause land and ocean at the same latitudes to heat and cool differently, identifying how these factors affect temperatures.

☑ Summarize some of the main patterns in the global distribution of average, maximum, and minimum temperatures.

2.17

How Are Variations in Insolation Expressed Between the North and South Poles?

VARIATIONS IN INSOLATION, both as a function of latitude and from season to season, help explain many aspects of our world—average temperatures, hours of daylight, type of climate and weather, type of landscape, and overall livability of a place. For a transect down the west coasts of the Americas, from the Arctic to the Antarctic, we examine the average monthly amounts of insolation, length of day, and temperature, as a way to connect concepts in this chapter with actual places. Examine the photographs, graphs, and text for each place, and think about what explains the patterns for that place and the variations from one place to the next. For each place, the graph on the left shows variation in insolation from month to month at the top of the atmosphere, whereas the graph on the right shows average number of daylight hours (red boxes) and average monthly temperature (blue curve).

02.18.a1 ANWR, AK

Northern Alaska and Canada

The North Pole is located in the Arctic Ocean, but parts of Alaska and Canada are north of the Arctic Circle (66.5° N). Compare the graphs to the right, which show the monthly variation in insolation (first graph), number of daylight hours (bar graph) and average temperatures (line graph). Note that during parts of winter there is no daylight.

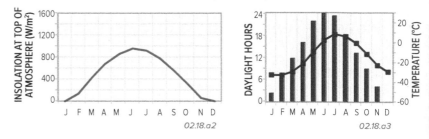
02.18.a2 02.18.a3

CAMBRIDGE BAY, NUNAVUT, CANADA

02.18.a4 British Columbia, Canada

Pacific Northwest

The northwestern part of the mainland U.S. and adjacent parts of British Columbia, Canada, straddle the famous 49th parallel (49° N latitude). They are squarely in the mid-latitudes. Like most of the places on this page, the region is near the ocean, so its temperature variations are moderated somewhat by the waters of the adjacent ocean.

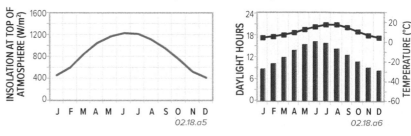
02.18.a5 02.18.a6

VANCOUVER, BRITISH COLUMBIA, CANADA

02.18.a7 Baja, Mexico

California Sur, Baja Mexico

Baja California, part of Mexico, is a desert peninsula bordered by the Pacific Ocean to the west and the Gulf of California to the east. La Paz, the capital of Baja California Sur, is at a latitude of 24° N, just north of the Tropic of Cancer (23.5° N). Note that the graphs for Baja and the two previous places display a maximum centered on June to August (summer).

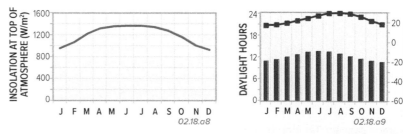
02.18.a8 02.18.a9

LA PAZ, BAJA CALIFORNIA SUR, MEXICO

02.18.a10 Ecuador

Ecuador

The South American country of Ecuador is named for its position on the equator. It is on the west coast of the continent and contains parts of the Andes, Amazon rain forest, and the Galápagos Islands, famous for their active volcanoes and unusual animals. Note the pattern on the length-of-day bar graph and the minor variation in monthly insolation for near the equator.

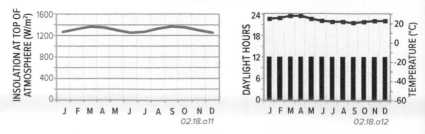
02.18.a11 02.18.a12

GALÁPAGOS ISLANDS

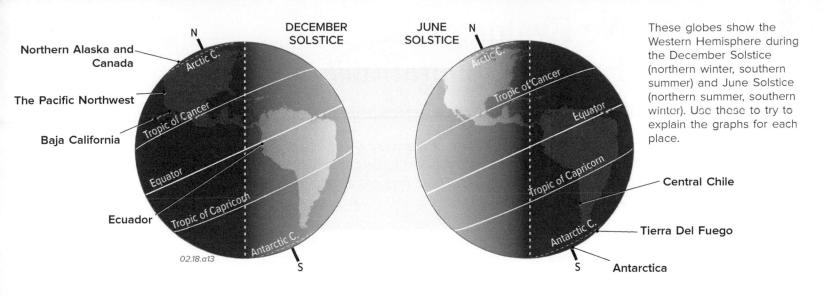

Northern Alaska and Canada

The Pacific Northwest

Baja California

Ecuador

DECEMBER SOLSTICE

JUNE SOLSTICE

Central Chile

Tierra Del Fuego

Antarctica

These globes show the Western Hemisphere during the December Solstice (northern winter, southern summer) and June Solstice (northern summer, southern winter). Use these to try to explain the graphs for each place.

02.18.a13

Central Chile

02.18.a14 Santiago, Chile

About halfway down the western coast of South America is the central part of Chile. The country's capital of Santiago is at a latitude of 33.5° S, south of the Tropic of Capricorn. It is inland and higher than the coast. Note that in the Southern Hemisphere, the graphs of insolation, hours of daylight, and temperatures now have June through August troughs rather than peaks. You know why, don't you?

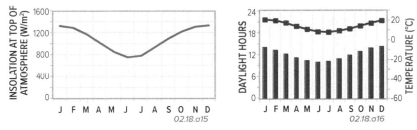

SANTIAGO, CHILE

02.18.a15

02.18.a16

Tierra del Fuego

02.18.a17 Ushaia, Argentina

Tierra del Fuego is the southernmost tip of South America, in southern Chile and Argentina. Ushuaia, Argentina, at a latitude of 54.8° S, is called the southernmost city in the world. It is not within the Antarctic Circle, so every day has some daylight and darkness. The southern tip of South America, surrounded by cold seas, is a frigid, stormy place.

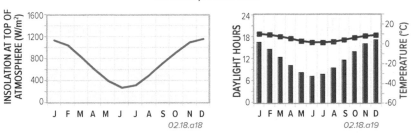

USHUAIA, ARGENTINA

02.18.a18

02.18.a19

Antarctica

02.18.a20 Antarctica

Antarctica, centered over the South Pole, consists of vast plains of ice with some impressive mountain ranges. The main part of the continent is within the Antarctic Circle, so it has months of darkness. The South Pole has continuous darkness from April to September (the southern winter), but 24 hours of sunlight from October to February. It is always a very cold place.

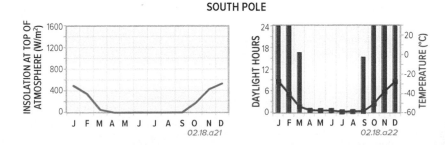

SOUTH POLE

02.18.a21

02.18.a22

Before You Leave This Page

✓ Sketch the general patterns of average insolation, number of daylight hours, and temperatures for three sites: one in the Northern Hemisphere, one in the Southern Hemisphere, and one near the equator. Explain the patterns in terms of latitude.

How Do We Evaluate Sites for Solar-Energy Generation?

SOLAR ENERGY IS A RENEWABLE SOURCE of light, heat, and electricity. Solar energy is typically collected with a solar panel, which can generate electricity or can heat air, water, or some other fluid. You have an opportunity to evaluate the solar-energy potential of five sites in South America, using concepts you have learned about insolation in this chapter. To do this, you will use Sun angles at each site to determine how directly sunlight will strike solar panels at different times of the year. You will then consider some regional factors that influence how much insolation reaches the ground. From these data, assess how efficient each site will be at generating solar energy.

Goals of This Exercise:

- Understand the importance of latitude in constraining the maximum possible amount of insolation striking the surface, to help evaluate the solar-energy potential of a site.
- Use factors controlling regional amounts of insolation, along with characteristics of the geographic setting of a site, to evaluate how much insolation is likely to actually arrive at Earth's surface, the key factor in the solar-energy potential of each site.

1. The globe below is a view centered on South America, and it shows the South Pole and adjacent parts of the Southern Hemisphere. It locates five possible sites you will consider for their solar-energy potential: (1) the Galápagos Islands of Ecuador; (2) Macapá, a small settlement in Brazil; (3) La Serena, a coastal town of Chile; (4) near Mar Chiquita, in central Argentina, and (5) Ushuaia, Argentina. Some characteristics of each site are described below.

Site	Lati-tude	Sun Angle		
		Equi-nox	Dec. Sol.	June Sol.
Galápagos	0°	90°	66.5° S	66.5° N
Macapá	0°	90°	66.5° S	66.5° N
La Serena	30° S	60°	82.5°	36.5°
Mar Chiquita	30° S	60°	82.5°	36.5°
Ushuaia	~60° S	30°	53.5	6.5°

2. The Galápagos are a series of islands in the Pacific Ocean, far west of South America. They are known for their exotic animal species, made famous by the visits of Charles Darwin. Although surrounded by the ocean, many parts of the islands receive little rain.

3. La Serena is a popular tourist site along the Pacific coast of Chile. It has a cool, desert climate, caused in part by a cold ocean current (the Humboldt Current) that limits the amount of moisture in the air. It is on the fringe of the Atacama Desert, the driest place on Earth (the tan strip along the coast to the north of 30° S on the globe).

4. Ushuaia, the southernmost city in the world, is near Cape Horn. Cape Horn is surrounded on three sides by ocean, some of which is coming from near Antarctica and is very cold. As a result, Ushuaia has a relatively cold and humid climate. It is close to 55° S latitude, but for this exercise we will consider it to be at 60° S to simplify our calculations.

30° N

Macapá

0°

Galápagos

Mar Chiquita

30° S

La Serena

Ushuaia

02.19.a1

7. From the latitude of each site we can calculate the Sun angle at each site for an equinox and both solstices. These results are presented in the table above.

6. Macapá is a small community in Brazil, near the mouth of the Amazon River, just inland from the sea. It is next to the Amazon River and is part of the huge Amazon rain forest, which stretches from the Atlantic Ocean westward all the way to the foothills of the Andes. As part of the rain forest, Macapá is a humid, rainy place.

5. Mar Chiquita is a small inland lake on the plains of central Argentina, far from the oceans but at a relatively low elevation. The proposed site is north of the lake, far enough that the lake does not significantly affect the local climate. This part of Argentina is east of the Andes, the huge mountain range along the western coast of South America. The Andes block moisture from the west, causing the plains of Argentina to be relatively dry.

Procedures

1. Read descriptions of each site and consider how the geographic setting might impact the favorability of the site for solar-energy production.

2. For each site, use the Sun angles to look up the maximum percentage of solar energy that is theoretically available.

3. Consider how insolation varies with latitude and from season to season as an important consideration for the suitability of each site. Graphs from the previous two-page spread (the Connections spread) will be helpful here.

4. List the pros and cons of building a solar-energy facility at each site.

5. OPTIONAL EXERCISE: Your instructor may provide you with simple formulas for calculating Sun angle from latitude and have you determine Sun angles and solar favorability for other sites, some near where you live.

8. In considering the various sites, an important factor is that the solar panel will be horizontal. So we can use Sun angle as a measure of whether the sunlight is coming in perpendicular to the solar panel, the optimum orientation. Although this orientation works fine near the equator, at other latitudes solar panels are inclined at some other angle so as to maximize the amount of sunlight (keeping the panel perpendicular to the direction of sunlight, wherever possible).

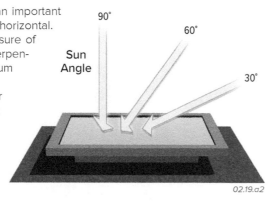

02.19.a2

9. Use the Sun angle for each site to determine the maximum percentage of energy that can be produced by the solar panel for that Sun angle, relative to the maximum amount it could produce if it were perfectly perpendicular to the Sun. Write down your results to guide your considerations in evaluating each site.

Percentage of Solar Energy Available	
Sun Angle	Percent
90°	100%
80°	98%
70°	94%
60°	87%
50°	76%
40°	64%
30°	50%
20°	34%
10°	17%
0°	0%

10. For each location, use your angle calculations to predict how the amount of insolation arriving at the top of the atmosphere will vary from season to season. Key times in the seasonal changes will be the solstices (December and June) and the equinoxes (March and September).

11. Since your calculations represent the amount of insolation reaching the top of the atmosphere, now consider how the climate of each site might decrease this amount, such as from excessive cloudiness. Use the globes below to complete this step. Finally, combine your angle calculations with results from the globes below to make a list showing the pros and cons of each site. Choose the best site and be able to defend your conclusions.

Downward Shortwave Radiation Flux This globe shows the amount of shortwave radiation (insolation) that actually reaches the surface. Red, orange, and yellow show larger amounts of insolation reaching the surface, whereas blue and purple show smaller amounts. The generally greater amounts of insolation in the tropics versus the low amounts near the South Pole show the effects of latitude. The deviations from the broad pattern, such as the green and bluish green low over the Amazon, are due to processes in the atmosphere that reflect, absorb, or scatter insolation. Use what you know about the geographic setting of each site to propose what processes explain the pattern on this globe. The large globe on the previous page shows the topography (mountains versus low, flat regions) and land cover (vegetation versus rocks). These variations affect factors like cloudiness and the amount of rain.

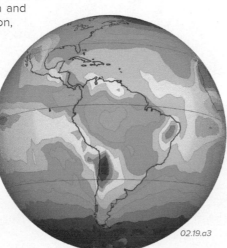

02.19.a3   02.19.a4

Outgoing Longwave Radiation Flux (OLR) This globe shows the amount of OLR emitted by Earth's surface. Such radiation is not useful for generating solar energy because it is longwave and, therefore, less energetic than shortwave radiation, and because it is going upward, not down. It is a useful indication, however, of processes going on in the atmosphere because it suggests, for example, the abundance of water vapor (a greenhouse gas) and clouds, which can absorb longwave radiation before it exits the atmosphere, thereby reducing OLR. On this globe, orange represents large amounts of OLR, whereas purple and blue represent much smaller amounts. What do the patterns indicate, and what are the implications of both globes for solar energy? Again, examine the large globe on the previous page for clues.

3 Atmospheric Motion

MOTION OF THE EARTH'S ATMOSPHERE has a great influence on human lives by controlling climate, rainfall, weather patterns, and long-range transportation. It is driven largely by differences in insolation, with influences from other factors, including topography, land-sea interfaces, and especially rotation of the planet. These factors control motion at local scales, like between a mountain and valley, at larger scales encompassing major storm systems, and at global scales, determining the prevailing wind directions for the broader planet. All of these circulations are governed by similar physical principles, which explain wind, weather patterns, and climate.

Broad-scale patterns of atmospheric circulation are shown here for the Northern Hemisphere. Examine all the components on this figure and think about what you know about each. Do you recognize some of the features and names? Two features on this figure are identified with the term "jet stream." You may have heard this term watching the nightly weather report or from a captain on a cross-country airline flight.

What is a jet stream and what effect does it have on weather and flying?

Prominent labels of H and L represent areas with relatively higher and lower air pressure, respectively.

What is air pressure and why do some areas have higher or lower pressure than other areas?

Distinctive wind patterns, shown by white arrows, are associated with the areas of high and low pressure. The winds are flowing outward and in a clockwise direction from the high, but inward and in a counterclockwise direction from the low. These directions would be reversed for highs and lows in the Southern Hemisphere.

Why do spiral wind patterns develop around areas of high and low pressures, and why are the patterns reversed in the Southern Hemisphere?

North of the equator, prevailing winds (shown with large gray arrows) have gently curved shapes. For most of human history, transportation routes depended on local and regional atmospheric circulation. These winds were named "trade winds" because of their importance in dictating the patterns of world commerce. The trade winds circulate from Spain southwestward, causing Christopher Columbus to land in the Bahamas rather than the present U.S.

What causes winds blowing toward the equator to be deflected to the west?

Prevailing winds from the north and south converge near the equator. This zone of convergence, called the *Intertropical Convergence Zone (ITCZ)*, is a locus of humid air and stormy weather.

What causes winds to converge near the equator, and why does this convergence cause unsettled weather?

Polar Cell

Polar Easterlies

60° N

Polar Front Jet Stream

L

H

30° N

Northeast Trade Winds

ITCZ

EQUATOR

TOPICS IN THIS CHAPTER

Polar Tropopause

Subtropical Jet Stream

30° N

Northeast Trade Winds

Hadley Cell

ITCZ

Tropical Tropopause

03.00.a1

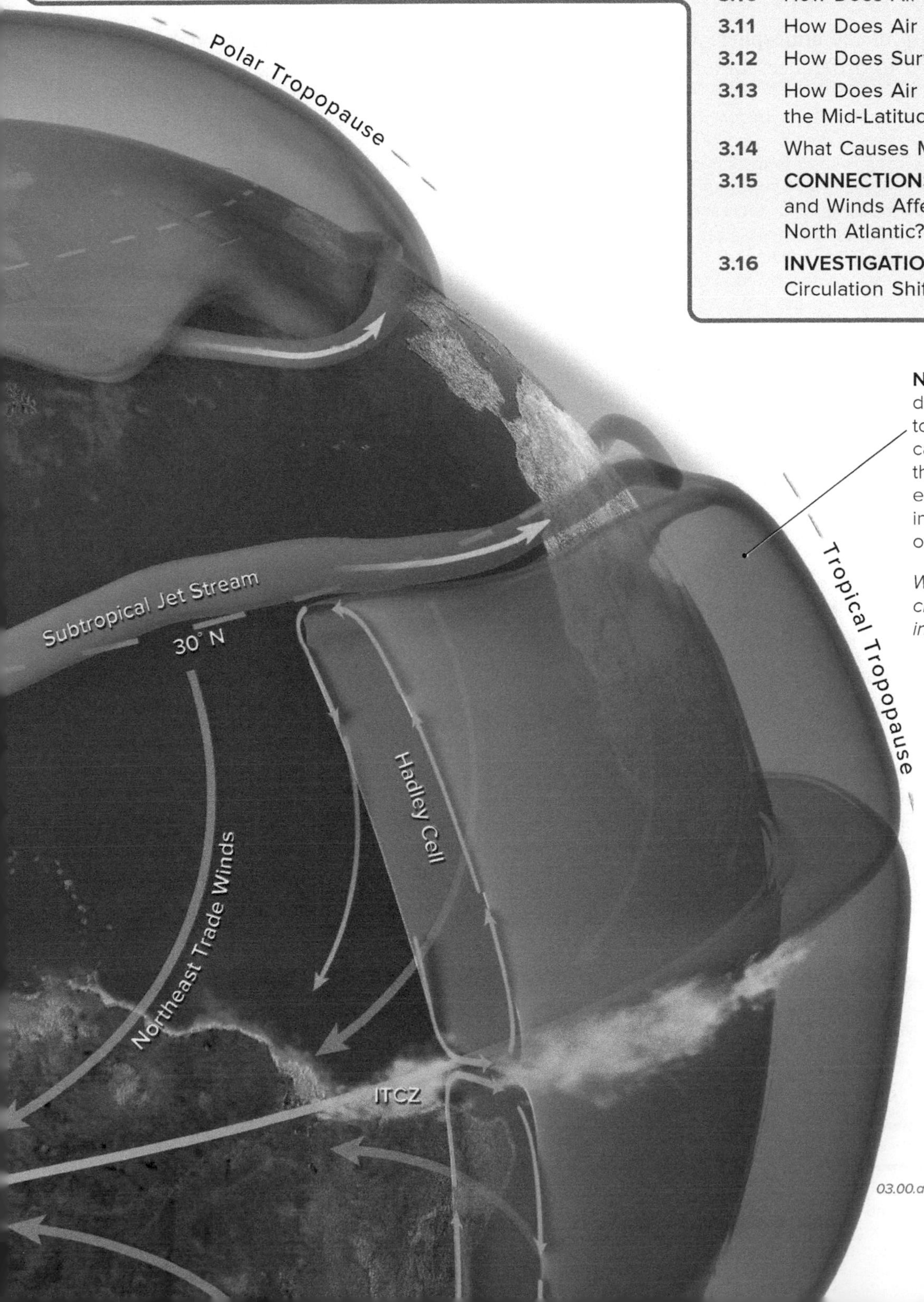

Near-surface winds interact with upward- and downward-flowing air higher in the atmosphere, together forming huge tube-shaped air circuits called *circulation cells*. The most prominent of these are *Hadley cells*, one of which occurs on either side of the equator. The circulation cells influence the altitude of the tropopause, the top of the troposphere.

What controls the existence and location of circulation cells, and how do the Hadley cells influence global weather and climate?

Motion in the atmosphere affects us in many ways. It controls short-term weather and long-term climate, including typical average, maximum, and minimum temperatures. The broad-scale patterns of air circulation, along with effects of local winds, cause winds to change direction with the seasons and from night to day. Regional air circulation affects the amount and timing of rainfall for a region, which in turn controls the climate, types of soils, vegetation, agriculture, and animals situated in an area. Winds determine which areas of the U.S. are more conducive to wind-power generation than others. The result of these global, regional, and local atmospheric motions is a world in which the tropics are not too hot, the polar areas are not too cold, and no areas have too little moisture for life.

3.0

3.1 How Do Gases Respond to Changes in Temperature and Pressure?

THE ATMOSPHERE CONSISTS LARGELY OF GASES, with lesser amounts of liquids, such as drops of water, and solids, such as dust and ice. By their nature, gases expand easily or contract in volume in response to changes in temperature and pressure. Variations in temperature and resulting changes in pressure are the main drivers of motion in the atmosphere.

A How Does a Gas Behave When Heated or Cooled?

The amount of insolation at the top of the atmosphere varies considerably from place to place and through time. These variations in insolation in turn lead to differences in temperatures, to which gases in the atmosphere respond.

03.01.a1 Namibia, Africa

1. Consider what happens when we want to make a hot air balloon rise (◄). Typically, a propane-powered burner heats ambient air, causing the air to expand in volume. This increase in volume inflates the balloon. Since the same amount of gas now occupies a much larger volume, the *density* of the heated air is less than the density of the surrounding air, so the balloon rises. So, as air increases in temperature, it tends to increase in volume and become less dense.

2. The figure below shows how a quantity of gas responds to either an increase in temperature (heating) or a decrease in temperature (cooling). The starting condition is represented by the cube of gas on the left.

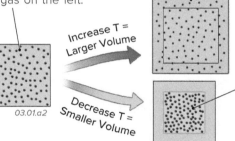

Increase T = Larger Volume

Decrease T = Smaller Volume

03.01.a2

3. An increase in the temperature of a gas means more energetic molecules, so a larger volume is needed to accommodate the same amount of gas.

4. If a gas cools, the molecules within it have less kinetic energy (motions) and can therefore be packed into a smaller volume. The gas has a higher density and will tend to sink.

5. This example shows that temperature and volume of a gas are directly related—in fact they are *proportional* if pressure is held constant. Such a proportional relationship means that if temperature is doubled, volume doubles too. If temperature decreases by half, volume does too. This specific relationship is called *Charles's Law*, which is one of the fundamental laws governing the behavior of gases, and it explains why a hot air balloon rises.

B What Happens When a Gas Is Compressed?

If a gas is held at a constant temperature but forced to occupy a smaller *volume*, the *pressure* of the gas increases. Pressure is proportional to the number of collisions of the molecules. If the same gas fills a larger volume, the collisions and amount of pressure both decrease. In both cases, if we instead change the pressure, the volume of the gas will adjust accordingly. A material, like a gas, that can be compressed, is said to be *compressible*.

1. Molecules of gas in the sealed container in the left canister below are under pressure, represented by the two weights resting on top. At some temperature, the molecules have a corresponding amount of energy, and some of the moving molecules are hitting the movable lid, resisting the downward force of the attached weight.

2. Removing a weight reduces the downward pressure on the gas. However, the gas retains its same average energy level (temperature) and therefore exerts the same upward force on the movable lid as before. The upward force from the gas molecules exceeds the downward force of the weight and so raises the lid, increasing the volume occupied by the gas. In this way, a decrease in pressure results in an increase in volume, if the gas does not change temperature.

Less Pressure = More Volume

3. Increasing the downward pressure by adding a third weight on the original canister causes the lid to slide down. This increase in pressure causes a decrease in volume. As the gas is compressed into a smaller volume, the number of the molecules impacting the lid increases. When this upward force from the gas molecules equals the downward force from the weight, the lid stops moving, and the volume and pressure of the gas stop changing.

4. The relationship between pressure and volume of a gas, under conditions of constant temperature, is *inversely proportional*—if pressure increases, volume decreases. If pressure decreases, volume increases. Either pressure or volume can change, and the other factor responds accordingly, changing in the opposite direction by a proportional amount. That is, if the volume is cut in half, the pressure doubles. If the volume doubles, the pressure is cut in half. This inversely proportional relationship between pressure and volume, under constant temperature, is called *Boyle's Law*.

03.01.b1

More Pressure = Less Volume

C How Are Temperatures and Pressures Related?

Since Charles's Law relates volume to temperature, and Boyle's Law relates volume to pressure, we might suspect that we can relate temperature and pressure. Combining Charles's Law and Boyle's Law leads to the *Ideal Gas Law*, which relates temperature, pressure, and density (mass divided by volume). Basic aspects of the Ideal Gas Law help explain the processes that drive the motion of matter and associated energy in the atmosphere.

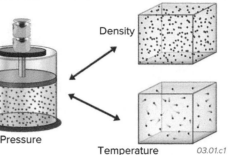

Density

Pressure

Temperature   03.01.c1

1. We can represent the Ideal Gas Law with a figure, with words, or with an equation. We begin with this figure (▶), which expresses the two sides of the equation. On one side of the equation (the left in this figure) is pressure. On the right side of the equation are density and temperature. The Ideal Gas Law states that if we increase a variable on one side of the equation (like increasing pressure), then one or both of the variables on the other side of the equation have to change in the same direction—density or temperature have to also change, or perhaps both do.

2. Examine this figure and envision changing any one of the three variables (pressure, density, or temperature), and consider how the other two variables would respond to satisfy the visual equation.

3. What happens if pressure increases? If temperature does not change, then density must increase. If pressure increases but density does not change, then temperature has to increase. Alternatively, temperature and density can both change. This three-way relationship partly explains why temperatures are generally warmer and the air is more dense at low elevations, where the air is compressed by the entire weight of the atmosphere, than at higher elevations, where there is less air. Higher pressure often results in higher temperatures.

4. What does the relationship predict will happen if a gas is heated to a higher temperature? If the density does not change, the pressure exerted by the gas on the plunger must increase. If the pressure does not change, the density must decrease. This is because density and temperature are on the same side of the equation, so an increase in one must be matched by a decrease in the other—if the other side of the equation (pressure) does not change. The relationship indicates that heated air can become less dense, which allows it to rise, like in a hot air balloon.

5. The Ideal Gas Law can also be expressed by the equation to the right:

$$P = R\rho T$$

where P is pressure, R is a constant, ρ is density (shown by the Greek letter rho), and T is temperature. Note how this equation roughly corresponds to the figure above.

D How Can Differences in Insolation Change Temperatures, Pressures, and Density in the Atmosphere?

The way gas responds to changes in temperature and pressure is the fundamental driver of motion in the atmosphere. Since temperature changes are largely due to insolation, we can examine how insolation affects the physical properties of gas and how this drives atmospheric motion.

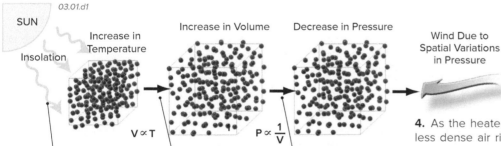

03.01.d1

SUN

Insolation

Increase in Temperature

$V \propto T$

Increase in Volume

$P \propto \dfrac{1}{V}$

Decrease in Pressure

Wind Due to Spatial Variations in Pressure

5. In this way, the response of gas to changes in temperature, pressure, and density (or volume), as expressed by the gas laws, is the primary cause of motion in the atmosphere. Variations in insolation cause changes in temperature, pressure, and density, which in turn cause air to move within the atmosphere.

1. The Sun is the major energy source for Earth's weather, climate, and movements of energy and matter in the atmosphere and oceans. In the figure above, insolation strikes Earth's surface (land or water), which in turn heats a volume of gas in the overlying atmosphere.

2. The increase in temperature results in expansion of the gas because of the increased kinetic energy of the molecules in the gas; expansion is an increase in volume. If the same number of gas molecules occupy more volume, the density of the air decreases (the air becomes less dense).

3. The increase in volume can result in a decrease in pressure (less frequent molecular collisions). As a result, the air mass is now less dense than adjacent air that was heated less. The more strongly heated and expanded air rises because it is less dense relative to surrounding air (which was not heated as much and so is more dense).

4. As the heated, less dense air rises, adjacent air flows into the area to replace the rising air. The end result is vertical and lateral movement of air—vertical motion within the rising air, and lateral motion of surrounding air toward the area vacated by the rising air.

Before You Leave This Page

✓ Sketch and explain why a gas under constant pressure expands and contracts with changes in temperature.

✓ Sketch and explain why a gas at a constant temperature expands or contracts with changes in pressure.

✓ Sketch and summarize the Ideal Gas Law and how the three gas laws explain motion in the atmosphere due to variations in insolation.

3.1

3.2 What Is Air Pressure?

PRESSURE OF GASES WITHIN THE ATMOSPHERE is highly variable, both vertically and laterally. These variations in pressure determine the nature and direction of atmospheric motions. If one place in the atmosphere has higher pressure than another place, this imbalance of pressure (and therefore also atmospheric mass) tends to be evened out by the flow of air. How do we describe and measure pressure, and how do we use these measurements to understand or even predict the flow of air?

A What Is Pressure?

Pressure is an expression of the force exerted on an area, usually from all directions. In the case of a gas, pressure is related to the frequency of molecular collisions, as freely moving gas molecules collide with other objects, such as the walls of a container holding the gas. It is such collisions that keep a balloon, soccer ball, or bicycle tire inflated.

1. Molecules of gas in a sealed glass container move rapidly in random directions, and some strike the walls of the container. The force imparted by these collisions is pressure. The more collisions there are, the more pressure is exerted on the walls of the container.

03.02.a1

2. If we push down on the lid of the container, the same number of molecules are confined into a smaller space. Lower parts of the container walls are now struck by a greater number of the more closely packed gas molecules, so the pressure is greater. Decreasing the volume of a gas increases its pressure, consistent with Boyle's Law.

3. What happens if we put a weight (▶) on top of the lid (center container) and then either cool or heat the gas in the container?

4. If we cool the container by placing it in ice, the molecules become less energetic and so strike the walls and lid of the container less often — the gas pressure decreases and the lid moves down.

03.02.a2

5. If we instead heat the container, the gas molecules become more energetic and strike the walls and lid of the container more often — the gas pressure increases and lifts the lid.

---

6. The equation to the right illustrates what pressure actually measures and the units we use to describe it. The units of pressure are used throughout this book in describing weather, climate, and the flow of water.

$$\text{Pressure} = \frac{\text{Force}}{\text{Area}} = \frac{\text{Mass} \times \text{Acceleration}}{\text{Area}} = \frac{\text{kg} \cdot \text{m sec}^{-2}}{\text{m}^2} = \frac{\text{Newtons}}{\text{m}^2} = \text{Pascals} \quad \textbf{\textit{100 Pascals = 1 millibar (mb)}}$$

7. Pressure is a force exerted on a given surface area.

8. According to Newton's second law, force is the product of mass and acceleration.

9. The unit of mass is the kilogram (kg), acceleration is in meters/second per second (m/s²), and area is in square meters (m²).

10. A force of one kg·m/s² is called a *Newton*. Pressure, measured in *Pascals*, is an expression of the number of Newtons of force exerted on a square meter of surface.

11. The air pressure at Earth's surface is many Pascals, so we express pressures in a larger, related unit called a *bar*, or in *millibars* (1/1000 of a bar).

B How Is Air Pressure Measured?

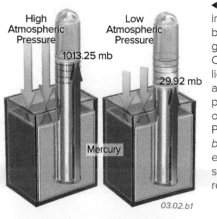

High Atmospheric Pressure
1013.25 mb

Low Atmospheric Pressure
29.92 mb

Mercury

03.02.b1

◀ We can measure air pressure with an instrument called a *barometer*. The barometer shown to the left is a sealed glass tube fixed in liquid mercury. Changes in air pressure cause the liquid level in the tube to rise or fall, allowing the measurement of relative pressure. Such barometers have units of inches (or centimeters) of mercury. Pressure is also reported in units of a *bar*, with one bar being approximately equal to the average air pressure at sea level. Modern digital instruments record pressure in millibars.

▶ Meteorologists measure air pressure at vertical heights in the atmosphere using hydrogen- or helium-filled balloons, like this one. An instrument package called a *radiosonde* is suspended from the balloon. Sensors measure pressure, temperature, humidity, and position (using a GPS) as the balloon ascends, and these measurements are transmitted via radio waves to a central computer. Wind speed and direction are inferred from successive positions of the radiosonde. The balloon eventually pops and the radiosonde parachutes to the ground.

03.02.b2

C How Does Air Pressure Vary Vertically?

1. Air pressure in the atmosphere is not constant. The largest variation is vertically, with an abrupt decrease in pressure upward from near the surface. The red curve on this figure shows how the air pressure, measured in millibars (mb), decreases from Earth's surface to the top of the atmosphere.

2. This diagram depicts the main layers of the atmosphere (troposphere, stratosphere, etc.) and highlights some of the features observed in each part, such as auroras in the thermosphere, shooting stars that mostly burn up in the mesosphere, and the restriction of most clouds and weather to the troposphere. The top of the troposphere is the tropopause.

3. Colors along the left edge of the diagram convey temperature variations within and between the atmospheric layers. These vertical temperature variations affect the density of the air, impacting air motions caused by the Sun heating the Earth's surface.

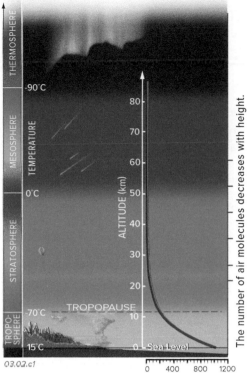

03.02.c1

4. In the thermosphere and mesosphere, gas molecules are relatively sparse and temperatures are low (−90°C at the thermosphere-mesosphere boundary). As a result of the sparseness of molecules, air pressures are very low (less than one millibar).

5. The abundance of gas molecules increases down into the stratosphere, and this is accompanied by an increase in air pressure (the bending of the red curve to the right as it goes downward). Air pressures at the base of the stratosphere (the tropopause) have increased to about one-fifth of pressures measured at sea level.

6. The pull of Earth's gravity holds most gas molecules close to Earth's surface, in the troposphere. Air pressure increases downward in the troposphere because of a greater abundance of molecules downward and the larger total number of molecules pressing down from the layers above. The highest air pressures are close to the surface, and at the lowest elevations. Sea level is the reference level for air pressure, with an average pressure of 1,013 mb (a little over 1 bar).

D How Does Air Pressure Vary Laterally?

1. Air pressure also varies laterally, from area to area, and from hour to hour, and these variations are typically represented on maps, like the one shown here. Such maps either show the pressure conditions at a specific date and time or show pressure values averaged over some time period, like a month or a year. To allow us to compare different regions and to see the larger patterns, the map uses pressure values that are corrected to sea level, or their sea-level equivalent. In this way, we eliminate the effects of differences in elevation from place to place.

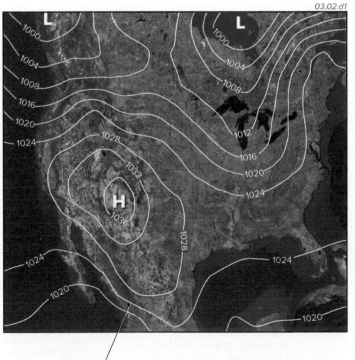

03.02.d1

3. Most maps of air pressure feature the large capital letters H and L. An H represents an area of relatively higher pressure called a high-pressure area or simply a *high*. An L represents a low-pressure area, commonly called a *low*. An elongated area of high pressure can be called a *ridge* of high pressure and an elongated area of low pressure is a *trough*.

4. The map patterns change with time, corresponding to changes in air pressure. Patterns typical for a region also change from season to season.

2. Such maps of air pressure contain numbered lines, called *isobars*, that connect locations with equal pressure. If you could follow an isobar across the countryside, you would follow a path along which the pressure values, once corrected to their sea-level equivalents, would be equal. Successive isobars are numbered to represent different values of air pressure, usually in millibars (e.g., 1,024), and there is generally a constant difference in pressure between two adjacent isobars (a 4 mb difference on this map). Note that isobars do not cross, but can completely encircle an area.

Before You Leave This Page

☑ Describe the concept of pressure and how it responds when a gas is subjected to changes in temperature and volume.

☑ Explain how we measure pressure and what units we use.

☑ Sketch and explain how air pressure varies vertically in the atmosphere.

☑ Describe what a map of pressure shows, and explain the significance of an isobar, H, and L on such a map.

3.2

3.3 What Causes Pressure Variations and Winds?

THE MOVEMENT OF AIR IN THE ATMOSPHERE produces *wind,* or movement of air relative to Earth's surface. Circulation in the atmosphere is caused by pressure differences generated primarily by uneven insolation. Air flows from areas of higher pressure, where air sinks, to areas of lower pressure, where air rises.

A How Do We Measure the Strength and Direction of Wind?

Wind speed and direction are among the most important measurements in the study of weather and climate. On short timescales, wind can indicate which way a weather system is moving and the strength of a storm. When considered over longer timescales, winds indicate general atmospheric circulation patterns, a key aspect of climate.

03.03.a1

1. Wind directions can be assessed as easily as throwing something light into the air and tracking which way it goes, but it is best done with a specially designed device, called an *anemometer* (◄), that can measure the wind speed and direction. Wind speed is expressed in units of distance per time (km/hr) or as *knots*, which is a unit expressing nautical miles per hour. One knot is equal to 1.15 miles/hr or 1.85 km/hr.

2. Wind direction is conveyed as the direction *from which the wind is blowing*. Wind direction is commonly expressed with words (►), such as a northerly wind (blowing from the north). It can instead be described as an *azimuth* in degrees clockwise from north. In this scheme, north is 0°, east is 090°, south is 180°, and west is 270°. Go ahead and write these numbers on the appropriate place on this figure.

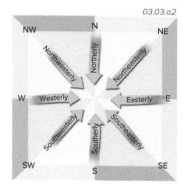

03.03.a2

3. The atmosphere also has vertical motion, such as convection due to heating of the surface by insolation. A local, upward flow is an *updraft* and a local, downward one is a *downdraft*.

B What Causes Air to Move?

Air moves because there are variations in air pressures, in density of the air, or in both (recall that pressure and density are related via the Ideal Gas Law). Such pressure and density variations are mostly caused by differential heating of the air (due to differences in insolation) or by air currents that converge or diverge. The atmosphere is not a closed container, so changes in volume (i.e., air being compressed or expanded) come into play. These volume changes can make air pile up or spread out, resulting in variations in air pressure.

1. Movement of air occurs to equalize a difference in air pressure between two adjacent areas (►), that is, a *pressure gradient*. Air molecules in high-pressure zones are packed more closely together than in low-pressure zones, so gas molecules in high-pressure zones

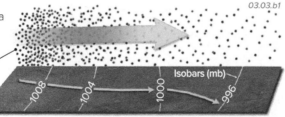

Isobars (mb)
1008 · 1004 · 1000 · 996

03.03.b1

tend to spread out toward low-pressure zones. As a result, air moves from higher to lower pressure, in the simplest case (as shown here) perpendicular to isobars.

2. High-pressure zones and low-pressure zones can be formed by atmospheric currents high in the atmosphere that converge or diverge (▼). *Converging* air currents compress more air into a smaller space, increasing the air pressure. *Diverging* air currents move air away from an area, decreasing pressure. Forces associated with converging and diverging air are called *dynamic forcing*.

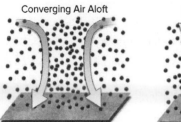 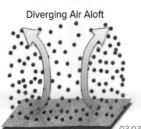

Converging Air Aloft   Diverging Air Aloft

03.03.b2

3. Most variations in air pressure and most winds, however, are caused by thermal effects, specifically differences in insolation from place to place. This cross section (►) shows a high-pressure zone (on the left) caused by the sinking of cold, high-altitude air toward the surface. In the adjacent low-pressure zone (on the right), warmer near-surface temperatures have caused air to expand, become less dense, and rise, causing low pressure. Near the surface, air would flow away from the high pressure and toward the low pressure. Different air currents would form higher, in the upper troposphere, to accommodate the sinking and rising of the air.

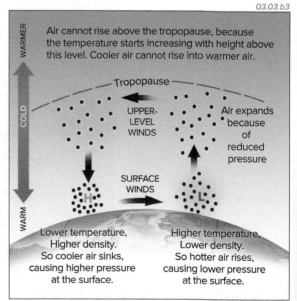

03.03.b3

Air cannot rise above the tropopause, because the temperature starts increasing with height above this level. Cooler air cannot rise into warmer air.

WARMER · COLD · WARM

Tropopause

UPPER-LEVEL WINDS

Air expands because of reduced pressure

SURFACE WINDS

H   L

Lower temperature, Higher density. So cooler air sinks, causing higher pressure at the surface.

Higher temperature, Lower density. So hotter air rises, causing lower pressure at the surface.

C   What Forces Result from Differences in Air Pressure?

Differences in air pressure, whether caused by thermal effects or dynamic forcing, produce a *pressure gradient* between adjacent areas of high and low pressure. Associated with this pressure gradient are forces that cause air to flow. Pressure gradients can exist vertically in the atmosphere or laterally from one region to another.

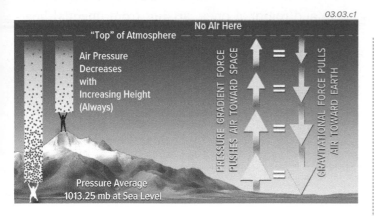

03.03.c1

3. Lateral variations in air pressure also set up horizontal pressure gradients, and a pressure-gradient force directed from zones of higher pressure to zones of lower pressure. On the map below, the pressure-gradient force acts to cause air to flow from high pressure toward lower pressures, as illustrated by the blue arrows on the map.

4. Places where isobars are close have a steep pressure gradient, and a strong pressure-gradient force, so movement of the atmosphere (i.e., winds) will generally be strong in these areas.

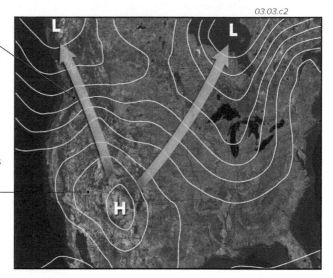

03.03.c2

5. Places where isobars are farther apart have a more gentle pressure gradient, thus a weak pressure-gradient force, and generally lighter winds. Although winds tend to blow from high to low pressure, other factors, such as Earth's rotation, complicate this otherwise simple picture, causing winds patterns to be more complex and interesting.

1. Elevation differences cause the largest differences in air pressure. At high elevations, there is less atmospheric mass overhead to exert a downward force on the atmosphere. As a result, density decreases with elevation, and air pressure does too.

2. These vertical variations in air pressure cause a pressure gradient in the atmosphere, with higher pressures at low elevations and lower pressures in the upper atmosphere. This pressure gradient can be thought of as a force directed from high pressures to lower ones. This *pressure-gradient force* is opposed by the downward-directed force of gravity, which is strongest closer to Earth's surface.

D   How Does Friction Disrupt Airflow?

As is typical for nature, some forces act to cause movement and other forces act to resist movement. The pressure-gradient force acts to cause air movement, where friction acts to resist movement.

1. Friction occurs when flowing air interacts with Earth's surface.

2. Wind is slowed near the surface because of friction along the air-Earth interface, as represented in this figure by the shorter blue arrows low in the atmosphere. As the air slows, it loses momentum (which is mass times velocity). Some momentum from the moving air can be transferred to the land, such as when strong winds pick up and move dust or cause trees to sway in the wind. It is also transferred to surface waters, causing some currents in oceans and lakes and forming surface waves.

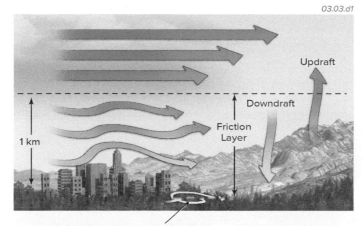

03.03.d1

3. Friction with Earth's surface, whether land or water, also causes the wind patterns near the surface to become more complicated. On land, air is forced to move over hills and mountains, through valleys, around trees and other plants, and over and around buildings and other constructed features. As a result, the flow patterns become more curved and complex, or *turbulent*, near the surface, with local flow paths that may double back against the regional flow, like an eddy in a flowing river. Friction from the surface is mostly restricted to the lower 1 km of the atmosphere, which is called the *friction layer*.

4. Stronger winds occur aloft, in part because these areas are farther from the frictional effects of Earth's surface. Some friction occurs internally to the air, even at these heights, because adjacent masses of air can move at different rates or in different directions. Friction can also accompany vertical movements in updrafts and downdrafts.

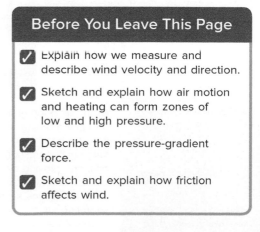

Before You Leave This Page

✓ Explain how we measure and describe wind velocity and direction.

✓ Sketch and explain how air motion and heating can form zones of low and high pressure.

✓ Describe the pressure-gradient force.

✓ Sketch and explain how friction affects wind.

3.3

3.4 How Do Variations in Temperature and Pressure Cause Local Atmospheric Circulation?

PRESSURE GRADIENTS INDUCE FLOW at all scales, including local and regional ones, arising from unequal heating by insolation and latent heat, by differing thermal responses of land versus sea, and even from the construction of large metropolitan areas. Such circulations contribute to the climate of a place, particularly in the absence of more powerful circulations.

A What Causes Breezes Along Coasts to Reverse Direction Between Day and Night?

People who live along coasts know that gentle winds sometimes blow in from the sea and at other times blow toward the sea. The gentle wind is called a *sea breeze* if it blows from sea in toward the land and a *land breeze* if it blows from the land out to sea. Such breezes are mostly due to differences in the way that land and sea warm up during the day and cool down at night.

The Sea Breeze (Daytime)

1. During daytime hours, land heats up significantly, particularly in summer. Land, which has a lower specific heat than water, heats up more overall and more rapidly than water does. The hot air over the land surface rises, inducing a local low-pressure area over the land.

03.04.a2 Frankfort, MI

2. At the same time, air over the water body is chilled by the relatively cool water temperatures and by cooling associated with evaporation. The relatively cool air over the water sinks, inducing a local high-pressure area over the water.

03.04.a1

3. The difference between the low pressure over land and the high pressure over the water represents a pressure gradient. The associated pressure-gradient force pushes air near the surface from higher pressure to lower pressure. This flow is an *onshore breeze* or *sea breeze* that feels cool to people on the beach.

5. In this photograph (◀), heating of the land causes air over the land to rise, drawing in moist air from the adjacent water. Rising of the moist air along the coast forms scattered clouds, but it can draw in much thicker masses of clouds, forming coastal fog and overcast skies.

4. Air aloft moves in the opposite direction, from land to sea. This is a response to an upper-level pressure gradient caused by the "extra" air rising over land and less upper-level air over the sea, due to air sinking.

The Land Breeze (Nighttime)

1. At night, land cools significantly, particularly if there isn't much cloud cover. The relatively cool air over the land surface sinks, inducing a surface high-pressure area. Sometime in the evening hours, the surface low that develops over land during daytime weakens and becomes a high.

03.04.a4 Philippines

2. The water body doesn't cool as much at night, partly because of water's higher specific heat capacity and because of mixing. So air over the water stays relatively warm compared to the air over the land. Therefore, this air rises, generating relatively low pressure over the water body.

03.04.a3

3. The difference between the high pressure over land and the low pressure over the water is a pressure gradient. The pressure-gradient force pushes near-surface air from higher to lower pressure, from the land to the water as an *offshore breeze* or *land breeze*.

6. In this photograph (◀), an early morning offshore flow, from the land to the sea, pushes moist air offshore, moving clouds out to sea and causing clear conditions along the shoreline.

4. Aloft, the pressure-gradient force causes the air that rose over the low to return to land to replace the air that sank to create the surface high.

5. The strength of sea and land breezes is proportional to the temperature gradient. The land breeze circulation strengthens through the night, begins to weaken at sunrise, stops sometime in the morning, and then reverses to a sea breeze, when it strengthens until peaking in the afternoon.

B How Does Topography Cause Pressure Gradients?

On land, topography can cause pressure gradients and winds that blow in different directions during the day versus at night. These are primarily related to differences in the way land and the atmosphere heat up and cool down, and to temperature differences as a function of elevation.

1. The Sun heats the surface more efficiently than it heats the atmosphere. During the day, the land heats up and in turn heats the adjacent air. Such heated air from lower elevations is especially warm and rises (▶), causing upslope winds, known as a *valley breeze*, or *anabatic wind*. Air that is farther from the surface, aloft in the center of the figure shown here, is less heated and so sinks to replace the rising air. Upslope winds are most obvious when winds from other types of circulation, like storms, are weak.

03.04.b1

03.04.b2

2. At night, the situation is reversed (◀). With no insolation to warm the surface, the surface will cool more rapidly than the air above it, as it emits long-wave radiation to the atmosphere. Thus, the land and adjacent air at higher elevations cool significantly, and the resulting relatively cool surface air will move downslope, producing a *mountain breeze*, or *katabatic wind*. The cool air displaces warmer air rising in the valley and flowing upward, where it can replenish the air lost at higher elevations. The end result is that valley bottoms can be substantially colder at night than areas at a slightly higher elevation.

3. Pollution in rugged terrain is mostly produced in the valleys, where people are more likely to live, so the daytime valley breezes push polluted air upslope. But if the pollution is too intense or the slopes are too high and steep to push the air over the ridges and peaks before nightfall, the mountain breeze will send the polluted air back down into the valley to join the new polluted air produced during the next day. The cycle repeats day after day. Persistent high pressure (with its sinking air) over the area can make the situation even worse. Some of the poorest air quality exists in valleys surrounded by higher terrain, such as Mexico City.

C How Do Urban Areas Affect Pressure Features?

Human development of land can cause pressure gradients, as natural ground cover is replaced by buildings, asphalt, and concrete. These human-produced materials have different properties from those of natural materials, and the resulting temperature differences can cause pressure differences.

1. Urban areas, such as cities and towns, are usually warmer than the surrounding rural environments, for several reasons. Urban areas often lack trees for shade and may have less standing water for latent heating rather than sensible heating. They contain waste heat from engines, street lights, and fireplaces, along with urban building and road materials that absorb intense heat. The urban area causes the air to heat up more than air in the less developed surroundings. The warmer air over the city is less dense and rises, leaving behind a local low-pressure area.

2. The outlying areas are generally cooler because they are more likely to contain vegetation for shading and typically have more near-surface water, which absorbs insolation for latent heat rather than sensible heating. Outlying areas also have less human-generated waste heat.

3. This phenomenon is known as an *urban heat island* (UHI). Urban-rural temperature differences usually peak in the evening hours, after a day of surface heating and busy human activities that generate the urban heating. The UHI is responsible for some of the modern observed increase in surface temperatures.

03.04.c1

4. This urban-caused circulation of air can redistribute heat and pollution from the city center to outlying suburban and rural areas. Polluted air can be lifted out of the urban area, flow laterally, and descend into outlying areas. Higher pressure in the outlying areas may trap pollution near the ground. As with the sea breezes and valley breezes, the urban heat island can be diminished by storms and other strong air-circulation patterns.

Before You Leave This Page	☑ Sketch, label, and explain the formation of sea and land breezes.
	☑ Sketch, label, and explain the formation of valley and mountain breezes.
	☑ Sketch, label, and explain what an urban heat island is and how it can cause the flow of air and pollution.

3.4

3.5 What Are Some Significant Regional Winds?

DIFFERENCES IN AIR PRESSURE cause a variety of regional to local wind conditions, such as those associated with storms, which are discussed in the chapter on weather. Some local winds are not so much related to weather systems as they are to differences in pressure that tend to occur at certain times of the year or after the establishment of an area of high pressure. These local to regional winds have interesting names, like Chinook winds or Santa Ana winds, and can have profound impacts on people.

A What Is a Chinook Wind and Where Do They Form?

The term *Chinook* originated in the Pacific Northwest and can refer to several types of winds. The most common usage is for a warm, dry wind that blows down the flanks of a mountain range. Chinooks are so warm and dry they are called "snow eaters," for the way in which they can cause a sudden melting of snow and ice on the ground. The onset of a Chinook wind can cause a sudden rise in temperatures, especially during the winter. A Chinook in Loma, Montana, caused temperatures to rise from −48°C (−59°F) to 9°C (49°F) within a 24-hour period, the most change recorded for a single day in the U.S. In Spearfish, South Dakota, a Chinook off the adjacent Black Hills caused temperatures to rise 27 C° (49 F°) in two minutes, the world's fastest rise in temperature ever recorded!

1. This figure depicts the formation of a *Chinook wind*. The process begins when winds push moist air against the *windward* side of a mountain, where windward refers to the side from which wind is blowing. As the moist air rises up the mountain, it cools, causing the formation of clouds, a process that also releases latent heat. The heat warms the air, which continues rising toward the mountain peaks.

03.05.a2

COMMON SITES OF CHINOOK WINDS

Eastern Side of Cascade Range

Eastern Side of Rocky Mountains

Eastern Side of Sierra Nevada

Windward Side

Leeward Side

Chinook Wind

03.05.a1

2. Once the air reaches the peak, it begins flowing down the other side—the *leeward* side, of the mountain (the side opposite the windward side). As the air descends, it continues drying out and is compressed and heated. It was also warmed from the release of latent heat on the windward side. The warm, dry air descends from the mountain and spreads across the adjacent lowland, forming a Chinook wind.

3. This map shows the locations where Chinook winds are relatively common. As expected, Chinooks occur on the leeward side of mountain ranges (prevailing winds are from the west to east in this region), in Alaska, and elsewhere.

B How Do Katabatic Winds Affect Polar Regions?

1. On the previous two pages, we introduced the term *katabatic wind* for a wind that blows downslope, forming a cool *mountain breeze*. More regional and pronounced katabatic winds affect Antarctica and Greenland, both of which have a high central landmass surrounded by ocean. Air over the middle of the landmasses is very cold and so also very dense, flowing off the central topographic highs and down the icy slopes.

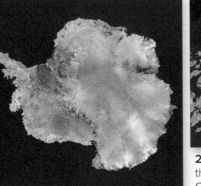

03.05.b1

03.05.b2

03.05.b3 Antarctica

2. These two perspective views show that Antarctica (on the left) and Greenland (on the right) both have a broad, high area centered in the middle of the ice. Katabatic winds blow down off these high areas in all directions. These winds are especially strong where they are channeled down valleys, such as the famous Dry Valleys of Antarctica, so named because strong katabatic winds have stripped most ice and snow off the land surface.

3. Katabatic winds in Antarctica generally involve cold but dry air, but they can interact with clouds along the coast, sometimes creating stunning effects as the cold air and clouds spill off the highlands, similar to the scene in the photograph above.

C How Do Santa Ana Winds Affect Southern California?

Winds in Southern California typically blow from west to east (that is, they are westerlies), bringing relatively cool and moist air from the Pacific Ocean eastward onto land, especially in areas right along the coast, like Los Angeles and San Diego. These coastal cities also often have onshore sea breezes during the day and offshore land breezes at night. At other times, however, regional winds, called *Santa Ana winds*, blow from the northeast and bring dry, hot air toward the coast, causing hot, uncomfortable weather and setting the stage for horrendous wildfires.

Setting of Santa Ana Winds

1. *Santa Ana winds* are regional winds that blow from the northeast (▶), typically developing during spring and fall, when high pressure forms over the deserts of eastern California and Nevada. Circulation of air associated with the area of high pressure pushes winds south and westward, toward the coast, in marked contrast to the normal onshore flow. This air is coming from the Mojave Desert and other desert areas to the north and east, so it is very dry.

03.05.c1

2. Santa Ana winds from the Mojave Desert are partially blocked by the mountains on the northern and eastern sides of Los Angeles. The winds spill through mountain passes, such as Cajon Pass northeast of Los Angeles, and are funneled down the canyons and into the Los Angeles basin. The funneling effect causes winds to be especially strong within the canyons. As the air moves from higher deserts (to the northeast) down toward Los Angeles, the air compresses and heats up. As a result, during an episode of Santa Ana winds, the coastal areas of Southern California experience much hotter and drier weather conditions than are normal. Due to this behavior of air flowing from high to low areas, Santa Ana winds are considered to be a type of katabatic wind.

Effect of Santa Ana Winds on Wildfires

03.05.c2 San Diego area, CA

1. The hills and mountains of coastal Southern California receive enough precipitation to be covered with thick brush (▲), such as oak, or by forests at higher elevations. During a Santa Ana wind, the fast winds dry out the brush, trees, and other vegetation, making it prone to wildfires.

2. The hills and mountains of Southern California experience some of the most spectacular but devastating wildfires of any place on the planet (▶). Santa Ana winds push these fires southwestward, toward the cities and through neighborhoods in the foothills. Wildfires associated with Santa Ana winds can burn thousands of homes, causing hundreds of millions of dollars in damage. Pushed by the strong winds down the canyons, the fast-moving fires can cause the deaths of firefighters and people who did not evacuate in time.

3. This amazing image from NASA (▶) combines a satellite image of Southern California and adjacent states with the locations of fires (shown in red), as determined by processing a different kind of satellite data. The smoke produced by a number of fires trails off across the Pacific Ocean, clearly showing Santa Ana winds blowing from the northeast. These fires killed 9 people and injured dozens of others, destroyed more than 1,500 homes, and burned more than 2,000 km² of forest, brush, and neighborhoods. The region was declared a federal and state emergency, as more than one million people were evacuated, the largest such evacuation in California history. The especially large amount of destruction from these fires was due to the combination of strong Santa Ana winds and a prolonged drought that had dried out the natural vegetation.

03.05.c3

03.05.c4

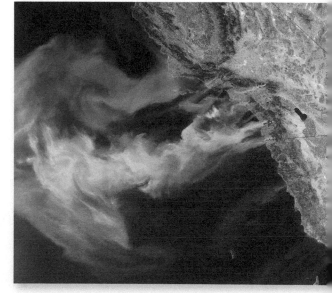

Before You Leave This Page

✓ Sketch and explain the origins of a Chinook and of a katabatic wind.

✓ Sketch and summarize the origin and setting of Santa Ana winds and why they are associated with destructive wildfires.

3.5

3.6 How Do Variations in Insolation Cause Global Patterns of Air Pressure and Circulation?

SEASONAL AND LATITUDINAL variations in insolation cause regional differences in air pressure, which in turn set up regional and global systems of air circulation. These circulation patterns account for many of the characteristics of a region's climate (hot, cold, wet, dry), prevailing wind directions, and typical weather during different times of the year. Here, we focus on vertical motions in the atmosphere resulting from global variations in insolation and air pressure.

A What Pressure Variations and Air Motions Result from Differences in Insolation?

1. On this figure, the top graph plots the average amount of insolation striking the top of the atmosphere as a function of latitude. On the surface below, the large letters represent high- and low-pressure zones. Arrows show vertical and horizontal airflow and are color coded to convey the overall temperature of air.

2. The maximum amount of insolation striking the Earth is along the equator and the rest of the tropics. This heats up the air, causing the warm air to expand and rise. The expansion and rising results in a zone of surface low pressure (L) in equatorial regions. The upward flow of air helps increase the height of the tropopause over the equator, as shown by the dashed line.

3. Surface winds flow toward the low pressure to replace the rising tropical air. The rising tropical air cannot continue past the tropopause, so as it reaches these heights it flows away from the equator (EQ) to make room for more air rising from below. This upper-level air descends in the subtropics, near 30° latitude, where it forms a zone of high pressure (H).

4. The amount of insolation decreases away from the equator, with a relatively sharp drop-off across the mid-latitudes (30° to 60° latitude). Descending air and high pressure in the subtropics causes surface air to flow toward higher latitudes (to the right in this diagram).

6. This global pattern of rising and sinking air and resulting low and high air pressures dominates the motion of Earth's atmosphere. The pattern results from variations in insolation and is compensated by horizontal flows, both near the surface and in the upper atmosphere.

5. The amount of insolation reaches a minimum near the poles. Air near the poles is very cold and dense, sinking to form a zone of high pressure near the surface. The descending air flows away from the pole, toward 60° latitude. In the upper atmosphere, air flows toward the poles to replenish the air that sank.

03.06.a1

Maximum

AMOUNT OF INSOLATION (W/m²)

Minimum

EQ   30°N   60°N   POLE

Troposphere

Vertical Motion

Surface Winds

Air Pressure

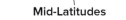

L   H   L   H

EQ   30°N   60°N   POLE

Equator	Subtropics	Mid-Latitudes	Poles

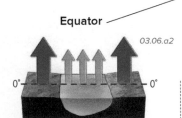

03.06.a2

0° — 0°

03.06.a3

30°

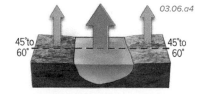

03.06.a4

45° to 60°   45° to 60°

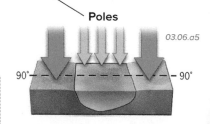

03.06.a5

90° — 90°

7. In equatorial regions, which have an energy excess, the rising warm air produces low air pressure. The land warms faster than the sea, so the air is even warmer and the air pressure is correspondingly lower, over the landmasses. Rising warm air and low air pressure also prevail over the equatorial oceans, although temperatures tend to be cooler and pressures are not quite so low.

8. In the subtropics, at approximately 30°, cool air descends from the upper troposphere and causes higher-than-normal air pressure. In the subtropics, oceans are usually cooler than the surrounding continents. Air descending over the cooler oceans will warm up less than air descending over the warmer continents, and so air pressures are particularly high over the oceans.

9. Air in the mid-latitudes is forced to ascend by surface air converging from the poles and subtropics. This zone has an energy deficit, and oceans retain their energy and are generally warmer than continents. Air forced to rise above the oceans is therefore warmer than that rising over the continents. As a result, the air over the oceans is more likely to rise, and the associated low-pressure zone tends to be stronger over the oceans than over the land.

10. The cold, dense, descending air over the poles produces a zone of high air pressure. This zone has a large energy deficit. In such zones, continents lose their energy faster, and are therefore cooler, than the adjacent polar oceans, so air pressures are extremely high over the cold continents. The slightly warmer temperature of the adjacent oceans somewhat diminishes the high air pressures.

B How Does Sea-Level Air Pressure Vary Globally?

Average Annual Air Pressure

1. This map shows sea-level equivalent air pressure averaged for 1981 to 2010. Observe the main pattern and compare these patterns with the figures on the previous page. Can you explain the larger patterns on this map?

2. Two belts of high pressure (shown in light gray) encircle the globe at about 30° N and 30° S (the subtropics). Between these two is a belt of lower pressure (shown in medium gray) in the tropics, straddling the equator. The equatorial low pressure and flanking high-pressure zones are due to the large air current that rises in the tropics and descends in the subtropics.

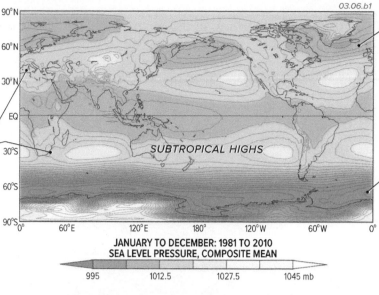

JANUARY TO DECEMBER: 1981 TO 2010
SEA LEVEL PRESSURE, COMPOSITE MEAN
995   1012.5   1027.5   1045 mb

3. A set of low-pressure areas (dark or gray) occurs near 60° N. Note that the lows are best developed in the oceans, and are poorly developed on land.

4. A prominent air-pressure feature on this map is a belt of extremely low pressure (shown in dark gray) in the ocean just off Antarctica. This belt is so well developed in the Southern Hemisphere because of the abundant ocean surface, uninterrupted by continental landmasses at this latitude (60° S).

5. An intense high-pressure belt (very light gray) occurs over continental Antarctica, in contrast to the oceanic Arctic.

January Air Pressure

6. Patterns of air pressure change with the seasons, following seasonal changes in insolation patterns. In January, one of the most prominent air-pressure features is a high-pressure area over Siberia, Russia, called the *Siberian High*. As discussed later, this high pressure helps drive the monsoon that affects much of southern Asia.

7. The broad belt of tropical low pressure moves toward the Southern Hemisphere, following the direct rays of the Sun (remember that January is in the southern summer). The migration is particularly noticeable over the hot land surfaces, like Australia.

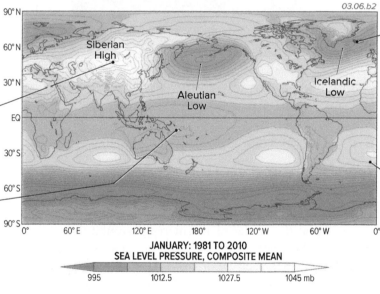

JANUARY: 1981 TO 2010
SEA LEVEL PRESSURE, COMPOSITE MEAN
995   1012.5   1027.5   1045 mb

8. A large area of low pressure, called the *Icelandic Low*, strengthens over the northern Atlantic Ocean, wrapping around Greenland. As ocean waters retain their heat better at this cold time of year, a similar low, the *Aleutian Low*, develops in the northern Pacific Ocean west of Canada.

9. Elongated highs (shown in light gray) occur over the oceans in the Southern Hemisphere subtropics, but not over the adjacent continents. The highs are enhanced by the relatively cool oceans in this region. Farther south is the pronounced belt of low pressure in the cold oceans that encircle Antarctica.

July Air Pressure

10. Air-pressure patterns change markedly by July, the northern summer. In the Northern Hemisphere, the Siberian High has dissipated as warm air over interior Asia rises. Pressure gradients across the Northern Hemisphere are weaker in July than in January, because the equator-to-pole energy gradient isn't as steep in the summer as it is in the winter.

11. Typical patterns remain in the Southern Hemisphere, with belts of high pressure in the subtropics, flanked to the south by a continuous belt of very low pressure across the southern oceans. High pressure strengthens over the main landmass of Antarctica during the prolonged darkness of the southern winter.

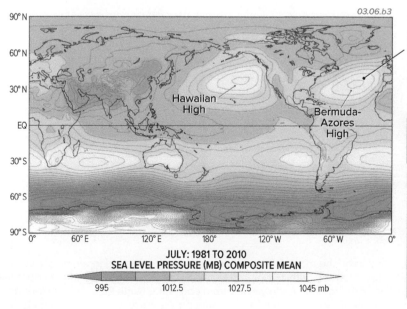

JULY: 1981 TO 2010
SEA LEVEL PRESSURE (MB) COMPOSITE MEAN
995   1012.5   1027.5   1045 mb

12. Bullseye-shaped high-pressure areas strengthen over the oceans, one over the central Atlantic (the *Bermuda-Azores High*) and another in the Pacific (*Hawaiian High*).

Before You Leave This Page

☑ Sketch and explain global air-circulation patterns, summarizing how they differ over land versus oceans.

☑ Summarize the main patterns of air pressure, identifying key features.

3.6

3.7 What Is the Coriolis Effect?

THE PRESSURE-GRADIENT FORCE drives airflow in the atmosphere, but winds do not blow in exactly the direction we would predict if we only consider pressure gradients. All objects—whether air masses, ocean waters, or airplanes—moving across the surface of the Earth display an apparent deflection from the intended path. The cause of this deflection is the *Coriolis effect*. Why does this apparent deflection occur?

A What Is the Coriolis Effect?

1. The *Coriolis effect* refers to the apparent deflection in the path of a moving object in response to rotation of the Earth. The easiest way to envision this is by considering air that is moving from north to south or south to north. Earth's atmosphere, including any moving air, is being carried around the Earth by rotation.

2. The blue arrows show how much distance the surface rotates in an hour. The arrows are longer near the equator, indicating a relatively long distance that these areas have to travel, and therefore faster velocities. The distances traveled and the linear velocities gradually decrease toward the poles.

3. At the poles, the distance traveled and velocity are both zero—the surface has no rotation-related sideways velocity. In 24 hours, an area directly at the pole would simply spin 360°, whereas an area at the equator would have moved approximately 40,000 km (the circumference of the Earth).

03.07.a1

4. As air moves toward the poles, it possesses the eastward momentum that it had when it was closer to the equator. So, it appears, from the perspective on Earth's surface, to be deflected to the right (to the east).

5. The opposite occurs as air moves toward the equator and encounters areas with a faster surface velocity. The air appears to lag behind, deflecting to the west as if it were being left behind by Earth's rotation. Note that in the Northern Hemisphere, air deflects to the right of the flow (not necessarily to the right as you look at it on a map), irrespective of which way it is moving (toward the pole, away from the pole, or in some other direction).

6. In the Southern Hemisphere, air moving toward the pole travels from faster rotating areas to slower ones, so it appears to be rotating faster than the surface—it deflects to the left.

B Why Do Moving Objects Appear to Deflect Right or Left on a Rotating Planet?

To visualize in a different way why moving objects on a rotating planet appear to deflect left or right, examine these overhead views of a merry-go-round that is rotating counterclockwise (in the same way as Earth when viewed from above the North Pole). One person located at the center of the merry-go-round throws a ball to a second person standing near the outside edge of the merry-go-round. The path of the ball can be measured relative to two frames of reference: the two clumps of trees, which are fixed in our perspective, or from the children on the merry-go-round, which is moving.

1. The person at the center of the merry-go-round slowly tosses a purple ball toward an outer person, in the direction of the upper two trees. The intended path of the ball is shown by the yellow arrow. The outer part of the merry-go-round moves faster than the center.

03.07.b1

2. After a short time, the ball is heading toward the two trees, but, relative to the intended path (yellow arrow) or from the perspective of the thrower, the purple ball seems to be veering away to the right, because the thrower rotated.

03.07.b2

3. With each passing time period, the intended receiver moves farther away from the ball as the ball goes toward the two trees. As viewed from the thrower, the ball deflected to the right relative to the intended path.

03.07.b

4. In the last figure, the ball's path traced upon the moving framework of the merry-go-round (open purple circles) reveals an apparent deflection to the right (shown with a dashed red line) of the intended path. However, relative to the fixed reference of the upper two trees, the ball has actually followed a straight line. This view is similar to one of the rotating Earth viewed from above the North Pole. The thrower and receiver are two locations at different latitudes, and the ball represents an air mass moving from the slow-moving pole toward the faster-moving equator.

C Will Deflection Occur if Objects Move Between Points on the Same Latitude?

A similar deflection occurs if an object moves parallel to latitude on a rotating planet. To visualize why this is so, we return to the merry-go-round, which is still rotating counterclockwise, like Earth viewed from above the North Pole. As before, it is key to consider movements in terms of a fixed reference frame and a reference frame that is moving.

1. The person throwing the ball is on the outside of the merry-go-round along with the receiver. The intended path of the ball is shown by the yellow arrow. Since the players are the same distance out from the center, they are moving at the same rate.

03.07.c1

2. After a short time, the ball is heading along its original path (the purple path) relative to the upper two trees (the fixed reference frame). In the intervening time since the throw, however, the thrower and receiver have both moved (a moving reference frame).

03.07.c2

3. From the moving frame of reference of the thrower, the ball appears to be deflected to the right of the intended path, with the deflection shown by the orange dashed line.

03.07.c3

4. This example represents the apparent deflection of air (or any other object) moving *parallel to latitude*. So regardless of whether objects are moving in the north-south or the east-west directions, the objects appear to be deflected from their intended path. Moving objects have an apparent deflection to the right of their intended path in the Northern Hemisphere and to the left in the Southern Hemisphere. This left or right deflection due to the Coriolis effect accounts for the directions of prevailing winds, the paths of storms, and the internal rotation within hurricanes.

D What Affects the Strength of the Coriolis Effect?

Since the Coriolis effect is related to the rate at which sites of the surface move during rotation of the Earth, we would suspect the strength of the effect may vary with latitude. It is also influenced by how fast objects are moving.

03.07.d1

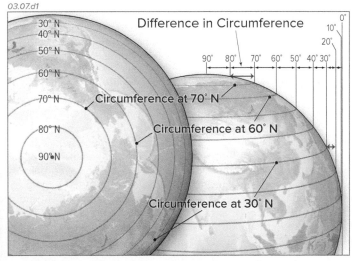

Difference in Circumference

Circumference at 70° N
Circumference at 60° N
Circumference at 30° N

1. When viewed from above the poles, the parallels of latitude constitute a series of concentric circles increasing in circumference from the poles to the equator. Moving from one latitude to another, like from the pole to 80° N, the percentage increase in circumference is much greater at high latitudes than nearer the equator. Note the difference in circumference for every 10° difference in latitude. Thus, the Coriolis effect is greatest at high latitudes, where the velocity of the moving reference frame changes most rapidly relative to the moving object.

3. The Coriolis effect is stronger for an object with a large velocity. In the case of a rotating storm, the deflection can be related to movement of the entire storm across Earth's surface, rotations within the storm, and other motions.

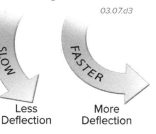

SLOW — Less Deflection

FASTER — More Deflection

03.07.d3

03.07.d2 Hurricane Irene 2011

2. The Coriolis effect is expressed daily in many ways, including the shape of storms as viewed by satellites and featured on the daily weather report, the rotation of hurricanes (◄), and the changes in wind directions as a large storm approaches and then exits your town.

Before You Leave This Page

☑ Describe the Coriolis effect and explain the direction of apparent deflection of moving objects in the Northern and Southern Hemispheres.

☑ Explain why a moving object on a rotating reference frame shows an apparent deflection.

☑ Explain the factors that control the strength of the Coriolis effect.

3.7

3.8 How Does the Coriolis Effect Influence Wind Direction at Different Heights?

PRESSURE GRADIENTS INITIATE MOTION in the atmosphere, but the actual direction in which the air moves is greatly influenced by the Coriolis effect. Close to the surface, where friction with the planetary surface is greatest and wind velocities are lowest, the pressure gradient dominates. Higher in the atmosphere, winds have higher velocities and the Coriolis effect becomes more important. The resulting patterns of airflow can dominate weather systems.

A Why Do Wind Speed and Direction Change with Height Near the Surface?

The direction in which air moves is determined by three factors: (1) the pressure-gradient force (winds blow from higher pressure toward lower pressure); (2) the Coriolis effect, which appears to deflect objects moving across Earth's surface, and (3) friction with Earth's surface, which cannot change direction by itself but can interact with the other two forces to change the speed and direction of wind. Friction becomes less important upward.

1. This figure shows how wind direction can change upward, from the surface to well above the friction layer. Winds are fastest higher up and progressively slower down closer to the surface. This example is for a site in the Northern Hemisphere.

2. The pressure-gradient force, caused by the pressure gradient between a high (H) and a low (L), causes winds right above the surface to blow from higher pressure to lower pressure, in an attempt to even out the pressure gradient. Even though friction is strong near the surface, it merely slows the wind; it does not by itself change the wind's direction.

03.08.a1

4. At some height, the deflection will be such that the winds flow *parallel* to the pressure gradient—no longer from higher to lower pressure. Further turning cannot happen because if it did, the flow would be going against the pressure gradient, from lower pressure toward higher pressure. The height at which this flow neither toward nor away from lower pressure (i.e., parallel to isobars) occurs is called the *geostrophic level*. It typically occurs when air pressure is about half that at the surface (i.e., the 500 mb level).

3. Higher up, friction is decreased, so winds are stronger. Since the Coriolis effect is proportional to wind speed, it begins deflecting air to the right (or to the left in the Southern Hemisphere). This causes successively more and more deflection with height.

5. To illustrate how wind direction can change with altitude, examine what commonly happens to a balloon as it ascends from the surface. The red arrows indicate wind direction and the inset maps show isobars between high- and low-pressure areas. Follow the balloon's progress from the surface upward, starting with the text below.

6. Right after the balloon takes off, in Position 1 on this figure, the pressure-gradient force and friction are the primary factors that affect wind speed. The balloon will go from higher pressure to lower pressure, across the pressure gradient as it rises. Friction slows the circulation in this lowest part of the friction layer, which is called the *surface boundary layer*.

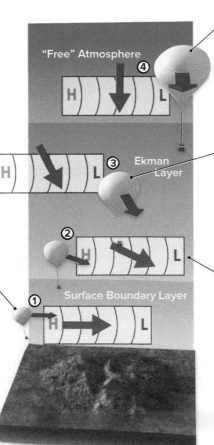

"Free" Atmosphere

Ekman Layer

Surface Boundary Layer

03.08.a2

9. Eventually, the balloon will ascend to a height (1–2 km above the surface) at which friction becomes negligible (above the *friction layer*). At that point, only the pressure-gradient force and the Coriolis effect are important in dictating flow. The wind speed is faster because of the reduced friction, so the Coriolis effect continues to pull the balloon to the right until it no longer flows toward lower pressure. Instead, the wind is perpendicular to the pressure-gradient force and parallel to isobars. This type of wind, flowing parallel to the isobars, is called a *geostrophic wind*.

8. As the balloon continues to rise, near Position 3, it experiences less and less friction, so the winds strengthen with height. This makes the Coriolis deflection even stronger with increasing height. The winds and the balloon move even less from higher pressure to lower pressure, instead turning at an angle to the isobars. This layer characterized by a turning of the winds with height (shown in Positions 2 and 3) is called the *Ekman layer*. It exists from about 100 m to about 1–2 km above the surface. The Ekman layer is a part of the friction layer.

7. In Position 2, the balloon has risen high enough from the surface that friction becomes weaker. This weakening of friction speeds up the balloon, and it also increases the impact of the Coriolis effect, because faster winds cause an increasing rightward deflection. This causes the balloon to move somewhat to the right of the direction that the pressure-gradient force would otherwise take it. The balloon therefore begins to travel more to the right as it rises.

03.08.a3 Namibia

B How Do We Depict Upper-Level Wind Patterns?

1. This map shows another way to represent pressure in the atmosphere, by contouring the height (in meters) at which 500 mb of pressure is reached. The contour lines represent lines of constant height and are called *isohypses*. Surface pressure is about 1000 mb, so the 500 mb level represents the height bounding the lower half of the atmospheric mass. This height is also generally considered to be the lowest level at which geostrophic winds occur. The higher the 500 mb level occurs, the higher the pressure. These "constant pressure surface maps" are used by the National Weather Service and other organizations in making weather maps, because of the need for constant pressure surfaces in the aviation industry.

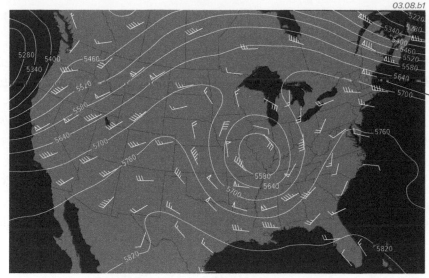
03.08.b1

2. This map also has symbols, called *wind barbs*, that show the direction and velocity from which the wind is blowing. The barb is parallel to wind direction, and each flag on the barb represents 50 knot (kt) winds (two flags indicate winds of 100 knots). Each full hash mark designates 10 kt, and each half-barb represents 5 kt. The wind barbs on this map show that upper-level winds generally flow parallel to the isohypses (they are geostrophic).

C Which Way Does Air Flow Around Enclosed High- and Low-Pressure Areas Aloft?

Aloft, in upper parts of the troposphere, well above the friction layer, wind directions are dominated by the pressure-gradient force and the Coriolis effect and so are typically geostrophic. As a result, air circulation can be nearly circular around upper-level lows and highs. In situations in which the flow is nearly circular, the low or high pressure typically lasts for a longer period of time than in cases where the flow is not enclosed as tightly, such as in an upper-level *ridge* or *trough* of pressure. Closed circulations at the 500 mb level or above suggest a strong circulation that is not likely to dissipate soon. The figures below are map views.

Northern Hemisphere

03.08.c1

03.08.c2

For low pressure, the pressure-gradient force pushes air into the low from all directions, but at high altitudes the Coriolis effect deflects this air until it parallels the isohypses. In the Northern Hemisphere, the deflection into a low is to the right, so the upper-level flow goes *counterclockwise* around the enclosed low-pressure area. This type of curved flow is called *gradient flow*.

For a high-pressure area, the pressure-gradient force pushes air out of the high in all directions, but the Coriolis effect deflects winds to the right, parallel to the isohypses. Due to this deflection, the gradient flow goes *clockwise* around the high-pressure area.

Southern Hemisphere

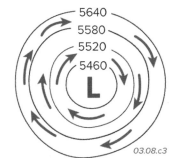
03.08.c3

03.08.c4

Similar processes occur around lows in the Southern Hemisphere. Air pushed into the low from all directions is deflected to the left by the Coriolis effect, until it flows parallel to the isohypses. The gradient flow therefore goes *clockwise* around a low in the Southern Hemisphere.

In the Southern Hemisphere, deflection of air moving out from a high is to the left, causing the gradient flow to be counterclockwise around a high. In both hemispheres, pressure gradients tend to be weaker around highs than around lows, because latent heat released by rising air in lows allows for stronger pressure gradients and faster winds.

Before You Leave This Page

☑ Sketch and explain why wind directions change from the surface to higher levels in the troposphere.

☑ Explain what a geostrophic wind is and how you would recognize one on a map that represents air pressure.

☑ Sketch upper-level flow around low-pressure and high-pressure areas in the Northern and Southern Hemispheres.

3.8

3.9 How Do the Coriolis Effect and Friction Influence Atmospheric Circulation?

THE CORIOLIS EFFECT AND FRICTION affect the patterns of air movement set in motion by pressure gradients. These phenomena influence wind direction from local scales, affecting the rotation and wind patterns of individual storms, to global scales, affecting wind patterns of the entire planet. How do pressure gradients, the Coriolis effect, and friction explain wind patterns at such diverse scales? These two pages provide an overview of global and regional patterns of air circulation, with more in-depth coverage to follow in subsequent sections.

A What Influences the Patterns of Airflow Around Low- and High-Pressure Areas?

Once the pressure-gradient force starts air moving, the air is affected by the Coriolis effect and, if close to Earth's surface, is acted on by friction with the surface. The results of these interactions produce distinctive and familiar patterns of wind and circulation of air around areas of low and high pressure.

Low- and High-Pressure Areas

03.09.a1

1. In an area of *low pressure*, the pressure-gradient force pushes air laterally into the low from all directions. If the Earth were not rotating and therefore had no Coriolis effect, this simple inward flow pattern would remain intact. Friction would likewise not perturb this pattern because friction does not change the direction of airflow, only the speed. Near the center of the low, the converging winds force some air upward, higher into the atmosphere.

03.09.a2

2. In an area of *high pressure*, the pressure-gradient force pushes air laterally out in all directions. As with a low pressure, this outward flow of air is influenced by the Coriolis effect and by friction near Earth's surface. As air flows outward from the high, it is replaced by air flowing down from higher in the atmosphere.

Northern Hemisphere

03.09.a3

3. Earth does rotate, so the Coriolis effect, represented in this figure by green arrows, deflects the inward-flowing air associated with a low-pressure area. This deflection causes an inward-spiraling rotation pattern called a *cyclone*. In the Northern Hemisphere, the Coriolis deflection is to the right of its intended path, causing a *counterclockwise* rotation of air around the low-pressure zone. This counterclockwise rotation is evident in Northern Hemisphere storms.

03.09.a4

4. In a Northern Hemisphere high-pressure area, the Coriolis effect deflects winds to the right, but friction again slows the winds. This results in an outward-spiraling rotation pattern called an *anticyclone*. In the Northern Hemisphere, circulation around an anticyclone is *clockwise* around the high-pressure zone.

Southern Hemisphere

03.09.a5

5. For a low-pressure area in the Southern Hemisphere, the air flowing inward toward a low is deflected to the left by the Coriolis effect. This causes air within a cyclone in the Southern Hemisphere to rotate *clockwise*, opposite to what is observed in the Northern Hemisphere. In either hemisphere, friction causes wind patterns within cyclones to be intermediate between straight-inward winds and circular gradient winds.

03.09.a6

6. For a high in the Southern Hemisphere, the air flowing outward from the high is deflected to the left by the Coriolis effect. This causes the air within an anticyclone in the Southern Hemisphere to rotate *counterclockwise*, opposite to the pattern in the Northern Hemisphere. Therefore, cyclones and anticyclones can rotate either clockwise or counterclockwise, depending on hemisphere.

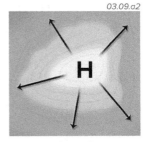

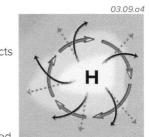

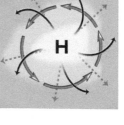

03.09.a7

7. This satellite image (◄) shows a cyclone near Iceland (Northern Hemisphere). It is characterized by a circular rotation of strong winds around the innermost part of the storm. It displays a distinctive inward-spiraling flow of the clouds and air around an area of low air pressure. Within a cyclone, air moves rapidly down the pressure gradient toward the very low pressure at the center of the storm. In the Northern Hemisphere, the air is deflected to the right of its intended path, resulting in an overall counterclockwise rotation of the storm. The distinctive spiral pattern results from a combination of the pressure-gradient force (which started the air moving), the Coriolis effect, and friction with Earth's surface. A hurricane, which is a huge tropical cyclone, shows a similar pattern of rotation around the center of the storm—the eye of the hurricane.

B What Factors Influence Global Wind Patterns?

Insolation warms equatorial regions more than the poles, setting up pressure differences that drive large-scale atmospheric motion. Once set in motion, the moving air is acted upon by the Coriolis effect and by friction with Earth's surface. These factors combine to produce curving patterns of circulating wind that dominate wind directions, climate, and weather.

Rotation and Deflection

1. Earth is a spinning globe, with the equatorial region having a higher velocity than polar regions. As a result, air moving north or south is deflected from its intended path by the Coriolis effect.

2. Equatorial regions preferentially heat up. As Earth rotates, the Sun's heat forms a band of warm air that encircles the globe and is re-energized by sunlight each day.

3. Warmed equatorial air rises and then flows north and south, away from the equator at upper levels of the troposphere. Air at the surface flows toward the equator to replace the air that rises. The Coriolis effect deflects this equatorial-flowing surface wind toward the west (to the left in the Southern Hemisphere and to the right in the Northern Hemisphere).

4. These flows of air combine into huge, tube-shaped cells of circulating winds, called *circulation cells*. Very fast flows of air along the boundaries of some circulation cells are *jet streams*.

Prevailing Winds

5. Wind direction is referenced by the direction from which it is coming. A wind coming from the east is said to be an "east wind." A wind that generally blows from the east is an *easterly*.

6. Polar regions receive the least solar heating and are very cold. Surface winds move away from the poles, carrying cold air with them. *Polar easterlies* blow away from the North Pole and have a large apparent deflection toward the west because the Coriolis effect is strong at high latitudes.

7. *Westerlies* dominate a central belt across the U.S. and Europe, so weather in these areas generally moves from west to east.

8. *Northeast trade winds* are easterlies, blowing from the northeast. They were named by sailors, who took advantage of the winds to sail from the so-called Old World to the New World.

9. *Southeast trade winds* blow from the southeast toward the equator. Near the equator, they meet the northeast trade winds in a stormy boundary called the *Intertropical Convergence Zone (ITCZ)*.

10. *Westerlies* also occur in the Southern Hemisphere and are locally very strong because this belt is mostly over the oceans and has few continents to generate additional friction to disrupt the winds.

11. *Polar easterlies* at the surface flow away from the South Pole and deflect toward the west but are mostly on the back side of the globe in this view. The Coriolis effect causes this large apparent deflection due to the very high latitudes.

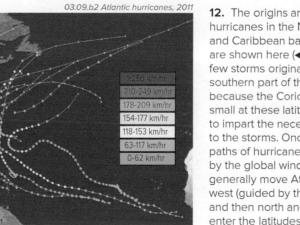

03.09.b2 Atlantic hurricanes, 2011

>250 km/hr	
210-249 km/hr	
178-209 km/hr	
154-177 km/hr	
118-153 km/hr	
63-117 km/hr	
0-62 km/hr	

12. The origins and paths of hurricanes in the North Atlantic and Caribbean basins in 2011 are shown here (◄). Note that few storms originate in the far southern part of this map, partly because the Coriolis effect is so small at these latitudes that it fails to impart the necessary rotation to the storms. Once formed, the paths of hurricanes are steered by the global wind patterns, which generally move Atlantic hurricanes west (guided by the trade winds) and then north and east once they enter the latitudes of the westerlies.

Before You Leave This Page

✓ Sketch, label, and explain the combined effect of the pressure-gradient force, Coriolis effect, and friction, and how this is expressed in flow around anticyclones and cyclones in both the Northern and Southern Hemispheres.

✓ Locate and name on a map the main belts of global winds.

3.9

3.10 How Does Air Circulate in the Tropics?

TROPICAL CIRCULATION is driven by the intense solar heating of land and seas near the equator. The heated air rises and spreads out from the equator, setting up huge, recirculating cells of flowing air. The rising air results in a belt of tropical low pressure, and where the air descends back toward the surface is a belt of subtropical high pressure. What determines where the rising and sinking occur, and how does the Coriolis effect influence this flow?

General Circulation in the Tropics

1. Examine the large figure below and note the main features. What do you observe, and can you explain most of these features using concepts you learned from previous parts of the chapter? Tropical areas are known for their lush vegetation (▶), which in turn is due largely to relatively abundant and consistent insolation, warm temperatures, and abundant rainfall. After thinking about these aspects, read the rest of the text.

03.10.a2 Kakadu World Heritage Site, Australia

2. At the surface, winds generally converge on the equator from the north and south. The south-flowing winds in the Northern Hemisphere are apparently deflected to the right relative to their original path, blowing from the northeast. These winds are called the *northeast trade winds* because they guided sailing ships from the so-called Old World (Europe and Africa) to the New World (the Americas).

3. A belt of high pressure occurs near 30° N and 30° S, where air descends to the surface of the Earth. This air rose in the low pressure located near the equator, as a result of excess heating.

4. The rising and descending air, and the related high- and low-pressure areas, are linked together in a huge cell of convecting air—the *Hadley cell*. One Hadley cell occurs north of the equator and another just south of the equator.

5. Note that the Hadley cell extends to approximately 30° north and south of the equator, so it generally encompasses all the tropics and some distance beyond.

6. In the Southern Hemisphere, winds blowing toward the equator are deflected to the left (west), resulting in winds blowing from the southeast, forming the *southeast trade winds*.

60°N

30°N

Counterclockwise Earth Rotation

Northeast Trade Winds

Northeast Trade Winds

EQ

Hadley Cell

30°S

Southeast Trade Winds

Southeast Trade Winds

60°S

Hadley Cell

03.10.a1

Formation of Hadley Cells

1. Insolation, on average, is most intense near the equator, in the tropics. The position of the overhead Sun migrates between the Tropic of Cancer and Tropic of Capricorn from season to season. The Sun-heated air rises from the tropics, forming a belt of low pressure at the surface. As the warm, moist air rises, the air cools somewhat, forming clouds; this accounts for the typical cloudiness and haziness of many tropical areas. Condensation of drops further heats the air, aiding its rise.

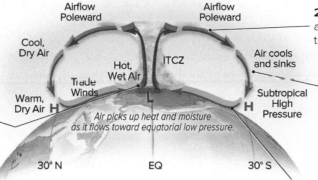

03.10.a3

2. After rising, this air spreads out poleward as it approaches the upper boundary of the troposphere (the tropopause).

3. Once the upper-level flow reaches about 30° N and 30° S latitude, it sinks, both because it begins to cool aloft and due to forces arising from the Earth's rotation. This sinking air dynamically compresses itself and the surrounding air, producing the subtropical high-pressure systems.

4. Once near the surface, the air flows back toward the equator to replace the air that rose. The flow from the two hemispheres converges at the ITCZ.

Influence of the Coriolis Effect

5. As the air flows toward the equator in each hemisphere from the subtropical high to the ITCZ, the Coriolis effect pulls it to the right (in the Northern Hemisphere) or left (in the Southern Hemisphere) of its intended path, as shown by the arrows on the left side of this diagram. The Coriolis effect is weak near the equator, however, so the deflection is only slight. The result is surface air flowing from northeast to southwest in the Northern-Hemisphere tropics (the northeast trade winds) and from southeast to northwest in the Southern-Hemisphere tropics (the southeast trade winds).

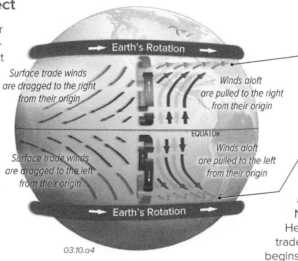

03.10.a4

6. In the Northern Hemisphere, as the air flows poleward after rising at the ITCZ, the weak Coriolis effect also pulls the air slightly to the right of its intended path. The result is that some of the upper-level air moves from southwest to northeast at the top of the Northern Hemisphere Hadley cell.

7. In the Southern Hemisphere, the Coriolis effect deflects the upper-level winds to the left of their intended path, causing a northwest-to-southeast flow at the top of the Southern Hemisphere Hadley cell.

8. As the seasons progress, the set of Hadley cells and the ITCZ migrate—to the Northern Hemisphere in Northern-Hemisphere summer and to the Southern Hemisphere in Southern-Hemisphere summer. If the trade-wind flow crosses the equator, the Coriolis deflection begins to occur in the opposite direction, and the winds can reverse direction (not shown).

Seasonal Variations in the Position of the Intertropical Convergence Zone

9. As the overhead Sun shifts north and south within the tropics from season to season, the ITCZ shifts, too. In the northern summer, it shifts to the north. The typical June position of the ITCZ is the reddish line on the figure below, and the December position is the blue line.

10. The ITCZ generally extends poleward over large landmasses in the hemisphere that is experiencing summer. This larger shift over the land than over the oceans is because of the more intense heating of land surfaces.

11. Unlike the ITCZ, the subtropical high pressure doesn't exist in a continuous belt around the Earth. The ocean-covered surfaces support high pressure better than land surfaces because land heats up too much at these latitudes, especially in summer. The heated air over the land rises, counteracting the tendency for sinking air in the Hadley cell. So the subtropical high pressure tends to be more vigorous over the oceans.

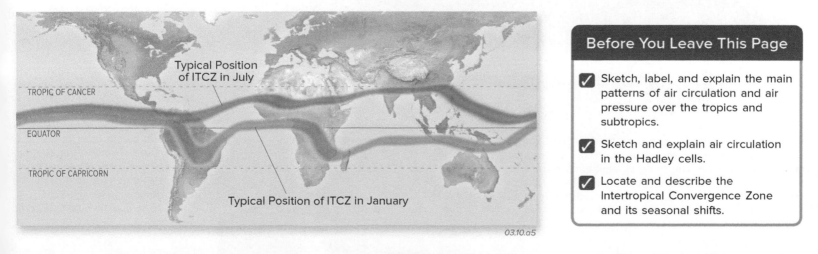

03.10.a5

Before You Leave This Page

☑ Sketch, label, and explain the main patterns of air circulation and air pressure over the tropics and subtropics.

☑ Sketch and explain air circulation in the Hadley cells.

☑ Locate and describe the Intertropical Convergence Zone and its seasonal shifts.

3.10

3.11 How Does Air Circulate in High Latitudes?

POLAR REGIONS RECEIVE LITTLE INSOLATION compared to the rest of Earth. As a result, the poles are very cold places that experience winter darkness for months at a time. Air circulation around the poles results from this relative lack of solar heating and also the proximity to the axis of rotation for the planet. The encroachment of polar air away from the poles can cause nearby areas to experience very cold temperatures. Airflow away from the poles results in a belt of relatively stormy weather near 45° to 60° N and 45° to 60° S.

General Circulation at High Latitudes

1. Examine the large figure below and observe the main features near the poles. Note the circulation directions near the surface versus those aloft. After you have made your observations, read the rest of the text.

2. Cold, dense air sinks near the North Pole. As it nears the surface, it then flows outward, away from the poles (to the south).

3. As the air flows south, it is deflected to the right by the Coriolis effect, which is very strong at these latitudes. As a result, surface winds generally encircle the North Pole, blowing in a clockwise direction when viewed from above the pole (▶). In the small globe to the right, the golden arrows show Earth's rotation and light-yellow arrows show surface winds.

03.11.a2

4. The south-flowing air eventually begins to heat up and rise, usually somewhere between 60° and 45° latitude. This rising air causes a series of low-pressure areas at the surface, called the *subpolar lows* (L on this figure). Once the air rises to its maximum height, the flow turns back to the north, completing a circulating cell of cold air—the *polar cell*. The polar cell is represented here by the large blue arrows, with air rising near 60° N and descending at the pole.

03.11.a1

03.11.a4 *Icebergs off the Antarctic Peninsula, Antarctica*

6. As near the North Pole, cold air flowing away from the South Pole eventually heats up enough to rise, producing a belt of low pressure. The rising air aloft turns south and descends back near the pole, completing the polar cell. The polar cell involves very cold air at such high latitudes, causing the land to largely be covered year-round in ice and snow (▶).

03.11.a3

5. A similar situation occurs around the South Pole, where surface air circulates around the pole, but in a counterclockwise direction when viewed from below the South Pole. These circular winds from the east, *polar easterlies*, are in response to the Coriolis effect, which is in turn caused by rotation of the Earth and enhanced by the comparative lack of surface friction with the ocean surfaces that dominate these latitudes. Remember that this view is from below the South Pole, a different perspective than you are used to.

Circulation Around the Poles

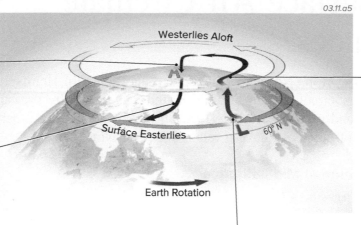

03.11.a5

1. The very cold air over the poles is so dense that it has a tendency to sink vigorously to the surface, creating high surface pressure—*polar highs*. The air then moves equatorward, because that is the only direction it can go from the pole.

2. The Coriolis effect is very strong at high latitudes, so the air deflects strongly and circulates around the pole, as shown here for the North Pole. Around the North Pole, the surface winds moving south deflect to the right of their intended path and so blow from the east—they are *polar easterlies*.

4. At upper levels, the return flow of air northward toward the pole is also deflected to the right of its intended path (in the Northern Hemisphere). As it is turned to the right, it blows from the west (a westerly flow aloft). So not only is air flowing away from the pole near the surface and toward the pole aloft, the surface and upper-level airflows are rotating in opposite directions (clockwise near the surface, counterclockwise aloft). This is difficult to capture in a single perspective, which is why the polar flow is represented on this page with several figures.

3. As the air flows away from the pole, it warms and rises, producing low surface pressure—subpolar lows. In the winter, the subpolar lows are particularly intense over water bodies because the water is relatively warm at that time of year, relative to air elsewhere at these latitudes. The water warms the air, allowing it to rise.

Northern Hemisphere

5. The figure below shows the Northern Hemisphere polar cell, as viewed directly down on the North Pole. The slightly faded arrows depict surface flows (easterlies), whereas the brighter arrows show upper-level flow (westerlies). Color gradations on arrows indicate whether air is warming (blue to red from tail to head) or cooling (red to blue).

6. High surface pressure is present at the pole, but shifts slightly in position from season to season. Low-pressure zones (the subpolar lows) occur over the adjacent oceans. The subpolar low in the Atlantic is the *Icelandic Low*. Another subpolar low, on the opposite side of the North Pole, occurs over the northern Pacific Ocean, and is the *Aleutian Low*, named for the Aleutian Islands west of mainland Alaska.

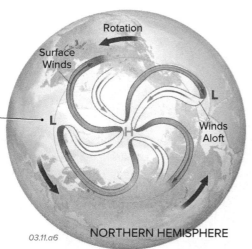

03.11.a6   NORTHERN HEMISPHERE

Southern Hemisphere

7. The polar cell in the Southern Hemisphere, shown below, is over the South Pole. Unlike the polar cell in the Northern Hemisphere, this one is centered over land—Antarctica. Antarctica is surrounded by uninterrupted oceans. The entire region is a very cold place, so even the air that is shown as warming is still very cold.

8. Surface winds flowing away from the pole are deflected to the left of their intended path, and so circulate counterclockwise around the pole when viewed from below (polar easterlies). Winds aloft move toward the pole and are deflected left of their intended path, flowing clockwise, in the opposite direction from the surface winds. The outward flowing surface air is balanced by the inward flowing air aloft. A continuous belt of low pressure occurs over the ocean.

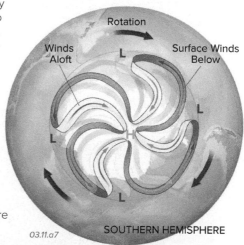

03.11.a7   SOUTHERN HEMISPHERE

Differences Between the North and South Polar Areas

The general patterns of atmospheric flow are similar for the North and South polar regions, except for the reversed directions of circulation (clockwise versus counterclockwise) related to different deflection directions for the Coriolis effect in the Northern and Southern Hemispheres. One striking difference, however, is the difference between the land-sea distributions around the two poles. The North Pole is over ocean, although this ocean (the Arctic Ocean) is mostly frozen most of the year. The Arctic Ocean is nearly enclosed by the surrounding continents (North America, Asia, and Europe).

In contrast, the South Pole is over land, the frozen continent of Antarctica, and this continent is completely surrounded by open ocean. Differences in the responses of land and water to warming, cooling, and latent heat account for some difference in the specifics, such as having two main low-pressure areas in the oceans adjacent to the north polar region, but a continuous belt of low pressure in the ocean around Antarctica.

Before You Leave This Page

✓ Sketch, label, and explain the main patterns of air circulation and air pressure in the polar regions.

✓ Sketch and explain the polar cells, summarizing why circulation occurs.

✓ Explain the differences between polar circulation and pressures in Northern and Southern Hemispheres.

3.11

3.12 How Does Surface Air Circulate in Mid-Latitudes?

THE MID-LATITUDES are regions, in the Northern Hemisphere and Southern Hemisphere, that lie between the tropics (23.5°) and polar circles (66.5°). Air circulation in the mid-latitudes is driven by pressures set up by circulation in the adjacent tropics and polar regions, and by the Coriolis effect. Surface winds within most of the mid-latitudes blow from west to east (westerlies), but the subtropics can have a relative lack of wind.

03.12.a2 Banff, Alberta, Canada

General Circulation at Mid-Latitudes

1. Examine the large figure below and observe the main features in the mid-latitudes, especially the region between 30° and 60° latitude (the dashed lines) in both the Northern and Southern Hemispheres. The mid-latitudes are characterized by distinct seasons and changing weather patterns, and from forested areas (◄) to deserts. After observing the features on the large globe and reflecting on implications for weather, read the rest of the text.

2. Cold surface air flowing away from the poles warms and ascends along the poleward edge of the mid-latitudes. This rising air causes storminess and low pressure: the *subpolar lows* (L on this figure). This rising air is part of the cold *polar cell*.

3. In the Northern Hemisphere, air from the south flows toward the subpolar lows and away from high pressure in the subtropics (near 30–35° latitude). The Coriolis effect is relatively strong in the mid-latitudes and deflects this flow to the right, forming a belt of wind blowing from west to east—the *westerlies*.

4. Along the edge of the mid-latitudes, in the subtropics, air in the descending limb of the Hadley cell causes high pressure. These regions near 30° N and 30° S—the *horse latitudes*—can have weeks without wind, which posed a hazard to early sailing ships and their cargo.

5. The top part of the graph below shows how the combination of a northward-directed pressure-gradient force and eastward-directed Coriolis effect results in the westerlies in the Northern Hemisphere. The Coriolis effect strengthens with increasing latitude, turning winds even more to the east as they move north.

03.12.a1

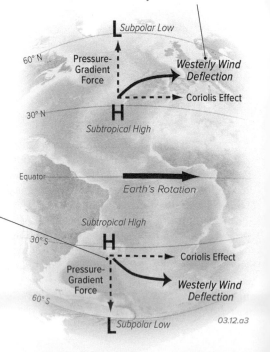

6. In the Southern Hemisphere, air is driven south by high pressure in the subtropics. As this air flows south, it is deflected to the east by the Coriolis effect. The result is again west-to-east flow in the middle latitudes, forming a Southern Hemisphere belt of westerlies. Note that in both the Northern and Southern Hemispheres, the Coriolis effect is deflecting flows in the mid-latitudes toward the east, in the direction the Earth is rotating.

7. The Southern Hemisphere has a belt of low pressure (the subpolar lows) caused by rising air in the polar cell. Air flows toward these lows and away from the subtropical highs, deflecting to the east and becoming westerlies. These westerlies are especially strong (there are no wind-blocking continents) and so are called the *roaring forties* (between 40° S and 50° S latitude).

03.12.a3

Circulation Around Highs and Lows in the Northern Hemisphere

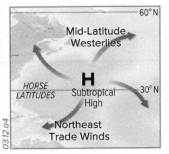

1. In the Northern Hemisphere, high pressure commonly occurs in the subtropics, along the horse latitudes. Winds around the highs are deflected in a clockwise direction by the Coriolis effect. Winds north of the high reinforce the westerlies, whereas winds to the south reinforce the easterly trade winds. As winds spin around the high, they push cold air equatorward and warm air poleward.

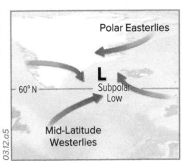

2. Circulation around lows, including subpolar lows, is counterclockwise in the Northern Hemisphere. The subpolar lows reside between polar easterlies to the north and westerlies to the south (in the mid-latitudes). The rising air and shearing between opposite-directed winds causes windy, stormy conditions.

Circulation Around Highs and Lows in the Southern Hemisphere

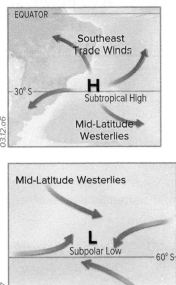

3. In the Southern Hemisphere, high pressure around the subtropics is accompanied by counterclockwise circulation. The highs are situated between southeast trade winds to the north and westerlies in the mid-latitudes to the south. As in the Northern Hemisphere, circulation around pressure systems redistributes warm and cold air and moisture.

4. Lows in the Southern Hemisphere, such as the subpolar lows, have clockwise circulation. The lows lie between the westerlies and polar easterlies, and this latitudinal belt is extremely windy and stormy. At this latitude there are few large landmasses to disrupt the winds.

Migration of Pressure Areas in the Mid-Latitudes

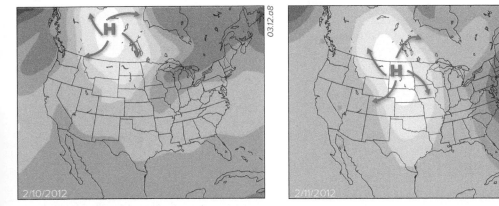

5. Embedded within the mid-latitude westerlies are traveling, spinning circulations of high and low pressure, much like a swirling flow that is embedded within the overall current in a stream. The four figures here show the movement of a high-pressure area (an *anticyclone*) guided by the mid-latitude westerlies. Anticyclones generally move equatorward as they migrate eastward, as shown in this series of daily images. Traveling low-pressure systems called *cyclones*, not shown here, tend to move poleward as they travel from west to east across the mid-latitudes; they are associated with stormy weather. As either type of pressure migrates through an area, it causes the wind direction to change relatively suddenly. As a result, wind and other weather patterns change more quickly in the mid-latitudes than in any other part of the Earth.

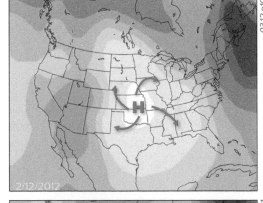
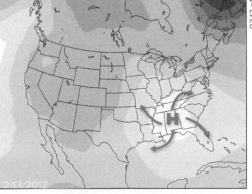

Before You Leave This Page

✓ Sketch, label, and describe the general circulation in the mid-latitudes, in both the Northern and Southern Hemispheres, and explain the processes that cause air in the mid-latitudes to circulate generally from west to east.

✓ Sketch, label, and explain the circulation around areas of high and low pressure, and how they relate to the prevailing winds.

✓ Summarize the overall movement of mid-latitude cyclones and anticyclones.

6. Note how the winds change in any area across which the pressure area migrates. In the case shown here, an area has northwesterly winds as the high pressure approaches from the northwest, followed by southeast winds once the high has passed to the east.

3.12

1000   1015   1025   1040mb

3.13 How Does Air Circulate Aloft over the Mid-Latitudes?

SURFACE WINDS IN THE MID-LATITUDES are generally from west to east in both hemispheres. Higher in the troposphere, the main features are two currents of fast-moving air—jet streams—that encircle the globe near the boundaries of the mid-latitudes. What factors determine the direction and speed of airflow aloft in the mid-latitudes, such as in the jet streams?

A What General Circulation Occurs Aloft over the Westerlies?

The main direction of surface air in the mid-latitudes is as curving arcs that move poleward away from the subtropics and then bend increasingly to the east, becoming westerlies. What happens aloft? The figure below shows the setting of upper-level airflows in the mid-latitudes, between the polar cell and the Hadley cell. The pole is to the left.

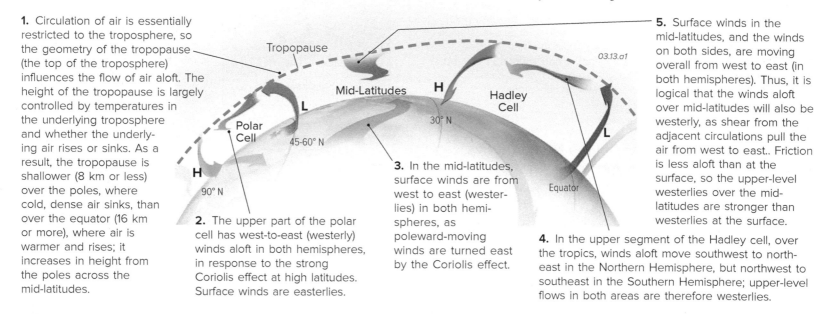

1. Circulation of air is essentially restricted to the troposphere, so the geometry of the tropopause (the top of the troposphere) influences the flow of air aloft. The height of the tropopause is largely controlled by temperatures in the underlying troposphere and whether the underlying air rises or sinks. As a result, the tropopause is shallower (8 km or less) over the poles, where cold, dense air sinks, than over the equator (16 km or more), where air is warmer and rises; it increases in height from the poles across the mid-latitudes.

2. The upper part of the polar cell has west-to-east (westerly) winds aloft in both hemispheres, in response to the strong Coriolis effect at high latitudes. Surface winds are easterlies.

3. In the mid-latitudes, surface winds are from west to east (westerlies) in both hemispheres, as poleward-moving winds are turned east by the Coriolis effect.

4. In the upper segment of the Hadley cell, over the tropics, winds aloft move southwest to northeast in the Northern Hemisphere, but northwest to southeast in the Southern Hemisphere; upper-level flows in both areas are therefore westerlies.

5. Surface winds in the mid-latitudes, and the winds on both sides, are moving overall from west to east (in both hemispheres). Thus, it is logical that the winds aloft over mid-latitudes will also be westerly, as shear from the adjacent circulations pull the air from west to east.. Friction is less aloft than at the surface, so the upper-level westerlies over the mid-latitudes are stronger than westerlies at the surface.

B Is There a Circulation Cell in the Mid-Latitudes?

1. If there were a cell of circulating air in the mid-latitudes, as there is in the tropical and polar latitudes, then the sinking from the poleward edge of the Hadley cell would cause shear that would also induce sinking at the equatorward edge of the mid-latitudes.

2. Likewise, the rising motion at the equatorward edge of the polar cell (the subpolar lows) would induce rising next to the poleward edge of the mid-latitudes. This rising motion, coupled with the sinking next to the Hadley cell, would logically set up upper-level circulation over the mid-latitudes, with air moving from the pole toward the tropics. This hypothetical mid-latitude circulation system is known as the *Ferrel cell*.

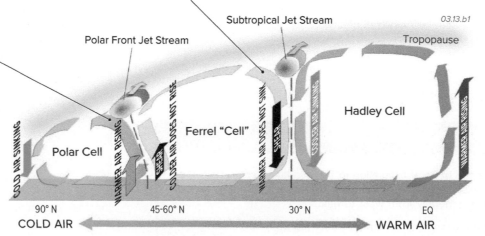

3. Descending air near the subtropical highs flows poleward along the surface and is turned toward the east by the Coriolis effect, contributing to the mid-latitude westerlies. Aloft, some rising air at the subpolar lows moves equatorward.

4. However, other factors cause the Ferrel cell to be developed very weakly or not at all. Thermal properties, which result in differences in air density, also dictate whether air will rise. Air on the subtropical side of the Ferrel cell (near 30° latitude) is relatively warm, and so it should be rising, not sinking. Air on the polar side of the cell is colder, and so it should be sinking, not rising. In other words, the thermal effects of the Ferrel cell's warmer air rising and cooler air sinking somewhat offset the dynamic effect of the shear from the adjacent Hadley and polar cells. The result is the near absence of a circulation cell in the mid-latitudes.

C What Are Jet Streams and How Do They Form?

Fast-moving, relatively narrow currents of wind, called *jet streams*, flow aloft along the boundaries of the mid-latitude air currents. One jet stream is located along the edge of the polar cell and another is along the edge of the tropical Hadley cell. How do such jet streams form and what controls their locations?

1. Jet streams are strong air currents that are long and relatively narrow, resembling a ribbon or tube wrapping around the planet. This figure shows two jet streams: the *polar front jet stream* and the *subtropical jet stream*. They are shown for the Northern Hemisphere, but equivalents are also present in corresponding positions in the Southern Hemisphere. The jet streams are typically a few hundred kilometers wide and about 5 km thick, and are located high in the troposphere, near the tropo-pause. They generally follow paths that curve or meander and can locally split into sepa-rate strands that eventually rejoin. The four jet streams all blow from west to east as a result of Earth's rotation.

2. The subtropical jet stream circles around the globe at about 30° latitude (in both hemispheres), near the boundary between the Hadley cell and westerlies aloft in the mid-latitudes.

3. In both the Northern and Southern Hemispheres, the polar front jet stream encircles the globe near the edge of the polar cell. It shifts position north and south from time to time, but typically resides between 45° and 60° latitude. As it shifts farther away from the pole, it brings cold air into the mid-latitudes, like the center of North America.

4. In addition to influencing weather, jet streams affect air travel in good and bad ways. A plane flying in the same direction as the jet stream (to the east) moves faster than normal, but a plane flying against the jet stream (west) is slowed. Jet streams also cause clear-sky turbulence on flights.

03.13.c1

D What Are Rossby Waves and How Do They Relate to Jet Streams?

1. Jet streams do not track around the planet along perfectly circular routes, but instead typically follow more irregular curved paths, like along the thick blue line in the polar projection below. These meandering paths, when viewed in three dimensions, resemble curving waves, and are called *Rossby waves* after the scientist who discovered them. The lines on this map, which is viewed directly down on the North Pole, are isohypses with dark colors showing local areas of low heights (low pressure). Notice that these are concentrated in the subpolar areas.

2. Where the polar jet stream curves toward the pole, this part of the jet stream is called a *ridge*. Ridges bend poleward.

3. Where the polar jet stream curves away from the pole, we call that type of curve a *trough*. Troughs bend equatorward.

4. The size and position of ridges and troughs is not constant, but changes over days, weeks, and months. Troughs can get "deeper" or "shallower," while ridges get "stronger" or "weaker."

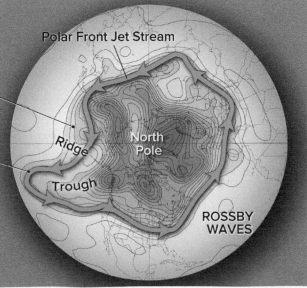

03.13.d1

5. Bends in the polar jet stream (Rossby waves) have a great influence on weather in nearby parts of the mid-latitudes. When a trough becomes more accentuated and shifts farther away from the pole (i.e., it deepens), it allows cold Arctic air to extend farther away from the pole, sometimes called a "polar vortex" by popular media. Also, as discussed in the weather chapter, Rossby waves influence surface high and low pressures near the polar jet and help strengthen, weaken, and guide mid-latitude cyclones as they migrate across the surface. Accordingly, we return to Rossby waves later.

Before You Leave This Page

☑ Describe the flow of air aloft over the mid-latitudes.

☑ Sketch, label, and explain the two types of jet streams.

☑ Sketch and explain Rossby waves, including a trough and ridge.

3.13

3.14 What Causes Monsoons?

A COMMON MISCONCEPTION is that the word "monsoon" refers to a type of rainfall, but the word actually refers to winds that reverse directions depending on the season. One of these seasonal wind directions typically brings dry conditions and the other brings wet conditions. Monsoons impact a majority of the world's population.

A What Are the Features of the Asian Monsoon?

One way to characterize a monsoon is to compare maps showing wind directions for different times of the year. Such maps can then be compared to rainfall records to determine which seasonal wind directions bring dry conditions and which ones bring wet conditions. The maps below show climatological wind conditions, averaged over three decades, for two different months—January and July. Arrows show wind directions, and shading represents pressure at sea level, with light gray being high and dark gray being low. Examine the patterns of circulation for each month and then compare the patterns between the months.

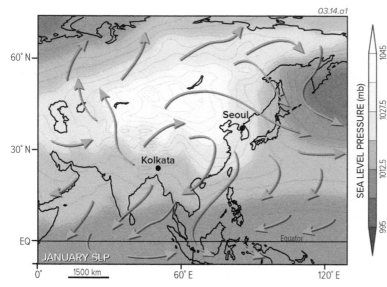

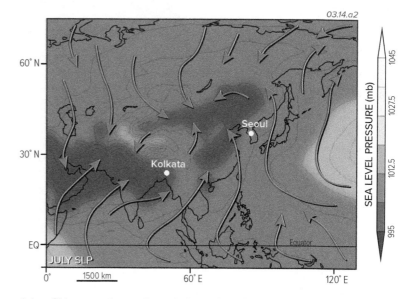

January—This map shows typical wind conditions for Asia during January. In the center of the map, winds define a region where flow is clockwise and outward, centered on the light-colored area of high pressure (an anticyclone). This high-pressure area, the *Siberian High*, forms from cold, sinking air over Siberia. This circulation brings very dry air (from the cold interior of the continent) from the north over southern Asia and from the northwest across eastern Asia. We would predict from these wind patterns that little precipitation would occur in much of Asia at this time.

July—This map shows that wind conditions for the same region during July are totally different than they are for January. Circulation that marked the high pressure is gone, replaced by an area of inward and counterclockwise flow over Tibet (north of Kolkata). In the Northern Hemisphere, this pattern of circulation is diagnostic of a low-pressure area, which in this case is caused by warm, rising air that accompanies warming of the Asian landmass. This low is called the *Tibetan Low*. This circulation brings very humid air from the southwest over southern and southeastern Asia. How do you think this circulation affects rainfall?

Seasonal Variation in Precipitation

Observe these graphs showing average monthly precipitation amounts for two very different parts of Asia: Kolkata, one of the largest metropolitan areas of India, and Seoul, the capital of South Korea. For both cities, notice prominent precipitation peaks that occur during the summer—the wet season. The increase in precipitation during the wet season results from the flow of moist air from oceans onto land, toward the Tibetan Low. The dry season, during the winter months, reflects the flow of dry air from the land, flowing outward from the Siberian High.

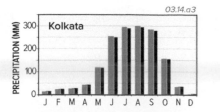

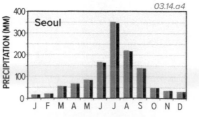

Effect of the Monsoon on Vegetation

These satellite images show increased vegetation due to monsoon-related rains along the western coast of India. The left image is during the dry season, when wind patterns bring in dry air. The right image is from the end of the monsoon. Note the increase in plant cover (green areas) during the monsoon-caused rainy season.

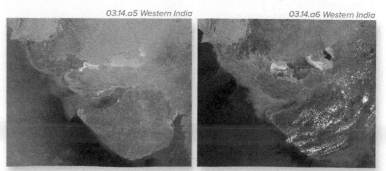

03.14.a5 Western India

03.14.a6 Western India

B What Other Regions Experience Monsoon Circulations?

West Africa

January—In January, near-surface winds in West Africa largely flow from the northeast, bringing in dry air from inland areas, including the Sahara Desert, and carrying it southwest to coastal areas and farther offshore. Such *offshore flows* generally result in dry weather.

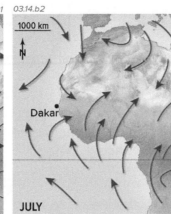

JANUARY

JULY

July—A shift in wind direction in July brings moist ocean air from several directions onto the very hot land where air has risen. This change in wind direction causes enormous differences in precipitation, as shown by the graph below for Dakar, Senegal. Along with the increase in precipitation comes an increase in the amount of vegetation. In Dakar and much of the region, precipitation is nearly nonexistent in January and adjacent months.

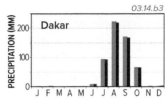

Northern Australia

January—In January (the southern summer), winds over northern Australia bring moist air from the ocean onto the heated land surface.

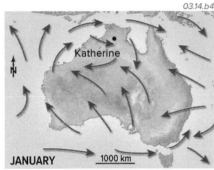

JANUARY

JULY

July—The wind shifts by July (winter) as the land surface cools, creating higher pressure over the land. This causes a large drop in precipitation, as shown by the graph below for Katherine, Australia. The monsoon flow in July results in little rain.

Southwestern U.S.

January—Southwestern North America, most of which is desert, has a less dramatic, but still important monsoon effect. In the winter months, winds blow from various directions, and winter precipitation in this region is from brief incursions of cold, wet air (i.e., cold fronts) from the northwest.

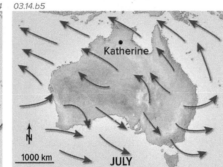

JANUARY

JULY

July—During the late summer months, heating of the land surface and the resulting low pressure causes a shift in winds. Winds from the south bring moist air northward from the Gulf of Mexico and Gulf of California, and summer thunderstorms form when this air interacts with the heated land. These summer thunderstorms cause precipitation to peak in August, as shown by the graph below for Tempe, Arizona. Note the different scale needed to show the relatively small amounts of precipitation in this desert area versus the previous ones. Nearly as much precipitation falls in the winter from the cold fronts.

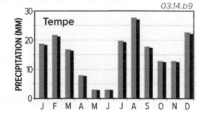

The Effect of Monsoons on Cultures

Monsoons greatly influence the lives of people living in regions with seasonal shifts in wind. The main effects of a monsoon are seasonal variations in precipitation, which in turn affect water supplies, amount of vegetation, and overall livability for some normally dry landscapes. Many cultures plan their activities around these seasonal changes, conserving water during the dry season and taking advantage of the plentiful water during the wet season.

The monsoon pervades the psyches of people in southern and southeastern Asia, especially the region from India to Vietnam, in ways not fathomable to most North Americans or Europeans. The influence on agriculture, including the cultivation of rice, and on flooding and other natural hazards is obvious, but the monsoon also appears in literature, art, music, architecture, and nearly every other aspect of culture. Ceremonies commonly mark the anticipated start of the monsoon. In years when the monsoon rains arrive later than usual, people become very concerned that harvests will suffer. The date of the onset of the monsoon rains varies by location, but generally proceeds from south to north with the onset in April and May.

Before You Leave This Page

☑ Explain what causes a monsoon, using examples from Asia, West Africa, northern Australia, and southwestern North America.

☑ Describe some of the effects of shifting monsoonal winds.

3.14

3.15 How Have Global Pressures and Winds Affected History in the North Atlantic?

INTERCONTINENTAL TRAVEL AND TRADE have relied upon moving currents in the air and oceans. Before the 20th century, when transoceanic travel and shipping relied on wind power, global winds, such as the trade winds and westerlies, dictated which directions of travel were possible at different latitudes. The directions of global winds therefore greatly influenced the exploration and colonization of the Americas, and traces of that influence can be detected in past and present cultures. The dashed, colored lines in the larger central globe spanning these two pages represent the paths of explorers or trade routes, as described below.

This small globe (▶) shows the main belts of global winds, controlled by the pressure-gradient force and Coriolis effect. Can you locate and name the main wind belts? In the Northern Hemisphere, trade winds (shown in red) blow from northeast to southwest, allowing wind-powered sailing ships to travel from western Africa to Central America and the northern part of South America.

03.15.a2

The polar easterlies (shown in blue) likewise allowed westward travel, but under cold and stormy conditions.

In contrast, westerlies (shown in purple) allowed a return trip from North America to Europe, but only in the mid-latitudes. Ships commonly had to travel north or south along coasts to catch a prevailing wind going in the direction of intended travel across the Atlantic Ocean.

03.15.a3 Cotton plantation along the Mississippi River

◀ Plantations in the southeastern U.S. depended upon African slave labor to cultivate and harvest cotton, tobacco, and tropical agricultural products. In contrast, the economy of the northeastern U.S. was based more upon manufacturing and the refining of products grown on the southern plantations, with the final refined product being exported to western Europe. The traditional cultural divide between the northern states and southern states is the Mason-Dixon line (near 40° N), which corresponds roughly to the global climatological divide between the zones of excess climatic energy to the south and an energy deficit to the north. The excess and deficit energy settings controlled which crops could be grown in the two regions and allowed different kinds of activities by impacting the climate. This climate-energy boundary contributed to the political and social attitudes that existed then and today.

03.15.a4 Sugar plantation

Slaves and various provisions were brought from Africa on ships utilizing the trade winds on either side of the equator, winds that provided a direct conduit of trade from western Africa to equatorial parts of the Americas. Most slaves were put to work on labor-intensive tasks, such as agriculture, in and around the tropical regions of the Caribbean (◀). Their descendants still exert a great influence over the distinct culture of the islands and coastal areas in the region.

03.15.a5 Sir Francis Drake

After Portugal and Spain agreed to divide the "new" world into two separate spheres of influence, the Spanish encountered difficulty in transporting products and plunder from the Pacific empire across Portuguese territories, especially Brazil. Spanish goods were brought to Caribbean ports or were transported across the isthmus of Central America, before being stockpiled temporarily in Spanish ports in the tropics, including Havana, Cuba. Treasure-bearing sailing ships departing tropical ports had to sail north up the coast of Anglo-America (yellow paths on the large globe) in order to reach the mid-latitude westerlies, which then carried the ships east to Spain. This coast provided an ideal safe haven for English privateers, like Sir Francis Drake (◀), who captured many ships and their content, enriching the privateers and England in the process, and laying the financial foundation for the British Empire.

03.15.a6 Remains of Viking settlement, Iceland

◄ The Viking explorer Leif Erickson probably visited North America about 500 years before Christopher Columbus. Moving westward from Scandinavia, the Vikings progressively colonized Iceland and Greenland before journeying across the Davis Straits to Baffin Island and the coast of Labrador (blue path on the globe). The remains of a Viking settlement have been found on the northernmost tip of Newfoundland at L'Anse Aux Meadows. Viking ships would have employed the polar easterlies and associated ocean currents to aid their westward migration. Ironically, a shift to a much colder climate in the late 15th century caused the extinction of the settlements in Greenland at exactly the same time as Columbus was embarking on his explorations.

03.15.a7 Titanic

▲ When the steam-powered *Titanic* left Southampton, England (51° N), for New York City (41° N) in April 1912, global winds and currents were no longer such a strong determinant of oceanic travel routes. The planned route (red path on the globe), however, crosses a treacherous part of the North Atlantic Ocean—the region with stormy weather near the subpolar lows. The *Titanic* struck an iceberg brought south by global winds and ocean currents, sinking at 42° N off the east coast of Canada.

The prosperity of many trading ports in western Europe and the Americas (▼) during the early 19th century was based on various types of interconnected trade that followed a triangular path (green path on the globe). Manufactured goods from the emerging European industrial economies were loaded for trade with West Africa. There, African slaves were taken on board and carried west by the trade winds to the Americas. Once in the Americas, the slaves were traded for the products of the tropical plantation system. Using the westerlies, the raw goods were returned to western Europe for consumption or industry. The triangular path was designed to maximize favorable winds and to avoid the subtropical high pressures and calm winds of the horse latitudes.

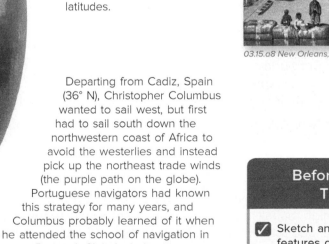

03.15.a8 New Orleans, LA

Departing from Cadiz, Spain (36° N), Christopher Columbus wanted to sail west, but first had to sail south down the northwestern coast of Africa to avoid the westerlies and instead pick up the northeast trade winds (the purple path on the globe). Portuguese navigators had known this strategy for many years, and Columbus probably learned of it when he attended the school of navigation in Sagres, Portugal. Global wind patterns therefore controlled Columbus's path and explain why in all likelihood, he first stepped ashore in the Bahamas (24° N), much farther south than Cadiz, where he started.

03.15.a1

Before You Leave This Page

☑ Sketch and describe the main features of atmospheric circulation in the northern Atlantic Ocean and how they influenced early travel.

3.16 What Occurs During Seasonal Circulation Shifts?

GLOBAL ATMOSPHERIC CIRCULATION responds directly to insolation. As the Sun's direct rays migrate seasonally, belts of winds, such as the westerlies, migrate too. In this investigation, you will examine the general circulation of the atmosphere, as expressed by data on air pressure, wind velocity, and cloud cover for two months with very different seasons—January and July.

Goals of This Exercise:

- Identify major patterns in air pressure, wind velocity, and cloud cover for each season.
- From these data, identify the major features of the global atmospheric circulation in each season.
- Assess and explain the degree of seasonal movement of these circulation features.

When examining broad-scale patterns of the Earth, such as global circulation patterns, a useful strategy is to focus on one part of the system at a time. Another often-recommended strategy is to begin with relatively simple parts of a system before moving to more complex ones. For this investigation, you will infer global patterns of air circulation by focusing on the Atlantic Ocean and adjacent lands (▶).

This globe is centered on the central Atlantic, and its top is slightly tilted toward you to better show the Northern Hemisphere. As a result of this tilt, Antarctica is barely visible at the bottom of the globe. The colors on land, derived from satellite data, depict rocks and sand in tan and brown. Vegetation is in various shades of green, with the darkest green indicating the thickest vegetation (usually forests). Shallow waters in the Caribbean region (on the left side of the globe) are light blue.

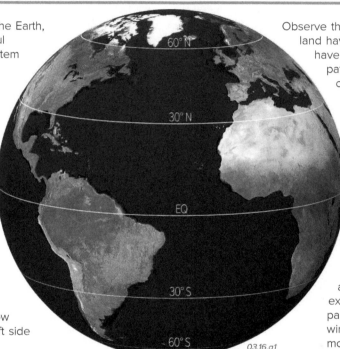

03.16.a1

Observe the entire scene, noting which areas on land have the most vegetation and which ones have the least. Compare these vegetation patterns with patterns of atmospheric circulation and air pressure, like subtropical highs and the Intertropical Convergence Zone (ITCZ).

Consider what directions of prevailing winds would occur in different belts of latitude. For example, where in this globe are the two belts of trade winds (one north and one south of the equator)? How about the mid-latitude belts of westerlies in each hemisphere? Consider how these winds might blow moisture-rich air from the ocean onto land. After you have thought about these aspects, read the procedures below and examine the globes and text on the next page, which highlight average air pressure, wind velocity, and cloud cover for two months—January and July.

Procedures

Complete the following steps on a worksheet provided by your instructor or as an online activity.

1. Study the two globes showing air pressure (on the next page), and note areas with high and low pressure. Locate the Icelandic Low and Bermuda-Azores High, and determine for which season each is strongest or if there is not much difference between the seasons. Then, locate a belt of low pressure near the equator and the adjacent belts of subtropical highs on either side. Mark and label the approximate locations of these features on the globe on the worksheet.

2. Next, examine the two globes that show wind velocity. In the appropriate place on the worksheet, draw a few arrows to represent the main wind patterns for different regions in each month. Label the two belts of westerlies and the two belts of trade winds. If the horse latitudes are visible for any hemisphere and season, label them as well. Mark any somewhat circular patterns of regional winds and indicate what pressure feature is associated with each.

3. Examine the two globes that show the average cloud cover for each month. From these patterns, label areas that you interpret to have high rainfall in the tropics due to proximity to the ITCZ or low rainfall due to position in a subtropical high. Examine how the cloud patterns correspond to the amount of vegetation, pressure, and winds.

4. Sketch and explain how the different features of circulation and air pressure change between the two months. Answer all the questions on the worksheet or online. OPTIONAL EXERCISE: Your instructor may have you write a short report (accompanied by a sketch) summarizing your conclusions and predicting how seasonal shifts would affect people.

Air Pressure

These two globes show average air pressure over the Atlantic and adjacent land areas during the months of January and July. Lighter gray indicates relatively high pressures, whereas darker gray indicates low pressures. The lines encircling the globe are the equator and 30° and 60° (N and S) latitudes.

Observe the main patterns on these two globes, noting the positions of high pressure and low pressure and how the positions, shapes, and strengths change between the two seasons. Then, complete the steps described in the procedures section.

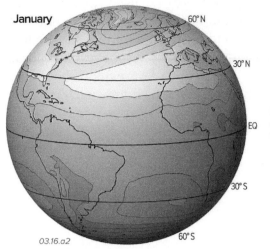

03.16.a2

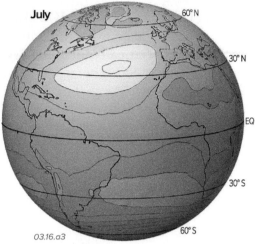

03.16.a3

Wind Velocity

These globes show average wind velocities for January and July. The arrows show the directions, while the shading represents the speed, with darker being faster. In this exercise, the directions are more important than the speeds, but both tell part of the story.

Observe the patterns, identifying those that are related to global circulation (i.e., westerlies) versus those that are related to more regional features, such as the Bermuda-Azores High and the Icelandic Low. Note also the position of where winds converge (ITCZ) along the equator and how this position changes between the two months.

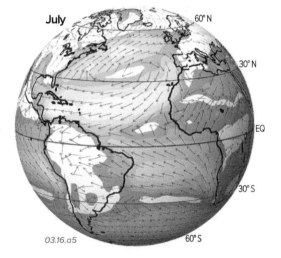

03.16.a4

03.16.a5

Cloud Cover

The clouds that form, move above Earth's surface, and disappear can be detected and tracked with satellites, shown here for January and July of a recent year. On these globes, light colors that obscure the land and ocean indicate more abundant clouds (and often precipitation), whereas the land shows through in areas that average fewer clouds.

Observe the patterns, noting which areas are cloudiest and which ones generally have clear skies. Relate these patterns of clouds to the following: amount of vegetation on the land, air pressure for that month, and average wind directions. Answer all questions on the worksheet or online.

03.16.a6

03.16.a7

3.16

4 Atmospheric Moisture

MOISTURE IN THE ATMOSPHERE, in the form of water vapor, liquid water, and ice, controls most aspects of our weather and climate. Moisture moves back and forth from Earth's surface to the atmosphere and, once in the atmosphere, is transferred vertically and laterally by moving air. Atmospheric moisture is expressed as clouds, precipitation, storms, weather fronts, and other phenomena. In this chapter, we explore how and where moisture occurs in the atmosphere, what happens when moisture moves with air currents, and how such motions are expressed as clouds, precipitation, and other aspects of weather.

Examine this central figure and observe the various features. Try to identify which features involve water in one form or another. Next, consider how moisture can move from one place to another, and from one state of matter (gas, liquid, solid) to another. After you have done this for each main feature shown, read the text surrounding the figure.

04.00.a2 Cayman Islands

Our world has various types of clouds, some of which are shown here in the large figure and the photograph above. Some clouds are thin and wispy, whereas others are tall and puffy. Some are high in the troposphere, while others are close to Earth's surface. Some clouds are associated with precipitation (rain, snow, hail, or sleet), but most are not.

What is in a cloud, how do we classify and name different types of clouds, and what do different types of clouds tell us about what is going on in the atmosphere?

04.00.a3 Plush, OR

Condensation

Precipitation

Evapo-transpirati

Evaporation

Runoff

Most clouds and precipitation are caused by cooling of moist air. Such cooling generally occurs when air rises into cooler parts of the atmosphere. Cooling of moist air can also occur when warmer air interacts with cooler land or water. As moist air cools, water vapor can change into a liquid (water drops) through the process of *condensation*. If cold enough, water vapor can instead form ice crystals through the process of *deposition*. Also, water drops and ice crystals can form from one another. In any case, the water drops and ice crystals are what form clouds and, under the right conditions, cause various types of precipitation (◄).

04.00.a1

For each place where a cloud is shown in the large figure, what are some possible reasons why the air might be cooling at that location?

TOPICS IN THIS CHAPTER

The amount of moisture in the atmosphere varies laterally from place to place and vertically between different heights in the atmosphere. We commonly use the term *humidity* to express how much moisture is in the atmosphere.

In this figure, predict which areas would have more humidity compared to other areas. How would this extra humidity be expressed in the atmosphere, and how would increased humidity affect your weather?

Insolation

Evaporation

Evaporation

Ocean Currents

04.00.a4 Costa Rica

Water moves from the surface to the atmosphere through the process of *evaporation*, where liquid water becomes water vapor. Most evaporation occurs over the oceans, but significant amounts also occur over lakes, streams, and other surface water. Water also gets into the atmosphere from moisture released by moist soil and by plants through their leaves to the atmosphere, a process called *transpiration*. The combination of plant transpiration and evaporation from Earth's surface is *evapotranspiration*. The released water vapor is invisible but can condense into visible mists (◄).

On the large figure, examine the surface, both the water and land, and identify places where moisture could move from the surface to the atmosphere.

Expressions of Moisture

The large figure on these two pages shows a wide variety of ways in which moisture is expressed on the surface and in the atmosphere. Water is easily identifiable in the large body of water (ocean) on the right, in the two lakes on land, and in the streams. This water is shown in liquid form, but under cold-enough circumstances there could also be ice, such as if the lakes were at least partially frozen or if the ocean had some sea ice (frozen seawater) or icebergs (large floating chunks of ice). We can easily recognize water in its solid form (ice) in the snow on the high mountains. There would also be a zone of especially moist air (more water vapor) over and adjacent to the ocean, lakes, streams, and wet parts of the land, but we cannot see it because water vapor is not visible.

Even less obvious is the water contained in the green plants and the underlying soil. This moisture moves from these sites to and from the atmosphere, such as rain that falls from the clouds, soaks into the soil, and is taken up by plants, which can then release this moisture back into the atmosphere via transpiration.

Moisture in the atmosphere is expressed in the various types of clouds. Most clouds are visible to us because they contain tiny drops of liquid water—even if there is no rain associated with the cloud. The wispy, higher-level clouds consist of ice crystals, which are easily blown and streaked out by the strong upper-level winds. The higher parts of some of the taller lower-level clouds would also contain ice, which can remain aloft or eventually accumulate into such large particles that it falls to the surface as snowflakes or hail. In all, many aspects of our planet owe their existence to moisture in the atmosphere and to the exchange of moisture between the atmosphere and the surface.

4.0

4.1 How Does Water Occur in the Atmosphere?

THE PRESENCE AND ABUNDANCE OF WATER in the atmosphere are a fundamental control of weather and climate, which both have a profound influence on our lives. The molecular structure of water causes it to have special properties that we can observe every day and that are important to life on Earth. In what forms does water occur in the atmosphere, and how did the water get there?

A How Is Water Expressed In and Near the Atmosphere?

Examine this photograph of the Himalaya in Tibet (▶) and identify all the places where there are visible expressions of water. Are there places where water is likely to be present but not visible? Ponder this for a moment before reading on.

04.01.a1 Himalaya, Tibet

Water in the atmosphere, as on the Earth's surface and in its subsurface, occurs in three forms: as a gas (water vapor), a liquid (liquid water), and a solid (ice). In this scene, liquid water in the atmosphere is expressed as tiny drops in the clouds and as any raindrops falling from the clouds. Ice crystals, expressed as snow, are on the surface, but are also present as small crystals in some of the clouds. Water vapor is also present, but is not obvious—it always occurs as an invisible gas. Clouds, like the rest of the atmosphere, contain water vapor but only the airborne drops and ice crystals are visible.

Tiny water drops also form fog and mist, such as on the Galápagos Islands (▶). Such moisture sustains unusual plants that extract moisture directly from the atmosphere or from the liquid water (dew) that forms on leaves and other hard surfaces, including on the back of the Giant Tortoise (~1 m across) grazing on vegetation sustained by the mist.

04.01.a2 Galápagos Islands

B What Are Some Important Properties of Water?

Recall that water (H_2O) is a molecule composed of one oxygen atom strongly bonded to two hydrogen atoms. The asymmetrical arrangement of the hydrogen atoms causes the molecule to have a positive charge on the side with the hydrogen atoms and a negative charge on the opposite side, near the oxygen atom. A molecule with this charge distribution is said to be *polar*, and the polar nature of water is why it is such a good solvent (can dissolve other substances). This polar character has many other implications.

Hydrogen Bond

04.01.b1

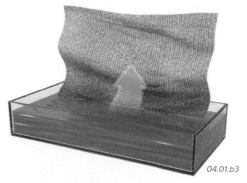

04.01.b3

Surface tension allows water to attach itself to other objects, such as this damp cloth (▲). Note that the moisture has climbed up the cloth, higher than the level of the water. This ability of water to travel upward within small spaces is called *capillary action*. Capillary action is important in the upward motion of liquid water in soil, drawing soil water up toward the surface where it can evaporate. It is also important in plants, allowing water to rise from the roots through the branches and into the leaves.

The positive side of one water molecule is attracted to the negative side of an adjacent water molecule, forming a weak bond (i.e., a *hydrogen bond*) that tends to keep adjacent water molecules together. This attraction is what causes water to tend to stay together as a discrete drop, rather than flowing away (▶). This tendency for water to stay together with a discrete outer surface is called *surface tension*. Surface tension has to be overcome during evaporation, because it tends to hold water molecules within the liquid rather than letting them escape into the air. Surface tension also must be overcome to allow small drops of water to form in the atmosphere, as in clouds, and to allow these small drops to combine into larger drops, forming a raindrop.

04.01.b2

C How Do Water Molecules Move Between Liquid Water and Water Vapor?

Water molecules move between the three states—gas, liquid, and solid—in response to changes in the energy of the system, mostly to changes in thermal energy from insolation (or lack thereof). What actually occurs during such changes in state at the level of individual molecules? Here we take a closer look at the movement of water molecules between liquid and vapor, focusing on how the energy levels of the molecules instigate change.

1. This figure shows a totally closed container of water and air, with water molecules in each state color coded for their energy levels, with yellow, orange, and red representing highest energy levels, and green, blue, and purple representing lower energy levels. Based on the colors in the container, molecules of water vapor (in the air) tend, on average, to have higher energy levels than water molecules in the liquid.

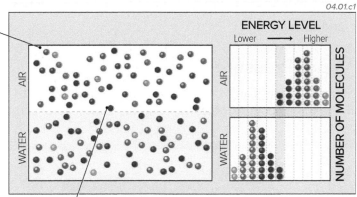

04.01.c1

3. The graphs (histograms) on the right side of the figure display the frequency of molecules at various energy levels in the air (upper histogram) and water (lower histogram). In general, energy levels are higher in the air, as expressed by the peak of the air histogram being farther to the right (toward higher energy levels) than the peak for the water (liquid) histogram. This higher energy of the vapor is largely because of the *latent heat of vaporization*, which is energy the molecules gained primarily from insolation as they moved from liquid to vapor states. The critical energy level where changes in state (evaporation or condensation) occur is indicated by the dashed, vertical gray line common to both graphs. This system is in *equilibrium* when as many molecules are changing from liquid to gas as are moving in the opposite direction.

2. Note, however, that some molecules in the air have similar energy levels (purplish red on this figure) to those in the water. If the energy levels of vapor molecules are sufficiently low, these low-energy molecules will condense and join the liquid. Similarly, some molecules of the liquid will have sufficient energy to evaporate and pass into the vapor phase.

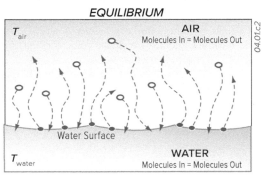

04.01.c2

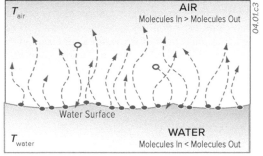

04.01.c3

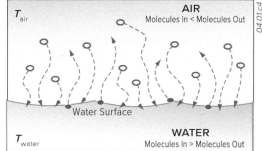

04.01.c4

4. Due to the overlap in energies of molecules in the liquid and vapor, molecules are constantly moving from one state to the other. This figure illustrates that some molecules in the liquid attain high enough energy states to escape the water surface and become a molecule of water vapor (evaporation). In contrast, some vapor molecules will drop low enough in energy levels that they will join the liquid (condensation). In equilibrium there is an equal exchange of water molecules between the liquid and vapor.

5. If the entire system is at higher overall energy levels, such as when the water and air are heated by the Sun or on a stove, many molecules in the liquid become more energetic, reaching energy levels high enough to allow them to escape into the air (i.e., evaporate). Fewer gas molecules in the air have low enough energy levels to condense into liquid. As a result of an increase in the energy of the system, increased evaporation causes the number of gas molecules (water vapor) to increase, while the liquid water loses mass.

6. If the system has lower overall energy levels, as when it is cooled, more gas molecules condense into liquid, while fewer molecules in the liquid evaporate. Since more water molecules condense onto the water surface than evaporate from it, the liquid gains mass. In contrast, the gas loses water molecules, causing a decrease in the amount of water vapor in the air. This transfer from vapor to liquid occurs when water drops condense on the outside of a cold glass or beverage can.

Movement of Water Molecules Into and Out of Ice

Similar processes occur when water molecules move between ice (the solid state) and liquid and gaseous states. When ice and liquid water are in contact, the energy levels of some molecules in the ice will overlap with some of those in the liquid. Some molecules will move from ice to water (*melting*) and others will move in the opposite direction (*freezing*). Likewise, when ice and vapor are in contact, some molecules in the vapor become solid ice (*deposition*), whereas some molecules in the ice move into the vapor (*sublimation*). How many move in each direction, and the resulting gains and losses of mass, depend on the overall energy level of the system. If there is equilibrium, the same number of molecules will move in opposite directions, but generally the ice, liquid, or vapor is losing mass to one of the other states.

Before You Leave This Page

✓ List some ways water is expressed in the environment.

✓ Sketch and describe the molecular structure of water and how it imparts important properties to water.

✓ Sketch and describe how overlap of energies causes molecules to move between liquid and vapor phases.

4.1

4.2 What Is Humidity?

THE AMOUNT OF WATER VAPOR in the air is referred to as *humidity*. Humidity is something we can sense, affecting whether the air feels humid or dry. We are most familiar with one measure of humidity — *relative humidity*, a term commonly used on daily weather reports. There are other measures of humidity, some of which are more useful for comparing the amount of moisture between different elevations, times, and regions. Understanding humidity leads to a much better understanding of weather and climate.

A What Are Humidity and Vapor Pressure?

The atmosphere is nearly all nitrogen and oxygen, but it contains a small but variable (<1–4%) amount of water vapor (and other gases). The term *humidity* conveys the amount of water vapor in the atmosphere, and the amount of humidity can be represented in several ways.

Humidity and Vapor Pressure

1. Imagine two cubes, both the same size and partially filled with water but mostly filled with air (▶). In both cubes, some water molecules have evaporated from the liquid, becoming water vapor in the air. The amount of water vapor in the air is the humidity of the air. In this figure, water vapor is represented by small blue dots in the air.

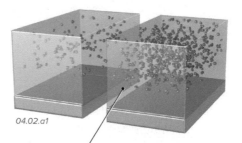

04.02.a1

2. In the example shown here, the cube on the right has more water vapor than does the left cube, so the air in the right cube has a higher humidity than does air in the left cube. If we captured and weighed all the water vapor in either cube, we could determine how much water vapor was in that volume of air. Such a measure, called *absolute humidity*, is in units of mass per volume, such as grams per cubic meter (g/m³). Since this value changes as air expands or compresses (changes in volume), as commonly occurs in the atmosphere, absolute humidity is not a very useful measure in understanding the atmosphere.

3. Recall that the weight of the atmosphere pushing down causes atmospheric pressure. Some amount of this pressure is due to the weight of the water vapor that is part of the atmosphere, and this provides us with another way to represent humidity. The amount of pressure contributed by the water vapor is, not surprisingly, called the *vapor pressure*. Vapor pressure is expressed using the same units as atmospheric pressure, such as millibars (mb). For a given volume of air, if there are more water vapor molecules present, then the vapor pressure is higher. In meteorology, vapor pressure is represented by the small letter *e*.

Water-Vapor Capacity

4. Air can include only a limited amount of water vapor, and the maximum amount it can include is called its *water-vapor capacity* (commonly depicted by e_s, where the *s* stands for saturation). The water-vapor capacity of air varies with the temperature of that air, as shown in the graph below. The blue curve represents the maximum amount of water vapor the air can contain at different temperatures, with the water-vapor capacity increasing exponentially with increasing temperature — warmer air has a greater capacity for water vapor than does cooler air. Water-vapor capacity can be expressed in millibars (mb).

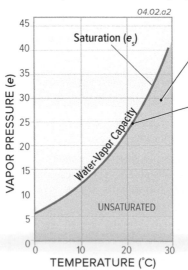

04.02.a2

5. If the conditions of temperature and vapor pressure for some volume of air plot in the shaded area on this graph, the air can hold even more water vapor — the air is *unsaturated*. These are the conditions we normally experience when the weather is fair (not raining or snowing).

6. If the conditions plot directly on the blue line, the air has as much water vapor as possible — it is *saturated* with respect to water vapor. Under these conditions, the water vapor begins to form droplets via the process of condensation (or ice crystals via the process of deposition if under colder conditions). In other words, when the atmosphere reaches saturation, drops of water form, such as in clouds and fog, perhaps followed by precipitation. In certain settings, conditions can be slightly above the blue line, and the air is said to be *supersaturated*.

Relative Humidity

7. A common way to convey humidity is to express it as a ratio of how much water vapor is in the air relative to the maximum amount of water vapor that is possible (i.e., the water-vapor capacity). In the graph below, the green dot represents a vapor pressure of 20 mb and a temperature of 29°C. At that temperature, saturation would occur at about 40 mb, so the vapor pressure (humidity) represented by the green dot is half, or 50% of the maximum possible. The atmosphere is unsaturated in this case. This percentage, representing the observed vapor pressure divided by the maximum possible vapor pressure (i.e., water-vapor capacity), is the *relative humidity*, the term we hear on daily weather reports. Relative humidity varies from single digits (e.g., 5%) in very dry places, such as dry seasons in a desert, to 90% to 100% in very humid conditions. We can feel the difference!

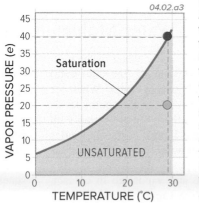

04.02.a3

B   What Is Specific Humidity?

Relative humidity helps describe how humid the air feels to us and indicates how close the air is to saturation, but relative humidity varies with temperature, even if the amount of water vapor in the air has not changed. Therefore, we also use another measurement of humidity, called *specific humidity*, which compares the mass of water vapor in a body of air to the total mass of that air. An advantage of this approach is that a specific-humidity measurement is not affected by changes in temperature, pressure, and volume that occur when air masses move, such as up or down in elevation. Specific humidity, therefore, is a convenient way of comparing the amount of moisture at different elevations, at different times, and from region to region.

1. Specific humidity is expressed as the ratio of the mass of water vapor in some body of air to the total mass of that air. Specific humidity is represented by a lowercase q and is calculated in the example below.

2. This cube (▶) contains molecules of water vapor (shown as blue dots) dispersed through a mass of atmosphere (filling the rest of the cube). In this example, the mass of all the water vapor is 12 grams (g), and the total mass of the air (including the water vapor) is 1 kilogram (kg). The specific humidity is therefore the following ratio:

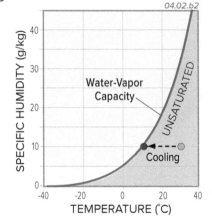

04.02.b1

mass of water vapor/mass of air = specific humidity (q)

12 g / 1 kg = 12 g/kg

3. The calculated specific humidity of 12 g/kg is a typical value for the air (1.2%). We use the units of g/kg because the amount of water vapor is small compared to the total amount of air.

4. Note that since specific humidity is calculated with the *masses* of water vapor and air, it is not directly dependent on the temperature, even though places with different temperatures often have very different specific humidities. It is also largely independent of changes in volume (and therefore pressure)—even if we shrink the cube it will still contain the same ratio of the mass of water vapor to total mass of air. The specific humidity of an air mass therefore changes only slightly as that air warms, cools, or changes pressure as it moves in the atmosphere. The specific humidity can only be changed by adding or subtracting water vapor, such as adding water vapor through evaporation of surface waters or by losing water vapor through the formation of precipitation.

Specific Humidity and Water-Vapor Capacity

5. This graph plots specific humidity and temperature, with the blue line showing conditions where air is saturated with water vapor. The curve has the same shape as the curves on the other page because it also represents water-vapor capacity. It curves strongly upward with higher temperatures. The curve does not show how specific humidity of an air mass varies with temperature, only the conditions under which it becomes saturated. The shaded field represents conditions where air is unsaturated.

6. At low temperatures, represented by the left part of the graph, much less water vapor can exist than when air is warmer (the right part of the graph).

04.02.b2

Graph: SPECIFIC HUMIDITY (g/kg) vs TEMPERATURE (°C). Water-Vapor Capacity curve, UNSATURATED region shaded, Cooling arrow.

7. The green dot represents air that has a specific humidity of 10 g/kg and a temperature of 30°C. If we cool the air (moving left on the graph), the air can reach saturation (the red dot), but note that the specific humidity does not change, because it is a ratio of masses. Seen in this way, the air's ability to hold more moisture does change with temperature (the blue curve), but specific humidity tells us the concentration of water vapor in the air, independent of temperature. For this reason, specific humidity is very useful for comparing the moisture content of air in different settings. If air is unsaturated, it allows more evaporation, and the amount of evaporation it allows increases with higher temperatures.

Comparing Different Measures of Humidity

8. The table below lists water-vapor capacity, specific humidity, and relative humidity for air at four different temperatures. Note that the specific humidity does not change as the air cools, but the relative humidity increases until it reaches 100% at 10°C (i.e., the air becomes saturated).

Temperature (°C)	Water-Vapor Capacity (g/m³)	Specific Humidity (g/kg)	Relative Humidity
40	50	10	20%
30	28	10	36%
20	15	10	67%
10	10	10	100%

Before You Leave This Page

✓ Describe humidity, vapor pressure, and relative humidity, using sketches or graphs.

✓ Describe specific humidity, how it is calculated, and whether it changes with temperature.

✓ Contrast relative and specific humidity and when each is useful.

4.2

4.3 How Does Specific Humidity Vary Globally and Seasonally?

HUMIDITY OF THE AIR VARIES greatly from region to region and between different altitudes in the atmosphere. It also varies from one time to another, such as between different seasons. To compare different regions, altitudes, and seasons, we generally use specific humidity, which expresses the amount of water vapor in the air, independent of variations in temperature. From what you know about the Earth, which regions do you think have the highest specific humidity and which ones have the lowest? How do you think these will vary between seasons?

A How Does Specific Humidity Vary Globally?

Specific humidity offers some advantages for comparing moisture of various places with different temperatures. The globes below depict specific humidity averaged over an entire year. On all the globes and maps on these next two pages, blues and greens represent higher values of specific humidity, whereas tan represents low values. Observe the globes below, identifying areas that have unusually high or low specific humidity, and consider possible explanations for these humid and dry places, respectively.

04.03.a1

1. The main pattern that emerges from inspection of these globes is that specific humidity varies primarily as a function of latitude—areas near the equator have much higher specific humidity than areas farther north and south (toward the poles).

2. A belt of high specific humidity straddles the equator, coinciding with the tropics. This region receives, on average, the most insolation, which in turn causes a relatively high amount of evaporation and higher content of water vapor in the air. Also, warmer air has a relatively high capacity for water vapor (recall the blue curves from the previous pages).

3. The lowest values are near the poles, where cold temperatures and a cover of ice over the surface limit the amount of evaporation. Also, cold air has a lower capacity for water vapor. For all these reasons, there is less water vapor in polar regions than at lower latitudes.

4. The patterns of specific humidity are less complicated over the oceans than over land. The patterns over the oceans do not exactly follow latitude, mostly because of the influence of ocean currents that move warm and cold water north and south along the edges of continents. In the Atlantic Ocean, the warm Gulf Stream Current brings warm water northward along the East Coast of North America. The northward flow of warm water is accompanied by a northward expansion of moderate humidities toward Europe.

5. Patterns of specific humidity are more complicated on land, mostly reflecting the influence of topography, especially large mountain belts that interfere with prevailing winds, such as the Andes of South America.

SPECIFIC HUMIDITY (g/kg)

| 3 | 6 | 9 | 12 | 15 | 18 | 21 | 24 | 27 | 30 | 33 |

6. Very low specific humidity also characterizes land areas along the subtropics, including the northern part of Africa. This region is the site of the Sahara Desert, the world's largest desert. We commonly associate deserts with dry air, and this is indeed reflected in the low values of specific humidity. The low humidity continues eastward across deserts of the Arabian Peninsula and onto southern Asia.

7. In the ocean west of southern Africa is another bend in the patterns of humidity. Lower specific humidities extend to the north offshore of the west coast of southern Africa and then bend westward into the South Atlantic Ocean. Think about what might cause this before reading on.

8. As you probably surmised, this pattern off southern Africa reflects another ocean current, but this time a cold current that brings cold water north, accounting for the lower humidity.

9. The annual average of specific humidity represented on these globes does not tell the whole story. In some regions, there are huge seasonal variations in humidity as the prevailing wind directions shift, such as in association with a monsoon. Before we explore these seasonal variations on the next page, think about how the large patterns on both these globes might change from January (the northern winter and southern summer) to July (northern summer and southern winter).

04.03.a2

B How Does Specific Humidity Vary from Season to Season?

Specific humidity varies from one season to the next, as represented by the two maps below, which show average specific humidities for the months of January and July. On both maps, note the values for the region where you live, and think about how this relates to the typical weather you experience at these two times of year.

January

1. In winter, specific humidity is very low at the poles, because there is so little energy available to evaporate surface water, most of which is frozen anyway.

2. In the mid-latitudes, specific humidity values are fairly low in winter, largely because the moderately cool temperatures limit the amount of evaporation.

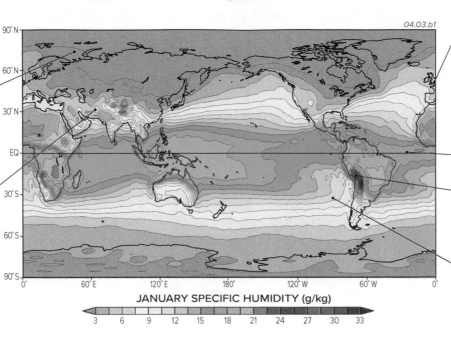

JANUARY SPECIFIC HUMIDITY (g/kg)

3   6   9   12   15   18   21   24   27   30   33

3. Where water is significantly warmer than the air above it, such as in the British Isles and over the mid-latitude oceans, the warm, moist air adjacent to the water will rise more because it is less dense. This increased air turbulence allows for more water to evaporate, thereby increasing the humidity.

4. The most humid regions are tropical and equatorial surfaces.

5. Large topographic features, like the Andes Mountains of South America, influence regional patterns as the southeast trade winds push moist air up against the east side of the Andes.

6. The Southern Hemisphere is mostly water, so specific humidity is generally high, particularly in January, the heart of the Southern Hemisphere summer.

July

7. The overall pattern is fairly similar for July, even though winter and summer have switched between the Northern and Southern Hemispheres. As in January, the highest values of humidity are mostly in the tropics and the lowest values are near the poles.

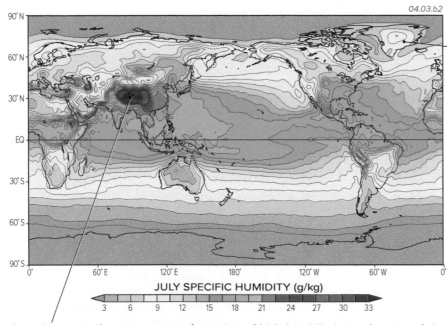

JULY SPECIFIC HUMIDITY (g/kg)

3   6   9   12   15   18   21   24   27   30   33

9. In July, during the Northern Hemisphere summer, humid air reaches farther northward in response to the increased summer temperatures and the resulting increase in evaporation.

10. At this time, during the Southern Hemisphere winter, desert regions of the Southern Hemisphere, such as central Australia and southern Africa, have much lower humidity than in January. This is due to the sinking motion from subtropical high-pressure areas, which brings drier, subsiding air from aloft down to the surface.

11. Because the Southern Hemisphere has much less land than the Northern Hemisphere, it appears more similar between January and July than does the Northern Hemisphere.

8. A striking difference from January is the appearance of a region of high humidity in southeastern Asia. This is related to the Asian monsoon, where a seasonal change in prevailing wind direction brings warm, moist air from the adjacent oceans northward into Asia, spreading humidity across India, China, Korea, Japan, and large areas of southeastern Asia.

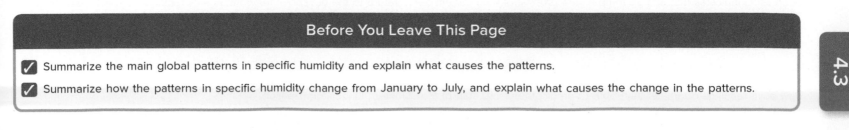

Before You Leave This Page

☑ Summarize the main global patterns in specific humidity and explain what causes the patterns.

☑ Summarize how the patterns in specific humidity change from January to July, and explain what causes the change in the patterns.

4.3

4.4 What Is the Dew Point?

ANOTHER USEFUL MEASURE OF HUMIDITY is the *dew point*, the temperature to which a volume of air must be cooled to become saturated with water vapor. If the air temperature is at the dew point, the air is so saturated with water vapor that vapor begins to condense as drops of liquid water, such as drops in clouds or rain, or as water drops (dew) on solid surfaces. When you wake in the morning and everything outside is covered with moisture (dew), it signifies very humid air and that night-time temperatures cooled to the dew point.

A How Is the Dew Point Expressed?

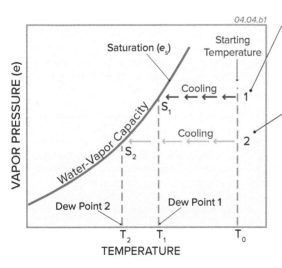
04.04.a1 Colorado

◄ *Dew* is expressed as drops of liquid water that condense out of the atmosphere and onto plants, rocks, walls, or any solid surface. It typically forms at night, in response to the air cooling. If temperatures are at or below freezing, earlier formed drops of dew can freeze or ice can form directly from the air (deposition). In any case, drops of dew or crystals of ice generally disappear rapidly once the Sun rises and air temperatures begin to climb.

▶ To explore the formation of dew, examine the sealed glass container on the left. It contains liquid water and water vapor (represented as blue dots). The container is warm enough that all the available water vapor can exist in the air without condensing—that is, the air is unsaturated.

04.04.a2

If we cool that same container with ice cubes, the temperature of the water and air will decrease. Recall that cold air has lower water-vapor capacity than warm air, so as the air cools it moves closer to saturation. At some temperature, the cooled air reaches saturation and can no longer contain so much vapor. As a result, the water begins to condense as drops on the inside of the container, forming dew. As we know from our experience with cans or glasses of an icy beverage, drops of dew can also form on the outside of the container, especially if the surrounding air has high humidity.

B What Is the Dew-Point Temperature?

1. To better visualize why cooling can cause the formation of dew, we return to a familiar graph, one that plots vapor pressure (e) versus temperature (▼). The curved blue line marks conditions where air is saturated in water vapor; it is labeled as e_s, where the s indicates saturation.

04.04.b1

2. We begin with an air mass at a certain temperature (T_0) and vapor pressure, marked as position 1, well within the unsaturated field of the graph. As this air cools, its position on the graph moves horizontally to the left (temperature is changing but vapor pressure is not), as indicated by the reddish arrows. It cools until it reaches saturation at position S_1, at which point the air has a relative humidity of 100%. The temperature at which saturation occurs is the *dew-point temperature* and can be read from the graph by drawing a vertical line down from S_1 to the temperature axis (T_1).

3. A second air mass, at position 2, starts at the same temperature (T_0) but has less water vapor in it (lower vapor pressure) than does air mass 1. If the air cools, as indicated by the green arrows pointing to the left, it also approaches the saturation curve. In order to reach saturation (at point S_2), however, it has to cool more than air mass 1. Air mass 2 has less water vapor than air mass 1, and as a result has a lower dew-point temperature (T_2).

4. To generalize, humid air at a given temperature has a higher dew-point temperature than does less humid air at the same temperature. If air is very humid, for example with a relative humidity of 80% to 90%, it may need to cool only a few degrees to reach its dew point. In contrast, air that is very dry (like the 5% to 10% relative humidities common in deserts) has to cool significantly to reach its dew point, and on most nights does not.

5. The difference between current air temperature and the dew-point temperature is known as the *dew-point depression*. A large dew-point depression indicates that a large temperature drop—from the current temperature to the dew-point temperature—is required to reach saturation. Thus, large dew-point depressions suggest dry air, whereas small dew-point depressions suggest humid air and that only a minor drop in temperature will cause saturation. Dew-point depression is used in some agricultural and forestry applications and tells us how much temperatures might drop overnight, since air temperatures at night generally do not cool much below the afternoon's dew point.

6. Using dew-point temperature as an index of moisture overcomes most of the difficulties of using relative humidity, because dew-point temperature does not change as much through the course of a "normal" day as relative humidity does. In the Arizona desert, weather forecasters use dew-point readings of greater than 55°F to predict the official start of the summer thunderstorm season.

C Which Parts of the U.S. Have the Highest and Lowest Dew Points?

Dew-point temperature is an important measure of the humidity of air, as well as a predictor of dew and precipitation. The maps below show average monthly dew-point temperatures and air temperatures across the conterminous U.S. for January and July. Blue colors represent low (cold) dew-point temperatures, whereas orange and red indicate high dew-point temperatures. While observing and comparing the two maps, identify areas that are unusually high or low and consider possible explanations for the patterns you observe. Find the values for the area where you live or would like to visit and think about how this relates to the climate there and how humid it feels at different times of the year.

January

1. Extremely low dew-point temperatures, as characterize the northern states in January, generally mean that the air is so dry that clouds do not form readily. Without cloud cover, night-time radiational cooling is intense, making January temperatures even more frigid.

2. Topography, such as mountain ranges in western North America, block moisture from penetrating to some regions, leading to low dew-point temperatures in the desert Southwest.

3. This map (▶) shows normal January air temperatures, averaged over the entire month. The warmest areas are in the Southwest and South Florida.

AVG. DEW-POINT TEMPERATURE (JANUARY)
04.04.c1

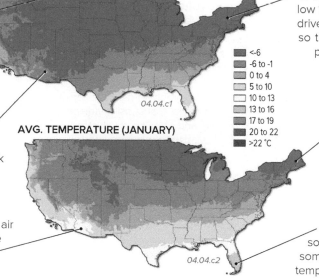
AVG. TEMPERATURE (JANUARY)
04.04.c2

<6
-6 to -1
0 to 4
5 to 10
10 to 13
13 to 16
17 to 19
20 to 22
>22 °C

4. January dew-point temperatures are extremely low in northern areas. Little insolation is available to drive the evaporation process under such conditions, so there is limited moisture in the air. When dew-point temperatures are below 0°C, they are called "frost-point" temperatures, and ice forms instead of dew.

5. Dew-point temperatures in eastern North America generally are not far below the air temperature, so air there does not have to cool very much to become saturated. Thus, clouds form even if there are no weather fronts. This is why many eastern states have overcast days in the winter.

6. January temperatures in the Southeast, especially southern Florida, are more moderate, so there is enough energy available to evaporate some water. This results in relatively high dew-point temperatures compared to the rest of the country.

July

7. Summer dew-point temperatures are much higher than in the winter. Increased insolation allows more energy to drive up temperatures and cause more evaporation, which increases humidity.

8. Areas in rain shadows of the moisture sources, such as the Intermontane West, have dew points that remain far below their air temperatures shown on the lower map. A slightly larger amount of moisture flows into southern Arizona from the Gulf of California in the summer, as represented by the small orange spots on the southwestern edge of this map.

9. Summer dew-point temperatures are high in the Southeast, as depicted by the orange colors that fringe the Gulf Coast in the dew-point map for July. This high humidity, combined with high summer air temperatures, creates an oppressive summer climate. Strong clockwise flow around the high-pressure system near Bermuda (the Bermuda-Azores High) pumps moisture from the Gulf of Mexico, Caribbean Sea, and Atlantic Ocean deep into the heart of North America. There are no topographic barriers that would interfere with the movement of this moist air inland, so moisture from the ocean flows onto land, raising dew-point temperatures.

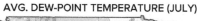
AVG. DEW-POINT TEMPERATURE (JULY)
04.04.c3

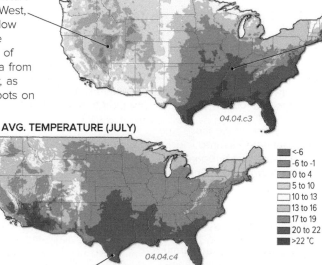
AVG. TEMPERATURE (JULY)
04.04.c4

<6
-6 to -1
0 to 4
5 to 10
10 to 13
13 to 16
17 to 19
20 to 22
>22 °C

10. Notice the steep gradient in dew-point temperature across the Great Plains, from moderately low dew points in the plains of eastern parts of New Mexico, Colorado, Wyoming, and Montana, to much higher dew points toward the Mississippi River. This gradient mostly reflects the northward flow of moisture from the Gulf of Mexico.

11. We can use the combination of these two maps to identify which parts of the country are hot and humid during the summer and those that are cool and less humid—useful in planning a summer vacation!

Before You Leave This Page

✓ Describe what the dew point represents, using a graph to show how cooling can cause air to reach its dew point (saturation).

✓ Describe the general patterns of dew point in the U.S., identifying some of the factors that contribute to high and low dew-point variations from place to place and winter to summer.

4.4

4.5 What Happens When Air Rises or Sinks?

THE ATMOSPHERE IS A DYNAMIC ENVIRONMENT, with upward, downward, and sideways motion. What happens to air that rises and encounters different temperatures and pressures? What happens when air sinks? The vertical motions of air cause clouds, precipitation, and many other weather phenomena.

A At What Rate Does an Air Parcel Cool As It Rises?

To explore the vertical motions of air, it is useful to track what happens to a discrete mass of air, called a *parcel* (used in the same way we refer to a lot of subdivided land as a "land parcel"). As we might expect, air changes temperature as it rises or sinks, largely in response to changes in air pressure and volume that accompany vertical motions through the air. Let's examine what happens as we follow a rising parcel of air, beginning with text at the bottom left (number 1).

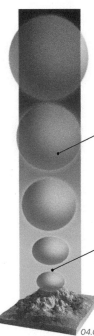

3. When there is no energy exchanged between the parcel and its surroundings, as in the case shown here, the process is said to be *adiabatic*. When such an exchange does occur, the process is *diabatic*. Increasing insolation of the surface as noon approaches is an example of diabatic heating.

2. As the air rises higher, the air pressure continues to decrease. If the parcel does not acquire or lose additional energy with its surroundings, then the decrease in pressure is accompanied by (1) an increase in volume (as shown by the increasing size of the air parcel), and (2) a decrease in temperature.

1. If the parcel of air starts near the ground, its volume is confined by the weight of the overlying atmosphere, which exerts significant pressure. If the air begins rising, there is less air on top, which results in a decrease in air pressure.

04.05.a1

4. If the rising motion can be assumed to be adiabatic, we can calculate how much a rising air parcel cools for a given rise in elevation, a rate called the *lapse rate*. Lapse rate is in units of degrees per vertical distance, usually in C°/km.

5. This graph (▶) plots the lapse rate for air that is rising adiabatically and remains unsaturated. The lapse rate is calculated simply as the change in temperature for the change in height, and reported as a ratio (C°/km). In the case shown here, the air cools 10 C° in one kilometer, or 10 C°/km.

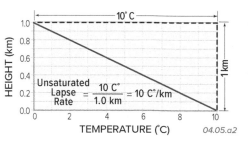

$$\text{Unsaturated Lapse Rate} = \frac{10\ C°}{1.0\ km} = 10\ C°/km$$

04.05.a2

6. If any parcel of air *rises* adiabatically and does not become saturated during its ascent, it always cools at 10 C°/km. This constant lapse rate, called the *unsaturated adiabatic lapse rate*, applies everywhere adiabatically rising air remains unsaturated. By convention, positive lapse rates are for temperatures that cool with height, as shown by the orange line on the graph above.

7. If a parcel *descends* adiabatically, it warms at the same rate that it cooled on ascent—10 C°/km. An adiabatically descending parcel of air will remain unsaturated, because the slightest bit of warming will increase the water-vapor capacity and therefore decrease the relative humidity. If the relative humidity is below 100%, the air is unsaturated.

B How Much Does a Saturated Air Parcel Cool as It Rises?

1. A different situation occurs if the atmosphere is saturated. As a saturated parcel of air rises, the cooling decreases the water-vapor capacity of the air and, since the air is already saturated, the moisture must come out of the vapor via condensation, freezing, or deposition, as illustrated in the diagram below.

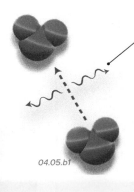

2. As water vapor molecules condense, they release latent heat that warms the local environment. This warming counteracts some of the cooling caused by the expansion of the parcel during adiabatic ascent. As a result, a rising parcel of saturated air cannot cool as much with increasing height as does unsaturated air. The lapse rate for saturated air will be less than the unsaturated adiabatic lapse rate.

04.05.b1

3. The rate at which saturated air cools with adiabatic ascent is called the *saturated adiabatic lapse rate*. This rate depends on temperature because at higher temperatures more water vapor is available for condensation, freezing, or deposition. The figure below plots unsaturated adiabatic lapse rate in red and the saturated adiabatic lapse rate in blue, for different starting temperatures on the surface (on the bottom of the graph).

4. The slope of the lapse rate for unsaturated air (red lines) remains the same, regardless of starting temperature. The lapse rates for saturated air (blue lines), however, vary for different temperatures, as reflected by different slopes of the blue lines.

5. For any starting temperature, the blue line (saturated lapse rates) consistently plots to the right of the red line (unsaturated rate). This means that rising unsaturated air cools at a faster rate than saturated air (which is warmed by latent heat).

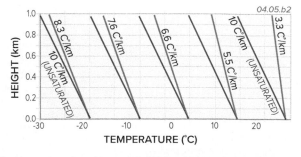

04.05.b2

6. Note that the slopes for the unsaturated and saturated air are only slightly different for low temperatures (not much water vapor), but become more different at higher temperatures (more water vapor). This means that with increasing starting temperatures (farther right on the graph), warmer saturated air parcels cool less with height.

C What Is the Lapse Rate of Air That Is Not Rising or Sinking?

Air adjacent to a rising parcel of air can have a totally different lapse rate than the rising air, and these rates can vary significantly from day to day and place to place, depending on the local weather, climatic setting, and other factors.

1. The temperature change with height for air that is not rising or sinking is known as the *environmental lapse rate*. The environmental lapse rate can be quite variable. On this graph, each green line represents the environmental lapse rate under certain atmospheric conditions. The different lapse rates are expressed as different slopes (inclination of the lines). When dealing with lapse rates, physical geographers call a lapse rate "steep" if the parcel cools a lot with height (as with the left line),

even though that line does not look steep on this kind of graph. Since an environmental lapse rate is for air that is not rising or sinking, it represents the ambient conditions.

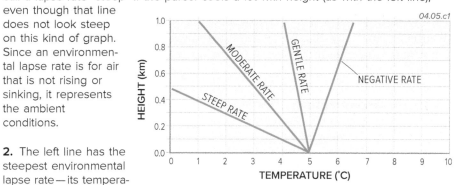

2. The left line has the steepest environmental lapse rate—its temperature cools the most rapidly with height. The air temperature starts at 5°C at the surface but decreases 5 degrees (Celsius) by a height of 0.5 km. This yields an environmental lapse rate of 5 C°/0.5 km, or 10 C°/km.

5. The rightmost line has a *negative lapse rate*. It depicts a situation where temperature *increases* with increasing height. In such situations the environmental lapse rate is said to be *negative* because the temperature cools negatively (i.e., warms) with increasing height. A situation with a negative environmental lapse rate is called a *temperature inversion*.

4. The third line (gentle rate) represents an even gentler lapse rate. This air starts at 5°C at the surface, but the temperature only decreases to 4°C one kilometer higher in the air. This equates to an environmental lapse rate of 1 C°/km. The gradual nature of this change is why this type of rate is called a gentle lapse rate (even though the line looks steep on this graph).

3. The second line from the left has a moderate (less steep) lapse rate. For this lapse rate, temperature decreases more slowly upward in the atmosphere. The air is 5°C at the surface, but temperatures cool to 1°C by one kilometer higher. The change in temperature is therefore 4 C° between the surface and one kilometer in altitude, yielding an environmental lapse rate of 4 C°/km.

D What Do the Different Types of Lapse Rates Indicate About the Stability of Air?

The three types of lapse rates—unsaturated adiabatic lapse rate, saturated adiabatic lapse rate, and environmental lapse rate—each indicate something about conditions in the atmosphere. By comparing the three types of lapse rates, we can predict whether air will rise, sink, or not move vertically. That is, we can tell how stable the air mass is. The rising of air due to its lower density than air around it is called *free convection*, or simply *convection*.

1. This graph compares different types of lapse rates for an adiabatically moving air parcel that starts from the surface at 10°C. The unsaturated adiabatic lapse rate is shown by the red line, and the saturated adiabatic lapse rate is depicted by the blue line. In green are three different environmental lapse rates with varying steepness.

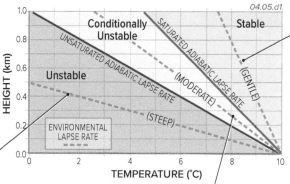

2. The green line in the lower left part of the graph has a steep environmental lapse rate, steeper than either the unsaturated or saturated lapse rates. If an air parcel starts to rise, it changes temperature according to one of the adiabatic lapse rates (red line or blue line, depending on whether it is unsaturated or saturated). Once either type of air has risen even a small amount, it will be warmer and less dense than the surrounding environment at a given height, so it will tend to continue rising. This will occur for either unsaturated or saturated adiabatically rising air if the environmental lapse rate falls within the orange region on this graph. In such situations, any rising air parcel will remain warmer than its surroundings and so will tend to continue rising, producing conditions that are called *unstable* (where air is rising).

3. A different situation occurs if the environmental lapse rate is moderate (the middle green line), plotting between the unsaturated and saturated lapse rates. In this case, if the adiabatically rising air is saturated (blue line), it will remain warmer than its surroundings (green line) and will tend to continue rising. If it is unsaturated, however, it will become cooler than its surroundings (green line is to the right of the red line in this case) and tend to stop rising. For this reason, the situation represented by the yellow area is termed *conditionally unstable*. In this situation, the addition of moisture to the air increases the likelihood for rising motions and other unstable conditions.

4. The rightmost green line represents a gentle environmental lapse rate, where the ambient air cools relatively slowly with increasing height. As a result, an adiabatically rising air parcel (shown by the red or blue lines) would be cooler than the surrounding environment. In this case, the air parcel would cease rising because it is cooler and has a higher density than the surrounding air. At any given height, it will more likely sink, a condition known as a *stable atmosphere*. This will happen any time the environmental lapse rate is more gentle than the saturated rate (the green-shaded area). If the environmental lapse rate is negative (sloping up to the right), it forms a temperature inversion—a very stable atmospheric setting.

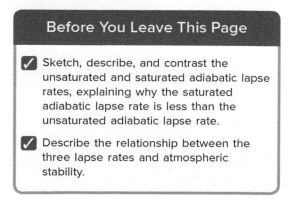

Before You Leave This Page

☑ Sketch, describe, and contrast the unsaturated and saturated adiabatic lapse rates, explaining why the saturated adiabatic lapse rate is less than the unsaturated adiabatic lapse rate.

☑ Describe the relationship between the three lapse rates and atmospheric stability.

4.6 How Does the Surface Affect the Rising of Air?

EARTH'S SURFACE CONSISTS OF A VARIETY OF MATERIALS, including bare rock, soils, forests, cities, water, and ice. Each of these materials responds differently to insolation and to changes in temperature and humidity of the adjacent air. In turn, these surface materials can affect the temperature and humidity of that air. On land, these materials are referred to as *land cover*. Some changes in land cover are natural and others are caused by humans.

A How Does Human Development Impact Local Atmospheric Stability?

The amount of energy reflected and absorbed by different land surfaces varies and depends on several factors, including the type of land cover. Human modification of the land surface, especially changes in land cover, can affect the temperature of the land and overlying air. This can cause the environmental lapse rate to change, influencing whether air rises and affecting the stability of the local atmosphere.

Deforestation

1. Various human activities greatly affect the surface of the Earth and the type of land cover. One ongoing problem is *deforestation*, where native forests and other vegetation are cut down and removed as part of development. In deforestation, dark forest cover and other natural vegetation are often replaced with less heavily vegetated farms with abundant bare soil. The loss of plant cover commonly increases surface temperature by reducing shading and because insolation goes into heating the ground, rather than into evaporating water that was transpired by the forest.
In developed land, vegetated areas are removed and drainage systems remove standing water. Thus, after development, more of the incoming energy is used in sensible heating and less is stored as latent heat.

04.06.a1 West Virginia

Urbanization

2. Another major change in land cover occurs during *urbanization*, where natural lands, farms, and parks are replaced by asphalt, concrete, and buildings during the growth of cities and towns. Many aspects of development, including urbanization, cause a change in albedo (what percentage of insolation is reflected back into the atmosphere) and in the heat-retention characteristics of the land cover. If darker, rougher-textured materials are replaced with lighter colored, smoother-textured ones, the albedo increases, and more insolation is reflected off the surface. Often, however, dark asphalt and other materials more effectively absorb insolation, heating up more during the day and giving back this heat in the evening, raising city temperatures, forming a warmer overall environment, called an *urban heat island* (UHI).

04.06.a2 Houston, TX

Effects of Development on Atmospheric Stability

3. Changes in land cover due to development can result in changes to surface temperatures and therefore to environmental lapse rates. In the figure below, an area of land originally consisted of natural vegetation, with trees and grass, and a small pond. After development, the same area was covered with asphalt, concrete, and bare dirt.

4. Deforestation and urbanization both occurred in this area, so we predict that more insolation will become sensible heat and less will be latent heat.
This should cause more heating of the land surface, resulting in higher surface temperatures, both during the day and especially at night, as part of UHI. If the surface temperature increases but the temperature higher in the atmosphere does not change as much, there will be a steeper environmental lapse rate—a larger decrease in temperature from the surface upward to some height.

Before Development   After Development
04.06.a3

5. We can represent these changes with a graph (▶) showing changes to surface temperatures and environmental lapse rates. The graph shows the environmental lapse rate in green and the unsaturated adiabatic lapse rate in red, for times before develop-ment (Time 1, on the left) and after development (Time 2, on the right). Examine this graph, note the changes between the two times, and consider what the implications might be for this somewhat extreme example (temperature changes this large due to development are rare).

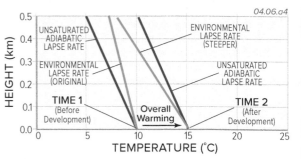
04.06.a4

6. The most obvious change is a shift to the right of the starting points of the two lapse-rate lines for Time 2 (after development). This shift reflects an increase in the local average temperature. Another change is a steepening of the environmental lapse rate, because the surface warmed up a lot, but air farther up did not warm as much; thus, there is a larger change in temperature from the surface to that height (a steep rate). Finally, comparing the slopes of the lapse rates indicates that the atmosphere was stable (rising air was cooler than ambient air) before development but is unstable (rising air is warmer than ambient air) after development. This can change the amount of lifting (vertical motion), which may enhance the likelihood of precipitation if sufficient moisture is present.

B   How Do Water and Ice on the Surface Affect Stability?

Water has different thermal properties than land. Water has a higher *specific heat capacity* than air, meaning it can hold more heat. Water has more *thermal inertia*, which means that it reacts more slowly to changes in temperature imposed on it by the Sun, atmosphere, or land. In general, water surfaces stay cooler than inland areas in summer and remain warmer than inland areas in winter. Warm ocean currents and cold currents can influence the temperatures and stability profile of the overlying atmosphere. Ice and snow also affect surface temperatures, lapse rates, and atmospheric stability.

Warm Ocean Current

1. Warm ocean currents generally flow from lower latitudes into higher latitudes, bringing in water that is warmer than that outside the current (▶). Consider what a warming of the ocean surface due to a warm current might do to conditions in the overlying atmosphere.

2. This figure (▶) shows the environmental lapse rate and the unsaturated adiabatic lapse rate for the atmosphere over a warm ocean current. Near-surface air temperatures are relatively warm because they are heated by the underlying warm water.

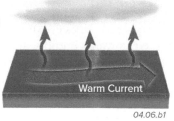

04.06.b1

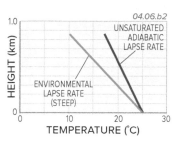

04.06.b2

Cold Ocean Current

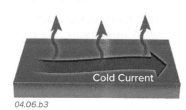

04.06.b3

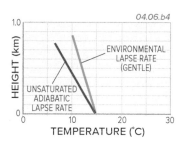

04.06.b4

4. Cold ocean currents generally bring colder water from higher latitudes into warm environments in lower latitudes (◀). We would predict that a cold current would cool the air near the surface, but what other effects will this have?

5. This figure (◀) shows the same types of lapse rates for the atmosphere over a cold ocean current. Near-surface air temperatures are relatively colder because of the underlying cold water, as reflected in a shift of both curves to the left on this graph compared to the previous graph for the warm current.

3. The warm water has less of an effect higher in the atmosphere, so the temperature difference between the near-surface air and higher air is relatively large. This results in a steep environmental lapse rate. The slope of the unsaturated adiabatic lapse rate is not influenced by the water temperature (it is always the same, everywhere). If the environmental lapse rate plots to the left of the unsaturated adiabatic lapse rate, adiabatically rising air will remain warmer than its surroundings and will continue to rise, causing an unstable atmosphere. The rising, moist air can form clouds or rain.

6. The cooling of near-surface air results in a relatively small difference between near-surface and higher air. The environmental lapse rate is therefore gentle (which plots as a steeper line on this type of graph). The relatively gentle environmental lapse rate is now to the right of the unsaturated adiabatic lapse rate, indicating that air rising adiabatically will become cooler (and more dense) than the surrounding air, so it will cease rising. As a consequence, the atmosphere over a cold current tends to be stable. This stability and the limited water-vapor capacity of cold air will tend to inhibit vertical motion, which will restrict vertical cloud development and precipitation.

Snow and Ice on the Surface

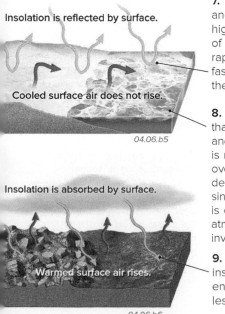

04.06.b5

04.06.b6

7. Snow and ice have a major impact on atmospheric temperatures, and not just because they are cold. Snow and ice both have very high albedo, as high as 90%, which is much higher than the albedo of rocks, plants, and other land cover. Also, snow and ice can appear rapidly, such as during a snowstorm, and can melt away nearly as fast. This appearing and disappearing can result in major changes in the thermal character of the surface over quite short timescales.

8. These two figures illustrate some differences between surfaces that are covered by snow and ice, versus those that are not. If snow and ice cover the ground and any frozen lakes, most of the insolation is reflected, keeping the surface cold. Cold air is dense, so the air over an ice or snow surface tends to sink. This increases the tendency for high pressure to exist over the ice-snow surface. The sinking air can trap cold air near the surface. The air near the surface is cooled by the ice and snow, but the air above it is not, so the atmosphere is stable. Such conditions can result in a temperature inversion, a very stable condition.

9. Once the snow and ice melt, the albedo decreases, allowing insolation to warm the surface. The warmer surface warms the air, ending the inversion, changing the lapse rate, and making the air less stable.

Before You Leave This Page

☑ Explain how human development can affect local surface temperatures and atmospheric conditions.

☑ Sketch and explain how warm and cold ocean currents can affect near-surface atmospheric stability and the amount of clouds and precipitation.

☑ Summarize how snow and ice on land and water can influence atmospheric stability and what happens when the snow and ice melt.

4.6

4.7 What Mechanisms Can Force Air to Rise?

AIR RISES FOR VARIOUS REASONS, some caused by differences in density between the air and its surroundings, and others as a result of externally imposed factors, like a mountain. If an air parcel rises because atmospheric conditions are unstable, *free convection* results, via the processes described in the previous two spreads. If air is forced to rise due to external factors, it is called *forced convection*. Forced convection can occur due to four factors: the orographic effect, low-level convergence, upper-level divergence, and frontal lifting.

A How Do Mountains Affect Rising and Sinking of Air?

Laterally moving air masses interact with topographic features on the land surface, commonly being deflected up and over mountain ranges, ridges, and isolated mountains. As the air is forced upward, it cools and can begin to form clouds and perhaps precipitation. This type of interaction between topography and moving air is called an *orographic effect*.

Interactions between Moving Air and Topography

04.07.a1

04.07.a2

04.07.a3

1. As laterally moving air encounters friction with the surface, the wind is slowed down and faster moving air behind it begins to pile up, forcing some air to rise. A common scenario for this involves a sea breeze, where winds blowing across a large water body slow down as they encounter the more irregular land surface. This can result in clouds forming along coastal mountains.

2. Mountainous areas represent obstructions to low-level winds. As air encounters topography, it slows down, piles up, and rises due to an *orographic effect*. As the air rises, it cools by adiabatic expansion, forming a cloud if it cools to the dew-point temperature. This orographic effect is why thunderstorms are usually more common over mountain peaks than over the adjacent valleys.

3. During daytime, the Sun heats the surface more efficiently than the air above it. As the surface of a mountain warms, the adjacent air warms, which then flows upslope. When winds flowing up opposite sides of the mountain converge at the summit, they rise more. This is yet another reason why cloud formation seems to occur preferentially over mountains.

The Orographic Effect and Adiabatic Temperature Changes

4. Movement of air up, over, and down the flanks of a mountain results in interesting changes in temperature of the moving air. In the figure shown here, air is moving from left to right across the area. Moist air on the left (windward) side of the mountain encounters the flank of the mountain and is forced to move upslope. Once the air reaches the top, the air flows down the right (leeward) side. Examine this setting and think about what happens to the temperature and saturation of the air on its way up and down.

04.07.a4

7. By the time the air reaches the top, it has been warmed by latent heat released during condensation that formed the clouds and possibly precipitation. As the air moves down the other side (the leeward side), it will compress and warm, and the air will become unsaturated (warm air has higher water-vapor capacity than cold air). It will remain unsaturated as it descends, thereby causing a drying effect on this side of the mountain. As it descends, it warms according to the unsaturated adiabatic rate of 10 C°/km. As a result of this process, the air at the leeward side will be warmer than the air at the same elevation on the windward side.

5. As the air encounters the base of the mountain, it begins moving up the slope. It is initially unsaturated, so as it rises it cools according to the unsaturated adiabatic lapse rate, which is 10 C°/km everywhere. As the air cools, it approaches its dew point (the temperature at which the air becomes saturated with respect to water vapor).

6. As the air flows upslope, at some elevation it has cooled enough that it becomes saturated (reaches the dew point), and at this level clouds form. The height at which clouds begin to form is called the *lifting condensation level*. As the air, now saturated, continues rising, it now cools at the saturated adiabatic lapse rate, which is somewhat less than the unsaturated adiabatic rate. With favorable conditions, the cloud may begin to precipitate.

B How Do Convergence and Divergence of Air Cause Rising?

As air moves laterally across the surface of the Earth and at higher levels, it can encounter obstacles that impede continued motion. In addition to obstacles such as mountains, the obstacle may be other air masses moving in an opposing direction.

Convergence of Low-Level Air Masses

04.07.b1

04.07.b2

1. Low-level convergence (▲) occurs when winds from two systems are moving on a collision course. The easiest way for the colliding air to move is upward. In such cases, the air parcel must rise whether it is stable or unstable. If the air reaches its dew point as it ascends and cools, water vapor condenses into tiny drops that form clouds and perhaps precipitation.

2. The best example of a location where low-level convergence is important at continental scales is the *Intertropical Convergence Zone* (ITCZ), which encircles the Earth, more or less centered on the equator. In the satellite image above (▲), the ITCZ appears as the belt of moist air and clouds shown in red for its position during the northern summer and blue for its position during the southern summer. Along the ITCZ, the northeast trade winds, coming from north of the equator, converge with the southeast trade winds coming from south of the equator. The two sets of trade winds converge at the ITCZ, causing cloud cover as moist air is uplifted to a height at which it cools to its dew-point temperature.

Upper-Level Convergence and Divergence

3. Convergence can also occur from the way that upper-level air circulates around bends in the polar front jet stream, as illustrated in the figure to the right. These bends in the jet stream are called *Rossby waves*.

4. In upper levels of the troposphere, air that turns around ridges of high pressure will tend to move slightly faster than air around low-pressure troughs. This is because air turning around ridges moves to the right in the Northern Hemisphere (as shown in this example), in the same rotation direction as the Coriolis effect. The two forces work together on this type of bend.

5. Around troughs, air turns to the left (in the Northern Hemisphere), in opposition to the Coriolis effect. Along these bends, the Coriolis effect works against the flow of air, and the air moves slightly more slowly.

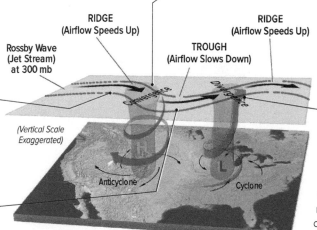
04.07.b3
(Vertical Scale Exaggerated)

RIDGE (Airflow Speeds Up)
RIDGE (Airflow Speeds Up)
TROUGH (Airflow Slows Down)
Rossby Wave (Jet Stream) at 300 mb
Convergence
Divergence
Anticyclone
Cyclone

6. The slowing down of air as it moves from the upper-level ridge of high pressure to the upper-level trough of low pressure causes air to *converge* aloft on the ridge-to-trough side of the Rossby wave. This piled-up air then sinks and warms adiabatically, causing high pressure near the surface.

7. On the opposite side of the trough, the air speeds up as it turns into the next ridge. This speeding up causes the air to *diverge* and spread apart. Air from below flows into this zone of divergence, causing low pressures near the surface. This promotes rising of the near-surface air. All of the directions of rotation shown here are opposite in the Southern Hemisphere.

C How Do Fronts Induce Rising Motion?

Whenever two air masses with significantly different temperatures meet (▶), the warmer air will be forced up and over the colder air because the warm air is less dense. This process is called *frontal lifting*. As this warm air rises, it may form clouds or begin to precipitate, as shown by colors on this weather map (▶) near a weather front marked by the blue line. The teeth on the line indicate which way the front is moving. The weather associated with fronts is presented in the next chapter.

Warm Air
Cool Air
04.07.c1

04.07.c2

Before You Leave This Page

✓ Sketch and explain how interactions between moving air masses and topographic features can cause air to rise, form clouds, precipitate, and dry out an area.

✓ Describe how lower-level and upper-level convergence and divergence can cause air to rise.

✓ Explain how fronts can force air to rise.

4.7

4.8 What Do Clouds Tell Us About Weather?

CLOUDS ARE ACCUMULATIONS of liquid water and ice suspended in the air. The types and amounts of clouds vary from place to place, from time to time, and from season to season. What are the different types of clouds, and how does each type form? Clouds provide clues not only about the amount and distribution of moisture at a given place and time, but also about conditions in the atmosphere, specifically the stability of the atmosphere and the type of convection that is occurring. This, in turn, informs us about important aspects of our weather.

We classify clouds based on three main factors: their form, their altitude, and whether they are associated with precipitation. Each of these aspects is summarized below, and examples of the main types of clouds are pictured in the large central figure.

Cloud Form

1. The three main forms of clouds are cumuliform, stratiform, and cirriform. *Cumuliform* clouds generally are taller than they are wide, or at least have a lumpy appearance. "Cumulus" in Latin means "a heap," and forms the root of words like accumulation. *Stratiform* clouds have the form of one or more layers (stratus means "to stretch" in Latin) that commonly form a continuous cover across the sky, *Cirriform* are feathery, wispy clouds, named after cirrus, the Latin word for a "lock of hair."

Cloud Altitude

2. We also subdivide clouds as to whether they are relatively low in the troposphere (low clouds), are near the top of the troposphere (high clouds), or are at an intermediate elevation (called mid-level clouds or designated with the prefix "alto"). For example, a mid-level cumuliform cloud is referred to as an altocumulus cloud. Some clouds extend from low levels up to high ones.

Precipitation

3. Clouds are composed of drops of liquid water, crystals of ice, or commonly some of each. Most drops and crystals are held aloft within the cloud because the buoyancy forces from unstable rising air exceed the gravitational force that pulls them toward Earth's surface. When the drops or crystals become too large to stay aloft, or the rising force weakens, the drops and crystals fall to the surface as rain, snow, hail, or some other type of precipitation. Clouds experiencing precipitation are designated with the prefix "nimbo" or the suffix "nimbus." A stratiform cloud that is precipitating is a nimbostratus cloud, and a cumulus cloud that is precipitating is a cumulonimbus. Even with all the possible combinations of words for form, height, and precipitation, there are only about ten main types of clouds, which are illustrated here.

Cumuliform Clouds

4. *Cumulus clouds* all have a somewhat puffy aspect but have different appearances, depending on their altitude and how unstable the air is (i.e., how strongly the air is rising). Cumulus clouds are produced by air rising generally under its own impetus (free convection), and indicate that air is unstable to some degree.

04.08.a2

6. Mid-level clouds (▲) contain both liquid water and ice particles, with the level at which ice can be present being lower in winter. Cumuliform clouds at these levels are termed *altocumulus*. They are puffy and only somewhat wispy.

04.08.a3 Durango, CO

7. Typical clouds (▲) on many days are low-level clouds with a puffy (cumuliform) appearance, or cumulus clouds, which can be widespread across the sky or localized over mountain peaks. When the atmosphere is only slightly unstable and water vapor is not too abundant, cumulus clouds are not very tall. Such clouds are typical during morning hours before surface heating causes air to rise more rapidly, and at times when conditions are only slightly unstable.

04.08.a4

5. Clouds high in the atmosphere (▶), are composed almost entirely of ice. When such high-level clouds are partly cumuliform (lumpy) and partly cirriform (wispy), they are *cirrocumulus clouds*.

04.08.a5

8. Cumulus clouds can occur in discrete layers (▶), called *stratocumulus clouds*. Like normal cumulus clouds, stratocumulus clouds are typically low clouds that are composed entirely of tiny drops of liquid water.

9. If moisture is abundant and the atmosphere is very unstable, strong lifting in the atmosphere can form cumulus clouds with great vertical extent (▲). Tall cumulus clouds associated with precipitation are *cumulonimbus clouds*, and commonly are accompanied by lightning. A cumulonimbus cloud contains liquid water near its base but can contain ice particles near its top, permitting hail to form. If well developed, such clouds develop an anvil shape at the top that points downwind, as the stronger upper-level winds push the moisture and storm cloud sideways.

Cirriform Clouds

10. *Cirrus clouds* are high in the troposphere and have a distinctive wispy appearance (▶), like the tail of a horse. At these levels, air temperatures are well below freezing, so these clouds are composed of ice crystals, which are easily blown and "smeared" about by the strong, upper level winds. The ice refracts (bends) the light, enhancing the wispy, shimmery appearance of most cirrus clouds. Cirrus clouds often indicate that stormier weather is approaching, usually within several days.

11. A high flying jet can leave a linear streak of clouds behind it, called a *contrail* (short for condensation trail). These form when water vapor produced by the engines encounters the much colder outside air, causing condensation of water drops or the formation of ice crystals. Once formed, contrails can be dispersed and sheared by the wind. They also contribute to the formation of cirrus clouds.

04.08.a7

Stratiform Clouds
04.08.a8

12. Stratiform clouds have a layered or spread-out aspect (▲). They are the clouds we observe when skies are overcast, but they may obscure just a part of the sky. Such layers of clouds form when air is forced upward (forced convection), such as by surface convergence or movement of a frontal boundary. Stratiform clouds are oriented laterally rather than vertically, implying that vertical motion is resisted by the atmospheric conditions.

04.08.a9 Grand Junction, CO

13. Stratiform clouds in the upper troposphere are high enough to be composed solely of ice and so have a somewhat wispy (cirriform), sheetlike appearance (▶). Accordingly, they are called *cirrostratus*.

04.08.a10 Yampa River, CO

14. Mid-level stratiform clouds contain both ice and liquid water and can be expressed as high overcast. These mid-level clouds are termed *altostratus*. Stratus clouds can form at several levels (▶).

04.08.a1

04.08.a11 Maine

16. A stratiform cloud that is precipitating is a *nimbostratus* cloud, as shown here (◀). This is a typical cloud associated with regional storms. A low-level stratiform cloud that is not precipitating is simply called a *stratus* cloud. Stratus can form a low blanket of clouds that causes overcast conditions.

15. *Fog* is a cloud that hugs the ground surface. It is described in detail in the next two spreads.

Before You Leave This Page

☑ Sketch each of the main types of clouds, explaining for each type the reasons for its appearance and how it forms.

4.8

4.9 What Conditions Produce Fog?

FOG IS SIMPLY A CLOUD at ground level, so the same conditions that create a cloud also produce fog. Specifically, fog is produced by cooling of the air, by increasing the humidity of the air, or some combination, with the end result that water vapor content in the air reaches saturation. Fog forms in several distinct settings, each of which causes air at or near ground level to become saturated.

A What Is Fog?

Fog is a ground-level cloud that obscures visibility, such as of this lighthouse (◄). It is most common along coastlines, over large water bodies, and in certain low areas that have the right setting to produce fog, but fog can form nearly anywhere, under the right conditions.

04.09.a1 Marshall Point, ME

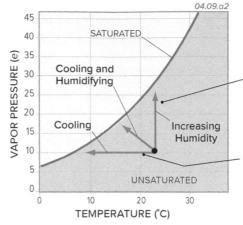

04.09.a2

Fog indicates that the relative humidity has reached 100% and the air temperature has reached the dew-point (or frost-point) temperature, causing saturation. One way in which air can reach saturation is by increasing the moisture content until vapor pressure reaches the water-vapor capacity, moving up on this graph (◄).

Another way to reach saturation is by cooling the air until it reaches the dew-point (or frost-point) temperature, represented on this graph as moving to the left.

A combination of cooling and increasing humidity can also cause saturation, as indicated by the upward sloping arrow on this graph.

B What Conditions Can Add Enough Humidity to the Air to Form Fog?

Fog can form where enough moisture is added to the air to cause saturation (100% relative humidity), but more commonly fog forms where the addition of moisture occurs simultaneously with the chilling of the air. Addition of moisture to the ambient air may happen when a nearby source of water for evaporation exists, or when precipitation evaporates as it falls through drier air.

Evaporation Fog

04.09.b1 Central Idaho

1. When a warm water body underlies colder air (◄), the environmental lapse rate is steep and the atmosphere is unstable. As the unstable air rises, it can incorporate water molecules that evaporate from the water body, increasing the humidity of the air. The increase in humidity and cooling causes the rising air to reach saturation, forming a type of fog called an *evaporation fog*. The instability causes an evaporation fog to rise in vertical columns, as shown in the photograph to the left. This phenomenon is common in fall, when water bodies are likely to remain warm from the long summer, but air above them begins to cool.

Valley Fog Produced by Increased Humidity

2. In rugged terrain, cold nocturnal (nighttime) downslope winds can trap atmospheric moisture in valleys. If moisture is added to the valley air, such as from wetlands, irrigated fields, canals, and lakes, the humidified and cooled air can reach the dew point, forming a *valley fog*. Perhaps the most famous valley fog in the U.S. is the tule fog of the Central Valley of California, the white area in the center of this regional satellite image (►).

04.09.b2

Precipitation Fog

3. Fog can also form in association with precipitation, such as occurs along a warm front. A warm front is a mass of warm air that moves laterally, trying to displace cold air in its path. Along a warm front, warmer, less dense air is forced to rise at a gentle angle over more dense, colder air.

4. As the moist, warm air rises over the cold air, it is cooled and can reach saturation, forming clouds and rain. The released rain then begins to fall through the underlying cold air and toward the cold surface.

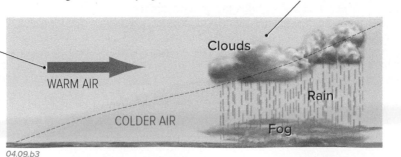

04.09.b3

5. If the underlying colder air is dry, it can cause some or all of the falling raindrops to evaporate, increasing the humidity of the cold air. Since cold air has a lower water-vapor capacity than warm air, the increase in humidity can cause the cold air to reach its dew point, forming fog and other low-level clouds. Rain that reaches the ground can also evaporate, increasing the humidity of the near-surface air, potentially forming more fog. In both cases, evaporation of precipitation is causing a *precipitation fog*.

C In What Settings Does Enough Near-Surface Cooling Occur to Form Fog?

Fog can also form where moist air is cooled enough to reach its dew-point temperature. This can occur where moist air moves over a colder part of Earth's surface, such as over snow or over a cold ocean current. It also occurs when moist air cools if it moves vertically. Several types of fog form primarily from cooling of moist air.

Radiation Fog

1. Although the term "radiation fog" sounds like something in the latest zombie movie, it is actually related to cooling of the ground surface by radiant heat loss. If skies at night are clear (i.e., cloud cover is sparse), and winds are light or calm, then the longwave radiation loss from the surface to the atmosphere will be relatively efficient, causing the surface to cool significantly.

2. Under these conditions, especially if near-surface air is humid, the surface temperature can drop to the dew point (or frost point) and cool the immediately overlying air, causing it to become saturated in water vapor. The condensation of water forms fog, in this case called a *radiation fog*. If the surface cools substantially, a temperature inversion will occur (colder air below warmer air), stabilizing the atmosphere and keeping the fog in place. Sometimes such fog can become concentrated in valleys and other topographically low places, therefore representing another way to form a valley fog.

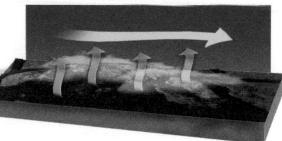

04.09.c1

Advection Fog

3. When warm air flows over a colder surface, the air loses some of its energy to the surface and therefore cools. If it cools to the dew point or frost point, the moisture in the air will condense and produce fog, as occurred when moist air moved over this glacier (▶). The movement of air laterally is called advection, so a fog produced in this manner is an *advection fog*.

04.09.c2 Athabasca Glacier, Alberta, Canada

4. A similar process produces fog along many coastlines. If moist air over an ocean or large lake blows onshore, it can encounter ground that is colder than the body of water.

04.09.c3

The cold ground lowers the temperature of the moist air, often enough to cause saturation and an advection fog. Advection fogs also form where warm, moist air blows over a cold ocean current (not shown). Advection fog is common along coastal California, which has a cold ocean current, called the California Current, offshore, bringing cold waters south along the coast.

Upslope Fog

6. As air is forced uphill by the prevailing local wind, it will cool as it rises into lower atmospheric pressures. At some level above the surface, the air may cool to the dew-point (or frost-point) temperature, and a cloud will form. On the mountain slope, the cloud forms on the ground, so it is a fog (▼). This type of fog is commonly called an *upslope fog* to indicate that the fog is caused by the upslope motion of air.

04.09.c4

04.09.c5 Cumberland Gap, KY

Valley Fog Caused by Descending, Cold Air

5. *Valley fog* can form in various ways, some already described. In mountainous terrain, air near the surface cools at night because of longwave radiative loss. This air will flow downhill because it is more dense than warmer air. It may remain colder than the surrounding air, even as it warms adiabatically on descent. If so, the descending air may chill the ambient air in the valley. If the ambient air is cooled to the dew-point or frost-point temperature, it will become saturated and form a valley fog.

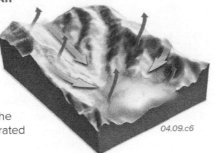

04.09.c6

Before You Leave This Page

☑ Sketch and describe three ways that increasing humidity, cooling, or some of both, can form fog.

☑ Sketch and describe four settings in which near-surface air can be cooled enough to form fog.

4.9

4.10 Where and When Is Fog Most Likely?

CERTAIN SETTINGS AND CONDITIONS are conducive to the formation of fog, so we might anticipate that some regions will have more fog than others. Also, as temperatures and humidity change with the seasons, some times of the year are likely to be foggier than other times. Considering all the places you know, which ones have more fog, and why do you think this is so?

A In Which Settings Is Fog Likely?

Formation of fog occurs when the combination of added moisture and cooling temperatures cause the air to reach saturation (100% relative humidity). What settings would most likely have fog?

1. Examine this diagram and try to identify any factors that you think could influence the amount of fog. Then read the different blocks of text, which will describe some of these factors—you may identify more than are described.

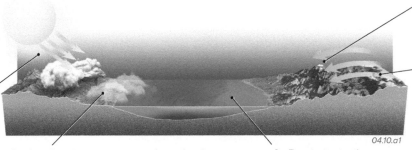

04.10.a1

2. An important factor is the climate of a place, especially the temperatures and humidity levels that are typical. Regions at lower latitudes will generally receive more direct insolation than regions at higher latitudes, and this variation greatly affects the temperature, which in turn controls how much moisture the air can evaporate.

3. Another important control on the frequency of fog will be the typical humidity of a place. Is it a humid place, commonly covered with clouds, or a drier, inland desert that commonly has clear skies?

4. Proximity to the ocean or other large bodies of water will clearly be a major factor in fog formation, since such water bodies can contribute abundant moisture to the air, if they are warm enough. Water also warms up and cools down more slowly than land, so it tends to help moderate any temperature swings of the adjacent land (cooler days and warmer nights than land farther away from water).

5. Prevailing wind directions are important since they can bring moist or dry air into an area. Especially important is whether breezes along a shoreline blow onshore (inland) or offshore (toward the water).

6. Topography and elevation are clearly key factors, since they control temperatures and influence local wind directions. Mountain slopes have upslope winds at some times and downslope winds at others. They also interact with the prevailing winds.

B What Regional Settings Are Favorable for the Formation of Fog?

1. This map shows major ocean currents and land cover. Warm ocean currents are in red and cold ocean currents are blue. On the land, green colors show abundant vegetation, tan colors show areas with less vegetation, and white areas are covered by ice and snow. Examine this map and think about how these ocean currents and regional ground-cover patterns could influence the frequency of fog. Areas of fog are shown schematically as patches of clouds.

2. Radiation fogs are common in inland areas, which often experience clear skies and a significant drop in temperature at night. Radiation fog can persist when there is a temperature inversion to stabilize the atmosphere and hold the fog in place, low to the ground.

3. Latitude plays a major role in whether fog forms because it influences regional temperatures, humidity, wind directions, ocean currents, and the types of storm systems.

8. Ice and snow on the land and on frozen oceans greatly cools the adjacent air, causing intense temperature inversions over the polar ice caps. These inversions lead to widespread radiation fog, especially beneath the polar highs.

7. Fog is frequent along the windward sides of mountain ranges, such as those along the western side of North and South America. Near these mountains, fog is more common where moisture is abundant, and this is controlled by regional patterns of winds and ocean currents.

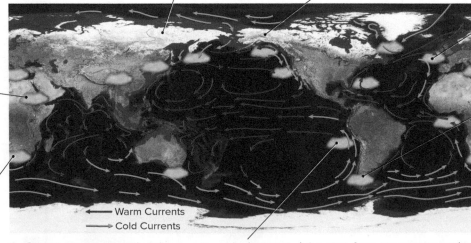

— Warm Currents
— Cold Currents

04.10.b1

6. Great Britain is very well known for its cold and damp advection fogs.

5. Over the open oceans, evaporation fogs are common as water evaporates and rises into overlying cooler air.

4. Ocean currents greatly influence fog development. Advection fogs are common where cold ocean currents extend into otherwise warm areas near the coast, such as along the western coasts of South America and southern Africa. Sometimes these fogs provide the major source of moisture in desert areas, such as the Atacama Desert of Chile and the Namib Desert of Namibia.

C Which Parts of the U.S. Have the Most and Least Fog?

Mean Number of Fog Days Per Month (average for entire year)

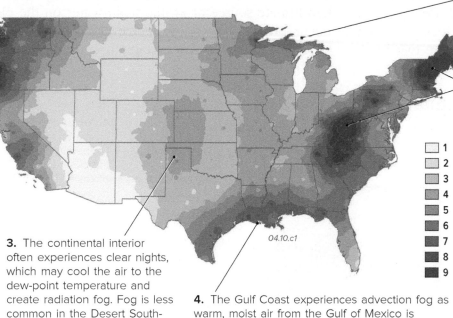

04.10.c1

1. This map shows the average number of days of fog experienced by areas in the conterminous 48 states, with darker colors indicating more fog. Observe the patterns on this map, noting how common fog is in the area where you live or visit.

2. The Pacific Northwest frequently has fog. As saturated marine air is pushed by the mid-latitude westerlies toward the Pacific Northwest, it gets chilled from beneath by a cold ocean current, causing an advection fog. Other fog is caused by air moving upslope on the western sides of the Cascades and other mountains. San Francisco experiences advection fog for the same reason, and the mountains surrounding the bay cause the fog to be funneled and focused in the Bay Area.

3. The continental interior often experiences clear nights, which may cool the air to the dew-point temperature and create radiation fog. Fog is less common in the Desert Southwest, such as in Arizona, which often lacks abundant moisture.

4. The Gulf Coast experiences advection fog as warm, moist air from the Gulf of Mexico is advected onshore due to clockwise circulation around the Bermuda-Azores high-pressure area.

5. The Great Lakes and other large bodies of water support evaporation fog, particularly in fall.

6. In New England and parts of the Appalachian Mountains, fog often occurs as cyclones and anticyclones over the Atlantic push moist and relatively warm maritime air over the colder continent. This creates advection and precipitation fog. Cyclones and anticyclones also produce upslope fog farther south, along the flanks of the Appalachian Mountains. Valley fogs are also common in this region.

Legend (1–9)

Seasonal Variations in the Number of Fog Days Per Month

Winter

7. In January (▶), advection fogs occur frequently in the Pacific Northwest and other coastal areas. These are caused by relatively warm air over the oceans moving over much colder land, which cools the air and leads to the formation of fog.

Spring

8. By March, fogs become less frequent across many parts of the country (▼). Still, radiation fogs tend to be fairly common across the middle of the country under clear nighttime skies. The Gulf Coast experiences frequent fog caused by moist air from the Gulf of Mexico.

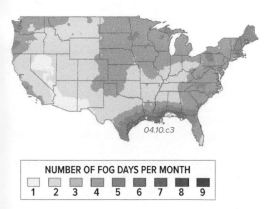

04.10.c3

04.10.c2

04.10.c4

Summer

9. Fog is least common in most of the U.S. in summer (▲). Even though the vapor pressure is high, higher temperatures mean that it is difficult for the air to reach saturation. Fog is most common in New England and in mountainous areas, such as West Virginia, where fogs are largely attributable to upslope, precipitation, and radiation fogs.

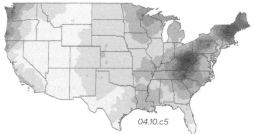

04.10.c5

Fall

10. Fall (▲), like summer, is a time of less frequent fog, except perhaps for upslope fogs in the Appalachians. In addition, downslope winds at night can trap moisture in the adjacent valleys. In subsequent months, the patterns characteristic of winter begin to become reestablished.

NUMBER OF FOG DAYS PER MONTH
1 2 3 4 5 6 7 8 9

Before You Leave This Page

✔ Sketch and describe some factors that influence the frequency of fog.

✔ Summarize how these factors affect the global, regional (U.S.), and seasonal distributions of fog.

4.10

4.11 How Does Precipitation Form?

THE PROCESS OF PRECIPITATION is vital to life on Earth, helping to redistribute water from the oceans to the atmosphere to the land. Precipitation is the ultimate source of all the fresh water on the planet, which we depend on in our daily lives. How does precipitation occur? What is going on inside clouds that forms raindrops, snowflakes, and hail? It turns out there are two important mechanisms by which precipitation droplets form, one of which is somewhat surprising.

A What Is Precipitation?

1. Precipitation is the process whereby liquid droplets of water (raindrops), solid bits of ice (snowflakes and hail), or some combination of these fall from the sky. Examine this figure and consider all the processes that have to occur to cause rain, snow, or hail.

2. The cycle begins with evaporation of water in the oceans and other parts of Earth's surface, a process that puts water vapor into the atmosphere. Next, the water vapor forms the various types of clouds, which can contain tiny drops of water, ice crystals, or some of each.

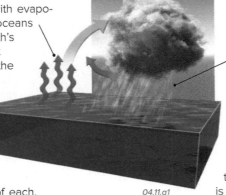
04.11.a1

3. For precipitation, some processes are occurring within the cloud that make the water droplets or bits of ice heavy enough that the pull of gravity can overwhelm the buoyancy forces (atmospheric instability) that uplift air within the cloud (the rising air is how most clouds form).

4. Whether a cloud contains drops of liquid, ice crystals, or some combination depends primarily on the temperature of the cloud. If ambient temperatures near the cloud are warm, the cloud will contain mostly drops of liquid. Temperatures decrease upward within a cloud, however, so the upper levels can contain ice, while the lower levels contain drops.

5. Under cold ambient conditions, a cloud contains mostly ice throughout its vertical extent. Clouds at intermediate ambient temperature, or intermediate altitudes, will contain a mix of drops and ice.

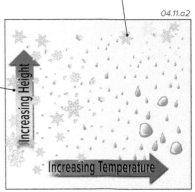
04.11.a2

Increasing Height

Increasing Temperature

B How Do Water Droplets Form and Grow?

1. An important factor in how water droplets in a cloud grow to become raindrops is the immense differences in size between the various players. Raindrops are huge compared to the size of water droplets that form a cloud, as shown by their relative sizes in the properly scaled diagram below.

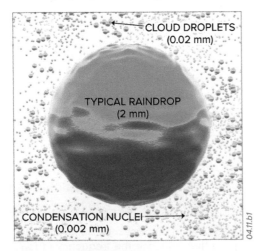

CLOUD DROPLETS (0.02 mm)

TYPICAL RAINDROP (2 mm)

CONDENSATION NUCLEI (0.002 mm)

04.11.b1

2. It is energetically difficult for the tiny water droplets in clouds to just form by themselves, but it is easier for them to form if they condense around even tinier particles, such as dust, salt, and smoke. Due to this role, such particles are called *condensation nuclei*.

3. The figure to the right illustrates what can happen to a moving water drop that interacts with smaller cloud droplets around it.

4. Some larger drops form when liquid water droplets in clouds merge and grow to a size that can be pulled down by gravity. A water droplet begins to fall as soon as the downward-directed gravitational force exceeds the upward-directed buoyancy force caused by instability. This occurs sooner for larger droplets than for smaller droplets, and the larger drop overtakes the smaller droplets on their descent, making the falling drop even larger. Eventually it falls as a raindrop. The growth of a drop in this manner is called *collision-coalescence*.

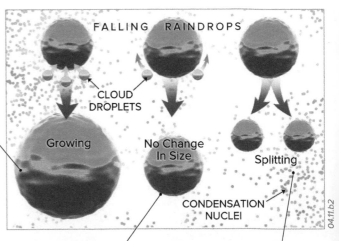

FALLING RAINDROPS

CLOUD DROPLETS

Growing   No Change In Size   Splitting

CONDENSATION NUCLEI

04.11.b2

5. In other cases, the smaller droplets simply slide past the falling drop, because the collision is not enough to break the surface tension that tends to keep each drop intact. In this case, the falling drop will not increase in size as it falls. This situation is actually very likely under many atmospheric conditions.

6. Alternatively, wind resistance can reshape the falling drop until it becomes easier to break the drop into separate drops than to retain its contorted shape. In this way, a falling raindrop becomes smaller, which can also occur if water molecules on the outside of the drop simply evaporate into the surrounding air.

C How Do Ice, Liquid, and Vapor Interact Within a Cloud?

Many clouds are in environments in which the temperature is below freezing, even in summer. Exceptions are clouds in the tropics and those low in the atmosphere, or a cloud being warmed by latent heat during precipitation. Even when temperatures are below freezing, some liquid water droplets exist. This sets up a situation in which water exists in all three phases—vapor, liquid, and ice (solid). The interaction between these three coexisting phases allows precipitation to form much more easily than would otherwise be possible if only vapor and liquid were present.

1. The diagram to the right shows the familiar saturation curve for water vapor, but focuses on conditions slightly above and slightly below freezing. At temperatures below freezing, the saturation curve for water vapor near ice (the dashed line) diverges from that for water vapor and liquid (the solid line). The divergence of the two curves begins at the freezing point (0°C) because ice can't easily exist above freezing.

2. The area between the two curves, shaded orange, depicts temperatures at which air is unsaturated with water vapor relative to a liquid but above saturation for air next to ice in the cloud. Another way to state this is that saturation is reached with slightly less water vapor when air is in contact with ice in the cloud than when it is in contact with liquid water, even at the same temperature.

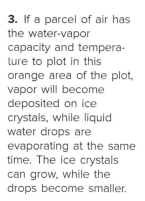

04.11.c1

3. If a parcel of air has the water-vapor capacity and temperature to plot in this orange area of the plot, vapor will become deposited on ice crystals, while liquid water drops are evaporating at the same time. The ice crystals can grow, while the drops become smaller.

---

4. This difference in saturation levels aids in the formation of precipitation. If ice, water vapor, and drops of liquid water coexist at the conditions specified by the orange field in the previous figure, then the air will be unsaturated next to the liquid drop. As a result, water molecules will evaporate from the drops, increasing the relative humidity of the surrounding air.

5. As the liberated water vapor molecules diffuse into the air, they cause an increase in water vapor adjacent to a nearby snowflake. The air near the snowflake therefore becomes saturated or supersaturated with respect to the ice. This in turn causes water vapor molecules to be deposited on the snowflake, enlarging it.

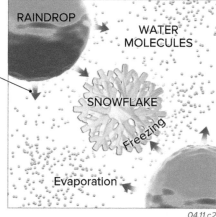

04.11.c2

6. Depending on the air temperature, the ice can remain as a solid or if temperature increases slightly, the ice can melt, going directly from ice to liquid. In this way, liquid water becomes water vapor, which then becomes ice, which then becomes drops, an easier path (under many conditions) than trying to overcome surface tension to grow drops or the difficult energetics of nucleation required to make new drops. This process of precipitation is often called the *Bergeron process* after one of the scientists who discovered it.

How Cloud Seeding Works

Technology exists to "help" precipitation occur more efficiently in drought-stricken areas. Strategies generally involve injecting particles into clouds to enhance droplet or ice crystal growth, either via a ground-based delivery system or more commonly airplanes that fly through the clouds, as shown here. One strategy uses dry ice to cool the cloud to temperatures so low that vapor deposits onto ice more readily. A second strategy is to inject into the cloud microscopic solid particles that act as condensation

04.11.t1

nuclei. The idea is that water vapor can condense or deposit much more easily when it has something to "hold onto" during the phase change. Silver iodide has often been used for this purpose because its crystalline structure maximizes opportunities for vapor to attach to it. These cloud seeding efforts have achieved mixed results. In some cases, precipitation has been enhanced or shifted to other areas. Interventions into natural processes, however, often have unintended consequences, such as, in this case, silver-iodide

pollution from the chemicals used to seed the cloud. Unintended consequences are always a concern when trying to affect a natural system.

Before You Leave This Page

✓ Explain what precipitation is and how the size of a drop can change.

✓ Sketch the water vapor capacity curve as a function of temperature, with respect to liquid water and ice, and describe how this leads to a Bergeron process for precipitation.

✓ Sketch and explain cloud seeding.

4.11

4.12 How Do Sleet and Freezing Rain Form?

RAIN AND SNOW ARE THE MOST COMMON forms of precipitation, but other types of precipitation are also important. The term *sleet* is used for a mixture of snow and rain or for precipitation as small partially frozen pellets that are too small to be called hail. *Freezing rain* is precipitation that reaches the ground as raindrops, but that immediately freezes upon contact with cold objects on the surface, such as the ground, roads, plants, and power lines. Can you envision the situations that could cause sleet or freezing rain?

A Under What Conditions Do Sleet and Freezing Rain Form?

Sleet and freezing rain are often confused, and they have several similarities. Both result from snow and ice pellets that melt on their way down to the surface as they pass through a layer that is warmer than freezing.

Sleet

1. This graph shows a hypothetical temperature profile (the green line) that would allow sleet to form. Note that this temperature profile shows a subfreezing layer close to the surface, overlain by a zone of warmer temperatures (just above freezing)—a temperature inversion.

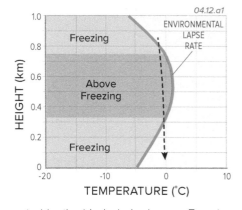

04.12.a1

Freezing Rain

3. This similar graph shows the conditions needed for freezing rain. As for sleet, the temperature profile displays a temperature inversion, but this time with a thinner zone of subfreezing temperatures near the surface. As for sleet, there is a temperature inversion with warmer temperatures (above freezing) above the near-surface air.

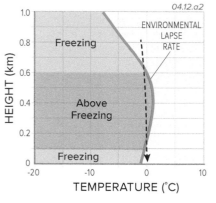

04.12.a2

2. Falling precipitation is represented by the black dashed arrow. Freezing temperatures are present higher in the atmosphere, in the upper part of the graph. As snow falls, it crosses from freezing conditions through the zone of above-freezing temperatures and then back into freezing ones some distance above the surface. As a result, the snow will at least partially melt on the way down, forming a drop of water that then partially refreezes before it reaches the surface. This creates a tiny pellet that bounces on impact with the surface. This type of precipitation is called *sleet.*

4. As snow falls through these conditions, it will at least partially melt when it encounters the mid-level interval of warmer-than-freezing air. As the drop continues falling, it again encounters freezing temperatures, but not until it is at or right above the surface. As a result, the drop does not have enough time to refreeze, and so hits the surface as rain that sticks to the ground instead of bouncing like sleet. Upon encountering the cold ground and objects on the surface, the drop immediately freezes, forming coatings of ice on the ground and objects.

B Where Do Sleet and Freezing Rain Form?

The scenario that provides the best opportunity for the temperature inversion required for the formation of sleet and freezing rain is ahead of the warm front, as shown in the cross section below.

1. A warm front is a mass of relatively warm air that moves across the surface, following the retreat of cold air in its path. In the cross section shown here, the warm air is on the left side (on top) of the frontal boundary and cold air is on the right side of (below) the boundary.

2. The warm air is less dense than the cold air, so it is wedged up and over the colder air mass. If warm air resides above colder air, it creates a temperature inversion. As the warm air is forced to rise, it cools, forming clouds and rain above and along the boundary.

6. Snow can be associated with a warm front if the warm air rises so high off the ground that it is below freezing. In this case, the entire profile remains below freezing, allowing snow to form aloft and reach the ground.

5. Even farther back from the surface warm front, rain produced aloft must pass through a thicker column of cold air. Thus, raindrops have enough time to freeze on the way down, forming sleet.

3. Close to the surface position of the front, rain produced along the front only falls through a short vertical interval of cold air. If raindrops pass through this zone without refreezing, they will reach the ground as normal rain. This is the first type of precipitation that announces the arrival of the warm front.

4. Farther back from the surface location of the front, the ground may still be at freezing temperatures. In this case, the rain freezes upon encountering the ground and other cold objects on the surface, producing freezing rain. As the warm front passes over an area hovering near freezing temperatures, therefore, normal rain follows freezing rain.

C What Regions Have a Higher Frequency of Freezing Rain?

1. This map of central North America (►) shows the average annual number of days with freezing rain. Examine this map for regional patterns and try to identify if the place where you live or visit has frequent freezing rain—you probably already know this answer, because freezing rain is not easily forgotten. A map of the frequency of sleet would show a similar distribution. Since latitude and elevation largely control temperature, they also influence where freezing rain occurs.

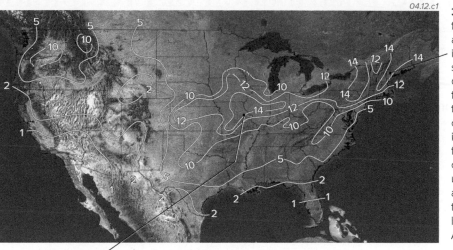
04.12.c1

3. Freezing rain is also frequent in the highland areas of the Northeast, including New England and some parts of the Appalachian Mountains. The freezing rain and sleet are the result of the same type of cyclonic storm system as in the Midwest. Sleet and freezing rain are also common in other humid uplands in mid-latitudes around the world, such as the Alps, Japanese highlands, and parts of the Andes of South America.

2. On this map, freezing rain is frequent in the Midwest, exhibiting a pattern that follows the typical paths of storms as they move eastward across the country. Such storms have a northward (counterclockwise) flow along their eastern, leading edge, bringing relatively warm air from the south that in many cases forms a warm front.

D How Does Terrain Contribute to the Distribution of Sleet and Freezing Rain?

Movement of air against mountains can also cause freezing rain and sleet. When a cold air mass is next to a mountain range, the circulation associated with it will cause air to move upslope as it encounters higher terrain, as shown here (►). The movement of air can be in response to larger circulation systems, such as a cyclone.

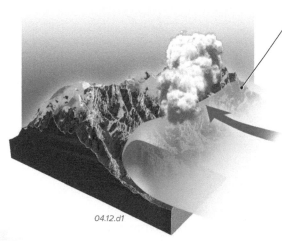
04.12.d1

A cold air mass beside the mountain is dense, while the air wedged above it may be warmer, leading to a temperature inversion. While the inversion would also stabilize the local atmosphere above the cold surface air mass, the movement upslope by the circulation around the pressure system may result in forced convection that produces rain at the lowest elevations, freezing rain at higher elevations, sleet farther upslope, and snow at the highest elevations. Mountains that block flow in this way are sometimes referred to as "cold air dams," as shown in the photograph below. Mountain-caused freezing rain, sleet, and snow are frequent next to the Rocky and Appalachian Mountains in winter.

04.12.d2

The Impacts of Freezing Rain

Freezing rain can cause extensive damage and poses severe hazards to animals, plants, and any object on the surface. Freezing rain can coat sidewalks and roadways with a layer of dangerously slippery ice, causing pedestrians to fall and automobiles to skid out of control. Freezing rain resembles rain as it falls, so people may believe that they are only driving in rain when in fact they are driving on an icy road. Even the rain that falls on windshields gives no clue that it will freeze on impact with the surface, because automobile defrosters and friction created by windshield wipers prevent freezing that might

alert the driver. The fact that sleet bounces informs motorists to its presence as ice, making it less dangerous.

Freezing rain adds considerably to the weight of trees, especially in fall or early winter, when leaves are still attached to tree limbs and branches. This may cause limbs to fall, perhaps onto power lines, which are themselves weighed down considerably by freezing rain. On average, freezing rain causes more deaths and property damage per hour than any other type of storm.

04.12.t1

Before You Leave This Page

☑ Sketch the temperature profiles that would lead to sleet and freezing rain.

☑ Sketch a warm front, showing where sleet and freezing rain are most likely to form.

☑ Characterize the spatial distribution of sleet and freezing rain in the U.S.

☑ Explain the impacts of freezing rain.

4.12

4.13 What Is the Distribution of Precipitation?

THE AMOUNT OF PRECIPITATION varies from region to region, season to season, and day to day. The daily variations are related to short-term changes in weather, but variations between regions and seasons reflect differences in the overall climatic setting, such as differences in latitude, prevailing wind directions, proximity to large water bodies, any nearby ocean currents, elevation of the land, and countless other factors. We explore some of these by examining regional variations in precipitation over North America and the world.

A How Does Precipitation Vary Across North America?

January

1. Examine this map, which shows average amounts of precipitation for North America for January, averaged over several decades. In northern and high-elevation regions, much of this precipitation is as snow. Note any regional patterns and the precipitation that is typical for where you live.

2. Heavy rainfall along the coast of Canada and the Pacific Northwest is due to proximity to the ocean combined with orographic effects that cause moisture-laden Pacific air to move upslope under the influence of the mid-latitude westerlies.

3. Inland from the western mountains, precipitation totals decline dramatically. These areas are blocked from oceanic moisture sources by mountains both to the west and to the east.

4. Precipitation decreases inland away from moisture sources and northward with lower temperatures. Cold air has lower water-vapor capacity and therefore will not produce clouds with a sufficient amount of moisture to allow for much precipitation.

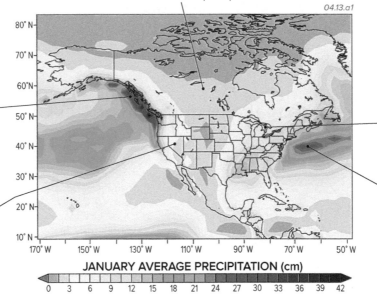

04.13.a1

JANUARY AVERAGE PRECIPITATION (cm)

0 3 6 9 12 15 18 21 24 27 30 33 36 39 42

5. The Great Lakes are much warmer than the overlying air in winter. This generates a steep environmental lapse rate, which makes an adiabatically moving air parcel likely to be warmer than its surrounding environment. Thus, the lakes destabilize the atmosphere, increasing local "*lake effect snow*" totals downwind.

6. Precipitation in both the Atlantic and Pacific oceans at this time of year is focused near the boundary between cold air and warm air, in part because these regions experience frequent passage of fronts.

July

7. This map shows precipitation totals for July, the middle of summer. Note the regional patterns and then compare this map with the one for January.

8. July totals are higher than in January in most of North America, except along the West Coast. In most areas, local surface heating destabilizes the atmosphere, but the cold California Current off the West Coast negates this effect in that region.

9. Summer precipitation totals are low in western North America and over the adjacent Pacific Ocean, even in places that are relatively wet in winter. This is because of the drying effects of sinking air on the eastern side of the Hawaiian high-pressure feature, which makes its closest approach to North America in summer.

10. July totals are much higher than those in January in the far north. The warmer air has more energy to evaporate water, producing water vapor that can then precipitate. Also, the region is much closer to the cold-warm boundary in summer than in winter, facilitating frontal lifting.

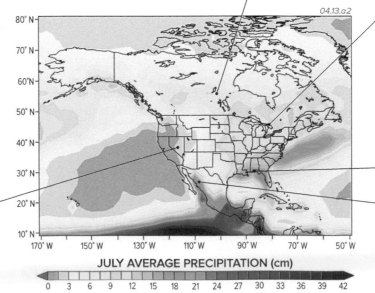

04.13.a2

JULY AVERAGE PRECIPITATION (cm)

0 3 6 9 12 15 18 21 24 27 30 33 36 39 42

11. The Great Lakes are cool relative to the overlying air in July. This weak environmental lapse rate (or even an inversion) stabilizes the atmosphere, the opposite of winter conditions, limiting the amount of precipitation.

12. In summer the Bermuda-Azores high-pressure area reaches its northernmost extent. The clockwise circulation around this feature supplies warm, moist air into interior North America.

13. The prong of higher precipitation that occurs in western Mexico is related to a northward flow of moisture as part of the "Arizona monsoon." Near the mountains, intense surface heating and orographic lifting cause summer thunderstorms.

B How Do Annual Precipitation Totals Vary Globally?

The globes below show average annual precipitation at a global scale. Examine the regional patterns and consider what factors might be causing them. You may want to refer back to the globes showing specific humidity earlier in this chapter. An area that most people consider to be "wet" generally has high precipitation and high humidity. How dry an area seems is a function of low precipitation, low humidity, and evaporation rates that exceed precipitation rates.

1. Globally, precipitation is concentrated in equatorial oceanic areas, expressed as a belt of blue and green colors near the equator. Precipitation generally decreases poleward, inland, and downwind from major mountain ranges.

2. The tropical rain forests of Indonesia, along with adjacent parts of the Indian and Pacific oceans, are among the wettest places on Earth. They are part of an east-west belt of high precipitation that roughly follows the Intertropical Convergence Zone (ITCZ). Along the ITCZ, convergence of moisture-laden air carried by the northeast and southeast trade winds causes lifting and precipitation. The Hadley cell migrates with the seasons, so the position and influence of the ITCZ do, too.

8. The east coasts of mid-latitude continents are generally moderately wet, as cold and warm air meet frequently, producing frontal precipitation. In summer, convective precipitation supplements the totals.

7. The idea that Siberia receives extremely large amounts of snowfall is a myth. The intense Siberian High is driven by cold, dense air, and it produces a sinking motion. The cold surface also acts to decrease the environmental lapse rate, making it more likely that an adiabatically moving air parcel will remain colder than its surrounding environment, enhancing atmospheric stability.

60° N
30° N
Equator
30° S
60° S
04.13.b1   *04.13.b2*   *04.13.b3*

3. Locations in the subtropics (30° N and S) tend to be very dry. This belt of latitudes contains many of the world's deserts, such as the Sahara and those of the Arabian Peninsula, southwest Africa, central Australia, and the American Southwest. Even though warm air would ordinarily tend to rise in these warm areas, this effect is counteracted by descending air in the downward-directed part of the Hadley cells. Some of these areas are especially dry because they are in the rain shadow of major mountain ranges. Exceptions to the dry subtropics occur along the east coast of continents (for example China), where there is an abundance of warm, moist air in association with warm surface ocean currents.

4. Polar areas are generally too cold to have much water in the atmosphere. Limited moisture, combined with the sinking of the intense cold air, opposes vertical cloud development. Furthermore, these areas are typically far from boundaries between warm and cold air, which can produce warm and cold fronts and their attendant storms. As a result, places near the poles receive very little precipitation.

5. The Amazon basin, with extensive tropical rain forests, is another region of high precipitation. This is due to the ITCZ and to orographic lifting of the moisture-laden trade winds over the eastern flanks of the Andes Mountains. The southwestern coast of South America has the driest deserts in the world.

6. Relatively little was known about precipitation over the oceans until recent years when satellite data became available. Oceanic patterns are similar to continental ones, except for the absence of complications caused by topography.

ANNUAL AVERAGE PRECIPITATION (cm)
0   36   73   109   146   183   219   255   292   328   365   402   438   474   511

How We Measure Precipitation

Rainfall is measured using a standard cylinder with a circumference that is 1/10 of the opening. The amount of rainfall can be read from graduated markings on the cylinder or from a measuring stick with hashmarks exaggerated 10 times. This allows for the observation of rainfall totals to the hundredths of an inch. Some rain gauges, like the one shown here, contain two cylinders, one for measuring typical rainfall amounts and a second larger cylinder used when extreme amounts of precipitation cause the first cylinder to overflow. When it snows, the snow is

04.13.t1

allowed to melt and the amount of liquid can be measured as described earlier in this paragraph. Snow typically has an average density of 8% that of water, so the equivalent of one centimeter of rain will produce, on average, more than 12 cm of snow (an inch of rainfall will produce a foot of snow), but this ratio can vary widely, depending on local conditions.

Before You Leave This Page

✓ Summarize factors influencing the distribution of precipitation in North America during January and July.

✓ Summarize the main patterns of global precipitation, and their main causal processes.

4.13

4.14 How Can Moisture Extremes Be Characterized?

THERE ARE GREAT EXTREMES in the distribution of moisture, with some regions being very dry for an extended period of time, as during a drought, whereas other regions receive huge amounts of precipitation in a short period of time. The cause of these extremes in moisture can be broad regional shifts in the climate or more localized shifts, such as a slight alteration in wind direction that changes local patterns of orographic lifting. Extremes of precipitation, whether too much or too little, are of great concern to physical geographers, as these extremes have the greatest impacts on people. Here, we describe several types of extreme events, how we document such events, and how we assess the risks for either too little or too much moisture.

A What Causes a Drought and How Can We Quantify It?

Drought is an unusually extended period when losses of surface water to evaporation and transpiration from vegetation in the area are uncharacteristically higher than the addition of water to the surface by precipitation. Drought places stress on plants and animals and can cause normally green and fertile land to dry up into eroded, dusty plains. Drought can cause water shortages by drying up streams and lakes, and by decreasing flow into groundwater. Three causes of drought are described below, but others are also common.

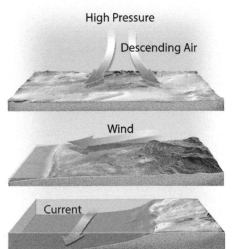

Drought can be caused by weather patterns that are atypical for an area. Persistent high pressure, with its sinking air (◄), can cause a region to be drier than normal. The map on the right (►) shows a large area of high pressure over the Desert Southwest. If such high pressure is persistent, the region can shift into drought conditions.

An unusual change in wind direction (◄), such as one that brings dry air from the land over an area that normally experiences an onshore flow of moist air from the ocean, can cause a drought.

A temporary change in the direction or strength of an ocean current (◄) can last multiple years, causing a temporary change in climate, possibly including drought if a cold current results in less moisture in the air.

04.14.a1

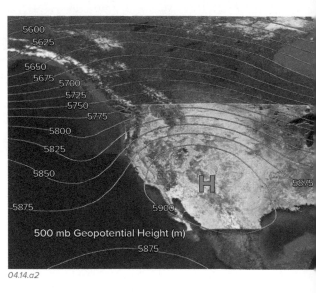

500 mb Geopotential Height (m)

04.14.a2

B How Can We Quantify the Risk of Drought?

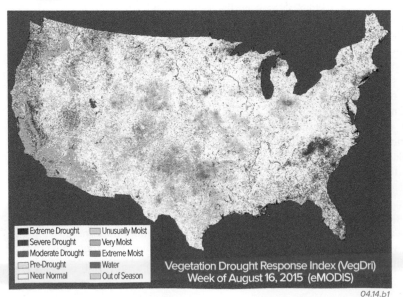

Vegetation Drought Response Index (VegDri)
Week of August 16, 2015 (eMODIS)

Legend:
- Extreme Drought
- Severe Drought
- Moderate Drought
- Pre-Drought
- Near Normal
- Unusually Moist
- Very Moist
- Extreme Moist
- Water
- Out of Season

04.14.b1

Satellite data, such as this eMODIS image (◄), are used to create drought indices based on vegetation health and other biophysical properties.

The numerically based *Palmer Drought Index* (PDI) uses climatic and soil moisture data to quantify the severity of the drought. Status is expressed relative to normal at that location and time of year, such as the values shown on the map here (►).

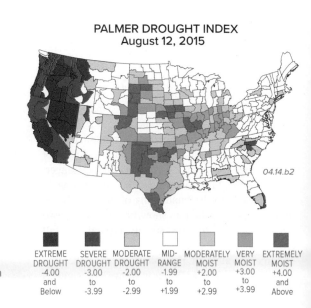

PALMER DROUGHT INDEX
August 12, 2015

04.14.b2

EXTREME DROUGHT	SEVERE DROUGHT	MODERATE DROUGHT	MID-RANGE	MODERATELY MOIST	VERY MOIST	EXTREMELY MOIST
-4.00 and Below	-3.00 to -3.99	-2.00 to -2.99	-1.99 to +1.99	+2.00 to +2.99	+3.00 to +3.99	+4.00 and Above

C How Can We Quantify and Depict the Risk of an Extreme Precipitation Event?

An important aspect to consider in planning land use or other activities is how often events of extremely high or low precipitation are likely to occur. We can calculate the average length of time that we would expect to see between events that equal or exceed a given magnitude, the *return period*. This length of time, also called a *recurrence interval*, allows geographers, land-use planners, farmers, and others to assess the risk of extreme events and to plan accordingly.

1. To calculate a return period, the first step is to specify the magnitude and duration of a hypothetical extreme event of interest. The magnitude is the amount of precipitation. The duration is the length of time that was required for that amount of precipitation to fall. We then use historical data to find where an event of this magnitude ranks with other events of similar duration. The results are entered into the equation below.

$$\text{Return Period (RP)} = (\text{number of years of record} + 1) / (\text{rank of event})$$

2. To illustrate this concept, we present a simple example. If 8.4 cm of precipitation fell over a 12-hour period at a given place, we could estimate its return period by finding its rank among the largest 12-hour rainfall totals in the 50 years of historic records at that location. If our 12-hour total ranks 30th among all 12-hour totals at the location (in the 50 years during which measurements were taken at that site), the return period would be 1.7 years. This means that we would expect 8.4 cm of precipitation to fall within a 12-hour period at that place once every 1.7 years, on average. Enter these numbers in the equation above and see what you get.

3. Using these maps, we could expect that within a 24-hour period in central Texas, about 8 cm of precipitation would fall once every 2 years, 13 cm would fall once every 10 years, and about 23 cm would fall once every 100 years, on average.

4. There are two inherent dangers with such approaches. First, it is extremely dangerous to extrapolate far beyond the observed historic data. How can we know what a 100-year event is if we only have 50 years of data? And second, people often forget that these numbers are just *average* return periods. If a 5-year event happened last year, that does not mean that another 5-year event will not occur again this year; it simply means that the magnitude would be *expected* to occur once every 5 years, on average.

Maps of Return Period

04.14.c1

2-YEAR 24-HOUR RAINFALL

2.5   5.0   7.6   10.1   12.7   17.8   22.9   27.9   33.0   38.1 cm

04.14.c2

10-YEAR 24-HOUR RAINFALL

2.5   5.0   7.6   10.1   12.7   17.8   22.9   27.9   33.0   38.1 cm

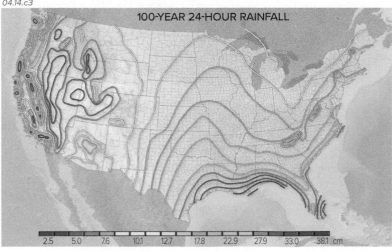

04.14.c3

100-YEAR 24-HOUR RAINFALL

2.5   5.0   7.6   10.1   12.7   17.8   22.9   27.9   33.0   38.1 cm

5. We can produce maps of the magnitude of rainfall events for various return periods and durations. Three are shown here for a 24-hour duration, including one for a 100-year return period. The "100-year storm" is often of interest because many structures are designed to withstand the effects of a 100-year storm. Examine these maps for the area in which you live or have some interest. Consider how such maps might be used, such as when planning land use near a river or how high to build a new bridge.

Before You Leave This Page

☑ Describe the concept of drought, and sketch and explain three possible causes.

☑ Explain the concept of a return period, and explain generally how one is calculated.

4.15 What Caused the Recent Great Plains Drought?

PARTS OF THE GREAT PLAINS, stretching from Texas to Montana, experienced severe drought in 2012, resulting in huge crop losses and other problems. These losses occurred in spite of recent technological advances, such as improved soil and water management practices and the use of drought-resistant varieties of crops. Some effects, such as crop loss, resulted directly from the event, but others, such as less food for livestock, were indirect results of the drought. What caused this drought, and what were some primary and secondary impacts? This drought nicely illustrates the onset of such conditions over a relatively short time frame, whereas other, more recent droughts, such as one in California, are more persistent.

A What Was the Meteorological Setting of the Drought?

Examine each pair of maps. The top maps show averages of the daily pressure (as represented by the height of the 500 mb pressure) for May and July of 2012. The bottom pair of maps shows specific humidity for the same two months. Compare the air pressure maps to the specific humidity maps and consider the relationship between the patterns of air pressure and humidity, if any exist.

Air Pressure

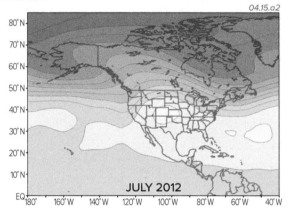

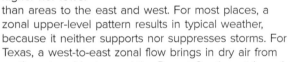

04.15.a1

04.15.a2

1. In May (▶), the air pressure patterns and upper atmosphere wind directions over the lower 48 states were essentially west-to-east, a condition meteorologists call a *zonal flow*. Pressures over the Great Plains were neither much higher nor lower than areas to the east and west. For most places, a zonal upper-level pattern results in typical weather, because it neither supports nor suppresses storms. For Texas, a west-to-east zonal flow brings in dry air from northwestern Mexico and the Desert Southwest (e.g., Arizona).

5440   5600   5720   5880
500mb GEOPOTENTIAL HEIGHT (m)

2. By July (◀), a strong ridge of high pressure developed over the Great Plains states, with the entire southern U.S. under high pressure (high 500 mb heights). The sinking air associated with this high pressure caused abundant adiabatic warming and compression, and little cloudiness and precipitation. High pressure also blocked or deflected storm tracks away from the region. As a result of these factors, drought extended across most of the central U.S., including much of the Great Plains.

Specific Humidity

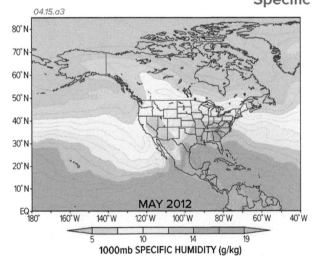

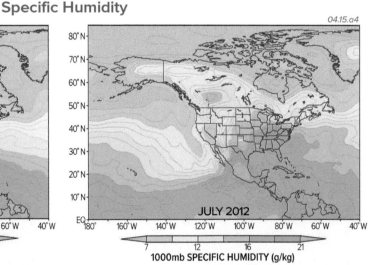

04.15.a3

04.15.a4

3. These two maps (▶) show specific humidity (the mass of moisture in a mass of air) for the same two months as shown above. Darker shades of green represent more moisture, whereas lighter shades indicate drier air. During May (the map on the left), dry air in the Desert Southwest and in the deserts of adjacent northern Mexico spread into west Texas, but moderately humid conditions existed in much of the Great Plains, such as in Kansas and Nebraska.

5   10   14   19
1000mb SPECIFIC HUMIDITY (g/kg)

7   12   16   21
1000mb SPECIFIC HUMIDITY (g/kg)

4. The specific-humidity map for July (◀) is not too different from the one for May. The clockwise flow around the Bermuda-Azores High, which makes its northernmost approach in July, has advected moisture into the interior of North America. A conclusion you can reach from comparing the four maps on this page is that the air-pressure conditions were very different, but the amount of moisture in the air was similar, although the patterns differ in detail.

B What Were the Patterns of Drought in the Middle of 2012?

The two maps below show the Palmer Drought Index for the same two months as depicted on the previous page. Examine the patterns of drought on these maps and compare them to the maps of air pressure and specific humidity. From comparing each set of maps, think about whether the drought in the Great Plains was mostly caused by a lack of humidity or by upper-level air pressure and patterns of flow.

On these maps, areas experiencing severe and extreme drought are shown in orange and brown, respectively.

Areas in light yellow are experiencing mid-range conditions, meaning not overly dry or overly moist. In May (▶), areas of the Great Plains with extreme drought are mostly restricted to isolated areas in northern Texas, eastern New Mexico, and especially southeastern Colorado. In the central and northern parts of the Great Plains, from Oklahoma to North Dakota (a north-south strip through the center of the map), conditions were in the mid-range. Note that California was also in a drought, which persisted for a number of years.

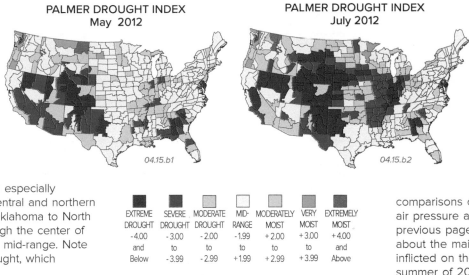

PALMER DROUGHT INDEX May 2012

04.15.b1

PALMER DROUGHT INDEX July 2012

04.15.b2

EXTREME DROUGHT -4.00 and Below	SEVERE DROUGHT -3.00 to -3.99	MODERATE DROUGHT -2.00 to -2.99	MID-RANGE -1.99 to +1.99	MODERATELY MOIST +2.00 to +2.99	VERY MOIST +3.00 to +3.99	EXTREMELY MOIST +4.00 and Above

By July (◀), extreme and severe drought had spread across more of northern Texas, eastern New Mexico, and into parts of Kansas, Nebraska, Wyoming, and the Dakotas. Nearly all of the Great Plains was experiencing at least moderate drought (dark yellow, orange, or brown). From your comparisons of these maps with those of air pressure and specific humidity (on the previous page), what do you conclude about the main cause of the drought inflicted on the Great Plains in the summer of 2012?

Primary and Secondary Impacts of Drought

In considering the effects of drought, some effects, such as drying out of agricultural fields and dying of crops, are due directly to the drought; these are *primary impacts*. Other effects are related to drought, but in a more indirect way, like an increase in the price of seeds that resulted from the initial dying of crops; such effects are *secondary impacts*. Here, we examine the primary and secondary impacts of the 2012 drought, focusing on Texas, but similar consequences of the drought were felt in other states.

An obvious primary effect of the drought was damage to crops due to the extreme dryness of the soil and lack of rainfall. In Texas, half of the cotton crop was lost, and the corn harvest decreased by 40%. In a similar manner, it was estimated that up to 10% of the trees in Texas were damaged. Wildlife populations dwindled as their habitats were degraded. Ranching likewise suffered the effects of the drought, with a sharp decrease in the number of cows. Such losses were largely caused by a diminished supply of drinking water or an increase in water contamination as ponds and other sources of surface water dried up. Throughout the drought-stricken region, the supplies of surface water and groundwater decreased, with reservoirs estimated as 40% below normal.

There were also secondary impacts. The dry conditions caused grass and brush to dry out (a direct impact), causing a series of wildfires that burned over 3 million acres in Texas, causing more than $750 million in damage to homes (a secondary impact). Trees and plants stressed by the drought became more vulnerable to later disease and pest infestations. High temperatures and low water levels (primary impacts) led to low amounts of dissolved oxygen in some lakes, ponds, and streams, resulting in fish dying (a secondary impact). Reduced seed production altered the food chain, even for species that weren't affected by the drought directly. Shrinking supplies of agricultural products resulted in higher prices.

Other secondary impacts are even less obvious. Transportation was delayed or halted along highways because of wildfires. Tourism decreased as fewer people visited lakes because the low water levels left an unpleasant and potentially dangerous situation for boaters. Also, the amount of recreational fishing declined for the same reason and because of the previously mentioned fish kills. Reduced water supplies in cooling reservoirs used by power-generation plants caused a reduction in power output, at the same time that energy demand peaked due to the higher temperatures (more

air conditioning needed). This resulted in rolling blackouts, providing an excellent example of a secondary impact. Geographers also use the concepts of primary and secondary impacts for other types of meteorological events, such as hurricanes and flooding. These terms are also considered in the context of global and regional changes that result from climate change and other events. Most geographers view problems from global, regional, and local perspectives.

Before You Leave This Page

✓ Summarize the patterns of upper-level air pressure and specific humidity that accompanied the 2012 drought in the Great Plains.

✓ Distinguish between primary and secondary impacts of an event, like drought, providing examples of each type of effect from the 2012 drought in Texas.

4.15

4.16 | How Do Global Patterns of Humidity, Water Vapor, and Precipitation Compare?

THE ABUNDANCE OF ATMOSPHERIC MOISTURE varies considerably from place to place on Earth. What are these variations and what causes them? In this exercise, you will observe and compare four types of global data: specific humidity, amount of water vapor, average precipitation, and what material is on the surface (land cover). You will describe and explain the patterns of atmospheric moisture and how these are reflected in the land cover.

Goals of This Exercise:

1. Describe regional patterns of atmospheric moisture, as measured using specific humidity, water vapor content, and precipitation, and then propose explanations for these patterns.

2. Explain how the patterns of atmospheric moisture are reflected in variations in land cover on satellite images.

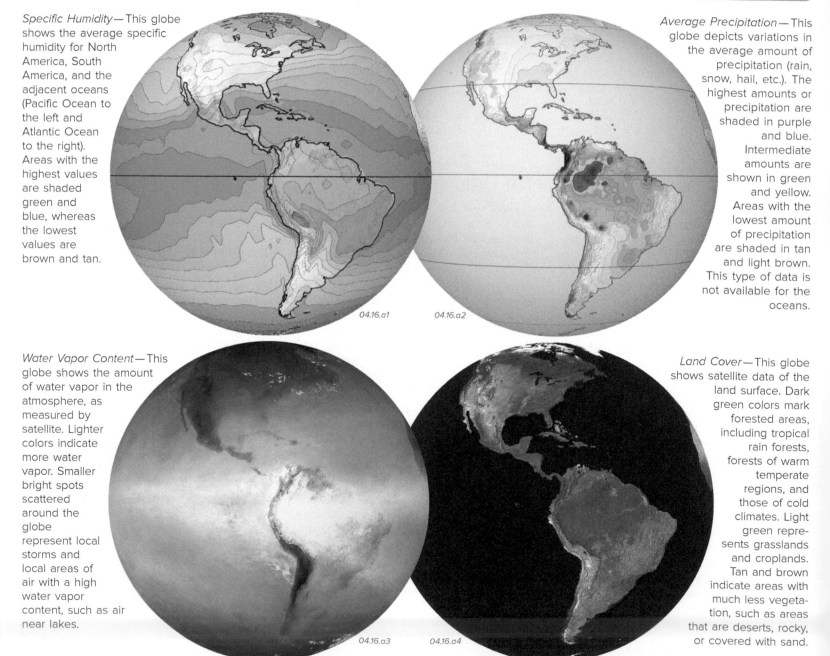

Specific Humidity—This globe shows the average specific humidity for North America, South America, and the adjacent oceans (Pacific Ocean to the left and Atlantic Ocean to the right). Areas with the highest values are shaded green and blue, whereas the lowest values are brown and tan.

Average Precipitation—This globe depicts variations in the average amount of precipitation (rain, snow, hail, etc.). The highest amounts or precipitation are shaded in purple and blue. Intermediate amounts are shown in green and yellow. Areas with the lowest amount of precipitation are shaded in tan and light brown. This type of data is not available for the oceans.

Water Vapor Content—This globe shows the amount of water vapor in the atmosphere, as measured by satellite. Lighter colors indicate more water vapor. Smaller bright spots scattered around the globe represent local storms and local areas of air with a high water vapor content, such as air near lakes.

Land Cover—This globe shows satellite data of the land surface. Dark green colors mark forested areas, including tropical rain forests, forests of warm temperate regions, and those of cold climates. Light green represents grasslands and croplands. Tan and brown indicate areas with much less vegetation, such as areas that are deserts, rocky, or covered with sand.

04.16.a1 04.16.a2

04.16.a3 04.16.a4

Procedures

For each region (the Americas on the left page and Africa on the right page), complete the following steps. Write your observations and interpretations on the worksheet or answer questions online.

1. Observe the larger patterns on each globe. Where are the amounts of any factor highest, where are they lowest, and is there a general pattern for this part of the globe? Note any regions that do not appear to fit the regional pattern. Record your observations on the worksheet.

2. Propose interpretations for the large-scale features, such as variations north and south or east to west. For example, why do certain areas have more moisture? Propose explanations for local patterns, such as areas that differ somewhat from the regional patterns. What features are associated with these local patterns? Record your observations on the worksheet.

3. Compare the four globes, proposing explanations for why the patterns on one globe are consistent with those on the other globes. Pay special attention to how the amounts of precipitation and the kind of land cover relate to the two measures of atmospheric moisture (specific humidity and water vapor). Can you then explain some of the large-scale features of our planet? Write your observations and interpretations on the worksheet or answer questions online.

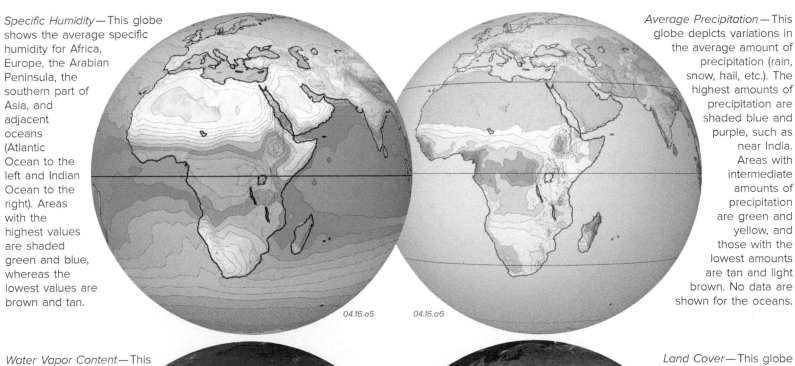

Specific Humidity—This globe shows the average specific humidity for Africa, Europe, the Arabian Peninsula, the southern part of Asia, and adjacent oceans (Atlantic Ocean to the left and Indian Ocean to the right). Areas with the highest values are shaded green and blue, whereas the lowest values are brown and tan.

Average Precipitation—This globe depicts variations in the average amount of precipitation (rain, snow, hail, etc.). The highest amounts of precipitation are shaded blue and purple, such as near India. Areas with intermediate amounts of precipitation are green and yellow, and those with the lowest amounts are tan and light brown. No data are shown for the oceans.

04.16.a5   04.16.a6

Water Vapor Content—This globe shows the amount of water vapor in the atmosphere, as measured by satellite. Lighter colors indicate more water vapor. Smaller bright spots scattered around the globe represent local storms and local areas of air with a high water vapor content, such as over the ocean near Antarctica.

Land Cover—This globe shows satellite data of the land surface. Dark green marks forested areas, including tropical rain forests. Light green shows grasslands and croplands. Tan and brown indicate areas with much less vegetation. In northern Africa and in the Arabian Peninsula, brown areas are rocky, whereas tan areas are mostly covered with sand (e.g., Sahara Desert).

04.16.a7   04.16.a8

4.16

5 Weather Systems and Severe Weather

OUR ATMOSPHERE IS DYNAMIC, featuring various types of weather systems, some relatively benign and beneficial, while others, like hurricanes and tornadoes, are dangerous and destructive. What types of weather do we experience, what causes weather, and what determines if a weather system is relatively mild or severe? In this chapter, we explore the main types of weather systems on Earth, including thunderstorms, hailstorms, tornadoes, and hurricanes.

05.00.a1 Eastern Colorado

Powerful tornadoes are among nature's most spectacular but frightening phenomena. A tornado forms beneath a thunderstorm as a swirling vortex of winds that can reach speeds of several hundred kilometers per hour. Tornadoes can have a tapered, almost delicate shape, or can be thick, dark, and ominous columns filled with debris.

How do tornadoes form, where do they occur, and how are they related to thunderstorms?

05.00.a2 Pickens County, AL

Tornadoes rampaged through the southeastern U.S. in April 2011, as part of the largest outbreak of tornadoes in history. Between April 25 and April 28, 358 tornadoes tore across Alabama and nearby states, completely destroying some houses, like the ones above, and killing more than 300 people.

How strong do tornadoes get, where do most occur, and what times of the day and year are they most likely to develop?

05.00.a3 Tuscaloosa, AL

The tornadoes carved swaths of destruction, totally leveling parts of neighborhoods (▲), while leaving nearby houses essentially undamaged, except for uprooted and toppled trees, downed power lines, and broken windows. In addition to the destruction caused by the tornadoes, deaths and damage resulted from hail, lightning, and flooding.

How do hail and lightning form, and are they related to one another?

05.00.a4

The tornado outbreak of late April 2011 was part of a large weather system, shown in the satellite image above. A swirling storm centered over the upper Midwest (in the upper part of the image) has a distinctive comma shape, with a long tail curving to the south and southwest. The tornadoes formed in the cluster of bright clouds to the southeast—out in front—of the main system.

What is a weather system, how does one form, and why do some acquire a curved shape, like a comma, or a coil?

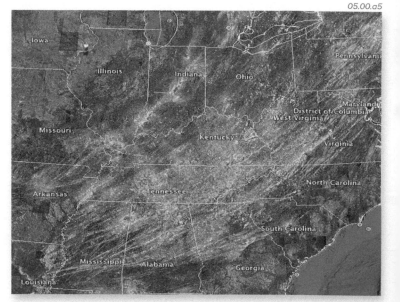

05.00.a5

Weather radar systems tracked each tornado as it moved across the region, as shown by this map depicting colorful lines (red signifies a very strong tornado) of tracks for each tornado. Some tornadoes stayed on the ground for hours and crossed multiple states. Radar images are a mainstay of daily weather reports.

How do radar systems work, what do they measure, and how do such data help us anticipate a storm's path and severity? Once a tornado or other severe storm is identified, how is the public notified?

TOPICS IN THIS CHAPTER

05.00.a6

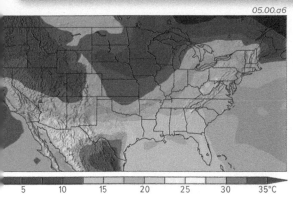

A map of temperatures for the day of peak tornado activity (◄) shows a boundary between warm air (shown in green and yellow) in the Southeast against cold air (shown in purple and blue) in the upper Midwest. A map of specific humidity (►) for that day indicates that the warm air was also humid, but the cold air was relatively dry.

Why are some masses of air warm, while others are cold? How are boundaries between warm and cold air related to thunderstorms and tornadoes?

05.00.a7

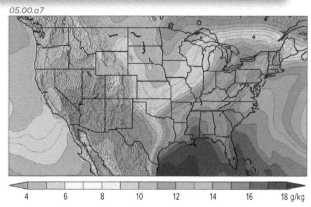

05.00.a8

Surface wind directions during this time (◄) show distinct patterns, with winds blowing from the south to the north in the area affected by the tornadoes (such as Alabama and Georgia). This wind direction brought warm, moist air northward. At higher levels, the winds turned to the right, moving storms from southwest to northeast. The colors on this map indicate wind speeds, with yellow and green representing the fastest winds.

Do regional wind patterns help us predict which way a storm will move?

Heavy precipitation was concentrated in the same region as the tornadoes (◄). In this map, purple and blue indicate the largest amounts of precipitation, which for these storms was as much as 40–50 cm (15 to 20 in.). The heavy rains, falling in a relatively short time period, produced severe flooding that destroyed roads and buildings and caused additional deaths.

What types of storms can produce very heavy precipitation?

05.00.a9

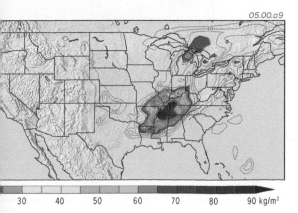

What Is Weather?

When we use the term *weather*, we are referring to *conditions in the atmosphere at some specific time and place*, whether it is right now, an hour ago, or sometime next week. Weather refers to the temperature, humidity, and windiness, and whether it is fair or there is precipitation. It also refers to whether an area is experiencing severe storms, such as a thunderstorm, tornado, hurricane, or snowstorm, or is in a time of exceptional dryness. Weather refers to the various aspects featured on these two pages, and much more! Examine the four maps on this page and figure out what the weather was like on April 27, 2011, where you live.

Weather affects all of our lives, from deciding what clothes to wear, what week to plant crops, or how soon to take shelter from some type of severe weather. In this chapter, we explore the fundamental processes that control weather and then examine the main types of weather phenomena, including rain and snowstorms, freezing rain, thunderstorms, hail, lightning, hurricanes, and, of course, tornadoes. Along the way, we explain features you can observe, like the types of clouds, to understand and predict the weather around you.

5.0

5.1 Why Does Weather Change?

THE ATMOSPHERE IS SO THIN that a single regional mass of cold or warm air can dominate the entire vertical extent of the troposphere. With time, such cold or warm air masses can expand or shrink, become colder or warmer, and can remain stationary or move great distances. In the narrow zone where the two air masses meet, weather conditions can change drastically across short distances and times.

A What Are Some Fundamental Controls of Weather?

Weather is controlled by the atmospheric processes described in previous parts of this book, and in fact those processes were described to get to this point—an understanding of weather and climate. Weather is a response to variations of insolation and resulting energy imbalances, to large-scale motion in the atmosphere, and to the important role of atmospheric moisture, including latent heat. Here, we briefly review the global-scale circulation patterns. Only the Northern Hemisphere is shown below, but similar patterns are also present in the Southern Hemisphere.

The most important features that control weather are the huge circulation cells that essentially divide each hemisphere into thirds (by latitude, not by area). Areas at high latitudes, poleward of approximately 60°, are under the effect of the *polar cell* and *polar easterlies*. High surface pressure dominates the pole.

The edge of the polar cell is marked by the *polar front jet stream*, a belt of high-speed winds that encircle the globe from west to east, high above the surface. The surface in this area has changeable and stormy weather due to rising air and shifting of the jet stream. Bends in the polar front jet stream are called *Rossby waves*, which exert a strong control on many weather systems.

A second jet stream, called the *subtropical jet stream*, encircles the globe (also west to east) high above the surface near 30° latitude. Between this jet stream and the equator, near-surface winds blow from the northeast, forming the *northeast trade winds*. Along the *equator*, high amounts of insolation-related heating cause rising air and associated low pressure. Winds converge from both sides of the equator, forming the *Intertropical Convergence Zone* (ITCZ).

Between the two jet streams, in the mid-latitudes and part of the subtropics, the general circulation is from the west—the *westerlies*. There is not a well-established, upper-level circulation cell over this region.

Between the subtropical jet stream and ITCZ is the *Hadley cell*, fed by rising air in the ITCZ. The cell descends in the subtropics, causing the *subtropical highs* (high pressure), with their associated normally dry conditions.

05.01.a1

B What Is an Air Mass?

05.01.

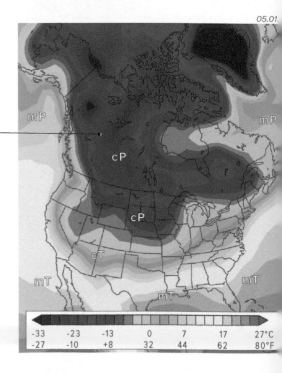

The pattern of global circulation helps form large bodies of air, called *air masses*, that have relatively uniform temperatures and moisture over large horizontal distances. We identify different types of air masses according to their temperature, moisture, and geographic characteristics.

The air-temperature map to the right depicts a cold air mass, represented by purple and blue, over the northern part of North America, including the northern Great Plains. A warmer air mass exists farther south, covering most of the southern and eastern U.S. The boundary between the two air masses is along the steep color gradient (a large change in color over a short distance) between the purple and green areas.

	Dry	Humid
Cold	**A, cP:** Stable, often with surface inversions; fair weather	**mP:** Unstable, often with cold air overlying warmer air adjacent to oceans
Warm	**cT:** Unstable, with hot surface and cooler air aloft, but lack of moisture	**mT, E:** Often unstable, with abundant moisture, especially E

The classification system for air masses uses a two-letter abbreviation. The first letter is lowercase and signifies the moisture characteristics, where "c" is for continental (dry) and "m" is for "maritime" (humid). The second letter is uppercase and refers to the temperature characteristics: "A" for Arctic (or Antarctic), "P" for polar (somewhat warmer than Arctic), "T" for tropical, and "E" for equatorial. An "A" air mass is inherently continental (dry), because water in such areas is often frozen. An "E" air mass is inherently maritime, because so much of the air at equatorial latitudes is over or near a source of water.

| -33 | -23 | -13 | 0 | 7 | 17 | 27°C |
| -27 | -10 | +8 | 32 | 44 | 62 | 80°F |

Continental Air Masses

Hot, dry air masses, typically occurring over land, are continental tropical (cT) air masses. Note that cT air masses occur in the Desert Southwest of North America, the Sahara and Arabian Deserts, Tibet, and central Australia. In areas under the influence of cT air masses, conditions can be oppressively hot and dry for many months during the year.

Arctic or Antarctic (A) air masses are formed near the North and South Poles. Not only are they extremely cold, but they are also very dry because the water-vapor capacity is low in such cold environments. Cold, dense air has a tendency to sink, and therefore, these air masses reinforce the region of polar high pressure (polar high) that resides over each pole. Surface winds near the pole move east-to-west (easterlies), so the surface A air masses tend to be steered westward as they drift away from the poles. The paths of A air masses can also be affected by curves in the upper-level polar front jet stream (Rossby waves), which can steer an A air mass toward the northern U.S. or into central Eurasia. These are the "Arctic blasts" that bring very cold air into the northern U.S.

Continental polar (cP) air masses are located farther away from the poles than A air masses. They are cold and dry, but less so than A air masses. cP air masses form over cold, inland areas, like northern Canada, Siberia, Mongolia, and north-central Europe. The southern edge of cP air masses is the polar front. In the winter, cP air masses can move significant distances equatorward, bringing cold air into the U.S. and southern Eurasia. In the summer, cP air masses are not as cold and do not reach as far south.

05.01.b2

Mollweide Projection

Maritime Air Masses

Due to their proximity to the equator, *equatorial (E) air masses* are very warm and moist. The high temperature and moisture support instability to great heights in the atmosphere, resulting in tropical clouds and storms. Storm development is aided by the general circulation of the atmosphere, which in equatorial regions promotes rising of the air, especially along the ITCZ.

Maritime tropical (mT) air masses form over warm, oceanic regions. mT air masses are warm and moist, and they have influence over much of the Earth because so much of the world (particularly in the low latitudes) is covered by ocean. For example, mT air tends to make most of the Atlantic Ocean and adjacent eastern U.S. warm and humid in summer. mT air masses are commonly near the subtropics.

Maritime polar (mP) air masses are cool and humid. They are not as cold as cP air masses because oceanic regions are not as cold as continental areas at the same latitude, particularly in winter. mP air masses are a cold, damp influence in places like Seattle and most of western Europe, as the westerlies push the cool, damp air eastward from their oceanic source region.

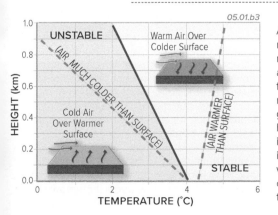

05.01.b3

Any air mass acquires characteristics of the region where it forms—its *source region*. Air masses that form inland or over frozen ocean areas (e.g., cP) are dry, whereas air masses that form over oceans (e.g., mT) are humid. Air masses that form in the low latitudes are generally warmer than those that form over high latitudes. When an air mass migrates from its source region across a different region (◄), its temperature and moisture will be modified, which can in turn affect the lapse rate, stability of the air, and how much mixing occurs within the air mass.

Before You Leave This Page

✓ Summarize the characteristics of each air mass type, where it is typically formed, and how it influences a region's climate.

✓ Explain factors that influence the rate at which an air mass acquires its characteristics and is modified as it migrates.

5.1

5.2 What Are Fronts?

THE NARROW ZONE separating two different air masses is called a *front* and is often the site of rising atmospheric motion. Whenever different air masses meet along a front, the less dense, warmer air mass is pushed up over the more dense, colder one. If the rising air cools to its dew-point temperature, cloud formation will begin, perhaps followed by precipitation. There are several types of fronts, depending on the manner in which one air mass is displacing the other and whether the front is moving. Fronts are associated with many of the storms most regions experience.

A What Processes Occur Along Cold and Warm Fronts?

1. The map below shows several fronts as decorated lines, among the areas of high pressure (H) and low pressure (L). The thin gray lines are isobars (lines of equal pressure), and the background of the map is a satellite image showing cloud cover. Examine each of the features on this map and note any correspondence between the fronts, high and low pressure, and cloud cover.

2. The blue lines represent *cold fronts*, the leading edges of cold air masses. Blue triangles along a cold front indicate the direction that the cold air mass is pushing. The east-facing triangles on the long cold front in the center of the U.S. indicate that the cold front is pushing eastward. Cold fronts are usually associated with *cP* but can also occur along the leading edge of an *A* or *mP* air mass.

3. The red lines represent *warm fronts*, the leading edges of warm air masses. The red semicircles along a warm front indicate the direction in which the warm air mass is moving. The warm front over northwestern North America is moving east. Warm fronts are usually associated with *mT* or *cT* air masses.

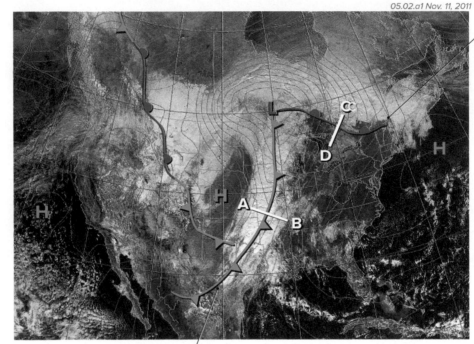

05.02.a1 Nov. 11, 2011

4. Let's examine the weather in the eastern half of the U.S., beginning with the long cold front. As the front and associated mass of cold, dense air move east, the less dense warmer air that is east of the front is displaced and forced upward.

6. To the north is a north-moving warm front. Air south of the warm front (like at "D") is warm and less dense, so it gets wedged over the more dense, cooler air north of it (at "C"). The gentle slope of this rising motion causes a shield of stratiform clouds over eastern Canada.

5. Warmer air ahead of the cold front (such as at B) moves up and over the colder air, creating cumuliform cloud cover. The air west of the cold front (at "A" for example) is cold and dense, so it tends to sink, associated with increasing surface pressure (H). The pressure gradient is very steep, as reflected by closely spaced isobars near the cold front.

Cold Fronts

7. The figure below shows a cross section through a cold front along line A-B on the map above.

The cold air mass displaces the warm air mass. Cold air is more dense and forces the warm air upward. If the rising warm air cools to its dew-point temperature, cloud cover and perhaps precipitation occur along the front.

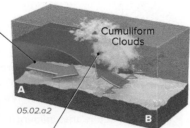

Cumuliform Clouds

05.02.a2

Warm Fronts

8. The figure below shows a cross section through a warm front along line C-D on the map above.

As the warm air mass follows and displaces the cool air mass, it slides up over the colder air. If the rising air cools to the dew-point temperature, stratiform clouds and perhaps precipitation occur along the warm front, but spread out over a significant width.

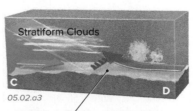

Stratiform Clouds

05.02.a3

Comparing Warm and Cold Fronts

9. Compare the cold-front and warm-front figures, observing similarities and differences, then read the text below.

The slope at which a warmer air mass is pushed upward along a cold front is much steeper than the slope at which it is forced upward along a warm front. This tends to support cumuliform cloud formation along cold fronts, but stratiform cloud cover along warm fronts. As a result, precipitation along cold fronts tends to be intense, localized, and short-lived, whereas precipitation along warm fronts tends to be light, widespread, and long-lived. Use these tips the next time a weather front crosses your hometown.

B What Is a Stationary Front?

Sometimes, for a period of several hours or days, neither the cold air mass nor the warm air mass is displacing the other. There is a temporary stall in motion of the front, and so such a stalled front is a *stationary front*. On a weather map, a stationary front, such as occurs here along the Atlantic Coast of North America, is represented by alternating blue and red line segments with alternating blue triangles and red semicircles. The blue triangles point away from the cold air, and the red semicircles point away from the warm air, as shown in the cross section along E-F on the map. Stationary fronts are associated with stratiform and cumuliform clouds.

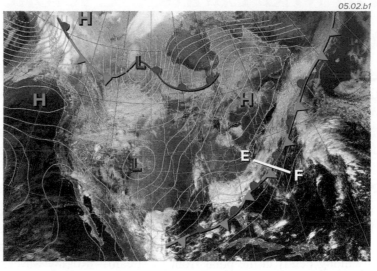
05.02.b1

05.02.b2

On this stationary front, the blue triangles point east, so the cold air is trying to push east and must therefore be on the opposite side of the front (on the west side). The warm air is therefore on the east side. A band of clouds and precipitation is along the front, mostly on the cold air side (west of the front). Since a station-ary front is not moving, prolonged precipitation can occur as the front sits over the same place for an extended time.

C What Types of Weather Are Characteristic of Cold, Warm, and Stationary Fronts?

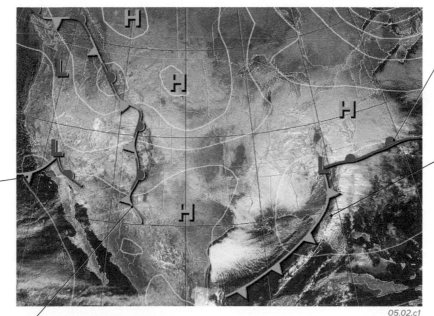
05.02.c1

1. On this map, examine the fronts, locations of high and low pressure, and the underlying satellite image showing the distribution of clouds. Find a cold front, warm front, and stationary front and for each one sketch a simple cross section across the front, showing which way it is moving. After you have done this, read the rest of the text around the figure.

2. For the storm system in the Southwest, the blue line marks a cold front that is moving to the south. The red line is a warm front moving to the northeast, but a warm front slopes opposite to the way it is moving (see the cross section on the previous page), to the southwest in this case.

4. In the East is an east-west oriented red line that marks a warm front. The semicircles point north, so this is the way the front is moving.

5. In the Southeast, the blue line is a cold front. The cold air is on the northwest side, and the front is moving to the southeast. The warm front and cold front join at an area of low pressure, marked by the red L. Symbols for four types of fronts are shown below. We discuss occluded fronts later.

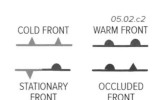
05.02.c2

COLD FRONT   WARM FRONT

STATIONARY FRONT   OCCLUDED FRONT

3. The red and blue line along the mountains in the west-central part of the country is a stationary front. The cold air is trying to move to the west, so it must be on the opposite side (east). Based on the cross section above, the front slopes away from the cold air.

6. This figure (▶) focuses on the low-pressure area in the East where the cold front (blue) meets the warm front (red). In this situation, the two fronts divide the region into three fairly distinct areas, called *sectors*. We reference each sector by the tempera-ture of the air in that region relative to the two other sectors. The *cold sector* is behind the cold front and naturally has the coldest air.

05.02.c3

7. The region north of the warm front is intermediate in temperature and is referred to as the *cool sector*.

8. The air in front of the cold front is the warmest and so is called the *warm sector*. The counterclockwise rotation around the low is bringing air north from lower latitudes and from oceans in the warm sector. Clouds along the cold front are clumpy and probably cumuliform, but clouds in other areas are less clumpy and so are probably mostly stratiform.

Before You Leave This Page

✓ Identify cold, warm, and stationary fronts on a weather map and suggest what types of air masses could be on each side of the front.

✓ Sketch a cold, warm, and stationary front in cross section, including typical cloud types and the weather likely to be associated with each front.

5.2

5.3 Where Do Mid-Latitude Cyclones Form and Cross North America?

FRONTS generally do not exist in isolation, but typically form, migrate, and fade away as part of a larger system called a *mid-latitude cyclone*. Cold and warm fronts usually trail from a central core of low pressure—a *cyclone*. While popular culture uses the term "cyclone" to refer to a tornado or other form of wind storm, a cyclone is really any enclosed area of low pressure. Mid-latitude cyclones migrate across Earth's surface guided by large-scale atmospheric circulation, like the polar front jet stream and the westerlies.

A What Surface and Upper-Level Conditions Form Mid-Latitude Cyclones?

Cyclones originate in various places as long as conditions in the lower and upper atmosphere are favorable for cyclogenesis (formation of a cyclone). Commonly, however, they develop downwind of mountain ranges or just offshore of cold land. In addition to forming near certain surface conditions, cyclogenesis is favored by certain types of flow in the upper atmosphere, especially along bends in the polar front jet stream—Rossby waves.

Leeward Side of Mountains

As air moves downslope on the lee side, it becomes stretched vertically as the depth of the atmosphere increases. In doing so, it acquires an increasing tendency for counterclockwise spin, or *vorticity*, wherever cold and warm air masses meet. In the U.S., westerlies cause cyclones to form east of large mountain ranges, such as the Rocky Mountains.

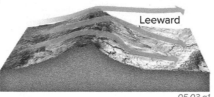

Leeward

05.03.a1

Offshore of Cold Land

Cyclones also form over water near cold land. This can occur during the colder times of the year, when the coastal land has cooled more rapidly than the adjacent water. Rising air over the ocean forms an area of low pressure just offshore. This, plus winds blowing toward the pole, causes the cold-warm interface along the coast to initiate potentially violent storms because of the abundance of moisture.

05.03.a2

Polar Front Jet Stream and Rossby Waves

1. A jet stream can develop bends, much like a meander in a river. In the Northern Hemisphere, a trough in the polar front jet stream is a bend to the south and a ridge is a bend back to the north. As the fast-moving air passes from a trough to a ridge it usually accelerates.

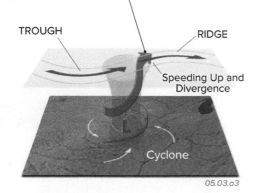

TROUGH   RIDGE

Speeding Up and Divergence

Cyclone

05.03.a3

2. As the air spreads out laterally, the decrease in pressure sets up an upward flow of air from lower in the atmosphere to the part of the jet stream that is diverging (spreading out). As a result, an area of low pressure forms at the surface and begins rotating, forming a cyclone. Times when the Rossby waves are most amplified (curved) are favorable for the formation of mid-latitude cyclones.

3. The map on the right shows how Rossby waves appear on a map showing upper-level air pressures. To show upper-level pressures, we commonly use a *constant-pressure map*, which plots lines, called *isohypses*, to represent the height (in meters) that you would have to go up in the atmosphere in order to reach a certain pressure. Higher heights correspond to higher pressure aloft. In this map, darker is low heights (low pressure) and light shading is high heights and high pressure. The dark, nearly circular feature is an area of low pressure, or a *trough*.

4. This map shows isohypses for 300 mb of pressure, high enough that air at this height only experiences approximately 30% of the atmospheric pressure compared to sea level (an average of 1,013 mb). A 300 mb pressure is up about 9,000 m (9 km), at the heights where the jet streams operate. The scale for the map is given here (▶), with heights in meters.

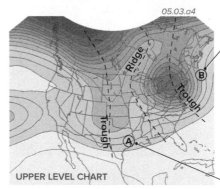

05.03.a4

Ridge

Trough

Trough

B

A

UPPER LEVEL CHART

GEOPOTENTIAL HEIGHT AT 300M

8950 9200 9400 9650m
05.03.a5

5. Cyclone B is on the flank of the deep trough of low pressure, on the trough-to-ridge side of a Rossby wave. In this position, an upward flow of air into the upper atmosphere helps strengthen the cyclone below. The cyclone will move to the northeast parallel to the isohypses, continuing to be strengthened.

6. Note that the pressure gradient is weaker near cyclone A. There is some lateral spreading (divergence) of the air, indicated by the shape of the isohypses, north and east of A, where the isohypses spread apart. Remember that air at these heights generally flows parallel to lines of equal pressure (equivalent to isohypses), so if isohypses diverge or converge, the airflow does too. The divergence at the 300 mb level helps draw air up from the surface, strengthening the cyclone. In the Northern Hemisphere (the case shown here), air overall flows from west to east (left to right).

B | What Are Common Tracks for Mid-Latitude Cyclones Across North America?

There are several common origins and paths of cyclones, influenced by the position of mountains and coastlines. Once formed, cyclones are guided by large-scale patterns of atmospheric circulation and so follow similar paths—or *storm tracks*—across the surface. The combinations of specific sites of formation and common tracks produce storms with similar characters, paths, and associated weather, like those shown below for North America.

1. Low pressure off the west coast of North America, in the Gulf of Alaska, can spawn cyclones that bring cold, moist air from the ocean inland across Canada and the Pacific Northwest. These storms can deliver heavy rain along the coast and snow at higher elevations. Due to orographic effects, precipitation is especially heavy along west-facing slopes of coastal mountains.

2. Cyclones also form on the leeward sides of major mountain ranges, like the Canadian Rockies in British Columbia and Alberta. A storm formed here commonly incorporates cold Arctic air and is called an *Alberta Clipper.* Similar storms form east of the southern Rocky Mountains in or near Colorado, but are not as cold.

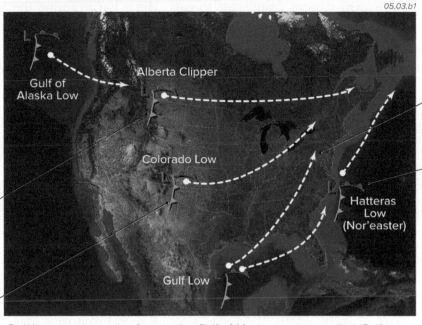

05.03.b1

5. Note that although the storm tracks come close together near New England, when the storm systems arrive, they are often older, having spent most of their latent energy before they track to the New England area. So, New England has many cloudy days but less severe weather than places experiencing less mature storms.

4. During winter, low pressure commonly occurs over the Atlantic Ocean, often near Cape Hatteras in the Outer Banks of North Carolina. Storms formed at this general area, called *Hatteras Lows,* move northward where they are also called *Nor'easters.* They are challenging to forecast because a slight variation in track can mean the difference between a foot or more of snow and none at all in the heavily populated areas of the Northeast.

3. Wintertime lows also form in the Gulf of Mexico, and are called *Gulf Lows.* They arise because of the contrast between the cold land and the waters of the Gulf, which remain fairly warm.

---

6. The two maps below show a surface cyclone, located at the low-pressure area labeled "L." The bottom map is two days later than the top map. The lines are 300 mb isohypses and are color coded to indicate their position with regard to the upper-level trough in the center of the top map. On these maps, note the correspondence of the surface cyclone and upper-level patterns.

7. The maps below show weather maps of the same cyclone (and an anticyclone to the west as the blue "H"), with fronts depicted using standard symbols for warm, cold, stationary, and occluded fronts. The lines represent surface air pressures.

8. The region shown is governed by west-to-east winds (westerlies) and movement of the Rossby waves, so the cyclone and anticyclone both generally move toward the east. However, the cyclone turns northeast and the anticyclone turns southeast, both moving parallel to the upper-level isohypses (left pair of maps). Mid-latitude cyclones will generally track northeastward, following the trough-to-ridge side of the Rossby wave. Accelerating and diverging upper-level winds on these sides will move the surface cyclone toward the northeast and strengthen it. Surface anticyclones often move southeastward, where sinking motion is supported.

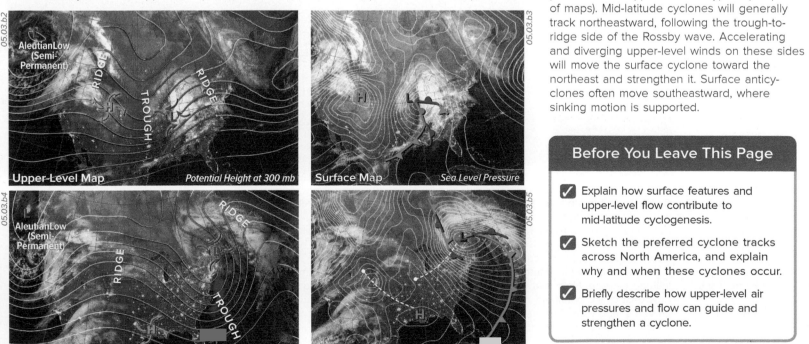

Before You Leave This Page

☑ Explain how surface features and upper-level flow contribute to mid-latitude cyclogenesis.

☑ Sketch the preferred cyclone tracks across North America, and explain why and when these cyclones occur.

☑ Briefly describe how upper-level air pressures and flow can guide and strengthen a cyclone.

5.3

5.4 How Do Mid-Latitude Cyclones Move and Evolve?

ONCE FORMED, MID-LATITUDE CYCLONES migrate across the surface and commonly evolve through a series of steps, due to the way that winds circulate around the area of low pressure. Circulation around the cyclone will steer the trailing fronts in a counterclockwise direction (in the Northern Hemisphere), while west-to-east motion of the westerlies and Rossby waves shift the entire storm system from west to east. Here, we examine the life stages of a mid-latitude cyclone as it moves west-to-east across North America.

A What Are the Three Stages in the Life Cycle of a Mid-Latitude Cyclone?

Formation (Cyclogenesis)

1. The map on the left shows several fronts as decorated lines, areas of high and low pressure, and isobars as thin lines. A mid-latitude cyclone (Cylone A on the map) in late September begins when a relatively cold air mass and a warmer one meet along a frontal boundary. Here, cold and warmer air masses meet along a stationary front near Cyclone A. Another cyclone (B) is near New England.

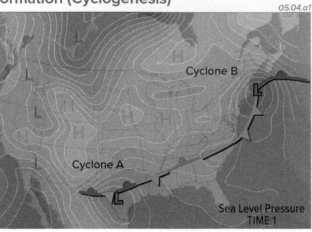
05.04.a1

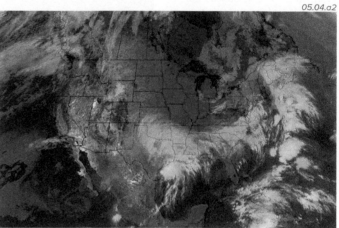
05.04.a2

2. At some point along this boundary, a small amount of surface convergence occurs because of local circulation features that push one air mass into the other, because of topographic influences, or by some other local mechanism. This convergence causes the low pressure to intensify.

3. This satellite image shows clouds and precipitation at the same time as represented by the weather map on the left. Note that the clouds and precipitation are associated with both cyclones on this map (Cyclone A over Texas, and Cyclone B over New England).

Maturity

4. At a later time, the low pressure associated with Cyclone A has moved east and intensified with a stronger pressure gradient, causing more surface air to move toward it. As this air begins to converge from all directions, the Coriolis effect causes the air to have an apparent deflection to the right. This results in counterclockwise flow that pushes cold air southward, turning one segment of the stationary front into an eastward-advancing cold front.

05.04.a3

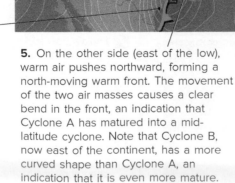

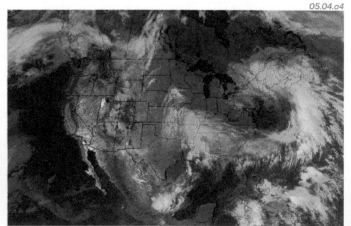
05.04.a4

5. On the other side (east of the low), warm air pushes northward, forming a north-moving warm front. The movement of the two air masses causes a clear bend in the front, an indication that Cyclone A has matured into a mid-latitude cyclone. Note that Cyclone B, now east of the continent, has a more curved shape than Cyclone A, an indication that it is even more mature.

6. Both cyclones are visible in the satellite image above. Although not obvious on this image, Cyclone A now has distinct cold, warm, and cool sectors of air. The "cold" sector is behind (west of) the cold front, whereas the "warm" sector is behind (south of) the warm front. The cool sector is in front of (north of) the warm front—neither the warm front nor the cold front have passed through this sector. On the satellite image, dense cumuliform clouds (along the south edge of the image) form along the cold front, while indistinct stratiform clouds form along the warm front and in the cool sector. Advancing cold air can easily displace warmer air, but the warmer air has more difficulty moving the cool air; with time the cold front advances faster than the warm front, as in Cyclone B.

Occlusion

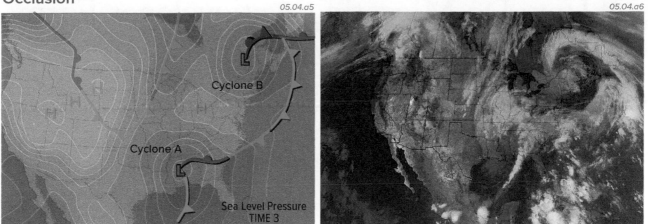

05.04.a5

05.04.a6

7. With time, the westerlies push both weather systems toward the east or northeast. Eventually, the cold fronts will catch up with, and even overtake, the warm fronts as both flow counterclockwise around the low-pressure areas. When this happens, the warmer air mass is lifted above the surface, because it is less dense than both the cold air and the "cool" air. This process is called *occlusion*, and a front that has experienced occlusion is an *occluded front*. In occlusion, the moisture is moved counterclockwise around the low, so that the heaviest precipitation may be northwest of the low.

8. Occluded fronts, like that near cyclone B on the map above, are purple on weather maps with the triangles and semicircles pointing in the direction that the system is moving. Though many mid-latitude cyclones never occlude, occlusion often, but not always, signals the end of the mid-latitude cyclone's life, because by this time the cold air has warmed and the warm air has cooled. Some occluded fronts can be sustained for some time.

OCCLUDED FRONT

9. Notice on the satellite image that the sky over the occluded front is filled with clouds. The cumuliform cloudiness associated with the cold front more or less merges with the stratiform clouds associated with the warm front. Occluded fronts, however, can have a wide variety of cloudy features.

10. This figure (►) contains three cross sections that capture different stages in the process of occlusion. Occlusion begins nearest to the low-pressure core because there the cold front has the least distance to catch up to the warm front, and it proceeds away from the low over time. To visualize what occurs, move through the three cross sections, which represent different places, but also successive stages in the occlusion process.

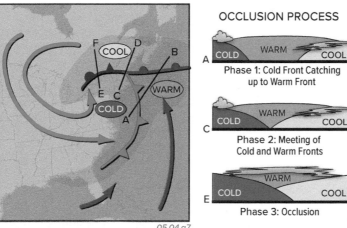

OCCLUSION PROCESS

Phase 1: Cold Front Catching up to Warm Front

Phase 2: Meeting of Cold and Warm Fronts

Phase 3: Occlusion

05.04.a7

11. In cross section A–B, the cold front is moving to the east and the warm front is moving to the north, but the cold front is moving faster and so it is getting closer to the warm front.

12. Here, the cold front has caught up with the warm front, trapping the warm, moist air above the mass of cold air and the mass of cool air, which are in the process of joining together.

13. At segment E–F, the two air masses are fully beneath the warm air, lifting it higher. As the warm air is forced up, it cools, forming clouds and precipitation. The air north of the warm front is cool but not cold because it is blocked from receiving a fresh supply of cold air from the air mass to the west.

B   How Do the Great Lakes Interact with Cyclones?

1. The Great Lakes in the north-central part of the U.S. are famous for interacting with winter storms, producing intense snowfall called *lake-effect snows*. The largest amounts of lake-effect snow occur after mid-latitude cyclones and associated cold fronts have moved eastward across the Great Lakes, and higher pressure behind them begins to approach the area. The circulation over the water causes moist air to flow onto land and dump huge amounts of snow on land south and east of the lakes, including the cities of Syracuse and Buffalo in upstate New York.

5.04.b1

Moist Air

2. As a cyclone moves across one of the lakes, it encounters water temperatures that are somewhat warmer than the air or the surrounding land. This is largely due to water changing temperature more slowly than air or land from fall to winter. In winter, especially early winter, the Great Lakes are much warmer than the air above them or the land beside them. This causes steep environmental lapse rates, destabilizing the atmosphere and strengthening a cyclone passing over the Great Lakes.

3. Having flowed across the Great Lakes, the moist, destabilized air is pushed inland by an approaching anticyclone, depositing snow on adjacent downwind land, especially where topography, such as the mountains of upstate New York, cause orographic enhancement.

Before You Leave This Page

✓ Sketch and describe the three stages in the life cycle of a mid-latitude cyclone (cyclogenesis, maturity, and occlusion).

✓ Sketch and explain how and why the process of occlusion occurs, using a map or cross section to illustrate the progression with time.

✓ Sketch and describe what causes lake-effect snows in the Great Lakes region.

5.5 How Do Migrating Anticyclones Form and Affect North America?

ANTICYCLONES ARE ROTATING SYSTEMS of high pressure. Most migrate across Earth's surface, guided by broad-scale atmospheric circulation, but some stay or reappear in the same general area year after year. As migrating anticyclones approach, pass over, or leave an area, they cause changes in wind direction, temperature, and moisture. They are typically associated with non-stormy weather, but they sometimes cause fog and an increase in smog. They can block the motion of cyclones or guide them one way or another, and they influence our weather in other ways.

A Where Do Anticyclones Occur?

There are two types of anticyclones: *semipermanent anticyclones* that remain in the same general area for entire seasons and year after year, and *migrating anticyclones*. Both types are zones of high pressure associated with clockwise rotation in the Northern Hemisphere and counterclockwise rotation in the Southern Hemisphere, opposite to the rotation of cyclones. Unlike cyclones, no fronts trail from anticyclones, because the sinking air doesn't allow for cooling, cloud formation, and the release of latent energy needed for generating strong winds. Pressure gradients around anticyclones are generally weak, as are winds. Where cyclones tend to bring stormy, unsettled weather, anticyclones bring atmospheric stability and fair weather.

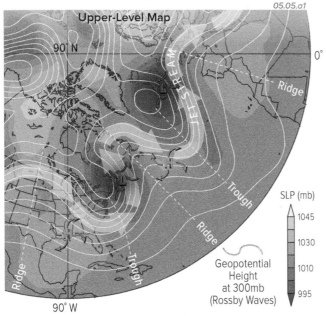

05.05.a1

This map shows two surface anticyclones, each shown as a blue "H," and two surface cyclones, each marked by a red "L." The polar front jet stream is represented by large blue arrows. The lines are isohypses at the 300 mb level, and so depict *upper-level* pressures. The gray shaded areas show *surface* air pressures, with darker gray indicating lower pressures. Each cyclone is associated with an area of low *surface* air pressure.

The cyclones are beneath the east sides of the upper-level troughs (where air is diverging on the trough-to-ridge side of a Rossby wave). Note the positions of the anticyclones. In both cases, the anticyclones are beneath the east side of an upper-level ridge as defined by where the 300 mb isohypses bend toward the pole. That is, the anticyclones are beneath the ridge-to-trough side of a Rossby wave.

Since the positions of the surface anticyclones are set to some degree by the position of the Rossby waves, they can remain relatively fixed as long as the ridges in Rossby waves remain in about the same location. In this way, the positions of some anticyclones remain broadly fixed year after year, but they may strengthen, weaken, or shift laterally from season to season. These are the semipermanent anticyclones, and they include some familiar high-pressure areas, like the Bermuda-Azores High, the Hawaiian High, and the Siberian High. Just as there are these semipermanent anticyclones, there are *semipermanent cyclones*, like the Aleutian Low and the Icelandic Low. Which named pressure feature is the northernmost low-pressure zone shown on this map?

B Where Do Mid-Latitude Anticyclones Commonly Form?

High-pressure zones associated with anticyclones form in different settings than cyclones. Once formed, most mid-latitude anticyclones migrate in the direction of large-scale atmospheric circulation (west to east), all the while rotating clockwise (in the Northern Hemisphere). Many anticyclones result from convergence in Rossby waves.

Where upper-level air moves from the ridge to the trough of a Rossby wave, the air converges and slows down slightly. This in turn forces some of the air downward, toward the surface, causing an area of surface high pressure below the convergence. The air around this high pressure rotates, clockwise in the Northern Hemisphere and counterclockwise in the Southern Hemisphere. In this manner, a surface anticyclone forms below the ridge-to-trough part of a Rossby wave. As the air descends, it heats up adiabatically.

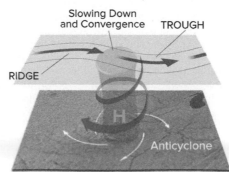

05.05.b1

Once formed, the mid-latitude anticyclone will migrate across the surface, guided by regional wind directions, such as the westerlies, and by the upper-level air circulation patterns. As we observed previously, anticyclones commonly move parallel to the isohypses, moving east and southeast across the U.S. Zones of high pressure can be formed in other ways, such as by descending air in the subtropical limb of the Hadley cell, forming subtropical high-pressure areas.

C How Do Anticyclones Typically Migrate Across North America?

Migrating mid-latitude anticyclones can cross the entire continent, driven to the east by the prevailing westerlies. They form and are likely to grow beneath the ridge-to-trough side of the Rossby wave, and they tend to migrate beneath this side of the wave as well, in a general east-southeast direction.

1. The maps below track the path of a surface anticyclone (shown with a blue "H" and clockwise rotation) during the course of several days. The shades of gray represent surface air pressure, with lighter gray indicating high pressure and darker gray indicating lower pressures. The yellow lines are isohypses at the 300 mb level, depicting upper-level conditions. The map also shows several surface cyclones (each as a red "L"), but our focus here is on the anticyclone. The legend is to the right (▶).

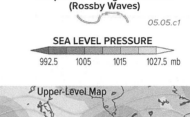

Geopotential Height at 300mb (Rossby Waves)
05.05.c1

SEA LEVEL PRESSURE
992.5   1005   1015   1027.5 mb

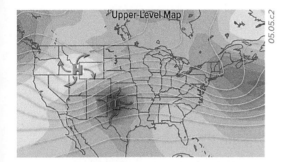
05.05.c2

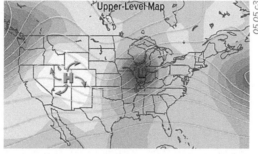
05.05.c3

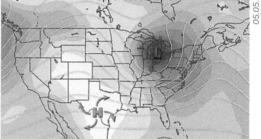
05.05.c4

2. In this first map, an anticyclone is located over western Wyoming, and a cyclone lies just to its southeast. There is a strong pressure gradient between the two, as implied by the rapidly changing shades of gray. This results in strong 300 mb winds between the two features and the tightly packed isohypses in the area.

3. On this second map, the anticyclone has moved south, into the Southwest. The clockwise circulation around the high causes a different wind direction on each side of the high. High pressure generally means clear skies, especially in this relatively dry region of the continental interior.

4. In this map, the anticyclone has moved to south Texas as the ridge overlying it flattened out. On this and the previous map, the anticyclone is on the east flank of an upper-level ridge—that is, the surface anticyclone is beneath the ridge-to-trough side of a Rossby wave.

D What Changes in Weather Occur When an Anticyclone Migrates Over an Area?

An anticyclone is characterized by descending air, so it tends to bring stability to the atmosphere, resulting in clear skies and fair weather, or fog under certain circumstances.

1. Air descending (▶) in the atmosphere is generally cooler, but warming adiabatically. In addition, the clear skies allow for more insolation during the day, warming the land and air. The same clear skies also allow much of this heat to radiate into space at night, resulting in cool nights and a large temperature contrast between day and night.

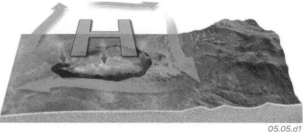
05.05.d1

3. The clockwise circulation can pin cold air against mountains when the cold air is too dense to move uphill. This often occurs on the western side of the Appalachian Mountains, and, if temperatures are cold enough, often creates temperature inversions with sleet, freezing rain, and snow.

2. The moisture in the surface air may be cooled and trapped by this descending motion, causing fog in certain settings. The downward flow can trap smog and other air pollution near the surface, especially if there is a temperature inversion.

4. As an anticyclone approaches an area (▶), winds will change direction depending upon the direction of approach (usually from the west or northwest). In front of an eastward progressing high (on the right in this view), cold air blows in from the north. After the high passes (like on the left in this view), warm winds blow from the south. North or south of the high, winds blow east or west. By employing this pattern, we can use wind directions to assess the larger weather picture and anticipate if a high-pressure zone is approaching or has passed. This approach works for cyclones, too, but the wind directions are opposite.

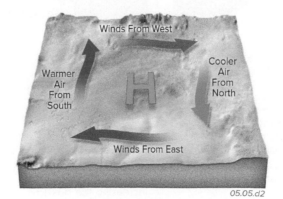
05.05.d2

Before You Leave This Page

✓ Sketch and explain the relationship between mid-latitude anticyclones and upper-level features, including Rossby waves.

✓ Summarize how most anticyclones migrate across North America.

✓ Describe the weather associated with anticyclones and how wind directions change when an anticyclone migrates over an area.

5.5

5.6 What Conditions Produce Thunderstorms?

THUNDERSTORMS ARE COLUMNS of moist, turbulent air with variable amounts of rain, strong wind, lightning, and hail. They are perhaps the most fundamental of all organized weather systems. They can provide needed rainfall for crops, but they can be accompanied by severe weather. We begin by examining the formation, growth, and decay of an individual thunderstorm that is independent of larger storm systems—the *single-cell thunderstorm*—and then explore the formation of thunderstorms that occur in clusters or in large, consolidated masses.

A What Stages of Development Characterize Single-Cell Thunderstorms?

Cumulus Stage

05.06.a1

1. The initial development of a single-cell thunderstorm—the *cumulus stage*—occurs when the surface heats more rapidly than the atmosphere above it. This causes the heated near-surface air to rise relative to adjacent air, forming an *updraft*. Usually, such conditions begin to appear in mid-to-late morning, particularly in summertime. If the air is moist, one or more isolated *cumuliform* clouds will begin to form, starting out as a cumulus cloud. No rain is falling during this stage—the cloud is still growing, as indicated by the puffy, cauliflower-shaped top. Most developing cumuliform clouds never get past the first (cumulus) stage, either because the atmosphere is not sufficiently unstable, or is not moist enough, or both.

Mature Stage

05.06.a2

2. The beginning of rainfall signals the onset of the next stage—the *mature stage*. At this stage, the updrafts that allowed the cloud to grow in the cumulus stage are accompanied by downdrafts induced by falling precipitation. Most *lightning* occurs during this stage. If the thunderstorm is too tall, its top can flatten against the tropopause, where the very stable conditions in the temperature inversion of the stratosphere forbid higher vertical development. The strong upper-level winds produce an anvil-shaped cloud top. Evaporation of some cloud water drops cools the downdrafts, particularly on hot summer afternoons. Upon hitting the surface, the downdraft will move laterally in the direction of winds that are pushing the storm, producing cool wind gusts—the *gust front* we feel before a storm.

Dissipating Stage

05.06.a3

3. Eventually, enough rain falls to cool the surface, removing the unstable conditions that caused rising motions. This *dissipating stage* is characterized by continued rainfall until much of the moisture is rained out of the cloud. At this point, the cloud begins to disappear, taking on a ragged, feathery appearance, as it mixes with dry air from outside the cloud and evaporates into the surrounding air. Such mixing in of other air is called *entrainment*. Water in the cloud that does not precipitate may eventually evaporate into the surrounding drier air.

4. The entire life cycle of an individual single-cell thunderstorm (▶) is on the order of 1–2 hours, but some last much longer, and several single-cell thunderstorms can occur in sequence. Unless they occur in association with larger systems like a mid-latitude cyclone, single-cell thunderstorms are far more likely to occur in summer than in winter. They are also far more likely to occur sometime between late morning and late afternoon than during any other time of day. Longer-lived and more severe thunderstorms can occur if conditions are especially favorable, such as abundant moisture or some other atmospheric phenomenon that enhances and strengthens the thunderstorm. Such thunderstorms can cause extensive damage, as described on subsequent pages.

05.06.a4 Wind River Mtns., WY

Cloud Mergers

05.06.a5

5. A single-cell thunderstorm can form or grow when two cumuliform clouds are forced together by winds. In this figure, adjacent fair-weather cumuliform clouds are supported by uplift associated with an unstable atmosphere, but no cloud has enough moisture and uplift to be dangerous by itself. When the clouds collide, the instability and associated uplift beneath the collision point dictate that the moisture from both clouds must move upward. This causes very rapid vertical development and a sudden storm that may be difficult to anticipate.

Upper-Wind Effects

05.06.a6

6. A thunderstorm that forms beneath strong winds aloft, such as the polar front jet stream, has an increased likelihood for strong development or intensification. Any increase in horizontal wind speed with height is called *vertical wind shear*. The fast winds in the top part of a tall thunderstorm, driven by the jet stream, may allow some of the rain from the storm to fall in areas separated from the updrafts. This reduces the possibility that the downdrafts associated with the falling precipitation will get in the way of, and cool, the updrafts that are sustaining the system. If vertical wind shear allows the precipitation to remain separated from the zone of uplift, the surface remains warm, allowing updrafts and overall atmospheric instability to continue for an extended time. This in turn allows the thunderstorm to persist and possibly grow stronger, increasing the possibility of more severe weather.

B What Conditions Form Large Thunderstorms or Clusters of Thunderstorms?

Thunderstorms can also occur in clusters, called a *multi-cell thunderstorm*, or can coalesce into a much larger single storm, called a *supercell thunderstorm*. Clusters of thunderstorms can form along a front or when the individual storms are sufficiently close to reinforce and strengthen one another. In either case, the thunderstorms can cause severe weather.

1. This figure (▶) shows two thunderstorms that are close enough to affect one another. In well-organized thunderstorms, downdrafts from one thunderstorm can contribute to updrafts in an adjacent thunderstorm. This happens because the downdraft of cold, dense air spreads out when it hits the surface, pushing warmer air upward, thereby strengthening a nearby thunderstorm or even triggering a new one. The downdrafts and resulting outflow can cause warm air ahead of the outflow (A) to be displaced upward to such an extent that it triggers another thunderstorm, or even

05.06.b1
Enhanced Uplift at "A" Causes Cloud Growth

Ⓐ

05.06.b3
cP
cT
mT

a row of thunderstorms, as shown in the radar image to the left. These types of systems, where downdrafts coming from a line of storms produce another line of storms, proceed in a "conveyor belt" fashion and can affect huge regions. These types of storms are called *multi-cell thunderstorms*.

05.06.b2 Binghamton, NY

2. Lines of thunderstorms commonly form along cold fronts, especially where the two air masses have very different moisture characteristics. This often happens as hot, dry (*cT*) air masses from the West interact with warm, moist (*mT*) air masses from the Southeast (◀). This interaction can be driven by movement of a cold front and mid-latitude cyclone, but both of these air masses will be in the warm sector of the cyclone, as shown here. Such air masses meet at a typically north-south boundary called the *dry line*. Along the dry line, the more humid air mass will be lifted up and over the drier air mass, assuming that both are at the same temperature, because humid air is less dense than dry air. The resulting storms can have an ominous front (▶), a *squall line,* and so are called *squall-line thunderstorms*. They can be very severe due to the abundant moisture in the humid air mass, often even more severe than the storms generated by the subsequent passage of the cold front.

05.06.b4 Belleville, KS

3. Some single-cell thunderstorms develop into tall and powerful *supercell thunderstorms* (▶). Like other large thunderstorms, supercell thunderstorms commonly have a flat-topped, anvil shape. The long point of the anvil generally indicates the direction the storm is moving (left to right in the figure below), because the strong upper-level winds push the entire surface storm in the same direction that they push its top. The anvil indicates the presence of vertical wind shear (increasing wind speed with height), which assists storm development by enhancing uplift and allowing precipitation to fall away from the updrafts.

4. Intense rain and hail form within supercells and fall toward the ground, often in brief but heavy downpours. Note that hail can form in other types of storms, too. *Lightning* and *thunder* are also present in most supercells. Lightning may discharge in the air or reach the ground.

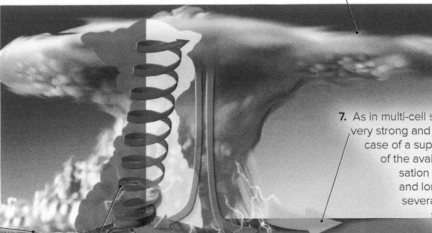

05.06.b5

05.06.b6 Humboldt, NE

7. As in multi-cell storms, outflow from downdrafts can be very strong and can be the catalyst of other storms. In the case of a supercell, the system is so strong that much of the available latent energy released in condensation and deposition is consumed in the large and long-lasting supercells. Such storms can last several hours, be 20–50 km in diameter, and support more catastrophic severe weather than other systems.

5. Supercells feature a rotating vortex called a *mesocyclone*. All supercells have at least one mesocyclone, and some supercells may support several. The rotating vortex is interpreted to have started out horizontal, where it was formed by horizontal shearing of the wind against the surface. Later, updrafts within the thunderstorm distort the vortex into a steep orientation.

6. Subsequent *updrafts* can spiral into the mesocyclone, strengthening it by tightening its rotation and ultimately leading to formation of a tornado. About one-third of mesocyclones form tornadoes.

Before You Leave This Page

✓ Sketch and explain the processes that occur during the life cycle of a single-cell thunderstorm, and what can enhance them.

✓ Describe the formation of either multi-cell or supercell thunderstorms.

5.6

5.7 Where Are Thunderstorms Most Common?

THUNDERSTORMS CAN OCCUR as isolated single-cell storms, forming over mountains and other local features, or can be embedded in larger weather systems, like mid-latitude cyclones or multi-cell thunderstorms. By knowing how thunderstorms form, can you predict where they should be most common?

A How Does Thunderstorm Frequency Vary Globally?

Space-based detections of lightning can be used to determine the global distribution of thunderstorms. Optical sensors on satellites use high-speed cameras to look into the tops of clouds and detect flashes of lightning. Observe the map below, which shows the average yearly counts of lightning flashes per square kilometer based on NASA satellite data.

1. These data represent lightning observed over 8 years. Places with the highest frequency of lightning strikes are red or black. Places where fewer than one flash occurred (on average) each year are white, gray, or purple, and places with moderate numbers of flashes are green and yellow. Observe the main patterns before you read the associated text.

2. The globes show that lightning strikes are much more common on land than over the oceans. Large landmasses can heat up during the day and cause free convection, resulting in thunderstorms. Above the oceans, thunderstorms are more frequent near coastlines than in areas far from a coast; this reflects the sharp contrasts in air masses that occur along coastlines.

05.07.a1

4. North America exhibits frequent thunderstorms, reflecting a supply of warm, moist air from the Gulf of Mexico, abundant solar heating of the land, and the frequent passages of fronts. Europe commonly lacks much contrast between air masses, as the westerlies push a continual parade of *mP* air masses across the continent. Also, the mountains mostly trend east-west, minimizing the interaction between very hot and very cold air masses. As a result, Europe has fewer thunderstorms than might be expected, considering the frequency of frontal passages and rainy days.

3. Most of the 2,000 or so thunderstorms occurring at any time in the world are located within the Intertropical Convergence Zone (ITCZ), especially in the afternoon. Abundant tropical moisture, strong surface heating, and vigorous trade-wind convergence are responsible. Note the high density of thunderstorms in central Africa, attributable to the ITCZ and a large continent. ITCZ-related thunderstorms are also abundant in southern Central America, Southeast Asia, and Indonesia (northwest of Australia).

05.07.a2

05.07.a3

5. Polar regions, including Antarctica, are too cold to have moisture sufficient for the condensation and deposition required to support thunderstorms, even in the summer. Both poles are among the places least likely to experience thunderstorms and lightning.

LIGHTNING FLASHES/KM²/YEAR

| 0.1 | 0.4 | 1.4 | 5 | 20 | 70 |

05.07.a4

Seasonal Variations

6. These maps show the variations in thunderstorm-associated lightning between summer and winter. The top map is for winter in the Northern Hemisphere and summer in the Southern Hemisphere. The bottom map shows the opposite (northern summer; southern winter). Examine these maps and compare the number of thunderstorms in summer versus winter for different regions.

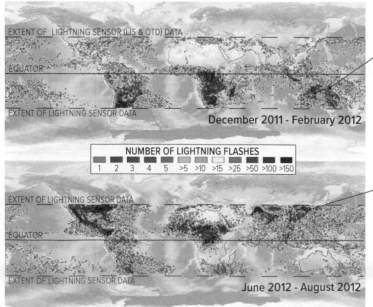

EXTENT OF LIGHTNING SENSOR (LIS & OTD) DATA

EQUATOR

EXTENT OF LIGHTNING SENSOR DATA

December 2011 - February 2012

NUMBER OF LIGHTNING FLASHES

| 1 | 2 | 3 | 4 | 5 | >5 | >10 | >15 | >25 | >50 | >100 | >150 |

EXTENT OF LIGHTNING SENSOR DATA

EQUATOR

EXTENT OF LIGHTNING SENSOR DATA

June 2012 - August 2012

7. From December through February, there are many more thunderstorms south of the equator than north of it. Note, for example, the difference in Australia and Southeast Asia between the two maps. During this time, thunderstorms are also abundant in southern Africa and in South America.

8. During June through August, the locus of thunderstorm activity shifts to the Northern Hemisphere, where it is summer. The difference in thunderstorm activity between the two maps is especially noticeable in North America and Central America. Thunderstorms in mainland Asia are tied closely to the summer monsoon, when surface heating combines with moist onshore flow to create strong convection. Central Africa, straddling the equator, reports thunderstorms during both seasons. Sinking air in the descending limb of the Hadley cell, combined with limited moisture, suppresses the development of thunderstorms in the Sahara Desert and Arabian Desert.

B How Does Thunderstorm Frequency Vary Throughout the Continental U.S.?

1. This map shows the frequency of thunderstorms in different parts of the U.S., as expressed by the number of days per year that an area has thunderstorms.

2. Few thunderstorms occur on the Pacific Coast, even though there are many rainy days. Surface heating is seldom strong enough to create free convection. The difference between air masses across frontal boundaries is too slight to generate vigorous uplift. Cold air masses are not bitterly cold, and warm air masses are not too hot due to the moderating influence of the Pacific Ocean. Orographic uplift can generate an occasional thunderstorm, but even these are not too frequent as most of the coastal ranges are not very high and the highest ones are farther north, next to cold ocean waters.

3. Thunderstorms are somewhat more common elsewhere in the West, even though moisture is relatively scarce. More thunderstorms occur on the peaks of the highest mountains, where surface daytime heating moves air upslope from all sides, converging at the peaks and forcing additional uplift.

4. In the Midwest and Northeast, thunderstorms increase in frequency from north to south. Water-vapor capacities in the northern states are so low most of the year that they limit the supply of atmospheric moisture available for release as latent heat during condensation, which is essential to generate severe storms.

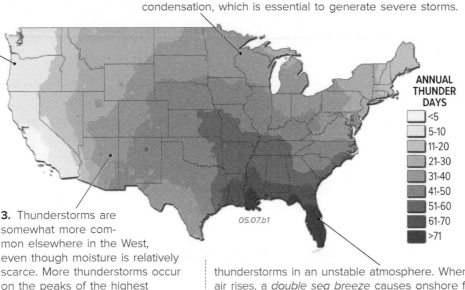

05.07.b1

ANNUAL THUNDER DAYS
- <5
- 5-10
- 11-20
- 21-30
- 31-40
- 41-50
- 51-60
- 61-70
- >71

5. The Southeast has the highest frequency of thunderstorms in the U.S., with the Florida peninsula experiencing the most thunderstorms in the country. Winter and early spring thunderstorms mostly originate from cold fronts trailing from Gulf Lows and Colorado Lows. Spring, summer, and, to some extent, fall thunderstorms occur because of convective afternoon thunderstorms in an unstable atmosphere. When the surface heats up and air rises, a *double sea breeze* causes onshore flow from both the Gulf of Mexico and the Atlantic to fill in the low pressure over Florida. Late summer and fall also bring tropical cyclone-induced thunderstorms. The region has a combination of warm seawater, warm air, a sea-land contrast, and a position near storm tracks—a perfect recipe for thunderstorms.

Seasonal Variations

6. These maps show the frequency of thunderstorms for different months. In March and April, thunderstorms occur inland along the Lower Mississippi Valley because cold and warm air masses come into contact with each other at strong cold fronts, accompanied by moisture from the Gulf of Mexico.

7. During May, the peak in Texas results from the combination of early summer convection on some days with late season cold-front activity on others.

8. In June, the location of thunderstorm activity shifts farther north and inland, onto the Great Plains. Land surfaces are starting to warm up, but cold air is nearby in Canada and sometimes flows into the plains. Cyclonic activity enhances the likelihood of thunderstorms in the central U.S. Thunderstorm activity begins to pick up in Florida, with the arrival of summer.

MAR
SEP
APR
AUG
MAY
JUL
JUN

05.07.b2

THUNDER DAYS
- <0.5
- 0.5–2.4
- 2.5–6.4
- 6.5–8.4
- 8.5–2.4
- 12.5–5.4
- >15.5

9. The September peak in Florida is the result of tropical cyclones (of strengths varying from tropical waves to hurricanes), enhancing the convectional thunderstorm activity. Hurricane season peaks in August and September. The rest of the southeastern U.S. has the same ingredients for thunderstorm formation as Florida, except for the double sea breeze.

10. Most of the U.S. sees a summer peak in thunderstorm days. By July, surface heating allows for destabilization of the atmosphere during afternoon hours. Summer cold fronts supplement these convectional thunderstorms in the Northeast and north-central states, while tropical cyclones can add to the frequency of thunderstorms in the Southeast. Increased thunderstorm activity in the Southwest and intermountain states of the West largely reflects the onset of the Southwest monsoon.

Before You Leave This Page

☑ Summarize variations in thunderstorm frequency around the world, explaining the factors influencing the frequency and seasonality.

☑ Describe which parts of the U.S. experience frequent thunderstorms, explaining the factors influencing the frequency and seasonality.

5.8 What Causes Hail?

HAIL IS A BALL OF ICE that forms under freezing temperatures within a cumulonimbus cloud and that subsequently falls toward the surface. Large hail that reaches the surface can do so at high-enough speeds and with enough force to smash windows, dent cars, and destroy entire fields of crops. How does hail form, what controls the size of individual chunks of hail, and where and when is hail most common?

A How Does Hail Form?

Most of the largest hail forms in association with a supercell thunderstorm, but hail can also occur with other types of thunderstorms. The photograph to the right shows pea-sized pieces of hail, or *hailstones*, and the figure below illustrates the interesting way in which hail forms. Examine the figure and then begin reading clockwise from the lower left as we follow a single hailstone through the storm. Hail, lightning, and thunder are commonly associated with one another, so the thunderstorms that cause these phenomena are commonly called *thunderheads*.

05.08.a2 South Mtns., AZ

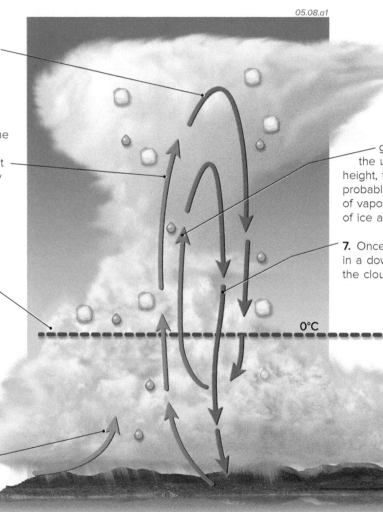
05.08.a1

4. Eventually the hailstone becomes influenced by a nearby downdraft, shown here as the beginning of a downward-arcing flow path.

3. As the central part of the storm continues to rise, more vapor deposition occurs and the hailstone grows. The hailstone will continue rising as long as it is not too heavy to be lifted by the updraft.

2. Once air rises high enough that it enters the subfreezing part of the cumulonimbus cloud (above the red line on this figure), the Bergeron process causes water vapor (not shown) to deposit onto an ice particle, causing it to grow. The resulting hail begins as a small, more-or-less-spherical object.

1. Supercell thunderstorms and other strong cumulonimbus clouds have strong updrafts, capable of lifting raindrops, chunks of ice, and even airplanes unfortunate enough to fly into the storm. Such updrafts are associated with the mature stage of a thunderstorm, when air rises against gravity because of a very unstable atmosphere.

0°C

5. If the downdraft causes the hailstone to be carried below the freezing height (the red line), the outer surface of the ice can partially melt. At this stage, a core of ice is coated with a layer of liquid water. While temperatures remain above freezing, more liquid water can coat the hailstone.

6. Before the hailstone can fall to the ground, it may get into another updraft. After the updraft lifts the hailstone above the freezing height, the coating of liquid water, most of which was probably melted ice, can refreeze. More deposition of vapor onto ice occurs, accumulating another ring of ice around the hailstone.

7. Once again, the developing hailstone gets caught in a downdraft. After falling to the warmer parts of the cloud (below the red line), more melting occurs.

8. This cycle can repeat many times—freezing and deposition of ice (via the Bergeron process) onto the hailstone above the freezing level, alternating with melting of an outside layer of ice when the hailstone is below the freezing level. Each trip up and down results in another layer of ice added to the hailstone. Eventually the developing hailstone becomes too heavy to be lifted by the updraft, and gravity pulls it to the surface. The more vigorous the updraft and the greater the instability, the larger the hailstone will be before it is finally pulled to the ground by gravity. The larger the hailstone and the faster it hits the ground, the more force it has and the more damage it can cause.

9. A number of factors influence the process of hail formation. For example, the freezing level is higher in summer than in winter. Hail formation is most likely at the time of year when the cloud extends substantially both above and below freezing. If a cloud is entirely warmer than freezing (which is unlikely), it will only produce rain. If the entire cloud is below freezing temperatures, it will produce snow. Other important factors include the amount of moisture, vertical extent of the cloud, the strength of the updrafts and downdrafts, and the temperature contrast between air masses across a front.

B Where and When Does Hail Tend to Occur?

Hail is associated with thunderstorms, so its distribution across the globe and in the U.S. is similar to the distribution of thunderstorms. As in the case of most other forms of severe mid-latitude weather, the U.S. experiences more than its share of hail compared to other locations on Earth. These maps show the geography and seasonality of hail in the contiguous U.S. Observe the general patterns shown on the large map below, and then examine the patterns for a certain month in the series of smaller maps. Try to explain the main patterns before reading the associated text.

1. This map depicts the average number of days per year when a county experiences hail. From the legend, we can see that some areas average less than one day of hail per year, whereas others average more than 10.

2. Frequencies are low on the Pacific Coast, where contrasts between air masses are typically weak. Frequencies are also low in most of the interior of the West, where the air is usually dry.

3. Frequencies are slightly higher in a north-south band through Arizona, which experiences thunderstorms during a summer monsoon.

4. Hail occurs most frequently over the Great Plains, in the center of the map, in a wide, north-south band stretching from central Texas to the Canadian border (and beyond). The average size of hailstones is larger over the Great Plains than anywhere else in the U.S., and perhaps anywhere else in the world. Most damage from hail is from larger hailstones (e.g., >5 cm), and the Great Plains often experiences such hail sizes.

MEAN ANNUAL NUMBER OF HAIL DAYS (60 Year Average)

Legend:
- 1–2
- 2–4
- 4–7
- 7–15

05.08.b1

5. Frequencies in the eastern U.S. are intermediate and spotty in distribution. The size of hailstones generally becomes smaller toward the coast.

6. Florida and areas along the Gulf Coast have frequent thunderstorms, but low to intermediate frequencies of hail. Here, thunderstorms are seldom tall enough and well-organized enough to form hail. Furthermore, temperature contrasts between air masses are usually not as great as over the Great Plains.

Seasonal Variations

7. The four maps below show hail frequencies for four different months, each representative of a season (winter, spring, summer, and fall, from left to right). How does hail vary from season to season?

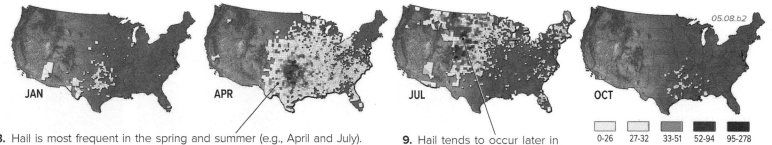

JAN   APR   JUL   OCT

05.08.b2

TOTAL NUMBER OF HAIL DAYS (60 Year Average)

- 0-26
- 27-32
- 33-51
- 52-94
- 95-278

8. Hail is most frequent in the spring and summer (e.g., April and July). In spring, the surface warms more rapidly than the overlying air, which must be heated by the surface itself. Successively higher locations in the atmosphere require longer to heat. This sets up a steep environmental lapse rate, in which the surface can be quite warm while the atmosphere is still cold from winter. This destabilizes the atmosphere, setting the stage for the cumulonimbus clouds that are required for hail to form. In fall, the pattern is reversed as the surface cools more rapidly than the air above it, and the atmosphere is seldom so unstable.

9. Hail tends to occur later in the calendar year in the northern states than in the southern states. This is no surprise because the boundary between the cold and warm air masses migrates northward as summer approaches. In fall (October) and winter (January), the surface is simply too cold in most of the U.S. to support the atmospheric instability needed for frequent cumulonimbus clouds and intense uplift.

Time-of-Day Variations

10. This graph plots average frequency of hail versus time of day for five cities. What do you observe? All of these cities show a strong tendency for hail in the afternoon, when the ground has warmed enough to destabilize the atmosphere. These patterns are very similar to those for severe thunderstorms.

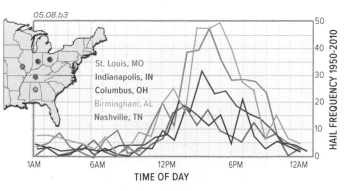

05.08.b3

St. Louis, MO
Indianapolis, IN
Columbus, OH
Birmingham, AL
Nashville, TN

HAIL FREQUENCY 1950-2010

TIME OF DAY

1AM   6AM   12PM   6PM   12AM

Before You Leave This Page

☑ Sketch and explain the formation of hail within a cumulonimbus cloud, identifying all essential components.

☑ Explain the factors contributing to the geography, seasonality, and time-of-day variations for hailstorms in the continental U.S.

5.8

5.9 What Causes Lightning and Thunder?

LIGHTNING IS A DANGEROUS but intriguing weather phenomenon. It results from electrical currents within a storm cloud, releasing tremendous amounts of energy that can strike the ground. In the U.S., more than 80 deaths per year are due to lightning. Although this number is much less than those killed simply by extreme heat and extreme cold, lightning is responsible for more deaths than any other type of severe storm in the U.S.

A How Is Lightning Generated?

Lightning is a sudden and large discharge of electrical energy generated naturally within the atmosphere, typically expressed as a bolt of electricity that flashes briefly and then is gone. Most lightning occurs as *cloud-to-cloud* lightning or occurs within a single cloud, but about 25% of lightning strikes the ground (called *cloud-to-ground lightning*). Some lightning is less discrete, appearing to form diffuse sheets, called *sheet lightning*.

05.09.a1

1. Most lightning occurs as a single discrete bolt that can be relatively straight or extremely irregular with many obvious branches (◀). Lightning is associated with cumulonimbus clouds, the kind that form thunderstorms (▶). Ice particles in the upper part of such clouds have a positive electrical charge, but the charge is mostly negative in the lower cloud where there is no ice. This charge separation is believed to happen because of the Bergeron process and different physical properties of ice particles and liquid water, but the exact process is unknown. The charge imbalance between the top and bottom of the cloud sets in motion a specific sequence of events, depicted in the series of figures below.

05.09.a2 west central IL

2. As a result of this charge imbalance, there is an excess of positive charges in the top of the cloud relative to the bottom of the cloud. Usually the Earth has a negative charge too, but as the cloud passes over a place, the negatively charged cloud base induces a positive charge directly below, and for several kilometers around, the cloud.

3. Recall that conduction requires energy transfer from one molecule to another, and therefore is most efficient in metals and other conductive materials. Air is a poor electrical conductor, so the flow of electricity through air cannot occur until the electrical potential becomes great enough to overcome the insulating effect of air.

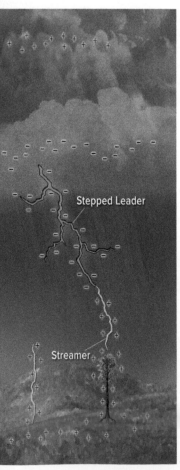

4. Lightning starts with negative charges near the base of the cloud. The negative charges begin to form an invisible channel of ionized molecules through the air, from the base of the cloud toward the positively charged ground. As the channel develops, it separates into a series of branches, all still invisible, and repeatedly starts and stops, advancing in a zigzag path, resembling a staircase. For this reason, it is called a *stepped leader*.

5. As the negatively charged stepped leader approaches the ground, positively charged streams of energy, called *streamers*, move upward to join the leader, commonly reaching up from the highest object on the ground, such as a tree, pole, building, or other tall object.

6. When the streamer makes contact with the stepped leader, it establishes a pathway of ionized molecules through which electricity can flow between the cloud and the ground.

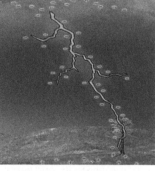
05.09.a4

7. Once the streamer and stepped leader meet, an electrical connection is established and electrons flow from the base of the cloud to the ground, but this flow is still invisible.

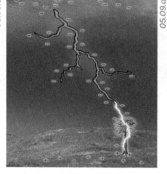
05.09.a5

8. As the negative charge approaches the ground, the positive charge begins to flow upward in what is called a *return stroke*. The return stroke begins near the surface.

9. The return stroke is nearly instantaneous, lighting up the sky with the discharge of energy. This is the main, bright flash we observe and recognize as lightning.

10. Once the electrical pathway is established, negative charges flow from the cloud to the ground, continuing to light up the sky with no obvious interruption after the return stroke. In most cases, there is then a short pause, after which a new negatively charged leader, called a *dart leader*, starts down the established channel from the cloud to the ground. The discharge of the dart leader causes another flash of light, distinctly after the initial flash. Lightning strikes can have several or several dozen flashes from dart leaders.

05.09.a3   05.09.a7

Thunder

11. *Thunder* results from rapid heating and expansion of the air along the path of the lightning bolt (◄). Around a lightning discharge, the air temperature approaches 15,000–30,000°C, which is much hotter than the surface of the Sun. The extreme heat causes this air to expand quickly away from the lightning stroke, as shown by the red arrows on this figure.

12. The expansion creates a momentary vacuum around the stroke. Almost instantly after the expansion, air rushes in from all sides toward the location where the lightning stroke passed. This is depicted by the green arrows. The sound of the matter in the various air streams colliding is what makes the sound of thunder.

05.09.a8

Using Thunder to Estimate Distances to Storms

13. Lightning travels at the speed of light. At about 300,000 km/s (186,000 mi/s), it can be considered instantaneous. But thunder only travels at the speed of sound (1,225 km/hr, or 760 mi/hr). Since there are 3,600 seconds in an hour, this means that the sound of thunder travels about 1.7 km or 1 mile in 5 seconds. So if you start counting when you see lightning, and can count to five before you hear the thunder, you are one mile away from the lightning. But beware: lightning does not necessarily flow straight down the middle of a storm. The strike could be farther ahead or behind the cloud from which it originated.

Elves, Sprites, and Jets

14. Intense lightning can cause a flat disk of dim reddish light, called an *elve*, to form about 96 km above the Earth. The light of an elve radiates outward in every direction, spreading over a huge area of sky.

15. Immediately after a very energetic bolt of lightning strikes the ground, ghostly red lights, called *red sprites*, may shoot straight up from the top of a thunderhead. Some red sprites soar up to 100 km into the atmosphere.

16. *Blue jets* are dim, blue streaks of light. They look like quick puffs of smoke that burst out of a thunderhead, arc upward, and then fade away. Blue jets can climb as high as 30 km into the atmosphere.

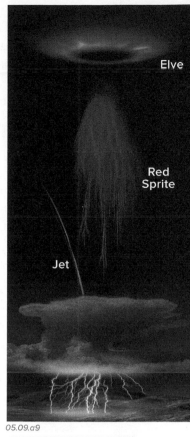

Elve

Red Sprite

Jet

05.09.a9

Safety Tips Around Lightning

Lightning is extremely dangerous. The map below shows the number of lightning-related fatalities for each state from 1959 to 2014. How dangerous are states with which you are familiar? What factors contribute to this spatial pattern? How at risk are you for lightning?

Conduction of electricity is easiest in solids and most difficult in gases. Therefore, direct contact with any solid object, particularly a tall one that conducts electricity well (such as a tall metal post), is extremely dangerous in a thunderstorm. Tall objects minimize the distance that the electrical energy needs to travel through a gas (air) during lightning.

A house or other enclosed building offers the best protection from lightning. Open shelters on athletic fields, golf courses, and picnic areas provide little or no protection from lightning. When inside a building, stay away from windows and doors, and avoid contact with plumbing, corded phones, and electrical equipment. Basements are generally safe havens, but avoid contact with concrete walls that may contain metal reinforcing bars. A car is a relatively safe place, because the metal body and frame carry the electrical current if struck.

If no safe shelter is available, stay away from trees and other tall objects, and crouch down with your weight on your toes and your feet close together. Lower your head and get as low as possible without touching your hands or knees to the ground. Do not lie down!

Consider the 30-30 rule: if the time between seeing a flash and hearing the thunder is 30 seconds or less, the lightning is close enough to hit you, so seek shelter immediately. After the last flash of lightning, wait 30 minutes before leaving a shelter.

05.09.t2

WHEN THUNDER ROARS **GO INDOORS!**

NUMBER OF LIGHTNING FATALITIES BY STATE, 1959–2014

Alaska - 0
D.C. - 5
Hawaii - 0
Puerto Rico - 33

05.09.t1

STATES RANKED BY FATALITY RATE PER MILLION PEOPLE

1–10	11–20	21–30	31–52

Before You Leave This Page

✓ Sketch and explain how and why lightning and thunder form, and describe the factors that set the stage for their formation.

✓ Summarize some strategies for remaining safe from lightning and the reason for using each strategy.

5.9

5.10 What Is a Tornado?

A TORNADO IS THE MOST INTENSE type of storm known, expressed as a column of whirling winds racing around a vortex at speeds up to hundreds of kilometers per hour. The May 1999 tornadoes in Oklahoma City, Oklahoma, were estimated to have wind speeds over 500 km/hr. Fortunately, most tornadoes are less than several hundred meters wide, so they usually only damage narrow areas, but larger tornadoes can remain on the ground for hours and create a path of near-total destruction across one or more states.

A What Is a Tornado and How Does One Form?

A *tornado* is a naturally occurring, rapidly spinning vortex of air and debris that extends from the base of a cumulonimbus cloud to the ground. Tornadoes typically have diameters of tens to hundreds of meters, but the largest ones can be 2–3 km wide. Most individual tornadoes remain on the ground for minutes to tens of minutes and have paths on the ground of less than a few kilometers. Some very rare, long-lived tornadoes track over 100 km, last a few hours, and destroy everything in their direct path.

05.10.a1

05.10.a2

1. This photograph (◄) shows the characteristics of a well-developed tornado, with the classic shape of a downward-tapering funnel. The tornado is visible because of the condensation of moisture, expressed as water drops, within the vortex. A brown cloud of debris picked up from the ground typically surrounds the base of the tornado, extending out farther than the vortex of strongest winds. As in the movies, if you are within the debris field, you are in trouble. Some tornadoes have wider and less elegant shapes, like a stubby, thick pillar.

Formation of a Tornado

2. The formation of a tornado is complicated and not fully understood. Different sets of circumstances are present in different tornadoes, but tornado formation generally begins with horizontal winds moving across the surface. Wind speeds are usually higher aloft and decrease downward due to friction with the surface. This difference in wind speeds causes a *vertical wind shear* within the wind column. Wind shear can result in a rotating, subhorizontal vortex (▲). Vertical wind shear generates a type of rotation known as *mechanical turbulence*, a normal occurrence, which by itself does not lead to severe weather.

05.10.a3

3. In an unstable atmosphere, rising air (an updraft) can perturb the rotating tube of turbulent air (◄). With enough lifting, the tube will become more vertically oriented (►). As the tube is lifted, it generally splits into several short segments, some of which may even be rotating in opposite directions.

05.10.a4

4. Eventually a segment of mechanical turbulence may become vertical, at which point it merges with the original and unstable updraft. Once it spans the distance from the cloud to the ground, it is a *tornado*. If a funnel-shaped cloud does not reach the ground, it is a *funnel cloud*. In the Northern Hemisphere, the parent storm system often rotates counterclockwise, so the tornado also spins counterclockwise, except in a few rare cases. As a tornado tightens its rotation, its wind speeds increase, making it a stronger, more powerful, and potentially more dangerous storm.

5. This figure (►) shows the essential aspects needed to form a tornado. The main ingredient is a well-developed cumulonimbus storm cloud, especially one associated with a supercell thunderstorm. A favorable regional setting is another prerequisite, such as a mid-latitude cyclone that puts cold air adjacent to much warmer, moister air. The position and motion of the cyclone are in turn controlled by such factors as the position of the polar front jet stream.

6. Another necessary ingredient is vertical wind shear near the surface, caused by higher speed winds aloft and slower winds near the surface. Mountains and other topographic irregularities tend to disrupt a regular wind-shear pattern, which is one reason why tornadoes are more common in areas that have relatively low relief (little topography) than in areas with mountains.

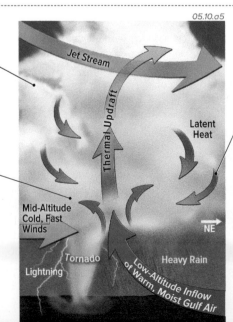
05.10.a5

Jet Stream
Thermal Updraft
Latent Heat
Mid-Altitude Cold, Fast Winds
NE
Tornado
Heavy Rain
Lightning
Low-Altitude Inflow of Warm, Moist Gulf Air

7. Formation of a tornado—and its host thunderstorm—requires an unstable atmosphere. Without it, there will not be enough lifting to form the cloud and to realign a vortex into a vertical orientation.

8. A storm must have strong updrafts in order to lift the horizontally spinning tube of air. The host storm must also possess strong internal rotation. Such rotation, which can be observed on weather radar, is one aspect weather forecasters use to gauge whether or not a severe thunderstorm is likely to spawn tornadoes. Storms that spawn tornadoes commonly have a *hook echo* pattern on radar (▼).

05.10.a6 Lacombe, LA

B What Weather Systems Produce Tornadoes?

Tornadoes are related to strong thunderstorms, which form in a number of settings. Most tornadoes occur in association with multi-cell or large supercell thunderstorms, which are generally associated with cold fronts. Others occur embedded in tropical cyclones or with more isolated single-cell thunderstorms.

Associated with Cold Fronts

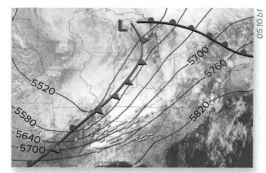

05.10.b1

Most catastrophic tornadoes are associated with multi-cell or supercell thunderstorms of the Great Plains and Southeast, usually in the warm sector of cold fronts. On the map above, a cold front (in blue) stretches south from a cyclone (marked by an "L") over the Great Lakes. In front of the cold front are yellow lines showing the tracks of torna-does. These tornadoes formed in associa-tion with squall-line thunderstorms along the dry line. Such tornadoes would more likely occur in the afternoon and early evening when thunderstorms are most active.

Tropical Cyclones

05.10.b2 Hurricane Isaac, 2012

Tropical cyclones, like hurricanes, spawn tornadoes too, but a large percentage of these are not terribly strong because the upper-air pattern favored for tornado development is obliterated by tropical cyclone circulation. In the Northern Hemisphere, tornadoes are most common in the northeastern quadrant of a storm, such as embedded in the thicker swirls of clouds in the northeastern (upper right) part of Hurricane Isaac (▲). Tornadoes closer to the center of the tropical cyclone may occur at any time of day and are generally weaker than those farther from the center, which mostly occur in the afternoon.

Single-Cell Thunderstorms

05.10.b3

Single-cell thunderstorms can spawn tornadoes, but the tornadoes are generally small and weak, in part because the storm system is itself typically not very large or powerful. Tornadoes formed in this setting are generally restricted to the afternoon, when surface heating destabi-lizes the atmosphere and causes a peak in thunderstorm activity. Tornadoes formed in any setting can have a distinctive, sharp, low-forming cloud, called a *wall cloud* (▲), indicating that a tornado may be near. Storm chasers (people who purposely try to get close to a storm to observe and photograph it), are on the lookout for wall clouds.

C How Are Tornadoes Classified and What Type of Damage Do They Cause?

Tornadoes are classified using the *Enhanced Fujita scale* (EF-Scale), which is based on estimated wind speed as evidenced by damage, shown in the table below. The amount of damage a tornado does is controlled by its strength (i.e., its EF rating), how long it stays on the ground, and what type of features are in the way (e.g., a city versus farmland).

This tornado is an awe-inspiring sight, but such thinly tapered tornadoes are typically not the strongest. The blue sky in the back-ground suggests this may be related to a single-cell thunderstorm. It is passing over agricultural fields, so it may cause relatively limited damage, except to crops.

05.10.c1 Colorado

This photograph illustrates the type of near-total destruction that tornadoes can cause. The wood-frame houses in the foreground were completely disassem-bled by the strong winds within a tornado, while the houses in the background are still standing (kind of) but suffered major damage.

05.10.c2

Enhanced Fujita Scale			
Strength	3-Second Wind Gust		Examples of Damage
	(km/hr)	(mi/hr)	
EF0	105–137	65–85	Branches/small trees fall
EF1	138–178	86–110	Mobile homes overturned
EF2	179–218	111–135	Roofs blown off houses
EF3	219–266	136–165	Walls blown off houses
EF4	267–322	166–200	Strong frame homes leveled
EF5	>322	200+	Everything obliterated

Before You Leave This Page

☑ Sketch and explain what a tornado is and how one forms.

☑ Describe the weather systems that produce tornadoes.

☑ Summarize the Enhanced Fujita classification of tornado damage.

5.10

5.11 Where and When Do Tornadoes Strike?

THE INTERIOR OF NORTH AMERICA has by far the most frequent tornado activity in the world. Tornadoes also occur, but far less frequently, in Europe, in the southeastern parts of South America and South Africa, and in a few other regions. Large parts of some continents, like all of North Africa and northern Asia, do not report any tornadoes. Where and when should we expect tornadoes in the U.S., and what causes the huge disparity between the central U.S. and most of the rest of the world?

A How Does Tornado Frequency Vary Across the Continental U.S.?

The continental U.S. is the site of numerous tornadoes each year, especially in a north-south belt through the middle of the country, mostly corresponding to the Great Plains. Most of these tornadoes occur in the warm sector of a mid-latitude cyclone, ahead of an associated cold front that has great thermal contrasts between the warm and cold sectors. Such situations are most common in the mid-latitudes.

1. Examine this map, which shows the frequency of tornadoes by county across the continental U.S. From this map, three areas experience the highest incidence of tornadoes, one in the center of the country, one along the Gulf Coast, and one in Florida. The patterns reflect the typical types of fronts and other weather activity that occur in the different regions. The geography of North America also contributes to the high frequencies. The major mountain ranges are oriented from north-to-south rather than east-west, which allows warm air and cold air to meet in the center of the continent.

3. Florida experiences frequent tornadoes, many forming in association with single-cell thunderstorms or with tropical cyclones. These are mostly weak tornadoes and are overall not as deadly. For tornadoes related to tropical cyclones, people have already evacuated from less sturdy structures because of earlier warnings about the approaching larger cyclone.

TOTAL NUMBER OF REPORTED TORNADOES FROM 1950-2012

- 0
- 1–30
- 31–60
- 61–120
- 121–180
- 181–220
- 221–252

2. The region of most tornadoes is along a north-south belt, running from north Texas northward to the Dakotas and into Canada—this is justly called *Tornado Alley*. Many of these tornadoes are strong, but some storms lack enough moisture to produce violent tornadoes.

4. Another concentration of tornadoes appears along the Gulf Coast and adjacent inland areas—a region called *Dixie Alley*. Dixie Alley tornadoes are often more deadly than their Tornado Alley cousins, in part because they are often hidden in thunderstorms amid abundant Gulf of Mexico moisture.

Seasonal Variations

5. The maps below show frequencies of tornadoes for four different months, each representing a season, from 1950–2010. Compare the frequencies in different seasons and try to explain regional patterns.

6. In winter, tornadoes are rare except in Dixie Alley and Florida—areas near relatively warm waters of the Gulf of Mexico. Boundaries between warm and cold air masses are commonly located in this part of the Southeast, more so than anywhere else at this time of year.

7. In spring, the land starts to warm but warming of the atmosphere lags behind, resulting in an unstable atmosphere. The polar front jet stream tends to be strong and displaced southward at this time, which gives upper-level support for tornadoes in the form of wind shear. Tornadoes become more common.

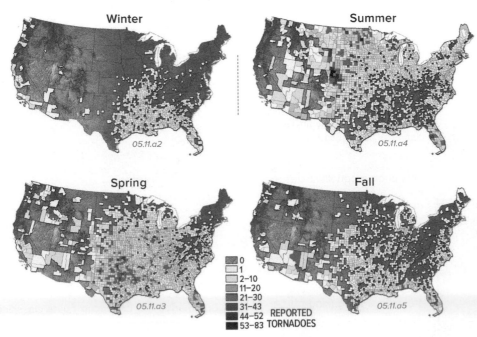

REPORTED TORNADOES
- 0
- 1
- 2–10
- 11–20
- 21–30
- 31–43
- 44–52
- 53–83

8. Warming continues in the summer, and the boundary between warm and cold air migrates northward. Tornado activity follows into the northern Great Plains and areas to the east. In addition to increased surface heating and convection, moisture flows in from the south.

9. The frequency of tornadoes decreases in fall because lingering summer warmth in the atmosphere remains while the land begins to cool quickly. This stabilizes the atmosphere. Tropical cyclones as a source of tornadoes remain a threat along the Gulf Coast and Florida.

B How Does Tornado Frequency Vary Globally?

Tornadoes occur in other parts of the world, but are less frequent than in the tornado-prone parts of the U.S. Examine the globes below, where each dot represents an area where tornadoes occur, and think how it compares to the previous one displaying the global distribution of thunderstorms, as expressed by lightning measurements.

1. The U.S. clearly experiences the highest frequency of tornadoes. Tornadoes mostly occur in the Great Plains and the Southeast.

2. Tornadoes also occur in Hawaii (not shown but close to the western edge of the globe) and also in the prairies of southern Canada and near the Great Lakes.

3. Unlike North America, Europe has large, east-west-oriented mountain ranges that inhibit the meeting of very warm and very cold air. Therefore, tornado frequencies are lower in Europe than in North America. Nevertheless, a mid-latitude location, a high population density, and sophisticated detection abilities ensure that most tornadoes that form are sighted.

4. Central and northern Asia report few tornadoes, but part of this may be due to under-reporting in the vast, less populated areas. Tornadoes occur in the Indian Subcontinent, where strong storms accompany the inland flow of marine moisture during the monsoon.

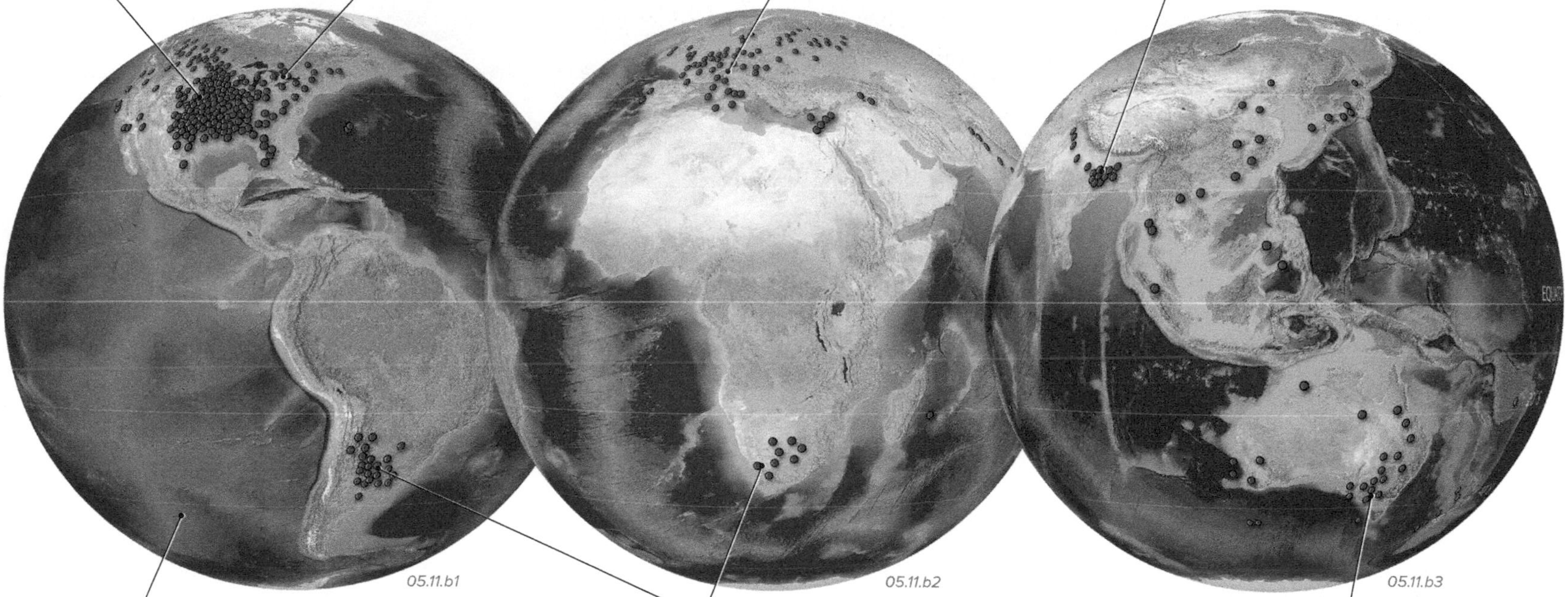

05.11.b1   05.11.b2   05.11.b3

5. Few tornadoes are known to occur over open seas because surface heating is far less intense over oceans than over land. The apparent lack of tornadoes is partly due to a lack of radar coverage over the oceans to detect them. Regions covered by sophisticated detection systems, such as the U.S. and Canada, tend to report more tornadoes.

6. The lowlands of mid-latitude South America and Africa also have some tornadoes. Recall the maps showing lightning-indicated thunderstorms here.

7. The high frequency of tornadoes in south Asia and parts of Australia is largely attributable to their proximity to a supply of warm, moist air. These areas also showed a high density of activity on the lightning-strike map.

Time-of-Day Variations

05.11.b4

168 Tornadoes in the Area of St. Louis, Missouri

TIME OF DAY

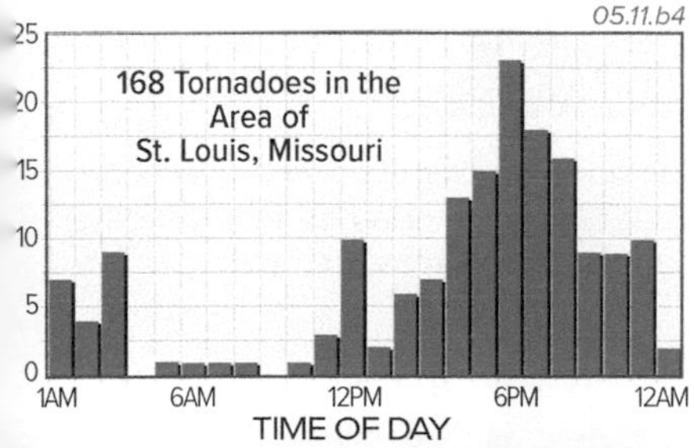

05.11.b5 Salina, KS 2012 (EF4)

8. This graph shows the average frequency of tornadoes versus time of day, using 60 years of data for St. Louis, Missouri. Here, tornadoes typically occur in the afternoon but continue into the evening and night. These nocturnal tornadoes are often the most deadly, for two reasons. First, people are asleep or otherwise less likely to be alert for tornado watches and warnings. Second, nocturnal tornadoes are proportionately more likely to occur in association with storm systems that are well organized, long-lived, and strong, such as multi-cell storms. Other locations in Tornado Alley, Dixie Alley, and Florida display similar temporal patterns.

9. The tendency for tornadoes to strike in the afternoon and early evening partly accounts for the relatively dark and foreboding aspect (◄) of most photographs of tornadoes—the Sun is low in the sky or has started to set. The thick, dark clouds help too, especially when the tornado occurs in multi-cell or supercell thunderstorms.

Before You Leave This Page

☑ Sketch or summarize the frequency and seasonality of tornadoes across the U.S., identifying the main factors that explain the observed patterns.

☑ Briefly describe the factors contributing to variations in tornado frequency around the world.

☑ Draw a simple graph of tornado frequency by time of day, and explain the reasons for the pattern.

5.11

5.12 What Are Some Other Types of Wind Storms?

SEVERE WEATHER INCLUDES VARIOUS WIND STORMS in addition to tornadoes and cyclones. Some of these other storms—with interesting names like waterspouts, microbursts, haboobs, and derechos—can be extremely dangerous. What are their characteristics, how do they form, and where do they occur?

Waterspouts

1. A *waterspout* is a rotating, columnar vortex of air and small water droplets that is over a body of water, usually the ocean or a large lake. There are two kinds: tornadic waterspouts and fair-weather waterspouts. Waterspouts are usually less intense than tornadoes, with smaller funnel diameters, weaker wind speeds, shorter lives, and shorter path lengths.

2. Some waterspouts resemble tornadoes (▶), and some are actual tornadoes that either formed over the water or formed over land and later moved over the water. These large, storm-related tornadic waterspouts form in the same way tornadoes on land form—strong updrafts associated with a powerful thunderstorm, such as a

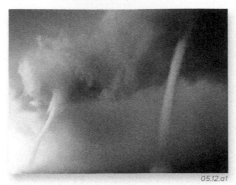

05.12.a1 05.12.a2

supercell thunderstorm. Such waterspouts, like any tornado, are dangerous, but are generally not as powerful because the cooler surface temperatures over water limit the atmospheric instability and the amount of lifting. This type of tornadic waterspout can be recognized by its association with a thunderstorm and is much larger and stronger than a fair-weather waterspout.

3. The other type of waterspout is associated with normal cumulus clouds, and not typically with thunderstorms. Since many of these form when it is not raining, they are called fair-weather waterspouts. A fair-weather waterspout is a rotating column of air that starts from wind shear near the water that gets lifted upward by mild updrafts that are also forming the overlying cumulus cloud. The water in the spout was not picked up from the underlying water, but is condensation produced as the air is rising and cooling. This type of waterspout is most common in the tropics, such as in the Florida Keys, but it can also form in mid-latitudes, like on the Great Lakes.

Derechos

4. A *derecho*, whose name is derived from the Spanish word for "straight," is a regional wind storm characterized by strong, nonrotating winds. The winds can advance across huge areas and cross several states in a matter of hours. Wind velocities can be hurricane-strength and can damage trees, power lines, and other structures.

5. This map (▶) shows a series of radar images (strongest echoes in red) capturing the progression of a single derecho that started near the Great Lakes (in the upper left) and moved east with time. As the derecho moved, it spread, forming a radar

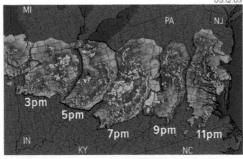
05.12.a3

echo in the shape of a bow (like an archer's bow). Such a *bow echo* is characteristic of a derecho. Note how rapidly this derecho traveled, moving from Lake Michigan to the Atlantic Ocean in about 10 hours, at an average speed of about 150 km/hr.

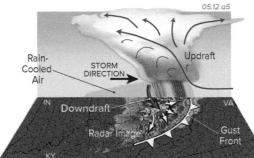
05.12.a5

7. Derechos are caused by downdrafts (red arrow in this figure) from a band of thunderstorms in a multi-cell storm, where a downdraft from one thunderstorm forms or strengthens an adjacent thunderstorm. Such downdrafts can be sustained for a long time and over long distances, forming a derecho. Embedded within a derecho may be other types of wind storms, including tornadoes, squall lines, and microbursts.

6. On this map (▶), contours depict the number of derechos experienced in different areas. The Ohio Valley, Upper Midwest, and Great Plains of the U.S. may experience more derechos than any place in the world. Summer is the most common season for derechos. They are long lived and can occur at any time of day or night.

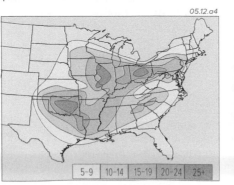
05.12.a4

| 5-9 | 10-14 | 15-19 | 20-24 | 25+ |

05.12.a6 Oklahoma

8. Derechos can be accompanied by a very distinctive and ominous type of cloud called a *shelf cloud*, a spectacular example of which is shown in this photograph (◀). Similar-appearing clouds can form in other ways, such as near tornadoes and dry lines.

Microbursts and Other Powerful Downdrafts

9. *Microbursts* are brief but strong, downward-moving winds (downbursts) that can flatten trees or structures. Unlike tornadoes, microbursts produce winds that move in a straight line, rather than rotating, so sometimes people refer to microbursts as *straight-line winds*. A microburst can be classified as being wet or dry, depending on whether it is embedded within a thunderstorm. Microbursts can be more severe than some tornadoes, and although only lasting a few seconds, are a major aviation hazard.

05.12.a7 Austin, TX

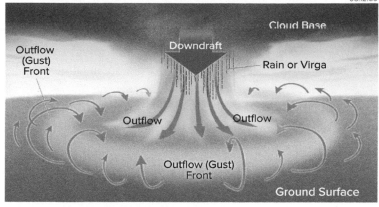
05.12.a9

10. This photograph (◄) shows a wet microburst within a thunderstorm. It is expressed as a downward burst of moist air that, in this case, is carrying an increased amount of rain. Note how the rain is being spread out in all directions as it nears the ground. Microbursts generally affect a relatively small area, but they can extend over areas as large as several counties. In either case, their duration is short (measured in seconds or minutes).

11. Dry microbursts are downdrafts, generally not associated with rain. Under these conditions, they may pick up dust and loose materials, moving them as a small dust storm, or *haboob* described below. Both wet and dry microbursts can be strong enough to cause significant damage, with or without dust.

12. Under some circumstances, a diffuse column of raindrops extends down from a cloud but evaporates before the drops reach the ground. This phenomenon, called *virga*, can occur for a variety of reasons, including downdrafts. Where associated with downdrafts, the descending air can compress and heat up as it is pushed downward, helping evaporate the water drops before they reach the ground.

05.12.a8 Hermosa Cliffs, CO

13. The figure above illustrates the setting in which most microbursts occur. Late in the life cycle of a thunderstorm, the lifting power of the convecting air begins to be overwhelmed by the downward force of precipitation and downdrafts. These cool the lower parts of the cloud, the air below, and the surface, limiting the heat available for additional convection. At this point, the thunderstorm can start to collapse (the stage shown in the figure above). This can cause a very strong downdraft, which upon reaching the surface causes a microburst. Microbursts can move out from a dying thunderstorm in several directions or mostly in one direction. The tell-tale sign of a microburst is a "starburst" damage pattern (►), where features are blown radially, rather than a pattern of damage that indicates rotation around a central core (as in a tornado).

05.12.a10 Burnett County, MN

Haboobs and Dust Devils

14. A name applied to severe downdraft-related, nonrotating dust storms is *haboob* (▼), derived from an Arabic word for "blasting." These are formed in a similar way to microbursts, at the collapse of a dying thunderstorm. They can pick up dust from desert areas and plowed fields, producing a thick cloud of dust that advances rapidly across the landscape. They can damage windows, cause respiratory issues, and create very hazardous driving conditions as visibility drops to a few meters in a matter of seconds.

05.12.a12 Phoenix, AZ

15. Often overlooked as a hazardous weather phenomenon is a *dust devil* (▲), also called a whirlwind or simply a dust storm. Like other rotating storms, these develop when the atmosphere is extremely unstable and upper-level patterns encourage rising motion. The upward motion at the surface creates strong pressure gradients, which strengthen the system seemingly uncontrollably. Eventually, the energy is expended, friction slows the wind, and the dust devil dissipates.

16. Dust devils and haboobs form in arid regions, so their strength is limited by the generally small latent energy released in condensation and deposition. But the thick dust, in addition to reducing visibility, can remove topsoil and erode the landscape.

05.12.a11 Richland, WA

> ### Before You Leave This Page
>
> ☑ Describe two types of waterspouts and settings in which each forms.
>
> ☑ Sketch and describe a derecho, explaining how one forms and the type of associated thunderstorm.
>
> ☑ Sketch and explain how most microbursts form and summarize their varied manifestations.
>
> ☑ Summarize how haboobs and dust devils form.

5.12

5.13 What Is a Tropical Cyclone?

HURRICANES AND OTHER TROPICAL CYCLONES are some of nature's most impressive spectacles. These immense seasonal storms form over warm, tropical waters and can cause extremely heavy precipitation, all the while deriving energy from latent heat released by condensation and deposition. In the Atlantic and eastern Pacific, the most intense type of tropical cyclone is called a *hurricane* while in the western Pacific and Indian Oceans, the terms used are *typhoon* and *cyclone*, respectively. Tropical cyclones take days to traverse an ocean and spread energy as they move across the globe. How do such storms form, and how are they sustained?

A What Are the Characteristics of a Tropical Cyclone?

Tropical cyclones, hurricanes, and *typhoons* are characterized by low atmospheric pressure, large areas of strongly rotating winds, locally elevated sea levels near the storm, high wind-driven waves, coastal flooding and erosion, and flooding along streams as the storm passes inland.

1. Tropical cyclones are huge, circulating masses of clouds and warm, moist air. They are zones of low atmospheric pressure that cause air to rise and condense, creating locally intense rainfall.

2. Warm tropical ocean waters fuel tropical cyclone formation. Water vapor evaporated from the sea surface mixes with the air, rises, cools, and produces clouds, releasing energy during condensation, which warms the air around it. This energy further enhances the upward motion, drawing more surface moisture to replace the rising air. In this way, tropical cyclones, once formed, provide their own fuel (energy).

3. If the tropical cyclone encounters additional warm water in its path, extra energy is input to the storm, winds increase, and the storm grows in strength and size. A tropical cyclone dissipates when it passes over land, cold water, or mixes with much drier air.

8. In the Northern Hemisphere, the right front quadrant is the most destructive part of the storm. This is the zone where the counterclockwise winds push the waves, called a *storm surge,* onshore with additional momentum.

05.13.a1

7. The strongest winds and most severe thunderstorms are usually found within the *eyewall* — the area immediately surrounding the eye.

6. Dry air flows down the center of the storm, compresses, and evaporates clouds, forming a cylinder of relatively clear, calm air, called the *eye.* Sinking formed this eye, a hole down into the clouds, in Hurricane Ike in 2008 (▼).

0.

4. Differences in atmospheric pressure between the surrounding environment and the center of the tropical cyclone cause sea level to bow up by several centimeters. This effect is greatly exaggerated in the figure above to make it more visible. This mound of water can add to flooding problems when the cyclone reaches land, an event called *landfall.*

5. Like all enclosed low-pressure zones (cyclones), the Coriolis effect causes tropical cyclones to spiral counterclockwise in the Northern Hemisphere and clockwise in the Southern Hemisphere. The storm's overall track is steered by air pressures and currents high in the atmosphere. The track is greatly influenced by ridges and other areas of high pressure, which can block or deflect a cyclone, either toward or away from land.

B Why Do Tropical Cyclones Rotate?

Tropical cyclones, like mid-latitude cyclones, are a result of multiple forces that influence where and how a cyclone moves. The pressure-gradient force acts to move air into the lower pressure from all sides, but this flow of air is deflected by Earth's eastward rotation—the Coriolis effect. The Coriolis effect is fairly weak at low latitudes but increases with the linear velocity of an object, like fast-moving winds associated with a tropical cyclone. This figure is a labeled satellite image of Hurricane Irene in 2011, a very large hurricane that caused extensive damage in the Caribbean, Gulf Coast, and the East Coast of the U.S.

05.13.b1

The Coriolis effect deflects moving air to the right of its trajectory in the Northern Hemisphere (as shown by the red arrows on this figure) and to the left of its trajectory in the Southern Hemisphere. This produces a vortex shape and rotation of the entire cyclone system, with the now-familiar counterclockwise spin in the Northern Hemisphere and a clockwise spin in the Southern Hemisphere.

Earth's Rotation

C Where and How Are Tropical Cyclones Formed?

Tropical cyclones begin in the tropics as relatively weak disturbances in normal patterns of air pressures and wind directions. Few disturbances develop further, but some are dramatically strengthened by the environments through which they pass, attaining low-enough pressures and high-enough wind speeds to become tropical cyclones. Here, we examine a common scenario that generates tropical cyclones over the Atlantic Ocean.

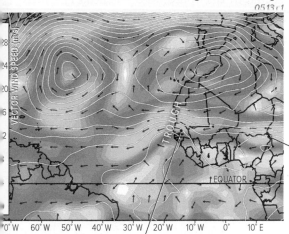

1. Tropical cyclones in the Atlantic Ocean often begin over or near northwest Africa in association with a west-flowing stream of air in the northeast trade winds called the *African Easterly Jet* (AEJ), shown on the map to the left. The lines on the map are isohypses at the 700 mb level, and colors indicate wind speeds at a slightly higher level (red, yellow, and green are fastest). The relatively fast winds (green belt) of the AEJ flow across the arid regions of Africa. Note that the jet has wavelike bends north and south, similar to those observed for the polar front jet (Rossby waves).

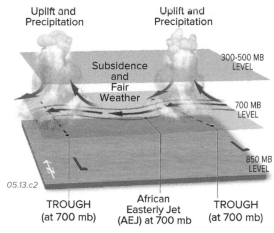

05.13.c2

2. The waves in the AEJ are termed *easterly waves* because the bends propagate from east to west. Since lower pressure is on the equatorward side of the AEJ closer to the Intertropical Convergence Zone (ITCZ), the poleward bulge of the wave describes an area of low pressure, called a *trough*. Like those in the mid-latitudes, troughs are associated with low pressure, lifting air, and cloudy or stormy weather. But unlike those in the mid-latitudes, tropical troughs bend poleward. If a westward-traveling easterly wave encounters conditions favorable for further strengthening, such as warm sea temperatures, it can incrementally grow into a tropical depression, tropical storm, and hurricane.

4. This figure (▲) shows that the east side of a northward bend in the AEJ (a trough) is associated with lifting and unsettled weather; the west side experiences fair weather. The explanation for this is similar to that for Rossby waves. If the storm can survive its trip through the west side of the wave, it may intensify afterwards.

3. This color-enhanced satellite image (▶) shows a belt of thunderstorms (in red) associated with westward-moving easterly waves and tropical depressions. The easternmost wave (on the right side of the image) is still over Africa, where it formed, but ones to the left have been carried westward over the Atlantic Ocean by the west-blowing trade winds, which dominate surface winds in this region. As the thunderstorms move over the warm waters of the Atlantic, they can strengthen.

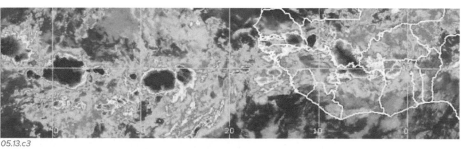

05.13.c3

Site of Origin and Tracks

5. A tropical cyclone forms under specific conditions limited to certain regions and then travels along a path called a *storm track*. These globes (▼) display global storm tracks of tropical cyclones in various parts of the world, colored according to strength. Tropical storms originate over warm water and mostly travel between 10° and 15° latitude and the subtropics—they cannot cross the equator. Note a general lack of tropical cyclones right along the equator, due to the absence of the Coriolis effect there. The locus of activity shifts seasonally, following the migration of the overhead Sun and the regions of the warmest water.

6. Tropical cyclones in the Atlantic Ocean, the ones that affect the Gulf and East Coast of the U.S., mostly start west of Africa (related to the AEJ) from June 1 to November 30, the *Atlantic Hurricane season*.

7. Tropical cyclones, including hurricanes, also form in the tropical Pacific Ocean, but typically begin earlier, from May 15, and go through November.

8. Tropical cyclones in the western Pacific and the northern Indian Ocean mostly form between June and November but can occur at any time of year.

9. Most in the Southern Hemisphere form from January to March, the southern summer. Cyclones rotate clockwise and turn south and southwest in the Southern Hemisphere but rotate counterclockwise and turn north and northeast in the Northern Hemisphere, reflecting the different deflections of the Coriolis effect on either side of the equator.

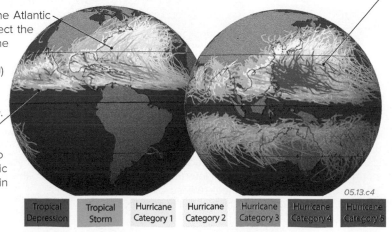

05.13.c4

| Tropical Depression | Tropical Storm | Hurricane Category 1 | Hurricane Category 2 | Hurricane Category 3 | Hurricane Category 4 | Hurricane Category 5 |

Before You Leave This Page

☑ Sketch, label, and explain the features and rotation of a tropical cyclone.

☑ Sketch and explain how an easterly wave can initiate a tropical cyclone, and describe where tropical cyclones originate, including seasonal variations.

5.14 What Affects the Strength of a Tropical Cyclone?

AS A TROPICAL CYCLONE MOVES, it responds quickly to changes in the environment it encounters. In some cases, environmental changes will cause a tropical cyclone to strengthen, perhaps becoming a hurricane or becoming a stronger hurricane. In other cases, the new environmental conditions cause a cyclone to weaken, dissipate, and eventually disappear. Consider the conditions under which a tropical cyclone forms and what factors might strengthen or weaken it, such as sea-surface temperatures and whether they are over water or land.

A What Conditions Strengthen or Weaken a Tropical Cyclone?

Changes in Water Temperature

05.14.a1

05.14.a2

05.14.a3

1. These figures track the changes that occur when a tropical cyclone migrating across the surface encounters different conditions. The initial cyclone (shown above) is still growing, as expressed by its cumuliform top. The water below is just warm enough to sustain growth.

2. If the cyclone encounters even warmer sea temperatures, it may strengthen. The warmer water adds heat and additional moisture to the cyclone, moisture that will later generate even more heat during condensation (latent heat), a key factor in cyclone growth.

3. If the cyclone instead encounters cooler water, as when an Atlantic hurricane turns north in mid-ocean, it can lose thermal energy to the water. Also, the cooler water will not contribute much moisture, and can even remove moisture by cooling the air.

Encounters with Other Air Masses

4. If a cyclone advances into, or incorporates, a wetter air mass, the additional moisture can strengthen the cyclone. The incorporation of more moisture is unlikely as an air mass supporting a tropical cyclone is often at or near saturation already.

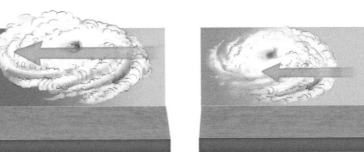
05.14.a4   05.14.a5

5. More commonly, a cyclone advances into, or incorporates drier air. As the dry air is mixed in (*entrained*), cloud development ceases and cloud cover thins, eventually disappearing due to evaporation into the drier air. The loss of moisture robs latent heat from the storm, and this is how many cyclones dissipate.

Encounters with Land

6. A migrating cyclone will often cross onto land. Once it is over land, the cyclone no longer has an unlimited supply of empowering moisture. Also, the land is dry and/or cold, so it will remove moisture and/or energy from the storm.

05.14.a6

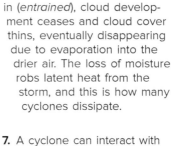
05.14.a7

7. A cyclone can interact with surface topography. This increases friction, slowing the winds. Mountains also remove large amounts of moisture in the form of precipitation due to orographic effects. Such settings are among the main causes of cyclone-related flooding and fatalities.

Upper-Level Wind and Pressure

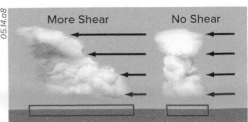
05.14.a8

8. Vertical wind shear is the difference in wind speed or direction between the upper and lower atmosphere. While strong vertical wind shear assists in extra-tropical thunderstorm development, it generally tears apart or dissipates tropical cyclones. Fast winds in the upper troposphere will shear off the top of the developing tropical cyclone, as the near-surface trade winds push the base of the storm in the opposite direction. Note that vertical wind shear can spread out a storm's latent heat release over a larger area (as represented by the red boxes), limiting buildup of the storm.

9. Although not illustrated here, *upper-level pressures* can act to strengthen or weaken a tropical cyclone, like they do for a mid-latitude cyclone. They also influence the path of the storm.

B   What Types of Damage Are Associated with Tropical Cyclones?

05.14.b1 Mississippi

05.14.b2 Long Beach, NY

05.14.b3 Gulfport, MS

1. Much damage from hurricanes and other tropical cyclones is from the high winds (▲), which can reach hundreds of kilometers per hour. Also, tornadoes and other strong local winds embedded within the larger storm can cause additional localized damage.

2. In addition to being subjected to strong winds, shorelines are afflicted by rough surf associated with the very large waves. The waves can erode parts of the beach, making houses even less protected, dumping sand far into neighborhoods (▲).

3. The coastal flooding associated with tropical cyclones also inundates low-lying areas along the coast. Much damage and death, however, also result from the flooding that occurs farther inland from streams having waters swollen by very high amounts of rainfall.

4. Recall that tropical cyclones form a mound of water beneath them, in response to the very low air pressure. When the storm approaches shallow coastal waters, this mound forms a wave, called a storm surge (▼), that rises as it approaches the shore. A storm surge can be devastating, flooding coastal areas with over 9 m (30 ft) of water. If the storm surge arrives at the same time as high tide, it is superimposed on the already-high water level. Also, the strong winds

05.14.b5 Staten Island, NY

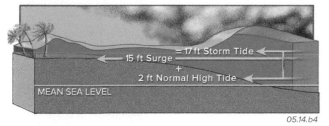

= 17 ft Storm Tide
15 ft Surge
+
2 ft Normal High Tide
MEAN SEA LEVEL

05.14.b4

around cyclones push water in front of them, so the amount of storm surge can increase if a shoreline is in the direct path of the winds (northeast side of storms hitting the U.S.). Storm surges can bring marine objects onshore, including ships (▶).

Strengths of Tropical Cyclones

Meteorologists and newscasters commonly refer to a "category 3 hurricane," but what does this term actually indicate? This table shows how we categorize the strength of tropical cyclones.

The least energetic type of tropical cyclone considered dangerous is a *tropical depression*, so named because it is an area of low pressure. A tropical depression has at least one closed isobar on an air-pressure map, and successive tropical depressions are numbered sequentially

in a given ocean basin each year (1, 2, 3, etc.). Once a tropical cyclone gains wind speeds of 64 km/hr (39 mi/hr), it is called a *tropical storm* and given a name. The first storm of the season in a given ocean basin receives a name that begins with an A, the second begins with B, and so on. If a named storm strengthens to wind speeds of 119 km/hr (74 mi/hr), it is termed a hurricane, typhoon, or cyclone, depending on the part of the world where it occurs. Atlantic and eastern Pacific hurricanes are then further

classified according to their intensity on the *Saffir-Simpson scale* (named after the two people who devised it). The weakest hurricane is a category 1, and the strongest is a category 5. The amount of expected damage increases non-linearly with storm strength (e.g., a category 4 storm generally produces more than four times the damage of a category 1 storm). Storms of Category 3–5 are often called "major" hurricanes, typhoons, or cyclones.

Tropical Storm Category	Maximum Sustained Winds	Anticipated Storm Surge	Examples
Depression	37–63 km/hr (23–38 mi/hr)		
Storm	64–118 km/hr (39–73 mi/hr)	> 1.0 m (> 3 ft)	Allison (2001)
Category 1	119–152 km/hr (74–95 mi/hr)	1.0–1.7 m (3–5 ft)	Irene (2011); Sandy (2012)
Category 2	153–176 km/hr (96–110 mi/hr)	1.8–2.6 m (6–8 ft)	Frances (2004); Ike (2008)
Category 3	177–209 km/hr (111–130 mi/hr)	2.7–3.8 m (9–12 ft)	Katrina (2005); Wilma (2005)
Category 4	210–250 km/hr (131–155 mi/hr)	3.9–5.6 m (13–18 ft)	Galveston (1900); Charley (2004)
Category 5	> 251 km/hr (> 155 mi/hr)	> 5.6 m (>18 ft)	Andrew (1992); Camille (1969)

Before You Leave This Page

☑ Sketch and explain the conditions that strengthen a tropical cyclone.

☑ Summarize the types of damage associated with tropical cyclones.

☑ Explain how we categorize tropical cyclones.

5.15 How Are Weather Forecasts Made?

THREE INGREDIENTS ARE NECESSARY for weather—energy, motion (both horizontal and vertical), and atmospheric moisture. The physical processes, such as energy transformations during phase changes of water (between solid, liquid, and gaseous states) are well understood, but several problems limit accurate weather forecasting. First, the equations describing the various transformations across three-dimensional space are complicated, requiring the highest powered computers. Data that are input into the models are often inadequate in quantity and quality. Also, it is impossible to know all the initial boundary conditions or other complexities.

A What Are the Sources of Data for Weather Forecasting Models?

05.15.a3

05.15.a1

1. Airborne Instruments—Every 12 hours, weather balloons (◄) are launched synchronously at about 800 sites around the world. These balloons are equipped with computerized sensors that report the vertical profiles of temperature, pressure, humidity, wind speed and wind direction, and sometimes other data, like ozone concentrations. The balloon ascends until it eventually bursts above the 200 mb level, and the instrument floats down on a parachute, to be used again.

2. Satellite Data—These provide estimates of temperature, humidity, wind speeds, and other variables. They may be the only data available for the oceans and other sparsely inhabited parts of the Earth. Some weather satellites (▶) trace paths around the world over periods of several hours, while others remain over the same location, following the Earth as it rotates on its axis.

3. Radar—Meteorologists rely extensively on radar, some of which is collected from ground and some from satellites. Doppler radar stations (◄) emit microwaves that are reflected back to the receiver from water droplets. Differences in wave frequency reveal the direction of movement, both of the water droplets and of the circulation steering them.

4. Marine Data—Some marine areas have moored buoys (▶) that measure air and sea temperature, pressure, wind speed and direction, and wave speed and height. Portable buoys and a variety of floating objects are used to track ocean currents, although similar data can be collected by satellites. Some buoys descend to measure water temperatures at different depths, then rise to the surface to relay their information.

05.15.a2 Underwood, SD

05.15.a4

5. All of these data are reported telemetrically to a central receiving station in suburban Washington, D.C., for processing. The received data are first checked for obvious errors and to make sure the sensors are working correctly. After this first crucial step, data from the different sources are spatially averaged and interpolated to make the huge data sets a bit more manageable. For example, the atmosphere is commonly depicted as a series of points, each representing the center of a cube of air, with a three-dimensional array of virtual cubes encircling the Earth and stacked dozens high. The interpolated data are most accurate where derived from closely spaced, actual measurements, and much less accurate for regions or levels of the atmosphere with few actual measurements.

6. Upper-level data used to determine the value of the center point for cubes higher in the atmosphere are far less abundant than surface data, so the accuracy of the interpolation typically decreases upward. The smaller the cubes, the higher the spatial resolution, but smaller cubes also increase the computation time. Maps for various vertical levels, called *analysis products* (as opposed to *forecast products* used to predict weather) show the values of the interpolated data. These maps are produced for the U.S. by the National Weather Service and for the world by meteorologists affiliated with the World Meteorological Organization.

B What Equations Are Used in Weather Forecasting Models?

1. All weather forecasting models are based on only seven equations. Three of the equations describe atmospheric movement and are manifestations of Newton's Laws of Motion. These equations represent velocity or acceleration in three dimensions—west-east, north-south, and up-down in the atmosphere (▶).

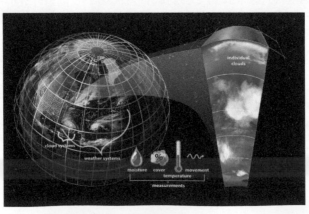

05.15.b1

2. A fourth equation accounts for density changes resulting from the piling up or spreading out of air in each cube. The fifth equation describes the rate of change of atmospheric moisture and accounts for changes of state of water in each cube. The sixth equation describes the rate of change of incoming energy (diabatic heating) from the Sun, the energy that powers all motion. Finally, the seventh equation is the Ideal Gas Law, which describes the relationship among temperature, pressure, and density in each cube, and how these intrinsic attributes change in response to all motions and all the other processes.

C How Are the Input Data Used in Computer Models?

1. Once the data have been collected, checked for accuracy, and interpolated into a three-dimensional grid, like the one in the figure below, the next step is to use the data as inputs into models of how our atmosphere works. To do this, the variables at the center point of each cube are input into the seven equations. Each equation (except for the Ideal Gas Law) calculates a *rate of change* of the variables in that equation. For example, an equation might calculate how fast the atmospheric velocity changes (i.e., accelerates or decelerates) in a west-to-east or south-to-north direction. Or, we could model how fast atmospheric density or moisture changes due to some change in the environment. Each step in the model forecasts these rates of change for a relatively short interval of time (usually minutes). This is called the *time step*.

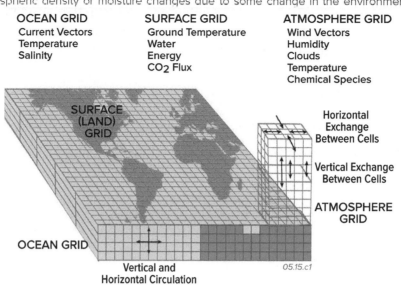

OCEAN GRID
Current Vectors
Temperature
Salinity

SURFACE GRID
Ground Temperature
Water
Energy
CO_2 Flux

ATMOSPHERE GRID
Wind Vectors
Humidity
Clouds
Temperature
Chemical Species

SURFACE (LAND) GRID

Horizontal Exchange Between Cells

Vertical Exchange Between Cells

ATMOSPHERE GRID

OCEAN GRID

Vertical and Horizontal Circulation

05.15.c1

2. In this way, answers to questions we have about weather and climate come from solving the seven basic equations simultaneously at each centerpoint. A model takes into account the values of each variable within a cube and in adjacent cubes (including the ones above and below it), for the thousands of cubes. Then, the model generates a forecast of the atmospheric conditions, starting at some specific time.

3. To forecast hours or days into the future, a model produces a series of forecasts, each representing one short time step into the future. Each incremental step uses the previous step's forecast (i.e., the "answers" for each cube) as a starting input for the next forecast. Many iterations of the model are run in order to eventually generate forecasts for hours to days into the future.

4. Any inaccuracies in input data cause errors in the forecast from the start. Also, with each time step, the forecast will not be 100% accurate, because of an incomplete understanding of the physics involved in exchanges between the atmosphere and the land or ocean surface, or because processes smaller than the grid are at work. Given the complexity of the weather forecasting problem, especially the diverse set of spatially varying parameters, it is amazing that forecasts are as accurate as they are.

How Meteorologists Design, Use, and Intervene in Models

Even though computer models make various types of calculations and forecasts, humans are needed to determine which model output is most trusted in a given situation. For example, a meteorologist's experience and intuition may tell her that—at a particular time of year in a particular type of meteorological setting—one model is more successful at predicting precipitation than another model. Since weather forecasts can have profound implications for people's lives, such as whether to evacuate or take shelter, it is important to keep a human "in the loop."

Meteorologists determine what data and assumptions go into a model, and they design a model to answer a question at a specific temporal or spatial scale. Some models have shorter time steps (i.e., better temporal resolution) than others. This can be an advantage since a prediction is more likely to be accurate if it is made over a shorter time period. In addition, a poor temporal resolution could allow fast-forming weather systems to occur without being taken into account in subsequent calculations. Better temporal resolution, however, comes at the expense of computation time.

Likewise, models differ in their spatial resolution (including the number of vertical levels included), with finer resolution giving more precision, but requiring more computational power. The size of the area for which the forecast is generated is an important consideration. Modeling larger spatial domains gives more assurance that storm systems or other features will not creep into the area of interest undetected, but larger areas also require more computations.

Finally, some models express motion not in terms of what is happening at a series of three-dimensional gridpoints, but as a series of waves—these are called *spectral models*. They can be very accurate and may represent the "wave of the future" in weather forecasting models, particularly for longer forecasts.

In addition to collecting data, designing accurate data-collection strategies, and developing and interpreting models, atmospheric scientists research the physics and feedbacks involved in complex weather systems, with a goal of refining the models so that forecasts and other analyses are increasingly accurate. Accurate portraying of these complex systems is generally more important in models for predicting long-term climatic changes than for predicting weather.

Before You Leave This Page

✓ Explain the sources of data for weather forecasting models.

✓ Describe the seven basic equations in all weather forecasting models and what each one represents.

✓ Summarize the various trade-offs between features offered by different weather forecasting models.

✓ Explain the role of meteorologists in weather forecasting, including the use of models.

5.16 How Are We Warned About Severe Weather?

IN 1988, THE U.S. NATIONAL WEATHER SERVICE (NWS) began an aggressive program to modernize its capabilities in severe weather detection, forecasting, and communication, in support of its mission to protect life and property. As this program has been implemented, weather forecasts have shown amazing improvement, particularly in severe weather advisories. The centerpiece of this modernization was the installation of Doppler radar at each of its 122 locations in the U.S. and its territories. Weather-service organizations in other countries have made similar progress in deploying instruments and improving forecasts.

A How Does Doppler Radar Work?

Doppler radar is valuable because it tells us not only about the movement of storm systems (as conventional radar does), but also about mass and movement within individual storm clouds. We can observe real-time Doppler radar images of storms, tracking their progress on weather broadcast and cable channels, websites, and even apps on our phones, tablets, and game consoles.

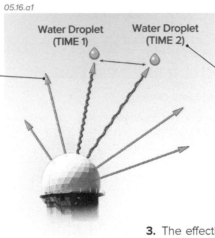

05.16.a1

Doppler Radar Facilities

1. In a Doppler radar facility, the transmitter emits microwave energy in all directions. When the radiation encounters an object, it is reflected in various directions, including back toward the facility. In accordance with the *Doppler effect*, the wavelength of the return signal increases as the reflecting particle moves away from the transmitter and decreases as it moves toward the transmitter. Using this principle, Doppler radar can calculate the directions and velocities of water drops in clouds and of raindrops, snowflakes, and other forms of precipitation.

2. A receiver at the same facility measures radar signals reflected by solid and liquid particles in the atmosphere in all directions from the facility. The system can calculate the distance that particles have moved between two time periods, here labeled Time 1 and Time 2. By knowing the distance and how long it took to travel that distance, we can calculate the *velocity*. By knowing the velocity at different times, we can calculate *accelerations*. Doppler radar can keep track of the motions of millions of particles through time, deciphering an amazingly complicated array of signals. The resulting data and images vividly depict the complex motions of atmospheric masses. By analyzing the full assemblage of changes in wavelength associated with each reflection, we can measure the motion of every part of the system.

3. The effective range of a Doppler radar system is about 460 km (285 miles) for detecting the location and movement of precipitation, and about 240 km (150 miles) for detecting motion within a weather system.

Data from Doppler Radar

4. Doppler radar can detect motion within single- and multi-cell thunderstorms and tropical cyclones, and the motion of the storms themselves, with greater precision and timeliness than other tools. This greatly improves forecasting of the likely track of a system, particularly for smaller storms that may go undetected by satellite. Many of the maps used to illustrate this chapter are Doppler radar images.

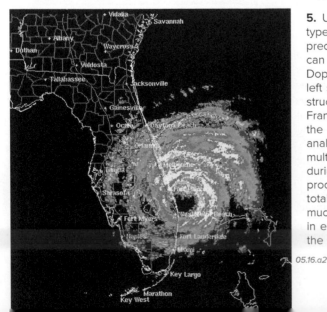

05.16.a2

5. Using Doppler radar, the type and intensity of precipitation from the system can also be estimated. This Doppler radar image to the left shows details of the structure within Hurricane Frances (2004), including the coiled shape. By analyzing such data for multiple time increments during a storm, maps can be produced showing storm totals, which represent how much total precipitation fell in each area as a result of the entire storm.

6. Doppler can also detect rotating motion within storms such as in mesocyclones, tornadoes, and microbursts. Many tornadoes touch down sometime after a mesocyclone begins forming and rotating, so meteorologists at the NWS can often issue a tornado warning more than 30 minutes before a predicted tornado actually touches down. Although Doppler cannot yet distinguish mesocyclones that will spawn a tornado from those that will not, this capability will likely be available sometime in the near future. Software systems at NWS automatically alert meteorologists of zones of Doppler-detected wind shear in the atmosphere, a potentially life-saving feature.

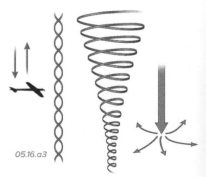

05.16.a3

B What Are Some Limitations of Doppler Radar?

05.16.b1

1. Like any technology, Doppler radar (◀) has certain limitations. Stationary objects, such as buildings and trees, reflect the Doppler radar beam, so these are part of the signal received by the receiver. Reflections from these non-weather objects are considered *noise* or *false echoes*. Objects causing false echoes are collectively referred to as *ground clutter*. Most of these objects are not moving, so they are relatively easy to recognize on the radar signal. They are filtered out using computing software that assumes that echoes that do not move for a predetermined period of time must not be precipitation. Ground clutter remains a problem, however, particularly in urban areas near the transmitter. In cities with numerous tall buildings, ground clutter can be so pervasive that data close to the city need to be masked out, producing radar plots that have a central circular area with no displayed data. You have probably seen these on weather broadcasts for larger cities.

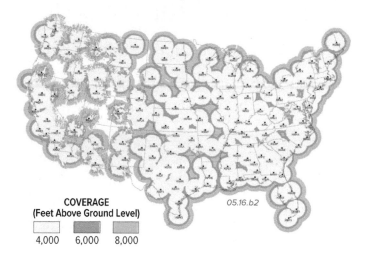

COVERAGE
(Feet Above Ground Level)

| 4,000 | 6,000 | 8,000 |

05.16.b2

2. Birds are another source of false echoes. Echoes from birds are somewhat more difficult to isolate on the radar signal because, unlike buildings, birds move and are in the air, not on the ground. Also birds come in various sizes and can be locally numerous, especially during migrations.

05.16.b3

3. Another limitation of Doppler radar is that some parts of the country are either not near a radar site or are shielded by topography, ground clutter, or even the curvature of the Earth. The network of Doppler radar stations across the U.S. (▲) generally provides adequate coverage, but there are some gaps. Similar networks of radar installations are present in southern Canada, most of Europe, eastern Asia, South Africa, Australia, and New Zealand. Many countries are not so well covered, if at all.

C How Does the National Weather Service Alert the Public About Severe Weather?

1. The NWS has local offices spread across the country (▶). Local offices collect and analyze data, monitor the weather conditions, and relay this information into the vast computer network of the NWS.

2. In the U.S., the NWS Storm Prediction Center (SPC) in Norman, Oklahoma, analyzes these data and issues storm *watches* when conditions in an area are favorable for life-threatening thunderstorms, tornadoes, winter storms, and strong winds. Watches usually cover a broad area, sometimes larger than an average-sized state. Flood watches are issued by the NWS in cooperation with one or more of the 14 National River Forecast Centers.

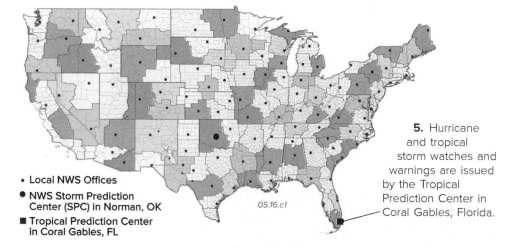

• **Local NWS Offices**

● **NWS Storm Prediction Center (SPC) in Norman, OK**

■ **Tropical Prediction Center in Coral Gables, FL**

05.16.c1

5. Hurricane and tropical storm watches and warnings are issued by the Tropical Prediction Center in Coral Gables, Florida.

3. By contrast, a *warning* of any of these severe weather types indicates more imminent conditions than a watch. The NWS issues warnings as bulletins that specify the location of the observed severe weather, its anticipated path, and the time that it is expected to occur at various locations in the warning zone. Except for the case of hurricanes, warnings are issued by the local NWS office. Warnings usually cover a much smaller area of predicted damage, such as a county or two, and also have a more focused time when damage is expected. A watch may be in effect for several hours, but a warning can be about an event that potentially will occur minutes after the warning is issued.

4. As soon as a severe weather *watch* is issued for your area, move to a sturdy shelter and stay alert to the weather by listening to the radio or watching TV. During a *warning*, immediately seek the safest location within that shelter by moving away from windows and toward the most reinforced parts of the structure, such as a bathroom, basement, or interior hallway or closet. Mobile homes are inadequate shelters in any type of severe weather.

Before You Leave This Page

☑ Sketch and explain the principles behind Doppler radar, the types of data and interpretations it produces, and its limitations.

☑ Explain the difference between a severe weather "watch" and a "warning" issued by the NWS.

5.17 What Happened During Hurricane Sandy?

SEVERE STORMS are extremely dangerous, capable of killing thousands of people and inflicting incomprehensible damage. Severe weather—and the more benign varieties—develop, operate, move, and dissipate according to fundamental principles presented in this chapter. Here, we examine Hurricane Sandy, which in 2012 tracked up the East Coast of the U.S., before coming onshore in New Jersey.

05.17.a1

05.17.a2

1. Hurricane Sandy began as a storm in the tropics (tropical cyclone) and became a hurricane that skirted the East Coast of the U.S. before coming onshore in New Jersey on October 29, 2012, as what many in the media and public called "Superstorm Sandy." On this satellite image (◄), Sandy appears as a swirl of clouds off the East Coast. Clearly visible is a well-defined center of the storm—the eye of the hurricane. To the west of Sandy is a long band of clouds, stretching from Florida to New England. What type of weather system do you think these clouds represent? Sandy later merged with this second weather system, which consisted of a cold front associated with a mid-latitude cyclone centered over New England. Sandy was weakening from hurricane status to tropical storm status as it came on shore, but the interactions with the second weather system had catastrophic results: over $75 billion in damages and nearly 150 deaths, about half in the U.S.

2. The superstorm was large compared to typical hurricanes, with winds of 65 km (40 miles) per hour extending across an area more than 1,500 km (900 miles) in diameter. In this satellite image (▲), clouds associated with the storm cover the entire northeastern U.S. and large areas of adjacent Canada and the Atlantic Ocean.

05.17.a3

05.17.a4 Oceansat-2

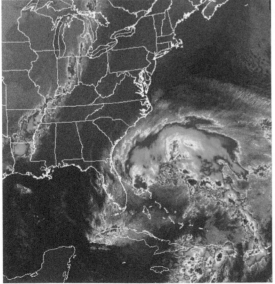

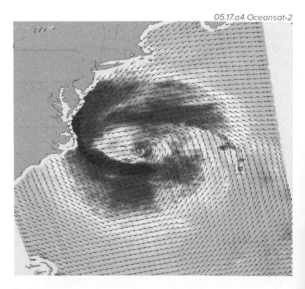

3. Enhanced infrared satellite images, like this colorful map (◄), are a mainstay of daily weather reports. In this map, clouds are shown in light gray where they are relatively thin but blue, green, yellow, and red where thicker, with red being the thickest. We can extract estimates of cloud thicknesses from an infrared image because the temperature of the top of a cloud is mostly a function of how high it is—colder typically means a higher top, greater overall thickness, and more precipitation. Examine this scene and consider what each feature is, why it has the shape it does, and which way it is likely to be moving. Also think about what processes are occurring in each place.

4. Sandy, still a hurricane at this time, is the large area of intense clouds and rainfall on the right side of the image (▲). The storm has a coiled shape, reflecting rotation around cyclones due to the Coriolis effect. It is moving north, as many storms do in this part of the Atlantic. Note a second line of clouds and precipitation to the west on land. These clouds formed along a cold front, a boundary between cold air to the west and warm air to the east. Sandy is within the *warm sector* of the storm. On this image, the different shades of gray on either side of the front reflect the temperature contrast between the two air masses, colder (lighter gray) to the west and warmer (medium gray) to the east.

5. The image above shows wind speeds near the ocean surface as measured by radar on a satellite orbiting Earth. Wind speeds exceeding 65 km (40 miles) per hour are yellow, those above 80 kph (50 mph) are orange, and those above 95 kph (60 mph) are dark red. At the time shown in this image, strong winds along the western side were striking the Outer Banks of North Carolina, which is an especially vulnerable area for flooding due to storm surges associated with hurricanes.

Movement and Evolution of Hurricane Sandy

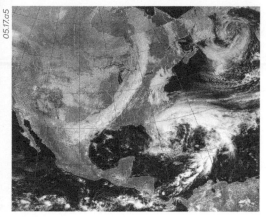

05.17.a5

05.17.a6

05.17.a7

6. Hurricane Sandy originated in the tropics and then migrated northwest toward the North American continent before turning north and tracking along the East Coast. This sequence of three satellite images shows the evolution of Sandy with time, with the earliest image on the left and the latest image on the right. In the earliest image (▲), Sandy is leaving the tropics as a somewhat dispersed, poorly organized storm. Note the curved cold front crossing the center of the U.S.

7. In this second image (▲), Sandy has moved farther north, to a position off the Outer Banks. It has become more tightly coiled and better organized, which causes the hurricane to intensify and the destructive winds to speed up. As the hurricane and the mid-latitude cyclone to the west get closer, they begin to interact. Note that the southern "tail" of the cold front is curving toward the east, as it gets drawn into the counterclockwise circulation around the hurricane.

8. In this final image (▲), Sandy and the cold front have collided, causing exceptional storminess across the Northeast. Many hurricanes turn northeast and head out to sea as they come up the coast, but Sandy turned northwest, directly into New Jersey. This unusual track was caused by an area of high pressure located over easternmost Canada, which blocked Sandy from moving northeast. The interaction between Sandy and the second storm made the situation much worse.

Hurricane Sandy

Hurricane Sandy began as a tropical disturbance in the Caribbean. The system strengthened considerably because it remained over very warm waters and was in a region with limited vertical wind shear. Sandy evolved into a named tropical storm and then a hurricane. The storm moved over Jamaica and Cuba, where it was weakened due to interactions with the land. It was downgraded to a tropical storm as it moved into the Atlantic. Once over open, warm waters of the Gulf Stream, it was reinvigorated, partly because of the approaching second storm to the west.

Several aspects of Sandy were very unusual, accounting for the excessive amount of damage. First, Sandy was a huge hurricane by any standards, affecting essentially the entire eastern half of the U.S. in one way or another, even though its wind speeds were not very strong by comparison to other hurricanes. A second factor was the way Sandy interacted with the mid-latitude cyclone to the west. Sandy formed fairly late in the hurricane season, and the front, with its accompanying very cold air, was fairly early for that type of winter storm. The interaction between the tropical-derived moisture and very cold air caused extremely heavy precipitation, including snow, ice, and blizzards, aspects we do not typically associate with hurricanes. Third, Sandy struck a part of the coast that does not experience frequent hurricanes, so some structures, especially those along the beach, were not built to withstand such storms.

The three photographs included here represent a small sample of the destruction caused by Sandy. The aerial image to the left shows damage to the Mantoloking area of New Jersey, including destruction of a highway and critical bridge, as well as severe damage to houses, especially those facing the ocean (on the right side of the aerial photograph). The two photographs on the right are from Coney Island, New York, and the coast of New Jersey. Consider what factors resulted in such severe damage to all three sites.

05.17.t2 Coney Island, NY

05.17.t3 New Jersey coastline

05.17.t1 Mantoloking, NJ

Before You Leave This Page

✓ Summarize how Hurricane Sandy formed, how it changed over time, and how it differed from a typical hurricane.

5.17

5.18 Where Would You Expect Severe Weather?

YOU ARE A METEOROLOGIST at the National Weather Service and are responsible for providing an overview of weather conditions across the U.S. To do this, you will study maps showing weather conditions at the surface and aloft and then determine where and how different air masses are interacting. In addition, you will identify areas where some type of severe weather is possible and identify what type of severe weather is likely.

> **Goal of This Exercise:**
> - Apply principles from this chapter to evaluate the weather conditions and make a forecast for where severe weather is most likely to occur.

Initial Procedures

Follow the steps below, entering your answers for each step in the appropriate place on the worksheet or online.

1. Observe and characterize the main weather features, including cloud cover, types of air masses, fronts, air pressure, moisture, Rossby wave ridges and troughs, and the polar front jet stream.

2. Consider all the surface and upper-level data to assess the likelihood of severe weather, identifying areas where different types of severe weather were likely to occur on two successive days.

These two pages display different types of maps, each showing certain weather features. All of the maps cover the main part of North America, but they do not cover precisely the same area.

Surface Conditions

These satellite images show the conditions for two successive days (Day 1 on the left, Day 2 on the right). Observe these images and propose what weather features (such as storms and fronts) each shows. Then, compare the two images to determine how each feature is moving. Recall that most weather systems in this region move from west to east, guided by the westerlies. Complete this phase before continuing on to examine other maps. Copies of these maps are on the worksheet.

DAY 1　　　　　　　　　　　　　　　　　　　　　　　　　　**DAY 2**

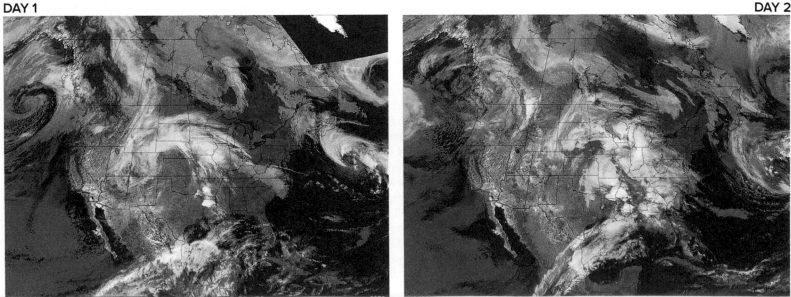

05.18.a1　　　　　　　　　　　　　　　　　　　　　　　　　　05.18.a2

Temperature

These maps show the average air temperature at the surface for each day. Red and orange show the warmest temperatures, purple and blue show the coldest, and green and yellow depict intermediate temperatures. Identify the boundaries between air masses of different temperature and propose what types of fronts those boundaries represent.

Atmospheric Moisture

These maps represent the amount of water in a column of atmosphere (in kg/m²) that would fall as precipitation if all of the water vapor condensed. Darker areas have abundant water and lighter ones have less water available to precipitation. Using the temperature and moisture maps, identify different types of air masses (e.g., cold and dry, versus warm and humid).

Conditions Aloft

These maps represent air pressures with 300-mb isohypses (white lines), so they represent upper-level conditions. Arrows show the general directions of airflow, which in these upper levels are parallel to the isohypses. Colors show estimated wind speeds, with red and orange representing the fastest speeds, purple and white showing the slowest, and green and yellow depicting intermediate wind speeds. Based on the high wind speeds and close

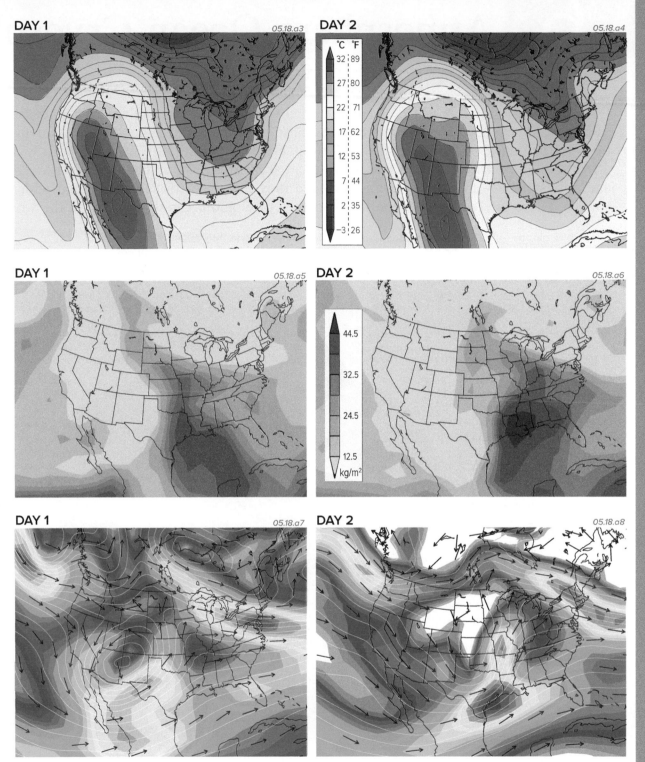

spacing of the isohypses, where is the polar front jet stream? Which areas have upper-level support for storm development based on likely positions of upper-level divergence in the Rossby waves?

Final Procedure

3. Using the various data presented here, divide the map area into different regions, each characterized by a particular air mass. Describe the types of boundaries that are likely between these air masses, based on their location and direction of movement. Shade in regions where severe weather was possible for Day 1 and for Day 2. Your instructor may have you answer questions on a worksheet or online.

5.18

6 Atmosphere-Ocean-Cryosphere Interactions

OVER 70% OF THE PLANET'S surface is covered by oceans, which exchange energy and moisture with the overlying atmosphere. Oceans move in response to three main factors—winds moving above them, spatial variations in the density of water, and the Coriolis effect. In addition to responding to wind directions, oceans in turn affect the temperatures and pressures of the adjacent atmosphere. These two major systems of the Earth—oceans and atmosphere—are closely connected, and they interact with each other and with the planet's ice—the cryosphere. These interactions are the major causes of climatic variability around the world.

These two globes represent temperatures during two time periods, one from 1964–1976 and one from 1978–1990. Specifically, the globes show how average land temperatures and sea-surface temperatures (SST) during these two time periods departed (differed) from long-term averages. For each globe, the average temperature for that time period was compared to average long-term temperatures. If the temperatures during that time were warmer than the long-term average, the values are positive (shown in red, orange, and yellow). If the temperatures during that time period were colder than the long-term average, the values are negative (shown in purple, blue, and green). Referencing a data set to long-term averages produces a comparison we call a "departure" or "anomaly." Observe the two globes for patterns and for changes between the two time periods.

What patterns do you observe and what happened to cause such different patterns over time?

Temperature Departure from Average (1964–1976)

Northwestern North America, especially Alaska, was colder than normal for those locations, as shown by the purple and dark blue colors over land, from 1964 to 1976. The adjacent waters of the Pacific were also relatively cold, but only along the western coast of North America. The colder than normal temperatures continued south into southwestern North America and adjacent waters. In contrast, waters farther out in the Pacific were warmer than usual, as depicted by the large region colored red and orange.

Temperature Departure from Average (1978–1990)

A shift in regional temperatures occurred around 1977, as shown by the different patterns between the globes. Alaska and adjacent waters of the eastern Pacific warmed significantly, as reflected by red and orange areas over Alaska. In contrast, cooling occurred in the middle of the northern Pacific, as indicated by the blue and green colors. The patterns in these two globes are nearly opposites, representing two common modes in temperatures in this region: a *cool phase*, in which SST are cooler than usual in the eastern north Pacific near Alaska and warmer than usual in the mid-north Pacific, and a *warm phase*, with the opposite pattern. The switch from one mode to another is named the *Pacific Decadal Oscillation* (PDO).

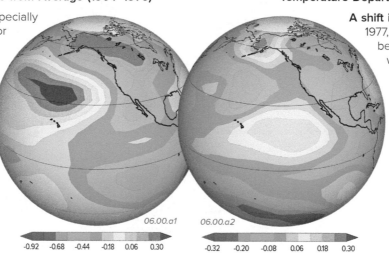

06.00.a1   06.00.a2

-0.92 -0.68 -0.44 -0.18 0.06 0.30

-0.32 -0.20 -0.08 0.06 0.18 0.30

What causes warming or cooling of sea-surface temperatures, and how can one part of the ocean be colder than normal while another part is warmer than normal?

Why are the patterns on these two globes the opposite of each other, and what causes these shifts, as part of the PDO?

Regional changes in SST can be quantified by statistically comparing SST in the eastern and western Pacific and calculating an index (called the *PDO index*) that represents the difference. When the waters off Alaska are warmer than those in the central Pacific, the PDO index is positive. The top graph to the right plots the PDO index for several years and shows a major shift in climate at about 1977—the same temperature shift represented by the difference in the two globes. Red bars represent the warm phase.

What is an ocean oscillation, and do similar oscillations occur in other parts of the Pacific or even in other oceans?

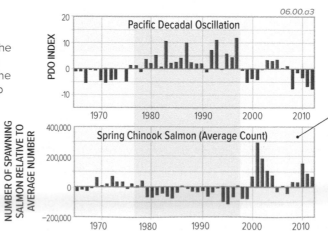

06.00.a3

Pacific Decadal Oscillation

PDO INDEX

20
10
0
-10

1970   1980   1990   2000   2010

NUMBER OF SPAWNING SALMON RELATIVE TO AVERAGE NUMBER

400,000

Spring Chinook Salmon (Average Count)

200,000

0

-200,000

1970   1980   1990   2000   2010

The Pacific Decadal Oscillation, including its warm and cool phases, was first recognized by a fisheries scientist trying to explain year-to-year variations in how many salmon return to their spawning grounds in streams along the Pacific, as shown in the bottom graph. Comparison of the graphs reveals a striking correlation—the number of spawning salmon is higher than average during cool phases of the PDO and less during warm phases.

What causes these warm and cool phases in an ocean?

TOPICS IN THIS CHAPTER

06.00.a4 Mount Baker, WA

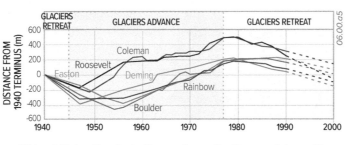

This shift in climate in the northern Pacific correlates with a change in glaciers on Mount Baker (◄), one of the large volcanic mountains in the Cascade Range. As shown in the graph above, glaciers were advancing before this shift (during the cool phase) and retreating after it (during the warm phase). As shown on the graph of PDO index on the previous page, sometime around 1998 the PDO went back into a cool phase, heralding the return of cooler temperatures.

Do ocean oscillations affect the amount of precipitation and global amounts of ice?

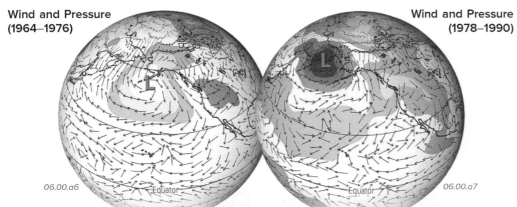

Wind and Pressure (1964–1976)

Wind and Pressure (1978–1990)

06.00.a6 06.00.a7

Shifts in the PDO are accompanied by changes in winds and air pressure. The two globes above show anomalies (departure from average) in wind directions (arrows) and surface air pressure (shading, where dark gray is lower pressure). During a PDO cool phase (left globe), the Aleutian Low is very weak and the counterclockwise winds weaken (clockwise anomalies). This gives the Alaska coast the colder temperatures during the cool phase. During a PDO warm phase (right globe), the Aleutian Low is much lower, and stronger counterclockwise winds around the Aleutian Low bring warm SST to Alaska's coast. Since wind directions are controlled largely by differences in air pressure, the patterns of air pressure also differ between these two phases.

How do changes in SST relate to changes in air pressure and wind direction? Do the oscillations cause the wind patterns or vice versa, or are both related to changes in air pressure?

The PDO and Climate

In 1977, a major shift occurred in the patterns of sea-surface temperatures (SST) in the North Pacific. Before that time, waters in the Gulf of Alaska had been relatively cool, while waters in the central north Pacific had been warmer than normal. After 1977, the patterns switched, with warmer than normal waters along the coasts of Alaska and British Columbia, but colder than normal waters in the central north Pacific. This change, sometimes called the *Great Pacific Climate Shift,* had profound implications for conditions on land in western North America and in the Pacific. This shift correlated with changes in glaciers, temperatures on land, and the number of spawning salmon.

In the past several decades, this climate shift has been recognized as part of a series of cyclical changes in broad-scale patterns in air pressure, wind directions, and SST, where the Pacific Ocean has two dominant modes: a *warm mode* and a *cool mode.* From data collected over the last century, scientists have been able to establish that the Pacific Ocean stays in the warm mode for several decades before switching to the cool mode for the next several decades: thus the "decadal" part of the Pacific Decadal Oscillation name. The term "oscillation" refers to a climatic condition, such as air pressure or SST, that shifts back and forth between two distinct modes or phases. The Pacific Decadal Oscillation is but one such oscillation. Similar oscillations, described in this chapter, occur in the equatorial Pacific, Atlantic Ocean, and Indian Ocean, each greatly impacting regional climate.

6.0

6.1 What Causes Ocean Currents?

SURFACE WATER OF THE OCEANS circulates in huge currents that generally carry warmer water toward the poles and colder water toward the equator. Extremely large quantities of energy are stored in the uppermost 100 meters of the global oceans, and surface ocean currents redistribute this energy from one part of the ocean to another, carrying it from zones of excess energy to those with an energy deficit. What causes ocean currents, and can we make some general predictions about which way ocean currents would likely flow?

A What Causes Surface Waters of the Ocean to Move and Which Way Do They Go?

Various processes cause surface water to move, but the primary driving force is wind. Wind blowing across a body of water transfers some of its momentum onto the surface of the water, causing it to move. On a smaller scale, this causes waves, and on a global scale, this sets in motion large currents that can traverse an entire ocean.

1. As wind blows across water, it pushes against the surface, causing the uppermost water to begin flowing.

2. This water in turn transfers some of its momentum to waters just below the surface, setting up an overall shearing of the water column, with the strongest shear force (depicted by arrows) at the surface.

3. The effects of wind stress, and therefore current velocity caused by the wind, decrease with depth. They vanish at a depth of about 100 m, a depth known as the *null point*.

06.01.a1

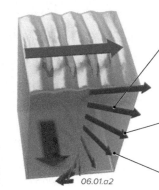

06.01.a2

4. If the planet were not rotating, water above the null point would flow parallel to the direction in which the wind is blowing. On Earth, however, as soon as the water starts moving it is subject to the Coriolis effect, and so it has an apparent deflection away from its initial direction, turning to the right in the Northern Hemisphere.

5. As the effects of the shearing force decrease downward, toward the null point, the direction of flow turns farther to the right with depth. In this way, the direction and velocity of flow, when plotted as a series of arrows, define a downward spiraling shape—an *Ekman spiral*.

6. As a result of these competing influences, the overall direction of flow of the uppermost ocean, when averaged between the surface and the null point, is at an angle to surface winds.

B Which Way Do Surface Winds Blow Against the Ocean Surface?

If wind patterns drive the flow of the upper part of the ocean, we should be able to anticipate the directions of ocean currents by considering the directions of wind.

1. This globe reviews the main belts of winds encircling Earth. Examine the globe and consider what pattern of ocean currents might result from these wind directions. Try this before reading on.

2. As you thought about likely directions of currents, you probably realized that eventually any current that flows east or west will at some point run into a continent—the continents are in the way, so an ocean current must turn to avoid the continent.

3. Recall that winds are generally not very strong along the equator, the so-called *doldrums*. They are also very weak in areas of descending air in the subtropics, at 30° N and 30° S latitude—the *horse latitudes*.

4. Another factor to consider is that the apparent deflection of moving objects by the Coriolis effect is opposite across the equator. The apparent deflection is to the right north of the equator and to the left south of it.

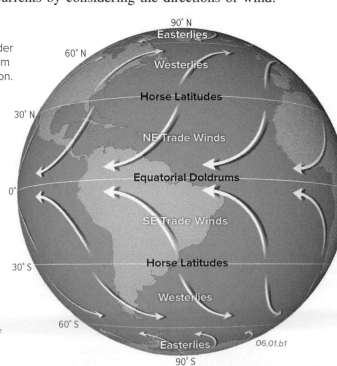

06.01.b1

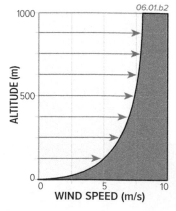

06.01.b2

5. This graph (▲) shows the general relationship between wind speed and altitude. Winds are slower near the surface, and they increase upward. Winds slow down near the surface because some of their energy of motion (kinetic energy) is transferred from the wind to the surface, producing waves and currents. As there is over land, the atmosphere has a friction layer over water.

C What Is the Anticipated Pattern of Global Surface Ocean Currents?

Given what we know about the global distribution of surface pressure cells and the direction of surface winds, it is reasonable to make a first approximation of the direction of surface ocean currents as a series of interacting circular motions, or *gyres*, like the ones in the stylized ocean shown below.

1. A key characteristic of this hypothetical planet is that the central ocean extends from pole to pole, so it crosses all of the regional wind belts, from the equatorial doldrums to the polar easterlies. The adjacent continents also nearly extend from pole to pole, effectively confining the ocean on both sides.

2. Surface waters flowing from a lower latitude (closer to the equator) toward a higher latitude (closer to the poles) transport water with greater stored energy than expected at that latitude and are considered "warm" currents. Those flowing in the reverse direction, from higher to lower latitudes, have less stored energy than expected and are termed "cold" currents. The terms *warm* and *cold* should only be considered relative to the environment through which they are passing and are indicative of whether they are bringing energy to or removing it from an environment. Accordingly, warm currents are colored red and cold currents are colored blue. Note that a single gyre brings warm water to the coast of one continent but cold water to the coast of the other continent, as described below.

East Coast of Continent

45° N–60° N: Cold surface waters originating from north polar regions.

25° N–35° N: Warm surface waters originating from equatorial regions and moving northward.

25° S–35° S: Warm surface waters originating from equatorial regions. Note that this happens with winds from the opposite direction compared to the Northern Hemisphere.

45° S–60° S: Cold surface waters originating from south polar regions. Note that this happens with winds from the opposite direction compared to the Northern Hemisphere.

West Coast of Continent

45° N–60° N: Warm surface waters originating from subtropical regions and moving northward.

25° N–35° N: Cold surface waters originating from mid-latitude regions.

Equatorial Latitudes: Mostly warm waters flowing east to west.

25° S–35° S: Cold surface waters originating from mid-latitude regions and flowing north.

45° S–60° S: Warm surface waters originating from subtropical regions.

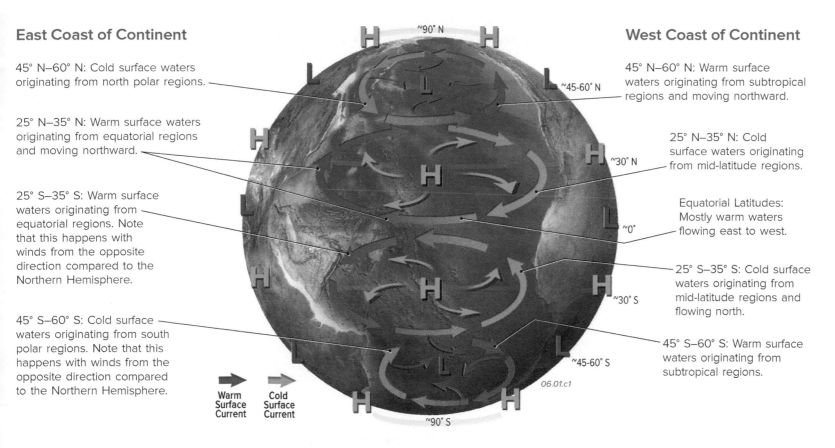

Warm Surface Current   Cold Surface Current

06.01.c1

To remind us that generalized models are just that, we include a numeric model of actual ocean currents, whose squiggly lines bring to mind the style of painter Vincent van Gogh (▼).

06.01.c2

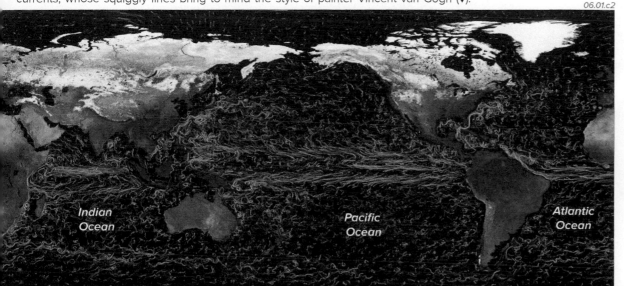

Indian Ocean   Pacific Ocean   Atlantic Ocean

Before You Leave This Page

☑ Sketch and explain how motion is transferred from the atmosphere to the ocean surface and how the Ekman spiral results.

☑ Sketch and explain the anticipated directions of surface ocean currents in a simple one-ocean, two-continent global model, identifying warm and cold currents.

6.1

6.2 What Is the Global Pattern of Surface Currents?

OCEAN CURRENTS on Earth follow the main patterns we would predict from our understanding of global wind directions, but irregular shapes of the continents cause important complexities, such as isolating some protected seas from the larger circulation patterns. The global figure below depicts a summary of major surface currents and gyres in the oceans. The smaller globes and accompanying text each highlight major currents in different regions. The major currents have profound implications for the climate, weather, ecology, and the lives of the inhabitants, so we recommend that you take some time to learn their names.

1. This large map shows the locations, flow directions, and characteristics of ocean currents on a global scale. Warm currents are shown in red, and cold currents are shown in blue.

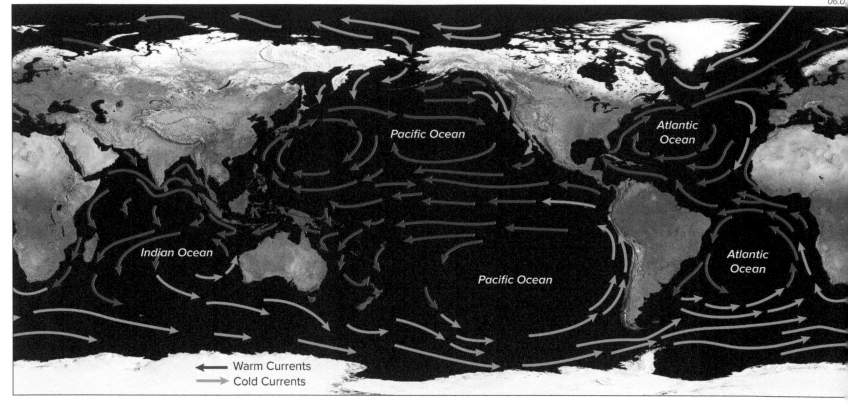

06.0.

Indian Ocean

2. Most of the Indian Ocean lies south of the equator. Surface circulation in the Indian Ocean is dominated by the flow around the southern subtropical gyre, consisting of (1) the warm, south-flowing *Mozambique Current* east of Africa, (2) the cool, north-flowing *Western Australian Current*, and (3) a west-flowing segment of the gyre along the equator to complete the loop. South of 40° S, waters become entwined with the *Antarctic Circumpolar Current*. Surface flow patterns vary seasonally according to the winds of the strong Asian monsoon and other factors.

06.02.a2

Western Pacific Ocean

3. The largest ocean basin on Earth, the Pacific basin, displays many of the characteristics expected from the model. It contains two subtropical gyres, one in the Northern Hemisphere and one in the Southern Hemisphere.

4. Northwestern Pacific: Water enters the western Pacific driven by westward flow on either side of the equator. As waters north of the equator approach the Philippines and mainland Asia, they turn north, carrying warm water past southern China and Japan in the *Kuroshio Current*. At about 40° N, this flow then turns east as the *North Pacific Current*.

5. Southwestern Pacific: South of the equator, the west-flowing current turns south down the east coast of Australia as the warm *East Australian Current*. Farther south, it becomes enmeshed with the strong *West Wind Drift*, driving eastward toward South America.

06.02.a3

Northern Pacific Ocean

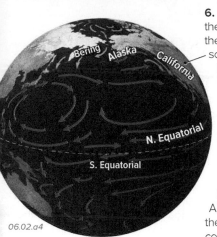

06.02.a4

6. The northern subtropical gyre in the Pacific produces a cold current, the *California Current*, that flows southward, down the west coast of North America. A portion of the east-moving, mid-latitude water turns north as it approaches northwestern North America, turning northward along the coast of northern British Columbia and Alaska as the *Alaska Current*, bringing somewhat warmer waters to these northerly latitudes. As it moves west, the Alaska Current ultimately mixes with the cold waters of the Bering Current coming from the Arctic Ocean.

Southeastern Pacific Ocean

7. The cold waters of a current (the West Wind Drift) flowing east from Australia turn north along the west coast of South America. During this long track across the southern Pacific, the waters turn cold and form a very pronounced cold current, called the *Humboldt Current* (also locally known as the *Peru* or *Chile Current*). As the Humboldt Current approaches the equator, it turns back to the west, toward Australia, to complete the circuit of this huge gyre. The west-flowing segment is the *South Equatorial Current*.

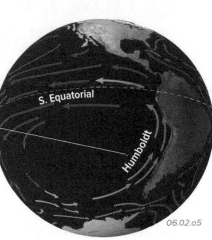

06.02.a5

Northern Atlantic Ocean

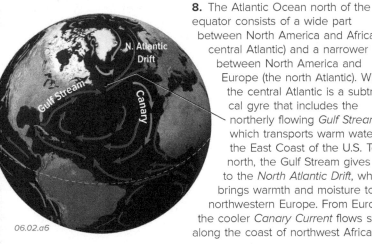

06.02.a6

8. The Atlantic Ocean north of the equator consists of a wide part between North America and Africa (the central Atlantic) and a narrower part between North America and Europe (the north Atlantic). Within the central Atlantic is a subtropical gyre that includes the northerly flowing *Gulf Stream*, which transports warm water up the East Coast of the U.S. To the north, the Gulf Stream gives way to the *North Atlantic Drift*, which brings warmth and moisture to northwestern Europe. From Europe, the cooler *Canary Current* flows south along the coast of northwest Africa.

Southern Atlantic Ocean

9. The part of the Atlantic Ocean south of the equator, between South America and Africa, is the south Atlantic. Within this region is the Southern Hemisphere gyre, which includes the cold *Benguela Current*, which flows north along the western coast of southern Africa. After flowing west, the gyre turns south along the eastern coast of South America, becoming the warm *Brazil Current*. To the south, this current is deflected to the east by its interactions with the *Antarctic Circumpolar Current*, flowing eastward.

06.02.a7

Arctic Ocean

06.02.a8

10. Unlike its southern equivalent, the northern polar region, north of 80° N, is occupied entirely by ocean—the Arctic Ocean. The major outlets for the Arctic Ocean include the Pacific basin by means of the cold *Bering Current*, and the Atlantic via the *Labrador* and *Greenland currents*, which bring cool water down the east coast of Canada and Greenland. A major flow of warm ocean waters into the Arctic Ocean is accomplished by the North Atlantic Drift flowing northward along the coast of northwestern Europe.

Southern Ocean

11. Ocean waters surrounding Antarctica are regarded as the southern parts of the Pacific, Indian, and Atlantic oceans. No land interruptions, other than the southern tip of South America and New Zealand, exist in this region. The entire polar zone is occupied by continental Antarctica. Ocean-continent contrasts here are parallel to latitude. Strong pressure gradients and relative lack of land interruptions induce extremely strong westerly winds at these latitudes—known by sailors as the "roaring forties" and the "screaming sixties." Circling around Antarctica is the strong West Wind Drift, also called the *Antarctic Circumpolar Current.*

06.02.a9

Before You Leave This Page

✓ Sketch and explain the main patterns of ocean currents in the Northern Hemisphere versus those in the Southern Hemisphere, noting similarities and differences with a simple one-ocean model.

6.3 How Do Sea-Surface Temperatures Vary from Place to Place and Season to Season?

SEA-SURFACE TEMPERATURES (SST) vary greatly, from bathwater warm to slightly below freezing. Early data on SST were collected from ships, but since the 1970s, satellites have collected voluminous SST data, documenting variations in temperature from region to region, season to season, and decade to decade. SST data have become even more important as climatologists investigate the causes and consequences of global warming and other types of climate change. How do SST vary on Earth, and what do such data tell us?

Average Sea-Surface Temperatures

1. These two globes both show SST averaged over the entire year and over several decades. Red and orange represent the warmest temperatures, purple and blue indicate the coldest temperatures, and green and yellow show intermediate temperatures.

2. Observe the global patterns and propose possible explanations based on how insolation varies from place to place. Then, examine the smaller patterns in the context of ocean currents (their position, direction of motion, and warm-versus-cold character). After you have done this for both globes, read the associated text.

3. The most obvious pattern in the SST data, not surprisingly, is the large temperature change between tropical latitudes and polar regions. Polar regions are green, blue, and purple (from cold to coldest) on this globe—they are the coldest seas on Earth.

4. Note, however, that the warmest seas (in red) are not necessarily along the equator. If you expected the equator to be warmest, how could you explain this apparent discrepancy? Part of the explanation is that seas at the equator are overlain by rising air and so commonly have cloud cover that reflects some insolation back into space. Also, some of the warmest seas are in the subtropics, where descending air dries out the surface, limiting cloud cover and permitting more sunlight to reach the sea.

5. An aspect to consider is whether an area of sea has a wide-open connection with open ocean. If so, that water can mix with or move to parts of the ocean where the waters are deeper and colder. On this globe, the warmest waters are in the relatively shallow parts of the Caribbean Sea and Gulf of Mexico, areas whose connection with the Atlantic Ocean is somewhat hindered by the many islands and peninsulas, so the Sun-warmed water mostly remains in those regions.

06.03.a1

-29 -25 -21 -17 -14 -10 -6 -2 2 6 10 14 17 21 25 29

6. This globe shows most of the Indian Ocean (on the left) and about half of the Pacific Ocean (on the right). What patterns do you observe on this side of the globe, and what are some differences between this side and the other side of the planet, as shown in the top globe?

7. The semi-enclosed, shallow seas between the mainland of southeastern Asia and the islands of Indonesia and the Philippines are relatively warm, as is the Indian Ocean. As discussed later in this chapter, SST in this region, while always warm, can vary somewhat in response to an important ocean oscillation called the El Niño-Southern Oscillation, or ENSO for short. More on this important oscillation later.

8. Patterns are more regular in the middle of the Pacific Ocean, away from the complicating effects of continents, which influence weather patterns and steer ocean currents along their coasts. The SST data form bands of colors that mostly follow along lines of latitude. The patterns are more complicated closer to the continents, such as near Japan.

9. The southern parts of the Indian Ocean and Pacific Ocean have similar, latitude-parallel gradations. Again, this pattern largely reflects the lack of complicating landmasses. Only the southernmost tip of South America (not shown), Tasmania (the island south of mainland Australia), and the South Island of New Zealand extend south of 40° S latitude. Between these places and Antarctica, the ocean continues unimpeded.

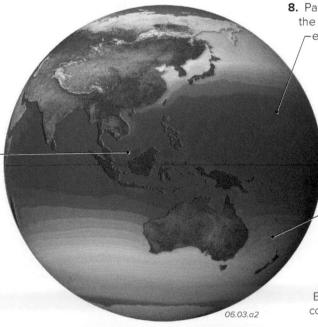

06.03.a2

January Sea-Surface Temperatures

10. These two globes focus on the Atlantic Ocean, but at two different seasons. The left globe shows average SST for January, which is winter in the Northern Hemisphere but summer in the Southern Hemisphere. The right globe shows the same data for July. Observe the main patterns and compare the two globes.

11. In January, the colors of SST shift slightly but noticeably to the south, following the shift of maximum insolation as summer temperatures increase in the Southern Hemisphere. Also, whichever hemisphere is having summer experiences an overall expansion of warm SST.

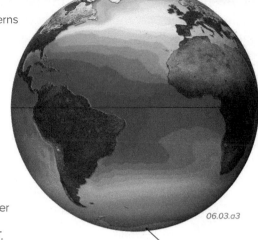

06.03.a3   06.03.a4

12. In January (the southern summer), the waters around Antarctica are not as cold as they are in July (the southern winter; the right globe). The opposite pattern occurs near the North Pole—waters there are colder in January, the northern winter, than during July, the northern summer.

July Sea-Surface Temperatures

13. Compare the patterns of SST on the two sides of the ocean. On the western side of the ocean, next to the Americas, warm waters spread relatively far to the north and south, away from the equator. This larger expanse of warm waters reveals the effects of ocean currents, like the Gulf Stream, moving warm water away from the equator, along the eastern coasts of North and South America. Warm waters of the Gulf Stream continue toward Europe.

14. Compare this with the pattern on the eastern side of the ocean, adjacent to Europe and Africa. Here, relatively cooler waters reach farther toward the equator. This pattern is especially obvious along the west coast of southern Africa, where the north-flowing Benguela Current brings cold water northward along the coast. A similar current, the Canary Current, brings cold water south along the coast of northwestern Africa. As a result of the opposite flows on the east versus west sides of the Atlantic Ocean, the width of warm water is much less near Africa than it is near South America and the Caribbean Sea.

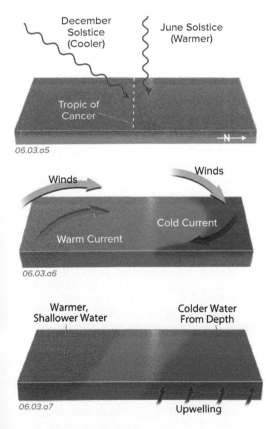

06.03.a5

06.03.a6

06.03.a7

Factors Influencing Sea-Surface Temperatures

15. From the four SST globes on these pages, the main pattern that emerges is the strong correlation of warm waters to low latitude. In the tropics, the overhead rays of the Sun strongly heat the oceans, especially during summer for that hemisphere (◄). Much less solar heating occurs near the pole; none for much of the winter.

16. Another influence is ocean currents, such as the Gulf Stream (◄), which mostly move warm water away from the equator on the western side of oceans from about 20° to 50° north and south. In contrast, cold currents flow toward the equator on the eastern sides of oceans, as occurs with the Benguela Current off Africa, the Humboldt Current along the western side of South America, and the California Current down the western side of North America. Currents largely reflect regional wind patterns.

17. A third important factor is the flow of cold, deep waters toward the surface, or *upwelling* (◄). Cold-water upwelling occurs along the California coast, off the western coast of South America, and in other places. Offshore winds push water away from the coast, causing deeper water to upwell to replace this water. Upwelling leaves only a subtle imprint on global temperature maps, but it strongly influences the biological productivity of the ocean and the activities in many coastal communities.

18. SST can greatly affect adjacent land. This photograph (►) shows desert along the western coast of southern Africa with the Atlantic Ocean in the distance. The dryness is partly due to the cold, stabilizing Benguela Current offshore and to prevailing winds that blow offshore.

06.03.a8 Namib Desert, Namibia

Before You Leave This Page

✓ Summarize the main patterns in sea-surface temperatures as observed using average temperatures.

✓ Describe and explain the main differences between SST in January and those in July, giving specific examples.

✓ Sketch and explain some factors that affect SST, and how ocean currents affect SST of western versus eastern sides of an ocean, providing examples from the Atlantic.

6.3

6.4 What Causes Water to Rise or Sink?

MOVEMENT WITHIN THE OCEANS arises from other factors in addition to shearing induced by the wind. Such factors include variations in the density of water in response to changes in its temperature and salinity. Changes in temperature and salinity occur from flows of fresh water into and out of the oceans as a result of precipitation, evaporation, ice formation and melting, and inputs from rivers. The temperatures of ocean waters and the relative amounts of fresh water control the density of water, and these differences in density can alter the circulation of ocean waters.

A How Does the Temperature of Water Control Its Density?

Like most substances, water changes its density when subjected to increases or decreases in temperature. Unlike other substances, however, water exhibits some peculiar density changes as it approaches freezing and begins to solidify into ice.

Polarity of the Water Molecule

1. A key attribute of water that controls its behavior is its polarity. Recall that in a water molecule, the two hydrogen atoms are on one side of the oxygen, giving the molecule a polarity, which causes water molecules to be attracted to charged atoms (*ions*). This allows water to interact with other chemical substances, such as salt.

06.04.a1

Hydrogen Bonding

2. Water molecules are attracted to each other as well as to ions. In water, a bond called a *hydrogen bond* forms between one molecule's hydrogen atom and another molecule's oxygen atom.

06.04.a2

3. Hydrogen bonds form when the polarities of two adjacent water molecules attract one another. The hydrogen bond is responsible for some of water's unique properties (e.g., viscosity and surface tension).

Density of Water as a Function of Temperature

4. This graph (▶) shows how the density of fresh water changes with temperature. The curve on the right side shows how liquid water changes density in temperatures slightly above freezing. The vertical break marks 0°C, the temperature at which fresh water freezes. The angled line on the left shows how ice changes density with temperature. Examine the main features of this graph before we explore it further. Read clockwise around the graph, beginning on the other side.

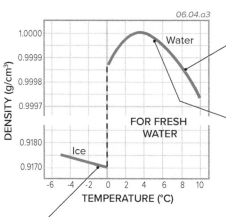

06.04.a3

5. At temperatures above 4°C, water, like most substances, becomes more dense as it cools, as shown by moving from right to left up the sloping line on the right side of the graph. With cooling temperature, individual molecules become less energetic and can occupy a smaller volume, thereby increasing the water's density. Fresh water reaches its maximum density (1 g/cm³) at a temperature of approximately 4°C (39°F).

6. At a temperature of 4°C, however, fresh water begins to decrease in density, as is shown by the downward bend in the graph between 4°C and 0°C. This decrease in density occurs because, at temperatures below 4°C, the strength of the hydrogen bonds starts to exceed the disruptive random motions of the less energetic molecules, and the molecules start to organize themselves into a more regular arrangement that occupies more volume and decreases density.

7. As ice (the solid state of water) forms, the crystalline structure of the hydrogen bonds dominates, placing adjacent molecules about 10% more distant, on average, from their nearest neighbors than when in the liquid state at 4°C. This causes the density of ice to drop to about 0.917 g/cm³ (at 4°C, liquid has a density of 1.00 g/cm³) This decrease in density when water freezes causes ice to float on water—less dense materials, like wood and ice, float on a denser liquid, like water. Our world would be a very different place if ice were denser than water. Ice would begin to grow upward from the ocean floor over time until it occupied a very large percentage of the oceans.

Sinking and Rising of Water as a Function of Temperature

8. These density changes of water as a function of temperature have huge implications for our natural world, many of which will be discussed later in this chapter and throughout the book.

9. At temperatures between freezing and 4°C, slightly warmer waters are denser than cooler waters (if they have the same salinity), and will therefore sink, helping to redistribute warmth downward in the ocean. Once ice forms, it is even less dense than warmer waters, so it floats. In most substances, the solid state is denser than the liquid one, but not for water. Ice is approximately 9% less dense than liquid water.

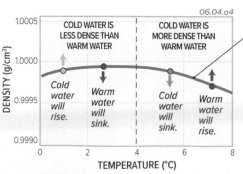

06.04.a4

10. At temperatures above 4°C, water behaves like most other substances, becoming less dense as it warms. At temperatures above 4°C, cooler water is denser than warmer water and will therefore sink.

B How Does Change in Salinity Affect the Density of Water?

How Salt Occurs in Water and Changes Its Density

1. We all know that water can dissolve some solid materials, like salt, but how does water do this? When we add salt crystals, like the one represented by the array of chlorine (Cl) and sodium (Na) atoms in the center of this block, to a pot of water, each crystal is surrounded by water molecules. In the salt mineral halite, sodium atoms have loaned an electron and so have a positive charge (Na⁺). Such positive ions are called *cations*. Chlorine has gained an electron and so has a negative charge (Cl⁻). Such negative ions are *anions*.

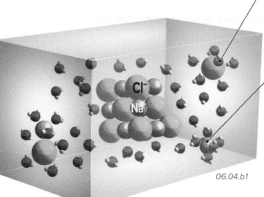

06.04.b1

2. The negatively charged chlorine anion is attracted to the positive (H) end of the water molecule. If this attraction is strong enough, it can pull the chlorine away from the halite crystal and into the water.

3. In a similar manner, sodium in the salt crystal is a positively charged cation (Na⁺), and thus is attracted to the negative side of any adjacent water molecule. This attraction can pull the sodium ion away from the salt crystal and into the water.

4. Once dissolved in water, the positively charged sodium cation (Na⁺) will be surrounded by the negative sides of water molecules, limiting its ability to rejoin the salt crystal. The negatively charged chlorine anion (Cl⁻) is likewise surrounded by the positive side of the encircling water molecules.

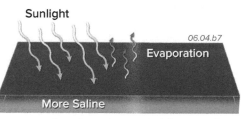

06.04.b2

5. As the salinity of water increases so does its density, in a reasonably linear fashion. Elevated saline content in water increases its density, so the saline water will sink. Less saline water will rise. The exact nature of the relationship between salinity and density also depends upon temperature.

Decreasing the Salinity of Water

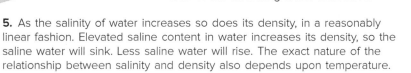

Sunlight

06.04.b3

Less Saline

06.04.b4

River

Less Saline

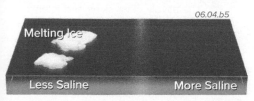

06.04.b5

Melting Ice

Less Saline   More Saline

6. One way to decrease the salinity of water is to dilute it with fresh water, such as by adding precipitation to the body of water.

7. Dilution of salt content also occurs from the introduction of fresh water, as from a stream. Most rivers contain some dissolved salt, but much less than the sea.

8. Ice consists of fresh water because it does not easily incorporate salt into its crystalline structure when it freezes. When ice melts, it adds fresh water to the body of water, diluting the salt content of the water and decreasing the salinity. Ice melting on land likewise provides abundant fresh water to streams, most of which drain into some body of water.

06.04.b6 *Lemaire Channel, Antarctica*

9. Several types of ice occur in water. The most familiar type is an *iceberg* (◄), which is a piece of ice broken off a glacier that flowed into a sea or lake. The second type is *sea ice* (not shown), which forms from freezing of seawater into a smoother layer of ice coating part of the sea.

Increasing the Salinity of Water

Sunlight

06.04.b7

Evaporation

More Saline

10. The most obvious way to increase the salinity of water is through evaporation, which removes water molecules but leaves the salt behind. The water vapor contains no salt and can be carried far away by motion in the atmosphere, eventually producing freshwater precipitation as rain, snow, sleet, or hail. This fresh water forms the streams and glaciers on land.

ICE-FORMING   06.04.b8

Ice Shelf

More Saline   Less Saline

11. Salinity also increases when ice forms on the sea. Fresh water is removed during freezing, leaving behind the salt in the water. As a result, the salinity of seawater increases below and adjacent to sea ice. If the sea ice melts, as can occur during the summer, it introduces the fresh water back into the ocean. We explore further the relationship between salinity, density, and temperature in the next two pages.

Before You Leave This Page

☑ Sketch a water molecule and explain why it has polarity.

☑ Sketch and explain the relationships between the temperature of water and its density, including changes at 4°C and 0°C.

☑ Describe how water dissolves minerals, and how salinity affects the density of water.

☑ Sketch and explain three ways to decrease the salinity of water and two ways to increase the salinity of water.

6.5 What Are the Global Patterns of Temperature and Salinity?

THE OCEANS VARY IN SALINITY from place to place and with depth. Spatial variations are due to differences in the amount of evaporation and formation of ice, which increase salinity, versus the amount of precipitation, input from streams, and melting of ice, all of which add fresh water that dilutes the salinity. Also, saline waters are denser than fresh water, so there are changes in salinity as a function of depth within the oceans.

A What Is the Observed Pattern of Sea-Surface Salinities?

The oceans vary in temperature and salinity, and therefore also in density, laterally across the surface and from shallower to greater depths. The maps below summarize the global variations in SST and surface salinity. Examine the patterns on each map and then compare the patterns between the two maps, region by region.

1. This map (▶), now familiar to you, shows typical SST, with red and orange being warmest and blue being coolest. Note where the warmest seawater is located, and think about what implication this might have for salinity. Do you think this warm water will be more or less saline than waters that are somewhat cooler, like those shown in yellow on this map?

2. The dominant feature of global variations in SST is the equator-to-pole contrast arising from the supply of insolation and the amount of energy available to heat the water.

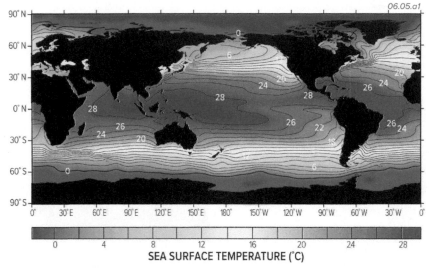

SEA SURFACE TEMPERATURE (°C)

3. Where are the coldest waters located? Naturally, they are near the poles. The cold end of the temperature scale below the map indicates that some of the coldest areas are at or even below freezing. We would expect ice to be present in these areas. The ice could be forming and increasing the salinity of the seawater left behind, or it could be melting and introducing fresh water into the sea, decreasing the salinity. We would predict that ice would influence salinity.

----

4. This map (▶) shows salinities of the ocean surface on the same day, with red areas having the greatest salinities and blue and purple having the smallest. First, observe the main patterns. Then, think about how you might explain these patterns by considering the processes described on the previous two pages, such as the amount of evaporation and cloud cover. After you have done these, read on.

5. The lowest salinities, represented by the regions in blue and purple, are mostly at high latitudes, such as areas closer to the poles. We can largely explain these low salinities by high-latitude regions being cold, with low amounts of insolation, and therefore low amounts of evaporation. Note, however, that some areas of low salinities are well away from the poles. One low-salinity region is in the very warm water that is southeast of Asia; therefore, the warmest areas of the ocean are not necessarily those with the highest salinity.

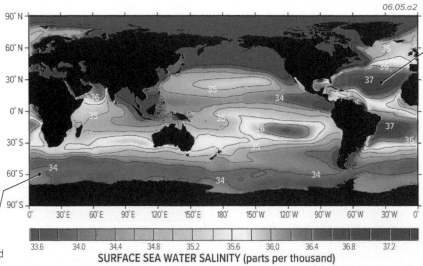

SURFACE SEA WATER SALINITY (parts per thousand)

7. High salinities occur in the subtropics, coinciding with locations of the subtropical high pressures. These areas have relatively clear skies and intense sunlight, which together cause high rates of evaporation. Also, the subtropics have low amounts of precipitation and dilution. The Mediterranean is very saline because it experiences high evaporation rates and has a limited connection to the Atlantic, with few high-discharge rivers flowing into it. Also note the presence of some moderate-salinity water in the northern Atlantic Ocean, near Greenland.

6. Low salinities away from high latitudes occur (1) where large rivers, like the Mississippi, Amazon, and Ganges-Brahmaputra rivers, deliver fresh water to the sea, (2) where cold currents, like the California Current and Humboldt Current, bring less saline waters away from the pole, and (3) in areas of high precipitation, like west of Central America and in the previously mentioned warm-water area southeast of Asia. Find all of these areas on the map.

B How Do Temperature and Salinity Vary with Depth?

The oceans also vary in temperature and salinity as a function of depth. Seawater is warmest near the surface, where it is warmed by the Sun. The distribution of salinity is similar, with the most saline waters near the surface, where there is evaporation. Salinity has more complex patterns too, because saline waters are more dense than less saline water and so they can sink. We present the temperature and salinity data in two pole-to-pole cross sections (North Pole on the right). The black, jagged shape at the bottom is the vertically exaggerated seafloor.

1. This pole-to-pole cross section (taken at 20° W longitude) shows Atlantic Ocean temperatures from the surface to the seafloor. Warmer surface waters overlie colder, deep ones. When this occurs, less dense water overlies more dense water, a highly stable arrangement.

---

3. This cross section shows salinity (measured in a unit of salinity called PSU) along the same line, with the North Pole on the right. Examine the patterns on this cross section and compare it with the temperature cross section above. Predict which way some of the water is moving at depth.

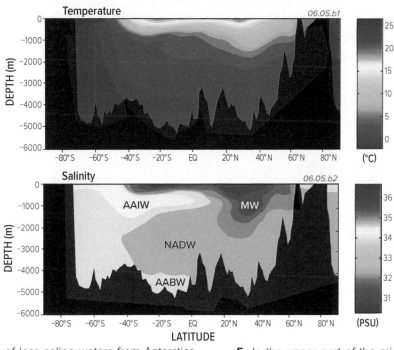

2. Only at polar latitudes, both in the far north (right) and far south (left), are surface and deep ocean water temperatures similar. When shallow and deep waters are similar in temperature and density, they can mix more easily, bringing deep water up toward the surface and causing shallow water to sink. Such mixing of waters from different depths is called *overturning* or *turnover*.

---

6. Last, but not least, a key feature is an extensive tongue of saline polar waters that descends near Iceland near 60° N to 70° N and then extends well south of the equator as the *North Atlantic Deep Waters* (NADW). As we shall explore shortly, this flow of saline waters down into the depths of the ocean and to the south has a major role in moderating our climate.

4. In this cross section, a tongue of less saline waters from Antarctica, called the *Antarctic Intermediate Waters* (AAIW), forms a wedge below saline waters near the surface with somewhat more saline waters at depth. Along the bottom are some moderately saline, very cold and dense waters known as *Antarctic Bottom Waters* (AABW).

5. In the upper part of the middle of the cross section is a relatively thin veneer of extremely saline waters that formed from evaporation and limited rainfall in the subtropics. A thicker batch of salinity at 40° N marks saline waters (MW) exiting the Mediterranean Sea, which also explains the increased temperatures at 30° N to 40° N.

C What Are the Regional Relationships Among Temperature, Salinity, and Density?

1. To explore how temperature, salinity, and density vary from region to region, this graph plots temperature on the vertical axis and salinity on the horizontal axis. Contours crossing the graph represent differences in density in parts per thousand, as calculated from the temperature and salinity. Greatest densities are in the lower-right corner. The yellow star represents the mean temperature and salinity of the world's oceans. Examine this graph and consider why a region plots where it does.

2. Equatorial waters are warm due to high insolation and are less saline than average seawater due to heavy tropical rainfall. Warm temperatures and low salinity produce low densities (i.e., numbers on the contours).

3. Mid-latitude waters are moderate in temperature and salinity. Compared to the average ocean, they are cooler because they receive less insolation than average, but less saline because precipitation generally exceeds evaporation.

4. Ice-free polar regions, predictably, have relatively cold seawater. They are also slightly less saline due to lower amounts of insolation and associated evaporation at high latitudes.

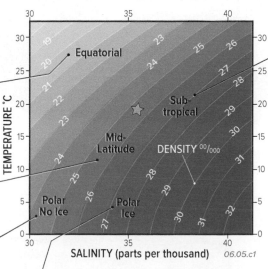

5. In polar zones where sea ice is actively forming, ice removes some fresh water from seawater, leading to greater salinities.

6. Subtropical waters are cooler than equatorial ones, but more saline, because surface evaporation exceeds precipitation. The salinity in particular leads to higher densities of these waters. The most saline waters in the world are in parts of the subtropics, especially where an area of the sea is mostly surrounded by continents and has somewhat restricted interchange with the open ocean. Such very saline waters are in the Red Sea and Mediterranean Sea.

Before You Leave This Page

✓ Summarize the global patterns of ocean salinity and temperature, how different regions compare, and what factors cause the larger patterns.

✓ Summarize how salinity and temperature vary as a function of depth, identifying the major distinct types and names of water at depth.

6.5

6.6 What Processes Affect Ocean Temperature and Salinity in Tropical and Polar Regions?

THE EXTREMES IN SEAWATER TEMPERATURE AND SALINITY occur at high and low latitudes. The warmest and most saline waters occur in the tropics and especially the subtropics. Very cold, less saline waters occur near the poles, away from where ice is actively forming. Here, we examine the processes occurring in some of the warmest and coldest parts of the ocean to illustrate how these important end members affect the planet's oceans and the planetary ocean-atmosphere-cryosphere system.

A What Are Warm Pools and How Are They Formed?

1. Within Earth's oceans are regions where the water temperatures are higher than in adjacent regions, and these warm regions persist, or re-form, year after year. These regions of warm seawater are called *warm pools*. The figure here illustrates how they are formed and maintained.

2. Excess insolation (compared to outgoing longwave radiation) generates warm air and warm sea-surface temperatures (29–30°C). Abundant precipitation falls on the surface of the tropical and equatorial oceans where, despite the high temperatures, precipitation exceeds losses to evaporation.

3. Warm, rising air carries large quantities of water vapor into the troposphere. It cools adiabatically but more slowly than the air around it and therefore continues to rise due to instability, forming very deep convection.

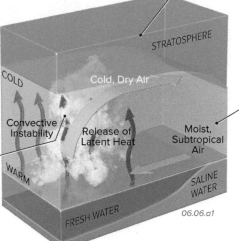

06.06.a1

4. Deep convection over tropical and equatorial regions leads to the release of latent heat and an expansion of the troposphere, causing the tropopause to be higher (17–18 km) over the tropics than over the rest of the planet.

5. Air at the tropopause and in the lower stratosphere is very cold (as low as −85°C) with extremely low moisture content (4 parts per million). The rising air is no longer less dense than air around it since the stratosphere has a temperature inversion. Thus, air moves laterally toward the poles (the yellow arrow) as if along an invisible ceiling.

6. Air moves horizontally across the ocean surface to replace the unstable rising equatorial air, bringing water vapor evaporated from the subtropical regions, where evaporation exceeds precipitation. Once in the tropics, the air rises, condenses, and releases latent heat, which was derived from the subtropics. The tropics therefore accumulate excess energy, to be released in this restricted area of rising air and instability. The ocean warms, forming a warm pool.

Location and Influence of Warm Pools

7. These two globes show SST, but they focus on areas of warm pools, which occur in the warmest waters, those shaded red and orange.

8. Earth's warmest and largest warm pool is near Southeast Asia, broadly between China and Australia. The warm pool is centered on the shallow seas that are partially enclosed by the islands of Indonesia. This warm pool is commonly called the *Western Pacific Warm Pool*. Warm waters in this region also extend into the adjacent Indian Ocean.

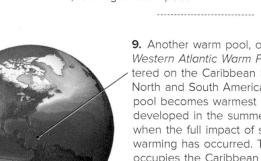

06.06.a2   06.06.a3

9. Another warm pool, often called the *Western Atlantic Warm Pool*, is centered on the Caribbean Sea, between North and South America. This warm pool becomes warmest and best developed in the summer and fall when the full impact of summer warming has occurred. The warm water occupies the Caribbean Sea, Gulf of Mexico, and adjacent parts of the western Atlantic Ocean. The shapes of North America and South America enclose the Caribbean part of this warm pool on three sides.

06.06.a4 Raja Ampat, Indonesia

10. The warm waters of the Western Pacific Warm Pool sustain rich and diverse life. These seas are famous for their coral reefs and clear, warm waters, stretching from Indonesia (◄) to east of the Philippines, in the island republic of Palau (►).

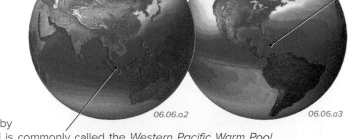

06.06.a5 Palau

11. The Western Atlantic Warm Pool forms the warm waters of Caribbean vacation spots, like the Florida Keys, the Cayman Islands, and beaches near Cancún, Mexico (►).

06.06.a6 Akumal, Mexico

B   What Processes Occur Around the Poles?

1. In terms of temperatures and humidity, the poles are the exact opposites of a warm pool. Here, we explore the processes near the poles, using a simple example where the entire polar region is ocean (no land).

2. Cold air aloft moving poleward is extremely dry. Near the pole, it subsides and warms adiabatically. The combination of extremely low moisture contents and warming of the sinking air results in very few clouds.

3. The lack of water vapor and clouds encourages the loss of outgoing longwave radiation. When combined with the sinking and warming air, this creates a very stable atmosphere due to a near-surface temperature inversion (warmer air over cooler air), which inhibits cloud formation and precipitation.

4. Ice floats on the surface of the ocean, due to the unusual physical properties of water. The high albedo (reflectance) of snow and ice reflects much of the already limited insolation back toward space, while the absence of clouds and water vapor ensures considerable heat loss through longwave radiation back toward space. As a result, the surface remains cold.

7. At the same time, cold polar air blows outward across the surface away from the pole, ultimately replacing the warm rising air at the equator. In this way, the oceans and atmosphere are both moving energy and material toward and away from the poles.

6. As the dense, polar waters descend, surface waters from lower latitudes flow toward the poles to replace them. This flow of dense saline waters downward, and their replacement by poleward surface flows of warmer water, sets up circulation involving both the shallow and deep waters.

5. The formation of ice on the surface of the polar oceans increases the salinity of the underlying waters, turning them into *brine*—very saline water. This and the very cold water temperatures increase the density of the surface waters to the point that they attain a higher density than that of the deep ocean waters below them, causing surface water to descend.

06.06.b1 — STRATOSPHERE / Dry / Stable / COLD POLAR AIR / High Surface Albedo / ICE / COOLING SALINE WATER

06.06.b2 Antarctica

06.06.b3 McMurdo Sound, Antarctica

06.06.b4 Hekla Volcano, Iceland

8. Ice shelves of Antarctica (◄) gain ice in three main ways, all of which result in ice that is fresh, not saline: (1) freezing of seawater below but excluding salt (called *brine exclusion*), (2) flow of glaciers from land to sea, and (3) snowfall and windblown snow on top. Brine exclusion makes adjacent water more saline, but melting of ice will release fresh water, decreasing the salinity. The resulting salinities indicate the relative contributions of freezing versus melting. The photograph to the right is of an ice shelf in Antarctica, but taken underwater, looking up at its underside, with a cooperative jellyfish for scale.

9. Iceland is a large island in the northern Atlantic Ocean. It is cloudier and stormier than most Arctic places, however, because of the effect of the Gulf Stream, which brings warmer water and moist air farther poleward than is typical.

10. Many scientists believe that this global-scale connection between the poles—along with feedbacks between the oceans, atmosphere, and cryosphere—is responsible for the abrupt changes (vertical gray lines) in climate recorded by ice cores from Greenland and Antarctica (▲). Rapid drops in Greenland temperatures are strongly associated with simultaneous Antarctic warming. These changes occur over short geologic time periods of one to two thousand years, but not as fast as Hollywood would have us believe.

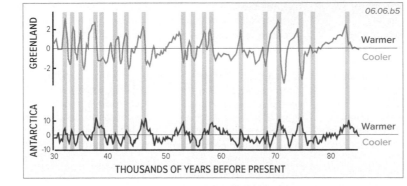
06.06.b5 — GREENLAND / Warmer / Cooler / ANTARCTICA / Warmer / Cooler / THOUSANDS OF YEARS BEFORE PRESENT

Before You Leave This Page

- [✓] Sketch and explain processes that form low-latitude warm pools, and identify where they are located.

- [✓] Sketch and explain processes that occur in polar areas and how they affect the salinity of water.

6.6

6.7 How Are the Atmosphere, Oceans, and Cryosphere Coupled?

THE OCEAN SURFACE marks the boundary between Earth's major systems—oceans and atmosphere. Both systems move mass and energy laterally and vertically, primarily in response to variations in density and to the equator-to-pole energy gradient. Deep-flowing abyssal waters are generally cooler and denser than surface waters, but they rise in some places and sink in others. The global, three-dimensional flow of matter and energy between the ocean, atmosphere, and cryosphere is a fundamental determinant of global climate.

A How and Why Do Temperature, Salinity, and Density of Seawater Vary?

The oceans vary in temperature and salinity, and therefore also in density, laterally across the surface and from shallower to greater depths. The figure below summarizes these variations from equator to pole, and between upper and lower levels of the ocean. Begin with the text above the figure, which describes the atmospheric processes imposed on the underlying sea, and then read how the sea responds to these changes.

1. Atmospheric motions result from the energy gained by insolation into the lower atmosphere, and by radiative cooling at higher levels. Warm, rising air is located over the seasonally migrating Intertropical Convergence Zone (ITCZ), centered at the equator. As the warm air rises in these regions it cools and produces condensation, clouds, and precipitation.

2. In the subtropics, air is dominantly sinking in the descending limb of the Hadley cell. This forms semipermanent subtropical anticyclones, where high pressure causes warming and limits the cloudiness and precipitation.

3. Rising air is also common at the poleward edge of the mid-latitudes. Upward atmospheric motions are caused by the rising limb of the polar cell and differences in density between cool polar and warm subtropical air masses, leading to low pressure, clouds, and frontal precipitation. These latitudes receive less insolation than subtropical ones, which, together with the higher albedo of the clouds, reduces evaporation.

4. Near the poles, descending air, combined with the limited insolation, keeps polar regions cold and relatively dry. This in turn limits the amount of evaporation and causes a noticeable absence of precipitation.

5. Clouds in equatorial and tropical regions allow more precipitation than local evaporation, causing a decrease in salinities near the surface. The warmest, less saline waters have low density, forcing waters of intermediate density beneath them immediately around the equator.

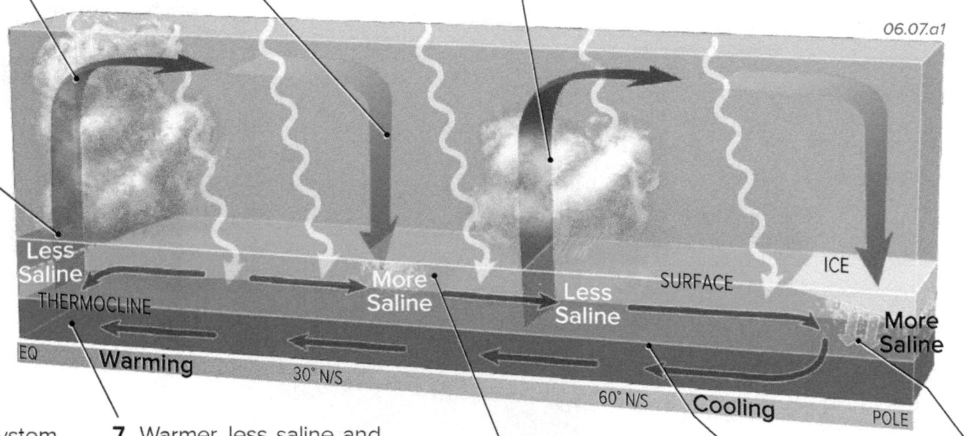

06.07.a1

11. As cold and saline waters near the pole sink deep into the oceans, they begin to flow back toward the equator as *abyssal waters*. In contrast, the flow of surface waters (i.e., those in the photic zone) is generally toward the poles. Together, the poleward-flowing surface waters and equator-flowing deep waters form a circuit of flowing water that acts to "overturn" waters of the ocean.

6. Unlike the atmospheric system, the oceanic system is heated from above. Insolation is absorbed in the upper layers of the oceans but does not penetrate far beyond a depth of at most 100 m. This upper zone is warmer and less dense, and is known as the *photic zone*. Deep ocean waters below these depths are consistently at between −1°C and 4°C throughout the world's oceans. These waters stay in liquid form at barely subzero temperatures because the salinity of the water decreases the freezing temperature (like antifreeze in your car) and because of the pressure exerted by the great depths of overlying ocean waters.

7. Warmer, less saline and less dense waters "sitting" on top of cooler, more dense waters is an extremely stable arrangement which discourages mixing between the two sets of water. The transition from the photic zone to deep ocean waters is usually marked by a very rapid change in density and in temperatures—a boundary called the *thermocline*. This figure represents the thermocline as the boundary between lighter blue (warmer) waters and darker blue (colder) waters.

8. Much insolation reaches the ocean surface in the subtropics, in part due to high Sun angles and a general absence of clouds. The combination of high evaporation and low precipitation leads to these subtropical waters becoming more saline. A marked vertical gradient in water density is a *pycnocline*. The pycnocline commonly coincides with the thermocline, so it can be called a *thermopycnocline*.

9. In the mid-latitudes, lower insolation and the higher albedo of the clouds reduces evaporation, which, in combination with increased frontal precipitation, leads to a freshening of surface waters.

10. The formation of sea ice in polar latitudes preferentially extracts fresh water from seawater, leaving a concentration of dissolved salt in the very cold ocean waters. The combination of cool polar sea temperatures and high salinity produces surface waters of sufficient density to descend.

B What Is the Thermohaline Conveyor?

The flow of deep water and surface water combine to form a global oceanic circulation system, commonly called the *thermohaline conveyor* because it is driven by differences in water temperatures and salinities. This circulation system is among the most important on Earth. What is it, where is it, and how does it operate? The best way to understand it is to follow water through the system, beginning in the Caribbean Sea.

1. Ocean waters in the tropical Atlantic Ocean and Caribbean Sea, along with those in the Gulf of Mexico, heat up before entering the subtropics where their salinity increases through evaporation. Cold waters from the *Greenland Sea (1)* and the *Labrador Sea (2)* cool these salty waters in the North Atlantic around Iceland and the Maritime Provinces of Canada, making them sufficiently dense to descend deep into the Atlantic as the North Atlantic Deep Waters (NADW).

2. At depth, the NADW flow south through the Atlantic Ocean, threading their way between Africa and the Americas, crossing the equator, and turning east when they encounter Antarctica.

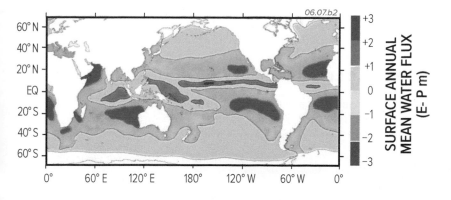

06.07.b1

5. Neither the Indian nor Pacific Ocean basins extend sufficiently poleward to generate their own northern deep ocean waters, but deep waters upwell in the *Indian Ocean (5)* and join the surface flow.

4. Once in the Pacific, the deep waters turn north, again crossing the equator. The northward-moving deep waters are ultimately forced to rise, or upwell, at the Pacific basin's northern limit, south of *Alaska (6)*. Surface currents move these waters back south, into the tropical and equatorial regions, ultimately returning to the Atlantic south of Africa and completing the loop to the Gulf Stream.

3. Extensive ice shelves extending from Antarctica over the adjacent *Weddell Sea (3)* and *Ross Sea (4)* exclude salt water during their formation, causing the nearby waters to be very cold and saline. These Antarctic Bottom Waters (AABW) sink to join the NADW in moving eastward. Some deep waters branch northward into the Indian Ocean basin, while the remainder continues into the Pacific basin, mixing along the way and becoming slightly warmer and less dense.

---

6. There is concern about how climate change could impact the operation of the thermohaline conveyor. To evaluate this issue, we can consider how climate change could impact global temperatures, and how these in turn would affect the amounts, distribution, and movement of water due to changes in evaporation, rainfall, and runoff. These factors affect temperatures, salinity, and density of water, which are critical in driving the thermohaline conveyor.

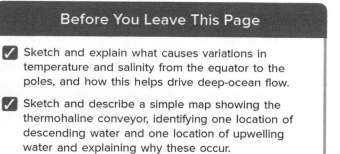

06.07.b2

7. This map (◄) shows evaporation (E) minus precipitation over the oceans (Pm), expressing the flux of water from the surface to the atmosphere. Surface areas in red are losing the most water, and surface areas in purple are gaining water. Areas along the equator gain water from the large amounts of precipitation, whereas excess evaporation in the subtropics causes the surface to lose water. These fluxes are accompanied by changes in salinity of seawater. Added rainwater decreases the salinity, whereas evaporation increases salinity. Changes in salinity and temperature can influence the thermohaline conveyor, which in turn influences climates in North America, Europe, and other regions.

8. This map (►) illustrates the pathways and changing water temperatures of the Gulf Stream as it works its way up the East Coast of North America. The stream is warm (orange) in the south where it starts, but cools (yellow to green) as it gets farther north and east, toward Europe.

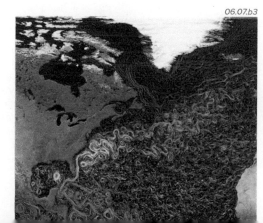

06.07.b3

Before You Leave This Page

✓ Sketch and explain what causes variations in temperature and salinity from the equator to the poles, and how this helps drive deep-ocean flow.

✓ Sketch and describe a simple map showing the thermohaline conveyor, identifying one location of descending water and one location of upwelling water and explaining why these occur.

6.7

6.8 | What Connects Equatorial Atmospheric and Oceanic Circulation?

EQUATORIAL AND TROPICAL REGIONS are areas of greatest excess energy. In the same way that the Hadley cells establish connections between latitudes, there are broad-scale, *east-west* connections within the equatorial atmosphere-ocean system. These connections are a major cause of global climate variability.

A | How Do Continents Impact Ocean Circulation?

The equator crosses two broad areas of land—South America and Africa—and a large number of islands of the maritime southeast Pacific, including those in Indonesia. These various landmasses block east-west flow in the oceans, turning ocean currents to the north or south. They also cause deep ocean currents to upwell along their flanks.

1. If there were no equatorial or tropical continents, the world would be circumscribed by a fairly homogeneous band of warm ocean water. The trade winds, north and south of the equator, would exert a small stress on the surface of the ocean, causing a net westward movement of the waters.

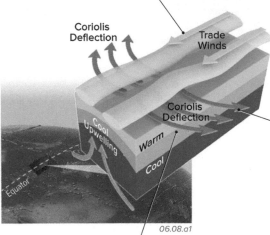

06.08.a1

2. The convergence of winds produces a slight mound of water, which then spreads north and south of the equator. Once in motion away from the equator, the Coriolis effect causes the moving waters to deflect from their intended path, causing warmer surface waters to flow away from the equator in both directions—to the northeast, north of the equator, and to the southeast, south of the equator.

3. As surface waters move away from the equator, deeper and cooler water beneath the equator must rise, or upwell to replace the dispersing surface water. This results in slightly cooler water along the equator than to either side.

4. Continents and other landmasses are in the way of this westward flow of tropical ocean water, causing the warm water to accumulate, or pool, along the eastern side of the land, as expressed on this SST map (▼).

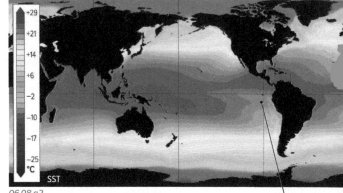

06.08.a2

5. Islands and peninsulas between Australia and Asia block westward movement of warm surface waters, causing a ponding, or a warm pool, in the western equatorial Pacific.

6. Cooler waters, an expression of equatorial upwelling, are visible in the eastern Pacific.

B | How Do Warm Pools Influence Atmospheric Circulation Above Oceans?

1. Once formed along the eastern side of a continent, a warm pool can set up an east-west circuit of flowing air through the tropics.

2. Warm, moist air rises above the Western Pacific Warm Pool, where temperatures regularly exceed 28°C. As the air rises, it cools adiabatically, producing condensation, clouds, and intense rainfall, as well as releasing large quantities of latent heat into the troposphere immediately above the warm pool.

3. Unable to penetrate vertically beyond the temperature inversion at the tropopause, the air moves horizontally away from the warm pool. As it moves, it mixes with the surrounding cool air and cools.

4. Having cooled sufficiently, the air descends over the eastern Pacific, warming adiabatically.

5. This cooler (but warming) and denser air from above turns west and passes back over the surface as part of the trade winds, warming and evaporating moisture from the ocean surface as it proceeds westward.

6. This pattern of east-west equatorial atmospheric circulation, initiated by warm air rising above warm pools, is known as the *Walker cell circulation*. In 1923, Sir Gilbert Walker first reported "great east-west swayings" of the near-equatorial flow, in the manner represented on this figure.

06.08.b1

C Do Similar Circulation Patterns Exist over Other Oceans?

1. Warm pools are pronounced features of the global atmosphere-ocean system, as displayed on this SST map. Trade winds convey warm waters westward through the equatorial regions. In the Atlantic, this movement is blocked by South and Central America, ponding warm water off northern South America and in the Caribbean.

2. As shown in the wind-circulation patterns below the SST map, the Walker-like cell in the Atlantic begins with warm, rising air over the Amazon basin of South America—a hot, humid area of rain forest flanking the Amazon River and its tributaries. To the west are the Andes along western South America, and the descending limb of the Pacific Walker cell, both of which restrict flow to the west from the Amazon.

5. The Indian Ocean differs, as its warm pool is located at its eastern edge. This reflects the influence of the adjacent larger Pacific warm pool and the porous barrier between the two. The break between the Pacific and Indian Ocean basins at equatorial latitudes is discontinuous, marked by the islands and straits of Indonesia and maritime Southeast Asia (collectively referred to as Australasia), through which warm waters may pass. Also, the Indian Ocean is unique in not extending beyond about 20° N. The large continental mass gives rise to the Asian monsoon system, whose summer winds are from the southwest, blowing waters back to the east. Low atmospheric pressure is experienced at the surface above the warm pools, and higher pressure is beneath areas where the air is descending.

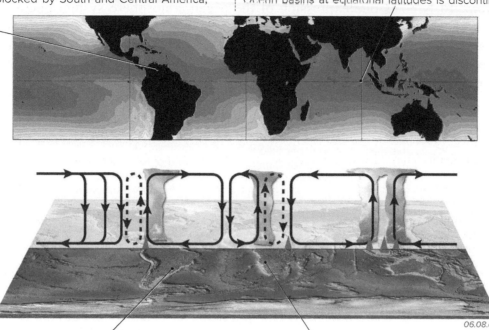

06.08.c1

3. The Atlantic Ocean is narrower than the Pacific, so the pattern of circulation is not as wide. Air rises above the Caribbean warm pool and the extensive wetlands and rain forests of the Amazon lowlands, resulting in intense tropospheric convection, second only in its intensity to that of the Western Pacific Warm Pool. Air in the Atlantic cell descends in the eastern Atlantic and over the western side of North Africa, enhancing the regional dryness.

4. Heating and evapotranspiration from the moist forested surface of the Congo basin produces another zone of rising air and convection over this region, drawing moisture from the Atlantic deep into central Africa. The resulting Walker cell is narrow, being constrained on the west by the Atlantic cell and on the east by the Indian Ocean circulation and air motions in East Africa.

How These Circulation Systems Were Discovered

At the end of the 19th and into the early 20th century, India suffered several catastrophic droughts responsible for the deaths of millions. These droughts are displayed on the graph below, which shows departures of the total rainfall received annually in India from the summer monsoon compared to long-term averages (1870–2000).

Sir Gilbert Walker was head of the British Indian Meteorological Department during the droughts of 1918 and 1920. Walker was interested in understanding the causes of drought, which naturally led him to investigate regional wind patterns and ocean temperatures, both of which are critically important to the Indian monsoon that is so crucial to the lives of the people of India and adjacent countries. Walker began publishing his research just as severe droughts ceased. Thus, his contribution was undervalued in his lifetime, as reflected in his 1959 obituary which states sardonically, "Walker's hope was to unearth relations useful for . . . a starting point for a theory of world weather. It hardly seems to have worked out like that."

Only in the late 1960s, as drought became a problem again, was the significance of Walker's discovery fully recognized, and this principal causal mechanism of equatorial circulation and climatic variability was named in his honor.

As we will learn in following pages, the equatorial circulation pattern recognized by Walker has major implications for regional climates on both sides of the Pacific. This is an excellent example of how investigation of a problem (the monsoon in India) led to insights with broader and unanticipated implications in other realms.

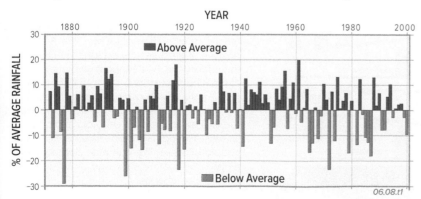

06.08.t1

Before You Leave This Page

- ✓ Sketch and explain the concept of equatorial upwelling.
- ✓ Sketch the warm pools in each of the major ocean basins straddling the equator, and explain their distribution.
- ✓ Sketch and explain the Walker cell over the equatorial Pacific Ocean, and similar cells elsewhere.

6.8

6.9 What Are El Niño and the Southern Oscillation?

THE MOST WELL KNOWN AND PUBLICIZED cause of climate variability, floods, droughts, hurricanes, heat waves, and landslides is a change in the strength of winds and ocean currents west of South America in what has become known as *El Niño*. El Niño is one expression of the ocean-atmosphere system operating over and within the equatorial Pacific Ocean. Phenomena associated with El Niño are more correctly termed *ENSO* because they involve oceanic (El Niño, *EN*) and atmospheric (Southern Oscillation, *SO*) components, which are linked but respond at different rates. This difference in speed of response, combined with the linkages between the two systems, leads to the intrinsically erratic nature of ENSO.

A What Is El Niño?

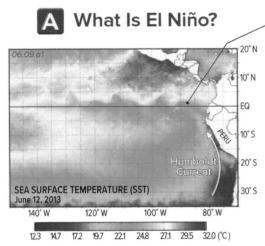

SEA SURFACE TEMPERATURE (SST)
June 12, 2013

12.3 14.7 17.2 19.7 22.1 24.8 27.1 29.5 32.0 (°C)

The waters off the coast of northern Peru are generally cold, as shown in this map of SST (◄), because the north-flowing Humboldt Current brings very cold water from farther south. The cold character of these waters is intensified by the upwelling of cold water near the shore. This upwelling occurs because warmer surface waters are pushed westward, away from shore, by dry southeasterly trade winds descending over the Andes and by complex sea motions. The cold upwelling limits evaporation, producing a dry desert coastline, but it brings to the surface nutrients that nourish plankton and cold-water fish, such as anchovy.

Fishermen noted that the waters began warming (►) toward the end of November and early in December (Southern Hemisphere summer), heralding the beginning of the Christmas season. They named this local phenomenon "El Niño," in reference to the boy child or "Christ child." Occasionally, as in 1972–73, 1982–83, 1997–98, and 2015–16, the warming is exceptionally strong and persistent, referred to as a strong *El Niño* pattern. This brings abundant rains to the region, with flash flooding inundating extensive areas. The anchovy and accompanying ecosystem of birds and sea mammals disappear, and the fishing industry is decimated.

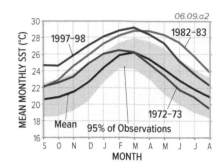

B What Is the Southern Oscillation?

1. Associated with changes in the strength of El Niño are fluctuations in air pressures in the Pacific. This map (►) shows surface air pressures for part of the Pacific Ocean, centered on the equator. Low pressures are darker, whereas high pressures are lighter in color. Sir Gilbert Walker noticed that when air pressure increased in Darwin, Australia, air pressure typically decreased in Tahiti, a cluster of islands farther east in the central Pacific. When pressure decreased in Darwin, it increased in Tahiti. Walker termed this the "great east-west swaying" of the equatorial atmosphere—the *Southern Oscillation*. Even though it is a tropical oscillation, it seemed "Southern" to Walker, so it retained this misnomer.

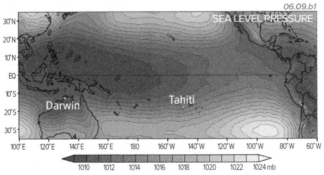

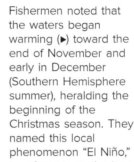

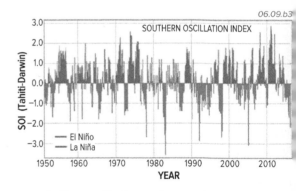

3. To quantify these changes in pressure, Walker created a *Southern Oscillation Index* (SOI), which represents the difference in air pressure between Tahiti and Darwin. In its simplest form, it is calculated by subtracting the air pressure measured at Darwin from that measured at Tahiti (▲). If air pressure is higher at Tahiti than at Darwin, the SOI is positive. If air pressure is higher at Darwin than Tahiti, the SOI is negative. Patterns of alternating fluctuations are apparent in the plot above, which compares pressures to the long-term mean. The graph shows that the SOI switches back and forth from positive to negative modes, generally every couple of years. We can anticipate that as air-pressure patterns change, wind directions should also change.

2. Walker envisioned these coordinated fluctuations in air pressures—the Southern Oscillation—as the ascending and descending limbs of a Walker cell over the Pacific. These fluctuations can be portrayed by a simple system (►), where a change in the air pressures in one limb of the cell is matched by an opposite change in pressure at the other limb. For the Walker cell in the Pacific, conditions at Darwin, Australia, represent the ascending limb (near the warm pool, on the left) and those at the Pacific island of Tahiti represent the descending limb (on the right).

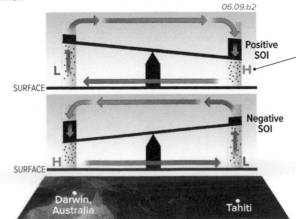

C How Is the Southern Oscillation Linked to World Weather?

The Southern Oscillation represents a major shift in air pressure, winds, and other factors in the equatorial Pacific, but Walker recognized that changes in the Pacific correlated with changes in weather elsewhere in the world. To investigate these changes, he hand-plotted air-pressure measurements on a world map (top map below). There is a remarkable correspondence between one of Walker's hand-calculated maps and output from a supercomputer (bottom map).

1. Walker's map (▶), drawn in 1936, showed which parts of the globe had air pressures that correlated with the Southern Oscillation Index (SOI), based on observations from only 50 weather stations. Numbers on the map indicate the strength of the relationship, or correlation. Relationships within the area defined by the solid lines are (statistically) significant. The dotted line separates areas in which pressures fluctuate in the same way as Tahiti (positive numbers) from those in which fluctuations are in the opposing direction (negative numbers).

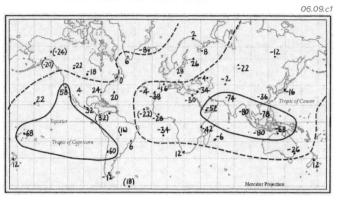
06.09.c1

2. Even this early map demonstrates that extensive areas over Australasia and the Indian Ocean, as far as India and Africa and portions of the eastern Pacific, display significant and opposing relationships to shifts in air pressure represented by the SOI.

3. Today we still use Walker's idea of the Southern Oscillation, except Walker used a different sign convention for this particular kind of map. This modern map (▶), based on the National Center for Climate Prediction's global climate model, displays correlations in tenths.

4. Positive associations, represented by shades of light gray, occur in the eastern Pacific, stretching from southern Chile to California. When the index is positive (reflecting higher pressures at Tahiti), these areas will also experience higher pressures and probably less rain. When the index is negative (higher pressures at Darwin), they will experience lower pressures and, in all likelihood, more rainfall.

06.09.c2

May to Apr: 1958 to 2008: Surface Sea Level Pressure
Seasonal Correlation w/ May to Apr SOI

−1   −0.7   −0.4   −0.1   0.1   0.4   0.7   1

5. Negative values are extensive, covering Australasia and Maritime Southeast Asia, and stretching across the Indian Ocean as far as India and Africa. These areas will have lower pressures and probably more rain when the SOI is positive (higher pressures at Tahiti), but high pressures and less rain when the index is negative.

6. The dynamics of ENSO are intimately related to the nature of the thermocline across the equatorial Pacific. Recall that oceans have an upper surface layer that is generally impacted by insolation, with temperatures decreasing with depth. The transition between the warmer and cooler waters occurs along a clearly defined layer where temperatures cool rapidly with depth (the thermocline). Because water temperature and density are related, this thermocline corresponds to a similar marked change in density.

Surface

DEPTH

Warmer

THERMOCLINE

Cooler

06.09.c3

TEMPERATURE

7. During an El Niño, the thermocline moves deeper in the eastern Pacific (near South America), a further expression of the abundance of warm water near the surface and an associated decrease in the number of cold-water fish caught in the region. The following two pages describe the different phases and manifestations of ENSO, including the response of the thermocline.

06.09.c4 Sulawesi, Indonesia   06.09.c5 Isla Isabela, Galápagos   06.09.c6 Central Ecuador

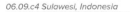

8. As recognized by Walker, the effects of ENSO are widespread across the Pacific, affecting weather from the westernmost Pacific (e.g., Sulawesi) to South America (e.g., Ecuador), and all the islands in between (e.g., Galápagos). These events affect sea and land temperatures, air pressures, wind directions, rainfall amounts and seasonality, and other important aspects of weather.

Before You Leave This Page

✓ Describe what the term El Niño explains and how it is expressed.

✓ Describe the Southern Oscillation in terms of the Walker cell circulation and the general relationship between pressures at Darwin and Tahiti.

✓ Describe some areas of the world where weather is affected by the Southern Oscillation.

6.9

6.10 What Are the Phases of ENSO?

THE ATMOSPHERE-OCEAN SYSTEM in the equatorial Pacific is constantly changing. Although each year has its own unique characteristics, certain atmosphere-ocean patterns repeat, displaying a limited number of modes. We can use surface-water temperatures in the eastern equatorial Pacific to designate conditions as one of three phases of the *El Niño-Southern Oscillation* (ENSO) system—neutral (or "normal"), warm (*El Niño*), and cold (*La Niña*).

A What Are Atmosphere-Ocean Conditions During the Three Phases of ENSO?

El Niño and La Niña phases represent the end-members of ENSO, but sometimes the region does not display the character of either phase. Instead, conditions are deemed to be neither and are therefore assigned to the *neutral phase* of ENSO. To understand the extremes (El Niño and La Niña), we begin with the neutral situation.

Neutral Phase of ENSO

1. Warm, unstable, rising air over the western equatorial Pacific warm pool produces low atmospheric pressures near the surface.

2. Walker cell circulation in the equatorial troposphere brings cool, dry air eastward along the tropopause.

3. Cool, descending air over the eastern equatorial Pacific produces dominantly high atmospheric pressure at the surface and stable conditions in the atmosphere.

9. The warm, moist air above the warm pool rises under the influence of low pressures, producing intense tropical rainfalls that maintain the less saline, less dense fresh water on the surface of the warm pool.

8. Warm waters blown to the west not only depress the thermocline to about 150 m below the surface, but also physically raise the height of the western equatorial Pacific compared to the eastern Pacific.

4. Easterly trade winds flow over the Andes mountain range and then continue to the west across the ocean, pushing west against the surface waters along the coast of South America. The easterlies continue propelling the warm water westward toward Australia and southeast Asia, allowing the waters to warm even more as they are heated by insolation along the equator.

5. Westward displacement of surface waters, and offshore winds, induces upwelling of cold, deep ocean waters just off the coast of western South America. Abundant insolation under clear skies warms these rising waters somewhat, so there is no density-caused return of surface waters to depth.

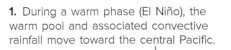

Western Pacific · Eastern Pacific · Thermocline · 06.10.a1

7. In the western Pacific, surface waters are warm (over 28°C) and less saline because of abundant precipitation and stream runoff from heavy precipitation that falls on land. The warm surface waters (the warm pool) overlie cooler, deeper ocean water—a stable situation.

6. The thermocline slopes to the west, being over three times deeper in the western Pacific than in the eastern Pacific. This condition can only be maintained by a series of feedbacks, including the strength of the trade winds.

Warm Phase of ENSO (El Niño)

1. During a warm phase (El Niño), the warm pool and associated convective rainfall move toward the central Pacific.

2. El Niño conditions are also characterized by weakened Walker cell circulation over the equatorial Pacific. This is expressed by decreased winds aloft and by a reduction in the strength and geographic range of the easterly trade winds near the surface.

6. For Australia, Indonesia, and the westernmost Pacific, El Niño brings higher atmospheric pressures, reduced rainfall, and westerly winds. The warm pool and associated convective rainfalls move toward the central Pacific, allowing cooler surface waters in the far west.

5. Changes in the strength of the winds, in temperatures, and in the movements of near-surface waters cause the thermocline to become somewhat shallower in the west and deeper in the east, but it still slopes to the west.

3. Upon reaching South America, the cool air descends over equatorial parts of the Andes, increasing atmospheric pressure, limiting convectional uplift, and reducing associated rainfall in Colombia and parts of the Amazon.

4. Weakening of the trade winds reduces coastal upwelling of cold water, which, combined with the eastern displacement of the descending air, promotes a more southerly location of the ITCZ in the Southern summer and increased precipitation in the normally dry coastal regions of Peru and Ecuador.

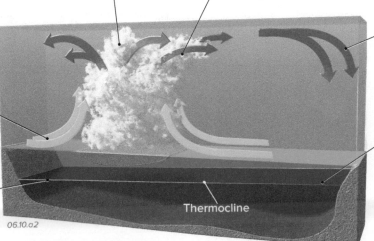

Thermocline · 06.10.a2

Cold Phase of ENSO (La Niña)

1. In many ways, the cold phase of ENSO (La Niña) displays conditions opposite to an El Niño, hence the opposing name.

2. During a cold phase of ENSO (La Niña), Walker cell circulation strengthens over the equatorial Pacific. This increases winds aloft and causes near-surface easterly trade winds to strengthen, driving warmer surface waters westward toward Australasia and Indonesia.

8. The region of equatorial rainfall associated with the warm pool expands and the amount of rainfall increases.

3. Enhanced easterly trade winds bring more moisture to the equatorial parts of the Andes and to nearby areas of the Amazon basin. Orographic effects cause heavy precipitation on the Amazon (east) side of the mountain range (not shown).

7. In the western Pacific, strong easterlies push warm waters to the west where they accumulate against the continent, forming a warmer and more expansive warm pool. In response, the thermocline of the western equatorial Pacific is pushed much deeper, further increasing the slope of the thermocline to the west.

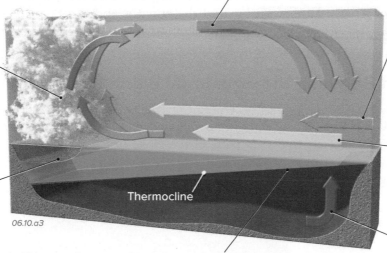

Thermocline

06.10.a3

4. Partially depleted of moisture and driven by stronger trade winds, dry air descends westward off the Andes and onto the coast. The flow of dry air, combined with the descending limb of the Walker cell, produces clear skies and dry conditions along the coast.

6. The upwelling near South America raises the thermocline and causes it to slope steeper to the west. Cold water is now closer to the surface, producing favorable conditions for cold-water fish.

5. As surface waters push westward and the Humboldt Current turns west, deep waters rise (strong upwelling). The resulting cool SST and descending dry, stable air conspire to produce excessive drought in coastal regions of Peru.

B How Are ENSO Phases Expressed in Sea-Surface Temperatures?

As the Pacific region shifts between the warm (El Niño), cold (La Niña), and neutral phases, sea-surface temperatures (SST), atmospheric pressures, and winds interact all over the equatorial Pacific. These variations are recorded by numerous types of historical data, especially in SST. The globes below show SST for the western Pacific (near Asia) and eastern Pacific (near the Americas) for each phase of ENSO—neutral, warm, and cold. The colors represent whether SST are warmer than normal (red and orange), colder than normal (blue), or about average (light).

Neutral Phase of ENSO

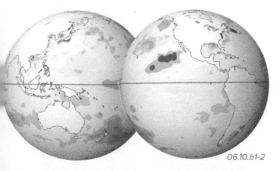

06.10.b1-2

Warm Phase of ENSO (El Niño)

06.10.b3-4

Cold Phase of ENSO (La Niña)

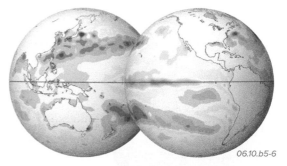

06.10.b5-6

During the *neutral phase* of ENSO, SST along the equator in the Pacific are about average, with no obvious warmer or colder than normal waters near the Western Pacific Warm Pool (left globe) or South America (right globe). An area of warmer than normal SST occurs southwest of North America, but this is not obviously related to ENSO.

During the *warm phase* of ENSO, a belt of much warmer than normal water appears along the equator in the eastern Pacific, west of South America. This warm water is the signature of an El Niño, causing the decrease in cold-water fishes. SST in the western Pacific are a little cooler than average, but an El Niño is most strongly expressed in the eastern Pacific (right globe).

During the *cold phase* of ENSO (a La Niña), a belt of colder than normal water occurs along the equator west of South America, hence the name "cold phase." The western Pacific (left globe), however, now has waters that are warmer than normal. These warm waters are quite widespread in this region, extending from Japan to Australia.

Before You Leave This Page

☑ Sketch and explain atmosphere-ocean conditions for each of the three typical phases of ENSO, noting typical vertical and horizontal air circulation, sea-surface temperatures, relative position of the thermocline, and locations of areas of excess rain and drought.

☑ Summarize how each of the three phases of ENSO (neutral, warm, and cold) are expressed in SST of the equatorial Pacific Ocean.

6.10

6.11 Do Impacts of ENSO Reach Beyond the Tropics?

WHILE THE IMMEDIATE IMPACTS OF ENSO are restricted to regions near the equatorial Pacific, shifts between different phases of ENSO cause climate variability well beyond the equator and the Pacific. When atmospheric conditions in one region affect a distant region, these distant associations are called *teleconnections*, which in this case are caused by interactions with the subtropical and polar front jet streams.

A What Is the Connection Between the Equatorial Warm Pool and the Extra-Tropics?

The effects of an ENSO can spread out of the tropics because air near the equator rises and flows northward as part of the Hadley cell. Motions in the Walker and Hadley cells occur simultaneously, so equatorial air can be incorporated into both circulation cells, spreading ENSO effects to distant parts of the globe (a teleconnection). To explore how this teleconnection operates, we follow a packet of the warm air that was produced in the Western Pacific Warm Pool during a cold-phase (La Niña) event. Begin reading on the lower left of the figure below.

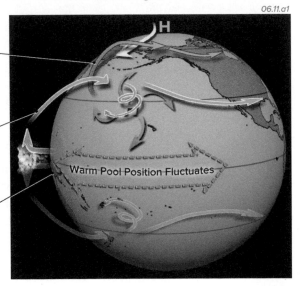

06.11.a1

3. Warmer air flowing poleward from the subtropics is forced to rise over the cold polar air, along the polar front. This poleward-moving air also has an apparent deflection to its right (in the Northern Hemisphere), joining the rapidly east-flowing polar front jet stream.

2. When the rising air reaches the upper levels of the troposphere, it is transported poleward in the Hadley cell. It has an apparent deflection to the right (due to the Coriolis effect) and becomes part of the rapidly eastward-moving subtropical jet stream. Eventually, the air descends in the subtropics near 30° N latitude.

1. During a La Niña phase, excess warm water accumulates in the Western Pacific Warm Pool. As this excess energy transfers from the ocean to the atmosphere, convection begins to carry the warm, moist air upward.

Warm Pool Position Fluctuates

4. The different phases of ENSO are distinguished by the position of the western warm pool (west for La Niña, farther east for El Niño). As this position changes, moist tropical air is fed into the jet streams in different locations along their paths, affecting the subtropical and polar jets, particularly "downstream" (east) in the jet streams (in both hemispheres). Rising air from the warm pools also flows south into the Southern Hemisphere, affecting weather there as well.

ENSO and Patterns of Atmospheric Pressure

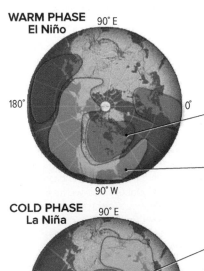

WARM PHASE
El Niño 90° E

180°   0°

90° W

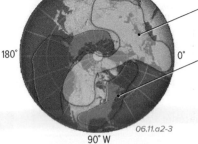

COLD PHASE
La Niña 90° E

180°   0°

06.11.a2-3

90° W

5. These two views (◄) of the Northern Hemisphere summarize changes in atmospheric pressure that typically accompany the warm and cold phases of ENSO. For each phase, areas in blue typically have atmospheric pressures that are lower than normal and those in red typically have higher than normal atmospheric pressures. The colors do not indicate absolute pressure, just deviations from normal (anomalies).

6. During a warm phase (El Niño), areas of higher than normal pressure are typically centered over Hudson Bay, the north-central Pacific, southwestern Europe, and western Africa.

7. At the same time, lower than normal pressures form in the eastern Pacific, Mexico, and across the southern U.S.

- -

8. During a cold phase (La Niña), the patterns of atmospheric pressure are nearly reversed from those of an El Niño. Lower than normal pressures extend across much of southern Europe, Greenland, and western North America.

9. During a La Niña, higher than normal atmospheric pressures are situated along the East Coast of the U.S. and cover most of the eastern Pacific. Recall that high pressure is associated with dry, sinking air, whereas low pressure is associated with rising air and unsettled weather. We would predict such conditions in the red (high-pressure) and blue (low-pressure) areas on these maps.

ENSO and Jet Streams

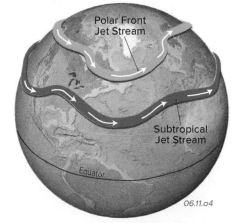

Polar Front
Jet Stream

Subtropical
Jet Stream

Equator

06.11.a4

10. The polar front and subtropical jet streams flow in a continuously changing wave-form around the globe, with the positions of the waves governed by locations of high and low pressure. Changes in the strength and location of the global pressure cells, as occur during ENSO events, therefore modify the usual pattern of the jet streams. The jet streams in turn control the typical paths of weather-generated storms and other weather patterns.

B What Are the Typical Responses to ENSO Around the Pacific Basin?

ENSO greatly affects weather conditions over the Pacific and in adjacent continents. The four maps below summarize the weather conditions that result from ENSO during summer and winter. Note that due to the increased energy gradient between the equator and the poles during winter, the "winter" hemisphere frequently shows greater changes by ENSO phase than the "summer" hemisphere.

Warm Phase (El Niño)

1. During the warm (El Niño) phase, a weakening of the Walker cell and easterlies (trade winds) shifts the Pacific warm pool eastward and limits upwelling next to South America.

2. As the Western Pacific warm pool shifts eastward, dry conditions spread across the far western Pacific (yellow). If prolonged, these conditions cause drought in this area.

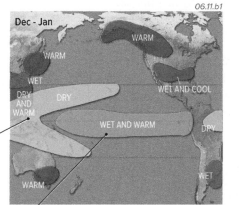

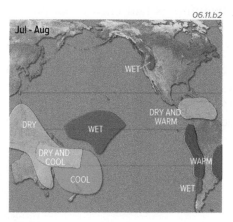

5. In July and August during El Niño, the central Pacific stays rainy and the western Pacific stays relatively dry.

6. During these months, effects of El Niño diminish in the Northern Hemisphere (where it is summer) but intensify in the Southern Hemisphere (where it is winter and jet streams are more robust).

3. The locus of precipitation follows the warm pool into the middle of the Pacific. The weakened easterlies also allow more December–January precipitation along the western coast of South America.

4. In North America, ENSO-related alteration of the path of the polar front in the winter affects the northern part of the continent, while changes in the subtropical jet stream affect conditions over the southern part of the continent.

7. The position of the southern subtropical jet stream controls the paths of precipitation-bearing frontal systems in the mid-latitudes of the Southern Hemisphere. Central Chile is most sensitive to shifts in the passage of these winter storms. The quantity of snow falling in the Andes controls water available for irrigation in Chilean and Argentinian rivers during the next dry summer season.

Cold Phase (La Niña)

8. During December–January of a cold (La Niña) phase, the Walker cell and easterlies intensify, pushing warm water westward into the Western Pacific Warm Pool, causing wet conditions across that region.

9. The westward ocean currents cause upwelling along the western coast of South America. The presence of cold waters in the eastern equatorial Pacific causes dry and cool conditions there. The position and intensity of the ITCZ changes, affecting the strength of trade winds, regional precipitation, and tropical storms in the Caribbean region.

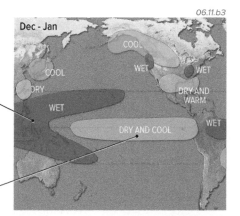

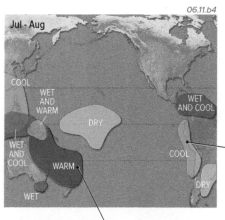

11. In July and August during La Niña, the western Pacific is warm and wet because of the presence of an extensive and strong warm pool. The cooler central Pacific stays relatively dry. Along the west coast of South America, the appearance of excessively cold water cools the coast during the southern winter, decreasing rainfall.

10. Effects of La Niña are abundant in North America. These include cooler and wetter than normal conditions next to the Gulf of Alaska, but dry and warm conditions in the southeast U.S.

12. In the Polynesian islands and northeastern Australia, warm, moist equatorial air clashes with cooler sub-Antarctic air along a northwest-southeast zone (shown in orange). Changes in the strength of the subtropical pressures shift this precipitation back and forth like a large door hinged over northeastern Australia.

Before You Leave This Page

☑ Sketch and explain the atmospheric connection of the Western Pacific Warm Pool and the subtropical and polar front jet streams.

☑ Explain how warm (El Niño) and cold (La Niña) phases of ENSO can affect the position of jet streams and other seasonal weather patterns, using a couple of examples from North America and one from elsewhere in the world.

6.12 How Does an El Niño Start and Stop?

THE ENSO PHASES, EL NIÑO AND LA NIÑA, have profound impacts throughout the Pacific basin and neighboring continents. What begins as a local adjustment in sea level and depth of the thermocline in the western Pacific can spread across the equatorial Pacific, triggering an El Niño. Once established, an El Niño directly impacts weather in locations along the western coast of Central and South America and beyond. For these reasons, much effort has gone into devising methods to predict the beginnings and ends of El Niño and La Niña phases.

A What Are the First Signs That an El Niño May Begin?

The continuously and gradually shifting relationship between the atmosphere and ocean means that no true "starting point" exists for the onset of an ENSO-related event, such as the beginning of a warm phase (El Niño). The first obvious sign of El Niño is a temporary weakening of the trade winds just east of the International Date Line (IDL). Another is an increase in atmospheric pressure in the westernmost Pacific, as measured in Darwin, Australia, relative to islands farther east in the Pacific, such as Tahiti. Recall that these were the two sites used to define the Southern Oscillation Index (SOI).

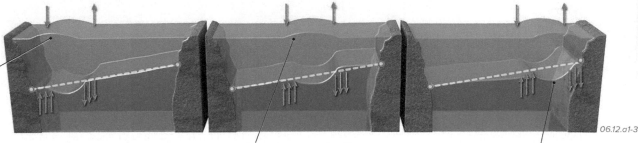

06.12.a1-3

1. During the warm phase of ENSO (i.e., an El Niño), easterly trade winds that normally blow water westward, piling up warm water in the Western Pacific Warm Pool, weaken. This causes a mound of warm water, or a wave, hundreds of kilometers wide to start to move to the east, traveling at about 3 m/s. This type of wave is called a *Kelvin wave*.

2. As the mound of warm water moves east along the equator, it raises sea level by a few centimeters and causes more noticeable warming of the SST. The additional thickness of warmer water depresses the thermocline beneath it, while cooler water rises to the west, elevating the thermocline in areas through which the wave has passed.

3. The combined surface and subsurface perturbations proceed eastward, with the wave moving faster than any individual water molecule. Once set in motion, the wave continues eastward, mostly unaffected by surface winds (lower parts of the wave are too deep). The Kelvin wave moves east until it reaches the shallow seafloor along the west coast of South America.

4. A Kelvin wave is detectable from slight increases in sea level as the wave passes islands and buoys in the Pacific. The slight rise of sea level is also measurable by satellites. This image (▶) shows not SST, but height of the sea surface (like a topographic map), with red being higher and blue being lower. A Kelvin wave is imaged here as the area of high sea levels (red).

Motion of a Kelvin Wave

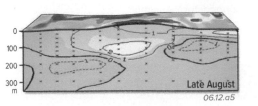

06.12.a4

5. An east-west-oriented mound of surface or subsurface waters might be expected to move poleward (parallel to the yellow arrows), away from the areas of higher sea level, dissipating the wave. However, waters that move away from the equator eventually have an apparent deflection from the Coriolis effect. This deflection bends them back toward the equator. As the Kelvin wave moves east, it therefore remains concentrated along the equator by the Coriolis effect and also by the convergence of trade winds along the ITCZ.

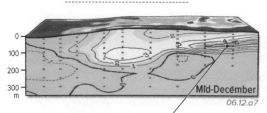

06.12.a5 06.12.a6 06.12.a7

6. This sequence of block diagrams tracks the eastward movement of a Kelvin wave, as recorded by anomalous water heights (top of the block) and temperatures at depth (front of the block). Yellow, orange, and red represent higher than normal heights and temperatures.

7. The mound of warm water starts in the west and moves eastward along the equatorial Pacific.

8. The warmer than normal water is concentrated at depth along the thermocline, which is inclined upward toward the east. As the wave moves east, cooler waters (shown in blue and purple) flow in behind it in the west.

B What Happens After the Kelvin Wave Hits South America?

The Kelvin wave moves east until it is confronted with the barrier of continental South America. At that point, some waters spread out along the coast, while some are reflected westward back across the Pacific.

1. The colors on the globe and the top surface of each block diagram again show height of the sea surface, not SST. Sea levels are anomalously high in the red areas, whereas they are anomalously low in blue and purple areas. The cross sections on the sides of the blocks indicate temperatures of the sea at various depths, with red being warmest and blue being coldest. They show a progression of what occurs as the Kelvin wave bounces westward of South America.

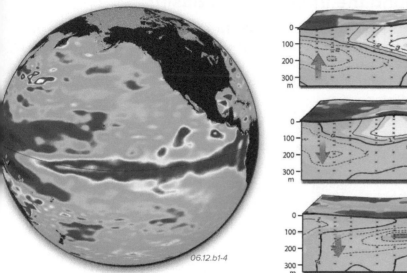

06.12.b1-4

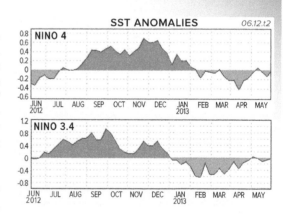

2. The orange and yellow colors in shallow depths represent the warm remnants of the Kelvin wave, which have started moving west by the time shown here (i.e., the front of the wave has already bounced off South America). The warm waters spread along the coasts to the north and south of where the nose of the wave first impacted the coast. In this way, the oceans west of Peru and Chile have warmed—the signature of El Niño.

3. Between the top cross section and the middle cross section, the warm waters moved westward and started dispersing, becoming less distinct. As the warm waters move westward, note that colder waters (the blue and green at depth) begin to extend eastward beneath the warm waters.

4. As the warm water spreads out away from the equator, cold waters from depth upwell into the area vacated by the spreading waters—causing the termination of El Niño. Eventually, warmer than usual conditions prevail in the western Pacific and cooler waters return to the east, the exact opposite of El Niño and the signature of La Niña, the cold phase of the ENSO phenomenon.

Predicting El Niño

The severe effects of El Niño on the commercial fishing industry of Peru and nearby countries, and the wide-reaching climatic impacts, have led governments to invest significant resources toward studying and predicting ENSO events, especially El Niño. In recent decades, warm phases of ENSO (El Niño) have displayed a tendency to reoccur about once every 5–7 years and to last 12 to 18 months. A severe El Niño in 1982–1983 caused much damage to the economy of Peru, prompting more detailed study of the phenomenon. As a result of these efforts, scientists predicted major El Niño events in 1986–1987, 1997–1998, and 2015–2016, allowing the affected countries to prepare for the crisis.

The types of data considered in these predictions include changes in atmospheric pressure (e.g., SOI), SST (measured by ships, buoys, and satellites, shown in the map to the right), observed decreases in the trade winds, and relative changes in sea level between sites in the western Pacific and those farther east. The various types of data are processed and modeled using supercomputers because the number of variables and number of

computations are both substantial. Some models try to simulate nature by using complicated equations of atmospheric and oceanic motion and energy transfer, and others try to identify the "fingerprint" of an ENSO event from records of past events. A common forecasting tool is to calculate and plot graphs of SST anomalies over time (as in the graph to the right), for different regions of the Pacific (numbered NINO 1, 2, 3, and 4). NINO areas can be considered together, such as NINO3-4 for a combination of the NINO 3 and NINO 4 regions. It is an evolving and exciting field of study, with important implications for our understanding of regional and global climates and how they change.

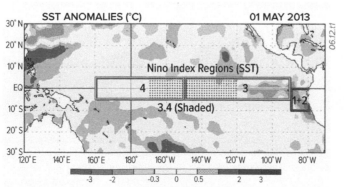

Before You Leave This Page

☑ Sketch and explain how a Kelvin wave develops, moves east, and contributes to an El Niño event when it encounters South America.

☑ Explain what happens to the warm waters when the Kelvin wave hits South America and how subsequent events lead to the demise of El Niño.

☑ Summarize some approaches and types of data used to predict El Niño.

6.13 Do Other Oceans Display Oscillations?

ENSO REMAINS THE GREATEST CAUSE of climatic variability on a global scale, but oscillations have been identified in other ocean basins and shown to have regional impacts. Whether these phenomena are linked or "coupled" with ENSO is still debated, but they may be responsible for long-term variations in our climate. One such oscillation occurs in the Indian Ocean, and another occurs in the Atlantic Ocean, both of which are described here. A third oscillation occurs in the Pacific (the PDO) and was presented at the beginning of the chapter.

Is ENSO Linked to Climate Variability Around the Indian Ocean Basin?

The Walker circulation cell over the Pacific shares a common ascending limb with a similar cell over the Indian Ocean. It is reasonable to assume, therefore, that atmosphere-ocean interactions in both may be linked, like two giant atmospheric gears that are meshed over the warm pool in the western Pacific and eastern Indian Ocean. Changes in the strength or location of that warm pool likely has consequences for areas bordering the Indian Ocean basin.

1. The atmosphere over the Indian Ocean features a Walker-like cell, with ascending air over the warm pool on the eastern side of the ocean, near Indonesia, northern Australia, and Australasia. This convectively rising air produces clouds and precipitation in these areas.

2. The rising air flows west once it nears the tropopause and then descends over the western equatorial Indian Ocean, off the coast of Africa.

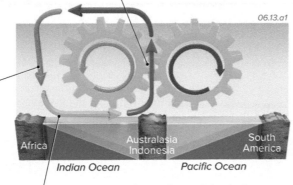

06.13.a1

3. The descending air then flows east across the surface of the Indian Ocean, to once again join the ascending limb over Australasia and Indonesia. This circulation is called the *Indian Ocean Dipole* (IOD).

4. We study variations in circulation for this region by examining differences in SST and pressures between an eastern area near Sumatra (near the ascending cell) and a western area off the coast of Africa (under the descending cell). These two locations are shown on the SST map below, on which orange shows areas that are warmer than average and blue and green shows areas that are colder.

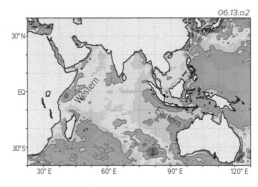

06.13.a2

5. This map shows SST for 1997, a year of a strong warm phase of ENSO (El Niño). In the Indian Ocean, this event produced cooler water in the area of Indonesia (blue and green) as the Western Pacific Warm Pool moved toward the central Pacific.

Response to a Warm-Phase ENSO

6. In 1997, in association with a strong warm-phase ENSO (El Niño off South America), the Western Pacific Warm Pool shifted eastward into the central Pacific. This also caused an eastward shift in the position of the ascending air, which in turn determines where the Pacific Walker cell meshes with the IOD.

7. This shift reduced the amount of rising air over Indonesia and moved the descending limb of the cell eastward, away from the coast of Africa (more over the Indian Ocean).

8. The weakened, displaced cell caused the southeast trade winds to strengthen and blow warmer surface waters to the west, piling them up along the coast of East Africa and limiting the amount of upwelling that normally characterizes that area. Cooler water was upwelled west of Indonesia to replace the water blown to the west. The configuration of warmer-than-normal waters in the west and cooler ones in the east, a characteristic of the El Niño phase, is termed a *positive IOD*.

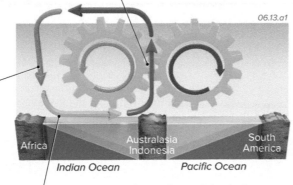

06.13.a3

Response to a Cold-Phase ENSO

9. During 1998, a cold-phase ENSO (La Niña) prevailed, bringing unusually warm waters to Australasia. The warm waters control the position of the ascending cell of the Indian Ocean dipole, which in turn affects conditions over the entire Indian Ocean.

10. As the circulation cell migrated westward, descending air returned to the coast of Africa. The southeast trade winds weakened as they were combatted by the west-to-east surface Walker circulation. Winds pushed warmer water toward Indonesia, allowing colder water to upwell along the coast of equatorial East Africa, an area rich in marine life. This configuration, with cooler waters to the west and warmer ones in the east, is termed a *negative IOD*.

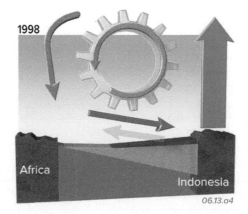

06.13.a4

B What Is the North Atlantic Oscillation?

Air pressures in the North Atlantic are dominated by two semipermanent features—the *Icelandic Low* to the north and the *Bermuda-Azores High* to the south. Each pressure zone is stronger at some times than others, and the difference in pressure between the two varies as well. This variability in pressure difference between the two pressure centers is known as the *North Atlantic Oscillation* (NAO).

1. The *Icelandic Low* is a region of low atmospheric pressure that resides over the water of the North Atlantic Ocean, generally near Iceland. Circulation around the low is counterclockwise and inward.

2. The *Bermuda-Azores High* is a region of high pressure located farther south in the Atlantic, typically in the vicinity of the Azores islands west of Spain. Circulation around the high is clockwise and outward. The position and strength of the high vary somewhat with time, but the high stays in this general vicinity.

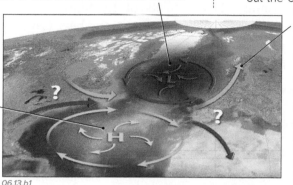

06.13.b1

3. The opposite rotations of the two pressure areas act to funnel material, such as a storm, in from the west, accelerate it, and eject it out the other side. Material leaving the "funnel" on the east side can turn to the northeast if it is mostly under the influence of the counterclockwise, inward rotation of the low, or to the southeast if it is instead influenced by the clockwise and outward rotation of the high.

4. The *NAO index* is expressed quantitatively as the *difference in pressure* between the Bermuda-Azores High and the Icelandic Low. When the difference in pressure is greater than normal, the NAO is said to be in a *positive phase*. When the pressure difference is less than normal, the NAO is in its *negative phase*. Like the other indices, we commonly plot the index as a function of time to discern patterns of weather and climate (not shown).

Weather Impacts of a Positive NAO

5. The relative strength and positions of the high and low control the position and strength of the polar front jet stream over eastern North America, the storm tracks across the North Atlantic, and the trajectory of the jet stream and storms across northwestern Europe, the Mediterranean, and North Africa. This top map shows the typical winter impacts of a positive NAO, when there is a large pressure difference between the low and high.

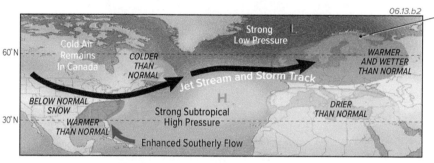

06.13.b2

6. Northwest Europe and European Russia experience warmer, wetter winters, while southern Europe, North Africa, and the Middle East suffer drought in what is usually the wet season.

7. When the NAO index is positive, the polar front jet stream and associated storms are steered northward, leading to warmer winters in the northern U.S. and unusual tropical air over the Southeast.

Weather Impacts of a Negative NAO

8. A negative NAO index occurs when there is less pressure difference (i.e., less of a pressure gradient) between the high and low. The Bermuda-Azores High is not as strong and so is less able to block or guide storm systems compared to the way it does under a positive NAO. In the winter, the weakened high pressure allows the polar front jet stream and storm tracks to move southward, permitting the passage of cold Canadian air over the eastern U.S. and resulting in greater than normal snowfall. Once over the Atlantic, the jet stream and storm track continue in an eastward direction toward Europe and Africa.

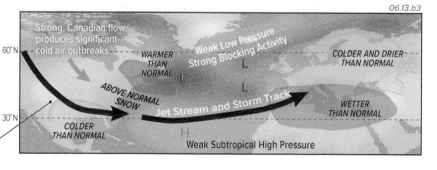

06.13.b3

9. During a negative NAO, cold, dry continental air dominates Russia and northwest Europe, limiting the amount of clouds and precipitation. The lack of clouds enables a large loss in radiant heat, keeping the region cold.

10. A southerly position of the jet stream and storm track bring moist Atlantic air over southern Europe, the Mediterranean, and North Africa.

Before You Leave This Page

✓ Sketch and explain the Indian Ocean Dipole, including changes in winds and water temperature during positive and negative phases.

✓ Explain the North Atlantic Oscillation, how the positions of the Icelandic Low and Bermuda-Azores High guide passing storms, and how positive and negative phases of the NAO impact winter temperatures and precipitation on either side of the North Atlantic.

6.14 What Influences Climates Near the Southern Isthmus of Central America?

THE SOUTHERN ISTHMUS OF CENTRAL AMERICA is geographically and climatically one of the most interesting places in the world. It is a narrow strip of land that connects Central America and South America and that separates very different waters of the Pacific and Atlantic Oceans. The climate of the region reflects intricately changing wind patterns and atmosphere-ocean interactions, including a migrating ITCZ, effects of El Niño–Southern Oscillation (ENSO), and mountains with rain shadows on different sides at different times. The interactions between these various components lead to unusual climates and large variability within a single year and from year to year. The region illustrates connections between winds, ocean currents, SST, weather, and flooding. The details are less important than the general relationships.

1. The region around the Isthmus of Panamá includes the southward-tapering tip of Central America, including Costa Rica and Panamá, and the wider, northwest end of South America, including the country of Colombia. The spine of the isthmus is a volcanic mountain range in Costa Rica, but the Panamá part is lower (hence the location of the Panamá Canal). Colombia contains the north end of the Andes mountain range.

06.14.a1

2. West of the land is the Pacific Ocean, and to the east is the Caribbean Sea, a part of the Atlantic Ocean. The Caribbean Sea is bounded on three sides by land—North America to the north, South America to the south, and Central America, including the Isthmus of Panamá, to the west. On the east, the Caribbean is open to the main part of the Atlantic through gaps between individual, mostly small islands, of the Greater and Lesser Antilles. As a result of the Caribbean being partly enclosed and being within the tropics and subtropics, the waters of the Caribbean are generally warmer than those of the Pacific.

Regional Wind Patterns and Positions of the ITCZ

3. The isthmus and the adjacent parts of South America are north of the equator, but in tropical latitudes, centered around 10° N.

4. Crossing the region in an approximately east-west direction is the Intertropical Convergence Zone (ITCZ), which has a seasonal shift north and south. In September it is as far north as 13° N.

5. In January, the ITCZ shifts to the south to near 5° N, following the seasonal migration of direct-overhead insolation to the south.

06.14.a2

6. Much of the region is dominated by the northeast trade winds that blow from the Caribbean Sea, across the isthmus. Embedded in the trade winds is a low-altitude, fast-flowing current of air called the Caribbean Low-Level Jet (CLLJ).

7. In most tropical regions, the ITCZ is north of the equator during the northern summer (e.g., July) and south of the equator during the southern summer (e.g., January). In this region, however, the ITCZ stays in the Northern Hemisphere because of the cold Humboldt Current farther south. Related to this position are southwesterly winds that originated in the Southern Hemisphere as the southeast trades and are deflected back northeastward when they cross the equator as they flow toward the ITCZ. They reach farther north during times when the ITCZ is farther north.

Wind Patterns and Weather During the Winter

8. From November to April, while the ITCZ is displaced to the south, the northeast trade winds and the CLLJ flow southwest across the isthmus, carrying warm, moist air from the Caribbean Sea. The Caribbean is part of the Western Atlantic Warm Pool, similar in origin and setting to the warm pool of the Pacific. The Atlantic warm pool forms when trade winds blow water westward until it piles up against Central America. When the moist air is forced up over the mountains of the isthmus at Costa Rica, heavy orographic rains fall on the Caribbean flanks of the mountains, while the Pacific coast experiences a dry season from the rain-shadow effect.

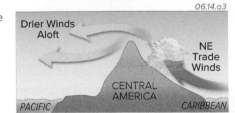

06.14.a3

Wind Patterns and Weather During the Summer

06.14.a4

9. During the opposite time of year, from May to October, the Pacific coast of Costa Rica experiences its rainy season. This corresponds to the northernmost migration of the ITCZ across the region, which allows the southwesterly winds to bring in equatorial Pacific moisture. As the moist Pacific air is forced eastward over the mountains, orographic rain falls on the west side. At the same time, the rising air in the ITCZ is over the area, producing heavy rain on a nearly daily basis.

Regional Wind Patterns and Positions of the ITCZ

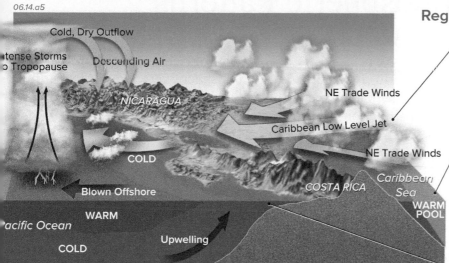

06.14.a5

10. In July and August, the rainy season of western Costa Rica is temporarily interrupted by a dry spell of several weeks called the "Veranillos" (little summer). This dry spell occurs when the Caribbean Low-Level Jet (CLLJ) reaches its maximum intensity. The jet accelerates through the topographic break in the mountain belt between Costa Rica and Nicaragua.

11. Recall that the Caribbean Sea is part of the Western Atlantic Warm Pool, so it has very warm water during July and August (during the northern summer). As a result, the trade winds and CLLJ bring moist air across the region. As this air climbs the eastern side of the mountains, much of the moisture is lost to precipitation, yielding a rainy season on this eastern (Atlantic) side of the country during these months.

12. During July and August, the vigor of the trade winds and CLLJ blow surface waters of the Pacific Ocean offshore, causing upwelling of cooler water in the Pacific at the Costa Rica/ Nicaragua border and a temporary suppression of the convective activity in the ITCZ. This causes the short dry season (Veranillos) on the western (Pacific) side of the country.

13. Warm, moist air farther offshore produces deep tropical convection, and an outflow of cooler, drier air descends over northwestern Costa Rica, preventing convection over the upwelling.

ENSO Warm Phase

ENSO Cold Phase

14. The weather patterns are already very interesting and unusual, but this region also falls under the influence of ENSO. These two globes show SST during ENSO warm and cold phases, relative to the long-term averages. Red and orange show SST above their typical value for a region, whereas blue shows SST that are colder than normal for that area. What patterns do you observe near Central America?

15. During a warm phase of ENSO (El Niño), warmer waters appear in the eastern equatorial Pacific, as shown by the red colors along the equator. Note that Pacific waters west of the isthmus are warmer than usual, but waters of the Caribbean are colder than normal. The warmer than usual waters of the equatorial Pacific are a site of rising air, drawing the ITCZ south and west of its usual position. The displaced ITCZ reduces rainfall along the Pacific flank of Costa Rica and enhances the Caribbean trade winds. This increases rains along the Caribbean slope, while the Pacific slope suffers from an enhanced rain-shadow effect and an extended Veranillos.

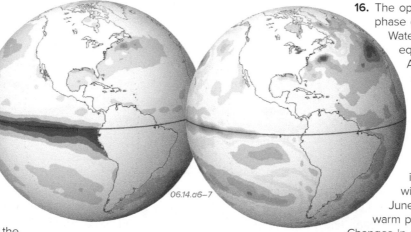

06.14.a6–7

16. The opposite occurs during a cold phase of ENSO, as shown in this globe. Waters are colder than normal in the equatorial Pacific west of South America, but are more typical offshore of Central America. The waters of the Caribbean, however, are warmer than typical. The changes in water temperatures and air pressures between the warm and cold phases of ENSO cause changes in the directions and strengths of winds. For instance, the CLLJ in June through August is stronger in warm phases than in cold phases. Changes in position of the ITCZ and strengths of winds cause differences in the risk of flooding, as depicted in the graphs below.

Before You Leave This Page

☑ Sketch and label the main wind features that affect this region, explaining how each one forms.

☑ Sketch and explain a cross section for each of the main settings for rainfall in the isthmus and in nearby South America.

☑ Summarize some changes that occur when the region is in a warm or cold phase of ENSO.

17. These two graphs (▶) show the risk of flooding during warm and cold ENSO events. Flooding on the Pacific flank (left graph) corresponds to the northward migration of the ITCZ during June through September. In warm phases (red bars), this migration is reduced, the CLLJ intensifies, and the Veranillos expands. In contrast, rivers on the Caribbean flank (right graph) generally flood when the CLLJ intensifies (JA and NDJ). Flooding intensifies in the Caribbean during warm phases (red bars) and in the Pacific during cold phases.

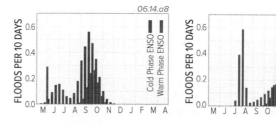

06.14.a8

06.14.a9

What Oceanic and Atmospheric Patterns Are Predicted for a Newly Discovered Planet?

PLANET W is a newly discovered planet that is similar to Earth. It has oceans, an atmosphere very similar to Earth's, and ice at both poles. Since no astronaut has yet ventured to the planet, we currently only know things we can observe from a distance, specifically the physical geography—the distribution of land, oceans, and sea ice, as well as the larger features on land. To guide future expeditions, you will use the known physical geography to predict the global patterns of wind, ocean currents, and atmosphere-ocean-cryosphere interactions.

Goals of This Exercise:

- Observe the global distribution of continents and oceans on Planet W.
- Create one or more maps showing the predicted geometries of global winds, ocean currents, and major atmospheric features, like the ITCZ and polar front.
- Draw cross sections that portray the global patterns, predicting the locations of warm pools, zones of high and low precipitation, and possible sites for ENSO-type oscillations.
- Use your predictions to propose two habitable sites for the first two landing parties.

Main Procedures

Follow the steps below, entering your answers for each step in the appropriate place on the worksheet or online.

1. Observe the distribution of continents and oceans on the planet. Note any features that you think will have a major impact on the climate, weather patterns, and other aspects of atmosphere-ocean-cryosphere interactions. Descriptions of some features are on the next page.

2. Draw on the worksheet the predicted patterns of atmospheric wind circulation (trade winds, westerlies, etc.), assuming that the patterns are similar to those on Earth.

3. Draw on the worksheet the likely paths of ocean currents within the central ocean. Identify the segments of currents that will be warm currents, cold currents, or neither.

4. Draw on the worksheet the likely locations of warm pools and shade in red any area on land whose climate will be heavily influenced by the presence of this warm pool. Shade in parts of the ocean that are likely to be more saline than others.

5. From the pattern of winds and ocean currents, identify which side of a mountain range would have relatively high precipitation and which side would have lower precipitation because it is in a rain shadow. Color in these zones on the worksheet, using blue for mountain flanks with high precipitation and yellow for areas of low precipitation. Note, however, that precipitation patterns along a mountain range can change as it passes from one zone of prevailing winds into another, or if wind directions change from season to season.

Optional Procedures

Your instructor may also have you complete the following steps. Complete your answers on a sheet of graph paper.

6. Draw four cross sections across the central ocean, extending from one side of the globe to the other. Of these cross sections, draw one in an east-west direction across each of the following zones: (1) between the equator and 30° (N or S); (2) within a zone of westerlies (N or S); and (3) within the south polar zone, south of 60° S. The fourth cross section can be drawn at any location and in any direction, but it should depict some aspect of the oceanic and atmospheric circulation that is not fully captured by the other cross sections. On each cross section, draw the surface wind patterns, zones of rising or sinking air, and possible geometries of a Walker cell or other type of circulation, where appropriate. Be prepared to discuss your observations and interpretations.

7. Draw a north-south cross section through the oceans, showing shallow and deep flows that could link up into a thermohaline conveyor. Describe how this might influence the climate of the entire planet and what might happen if this system stops working.

Some Important Observations About the Planet

A. The planet rotates around its axis once every 24 hours in the same direction as Earth. Therefore it has days and nights similar to Earth's. It orbits a sun similar to our own, at about the same distance. Its spin axis is slightly tilted relative to the orbital plane by approximately the same amount as on Earth, so this planet has seasons. The tilt is not shown in this view of the planet.

B. The planet has well-developed atmospheric currents, with wind patterns similar to those on Earth.

C. The oceans are likewise similar to those on Earth. They are predicted to have ocean currents that circulate in patterns similar to those on Earth. The oceans are predicted to have variations in sea-surface temperature, salinity, and therefore density, but no data are available for these aspects.

D. The planet has ice and snow near both poles, some on land (glaciers and ice sheets) and some over the ocean (sea ice).

1. A continent, simply called Polar Continent for now (you can name it if you want), is over the North Pole, but slightly off to one side. The western part of the continent extends south of 60°, but there are no data about whether this part has ice. There appear to be ice shelves and sea ice surrounding part of the continent.

3. There is a gap between Polar Continent and those continents to the south, providing a relatively land-free zone through which the ocean can circulate. Consider how this gap might influence speeds of winds and ocean currents here.

4. A nearly continuous mountain range runs along the coast of Eastern Continent.

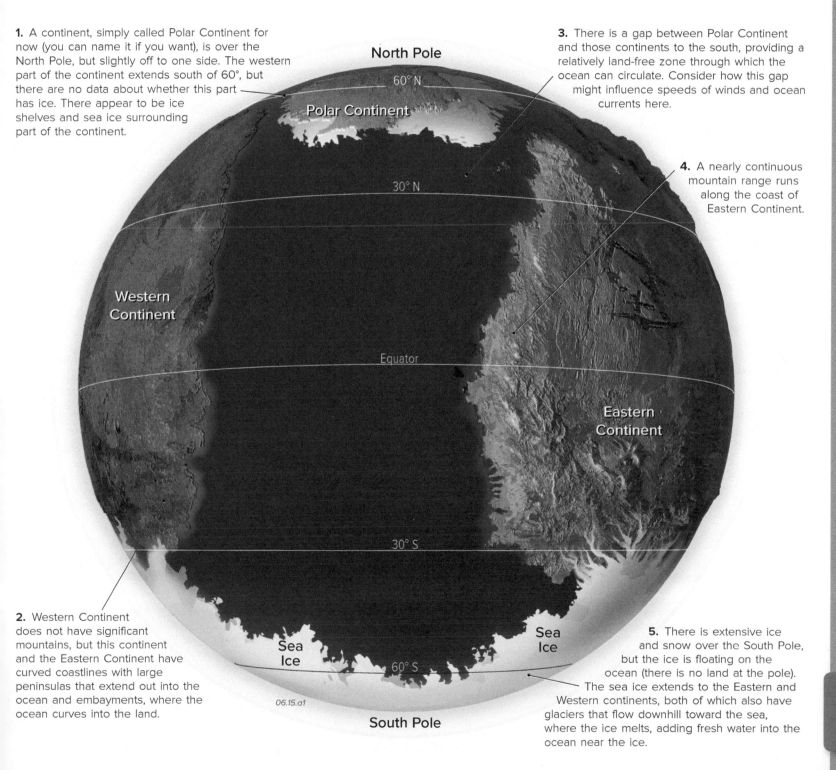

06.15.a1

2. Western Continent does not have significant mountains, but this continent and the Eastern Continent have curved coastlines with large peninsulas that extend out into the ocean and embayments, where the ocean curves into the land.

5. There is extensive ice and snow over the South Pole, but the ice is floating on the ocean (there is no land at the pole). The sea ice extends to the Eastern and Western continents, both of which also have glaciers that flow downhill toward the sea, where the ice melts, adding fresh water into the ocean near the ice.

Climates Around the World

CLIMATE IS THE LONG-TERM PATTERN of weather at a place, including not just the average condition but also the variability and extremes of weather. It is largely controlled by an area's geographic setting, especially latitude, elevation, and position relative to the ocean. Climate is also influenced by atmospheric pressure and moisture, prevailing and seasonal wind directions, and nearby ocean currents—in other words, everything we have explored so far in this book. Observed in this context, climate is the culmination of the various factors that move energy, air, and water (in all three states) in Earth's climate system.

The climate of a region can be described, measured, and represented in many ways, including maps of annual average temperature and precipitation, like the two globes below (temperature is on the left). To examine long-term climate, as opposed to short-term weather, we consider data averaged over years or decades. Climate is what we expect; weather is what we get. Remember that climate includes not only average values but also extremes and variability about the averages.

What is the difference between weather and climate, and how are the two related?

Maps of climatic variables, like the ones below, visually represent large amounts of data for the land, sea, or atmosphere. For numerous measures, like temperature and precipitation, maps can depict annual averages, values for a particular month (perhaps the warmest month) as averaged over decades, or some measures of seasonal extremes or variability from one decade to another.

What are some regional and global patterns in average temperature and precipitation on the two globes below?

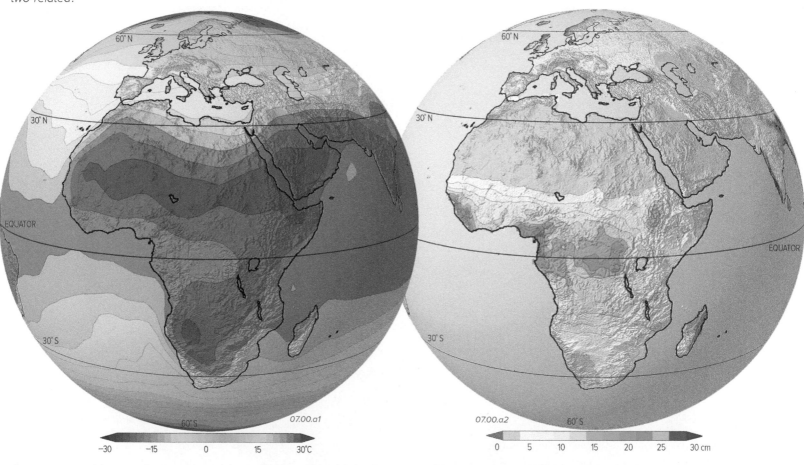

07.00.a1

−30   −15   0   15   30°C

07.00.a2

0   5   10   15   20   25   30 cm

Average annual temperatures vary greatly across the planet (▲). In the map above, high average temperatures are in red and orange, and cooler ones are in green and blue. What patterns do you observe? Much of Africa, but especially the central and northern parts, are very warm, much warmer than Europe. As anyone who has traveled to these places knows, they have very different climates. Africa straddles the equator, so some parts of it are likely to be tropical, whereas Europe is situated well north of the equator. The hottest areas, however, are in the subtropics.

Do all areas with the same average temperature have the same climate? Are warm temperatures, by themselves, enough to define an area as tropical?

The average amount of precipitation (▲) is another factor most people consider when talking about the climate of an area. We might say that Scotland, the northern part of the British Isles is a "cold, rainy place," but the Sahara of northern Africa is an "extremely hot and dry place." With such statements, we are expressing something about the average temperature or amount of precipitation but not what occurs on any given day (which is weather). Also, many regions have distinct wet and dry seasons, an important aspect of climate.

How can averages and variability in temperature and precipitation be combined to represent the climate of a region?

TOPICS IN THIS CHAPTER

By considering temperature and precipitation, both in terms of their averages and in terms of average maxima, average minima, and seasonal variations, geographers developed several climate classification systems, including the one portrayed on this globe (▶). The colors and letters indicate what type of climate characterizes an area. Examine the color patterns and letters on this globe and compare these to the globes showing average annual temperature and precipitation (on previous page). What correlations do you note?

How do we classify climates, and what do the letters assigned to a region indicate about its climate?

Oceans have characteristic climates, just like land areas. Note on this map that the climate of a land area generally continues out across an adjacent body of water, whether it is the Atlantic and Indian oceans or the Mediterranean Sea. Evaporation of water in the oceans provides energy for weather systems and other dynamic processes (like flooding), so assessing the climate of the oceans is important in understanding Earth.

On land, we consider types of plants in an area when talking about its climate, but how do we assign a climate zone to part of an ocean?

The climate of a region is influenced by many factors discussed in this book, including latitude and its relationship to Sun angle, elevation, topography, the locations of semipermanent pressure features, prevailing wind directions, common storm tracks, sea-surface temperatures, ocean currents, humidity, atmospheric oscillations, and teleconnections. These and other factors act in concert to produce the characteristic climates around the world, such as these in Africa and Europe.

How do Sun angle, topography, pressure features, humidity, storm tracks, ocean currents, and other factors interact to produce the observed variations in climate?

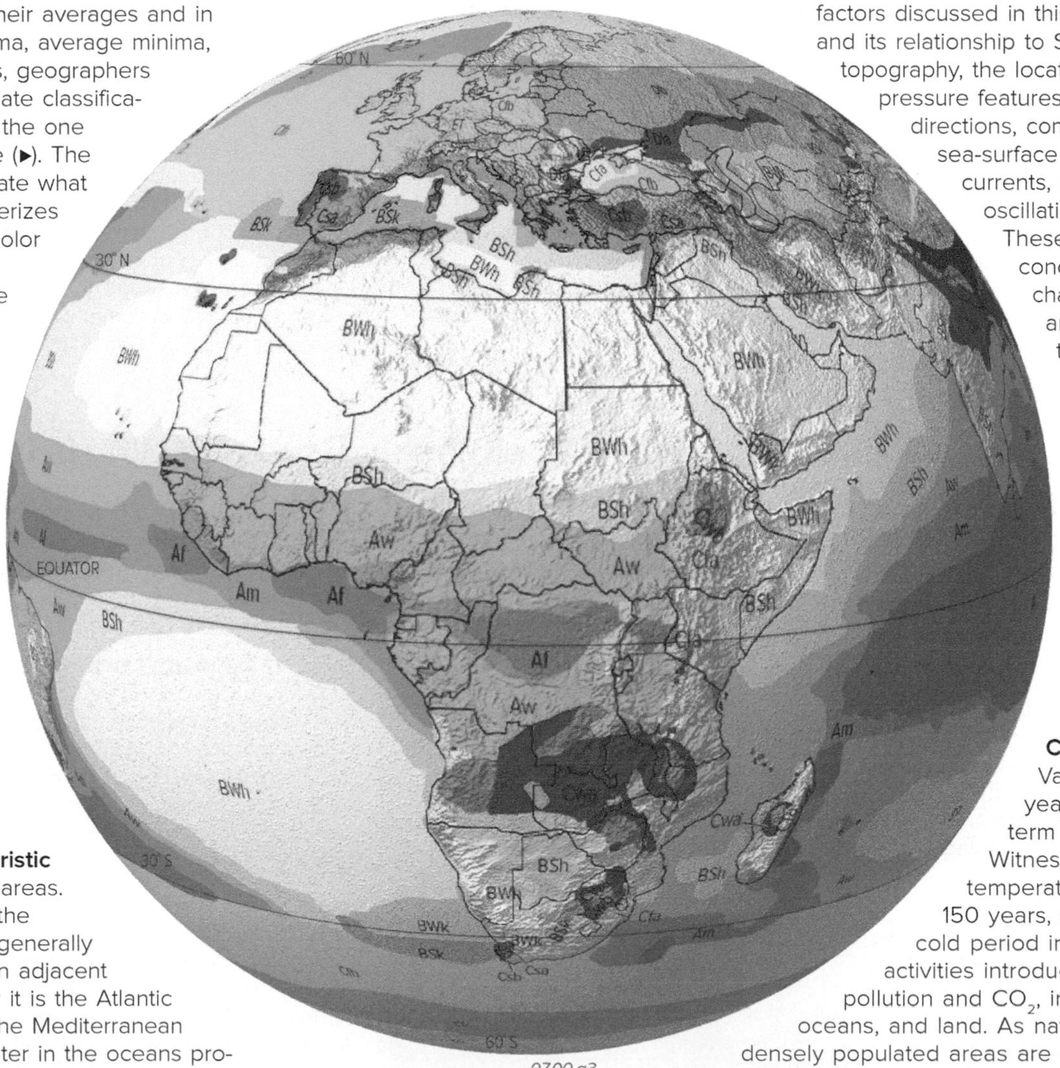

Climate is not static. Variations happen from year to year, and longer-term changes also occur. Witness the rise of global temperatures in the last 150 years, since the end of a very cold period in the 1800s. Human activities introduce materials, such as pollution and CO_2, into the atmosphere, oceans, and land. As natural lands or less densely populated areas are converted into cities and towns, long-term averages of temperature, precipitation, cloud cover, and local winds are affected—that is, urbanization changes the climate.

To what extent are the observed increases in global temperatures during the past century and a half attributable to human activities?

07.00.a3

7.0

7.1 How Do We Classify Climates?

CLASSIFICATION IS THE PROCESS of grouping similar items together and separating dissimilar items. Items that are similar in some ways can be different in others, and items that are in different groups can have certain similarities. Classifications allow us to examine general patterns, with the caveat that interpretations derived from classified groups are best done cautiously. Here, we examine a well-established method of classifying climates.

A What Is the Purpose of Classifying Climates?

Climates are classified to let us observe broad patterns and to simplify communication about the characteristics of a region. Historically there have been thousands of meteorological stations collecting weather and climatic information, and in the past few decades this wealth of knowledge has expanded with the addition of large quantities of remotely sensed information. Classification helps us to seek useful generalizations from this enormous amount of information. Climate can be classified in many ways, with the exact nature of the classification depending upon the research question that is of interest.

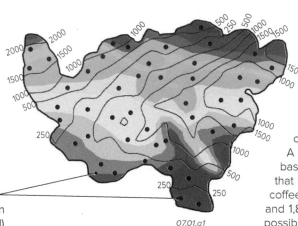

07.01.a1

1. Colors on this map of an imaginary country represent mean annual temperatures (reds above 28°C, blue and purple below 0°C). The black contours represent mean annual precipitation in millimeters. Meteorological stations are shown as dots. Examine the patterns on this map and think about how you might classify different parts of the country (e.g., cold and wet).

2. The southernmost areas are hot, which is an important fact if we were investigating an outbreak of a heat-related sickness. But we might classify stations in the southeast differently than those in the southwest because of the differences in precipitation (and therefore in humidity and perhaps in mold).

3. A study of the timing and magnitude of floods might focus on (1) regions where winter temperatures reach below freezing, opening up the possibility of flooding due to spring snowmelt, and (2) the amount of winter precipitation, which controls how much snow accumulates.

4. Most crops have fairly well-defined climatological requirements for their growth. A climate classification could be developed based on "optimal," "fair," or "poor" for growing that crop. For example, the best conditions for coffee are average temperatures of about 20°C and 1,800–2,800 mm of precipitation. Areas for possible coffee cultivation in this country would be very limited.

B What Climate Classification System Is Most Widely Used and Why Is It Useful?

Vladmir Köppen was a botanist interested in the global distribution of vegetation types. He surmised that the annual temperature and precipitation regimes determined the types of natural vegetation. Other influences on plant growth, such as which way a slope faces, the number of consecutive days without precipitation, and the types of soils, have more localized effects and so were not considered by Köppen.

On Köppen's maps, the lines separating different types of climate also separate different realms of natural vegetation. For each climate type, Köppen carefully selected thresholds of temperature and precipitation that preserved the boundaries for that characteristic assemblage of vegetation. The original scheme underwent modification as more information on global climates and vegetation became available. The resulting *Köppen climate classification* is the most widely used system; an example is portrayed on this climate map of South America (▶). Notice how the patterns of colors correlate with latitude, elevation, proximity to the ocean, and other aspects of position within the continent. Some of the boundaries would be recognizable on a satellite image.

07.01.b1

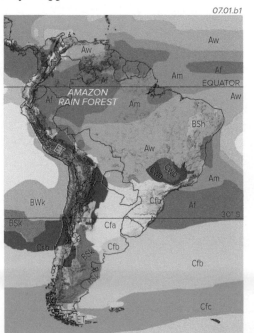

In reality, boundaries between climatic types are not static from year to year and not as clearly delineated as the map would suggest. Instead a boundary between vegetation and climate types usually occurs across a broad transition zone, known as an *ecotone*. In South America, an ecotone occurs between the Amazon rain forest and mountain climates of the Andes.

Köppen's system was designed to delineate vegetation realms, so it overlays well with worldwide natural vegetation zones—*biomes*. Both climate and vegetation are major controls on the types of soil that develop, leading to a notable similarity between the Köppen classification and maps of world soils. These linkages between climate, vegetation, and soils ensure that this classification is very useful in understanding the physical geography of our continents at planetary and regional scales. More recent satellite data availability over oceans allows us to compute the climate type over the ocean. Köppen was interested in vegetation realms, so he restricted his classification to the land.

C How Is the Köppen Classification System Organized?

The Köppen climate classification system has five major categories, represented by the first (capital) letter of the labels on the South America map. These are the following: A–*tropical climates*, B–*arid climates*, C–*temperate mid-latitude climates*, D–*harsh mid-latitude climates*, and E–*polar climates*, each represented by a photograph below. The five categories are further subdivided using a succession of criteria of temperature and the availability of water, as illustrated by the flow chart below. In the A, C, and D climate types, the second letter is lowercase and indicates whether the climate is wet year-round (f), has a dry summer (s), has a dry winter (w), or experiences a monsoon (m). The second letter is uppercase for arid and polar climates. Subsequent pages list key characteristics of each type of climate, but follow the flow chart below first.

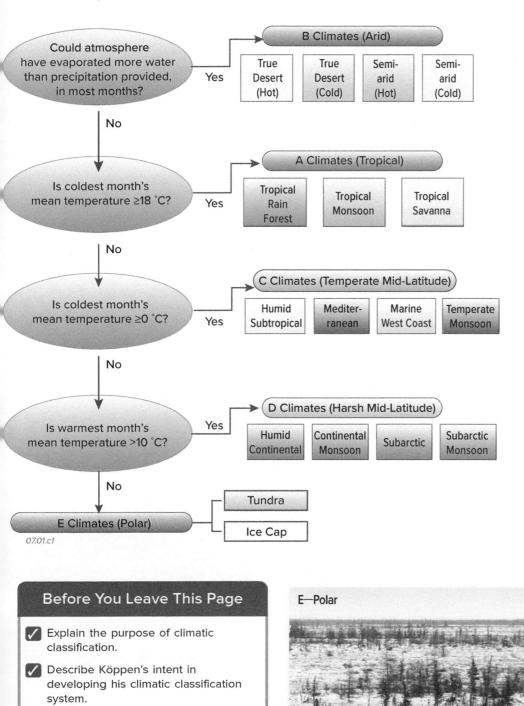

Could atmosphere have evaporated more water than precipitation provided, in most months? — Yes →

B Climates (Arid)

| True Desert (Hot) | True Desert (Cold) | Semi-arid (Hot) | Semi-arid (Cold) |

No ↓

Is coldest month's mean temperature ≥18 °C? — Yes →

A Climates (Tropical)

| Tropical Rain Forest | Tropical Monsoon | Tropical Savanna |

No ↓

Is coldest month's mean temperature ≥0 °C? — Yes →

C Climates (Temperate Mid-Latitude)

| Humid Subtropical | Mediter-ranean | Marine West Coast | Temperate Monsoon |

No ↓

Is warmest month's mean temperature >10 °C? — Yes →

D Climates (Harsh Mid-Latitude)

| Humid Continental | Continental Monsoon | Subarctic | Subarctic Monsoon |

No ↓

E Climates (Polar)

| Tundra |
| Ice Cap |

07.01.c1

B—Arid

07.01.c3 Namibia

A—Tropical

07.01.c2 Grand Cayman Islands

C—Temperate

07.01.c4 Germany Valley, WV

E—Polar

07.01.c6 Churchill, Canada

D—Harsh Mid-Latitude

07.01.c5 Northern British Columbia, Canada

Before You Leave This Page

✓ Explain the purpose of climatic classification.

✓ Describe Köppen's intent in developing his climatic classification system.

✓ Discuss the five main categories of climates in the Köppen climatic classification system.

7.1

7.2 Where Are Different Climate Types Located?

THE KÖPPEN CLIMATE CLASSIFICATION was defined for land areas and generally is depicted as a map covering a large area and portraying the regional or global distributions of different climate types. The map on these two pages covers the entire planet and is an *Extended Köppen Classification*, because we have extended the land-defined Köppen classification to encompass similar climatic conditions over the oceans. This allows us to better explore how different climates relate to regional atmospheric features, like high pressure in the subtropics. Study the patterns of different climates at the scale of the entire planet and then focus on a single continent or part of a continent. Refer to tables on the next two-page spread for a key to the labels. What is the label for the place where you live?

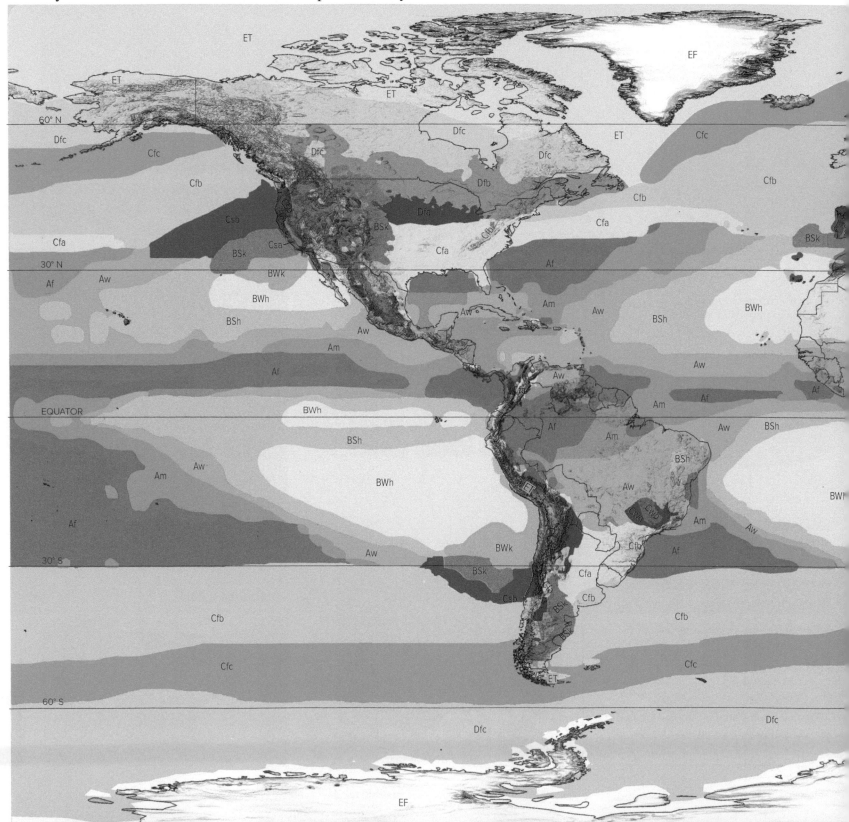

Group Letter	Name	Examples (West to East)
A	Tropical	Central America, Caribbean, Amazon, Congo, Indonesia
B	Arid	Southwest U.S., Sahara, Arabia, Tibet, central Australia
C	Temperate Mid-Latitude	Southeast U.S., NE Argentina, Mediterranean basin, Japan
D	Harsh Mid-Latitude	Southern Canada, northeastern Europe, Siberia
E	Polar	Northern Alaska, Arctic, Greenland, Antarctica

Before You Leave This Page

✔ Summarize where on the planet the main climate groups are most common (A—tropical, B—arid, C—temperate mid-latitude, D—harsh mid-latitude, and E—polar); explain how their locations compare to latitude, regional topography, and position in the interior of a continent versus next to the ocean.

07.02.a1

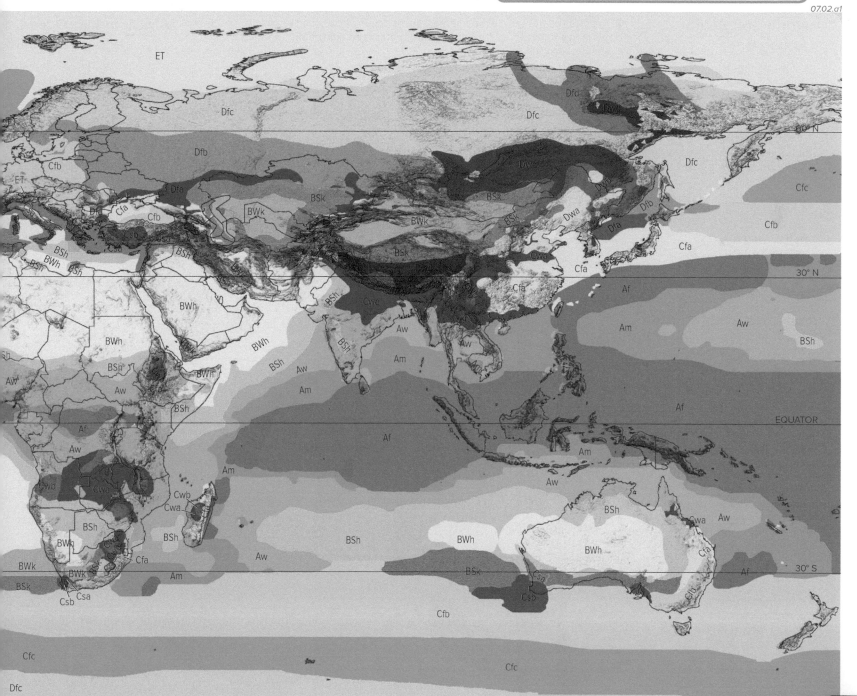

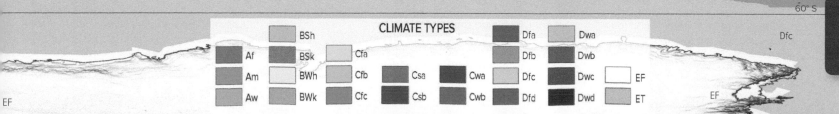

CLIMATE TYPES

	BSh		Dfa	Dwa
Af	BSk	Cfa	Dfb	Dwb
Am	BWh	Cfb	Csa	Cwa	Dfc	Dwc	EF
Aw	BWk	Cfc	Csb	Cwb	Dfd	Dwd	ET

7.3 What Are the Most Common Climate Types?

EACH TYPE OF CLIMATE, as defined by the Köppen climate classification, has specific characteristics that distinguish it from any other climate type. Once a climate is assigned to one of the five main groups (A–E), it is then categorized by factors such as whether the summer is hot, warm, or cool, and whether most precipitation falls in the summer, winter, throughout the year, or during a monsoon. The table on the right-hand page provides a reference for characteristics of each climate type. Four globes present the same information as the flat map on the previous two pages. Carefully observe each globe, guided by the table, and then read the text blocks in clockwise order around the globes.

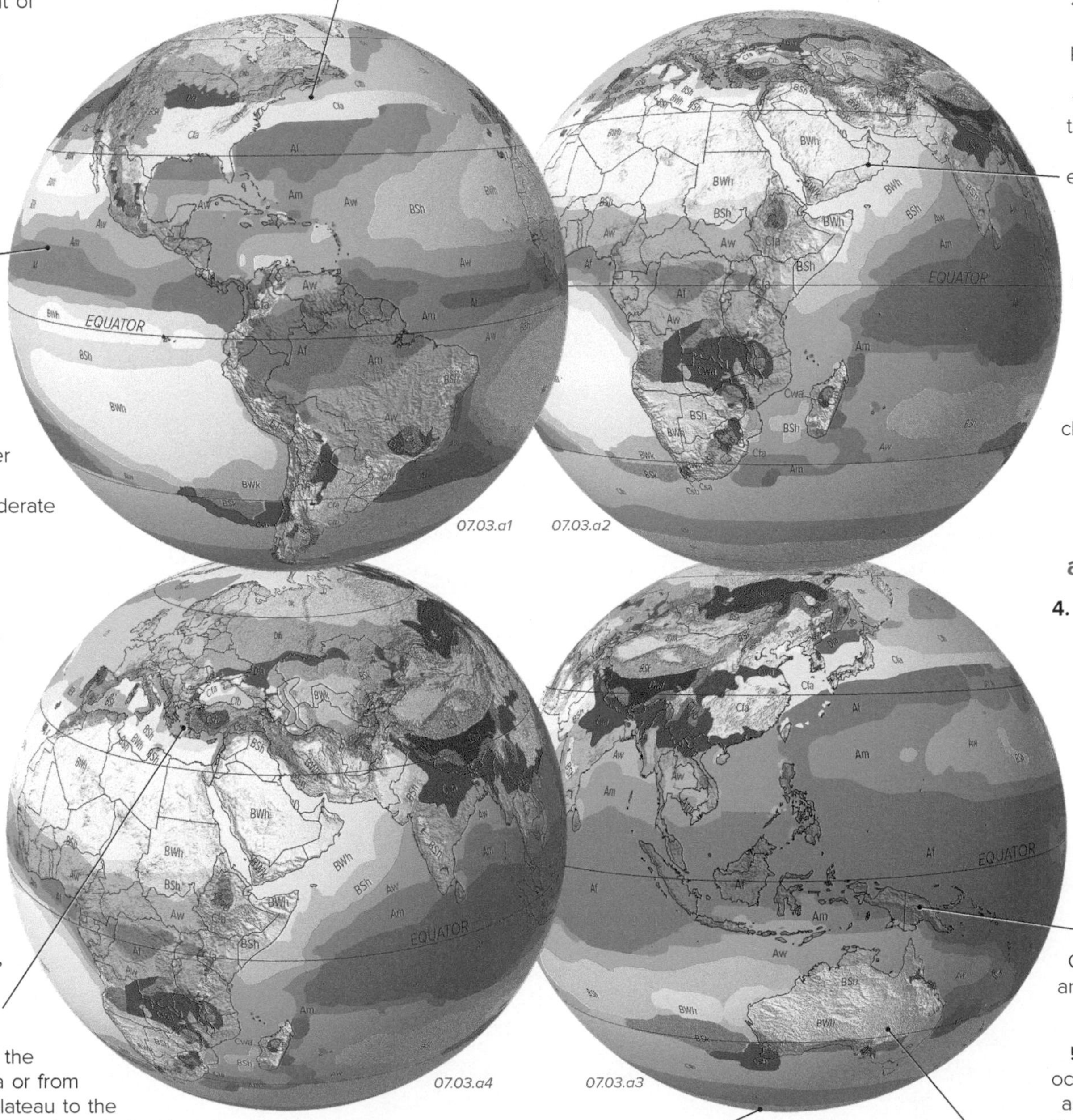

North and South America

1. The arrangement of climates is least complicated in the oceans, away from the land. Here, climate types are arranged in belts that roughly parallel latitude, progressing from A-type (tropical) near the equator to B, C, and E toward the poles. Some climate types, such as D-types (harsh mid-latitude), are more common over land than over the oceans, which moderate the temperatures.

2. Departures from this simple arrangement occur where ocean currents displace warm and cold water, and therefore climate types, north and south, like along the Gulf Stream.

Africa, Indian Ocean, and Antarctica

3. B-type climates are those where precipitation is not abundant enough to replenish water that could possibly be lost through evapotranspiration (*potential evapotranspiration*). B-type climates generally occur in the dry, sinking air that characterizes the subtropics. Note that B-type climates also occur over the ocean.

07.03.a1 *07.03.a2*

Europe and Western Asia

7. The patterns of climate types are most complex on land, especially in areas with large amounts of high relief and variations in topography. Steep gradients in climate types, as in western Asia, commonly reflect steep topographic gradients, such as those northeast of the Mediterranean Sea or from the high Tibetan Plateau to the tropical lowlands of southern India. Note that D-type climates over landmasses are largely restricted to the Northern Hemisphere, as in Siberia, but an oceanic belt of D-type climates also encircles Antarctica (not shown in this view).

Eastern Asia and Australia

4. As shown by the Indian and Pacific oceans, most A-type climates occur over the ocean, and these conditions also encompass islands and adjacent landmasses, as in the region of Papua New Guinea, Indonesia, and the Philippines.

5. Climates in the ocean are generally asymmetric around continents, as on either side of Australia, reflecting the asymmetrical circulation of ocean currents, which bring cooler water and stable atmospheric conditions along one side and warmer water with unstable conditions along the opposite side.

6. E-type (polar) climates are restricted to Antarctica and the surrounding ocean and to the Arctic Ocean and the surrounding land.

07.03.a4 *07.03.a3*

Common Climate Types

First Letter	Second Letter	Third Letter	Climate Type	Descriptive Name	Description of Climate
A Tropical Climates	f		*Af*	Tropical Rain Forest	Generally hot with no cold season; wet year-round
	m		*Am*	Tropical Monsoon	Generally hot with no cold season; wet for most of the year
	w		*Aw*	Tropical Savanna or Tropical Wet-Dry	Generally hot with no cold season; wet for about half of the year
B Arid Climates	S	h	*BSh*	Hot Steppe	Semiarid; annual average temperature exceeds 12°C (54°F)
	S	k	*BSk*	Cold Steppe	Semiarid; annual average temperature is below 12°C (54°F)
	W	h	*BWh*	Hot Desert	Arid; annual average temperature exceeds 12°C (54°F)
	W	k	*BWk*	Cold Desert	Arid; annual average temperature is below 12°C (54°F)
C Temperate Mid-Latitude Climates	f	a	*Cfa*	Humid Subtropical	Wet year-round; long, hot summer and short, intermittent cold season
	f	b	*Cfb*	Marine West Coast	Wet year-round; warm summer and a cold but not severe winter
	f	c	*Cfc*	Marine Mild Summer	Wet year-round; mild summer and a cold but not severe winter
	s	a	*Csa*	Mediterranean	Wet in winter but dry in summer; long, hot summer and short, intermittent cold season
	s	b	*Csb*	Mediterranean	Wet in winter but dry in summer; warm summer and a cold but not severe winter
	w	a	*Cwa*	Temperate Monsoon	Wet in summer but dry in winter; long, hot summer and short, intermittent cold season
	w	b	*Cwb*	Temperate Monsoon	Wet in summer but dry in winter; warm summer and a cold but not severe winter
D Harsh Mid-Latitude Climates	f	a	*Dfa*	Humid Continental	Wet year-round; hot summer and severe winter
	f	b	*Dfb*	Humid Continental	Wet year-round; warm summer and long, severe winter
	w	a	*Dwa*	Continental Monsoon	Wet in summer but dry in winter; hot summer and severe winter
	w	b	*Dwb*	Continental Monsoon	Wet in summer but dry in winter; warm summer and long, severe winter
	f	c	*Dfc*	Subarctic	Wet year-round; short, mild summer and very long, severe winter
	f	d	*Dfd*	Subarctic	Wet year-round; very short, cool, intermittent summer, and very long, extremely severe winter
	w	c	*Dwc*	Subarctic Monsoon	Wet in summer but dry in winter; short, mild summer and very long, severe winter
	w	d	*Dwd*	Subarctic Monsoon	Wet in summer but dry in winter; very short, cool, intermittent summer, and very long, extremely severe winter
E Polar Climates	T		*ET*	Tundra	Cold, with no month having an average temperature above 10°C (50°F); freezes occur in every month
	F		*EF*	Ice Cap	Cold, with each month having an average temperature below freezing

First Letter: Climate Group; generally from A near the equator to E near the poles. Second letter generally describes precipitation amounts and seasonality.

Lowercase Second Letter (Groups A, C, and D): f = wet year-round; m = monsoon; s = dry summer; w = dry winter

Uppercase Second Letter for Group B: S = steppe; W = desert; Uppercase Second Letter for Group E: T = tundra; F = ice cap

Third Letter: h = hot; k = cold; a = hot summers; b = warm summers; c = mild summers; d = cool summers

Before You Leave This Page

☑ Summarize where on the planet the main climate groups (A–E) are most common, and explain how their locations compare to latitude, regional topography, and position in the interior of a continent versus next to the ocean.

☑ Summarize the factors considered (such as hot summers) when subdividing different types of climate within a main group.

7.3

7.4 What Is the Setting of Tropical Climates?

A-TYPE CLIMATES in the Köppen climatic classification system are "tropical," having consistently warm temperatures all year. Precipitation in these zones is primarily caused by the convergence of the trade winds along the Intertropical Convergence Zone (ITCZ), which shifts with the season to locations north and south of the equator. There are three types of A climates in the Köppen system—*Tropical Rain Forest*, *Tropical Savanna*, and *Tropical Monsoon*.

A Where Are the Various Types of Tropical Climates?

These globes show the distribution of each of the three types of Group A (tropical) climates. All three globes show all three types, but each type is discussed separately below the globes. Note that all three climate types are centered in the tropics, but they widen or narrow considerably from ocean to ocean and from continent to continent.

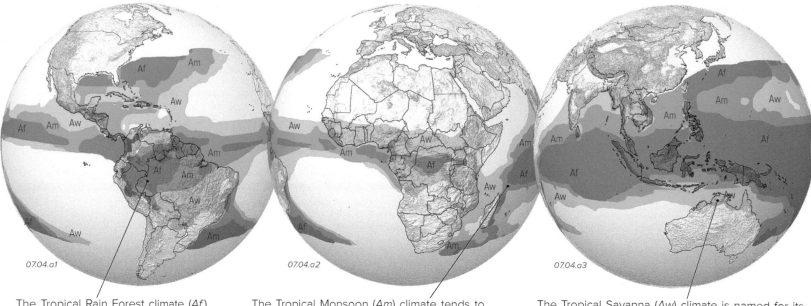

07.04.a1   07.04.a2   07.04.a3

The Tropical Rain Forest climate (*Af*) is characterized by abundant precipitation year-round, which supports lush vegetation. Notice that over land, this type is surprisingly limited in extent, but it is more widespread over the oceans.

The Tropical Monsoon (*Am*) climate tends to occur between *Af* and *Aw* climates, as in parts of the western Pacific and Indian oceans that have wet and dry seasons, typically due to a seasonal switch in the dominant wind direction. The short dry seasons seldom allow soils to dry fully, and so they support lush forests.

The Tropical Savanna (*Aw*) climate is named for its typical vegetation of grasslands and a few scattered trees. It occurs poleward of the other A-type climates and is therefore subject to the alternating influences of the subtropical highs and the ITCZ. Temperatures vary throughout the year somewhat more than in other A-type climates.

B Why Are Temperatures Consistent Throughout the Year in A-Type Climates?

A-type climates share a trait that we commonly associate with the word *tropical*—warm and relatively consistent temperatures throughout the year, a direct consequence of the way the Sun interacts with our tilted planet.

The noontime Sun is generally high over the tropics, as it migrates during the course of a year, from directly above 23.5° S latitude on the December Solstice to above 23.5° N latitude on the June Solstice, and back above 23.5° S again by the following December. As a result, it remains high over the region between these two latitudes (the tropics).

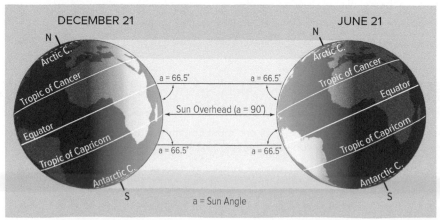

This setting causes the areas between the tropics to be heated with similar intensity throughout the year, unlike the seasonal variations that occur beyond the tropics. Also, the nearly overhead position of the Sun means the lengths of day and night vary little seasonally. Locations near the equator experience approximately 12 hours of sunlight throughout the year, but not all areas in climate Group A do because they extend away from the equator.

07.04.b1

C How Does Precipitation Differ Between the Three Types of Tropical Climates?

Tropical Rain Forest (*Af*)

07.04.c1 Palau

1. In a Tropical Rain Forest (*Af*) climate, precipitation exceeds losses through evaporation and transpiration in most months of the year. In addition to the lush vegetation, streams carry large volumes of water from the *Af* climates.

----

4. Trees become shorter and more widely spaced from climate *Af* to *Am* to *Aw*, from left to right in this figure (▶).

Tropical Monsoon (*Am*)

07.04.c2 Cambodia

2. For most months in a Tropical Monsoon (*Am*) climate, precipitation exceeds the needs of vegetation. The excess precipitation stored in the soil during these months allows rain forests to survive a few relatively dry months. Vegetation becomes slightly shorter and sparser in this zone compared to *Af*.

Tropical Savanna (*Aw*)

07.04.c3 Great Rift Valley, Kenya

3. The Tropical Savanna climate (*Aw*) is also known as a *Tropical Wet-Dry* climate because about half of the months receive abundant precipitation and the other half are very dry. The dry season is too long to support forests because the soil moisture becomes depleted during that time. Instead, the vegetation is grassland with scattered short, wide trees, such as acacia trees.

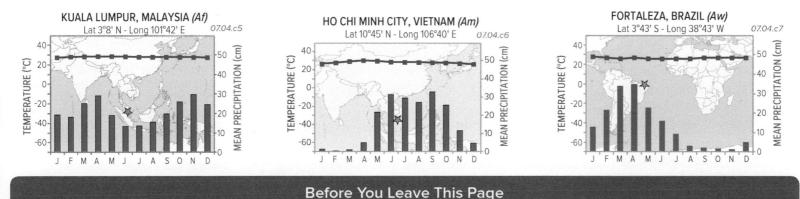

07.04.c4

Effects of the ITCZ

5. In an *Af* climate, the ITCZ brings rain most days throughout the year, although rainfall can be heavier in some months than others. *Af* climates remain under the influence of the ITCZ, even as it migrates north and south with the seasons.

6. For an *Am* climate, the ITCZ is a factor in most months, especially during the summer when it is typically overhead or very close. During low-Sun-season months, the ITCZ moves too far away to generate much precipitation.

7. In an *Aw* climate, the migrating ITCZ brings abundant summer rain. The low-Sun-season is dry and warm as these areas come under the influence of the subtropical highs.

Examples

8. Group A climates are all tropical, so there is very little seasonality in temperature. Precipitation patterns do vary among the three types, displaying some seasonality. *Af* climates are closest to the equator and so display the least variation in precipitation over the course of a year, whereas *Aw* climates are farthest away from the equator and so experience the greatest seasonal changes. *Am* climates are typically in between in position and character. The plots below are *climographs*, each showing temperature as a line graph and amounts of precipitation as a bar graph. Observe the patterns and compare them with the information above about the ITCZ.

KUALA LUMPUR, MALAYSIA (*Af*)
Lat 3°8' N - Long 101°42' E   07.04.c5

HO CHI MINH CITY, VIETNAM (*Am*)
Lat 10°45' N - Long 106°40' E   07.04.c6

FORTALEZA, BRAZIL (*Aw*)
Lat 3°43' S - Long 38°43' W   07.04.c7

Before You Leave This Page

☑ Summarize where Group A climates generally occur, and the typical positions of *Af*, *Am*, and *Aw* climates relative to one another.

☑ Explain why the temperature remains nearly constant throughout the year in A climates, particularly in Tropical Rain Forest climates.

☑ Explain the relationship between the position of the ITCZ and the geographic location of each zone, and how this affects any seasonality, or lack thereof, in the amount of precipitation (i.e., wet and dry seasons).

7.4

7.5 What Conditions Cause Arid Climates?

ARID CLIMATES OCCUPY a greater portion of Earth's land surface than any other climate category. They comprise Group B in the Köppen classification and are subdivided into desert climates and steppe climates, with the distinction being that deserts are more arid than steppes. Deserts and steppes are further subclassified into hot or cold categories—not all deserts or steppes are hot! In the Köppen classification of Group B climates, *potential evapotranspiration* is taken into account, in addition to precipitation; an area is classified as arid only if precipitation does not offset the potential loss of water through evaporation from the surface and transpiration through leaf surfaces.

A Where Are the Various Types of Arid Climates?

These globes show the distribution of the four types of arid climates—*Hot Desert* (BWh), *Cold Desert* (BWk), *Hot Steppe* (BSh), and *Cold Steppe* (BSk).

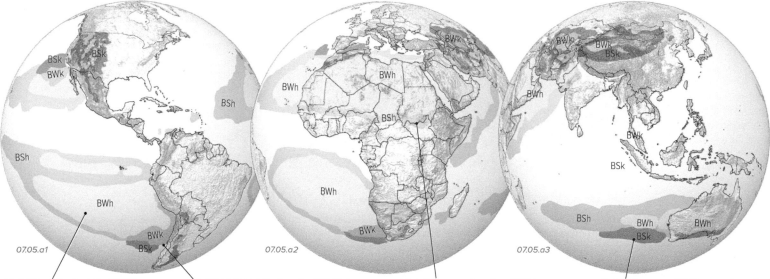

07.05.a1

07.05.a2

07.05.a3

A Hot Desert climate (BWh) covers huge areas of Africa (e.g., Sahara), the Arabian Peninsula, the interior of Australia, and parts of the American Southwest. It also extends over large areas of subtropical ocean.

A Cold Desert climate (BWk) is less common, occurring mostly in the interior of Asia (like southern Mongolia) and near cold ocean currents, like the Humboldt Current off the west coast of South America, as shown here.

Hot Steppe climates (BSh) generally surround the hot deserts and represent transitions to more humid climates, as across the Sahel region south of the Sahara and north and east of the hot deserts of Australia.

Cold Steppes (BSk) are abundant along the Great Plains and western interior of North America, in Tibet and other parts of central Asia, and in southern Australia. They cover relatively small parts of the ocean.

B What Controls Temperatures in Arid Climates?

07.05.b1 Big Maria Mtns., CA

07.05.b2 Sossusvlei, Namibia

1. Most deserts, like those in the American Southwest (◄), are in subtropical regions, where the noon Sun is generally high, so insolation arrives nearly perpendicular to the surface. At the latitudes of most arid climates, the noontime Sun is high for several consecutive months.

2. Many, but not all, arid climates lack clouds and precipitation for much of the year. If there is little humidity, insolation is transmitted through the atmosphere (◄) with a minimum amount being absorbed in the atmosphere before it reaches the surface, promoting warm days.

3. At night, longwave radiation is lost to space efficiently because of the lack of clouds and water vapor to absorb outgoing longwave radiation, so temperatures fall rapidly.

07.05.b3

Incoming Insolation

Outgoing Longwave Radiation

Sun Angle

4. Cold Steppe (BSk) and Cold Desert (BWk) climates have the relative dryness of their hot counterparts, but have lower temperatures, like in Tibet and Mongolia. Tibet is cold because it has high elevations, averaging 4,500 m (about 14,800 ft). The limited amount of precipitation relative to potential evapotranspiration produces a characteristic sparseness of plants (▼).

07.05.b4 Tibet

C What Factors Contribute to the Lack of Precipitation in Arid Climates?

An area designated as an arid climate (Group B) in the Köppen classification system must have the *potential* to lose more moisture through evapotranspiration (called potential evapotranspiration) than it receives in precipitation. This condition can result from a limited amount of precipitation and a high demand for water by having intense insolation (i.e., high potential evapotranspiration); it is always some combination of the following factors.

Descending Air Along the Subtropics

The main cause of deserts and other arid areas is their position along the subtropics, where the descending limb of the Hadley cell brings dry air and high pressure. Note how the arid lands in the Sahara and the Arabian Peninsula (▶) form an east-west band along subtropical latitudes, a manifestation of the Hadley-caused high pressure. The descending air generally means that even if the air is humid, it cannot rise easily, which limits formation of clouds and precipitation. Arid lands of Australia and the American Southwest occupy similar subtropical settings.

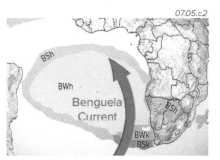
07.05.c1

Rain Shadow

An area can be dry because it is in the rain shadow of a mountain range. Southern and western sides of the Hawaiian Islands (▶) are on the rain-shadow side of the mountains. Moist air transported by the northeast trades moves up the northern and eastern slopes, causing heavy rain on those sides, but drying out the western, downwind sides.

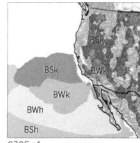
07.05.c3

Cold Ocean Currents

Other desert climate areas (*BWh* and *BWk*) and steppe climate areas (*BSh* and *BSk*) owe their existence partly or mostly to their proximity to cold ocean currents, which limit the amount of moisture in the air and promote atmospheric stability. One such area is the Namib Desert of southwest Africa (▶), adjacent to the cold Benguela Current (coming from nearer Antarctica). Another is the Atacama Desert along the southwest coast of South America, beside the cold Humboldt Current.

07.05.c2

Distance to Ocean

Many arid (Group B) areas are dry because they are in the interior of a continent, far from oceanic moisture sources, as in the case of the western interior of the U.S. (▶); they are also in the rain shadow of mountains along the West Coast.

07.05.c4

D What Are Some Examples of Climate Conditions in Arid Climates?

The climographs below convey temperatures (the lines) and precipitation (bar graphs) in areas designated Group B (desert and steppe climates). As we might expect, precipitation totals are very small, and what precipitation falls evaporates quickly in the dry air, leaving a parched landscape.

Desert Climates

Khartoum (▶) is an example of a hot desert in the subtropics. The region receives essentially no rain for most of the year. The meager July–August rains result from a slight influence of the northern edge of the ITCZ. Khartoum is near the tropics, so its temperature seasonality is dampened and temperatures remain high, showing limited range.

KHARTOUM, SUDAN (BWh)
Lat 15°36' N - Long 32°32' E   07.05.d1

Few people live in the world's cold deserts. This town in China (▶) lies on the fringe of the desert/steppe ecotone. Temperatures vary markedly and are very cold during the winter, but precipitation is relatively low. Most cold deserts have even less precipitation than shown here.

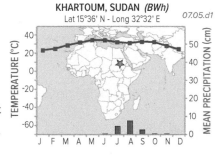
URUMQI, CHINA (BWk)
Lat 43°43' N - Long 87°38' E   07.05.d2

Steppe Climates

COBAR, AUSTRALIA (BSh)
Lat 31°29' S - Long 145°48' E   07.05.d3

TEHRAN, IRAN (BSk)
Lat 35°45' N - Long 51°45' E   07.05.d4

Steppe climates are more abundant and far more heavily populated than deserts. They are transition zones between desert climates and more humid climates. Precipitation in such environments is variable from year to year, with some years receiving desert-like precipitation totals and others experiencing precipitation more characteristic of humid climates. Like natural vegetation, human populations in these regions must be able to adapt to periodic dry spells and severe water shortages.

Before You Leave This Page

✓ Summarize the climate properties of Group B climates and where they generally occur.

✓ Sketch and explain why temperatures are high and precipitation is low in hot desert climates and other arid regions, providing examples.

7.5

7.6 What Causes Warm Temperate Climates?

TEMPERATE MID-LATITUDE CLIMATES, designated as Group C of the Köppen system, experience moderate temperatures and precipitation and so are sites of particularly intense human activity. Some temperate climates are relatively warm, because they occur mostly in subtropical latitudes. These include Humid Subtropical, Mediterranean, and Temperate Monsoon climates.

A What Is the Spatial Distribution of the Various Temperate Mid-Latitude Climates?

The globes below depict distributions of six temperate mid-latitude climate types. These include *Humid Subtropical* climate (*Cfa*), two *Mediterranean* climates characterized by dry summers (*Csa* and *Csb*), and three other climates in which precipitation is predominantly in summer due to continental monsoon effects, for which they receive the "w" or "winter dry" designation (*Cwa*, *Cwb*, and *Cwc*).

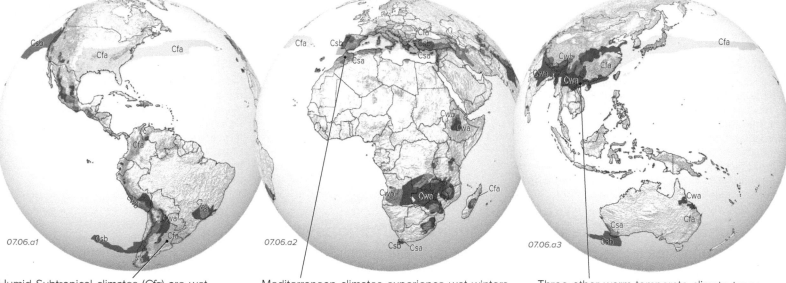

07.06.a1   07.06.a2   07.06.a3

Humid Subtropical climates (*Cfa*) are wet year-round and have hot summers with relatively short and intermittent cold seasons. They generally occur on the east coasts of continents and adjacent oceans—southeastern South America (shown here), southeastern U.S., eastern Asia (including much of Japan and part of China), and eastern Australia. Since *Cfa* climates are defined as having hot, rather than warm, summers, their marine extent is somewhat limited.

Mediterranean climates experience wet winters but dry and hot (*Csa*), or dry and warm (*Csb*), summers. These are pleasant climates for vacationers or retirees. In addition to partly encircling the Mediterranean Sea, they generally occur near the west coasts of continents and in adjacent oceans, including along the Pacific coast of the U.S., southwestern South America, southern Eurasia, and the southwestern tips of Africa and Australia. Coastal California is famous for its moderate, Mediterranean climate.

Three other warm temperate climate types experience large variations in precipitation between seasons, generally associated with shifts in wind direction—a *monsoon*. Places with Temperate Monsoon climates receive most of their rainfall in the summer but are dry in the winter, and they can have hot (*Cwa*), warm (*Cwb*), or very rarely, mild (*Cwc*) summers. They are best known from India to Southeast Asia, but they also occur in southern South America, southern Africa, and in northeastern Australia.

B What Affects Temperatures in Temperate Climates?

07.06.b1 Crete

1. When we think of a Mediterranean climate, we envision a place like Crete (◄), with warm summers, fair-weather cumulus clouds, and maybe a nearby ocean or sea. The pleasant temperatures are due largely to a location in or near the subtropics, where abundant insolation warms the land and water, which in turn warms the air.

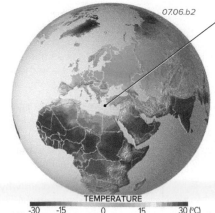

07.06.b2

TEMPERATURE
-30   -15   0   15   30 (°C)

2. On this globe showing average annual temperature, the Mediterranean (the nearly enclosed sea in the center) is a zone of transition, between very hot temperatures of the subtropics to the south and colder temperatures farther north. This characterizes a temperate climate—it is intermediate between tropical and harsh climates. Summers are warm to hot, because the Sun is high and day lengths are moderately long. Coastal areas have temperatures that are moderated by nearby bodies of water. Summers are influenced exclusively by tropical air masses, especially maritime tropical air masses and less often by continental tropical air masses. In winter, temperate climates are influenced by polar and tropical air masses, but the polar air masses are usually moderated by the time they reach these latitudes.

C What Affects Precipitation in Temperate Climates?

Temperate climates, especially those in or near the subtropics, are affected greatly by subtropical highs, like the *Bermuda-Azores High* in the north Atlantic and the *South Atlantic High*. When these high-pressure areas are strong and nearby, they suppress the formation of clouds and precipitation. Precipitation can occur more easily when the highs are farther away and when they are weaker.

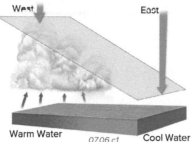

West   East

Warm Water   07.06.c1   Cool Water

1. Subtropical highs display a very important asymmetry—sinking motions descend to the surface on the eastern side (▶), which coincides with the *western side* of adjacent continents. However, the sinking does not extend to the surface on the western side, which corresponds to the *eastern side* of continents.

2. The "tilting" of the subtropical highs is caused primarily by the difference in ocean currents on each side of the ocean. Cold currents on the eastern sides of ocean basins stabilize the atmosphere and allow sinking of air to the surface. Warm currents on the western margins destabilize the atmosphere by adding energy to the surface, thereby encouraging rising atmospheric motion. Similar processes occur in both the Northern and Southern hemispheres, because in both settings cold currents are on the east side of the ocean and warm currents are on the west side.

3. Hadley cells migrate and change width seasonally, following the position of maximum insolation northward during northern summer and southward in southern summer. Since the subtropical highs represent the descending limb of the Hadley cell, they shift and change in intensity and size too, as shown here.

4. In both hemispheres, the subtropical highs expand and make their closest approach to regions of temperate climates during the summer (June in the Northern Hemisphere and December in the Southern Hemisphere). The strong sinking action on the eastern side of a high, when it makes its closest approach in summer, stifles rain on the west coasts of the continents. Mediterranean (*Csa, Csb*) climates are characterized by this lack of rain in the summer.

5. Because sinking associated with a subtropical high does not extend to the surface on its western sides, the east sides of continents are places with Humid Subtropical climates. Here, clouds can grow vertically and summer rain falls in places like southeastern South America or the southeastern U.S. On this figure, note that Mediterranean climates (in light and dark pink) are more common on the west coasts of continents, but Humid Subtropical climates (shown in light green) are more common along the east coasts of continents—a direct response to the asymmetry of the subtropical highs.

6. With the change in seasons, the Hadley cells migrate toward the opposite hemisphere (e.g., into the Southern Hemisphere during December, the northern winter). The subtropical high in the hemisphere experiencing winter also shifts toward the equator and becomes smaller and less intense, so it no longer blocks migrating weather systems or the southern advance of polar air masses. As a result, frontal precipitation from mid-latitude cyclones spreads across regions of temperate climate during the winter.

7. Thus, a Mediterranean climate owes its characteristic dry and warm or hot summer, but cooler, wet winter, to its subtropical latitude and to the interactions between migrating subtropical highs and ocean currents—a lot of things need to happen for a nice day on a California beach.

JUN   DEC

JUN   DEC

07.06.c2

Examples

ATHENS, GREECE (Csa)
Lat 37°58' N - Long 23°43' E   07.06.c3

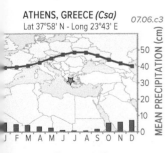

Athens, Greece, along the northern shore of the Mediterranean, experiences typical Mediterranean climate, with hot and very dry summers.

TAIPEI, TAIWAN (Cfa)
Lat 25°02' N - Long 121°38' E   07.06.c4

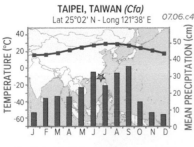

9. Taipei, Taiwan, has a Humid Subtropical climate, but winter precipitation is somewhat limited by wind patterns established after the end of the summertime Asian monsoon.

LUCKNOW, INDIA (Cwa)
Lat 26°50' N - Long 80°56' E   07.06.c5

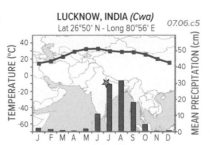

10. Lucknow, India, is in the region affected by the South Asian monsoon, with dramatic variations in rainfall during the year, earning its monsoon designation (*Cwa*).

Before You Leave This Page

☑ Summarize the three varieties of Group C climates (Mediterranean, Humid Subtropical, and Temperate Monsoon) and where they generally occur.

☑ Describe factors that influence temperatures of a temperate climate.

☑ Sketch and explain how migration of subtropical highs produces different climates on west versus east coasts of a continent.

7.7 What Are the Settings of Mid-Latitude Climates?

NON-ARID, MID-LATITUDE CLIMATES tend to experience a relatively even distribution of precipitation year-round. Some are dominated by maritime air masses that moderate temperature swings and result in relatively mild winters and summers; they are within Group C in the Köppen designation. Other mid-latitude climates involve more severe winters because they are dominated by continental air masses, and these fall into the *D* classification. Mid-latitude regions are affected by westerlies, and precipitation patterns reflect the role of mid-latitude cyclones.

A Where Are Marine West Coast and Humid Continental Climates Found?

The globes below portray the distributions of three mid-latitude climate types—*Marine West Coast* climate (*Cfb* and *Cfc*), *Humid Continental* climate (*Dfa* and *Dfb*), and *Continental Monsoon* climate (*Dwa* and *Dwb*). As you can observe, most are in cold places.

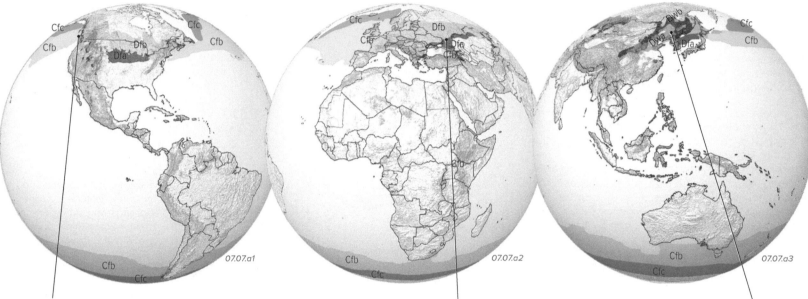

07.07.a1 07.07.a2 07.07.a3

The Marine West Coast climates (*Cfb* and *Cfc*) occur over or near oceans at moderately high latitudes, like the southwest coast of Canada. An abundant supply of moisture provides precipitation throughout the year. The influence of the ocean means that summers are warm (*Cfb*) or cool (*Cfc*), and winters are cold but not severe. Such climates occur in northern and southern parts of oceans and across most of western Europe.

Humid Continental climates (*Dfa* and *Dfb*) occupy the interior of continents, but only in the Northern Hemisphere—southern continents are not wide enough at these latitudes to allow these climates. Humid Continental climates are wet year-round and feature hot (*Dfa*) or warm (*Dfb*) summers but long and cold (*Dfa*) to severe (*Dfb*) winters. They are the climates of southern Russia and the upper Midwest and Great Lakes region of the U.S.

Continental Monsoon climates (*Dwa* and *Dwb*) are restricted in distribution, occurring mostly in Asia, in the interior of China and north of the Korean Peninsula. In both locations, precipitation varies greatly during the year, partly because of nearby monsoons. As the name implies, *Dwa* climates have dry winters and hot, wet summers. *Dwa* occurs farther north and has warm (not hot) summers.

Examples

07.07.a4 Vancouver, British Columbia

Mid-latitude climates experience long days in summer and short ones in winter. The influence of the ocean (◄) causes marine climates (*Cfb* and *Cfc*) to have less seasonality in temperatures than continental ones.

Humid Continental (*Dfa* and *Dfb*) climates are wet year-round, but they are farther from any oceanic influences and so are more seasonal (▼), experiencing warm or hot summers but much colder winters.

07.07.a5 Parnell Tower, WI

Climate	Distance from Ocean	Temperature Regime
Marine West Coast (warm or mild summer)	Immediately downwind of ocean, in mid-latitude westerlies	Warm to cool summers, mild winters
Humid Continental (hot or warm summer)	Far from the moderating influence of the oceans	Hot or warm summers, cold winters
Continental Monsoon (hot or warm summer)	May be near oceans, but circulation during part of year makes oceanic air a non-factor	Warm summers, dry winters

B What Affects Precipitation in Non-Arid Mid-Latitude Climates?

The mid-latitude westerlies bring a parade of mid-latitude cyclones to areas of Marine West Coast climates (*Cfb* and *Cfc*) and Humid Continental climates (*Dfa* and *Dfb*). This ensures abundant precipitation throughout the year in marine climates and adequate precipitation throughout the year in continental ones. Humid Continental climates of Asia experience little precipitation in winter, but the climates are not arid because potential evapotranspiration is low in winter there, so water is not removed readily from the surface. What other factors might affect the precipitation climatology of Marine West Coast and Humid Continental climates?

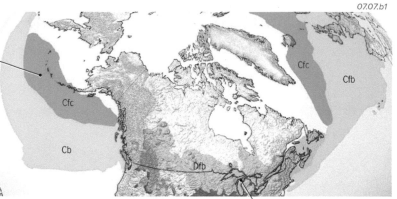

07.07.b1

Marine West Coast

1. The two mid-latitude marine climates (*Cfb* and *Cfc*) are either over relatively cool oceans or over land areas where winds bring oceanic air across a continent, as along the Pacific coast of Canada and in western Europe. Even though the subtropical high-pressure zones are far away, they still make their nearest approach to these regions in the summer and suppress precipitation somewhat, particularly on the southern fringes of the Marine West Coast climate. As a result, precipitation tends to peak in October (when the oceans are warmest) and December (when the subtropical high is farthest away) in Northern Hemisphere areas under these two marine climates.

Humid Continental

2. Humid Continental (*Dfa, Dfb*) climates receive precipitation throughout the year, including snow in the colder months. Summers are somewhat wetter than the rest of the year in most Humid Continental climates because of convective thunderstorms on hot summer afternoons and the high water-vapor capacity in warm air. In North America, the clockwise flow of Gulf of Mexico moisture around the expanded Bermuda-Azores High adds to the summer precipitation potential. In winter, the combination of moisture from the Great Lakes and bitter cold air results in heavy snowfall (lake-effect snows) on lands downwind of the lakes.

Influence of the Westerlies

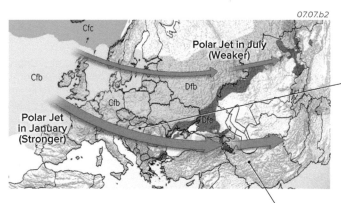

07.07.b2

3. The mid-latitude westerlies, including the polar front jet stream (shown here in idealized form), push maritime polar air masses from the Atlantic deep into the interior of Europe. This is possible in part because south of Scotland, Europe lacks north-south-oriented mountain ranges to confine the moist air to the coastal zones. As a result, most of northwestern Europe is in a Marine West Coast climate (*Cfb*).

5. The east-west oriented Alps often prevent Arctic and polar air masses from reaching southern Europe. Nothing prevents such air masses from penetrating farther southward in eastern Europe, so the Humid Continental climates extend farther south in this region. Mid-latitude cyclones that occur over these continental-climate areas produce less precipitation than might be expected because of the increased distance from the source of moisture (i.e., the Atlantic Ocean and Mediterranean Sea).

4. The effects of this moist air fade farther eastward across the continent, allowing continental air masses to exert more control. As a result, countries of eastern Europe, like Belarus and Ukraine, experience Humid Continental climates (*Dfb*).

Examples: West-to-East Changes Across Eurasia

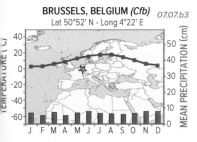

BRUSSELS, BELGIUM (Cfb) 07.07.b3
Lat 50°52' N - Long 4°22' E

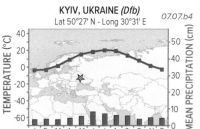

KYIV, UKRAINE (Dfb) 07.07.b4
Lat 50°27' N - Long 30°31' E

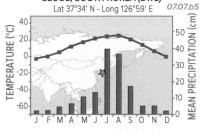

SEOUL, SOUTH KOREA (Dwa) 07.07.b5
Lat 37°34' N - Long 126°59' E

6. Brussels, Belgium, is very mild for its latitude. This is characteristic of Marine West Coast climates.

7. Kyiv (Kiev) Ukraine, farther east, has a Humid Continental climate (*Dfb*), with greater variations in temperature than Brussels.

8. Seoul, South Korea, has more precipitation seasonality than the *Dfb* climates because it is affected by the East Asian monsoon.

> ## Before You Leave This Page
>
> ✔ Summarize the varieties of mid-latitude climates (*Cfb, Cfc, Dfa, Dfb, Dwa,* and *Dwb*) and where they occur.
>
> ✔ Explain the factors that contribute to temperatures and precipitation seasonality for mid-latitude climates.
>
> ✔ Sketch and explain how and why climates vary across Eurasia.

7.7

7.8 What Causes Subarctic and Polar Climates?

SUBPOLAR AND ADJACENT CLIMATES occur at very high latitudes in both the Northern and Southern Hemispheres. Climates of polar regions, assigned to Group E in the Köppen system, are extremely cold and are subdivided into *Ice Cap* (*EF*) and *Tundra* climates (*ET*). Farther from the pole, the Tundra climate gives way to other *Group D* climates—the *Subarctic* climate (*Dfc, Dfd*), and *Subarctic Monsoon* (*Dwc, Dwd*) climate. The Subarctic and Group E climates are characterized by frigid temperatures and low precipitation totals.

A Where Are Subarctic, Tundra, and Ice Cap Climates Found?

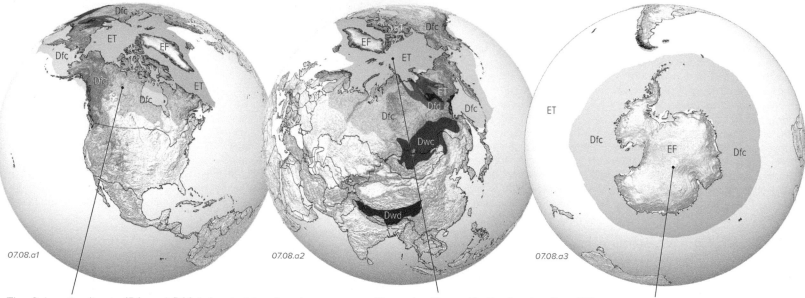

07.08.a1 07.08.a2 07.08.a3

The Subarctic climate (*Dfc* and *Dfd*) is located in a band away from the poles. The line separating it from more polar climates, the 10°C (50°F) isotherm in the warmest summer month, separates a climate that provides enough energy for trees to survive from one that does not. This boundary, called the *tree line*, is visible from satellites. The Subarctic climate occurs farther south at high elevations. A related subarctic climate, the Subarctic Monsoon climate (*Dwc* and *Dwd*) in Asia, receives even less winter precipitation than places with a Subarctic climate.

The polar (Group E) climates, Ice Cap (*EF*) and Tundra (*ET*), are centered around both poles (although only the North Pole is shown on this globe). In the Northern Hemisphere, the Ice Cap climate is restricted to Greenland, but the Tundra climate covers the Arctic Ocean, high Arctic islands, parts of the North Atlantic, and northernmost North America, Asia, and Europe.

Antarctica is mostly covered by ice and snow and so possesses an Ice Cap climate. The surrounding ocean is classified as a Subarctic climate, and true tundra (polar vegetation with no trees) occurs in some islands and other ice-free lands.

B What Causes Low Precipitation Amounts in Subarctic and Polar Climates?

1. Polar and adjacent Subarctic climates are characterized by low precipitation, except locally. This is largely because these are very cold places, as depicted on this globe of average annual temperatures. Very low temperatures have a number of implications for precipitation.

2. Very cold air has a low water-vapor capacity, so cold air can carry only limited amounts of water vapor. Without much water vapor, it is difficult to form clouds and even more difficult to generate precipitation. Also, much energy goes into latent heat during melting of ice and thawing of upper parts of the soil during the summer, so the surface doesn't heat up enough to generate convective precipitation.

3. For most of the year, subarctic and polar regions lie far from the zone where warm and cold air masses meet, so frontal precipitation is minimal. Tropical cyclones can never penetrate far enough poleward and inland to influence Subarctic, Subarctic Monsoon, and polar climates.

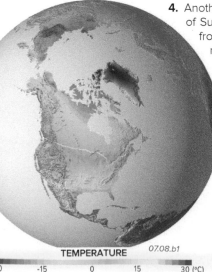

TEMPERATURE 07.08.b1
-30   -15   0   15   30 (°C)

4. Another limiting factor is that much of the region of Subarctic and polar climates is inland, or near frozen seas. Humidity flowing into the area must primarily originate over unfrozen seas away from the pole, but persistent high pressure over the pole causes wind to generally blow away from the poles, driving any moist air away. The high pressure and sinking air also limit atmospheric instability.

5. The most likely scenario to produce precipitation (especially over land) occurs in summertime, when the boundary separating cold and warm air masses—the polar front—makes its nearest approach to subarctic and polar latitudes. Marine areas often have precipitation peaks in fall, as the seasonal warming is delayed over oceans.

C What Controls the Temperature Regime in Subarctic and Polar Climates?

Limitations in Solar Heating

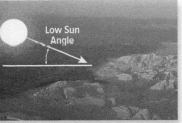

07.08.c1 Greenland

1. *Low Sun Angles*—The latitude of these climates dictates that Sun angles are very low during summer (◄) and below the horizon for much of winter (24 hours of darkness). A low Sun angle results in less sunlight per unit area and a long path length through the atmosphere, which increases the amount of insolation absorbed, reflected, and scattered in the air before it reaches the surface.

07.08.c2

2. *Seasonal Variation in Day Length*— Sunlight hours in winter are very short (◄), if they exist at all. Even though summer days have many hours (even 24 on some days inside the Arctic Circle), the low Sun angles prohibit the surface from warming much. Much of this energy is used to melt ice cover and to thaw the frozen soil layer, so less energy is available for sensible heating.

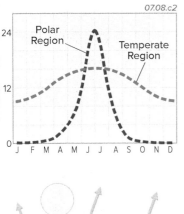

High Albedo

07.08.c3

3. *High Albedo*—Snow and ice surfaces (◄) have high albedo (i.e., reflect a high percentage of the incident solar energy), so a high percentage of the limited insolation is reflected directly back to space without any impact upon these areas. All these insolation-related factors maintain low polar and subpolar temperatures.

Atmospheric Circulation

4. *Continentality*—Northern polar oceanic areas are ice-covered for much, if not all, of the year (►). Extensive sea ice, which can surround the coastline and cover most of the Arctic Ocean, almost functions like a single large continent during some times of the year. The setting causes the North Pole to be characterized by intense continentality, with bitter low temperatures. The South Pole, centered over Antarctica, is a large, cold continent, with a fringe of sea ice that grows in the winter (e.g., July) and shrinks in the summer (December). The South Pole is higher in elevation than the North Pole and is by far the coldest place on Earth.

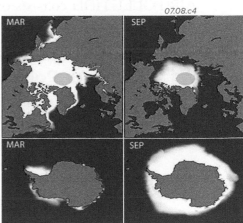

07.08.c4

5. *Atmospheric Circulation*— Strong upper-level westerlies circumnavigate the pole in each hemisphere (►). Air descending from aloft at the polar high is very cold, ensuring that Arctic and Subarctic climates are cold and dry. Arctic and Antarctic (A) air masses form near the poles and can migrate toward the equator as the westerlies dip equatorward in a trough, as is shown here over the eastern U.S.

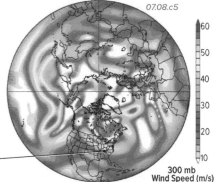

07.08.c5

300 mb
Wind Speed (m/s)

Examples

KRASNOYARSK, RUSSIA *(Dfc)*
Lat 56°1' N - Long 93°4' E

07.08.c6

6. Notice the extreme continentality in this example of a Subarctic climate in Siberia. These climates have the greatest annual temperature range on Earth. Precipitation is low but peaks distinctly in summer at most sites.

07.08.c9 Churchill, Canada

NUUK, GREENLAND *(ET)*
Lat 64°11' N - Long 51°44' E

07.08.c7

7. Greenland is mostly covered by an ice sheet, but a thin strip of coastal land is exposed. This site, Greenland's capital, is located on a peninsula and is influenced by relatively warm local ocean currents for its latitude. It gets more precipitation than most other places with a Tundra climate.

9. Tundra climates also include ice and snow, but they can have unfrozen water during the summer (◄).

10. Antarctica is mostly covered by ice, but rocks protrude through the ice in some mountains (►) and along the coast.

VOSTOK STATION, ANTARCTICA *(EF)*
Lat 78°28' S - Long 106°50' E

07.08.c8

8. Found throughout Antarctica, interior Greenland, and the tops of the highest mountains, Ice Cap climates have unimaginably low temperatures and very low precipitation. This site near the South Pole has never received measurable precipitation.

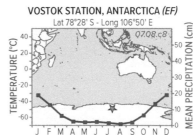

07.08.c10 Antarctica

Before You Leave This Page

✓ Describe or sketch the distribution and characteristics of polar and subarctic climates.

✓ Explain the major differences between Group D and Group E climates in the Köppen system.

✓ Summarize the factors that limit the amount of precipitation in Subarctic and polar climates.

✓ Explain what factors contribute to the very cold temperatures in Subarctic and polar climates.

7.8

7.9 How Does Air Quality Relate to Climate?

AIR POLLUTION consists of gases, liquids, and solids introduced into the atmosphere by human activities and deemed to be detrimental to humans and other creatures, plants, or other aspects of ecosystems. Most air pollution consists of noxious gases and liquids, car exhaust, and smoke and soot from industrial activities and fires.

A What Is Air Pollution and What Are Its Causes?

Air pollution can be in the form of molecules of gas, tiny drops of liquids, or solid particles that are small enough to be lifted into the air. The table at the bottom of the page lists some common air pollutants.

07.09.a1 Shanghai, China

07.09.a2 Wisconsin

07.09.a3

1. Much air pollution is in the form of gases, many of which are invisible. Some gases combine with other chemical compounds to produce visible smog (▲). Most gaseous air pollution consists of compounds of carbon, nitrogen, hydrogen, and oxygen.

2. Air pollution can also be tiny drops of liquid, including those in visible steam (▲) and more noxious liquids derived from sulfur dioxide (SO_2) and other chemicals. Steam is white to gray, but other chemicals and particles in the mix can turn the steam brown or black.

3. Air pollution can also consist of solid particles, including dust, whether it is from a dirt road or from the coal being hauled by this truck (▲). Tiny drops of liquids and solids in the atmosphere are called *aerosols*, but not all aerosols are caused by human activities.

07.09.a4   07.09.a5

07.09.a6 Amazon rain forest

4. Air pollution, whether it is as a gas, liquid, or solid, can be introduced into the atmosphere in various ways. One major source of air pollution is from automobile exhaust.

5. Large quantities of air pollution also arise from industrial activities, such as power-generating stations, factories, petroleum refineries, waste-water treatment plants, and mining.

6. Less industrial human activities, such as burning forests and other vegetation, also introduce gases, especially carbon dioxide (CO_2) and solid particulates, into the air.

Pollutant	Symbol	Characteristics, Source, and Effects
Carbon dioxide	CO_2	Colorless, odorless gas; produced by fossil fuels, burning of vegetation, and natural sources; greenhouse gas
Carbon monoxide	CO	Colorless, odorless gas; produced by combustion of petroleum products; decreases oxygen in blood
Methane	CH_4	Colorless gas with strong odor; produced by digestive processes of animals and production of petroleum; greenhouse gas
Nitrogen oxides	NO, NO_2	Colorless (NO) or yellow and brown (NO_2) gas; produced by combustion at high temperatures and pressures as in automobiles
Ozone	O_3	Unstable gas; produced by photochemical reactions in the atmosphere, as in smog; harmful to plant and animal respiration
Particulate matter	PM	Dust, ash, soot, salt; from surface disturbance, smokestacks, or volcanic activity; respiratory and other health issues
Sulfur oxides	SO_2, SO_3	Colorless gas with rotten-egg smell; combustion of sulfur-containing coal and petroleum, as well as natural sources (e.g., volcanoes); respiratory and other health issues, acid rain
Volatile organic compounds	VOC	Gases and liquids with chemical smells; industrial solvents and other uses, combustion of fossil fuels; various health issues; and involved in ozone formation

B What Are Some Hazardous Pollutants?

A family of atmospheric pollutants known as nitrogen oxides (NO_x) includes NO_2 (nitrites) and NO_3 (nitrates). These pollutants can have numerous environmental and epidemiological consequences once they enter the atmosphere, precipitate, and run off from land surfaces. Automobile engines are the primary producers of NO_x in the atmosphere.

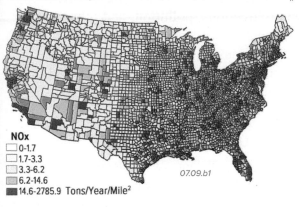

This map of NO_x concentrations by county (◄) shows that the largest metropolises and the sprawling Sunbelt cities that grew during the automobile era have the highest concentrations.

NOx
☐ 0-1.7
☐ 1.7-3.3
☐ 3.3-6.2
☐ 6.2-14.6
■ 14.6-2785.9 Tons/Year/Mile²

07.09.b1

Sulfur dioxide (SO_2) enters the atmosphere primarily through combustion of fossil fuels, particularly low-quality coal. SO_2 causes respiratory infections and diseases, and it damages vegetation and crops. In the 1990s and earlier, the coal-burning industrial regions in the eastern U.S. had the highest SO_2 concentrations in the country.

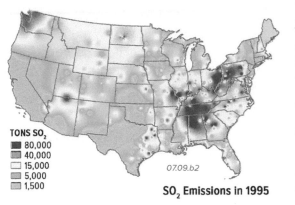

TONS SO₂
■ 80,000
■ 40,000
☐ 15,000
☐ 5,000
☐ 1,500

07.09.b2

SO₂ Emissions in 1995

Smaller red and orange spots in the Southwest, such as on the borders of Arizona, also mark the sites of coal-burning power plants, some of which have caused air pollution in the Grand Canyon and nearby parks.

C What Factors Influence the Severity of Air Pollution?

Many factors influence whether an area has relatively clear, pollution-free air or is heavily polluted. Air pollution starts with a *source* of pollution, but whether the pollution is dispersed (less concentrated but covering a larger area) or is concentrated, like in a single valley, depends on many topographic and atmospheric factors.

07.09.c1 Durango, CO

07.09.c2 Phoenix, AZ

07.09.c3 East Los Angeles, CA

Source—Air pollution comes from some type of source. If it comes from a single, relatively localized site, such as from a single smoke-stack (▲), it is a *point source*. If it comes from multiple sources it is a *non-point source*. How concentrated pollution is depends on whether it is a non-point or point source and how much pollution is actually emitted.

Atmospheric Conditions—Once air pollution is in the air, it is subject to various processes in the atmosphere, especially wind and vertical motion. Winds blow air pollution away from the source, perhaps clearing it from one side of a valley but concentrating it elsewhere (▲). Stable, sinking air, as during a temperature inversion, can trap pollution close to the source and the ground, and a lack of wind limits dispersal.

Topography—The shape of Earth's surface greatly influences winds close to the ground, so wind can trap or disperse air pollution. Shore-lines can also influence pollution distribution, because water and land respond differently to heating, cooling, and winds. Los Angeles, California, has both of these influences—high mountains that trap polluted air in the basin, and a nearby shoreline with inland sea breezes.

The Clean Air Act

Federal air quality legislation in the U.S. began in the 1960s. Initial attempts culminated with the *Clean Air Act of 1970*, which required that the federal Environmental Protection Agency (EPA) be created, along with state-run EPAs (sometimes with different names) in each state. The EPA was to monitor certain widespread (i.e., "criteria") pollutants nationwide with the goal of keeping the concentration of each below a "threshold" level deemed hazardous. Criteria pollutants are a family of solid aerosols known as PM_{10}, NO_2, SO_2, O_3, lead (Pb), and carbon monoxide (CO). The Clean Air Act was subsequently amended many times, generally with stricter standards and heavier penalties to communities for violating these standards. As a result, air quality has generally improved in the U.S. in the last 50 years. But air quality in many other regions of the world has deteriorated.

Before You Leave This Page

✓ Describe the three forms of air pollution and some typical sources.

✓ Describe the factors that influence the severity of air pollution, providing examples of each factor.

7.9

How Do Air Pollution and Urbanization Affect and Respond to the Local Climate?

THE CLIMATE OF A PLACE is a combination of the regional atmospheric conditions, such as those expressed by the Köppen climate classification, and local effects caused by topography, local wind directions, pollution, and other factors. These result in local variations, commonly called the *local climate* or the *microclimate*. The local climate is greatly influenced by how urbanized the area is. The constructed features of cities and towns influence how much insolation heats the surface, which in turn heats the air, and how this heat is retained and released. Cities affect their local climate by being warmer than rural areas, especially at night, a phenomenon known as the *urban heat island (UHI)*.

A   How Do Climate and Human Geography Affect the Distribution of Pollution?

Some emitted pollutants react with sunlight, water, or other contaminants to create a new pollutant—a *secondary pollutant*—that can be more harmful than the original. Climate and physical geography of Earth's surface play a direct role in the degree to which these pollutants are concentrated, but so do the spatial distribution of population density and activities of people, which are part of *human geography*. Sprawling cities and industrial areas are the largest sources of air pollution in the U.S., where air quality has improved steadily since passage of the original Clean Air Act in 1970.

Ozone Pollution

07.10.a1

Sunlight and *volatile organic compounds* can react with NO_x to generate secondary pollutants, including ozone (O_3). The same O_3 that protects us when it is in the stratosphere is toxic near the ground. It eats away things, such as leaf cuticles and human respiratory tissue. Sunny climates are the most

CONCENTRATIONS (Parts per Billion Volume)

0   5   10   15   20   25   30   40   50   60   70

susceptible to O_3 pollution, particularly in summer when the Sun angle is high and daylight hours are abundant. Atmospheric circulation moves this O_3 from one location and altitude to another.

Acid Rain

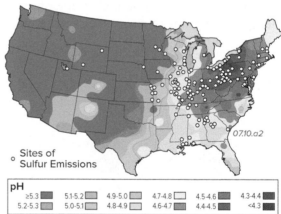

07.10.a2

○ Sites of
  Sulfur Emissions

pH
≥5.3 ■   5.1-5.2 □   4.9-5.0 □   4.7-4.8 □   4.5-4.6 ■   4.3-4.4 ■
5.2-5.3 □   5.0-5.1 □   4.8-4.9 □   4.6-4.7 ■   4.4-4.5 ■   <4.3 ■

Atmospheric water can react with SO_2 or NO_x to produce precipitation that is acidic (has a low pH). This map of average pH of precipitation in the U.S. shows that humid climates in industrial areas are particularly prone to acidic precipitation (*acid rain*).

B   What Is the Impact of Pollution on Energy and Precipitation?

1. Air pollution, in its gaseous, liquid, or solid forms, can interact with the shortwave radiation of insolation. This figure (▶) illustrates how air pollution operates within the atmosphere. When insolation interacts with some component of air pollution, energy is *reflected, scattered*, and *absorbed*. The reflected shortwave energy moves back to space, cooling the atmosphere more than if the pollutant didn't exist.

2. If the pollution is a gas, insolation is scattered in all directions equally. If it is a solid or liquid air pollutant—an *aerosol*—some of the energy is reflected, but much more of it is scattered forward (down to the surface). The downward-scattered radiation does heat the surface, but only to perhaps 90% of what it would have been heated without the aerosols being present.

3. Insolation may also be *absorbed* by the aerosol. This increases the temperature of the aerosol layer but prevents the surface below it from being heated as much. The effect of absorption, reflection, and scattering decreases the environmental lapse rate, stabilizing the atmosphere and reducing the likelihood of precipitation.

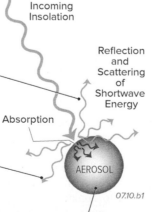

Incoming
Insolation

Reflection
and
Scattering
of
Shortwave
Energy

Absorption

AEROSOL

07.10.b1

4. This is not the whole story, however. Solid aerosols may also enhance precipitation by acting as *condensation nuclei*—solids that attract water and therefore form the center of a growing precipitation droplet. If condensation occurs, latent heat is released. If precipitation results, it makes the surface wetter and cooler, so the two effects partly cancel each other out.

5. Solid aerosols are often identified by their diameter, regardless of chemical composition. For example, PM_{10} stands for "particulate matter with diameters of ≤10 micrometers (μm)." Aside from the impacts on the local energy and water budgets, and therefore the local climate, aerosols impact the human respiratory system by causing difficulty breathing, cancers, and lung-tissue damage, particularly in the very young, the elderly, and people with preexisting respiratory illnesses. Smaller diameters tend to be more dangerous because they can penetrate deeper into the respiratory system.

C What Is an Urban Heat Island and What Causes One to Develop?

When a city is first built or enlarges, changes in land use and associated human construction replace whatever was there previously—natural lands, parks, undeveloped lots, agricultural fields, or some other kind of open space. Buildings, concrete, sidewalks, and glass windows have different albedos and thermal properties than the open space they replaced. As a result, the urban area becomes warmer than the surrounding rural area—the city is an urban heat island (UHI).

1. This figure illustrates some ways in which a UHI forms. Buildings and other constructed features interact in a complex way with incoming shortwave radiation, generally decreasing the albedo in some places, like from dark asphalt, and increasing it in others, such as from light-colored rooftops and reflective glass. Buildings that are close to one another can change the albedo and cause the radiation to reflect from one surface to another.

07.10.c1

4. Urban areas also have a high concentration of waste heat from industrial, domestic, and transportation sources, such as cars, furnaces, lighting, electrical devices of all sorts, and, in colder climates, burning fireplaces. Even air conditioners generate more heat than they remove, adding more waste heat to the local environment.

2. Building materials are designed to allow slow absorption of insolation during the hot daytime to prevent heat from penetrating the interiors of homes and other inside spaces. At night, when outdoor temperatures fall, the energy stored in the materials is slowly released to the atmosphere as longwave radiation. The buildings can hinder the escape of this longwave energy out to space, keeping the nighttime temperatures warmer.

3. Roads, gutters, and storm drains remove water from the surface quickly after rain falls, limiting the amount of water that infiltrates the ground. This prevents water from being evaporated and transpired at the site, and instead allows insolation to raise the temperature of the ground, rather than being converted into latent heat during evaporation. Also, urban areas have fewer tree-shaded areas and less standing water compared with rural areas, so this further allows direct heating of the ground.

5. The UHI is affected by the size of the city, amount and location of green space, and economic activities. Whereas an urban area is influenced by all of the factors described above, less-developed or more rural areas are relatively cooler overall, because they contain more open space, natural or cultivated vegetation, and surface water. Surface water helps limit warming because it results in more insolation being converted to latent heat rather than warming the surface (sensible heat). This map of the conterminous United States (▶) shows how much warmer different urban areas are compared to adjacent, less developed areas, such as grasslands, forests, and farms. The map was generated mostly using satellite data, including the type of land cover (e.g., concrete and asphalt versus natural vegetation). Can you identify what cities generate the largest UHI? Is there an urban heat island in the place where you live or go to school? Can you feel a difference between temperatures in the center of the city and those you experience when you travel just outside a city?

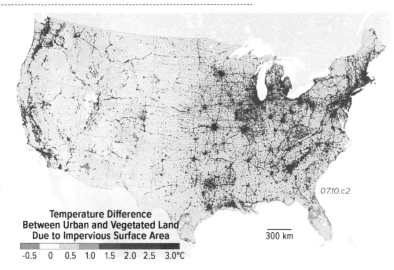

07.10.c2

Temperature Difference Between Urban and Vegetated Land Due to Impervious Surface Area

300 km

-0.5 0 0.5 1.0 1.5 2.0 2.5 3.0°C

Other Impacts Associated with Urbanization

Urbanization affects local environments in many other ways. In addition to increasing the temperature over the city (UHI), urbanization generally lowers the relative humidity because of higher temperatures and lack of surface water available to be evaporated. An exception to this occurs in some desert cities, like Phoenix, Arizona, where an abundance of irrigation canals, artificial lakes, golf courses, and watered lawns can raise humidities compared to the surrounding desert—an *oasis effect*.

Urbanization can also change wind speeds and directions. Wind speeds are generally lower in cities because buildings block winds and provide friction that slows the wind. Winds can be stronger locally if they blow parallel to streets, allowing them to be channeled between buildings, in what is called an *urban canyon*.

Since urbanization modulates temperature and pressure, it can also affect the amounts and spatial distribution of precipitation. For example, precipitation is generally greater downwind of a city because abundant particulates introduced into the air by urban activities provide nuclei for condensation. Fog can be reduced in cities by changes in temperature and humidity, but smog, which is related to pollution, is concentrated near cities.

Before You Leave This Page

✓ Describe how climate and human geography impact the severity of pollution.

✓ Describe the impacts of pollution, especially aerosols, on energy and precipitation.

✓ Explain what an urban heat island is, the features of a city that cause one to form, and other climatic impacts of urbanization.

7.10

7.11 What Is the Evidence for Climate Change?

OVER THE LAST 150 YEARS, people have measured atmospheric temperatures. This record, albeit short in geological terms, shows an overall increase in temperatures—*global warming*. There is currently much scientific and political discussion of this topic. What is the evidence that Earth's climate is changing?

A What Is Climate Change?

Earth's climate has been changing since the planet formed 4.55 billion years ago, varying over timescales of decades to millions of years. Climate change can include global trends in warming, cooling, precipitation, wind directions, and other related measures. Global warming means *increasing* global atmospheric and oceanic temperatures from some point in the past to the present, usually compared to an *arbitrary* mean global temperature (e.g., averaged from 1961 to 1990). Scientists examine various records of Earth's climate to investigate past changes. In addition to direct measurements, we infer past climatic conditions from other types of observations, called *proxy evidence*.

Thermometer Record

Thermometers provide a direct measurement of air temperature. This record shows an average variation in temperature for the last 140 years. According to this record, it appears that average air temperatures have increased over the last century. From the 1940s to the 1970s, the data show a relatively cool period.

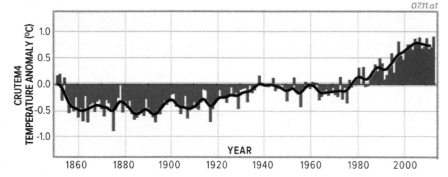

Temperature data collected in the last decade show less warming on a global scale since 1998, as shown by the downtick on the right end of the graph. Temperatures in the U.S. have been warmer overall, and global and U.S. temperatures remain well above the long-term average, at higher levels than have been recorded since we began recording temperatures.

Sea-Surface Temperatures

Another direct measurement of temperature is sea-surface temperature (SST), which is collected from buoys, ships, and more recently by satellites. Observe the SST graph below and then compare it to the air-temperature data to the left.

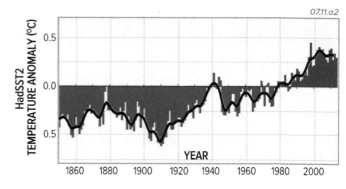

SST data show a relatively cold period prior to 1940, and strong overall warming from the 1940s to about 1998, when the rate of warming slowed but remained above the long-term average. These are the same trends displayed by the air-temperature data.

Glacier Length

Glaciers flow from areas of snow accumulation to lower elevations. The dynamics and energy flow of glacier movement and retreat are well understood. Most scientists interpret changes in the lengths of glaciers to be related to changes in atmospheric temperature and in the amount of precipitation. The combined

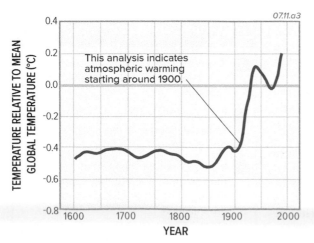

This analysis indicates atmospheric warming starting around 1900.

data from glaciers around the world are interpreted as a warming trend beginning before the turn of the 20th century. Note that changes in glacier length started around 1850, and much melting occurred prior to 1940. Recent warming has been accompanied by another increase in the rate of melting.

Ice Core Proxy

Continental glaciers on Greenland and Antarctica have yearly layers that record winter precipitation and summer dust accumulations. Scientists extract ice cores by drilling, and then they chemically analyze the gases and ice of these cores in refrigerated laboratories. Air trapped in tiny bubbles provides samples of the atmosphere (including CO_2 concentrations) back to at least 100,000 years ago. Oxygen and hydrogen *isotopes* in ice provide a proxy for temperature.

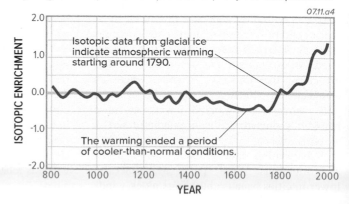

Isotopic data from glacial ice indicate atmospheric warming starting around 1790.

The warming ended a period of cooler-than-normal conditions.

Other Proxies

We can use other types of proxies, such as sediments deposited in lakes and oceans, minerals precipitated in caves, and tree rings. Tree-ring growth is partly dependent on climate. Some trees can grow for 300 years or more, and the thickness of tree rings of successive populations from temperature- and precipitation-sensitive forests can provide a climate record going back hundreds or thousands of years.

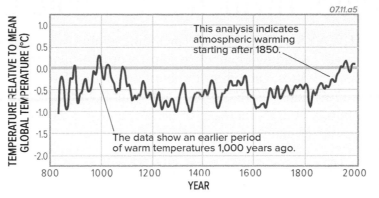

07.11.a5

This analysis indicates atmospheric warming starting after 1850.

The data show an earlier period of warm temperatures 1,000 years ago.

The record from a selection of other proxies shows an increase in air temperature between 1850 and 1998. Prior to this, global temperatures were lower, as part of a long cold period called the *Little Ice Age*. Before that, around 1,000 years ago, was a warm period, which is called the *Medieval Warm Period*.

Comparing Proxies

The National Academy of Sciences (NAS) published the curves in the chart to the right as part of a report to Congress. The report summarizes and compares the various temperature records shown on earlier graphs and some reconstructions produced by combining multiple types of proxies. The plot shows direct measurements, such as the temperature record, as well as the various types of proxy data, each in a different color. The curves have been smoothed to emphasize overall variations and show strong similarities in shape. Comparing the different types of data strengthens the case that global warming has occurred since the mid-1800s. The NAS concluded that Earth's atmosphere has warmed 0.6 C° in the last 100 years.

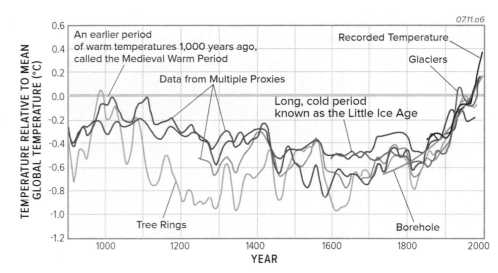

07.11.a6

An earlier period of warm temperatures 1,000 years ago, called the Medieval Warm Period

Data from Multiple Proxies

Recorded Temperature

Glaciers

Long, cold period known as the Little Ice Age

Tree Rings

Borehole

Using Satellites to Investigate Temperature Changes

Increasingly, satellites are used to measure and monitor changes in temperature and other expressions of climate with instruments designed to measure different aspects of change. For example, some satellites measure temperature, moisture, or cloud cover in the lower atmosphere, whereas others measure temperature, abundance of different gases, and attributes of higher parts of the atmosphere, like the stratosphere.

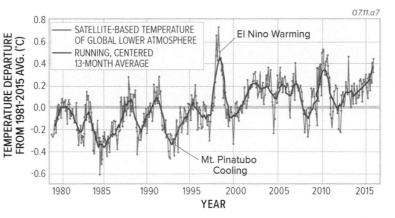

07.11.a7

SATELLITE-BASED TEMPERATURE OF GLOBAL LOWER ATMOSPHERE
RUNNING, CENTERED 13-MONTH AVERAGE

El Nino Warming

Mt. Pinatubo Cooling

This plot (◄) shows monthly averages of satellite measurements of temperatures in the lower atmosphere, along with a red curve fit through the data. The data show an overall warming trend, with shorter periods of cooling and warming, including a peak warm period around 1998, which coincided with a strong El Niño event.

Weather, Climate, and the Scientific Method

There is much discussion, and controversy, about one important aspect of climate change—*global warming*. As shown on these two pages, evidence for global warming since the mid-1800s is clear, but there is active debate about how much of the warming is attributable to human introduction of greenhouse gases (mostly CO_2) and how much is due to natural causes. Scientists are always testing, refining, or refuting new, old, or even widely accepted hypotheses—it is part of the scientific method. One misleading and nonscientific tactic used by members of the general public and media, on both sides of the global warming debate, is to point to a specific weather event, like a single hurricane or tornado, a Midwestern blizzard, or a lack of snow in some region, and say that this proves or disproves, or is a direct consequence of, global warming or the lack thereof. Such short-term events are *weather*, not *climate*, and have a complex and still incompletely understood relationship to climate change.

Before You Leave This Page

✓ Describe what climate change means and some ways it can be measured.

✓ Summarize the major lines of direct measurement and proxy evidence indicating global warming in the last 100 years.

✓ Explain conclusions using the combination of different temperature reconstructions.

7.11

7.12 What Factors Influence Climate Change?

CLIMATE CHANGE IS ALWAYS OCCURRING, including global warming since the mid-1800s. There are many natural causes of climate change, including changes in Earth's orbit around the Sun and changes in solar activity. Many scientists propose that human activities, including the burning of fossil fuels and the clearing of forests, contribute to climate change by releasing greenhouse gases to the atmosphere. Other factors may lead to *global cooling*, including ash from large volcanic eruptions and an increase in certain aerosols in the atmosphere. Here, we examine some of the factors that can influence our climate.

A   What Processes Influence Atmospheric Temperature Change?

Earth's surface temperatures are dominated by energy from the Sun. Insolation heats the oceans, land, and atmosphere, but several factors influence how much of this energy reaches the surface and how much is retained.

Interaction of Insolation with Earth's Atmosphere, Oceans, and Land

1. Nearly all of Earth's heating at the surface comes from insolation, which heats the atmosphere, land, and oceans. Most of this energy escapes eventually back into space in the form of longwave infrared energy. The rest is delayed by interactions with Earth, keeping the planet warm by a process called the *greenhouse effect*. The amount of insolation hitting Earth varies regularly, by a small amount, due to orbital fluctuations and changes in the Sun's energy output, as expressed by changes in sunspot activity. Sunspots are the darker areas that, on average, appear and disappear on the surface of the Sun in an 11-year solar cycle.

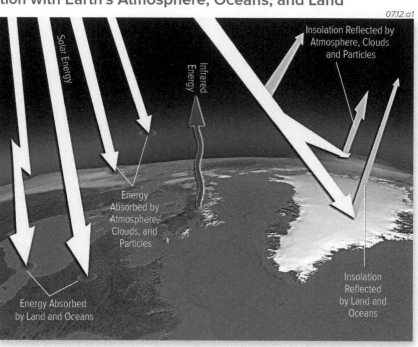

07.12.a1

2. Some insolation is *absorbed* by the atmosphere (shown as an orange disk in the figure), and some is *reflected* off the atmosphere. Much of the reflected insolation returns to space without heating Earth or its atmosphere, as depicted by the blue arrows.

3. Insolation is absorbed by clouds, by soot from burning, and by fine particles (aerosols), which are produced by volcanoes, industry, and automobiles. Some of this absorbed energy radiates back into space as *infrared* (longwave) *energy* (shown by the wavy red arrow). Clouds and particles also reflect some insolation.

4. Some insolation is reflected back to space from the land surface and oceans. Ice in continental glaciers is an effective reflector. As glaciers melt, darker land or ocean is uncovered. This increases the amount of energy that is absorbed by Earth and subsequently re-radiated back to the atmosphere.

5. Some insolation is absorbed by the land and the oceans, both of which then radiate infrared (longwave) energy back into the atmosphere. Some of this infrared energy is absorbed by atmospheric gases, such as water vapor (H_2O), carbon dioxide (CO_2), methane (CH_4), and nitrous oxide (N_2O), which are called *greenhouse gases*. Some portion of these gases is produced naturally and some is produced by human activities.

Orbital Variation

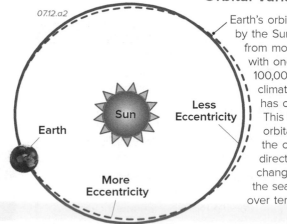

07.12.a2

Earth's orbit is affected almost exclusively by the Sun and Moon, causing it to cycle from more elliptical to more circular paths with one cycle completed over about 100,000 years. This influences Earth's climate, but the current global warming has occurred in less than 200 years. This is too short a time period for orbital changes to have caused all of the observed warming. The tilt and direction of Earth's axis of rotation also change, which changes the contrast of the seasons, but these effects occur over tens of thousands of years.

Variations in Solar Radiation

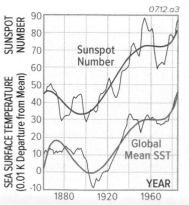

07.12.a3

The Sun is a candidate for causing climate change. Solar activity, as presented by the number of sunspots, clearly correlates to earlier changes in temperature (◄), but less so to recent warming. Cosmogenic isotopes in ice cores and tree rings (proxies) suggest that the Sun's energy emission has varied by only 1% over the last 1,000 years. Both data sets are relevant when evaluating the Sun's role in recent warming.

Greenhouse Gas Production

1. Several gases in Earth's atmosphere absorb infrared radiation emitted by Earth. This causes them to vibrate and heat up, and then to emit infrared radiation. This radiation can escape into space or be absorbed by other greenhouse gases, mostly lower in the atmosphere where the concentration of greenhouse gases is the highest. Since 1957, atmospheric scientists have collected air samples on the high peaks of Hawaii. The data for Mauna Loa are plotted on this graph (▶), which shows CO_2 content in the atmosphere as a function of time. The concentration of CO_2 in air, represented by the red line, has increased by 20% in the last 40 years. Most of this increase is attributed to humans, especially through the burning of fossil fuels.

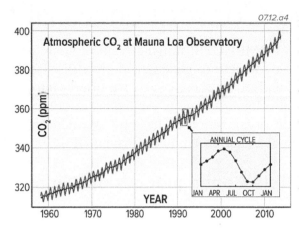

07.12.a4

2. Our oceans are a huge reservoir for dissolved gases and other chemical components, including CO_2. As the CO_2 of our atmosphere increases, a large amount of this goes into the oceans, where it affects seawater chemistry, especially the acidity. This *ocean acidification* and its impact on sea life lead to concerns about increases in atmospheric CO_2. Seawater can dissolve less CO_2 when it is warm, so it releases CO_2 back to the atmosphere as it warms, as is occurring now. So some increase in the atmospheric CO_2 could be a *result* of recent warming.

Greenhouse Gases and Temperature Change Records from Ice Cores

3. Long-term records of CO_2 and other greenhouse gases are contained within ice cores and can be traced back hundreds of thousands of years. Ice cores from Antarctica have been analyzed for CO_2, isotopic temperature, and isotopic age. The data, shown on this graph (▶), allow comparison between prior natural variations and changes in the last several hundred years.

4. How well do CO_2 concentrations in the atmosphere track temperature change? On this graph, the two trends are very similar. Graphs of other greenhouse gases, such as CH_4 versus temperature and N_2O versus temperature, show a similar correspondence. These large fluctuations in the past have natural causes.

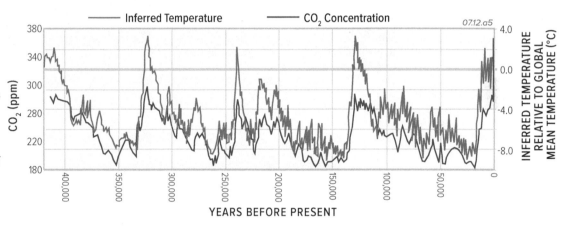

07.12.a5

5. Scientists have used data from ice cores and other proxies to interpret how atmospheric gases changed in the past few hundred years. These data indicate that levels of CO_2, CH_4, and N_2O have increased since the start of the Industrial Revolution of the last part of the 18th century. Many scientists infer that these increases in greenhouse gases are partly responsible for the recent increase in temperature, but there remains debate about how *much* our human activities have affected, or will affect, our climate.

Ocean Oscillations, Climate Change, and Climate Models

Recent discoveries have changed our view of the relationship between ocean currents and climate. Studies of the oceans have documented that separate ocean basins (e.g., the Pacific Ocean) display variations in temperature that occur over years or decades—*ocean oscillations*.

As discussed in the previous chapter, one long-recognized oscillation is the El Niño–La Niña effect and is called the El Niño–Southern Oscillation (ENSO for short), and another is the Pacific Decadal Oscillation (PDO). More recently, we have discovered other ocean oscillations, including the Atlantic Multidecadal Oscillation (AMO). As the name of the PDO and AMO imply, these oscillations

may last decades, whereas ENSO cycles last a year or two. Variations in sea-surface temperatures (SST) and atmospheric pressure can be used to calculate a numerical *index* for each oscillation, as a way to represent the changes. The graph here shows values for the Atlantic Ocean (AMO), relative to its long-term average. The AMO represents SST in the Atlantic basin—high numbers indicate warm SST. This

graph shows that Atlantic SST undergoes cycles that last several decades. These cycles also relate to periods of abundant or rarer tropical cyclones, and they also show some similar patterns to global temperatures. Because of such correlations, climate scientists are increasingly examining whether oscillations are primarily a cause or an effect of climate change.

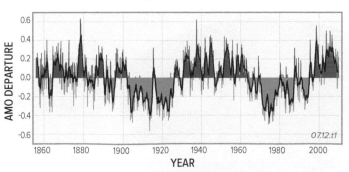

07.12.t1

Before You Leave This Page

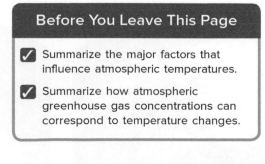

✓ Summarize the major factors that influence atmospheric temperatures.

✓ Summarize how atmospheric greenhouse gas concentrations can correspond to temperature changes.

7.12

What Are the Consequences of Climate Change?

CLIMATE CHANGE HAS MANY IMPACTS, ranging from obvious ones like an increase in global temperatures, to less obvious ones, such as a possible increase in malaria and other diseases. Some of these consequences are highly probable, whereas others are very speculative. On these pages, we briefly introduce some of the most likely results of climate change, specifically those related to global warming. Abundant information is available elsewhere, including reports by governmental and nongovernmental organizations. The most detailed and widely cited reports are those by the United Nations-sponsored Intergovernmental Panel on Climate Change (IPCC).

A What Are Some Possible Consequences of Climate Change to the Environment?

1. The figure below illustrates some consequences of global warming, but many more are possible. Examine the figure and think about what you know about the feature being depicted and how it might respond to an increase in average temperatures. Then read the text.

2. The most obvious result of global warming would be an *increase in global temperatures*. Computer models of climate indicate that many areas, especially those near the poles, will indeed increase in temperature, but other regions might get colder.

3. Higher temperatures would be predicted to cause more *melting* of snow and ice, resulting in glaciers melting back and becoming less extensive, as has been observed.

4. Higher temperatures and altered ocean circulation could lead to more drought in some places and even the expansion of desert areas, the process of *desertification*.

5. Changes in global temperatures and accompanying changes in precipitation patterns can affect the *distribution of communities* of plants and animals. In response to warming, such communities may shift to higher elevations or higher latitudes to stay within an optimal temperature range. Climate change is predicted to decrease the geographic range of some communities, while increasing the range of others. Included in this consideration are croplands, some of which will benefit from global warming, while others will suffer.

6. An increase in global temperatures is predicted to increase *evaporation* from warmer surface waters. This would increase both humidity and water-vapor capacity of the air. Warmer temperatures and more humidity may make the atmosphere less stable and increase *precipitation*. More precipitation, along with more melting of glaciers, would lead to more runoff from streams, with an associated increase in the amount of flooding. Increased runoff also means a larger influx of fresh water into the ocean.

7. Climate change has the potential to change the frequency and intensity of *severe weather.* For various reasons, global warming can either increase or decrease the frequency of certain types of severe storms. These are presented on the next page.

8. Global warming is predicted, and observed, to cause an increase in global sea-surface temperatures (SST) and a decrease in the amount of *sea ice* in the Arctic.

9. The various consequences have complex interactions, as between SST, stability, humidity, precipitation, runoff, and influx of fresh water into the ocean. Such interactions could result in other changes, like a change in ocean currents, including those involved in the *thermohaline conveyor*, an important moderator of global climate.

07.13.a1

10. The computer-generated images below show the decrease in the amount of sea ice in the Arctic from 1979 (left globe) to 2015 (right globe), which encompasses a period of global and regional warming. The amount of Arctic ice displays some year-to-year increases but shows an overall decrease with time.

11. Since the middle of the 1800s and before, sea level has been slowly increasing, rising 0.2 meters in 200 years. Currently, there is scientific debate about whether the rate of sea level rise is remaining the same or is accelerating with time.

1979

07.13.a2

2015

07.13.a3

Median extent of sea ice on September 11, 2015, for the 30-year period of 1981–2010.

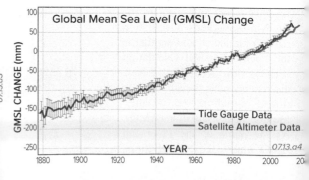

Global Mean Sea Level (GMSL) Change

GMSL CHANGE (mm)

— Tide Gauge Data
— Satellite Altimeter Data

YEAR   07.13.a4

1880 1900 1920 1940 1960 1980 2000 20

B How Might Climate Change Promote More Severe Weather?

In 2005, an unprecedented 28 named tropical storms formed in the Atlantic, and 19 formed in 2012. Yet we have also had many quiet years in recent times. Patterns such as these have led people to wonder whether we are moving into an era of more numerous and destructive tropical cyclones, with the increase attributed to climate change. What does the science say about these important questions? Observed increases in the frequency of all forms of severe weather are at least partially the result of Doppler radar and other improved remote-sensing technology, which allow us to detect storms more easily. One approach is to compile data on severe storms over time, to examine whether the frequency and severity of storms correlates with observations of climate change. But remember that simple correlation does not demonstrate causation.

1. Observational and modeling studies suggest that climate change will increase SST, which in turn might increase the intensity and duration of the strongest storms. Warming might also lengthen the severe tropical cyclone season. In theory, increased SST would allow an increase in evaporated water, which increases the amount of latent heat in the atmosphere. Since latent-energy release—when the evaporated water condenses or deposits—is a key ingredient for any type of severe weather, increased evaporation can result in more energy for storms. The end result is larger, taller clouds fueling more powerful storms.

2. Warmer sea and land surfaces are also likely to affect atmospheric stability. Recall that with warmer air near the surface, the environmental lapse rate will steepen. A steep environmental lapse rate enhances the ability of air to rise, increasing the instability of the atmosphere. In other words, warmer air near the surface will contribute to taller clouds as long as moisture is present.

3. Global warming could cause changes to upper-level circulation, which in turn could cause underlying areas to experience more frequent precipitation or drought. Changes in upper-level circulation patterns might make the upper levels of the atmosphere more or less favorable for cyclone formation. Specifically, if climate change displaces or decreases the flow of upper-level westerlies, fewer developing tropical cyclones may have their tops sheared off as they move westward. Alternatively, global warming could increase vertical wind shear, which might cause mid-latitude storms to be more severe. What do the data say?

07.13.b1

Frequency of Tropical Storms

4. The graph below plots the number of Atlantic storms for each year. The time periods represented on the graph correspond to the interval of time when global temperatures have demonstrably increased (although in detail, temperatures have gone up and down during several approximately 30-year cycles). Research using these and similar data suggests that short-lived Atlantic tropical cyclones may have become more frequent over the years, but moderate-duration storms have not.

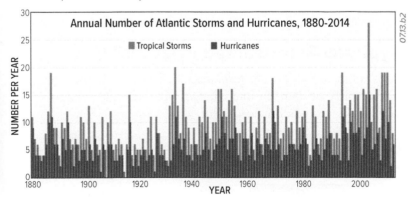

5. The research on this question is ongoing, with some studies predicting an increase in intensity, while others find a decrease in intensity. Other studies acknowledge that changes in tropical cyclones under climate change scenarios will be small in relation to other factors that create variability in tropical cyclones, such as ENSO.

Frequency of Tornadoes

6. Another question to evaluate is whether global warming has caused an increase in the number of mid-latitude cyclones or tornadoes. Some scientists have proposed that warming provides additional energy that might enhance the environment for such severe storms. Examine the graph below, which plots the number of tornadoes per year during the time when most warming occurred.

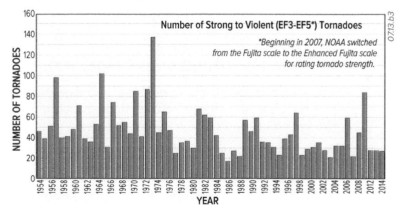

7. The data indicate that the number of strong-to-violent tornadoes in the U.S. has not increased over time. One possible explanation is that increasing surface temperatures would presumably be accompanied by increases in upper-level temperatures, leading to no net change in stability, as the adiabatically rising air and the air around it have both warmed.

Before You Leave This Page

✓ Sketch and explain some possible consequences of climate change, specifically global warming, to the environment.

✓ Discuss some ways that climate change could increase the frequency and strength of severe weather and some relevant data.

7.13

7.14 How Do We Use Computers to Study Climate Change?

COMPUTER MODELS are used to investigate issues related to climate, like global warming. The simplest climate models compute energy (temperature), water, or momentum for a small area over a limited time. The most complicated models use numerical methods and principles of physics to compute atmospheric conditions at many vertical levels across the entire Earth, for a long period of time. These are called *general circulation models* (GCMs). A key issue in climate modeling and prediction is understanding the nature of positive and negative feedbacks in the climate system.

A What Are GCMs and How Do They Approach Climate Problems?

Modeling Approach

1. As in weather forecasting models, which were discussed elsewhere, data are collected from a variety of sources, including aircraft, satellite imagery, balloon launches, and surface stations.

2. Data are quality controlled, to eliminate obvious errors.

3. Equations are run to simulate atmospheric variables at each point for a set time into the future.

4. The "answer" at each grid point becomes the starting point for the next iteration of calculations for another time into the future.

5. After a prediction far enough into the future is generated, the results are scrutinized and plotted on maps.

Modeling Procedures

6. The same general data sources are used for climate prediction as for weather forecasting. However, because small-scale weather features can create quite a bit of havoc on short time scales, it is generally more important to have a finer mesh of data for weather forecasting models. Both types of models are generally so complicated that they are run on supercomputers or clusters of networked, high-powered computers.

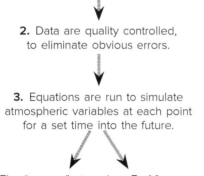

07.14.a1

7. Spatial analysis techniques are used to interpolate the data to a three-dimensional grid, with several vertical levels. The gridded data are run through the model, using the same seven basic equations that are used in weather forecasting models.

8. As in weather forecasting models, *spectral models* may be used. These use grid points initially, but then resolve atmospheric motion into a series of sine and cosine waves, which then become amplified or dampened based on the input of energy, matter, and momentum. The results of the spectral analysis then become re-interpolated to a grid.

B What Are Feedbacks and How Are They Typically Represented in GCMs?

Among the most important—and complicated—aspects of global climate models are *feedbacks*. A feedback is the way a system responds to a change in conditions, which in turn acts to amplify or dampen that change. Most climate scientists would agree that feedbacks are among the least understood parts of the global climate system.

1. These two figures were presented and discussed elsewhere, but they provide a useful review of feedbacks in the context of modeling. The left figure illustrates that we can think of natural phenomena as systems, containing matter and energy, with individual parts that interact in some ways more than others. In studying and modeling a system, we want to understand the *inputs* to some part of the system, in this case the input of moisture into the atmosphere from the water and ice (shown by the upward arrows). Then, we want to model, using physical principles, the *response* of this change to the system, such as to increased precipitation.

07.14.b1   07.14.b2

2. One way the system can respond to the imposed changes is through positive and negative feedbacks. A *positive feedback* amplifies those changes (causes more change in the same direction), while a *negative feedback* drives the system in the opposite direction, undoing some or all of the changes. In the figure shown here, a decrease in the amount of sunlight and an accompanying decrease in temperature cause an increase in the amount of snow and ice, which due to its high albedo reflects more insolation, causing even more cooling, a positive feedback.

C What Are Some Positive and Negative Feedbacks in the Climate System?

GCMs consider feedbacks, but they are difficult to understand and model because a chain reaction of effects may result from a single change. Shown below are some important feedbacks modeled in modern GCMs. Some of these feedbacks are not as straightforward, nor as well understood, as these simple figures imply.

Changes to the Surface

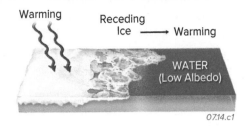

07.14.c1

1. Warming temperatures could melt more ice. Ice has a higher albedo than water, so replacing ice with water causes the surface to absorb more heat, causing even more warming in a positive feedback.

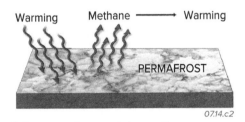

07.14.c2

2. Polar permafrost contains methane, a greenhouse gas. Warming the permafrost releases the methane, which causes more warming by absorbing more outgoing longwave radiation.

Water Vapor and Clouds

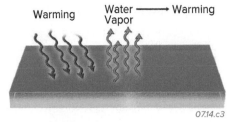

07.14.c3

3. Warming the oceans evaporates more water, the most abundant greenhouse gas. The extra water vapor could cause more warming by absorbing outgoing longwave radiation, which could evaporate more water vapor, which . . . you get the idea.

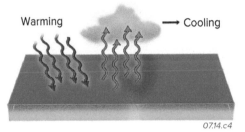

07.14.c4

4. But more water vapor should produce more low-level clouds, which reflect insolation. More water vapor and more clouds, therefore, could cause cooling, a negative feedback.

CO_2 In and Out of the Oceans

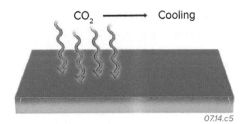

07.14.c5

5. Carbon dioxide (CO_2) goes in and out of the oceans, which hold over 50 times more CO_2 than the atmosphere. Any CO_2 that is emitted into the atmosphere but goes into the oceans reduces the rate of atmospheric warming.

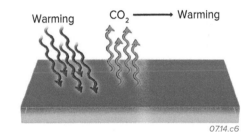

07.14.c6

6. If the oceans warm, however, CO_2 in the oceans may eventually be released back to the atmosphere through other processes, leading to more warming, a positive feedback.

The Importance of Feedbacks in Climate Models

To model past climate change and try to predict future climate change, GCMs and other climate models have to incorporate the effects of both positive and negative feedbacks, like those described above. The introduction of additional CO_2 into the atmosphere, such as by burning coal and other fossil fuels, has well-understood effects in the atmosphere, resulting in some amount of warming. Doubling the amount of CO_2 is predicted to cause approximately 1C° of warming, if there are no feedbacks. Such a modest temperature increase, by itself, might not be a cause for concern, since temperature fluctuations of this and much larger magnitudes have occurred many times in Earth's past, especially when viewed over geologic time scales.

A critical question involves how the Earth system will respond to such an increase in temperatures. If the warming results in a large amount of melting of ice, the resulting change in albedo can cause even more warming, a positive feedback. If the original warming causes the formation of more low-level clouds, the reflective nature of such clouds could be a negative feedback, limiting or even reducing the amount of warming. It is fair to say that if we don't understand feedbacks, it is impossible to model the behavior of Earth's complex climate system accurately.

For this reason, much scientific research has been and currently is investigating the feedbacks that are likely to have the largest influence on our climate. Scientists are studying the amount of ice in ice sheets and glaciers on land and in sea ice that forms from freezing of the surface of the ocean in polar climates. Some of this research can be done using satellites, but some is done by physically going onto ice-covered regions and observing and measuring what is happening. Atmospheric scientists are using satellites and other methods to measure the amount of water vapor at different levels in the atmosphere, since water vapor is also a greenhouse gas and is much more abundant than CO_2. If warming causes an increase in water vapor, the warming will be amplified, a positive feedback. As we better understand the feedbacks, we will better understand, and be able to model more accurately, the incredibly complex system that controls our climate.

Before You Leave This Page

✓ Explain what a GCM is, what kind of data it uses, and how the data are used in the model.

✓ Sketch and describe some of the major feedbacks incorporated in most GCMs.

✓ Explain the importance of feedbacks on GCMs and other climate models.

7.14

How Does the Climate System Sustain Life?

GLOBAL ENERGY, HYDROLOGIC SYSTEMS, AND BIOLOGIC SYSTEMS are interdependent, linked by processes, like photosynthesis, evapotranspiration, and albedo, through a system of complicated feedbacks. How does the climate stay at conditions favorable for life, despite the fact that it is always changing? Alternatively, can the biosphere help sustain a favorable environment through such self-regulating processes? This possibility is demonstrated in an Earth-like planet populated only by black or white daisies—*Daisy World*. Daisy World illustrates the connections between various components of the climate, including life.

A How Are Energy, Water, and Plants Linked in Daisy World?

1. On this hypothetical planet, Daisy World, there are two kinds of daisies—white daisies and black daisies (▶). The black daisies have a low albedo, so they absorb more insolation than do white ones. In hot or bright conditions, however, the dark color is a disadvantage because the black daisies retain so much insolation that they overheat. The white daisies reflect more insolation, and so they do better in hot environments, but not as well in cold ones, where they do not get enough energy. There are no "daisy-eaters," so the relative survival rates of the two daisies depend only on how they interact with the climate.

In too-cool times, black daisies out-compete white daisies.

Overheating

Daisy World

Initial Condition

In too-warm times, white daisies outcompete black daisies.

Overheating makes it too warm for black daisies.

Dynamic Equilibrium

Overcooling

Overcooling makes it too cold for white daisies.

07.15.a1

2. After observing the setup of Daisy World (◀), examine the figure below, which illustrates how energy, water, and the daisies interact to keep the climate within certain bounds of temperature and water vapor. The letter and number on each arrow refer to text adjacent to the figure. E arrows involve energy, W involves water, and B refers to the biosphere (daisies). Refer to chapters 2, 3, and 4 to review key topics discussed below.

Energy

E1. Insolation is identical in quantity and quality to that of Earth.

E2. Cloud albedo is similar to that on Earth, which ranges from 50% to 70%. The area of clouds varies, and is generated by the planetary water subsystem.

E3. The proportion of insolation reflected to space by clouds depends on the type and extent of clouds, which are controlled by the water subsystem.

E4. Insolation not reflected by clouds proceeds down toward the planetary surface.

E5. The surface is determined by planetary land cover (biosphere). The surface reflects a proportion of incident insolation back to space.

E6. Total energy absorbed by the biosphere is the sum of insolation (E5) and the re-radiated longwave radiation called *counter-radiation* (E10).

E7. Insolation returned to space, or planetary albedo, comprises energy reflected from changing cloud and land surface covers.

E8. Energy that does not leave via planetary albedo must exit as longwave radiation, controlled by the temperature of the planet and the efficiency at which emission occurs for the gas.

E9. Water vapor (water subsystem) is the only greenhouse gas in this imaginary world. Its quantity (clouds) therefore controls temperatures at the base of the clouds.

E10. Longwave radiation is radiated back to the surface by the atmosphere.

Water

W1. Like Earth, Daisy World is considered to be a closed system with respect to water, so no water leaves or enters the planet.

W2. Cloud cover is a function of the water vapor stored in the atmosphere (controlled by temperature and the energy subsystem), and the imbalance between evaporation inputs and precipitation outputs. When Precipitation (P) < Evaporation (E), cloud cover grows; when P > E, cloud cover shrinks.

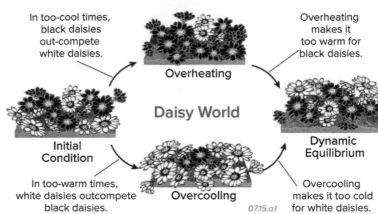

07.15.a2

W3. Evaporation is driven by latent heat flux (governed by the energy and biosphere subsystems), and by the quantity of water already in the atmosphere.

W4. Precipitation is positively related to cloud cover and to the average amount of precipitation.

W5. Temperature at the cloud base, which controls E10, is governed by surface temperature (biosphere subsystem) and the unsaturated or saturated adiabatic lapse rate.

W6. In reality, the adiabatic lapse rate is controlled by moisture content, defined by the planetary cloud cover.

Biosphere

B1. Black and white daisies share a constant mortality rate. Growth rates decline as temperatures above each daisy type diverge from a common optimum. In this way, the areas occupied by the two shades of daisy change with changes in the energy and water subsystems.

B2. Planetary surface temperature is a combination of temperatures over bare ground and black and white daisies, proportional to their planetary areas and albedos. Temperature rises as the number of black daisies increases, and vice versa.

B | How Do Biophysical Features on Daisy World Interact?

Variables in Daisy World—and on Earth—all interact with, and adjust to, one other. Some adjustments are almost instantaneous on an annual scale, like changes in planetary albedo, while other linkages, such as the daisy growth rate, take time to fully impact the system. This diagram illustrates some of the ways in which Daisy World could evolve through time, through positive and negative feedbacks between energy, water, and biology. In many natural systems the variables are constantly adjusting to one another because of time lags between the input and response.

Constraints to the Model

1. The air above black daisies is always warmer than above white daisies because of the black daisies' lower albedo (25%) and ability to retain insolation compared to that of white daisies (75%).

2. Planetary temperatures tend toward those associated with the prevailing daisy type (warmer over black daisies), although some of the planet is covered by bare ground.

3. As temperatures above each type of daisy approach or diverge from the optimum (22.5°C here), growth rates increase or decrease.

4. Mortality rates of all daisies remain fixed, thus proportions of the planet covered by each daisy do not change instantaneously. However, there is an accelerated changeover from white to black daisies as higher temperature above black daisies approaches the optimum, while temperatures above white daisies are getting colder, diverging from the optimum.

5. Higher proportions of black daisies reduce planetary albedo and rapidly warm the planet.

Feedbacks Within Daisy World

6. Suppose that an area has a higher proportion of black daisies. What do you think will happen to temperature and atmospheric moisture in this scenario?

7. The black daisies have a lower albedo, so they reflect less insolation back to space and retain more of the energy, heating up the surface. If water is available near the surface, a warmer planet may lead to more evaporation, more water vapor in the air, more clouds, a more active hydrologic cycle, and a higher proportion of the planet covered in clouds.

8. Greater cloudiness increases planetary albedo and reduces the insolation reaching the daisies. However, water vapor is a greenhouse gas, thus the presence of additional cloud cover, while increasing global albedo, also restricts cooling from the loss of outgoing longwave radiation.

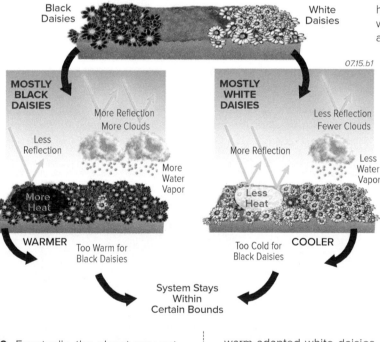

07.15.b1

9. Eventually, the planet may get too hot for the heat-absorbing black daisies, so more black daisies die, causing the planet to cool, which then favors the white daisies.

10. Suppose instead that an area has a higher proportion of white daisies. Predict what will happen to temperatures and atmospheric moisture in this scenario.

11. The white daisies have a higher albedo, so they reflect more insolation back to space, retaining less of the energy. This causes the surface to cool. A cooler planet has less evaporation, less water vapor, fewer clouds, and a less active hydrologic cycle. Less of the planet is covered in clouds, and there is less heat-trapping greenhouse gas. The proportion of white daisies continues to grow.

12. Following the peak abundance of white daisies, the high surface albedo (white daisies) and other factors cause a decrease in temperatures, decreasing the abundance of warm-adapted white daisies. At this point, the growth rate of white daisies declines at an ever-increasing rate, while that of the black daisies grows. A repetition of the cycle has then been initiated. Does Earth function in a similar way, with positive and negative feedbacks that keep the systems within certain bounds? We hope so!

Before You Leave This Page

✓ Describe the link between characteristics of the two types of daisies in Daisy World and how this can alter temperatures, atmospheric water, and the proportion of white and black daisies, at both local and global scales.

✓ Explain what happens with the hydrologic cycle, particularly evaporation and cloud cover, in Daisy World as the planetary temperature increases, and describe the effects on planetary energy budget and planetary temperatures.

✓ Describe the concept of a planet as a self-regulating system, operating within certain limits because of positive and negative feedbacks. You can build your explanation around Daisy World or around Earth.

7.16 What Climates and Weather Would Occur Here?

PLANET C is a hypothetical replica of Earth. It has oceans and an atmosphere very similar to Earth's. You will map the climatic zones on this planet, identifying cold and warm ocean currents, prevailing winds, potential locations of rain forests and deserts, places where the climate and topographic setting would be suitable for agriculture, and sites at risk for hurricanes.

Goals of This Exercise:

- Observe the general map patterns and photographs from different parts of the planet. Note any clues about likely weather and climate.
- Create a map showing probable climatic conditions, based on ocean currents and prevailing winds.
- Locate likely sites for rain forests, deserts, agricultural areas, and places where storms will strike land.

Procedures

Follow the steps below, entering your answers for each step in the appropriate place on the worksheet or online.

1. Observe the distribution of continents and oceans on the map below and on the larger version on the worksheet. Examine the photographs of the various areas and infer the environments and climates that are represented. Each photograph has a letter that corresponds to a letter on the map. Some photographs are on the next page.

2. Carefully examine the various types of data on the next page. For each data set, examine the climatic and weather implications for different parts of the planet.

3. Draw on the worksheet your interpretations of whether each ocean current is warm, cold, or neither, and whether prevailing winds will bring warmth or coolness, and cause dryness or precipitation.

4. Map the main climatic zones. Include the likely locations of rain forests, deserts, areas suitable for croplands, and areas at risk for hurricanes and other major tropical cyclones. Be prepared to discuss your observations and interpretations.

5. Your instructor may provide you with climographs so that you can assign Köppen climate types to different regions.

07.16.a2

07.16.a1

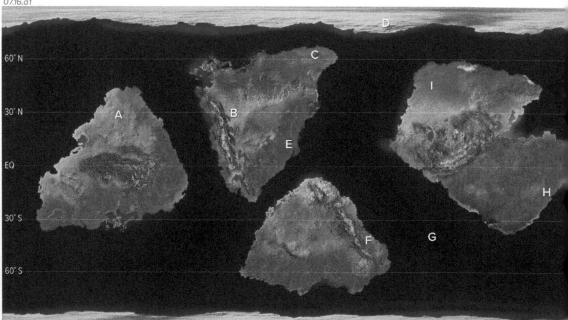

07.16.a3

07.16.a4

07.16.a5

07.16.a6

07.16.a7

Some Important Observations About the Planet

1. The planet rotates around its axis once every 24 hours in the same direction Earth does. Therefore, it has days and nights similar to Earth's. It orbits a sun similar to our own, at about the same distance. Its spin axis is slightly tilted relative to the orbital plane.

2. The oceans are likewise similar to those on Earth. They show similar variations in sea-surface temperature (SST) and have ocean currents that circulate around the edges of the main ocean basins and in gaps between the continents.

3. The planet has well-developed atmospheric currents, with wind patterns similar to those on Earth.

Preliminary Interpretation of Ocean Currents

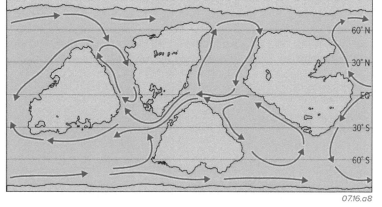

07.16.a8

This map shows the directions of ocean surface currents. From these data, identify which currents are probably cold, warm, or neither. Water temperatures will be key in inferring probable climates.

Preliminary Interpretation of Wind Data

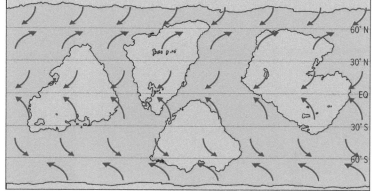

07.16.a9

These data show the directions of prevailing surface winds, which are similar to the patterns on Earth. Different wind directions occupy bands encircling the planet, parallel to the equator.

Ocean Water Temperatures and Sea-Surface Pressures

07.16.a10

This map depicts satellite data for sea-surface temperatures (SST) during late summer. The numbers give the temperature values in degrees Celsius (°C).

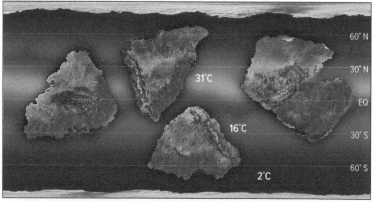

31°C

16°C

2°C

07.16.a11

07.16.a12

Climate Analysis

On the version of this map on the worksheet, show the probable locations of the following features by writing the associated letters on the map: rain forests (R), deserts (D), potential agricultural areas (A), and lands at highest risk for hurricanes (H). You can use colored pencils to shade in the extent of land associated with each letter. Your instructor may also have you assign likely Köppen climate types to different sites or regions. Complete this step on the worksheet.

07.16.a13

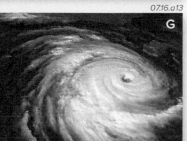

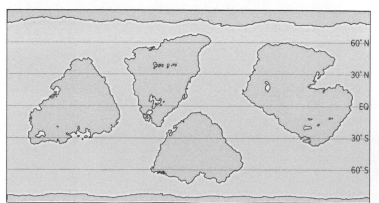

07.16.a14

CHAPTER
8 Water Resources

WATER IS ARGUABLY THE MOST IMPORTANT RESOURCE—life on Earth needs water to live and thrive. We are most familiar with *surface water,* water that occurs in streams, lakes, and oceans. Yet, the amount of fresh water in these settings is much less than the amount of fresh water that is frozen in ice and snow or that occurs in the subsurface as *groundwater.* This chapter is about surface water and groundwater and the important ways in which they interact.

08.00.a2 Shoshone, ID

The Snake River Plain, shown in this large satellite-based image, is a curved swath of low land that cuts through the mountains of southern Idaho. It contains a mixture of dry, sage-covered plains, water-filled reservoirs, green agricultural fields, and recent lava flows of dark-colored basalt, a volcanic rock. The Snake River winds its way through the plain. Most of Idaho's population lives on the Snake River Plain near the rivers and reservoirs.

The Big Lost River, Little Lost River, and adjacent streams that enter the plain from the north never reach the Snake River. Instead, the water from the rivers seeps into the ground between the grains in the sediment and through narrow fractures in the volcanic rocks. For this reason the rivers are called "lost."

Where does water that seeps into the subsurface go?

▲ **Within the Snake River Plain**, the Snake River has eroded a canyon down into layers of ledge-forming volcanic rocks and slope-forming sediment. The farmlands of the canyon bottom are on fertile sediment deposited by the river, and they receive water from rivers, springs, and rainfall, and from wells drilled to extract groundwater.

Lost River Area

Craters of the Moon National Monument

Snake River

Thousand Springs

Snake River

At Thousand Springs, large springs gush from the steep volcanic walls of the Snake River Canyon (◄). The canyon includes 15 of the 65 largest springs in the U.S., including those at Thousand Springs. The largest commercial trout farms in the U.S. use ponds fed by these springs.

What causes water from beneath the ground to flow to the surface as a spring, and where does the water in a spring come from?

08.00.a3 Thousand Springs, ID

TOPICS IN THIS CHAPTER

The Snake River winds through mountains and then flows west across the Snake River Plain. *Where does this river, flowing across such a dry plain, receive its water?*

08.00.a4 Jackson Hole, WY

The river begins its journey in Jackson Hole, Wyoming (▲), from streams that drain the Grand Tetons and nearby Gros Ventre mountain range. The relatively higher rainfall and snowmelt in these highlands sustain the river as it flows westward across the dry plains. Farther downstream, rivers and streams entering the plain from the east and south flow directly into the Snake River, increasing its flow.

Where does the water in rivers come from, and do most rivers gain or lose water from groundwater?

Yellowstone National Park

GRAND TETONS

Snake River

ke River

ocatello

25 km

08.00.a1

Disappearing Waters of the Northern Snake River Plain

Groundwater beneath the Snake River Plain is an essential resource for the region, providing most of the drinking water for cities and irrigation water for farms and ranches away from the actual river. Geoscientists study where this water comes from, how it moves through the subsurface, and potential limits on using this resource.

Some water enters the subsurface from the Big and Little Lost Rivers, which flow into the basin from the north and then disappear as their water sinks into the porous ground. Other groundwater comes directly from the main Snake River and from tributaries that enter the basin from the south and east. The largest influx of water to the subsurface is seepage from irrigated fields and associated canals.

The surface of the Snake River Plain slopes from northeast to southwest. The flow of groundwater follows this same pattern, flowing southwest and west through sediment and rocks in the subsurface. Groundwater derived from the disappearing rivers flows southwest, along the north side and center of the basin. The groundwater does not flow like an underground river but as water between the sediment grains and within fractures in the rocks. Where the Snake River Canyon intersects the flow of groundwater, water reemerges on the surface, pouring out at Thousand Springs. This region illustrates a main theme of this chapter—surface water and groundwater are a related, interconnected, valuable, and vulnerable resource.

Many lakes and farms are situated next to the Snake River. Farmers irrigate millions of acres of agriculture with surface water derived from reservoirs, lakes, streams, and groundwater pumped to the surface. Chemicals used by some farms cause contamination of groundwater and surface water.

What happens if groundwater is pumped from the subsurface faster than it is replaced by precipitation and other sources? What do we do if water supplies are contaminated?

8.0

8.1 Where Does Water Occur on the Planet?

WATER IS ABUNDANT ON EARTH, occurring in many settings. Most water is in the oceans, but is salty. Most fresh water is in ice and snow or in groundwater below the surface, with a smaller amount in lakes, wetlands, and streams. Water also exists in plants, animals, and soils and as water vapor in the atmosphere.

A From Where Did Earth's Water Come and Where Does It Occur Today?

Most water on Earth probably originated during the formation of the planet or from comets and other icy celestial objects that collided onto the surface, mostly early in Earth's history. Over time, much of this water moves to the surface, for example when magma releases water vapor during eruptions.

1. *Oceans*—Of Earth's total known inventory of water, an estimated 96.5% occurs in the oceans and seas as *saline* (salty) *water*. The remaining 3.5% is *fresh water* held in ice sheets and glaciers, groundwater, lakes, swamps, and other water features on the surface.

2. *Streams*—Streams are extremely important to us and are the main source of drinking water for many areas. They contain, however, only a very small amount of Earth's fresh water.

3. *Lakes*—Water occurs in lakes of various sizes. Many are freshwater lakes, but those in dry climates are saline or *brackish* (between fresh and saline). Lakes contain a majority of the liquid fresh water at Earth's surface, but most of this water lies in a few large lakes around the world.

4. *Swamps and Other Wetlands*—These wet places contain water lying on the surface and within the plants and shallow soil. They consti-tute about 11% of the liquid fresh water on the surface.

5. *Groundwater*—About 30% of Earth's total fresh water occurs as groundwater (water in the subsurface below the reach of tree and plant roots). Groundwater is mostly in small, open pores between sediment grains or within fractures that cut rocks. Most groundwater is fresh, but some is brackish or saline.

10. *Atmosphere*—A small, but very important, amount of Earth's water is contained in the atmosphere (0.001%). It occurs as invisible water vapor, as water droplets in clouds, and as rain, falling snow, and other types of precipitation.

9. *Glaciers*—Nearly 69% of Earth's fresh water is tied up in ice caps, glaciers, and permanent snow. A small amount also exists in *permafrost* and ground ice.

8. *Soil Water*—Earth's soils contain about as much water as the atmosphere (not much), but like water in the atmosphere, soil water is crucial to our existence.

7. *Biological Water*—Water occurs within the cells and structures of plants and animals. It is clearly important to us but represents an exceptionally small percentage of Earth's total water (0.0001%).

6. *Deep-Interior Waters*—An unknown, but perhaps very large amount of water is chemically bound in minerals of the crust and mantle. Some scientists think Earth's interior may contain more water than the oceans.

08.01.a1

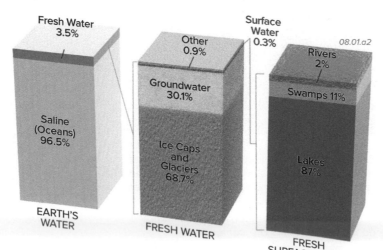

08.01.a2

11. These bar graphs show estimates of the distribu-tion of water on Earth's surface and the uppermost levels of the crust (▶). The left bar shows that the oceans contain 96.5% of Earth's total free water (water that is not bound up in minerals), but this water is saline. Only 3.5% of Earth's water is fresh (not saline). These graphs do not include water in Earth's atmosphere or deep interior.

12. The middle bar shows that most of Earth's fresh water occurs in ice caps and glaciers (◀). Almost all the rest is groundwater. Less than 1% occurs as liquid surface water in lakes, swamps, and streams.

13. The right bar shows where Earth's small percentage of fresh, liquid, surface water resides (◀). Most is in lakes, followed by swamps and rivers.

B How Does Water Move from One Setting to Another?

Water is in constant circulation on Earth's surface, moving from ocean to atmosphere, from atmosphere back to the surface, and in and out of the subsurface. The circulation of water from one part of this water system to another, whether above or below Earth's surface, is called the *hydrologic cycle*. From the perspective of living things, the hydrologic cycle is the critical system on Earth. It involves a number of important and mostly familiar processes. It is driven by energy from the Sun.

1. *Evaporation* — As water is heated by the Sun, some of its molecules become energized enough to break free of the attractive forces binding them together. Once free, they rise into the atmosphere as water vapor.

2. *Condensation* — As water vapor cools, such as when it rises, it condenses into a liquid or turns directly into a solid (ice, hail, or snow in deposition). These water drops and ice crystals then collect and form clouds.

3. *Precipitation* — When clouds cool, perhaps when they rise over a mountain range, the water molecules become less energetic and bond together, commonly falling as rain, snow, or hail, depending on the temperature of the air. Precipitation may reach the ground, evaporate as it falls, or be captured by leaves and other vegetation before reaching the ground.

4. *Sublimation* — Water molecules can go directly from a solid (ice) to vapor, a process called sublimation (not shown here). Sublimation is most common in cold, dry, and windy climates, such as some polar regions.

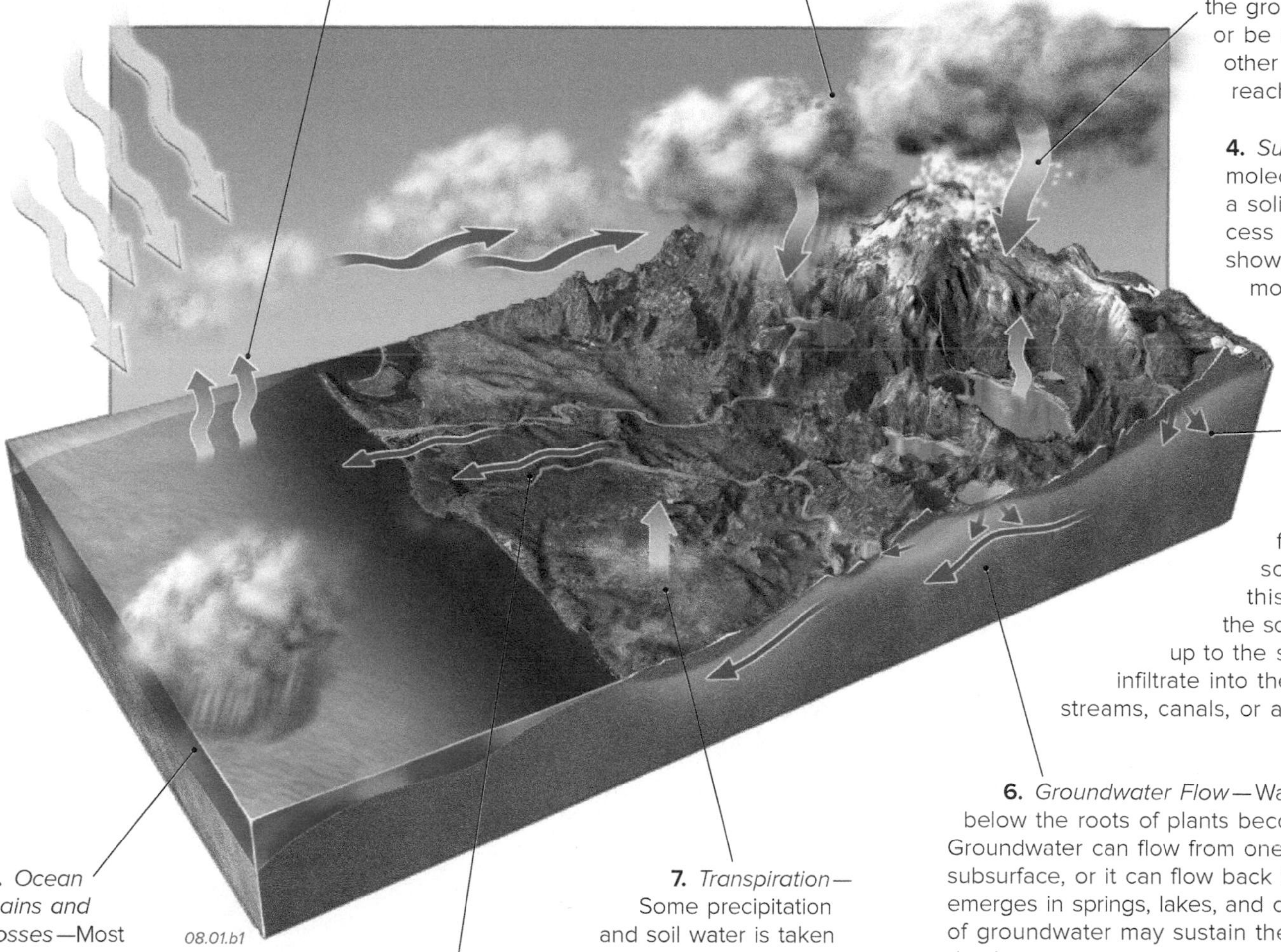

5. *Infiltration* — Some precipitation seeps into the ground, infiltrating through fractures and pores in soil and rocks. Some of this water remains within the soil, and some rises back up to the surface. Water can also infiltrate into the ground from lakes, streams, canals, or any body of water.

6. *Groundwater Flow* — Water that percolates below the roots of plants becomes groundwater. Groundwater can flow from one place to another in the subsurface, or it can flow back to the surface, where it emerges in springs, lakes, and other features. Such flow of groundwater may sustain these water bodies during dry times.

9. *Ocean Gains and Losses* — Most precipitation falls directly into the ocean, but oceans lose much more water to evaporation than they gain from precipitation. The difference is made up by runoff from land, keeping sea level close to a constant level.

08.01.b1

7. *Transpiration* — Some precipitation and soil water is taken up by root systems of plants. Through their leaves, plants emit water that evaporates into the atmosphere by the process of transpiration.

8. *Surface Runoff* — Rainfall or snowmelt can produce water that flows across the surface as runoff. Runoff from direct precipitation can be joined by runoff from melting snow and ice and by groundwater seeping onto the surface. The various types of runoff collect in streams and lakes. Most runoff is eventually carried to the ocean by streams where it can be evaporated, completing the hydrologic cycle.

Before You Leave This Page

☑ Summarize where most of Earth's total water resides.

☑ Sketch and describe the hydrologic cycle, summarizing the processes that shift water from one part to another.

8.1

What Is the Global Water Budget?

THE HYDROLOGIC CYCLE is key to many physical and biological processes on Earth as water moves in and out of different parts of the system. It also results in the conversion of solar energy to other forms of energy, such as when water that evaporated from the ocean later falls as rain, forms a river, and flows into the sea. Here, we summarize the amounts of water in different settings and the movement of that water from one setting to another. Knowing the specific amounts of water that is moving is much less important than understanding the processes and relative amounts.

A How Much Water Moves Between the Global Stores of Water?

1. When considering the motion of water in the Earth system, we want to know how *much* water a part of the system—called a *store*—holds, expressed as a volume (e.g., km³). We also measure how much water moves in and out of that store over some amount of time, which is a *flux*, in this case described with units of volume per time. The three main stores are the oceans, continents, and atmosphere (▶).

2. Annually 577,000 km³ of water evaporates and falls from the atmosphere. The daily quantity of energy involved in this process is approximately two million times the worldwide energy-generating capacity.

4. In terms of fluxes between the oceans and atmosphere, the ocean store would be reduced by 7% (84% minus 77%) annually.

ATMOSPHERE (12.9; ≈0.001%)

PRECIPITATION OVER OCEAN (77% of Total)

PRECIPITATION OVER LAND (23% of Total)

Difference = 7%

EVAPORATION OVER OCEAN (84% of Total)

Difference = 7%

EVAPORATION OVER LAND (16% of Total)

OCEANS (1,338,000; 97%)

GLOBAL RUNOFF (7% of Total Flux)

CONTINENTS (47,660; 3%)

UNITS IN 1,000 KM³

08.02.a1

5. Approximately 23% of global precipitation falls over continents. This is a little less than the percentage of the Earth's surface that continents occupy. However, only 16% of global evaporation comes from the continents—large areas of the continents have arid (dry) climates, and much continental water is stored as groundwater or as ice and snow in regions of low insolation.

6. In terms of fluxes, input of precipitation to the continents exceeds evaporation output by 7% annually. That is, every year, the continents gain water from their exchange with the atmosphere. Without further outputs, the stores of fresh water on the continents should be increasing.

7. The imbalances between the atmosphere-ocean and atmosphere-continent exchanges (difference in

3. The ocean holds about 97% of Earth's known water, and evaporation over oceans accounts for 84% of all global evaporation. About 77% of Earth's precipitation falls on the oceans. These high percentages emphasize the significance of oceans as moderators of global temperatures.

inputs and outputs) are redressed by the flow of excess waters from the continents to the oceans by all the rivers in the world—*global runoff*. Note that approximately two-thirds of all water falling on the continents leaves as evaporation and one-third leaves by streams. Evaporation from the global oceans is 12 times larger than the combined flows of all the streams in the world.

B What Factors Control How Quickly Water Moves Through Each Store?

The amount of water in the stores is of practical interest, but says little about the dynamic nature of the planetary surface. Movement of water between global hydrologic stores represents energy available to do work in shaping the Earth's surface. A measure of the "dynamism" of each store is the average length of time that a drop of water remains in that store—its average *residence time*.

1. Consider a box (a store) with 10 spheres (think water molecules), including one of particular interest in red. In each time period, one random sphere is removed (output) and a new sphere inserted (input). On average how many time periods will pass before the red sphere is selected (i.e., how long will the red sphere stay in the store)?

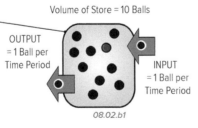

Volume of Store = 10 Balls

OUTPUT = 1 Ball per Time Period

INPUT = 1 Ball per Time Period

08.02.b1

Volume of Store = 10 Balls

OUTPUT = 2 Balls per Time Period

INPUT = 2 Balls per Time Period

08.02.b2

Volume of Store = 20 Balls

OUTPUT = 1 Ball per Time Period

INPUT = 1 Ball per Time Period

08.02.b3

2. Application of statistical theory indicates that the red sphere will most likely be removed (output) during the time period over which 10 spheres could have been output. It is less likely that the sphere would remain in the store longer than this. The average length of time that the sphere stays in the store is 10 time periods.

3. When the rates of inputs and outputs are increased (e.g., 2 spheres per time period), the chances of selecting the red sphere increase and the average time period declines to 5.

4. If the size of the store (number of spheres) is increased to 20, then the chance of selecting the red sphere at any time period declines and the average time period increases to 20.

5. By making a few assumptions, like the size of the system remaining constant, we can calculate the average residence time for a molecule if we know the volume of the store and the average rate of flow into and out of the store (average rate of movement). Such calculations are useful in evaluating local and global water budgets.

C What Are Average Residence Times in the Global Stores of Water?

1. Water moves at different rates through each different subsystem (e.g., evaporation from the ocean) and its stores (e.g., ocean and atmosphere). This figure depicts the average rates at which a water molecule moves from one store to another. This rate, in turn, indicates how long, on average, that molecule will stay in the store—its *residence time*. To calculate the residence time, we simply divide the total volume of the store by the flux leaving it.

2. The volume of water in the atmospheric store is very small (12,900 km³), yet 577,000 km³ of water enters and leaves it annually, yielding average residence times of approximately 0.02 yr (i.e., 12,900/577,000), or 8 days. Thus, once evaporated into the atmosphere from the planetary surface, a water molecule stays there on average a very short time. The atmosphere is therefore very responsive to changes in energy conditions at Earth's surface.

PRECIPITATION ONTO OCEAN (444/yr)

PRECIPITATION ONTO LAND (133/yr)

EVAPORATION FROM OCEAN (485/yr)

EVAPORATION FROM LAND (92/yr)

SURFACE RUNOFF (40/yr)

UNITS IN 1,000 KM³

08.02.c1

4. Water stored in and on the continents (47,660,000 km³) is smaller than the amount of water in the oceans but much larger than the amount in the atmosphere. The input is precipitation and the outputs are evaporation and runoff. The relatively limited spatial extent of continents causes the average annual fluxes to be considerably smaller than between ocean and atmosphere. The combination of smaller fluxes and the intermediate size of the continental store results in an average residence time of approximately 350 years. Recall that the vast majority of fresh water on continents is stored in groundwater and ice, both of which may "lock up" water for centuries. Here, we have ignored water that percolates into the soil and moves down to become groundwater.

3. Oceans are by far the largest store (1,338,000,000 km³). Inputs (precipitation and runoff) total 484,700 km³/yr, as does the output of evaporation. This yields an average residence time on the order of 3,000 years (1,338,000,000/484,700). The oceans are stratified vertically (warmer on top than on bottom) and evaporation varies geographically, but once water enters the ocean store it will likely stay there an average of several thousand years. Oceans therefore respond more slowly to changes, despite the fact that the atmosphere and oceans are strongly linked through evaporation and precipitation.

D How Are Fluxes of Water Linked to Shaping of Earth's Surface?

When we imagine the processes that reshape landscapes, we might think of events like powerful floods, scouring glaciers, turbulent mudflows, and blowing sand. All of these involve energy resulting from the movement of liquid water, ice, Earth materials, or the atmosphere. This type of energy is called *kinetic energy*—energy by virtue of motion. Also linked to the flux of water in some settings is *potential energy*—energy by virtue of being some height above a surface and capable of moving if allowed.

1. Examine this photograph of a balanced rock (▶), which obviously has the potential to fall sometime in the future—the rock has *potential energy* relative to the ground below (the datum). We can imagine how hard the rock would strike the ground when it fell, imparting a large force onto the ground it strikes. Clearly, this rock has a large amount of potential energy.

2. With the help of gravity, the rock's potential energy is converted into *kinetic energy* (the energy of motion) when the rock begins to fall. The resulting motion of the rock can perform *work*, for instance the crushing of stones on which the rock falls.

3. After the rock has fallen and is in its new, lower position, it has lost its potential energy relative to the ground surface (the datum). We could increase its potential energy if we used a crane to lift the rock back onto its pedestal. This would require an expenditure of energy by the crane, and that energy expenditure is now stored in terms of the greater elevation and therefore greater potential energy. Note that once the rock is on the ground it could be moved by running water, such as that released by the thunderstorm in the background. The relatively short average residence times of water (calculated in the section above) show that water can quickly evaporate from the ocean, be lifted by storms, and be pulled by gravity back to the Earth as precipitation, renewing the water's ability to move across the surface.

08.02.d1 Arches NP, UT

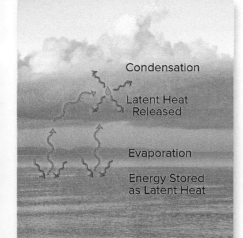
08.02.d2

Condensation

Latent Heat Released

Evaporation

Energy Stored as Latent Heat

4. By means of latent heat transfer and the evaporation of water, the Sun's energy is able to renew the potential energy of water molecules by lifting them tens of thousands of meters above the datum (sea level). Following condensation and the formation of precipitation, the molecules can fall on high-elevation topography to begin the journey downslope, picking up kinetic energy. The short residence time within the atmospheric store ensures that this process is repeated over and over again with a very high frequency.

Before You Leave This Page

☑ Sketch and explain the comparative magnitudes of the hydrologic fluxes between the global stores of water.

☑ Explain what is meant by the average residence time, and how the size of the store and magnitude of fluxes affect it.

☑ Explain potential and kinetic energy and how these could relate to the short average residence time of water in the atmosphere.

8.2

8.3 How Do We Evaluate Water Balances?

THE CONCEPT OF A WATER BALANCE is critical in understanding our world and relates to a broad range of topics, from evaluating the health of an ecosystem to gauging the sustainability of a city. Just as an accountant monitors the amount and flow of money in financial transactions, the water balance is an accounting of the amount of water in each store and its rate of movement through the hydrologic cycle. The *climatic water balance* allows us to estimate water supplies and flows. Computer models based on some simple equations are used to assist us in calculating the climatic water balance at a local place.

A How Can the Gain or Loss of Water at the Surface Be Estimated?

To determine the water balance, we need to know how much is entering the system (the input) and how much is lost from the system (the output). On the previous two pages, we did this at a global scale, determining how continents were gaining water through precipitation and losing an equal amount through evaporation and runoff—they were in balance. Equally important is determining the water balance at regional and local scales, but first we need to know the inputs and outputs.

08.03.a1

Input (Gain)

1. Water is gained at the surface by precipitation (P) or by runoff that enters the area, bringing in water that originated as precipitation from someplace else. Relatively simple instruments are usually used to measure P (in cm or inches), including a standard rain gauge (◄). Various factors cause measurements of P to be smaller than reality, including the effects of turbulence, water evaporating from inside the tube, and water drops that cling to the sides of the gauge and so are not measured.

2. When precipitation falls as snow, we can measure it with (1) a vertical rod marked with distances above the ground, (2) electronic height-measuring sensors, or (3) devices called snow pillows (►), which typically are several-meter-wide antifreeze-filled plastic sacks that measure the weight of snowpack and translate it into centimeters of water.

08.03.a2

3. More sophisticated tools, such as automated precipitation stations in the Automated Surface Observing System (ASOS) (◄), can measure and record P in real time. Radar is used to derive fast estimates of P, using a network of facilities. The resulting maps typically show highest rainfall amounts in orange and red. Satellites are also increasingly being used to measure P, especially in remote areas of the world.

08.03.a3

4. There is a vast network of varied weather stations across the U.S. and the more populated areas in the world (►).

08.03.a4

Output (Loss)

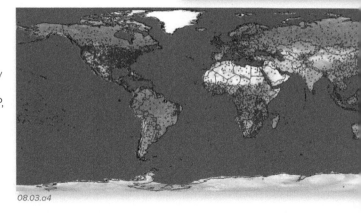

08.03.a5

5. Water is lost from the surface to the atmosphere through direct evaporation or through *transpiration*—evaporation of water that has moved from a plant's roots, up through the stem, and out from a leaf's surface. It is common to consider evaporated and transpired water together as a single process—*evapotranspiration*. The theoretical maximum rate at which water that can be evapotranspired from a wet surface, in centimeters or inches of precipitation over a given time period, is known as *potential evapotranspiration* (PE). PE is commonly estimated based on factors such as latitude, which determines Sun angle (solar intensity) and length of daylight hours, and temperature, because warmer air provides more energy to drive the evapotranspiration process. If P (precipitation) exceeds PE at a place for a certain period of time, there is "extra" water.

6. Other factors influence how much water is potentially or actually lost via evapotranspiration. For example, drier air and faster wind speeds both allow for more evapotranspiration. This globe (◄) shows average annual evapotranspiration values across the Americas. More water is lost from regions that have more surface water and plants, such as the Amazon rain forest of South America, the tropical regions of central America, and the Southeast U.S. Much less is lost in drier regions, such as in the Arctic and in deserts of the Southwest U.S.

B What Is the Influence of the Ratio of Precipitation to Potential Evapotranspiration?

When an area receives more precipitation than it can lose through potential evapotranspiration, the excess water must go somewhere. In contrast, there are many areas where potential evapotranspiration (PE) routinely exceeds the amount of precipitation (P), and these are classified as arid and are commonly deserts.

P Greater than PE

1. When rain falls on the surface or snow melts, the resulting surface water can go several places. It can *evaporate* directly back into the air, as would be common in an area with a high PE. Or, it could instead be taken up by plants, commonly through the roots, to later be released back into the air during *transpiration*.

2. The water could also accumulate in the soil. After the soil becomes saturated, any extra water is called *surplus* and begins flowing downhill as *surface runoff* in streams and in less constrained sheets of water flowing across the surface (*sheetflow*). The running water can leave the area, carrying the surplus water away, or can accumulate locally in lakes and wetlands. Such *standing water* can later evaporate back into the atmosphere.

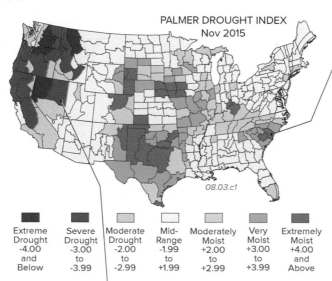

08.03.b1

3. Water that soaks into the soil is called *soil-water recharge*. This water exists in the part of the subsurface where much of the pore space is mostly filled with air rather than water—the *unsaturated zone*. Soil water can be pulled back up to the surface by roots or by capillary action, or it can continue deeper into the subsurface until it eventually enters a zone where water fills nearly all the pore spaces and fractures—the *saturated zone*— becoming groundwater. This is where most water occurs in the subsurface. The top of the saturated zone is the *water table,* shown as a dashed red line. As with groundwater, the "excess" water can flow away.

PE Greater than P

4. During periods when PE exceeds P, in addition to evaporating the precipitated water directly, the atmosphere will evapotranspire water that is stored in soil in an attempt to meet its demand for water PE. This is called *soil-water use*. Soil

08.03.b2 Baja California Sur, Mexico

can only hold a certain amount of water against gravity, termed its *field capacity*. A typical field capacity is about 15 cm (6 in) of equivalent precipitation. During times when PE exceeds P, the water stored in soil can be drawn down so much that the only soil water remaining strongly adheres to rock particles because of surface tension. Plants cannot easily get to this water, so this last amount of water storage in the soil is termed the *wilting point*. Like many other water-related measurements, wilting point is measured in centimeters or inches of precipitation. Once soil water reaches the wilting point, any further unmet demand for water—the part of PE that is not provided by P or by soil-water use—is called *deficit*.

C What Are Some Applications of the Climatic Water Balance?

1. Weekly, the U.S. Climate Prediction Center runs a climatic water-balance program at hundreds of weather stations, using the most recent measurements of P and calculations of PE. The model calculates surplus or deficit for each station. The calculated values are then compared to normal surplus or deficits for that station and that time of year, and the results are mapped as the *Palmer Drought Index* (▶). Drought occurs when deficit exceeds "normal" deficit for that location at that time of year.

08.03.c2 Yuma, AZ

PALMER DROUGHT INDEX
Nov 2015

08.03.c1

Extreme Drought −4.00 and Below	Severe Drought −3.00 to −3.99	Moderate Drought −2.00 to −2.99	Mid-Range −1.99 to +1.99	Moderately Moist +2.00 to +2.99	Very Moist +3.00 to +3.99	Extremely Moist +4.00 and Above

3. Places with abundant surplus may have other issues. Flooding may be a problem, and navigation on major rivers can be hazardous in such conditions. In extreme conditions, pressure on dams and levees can create major concerns. Places without large deficits or surpluses compared to "normal" are least likely to experience environmental stresses.

2. Places that are having a larger deficit than normal at that time of year for that location are considered to be in *drought*. In such places, forest fires may be a problem and planners may limit water use. Irrigation (◀) tries to supply enough water to keep agricultural soils above the wilting point, but this becomes more difficult and expensive during a drought.

Before You Leave This Page

✓ Explain various ways in which precipitation (P) is measured.

✓ Explain the concept of potential evapotranspiration and what happens when PE is greater than or less than P.

✓ Explain how we determine whether an area is experiencing a water deficit (drought) or a water surplus.

8.3

8.4 How Do Water Balances Vary Spatially?

A WATER-BALANCE DIAGRAM portrays the water balance of a place during any interval of time, identifying times when the area has a water-balance deficit versus times when it has an excess. Such diagrams allow us to better understand and manage our water resources and recognize limitations that water availability places on ecosystems and society. By comparing water-balance diagrams for different regions, we gain insight into how water resources vary as a function of climate and other factors.

1. These globes show the regional distribution of areas with a climatological water-balance surplus (purple and blue) and deficits (orange and brown). Examine the larger patterns on these globes and think about how the various climate types, such as arid and tropical, are expressed. Arranged around the globes are smaller graphs that show the climatological average water balance for each month within the year, each of which is a water-balance diagram. The diagrams begin with January (the "J" on the left) and then repeat January (the "J" on the right).

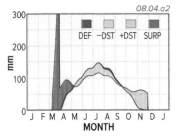
08.04.a2

2. On the globes, a blue-green area would be considered a "wet" climate, whereas the orange and yellow areas would be considered to have a "dry" climate. This does not mean that the wet climate would be rainy all the time or that a dry climate might not have some precipitation.

3. A water-balance diagram (◄) plots the average calculated water balance over time, like this one for Madison, Wisconsin. Light orange depicts times of soil-water use, darker orange indicates times of deficit, light blue shows soil-water recharge, and darker blue depicts surplus, which includes runoff, deep percolation to groundwater, and standing water. The two "DST" lines represent the estimated change in soil water from month to month (*DST* stands for "delta storage," the change in the amount of stored water). A negative DST (light orange) indicates that conditions are drying out relative to the month before, and a positive DST (light blue) indicates that conditions are getting wetter. Note how positive and negative DSTs can precede a time of surplus or deficit, respectively. This diagram for Madison, a Humid Continental climate, shows no climatological deficits and small surpluses, except when the spring thaw melts snow and ice, causing high runoff totals. Precipitation mostly arrives when it is needed and typically doesn't arrive in overwhelming amounts when it isn't needed. Why do you think no changes in storage are happening in the middle of winter? Need a hint? This is a liquid water balance!

08.04.a1

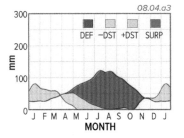
08.04.a3

4. The very long, dry summer in Mediterranean climates, such as Los Angeles (◄), causes large deficits during those months. The area receives much more precipitation in the winter, allowing for soil-water recharge from November into late winter and spring.

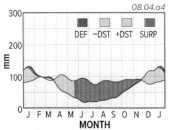
08.04.a4

5. Montevideo, Uruguay (◄), has a Humid Subtropical climate. Deficits are minimal even in the hot summer (which occurs in December through February) at this Southern Hemisphere location. Winter surpluses are kept relatively small by fairly high PE rates.

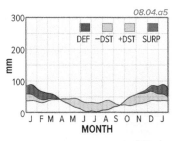
08.04.a5

6. Tierra del Fuego is the cold, southernmost tip of South America (▼). It has soil-water use or deficits most of the year (◄), except during the Southern Hemisphere winter (e.g., June and July).

Before You Leave These Pages

✓ Explain a water-balance diagram and why one might be useful in planning for wise use of water resources.

✓ Describe how the type of climate relates to water balance, as expressed in a water-balance diagram, such as in your town.

08.04.a6 Tierra del Fuego, Argentina

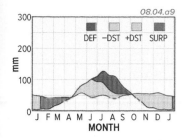

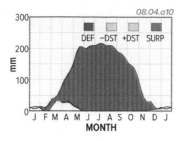

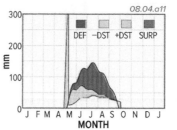

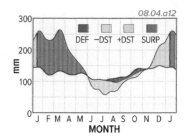

7. The Marine West Coast climates of northwestern Europe, represented by Paris (▲), have seasonal swings in water balance, with small deficits during summer. Precipitation rarely occurs in huge deluges and soils have a high field capacity, so recharge requires a long time.

8. Desert climates like that at Riyadh, Saudi Arabia (▲), show huge climatological deficits and little or no surplus. In this very dry region, water is often brought in artificially to supply the needs of people, animals, and crops.

9. Subarctic climates like in Yakutsk, Russia (▲), have odd water balance diagrams because there is no runoff during the long, frozen winter, but spring (May) snowmelt gives extremely high runoff totals. Long summer daylight hours create some deficit values.

10. Tropical Rain Forest climates, like Port Moresby, Papua New Guinea (▲), show long periods of surplus and only short periods of small deficits when the ITCZ moves farthest away. The warm temperatures and high rainfall amounts allow thick growths of rain-forest plants (▼).

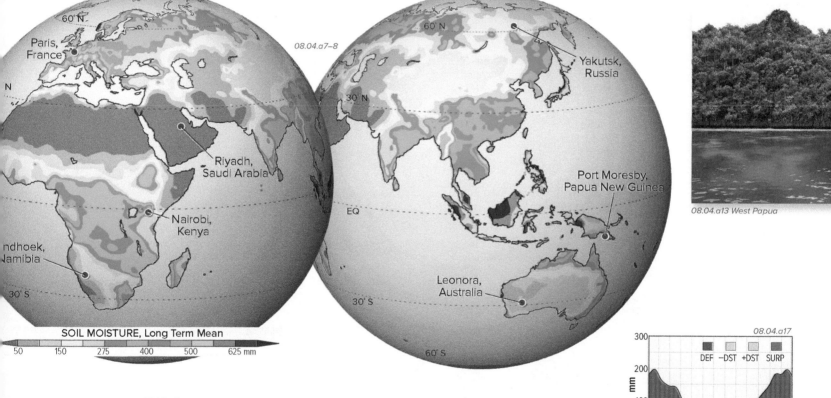

08.04.a7–8

SOIL MOISTURE, Long Term Mean

50   150   275   400   500   625 mm

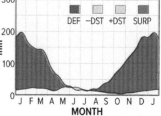

08.04.a13 West Papua

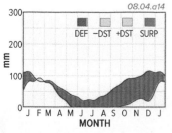

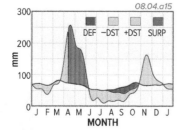

11. No climatological surpluses exist in the true desert climate of Windhoek, Namibia (▲), but deficits are smaller than might be expected because a cold ocean current keeps temperatures (and therefore PE) somewhat suppressed.

08.04.a16 Kenya

12. Tropical Savanna climates, as in Nairobi and the central parts of Kenya (▼), show double peaks in soil-water recharge (◄), which correspond to times of year when the ITCZ makes its closest approach. Soil-water use and deficits occur when the subtropical high approaches.

13. Leonora, in the Australian Outback (▲), has extremely high deficits, especially in the Southern Hemisphere summer (January). The lower temperatures, northward migration of the subtropical high, and occasional mid-latitude wave cyclone storminess in the Southern Hemisphere winter (June through August) permit a very brief period of soil-water recharge.

8.4

8.5 How Do We Use Freshwater Resources?

WE USE LARGE QUANTITIES OF WATER each day, for a variety of purposes, especially power generation and irrigation of farms. How much water does each of our activities consume, and where does the water come from?

A What Are the Main Ways in Which We Use Fresh Water?

Every 5 years, the U.S. Geological Survey conducts a detailed study of water use in the United States. The most recent USGS compilations show that we use freshwater in six or seven main ways, depending on how we classify the data. Water use in the United States is hundreds of billions of gallons per day, most of which is from surface waters. Examine the graph of freshwater usage below and think about the ways in which you use water.

08.05.a1

1. *Thermoelectric Power*—Electrical-generating power plants (◄) are the largest user of fresh water in the U.S., using 41% of the total amount of fresh water. Such power plants drive their turbines by converting water into steam and also use large amounts of water to cool hot components. Some of this water is from recycled sources, such as reclaimed sewage. Power plants are also the largest users of saline (salty) water.

2. *Irrigation*—Farms and ranches are the other large users of fresh water (►), using nearly 37% of fresh water. Farms use water from groundwater, streams, lakes, and reservoirs to irrigate grain, fruit, vegetables, cotton, animal feed, and other crops. Much of this water is lost through evaporation to the atmosphere before it can be used by plants. Downward-directed sprinkler systems cut down on water use by delivering water more directly to crops.

08.05.a2 Pasco, WA

3. *Public and Domestic Water Uses*—The third largest use of fresh water is by public water suppliers and other domestic uses. We consume water by drinking, bathing, watering lawns, filling artificial lakes (▼), and washing clothes, dishes, and cars. Much water from public water suppliers also goes to businesses. Most water for public and domestic use comes from streams and groundwater.

08.05.a3 Phoenix, AZ

FRESHWATER USAGE

Thermoelectric Power 41%

Irrigation 37%

Public and Domestic Use 12.6%

Industrial Uses 5.2%

Aquaculture 2.5%

Mining 1.1%

Livestock 0.6%

08.05.a4

08.05.a5 West Driefontane, South Africa

4. *Industrial and Mining Uses*—Fresh water and saline water are used extensively by industries, including factories, mills, and refineries (▲). Water is integral to many manufacturing operations, including making paper, steel, plastics, and concrete. Mining and related activities also use water in the extraction of metals and minerals from crushed rock.

08.05.a6

5. *Aquaculture*—According to the USGS study, we use approximately 2.5% of fresh water to raise fish and aquatic plants (◄). Most of this use is in Idaho, near the Thousand Springs area. Such water is not totally consumed—much is released back into the Snake River.

6. *Livestock*—Providing water for cows, sheep, horses, and other livestock accounts for only 0.6% of freshwater use (►), but much water is used for irrigation to raise hay, alfalfa, and other animal feed. Many ranches use small constructed reservoirs and ponds as the main water source for animals.

08.05.a7

B How Do We Use and Store Water?

08.05.b1 Grand Coulee Dam, WA

08.05.b2 Mississippi River, IA

08.05.b3 Black Hills, SD

1. *Electrical Generation*—The movement of surface water can generate electricity. To do this, we build dams that channel water through turbines in a hydroelectric power plant.

2. *Transportation*—We use many large waterways, such as the Mississippi River, as energy-efficient transportation systems to move agricultural products, chemicals, and other industrial products.

3. *Recreation*—People use surface water in lakes and streams for many types of recreation, including swimming, tubing, rafting, boating, and fishing. We also use fresh water to fill ponds, fountains, and swimming pools.

08.05.b4 Baltimore, MD

4. Surface waters are commonly stored in natural lakes and in constructed reservoirs behind earthen and concrete dams (◀). Water for drinking and other municipal uses can be stored in underground or above-ground storage tanks.

5. We construct canals, raised aqueducts, and large and small pipelines to move fresh water from one place to another (▶), especially to irrigate farms or to bring water to large cities. Pumps in water wells move groundwater to the surface.

08.05.b5 Tempe, AZ

C How Do We Refer to Volumes of Water?

Water-resource studies typically report volumes of water in one of three units: *gallons*, *liters*, or *acre-feet*. Gallons and liters are familiar terms, but the concept of an *acre-foot* of water requires some explanation.

1. How big is an acre? An acre covers an area of 4,047 m² (43,560 ft²). If a perfect square, an acre would be 64 m (210 ft) on a side. An acre is equivalent to 91 yards of an American football field. There are 640 acres in a square mile.

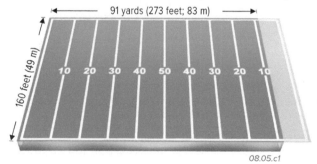
91 yards (273 feet; 83 m)
160 feet (49 m)
10 20 30 40 50 40 30 20 10
08.05.c1

2. An acre-foot of water is the volume of water required to cover an acre of land to a height of one foot. Imagine covering 91% of a football field (one acre) with a foot of water. An acre-foot is equivalent to about 326,000 gallons or more than 1.2 million liters of water.

Drinking-Water Standards in the U.S.

The U.S. Environmental Protection Agency (EPA) sets standards for safe drinking water. Nearly all *public* water supplies in the U.S. meet these standards, which can be found at *www.epa.gov*. These standards set a limit on the concentrations of selected contaminants in water. Small municipalities commonly have more trouble meeting these standards than large cities because of limited budgets for water analysis and for building and running facilities to remove contaminants. The EPA standards do not apply to private wells.

Many people prefer the taste of bottled water to public tap water, but there are generally no health reasons to buy bottled water so long as the public water provider meets all the federal, state, and local drinking water regulations. Commercially bottled water is monitored by the Food and Drug Administration (FDA) but is not monitored as closely as public water systems. The FDA requires a bottler to test its water source only once a year. Also, bottled water can cost as much as 1,000 times more than municipal drinking water.

Before You Leave This Page

✓ Describe ways we use fresh water, and which four uses consume the most.

✓ Describe how we use and store fresh water.

✓ Describe in familiar terms how much water is in an acre-foot.

✓ Describe what a drinking water standard is and who sets the limits.

8.5

8.6 How and Where Does Groundwater Flow?

MOST GROUNDWATER RESIDES in tiny pores between grains in sediment or in narrow fractures in rock. It flows beneath the surface in ways that are controlled by several key principles. The direction and rate of groundwater flow are largely controlled by the porous nature of the materials, the slope of the water table, and the geometry and nature of the subsurface rock. Some rock types allow easy groundwater flow, whereas others prevent significant movement.

A What Is the Geometry of the Water Table?

The *water table* defines the boundary between unsaturated and saturated rock and sediment. It usually is not a horizontal surface but instead has a three-dimensional shape that mimics the shape of the overlying land surface. The shape of the water table commonly has slopes, ridges, hills, and valleys. These features control which way groundwater flows.

1. In most environments, the water table has the same general shape as the overlying land surface but is more subdued. Where the land surface is high, the water table is also high. The similarity in shape between topography and the water table is less straightforward in some arid environments and in places where humans have pumped out groundwater faster than it can be replenished by precipitation.

2. The water table generally slopes from higher to lower areas. It is generally deeper below the surface under mountains than under lowlands, so its slope is less steep than that of the land surface. The shape of the water table is largely independent of the geometry of rock units through which the water table passes.

3. Groundwater just below the water table flows down the slope of the water table. In this example, it flows from left to right, from areas with a higher water table to areas with a lower water table. The blue arrows show flow directions of water right below the water table.

08.06.a1

4. Where the water table is horizontal, for example near this lake, groundwater may flow very slowly or not at all. Deeper water may flow in directions different from near-surface water.

6. Where the water table intersects the land surface, there may be lakes, wetlands, or a flowing stream. The stream in this figure occurs where the water table is at the surface. However, streams do not necessarily coincide with the water table, because some flowing streams are underlain by unsaturated materials, down into which water from the stream can seep.

5. The terminology used to describe features of a water table is derived from topography. A high part of the water table separating parts sloping in opposite directions is called a *groundwater divide*. Groundwater flows in opposite directions on either side of a groundwater divide into different *groundwater basins*.

B What Controls the Rate of Groundwater Flow?

1. The rate of groundwater flow is typically measured in meters per day, but can be much slower. Rate is primarily controlled by *permeability*, a material's ability to transmit fluid, which can vary greatly from one material to the next.

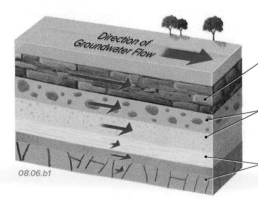

08.06.b1

2. The rate of groundwater flow is strongly controlled by the permeability of the rock type. In this diagram, flow is fastest in this highly permeable rock layer.

3. Flow is moderately fast in somewhat less permeable layers.

4. Flow is slowest in layers that have a very low permeability.

08.06.b2

5. The rate is controlled to a lesser extent by steepness of the water table because flow is driven by gravity.

Other factors being equal, water flows faster down a steep water-table slope and slower down a more gentle one. The slope of the water table is called the *hydraulic gradient*, and is measured in the same units as most gradients (amount of drop in some distance).

C   What Is an Aquifer?

An *aquifer* is a large body of permeable, saturated material through which groundwater can flow sufficiently to yield significant volumes of water to wells and springs. To be a good aquifer, a material must have high permeability.

1. The most common type of aquifer is an *unconfined aquifer* where the water-bearing unit is open (not restricted by impermeable rocks) to Earth's surface and atmosphere. Rainwater or surface water can seep unimpeded through the upper layers of rock and sediment into an unconfined aquifer.

2. A *confined aquifer* is separated from Earth's surface by rocks with low permeability. Here, a permeable aquifer is bounded above and below by layers of low-permeability rock.

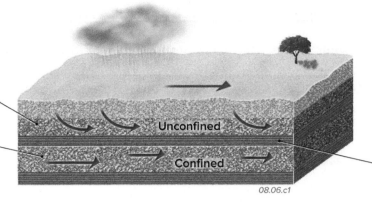

08.06.c1

3. A low-permeability unit, such as the thin gray layer in the middle, can restrict flow. An *impermeable* unit blocks flow completely. Such a unit is the opposite of an aquifer and is referred to as an *aquiclude*.

D   How Are Wells Related to the Water Table?

A well is a hole dug or drilled deep enough to intersect the water table. If the well is within an aquifer, water will fill the open space to the level of the water table. This freestanding water can be drawn out by buckets or pumps.

08.06.d1

This well has been drilled from the land surface downward past the water table. The aquifer is unconfined and has filled with water to the height of the water table.

08.06.d2

In dry seasons, or during periods of high groundwater use, some wells run dry. This occurs when the water table drops below the bottom of the well, which was not drilled deep enough into the aquifer.

08.06.d3

A *perched water table* sits above the main water table and generally forms where a discontinuous layer or lens of impermeable rock blocks and collects water infiltrating into the ground. Perched water bodies can make it difficult to predict the best site to drill an adjacent well and the depth of the water table in the new well.

Artesian Wells and Water

We often hear the word *artesian* in the context of bottled water or certain beverages. What does this term imply? Does it mean that the water is better tasting, more natural, or more healthy? The short answer to these three questions is no, or at least not necessarily.

The term *artesian* means that groundwater is in a confined aquifer and is under enough water pressure that the water rises some amount within a well. The water does not have to reach the surface for the well

08.06.t1

to be called artesian, but many artesian systems have enough pressure to force the water all the way to the surface, creating a well or spring. Although it is a catchy advertising term, the term *artesian* is not indicative of how the water tastes, whether it is more natural than other types of groundwater, or whether it is healthy. It only means that the groundwater is confined and under pressure, and as a result rises some amount in the well.

Before You Leave This Page

☑ Sketch and describe the typical geometry of the water table beneath a hill and a valley, showing the direction of groundwater flow.

☑ Summarize two factors that control the rate of groundwater flow.

☑ Sketch and describe the relationship of wells to the water table.

☑ Sketch and describe an unconfined, confined, and artesian aquifers.

8.6

8.7 What Is the Relationship Between Surface Water and Groundwater?

SURFACE WATER AND GROUNDWATER ARE NOT ISOLATED SYSTEMS. Rather, they are highly interconnected with water flowing from the surface to the subsurface and back again. Most groundwater forms from surface water that seeps into the ground, and some streams and lakes are fed by groundwater.

A How Does Water Move Between the Surface and Subsurface?

1. Surface water can soak into the subsurface and become groundwater if the surface material is permeable and the water table is deep enough so there is an unsaturated zone into which water can seep. Percolation of water into the groundwater system helps replenish or recharge water lost by wells, springs, or other parts of the system. Such replenishment is referred to as *groundwater recharge*.

08.07.a1

2. As long as topography does not intersect the water table, the groundwater will remain at depth, generally flowing toward low elevations.

3. Where the water table intersects the surface, groundwater can flow out onto the land. Such flow forms many springs and can add water to lakes and streams, keeping them from drying up. Seen in this context, a spring represents the interaction between surface topography and the water table, and whether groundwater forms a spring depends on the geometry of both. The water table is the more subdued of the two, so a spring is usually along a topographic low spot or a steeper part of a slope—the spring shown in this figure is at both a low spot and on a steeper part of the slope.

B What Causes Groundwater to Emerge as a Spring?

A *spring* is a place where groundwater flows out of the ground onto the surface. At most springs, the water table intersects the surface. This can occur in a variety of geologic settings, a few of which are summarized below. Some groundwater is heated by hot rocks, or perhaps magma, before coming to the surface, forming warm springs or hot springs. In rare cases, a hot spring that is near the boiling temperature of water can form a *geyser*, a kind of hot spring that intermittently erupts fountains or sprays of hot water and steam. Geysers can be spectacular displays, such as those by Old Faithful, a world-famous geyser in Yellowstone National Park of northwestern Wyoming and adjacent parts of Idaho and Montana.

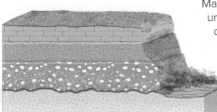

08.07.b1

Many springs are related to limestone aquifers. Where the saturated zone in the aquifer intersects the surface, water can flow out in a spring. In the spring to the right, water rushes out of dissolved fractures and other passageways in limestone, producing a deafening roar and earning the spring its name—Thunder Springs (▶). It is in the walls of a side canyon of the Grand Canyon, Arizona.

Many springs are related to boundaries between two different rock units. In the example shown here, a sequence of layered rocks sits on top of a less permeable rock type (the lowest unit). Groundwater flows down through the layered rocks until it encounters the hard, less permeable unit. It then flows laterally until it encounters a low point in the topography, emerging as a spring. Some springs are fairly subtle, not thunderous, and are called *seeps* (▶). This seep emanates from fractures within a granite, providing a local moist environment in a fairly dry region.

08.07.b2

08.07.b3 Grand Canyon, AZ

08.07.b4 Enchanted Rock SP, TX

C How Are Lakes Related to Groundwater?

Lakes can have various relationships to groundwater. Most lakes occur where the water table intersects the ground surface, but some have a different setting. Most wetlands represent the interaction between rainfall, surface water, and groundwater and may be nourished by groundwater flow.

1. Some lakes are perched above the water table. These lakes can be transient, lasting only a short time after precipitation. A perched lake can be permanent if the inflow of water into the lake, such as from runoff, is at least equal to the amount lost by outflow to the ground, by evaporation to the air, or by other means.

2. Most lakes mark where the water table intersects and rises above the land surface. A lake can be fed entirely or partially by inflow of groundwater.

3. Many lakes are along the bottoms of valleys where groundwater is commonly close to or at the surface. Such lakes may be nearly in equilibrium with the adjacent groundwater, neither gaining nor losing water.

4. Wetlands can form peripheral to lakes, commonly at the same level as the water table. Other lakes are perched on uplands that contain clay or other less permeable material close to the surface. The low permeability can trap precipitation and runoff, slowing the infiltration of water into the ground, forming a wetland from the ponded water.

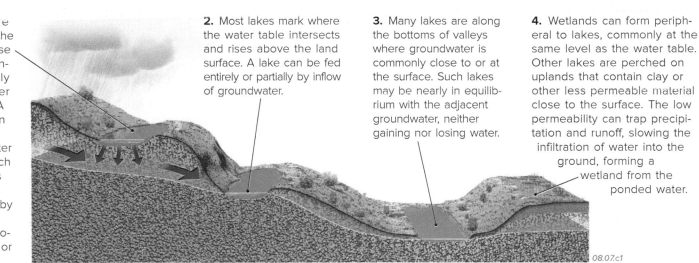

08.07.c1

D How Do Streams Interact with the Water Table?

Water in some rivers and smaller streams decreases to a trickle and entirely disappears farther down the drainage. In other cases, a stream will flow even though there has not been rain or snowmelt in a long time—what is the source of this water? These occurrences are a result of interactions of the stream with groundwater.

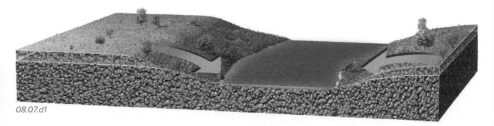

08.07.d1

1. Some streams and rivers are lower in elevation than the water table next to the stream, so groundwater flows into the stream or river. The blue arrows show the direction of groundwater flow below the water table. A part of a stream that receives water from the inflow of groundwater is said to be *gaining* or to be a *gaining stream.*

08.07.d3

3. Some losing streams disappear when they cross from hard, less permeable rocks to softer, more permeable materials. The water seeps into the ground, where it may continue to flow at a shallow depth in the loose sand and gravel in the basin.

08.07.d2

2. Other stream channels flow across an area where the water table is at some depth below the surface. The part of the stream that loses water from outflow to groundwater is said to be *losing* or to be a *losing stream.* The blue arrows show that groundwater below the water table flows down and away from the channel.

Before You Leave This Page

☑ Sketch and describe how the interaction of the water table with topography causes water to flow between the surface and subsurface.

☑ Sketch or describe what is required to form a spring and possible settings where this occurs.

☑ Sketch and describe ways that lakes and wetlands relate to groundwater.

☑ Describe gaining and losing streams and how a river or stream can lose its water entirely.

8.7

8.8 What Problems Are Associated with Groundwater Pumping?

THE SUPPLY OF GROUNDWATER IS FINITE, so pumping too much groundwater, a practice called *overpumping,* can result in serious problems. Overpumping can cause neighboring wells to dry up, land to subside, and gaping fissures to open across the land surface.

What Happens to the Water Table if Groundwater Is Overpumped?

Demands on water resources increase if an area's population grows, the amount of land being cultivated increases, or open space is replaced by industry. Groundwater is viewed as a way to acquire additional supplies of fresh water, so new wells are drilled or larger wells replace smaller ones when more water is needed.

1. A simple case illustrates the problems with overpumping. The two figures below show what happens when an unconfined aquifer is pumped, first by a small-volume pump and later by a larger pump. The topography of this area is fairly flat, there are no bodies of surface water, and a single type of permeable sediment composes the subsurface.

2. As people move into a nearby town, they drill a small well down to the water table to provide fresh water. The small well pulls out so little groundwater that the water table remains as it has for thousands of years, nearly flat and featureless. The well remains a dependable source of water because its bottom is below the water table.

Minor Groundwater Withdrawal

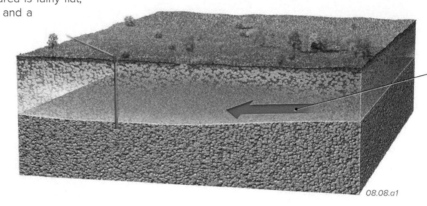

3. Across the entire area, groundwater flows from right to left, down the gentle slope of the water table. The blue arrow shows the direction of flow for groundwater in the saturated part of the aquifer, right below the water table. This arrow is drawn at the top of the water table so the arrow is visible.

08.08.a1

Increased Groundwater Withdrawal

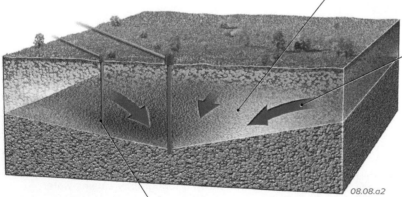

08.08.a2

4. As more people move into the surrounding area, they drill a larger well to extract larger volumes of water to satisfy the growing demand. The new, larger well pumps water so rapidly that groundwater around the well cannot flow in fast enough to replenish what is pumped out. This causes the local water table to drop, forming a funnel-shaped *cone of depression* around the well.

5. The direction of groundwater flow changes significantly across the entire area. Instead of flowing in one direction, groundwater now flows toward the larger well and into the cone of depression from all directions. The change in flow direction has unintended consequences. It may cause serious safety issues, since waste-disposal sites, such as landfills, are generally planned with the groundwater-flow direction in mind. The change in flow direction can bring contaminated water into previously fresh wells.

6. The original small well dries up because it no longer reaches the water table, which has been lowered by the larger well's cone of depression. A cone of depression is common around nearly all wells, but a large cone of depression, caused by overpumping of the aquifer, can have drastic consequences. It can dry up existing wells, change the direction of groundwater flow, and contaminate wells. In addition, overpumping can dry up streams and lakes, if they are fed by groundwater, or cause the roofs of caves to collapse because groundwater holds up the roof of some caves. Effects similar to those described above can also occur in a confined aquifer.

Before You Leave These Pages

✓ Sketch a cone of depression in cross section, describing how it forms and which way groundwater flows.

✓ Describe how a cone of depression can cause a well to become polluted.

✓ Sketch or describe some other problems associated with overpumping, including subsidence, fissures, and saltwater incursion.

B What Other Problems Are Caused by Excessive Groundwater Withdrawal?

Overpumping can cause the ground surface to subside if sediment within the underlying aquifer is dewatered and compacted. In certain settings, subsidence causes fissures to open on the surface.

Before Groundwater Pumping

1. Many areas have settings similar to this one: mountains composed of bedrock flank a valley or basin underlain by a thick sequence of sediment. Most water is pumped from beneath the sediment-filled basin and used by people in the valleys.

08.08.b1

2. Bedrock has interconnected fractures that give it some permeability, but it contains much less water than sediment in the basin.

3. The water table slopes from the mountains toward the basin, across the boundary between bedrock in the mountains and sediment beneath the valley.

After Groundwater Pumping

4. If we overpump groundwater, the water table can drop over much of the area. In some real-world cases it has dropped more than 100 m (~330 ft).

5. As the water table drops, the upper part of the original aquifer is now above the water table and has been dewatered. Sediment within and below the dewatered zone compacts because water pressure no longer holds open the pore spaces.

08.08.b2

6. Compaction of the sediment causes the overlying land surface to subside by several meters. Once the sediment compacts, most subsidence and loss of porosity are permanent and will not be undone if pumping stops and water levels rise again. Along the coast, such subsidence could lower an area below sea level.

7. The granite cannot compact, so open fissures develop across the land surface along the boundary between land that subsided (in the basin) and land that did not (in the mountains). The earth fissure here (▶) formed by this type of subsidence.

08.08.b3 Chandler, AZ

C How Can Groundwater Pumping Cause Saltwater Intrusion into Coastal Wells?

Some wells are by necessity near the coasts of oceans and seas. These wells have a special threat—overpumping can draw salt water into the well, a process referred to as *saltwater incursion* or *saltwater intrusion*.

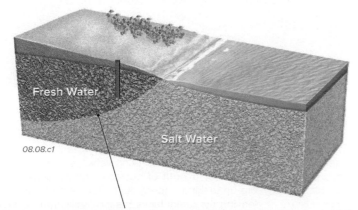
Fresh Water
Salt Water
08.08.c1

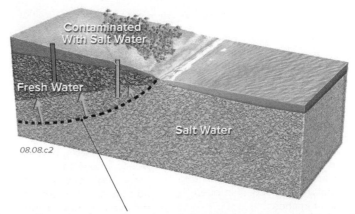
Contaminated With Salt Water
Fresh Water
Salt Water
08.08.c2

Along ocean coasts, fresh water commonly underlies the land, while groundwater beneath the seafloor is salty. Fresh water is less dense than salt water and forms a lens floating on top of salt water.

When wells on land are overpumped, the interface between fresh water and salt water moves up and inland (saltwater incursion). Wells closest to the coast will begin to pump salt water and will have to be shut down.

8.9 How Can Water Become Contaminated?

CONTAMINATION OF SURFACE- AND SUBSURFACE-WATER SUPPLIES is a major problem facing many communities. Some contaminants are natural products of the environment, whereas others have human sources, the direct result of our modern lifestyle. What are some main sources of water contamination?

1. Systematically examine each part of this figure, trying to recognize every potential source that could contaminate surface water and groundwater. Then read the accompanying text blocks.

2. Water contamination can have natural causes. Weathering of rocks releases chemical elements into surface water and groundwater—some of these elements are beneficial and others are not. Rocks, especially those that have been mineralized by hot fluids, may contain lead, sulfur, arsenic, or other potentially hazardous elements. Mining activities and natural erosion move mineralized rocks away from where they formed, further spreading these contaminants.

3. We use large amounts of petroleum and coal, which have to be discovered, extracted, transported, and processed. Any of these activities potentially cause pollution. Some of the worst disasters are leaks from pipelines and supertankers, and fires at refineries and storage tanks.

4. Old landfills are the repository for countless discarded items, many of which contain hazardous substances. Such items include diapers, old tires, lead batteries, toxic liquids from household or commercial use, mercury in compact fluorescent bulbs, and other garbage. If not properly sited and sealed from the environment, landfills can be major sources of pollution. Landfills along rivers, such as this one, can be breached by lateral erosion of channels. Supposedly impermeable linings beneath the landfill, if installed at all, can crack during settling and from daily landfill operations, allowing a toxic stew to seep into the underlying groundwater.

08.09.a1

5. One of the most basic types of contamination is human waste, which can end up in surface-water and groundwater supplies if proper sanitary procedures are not followed. Contamination of this sort comes from septic tanks, accidental spills from wastewater treatment facilities, or, in less affluent parts of the world, from waste disposal in open sewers and trenches.

6. Farms, ranches, and commercial orchards are contributors of chemical and organic contamination. Chemical contaminants include fertilizers that contain nitrates, insecticides to control pests, herbicides to combat invasive weeds, and defoliants to remove leaves before harvesting crops like cotton. Irrigated fields build up salts as water evaporates, and much of this gets carried into ditches by excess irrigation water. Animal waste, which contains harmful bacteria, hormones, and feed additives, is also a potential problem.

7. Gas stations can contaminate water because of leaks from underground storage tanks and spills that occur while filling vehicles. Gas stations frequently go out of business if they have to dig up leaking underground storage tanks, often leaving the leaky tanks behind. Spills from tanker trucks, railroad cars, and trucks delivering fuel from distribution hubs may cause water contamination if there is an accident.

8. To manufacture the items we use in our daily lives, factories use many different raw materials and chemicals. Plastic products, for example, are everywhere around us: containers for soda and bottled water, plastic bags for groceries and other purchases, and many parts of our cars. These plastics are mostly produced from petroleum, which must be refined and processed in refineries and plastic factories. Petroleum and various chemicals, along with the waste produced during the manufacturing process, can accidentally escape, as shown in this photograph, causing an industrial site to become heavily contaminated (▲). Liquid contamination may be pumped down "disposal wells," often ending up in the groundwater. Ponds intended for temporary storage can leak, contaminating surface water and groundwater. Fumes and particles emitted from smokestacks settle back to the ground or are washed down by rain and snowfall, possibly contaminating air, plants, buildings, soils, surface water, or groundwater.

08.09.a2 Russia

Water, Arsenic, and Bangladesh

Bangladesh, east of the main part of India, is a geographically challenged country. Much of it consists of lowlands that are easily flooded by high seas and the Ganges, one of the world's largest rivers.

One of Bangladesh's worst problems, however, is water contamination. For centuries, poor sanitation in this impoverished nation polluted the streams and other surface-water sources with cholera, dysentery, and other human- and animal-carried diseases. To provide a new source of water, people sank more than 10 million tube wells (created by pounding tubes into the soft sediment). Unfortunately, the sediment and groundwater have a high content of naturally derived arsenic, many times the recommended limit, causing arsenic poisoning on a scale never before seen. To help solve the problem, scientists from the U.S. and Bangladeshi governments have been sampling well waters, studying the surface and subsurface materials, drilling wells into a deeper aquifer (▼), and evaluating whether bacteria can be used to reduce arsenic concentrations.

Bangladesh 08.09.t1

9. Even if a community is careful with wastes, contamination can be carried into the area by streams that drain polluted areas upstream. Polluted surface water can seep into groundwater, and groundwater inflow can pollute streams. Soils can contaminate water, which then pollutes the next town downstream.

10. In the past, dry cleaners were sources of groundwater pollution because of the chemical solvents used to clean clothes without water. Such solvents have names from organic chemistry and commonly are referred to by their abbreviations, such as PCE for perchloroethylene ("perc" for short). Today, most such chemicals are no longer used.

11. Houses cause water pollution during the production of the materials used to build the house, from actual construction, and from day-to-day activities that include the use of fertilizer, termite treatment, and household pesticides. Oil and gas spilled from cars and other machines, along with oil improperly disposed of during do-it-yourself oil changes, can contaminate large volumes of fresh water.

12. We may be unaware of water contamination. Subsurface rock and sediment can contain hazardous natural substances, including metallic elements and radon. We may discover the contamination only if we drill into it, often because an unusual health issue appears in a local population.

Bangladesh 08.09.t2

Before You Leave This Page

☑ Describe the many ways that surface water and groundwater can become contaminated.

8.9

8.10 How Does Groundwater Contamination Move and How Do We Clean It Up?

WATER CONTAMINATION CAN BE OBVIOUS OR SUBTLE. Some streams and lakes have oily films and give off noxious fumes, but some contamination occurs in water that looks normal and tastes normal but contains hazardous amounts of a natural or human-related chemical or biologic component. How does contamination in groundwater move, how do we investigate its causes and consequences, and what are possible remedies?

A How Does Contamination Move in Groundwater?

As contamination enters groundwater, it typically moves along with the flowing groundwater. Contamination can remain concentrated, can spread out, or can be filtered by passage through sediment and rocks.

1. Groundwater contamination typically moves with the groundwater down the slope of the water table.

08.10.a1

2. Contamination from this septic tank will move to the right, away from the well. The direction of groundwater flow is clearly important in deciding where to put the septic tank relative to the well.

3. Contamination is drawn out parallel to the direction of groundwater flow.

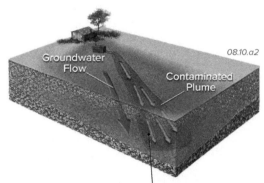

08.10.a2

4. Diffusion and mixing spread the contaminated zone as it migrates away from the source. Consequently, the shape of most contamination spreads out like smoke from a chimney and is called a *plume*.

5. Some contamination can be naturally filtered by materials through which the contaminated groundwater flows.

08.10.a3

6. Contamination from the septic tank on the left will be filtered by slow movement through sandy layers, whereas contamination from the septic tank on the right will flow rapidly away, unfiltered, through permeable layers.

B How Do We Investigate Groundwater Contamination?

Hydrologists are scientists who study the setting, movement, and quality of surface water and groundwater. They investigate groundwater and surface-water contamination using the same approaches they use for other types of water-related problems, plus a few extra strategies.

Substance	Concentration	Health Issues
Arsenic	0.01 mg/L	Cancer, numbness
Cadmium	0.005 mg/L	Kidneys, liver, lungs
Chromium	0.1 mg/L	Cancer, nasal issues
Lead	any	Kidneys, blood pressure
Trichloroethylene (TCE)	any	Cancer, kidneys

08.10.b1 Newcastle, England

2. Water contamination is fundamentally about chemicals and hazardous microbes, so geochemists collect *geochemical samples* that are analyzed either in the field or later by chemists in a laboratory. Some volatile organic compounds are detected using sensors that analyze soil gases given off by the soil.

1. Most surface-water and groundwater contamination is recognized by chemical analyses done by community water providers. In the U.S., water standards are set by the Environmental Protection Agency (EPA). This table lists the EPA drinking water standards for a few of the better known or more hazardous water contaminants. Values are in milligrams per liter (mg/L), which is equivalent to *parts per million* (ppm). A standard of 0.1 mg/L for chromium means that drinking water is above the limit if it contains more than about 1 atom of chromium for every 10 million molecules of water.

08.10.b2 Phoenix, AZ

3. Hydrologists conduct tests of an aquifer by pumping a well continuously at a specific rate and observing how that well and wells around it react during the pumping and after the pumps are turned off. This provides information about how fast groundwater and contamination might move.

C How Is Groundwater Contamination Tracked and Remediated?

Once groundwater contamination is identified, what do we do next? Hydrologists compile available information to compare the distribution of contamination with all relevant factors. One commonly used option to clean up, or *remediate*, a site of contamination is called "pump-and-treat." Some contamination can be mostly remediated, but remediation is much more expensive than avoiding the problem in the first place.

1. The first step to remediation is to understand the situation—what is the nature of the contamination, where is the contamination now, where did it come from, where is it going, and what are the geologic controls?

08.10.c1

2. In this area, contamination consists of chromium released by a chrome-plating shop. The water table slopes to the southeast, so this is the direction in which the upper levels of groundwater will flow. We predict that contamination will move in this same direction.

3. Chromium ions are carried away by groundwater flow and also diffuse through the water chemically, albeit at a slower rate. The combination of flow, diffusion, and mixing causes the contamination to spread out, forming a plume of contamination. There is no contamination up-flow (northwest) of the shop, but the plume of contamination will spread to the southeast.

4. To investigate the situation, we map elevations of the water table in meters to determine more precisely which way groundwater is flowing. In this case, the contours decrease in elevation to the southeast. Groundwater flows to the southeast, perpendicular to the contours (and toward lower elevation contours).

08.10.c2

5. We draw a second set of contours based on chemical analyses of the concentration of contamination, in this case chromium. For example, areas within the 5 mg/L contour have at least 5 mg/L chromium, and those within the 10 mg/L contour have at least 10 mg/L. The EPA limit for chromium is 0.1 mg/L, so these values are well above EPA standards.

6. From these maps, we can now determine where the contamination is, which way it is moving, and where it will go in the future (down the slope of the water table). If from interviews or historical records we can determine how long ago the contamination occurred, we can use simple calculations (distance/time) to get the rate of flow. We can also use computer simulations to model past and future movement.

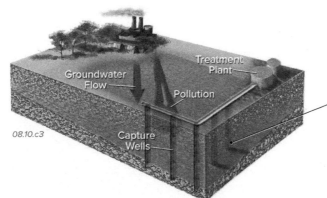

08.10.c3

7. Finally, we try to clean up the contamination. One strategy is to drill wells in front of the projected path of the contamination to contain, capture, and extract the contaminated water. Pumping brings contaminated water to the surface, where it is processed with carbon filters or other appropriate technology to separate the contaminant from the water. The cleaned water is typically reinjected into the ground, evaporated in evaporation ponds, or diverted into nearby streams.

A Civil Action

Woburn, Massachusetts, a small town 16 km (10 mi) north of Boston, was the site of a classic legal case involving groundwater contamination. The case was made famous in the book *A Civil Action* by Jonathan Harr and in a movie of the same name starring John Travolta.

The trouble began in the 1960s when the city drilled two new groundwater wells for municipal water supplies. The wells were drilled into glacial and river sediments that had filled an old valley. After the wells were installed, some residents complained that the water tasted odd and had a chemical odor. Over the next 20 years, residents began to show a high incidence of leukemia and other serious health problems. Chemical analyses showed that the groundwater was contaminated with trichloroethylene (TCE) and other volatile organic compounds. Local families filed a lawsuit against several chemical companies that were potentially responsible. The verdict remains complex, but the site is a classic example of the interaction of physical geography, water, health, and environmental law.

Before You Leave This Page

☑ Sketch a plume of contamination, showing how it relates to the source of contamination and the direction of groundwater flow.

☑ Describe some ways in which hydrologists investigate groundwater contamination.

☑ Sketch how chemical analyses define a plume of contamination and one way a plume could be remediated.

8.10

8.11 What Is Happening with the Ogallala Aquifer?

THE MOST IMPORTANT AQUIFER IN THE U.S. lies beneath the High Plains, stretching from South Dakota to Texas. It provides groundwater for about 30% of all cropland in the country, but it is severely threatened by overpumping. The setting, characteristics, groundwater flow, and water-use patterns of this aquifer connect many different aspects of water resources and illustrate their relationship to physical geography.

A What Is the Setting of the Ogallala Aquifer?

1. The *Ogallala aquifer* covers much of the High Plains area in the center of the U.S. The area outlined on this map represents the main part of the aquifer. The aquifer forms an irregularly shaped north-south belt from South Dakota and Wyoming through Nebraska, Colorado, Kansas, the panhandles of Oklahoma and Texas, and eastern New Mexico. The area underlain by the aquifer is one of the most agriculturally important regions of the world.

2. The Ogallala aquifer covers about 450,000 km² (174,000 mi²) and is currently the largest source of groundwater in the country. It provides 30% of all groundwater used for irrigation in the U.S. In 1980, near the height of the aquifer's use, 17.6 million acre-feet of water were withdrawn to irrigate 13 million acres of land. The water is used mostly for agriculture and rangeland. The main agricultural products include corn, wheat, soybeans, and feed for livestock.

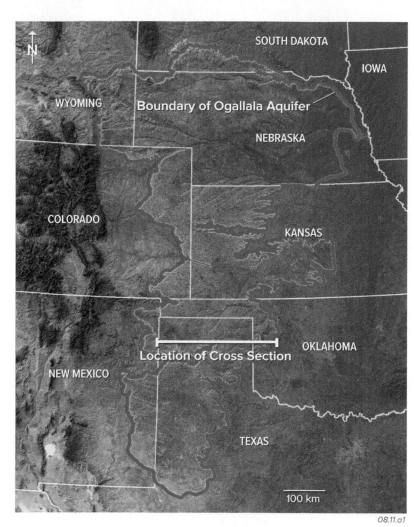

08.11.a1

3. The aquifer is named for the Ogallala Group, the main geologic formation in the aquifer. The formation was named in the early 1900s after the small Nebraskan town of Ogallala.

4. Much of the Ogallala Group consists of sand and other sediment (loose pieces of rock) carried by ancient streams and wind. The streams and wind spread the sediment over the landscape as a relatively continuous layer. This process ended when uplift and tilting of the Rocky Mountain region caused the streams to cut down into the landscape and carry their load of sediment farther east. Present-day streams continue to erode into the aquifer and drain eastward and southward, eventually flowing into the Gulf of Mexico.

The Aquifer in Cross Section

5. This vertically exaggerated cross section shows the thickness of the aquifer from west to east. It shows the aquifer in various colors; rocks below the aquifer are shaded bluish-gray. Note that the aquifer is at the surface and is an *unconfined aquifer*.

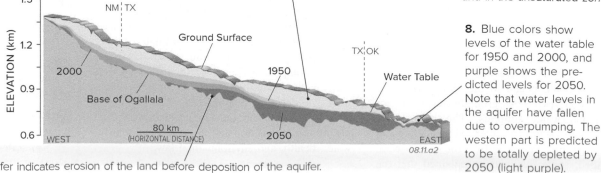

08.11.a2

7. The upper part of the aquifer (shaded yellow) is above the water table and in the *unsaturated zone*.

8. Blue colors show levels of the water table for 1950 and 2000, and purple shows the predicted levels for 2050. Note that water levels in the aquifer have fallen due to overpumping. The western part is predicted to be totally depleted by 2050 (light purple).

6. The irregular base of the aquifer indicates erosion of the land before deposition of the aquifer.

B Where Does Groundwater in the Aquifer Come from and How Is It Used?

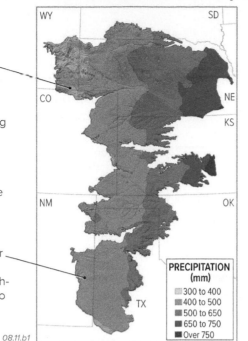

08.11.b1

1. Most of the water accumulating in the aquifer is from local precipitation. This map shows the amount of precipitation received across the area, with darker shades indicating more precipitation. The western part of the aquifer receives much less precipitation (rain, snow, and hail) than the eastern part.

2. Areas of the aquifer that receive the least precipitation—the south-western parts—are also those predicted to go dry by 2050.

3. This graph (▼) shows the water balance for the Ogallala aquifer. Water going into the aquifer is shown above the axis, whereas water being lost by the aquifer is below the axis. Some groundwater recharge occurs where water from precipitation seeps into the aquifer, especially in areas that receive higher amounts of precipitation, as either rain or snow.

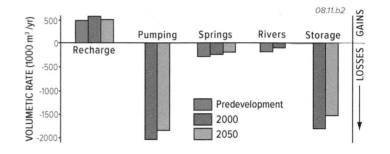

08.11.b2

4. The amount taken out of the aquifer by pumping, springs, and inflow into streams greatly exceeds the recharge, so most parts of the aquifer are being dewatered.

5. As the aquifer dewaters it compacts, which causes a decrease in porosity (open space) and a loss of pore space (in which to store water). This cannot be undone.

C How Has Overpumping Affected Water Levels in the Ogallala Aquifer?

Hydrologists employed by the U.S. government estimate that the aquifer contains 3.2 billion acre-feet of water! That is enough to cover the entire lower 48 states with 52 cm (1.7 feet) of water. How much has overpumping affected the aquifer's water levels, and what will happen to the region and to the country if large parts of the aquifer dry up?

1. This map shows the thickness (in meters) of the *saturated zone* within the aquifer. In some of its northern parts, more than 300 m (1,000 ft) of the aquifer is saturated with water, whereas less than 60 m (200 ft) remain saturated in the southern parts.

2. This map shows how many meters the water table dropped in elevation between 1980 and 2013 as a consequence of overpumping. The largest drops, exceeding 10 m, occurred in southwestern Kansas and the northern part of Texas. Compare this map to the one for precipitation.

3. Predictions—It is uncertain what will happen, but hydrologists are conducting detailed studies of key areas to try to predict what will happen in the next decades. Projections of current water use, combined with numerical models of the water balance, suggest that some parts of the aquifer will go dry by 2050. This will have catastrophic consequences for the local farmers, ranchers, and businesses, and for people across the country who depend on the aquifer for much of their food. Subsidence related to ground-water withdrawal and compaction of the aquifer will be an increasing concern. What do you think will happen to the region if this aquifer is partly pumped dry?

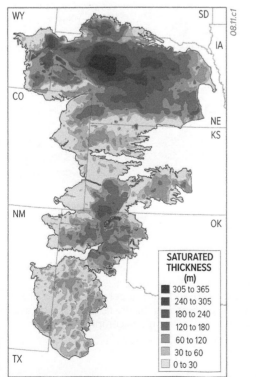

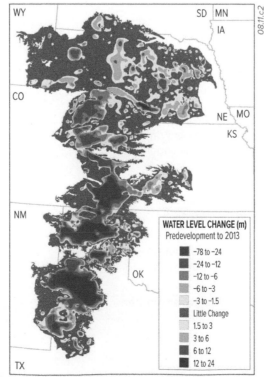

Before You Leave This Page

✓ Summarize the location, characteristics, and importance of the Ogallala aquifer.

✓ Summarize the water balance for the aquifer and how water levels have changed in the last several decades.

8.11

8.12 Who Polluted Surface Water and Groundwater in This Place?

SURFACE WATER AND GROUNDWATER in this area are contaminated. You will use the physical geography of the area, along with elevations of the water table and chemical analyses of the contaminated water, to determine where the contamination is, where it came from, and where it is going. There may be more than one source of contamination. From your conclusions, you will decide where to drill new wells for uncontaminated groundwater.

Goals of This Exercise:

- Observe the landscape to interpret the area's environmental setting.
- Read descriptions of various natural and constructed features.
- Use well data and water chemistry to draw a map showing where contamination is located.
- Use the map and other information to interpret where contamination originated, which facilities might be responsible, and where the contamination is headed.
- Determine a well location that is unlikely to be contaminated.

Procedures

Use the available information to complete the following steps, entering your answers on the worksheet or online.

1. This figure shows geologic features, streams, springs, and human-constructed features, including a series of wells (lettered A through P). Observe the distribution of rock units, sediment, rivers, springs, and other features on the landscape. Compare these observations with the cross sections on the sides of the terrain to interpret how the natural environment is expressed in different areas.

2. Read the descriptions of key features and consider how this information relates to the geographic setting, to the flow of surface water and groundwater, and to the contamination.

3. The data table on the next page shows elevation of the water table in each lettered well. On the worksheet, write the elevation of each well near its location on the map. Since groundwater flows from where the water table is higher to where it is lower, you can determine which way groundwater is flowing simply by assuming that it is flowing from higher numbers toward lower ones. On the map, draw arrows pointing in the general direction you think groundwater flows.

4. Use the data table showing concentrations of a contaminant, purposely unnamed here, in groundwater to shade in areas where there is contamination. Use darker shades for higher levels of contamination.

5. Use the groundwater map to interpret where the contamination most likely originated and which facilities were probably responsible. Mark a large X over these facilities on the map and explain your reasons in the worksheet.

6. Determine which of the lettered well sites will most likely remain free of contamination, and draw circles around two such wells.

1. The region contains a series of ridges to the east and a broad, gentle valley to the west. Small towns are scattered across the ridges and valleys. There are also several farms, a dairy, and a number of industrial sites, each of which is labeled with a unique name. Hydrologists studied one of these towns, *Springtown,* and concluded that it is not the source of any contamination.

2. A main river, called the *Black River* for its unusual dark, cloudy color, flows westward (right to left) through the center of the valley. The river contains water all year, even when it has not rained in quite a while. Both sides of the valley slope inward, north and south, toward the river.

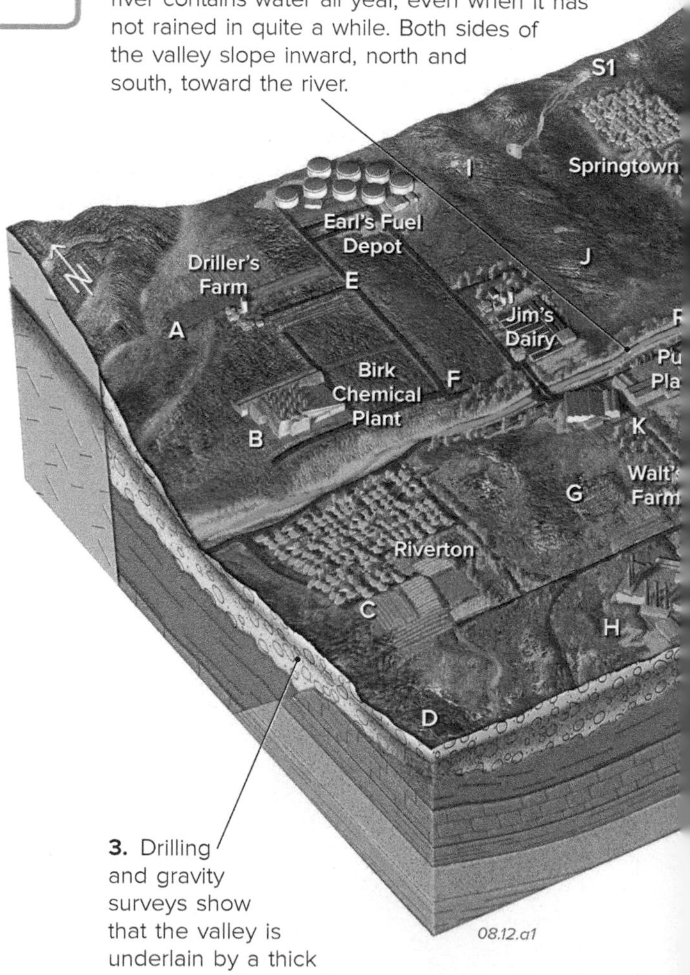

3. Drilling and gravity surveys show that the valley is underlain by a thick sequence of relatively unconsolidated and weakly cemented sand and gravel. Groundwater is mostly flowing in the sand and gravel, not in the more complicated rocks farther down.

08.12.a1

4. Bedrock units cross the landscape in a series of north-south stripes. One of the north-south valleys contains several large coal mines and a coal-burning, electrical-generating plant. An unsubstantiated rumor says that one of the mines had some sort of chemical spill that was never reported. Activity at the mines and power plant has caused fine coal dust to be blown around by the wind and washed into the smaller rivers that flow along the valley.

5. A north-south ridge is composed of Rock Unit 4. *Slidetown*, a new town on this ridge, is not a possible source of the contamination because it was built too recently. A few nice-tasting, freshwater springs issue from Rock Unit 4 where it is cut by small stream valleys.

08.12.a2

Sequence of Rock Layers

Rock Unit 9—Unconsolidated sand and gravel in the lower parts of the valley
Rock Unit 8—Permeable rock unit
Rock Unit 7—Impermeable, with coal
Rock Unit 6—Porous, cavernous limestone
Rock Unit 5—Impermeable rock unit
Rock Unit 4—Permeable rock unit
Rock Unit 3—Impermeable rock unit
Rock Unit 2—Moderately permeable with salty water
Rock Unit 1—Sparsely fractured granite; oldest rock in the area

6. The highest part of the region is a ridge of granite (Rock Unit 1) and layered rocks along the east edge of the area. This ridge receives quite a bit of rain during the summer and snow in the winter. Several clear streams begin in the ridge and flow westward toward the lowlands.

7. A company built a coal-burning power plant over tilted beds of Rock Unit 6, which contains many caves and sinkholes (pits where the ground has collapsed, usually into a cave). Rock Unit 6 is so permeable that the power plant has had difficulty keeping water in ponds built to dispose of wastewaters, which are rich in the chemical substances (including the contaminant) that are naturally present in coal.

8. The tables below list water-table elevations (Elev. WT) in meters and concentrations of contamination in milligrams per liter (mg/L) for each of the lettered wells (A–P). This table also lists the concentration of contamination in samples from four springs (S1–S4) and eight river segments (R1–R8). The location of each sample site is marked on the figure. Wells M, N, and P are deep wells, drilled into the aquifer of Rock Unit 6 at depth, although they first encountered water at a shallow depth. The chemical samples from these wells were collected from deep waters.

9. From mapping and other studies on the surface, geoscientists have determined the sequence of rock units, as described and listed in the upper right corner of this page. These studies also document broadly curving rock layers beneath the eastern part of the region. Note that contamination can flow through the subsurface, following permeable units, instead of passing horizontally through impermeable ones.

10. Based on shallow drilling, the water table (the top of the blue shading) mimics the topography, being higher beneath the ridges than beneath the valleys. Overall, the water table slopes from east to west (right to left), parallel to the regional slope of the land. All rocks below the water table are saturated with groundwater.

Labels on figure: Slidetown, R1, Cornhead Coal Mine, S4, R2, R3, N, R4, Coal Power Plant, R7, Kellogg, R5, CQ Coal Mine, Midnight Coal Mine, R6, P, S3, 200 m

Well	Elev. WT	mg/L
A	110	0
B	100	0
C	105	0
D	110	20
E	120	10
F	115	0
G	120	0
H	120	50

Well	Elev. WT	mg/L
I	130	30
J	130	0
K	120	0
L	130	0
M	150	50
N	150	0
O	140	0
P	150	0

Spring	mg/L
S1	50
S2	0
S3	0
S4	0

River	mg/L
R1	0
R2	20
R3	0
R4	0

River	mg/L
R5	0
R6	0
R7	5
R8	5

NATURAL LANDSCAPES HAVE VARIOUS EXPRESSIONS, including black lava flows, white sandy beaches, red cliffs, and gray granite hills. What causes these differences from one landscape to another, and how did the different kinds of materials form? We can observe and interpret the shape of a natural landscape—the *landform*—or the rocks, soils, and other earth materials that compose the landscape, as well as anything on top, such as vegetation. Here, we explore natural landscapes and the materials that compose them. We also examine the processes that form different types of earth materials and learn how we can recognize what types of materials are present simply from the appearance of a landscape.

09.00.a1

This perspective view (▶) shows satellite data superimposed over topography for southernmost California and adjacent Baja California, Mexico. The Peninsular Ranges, a forested mountainous area east of San Diego, are in greens and browns in the center of the image. The white line across the image, added for reference, marks the border between the United States and Mexico. Large parts of the mountains resemble a mound of granite boulders, as in the photograph below.

How does a landscape with this appearance form?

Peninsular Ranges

San Diego

UNITED STATES

MEXICO

10 km

09.00.a2 Jacumba, CA

The Peninsular Ranges contain many exposures of grayish-colored rocks, most of which are rocks similar to granite. When viewed up close (▼), the granite displays four different kinds of crystals: whitish, light pink, transparent gray, and black.

What are rocks made of, and what controls the color and other properties of a rock?

09.00.a3

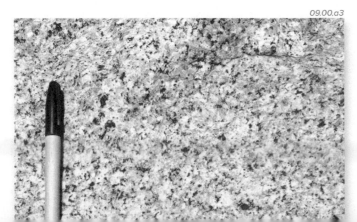

09.00.a4

San Diego County is a famous source of beautiful minerals, including tourmaline crystals (◀), that can be pink, purple, green, or all three colors.

What are crystals, how do they form, and where do we find them?

TOPICS IN THIS CHAPTER

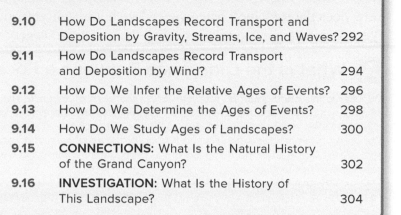

East of the Peninsular Ranges, the land drops down into the lowlands of the Salton Trough, which is characterized by sandy deserts, farmlands of the Imperial Valley, and several large, salty lakes, including the blue Salton Sea. The sand in the Salton Trough was eroded from the adjacent mountains and carried to the area by streams or strong winds.

What are most sand grains composed of?

Salton Sea

Salton Trough

Peninsular Ranges

The Peninsular Ranges

The area called the Peninsular Ranges is a broad, upland region that stretches 1,500 km from southernmost California and southward into the Baja Peninsula of Mexico. In this image, the mountains appear green because they are mostly covered by forests and other types of vegetation. The lowlands of the Salton Trough east (right) of the mountains receive much less rain and have a lighter color in this image because vegetation is sparse and sand and rocks cover the surface.

The contrast between the Peninsular Ranges and the Salton Trough illustrates some important aspects to consider when observing and thinking about landscapes. First, landscapes develop from the materials that are available. The mountains contain granite because that is the type of rock that was there, long before the mountains arose. The mountains would have a different appearance if they instead consisted of a material much weaker than granite and, in that case, might not even be mountains today. In addition to the types of materials, the geometry of the rock units and other materials can also impart a distinct style to the landscape. An appropriate place to begin thinking about landscapes is, therefore, to examine the repertoire of earth materials.

Once the materials are at the surface, they are acted on by the atmosphere, moisture in the soil, and other components of the environment that tend to break down rocks into loose materials. The kind of climate greatly influences this loosening process. Once loosened, these materials can be stripped off the land surface and transported away by running water, wind, and ice, eventually being deposited someplace else, as occurred for the sand and pebbles shown in the photograph below. This sequence of events is what shaped the mountains of the Peninsular Ranges and also deposited the resulting loose materials in the Salton Trough. These processes are still occurring, so the landscape continues to evolve.

09.00.a5 Rancho Vallecito, CA

09

9.1 What Materials Compose Landscapes?

WHAT MATERIALS MAKE UP THE WORLD around us? What do we see if we look closely at an exposure of rock? How does the rock look when viewed with a magnifying glass? Such properties of these rocks control landscape formation. We investigate these questions using the beautiful scenery of Yosemite National Park in California.

A What Is the Difference Between a Rock and a Mineral?

1. Observe this photograph of Yosemite Valley, the heart of Yosemite National Park. What do you notice about the landscape? This landscape is dominated by dramatic cliffs and steep slopes of massive gray rock perched above a green, forested valley. The valley is famous for waterfalls and for huge rock exposures. The appropriately named Half Dome is on the right side of the photograph. What would we see if we got closer to this landscape?

09.01.a1 Yosemite NP, CA

2. Closer examination reveals several different-colored crystals in the rock: whitish, pinkish, clear gray, and black. So what looks like a homogeneous gray rock from a distance is actually composed of different types of crystals. To better observe a rock at this scale, a geoscientist may collect a hand-sized piece, called a *hand specimen.* When examined more closely, the rock contains four different types of crystals with distinct appearances. The clear gray crystals all have similar chemical composition and physical properties, and so they represent one kind of solid substance called a *mineral.* The whitish crystals are a different kind of mineral, as are the black and pinkish crystals.

09.01.a2 Yosemite NP, CA

3. Rocks are composed of minerals, but what characteristics define a mineral? To be considered a mineral, a substance must fulfill all of the criteria listed below. A mineral is a naturally occurring, inorganic solid with an ordered internal (crystalline) structure and a relatively consistent chemical composition.

Natural

09.01.a3 Fluorite

4. A mineral must be natural. These crystals of fluorite (◄) grew naturally from hot water flowing through a rock. If a mineral must form naturally, then natural diamonds are minerals, but synthetic diamonds grown in the lab are not.

Inorganic

09.01.a4 Calcite

5. This crystal is inorganic and a mineral. Many shells have the same composition as the crystal, but they were made by clams and other creatures (they are organic); they generally are not considered to be minerals.

Solid

6. All minerals are solid, not liquid or gaseous. Ice, a solid, is a mineral, but liquid water is not, even though it has the same composition.

09.01.a5 Ice

09.01.a6 Calcite

Ordered Internal Structure

7. A mineral has an ordered internal *structure,* which means that atoms are arranged in a regular, repeating way. Such substances are considered to be *crystalline,* and they can form well-defined geometric crystals. This mineral (◄) is crystalline, and the shape of the crystals shows the internal arrangement of its atoms. If atoms are arranged in a random way, such as in a volcanic glass, it is not crystalline and not a mineral.

Chemical Composition

8. Minerals are homogeneous and so have specific chemical compositions that do not depend on the size of the sample that is analyzed. Table salt, which is the mineral *halite,* contains atoms of the chemical elements *sodium* (Na) and *chlorine* (Cl) in equal proportions, no matter how big or small the specimen. Most minerals have a specific chemical formula, like NaCl for halite.

09.01.a7 Salt

B What Are Some Common Minerals?

There are thousands of different minerals, but only a handful are very common and important to understanding Earth and its landscapes. The most common family of minerals is *silicate minerals*, which account for over 90% of the minerals in Earth's crust. In most silicate minerals, one silicon atom (Si) is bonded with four oxygen (O) atoms to form $(SiO_4)^{-4}$, which is the building block for the vast majority of silicate minerals. Other minerals do not contain silicon and so are called *nonsilicate minerals*; they include salt and the oxide minerals that cause many landscapes to be reddish.

Silicate Minerals

1. *Quartz* (▶) is a common mineral, with a formula of SiO_2. It is generally transparent to nearly white, but it can be many colors. It is a relatively hard mineral that commonly occurs as well-formed crystals, but it fractures irregularly, along smoothly curving surfaces.

09.01.b1

2. *Feldspar* is a very common mineral, coming in two varieties: *potassium feldspar* (▶) and *plagioclase*. Feldspars contain varying amounts of aluminum, potassium, sodium, and calcium. Feldspars are pink to cream-colored to gray.

09.01.b2

3. *Mica* refers to a family of minerals that breaks into flakes and sheets (▶) that are typically partially transparent and shiny because the flat surfaces reflect light. Micas contain potassium, aluminum, silicon, and oxygen and can be clear, silvery-gray, green, brown, or black.

09.01.b3

4. *Mafic minerals* are silicate minerals that, in addition to silicon and oxygen, contain magnesium and iron, such as in the mineral olivine (▶). These minerals are generally dark colored and include the common mineral groups *amphiboles, pyroxenes*, and dark micas.

09.01.b4

Nonsilicate Minerals

5. *Carbonate minerals* contain carbon and oxygen bonded with other elements. The most common carbonate mineral is *calcite* (▶), a common calcium-carbonate mineral $(CaCO_3)$. It may be almost clear but commonly has a cream to light-gray color. Other important carbonates include *dolomite*, which also contains magnesium.

09.01.b5

6. *Oxide minerals* contain oxygen bonded with a metallic element, like iron, copper, or titanium. *Hematite* (▶) is a common iron oxide (Fe_2O_3) that can be black, brown, silvery gray, or earthy red; it is the red color in rust and many red-rock landscapes. Another iron-oxide mineral is *magnetite* (Fe_3O_4), which is strongly magnetic.

09.01.b6

7. *Salt minerals* include several families of minerals, including *halite* (NaCl), which is the mineral in table salt. Halite grows and breaks into cubic shapes (▶) and generally forms from the evaporation of water. Other salt-like minerals include *gypsum*, a calcium-sulfate mineral that forms in environments similar to those in which halite forms.

09.01.b7

8. *Sulfide minerals* all contain sulfur bonded with a metallic element, like iron or copper. The most common sulfide mineral is *pyrite* (▶), an iron-sulfide mineral (FeS_2). Pyrite has a pale bronze to brass-yellow color (it is "fool's gold"). It commonly forms cube-shaped crystals. Other sulfide minerals contain copper, zinc, lead, and other metals.

09.01.b8

Clay Minerals

The term *clay* can refer either to a family of silicate minerals or to any very fine particles that are less than 0.002 millimeters in diameter. Clay minerals consist of sheets, similar to mica minerals, but the sheets are weakly held together, so they easily slip past one another, giving clays their slippery feel. When some clay minerals get wet, water pushes apart the weakly bonded sheets, causing the clay to expand. They also feel sticky when wet.

09.01.t1

Most clay minerals have light colors but may appear dark if mixed with other material (dark minerals or organic debris). Most clay minerals form when weathering breaks down rocks and minerals, especially feldspar, and they are an important component of soil. The fine grain size and low density of clay particles mean that they are easily picked up and transported long distances by wind and water. Clay can be deposited on land, such as on a floodplain or in a lake, but some makes it to the open ocean, where it finally settles to the ocean floor, forming submarine mud.

Before You Leave This Page

✔ Explain the relationship between rocks and minerals.

✔ Explain each characteristic that a material must have to be a mineral.

✔ List the common minerals and several characteristics of each.

9.1

9.2 How Do Rocks Form?

VARIOUS PROCESSES OPERATE on and within Earth to produce the variety of rocks we observe in landscapes. Many common rocks form in stream bottoms, beaches, or other familiar environments on Earth's surface. Other rocks form in less familiar environments, under high pressure deep within Earth, or at high temperatures associated with a volcano. To understand the different kinds of rock that can form, we explore the types of materials that characterize different modern-day environments.

A What Types of Sediments Form in Familiar Surface Environments?

Some types of rock form from the combination of loose materials that were brought together somehow in a location. Think about what you have observed on the ground in places such as lakes, streambeds, and mountainsides—probably mud, sand, and larger rocks. These loose materials are *sediment* and are formed by the breaking and wearing away of other rocks in the landscape. Although more hidden from us, sediment also occurs beneath the sea.

09.02.a1 Switzerland

09.02.a3 Granite, CO

1. *Glaciers* incorporate rock debris into their flowing, icy masses (◄). They carry a wide variety of sediment, from large, angular boulders to fine rock powder. They ultimately deposit the sediment along the edges of the melting ice.

2. *Steep mountain fronts* exhibit large, angular rocks that broke away from bedrock and moved downhill under the influence of gravity. Steep mountains may produce landslides and unstable, rocky slopes covered with angular blocks (►).

3. *Sand dunes* (►) are mostly sand, which has been shifted along the ground by wind. They contain sand because sand is too large for wind to move out of the area, and wind blows away smaller particles, like clays.

09.02.a4 Namibia

4. *Beaches* (▼) typically have waves, sand, broken shells, and rounded, well-worn stones. Some beaches are mostly sand, others are mostly stones, and some are mostly broken shells.

09.02.a2

09.02.a5 Naxos, Greece

6. *Stream channels* contain sand and bigger pebbles and cobbles, whereas low areas beside the channel accumulate smaller particles of silt and clay, which were carried farther from the stream channel during floods. Some streams flow into lakes, which have a muddy, clayey bottom, with sand around the lake shore.

5. In deeper water, the seafloor consists of mud and the remains of swimming and floating creatures that settle to the bottom. Seafloor closer to the land receives a greater contribution of sand and other sediment derived from the land. Streams and wind are especially effective in delivering this sediment from the land to the sea.

B What Types of Rocks Form in Hot or Deep Environments?

Some rocks form in environments that are unfamiliar to us and hidden from view, deep within Earth. Others form at very high temperatures associated with volcanic eruptions. Distinct families of rocks result from these rock-forming processes, which include solidification of magma, growing (precipitating) minerals from hot water, or the action of high temperatures and pressures that transform one type of rock into another type of rock. Read counterclockwise.

09.02.b2 Philippines

09.02.b3 Hawaii

1. In many volcanoes, molten rock (magma) flows onto the surface, creating *lava* that flows downhill or piles up around a vent (▶).

2. Explosive volcanoes erupt *volcanic ash* (▲), which can fall back to Earth and blanket the terrain or can surge rapidly and dangerously down the flanks of a volcano.

5. Distinctive rocks form when hot waters cool and grow minerals. This may occur beneath the surface or on the surface in hot springs (▼).

Magma

Heating of Rocks

3. Magma that does not erupt may cool and solidify in a *magma chamber*, forming granite or other rocks at depth. Heat from the magma chamber may bake adjacent rocks, changing them into different kinds of rocks.

Force

Force

09.02.b1

4. Deep within Earth where temperatures and pressures are high, forces can squeeze and deform rocks into new atomic arrangements and into new types of rocks. Under such force, solid rocks slowly flow, shear, and bend. Changing a rock by heat, pressure, or deformation is the process of *metamorphism*.

09.02.b4 Yellowstone NP, WY

Families of Rocks

The diverse environments shown on these pages produce many different types of rocks that, depending on the classification scheme, are grouped into three or four families. To interpret how rocks form, we observe modern environments and note the dominant types of sediment, lava, or other material. We infer that these same types of materials would have been produced in older, prehistoric versions of that environment. By doing this, we use modern examples to interpret ancient rocks and to understand how they formed. In this way, *the present is the key to the past.*

Sedimentary rocks form on Earth's surface, including the seafloor, mostly from loose sediment that is deposited by moving water, air, or ice. If loose sediment is buried, it can become consolidated into hard rock over time. Other types of sedimentary rocks form by growing minerals from water or by coral and other organisms that extract material directly from water.

Rocks formed from cooled and solidified magma are *igneous rocks*. These form when volcanoes erupt ash and lava or when molten rock crystallizes in magma chambers at depth.

Rocks changed by temperatures, pressures, or deformation are *metamorphic rocks*. Metamorphism can change sedimentary or igneous rocks, or even further modify preexisting metamorphic rocks. Finally, some rocks grow directly from hot water, and these are commonly considered to be a type of metamorphic rock because they result from fluids at high temperatures.

Before You Leave This Page

✓ Distinguish the three families of rocks by describing how each type forms.

✓ For each family of rocks, describe two settings where such rocks form and the processes that take place in each setting.

✓ Describe what we mean by "the present is the key to the past" and explain how it is used to interpret the origin of rocks and sediment.

9.3 What Can Happen to a Rock?

MANY THINGS CAN HAPPEN TO A ROCK after it forms. It can break apart into sediment or be buried deeply and metamorphosed. If temperatures are high enough, a rock can melt and then solidify to form an *igneous* rock. Uplift can bring metamorphic, igneous, and sedimentary rocks to the surface, where they break down into sediment. Examine the large figure below and think of all the things that can happen to a rock.

1. Weathering

A rock on the surface interacts with sunlight, rain, wind, plants, and animals. As a result, it may be mechanically broken apart or altered by chemical reactions via the process of *weathering*. Weathering creates sediment, which ranges from very fine clay to the large boulders shown here (▼).

09.03.a2 Central CO

2. Erosion and Transport

Rock pieces loosened or dissolved by weathering can be stripped away by *erosion* and moved away from their source. Glaciers, flowing water, wind, and the force of gravity on hillslopes can transport eroded material away.

TRANSPORT

09.03.a1

8. Uplift

At any point during its history, a rock may be *uplifted* back to the surface where it is again exposed to weathering. Uplift commonly occurs in mountains, but it can also occur over broad regions that lack mountains. Uplift is generally accompanied by deformation, metamorphism, and magmatism, which are normally grouped under the general term *tectonics*. Tectonics only requires deformation, but uplift, metamorphism, and magmatism are also commonly involved.

UPLIFT

6. Melting

A rock exposed to high temperatures may *melt* to produce a magma. Melting usually occurs at great depth within the Earth. The formation of magma, along with the associated processes, is called *magmatism*.

09.03.a3 Acadia NP, ME

7. Solidification

As magma cools, either at depth or after being erupted onto the surface, it will solidify and harden, a process called *solidification*. If crystals form during solidification, the process is *crystallization*, as illustrated by large, well-formed crystals (◄) that formed by slow cooling of magma at depth. Because it forms beneath the surface, granite cools very slowly and forms large crystals, and much later it is uplifted to the surface.

The Life and Times of a Rock — The Rock Cycle

This process, in which a rock may be moved from one place to another or even converted into a new type of rock, is the *rock cycle*. Scottish physician James Hutton first conceived of the rock cycle in the late 1700s as a way to explain the recycling of older rocks into new sediment. Most rocks do not go through the entire cycle, but instead move through only part of the cycle. Importantly, the different steps in the rock cycle can happen in almost any order. Steps are numbered on this page only to guide your reading and follow *possible* sequences of events for a single rock.

Suppose that uplift brings up a rock and exposes it at Earth's surface. Weathering dissolves and breaks the rock into smaller pieces that can be eroded and transported at least a short distance before being deposited. Under the right conditions, the rock fragments will be buried beneath other sediment or perhaps beneath volcanic rocks that are erupted onto the surface. Many times, however, sediment is not buried, but only weathered, eroded, transported, and deposited again. As an example of this circumstance, imagine a rounded rock in a streambed. When the flowing currents in the stream are strong enough, they pick up and carry the rock downstream, perhaps depositing it within or near the channel, where it may remain for years, centuries, or even millions of years. Later, a flood that is larger than the last one may pick up the rock and transport it farther downstream.

If the rock is buried, it has several possible paths. It can be buried to some depth and then uplifted back to the surface to be weathered, eroded, and transported again. Alternatively, it may be buried so deeply that it is metamorphosed under high temperatures and pressures. Uplift can bring the metamorphic rock to the surface.

If the rock remains at depth and is heated to even higher temperatures, it can melt. The magma that forms may remain at depth or may be erupted onto the surface. In either case, the magma eventually will cool and solidify into an igneous rock. Igneous rocks formed at depth may later be uplifted to the surface or they may remain at depth, where they can be metamorphosed or even remelted.

A key point to remember is that the rock cycle illustrates the possible things that can happen to a rock. Most rocks do not complete the cycle because of the many paths, interruptions, backtracking, and shortcuts a rock can take. The path a rock takes through the cycle depends on the specific natural events that happen and the order in which they occur. There are many possible variations in the path a rock can take.

3. Deposition

When kinetic energy associated with transportation decreases sufficiently, water, wind, and ice *deposit* their sediment. Sediment carried by streams can reside within or next to the channel or collect near the stream's mouth. The stream gravels in the photograph below are at rest for now but could be picked up and moved by a large flood. Some sediment is formed by the *ions* that are removed from water or by the actions of organisms.

09.03.a4 Tibet

4. Burial and Lithification

Once deposited, sediment can be buried and compacted by the weight of overlying material. Chemicals in groundwater can coat sedimentary grains with minerals and deposit natural cements that bind adjacent grains. The process of sediment becoming sedimentary rock is *lithification*.

UPLIFT

BURIAL

5. Deformation and Metamorphism

After a rock forms, strong forces can squeeze the rock and fold its layers, a process called *deformation*. If buried deeply enough, a rock can be heated and deformed to produce a metamorphic rock. The rock in the photograph to the left began as some other type of rock but was strongly deformed, metamorphosed, and converted into a metamorphic rock.

09.03.a5 Kettle Falls, WA

> ## Before You Leave This Page
>
> ✓ Sketch a simple version of the rock cycle, labeling and explaining, in your own words, the key processes.
>
> ✓ Describe why a rock might not experience the entire rock cycle.

9.3

9.4 What Are Some Common Sedimentary Rocks?

SEDIMENTARY ROCKS ARE THE MOST COMMON rock in many regions, forming many of the landscapes we encounter. They form when loose sediment, such as pebbles, sand, and clay, is deposited and then converted into a rock through the process of *lithification*. Different types of sedimentary starting material result in different types of sedimentary rocks.

A What Are Some Common Clastic Sedimentary Rocks?

Some sedimentary rocks are composed of rock and mineral pieces, called *clasts*, and are called *clastic rocks*. We describe and classify clastic sedimentary rocks based primarily on the sizes of clasts, along with other aspects such as clast roundness. Common clastic sedimentary rocks are shown below and described using these criteria of classification.

Gravel-Sized Clasts

1. *Conglomerate* has large, rounded clasts (pebbles, cobbles, or boulders) with sand and other fine sediment between the large clasts (◄). This conglomerate has well-rounded cobbles in a matrix of mostly quartz sand.

2 *Breccia* is similar to conglomerate except that the clasts are angular. Breccia usually has a jumbled appearance because most has a range of clast sizes, such as boulders in a mud-rich matrix, as shown here (►).

09.04.a1 Apache Trail, AZ

09.04.a2 Wickenburg, AZ

Sand-Sized Clasts

3. *Sandstone* consists of sand-sized grains (◄). It can contain some larger and smaller clasts, but it is mostly composed of sand. It generally has better-defined layers than conglomerate or breccia.

Mud-Sized Clasts

4. *Shale* consists mostly of fine-grained clasts (i.e., mud), especially very fine-grained clay minerals. The minerals are aligned, so the rock breaks in sheetlike pieces or chips, as displayed here (►).

09.04.a3 Durango, CO

09.04.a4 Boulder, CO

B How Do Clastic Sediments Become Lithified into Clastic Sedimentary Rock?

Compaction

1. As sediment is buried, increasing pressure pushes clasts together, a process called *compaction*. Compaction forces out excess water and causes sediments to lose as much as 40% of their volume. Originally loose sediment becomes more dense and more compact.

2. Sediments near Earth's surface, such as these sand grains, are a loose collection of clasts. The grains rest on one another but do not fit together tightly, so spaces, called *pore spaces*, exist between the grains. Pore spaces are generally filled with air and water.

3. As sand grains are buried, the weight of overlying sediment and other materials forces the grains closer together. The amount of pore space decreases as air or water is expelled, so the layer loses thickness. In this manner, sediment becomes compacted into sedimentary rock.

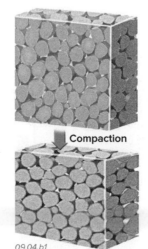

09.04.b1

Cementation

4. Even after sediment is compacted, adjacent clasts do not fit together perfectly. The remaining pore spaces are commonly filled with water that contains dissolved materials. The dissolved materials can *precipitate* (come out of solution) to form minerals that act as a natural cement that holds the sediment together, the process of *cementation*.

5. When sand grains and other sediment are deposited, abundant pore spaces exist between the grains, even after compaction. If these spaces are interconnected, as they are with sand grains, water can flow slowly through the sediments, carrying chemical components into or out of the sediment.

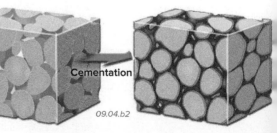

09.04.b2

6. As sediment is buried, water moving through pore spaces can precipitate minerals in the pore spaces. Such minerals are called *natural cement*, and they help turn the sediment into hard sedimentary rock.

C What Are Some Common Nonclastic Sedimentary Rocks?

Some sedimentary rocks are not composed of clasts and are therefore *nonclastic rocks*. Processes capable of producing nonclastic sedimentary rocks may be chemical, biological, or have both chemical and biological aspects. The photographs below show common nonclastic sedimentary rocks.

1. *Rock salt* refers to a rock composed of mostly the mineral halite (NaCl) or related minerals. Halite commonly precipitates as the water that dissolved it evaporates. Most table salt forms in this way.

09.04.c1 Paradox, CO

2. *Gypsum* refers to a type of mineral or a rock composed of that mineral. It mostly forms when seawater evaporates in tidal flats and narrow seas. It is the material in wallboard (sheetrock) in our houses.

09.04.c2 Eagle, CO

3. *Limestone*, made mostly of calcium carbonate (mostly calcite), forms when shells and coral skeletons are cemented together. Many limestones, like the one here, contain fossils, a record of past life.

09.04.c3 Superior, AZ

4. *Chalk* is a soft, very fine-grained limestone that forms from the accumulation of the calcium carbonate remains of microscopic organisms that float in the sea. Chalk forms the famous White Cliffs of Dover, England.

09.04.c4 Colorado Springs, CO

09.04.c5 Battle Mtn., NV

5. *Chert* is a silica-rich rock that forms in several ways. One way chert forms is in layers from the accumulation and compaction of tiny, silica-rich plankton shells that fall to the ocean bottom. It can also occur as layers and irregular masses within limestone that form from the mixing of water with different chemistries.

09.04.c6 Jasper Knob, MI

6. *Iron formation* is a rock composed of centimeter-thick layers of iron oxide (the minerals hematite and magnetite), iron carbonate, and iron silicate minerals, commonly intermixed with very fine-grained quartz. Most iron formations precipitated from seawater early in Earth's history. It is the main source of iron used in steel.

09.04.c7 Witbank, South Africa

7. *Coal* is formed from wood and other plant parts that have been buried, compacted, and heated enough to drive off most of their water and oxygen. Depending on the amount of heat and pressure, coal can be soft and dull or hard and shiny. There are different kinds of coal, including lignite, bituminous coal, and anthracite, in order of increasing quality.

Types of Natural Cement

There are four main types of natural cements that hold grains together: calcite, silica, clay minerals, and iron oxides. Other materials, such as gypsum, can function as cement but are less common.

Calcite ($CaCO_3$) is a common cement in sandstone and other sedimentary rocks. It holds grains together moderately well, but it is easily redissolved, so a calcite-cemented sandstone may become friable (crumbly).

Silica (SiO_2) acts as a cement in some sandstone and other sedimentary rocks. It forms a strong cement that can tightly bind grains, forming a tough, resistant rock.

Clay minerals can cement together larger grains, including sand. They may have been deposited with the sediment or formed from the alteration of feldspar or volcanic ash.

Hematite and other iron oxide minerals precipitate from water as a natural cement between the grains. Iron oxide minerals commonly give sediment deposited on land a reddish color, as displayed in the spectacular red-rock landscapes of the Desert Southwest.

Before You Leave This Page

- ✔ Describe the classification of common clastic sedimentary rocks.
- ✔ Describe what happens to clastic sediment as it becomes buried and converted into rock.
- ✔ Describe some common nonclastic sedimentary rocks.
- ✔ Describe the natural cements that are common in sedimentary rocks.

9.4

9.5 What Are Igneous Processes and Rocks?

IGNEOUS ROCKS FORM very differently from sedimentary rocks—they form by solidification of magma. Most igneous rocks have millimeter- to centimeter-sized crystals, but some have meter-long crystals, and others are composed not of minerals but of noncrystalline glass. Igneous rocks may contain holes, fragments, or compacted volcanic ash. In what settings do different igneous rocks form, and what is the character of the resulting rocks?

A In What Settings Do the Different Igneous Textures Form?

Igneous rocks display various textures, each of which indicates the environment in which the magma solidified. Magma can solidify at depth, erupt onto the surface as molten lava, or be explosively erupted as volcanic ash. Examine the figure below and think about where each texture in the photographs on these two pages might form.

09.05.a2 Northern NM

1. *Vesicles* are holes in volcanic rocks (◄) that form when gases dissolved in magma accumulate as bubbles at or near Earth's surface. Many lavas contain vesicles, and much of the material in volcanic ash forms when the thin walls between vesicles burst, shattering partially solidified magma into the sharp particles that compose volcanic ash.

09.05.a3 Inyo Mtns., CA

2. Some volcanic rocks are composed of angular fragments, a texture called *volcanic breccia*. These form in many ways, including from explosive eruptions of ash and rock fragments, from a lava flow that breaks apart as it partially solidifies while flowing, or from volcano-triggered mudflows and landslides.

09.05.a4 Greece

3. *Volcanic glass* forms when magma erupts on the surface and cools so quickly that crystals do not have time to form, as can occur in a lava flow or in volcanic ash.

4. Other igneous rocks cool slowly enough to grow crystals, with the size of the resulting crystals mostly revealing the time that was available for the magma to crystallize. A magma that is deeper needs more time to cool, and therefore solidify, than one nearer to the surface, so the size of crystals is mostly related to the depth, with larger crystals generally indicating greater depths. Examples of this variation in crystal size are shown below.

09.05.a1

9. Some *volcanic ash* erupts vertically in a column and settles back to Earth. This ash cools significantly before accumulating on the surface, so it consists of ash and other particles that are not strongly compressed together, as in the rock called *tuff* (►).

09.05.a5 Mt. Ta

10. Other volcanic ash erupts in thick clouds of hot gas and rock fragments, called *pyroclastic flows*, that flow rapidly downhill under the influence of gravity. The ash deposited by pyroclastic flows is very hot, and so most parts are compacted and welded together to some extent.

11. Some igneous rocks have a distinctive texture (called *porphyritic*) with larger crystals in a finer grained matrix (▼). Such rocks indicate that the magma cooled somewhat at depth to grow the larger crystals, but then rose and solidified on or closer to the surface, where it rapidly cooled to solidify the finer grained matrix.

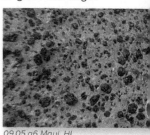

09.05.a6 Maui, HI

09.05.a7 Southern AZ

5. *Fine-grained* igneous rocks form if magma cools quickly and can only grow small crystals.

09.05.a8 South Africa

6. *Medium-grained* rocks have larger crystals (mm but not cm) that are easily visible to the unaided eye.

09.05.a9 Cave Creek, AZ

7. *Coarse-grained* igneous rocks form at greater depths, where slow cooling allows large crystals to grow.

09.05.a10 Grand Canyon, A

8. Some igneous rocks contain very large crystals (cm to m), and are called *pegmatite*.

B How Does the Composition of Igneous Rocks Vary?

Geoscientists organize igneous rocks according to the *sizes of crystals* and the *kinds of minerals* in a rock. Rocks with a light color and abundant quartz and feldspar are *felsic* rocks, whereas rocks that are dark and contain minerals rich in magnesium and iron are *mafic* or *ultramafic* rocks. *Intermediate* rocks are in between. The top two rows of photographs below show igneous rocks of different compositions and crystal size. The bottom two photographs show two other types of volcanic rocks.

	Felsic	Intermediate	Mafic

Coarsely Crystalline

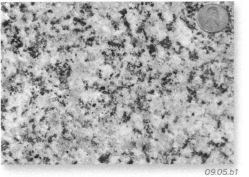

09.05.b1

1. *Granite* is a coarsely crystalline, light-colored igneous rock. The light color is due to the light-colored, felsic minerals—quartz and feldspar. Most granites also contain some dark or light-colored mica, some oxide minerals, and certain other minerals.

09.05.b2

2. *Diorite*, an intermediate rock, contains more mafic minerals than does granite. It is usually moderately dark from abundant dark silicate minerals, such as amphibole, and it also generally contains abundant plagioclase feldspar as the light-colored crystals.

09.05.b3 Selway, ID

3. *Gabbro* is a coarsely crystalline, mafic rock. It typically is dark and consists of dark-colored silicate minerals, like pyroxene. It can also contain light-gray feldspar. Some varieties are somewhat lighter colored, and some contain the greenish mineral olivine.

Finely Crystalline

09.05.b4 Shoshone, CA

4. *Rhyolite* is the fine-grained equivalent of granite. It is mostly a finely crystalline rock, but it can contain glass, volcanic ash, pieces of pumice, and visible crystals.

09.05.b5 Flagstaff, AZ

5. *Andesite,* the fine-grained equivalent of diorite, is commonly gray or greenish. Andesite commonly has visible crystals of cream-colored feldspar or dark amphibole.

09.05.b6 Grants, NM

6. *Basalt* is a dark mafic lava rock, equivalent in composition to gabbro. Most basalt is dark gray to nearly black, and can have some crystals and vesicles, as shown here.

Other

09.05.b7 Katmai, AK

7. *Pumice* is a volcanic rock containing many vesicles (holes). The holes are so abundant that most pumice floats on water. The solid material in pumice begins as volcanic glass, but over time it can convert into microscopic crystals.

09.05.b8 Superior, AZ

8. *Tuff* is a volcanic rock composed of a mix of volcanic ash, pumice, crystals, and rock fragments. If the particles of ash and pumice remain hot, they become compacted by overlying materials, becoming a hard, dense rock with flattened pumice, as shown here.

Before You Leave This Page

☑ Sketch an igneous system and show where the main igneous textures form.

☑ Describe or diagram the classification of some common igneous rocks, including some examples.

9.5

9.6 What Are Metamorphic Processes and Rocks?

METAMORPHISM INVOLVES CHANGING A ROCK that has become unstable, generally at elevated temperatures and pressures. A rock may be unstable because the minerals it contains are unstable, and as a result it develops new (metamorphic) minerals. It may be unstable because of the way the grains or layers are arranged, so it develops a new (metamorphic) texture. Alternatively, a rock may develop new metamorphic minerals and a new metamorphic texture.

A What Causes Metamorphism?

For a rock to be metamorphosed, it must be subjected to conditions of temperature, pressure, and fluid chemistry that make it unstable. This generally occurs when temperature (T) and pressure (P) increase, possibly accompanied by forces that cause the rock to *deform* or by an influx of deep fluids. There are two main kinds of metamorphism: *contact metamorphism* is caused by local heating by magma, typically without deformation, whereas *regional metamorphism* involves deformation along with heating over a broader region. Metamorphic temperatures can be as low as 100°C or can exceed 700–800°C, the temperature at which many crustal rocks begin to melt. Most metamorphism occurs at temperatures somewhere in between.

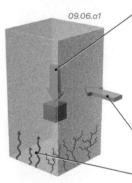

09.06.a1

1. *Pressure* increases with depth in Earth because rocks are buried more deeply. Higher pressures compress the rocks and, in combination with high temperatures, may cause some minerals to be unstable. Forces deep in Earth can cause rocks to move and deform.

2. *Temperature* increases with depth and near magma. An increase in temperature usually causes new minerals to grow or existing minerals to grow larger. It can also weaken the rocks, allowing them to deform. The crust contains abundant water and other fluids, shown here as blue in water-filled fractures. Such fluids interact with minerals and carry material into, through, and out of rocks.

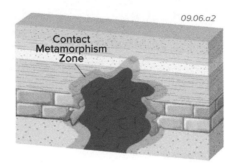

09.06.a2

Contact Metamorphism Zone

3. *Contact Metamorphism*—Rising magma efficiently brings thermal energy higher into the crust, heating the wall rocks. This is called contact metamorphism because it occurs near contacts (boundaries) of magma. Heating causes new minerals to grow or existing minerals to increase in size. In contact metamorphism, deformation may or may not occur with the heating.

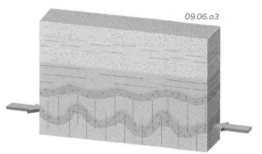

09.06.a3

4. *Regional Metamorphism*—In regional metamorphism, heating is often accompanied by enough force to cause deformation. The imposed forces can result from tectonics or burial. Regional metamorphism causes new minerals to grow and existing minerals to increase in size, while deformation during metamorphism generally results in the rock developing planar or linear metamorphic fabrics during metamorphism.

B What Processes Occur During Metamorphism and Deformation?

Many processes that operate during metamorphism change the minerals in a rock or change the arrangement of those minerals. Some metamorphic processes are related to heating or are chemical in character. Such processes cause grains to grow, recrystallize, redistribute themselves, or even dissolve in response to temperature, pressure, and any imposed forces. Some processes are physical and may deform or rotate individual crystals, grains, and layers, producing planar or linear metamorphic fabrics.

Chemical Processes

During metamorphism, movement of chemical elements can cause existing minerals to grow larger or can form new minerals. Minerals can grow in fairly random orientations, as occurs during contact metamorphism, which is due to an increase in temperature but not in forces. Or, minerals can grow with a preferred orientation, becoming aligned as in the example shown here. If minerals are arranged in lines, the rock has a *lineation*. If they are arranged in planes, the rock has a *foliation*. Many metamorphic rocks have a lineation and a foliation.

09.06.b1

Physical Processes

Metamorphism is generally accompanied by deformation of any original constituents of the rocks, like crystals or pebbles. Forces imposed on the rock (represented by blue arrows) result in deformation that can flatten grains and clasts that were initially somewhat spherical into shapes like pancakes or the thin, long top of a skateboard. If a rock is flattened during metamorphism, the deformed objects will define a foliation. If a rock is stretched in one direction, it acquires a lineation, commonly along with a foliation. Both physical and chemical processes occur during most metamorphism.

09.06.b2

C What Rocks Form When Sedimentary Rocks Are Metamorphosed?

There are many types of metamorphic rocks, and these rocks indicate diverse starting rock types and various conditions under which rocks can be deformed and metamorphosed. Some metamorphic rocks form when sedimentary rocks are subjected to a change in temperature, pressure, or both. Photographs below show what happens to three common sedimentary rocks—shale, sandstone, and limestone. Igneous rocks can also be metamorphosed with similar results, and metamorphic rocks can be metamorphosed several times under varying conditions of temperature and pressure.

Increasing Temperatures

Shale

09.06.c1 Larder Lake, Ontario, Canada

Slate

09.06.c2 Superior, AZ

Phyllite

09.06.c3 Hunt Valley, MD

Schist

09.06.c4 San Gabriel Mtns., CA

Gneiss

1. When a shale is metamorphosed at low to moderate temperature (less 200°C), it can develop cleavage (a type of foliation) and become *slate* (◄). Slates are dull (not shiny) and commonly dark.

2. At slightly higher temperatures, microscopic mica crystals give the rock a shiny aspect or sheen (◄). Such a rock is *phyllite*.

3. At higher temperatures (350°C), crystals of mica and other minerals become large enough to see. The resulting rock is a shiny rock called a *schist* (◄).

4. At even higher temperatures, chemical constituents are mobilized and light- and dark-colored minerals separate, forming a foliation and banded rock called *gneiss* (◄).

Sandstone

09.06.c5 Baraboo, WI

Quartzite

5. Most sandstones are predominantly quartz (SiO_2), a mineral that is stable over a wide range of temperature and pressure conditions. During metamorphism, quartz grains grow together and become so lightly bonded that fractures break across the grains rather than around them. This type of rock is a *quartzite* (▲). Quartzite is made of quartz, just like the original sandstone, and can preserve beds and other original sedimentary features.

09.06.c6 Big Maria Mtns., CA

Coarse Quartzite

6. With higher temperatures of metamorphism, the quartz in the rock begins to merge into larger crystals and can become a coarser grained quartzite (▲), in some cases with no individual grains left. Quartz is soluble and mobile in metamorphic fluids and so at high temperatures can be redistributed into quartz veins, which are common in metamorphic rocks.

Limestone

09.06.c7 Ios, Greece

Finely Crystalline Marble

7. Limestone consists mostly of calcite, a chemically reactive mineral. Low-temperature (<250°C) metamorphism of limestone causes calcite to recrystallize slightly, forming a finely crystalline *marble* (▲).

09.06.c8 Franklin Mtns., El Paso, TX

Impure Marble

8. At moderate temperatures (~400°C), marble becomes medium grained (▲). When metamorphosed, impurities in the limestone, like clay and chert, may produce various new metamorphic minerals, like these red garnets.

09.06.c9 Naxos, Greece

Coarse Marble

9. At higher temperatures, calcite crystals grow coarser to produce a coarsely crystalline marble (▲). The one shown here is nearly 100% calcite, but coarse marbles commonly also contain other minerals.

Before You Leave This Page

☑ Summarize causes of metamorphism, and describe or sketch a chemical and physical process that can accompany metamorphism.

☑ Describe the changes different sedimentary rocks undergo as they metamorphose and the metamorphic rocks they become.

9.7 How Are Different Rock Types Expressed in Landscapes?

SEDIMENTARY, IGNEOUS, AND METAMORPHIC ROCKS can each have a distinctive appearance in the landscape, often allowing us to recognize these rocks from a distance. With some practice, we can drive down a highway, observe the characteristics of a hill or mountain, and make an educated guess about what types of rocks are exposed. Here, we provide a brief introduction to interpreting the rock types of landscapes from a distance.

A How Are Sedimentary Rocks Expressed in the Landscape?

There are diverse types of sedimentary rocks, displaying a wide variety in colors, thickness of layers, and resistance to erosion. The unifying feature of most sedimentary rocks is the presence of visible layers. Each main type of sedimentary rock has certain characteristics that allow us to identify it from a distance.

Sandstone

09.07.a1 Moab, UT

1. Some sandstone layers are resistant to erosion and appear massive from a distance because they have little variation of grain size, as in the reddish cliff in this photograph. Such thick layers of sandstone generally were deposited by wind as sand dunes.

09.07.a2 Ruby Canyon, CO

2. Most sandstone has layers that differ from one another in color, grain size, or composition of the grains. Such layers can be parallel beds, as shown here, or cross beds centimeters to tens of meters high. Numerous layers mean many changes in conditions during deposition.

Fine-Grained Clastic Rocks

09.07.a3 Grand Junction, CO

3. Shale and other fine-grained clastic sedimentary rocks are relatively easily eroded rocks. Where exposed, these rocks typically form soft slopes covered by small, loose chips derived from weathering of the thinly bedded rocks.

09.07.a4 Petrified Forest, AZ

4. Shale and associated fine-grained rocks form another distinctive type of landform—*badlands*. Badlands have a soft, rounded appearance that exhibits the softness of the rocks. Badlands also have an intricate network of small drainages and eroded ridges carved into the soft rocks.

Limestone

09.07.a5 Chamonix, French Alps

09.07.a6 Austrian Alps

09.07.a7 Guilin, China

5. In some relatively dry climates, limestone and other carbonate rocks are relatively erosion-resistant. The rocks can form light-gray to dark-gray cliffs and steep slopes composed of beds that may vary slightly in thickness and color. Most limestone has visible layers, on either a small or large scale.

6. Limestone is very soluble, and so in very wet climates it dissolves and weathers quickly, especially along fractures and layers, forming caves, sinkholes, pits, and depressions.

7. Weathered and dissolved limestone forms distinctive landforms with caves, sinkholes, and pillars. Such landscapes formed by dissolution of limestone are called *karst terrain*.

Before You Leave This Page

✓ Describe the characteristics of some common sedimentary rocks, including their expression in landscapes.

✓ Describe the appearance of some common igneous rocks in landscapes.

✓ Describe some characteristics displayed by metamorphic rocks, as exposed in landscapes.

B How Are Igneous Rocks Expressed in the Landscape?

Igneous rocks form from magma, either from magma that solidifies below the surface, forming *intrusive* (plutonic) rocks, or magma that erupts onto the surface, forming *extrusive* (volcanic) rocks.

Granite makes up the bulk of the continental crust, and most granite is fractured. These fractures help speed up the weathering process, which wears away the edges and corners of fractures, resulting in rounded shapes in many granite exposures, a process called *spheroidal weathering*.

09.07.b1 Date Creek Mtns, AZ

Basaltic lava flows are the most common type of volcanic rock in many landscapes. Basalt flows are dark and spread out easily, forming distinct layers. Basaltic sequences generally have reddish zones from pyroclastic rocks that accompanied the eruption of lava, as shown in this photograph.

09.07.b2 Grand Coulee, WA

Volcanic rocks commonly have some sort of layers, and they also commonly have fractures perpendicular to the layers (▶). Depending on the rock compositions and the type of eruption that formed them, volcanic rocks can be various shades of gray, green, brown, tan, and cream colored.

09.07.b3 Copper Creek, AZ

09.07.b4 Columbia Plateau, WA

Many volcanic units, especially basaltic lava flows, have distinctive columnar fractures called *columnar joints*. Such joints form during the cooling and contraction of solidified igneous rocks. The size and orientation of the columns indicate how the rock cooled, but most columns are steep.

C How Are Metamorphic Rocks Expressed in Landscapes?

Rocks that have been changed (or metamorphosed) by increased temperature and pressure are metamorphic rocks. Different types of metamorphic rocks result from different starting rock types and different conditions under which rocks can be metamorphosed and deformed. These differences, in turn, are expressed by a somewhat variable appearance of metamorphic rocks in the landscape.

09.07.c1 Hunt Valley, MD

1. Metamorphic rocks can be shiny up close or even from a distance, especially if their mica minerals share a similar orientation (i.e., have a foliation) and reflect light. This schist (◀) is definitely shiny up close and from meters away.

3. Many metamorphic rocks have layers that form platy, jagged outcrops and tabular slabs of rock, as in the left photograph below. They can include numerous bodies of granite and other igneous rocks, some of which, as in the right photograph below, record how folded and deformed these metamorphic rocks really are (▼).

09.07.c3 Aurland Trail, Norway   09.07.c4 Grand Canyon, AZ

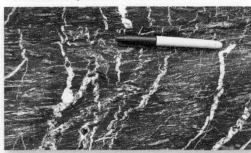
09.07.c2 San Juan Islands, WA

2. Metamorphic rocks commonly have mineral-filled fractures, called veins, like the quartz-filled veins shown here (◀). Such veins can contain other minerals and commonly represent materials mobilized by hot waters that accompany metamorphism.

9.8 What Controls the Appearance of Landscapes?

THE APPEARANCE OF A LANDSCAPE reflects the types of starting materials (rocks and loose sediment) and processes that act upon those materials, including precipitation, running water, wind, flowing ice, the pull of gravity, and the activities of plants, animals, and microbes. To understand the processes that shape landscapes, *geomorphologists*—physical geographers and geologists who study the spatial pattern, form, and evolution of landforms—often begin by observing a landscape to determine what is there. Most landscapes display a variety of features, such as different rock layers that we can distinguish by color, texture, and the way the rocks fracture.

A What Factors Influence the Development of Landscapes?

Examine the figure below and think about all the processes that might interact to produce a landscape.

1. A main control on the appearance of a landscape is the parent materials — what types of rocks, soils, and other materials are present. Some rocks are very hard and resistant to erosion, and these will tend to form peaks, cliffs, and steep slopes. Softer rocks will erode more easily, forming valleys or gentle slopes.

2. The orientation of a slope (i.e., which way the slope faces) has a strong control on the resulting landscape. In the Northern Hemisphere, south-facing slopes receive more sunlight than do north-facing ones, and so in some climates they will be drier and may have fewer plants. The relative lack of plants results in more soil erosion. This variation in slope appearance highlights the complex way all four spheres interact. If this figure shows an area in the Northern Hemisphere, are the north-facing slopes probably on the left or right side of the mountain range?

3. Climate has a great influence on the type of landscape that results. The amount and style of erosion are affected by the amount of rainfall and whether rainfall is spread out throughout the year or falls rapidly in a brief rainy season. Climate also influences the types and abundance of plants, which then determine whether soil develops and remains in place on the hillslope, held in place by roots. Another important climatic factor is whether the area experiences freezing temperatures, because when water in fractures freezes it expands, further loosening rocks that can then move downhill.

09.08.a1

B What Features Do Landscapes Display?

Observe this photograph, trying to identify distinct parts of the scene and then focusing on one part at a time. After examining the photograph, read the accompanying text.

09.08.b1 Canyonlands NP, UT

Color commonly catches our attention. These rocks are various shades of red, tan, and gray. If we were there, we would conclude that the rocks consist of consolidated sand and mud, and therefore are *sedimentary rocks.*

Another thing to notice is that this hill has different parts. A small knob of light-colored rocks sits on the very top.

Below the knob is a reddish and tan slope and a small reddish cliff.

There is a main, light-colored cliff, the upper part of which has a tan color and is fairly smooth and rounded.

Some parts of the cliff have horizontal lines that can be followed around corners of the cliff. These lines are the outward expression of *layers* within the rock. We call such layers in sedimentary rocks *beds* or *bedding*. These beds originally extended across the area prior to more recent erosion.

Lower parts of the cliff have a darker reddish-brown color and display many sharp angles and corners. Some of these corners coincide with vertical cracks, or *fractures*, that extend back into the rock. The reddish-brown color is a natural stain on the outside of the rocks.

Below the cliff is a slope that has pinkish-red areas locally covered by loose pieces of light-colored rock. A reasonable interpretation is that the loose pieces have fallen off the main cliff.

C What Are Some Strategies for Observing Landscapes?

Observe this photograph of Monument Valley, along the Arizona-Utah border, and try to recognize the types of features, such as layers and fractures, described on the previous page. After you have made your observations, read the text, which describes aspects to observe and some helpful strategies for looking at any landscape.

1. Most landscapes have a fairly complex appearance when viewed in their entirety, so a useful approach is to focus on one part of the landscape at a time. In the scene below, examine the left side of the image and compare it to the center. What similarities and differences do you observe?

2. Another approach is to let the landscape guide your observations from one part to another. In this scene, spend some time looking only at the cliff, and then focus on the reddish slope below the cliff. Next, pay attention to the piles of loose rocks that rest on the reddish slope.

3. Then, try focusing on one type of landscape feature at a time. In this photograph, start by concentrating on the fractures in the cliff. Are they steep, and are they evenly spaced? How do they affect the appearance of the cliff? Use the same approach to look at the ledges that cross the reddish slope.

09.08.c1 Monument Valley, AZ

4. Color is one of the first things we notice in any scene. Rocks, sediment, and soils have a range of colors depending on the composition of the materials and the environmental conditions imprinted on those materials. Some colors are integral to the rock, but others are a natural coating or stain on the outside surfaces of the rock. The rocks of Monument Valley are reddish brown to tan inside, but are locally coated by a darker brown coating.

5. Some rock types are more resistant to erosion than others and have more dramatic expressions in the landscape. Cliffs and ledges generally represent rock types that are hard to erode, as shown by the cliffs of hard sandstone in this photograph. Slopes or soil-covered areas contain weaker materials, such as the siltstone in the reddish slope and the loose wind-blown sand in the foreground.

8. To visualize different *components* of a landscape, draw a sketch that captures the main features but leaves out less important details. Compare the sketch below with the photograph above. Note how the sketch changes the way you look at the photograph.

7. The *shapes* of eroded rocks depend on the hardness of the rock, thickness of layers, spacing of fractures, and many other factors. Landscapes change over time, so shapes seen today will evolve into different shapes on timescales of years to millions of years.

6. Obvious features in many landscapes are *layers* in the rocks. The cliff represents a thick layer of sandstone, whereas the underlying slopes and ledges are the expression of dozens of layers. In this location, the layers are nearly horizontal, but layers can be tilted or even folded. These differences in orientation have a great impact on the appearance of the resulting landscape. To understand how layers influence the landscape, you first observe the layers and recognize how they are oriented.

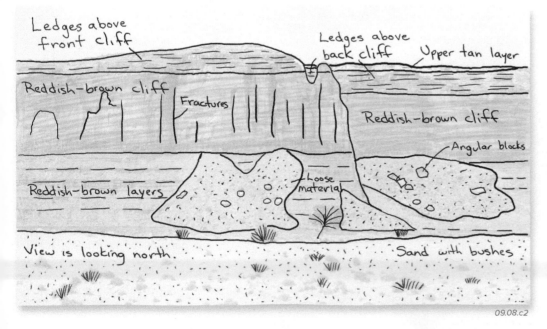

09.08.c2

Before You Leave This Page

✓ Summarize the different aspects or features you can observe in a landscape.

✓ Draw a simple sketch of a landscape photograph, identifying the main components, like those shown on these pages.

9.9 How Are Landscapes Weathered and Eroded?

WEATHERING AND EROSION begin with starting materials and then sculpt them into the landforms we observe. *Weathering* involves the on-site disintegration of rocks and other materials, and we subdivide it into two varieties — physical weathering and chemical weathering. *Erosion* is the scouring and stripping away of materials from a landscape, usually those that have been loosened by weathering. Together, weathering and erosion reshape a landscape and leave behind many signs of their actions.

A How Do Physical and Chemical Weathering Affect Materials in Landscapes?

Weathering involves physical and chemical processes that attack rocks on or near Earth's surface, loosening pieces and dissolving some material. Weathering can loosen material to form sediment. It can even produce soil on site. The type of weathering that occurs depends on the material that is weathered and the conditions during weathering. The processes of physical and chemical weathering are summarized below and are discussed in more detail in the chapter on soils.

Physical Weathering

1. *Physical weathering* is the physical breaking apart of rocks, soils, and other materials that are exposed to the environment. There are four major causes of physical weathering.

2. *Near-Surface Fracturing*—Many processes on or near the surface break material into smaller pieces. These include fracturing caused by rocks pulling away from a steep cliff. Fractures also occur when rocks expand as they are uplifted toward the surface and are progressively exposed to less pressure.

3. *Frost and Mineral Wedging*—Rocks, soils, and other materials can be broken as water freezes and expands in fractures. When the ice melts, the fractured pieces may become dislodged from the bedrock. Crystals of salt and other minerals that grow in thin fractures can also cause rocks to break apart.

4. *Thermal Expansion*—Rocks are heated by wildfires and by the Sun during the day. As rocks heat up, they expand, often irregularly, and may crack. This process probably plays a minor role in weathering, and scientists are currently debating its relative importance.

5. *Biological Activity*—Roots can grow downward into fractures and loose materials, prying them apart as the root diameter increases. Burrowing animals can transport and move rock and soil from depth to the surface where they are exposed to the elements, weathered, and eroded. Biological activity is especially efficient in soils and sediment, which are unconsolidated and therefore easy to move.

09.09.a1

Chemical Weathering

6. *Chemical weathering* includes several types of chemical reactions that affect rocks by breaking down minerals, causing new minerals to form, or by removing soluble material from the rock. Chemical weathering attacks both solid rock and loose rock fragments and produces ions in solution, loose grains and other pieces, and a covering of soil.

7. *Dissolution*—Some minerals, like calcite, are soluble in water, especially weakly acidic waters that are common in nature. These minerals, along with the rocks, sediment, and soil that contain them, can dissolve. The dissolved material may be carried away in streams, or groundwater, or be used locally by plants.

8. *Oxidation*—Some minerals, especially those containing iron, are unstable when exposed to Earth's atmosphere. These minerals can combine with oxygen to form oxide minerals, such as hematite, which compose the reddish and yellowish material that forms when metal rusts.

9. *Hydrolysis*—When silicate minerals are exposed to water, especially water that is somewhat acidic, the water reacts chemically with the minerals. This process commonly converts the original materials to clay minerals and produces leftover dissolved material that is carried away by the water. Hydrolysis is responsible for the formation of many clay-rich soils.

10. *Biological Reactions*—Decaying plants produce acids that can attack rocks, and some bacteria consume certain parts of rocks. These biological processes are different from the "Biological Activity" discussed above because they cause minerals to break down into their constituent elements.

09.09.a2

B How Does Erosion Operate and What Are the Results?

During erosion, material is removed from a landscape and transported away, and such wearing away of the surface is called *denudation*. Erosion removes material loosened by weathering, but it can also scour and move unweathered rock.

Gravity—Rocks, soils, and other materials are under the constant pull of gravity. If they reside on a flat surface, they likely will remain in place or move very slowly, but material on slopes (▶) can fall, slide, or tumble down the slope and accumulate farther down the slope or near the base.

09.09.b1 Seton Lake, British Columbia, Canada

Waves—Waves along shorelines, either along the ocean or the edge of a lake, constantly crash against the shore (◀). Wave action can carry away sand and other sediment and can erode into solid bedrock and undercut cliffs, causing even more material to fall onto the surf zone.

09.09.b4 Carmel, CA

Running Water—Streams use the power of running water and the scouring ability of sediment carried by the water to cause erosion. Running water can remove soil, sediment, and other loose materials from a landscape, or carve into solid bedrock (▶).

09.09.b2 Cascade Creek, Cascade, CO

Wind—Blowing wind can carry dust and fine sand grains, which act as sandpaper, scraping against and sanding away any materials exposed to the wind, including solid rock. The strange shapes here (◀) were scoured by blowing sand and ice (which also acts as a scouring agent).

09.09.b5 Dry Valleys, Antarctica

Ice—Glaciers are large, moving masses of ice on land that persist for many years. Along their sides and bases (▶) they scour away any loose material and even solid bedrock, causing ice in those parts to become darker because of all the eroded material.

09.09.b3 Glacier Bay, AK

Wind Erosion—Strong winds can remove so much material that the ground elevation is lowered (◀). This wind erosion can scour away soils, causing the land to be less productive. In this photograph, wind has sculpted soft sedimentary rocks into two pillars, remnants of the sedimentary layers that form the ridge in the background. This process is most efficient with soft and weak materials.

09.09.b6 Badlands NP, SD

The Effect of Erosion on Topographic Relief

Before You Leave This Page

✓ Distinguish the main processes of physical and chemical weathering.

✓ Describe the agents that cause erosion and some of the results.

09.09.b7 Sawtooth Mtns., ID

When a mountain forms and for some time afterwards, it is relatively steep as streams and glaciers carve down into the landscape faster than gravity and other processes can lower the peaks—relief is high.

09.09.b8 Baja California Sur, Mexico

Over time, weathering and erosion lower the peaks, lessen the slopes, and produce an overall softer-appearing, less rugged landscape. The steepest topographic features are typically removed first because they are most unstable.

09.09.b9 Cima Dome, southern CA

With even more time, erosion of bedrock can result in a gently sloping erosion surface. Such a surface is called a *pediment*, and it can form a gentle bedrock slope flanking a mountain or form a *pediment dome*, like the one above.

How Do Landscapes Record Transport and Deposition by Gravity, Streams, Ice, and Waves?

ONCE WEATHERED AND ERODED from a landscape, material can be transported farther away and later deposited. The main agents that transport earth materials are running water (streams), glaciers, waves in oceans and lakes, wind, and the ever-present force of gravity, which on its own moves materials down slopes and also drives or influences the other transport agents. Wind has some special characteristics and is discussed later.

A What Processes Transport and Deposit Sediment?

The same processes capable of erosion are also those that can simply continue transporting that material away. Gravity, streams, glaciers, waves, and wind are the main phenomena that transport material and are often called *agents of transport*. Each works in a different way, but each affects the material being transported.

Gravity on Slopes

1. All materials on Earth are subject to the downward pull of gravity. Material on slopes, especially on slopes that are relatively steep, can succumb to this force by falling, sliding, tumbling, or slowly creeping down the slope (▶). Gravity can be aided by moving water, mud, and ice.

09.10.a1 Seton Lake, British Columbia, Canada

09.10.a2 British Columbia, Canada

2. Material that has moved downhill under the force of gravity, called *colluvium*, may not be obvious where mostly covered by soil or vegetation. Colluvium underlies the forested hillsides in this photograph (◀) but is only exposed where the slope was undercut by an eroding stream.

Water and Mud on Slopes

3. Water and mud can transport material, including huge boulders (▶), down slopes and into stream valleys and other areas below. Such slurries of mud and other debris are called *debris flows* and, as shown in this view, can wreak havoc on anything in the way of the muddy torrent.

09.10.a3 Venezuela

09.10.a4 Death Valley, CA

4. Downhill, where a slope becomes less steep or adjoins a valley floor, the material being transported down the slope or in channels commonly accumulates at the bottom of a steep slope or at the base of the mountain, forming a cone-shaped feature (◀) known as an *alluvial fan*.

Glaciers

5. Glaciers are typically loaded with rock debris collected along the way. Much of this material was eroded from bedrock and consists of angular clasts of various sizes (▶). Glaciers also grind and pulverize material they transport, producing abundant particles of *silt* (small grains between sand and clay in size).

09.10.a5 Mendenhall Glacier, AK

09.10.a6 Athabasca Glacier, Alberta, Canada

6. A glacier deposits material along its base, sides, or front, usually all at the same time. Glacially deposited materials can be blanket-like, covering the ground somewhat evenly, but generally are more irregular (◀), forming curved ridges, hill-sized piles of sediment, or irregular terrain with pits and mounds.

Running Water in Streams

09.10.a7 Rampart Creek, Alberta, Canada

09.10.a8 Icefields Parkway, Alberta, Canada

09.10.a9 Animas River, Durango, CO

7. Flowing streams are a powerful agent of transport, carrying huge volumes of sediment from higher elevations to lower ones. Some streams (▲) carry a wide range of clast sizes, including sand, pebbles, and cobbles. Clasts may be carried only a short distance or hundreds to thousands of kilometers.

8. In landscapes, sediment recently deposited by streams typically forms ribbons or blankets of debris near or within the stream channel (▲). These features are transient, being covered and rearranged by the next large flow of water down the stream. They generally include partially rounded cobbles, pebbles, and sand.

9. Even streams that seem to be moving in a tranquil manner (▲) are transporting sediment. A tan or brown color to such streams indicates that the stream is carrying abundant fine sediment, mostly mud within the visible water and sand along the streambed. This sediment forms muddy *floodplains* next to the stream.

Waves Along Shorelines

10. Waves along a shoreline can be very powerful as they crash onto bedrock or beach sediments along the shore. Through this process, they pick up sediment, transporting it back and forth within a relatively limited area. Over time they transport material down the beach for long distances.

09.10.a10 Carmel, CA

09.10.a11 Naxos, Greece

11. Sediment moved by waves is mostly deposited along the beach. Depending on their setting and the types of materials that are available, beaches can be mostly covered by sand or by cobbles and other clasts, which became rounded when the waves pounded one clast against another.

B How Does Transport Affect the Size, Shape, and Sorting of Transported Material?

During transport by gravity, running water, ice, waves, or wind, clasts are subjected to tumbling, collisions with other clasts and with bedrock, and abrasion by fine sediment or ice. These actions may reduce the size of the clasts, increase their roundness, and sort the clasts by size. The distance that a clast has been transported is therefore a key factor to consider.

1. Bedrock exposed in mountains or along cliffs breaks off to form blocks the size of boulders and cobbles. Clasts near their source are usually large, angular, and poorly sorted (i.e., they have a wide range of clast sizes).

2. As boulders and cobbles are transported by gravity, streams, ice, or waves, their sharp corners break off because corners are the most exposed and weakest parts of a clast. The clasts become more rounded and smaller as pieces break off.

3. Far from their bedrock source, the clasts are worn into well-rounded pebbles and sand grains. River currents, beach waves, and wind separate clasts by size, eventually producing better-sorted sediments. Coarse materials are slowly left behind, and only the smaller clasts are carried far from the sediment source. Glaciers are an exception to this story because they continuously break rocks into angular clasts and can carry sediment long distances without doing any sorting.

More Rounding   More Sorting   Reduced Size

09.10.b1

9.10

9.11 How Do Landscapes Record Transport and Deposition by Wind?

WIND TRANSPORTS AND DEPOSITS SEDIMENT, but it is different from the other agents of transport because it derives its transporting energy from the atmosphere, not directly from gravity, so it can carry material uphill. Wind is limited to carrying material that is sand sized or smaller and produces a number of distinctive landforms, including sand dunes of various sizes and shapes. It also transports and deposits glacially derived silt, producing additional distinctive landforms.

A How Does Wind Transport Sediment?

Wind is generated by differences in air pressure and at times is strong enough to transport material, but only relatively small and lightweight fragments, like sand and clay. Transport of these materials by the wind is most efficient in dry climates, where there is limited vegetation to bind materials together and hold them on the ground.

1. Wind is capable of transporting sand and finer sediment, as well as lightweight plant fragments and other materials lying on the surface. It generally moves material in one of three ways and can deposit sediment in various settings, some of which are shown in photographs on these two pages.

2. Most materials on Earth's surface are not moved by the wind because they are too firmly attached to the land (such as rock outcrops), are too large or heavy to be moved, or are both.

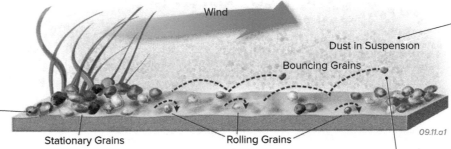

Wind

Dust in Suspension

Bouncing Grains

Stationary Grains

Rolling Grains

09.11.a1

5. Wind can pick up and carry finer material, such as dust, silt, and salt. This mode of transport is called *suspension*, and wind can keep some particles in the air for weeks, transporting them long distances, even across the oceans.

3. If wind velocity is great enough, it can roll or slide grains of sand and silt and other loose materials across the ground.

4. Very strong winds can lift sand grains, carry them short distances, and drop them. This process is akin to bouncing a grain along the surface and is called *saltation*.

Shape of Sand Dunes and Cross Beds

09.11.a5 Zion NP, UT

09.11.a2 Namibia, Africa

09.11.a3

6. The most obvious expression of material being transported and deposited by wind is in a sand dune (▲). Examine the features of this dune. The dune, like most others, is clearly asymmetrical, with a steep side facing to the right, down which sand is slipping—this is called the *slip face* and is on the downwind side (the *leeward side*). The left side of this dune has a more gentle slope, and is on the upwind side (the *windward side*). With time, the dune migrates from left to right, in the direction the wind blows.

7. When sand moves over a dune, it blows up the windward side and then is blown over or slides down the leeward side, depositing sand on the slip face. Over time, this produces a series of thin, curved layers on the downwind side of the dune, as shown above. Such beds are formed at an angle, and so are called *cross beds*. After cross beds are formed, the top of the dune can be blown off, truncating the cross beds (as in the lower part of the figure above). The image below shows truncated cross beds (lines) with a new layer of light-colored sand starting over the top.

09.11.a4 Namibia

8. In addition to being formed by wind, cross beds can also form from running water. Cross beds up to several meters high form within streams, deltas, and shorelines, but larger cross beds (▲) typically form in large sand dunes. The cross beds above are in a sandstone that was deposited 190 million years ago. There are several sets (thick layers) of cross beds, with the lower set being truncated at the base of the middle set. Cross beds in each set preserve the curved profile of the front of the dune. From their shape, the dunes tell us that the winds were blowing from left to right at this ancient time.

B   What Landforms Does Wind Produce?

Windblown sand and silt accumulate in some distinctive features, including sand dunes of various types, each of which forms under different wind conditions. Wind generates other landforms and smaller features, which are most obvious in deserts and other arid lands. Processes and features related to the wind are described as *aeolian*.

1. Some dunes have a crescent shape (▶), with tails pointing in the prevailing downwind direction (from left to right in this case). This common type of dune is a *crescent dune* or *barchan dune*. Each dune migrates in the direction of the tails, which marks the direction of the prevailing wind.

09.11.b1 Morocco

09.11.b4 Namibia

4. Dunes that are actively moving and being reshaped by the wind are called *active dunes*. Some dunes are not so active, allowing plants to take hold (◀), sometimes starting on the sheltered, leeward side. We commonly call such dunes *stabilized dunes*.

2. Other dunes are more linear or gently curved (▶) and can be many kilometers long. They are *longitudinal dunes* if they form parallel to prevailing winds or are *transverse dunes* if they form perpendicular to prevailing wind direction.

09.11.b2 Namibia

09.11b5 Namibia

5. Some winds are strong enough that sand is blown through, without accumulating in sizeable dunes. Winds can instead leave thin streaks of sand (◀) parallel to the prevailing wind. Such *wind streaks* can form on the downwind (leeward) side of small plants and stones.

3. Many dunes have more irregular, complex shapes, like these dunes (▶). If a dune has variably trending sand ridges radiating out from a central peak, it is a *star dune*, like these. Star dunes and other irregular dunes form where wind directions are highly variable over time.

09.11.b3 Namibia

09.11.b6 Tibet

6. *Loess* is wind-deposited silt and clay. Recently formed deposits can be a thin blanket over topography or thicker accumulations in valleys (◀). Loess is very common in parts of the Midwest United States, where it was the starting material for highly productive soils.

Landforms Influenced by the Wind

7. In many climates, especially deserts, rock surfaces develop a dark coating if left undisturbed for hundreds to thousands of years. This coating, called *rock varnish* or *desert varnish*, consists of iron-oxide and manganese-oxide materials, which are mostly derived from wind-blown dust. Rock varnish becomes darker the longer a rock is exposed at the surface, with very dark varnish requiring thousands of years. The darkness of varnish is therefore an indicator of how long that rock surface has been exposed. Rock varnish can be weathered or worn away, resetting the process. The dark varnished boulders shown here (▲) sat undisturbed on the surface for thousands of years, before parts of the varnish were scraped off by Native Americans to form artistic petroglyphs.

09.11.b7 Picture Rocks SP, AZ

09.11.b8 Granite Wash Mtns., AZ

8. In some settings, stones become concentrated on the surface through time, forming a feature called *desert pavement.* Over time, finer materials blow away, wash away, or move down into the soil, while pebbles and larger clasts remain on the surface or move up from just below the surface. If left undisturbed, the pavement becomes better developed over time, and exposed stones get coated with desert varnish, like the ones shown here. It takes more than 10,000 years to form a well-developed pavement with darkly varnished stones. Desert pavement can therefore be used as an indication of age of that surface.

Before You Leave This Page

☑ Sketch and explain how wind transports material.

☑ Sketch and describe how wind blows sand in a dune and how this forms cross beds.

☑ Summarize some common landforms formed by the wind or features in which wind is involved.

9.11

How Do We Infer the Relative Ages of Events?

TO DECIPHER THE HISTORY OF A LANDSCAPE, we use several strategies to determine the ages of different rock units, features, and events. The first strategy is to determine the age of one rock relative to another, using a series of commonsense approaches collectively called *relative dating*. We then try to assign actual numbers, in thousands to billions of years, to this relative chronology, using analytical dating methods, or *isotopic dating*. Also, *fossils* allow us to compare ages of different rock layers and to construct the *geologic timescale*. We start here with five main principles of relative dating.

Principle 1: Most Sediments Are Deposited in Horizontal Layers

Most sediments and many volcanic units are deposited in layers that originally are more or less horizontal, a principle called *original horizontality*. If layers are no longer horizontal, some event affected the layers after they formed. The few exceptions to the principle are small in scale and in special environments, such as the face of a sand dune or the undersea slopes of a delta.

09.12.a1 Goosenecks of the San Juan, UT

09.12.a2 San Juan River, UT

1. These canyon walls expose horizontal gray and reddish layers. These layers were deposited horizontally and have remained nearly so for 300 million years.

2. Just to the east, the same gray and reddish layers are *folded*. They are no longer horizontal, so something must have happened, like tilting associated with some type of deformation.

Principle 2: A Younger Sedimentary or Volcanic Unit Is Deposited on Top of Older Units

When a layer of sediment or a volcanic rock is deposited, any rock unit on which it rests must be older, a concept called the *principle of superposition*. This principle is illustrated below.

09.12.a3

3. A layer of tan sediment is deposited over older rocks.

09.12.a4

4. A series of horizontal red layers are then deposited over the first tan layer.

09.12.a5

5. A third series of layers is deposited last and is on top. In this sequence, the oldest layer is on the bottom and the youngest layer is on the top.

6. Observe all the different layers in this rock sequence (▶). The sediments were deposited and they lithified to form sedimentary rock long before the river eroded the canyon. Which exposed layer is oldest, and where would you look to find the youngest rock layer?

09.12.a6 Dead Horse Point, UT

7. Examine this sequence of rocks (▶) and predict which layers are oldest. It is most likely on the left, in the lowest part of the section. The rock layers here were tilted toward the right as the mountain to the left was uplifted. So before tilting, the rocks on the left were lowest.

09.12.a7 San Juan River, UT

Principle 3: A Younger Sediment or Rock Can Contain Pieces of an Older Rock

When a rock or deposit forms, it may incorporate pieces, or clasts, of older rock. A cobble eroded from bedrock and carried by a stream cannot exist unless the bedrock from which it formed was already there. The presence of clasts of an older rock in a younger rock clearly indicates the relative ages, even if you cannot see the two rock units in contact with one another.

8. The dark, lower basalt contributed clasts into an overlying layer of tan conglomerate. The conglomerate contains clasts of—and is therefore younger than—the basalt. The conglomerate also filled fractures in the basalt (▶).

09.12.a8 Lake Pleasant, AZ

9. A light-colored granite contains dark pieces, called *inclusions,* of older metamorphic rocks that fell into the magma. The metamorphic rocks, and their metamorphic layering, are contained within, and are older than, the granite (▶).

09.12.a9 Harcuvar Mtns., AZ

Principle 4: A Younger Rock or Feature Can Cut Across Any Older Rock or Feature

Many rocks are crosscut by *fractures* (joints and faults), so the rocks were there before the fractures formed. Veins and sheetlike bodies of magma can also invade into or across preexisting rock units, also showing *crosscutting relations.*

10. Several fractures cut across these limestone layers, so they formed after the rock already existed (▶). The fractures are said to be *crosscutting.*

09.12.a10 Little Colorado River, AZ

11. Light-colored sheetlike bodies (called *dikes*) of granite crosscut through darker igneous rocks. The crosscutting bodies are younger than the dark igneous rocks through which the granite cuts (▶).

09.12.a11 Santa Catalina Mtns., AZ

Principle 5: Younger Rocks and Features Can Cause Changes Along Their Contacts with Older Rocks

Magma comes into contact with preexisting rocks when it erupts onto the surface or solidifies at depth. In either setting, the magma may locally bake adjacent rock, or fluids from the magma may chemically alter nearby rocks. These changes, called *contact effects,* indicate that the magma is younger than the rocks that were altered.

12. A sheetlike body (dike) of basalt intrudes across a grayish sedimentary rock. Heat and fluids from the magma affected the older sedimentary rock, causing a reddish baked zone next to the basalt (▼). The baked zone shows that the basalt is the younger of the two rocks.

13. A lava flow or hot pyroclastic flow can bake and redden older underlying rocks, as shown here (▼). Sediments deposited on top of the volcanic unit long after the eruption will not be baked. A layer of magma injecting into the middle of a sequence of rocks can bake rocks both above and below.

09.12.a12 Bloody Basin, AZ

09.12.a13 Lewiston, WA

Before You Leave This Page

✓ Sketch and explain each of the five principles of relative dating, providing an example of each principle.

✓ Apply the principles of relative dating to a photograph or sketch showing geologic relations among several rock units, or among rock units and structures.

9.12

9.13 How Do We Determine the Ages of Events?

DETERMINING THE RELATIVE ORDER OF EVENTS is only one part of deciphering the history of a landscape. We also want to know when these events occurred, to assign ages in hundreds, thousands, millions, or billions of years. This is done by using fossils, some of which indicate a specific geologic time. We also use *isotopic dating methods*, most of which involve chemically analyzing a rock for the products of natural radioactive decay. The combination of relative dating, fossils, and isotopic dates allows us to reconstruct the chronology of a landscape.

A What Are Fossils and What Do They Indicate About the History of a Landscape?

Fossils are any remains, traces, or imprints of a plant or animal that are preserved in a rock or sediment. Fossils can provide information about how old the fossil-bearing layer is and in what environment the plant or animal lived.

09.13.a1 Caesar Creek SP, OH

09.13.a2 Hot Springs, SD

09.13.a3 Morocco, Africa

Most fossils are preserved hard parts, or parts that have been replaced by hard minerals, of marine organisms, including shellfish (▲) and coral. Other fossils are impressions of soft creatures or tracks left by creatures that walked, scurried, or crawled in mud.

Certain fossils are good indicators of age because they are distinctive and identifiable and represent creatures that only lived during a certain interval of time. The bones shown above are from mammoths and other large creatures that lived thousands of years ago.

Fossils can also provide information about the environment in which they lived. For example, if we find fossils of corals, crinoids (▲), and other marine organisms, we can be fairly certain that those rocks were deposited under water, probably in an ancient seaway.

B How Does Radioactive Decay Occur?

All atoms of any given element must have the same number of protons, but some differ in the number of neutrons they contain. Thus, different varieties of the same element may have different *atomic mass* (the combined mass from protons and neutrons); these varieties of the same element are called *isotopes*. Some isotopes are unstable through time, changing into a new element or isotope by the process of *radioactive decay*.

09.13.b1–3

1. This schematic figure shows atoms before any radioactive decay. These starting atoms are called the *parent atoms* or *parent isotopes*. Over time, some of the parent isotope will decay into a different element called the *daughter atom* or *daughter product*.

2. At a later time, half of the parent atoms (green) will have decayed into the daughter product (purple). The amount of time it takes for this to occur is called the *half-life*. After one half-life, there are an equal number of parent and daughter atoms.

3. After a time equal to another half-life has passed, half of the remaining parent atoms have decayed into daughter atoms. That is, after two half-lives, 3/4 of the parent atoms have decayed and 1/4 remain.

4. This table summarizes the type of radioactive decay shown in the figures above. If the number of parent atoms was initially 100, half of the parent atoms (50) will have decayed to atoms of the daughter product after one half-life. After two half-lives, only 25 parent atoms remain, alongside 75 daughter atoms.

	Before Any Decay	After One Half-Life	After Two Half-Lives
Atoms of Parent	100	50	25
Atoms of Daughter	0	50	75

5. Decay rates are different for different radioactive elements, but for any given isotope, the decay rate is always the same, being predictable and measurable in the laboratory. We, therefore, can calculate the length of time since the rock formed by measuring the ratio of parent atoms to daughter atoms in the rock. Dating rocks using radioactive decay is called *isotopic dating*.

C What Are the Main Subdivisions of the Geologic Timescale?

Early scientists discovered that the types of fossils change upward through sections of rock and that two different sites could have matching fossils. Using this principle, called *faunal succession*, they could interpret the relative age of two sequences of rock. They then recognized sequences of related layers across Europe and in North America and named different geologic time periods after places where rocks of that age are exposed. Eventually, we developed the geologic timescale, summarized in the figure below. The largest time intervals shown below are *eras*, and include the Paleozoic, Mesozoic, and Cenozoic Eras, from oldest to youngest. Boundaries between eras are marked by major changes in fossils, specifically the disappearance (extinction) of many species and families of creatures. Such major extinctions are referred to as *mass extinctions*. Geoscientists subdivided each era into several *periods*, shown below, with derivations of names of periods shown to the right of the column. *Ma* means millions of years before present.

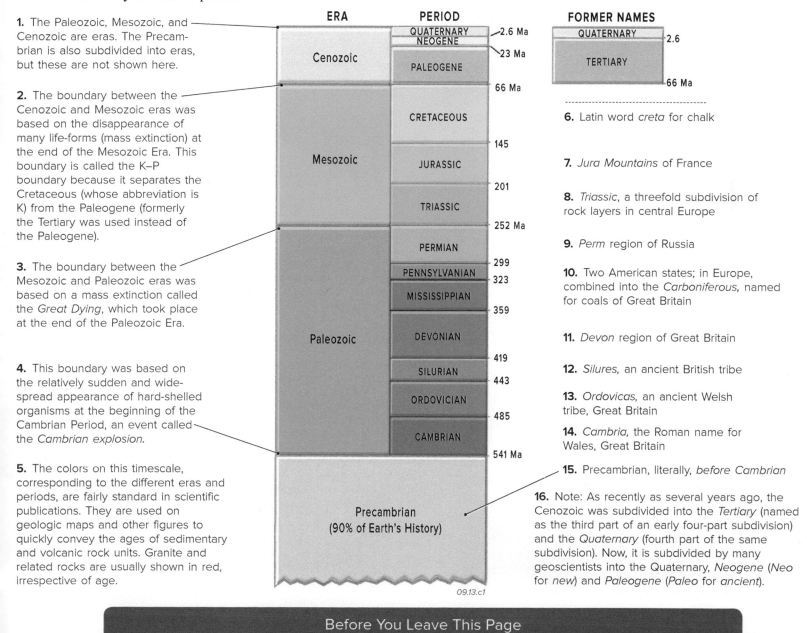

1. The Paleozoic, Mesozoic, and Cenozoic are eras. The Precambrian is also subdivided into eras, but these are not shown here.

2. The boundary between the Cenozoic and Mesozoic eras was based on the disappearance of many life-forms (mass extinction) at the end of the Mesozoic Era. This boundary is called the K–P boundary because it separates the Cretaceous (whose abbreviation is K) from the Paleogene (formerly the Tertiary was used instead of the Paleogene).

3. The boundary between the Mesozoic and Paleozoic eras was based on a mass extinction called the *Great Dying*, which took place at the end of the Paleozoic Era.

4. This boundary was based on the relatively sudden and widespread appearance of hard-shelled organisms at the beginning of the Cambrian Period, an event called the *Cambrian explosion*.

5. The colors on this timescale, corresponding to the different eras and periods, are fairly standard in scientific publications. They are used on geologic maps and other figures to quickly convey the ages of sedimentary and volcanic rock units. Granite and related rocks are usually shown in red, irrespective of age.

6. Latin word *creta* for chalk

7. *Jura Mountains* of France

8. *Triassic*, a threefold subdivision of rock layers in central Europe

9. *Perm* region of Russia

10. Two American states; in Europe, combined into the *Carboniferous*, named for coals of Great Britain

11. *Devon* region of Great Britain

12. *Silures*, an ancient British tribe

13. *Ordovicas*, an ancient Welsh tribe, Great Britain

14. *Cambria*, the Roman name for Wales, Great Britain

15. Precambrian, literally, *before Cambrian*

16. Note: As recently as several years ago, the Cenozoic was subdivided into the *Tertiary* (named as the third part of an early four-part subdivision) and the *Quaternary* (fourth part of the same subdivision). Now, it is subdivided by many geoscientists into the Quaternary, *Neogene* (*Neo* for *new*) and *Paleogene* (*Paleo* for *ancient*).

09.13.c1

Before You Leave This Page

☑ Summarize what fossils are and what they can tell us about a fossil-bearing sediment or rock.

☑ Explain how to determine how many half-lives have passed based on the ratio of parent to daughter atoms.

☑ Briefly summarize how the geologic timescale was developed.

☑ From oldest to youngest, list the four main geologic eras and the 13 periods.

9.13

How Do We Study Ages of Landscapes?

KNOWING THE RELATIVE AGES OF ROCKS AND STRUCTURES provides only one piece of the geologic story. We also need to understand when and how landscape features, such as mountains and valleys, formed.

A How Does a Typical Landscape Form?

Most landscapes have a similar history—rocks form and then are eroded. The histories of many regions typically include the deposition of a sequence of sedimentary layers, lithification into rocks, and later erosion of the rocks.

1. The sequence begins (▼) with deposition of a new sedimentary unit on top of preexisting metamorphic and igneous rocks. Most sediments, such as the top layer of sand shown here, are deposited as nearly horizontal layers.

09.14.a1

2. Through time, the depositional environment changes (◄) and a series of different sedimentary layers accumulates, with each younger layer being deposited on top.

09.14.a2

6. Observe this photograph and think about how the landscape formed before reading the text below the image. How many main rock layers do you observe? Use relative dating principles, like superposition, to infer the relative ages of the different rock layers you noted. Then, guided by the sequence of figures to the right, visualize how the area probably evolved over time.

09.14.a6 Grand View Point, Canyonlands NP, UT

3. Later, the layers are lithified. At some point, deposition stops, and all the layers that will be deposited are present (►). Weathering and erosion can begin.

09.14.a3

4. If the region undergoes broadscale uplift or the seas withdraw, leaving the area above sea level, the area can then be eroded by streams, glaciers, and wind. Erosion can more or less uniformly strip the entire land surface, removing the top layers. More likely, however, erosion will be faster in some areas, like along a stream cutting downward through softer rocks, as in a canyon (►).

09.14.a4

7. This canyon exposes five or six main sedimentary units and a number of smaller layers. There is a dominant light-colored cliff of sandstone. Layers below the sandstone are the oldest in this area, and red cliffs in the distance expose higher and younger layers. The far mountains are igneous intrusions that baked, and are younger than, the youngest layers in the cliffs. All the layers were deposited and lithified, and the intrusions were emplaced, before erosion began carving the canyon. During and after the initial canyon formed, the canyon was widened by continued erosion.

5. Erosion by a stream cuts downward, carving a deeper canyon. The canyon widens as small drainages erode as they flow into the main stream and as the steep canyon walls move downhill in landslides and slower movements. The combination of downcutting, widening, and development of subsidiary drainages, called *tributaries,* sculpts a deeper, wider, and more intricate canyon (►).

09.14.a5

B How Do We Infer the Age of a Landscape Surface?

To investigate when a landscape surface formed, we commonly try to find a rock unit or other geologic feature that was there before the surface formed or one that came after the surface already existed.

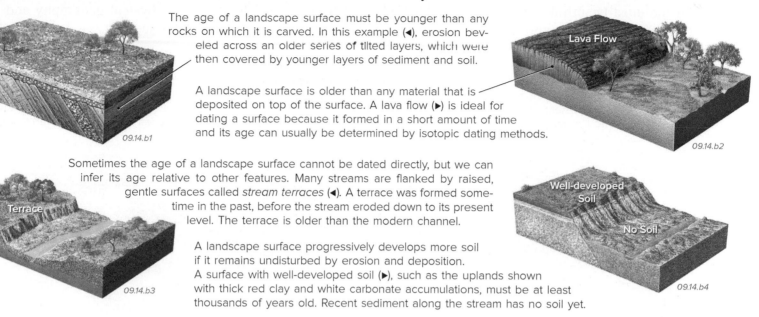

The age of a landscape surface must be younger than any rocks on which it is carved. In this example (◄), erosion beveled across an older series of tilted layers, which were then covered by younger layers of sediment and soil.

A landscape surface is older than any material that is deposited on top of the surface. A lava flow (►) is ideal for dating a surface because it formed in a short amount of time and its age can usually be determined by isotopic dating methods.

Sometimes the age of a landscape surface cannot be dated directly, but we can infer its age relative to other features. Many streams are flanked by raised, gentle surfaces called *stream terraces* (◄). A terrace was formed sometime in the past, before the stream eroded down to its present level. The terrace is older than the modern channel.

A landscape surface progressively develops more soil if it remains undisturbed by erosion and deposition. A surface with well-developed soil (►), such as the uplands shown with thick red clay and white carbonate accumulations, must be at least thousands of years old. Recent sediment along the stream has no soil yet.

09.14.b1
09.14.b2
09.14.b3
09.14.b4

C How Does Understanding Landscape Ages Help Us Evaluate Natural Hazards?

When assessing the potential for volcanic eruptions, earthquakes, flooding, and other natural hazards, we are interested in knowing what types of processes are involved with the hazard and *when* these events last occurred. We determine when these hazardous events last happened by applying principles of relative dating, landscape development, and isotopic dating.

Volcanic Eruptions

09.14.c1

Earth's surface contains many volcanoes and exposures of volcanic rocks. Knowing when a volcano last erupted, and which volcanic units are the most recent, help geoscientists evaluate the hazard of future eruptions. We try to determine the relative ages of the flows, how weathered and eroded each flow is, and the age from isotopic methods.

Earthquakes

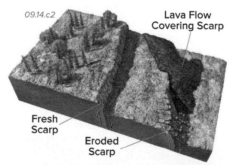

09.14.c2

Many recent earthquakes leave evidence as ruptures of the ground, forming a steep slope called a *fault scarp*. We study such scarps using relative dating, how degraded the scarp is from erosion, and isotopic methods. From this information, we can estimate how recently and how often earthquakes occurred.

Flooding

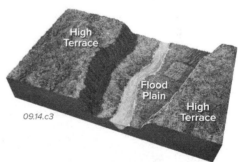

09.14.c3

We evaluate the potential for large floods using records of stream flow and from studying landforms beside the river. When streams flood, they often spill out onto their *floodplain*. Larger, more ancient floods may have covered higher stream terraces. By studying the terraces and their soil, we can estimate the sizes and ages of large, ancient floods.

Before You Leave This Page

☑ Sketch and describe the sequence of events represented in a typical landscape of flat-lying sedimentary rocks.

☑ Sketch and explain how you could infer the age of a landscape surface.

☑ Describe three ways that studying landscapes helps us better understand natural hazards and how to assess their likelihood.

What Is the Natural History of the Grand Canyon?

FOR SCENERY, THE GRAND CANYON HAS IT ALL. It contains some of the best exposed and studied, as well as one of the most beautiful, landscapes in the world. It is discussed in almost every physical geography and geology class because it so clearly expresses the history of events that formed this spectacular scenery.

1. This north view of the Grand Canyon region has satellite imagery superimposed on topography. Line A-B marks the location of a geologic cross section that represents a slice through the landscape. The Colorado River, which formed the canyon, flows from right to left, exits the canyon through high cliffs, and enters Lake Mead.

2. The Grand Canyon is eroded into the Colorado Plateau, a region of broad plateaus (high areas with mostly gentle relief) and deep canyons, which expose a mostly flat-lying sequence of Mesozoic and Paleozoic sedimentary rocks.

3. The river flows southwest across the area, cutting across nearly horizontal to locally tilted layers. The deepest part of the canyon is where the Colorado River erodes through the uplifted Kaibab Plateau.

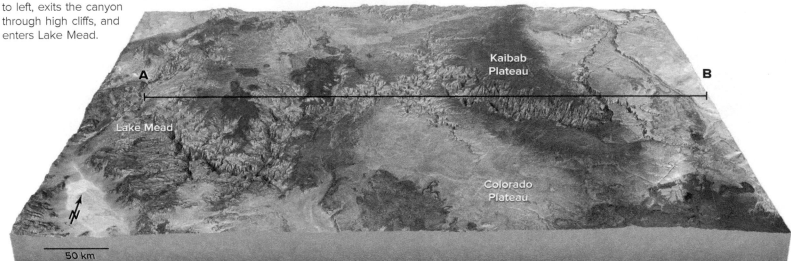

09.15.a1

4. Basalt flows cap some plateaus and predate formation of the main canyon. Based on isotopic dating methods, these rocks were dated to be 8 million years old.

5. The region has been cut into a series of blocks, with blocks to the west (left) being displaced downward relative to blocks on the east. These movements formed fault scarps that cut fairly recent lava flows; they also caused recent earthquakes.

6. Some lava flows poured down into the already-carved canyon, demonstrating that much of the canyon is older than 4 to 5 million years.

7. Sedimentary layers cap most plateaus and are warped over a few broad folds in the layers. Erosion was guided by, and so follows, the folded layers. Most erosion occurred by running water, but wind-related dunes and wind streaks are common.

8. Higher and younger sedimentary rocks are preserved on the down-folded sides of the folds and contain famous dinosaur tracks and petrified wood in the Painted Desert.

09.15.a2

9. The colorful walls of the canyon expose a flat-lying sequence of Paleozoic rocks. The different rock layers have different resistances and responses to erosion, creating the stair-stepped aspect of the upper canyon walls. The oldest layers are at the bottom.

10. The oldest rocks are 1.7 billion-year-old metamorphic rocks and granites. These hard rocks are exposed in the bottom of the canyon, where they form a steep and narrow inner canyon.

11. The steep canyon walls produce landslides that carve off pieces of the cliffs, sending them crashing downward toward the canyon bottom. This process, and other types of erosion, help widen the canyon over time, making it wide as well as deep. Floods and debris flows down tributaries move this material farther downhill.

12. Within the canyon walls are several surfaces, shown as squiggly lines, where erosion of underlying layers occurred before deposition of the overlying layers. This type of feature, called an *unconformity*, represents a time of erosion, not deposition. It is considered to be a gap in the record of events because no layers were deposited to indicate what occurred at this time.

Sequence of Rocks

09.15.a3

Late Paleozoic

Middle Paleozoic

Early Paleozoic

09.15.a4

Early Paleozoic

Upper Unconformity

Late Precambrian

Lower Unconformity

Precambrian Basement

Landscape History and Key Age Constraints

09.15.a5

13. The sequence of events in the Grand Canyon has been reconstructed using relative dating, fossils, and many different isotopic dating methods. The history of the landscape resulting from these studies is summarized below, which should be read from bottom to top (oldest to youngest).

20. *Modern Erosion*—The Colorado River and associated streams eroded down through the rock layers, millimeter by millimeter, over the last 6 million years. At the same time, the canyon gradually became wider as weathering, landsliding, stream action, and other processes eroded the canyon walls, causing them to retreat in both directions. It is the final sculpting of the canyon that makes it such a spectacular place. Weathering, erosion, and deposition are all occurring, so the view presented today is but one snapshot in the long movie represented by the history of the Grand Canyon.

19. *Deformation, Uplift, and Erosion*—The sedimentary layers largely have escaped deformation and remain nearly flat, except near a few faults and folds. Based on relative-dating methods, the folds formed well before the modern canyon was carved by the Colorado River. After the folds developed but before the canyon formed, many rock layers were eroded off the top of the layers in the canyon, but are still preserved elsewhere in the region. Fault scarps along the Hurricane fault show that some movement of the blocks is still occurring, indicating a moderate risk for earthquakes in this region.

18. *Deposition of Sedimentary Layers Exposed in the Canyon Walls*—After erosion carved the upper unconformity, various processes deposited the sedimentary layers that form the main walls of the canyon (shown in tan, blue, pink, orange, and bluish green). These sedimentary layers record a wide range of environments, including shallow marine, shorelines, streams, and a dune-covered desert. These rocks are dated with marine and nonmarine fossils as Paleozoic (250 to 550 million years ago). At times, the region was above sea level and being eroded, so no layers were deposited during these time periods, some of which lasted tens of millions of years.

17. *Tilting and Unconformity*—Layers in the Late Precambrian rocks were gently to moderately tilted and then beveled by erosion. This produced an upper *unconformity*, which represents as much as 500 million years of history.

16. *Late Precambrian Rocks and Unconformity*—In the Late Precambrian, sedimentary and volcanic rocks were deposited in horizontal layers across the upturned metamorphic layers and the lower unconformity. The lower parts of these late Precambrian rocks are dated by several isotopic methods at 1.1 billion years. Since the underlying basement rocks are 1.7 billion years old, the lower unconformity represents 600 million years of time not recorded by any rocks!

15. *Uplift and Erosion of the Oldest Rocks*—After the oldest rocks formed and were metamorphosed, they were uplifted and eroded over a period that lasted for hundreds of millions of years. Erosion beveled across the steep metamorphic layers. This erosion formed what is called the *lower unconformity*.

14. *Oldest Rocks*—Metamorphic rocks and granite in the bottom of the canyon represent the oldest events. They were formed, metamorphosed, and deformed to near-vertical orientations in the Precambrian, around 1.7 billion years ago.

The Percentage of Time that the Canyon Records

Although the canyon is a classic landscape with a thick sequence of numerous geologic formations, it represents a relatively small amount of geologic time. The oldest rocks are "only" about 1.7 billion years old, so the area contains no record for 2.8 billion years of early Earth history (4.5–1.7 billion years). Next, the two unconformities together cut out another 700 to 800 million years of history.

Even the Paleozoic sequence is missing more time than it records. The formations only represent five of the seven geologic periods in the Paleozoic (rocks of the Ordovician and Silurian Periods are not present), none of the formations span an entire period, and there are ancient erosion surfaces within the sequence. Mesozoic and Cenozoic rocks are largely absent in the canyon, so yet more time is unrepresented by rocks in the canyon walls.

What Is the History of This Landscape?

THIS DRAMATIC LANDSCAPE exposes various relationships that have been documented in the field and recorded as a series of short descriptions. Samples collected from the area were analyzed for their fossils. You will use this information to reconstruct the sequence and ages of events that produced features exposed in a landscape.

Goals of This Exercise:

- Observe the distribution of different rock types to characterize the sequence of layers and other features that are present in the landscape.
- Use descriptions of units and of key relationships between different features to infer the relative sequence of events.
- Summarize the history of the landscape.

Procedures

Use your observations to complete the following steps.

1. Observe the landscape to understand its various components.

2. From your observations and additional information, determine in what order the various components of the landscape formed. List the order, from oldest to youngest, on the worksheet, or online.

3. OPTIONAL: Your instructor may have you write a summary of the development of this landscape.

Field Notes

The units and features are described below. Each unit or feature has a letter assigned to it, but generally these do not indicate the order in which the features formed. Some letters were skipped so that some features would have letters that were easy to remember, such as V for the volcano.

Sequence of Layers—There is a sequence of sedimentary layers, most completely preserved on the right corner of the area. These units are lettered A–E, from bottom to top, in the order in which they formed. Most of the layers contain fossils of coral, shells, and other marine creatures.

Unit I—Finely crystalline igneous rock, forming a sheetlike dike that has baked units A, B, C, and G.

Feature F—Unit F is a fracture that has offset the units (such a fracture is called a fault). It cuts units A, B, and G. In a nearby area, the fault also cuts units C, D, and E. It does not cut the lava flow (unit L). Some other units are not near the fault.

Unit G—Coarse granite that is weathered near its contact with overlying unit A. This weathering occurred before unit A was deposited on top.

Units L and V—Unweathered lava flow (L) associated with a volcano (V). The lava flow rests on units G, A, and B. From observations nearby, it is younger than all the sedimentary layers, even the highest ones (such as unit E). Some of the higher layers were eroded away before the lava was erupted onto the surface.

Feature N—Narrow canyon.

Unit R—Partly consolidated river gravels with a thick, well-developed soil. Occupies ancient channels cut into older rocks. Contains fossils of land creatures.

This view shows a landscape with various rocks and features. There is a central plateau (high flat area) flanked by several mountains, a volcano, a canyon, and a number of lines and curved features that cross the landscape. The geometry of layers in the subsurface is shown on the sides of the block. An unconformity (buried erosion surface) is shown with a squiggly line, indicating some topographic relief along the erosion surface represented by the unconformity. A boundary where one unit was simply deposited on another is shown by a thin line, and a fault is marked by a thicker line.

1. A section of sedimentary layers forms a series of cliffs and slopes on three corners of the block. These are units A, B, C, D, and E, in order from oldest (lowest) to youngest (highest).

2. A dark igneous feature (I) is a dike that forms a linear wall across the landscape. It mostly is uninterrupted by other features, except for one obvious gap where it is cut by an ancient channel filled with unit R.

3. An older series of stream channels (R) cross the plateau and form low troughs in the topography. One channel goes all the way to the edge of the canyon, where it stops abruptly, evidently having been cut off. Along their lengths, the channels are partially filled by stream gravels and are characterized by well-developed, tan soils.

4. The top of one mountain in the area (right corner of this figure) exposes higher layers than are preserved elsewhere. If we wanted to study the sequence of sedimentary layers, this is the place we would go first.

09.16.a1

5. There is a cone-shaped volcano (V) surrounded by a black lava flow (L). Neither the volcanic deposits on the volcano nor the lava flow has developed any soil.

6. A fracture has broken and offset the rocks, and so is a fault (F). It forms an obvious line across parts of the area, but is not continuous. It is also shown in cross-section on the side of the block. It has not formed a fault scarp, but is expressed in the topography because it is the boundary between rock types that erode in slightly different ways. In a nearby area, the fault cuts the main sequence of layers, including layers A, B, C, D, and E.

7. The lowest unit in the area is a gray granite (G). It is mostly exposed in the bottom of the canyon, but underlies the rest of the area.

8. A narrow canyon (N) cuts through the area. The canyon is especially narrow in one segment where dark lava flows (L) have poured from the plateau and into the already-formed canyon.

9. Reconstruct the history using superposition, crosscutting relationships, and the relationship of different features to the landscape. Be systematic, focusing your attention on any pair of objects that are in contact. For example, does the fault cut the lava flow? Some objects may not be in direct contact with each other, but their relative ages can be determined by comparing their ages relative to some other feature.

Plate Tectonics and Regional Features

THE SURFACE OF EARTH IS NOTABLE for its dramatic mountains, beautiful valleys, and intricate coastlines. Beneath the sea are unexpected features, such as undersea mountain ranges, deep ocean trenches, and thousands of submarine mountains. In this chapter, we examine the distribution of these features, along with the locations of earthquakes and volcanoes, to explore the *theory of plate tectonics.*

These images of the world show large topographic features on the land, colored using satellite data that show areas of vegetation in green and areas of rocks and sand in tan. Colors on the seafloor indicate depths below sea level, ranging from light blue for seafloor that is at relatively shallow depths to dark blue for seafloor that is deep.

10.00.a1

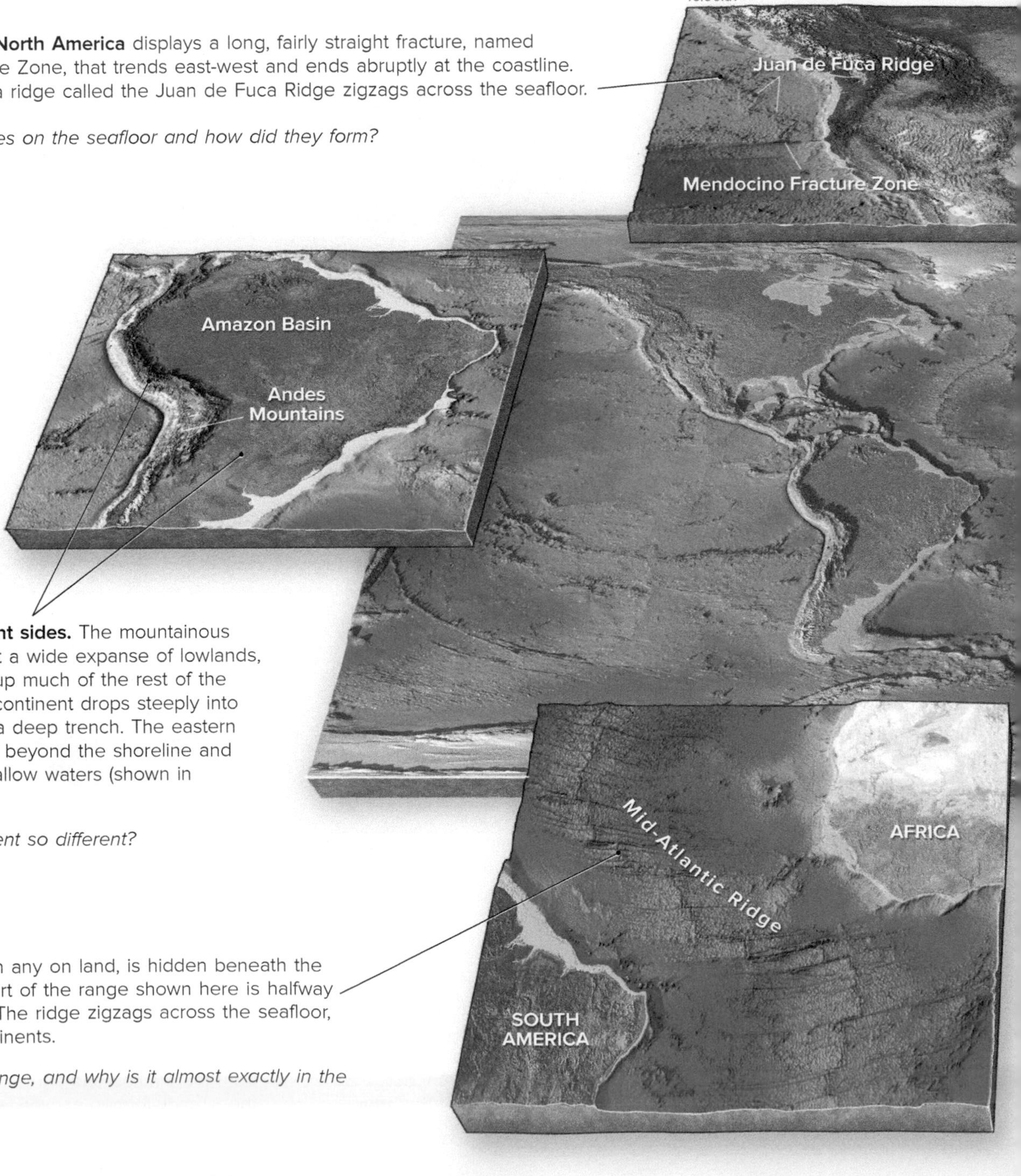

The seafloor west of North America displays a long, fairly straight fracture, named the Mendocino Fracture Zone, that trends east-west and ends abruptly at the coastline. North of this fracture, a ridge called the Juan de Fuca Ridge zigzags across the seafloor.

What are these features on the seafloor and how did they form?

Juan de Fuca Ridge

Mendocino Fracture Zone

Amazon Basin

Andes Mountains

South America has two very different sides. The mountainous Andes parallel the western coast, but a wide expanse of lowlands, including the Amazon Basin, makes up much of the rest of the continent. The western edge of the continent drops steeply into the Pacific Ocean and is flanked by a deep trench. The eastern edge of the continent continues well beyond the shoreline and forms a broad bench covered by shallow waters (shown in light blue).

Why are the two sides of the continent so different?

Mid-Atlantic Ridge

AFRICA

A huge mountain range, longer than any on land, is hidden beneath the waters of the Atlantic Ocean. The part of the range shown here is halfway between South America and Africa. The ridge zigzags across the seafloor, mimicking the shape of the two continents.

What is this underwater mountain range, and why is it almost exactly in the middle of the ocean?

SOUTH AMERICA

TOPICS IN THIS CHAPTER

The Tibetan Plateau of southern Asia rises four or five kilometers above the lowlands of India and Bangladesh to the south. The Himalaya mountain range with Mount Everest, the highest mountain on Earth, is perched on the southern edge of this plateau.

Why does this region have such a high elevation, and to what processes does it owe its existence?

Japan lies along the intersection of large, curving ridges mostly submerged beneath the ocean. Each ridge is flanked to the east by a deep trench in the seafloor. This area is well known for its destructive earthquakes and for Japan's picturesque volcano, Mount Fuji.

Do submarine ridges and trenches play a role in earthquake and volcanic activity?

The Arabian Peninsula provides much of the world's oil. East of the peninsula, the Persian Gulf has a shallow and smooth seafloor and is flanked by the world's largest oil fields. West of the peninsula, the Red Sea has a well-defined fissure-like feature down its center.

How did the Red Sea form, and what processes are causing its seafloor to be disrupted?

10.0

10.1 What Is Inside Earth?

UNDERSTANDING EARTH'S SURFACE requires knowing what is inside the Earth. You can directly observe the surface of Earth, but what is down below, in the subsurface? Earth consists of concentric layers that have different compositions. The outermost layer is the *crust*, which includes *continental crust* and *oceanic crust*. Beneath the crust is the *mantle*, Earth's most voluminous layer. The molten *outer core* and the solid *inner core* are at Earth's center.

A How Does Earth Change with Depth?

10.01.a1

10.01.a2

1. *Continental crust* has an average composition similar to this granite (◄). Continental crust, the thin, light-gray layer on the figure to the right, averages 35 to 40 km (20 to 25 mi) in thickness.

2. *Oceanic crust* exists beneath the deep oceans and has an average composition that is the same as basalt, a common dark lava rock (▼). Oceanic crust has an average thickness of about 7 km (4 mi), which is much thinner than can be shown here (the barely visible dark-gray layer).

10.01.a3 Grants, NM

3. The *mantle* extends from the base of the crust down to a depth of 2,900 km (1,800 mi). Much of the upper mantle is composed of the green mineral *olivine*, as exposed in the center of this rock (▼) brought to the surface in a volcano.

10.01.a4 Durango, Mexico

4. The lower mantle has a composition similar to the upper mantle, but it contains minerals formed at very high pressures. Nearly all of the mantle is solid, not molten. High temperatures cause some parts to be partially molten, while other parts flow because they are weak solids.

5. Based on studies of earthquakes, observations of meteorites, and models for the density of Earth, geoscientists interpret the *core* to consist of metallic iron and nickel, such as that observed in iron-nickel meteorites (◄). The outer core is molten, but the inner core is solid.

10.01.a5

Continental Crust

Oceanic Crust

Mantle

Upper Mantle

Lower Mantle

≈2900 km

≈5150 km

Outer Core

Inner Core

6370 km

B Are Some Layers Stronger than Others?

In addition to layers with different compositions, Earth has layers that are defined by strength and by how easily the material in the layers fractures or flows when subjected to forces.

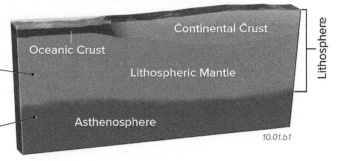

10.01.b1

1. The uppermost mantle is relatively strong and solidly attached to the overlying crust. The crust and uppermost mantle together form a rigid upper rigid layer called the *lithosphere* (*lithos* means *"stone"* in Greek), which averages about 100 km (about 60 mi) in thickness. The part of the uppermost mantle that is in the lithosphere is the *lithospheric mantle*.

2. The mantle directly beneath the lithosphere is mostly solid, but it is hotter than the rock above and can flow under pressure. This part of the upper mantle, called the *asthenosphere,* functions as a soft, weak zone under which the lithosphere moves. The word *asthenosphere* is from a Greek term for *"not strong."* In most regions, the asthenosphere is approximately 80 km to more than 150 km thick, so it can be deeper than 250 km.

C Why Do Some Regions Have High Elevations?

Why is the Gulf Coast of Texas near sea level, while the Colorado mountains are 3 to 5 km (2 to 3 mi) above sea level? Why are the continents mostly above sea level, but the ocean floor is below sea level? The primary factor controlling the elevation of a region is the thickness of the underlying crust.

1. The crust is less dense than the underlying mantle, and so it rests, or floats, on top of the mantle. The underlying lithospheric mantle is mostly solid, not liquid.

2. The thickness of continental crust ranges from less than 25 km (16 mi) to more than 60 km (37 mi). Regions that have high elevation generally have thick crust. The crust beneath the Rocky Mountains of Colorado is commonly more than 45 km (28 mi) thick.

3. The crust beneath low-elevation regions like the Gulf Coast of Texas is thinner. If the crust is thinner than 30 to 35 km (18 to 20 mi), the area will probably be below sea level, but it can still be part of the continent.

4. Most islands are volcanic mountains built on oceanic crust, but some are small pieces of continental crust.

5. Oceanic crust is thinner than continental crust and consists of denser rock than continental crust. As a result, regions underlain only by oceanic crust are well below sea level.

10.01.c1

Density and Isostasy

The relationship between regional elevation and crustal thickness is similar to that of wooden blocks of different thicknesses floating in water (▾). Wood floats on water because it is less dense than water. Ice floats on water because it is less dense than water, even though ice and water have the same composition. Thicker blocks of wood, like thicker parts of the crust, rise to higher elevations than do thinner blocks of wood. For Earth, we envision the crust being supported by mantle that is solid, unlike the liquid used in the wooden-block example.

This concept of different thicknesses of crust riding on the mantle is called *isostasy*. Isostasy explains most of the variations in elevation from one region to another, and it is commonly paraphrased by saying *mountain belts have thick*

crustal roots. As in the case of the floating wooden blocks, most of the change in crustal thickness occurs at depth and less occurs near the surface. Smaller, individual mountains do not necessarily have thick crustal roots. They can be supported by the strength of the crust, like a small lump of clay riding on one of the wooden blocks.

The *density* of the rocks also influences regional elevations. The fourth block shown here has the same thickness as the third block, but it consists of a denser type of wood. It therefore floats lower in the water. Likewise, a region of Earth underlain by especially dense crust or mantle is lower in elevation than a region with less dense crust or mantle, even if the two regions have similar thicknesses of crust. Temperature also controls the *thickness* of the lithosphere, and this affects a region's elevation.

10.01.t1

If the lithosphere in some region is heated, it expands, becoming less dense, and so the region rises in elevation. Thinner lithosphere also yields higher elevations because in this case less dense asthenosphere is replacing more dense lithosphere. So higher elevations result from thick crust, heated lithosphere, or thin lithosphere.

Before You Leave This Page

✓ Sketch the major layers of Earth.

✓ Sketch and describe differences in thickness and composition between continental crust and oceanic crust, and contrast the lithosphere and asthenosphere.

✓ Sketch and discuss how the principle of isostasy can explain differences in regional elevation.

10.1

10.2 What Are the Major Features of Earth?

OCEANS COVER AROUND 70% OF EARTH'S SURFACE. Seven continents make up most of the rest of the surface, and islands account for less than 2%. We are all familiar with the continents and their remarkable diversity of land-forms, from broad coastal plains to steep, snow-capped mountains. Features of the ocean floor, not generally seen by people, are just as diverse and include deep trenches and submarine mountain ranges. Islands exhibit great diversity, too. Some are large and isolated, but other islands form in arc shapes, ragged lines, or irregular clusters. What are the characteristics of each type of feature? In this chapter, we will explore how each of these features forms, so it is worth some time examining this map of the surface features of Earth.

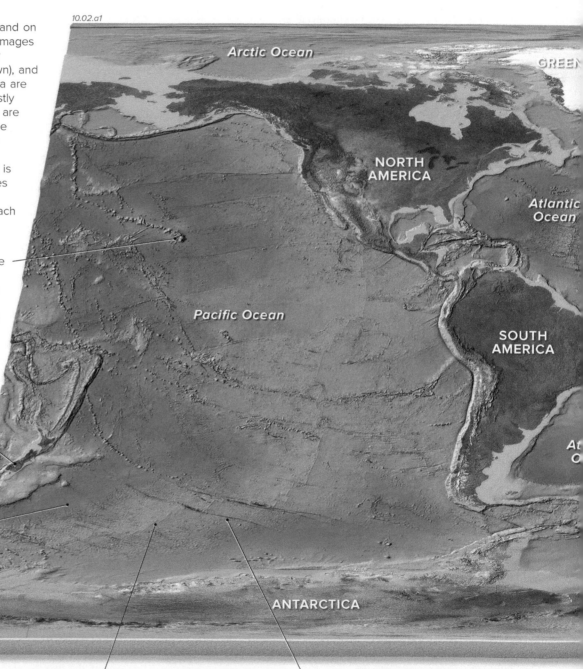

10.02.a1

1. This map shows large features on land and on the seafloor. The colors on land are from images taken by satellites orbiting Earth and show vegetated areas (green), rocky areas (brown), and sandy areas (tan). Greenland and Antarctica are white and light gray because they are mostly covered with ice and snow. Ocean waters are removed in order to reveal the slope of the seafloor. Colors of the ocean represent its depth, ranging from light blue where the seafloor is shallow to darker blue where it is deep. Examine this map and note the types of features you observe. Then, consider whether you have any ideas about how each type of feature forms.

2. Parts of the seafloor have mountains, the highest of which form islands, such as Hawaii. Most mountains on the seafloor do not reach sea level and are termed *seamounts*. Some islands, like Hawaii, and seamounts, are in long belts, which we refer to as *island and seamount chains*. Other islands and seamounts are isolated or form irregular clusters.

3. Some large islands, such as New Zealand, look like a small version of a continent.

4. Much of the ocean floor is moderately deep—3 to 5 km (9,800 to 16,000 ft)—and has a fairly smooth surface. Such a smooth, deep part of the seafloor is an *abyssal plain*.

5. A *mid-ocean ridge* is a broad, symmetrical ridge that crosses an ocean basin. Most ridges are 2 to 3 km (6,600–9,800 ft) higher than the average depth of the seafloor. One long ridge, named the *East Pacific Rise*, crosses the eastern Pacific and heads toward North America. Another occupies the middle of the Atlantic Ocean.

6. Cracks and steps cross the seafloor mostly at right angles to the mid-ocean ridges. Such a feature is an *oceanic fracture zone*.

7. Some continents continue outward from the shoreline under shallow seawater (light blue in this image) for hundreds of kilometers, forming submerged benches known as *continental shelves*. Which coastlines have broad continental shelves, like those surrounding Great Britain?

8. All continents contain large interior regions with gentle topography. Some continents have flat coastal plains, while others have mountains along their edges. Some mountains, like the Ural Mountains, are in the middle of continents.

9. Most continental areas have elevations of less than 1 to 2 km (3,300 to 6,600 ft). Broad, high regions, called *plateaus*, reach higher elevations, such as the Tibetan Plateau of southern Asia. Continents also contain mountain chains and individual mountains. Mount Everest, the highest point in the world, is almost 9 km (about 30,000 ft) in elevation.

Before You Leave This Page

☑ Identify on a world map the named continents and oceans.

☑ Identify on a world map the main types of features on the continents and in the oceans.

☑ Describe the main characteristics for each type of feature, including whether it occurs in the oceans, on continents, or as islands.

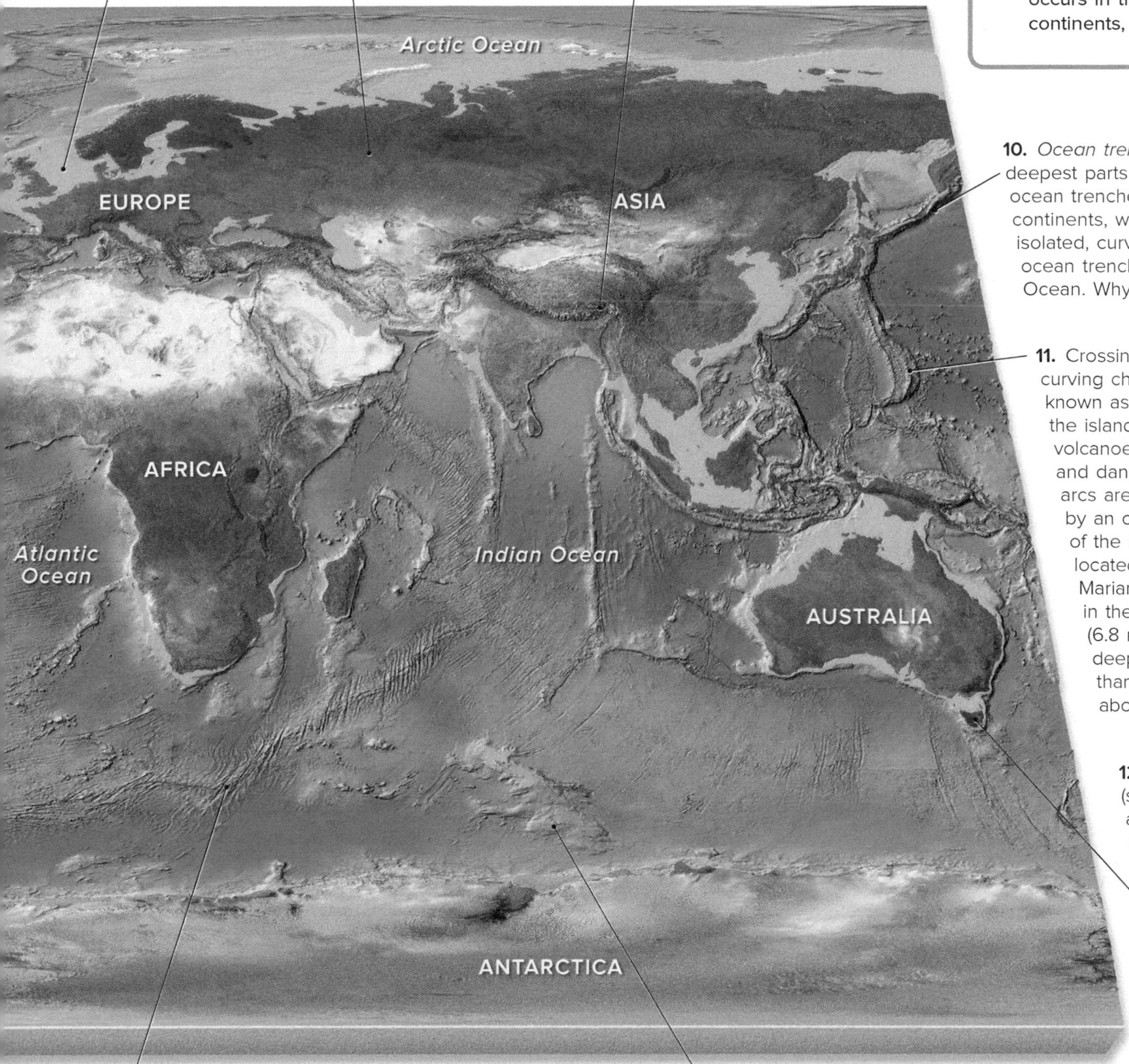

Arctic Ocean

EUROPE   ASIA

AFRICA

Atlantic Ocean

Indian Ocean

AUSTRALIA

ANTARCTICA

10. *Ocean trenches* make up the deepest parts of the ocean. Some ocean trenches follow the edges of continents, whereas others form isolated, curving troughs. Most ocean trenches are in the Pacific Ocean. Why are they here?

11. Crossing the seafloor are curving chains of islands, each known as an *island arc*. Most of the islands in an island arc are volcanoes, and many are active and dangerous. Most island arcs are flanked on one side by an ocean trench. Offshore of the Mariana island arc, located south of Japan, is the Mariana Trench, the deepest in the world at over 10.9 km (6.8 miles) deep—much deeper below sea level than Mount Everest is above sea level.

12. Some continents (such as South America) are flanked by an ocean trench, but other continents, such as Australia and Africa, have no nearby trenches.

14. Mid-ocean ridges and their associated fracture zones encircle much of the globe. In the Atlantic and Indian Oceans, they occupy a position halfway between the adjacent continents.

13. The oceans contain several broad, elevated regions, each of which is an *oceanic plateau*. The Kerguelen Plateau near Antarctica is one example, and another oceanic plateau lies northeast of Australia.

10.2

10.3 Why Do Some Continents Have Matching Shapes?

SOME CONTINENTS HAVE MATCHING SHAPES that appear to fit together like the pieces of a giant jigsaw puzzle. Alfred Wegener (1880–1930), a German physical geographer and meteorologist, observed the fit of these continents and tried to explain this and other data with a hypothesis called *continental drift*. Wegener argued that the continents were once joined together but later drifted apart. The hypothesis of continental drift was an important historical step that led to current theories that explain the distribution and shapes of the continents.

A Were the Continents Once Joined Together?

Fairly accurate world maps became available during the 1800s and scientists, including Alfred Wegener, noted that some continents, especially the southern continents, appeared to fit together. After considering many types of data, Wegener arrived at a creative explanation for this pattern.

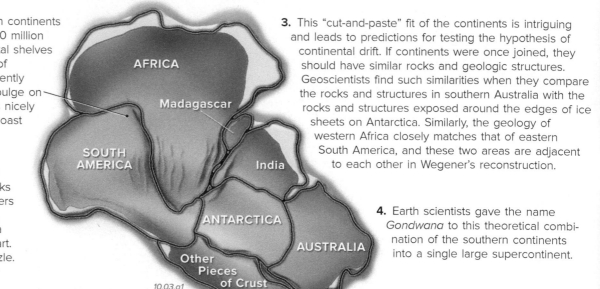

1. This figure shows how the southern continents are interpreted to have fit together 150 million years ago. In this figure, the continental shelves are included because they are parts of continents, even though they are currently underwater. In this arrangement, the bulge on the eastern side of South America fits nicely into the embayment on the western coast of Africa.

2. The fit of the continents and other supporting evidence preserved in rocks and fossils inspired Wegener and others to suggest that South America, Africa, Antarctica, Australia, and most of India were once joined but later drifted apart. Even Madagascar can fit into the puzzle.

3. This "cut-and-paste" fit of the continents is intriguing and leads to predictions for testing the hypothesis of continental drift. If continents were once joined, they should have similar rocks and geologic structures. Geoscientists find such similarities when they compare the rocks and structures in southern Australia with the rocks and structures exposed around the edges of ice sheets on Antarctica. Similarly, the geology of western Africa closely matches that of eastern South America, and these two areas are adjacent to each other in Wegener's reconstruction.

4. Earth scientists gave the name *Gondwana* to this theoretical combination of the southern continents into a single large supercontinent.

10.03.a1

B Is the Distribution of Fossils Consistent with Continental Drift?

Another piece of evidence supporting continental drift is the correspondence of the fossils of plants and land animals on continents now several thousand kilometers apart and separated by wide oceans.

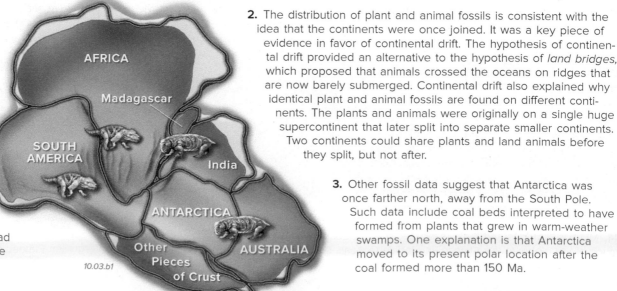

1. This figure illustrates that fossils of some land animals exist on several continents that are now separated by wide oceans. The animals lived more than 150 million years ago (abbreviated Ma) and are now extinct. These land animals could not swim across the wide oceans that currently separate the continents. Other key fossils linking the land areas of Gondwana are fossilized leaves of a seed-bearing plant that was widespread during late Paleozoic time (before 250 Ma).

2. The distribution of plant and animal fossils is consistent with the idea that the continents were once joined. It was a key piece of evidence in favor of continental drift. The hypothesis of continental drift provided an alternative to the hypothesis of *land bridges,* which proposed that animals crossed the oceans on ridges that are now barely submerged. Continental drift also explained why identical plant and animal fossils are found on different continents. The plants and animals were originally on a single huge supercontinent that later split into separate smaller continents. Two continents could share plants and land animals before they split, but not after.

3. Other fossil data suggest that Antarctica was once farther north, away from the South Pole. Such data include coal beds interpreted to have formed from plants that grew in warm-weather swamps. One explanation is that Antarctica moved to its present polar location after the coal formed more than 150 Ma.

10.03.b1

How Did Continental Drift Explain Glacial Deposits in Unusual Places?

Geoscientists studying continents in the Southern Hemisphere were puzzled by evidence that ancient glaciers had once covered places that today are close to the equator, and much too warm to have major glaciers.

1. This rounded outcrop in South Africa has a polished and scratched surface that is identical to those observed at the bases of modern glaciers. This observation is surprising because South Africa is currently a fairly warm and dry region without any glaciers.

10.03.c1 Kimberly, South Africa

10.03.c2 Kimberly, South Africa

2. Sedimentary rocks above the polished surface contain an unsorted collection of rocks of various sizes. Some of the rocks have scratch marks, like those seen near modern-day glaciers.

3. The scratch marks, or *striations*, on the polished bedrock surface tell geoscientists the direction that glaciers moved across the land as they gouged the bedrock. We interpret these scratch marks and other observations as evidence that glaciers moved across the area about 280 Ma.

4. The overall directions of glacial movement inferred from the scratch marks made it seem as if the glaciers had come from the oceans, something that is not seen today. Wegener discovered that these data made more sense when the continents were pieced back together into a larger, ancient continent, as shown in this illustration. According to this model, a polar ice cap was centered over South Africa and Antarctica 280 Ma, and the directions of glacial ice movement were those shown by the blue arrows.

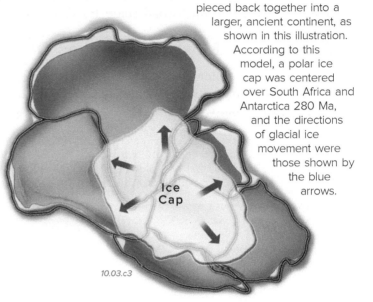

Ice Cap

10.03.c3

Old and New Ideas About Continental Drift

The hypothesis of continental drift received mixed reviews at first from physical geographers and other scientists. Geoscientists working in the Southern Hemisphere were intrigued by the idea because it explained the observed similarities in rocks, fossils, and geologic structures on opposite sides of the Atlantic and Indian Oceans. Geoscientists working in the Northern Hemisphere were more skeptical, in part because many had not seen the Southern Hemisphere data for themselves.

We now know that Wegener, with the evidence he considered, was on the right track. A crucial weakness of his hypothesis was that he could not explain how or why the continents moved. Wegener imagined that continents plowed through or over oceanic crust in the same way that a ship plows through the ocean. Scientists of his day, however, could demonstrate that this mechanism was not feasible. Continental crust is not strong enough to survive the forces needed to move a large mass across such a great distance while pushing aside oceanic crust. Because scientists of Wegener's time could show with experiments and calculations that this mechanism was unlikely, they practically

abandoned the hypothesis, in spite of its other appeals. The hypothesis probably would have been more widely accepted if Wegener or another scientist of that time had proposed a viable mechanism that explained how continents could move.

In the late 1950s, the idea of drifting continents again surfaced with the availability of new information about the topography, age, and magnetism of the seafloor. The magnetic data had largely been acquired in the search for enemy submarines during World War II. These data showed, for the first time, that the ocean floor (▼) had long submarine mountain

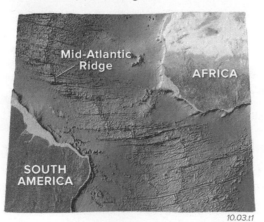

Mid-Atlantic Ridge

AFRICA

SOUTH AMERICA

10.03.t1

belts, such as the Mid-Atlantic Ridge in the middle of the Atlantic Ocean. Harry Hess and Robert Dietz, two geologists familiar with Wegener's work, examined the new data on ocean depths, and also new data on magnetism of the seafloor. Hess and Dietz both proposed that oceanic crust was spreading apart at underwater mountain belts, carrying the continents apart. This process of *seafloor spreading* rekindled interest in Alfred Wegener's idea of continental drift. Wegener's hypothesis morphed into the *theory of plate tectonics*, which is described later in this chapter.

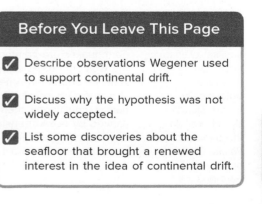

Before You Leave This Page

✓ Describe observations Wegener used to support continental drift.

✓ Discuss why the hypothesis was not widely accepted.

✓ List some discoveries about the seafloor that brought a renewed interest in the idea of continental drift.

10.3

10.4 What Is the Distribution of Earthquakes, Volcanoes, and Mountain Belts?

EARTHQUAKES AND VOLCANOES are spectacular manifestations of our dynamic Earth. Many of these are in distant places, but some are close to human settlements. The distributions of earthquakes and volcanoes are not random, but instead define clear patterns and show a close association with mountain belts and other regional features. These patterns show important, broad-scale Earth processes.

A Where Do Most Earthquakes Occur?

On this map, yellow circles show the locations of moderate to strong earthquakes that occurred over several recent decades. Observe the distribution of earthquakes before reading on. What patterns do you notice? Which regions have many earthquakes and which have few? Are earthquakes associated with certain types of features?

1. Earthquakes are not distributed uniformly across the planet. Most are concentrated in discrete belts, such as one that runs along the western coasts of North and South America.

2. Most earthquakes in the oceans occur along the winding crests of mid-ocean ridges. Where the ridges curve or zigzag, so do the patterns of earthquakes.

3. Earthquakes are sparse in some continental interiors but are abundant in others, like the Middle East, China, and near the Himalaya.

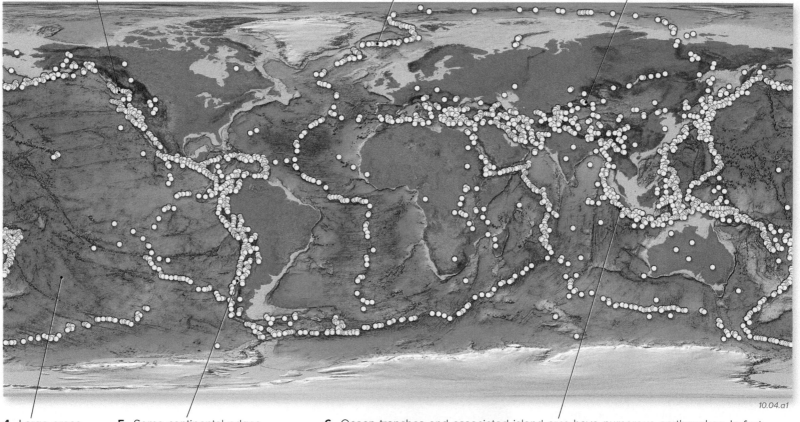

10.04.a1

4. Large areas of the seafloor, especially the abyssal plains, have few earthquakes. Volcanically active islands, like Hawaii, in the middle of the Pacific Ocean, do have earthquakes.

5. Some continental edges experience many earthquakes, but other edges have few. Earthquakes are common along the western coasts of South America and North America, and these edges also have narrow continental shelves flanked by ocean trenches. There are few earthquakes along the eastern coasts of the Americas, where the continental shelves are wide.

6. Ocean trenches and associated island arcs have numerous earthquakes. In fact, many of the world's largest and most deadly earthquakes occur near ocean trenches. Recent examples were the large earthquakes that produced deadly ocean waves in the Indian Ocean in 2004 and in Japan in 2011.

Before You Leave These Pages

☑ Show on a world relief map the major belts of earthquakes and volcanoes.

☑ Describe how the distribution of volcanoes corresponds to that of earthquakes.

☑ Compare the distributions of earthquakes, volcanoes, and high elevations.

B Which Areas Have Volcanoes?

On the map below, orange triangles show the locations of volcanoes that have been active in the last several million years. Observe the distribution of volcanoes and note which areas have volcanoes and which have none. How does this distribution compare with the distribution of earthquakes?

1. Volcanoes, like earthquakes, are widespread, but commonly occur in belts. One belt extends along the western coasts of North and South America.

2. Some volcanoes occur in the centers of oceans, such as the volcanoes near Iceland. Iceland is a large volcanic island along the mid-ocean ridge in the middle of the North Atlantic Ocean.

3. Volcanoes occur along the western edge of the Pacific Ocean, extending from north of Australia through the Philippines and Japan. Many are part of island arcs, associated with ocean trenches and earthquakes.

10.04.b1

4. Volcanic eruptions occur beneath the oceans, but this map shows only the largest submarine volcanic mountains. Volcanism is widespread along mid-ocean ridges.

5. Some volcanoes form in the middle of continents, such as in the eastern part of Africa and in China.

6. This map (▼) shows the topography of Earth's surface and seafloor, with high elevations in brown, low land elevations in green, shallow seafloor in light blue, and deep seafloor in dark blue. Using the three maps shown here, compare the distributions of earthquakes, volcanoes, and high elevations. Identify areas where there are (1) mountains but no earthquakes, (2) mountains but no volcanoes, and (3) earthquakes but no volcanoes. Make a list of these areas, or mark the areas on a map.

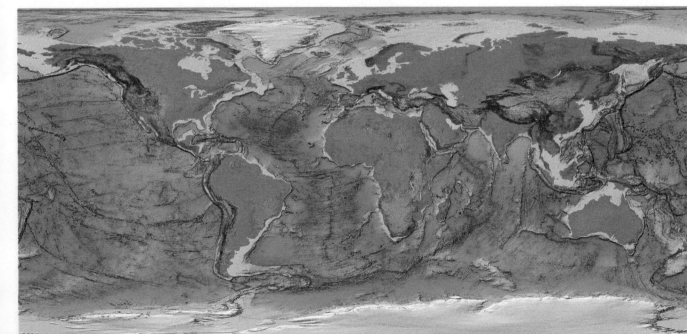

10.04.b2

10.4

10.5 What Causes Tectonic Activity to Occur in Belts?

MOST EARTHQUAKES AND VOLCANOES occur in belts around Earth's surface. Between these belts are vast regions that have comparatively little of this activity. Where are these belts of concentrated activity, and what explains this spatial distribution? What underlying processes cause these observed patterns? These and other questions helped lead to the *theory of plate tectonics*.

A What Do Earthquake and Volcanic Activity Tell Us About Earth's Lithosphere?

1. Examine the map below, which shows the locations of recent earthquakes (yellow circles) and volcanoes (orange triangles). After noting the patterns, compare this map with the lower map and then read the associated text.

2. On the upper map, there are large regions that have few earthquakes and volcanoes. These regions are relatively stable and intact pieces of Earth's outer layers. There are a dozen or so of these regions, each having edges defined by belts of earthquakes and volcanoes.

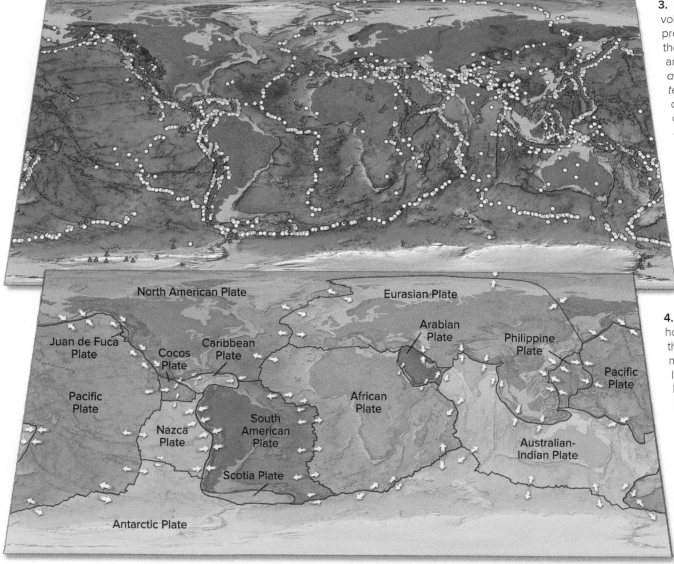

3. Earthquakes, volcanoes, and other processes that deform the crust and mantle are called *tectonic activity,* or simply *tectonics*. The belts of yellow and orange on the map are areas of *active tectonics*. The regions between the belts are relatively stable.

4. This lower map shows how we currently interpret the patterns on the upper map. Earth's strong outer layer, the *lithosphere,* is broken into a dozen or so fairly rigid pieces, called *tectonic plates*. This map shows names and boundaries of the larger plates. Spend some time learning the names and locations of the larger plates.

10.05.a1

5. Compare the two maps and note how the distribution of tectonic activity, especially earthquakes, outlines the shapes of the plates. Earthquakes are a better guide to plate boundaries than are volcanoes. Most, but not all, volcanoes are near plate boundaries, but many plate boundaries have no volcanoes.

6. Some earthquakes occur in the middle of plates, indicating that the situation is more complicated than a simple plate-tectonic model, in part because some parts of a plate are weaker than others. Forces can be transmitted through the strong parts, causing weaker parts to break and slip, generating an earthquake within the plate. Generally, most tectonic activity occurs near plate boundaries.

B How Do Plates Move Relative to One Another?

Plate boundaries have tectonic activity because plates are moving *relative to one another*. For this reason, we talk about the *relative motion* of plates across a plate boundary. Two plates can move away, toward, or sideways relative to one another, resulting in three types of plate boundaries: *divergent*, *convergent*, and *transform*.

Divergent Boundary

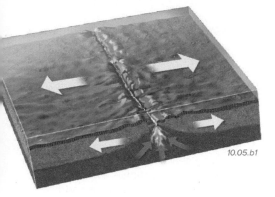

10.05.b1

Convergent Boundary

10.05.b2

Transform Boundary

10.05.b3

At a *divergent boundary*, two plates move apart relative to one another. In most cases, magma fills the space between the plates.

At a *convergent boundary*, two plates move toward one another. A typical result is that one plate slides under the other.

At a *transform boundary*, two plates move horizontally past one another, as shown by the white arrows on the top surface.

C Where Are the Three Types of Plate Boundaries?

This map shows plate boundaries according to type. Compare this map with the maps in part A and with those shown earlier in the chapter. For each major plate, note the types of boundaries between this plate and other plates it contacts. Then use the various maps to determine whether each type of plate boundary has the following features:

- Earthquakes
- Volcanoes
- Mountain belts
- Mid-ocean ridges
- Ocean trenches

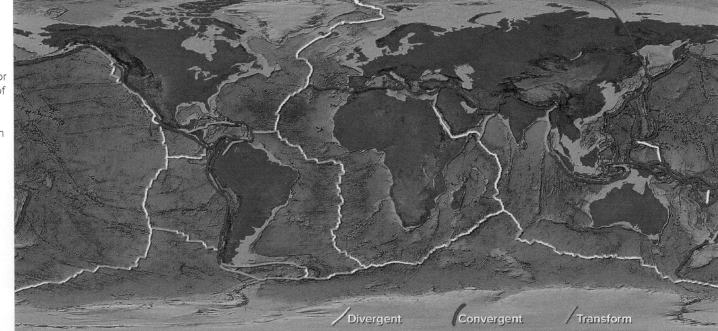

Divergent   Convergent   Transform

10.05.c1

Before You Leave This Page

✓ Describe plate tectonics and how it explains the distribution of tectonic activity.

✓ Sketch and explain the three types of plate boundaries.

✓ Compare the three types of plate boundaries with the distributions of earthquakes, volcanoes, mountain belts, mid-ocean ridges, and ocean trenches.

10.5

10.6 What Happens at Divergent Boundaries?

AT MID-OCEAN RIDGES, Earth's tectonic plates diverge (move apart). Ridges are the sites of many small to moderate-sized earthquakes and much submarine volcanism, and they record where two oceanic plates spread apart. On the continents, divergent motion can split a continent into two pieces, forming a continental rift and perhaps a new ocean basin as the pieces move apart.

A What Happens at Mid-Ocean Ridges?

Mid-ocean ridges are divergent plate boundaries where new oceanic lithosphere forms as two oceanic plates move apart. These boundaries are also called *spreading centers* because of the way the plates spread apart.

1. A narrow trough, or *rift*, runs along the axis of most mid-ocean ridges. The rift forms because large blocks of crust slip down as spreading occurs. The movement causes fracturing, resulting in frequent small to moderate-sized earthquakes.

2. As the plates move apart, solid mantle in the asthenosphere rises toward the surface. It partially melts as it rises, because decreasing pressures can no longer confine the rock as a solid. The molten rock (magma) rises along narrow conduits, accumulates in magma chambers beneath the rift, and eventually becomes part of the oceanic lithosphere.

3. Much of the magma solidifies at depth, but some erupts onto the seafloor, forming submarine lava flows. These eruptions create new ocean crust that is incorporated into the oceanic plates as they move apart.

4. Mid-ocean ridges are elevated above the surrounding seafloor because they consist of hotter, less dense materials, including magma. Lower density materials and thin lithosphere mean that the plate "floats" higher above the underlying asthenosphere. The elevation of the seafloor decreases away from the ridge because the rock cools and contracts, and because the less dense asthenosphere cools enough to become part of the denser lithosphere.

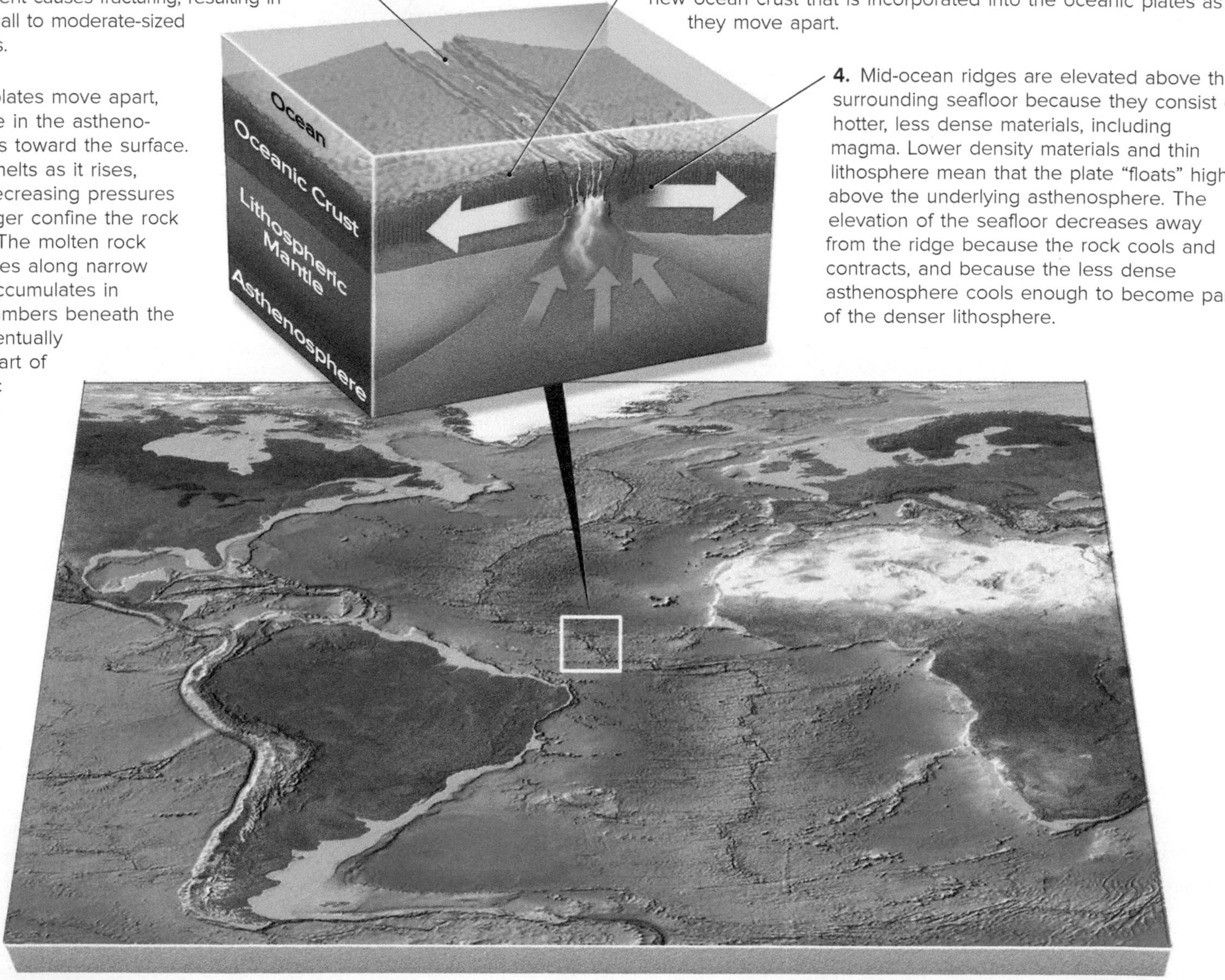

Ocean
Oceanic Crust
Lithospheric Mantle
Asthenosphere

10.06.a1–2

Before You Leave These Pages	☑ Sketch, label, and explain the features and processes of an oceanic divergent boundary.
	☑ Sketch and label the characteristics of a continental rift (i.e., a divergent boundary within a continent).
	☑ Sketch, label, and explain the stages of continental rifting, using East Africa and the Red Sea as example stages.

B What Happens When Divergence Splits a Continent Apart?

Most divergent plate boundaries are beneath oceans, but a divergent boundary may also form within a continent. This process creates a *continental rift,* such as the Great Rift Valley in East Africa. Rifting can lead to seafloor spreading and formation of a new ocean basin, following the progression shown here.

1. The initial stage of continental rifting commonly includes broad uplift of the land surface as mantle-derived magma ascends into and pushes up the crust.

----

3. Stretching of the crust causes large crustal blocks to drop down along faults, forming a *continental rift,* like in the Great Rift Valley. The downdropped blocks may form *basins* that can trap sediment and water, resulting in lakes.

----

5. If rifting continues, the continent splits into two pieces, and a narrow ocean basin forms as seafloor spreading takes place. A modern example of this is the narrow Red Sea, which runs between Africa and the Arabian Peninsula.

----

7. With continuing seafloor spreading, the ocean basin becomes progressively wider, eventually becoming a broad ocean like the modern-day Atlantic Ocean. The Atlantic Ocean basin formed when North and South America rifted away from Europe and Africa, following the sequence shown here.

----

2. The magma heats and can melt parts of the continental crust, producing additional magma. Heating of the crust causes it to expand, which results in further uplift.

----

4. Deep rifting causes solid mantle material in the asthenosphere to flow upward and partially melt. The resulting magma may solidify beneath the surface or may erupt from volcanoes and long fissures on the surface. The entire crust thins as it is pulled apart, so the central rift becomes lower in elevation over time. The Rio Grande Rift, which runs north-south through New Mexico, is another example of a continental rift in its early stages. An ancient example is the Midcontinental Rift, a mostly buried feature that underlies large parts of the north-central U.S.

----

6. As the edges of the continents move away from the heat associated with active spreading, the thinned crust cools and drops in elevation, eventually dropping below sea level. The continental margin ceases to be a plate boundary. A continental edge that lacks tectonic activity is called a *passive margin.*

----

8. Continental edges on both sides of the Atlantic are currently passive margins, lacking tectonic activity. Seafloor spreading continues today along the ridge in the middle of the Atlantic Ocean, so the Americas continue to move away from Europe and Africa.

10.06.b1–4

9. East Africa and adjacent seas illustrate the different stages of continental rifting. Here, a piece of continent has been rifted away from Africa, forming the Arabian Peninsula, and another piece is in the early stages of possibly doing the same (▶).

10. Early stages of rifting occur along the East African Rift, a long continental rift that begins near the Red Sea and extends into central Africa. The rift is within an elevated (uplifted) region and has several different segments, each featuring a downdropped rift. Some parts of the rift contain large lakes.

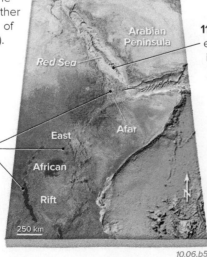

10.06.b5

11. The Red Sea represents the early stages of seafloor spreading. It began forming about 50 million years ago when the Arabian Peninsula rifted away from Africa. The Red Sea continues to spread and slowly grow wider.

12. The East African Rift is topographically lower than its flanks, but contains numerous active volcanoes, as represented by this mountain within the rift (▲). Volcanoes are especially common in the Afar region of Ethiopia, located where the rift joins the Red Sea. The Afar region is one of the most volcanically active regions on the planet.

10.06.b6 *East African Rift, Kenya*

10.6

10.7 What Happens at Convergent Boundaries?

CONVERGENT BOUNDARIES FORM when two plates move toward each other. Convergence can involve two oceanic plates, an oceanic plate and a continental plate, or two continental plates. Oceanic trenches, island arcs, and Earth's highest mountain belts form at convergent boundaries. Many of Earth's most dangerous volcanoes and largest earthquakes also occur along these boundaries. Convergent boundaries are among the most dangerous places on the planet!

A What Happens When Two Oceanic Plates Converge?

1. Convergence of two oceanic plates forms an *ocean-ocean convergent boundary.* One plate bends and slides beneath the other plate along an inclined zone. The process of one plate sliding beneath another plate is *subduction*, and the zone around the downward-moving plate is a *subduction zone.* Many large earthquakes occur in subduction zones.

2. An *oceanic trench* forms as the subducting plate moves down. Sediment and slices of oceanic crust collect in the trench. This sheared, scraped-off material generally remains completely submerged, but is exposed in a few islands, like Barbados, in the eastern Caribbean.

3. As the plate subducts, its temperature increases, releasing water from minerals in the downgoing plate. This water causes melting in the overlying asthenosphere, and the resulting magma is buoyant and rises into the overlying plate.

4. Some magma erupts, initially under the ocean and later as dangerous, explosive volcanoes that rise above the sea. With continued activity, the erupted lava and exploded volcanic fragments construct a curving belt of islands in an *island arc.* An example is the arc-shaped belt of the Aleutian Islands of Alaska. The area between the island arc and the ocean trench accumulates sediment, most of which comes from volcanic eruptions and from the erosion of volcanic materials in the arc.

5. Magma that solidifies at depth adds to the volume of the crust. Over time, the crust gets thicker and becomes transitional in character between oceanic and continental crust. Initially separate volcanic islands join to form more continuous strips of land, as occurred to form the island of Java in Indonesia.

Subducting Lithosphere

Asthenosphere

10.07.a1

B What Happens When an Oceanic Plate and a Continental Plate Converge?

1. The convergence of an oceanic and a continental plate forms an *ocean-continent convergent boundary.* Along this boundary, the denser oceanic plate subducts beneath the more buoyant continental plate.

2. An oceanic trench marks the plate boundary and receives sediment from the adjacent continent. Again, sediment and other material are scraped off the oceanic plate, forming a wedge of highly sheared material near the trench.

3. Volcanoes form on the surface of the overriding continental plate in the same way the volcanoes form in an ocean-ocean convergent boundary. These volcanoes erupt, often violently, producing large amounts of volcanic ash, lava, and mudflows, which pose a hazard for people who live nearby. Examples include large volcanoes of the Andes of South America and the Cascade Range of Washington, Oregon, northern California, and southern British Columbia.

4. Compression associated with the convergent boundary squeezes the crust for hundreds of kilometers into the continent. The crust deforms and thickens, resulting in uplift of the region. Uplift and volcanism may produce a high mountain range, such as the Andes.

5. Magma forms by melting of the asthenosphere above the subduction zone, due to fluids brought in by the subducted slab. The magma can solidify at depth, rise into the overlying continental crust before solidifying, or reach the surface and cause a volcanic eruption.

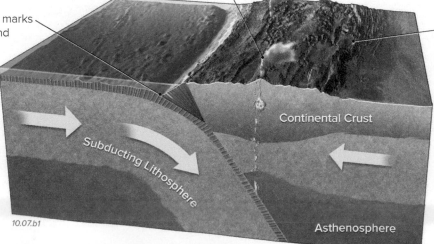

Subducting Lithosphere

Continental Crust

Asthenosphere

10.07.b1

C What Causes the Pacific Ring of Fire?

Volcanoes surround the Pacific Ocean, forming the *Pacific Ring of Fire*, as shown in the map below. The volcanoes extend from the southwestern Pacific, through the Philippines, Japan, and Alaska, and then down the western coasts of North and South America. The Ring of Fire results from subduction on both sides of the Pacific Ocean.

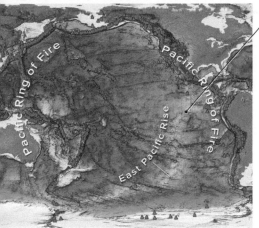

10.07.c1

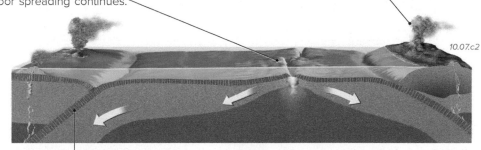

1. In the Pacific, new oceanic lithosphere forms along a mid-ocean ridge, the East Pacific Rise. The Pacific plate is west (left) of this boundary and several smaller plates are to the east (right). Once formed, new lithosphere moves away from the ridge as seafloor spreading continues.

2. Oceanic lithosphere of the Pacific plate and smaller plates subducts beneath the Americas, forming oceanic trenches on the seafloor and volcanoes on the overriding, mostly continental, plates.

10.07.c2

3. Subduction of oceanic lithosphere also occurs to the west, beneath Japan and island arcs of the western Pacific.

4. More oceanic plate is subducted than is produced along the East Pacific Rise, so the width of the Pacific Ocean is shrinking with time.

D What Happens When Two Continents Collide?

Two continental masses may converge along a *continent-continent convergent boundary*. This type of boundary is commonly called a *continental collision,* and it produces huge mountain ranges, like the Himalaya and Mount Everest in the photograph below.

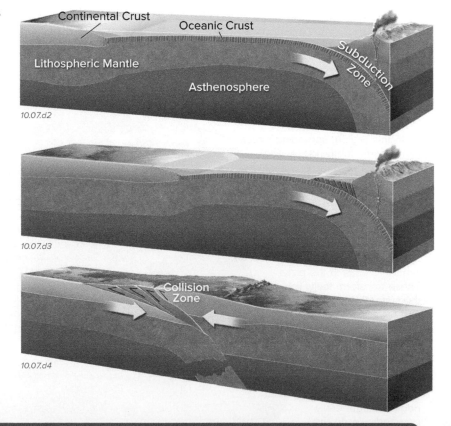

10.07.d2

10.07.d3

Collision Zone

10.07.d4

10.07.d1 Mount Everest

1. The large plate in the figure to the right is partly oceanic and partly continental. The oceanic part is being subducted to the right, under another continent at an ocean-continent convergent boundary.

2. As the oceanic part of the plate continues to subduct, the two continents approach each other. Magmatic activity occurs in the overriding plate above the subduction zone. The edge of the approaching continent has no such activity because it is not at a plate boundary yet.

3. When the converging continent arrives at the subduction zone, it may partially slide under the other continent or simply clog the subduction zone as the two continents collide. Because the two continents are thick and have the same density, neither can be easily subducted beneath the other and into the asthenosphere. Along the boundary, faults slice up the continental crust, stacking one slice on top of another. Continental collisions form enormous mountain belts and high plateaus, such as the Himalaya and Tibetan Plateau of southern Asia. The Himalaya and Tibetan Plateau are still forming today, rising at a rate of about 5 mm/yr, as the continental crust of India collides with the southern edge of the Eurasian plate.

Before You Leave This Page

☑ Sketch, label, and explain the features and processes associated with ocean-ocean and ocean-continent convergent boundaries.

☑ Sketch, label, and explain the steps leading to a continental collision (continent-continent convergent boundary).

10.7

10.8 What Happens Along Transform Boundaries?

AT TRANSFORM BOUNDARIES, PLATES SLIP HORIZONTALLY past each other along *transform faults*. In the oceans, transform faults are associated with mid-ocean ridges. Transform faults combine with spreading centers to form a zigzag pattern on the seafloor. A transform fault can link different types of plate boundaries, such as a mid-ocean ridge and an ocean trench. Some transform boundaries occur beside or within a continent, sliding one large crustal block past another, as occurs along the San Andreas fault in California.

A Why Do Mid-Ocean Ridges Have a Zigzag Pattern?

10.08.a1

1. To understand the zigzag character of mid-ocean ridges, examine how the two parts of this pizza have pulled apart, just like two *diverging* plates.

2. The break in the pizza did not follow a straight line. It took jogs to the left and the right, following cuts where the pizza was the weakest.

3. Openings created where the pizza pulled apart represent the segments of a mid-ocean ridge that are spreading apart. However, unlike a pizza, at a mid-ocean ridge, no open gaps exist because new material derived from the underlying mantle fills the space as fast as it opens, forming new oceanic crust.

4. The openings are linked by breaks, or *faults*, where the two parts of the pizza simply slide by one another. There are no gaps along these breaks, only *horizontal* movement of one plate sliding past the other. Arrows show the direction of relative motion. A fault that accommodates the horizontal movement of one tectonic plate past another is a *transform fault*. The spreading direction must be parallel to the transform faults and perpendicular to the spreading segments, so a zigzag pattern is required to allow a plate boundary to be curved.

Transform Faults Along the Mid-Ocean Ridge

5. Mid-ocean ridges, such as this one in the South Atlantic Ocean, have a zigzag pattern similar to the broken pizza.

6. In this region, spreading occurs along north-south ridges. The direction of spreading is east-west, perpendicular to the ridges.

7. East-west offsets are transform faults along which the two diverging plates simply slide past one another, like the breaks in the pizza. These transform faults link the spreading segments and have the relative motion shown by the white arrows.

8. Transform faults along mid-ocean ridges are generally perpendicular to the axis of the ridge. As in the pizza example, transform faults are parallel to the direction in which the two plates are spreading apart.

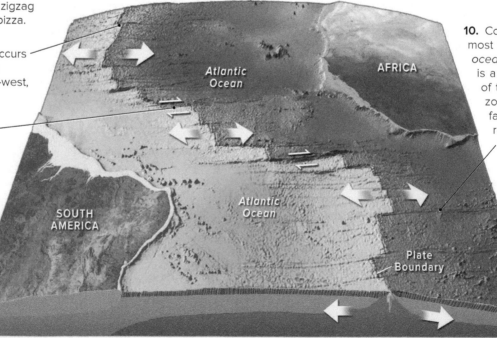

10.08.a2

10. Continuing outward from most transform faults is an *oceanic fracture zone*, which is a step in the elevation of the seafloor. A fracture zone is a former transform fault that now has no relative motion across it. It no longer separates two plates and instead is within a single plate. Opposite sides of the fracture zone have different elevations because they formed by seafloor spreading at different times in the past, so they have had different amounts of time to cool and subside after forming at the spreading center. Younger parts of the plate are warmer and higher than older parts, resulting in a step in elevation on the seafloor (an oceanic fracture zone).

9. The zigzag pattern of mid-ocean ridges shows the alternation of spreading segments with transform faults. In this example, the overall shape of the ridge mimics the edges of Africa and South America and so was largely inherited from the shape of the original rift that split the two continents apart.

B   What Are Some Other Transform Boundaries?

The Pacific seafloor and western North America contain several different transform boundaries. The boundary between the Pacific plate and the North American plate is mostly a transform boundary, with the Pacific plate moving northwest relative to the main part of North America.

1. The Queen Charlotte transform fault, shown as a long green line, lies along the edge of the continent, from north of Vancouver Island to southeastern Alaska.

2. The zigzag boundary between the Pacific plate and the small Juan de Fuca plate has three transform faults, shown here as green lines. These transform faults link the ridge segments that are spreading (shown here as yellow lines).

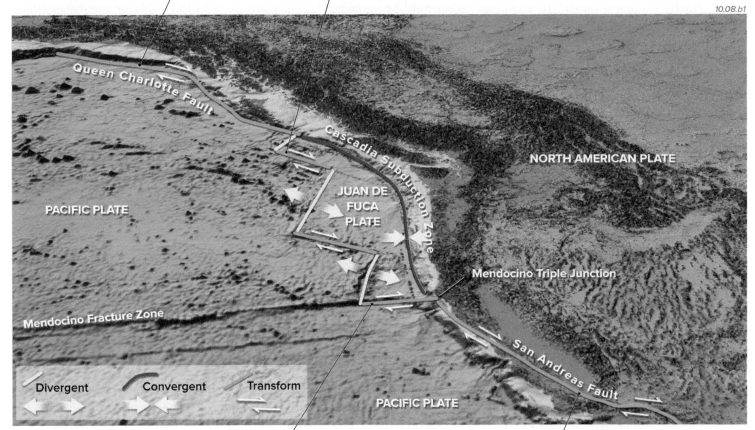

10.08.b1

3. The Mendocino fracture zone originated as a transform fault, but it is now entirely within the Pacific plate and is no longer active. Oceanic crust to the north is higher because it is younger than oceanic crust to the south.

4. A transform fault links a spreading center (between the Pacific plate and the Juan de Fuca plate) with the Cascadia subduction zone and the San Andreas fault. The place where the three plate boundaries meet is a *triple junction*. The Mendocino triple junction is the meeting place of two different transform faults and a subduction zone.

5. The San Andreas transform fault extends from north of San Francisco to southeast of Los Angeles. The part of California west of the fault is on the Pacific plate and is moving approximately 5 cm/yr to the northwest relative to the rest of North America. South of this map area, the transform boundary continues across southern California and into the Gulf of California.

Before You Leave This Page

☑ Sketch, label, and explain an oceanic transform boundary related to seafloor spreading at a mid-ocean ridge.

☑ Sketch, label, and explain the motion of transform faults along the west coast of North America.

6. Californians have a transform fault in their backyard, as well as many other smaller faults. In central California, the San Andreas fault forms linear valleys, abrupt mountain fronts, and lines of lakes. In the Carrizo Plain (▶), the fault is a linear gash in the topography. Some streams follow the fault and others jog to the right as they cross the fault, recording relative movement of the two sides. In this view, the North American plate is to the left, and the Pacific plate is to the right and is being displaced toward the viewer at several centimeters per year.

10.08.b2 Carrizo Plain, CA

10.8

Why and How Do Plates Move?

THE PROCESS OF PLATE TECTONICS circulates material back and forth between the asthenosphere and the lithosphere. Some asthenosphere becomes lithosphere at mid-ocean spreading centers and then takes a slow trip across the ocean floor before going back down into the asthenosphere at a subduction zone. Besides creating and destroying lithosphere, this process is the major way that Earth transports heat to the surface.

A What Moves the Plates?

How exactly do plates move? To move, an object must be subjected to a *driving force* (a force that drives the motion). The driving force must exceed the *resisting forces*—those forces that resist the movement, such as friction and any resistance from other material that is in the way. What forces drive the plates?

1. *Slab Pull*—Subducting oceanic lithosphere is more dense than asthenosphere, so gravity pulls the plate downward into the asthenosphere. Slab pull is a significant force, and a plate being subducted generally moves faster than a plate that is not being subducted. Subduction sets up other forces in the mantle that can work with or against slab pull.

2. *Ridge Push*—The mid-ocean ridge is higher than the seafloor away from the ridge because lithosphere near the ridge is thinner and hotter. Gravity causes the plate to slide away from the topographically high ridge and push the plate outward.

3. *Mantle Convection*—The asthenosphere, although a solid, is capable of flow. It experiences *convection*, where hot material rises due to its lower density, while cold material sinks because it is more dense. Hot material rises at mid-ocean ridges, cools, and eventually sinks back into the asthenosphere at a subduction zone. Convection also occurs at centers of upwelling mantle material called *hot spots*, and it can help or hinder the motion of a plate.

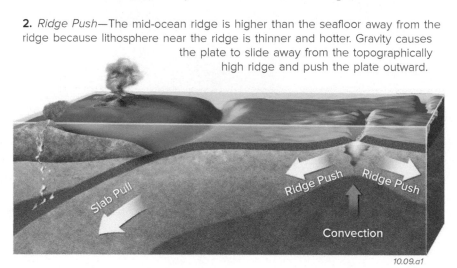

10.09.a1

B How Fast and in What Directions Do Plates Move Relative to One Another?

Plates move at 1 to 15 cm/yr, about as fast as your fingernails grow. This map, similar to one earlier in the chapter, shows velocities and relative motions along major plate boundaries, based on long-term rates. Arrows indicate whether the plate boundary has divergent (outward pointing), convergent (inward pointing), or transform (side by side) motion.

10.09.b1

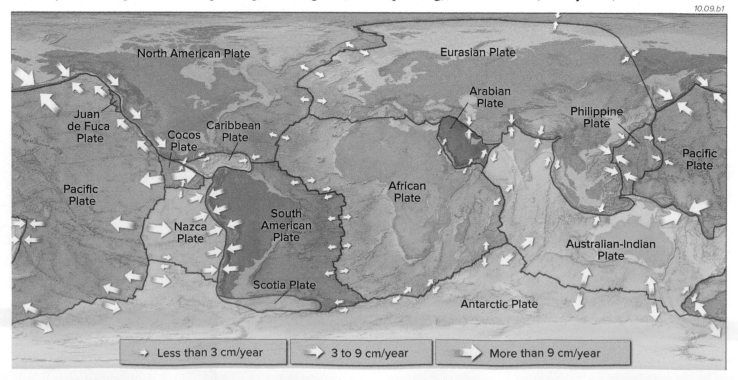

C Is There a Way to Measure Plate Motions Directly?

Modern technology allows direct measurement of plate motions using satellites, lasers, and other tools. The measured directions and rates of plate motions are consistent with our current concept of lithospheric plates and with the theory of plate tectonics. Examine this map and determine which way your home is moving.

1. Global Positioning System (GPS) is an accurate location technique that uses small radio receivers to record signals from several dozen Earth-orbiting satellites. By attaching GPS receivers to sites on land and monitoring changes in position over time, Earth scientists produce maps showing motions for each plate. Arrows point in the direction of motion, and longer arrows indicate faster motion.

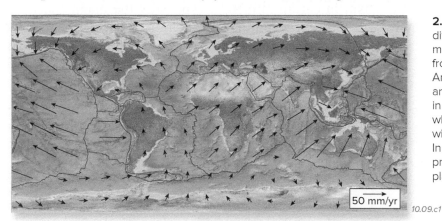

50 mm/yr

10.09.c1

2. Note the motions of different plates. Africa is moving to the northeast, away from South America. North America is moving westward and rotating counterclockwise in this view. The plate on which India rides is converging with Asia to form the Himalaya. In all, these motions match predictions from the theory of plate tectonics.

D Is the Age of the Seafloor Consistent with Plate Tectonics?

According to plate tectonics, oceanic crust forms from upwelling magma and spreading at a mid-ocean ridge. The oceanic crust then moves away from the ridge with further spreading. If this is so, the crust should be youngest near the ridge, where it was just formed, and should be progressively older away from the ridge. Also, oceanic crust near the ridge will not have had time to accumulate much sediment, but the sediment cover should thicken outward from the ridge.

1. Since 1968, ocean-drilling ships have drilled hundreds of deep holes into the seafloor. Geoscientists use drill cores, represented here by cylinders of rock and sediment, and other drilling results to measure the thickness of sediment and examine the underlying volcanic rocks (basalt). Geoscientists analyze samples of sediment, rock, and fossils to determine the age, character, and origin of the materials.

10.09.d1

◻ Sediments   ◼ Basalt

2. Drill core samples reveal that sediment is thin or absent on the ridge but becomes thicker away from the ridge. Age determinations on fossils in the sediment and from underlying volcanic rocks show that oceanic crust gets systematically older away from mid-ocean ridges. Such drilling results strongly support the theory of plate tectonics.

3. The map below shows the age of the seafloor, with letters marking the position of some mid-ocean ridges (R) and trenches (T). Purple represents the oldest areas (about 180 million years), and the darkest orange represents very young oceanic crust.

4. The youngest oceanic crust is near mid-ocean-ridge spreading centers (R), as we expect from plate tectonics.

5. The oldest oceanic crust in any ocean is the most distant from mid-ocean ridges. None is older than about 180 million years, because all older oceanic crust has been subducted (destroyed). The oldest seafloor is near passive margins (P).

10.09.d2

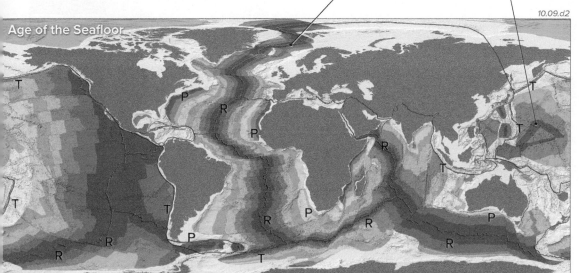

Age of the Seafloor

Before You Leave This Page

✓ Sketch and explain the driving forces of plate tectonics.

✓ Summarize the typical rates of relative motion between plates and describe one way we measure this.

✓ Predict the relative ages of seafloor from place to place using a map of the seafloor.

10.9

10.10 How Is Paleomagnetism Used to Determine Rates of Seafloor Spreading?

PALEOMAGNETISM IS THE ROCK RECORD OF PAST CHANGES in Earth's magnetic field. The magnetic field is strong enough to orient magnetism in certain minerals, especially the magnetic, iron-rich mineral *magnetite,* in the direction of the prevailing magnetic field. Magnetic directions preserved in volcanic rocks, intrusive rocks, and some sedimentary rocks provide an important way to determine the rates of seafloor spreading.

A What Causes Earth's Magnetic Field?

Earth has a metallic iron core, which is composed of a solid inner core surrounded by a liquid outer core. The liquid core flows and behaves like a *dynamo* (an electrical generator), creating a magnetic field around Earth.

1. The inner core transfers heat and less dense material to the liquid outer core. This transfer causes liquid in the outer core to rise, forming *convection currents.* These convection currents are limited to the outer core and are not the same as those in the upper mantle.

2. Movement of the molten iron is affected by forces associated with Earth's rotation. The resulting movement of liquid iron and electrical currents generates Earth's magnetic field.

10.10.a1

3. Earth's magnetic field currently flows from south to north (▶), causing the magnetic ends of a compass needle to point toward the north. This orientation is called a *normal polarity.*

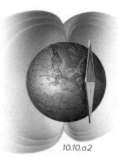

10.10.a2

10.10.a3

4. Many times in the geologic past, the magnetic field has had a *reversed polarity,* so that a compass needle would point south (◀). The switch between normal polarity and reversed polarity is a *magnetic reversal* and may only require as little as 100 years to occur.

B How Do Magnetic Reversals Help Us Infer the Age of Rocks?

The north and south magnetic poles have switched many times, typically remaining either normal or reversed anywhere from 100,000 years to a few million years. Scientists have constructed a magnetic timescale by isotopically dating sequences of rocks that contain magnetic reversals. This *geomagnetic polarity timescale* then serves as a reference to compare against other sequences of rocks.

10.10.b1

1. Scientists measure the direction and strength of the magnetism preserved in rocks with an instrument called a *magnetometer.* With this device, we can determine whether the magnetic field had a normal polarity or a reversed polarity when an igneous rock solidified and cooled or when a sedimentary layer accumulated.

2. This figure shows the series of magnetic reversals during the last 10 million years (▶), the most recent part of the Cenozoic Era. The abbreviation "Ma" stands for millions of years before present.

3. The timescale shows periods of normal magnetization (N) in black and those of reversed magnetization (R) in white. Variability in the spacing and duration of magnetic reversals produced a unique pattern through time. Geoscientists can measure the pattern of reversals in a rock sequence and compare this pattern to the magnetic timescale to see where the patterns match. This allows an estimate of the age of the magnetic rock or sediment. Other age constraints, including isotopic ages or fossils, are used to further refine the age of the magnetized rocks. The magnetic timescale is best documented for the last 180 million years because seafloor of this age is widely preserved, allowing measurements in many places.

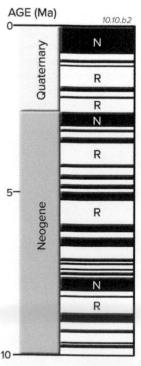

10.10.b2

AGE (Ma)

C How Are Magnetic Reversals Expressed at Mid-Ocean Ridges, and How Do Magnetic Patterns on the Seafloor Help Us Study Plate Tectonics?

In the 1950s, scientists discovered that the ocean floor displayed magnetic variations in the form of matching magnetic stripes on either side of the mid-ocean ridge. They interpreted the patterns to represent a magnetic field that had reversed its polarity, an idea that led to the theory of plate tectonics. Magnetic patterns allow us to estimate the ages of large areas of seafloor and to calculate the rates at which two diverging oceanic plates spread apart.

1. As the oceanic plates spread apart at a mid-ocean ridge, basaltic lava erupts onto the surface or solidifies at depth. As the rocks cool, the orientation of Earth's magnetic field is recorded by the iron-rich mineral magnetite. In this example, the magnetite records normal polarity (shown with a reddish color) at the time the rock forms. Rocks forming all along the axis of the mid-ocean ridge will have the same magnetic direction, forming a stripe of similarly magnetized rocks parallel to the ridge.

2. If the magnetic field reverses, any new rocks that form will acquire a reversed polarity (shown in white). Once the rocks have cooled, they retain their original magnetic direction, preserving the magnetic polarity in the seafloor. The new reversely magnetized rocks form a stripe along the mid-ocean ridge, and the previously formed, normally magnetized rocks have been split into stripes and have moved away from the ridge in opposite directions.

3. The magnetic poles have switched many times, and continued seafloor spreading produces a pattern of alternating magnetic stripes on the ocean floor. This pattern is strong enough to be detected by magnetic instruments towed behind a ship or a plane.

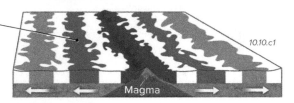

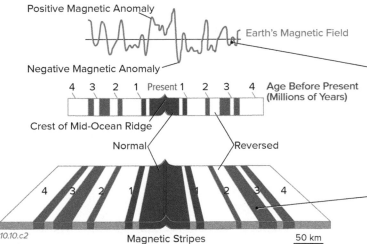

4. As magnetic instruments are towed behind a ship, the strength of the magnetic field is measured and plotted. Stronger measurements, representing rocks with normal magnetization, plot high on the graph and are called *positive magnetic anomalies*. The reverse magnetization of the rocks slightly weakens the measured magnetic signal and will plot low on the graph, forming a *negative magnetic anomaly*.

5. The seafloor patterns are compared with the patterns on the geomagnetic polarity timescale to assign ages to each reversal. We can simplify and visualize these data as reversely and normally magnetized stripes on the seafloor, as shown in this cross section.

6. We can calculate rates of seafloor spreading by measuring the width of a specific magnetic stripe in map or cross-section view and then dividing that distance by the length of time the stripe represents:

> *rate of spreading for stripe = width of stripe / time duration*

If a magnetic stripe is 60 km wide and formed over 2 million years, then the rate at which spreading formed the stripe was 30 km/m.y. This rate is equivalent to 3 cm/yr. Spreading added an equal width of oceanic crust to a plate on the other side of the mid-ocean ridge, so the total rate of spreading across the ridge was 60 km/m.y. (6 cm/yr), a typical rate of seafloor spreading. The rates determined from magnetic stripes nicely match those determined by GPS, producing strong support for plate tectonics.

7. The magnetic patterns on the seafloor, in addition to magnetic measurements on sequences of rocks and sediment on the seafloor and on land, demonstrate that Earth's magnetic field has reversed many times. Scientists are currently debating the possible causes of the magnetic reversals, with most explanations attributing reversals to chaotic flow in the molten outer core, which add to or subtract from the patterns caused by the dynamo, disrupting the prevailing magnetic field and causing a reversal. What do you think we might experience during a magnetic reversal in our lifetime?

Before You Leave This Page

✓ Describe how Earth's magnetic field is generated.

✓ Describe how magnetic reversals help with determining the age of rocks and sediment.

✓ Describe or sketch how magnetic patterns develop on the seafloor.

✓ Calculate the rate of seafloor spreading if given the width and duration of a magnetic stripe.

10.10

10.11 What Features Form at Oceanic Hot Spots?

SUBMARINE MOUNTAINS, called *seamounts*, rise above the seafloor. In some places, they reach the surface and make islands. In many cases, islands and seamounts form linear chains, as in Hawaii. The seafloor also has relatively high and broad areas that are oceanic plateaus. How are seamounts and oceanic plateaus formed?

A How Do Island and Seamount Chains Form?

Linear chains of islands and seamounts, and most clusters of islands in the oceans, have two key things in common: they were formed by volcanism and they are near sites interpreted to be above unusually high-temperature regions in the deep crust and upper mantle—*hot spots*. Hot spots are interpreted to represent the surface manifestation of hot masses of material rising from the deep mantle.

1. This figure shows how linear island and seamount chains can be related to a plate moving over a hot spot. At a hot spot, hot mantle rises and melts, forming magma that ascends into the overlying plate. The magma generated by a hot spot may solidify at depth or form a volcanic mountain on the ocean floor. If the submarine volcano grows high enough above the seafloor, it becomes a volcanic island. Each of the Hawaiian Islands consists of volcanoes. Geomorphologists consider the hot spot to be currently below or near the eastern side of the Big Island, near Kilauea volcano.

2. If the plate above the hot spot is moving relative to the hot spot, the volcanic island can move off the hot spot. As it does, that part of the plate cools and subsides, so volcanoes that start out as islands may sink beneath the sea to become seamounts. As an island sinks, erosion can bevel off its top, forming a flat-topped seamount (▶).

10.11.a2

3. As a plate moves over a hot spot, volcanism constructs a chain of volcanic islands and seamounts, each created when it was over the hot spot. According to this model, volcanoes above the hot spot may be erupting today, those close to but not above the hot spot are relatively young, and those farthest from the hot spot are older. If a plate is not moving or is moving very slowly, the hot spot forms a cluster of volcanic islands and seamounts instead of a linear chain. The Galápagos, a cluster of volcanic islands in the eastern Pacific, is interpreted to be above such a hot spot.

Hot Spot

10.11.a1

B How Do Oceanic Plateaus Form?

Some large regions of the seafloor rise a kilometer or more above their surroundings, forming *oceanic plateaus*. These plateaus are largely composed of flood basalts, like the seafloor in general. How do these form?

1000 km

10.11.b1

10.11.b2

10.11.b3

10.11.b4

1. This perspective shows the Kerguelen oceanic plateau, which rises above the surrounding seafloor in the southern Indian Ocean. The plateau is several thousand kilometers long, but it only reaches sea level in a few small islands. The small sliver of land showing in the lower right corner is part of Antarctica.

2. Geoscientists interpret oceanic plateaus as forming at hot spots, above rising hot mantle, called mantle plumes. The plumes travel through the mantle as solid masses, not liquids.

3. When the top of a plume encounters the base of the lithosphere, it causes widespread melting. Submarine basalts pour out onto the seafloor through fissures and central vents.

4. Immense volumes of basalt (as much as 50 million cubic kilometers) erupt onto the seafloor over millions of years. This volcanism creates a broad, high oceanic plateau.

C What Is the Distribution of Hot Spots, Linear Island Chains, and Oceanic Plateaus?

Hot spots have created many Pacific islands that we associate with tropical paradises and exotic destinations, like Tahiti and the Galápagos. Hawaii is the most famous island chain formed by movement of a plate over a hot spot, but several other linear island and seamount chains, in both the Atlantic and Pacific, formed in the same manner. We show oceanic and continental hot spots on the map below, but we discuss continental hot spots on the following two pages.

1. On this map, red dots show the locations of likely hot spots, many of which are located at the volcanically active ends of linear island chains. There is great debate, however, about which areas really are hot spots and how hot spots form.

2. The dark gray areas in the oceans represent linear island chains, clumps of islands, and oceanic plateaus, similar to this high area around Iceland, which is over a hot spot.

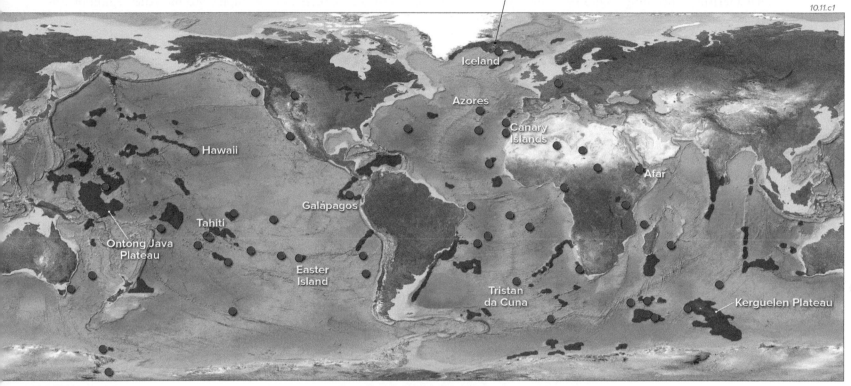

10.11.c1

3. The *Ontong Java Plateau* is the largest oceanic plateau on Earth, covering millions of square kilometers, nearly 1% of Earth's surface area. It formed in the middle of the Pacific Ocean 120 Ma and is no longer near the hot spot that produced it.

4. Volcanic islands near *Tahiti* define northwest-trending chains that are forming over several hot spots. In each chain, the islands to the northwest are older than those to the southeast, indicating that the Pacific plate is moving to the northwest relative to the underlying source of magma.

5. The *Galápagos* is a clump of volcanic islands west of South America. The western islands, shown in the satellite image to the lower left, are volcanically active and have erupted within the past several years. Eruptions build larger volcanoes and smaller volcanic cones, as shown in the photograph below.

6. *Tristan da Cunha,* a volcanic island in the South Atlantic Ocean, marks a hot spot just east of the Mid-Atlantic Ridge. Volcanism associated with the hot spot created a large submarine ridge (shown in gray) that tracks the motion of the African plate over the hot spot.

7. The *Kerguelen Plateau,* in the southern Indian Ocean, is the second largest oceanic plateau in the world. It mostly consists of basalt and was formed in several stages during the late Mesozoic (between 115 and 85 Ma).

10.11.c2

Western Galápagos Islands

10.11.c3 Galápagos Islands

Before You Leave This Page

✓ Describe or sketch how a hot spot can form a linear chain of islands and seamounts, or a cluster of islands, providing several examples.

✓ Describe how oceanic plateaus are interpreted to have formed, and provide some examples.

10.11

10.12 What Features Form at Continental Hot Spots?

A HOT SPOT WITHIN A CONTINENTAL PLATE is marked by high elevations, abundant volcanism, and continental rifting. Hot spots can facilitate complete rifting and separation of a continent into two pieces and can help determine where the split occurs. Several continental hot spots are active today.

A What Features Are Typical of Continental Hot Spots?

Hot spots are volcanic areas interpreted to be above rising mantle plumes. Continental hot spots are associated with certain characteristics, including high elevations, volcanism, and the presence of rifts. Two examples are the Afar region of East Africa and the Yellowstone region of the western U.S.

Afar Region, East Africa

1. Continental hot spots have high elevations largely because of heating and thinning of the lithosphere by a rising plume of hot mantle. Many geoscientists interpret the Afar region of eastern Africa to be located above a hot spot that is currently active.

2. The East African Rift is within the African plate. It may or may not evolve into a full rift that fragments the continent into two parts and that leads to seafloor spreading.

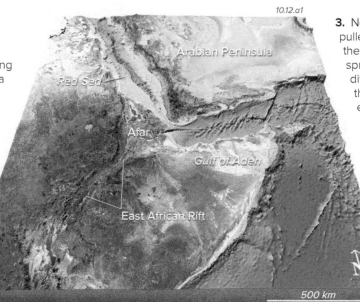

10.12.a1

3. Near the hot spot, the Arabian Peninsula has pulled away from Africa along the Red Sea and the Gulf of Aden. Beneath these seas, seafloor spreading generates new oceanic crust. The divergent plate boundary is along the rift down the middle of the Red Sea and then turns eastward, following the zigzag oceanic ridge through the Gulf of Aden.

4. The Red Sea, Gulf of Aden, and East African Rift meet in the Afar region of Ethiopia, branching off like three spokes on a wheel. The Afar region is among the most volcanically active areas on Earth and has experienced recent volcanic eruptions. Volcanism has been so prolific here that it has created a triangular area of new land in the corner of Africa from which the Arabian Peninsula pulled away.

Region Around Yellowstone National Park

5. Yellowstone is located in Wyoming, Idaho, and Montana and sits in a region that is higher in elevation than surrounding areas.

6. The Snake River Plain of southern Idaho is underlain by thick sequences of basalt and other volcanic rocks. It is the site of recent eruptions at Craters of the Moon National Monument.

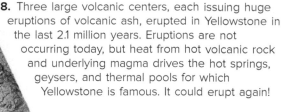

10.12.a2

8. Three large volcanic centers, each issuing huge eruptions of volcanic ash, erupted in Yellowstone in the last 2.1 million years. Eruptions are not occurring today, but heat from hot volcanic rock and underlying magma drives the hot springs, geysers, and thermal pools for which Yellowstone is famous. It could erupt again!

9. Yellowstone is interpreted to mark the present location of the hot spot, whereas the Snake River Plain records the track of North America as it moved southwest over the hot spot. Forming at the same time was the Basin and Range Province of Utah and Nevada, a broad continental rift adjacent to Yellowstone. The alternating valleys and mountains resulted from faults that uplifted some blocks of crust and downdropped others.

10.12.a3 Craters of the Moon, ID

7. Craters of the Moon National Monument (◀) in southern Idaho contains spectacular examples of recent, unweathered lava flows that are preserved in solid rock.

10. Yellowstone National Park, mostly in Wyoming, is named for the volcanic rocks turned yellowish (▶) by hot waters that accompanied the eruptions and continue to form the famous hot pools.

10.12.a4 Yellowstone NP

B How Do Continental Hot Spots Evolve?

Many continental hot spots underwent a similar sequence of events. They started with doming and ended with the formation of a new continental margin and a new ocean formed by seafloor spreading.

1. Hot spots mark where a mostly solid, hot mass rises, probably from the lower mantle, and encounters the base of the lithosphere. The rising material melts as a result of less pressure and also causes melting of nearby lithosphere.

2. As the upper mantle and crust heat up, a broad, domal uplift forms on the surface. Doming is accompanied by stretching of the crust, which commonly begins to break apart along three rifts that radiate out from the hot spot.

3. Some mantle-derived magma escapes to the surface and erupts as voluminous, very fluid basalts called *flood basalts*. Granitic-composition magmas form where mantle-derived magma causes melting of the continental crust.

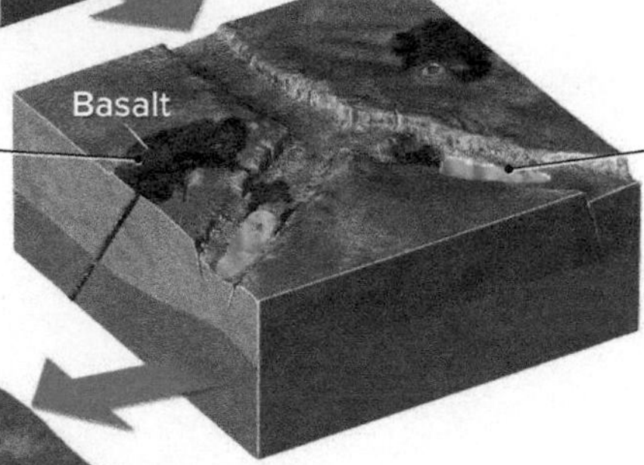

4. All three parts, or arms, of the rift are bordered by faults, which downdrop long fault blocks. The downdropped blocks form basins that contain lakes and are partially filled by sediment and rift-related volcanic rocks.

5. Complete rifting of the continent occurs along two arms of the rift. This results in a new continental margin and seafloor spreading in the new ocean basin. At the onset of spreading, the edge of the continent is uplifted because the lithosphere is heated and thinned due to the rifting.

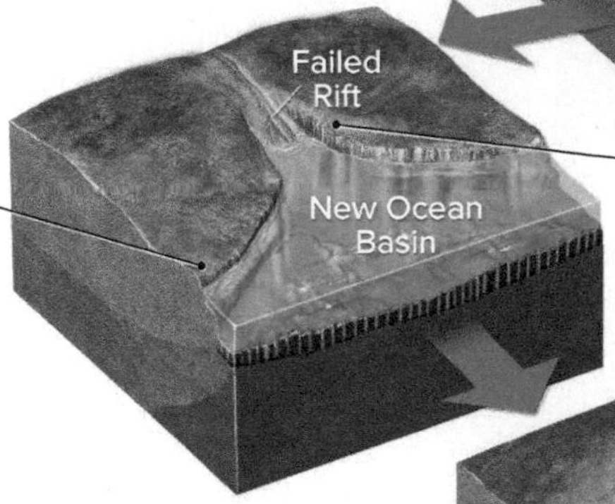

6. The third arm of the rift begins to become less active and fails to break up the continent into more pieces. This *failed rift* is lower than the surrounding continent and commonly becomes the site of major rivers.

7. As seafloor spreading continues, the generation of new oceanic lithosphere causes the mid-ocean ridge to move farther out to sea. The continental margin cools and subsides and is covered by marine sediment on the newly formed continental shelf. This continental margin is no longer a plate boundary and is now a passive margin.

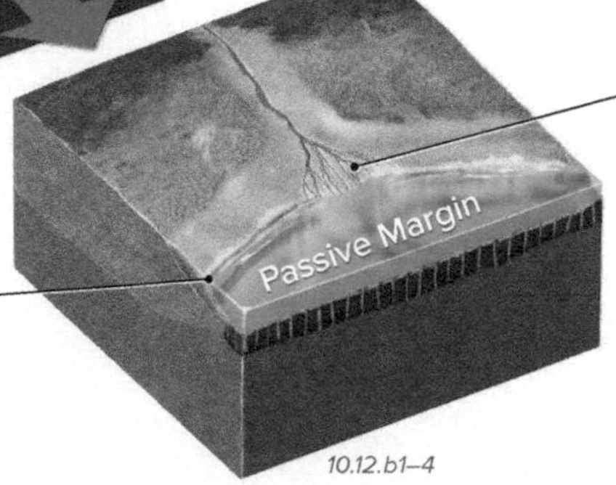

8. Sediment transported by streams down the failed rift will form a delta at the bend in the continent. This is currently occurring along the western coast of Equatorial Africa at the large inward bend in the coast (see the figure and text below).

10.12.b1–4

Hot Spots and Continental Outlines

Geoscientists conclude that hot spots have helped define the outlines of the continents by shaping the boundary along which continents separate from one another. The best example of this is the inward bend of the western coast of Africa near Cameroon. This bend occurs at the intersection of three arms of a rift, two of which led to the opening of the South Atlantic Ocean. The third *failed rift* cuts northeastward into Africa and is the site of several major rivers. Large eruptions of basalt (flood basalt) occurred along the rifts, and active volcanism near the failed rift may mark the location of a hot spot. This figure shows what the area may have looked like 110 Ma, when the continents first started to rift apart.

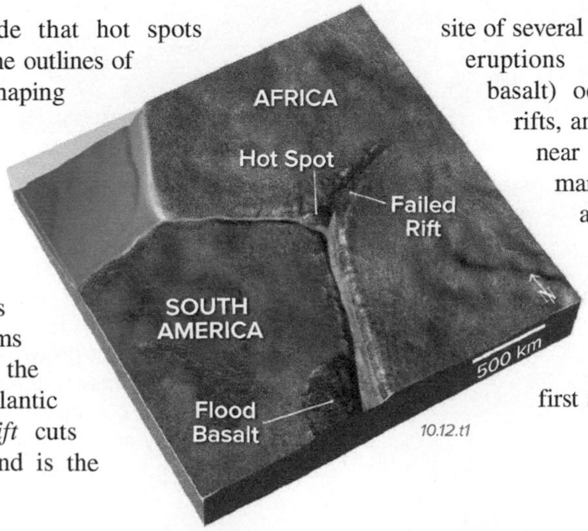

10.12.t1

Before You Leave This Page

☑ Summarize the features that are typical of continental hot spots, providing an example of each type of feature.

☑ Summarize or sketch how continental hot spots evolve over time.

☑ Describe or sketch how hot spots influence continental outlines, providing an example.

10.12

10.13 What Are Continents and How Do They Form?

CONTINENTS ARE AMONG THE LARGEST FEATURES of our planet, but what are they, how did they form, and does a continent remain the same size over long time periods? Continents have interesting histories, which at times involved splitting apart or joining together. Through geologic time, continents overall have tended to grow larger as they collide with and incorporate island arcs, oceanic plateaus, seamounts, and other tectonic features.

A What Is a Continent?

When people think of a continent, most envision its shape as seen by its present-day coastline. This outline, however, changes as sea level goes up or down, as it has many times in the distant past. We should therefore consider a slightly different definition of a continent—one that indicates a less transient outline.

1. The continent of Australia (▶) nicely illustrates the issues involved with defining the edge of a continent. The familiar outline of the continent is the shoreline, the boundary between the browns and greens of land versus the blues of the adjacent ocean on this map.

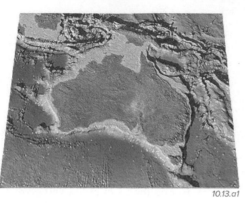
10.13.a1

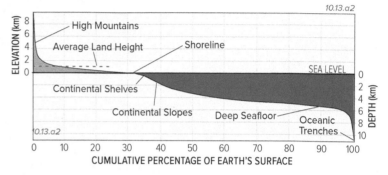
10.13.a2

10.13.a2

2. Notice, however, that surrounding the land is an area of relatively shallow ocean, represented on this figure by light blue. This blue shelf surrounding the continent is similar to the rest of the continent, except that it is barely under water. Earth scientists consider this shallow, continental shelf to be part of the Australian continent, putting the edge of the continent at the outer edge of the continental shelf, not at the coastline.

3. We can represent continents, and the rest of Earth's land and seafloor, in a graph that plots elevation versus the percentage of land or seafloor that is above that elevation (or depth). This graph, called a *hypsometric curve*, shows that somewhat less than 30% of Earth is continent at this time, with most of the rest being deep ocean. The major change between the continents and deep oceans occurs not at the shoreline, but across continental slopes. This graph illustrates that the continental shelves are more akin to the continents than to the deep seafloor.

B How Do Continents Form and Grow?

1. Examine this figure showing various tectonic features within and next to a central plate being consumed by subduction from both sides. If we predict where each feature will go as the plate motions continue, we expect that some of these features will collide with each other and with the continent to the right. When this occurs, they may be added to the front edge of the continent. A piece of exotic crust added to a continent is called a *tectonic terrane* (note the spelling).

2. Many terranes consist of volcanic and volcanic-related rocks that formed as island arcs. Island arcs often become terranes because they move across the ocean, potentially traveling long distances until they collide with, and become part of, another landmass, like a continent.

3. Some terranes have more continental characteristics, specifically thick granitic crust with continental sedimentary rocks. Such terranes generally represent pieces that were sliced or rifted off another continent and then tectonically transported until they collided with the edge of a continent.

4. Some terranes represent oceanic islands, oceanic plateaus, and other types of oceanic crust. Islands and plateaus have thicker-than-normal oceanic crust, so they are less likely to be subducted and more likely to be added to the edge of a continent. The oceanic plateau shown here is headed for the trench to the right and a collision with the edge of the adjacent continent.

5. For a terrane to become attached to a continent, it typically enters a subduction zone, where it is scraped off the subducting plate and tectonically added to the continent. Through this process of adding terranes along subduction zones, a continent grows. In fact, incorporating terranes along their edges is probably the main way most continents grow, and this process was very important in constructing the continents we have today.

10.13.b1

C What Are Common Features of Continents?

Many continents display a similar pattern, with a central region of older igneous and metamorphic rocks surrounded by a relatively thin veneer of younger, nearly flat-lying sedimentary layers.

1. Many continents, including North America, have a central region called a *continental shield*. A shield consists of relatively old metamorphic and igneous rocks, often simply called *crystalline rocks* or the *crystalline basement*. The rocks exposed in the shield represent the kinds of rocks that underlie much of the continent. In North America, the continental shield is mostly in Canada and is called the *Canadian Shield*.

2. Surrounding the shield is a broad region called the *continental platform*. It is characterized by nearly horizontal sedimentary rocks that were deposited on top of the older, lower rock. The sedimentary layers commonly have been gently warped and then eroded, which results in higher and lower rocks being exposed at the surface from place to place. In the U.S., the entire center of the country, between the Rocky and Appalachian Mountains, is a continental platform.

3. The boundary between the flat-lying platform sedimentary rocks and the underlying crystalline rocks is an old, buried erosion surface (an *unconformity*), which separates rocks with very different ages and histories.

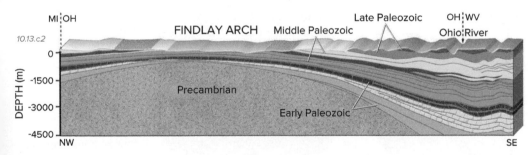

Continental Shield

Basin

Continental Platform

Coastal Plain

— Sedimentary Rocks

— Unconformity

— Crystalline Basement

10.13.c1

4. Sedimentary rocks in the interior generally have few major geologic structures, but some regions, such as the continental shield, have been gently warped upward, and others are warped downward, forming a low area called a *basin*. A large, ancient basin like this underlies nearly all of Michigan.

5. Most continents have at least one edge where elevation of the land slowly decreases toward the coast until it flattens out in a low plain barely above sea level. This low region is a *coastal plain* and is underlain by sedimentary units, many of which were deposited quite recently in geologic terms. In the U.S., the coastal plain warps around the southern and eastern edge of North America, from Texas to Florida to New Jersey.

10.13.c2

MI OH   FINDLAY ARCH   Middle Paleozoic   Late Paleozoic   OH WV   Ohio River

DEPTH (m)

0

-1500

-3000

-4500

Precambrian

Early Paleozoic

NW   SE

6. This geologic cross section across the state of Ohio (◄) is typical of the geology of central North America and of continental interiors in general. In this area, sedimentary layers dip gently off the flanks of a dome and are underlain by Precambrian basement. The section is vertically exaggerated, so true dips are less than shown here, and the thicknesses of the layers are greatly exaggerated.

In Suspect Terrain

California is the area many geoscientists think about when they study how continents grow from the addition of tectonic terranes. John McPhee's popular books, *In Suspect Terrain* and *Assembling California*, provide an accessible account of how terranes were recognized and how they added new real estate to North America. This map shows the various types of terranes added to the continent. The terranes include slices of oceanic crust and sediment, parts of island arcs, and a wedge-shaped mass of material, called an *accretionary prism*, that was scraped off oceanic plates that were subducted along a trench. The map does not show units that formed in place, such as granites in the Sierra Nevada.

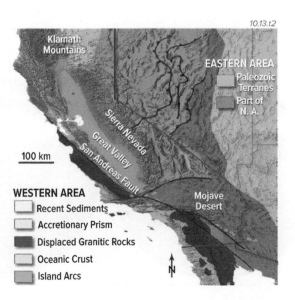

10.13.t2

Klamath Mountains

EASTERN AREA
Paleozoic Terranes
Part of N. A.

Sierra Nevada

Great Valley

San Andreas Fault

100 km

Mojave Desert

WESTERN AREA
☐ Recent Sediments
☐ Accretionary Prism
■ Displaced Granitic Rocks
☐ Oceanic Crust
☐ Island Arcs

N

Before You Leave This Page

☑ Sketch and describe a hypsometric curve and use it to explain how it helps us define the edge of a continent.

☑ Describe what a tectonic terrane is, the main tectonic settings in which terranes originate, and how terranes are added to crust.

☑ Summarize some common features of continents and how these are expressed in North America.

10.13

10.14 How Did the Continents Join and Split Apart?

CONTINENTS SHIFT THEIR POSITIONS over time in response to plate tectonics. They have rifted apart and collided, only to rift apart again. Where were the continents located in the past, and which mountains resulted from their motions? The figures below illustrate which continents were joined and how they separated. We start with 600 million years ago (Ma) and work forward to the present, as if watching a movie of Earth's history.

600 Ma: The Supercontinent of Rodinia

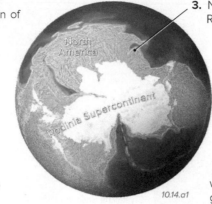

1. The images on these pages show one interpretation of where the continents were located in the past. These are artistic renderings based on available data, of the continents, mountains, and oceans. For most time periods, there are two views, one focused on the Western Hemisphere (images on the left) and one on the Eastern Hemisphere (images on the right), generally with some overlap. We begin here with a single image, centered on the South Pole.

2. Before the Paleozoic Era, in the last part of the Precambrian, all of the major continents were joined. This supercontinent is called *Rodinia*. Nearly all of the other side of the globe is a huge ocean.

10.14.a1

3. North America was in the initial stages of rifting from Rodinia. This rifting outlined the western margin of North America, but geologists are not certain which continent was adjacent to North America. Options include Australia, Antarctica, and Asia.

4. Large parts of Rodinia were near the South Pole, and there is evidence of widespread glaciation in these polar regions. Prior to this time, large parts of the Earth, more area than shown here, were probably covered with ice, a condition referred to as a "snowball Earth." This far back in the past, there are large uncertainties about how extensive the ice was and exactly how the continents were arranged, but geoscientists are trying to match the ages and sequences of rocks between different continents.

500 Ma: Dispersal of the Continents

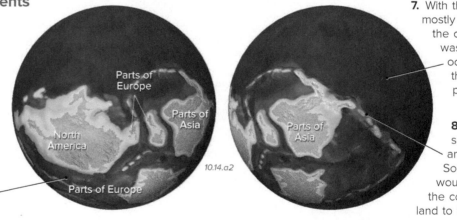

5. At 500 Ma, in the early part of the Paleozoic Era, North America and Europe were separate, moderate-sized continents that had not yet joined.

6. Antarctica, Australia, South America, and Africa were joined in the Southern Hemisphere, together forming the southern supercontinent of Gondwana, which was mostly located in the Southern Hemisphere (mostly out of view on these figures). Gondwana was separated from the northern continents by some width of ocean.

10.14.a2

7. With the continents still mostly clustered together, the other side of Earth was a single large ocean, much larger than the size of the present-day Pacific.

8. Island arcs surrounded Europe and parts of Asia. Some of the arcs would later collide with the continents, adding land to the continents.

370 Ma: Before Pangaea

9. In the middle of the Paleozoic Era, at 370 Ma, Europe and North America were fully joined but were not connected with Asia, which lay to the north. North America had collided with a microcontinent called Avalonia. This collision created mountains in what is now New England.

10. North America was approaching Africa and South America along a convergent margin. The continents were on a collision course as the intervening ocean became narrower over time.

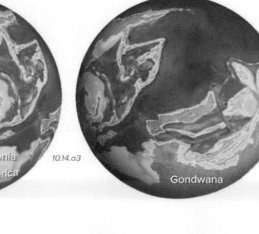

10.14.a3

11. Gondwana remained mostly intact, except for some slices of continental crust that probably were rifted away from the larger supercontinent. The microcontinent (Avalonia) that collided in New England was probably one of these rifted pieces, but it had broken away some time before 370 Ma, the time pictured here.

280 Ma: The Supercontinent of Pangaea

12. In the late Paleozoic, around 280 Ma, a continental collision between North America and the northern edge of Gondwana (South America and Africa) formed the Appalachian Mountains near the East Coast and the Ouachita Mountains (not labeled) in the southeastern U.S.

13. After this collision and a series of smaller collisions, all the continents were joined in a supercontinent called *Pangaea*. Pangaea is Greek, meaning "all Earth."

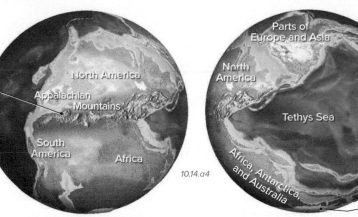

14. A wedge-shaped ocean, the *Tethys Sea*, separated Asia from landmasses farther to the south. Southern Africa, Australia, and Antarctica were close to the South Pole and so at this time were partly covered with ice.

150 Ma: Gondwana and Laurasia

15. At 150 Ma, in the Mesozoic, North America had separated from Africa and South America. The Atlantic Ocean now existed, and continents on either side of the Atlantic were moving away from each other due to seafloor spreading. The left globe is rotated so that the central Atlantic Ocean is in the center of the image.

16. During this time, North America was still joined with Europe and Asia, forming the northern supercontinent of *Laurasia*. South America had not yet rifted away from Africa.

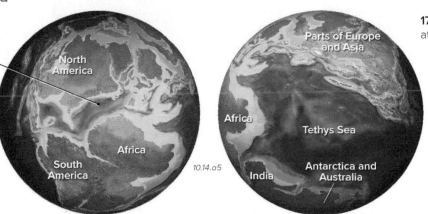

17. Antarctica was still attached to the southern tips of Africa and South America. India was attached to the northern edge of Antarctica. These continents were soon to be rifted apart, which would mark the end of *Gondwana*.

Present

18. This view shows the present-day configuration of continents and oceans. Examine these globes and think about the present-day plate boundaries, envisioning which way the continents are moving relative to one another. Use the current relative motions to predict where the continents were likely to have been in the recent past. Check your predictions by examining each set of previous globes as you step backward in time.

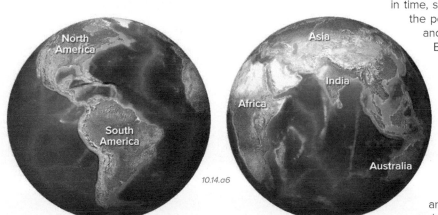

19. When you are done working backward in time, start at 600 Ma and track the position of every continent and watch each ocean open. By viewing this sequence, you are observing how our present-day world came to be.

20. Finally, think about how the position of the continents will change if present-day plate motions continue into the future. What collisions are yet in store for North and South America?

Before You Leave This Page

✓ Briefly summarize the general positions of the continents in the past, especially since 280 Ma.

✓ Identify times when the continents were joined in the supercontinents of Gondwana, Laurasia, Pangaea, and Rodinia.

10.14

10.15 Where Do Mountain Belts and High Regions Form?

MOUNTAIN BELTS AND OTHER HIGH REGIONS generally owe their high elevation to thick continental crust. Crust becomes thicker from deformation and additions from magmatism. In some cases, a region is higher than its surroundings due to processes originating in the mantle. Where are the world's main mountain belts and why did mountains form in these places?

A In Which Tectonic Settings Do Regional Mountain Belts Form?

Regional mountain ranges are hundreds or thousands of kilometers long. They are large enough that they can only be explained by major variations in the thickness and temperature of the crust and lithosphere. Most ranges occur near convergent plate boundaries or where there has been large-scale movement of material in the mantle.

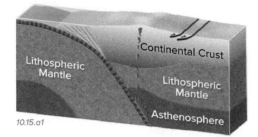

10.15.a1

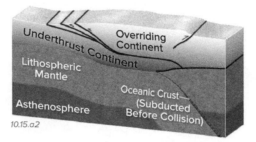

10.15.a2

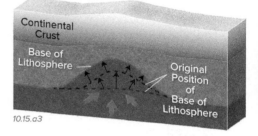

10.15.a3

Subduction Zones—Convergent margins are high in elevation because the crust is thickened by magmatic additions from the subduction zone and by crustal shortening. Also, in these regions, lithosphere is heated and replaced by less dense asthenosphere.

Continental Collisions—Collision zones have high elevations due to an increase in crustal thickness as one continent is shoved over another. In these settings, crustal thickening occurs by faulting, folding, and other forms of deformation.

Mantle Upwellings—Less dense asthenosphere can move upward into the lithosphere, causing regional uplift. This occurs near hot spots, plate boundaries, and in other settings; it is partly responsible for high elevations in parts of the western U.S.

B What Causes These Regions to Have High Elevation?

1. *Western Canada* has been a convergent margin for most of the last 100 Ma. Its mountain ranges overlie crust thickened from major faulting, from magmatic additions, and from collisions with island arcs and pieces of continental material.

2. The *Alps* mountain range of southern Europe is high because it has thick crust due to collisions between Europe and smaller continental blocks that came from the south.

3. The *Tibetan Plateau* and the *Himalaya* are extremely high because of very thick crust that resulted when the Indian continent collided with, and was partly shoved beneath, the continental crust of Asia.

10.15.b1

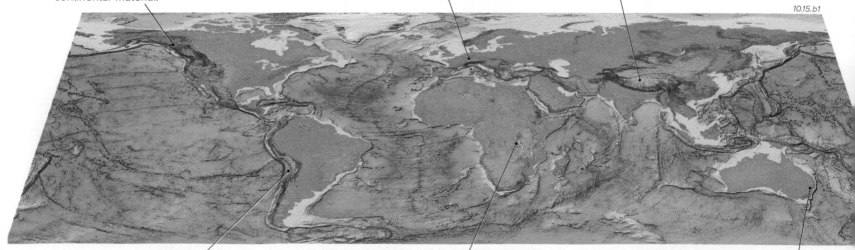

4. The *Andes* of South America are above a subduction zone. The underlying crust is hot and thick because of magmatic additions and crustal shortening.

5. The *East African Rift* is higher than most of Africa because of magmatic heating of the crust, thinning of the crust, and the presence of a hot spot related to mantle upwelling.

6. The *Great Divide Range* forms the eastern flank of Australia. There is currently no plate boundary here, and the age and cause of uplift are still being investigated.

C What Happens During the Erosion of Mountain Belts?

Mountains, once formed, are subjected to weathering and erosion. These processes wear mountains down but are countered by uplift related to *isostasy*. Uplift is driven by buoyancy due to the root of underlying thick crust.

Early Mountain Building

1. As a mountain belt forms, uplift is commonly faster than erosion, and the mountain becomes higher and more rugged over time. A high mountain belt results from uplift that is faster than erosion.

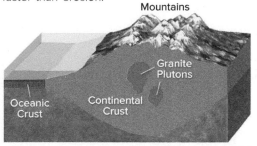

10.15.c1

2. As soon as the mountain belt starts forming, weathering and erosion begin to wear down, contributing sediment to streams. Sediment will be transported to adjacent low areas, called basins, perhaps in nearby oceans or on land.

Erosion and Isostatic Rebound

3. As material erodes from a mountain belt, there is less weight holding down the thick crustal root. The buoyant crust can uplift, a process called *isostatic rebound*.

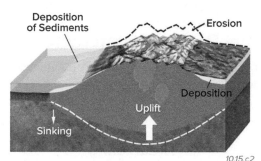

10.15.c2

4. Sediment derived from the mountain is deposited in nearby basins, on both the sea and continental sides. The added weight of the sediment depresses the crust in these regional basins, making room for more sediment.

Late Stages of Evolution

5. Erosion and isostasy cause rocks deep in the crust to be uplifted and exposed at the surface. As a result, many mountain belts expose granite and other deep rocks.

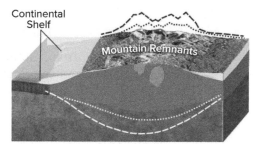

10.15.c3

6. Through simultaneous erosion and isostasy, the mountain is eroded down and the thick crustal root is gradually reduced in size. Material eroded from the mountain ends up in adjacent basins, increasing the crustal thickness beneath the basins.

D What Controls Regional Elevations in North America?

The vertically exaggerated topographic profile below illustrates how elevations vary from east to west across the U.S. It does not show the full thickness of crust, only the elevation of the land and depth of seafloor.

West Coast | Sierra Nevada | Basin and Range | Colorado Rockies | Great Plains | Mississippi River | Appalachian Mountains | East Coast

10.15.d1

1. Western North America is high mostly because crust was thickened along a convergent margin.

2. The moderately high elevation of the Basin and Range is largely due to hot, thin lithosphere.

3. Compression and shortening within the North American plate thickened crust in the Rocky Mountains. Additional uplift is due to a hot, thin lithosphere and upwelling asthenosphere associated with rifting.

4. Elevation decreases from the Great Plains toward the Mississippi River because in this direction the lithosphere becomes cooler and thicker.

5. The Appalachian Mountains were once a region of thick crust, due to the collision between North America and Africa. Much of this thickness has been lost due to erosion, and so the range has lost elevation over time.

Rule of Thumb for Elevations

Regional elevations are relatively low for regions with thinner crust, and relatively high for regions with thicker crust, but by how much? A rule of thumb is that increasing the thickness of the crust by 6 km will result in an increase in elevation of 1 km (about 3,300 ft). Here is an example from Arizona.

Phoenix sits at an elevation of 300 m (1,000 ft), whereas Flagstaff is at more than 2,100 m (7,000 ft). This difference is about 2 km, so the crust beneath Flagstaff should be 12 km thicker than the crust beneath Phoenix ($2 \times 6 = 12$). Scientific measurements show that the crust beneath Phoenix is about 28 km thick, whereas crust beneath Flagstaff is about 40 km thick. The difference is 12 km, the value we would predict from the rule of thumb.

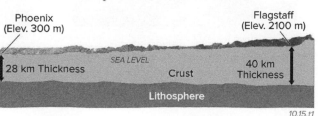

Phoenix (Elev. 300 m) — 28 km Thickness — SEA LEVEL — Crust — 40 km Thickness — Flagstaff (Elev. 2100 m) — Lithosphere

10.15.t1

Before You Leave This Page

✓ Sketch and explain the main tectonic settings of high regions, providing an example for each setting.

✓ Summarize the settings of the world's high mountains and plateaus.

✓ Explain how erosion and isostasy help expose deeply formed rocks in eroded mountain belts.

✓ Summarize differences in regional elevation across North America.

10.15

10.16 How Do Internal and External Processes Interact to Form Landscapes?

EARTH'S LANDSCAPES REVEAL THE INTERPLAY between internal processes—those originating from within the Earth—and external processes—those imposed on the Earth by its envelope of moving fluids (atmosphere and hydrosphere). Internal processes tend to be constructive, such as when they construct a volcano or cause uplift of a mountain range. External processes, especially erosion, tend to wear away parts of the landscape, but the eroded materials are deposited in other areas. Internal and external processes interact in some surprising ways.

A How Do Internal and External Processes Interact?

Internal processes arise from within the Earth and are manifested in phenomena such as volcanism, earthquakes, deformation, and mountain building. They are largely driven by the planet's internal heat. External processes, like weathering and erosion, occur on or near the surface and are mostly driven by gravity and the energy from the Sun and the resulting movement of air and moisture.

Tectonics

10.16.a1 Baja California Sur, Mexico

Most internal processes are expressed as tectonics. Volcanism and mountain building can build new landscape features, such as a volcano or mountain, or can modify existing landscapes by faulting and uplift (▲).

Weathering

10.16.a2 Baja California Sur, Mexico

Once rocks and other materials are exposed at the surface, weathering (an external process) begins to loosen pieces and round off corners (▲) and in general acts to disintegrate the rocks. It creates clasts that can be moved.

Erosion, Transport, Deposition

10.16.a3 Baja California Sur, Mexico

Clasts, clay, and other materials produced by weathering are then eroded away and transported by streams, waves, ice, wind, and gravity. Erosion, transport, and eventual deposition are all external processes.

B How Do Plate Tectonics Affect External Processes?

External processes can disintegrate and transport materials in a landscape, but tectonics can influence the rates at which different external processes are operating. The current arrangement and topography of the continents and ocean basins help guide wind and ocean currents that redistribute heat and moisture around Earth. Tectonics can change topography, which changes the local or regional climate, the rate of weathering, and other factors.

Mountains and Uplift

10.16.b1 Last Hope Sound, Chile

Mountain ranges intercept wind and water vapor, causing orographic effects that concentrate rainfall and cause rain shadows and other climatic effects. Elevation and moisture in turn affect rates of weathering and erosion.

Seas and Subsidence

10.16.b2 Portage, AK

Tectonics can slowly displace water from seas, flooding low parts of continents. Such slow regional flooding, or more rapid local earthquake-related tectonic subsidence (▲), can change local climates.

Volcanic Gas and Dust

10.16.b3 Augustine volcano, AK

Volcanic activity releases CO_2 and water vapor, which cause atmospheric warming. Volcanic ash and SO_2 gas from volcanoes reflect solar radiation, which may cause regional or global cooling.

C How Do Plate Movements Affect Climate and the Rates of External Processes?

As a tectonic plate moves, it changes position relative to Earth's climatic zones. Plate movements open and close channels that connect oceans, thereby changing ocean currents. Tectonics can uplift mountain ranges, form new ocean basins, or bring continents together. These changes can alter a region's elevation, temperature, wind flow patterns, and the amount and frequency of precipitation, all of which influence the rate of external processes.

1. To illustrate how tectonics can influence external processes, we use the three maps below to show the geography of North America at different times: 260 Ma, 75 Ma, and 20 Ma, from left to right.

2. Examine each map and think about how the configuration of land and water might change regional climates. Also, consider how the mountains would influence wind patterns and the amount of precipitation. In the first map, North America was just north of the equator so would experience northeast trade winds. In the other two maps, it was farther north and so would be under the influence of westerlies, as it is today. Only the general patterns are important here.

10.16.c1

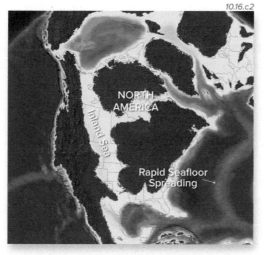
10.16.c2

10.16.c3

3. At about 260 Ma, North America was part of the supercontinent of Pangaea. The newly formed Appalachian Mountains were probably a huge, high mountain range, much like today's Himalaya. They blocked easterly trade winds, causing a rain shadow with deserts over the center of the continent, the same areas (e.g., upper Midwest) that receive abundant rainfall today.

4. At 75 Ma, shallow inland seas inundated the continent, bringing a warm, wet climate far inland, covering most of the continent. Tectonics along the west coast formed a large and fairly continuous belt of mountains inland from the coast. These changed the wind patterns and regional elevations, which affected the amounts, rates, and seasonality of rainfall, weathering, and erosion.

5. In the last 20 Ma, mountains along the West Coast, such as the Sierra Nevada and Cascades, increasingly blocked prevailing westerly winds, causing increased precipitation along the coast and the western slopes of the mountains, but a rain shadow with deserts inland. The patterns of humidity, rainfall, weathering, and erosion probably started to resemble those of the present day.

D How Do Internal and External Processes Influence the Steepness of Slopes?

In actively uplifting mountain belts, hillslopes indicate the interplay between internal processes, which are uplifting the mountains and causing slopes to be steeper, and external processes that are eroding the rising mountains and causing slopes to be less steep. An example from the south-facing slopes of the Himalaya nicely illustrates this interplay.

1. As mountains are uplifted, rocks at the surface are removed by water, ice, landslides, and the slower movement of material downhill. In some mountain ranges, erosion is almost as rapid as uplift and so limits the heights the mountains can achieve. Also, uplift of the mountains leads to regional climate changes, which may act to lower the range through increased erosion. In other areas, uplift seems to outpace erosion, so the mountains get higher and generally steeper.

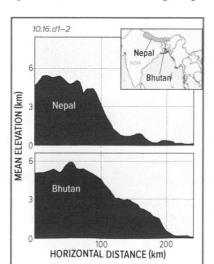

2. This graph is a topographic profile, showing the slope of the land. Mountain slopes of central Nepal have a curved topographic profile (◄) that may indicate a near balance between uplift and erosion.

3. Slopes in nearby Bhutan have straighter topographic profiles (◄), which may indicate that the mountains are uplifting faster than erosion can wear them down.

10.16

10.17 Why Is South America Lopsided?

THE TWO SIDES OF SOUTH AMERICA are very different. The western margin is mountainous, while the eastern side and center of the continent have much less relief. The differences are a result of the present plate boundaries and of the continent's geologic history during the last 200 million years. South America nicely illustrates many aspects of plate tectonics, including the connections between the tectonics of the land and seafloor. It also is an excellent example of how to analyze the major features of a region.

A What Is the Present Setting of South America?

The perspective view below shows South America and the surrounding oceans. Observe the topography of the continent, its margins, and the adjacent seafloor. See if you can find mid-ocean ridges, transform faults, ocean trenches, oceanic fracture zones, and other plate-tectonic features. From these features, infer the locations of plate boundaries in the oceans, and predict what type of motion (divergent, convergent, or transform) is likely along each boundary. Make your observations and predictions before reading the accompanying text in a counterclockwise direction.

1. The Galápagos Islands are located in the Pacific Ocean, west of South America. They consist of a cluster of about 20 volcanic islands, flanked by seamounts. Some of the islands are volcanically active and are interpreted to be over a hot spot. A mid-ocean ridge is also nearby.

9. The center of the South American continent has low, subdued topography because it is away from any plate boundaries. It is a relatively stable region that has no large volcanoes and few significant earthquakes. It is not tectonically active.

8. The Mid-Atlantic Ridge is here a divergent boundary between the South American and African plates. Seafloor spreading creates new oceanic lithosphere and moves the continents farther apart at a rate of 3 cm/yr. The ridge has spreading and transform segments.

2. The Andes mountain range follows the west coast of the continent and is the site of many dangerous earthquakes and volcanoes. A deep ocean trench along the edge of the continent marks where an oceanic plate subducts eastward beneath South America.

7. The eastern side of South America has a continental shelf that slopes gently toward the adjacent seafloor. There is neither a trench, nor any significant tectonic activity, nor any other evidence for a plate boundary. Instead, the continent and adjacent seafloor to the east are part of the same plate, and this edge of the continent is a passive margin.

3. The Pacific seafloor contains mid-ocean ridges with the characteristic zigzag pattern of a divergent boundary with offsets along transform faults. The Nazca plate lies north of this ridge, and the Antarctic plate is to the south.

6. The curved Scotia Island arc is related to a trench to the east and the westward subduction of the oceanic part of the South American plate.

10.17.a1

4. Many oceanic fracture zones cross the seafloor and were formed along transform faults, but they are no longer plate boundaries.

5. The southern edge of the continent is very abrupt and has a curving "tail" extending to the east. This edge of the South American plate is a transform boundary where South America is moving west relative to oceanic plates to the south.

B What Is the Geometry of the South American Plate and Its Neighbors?

This cross section shows how geoscientists interpret the configuration of plates beneath South America and the adjacent oceans. Compare this cross section with the plate boundaries you inferred in part A.

10.17.b1

1. At the Mid-Atlantic Ridge, new oceanic lithosphere is added to the African and South American plates as they move apart. As this occurs, the oceanic part of the South American plate, and the Atlantic Ocean, widen.

2. Along the eastern edge of South America, continental and oceanic parts of the plate are simply joined together along a passive margin. There is neither subduction, nor seafloor spreading, nor any type of plate boundary. As a result, the eastern continental margin of South America lacks volcanoes, earthquakes, and mountains.

3. A subduction zone dips under western South America, carrying oceanic lithosphere beneath the continent. Subduction causes large earthquakes and produces magma that feeds dangerous volcanoes in the Andes.

C How Did South America Develop Its Present Plate-Tectonic Configuration?

If South America is on a moving plate, where was it in the past? When did it become a separate continent, and when did its current plate boundaries develop? Here is one commonly agreed-upon interpretation.

1. Around 140 Ma, Africa and South America were part of the single large supercontinent of Gondwana. At about this time, a continental rift developed, starting to split South America away from the rest of Gondwana and causing it to become a separate continent. Hot spots (not shown) may have aided this rifting.

2. By 100 Ma, Africa and South America were completely separated by the South Atlantic Ocean. Spreading along the Mid-Atlantic Ridge moved the two continents farther apart with time. While the Atlantic Ocean was opening, oceanic plates in the Pacific were subducting beneath western South America. This subduction thickened the crust by compressing it horizontally and by adding magma, resulting in the formation and rise of the Andes mountain range.

3. Today, Africa and South America are still moving apart at a rate of several centimeters per year. As spreading along the mid-ocean ridge continues, the Atlantic Ocean gets wider. Earth, however, is not growing through time, and the expanding Atlantic Ocean is balanced by shrinking of the Pacific Ocean, whose oceanic lithosphere disappears into subduction zones along the Pacific Ring of Fire.

4. The photographs below contrast the rugged Patagonian Andes of western South America with landscapes farther east that have more gentle relief and are not tectonically active. The tectonics also significantly control the regional climates, with the Andes receiving abundant precipitation due to orographic effects, while the eastern lowlands are dry because they are downwind—in the rain shadow—of the Andes.

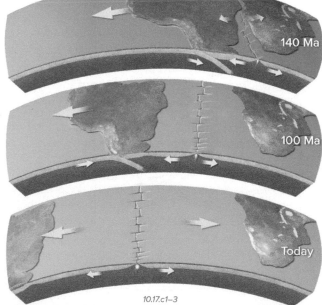

10.17.c1–3

10.17.c4 Cuernos del Paine, Chile

10.17.c5 Central Argentina

Before You Leave This Page

☑ Sketch and describe the present plate-tectonic setting of South America, and explain the main features on the continent and adjacent seafloor.

☑ Discuss the plate-tectonic evolution of South America over the last 140 million years.

What Is the Plate Tectonics of This Place?

AN UNDERSTANDING OF PLATE TECTONICS allows us to predict which places are at most risk from earthquakes and volcanoes. The most important things to know in this context are the locations and types of plate boundaries. In this exercise, you will examine an unknown ocean between two continents, make observations of the land and seafloor, and identify plate boundaries and other features. Using this information, you will predict the risk for earthquakes and volcanoes and determine the safest places to live.

Goals of This Exercise:

- Use the features of an ocean and two continental margins to identify possible plate boundaries and their types.
- Use the types of plate boundaries to predict the likelihood of earthquakes and volcanoes.
- Determine the safest sites for two cities, considering the potential for earthquakes and volcanic eruptions.
- Draw a cross section that shows the geometry of the plates at depth.

Procedures for the Map

This perspective view shows two continents, labeled A and B, and an intervening ocean. Use the topography on the land and seafloor to identify possible plate boundaries and then complete the following steps. Mark your answers on the map on the worksheet, which will be provided to you by your instructor in either paper or electronic form. Alternatively, your instructor may have you complete the investigation online.

1. Use the topographic features on land and the depths of the seafloor to identify possible plate boundaries. Draw lines showing the location of each plate boundary on the map in the worksheet. Label the boundaries as either divergent, convergent, transform, or no boundary present. Use colored pencils or different types of lines to better distinguish the different types of boundaries. Provide a legend that explains your colors and lines.

2. Draw circles [O], or use color shading, to show places, on land or in the ocean, where you think earthquakes are likely.

3. Draw triangles [▲] at places, on land or in the ocean, where you think volcanoes are likely. Remember that not all volcanoes form *directly on* the plate boundary; some form off to one side. For different plate-tectonic settings, consider where volcanoes form relative to that type of plate boundary.

4. Determine a relatively safe place to build one city on each continent. Show each location with a large plus sign [+] on the map. On the worksheet, explain your reasons for choosing these as the safest sites.

Oceanic Trench

CONTINENT A

Procedures for the Cross Section

The worksheet contains a modified version of this figure for you to use as a starting point for making a cross section. Add lines and colors to the front of the diagram to show the subsurface fractures. Use other figures in this chapter as guides to the thicknesses of the lithosphere and to the subsurface geometries typical for each type of plate boundary. Your cross section should only show features on the front of the block diagram, not features that do not reach the front edge. Your cross section should clearly:

1. Identify the crust, mantle, lithosphere, and asthenosphere, and show an accurate representation of their relative thicknesses.

2. Show the locations and relationships between lithospheric plates at any spreading center or subduction zone.

3. Include arrows to indicate which way the plates are moving relative to each other.

4. Show where melting is occurring at depth to form volcanoes on the surface.

10.18.a1

Coastal Plain

No Oceanic Trench

CONTINENT B

Continental Shield

11 Volcanoes, Deformation, and Earthquakes

EARTH'S SURFACE IS A REMARKABLE DIARY of the events and processes that shaped it. Magma generated at depth erupts to the surface in volcanoes, either as hot, glowing lava, or as spectacularly explosive eruptions of gas and volcanic ash. Tectonic forces break and bend solid rocks, frequently generating earthquakes that uplift mountains, lower valleys, and inflict destruction across entire regions. All landscapes contain clues—whether in the shape of a volcano or as small fractures exposed in a cliff—that record the processes and events that sculpt that landscape. What comes to mind when you think about a volcano, earthquake, or mountain?

11.00.a2 Mount Everest

11.00.a3 Tibet

◄ **Mount Everest** is the world's highest mountain, rising 8,848 m (29,029 ft) above sea level. It sits atop the Himalaya, the highest mountain range on Earth. On the northern side of the Himalaya is the Tibetan Plateau, the highest plateau on Earth. The photograph to the left was taken from the Tibet (north) side from one of the base camps where mountaineers begin their arduous and dangerous climb to the top. Traveling to the base camp, the climbers drive through mountain ranges composed of wildly folded rock layers (▶).

What processes produced such an incredibly high region, and are the folded layers somehow related to the same processes and events that uplifted the mountains and plateau?

11.00.a1

Mount Fuji on Honshu Island in Japan is a beautiful and symmetric mountain (▼). It is also an active volcano, which last erupted in 1708. From the shape of the volcano and the types of materials the volcano contains, we can tell that it is a type of volcano that has extremely explosive, dangerous eruptions. The Cascade Mountains of the Pacific Northwest contain the same type of dangerous volcano with similar, if less elegant, shapes.

What is a volcano, how do we recognize one, and what makes some volcanoes more dangerous than others?

Large earthquakes are common along the deep ocean trenches of the Pacific and Indian Oceans. The 2004 Indian Ocean earthquake unleashed a large wave called a *tsunami* that swept onto surrounding coastlines. The photograph below (▼) shows the city of Banda Aceh, Sumatra, after its destruction by the earthquake-generated wave. Most of this city of 320,000 people was stripped bare by the wave, and nearly a third of its inhabitants were killed.

What types of earthquakes are most likely to produce a tsunami?

11.00.a4

11.00.a5

TOPICS IN THIS CHAPTER

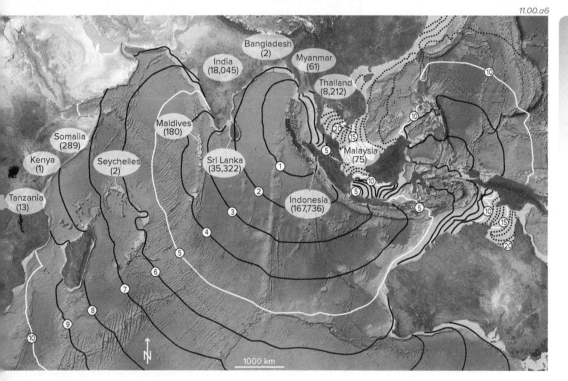

11.00.a6

Our Dynamic Planet

Volcanoes, earthquakes, and mountain ranges are vivid demonstrations of the dynamic nature of our planet. All are related to *internal processes*, those that originate deep beneath the surface and are powered mostly by Earth's internal heat. Many other features of natural landscapes— at the regional scale of a mountain range or at the local scale of a single exposure of rock—are related to these dynamic internal processes.

Volcanoes are a clear expression of this dynamism, bringing hot, molten rock from deep in Earth's crust or mantle and erupting it as magma on the surface. Different types of volcanoes erupt in characteristic ways, and often we can recognize the difference from a distance, just from the shape of the volcano or the color of the volcanic materials. The type of volcano determines how dangerous it is and what types of hazards it poses.

Most earthquakes occur when forces cause two parts of Earth's crust to shift abruptly relative to one another. Such movement can uplift a mountain range (e.g., the Himalaya), downdrop a valley, or horizontally slide one crustal block past another, as occurs along the San Andreas fault of California. If rock layers are soft enough, such as when they are warm and buried deeply, they may respond to the imposed forces by bending rather than breaking, forming folds, like those in Tibet. Many regions have it all—volcanoes, earthquakes, fractures, and folds. Here, we explore why.

The world's strongest earthquake in 40 years struck Indonesia on December 26, 2004. The magnitude 9.1 earthquake occurred west of Sumatra and was caused by movement on a fault, marked by the red line on this map. Yellow dots nearby show the locations of smaller, related earthquakes. The earthquake occurred beneath the ocean, abruptly uplifting a large region of seafloor and displacing overlying seawater that formed a massive wave—a tsunami—that spread across the Indian Ocean as a low wave, traveling at speeds approaching 800 km/hr (500 mi/hr)! The curved dark gray lines show a model of the wave's position by hour (numbers in small white circles). The tsunami increased in height as it approached the coasts of Indonesia, Thailand, Sri Lanka, India, east Africa, and various islands. Low coastal areas were inundated by as much as 20 to 30 m (65 to 100 ft) of water. Cities and villages, like Banda Aceh (◄), were completely demolished along hundreds of kilometers of coastline, leaving more than 220,000 people dead or missing. The numbers in ovals on the map above show the number of fatalities by location.

What causes earthquakes, where are they most likely to strike, and which ones are most hazardous?

11.00.a7 Banda Aceh, Indonesia

11.0

11.1 What Is a Volcano?

AN ERUPTING VOLCANO IS UNMISTAKABLE—glowing orange lava cascading down a hillside, molten fragments blasted into the air, or an ominous, billowing plume of gray volcanic ash rising into the atmosphere. But what if a volcano is not erupting? How do we tell if a mountain is a volcano?

A What Are the Characteristics of a Volcano?

How would you describe a volcano to someone who had never seen one? Examine the two volcano photographs below and look for common characteristics. Volcanoes are formed by magma, so the resulting rocks are igneous.

11.01.a1 Kilauea, HI

1. A volcano is a *vent* where magma and other volcanic products erupt onto the surface. The volcano on the left (the Pu'u 'Ō 'ō volcano in Hawaii Volcanoes National Park) produces large volumes of molten lava, whereas the one on the right (Kanaga volcano, Alaska) is erupting volcanic ash. Many geoscientists reserve the term *volcano* for hills or mountains that have been *constructed* by volcanic eruptions. Some eruptions do not produce hills or mountains, but we consider them to be volcanoes, too.

2. Most volcanoes have a *crater*, a roughly circular depression usually located near the top of the volcano. Other volcanoes have no obvious crater or are nothing but a crater.

3. Volcanoes consist of *volcanic rocks,* which form from lava, pumice, volcanic ash, and other volcanic materials.

4. Besides erupting from volcanoes that have the classic shape of a cone, magma erupts from fairly linear cracks called *fissures* and from huge circular depressions called *calderas.*

11.01.a2 Kanaga volcano, AK

5. Many volcanoes display evidence of having been active during the last several hundred to several million years, or even during the last several days. Such evidence can include a layer of volcanic ash on snow-covered hillslopes (left side of volcano shown above) or lava and ash flows that are relatively unweathered and that lack a well-developed soil. Over time, erosion degrades and disguises volcanoes and volcanic craters, making them less obvious.

B Is Every Hill Composed of Volcanic Rocks a Volcano?

If a landscape lacks most of the diagnostic features described above, it is probably not a volcano. Many mountains and hills are composed of volcanic rocks but are not volcanoes, and some volcanoes do not make mountains or hills.

1. This flat-topped hill (▼), called a *mesa*, has a cap of volcanic rocks, but it is not a volcano. It did not form over a volcanic vent. Instead, it is an eroded remnant of a once extensive lava flow that covered the region, as depicted in the figures to the right.

2. Lava erupts from a central volcanic vent or from a linear vent called a *fissure*. Once erupted, the lava spreads outward and cools into a solid rock.

3. Erosion removes the edges of the lava flow and works inward toward a central remnant.

11.01.b1 Hopi Buttes, AZ

4. The past location of the fissure is marked by a small, linear ridge of dark rocks that cross the landscape.

5. The lava flow is more resistant to erosion than underlying rocks and so forms a steep-sided, flat-topped mesa. It is a hill composed of volcanic rocks, but it is not over the vent and is not a volcano.

11.01.b2–4

C What Are Some Different Types of Volcanoes?

Volcanoes have different sizes and shapes and contain different types of volcanic materials. These variations are due to differences in the composition of the magmas and the style of the eruptions. There are four common types of volcanoes that are shaped like hills and mountains: *cinder cones, shield volcanoes, composite volcanoes,* and *volcanic domes.* Later in this chapter, we describe other types of volcanoes that are not hills or mountains. The types of volcanoes below are depicted at different scales, to allow the internal features of each to be shown relative to its size.

Cinder Cone

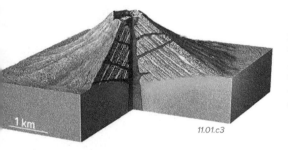

50 m

11.01.c1

1. *Cinder cones* (◀) are cone-shaped hills several hundred meters high, or higher, usually with a small crater at their summit. They are also called *scoria cones.* They contain loose black or red, pebble-sized volcanic *cinders,* along with larger volcanic *bombs;* most cinder is a type of basalt. Cinder cones generally occur as isolated features or in an area with numerous cones, but some form next to, or on the flanks of, composite and shield volcanoes.

Shield Volcano

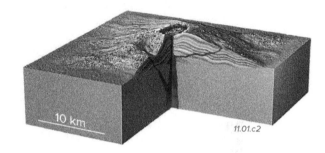

10 km

11.01.c2

2. *Shield volcanoes* (▶) have broad, gently curved slopes (shaped like an ancient battle shield) and can be relatively small (less than a kilometer across) or can form huge mountains tens of kilometers wide and thousands of meters high. They commonly contain a crater, or line of craters, and have fissures along their summit. Shield volcanoes consist mostly of lava flows of basalt, with smaller amounts of cinder and volcanic ash.

Composite Volcano

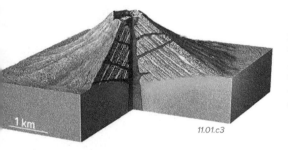

1 km

11.01.c3

3. *Composite volcanoes* (◀) are typically fairly symmetrical mountains thousands of meters high, with moderately steep slopes and a crater at the top. They may be large, but they are, on average, much smaller than shield volcanoes. Their name derives from the interlayering of various types of volcanic rocks and volcanic mudflows. They consist mostly of *andesite,* a fine-grained, typically gray, volcanic rock, but they can also contain other compositions of rocks. Composite volcanoes are a very dangerous type of volcano.

Volcanic Dome

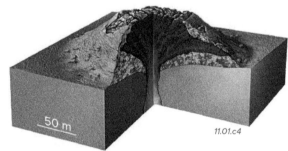

50 m

11.01.c4

4. *Volcanic domes* (▶) are dome-shaped features that may be hundreds of meters high. They consist of solidified lava, which can be highly fractured or mostly intact. Domes include some volcanic ash intermixed with rock fragments that are derived from solidified lava in the dome. They form where magma cannot easily flow, and so it piles up around a vent. Many domes are within craters of composite volcanoes or in larger volcanic depressions, called calderas, which are described later in this chapter.

The Relative Sizes of Different Types of Volcanoes

Volcanoes vary from small hills less than a hundred meters high to broad mountains tens of kilometers across. Although sizes vary quite a bit, we can make some generalizations about the relative sizes of the different volcano types.

The figure below illustrates that some types of volcanoes are larger than others. The volcanoes on this figure cannot be drawn to their true scale relative to one another because the largest shield volcanoes are so large that we cannot show them on the same drawing with small cinder cones. The figure does accurately show which volcanoes are the largest and which ones are the smallest.

Cinder cones and domes, which typically form during a single eruptive episode, are the smallest volcanoes. Shield volcanoes and composite volcanoes are much larger because they are constructed, layer by layer, by multiple eruptions. Shield volcanoes have more gentle slopes than cinder cones, domes, or composite volcanoes.

Scoria Cone
Dome
Small Shield
Composite Volcano
Large Shield

11.01.t1

Before You Leave This Page

✓ Sketch or describe the diagnostic characteristics of a volcano.

✓ Describe or sketch why not every hill composed of volcanic rocks is a volcano.

✓ Sketch and describe the four main types of volcanoes that construct hills and mountains.

✓ Sketch and describe the relative sizes of different types of volcanoes.

11.1

11.2 How Do Volcanoes Erupt?

THE DIFFERENT SHAPES OF VOLCANOES indicate differences in the style of eruption. Some eruptions are explosive and very dangerous, whereas others are comparatively calm and less dangerous. What causes these differences? The answer involves magma chemistry and gas content, both of which control how magma behaves near the surface. A key aspect of this behavior is the magma's resistance to flow, a parameter called *viscosity*. Rocks with high viscosity strongly resist flowing. An example of a non-volcanic substance with low viscosity is water, an example of a more viscous substance is honey.

A What Are Ways that Magma Erupts?

Magma may behave in several different ways once it reaches Earth's surface. It can erupt as a glowing mass of molten lava that issues from a vent and flows onto the surface. In contrast, some eruptions explosively throw bits of lava, volcanic ash, and other particles into the atmosphere, a process described as *pyroclastic* (meaning hot fragments). Both types of eruptions can occur from the same volcano, at the same time, or during different eruptions.

Eruptions of Lava

1. When magma erupts onto the surface and flows away from a vent (▶), it creates a *lava flow*. Erupted lava can be fairly fluid, flowing downhill like a fast river of molten rock. Some lava flows are not so fluid and travel only a short distance before solidifying.

11.02.a1 Kilauea, HI

2. A *lava dome* (▶) forms from the eruption of highly viscous lava, a lava that strongly resists flowing. The high viscosity of the lava causes the lava to pile up in a dome-shaped mass around the vent, instead of flowing away. Domes are often accompanied by several types of explosive eruptions.

11.02.a2 Mount St. Helens, WA

Pyroclastic Eruptions

3. Some explosive eruptions send molten lava into the air. A *lava fountain*, such as that shown here (▶), can accompany basaltic volcanism and results from a high initial gas content in a less viscous lava. The gas propels the lava and separates it into discrete pieces.

11.02.a3 Kilauea, HI

4. Other explosive eruptions eject a mixture of volcanic ash, pumice, and rock fragments into the air (▶). Airborne particles that are sand sized or smaller are *volcanic ash*. Ash mostly forms when bubbles blow apart bits of magma. Larger fragments are typically pumice, fragmented volcanic glass, and other shattered rocks.

11.02.a4 Redoubt volcano, AK

Two Different Types of Pyroclastic Eruptions from the Same Volcano

5. Augustine volcano in Alaska produces ash and pumice in two eruptive styles—an *eruption column* and a *pyroclastic flow*.

11.02.a5 Augustine volcano, AK

6. *Eruption Column*—Volcanic ash and pumice form when magma is blown apart by volcanic gases and can erupt high into the atmosphere, forming an *eruption column* (▲). The ash and pumice fall back to Earth as solidified and cooled pieces of rock. Finer particles of ash drift many kilometers away from the volcano and slowly settle to the ground.

7. *Pyroclastic Flow*—Some ash does not jet straight up but collapses down the side of the volcano as a dense, hot cloud of ash particles and gas (▼). This eruption style is a *pyroclastic flow* or simply an *ash flow*. A pyroclastic flow can be devastating because of its high speed (more than 100 km/hr) and high temperature (exceeding 500°C).

11.02.a6 Augustine volcano, AK

8. The two kinds of eruptions differ primarily because of gas content of the magma. An eruption column forms when large volumes of volcanic gas come out of the magma and overcome gravity to carry the cloud of ash and pumice up into the atmosphere.

9. A pyroclastic flow forms when the amount of gas is less and cannot support the eruption column, so the column rapidly collapses and flows downhill under the force of gravity.

11.02.a7

B How Do Gases Affect Magma?

1. To envision dissolved gas in magma, think about what happens when you open a bottle or can of soda. The liquid may have no bubbles until it is opened, at which time bubbles appear in the liquid, rise to the top, and perhaps cause the soda to spill out. The dissolved gas was always in the liquid, but it only became visible when you opened the top and released the pressure that held the gas in solution.

11.02.b1

2. Magma, like the soda, contains some dissolved gases, including H_2O (water vapor), CO_2 (carbon dioxide), and SO_2 (sulfur dioxide). These gases have a critical effect on eruption style and help the magma rise toward the surface.

3. As in this enlargement of the magma, confining pressure at depth keeps most of the gases in solution within the magma and keeps bubbles from forming.

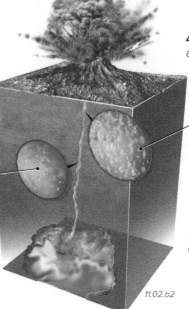

4. As the magma approaches the surface, pressure decreases and the gases cannot remain in solution. Bubbles of gas form in the magma. If enough bubbles form quickly, the expanding bubbles cause the magma to be more buoyant and help it rise toward the surface and erupt out of the volcano.

11.02.b2

C How Does Viscosity Affect Gases in Magma?

Viscosity, the resistance to flow, dictates how fast a magma can flow and how fast crystals and gas can move through the magma. When gas in a magma comes out of solution, movement of the resulting bubbles is resisted by the magma's viscosity. If the bubbles cannot escape, the magma is potentially more explosive.

11.02.c1 Mount St. Helens, WA

11.02.c2 Kīlauea, HI

More Viscous

◄ Felsic magmas—with abundant quartz and feldspar—contain a lot of silica, which forms in long, complicated chains, and so are relatively viscous. The high viscosity prevents gas from escaping easily. Gas builds up in the magma and, when it expands, greatly increases the pressure on the surrounding rock. This can cause explosive eruptions.

Less Viscous

▶ Mafic magmas—with abundant iron and magnesium—are less viscous, so gas bubbles can escape relatively easily. This can lead to a fairly nonexplosive eruption, such as this basaltic lava flow that flows smoothly downhill from the vent.

Composition, Viscosity, and Eruptive Style

Composition of magma is the main control on a volcano's eruptive style, shape, types of materials, and potential hazards. This is because composition, especially the amount, length, and linkage of silicate chains in the melt, controls viscosity and whether gas builds up in the magma. Composition of a magma also generally controls the temperature of the magma, with mafic (basaltic) magmas being hundreds of degrees hotter than a magma that is felsic (granitic) in composition. Temperature of a magma greatly influences its viscosity and therefore eruptive style.

Some magma has fewer linked silicate chains and so is relatively *less viscous* (flows more easily). Lower viscosity allows magma to flow from the volcano in a more fluid lava flow. If the magma is relatively fluid, it tends to spread out and form relatively gentle slopes, as in shield volcanoes. Explosive gases can build up somewhat

in low-viscosity magma, as demonstrated by lava fountains, but the resulting explosive eruptions are rather small and localized, scattering cinder and ash near the cinder cone.

Other magmas have more silicate chains, and the chains interfere with each other, restricting the flow of the magma and making it *more viscous*. Magma with a high viscosity does not flow easily across the surface, and so the erupted lava piles up close to the volcanic vent. As a result, this type of lava produces steep volcanic domes and steep composite volcanoes. High-viscosity magma can trap gas, leading to explosive pyroclastic eruptions of gas-propelled volcanic ash, pumice, and rock fragments. Such eruptions are associated with volcanic domes, composite volcanoes, and large volcanic calderas. Therefore, these volcanoes produce a mix of pyroclastic rocks and lava flows, mostly of felsic and intermediate composition. Composition

controls viscosity, temperature, eruptive style, the shape of the volcano, and the rock types that compose that volcano.

Before You Leave This Page

☑ Describe four ways that magma erupts.

☑ Describe the difference between an eruption column and a pyroclastic flow, and the role that gas plays in eruptive style.

☑ Explain how gas behaves at different depths in a magma and how it influences eruptive style.

☑ Describe how viscosity influences how explosive an eruption is.

11.2

11.3 What Volcanic Features Consist of Basalt?

BASALT IS THE MOST ABUNDANT VOLCANIC ROCK on Earth's surface. Dark-colored basalt covers large areas on every continent and forms the upper part of oceanic crust. When magma with a composition of basalt erupts, it can produce a variety of landforms. It forms cinder cones and associated dark lava flows, and it also produces shield volcanoes and lava flows that cover huge areas, called flood basalts. The type of volcanic feature that forms is largely controlled by the gas in the magma, the total volume of magma, and how fast the magma erupts.

A How Are Cinder Cones and Basalt Flows Expressed in Landscapes?

Basaltic magma has a relatively low viscosity compared to other magmas, and it erupts in characteristic ways. A basaltic eruption can form a lava flow or throw pieces of molten rock into the air. The lava can flow smoothly, like a hot, glowing stream, or it can partially solidify and break apart. Lava erupted into water forms distinctive pillow-shaped forms.

1. *Basaltic Eruptions*—At the beginning of many basaltic eruptions, gases carry bits of lava into the air, forming a lava fountain. The airborne bits of lava cool and then fall around the vent as loose pieces of cinder (also called scoria). The lava fountain may be followed by or accompanied by eruption of a basaltic lava flow.

11.03.a1 Hawaii

2. *Cinder Cones*—Pieces of cinder from the lava fountain gradually create a cone-shaped hill—a cinder cone. Ejected fragments can be as small as sand grains or as large as huge boulders. Cinder cones typically form in a short amount of time, from a few months to a few years, and generally are no more than 300 m (about 1,000 ft) high.

3. *Basaltic Lava Flows*—Fluid basaltic lava pours from the vent and flows downhill. Sometimes, as shown here, the lava fills up and overtops the crater in the cinder cone. At other times, a lava flow issues from cracks near the base of the cinder cone after most of the cone has been constructed.

Cinder Cones

11.03.a2 Sunset Crater, AZ

4. Most cinder cones begin with a conical shape (▲) and a central crater at the top of the cone. Young cinder cones have little soil or vegetation on them, and commonly are associated with dark, fresh-looking lava flows.

11.03.a3 Northern AZ

5. Over time, erosion wears away the summit of a cinder cone, making the cone into a rounded hill (▲) without a central crater. Erosion cuts into the slopes, and the slopes gradually build up a veneer of soil and plants.

Types of Lava Flows

11.03.a4 Hawaii

6. *Aa* lava (pronounced "ah-ah") is a type of rough-surfaced lava flow, formed when the lava breaks apart into a mass of jumbled, angular blocks of hardened lava that tumble down the front of the flow as it moves (▲). Aa forms a very rough, jagged surface.

7. *Pahoehoe* (pronounced "pa-hoy-hoy") is a type of lava flow that has an upper surface with small billowing folds that form a "ropy" texture. A pahoehoe lava flow is usually fed by a lava tube and grows as a series of lobes (▼). As the front of the flow solidifies, the lava breaks out and forms a new lobe, as shown here.

Pahoehoe lava moves relatively smoothly and easily compared to aa flows.

11.03.a5 Hawaii

Tubes and Pillows

11.03.a6 Hawaii

8. *Lava tubes* form when the surface of a lava flow solidifies to form an insulating roof over the hot, still-moving interior of the flow (▲). Lava flows insulated by lava tubes flow farther than lava flows on the surface because the lava stays hotter longer. If the tube drains, it becomes a curving, tube-shaped cave.

11.03.a7 San Juan Islands, WA

9. When fluid lava erupts into water, the lava grows forward as small, individual lumps that form rounded shapes called *pillows* (▶). Pillows are reliable evidence that lava erupted into water. Basaltic lava flows on the seafloor contain countless pillows.

B Q What Is a Shield Volcano?

Shield volcanoes have a broad, *shield-shaped* form and fairly gentle slopes when compared to other volcanoes, because their eruptions are dominated by relatively nonexplosive outpourings of low-viscosity lava, which can spread out from fissures and vents. Shield volcanoes form in various settings, but the largest ones, like those in Hawaii, formed at oceanic hot spots.

1. The Big Island of Hawaii consists mainly of three large volcanoes. Green areas are heavily vegetated, and recent lava flows are brown or dark gray.

2. Mauna Loa, the central mountain, is the world's largest volcano. It rises 9,000 m (29,500 ft) above the seafloor and is 4,170 m (13,680 ft) above sea level. From seafloor to peak, Mauna Loa is Earth's tallest mountain. Nearby Mauna Kea is an inactive shield volcano and the site of astronomical observatories.

3. Kilauea volcano, probably the most active volcano in the world, is on the southeastern side of the island. Recent lava flows (shown in dark grayish brown) flowed eastward, destroying roads and housing subdivisions.

11.03.b2 Mauna Loa, HI

4. In shield volcanoes, magma rises through a fracture and erupts onto the surface from a long fissure (▲). Large volumes of lava can flow out of the fissure, and escaping gas throws smaller amounts of molten rock into the air as a fiery curtain.

5. Eruptions also occur in more centralized vents, such as on Kilauea (▼). These vents are interpreted to overlie fissures.

03.b1

10 km

Mauna Kea

Kilauea

Mauna Loa

Fissure Eruption

11.03.b3 Kilauea, HI

6. The spine of Mauna Loa is a fissure from which lava flows erupted as recently as 1984. The fissure is the surface expression of one or more magma-filled fissures at depth. Recent activity on Mauna Loa has raised concerns about an eruption in the near future.

C Q What Are Flood Basalts and How Are They Expressed in Landscapes?

Flood basalts are basaltic lava flows covering vast areas that commonly occur in sequences several kilometers thick. They generally involve multiple eruption events, but individual lava flows can cover thousands of square kilometers and contain more than 1,000 cubic kilometers of magma! Flood basalts are fed by a series of long fissures.

A fissure forms when pressure from the magma pushes outward against the wall rocks, holding them apart while magma passes through. A wide fissure allows faster eruption rates,
11.03.c1
which result in a large volume of erupted magma that can remain hot and travel long distances. Narrower fissures restrict the eruption rates, leading to lower volume lava flows that cool off before they can travel as far.

11.03.c2 Palouse Canyon, WA

Some of the most famous flood basalts are in the Columbia Plateau of Washington, Oregon, and western Idaho. Cut into the plateau are canyons that expose multiple basalt flows, each forming a ledge or step in the canyon walls. Each flow represents a single eruption, separated by thousands of years in which not much happened. One basalt flow on the Columbia Plateau covers more than 130,000 km² (50,000 mi²) and probably erupted very quickly, perhaps in only several decades.

Before You Leave This Page

✓ Summarize the features of cinder cones and basalt flows.

✓ Describe the general characteristics of a shield volcano and how the eruptions occur.

✓ Describe flood basalts and explain why they can cover huge areas.

11.3

11.4 What Are Composite Volcanoes and Volcanic Domes?

COMPOSITE VOLCANOES FORM STEEP, CONICAL MOUNTAINS that are hard to mistake for anything other than a volcano. They are common above subduction zones, especially along the Pacific Ring of Fire. Composite volcanoes are extremely dangerous. Volcanic domes consist of smaller, dome-shaped masses of highly viscous lava that accumulated over and close to the associated volcanic vent—they are also very dangerous.

A What Are Some Characteristics of a Composite Volcano?

Composite volcanoes are constructed of interlayered volcanic material formed by lava flows, pyroclastic flows, falling ash, and volcano-related mudflows and landslides. Composite volcanoes, also called stratovolcanoes, erupt over long time periods, which explains their large size and complex internal structure.

11.04.a2 Mount St. Helens, WA

◄ **1.** *Eruption Column*—Composite volcanoes produce a distinctive column of pumice, ash, and gas that rises upward many tens of kilometers into the atmosphere. Coarser pieces settle around the volcano, but finer particles (volcanic ash) can drift hundreds of kilometers in the prevailing winds. An eruption column is one type of pyroclastic eruption.

Ash

11.04.a1

▶ **6.** *Pyroclastic Flows*— These are the most violent eruptions from the volcano. They form when the eruption column collapses downward as a dense, swirling cloud of hot gases, volcanic ash, and angular rocks. Pyroclastic flows are one of the main mechanisms by which these volcanoes are constructed.

11.04.a3 Mount Mayon, Philippir

2. *Shape*—Composite volcanoes display the classic volcano shape because most material erupts out of a central vent and then settles nearby. They have steep slopes because they form from small eruptions of viscous lava flows that pile up on the flanks of the volcano and help protect pyroclastic material from erosion. The shape represents one snapshot in a series of stacked volcanic mountains that have been built over time.

5. *Landslides and Mudflows*— Composite volcanoes can be large mountains that collect rain or snow. Rain and snowmelt mix with loose ash and rocks on the volcano's flanks, causing a volcano-derived mudflow called a *lahar* (▼). There is a high hazard for landslides and lahars because of the steep slopes, loose rocks, and abundant slippery clay minerals produced when hot water interacts with the volcanic rocks.

3. *Rocks*—Composite volcanoes consist of alternating layers of pyroclastic flows, lava flows, and deposits from landslides and mudflows. Most rocks are intermediate to felsic in composition. The volcanoes we see today formed during eruptions from long-lived vents and are built on and around earlier versions of the volcanoes. A large composite volcano can be constructed in tens of thousands of years, or a few hundred thousand years.

Lava

Pyroclastic Flows

Lahar

11.04.a4 Augustine volcano, AK

4. *Lava Domes and Flows*— Lava domes and flows can erupt from any level of a composite volcano. Lava may erupt from the summit crater or escape through vents on the volcano's sides or base. Lavas associated with composite volcanoes are moderately to highly viscous, and so they move slowly and with difficulty. The lava may break into blocks that fall, slide, or roll downhill, forming a tongue or apron (◄) of jumbled pieces from the lava.

11.04.a5 Mount St. Helens, WA

B What Are Some Characteristics of a Volcanic Dome?

Volcanic domes form when viscous lava mounds around a vent. Some volcanic domes have a nearly symmetrical dome shape, but most have an irregular shape because some parts of the dome have grown more than other parts or because one side of the dome has collapsed. Domes may be hundreds of meters high and one or several kilometers across, but they can be much smaller, too.

11.04.b1 Alaska

Dome

1. This rubble-covered dome (▲) formed near the end of the 1912 eruption in the Valley of Ten Thousand Smokes in Alaska. Volcanic domes commonly have this type of rubbly appearance because their outer surface consists of angular, broken pieces of the dome.

Growth of a Dome

2. Domes mostly grow from the inside as magma injects into the interior of the dome. This new material causes the dome to expand upward and outward, fracturing the partially solidified outer crust of the dome. This process creates the blocks of rubbly, solidified lava that coat the outside of the dome. The blocks and other pieces, which are as small as several centimeters across to as large as houses, commonly tumble or slide down the steep slopes on the side of a dome.

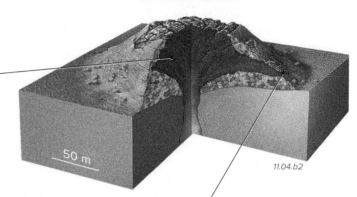
50 m
11.04.b2

3. Domes also grow as magma breaks through to the surface and flows outward as thick, slow-moving lava. As the magma advances, the front of the flow cools, solidifies, and can collapse into angular blocks and ash. Most domes display both modes of growth.

Collapse or Destruction of a Dome

4. Domes can be partially destroyed when steep flanks of the dome collapse and break into a jumble of blocks and ash that flow downhill as small-scale, but very dangerous, pyroclastic flows (▶).

11.04.b3   *11.04.b4*

5. Domes can also be destroyed by explosions originating within the dome (◀). These typically occur when magma solidifies in the conduit and traps gases that build up until the pressure can no longer be held.

Deadly Collapse of a Dome at Mount Unzen, Japan

Mount Unzen towers above a small city in southern Japan. The top of the mountain contains a steep volcanic dome that formed and collapsed repeatedly between 1990 and 1995. The collapsing domes unleashed more than 10,000 small pyroclastic flows (top photograph) toward the city below. In 1991, the opportunity to observe and film these small pyroclastic flows attracted journalists, volcanologists (scientists who study volcanoes), and other onlookers to the mountain. Unfortunately, partial collapse of the dome caused a pyroclastic flow larger than had occurred previously. This larger flow killed 43 journalists, volcanologists, and others and left a path of destruction through the valley (lower photograph). Note that damage was concentrated along valleys that drain the mountain.

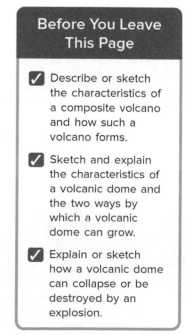
11.04.t1 Mount Unzen, Japan
11.04.t2 Mount Unzen, Japan

Before You Leave This Page

☑ Describe or sketch the characteristics of a composite volcano and how such a volcano forms.

☑ Sketch and explain the characteristics of a volcanic dome and the two ways by which a volcanic dome can grow.

☑ Explain or sketch how a volcanic dome can collapse or be destroyed by an explosion.

11.4

11.5 What Are Calderas?

CALDERA ERUPTIONS ARE AMONG NATURE'S most violent phenomena. They are also sometimes called "supervolcanoes" because they can spread volcanic ash over huge areas and erupt more than one thousand cubic kilometers of magma. As the magma vacates the magma chamber during an eruption, the roof of the chamber collapses to form a depression tens of kilometers across. The depression may fill with ash, lava flows, and sediment. On these two pages, we focus on large calderas, but smaller calderas can form on composite volcanoes or when magma withdraws beneath a shield volcano.

A What Is a Caldera?

A *caldera* is a large, basin-shaped volcanic depression, which typically has a low central part surrounded by a topographic escarpment, referred to as the *wall* of the caldera. The Valles Caldera of New Mexico nicely illustrates the important features of a caldera. It is relatively young and uneroded because it formed only two million years ago.

11.05.a1

1. This image shows satellite data superimposed on topography. The circular Valles Caldera contains a central depression, about 22 km (14 mi) across, surrounded by steep walls. The caldera formed when a huge volume of magma erupted from a shallow magma chamber, producing a large eruption column and pyroclastic flows.

2. Internally, the caldera contains a series of faulted blocks that have been down-dropped relative to rocks outside the caldera. Faulting and ground subsidence occurred at the same time as the main eruption, so relatively thick amounts of ash and other materials accumulated inside the caldera.

3. Small, rounded volcanic domes formed within the caldera after the main pyroclastic eruption. The domes were fed by fissures that tapped leftover magma. Some of this magma has solidified at depth, forming granite, but some may still be molten. The area has been explored for geothermal energy associated with the hot rocks at depth.

Valles Caldera

B How Does a Caldera Form?

The formation of a caldera and the associated eruption occur simultaneously—the caldera subsides in response to rapid removal of magma from the underlying chamber. The largest caldera eruptions produced volcanic ash layers more than 1,000 m thick.

11.05.b1

5 km

1. The first stage is the generation of felsic magma. The magma rises and accumulates in magma chambers that can be kilometers thick and tens of kilometers across. The chambers may be within several kilometers of the surface.

11.05.b2

2. Next, magma reaches the surface, either as lava flows or small pyroclastic eruptions. As the magma chamber loses material, the roof of the chamber subsides to occupy the space that is being vacated. Curved fractures allow crustal blocks to drop, outlining the edges of the caldera and providing many conduits for magma to reach the surface.

11.05.b3

3. The eruption forms columns and flows of pyroclastic material, much of which falls back into the caldera, creating a thick pile of ash. Landslides from steep caldera walls produce large blocks and smaller pieces that become part of the caldera deposit. Some ash escapes the caldera and covers surrounding areas.

11.05.b4

4. As the eruption subsides, magma rises through fissures along the edges or in the caldera, erupting onto the surface as volcanic domes.

C Did the Eruption of a Caldera Destroy a Civilization Near Greece?

Santorini, east of the Greek mainland, is a group of volcanic islands with geology that records a major caldera collapse only 3,500 years ago, about the time of the collapse of the Minoan civilization on the island of Crete.

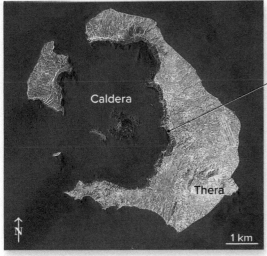
11.05.c1 Santorini, Greece

Caldera

Thera

1 km

1. This satellite view (◄) shows the islands of Santorini, including *Thera,* the largest island. Thera and the other islands encircle a submerged caldera that formed when the center of a larger volcanic island collapsed, leaving the modern islands as remnants. The steep cliffs around the caldera are eroded segments of the original wall of the caldera. The curving cliffs (►) expose volcanic layers, products of explosive volcanism that began one to two million years ago. Islands in the middle of the caldera were constructed by more recent eruptions.

11.05.c2 Santorini, Greece

2. The main eruption produced an ash column perhaps 40 km (25 mi) high, followed by pyroclastic flows. The erupted ash buried towns, now excavated (►), with up to 50 m of pumice and ash. Caldera collapse occurred as the eruption of ash emptied a large magma chamber.

11.05.c3 Akrotiri, Santorini, Greece

3. The collapse of the caldera evidently unleashed a large destructive wave that traveled southward across the sea, probably helping lead to the downfall of the Minoan civilization on the island of Crete. This destruction of the civilization on Santorini and collapse of the volcanic island into the sea may have started legends about the sinking of a landmass and city (Atlantis) into the sea.

D Why Is Yellowstone Caldera Considered to Be Dangerous?

Yellowstone is one of the world's largest active volcanic areas, occurring directly over a continental hot spot. Abundant geysers, hot springs, and other hydrothermal activity are leftovers from its recent volcanic history. During the last two million years, the Yellowstone region experienced three huge, caldera-forming eruptions. If a similar eruption occurred again, it would rain destructive ash over the Rocky Mountains and onto the Great Plains. Yellowstone is called a supervolcano.

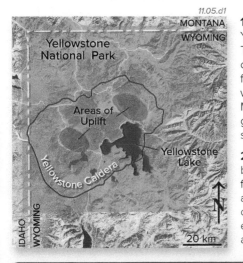
11.05.d1

MONTANA
WYOMING

Yellowstone National Park

Areas of Uplift

Yellowstone Lake

Yellowstone Caldera

Yellowstone Caldera

IDAHO
WYOMING

N

20 km

1. This image (◄) shows the outline of the youngest Yellowstone caldera, which formed 640,000 years ago. The boundaries of the caldera have been partially obscured by erosion, deposition of sediment, and lava flows that erupted after the caldera formed. Several areas within the caldera have been experiencing recent uplift. Magma and hot rocks at relatively shallow depth heat groundwater to form the spectacular geysers and hot springs for which Yellowstone is famous.

2. Ash from the three Yellowstone eruptions was carried by the wind and deposited over a huge area that extends from northern Mexico to southern Canada and as far east as the Mississippi River (►). A repeat of such an eruption could devastate the region around Yellowstone and cause extensive crop loss in the farmlands of the Great Plains and Midwest.

11.05.d2

Mount St. Helens Ash 1980

Yellowstone

Yellowstone Ash Bed 2

Yellowstone Ash Bed 1

Yellowstone Ash Bed 3

400 km

Before You Leave This Page

☑ Sketch and explain the characteristics of a caldera and the four stages by which one forms.

☑ Explain how a volcanic eruption destroyed Santorini.

☑ Describe what type of volcanic feature Yellowstone is, how it forms, and how a future eruption could impact the Great Plains.

11.6 What Hazards Are Associated with Volcanoes?

VOLCANIC ERUPTIONS can affect areas very close to the vent or can spread damaging volcanic ash over huge regions. The relative hazard depends on the type of eruption. Eruptions of pyroclastic flows have killed more than 30,000 people at a time, all within minutes. Surprisingly, volcanic eruptions can also cause huge floods.

A What Is Meant by a Hazard, a Risk, and a Disaster?

The terms *hazard*, *risk*, and *disaster* may seem more appropriate for a lesson about insurance, but geographers frequently apply these terms when discussing the effects that natural (and human-caused) events can have on humans and society. What is the difference between a hazard, a risk, and a disaster?

11.06.a2 Mount Rainier, WA

1. A *hazard* is the existence of a potentially dangerous situation or event, such as a potential landslide of a steep slope or a lava flow erupting from a volcano. The hazard in this photograph was a basaltic lava flow (◄).

2. *Risk* is an assessment of whether the hazard might have some *societal impact*, such as loss of life, damage to property, loss of employment, destruction of fields and forests, or implications for local or global climates. Remnants of destroyed houses, cars, and roads demonstrate that this area had a high risk for volcanic hazards.

11.06.a1 Kilauea, HI

3. The people living around Mount Rainier, including those in the cities of Tacoma and Seattle (▲), have a high risk from activity on the dangerous, composite volcano, especially from lahars. Some suburbs of Tacoma are built on or among deposits from lahars that flowed off the volcano within the last 1,000 years. If no people were living near this volcano, a hazard would still exist but there would be essentially no risk.

4. A *disaster* is an event that has occurred, often suddenly, resulting in severe injury, loss of lives, and damage to property. If the event has not yet occurred, it is a potential disaster.

B What Hazards Are Associated with Eruptions of Cinder, Ash, and Gas?

Basaltic eruptions can be deadly and destructive, but mostly to nearby areas. They can hurl lava and solid rock into the air and spew out dangerous gases. Fine ash ejected high into the air can cause damage that is more widespread.

Falling Objects

11.06.b1 Hawaii

1. During the formation of lava fountains (◄), most cinder falls back to Earth near the vent and piles up in a cinder cone. Hazards include being struck and burned by cinders and being struck by blobs of lava and other projectiles. Larger ejected pieces, called *volcanic bombs* (▼), pose a severe hazard for anyone close to the erupting cone.

11.06.b2 Flagstaff, AZ

Volcanic Ash from Cinder Cones

2. Sand-sized cinders and finer particles of ash can bury nearby structures and may cause breathing problems for people and livestock. In March and April of 2010, a shield volcano in Iceland, called Eyjafjallajökull, erupted large amounts of volcanic ash that drifted over Europe, shutting down most air travel and stranding hundreds of thousands of passengers. Eruption columns from cinder cones typically reach lesser heights of several kilometers and mostly impact areas nearby.

11.06.b3 Iceland

Gases

3. *Volcanic gases* are a significant hazard associated with many types of volcanoes. Gases such as carbon dioxide (CO_2) can cause asphyxiation. Other gases, including hydrogen sulfide (H_2S), cause death by paralysis. Gaseous sulfur dioxide (SO_2), hydrochloric acid (HCl), sulfuric acid (H_2SO_4), and fluorine compounds expelled during eruptions can destroy crops, kill livestock, and poison drinking water. Gas eruptions can kill hundreds of people.

11.06.b4 Krafla, Iceland

C What Are the Hazards Associated with Pyroclastic Eruptions?

Lava flows usually move slowly enough that people can get out of the way, but such flows can completely destroy any structures in their path. In contrast, pyroclastic flows race downhill and across the landscape at hundreds of kilometers per hour, incinerating and smothering any creature in their path. They can cause massive destruction. Eruption columns spread their damage over wider areas and over longer time periods.

1. On May 18, 1980, an earthquake triggered a massive landslide on the north side of Mount St. Helens in southwestern Washington state. The landslide reduced pressure on an underlying magma chamber, triggering a northward-directed blast that knocked over millions of trees and unleashed a pyroclastic flow. The pyroclastic flow swept downhill and across the landscape, burying and killing dozens of people and countless other living things in its path. This was followed immediately by a huge column of volcanic ash that rose 25 km (15 mi) into the atmosphere (◄). The ash was carried eastward by the wind and blocked sunlight as it settled back to Earth across a large area of Washington, Idaho, and Montana.

11.06.c1 Mount St. Helens, WA

2. After the main eruption, pyroclastic flows (◄) and other eruptions continued on Mount St. Helens until 1986. After a lull of several years, the volcano later built, destroyed, and rebuilt several new volcanic domes. The domes would grow at some times and collapse at others, unleashing smaller pyroclastic flows, like the one shown here. Activity subsided in 2008, but the volcano is still being monitored closely by instruments designed to detect changes in temperature, gas emissions, or steepness of slopes.

11.06.c2 Mount St. Helens, WA

3. Vesuvius is an active composite volcano near the city of Naples in southwestern Italy. In A.D. 79, a series of pyroclastic flows moved down the flank of the volcano, destroying the coastal towns of Pompeii and Herculaneum and killing more than 25,000 people. This image (►) is an artist's conception of an explosive Vesuvius eruption striking Naples, which covers most of the region shown. Naples and the surrounding area currently are home to over three million people. The area also has a significant hazard for lahars and landslides coming off the volcano, which looms over the city.

Herculaneum

Limit of Pyroclastic Flows

Pompeii

10 km

11.06.c3

4. The dashed red line marks the outward limit of pyroclastic flows from Vesuvius, but material from the eruption column covered a wider area. Note how much of the present city of Naples is within the area devastated by the eruption. Evidence indicates that the catastrophe began with earthquakes and the formation of an eruption column that deposited a layer of ash over Pompeii, killing some inhabitants. This was immediately followed by pyroclastic flows that raced down the mountainside and hit Pompeii. People were smothered, suffocated, subjected to thermal shock, or crushed.

Floods Associated with Volcanic Eruptions

Volcanic eruptions can be accompanied by several types of floods. A common type is a flood produced when a volcanic eruption melts a large amount of ice and snow, such as is common on the summits of some large composite volcanoes. The meltwater picks up unconsolidated volcanic ash, soil, and loose rocks on the steep slopes of the volcano, becoming a lahar. Lahars are among the main hazards from large volcanoes, especially for people who are in stream valleys downstream from the volcano. Lahars can also result from heavy rains during an eruption.

Catastrophic eruption-related floods occur in Iceland, a land of ice and fire, due to its location near the Arctic Circle and its position on top of a mid-ocean ridge and an oceanic hot spot. It has many glaciers, including a large ice sheet that covers 25% of the country. Beneath the ice are a half dozen shield volcanoes, several of which have erupted in the last several decades. When a volcanic eruption beneath an ice sheet melts the ice, it can release a catastrophic flood of meltwater. Such floods can be huge, carrying blocks of ice, rock, and other debris and causing widespread damage. Such floods carry so much sediment that they can cover large areas with a thick layer of mud, sand, and rocks.

Before You Leave This Page

☑ Explain how risk differs from hazard, and provide an example of each.

☑ Describe and compare the hazards associated with lava flows, lava fountains, volcanic gases, and various types of pyroclastic eruptions.

☑ Describe the eruptions that occurred at Mount St. Helens and Vesuvius, and what caused most deaths.

☑ Describe eruption-caused floods.

11.6

11.7 What Areas Have the Highest Potential for Volcanic Hazards?

IN SOME PLACES, THE RISK POSED by volcanic hazards is great. In others, it is inconsequential. Volcanic eruptions are much more likely in Indonesia than in Nebraska. Additionally, different types of volcanoes have different eruptive styles, so some volcanoes are more dangerous than others. What factors should we consider when determining which areas are the most dangerous and which are the safest? What information do we need to plan for an eruption?

A How Do We Assess the Danger Posed by a Volcano?

Potential hazards of a volcano depend on the type of volcano, which we can infer from its shape and rock types, and on its history. Examine the volcano below for clues about what type of volcano is present and how it might erupt. Think about how you would describe the types of eruptions that are possible. Then read on.

1. *Shape*—The shape of a volcano provides important clues about how dangerous the volcano might be. Volcanoes with steep slopes, such as composite volcanoes, are more dangerous because they involve potentially explosive, viscous magma and also are prone to landslides. Volcanoes that have relatively gentle slopes, such as most shield volcanoes, result from less explosive basaltic eruptions.

2. *Types of Volcanic Materials*—The types of materials on a volcano indicate the magma composition and style of eruption. A volcano is very dangerous if it contains deposits formed by pyroclastic flows. A volcano that erupted viscous or gas-rich magma is more dangerous than one that erupted less viscous magma, such as one that formed basalt. The volcano shown above is a steep, composite volcano composed of materials erupted as viscous lavas and pyroclastic flows.

11.07.a1 Augustine volcano, AK

3. *Age and History*—The age of a volcano is essential information. If the volcano has not erupted for a long time, maybe it is dormant (not active). The shape of a volcano, especially whether it still has a fresh-looking volcano shape or has been eroded, is one indicator of a volcano's age. Important clues are also provided by a volcano's history, if recorded, including oral histories from nearby people. Isotopic measurements on volcanic units can provide an accurate indication of a volcano's age. Studies of the sequence of volcanic layers, combined with isotopic ages, provide insight into how often eruptions recur. The volcano above clearly has recent activity, as expressed by the recent dark deposits covering the snow and the steam and other gases escaping from the summit dome.

B What Areas Around a Volcano Have the Highest Risk?

Once we have determined the type of volcano that is present, we consider other factors that help identify which areas near the volcano have the highest potential risk if people decide to live near the volcano.

11.07.b1

1. *Proximity*—The biggest factor determining potential risk is proximity— closeness to the volcano. The most hazardous place is inside an active crater. The potential risk decreases with increasing distance away from the volcano.

2. *Valleys*—Lava flows, small pyroclastic flows, and mudflows (lahars) are channeled into valleys carved into the volcano and surrounding areas. Such valleys are more dangerous than nearby ridges.

3. *Wind Direction*—Volcanic ash and pumice that are thrown from the volcano are carried farthest in the direction that the wind is blowing at the time of the eruption. Most regions have a prevailing wind direction, so a greater hazard of falling material exists in this direction from a volcano.

4. *Particulars*—Each volcano has its own peculiarities, and these influence which part of the volcano is most dangerous. Steeper parts of a volcano pose special risks, and one side of a volcano may contain a dome that could collapse and form pyroclastic flows. This image shows three small villages around a volcano. Is one village at greater risk than the others? Which one is in the least hazardous place, and what ideas led you to this conclusion?

C What Regions Have the Highest Risk for Volcanic Eruptions?

We can think on a broader regional scale about which regions are most dangerous. In North America, volcanoes are relatively common along the west coast and virtually absent east of the Rocky Mountains. Tectonic setting, especially proximity to certain types of plate boundaries or to a hot spot, is the major factor making some places more prone to volcanic hazards than others. This map shows the locations of volcanoes (orange triangles) that have been active in the last several million years.

1. The largest concentration of composite volcanoes is along the Pacific Ring of Fire. The volcanoes here form above subduction zones, in island arcs, and in mountain ranges along continental margins. Some subduction zone volcanoes erupt so vigorously that they form calderas.

2. Large volumes of fluid basaltic lava erupt on the seafloor at mid-ocean ridges. Such eruptions pose little risk to humans because almost all of these occur at the bottom of the ocean. The island of Iceland, where a mid-ocean ridge coincides with a hot spot, is an exception.

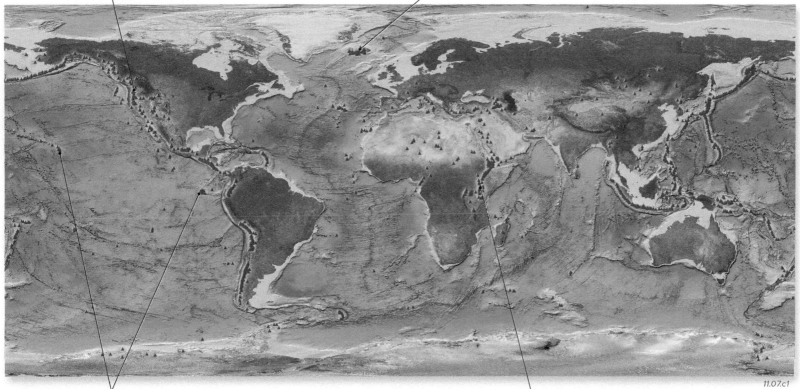

11.07.c1

3. Many shield volcanoes occur along lines of islands and submarine mountains in the Pacific and other oceans. Most of these linear island chains, and a few other clusters of islands, formed above hot spots. Hawaii and the Galápagos Islands are good examples. Shield volcanoes also occur in other settings, including on continents.

4. Some volcanic features, including basalt flows, cinder cones, and composite volcanoes, are in the middle of continents. Most of these form over hot spots or in continental rifts, such as the East African Rift.

Forecasts, Policy, and Publicity

Predicting volcanic eruptions is currently an imprecise science. There have been some fabulous successes and disappointing failures. Volcanologists have successfully predicted some eruptions by studying clusters of small earthquakes generated as magma rises through the crust, by measuring changes in the amount of gas released by volcanoes, and through other types of investigations. Some predictions led to evacuations that saved lives because government officials and citizens acted on the scientific evidence. Some predictions have been unsuccessful because an eruption that was considered possible or even likely did not occur. In other cases, predictions, policy, and publicity interacted in a bad way, with deadly results. In 1985, volcanologists working on Nevado del Ruiz, a composite volcano in the Colombian Andes, warned of an impending eruption. The city of Armero, with an estimated 29,000 inhabitants, lay in a valley that drained the steep volcano. Local government officials downplayed the risk and assured the citizens that there was no danger. A pyroclastic eruption occurred at night, melting snow and ice on the volcano and unleashing a lahar that moved at hundreds of kilometers per hour, engulfing most of Armero and killing more than 20,000 people.

Before You Leave This Page

☑ Summarize ways to assess the potential danger of a volcano based on its characteristics.

☑ Describe ways to identify which areas around a volcano have the highest risk.

☑ Describe how the plate-tectonic setting of a region influences its potential for volcanic hazards.

11.7

11.8 How Are Magmatic Conduits Exposed?

NOT ALL MAGMA REACHES THE SURFACE and erupts in a volcano. Instead, most magma solidifies at depth, forming bodies of igneous rock called *intrusions*. Some intrusions are small enough to be exposed on a single small hill or in a roadcut, whereas others comprise most of a mountain range. Intrusions can have a sheetlike, pipelike, or even lumpy geometry, and where exposed at the surface they form distinctive landscape features.

A What Features Represent Magma Chambers Within and Beneath Volcanoes?

Magma that erupts from volcanoes is fed through conduits that may be pipelike, sheetlike, or both. After the volcano erodes away, the solidified conduit can form a steep topographic feature called a *volcanic neck*. Larger, deeper magma chambers, when uplifted and exposed at the surface, can form irregularly shaped bodies of granite and related rocks.

1. Many volcanic necks form from the erosion of a volcano. In these figures (◄), a small volcano has been partially eroded, revealing a cross section through the volcano. A resistant and jointed volcanic conduit marks the center of the volcano.

2. This process formed the volcanic neck in this photograph (►), where erosion has cut through the middle of a volcano, exposing its internal architecture. As weathering and erosion wore down the volcano, they exposed the harder, more resistant rocks that solidified within the magmatic conduit that fed the volcano.

11.08.a1

11.08.a2 Mount Taylor, NM

3. Shiprock is a famous volcanic neck that rises above the landscape of New Mexico (▼). It consists of fragmented volcanic rocks. Shiprock and many other volcanic necks were not originally *inside* a volcano, but instead were magmatic conduits that formed well beneath the surface. The volcano above Shiprock was not a mountain, but a crater (pit) excavated by a violent explosion. The explosion occurred when magma ascending up a conduit encountered groundwater and generated huge amounts of steam, which expanded violently, causing an explosion (◄). After the volcanic eruption, erosion removed evidence of the crater and hundreds of meters of rock that once covered the area around the conduit.

11.08.a3 Shiprock, NM

11.08.a4

11.08.a5

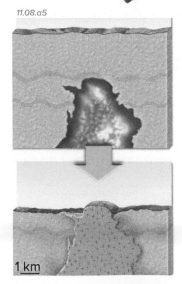

4. Some magma never reaches the surface, instead solidifying within a subsurface magma chamber (◄) and commonly forming irregularly shaped masses of granite or other *intrusive rocks* (i.e., igneous rocks that solidify below the surface). Such solidified magma chambers can remain in the subsurface or can be uplifted by tectonic processes and exposed by erosion. Once exposed, the irregular shape of the original magma chamber is expressed as an irregular outline and shape of the intrusive rock on the surface. The irregularly shaped exposure of granite to the right (►) occupied and solidified within a subsurface magma chamber and later was uplifted to the surface as part of a mountain range. In this example, the intrusion is more resistant to weathering and erosion than the rocks around it, so it forms steep, jagged peaks. Other intrusions are weaker than adjacent rocks and so form topographic low spots. If a number of intrusions cover an area of more than 100 km² (40 mi²), they form a *batholith*.

11.08.a6 Toiyabe Range, NV

1 km

B What Features Form When Magma Is Injected as Sheets?

Many small intrusions have the shape of thin or thick sheets, typically ranging in thickness from several centimeters to several tens or hundreds of meters. These form when underground forces allow magma to generate new fractures or to open up and inject into existing fractures. In some cases, magma squeezes between preexisting layers in the wall rocks, commonly between the horizontal layers of sedimentary rocks.

Dike

1. A *dike* is a sheetlike intrusion that cuts across any layers present in the host rocks (◄). Most dikes are steep because the magma pushes apart the rocks, commonly in a horizontal direction, as it rises vertically and fills the resulting crack to form a dike. Some dikes propagate vertically and others do so in a more or less horizontal direction, away from a volcanic conduit.

2. In this roadcut (blasted cut along a high-way), dikes of light-colored granite cut through an older, darker igneous rock (▶). Most of the dikes shown here are less than a meter thick.

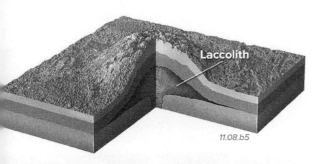
11.08.b1

11.08.b2 Franklin Mtns., TX

Sill

3. An intrusion that is parallel to layers in the host rocks is called a *sill* (◄). Most sills are gently inclined and form by pushing adjacent rocks upward rather than sideways.

4. These dark-colored sills intruded parallel to layers of light-colored, sedimentary wall rocks (▶). Like most sills, these contain steep fractures (columnar joints) formed by cooling of the sills after they solidified.

11.08.b3

11.08.b4 Salt River Canyon, AZ

Laccolith

5. In some areas, ascending magma encounters gently inclined layers and begins squeezing parallel to them, forming a sill. The magma then begins inflating a lump- or bulge-shaped magma body called a *laccolith*. As the magma chamber grows, the layers over the laccolith tilt outward and eventually define a dome (▼).

6. The Four Corners region of the American Southwest contains some of the world's most famous laccoliths, including these in the Henry Mountains of southern Utah. Here, the sedimentary layers bow up and over the laccoliths, forming distinctively shaped mountains (▶). The intrusions in the laccoliths solidified at depths of several kilometers (more than a mile) and were later uncovered by erosion.

11.08.b5

11.08.b6 Henry Mtns., UT

Before You Leave This Page

✓ Sketch and explain two ways that a volcanic neck can form and the way in which an irregularly shaped intrusion forms.

✓ Sketch and explain the difference between a dike and a sill, and explain why each has the orientation that it does.

✓ Sketch or describe the geometry of a laccolith and how you would recognize one in the landscape.

11.8

11.9 What Is Deformation and How Is It Expressed in Landscapes?

ANOTHER TECTONIC PROCESS IS DEFORMATION. Rock can be subjected to stress resulting from burial, tectonic activity, heating or cooling, and other processes. If the stresses are strong enough, they cause a rock to deform—move, rotate, change shape, or some combination of these. What kinds of stress affect rocks, what types of deformation can result from this stress, and how is deformation expressed in landscapes?

A What Are Force and Stress?

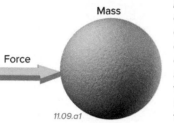

Force is a push or a pull that causes, or tends to cause, change in the motion of a body (◄). It is the amount of acceleration experienced by a mass (force = mass times acceleration).

11.09.a1

11.09.a2

The amount of force divided by the area where the force is applied is called the *stress*. Here (◄), the force from a metal weight is distributed evenly across the top of a broad, wooden pillar.

11.09.a3

If the same amount of weight is on a much thinner pillar (◄), the stress (force per unit area) on the pillar is greater. It might cause the pillar to splinter or break.

B How Do Materials Respond to Stress?

Rocks, sediment, and other materials may be subject to three types of stress: *compression*, *tension*, or *shear*. The way in which rocks respond to these stresses varies as a function of depth because rock strength changes with depth, as pressure and temperature increase downward in the crust. The examples below show how rocks respond to different types of stress under shallow levels of the crust, where materials are brittle and can fracture.

Type of Stress

| Compression | Tension | Shear |

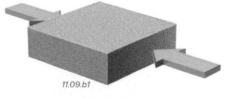

11.09.b1

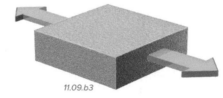

11.09.b3

11.09.b5

1. When stress pushes in on rock, the stress is called *compression*, shown by the inward-directed arrows above.

3. When stress is directed outward, pulling the rock, the stress is called *tension*. Tension is shown with stress arrows pointing away from the rock.

5. A third type of stress acts to *shear* the rock as if stresses on the edges of a block were applied in opposite directions.

Type of Structure Formed

11.09.b2

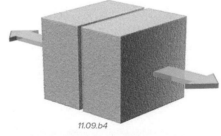

11.09.b4

11.09.b6

2. Compression in shallow levels of the crust can cause rocks to fracture and slip past one another, like pushing one block over another. In this figure, the rocks are responding to compression by trying to get out of the way. A fracture along which two rock masses slip past each other is a *fault*.

4. Tension can form fractures that help the rock stretch as it is pulled apart. Fluids, if present, can deposit minerals in the fracture, forming a *vein*. Tension may also cause slip along fractures, pulling one block past another (like in the figure to the left for compression, but with the blocks moving in the opposite direction).

6. Shearing in shallow parts of the crust usually forms a fault. Depending on the direction of shearing, one block can move up, down, or sideways relative to the block on the other side of the fault. In the case shown here, the two blocks are moving horizontally (sideways) relative to one another.

C How Do Materials Respond to Force and Stress?

Materials within Earth are subjected to forces from the weight of overlying rocks, from tectonic forces pushing or pulling on the rocks, from cooling and heating, and from pressurized fluids, such as water and magma. Just like the wooden pillar on the previous page, if a force is concentrated (i.e., high stress), a rock can break or otherwise deform. As a result, we normally talk about stress instead of force. These figures show stress with a blue arrow.

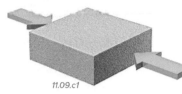

11.09.c1

1. A block of material may remain unchanged if subjected to only a small amount of stress (▶). If the imposed stresses are greater, three things can happen. The material may be *displaced* from one place to another, it may be *rotated*, or it may have its shape *modified*, or *strained*. All three responses may occur at the same time.

Displacement

2. In response to stress, a material may be moved, or displaced, from one place to another. During *displacement*, a material can behave as a rigid object or can change shape as it moves. In the photograph below, a thin sheet of light-colored granite has been displaced by movement along fractures (▼).

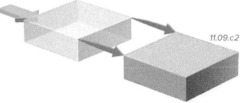

11.09.c2

11.09.c3 Tortolita Mtns., AZ

Rotation

3. A material may be rotated in response to stresses. Rotation can tilt the volume of material or spin it horizontally. The sedimentary layers in the photograph below were deposited as horizontal layers, but the layers have since been rotated. Rotation can be expressed by tilting, folding, or a partial spin of the material (▼).

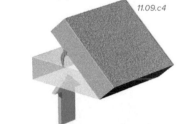

11.09.c4

11.09.c5 El Dorado Canyon, CO

Strain

4. A material can respond to stress by deforming internally—changing size or shape due to deformation. A change of size or shape is called *strain*. Stress is the cause, and strain is the effect. In the photography below, originally rounded pebbles in a rock were strained in response to stress squeezing the rock under moderately high temperatures (▼).

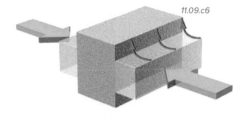

11.09.c6

11.09.c7 Granite Wash Mtns., AZ

Strength of Continental Crust at Depth

5. The strength of continental crust varies as a function of depth because temperature and pressure both increase downward (▼). At shallow levels of the crust, rocks deform by fracturing and other types of *brittle deformation*. Rocks in the upper crust become stronger with depth because increasing confining pressure acts to hold rocks together and makes slip along any fractures more difficult.

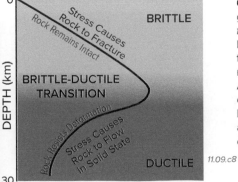

INCREASING STRENGTH ➡

Stress Causes Rock to Fracture
Rock Remains Intact

BRITTLE

BRITTLE-DUCTILE TRANSITION

Rock Resists Deformation

Stress Causes Rock to Flow in Solid State

DUCTILE

11.09.c8

DEPTH (km) 0 ... 30

6. Deeper, where pressure and temperature are greater, rocks may deform (flow) by *ductile deformation*. There is a gradational boundary, or transition, between the *upper brittle* and *lower ductile* parts of the crust. This typically occurs at a depth of approximately 15 km and temperatures of more than 300°C. At greater depths, the effects of temperature dominate over the effects of pressure, and rocks become progressively weaker, beginning to flow as a weak solid material. The strength of the crust decreases rapidly downward.

Before You Leave This Page

☑ Describe or illustrate the concept of stress.

☑ Sketch and explain the three types of stress, providing examples of the types of structures each forms.

☑ Sketch or describe the three ways that a material can respond to stress.

☑ Sketch and explain how crustal strength varies with depth.

11.9

11.10 How Are Fractures Expressed in Landscapes?

FRACTURES ARE THE MOST COMMON geologic structures. They range from countless small cracks visible in an outcrop (natural exposure of rock) to huge faults hundreds or thousands of kilometers long. What are the different types of fractures, how do the different types of fractures form, and how are fractures expressed in landscapes?

A In What Different Ways Do Rocks Fracture?

There are two main types of fractures: *joints* and *faults*. Joints and faults both result from stress but have different kinds and amounts of movement across the fracture.

Joints

11.10.a1

1. Most fractures form as simple cracks representing places where the rock has pulled apart by a small amount (◄). These cracks are called *joints* and are the most common type of fracture. They form under tension, where the rock is pulled apart.

2. These sandstone ledges (►) are cut by a series of near-vertical joints. The layers are not offset by the joints but are simply pulled apart by a very small amount.

11.10.a2 San Juan River, UT

Faults

3. A *fault* is a fracture where rocks have slipped past one another. Rocks across a fault can slip up and down, as shown here (►), or they can slip sideways or at some other angle. A fault displaces the rocks on one side relative to the other side.

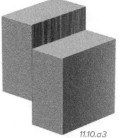

11.10.a3

4. The long fracture in the center of this photograph (◄) cuts across and offsets the rock layers. That is, the layers across the fault have been displaced relative to each other.

11.10.a4 Moab, UT

B What Stresses Form Joints?

A number of settings generate enough stress to cause joints, especially those related to tectonics, to cooling of hot rocks, and the removal of weight on top of buried rocks, each discussed below. In addition to these three causes, joints can result from stresses that accompany burial, slope failure (e.g., a rock slide), or when the weight of glaciers is removed by melting.

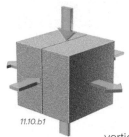

11.10.b1

1. The stresses that form joints arise from many sources, but they are mostly due to burial and to tectonic forces. Tectonic forces may push, pull, or shear the rock. These volcanic rocks (▼) are cut by vertical joints formed by tectonic stresses.

11.10.b3

2. Stresses build up as rocks get warmer or cooler. As some igneous rocks cool, they contract into polygon-shaped columns bounded by joints that commonly meet at 120° angles. The photograph below (▼) shows an example of such *columnar joints*.

11.10.b5

3. Stresses also arise during uplift of buried rocks, causing rocks to fracture due to reduced pressure. These joints, called *unloading joints*, form parallel to the surface and slice off thin sheets of rock, as in this granite (▼).

11.10.b2 South Fork, CO

11.10.b4 Iceland

11.10.b6 Estes Park, CO

C What Are the Main Types of Faults?

We classify faults based on the motion of one block relative to the other. There are three main types of faults: *normal faults,* *reverse faults,* and *strike-slip faults.* Black arrows show relative movement, and blue arrows show stress.

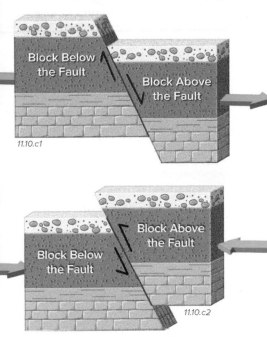

11.10.c1

Normal Fault—In a fault, if the block above the fault moves down relative to the block below the fault (◄), the fault is a *normal fault.* A normal fault forms when the rock units are pulled apart and lengthened, as for example by tension. The direction in which a fault (or layer) is inclined is called its *dip.* This normal fault dips to the right.

11.10.c2

Reverse Fault—If the block above the fault moves *up* relative to the block below the fault (◄), the fault is a *reverse fault.* A reverse fault forms as a result of *horizontal compression* and shortens the rock units in a horizontal direction. This reverse fault dips to the right.

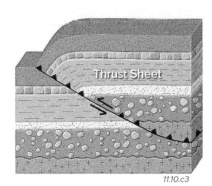

11.10.c3

Thrust Fault—A reverse fault that has a gentle dip (◄) is a *thrust fault.* The rock above the fault is called a *thrust sheet* and is pushed up and over the block below the fault. On maps and other figures, teeth indicate that a fault is a thrust.

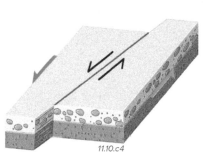

11.10.c4

Strike-Slip Fault—When rocks along a fault move with a side-to-side motion, parallel to the fault surface, the fault is a *strike-slip fault.* Relative motion is horizontal, offsetting the blocks laterally in one direction or the other.

D How Are Faults Expressed in Landscapes?

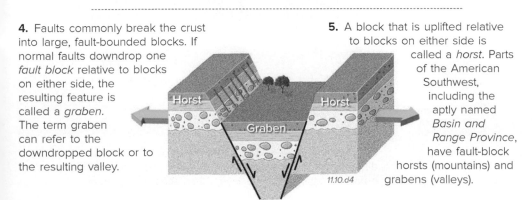

11.10.d1 Borah Peak, ID

11.10.d2 Echo Cliffs, AZ

11.10.d3 Buckskin Mtns., AZ

1. When fault movement offsets Earth's surface, as during an earthquake (▲), it can cause a step in the landscape, called a *fault scarp.* If a fault forms and later erosion wears away one side more than the other, the step in topography is called a *fault-line scarp.*

2. We commonly recognize faults because of *offsets* or abrupt *terminations* of layers. Also, rocks along faults are highly fractured and easily eroded, so they erode into linear topographic notches (▲). This fault truncates layers and forms a linear notch.

3. Up close, faults have other distinctive characteristics (▲). They can fold the truncated layers, juxtapose two very different rocks against one another, or display shattered and scratched rocks along the fault zone. This fault has all these diagnostic features.

4. Faults commonly break the crust into large, fault-bounded blocks. If normal faults downdrop one *fault block* relative to blocks on either side, the resulting feature is called a *graben.* The term graben can refer to the downdropped block or to the resulting valley.

11.10.d4

5. A block that is uplifted relative to blocks on either side is called a *horst.* Parts of the American Southwest, including the aptly named *Basin and Range Province,* have fault-block horsts (mountains) and grabens (valleys).

Before You Leave This Page

✓ Sketch and describe the formation and landscape expression of joints and faults.

✓ Sketch and describe three types of faults, showing the relative displacement and the type of stress.

✓ Sketch a horst and a graben.

11.10

How Are Folds Expressed in Landscapes?

DEFORMATION CAN FOLD ROCK LAYERS on such a grand scale that a single fold occupies an entire mountain range. Other folds are much smaller. Folds can form at depth and at the surface. The folded layers may be bent in gentle arcs or squeezed tightly into sharp angles. We classify and name folds based on their shape, in order to have a convenient way to describe what we observe in landscapes and in smaller exposures of rocks.

A What Is a Fold and What Are the Main Types of Folds?

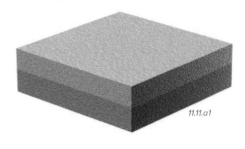

11.11.a1

1. Before folding, most rock layers are horizontal (◄) because most sedimentary and volcanic layers form with a more or less horizontal orientation.

2. Compression causes shortening, often accommodated by folding of the layers. When you scrunch up a rug, the folds (creases) are perpendicular to the direction of shortening, as shown here for rocks (▶). Compression can form folds, reverse faults, or both. It can occur in any tectonic setting but is most common along convergent plate-tectonic boundaries.

11.11.a2

11.11.a3 Tibet

Anticlines and Synclines

3. If rock layers warp up (◄), in the shape of an *A*, the fold is generally called an *anticline*. In an anticline, the *oldest* rocks are in the center of the fold. Within a fold, the place where the layers are mostly tightly bent or reverse their direction of tilt is the fold *hinge*. The hinge in this fold (◄) is along the top of the hill.

11.11.a4 Tibet

4. In the opposite case, where rock layers fold down (◄) in the shape of a *V* or *U*, the fold is generally called a *syncline*. In a syncline, the *youngest* rocks are in the center of the fold. The hinge of this fold is in the center, where the layers are mostly tightly folded and change direction. The tilted parts of the fold on either side of the hinge are called the *limbs* of the fold.

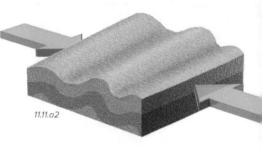

11.11.a5

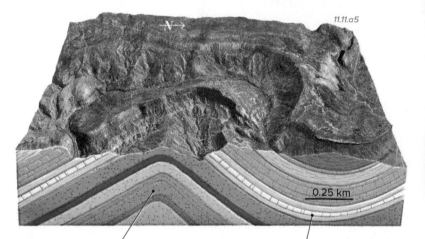

0.25 km

5. Synclines and anticlines occur together, usually as part of a series of folds. The upward fold of the layers on the left of the diagram is an anticline. For now, note how it has been eroded. More on this later.

6. The downward fold of layers on the right side of the diagram is a syncline. The beds that are inclined to the right in the center of the diagram are part of both folds. More specifically, they are a limb of both folds. Note also the landscape formed by the syncline.

Dome

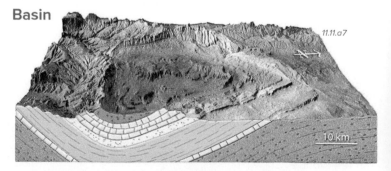

11.11.a6

20 km

7. Layers that are uplifted in a circular or elliptical area and dip away in all directions form a structural *dome*. Erosion exposes deeper and older rocks in the center of this dome.

Basin

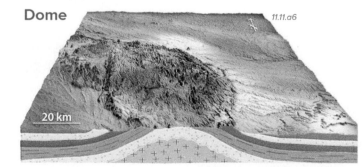

11.11.a7

10 km

8. A *basin*, formed by folding, is the opposite of a dome. Layers dip toward the center of the basin from all directions. The center of a basin contains younger layers than surrounding areas.

Monocline

11.11.a8 San Rafael Swell, UT

11.11.a9 Grand Junction, CO

9. In some folds, nearly flat layers bend down in one direction and then flatten out again. This type of fold is a *monocline,* a name that indicates that the fold has only one dipping segment. Monoclines can be tens of kilometers long or exposed in a small outcrop. Some monoclines have great names, such as the Coxcomb fold, Waterpocket fold, and Comb Ridge in Utah and Arizona. The monocline shown here is along the San Rafael Swell in central Utah.

10. This view shows a medium-sized monocline in sandstone layers. Horizontal layers on the left bend down in the center of the image (dip to the right) and then fold back to horizontal on the right side of the image.

B How Do Tilted and Folded Layers Erode?

11.11.b1 Dinosaur NP, CO

11.11.b2 Sideling Hill, MD

11.11.b3 Comb Ridge, UT

1. Some mountains, including this one (▲), are large folds, with the shape of the mountain mimicking the shape of the fold. In some cases the folding and formation of the mountain are occurring today, but generally folding occurred sometime in the past, followed by erosion that later uncovered an existing fold.

2. Commonly, erosion does just the opposite—it erodes the uplifted layers first, leaving behind a ridge of down-folded layers. In other words, the ridges contain synclines whereas the valleys were anticlines. This is very common in the Appalachian Mountains of Maryland (▲), Pennsylvania, and elsewhere.

3. Erosion can strip off easily eroded layers, but it slows upon encountering an underlying hard layer. Erosion of soft and hard layers can carve slopes (▲) parallel to the dipping (inclined) layers—this is known as a *dip slope.* A dip slope can follow planar dipping layers or gently curving ones. If a ridge has a dip slope on one side, it is called a *cuesta* or a *hogback,* like this one (►).

4. The Appalachian Mountains of the eastern U.S. are famous for their large folds and the way they were eroded. This map (▼) has a satellite image superimposed on topography for the area near the Susquehanna River in southeastern Pennsylvania. The image includes curving mountains and ridges (green) alternating with lowlands (pinkish brown). The river cuts downhill across the ridges and lowlands.

5. Some ridges and valleys are straight, but others curve across this distinctive region, known as the *Valley and Ridge Province.* The ridges and valleys represent the eroded hinges and limbs of folds. Can you identify examples of each? The hinges are curves and the limbs are more straight. Some anticlines are ridges and others are valleys; some synclines are valleys but others are ridges.

11.11.b4 Lyons, CO

11.11.b5

10 km

Before You Leave This Page

☑ Sketch a cross section of an anticline, syncline, and monocline, labeling the hinges and limbs.

☑ Sketch or describe a dome and a basin.

☑ Summarize or sketch the ways that folds are expressed in the landscape.

11.11

11.12 How Do Local Mountains and Basins Form?

VOLCANISM, DEFORMATION, AND OTHER PROCESSES can cause areas to be uplifted or to subside relative to adjacent regions. They can form huge *mountain ranges* hundreds of kilometers long, or they can also form mountains that are smaller, more local features. The distinction between regional mountain ranges and local mountains is important, because regional mountain ranges typically involve uplifted, thickened crust, while local mountains are small enough to simply rest upon—and be supported by—the crust. In a similar way, areas that have subsided (basins) can be regional features related to thin continental crust or local features related to smaller structures.

A How Does Volcanism Build Local Mountains?

A local mountain may be formed by a volcanic eruption that piles lava, ash, and cinder onto the crust. Such mountains vary in size from small cinder cones to large shield and composite volcanoes.

Volcanism creates mountains by piling volcanic materials on a preexisting surface. Some of the smallest volcanic mountains and hills are cinder cones. They are clearly local features, not requiring regional changes in the thickness of the underlying crust.

11.12.a1 Flagstaff, AZ   11.12.a2 Mount Hood, OR

Composite volcanoes consist of lava flows, variably compacted volcanic ash, and debris in mudflows and landslides. They commonly make lofty and steep mountains that have a typical volcano shape (◄). Many shield volcanoes are also large mountains, but they are not as steep.

B How Do Faults Form Local Mountains?

Local mountains can also arise through faulting. Thrust faults create mountains by thrusting one fault block up and over another. Normal faults also form local mountains, even though they stretch and thin the crust in a region.

Mountains Formed by Thrust Faulting

Thrust faulting will make a mountain if the overthrust block is uplifted faster than it is eroded, or if it is composed of hard, erosion-resistant rocks like granite (shown with the short-lines pattern).

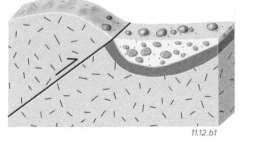
11.12.b1

Mountains Formed by Normal Faulting

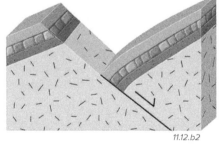
11.12.b2

During normal faulting, the block above the fault slips down, forming a basin. The other block remains high or is moved upward, and it can form a local mountain if it is not eroded away. Many linear mountains in the western U.S. were formed in this way.

C How Does Folding Form Local Mountains?

Another way to make local mountains is by folding. Folding can warp and uplift Earth's surface as well as the underlying rock layers. Uplift and erosion of a folded, hard layer can create a topographical high.

Folding can form mountains and hills by actively deforming the land surface and near-surface rocks. This upward-bending anticline formed very recently due to folding and uplift along the San Andreas fault in southern California.

11.12.c1 Mecca Hills, CA

11.12.c2 Mexican Hat, UT

In some mountains, a fold formed at depth millions of years ago, but more recently the fold was uplifted to the surface. Then, weathering and erosion—acting on the already folded layers—sculpted the landscape, guided by the geometry of the folded layers.

D In What Settings Do Local Basins Form?

A basin is an area that is lower than its surroundings (a topographic depression) that can accumulate sediment and volcanic rock. Local basins form in some of the same ways local mountains form. Most are related to faulting—specifically, the downdropping of one fault block relative to another.

Normal Fault Blocks

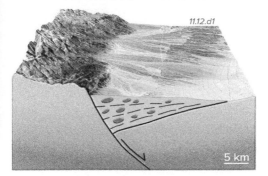

11.12.d1

5 km

Normal faulting can downdrop a block, forming a basin that fills with sediment. Such a mountain front starts out very steep and fairly straight, with a fault scarp along its base. Over time, erosion carves into the mountain, making it less steep and with a more irregular mountain front. Examples of this type of basin are in the Basin and Range of the Western U.S. and Connecticut River valley of New England.

Reverse and Thrust Faults

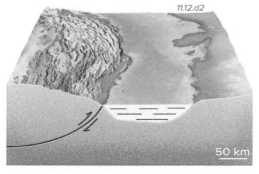

11.12.d2

50 km

Other basins form in front of a thrust fault, when the extra weight of the thrust sheet causes the crust to locally warp downward, forming a topographic depression (a basin). Such basins tend to be long, following the thrust fault laterally across the landscape. The Persian Gulf is this type of basin, as are many of the valleys in the Wyoming Rockies.

Strike-Slip Faults

11.12.d3

10 km

Basins can develop along a strike-slip fault if motion along the fault downdrops one block relative to another, while mostly shifting the two blocks horizontally past one another. An example of this type of basin is the Dead Sea of the Middle East—containing the lowest land elevations on Earth at 418 m (1,369 feet) below sea level.

E How Can Differential Erosion Form a Local Mountain?

Most variations in elevation of a landscape are due not to active deformation but to differences in the resistance of earth materials to weathering and erosion. Some materials are more resistant to erosion because they are harder, less fractured, or have some other protective features; these tend to form high or rugged parts of a landscape. Other materials are less resistant and so are more easily eroded away, tending to form low areas or soft-appearing slopes. When there are such variations in how materials are eroded, the process is called *differential erosion*.

11.12.e2 Moab, UT

1. Examine this diagram and observe how differential erosion resulted in the varied types of landscapes, then read the text.

2. Some mountains exist because they are composed of relatively harder rocks that resist erosion. In this scene, a hard granite forms a mountain that is surrounded by softer, more easily eroded layered rocks. A mountain or hill that remains when other rocks have been eroded down is an *erosional remnant*, like Stone Mountain in Georgia (▼), which is composed of hard and relatively unfractured granite.

1 km

11.12.e1

4. When layers are more gently inclined, differential erosion can form a landscape with alternating cliffs and slopes, like a series of steps. A resistant rock layer can protect softer rocks beneath from erosion, forming a local hill or mountain. Such a feature, if it has a nearly flat top, is a *mesa* (▲).

3. The geometry of layers influences the landscape. Here, tilted layers of different hardness were shaped by differential erosion into a series of curved ridges and valleys, like those in the Valley and Ridge Province of the eastern U.S.

11.12.e3 Stone Mountain, GA

Before You Leave This Page

✓ Describe how volcanism forms local mountains.

✓ Sketch and describe how thrust faulting, normal faulting, and folding can each build mountains.

✓ Describe three different ways in which faulting can form a basin.

✓ Describe how differential erosion can form a local mountain.

11.12

11.13 What Is an Earthquake?

AN EARTHQUAKE OCCURS WHEN ENERGY stored in rocks is suddenly released, typically in seconds. Most earthquakes are produced when stress builds up along a fault over a long time, eventually causing the fault to slip. Similar kinds of energy are released by volcanic eruptions, explosions, and even meteorite impacts. In special cases, human activities can lead to earthquakes by filling reservoirs or pumping fluids underground under high pressure.

A How Do We Describe an Earthquake?

When an earthquake occurs, it releases mechanical energy, some of which is transmitted through rocks as vibrations called *seismic waves*. These waves spread out from the site of the disturbance, travel through the interior or along the surface of Earth, and are recorded by scientific instruments at *seismic stations*.

1. The place where the earthquake is generated is called the *hypocenter* or *focus*. Most earthquakes occur at depths of less than 100 km (60 mi), and some occur as shallow as several kilometers. Earthquakes in subduction zones occur from shallow depths to as deep as 700 km (430 mi). The *epicenter* is the point on Earth's surface directly above where the earthquake occurs (directly above the hypocenter). If the seismic event happens on the surface, such as during a surface explosion, then the epicenter and hypocenter are the same.

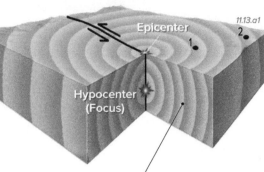

2. Seismic waves, once generated, spread in all directions. The curved bands show the peaks of waves radiating from the hypocenter. The intensity and duration of waves are measured by seismic stations (locations 1 and 2). In general, seismic stations closer to the hypocenter (station 1) will detect the waves sooner than those farther away (station 2).

3. An earthquake generates several different types of seismic waves, including *primary waves*, *secondary waves*, and *surface waves*. A primary wave, also called a *P-wave*, compresses the rock in the same direction as the wave propagates (▶) and can travel rapidly (about 6 km/s) through solids and liquids.

4. A secondary wave (▶), or *S-wave*, shears the rocks from side to side or up and down. This movement is perpendicular to the direction of travel. S-waves are slower (3.5 km/s) than P-waves and cannot travel through liquids, such as magma. *Surface waves* are similar to S-waves but travel along the surface of the Earth. All types of seismic waves help us study and understand earthquakes.

B What Causes Most Earthquakes?

Most earthquakes are generated by movement along faults. When blocks on opposite sides of a fault slip past one another abruptly, the movement generates seismic waves, while materials near the fault are pushed, pulled, and sheared. Slip along any type of fault can generate an earthquake.

Normal Faults

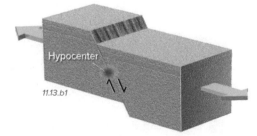

In a normal fault, the block above the fault moves down the dip of the fault relative to the block below the fault. The movement forms grooves on the fault surface. The crust is stretched horizontally, so earthquakes related to normal faults are common along divergent plate boundaries, such as oceanic spreading centers and in continental rifts.

Reverse and Thrust Faults

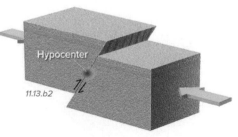

Many large earthquakes are generated along reverse faults, especially the gently dipping variety called thrust faults. In thrust and reverse faults, the block above the fault moves up with respect to the block below the fault. Such faults are formed by compressional forces, like those associated with subduction zones and continental collisions.

Strike-Slip Faults

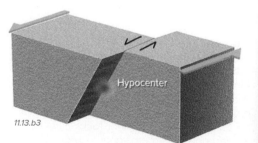

In strike-slip faults, the two sides of the fault slip horizontally past each other. This can generate large earthquakes. Most strike-slip faults are near vertical, but some have moderate dips. The largest strike-slip faults are along plate boundaries, like the San Andreas fault in California.

C How Do Volcanoes and Magma Cause Earthquakes?

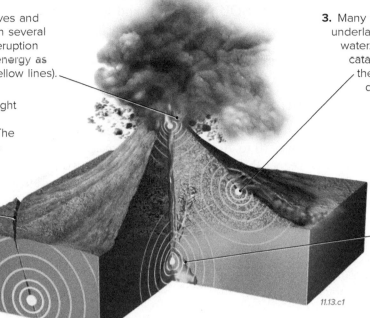

1. Volcanoes generate seismic waves and cause the ground to shake through several processes. An explosive volcanic eruption causes compression, transmitting energy as seismic waves (shown here with yellow lines).

2. Volcanoes add tremendous weight to the crust, and this loading can lead to faulting and earthquakes. The fault shown here, which caused an earthquake at depth, has faulted down the volcano relative to its surroundings. Fault movements and associated earthquakes would accompany caldera collapse, if it occurs.

3. Many volcanoes have steep, unstable slopes underlain by rocks altered and weakened by hot water. The flanks of such volcanoes can fall apart catastrophically, causing landslides that shake the ground as they break away and travel down the flank of the volcano. Numerous small earthquakes also occur as the rocks break, prior to the actual landslide.

4. As magma moves beneath a volcano, it can push rocks out of the way, causing earthquakes. Magma can push rocks sideways or open space by fracturing adjacent rocks and uplifting the Earth's surface. The emplacement of magma can cause a series of small and distinctive earthquakes, called *volcanic tremors*. All types of magma-related earthquakes are closely monitored by seismologists (scientists who study earthquakes), because they can warn us of an impending eruption.

11.13.c1

D What Are Some Other Causes of Seismic Waves?

Landslides

Catastrophic landslides, whether on land or beneath water, cause ground shaking. On the Big Island of Hawaii, lava flows form new land that can become unstable and suddenly collapse into the ocean. Seismometers at the nearby Hawaii Volcanoes National Park often record seismic waves caused by such landslides and by fractures opening up on land in response to sliding of the land toward the sea.

11.13.d1

Explosions

Mine blasts and nuclear explosions produce seismic waves measurable by distant seismic instruments. Monitoring compliance with nuclear test-ban treaties is done in part using a worldwide array of seismic instruments. These instruments recorded a nuclear bomb exploded by India in 1998. Seismic waves generated by a blast such as this are more abrupt than those caused by a natural earthquake.

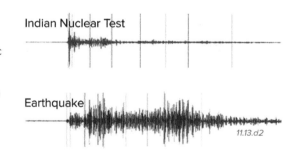

11.13.d2

Earthquakes Caused by Humans

Humans can cause earthquakes in several ways. Reservoirs built to store water fill rapidly and load the crust, which responds by flexing and faulting. After Lake Mead, behind Hoover Dam in Nevada and Arizona, was filled, hundreds of moderate earthquakes occurred under the reservoir between 1934 and 1944. Similarly, very shallow (less than 3 km deep) earthquakes occur near Monticello Reservoir in South Carolina. Worldwide, scientists have identified dozens of cases of earthquakes associated with dams. Most of the seismic activity occurs during the initial filling of a reservoir by water, which adds stress to underlying rocks.

Humans have also caused earthquakes by injecting wastewater underground via a deep well at the Rocky Mountain Arsenal northeast of Denver. This caused more than a thousand small earthquakes and two magnitude 5 earthquakes, which caused minor damage nearby. When the waste injection stopped and some waste was pumped back out of the ground, the number of earthquakes decreased. Similar earthquakes are currently being caused by the underground injection of wastewaters used to produce high-enough water pressures to fracture underground rocks (*hydraulic fracturing*, or "fracking"), in order to produce oil and natural gas from shales and similar rocks.

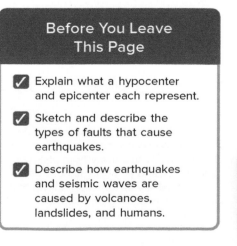

Before You Leave This Page

✔ Explain what a hypocenter and epicenter each represent.

✔ Sketch and describe the types of faults that cause earthquakes.

✔ Describe how earthquakes and seismic waves are caused by volcanoes, landslides, and humans.

11.13

11.14 Where Do Most Earthquakes Occur?

MOST EARTHQUAKES OCCUR ALONG PLATE BOUNDARIES, and maps of earthquake locations outline Earth's main tectonic plates. There are some regions, however, where seismicity (earthquake activity) is more widespread, reaching far away from plate boundaries and into the middle of continents. Where do earthquakes occur, and how can we explain the distribution of earthquakes across the planet?

A Where Do Earthquakes Occur in the Eastern Hemisphere?

This map and the one on the next page show the worldwide distribution of earthquake epicenters, colored according to depth. Yellow dots represent shallow earthquakes (0 to 70 km), green dots mark earthquakes with intermediate depths (70 to 300 km), and red dots indicate earthquakes deeper than 300 km. Examine these two maps and observe how earthquakes are distributed. Note how this distribution compares to other features, such as edges of continents, mid-ocean ridges, sites of subduction, and continental collisions.

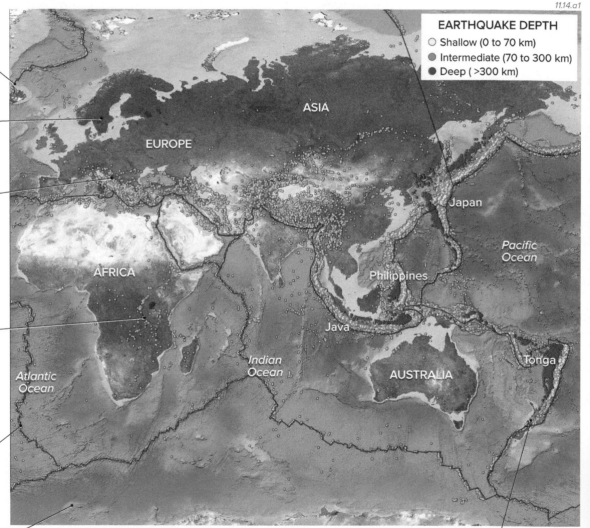

11.14.a1

EARTHQUAKE DEPTH
- ○ Shallow (0 to 70 km)
- ◒ Intermediate (70 to 300 km)
- ● Deep (>300 km)

1. Most earthquakes occur in narrow belts that coincide with plate boundaries. The belt of earthquakes north of Iceland marks a divergent plate boundary along a mid-ocean ridge.

2. Between the belts of earthquakes are some large regions with relatively few earthquakes, like the northern part of Europe.

3. A seismically active zone stretches along southern Europe and continues eastward into Asia. This activity follows a series of mostly convergent boundaries, including continental collisions, that are occurring from the Mediterranean Sea to Tibet. What kinds of faults are most likely dominant here?

4. A diffuse zone of seismic activity cuts across eastern Africa, following the East African Rift, a region of elevated topography, active volcanism, and faulted blocks. This region is a continental rift within Africa.

5. Mid-ocean ridges, such ones south of Africa, only have shallow earthquakes (only yellow dots on this map). In these locations, rifting and spreading of two oceanic plates produces faulting and magmatic activity, which in turn both cause earthquakes. Which type of fault is most likely dominant here?

6. Large regions of the ocean lack significant seismicity because they are not near a plate boundary. Some seismicity in the oceans occurs away from plate boundaries and is mostly related to volcanic activity or to minor faulting that accompanies cooling and subsidence of the oceanic lithosphere.

7. Seismicity is concentrated in the western Pacific, with the main zones of seismicity being associated with oceanic trenches and volcanic islands near Tonga, Java, the Philippines, and Japan. These zones run parallel to oceanic trenches and mark subduction zones. Worldwide, approximately 90% of significant earthquakes occur along subduction zones. Subduction zones have shallow, intermediate-depth, and deep earthquakes, and they are essentially the only places with deep and intermediate-depth earthquakes. Note that there is a consistent pattern of shallow earthquakes (yellow dots) close to the trench and progressively deeper earthquakes farther away (green and then red dots, moving away from the trench). What do you think causes this pattern? We address this topic on the next page.

B Where Do Earthquakes Occur in the Western Hemisphere?

The map below shows the Western Hemisphere, including North and South America and adjacent parts of the Pacific and Atlantic oceans. Observe the distribution of earthquakes, especially how earthquakes compare to the edges of continents, mid-ocean ridges, and sites of subduction. Do any occur where you live?

1. A belt of strong seismic activity occurs along the southern part of mainland Alaska and the Aleutian Islands to the west. This belt parallels an oceanic trench and contains shallow and intermediate-depth earthquakes. It marks a subduction zone where the oceanic Pacific plate subducts beneath Alaska. This belt is a continuation of the activity in Japan and the western Pacific (i.e., the Pacific Ring of Fire).

2. Earthquakes follow the west coast of North America and extend into the mountains of the West. These earthquakes reflect diverse types of faulting (strike-slip, normal, and thrust faulting), as well as volcanism.

3. Intense seismic activity follows the western coasts of Mexico, Central America, and South America. Included in this activity are deep and intermediate-depth earthquakes along subduction zones, especially the one beneath western South America. Shallow earthquakes are closer to the trench, and deep ones are farther away.

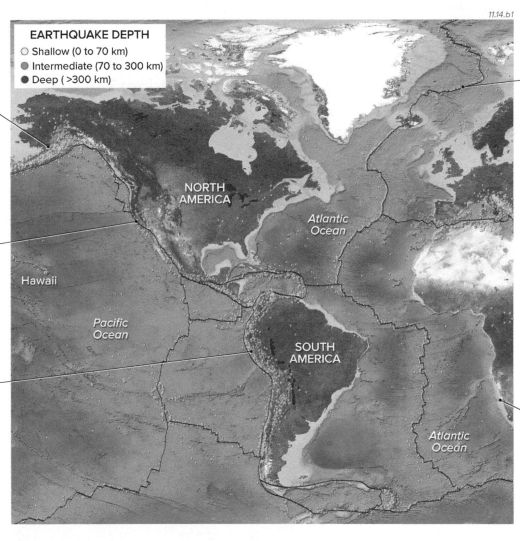

11.14.b1

6. A belt of shallow earthquakes follows the Mid-Atlantic Ridge, where the North and South American plates spread westward from the Eurasian and African plates. The pattern of earthquakes mimics the shape of the flanking continents, and the shape of the mid-ocean ridge is largely inherited from the time (200 to 80 million years ago) when these continents rifted apart.

7. Note the relative lack of seismicity along the west coast of Africa and east coasts of North and South America.

4. A deep oceanic trench flanks the western coast of South America, marking a subduction zone where oceanic plates subduct beneath the western side of the continent. Observe the pattern of earthquakes for this area on the large map above before examining the figure to the right.

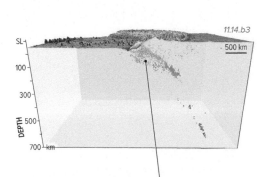

11.14.b3

5. In a side view, subduction-related earthquakes, shown as dots, are shallower to the west (near the trench) and deeper to the east, recording the descent of the oceanic plate. This pattern follows, and helps define, the position of the subducted slab, which is inclined from the shallow to the deep earthquakes.

Before You Leave This Page

☑ Summarize some generalizations about the distribution of earthquakes, especially the relationship to plate boundaries.

☑ Sketch and explain how you could recognize a subduction zone from a map showing earthquakes colored according to depth, and how you could infer which way the subduction zone is inclined.

11.14

11.15 What Causes Earthquakes Along Plate Boundaries and Within Plates?

DIFFERENT TECTONIC SETTINGS have different types of earthquakes. Earthquakes formed along a plate boundary generally record the relative movement along this boundary (divergent, convergent, or transform) or indicate other processes, such as magmatism, associated with the boundary. Other earthquakes occur in the middle of plates, for example those caused by continental rifting.

A How Are Earthquakes Related to Mid-Ocean Ridges?

Earthquakes are common along mid-ocean ridges, where two oceanic plates spread apart. Most of these earthquakes form at relatively shallow depths and are small or moderate in size. Some earthquakes result from spreading of the plates, whereas others record motion as the two plates slide by one another on transform faults.

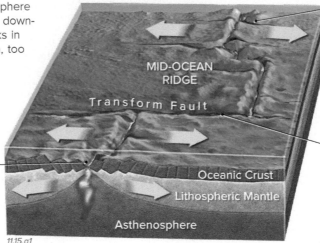

1. Seafloor spreading forms new oceanic lithosphere that is very hot and thin. Stress levels increase downward in Earth, but in mid-ocean ridges the rocks in the lithosphere get very hot at a shallow depth, too hot to fracture (they flow instead). As a result, earthquakes along mid-ocean ridges are relatively small and shallow, with hypocenters less than about 20 km (12 mi) deep.

2. Many earthquakes occur along the axis of the mid-ocean ridge, where spreading and slip along normal faults downdrop blocks along the narrow rift. Numerous small earthquakes also occur due to intrusion of magma into fissures.

3. As the newly created plate moves away from the ridge, it cools, subsides, and bends. The stress caused by the bending forms steep faults, which are associated with relatively small earthquakes.

4. Transform faults are a type of strike-slip fault, but one that is along a plate boundary. Near mid-ocean ridges, strike-slip earthquakes occur along transform faults that link adjacent segments of the spreading center. Largely because of the typically thin lithosphere, earthquakes along these oceanic transform faults are shallow. Shallow earthquakes are also typically small in magnitude.

11.15.a1

Labels in figure: MID-OCEAN RIDGE; Transform Fault; Oceanic Crust; Lithospheric Mantle; Asthenosphere

B How Are Earthquakes Related to Subduction Zones?

A subduction zone, where an oceanic plate underthrusts beneath another oceanic plate or a continental plate, undergoes compression and shearing along the plate boundary. It can produce very large earthquakes along and above the boundary.

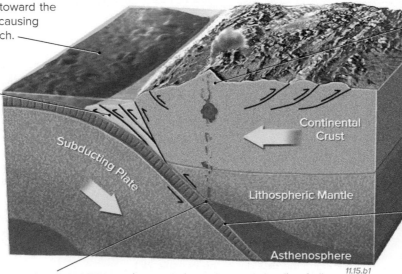

1. As the oceanic plate moves toward the trench, it is bent and stressed, causing earthquakes in front of the trench.

2. Larger earthquakes occur in the wedge of material that is scraped off the downgoing plate. Shearing within the wedge causes slip and earthquakes along numerous thrust faults.

3. Large earthquakes occur along the entire contact between the subducting plate and the overriding plate. The plate boundary is a huge thrust fault called a *megathrust*. Earthquakes along megathrusts are among the most damaging and deadly of all earthquakes. During large earthquakes, the megathrust can rupture upward all the way to the seafloor, displacing the seafloor and unleashing destructive waves in the ocean.

5. Earthquakes can also occur within the overriding plate due to movement of magma and from volcanic eruptions. Compressive stresses associated with plate convergence can cause thrust faulting behind the magmatic arc.

4. The downgoing oceanic plate is relatively cold, and so it continues to produce earthquakes from shearing along the boundary, from downward-pulling forces on the sinking slab, and from abrupt changes in mineralogy. Subduction zones are typically the only place in the world producing deep earthquakes, as deep as 700 km (430 mi). Below 700 km, the plate is too hot to cause earthquakes but flows slowly instead.

Labels in figure: Subducting Plate; Continental Crust; Lithospheric Mantle; Asthenosphere

11.15.b1

C How Are Earthquakes Related to Continental Collisions?

During continental collisions, one continental plate underthrusts beneath another. Collisions can be extremely complex, as different parts collide at different times and rates. Collisions cause large tectonic stresses that shear and cause faulting in a broad zone within the overriding and underthrusting plates. As a result, earthquakes are widely distributed along the collision zone, as represented by earthquakes in China and nearby parts of Asia.

1. Large thrust faults form near the plate boundary in both the overriding and underthrusting plates, causing large but shallow earthquakes. These earthquakes can be deadly in populated areas, such as India, Nepal, Pakistan, and Iran.

2. Large, deadly earthquakes are produced along the plate boundary, or megathrust, between the two continental plates.

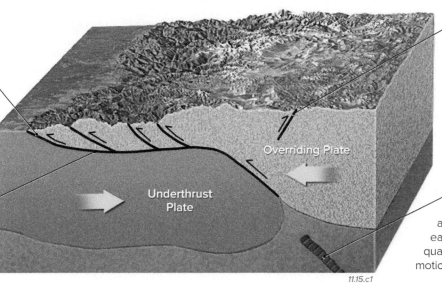

11.15.c1

3. Thrust faults also form within both continental plates, causing moderately large earthquakes. The immense stresses associated with a collision can reactivate older faults within the interior of either continent, as is presently occurring in Tibet and China. Strike-slip faults and normal faults may be generated as entire regions are stressed by the collision zone or are shoved or sheared out of the way.

4. The oceanic plate subducted prior to the collision is detached, so actual subduction and associated earthquakes stop. A few deep earthquakes have resulted from the sinking motion of such detached slabs.

D How Are Earthquakes Generated Within Continents?

In addition to continental collisions, earthquakes occur in other tectonic settings within continents. These settings include continental rifts, continental transform faults, magmatic areas, and reactivated preexisting faults.

1. Continental rifts generally produce normal faults, whether the rift is a plate boundary or is within a continental plate. The normal faults downdrop fault blocks into the rift (a graben), causing *normal-fault earthquakes.* Such earthquakes are typically moderate in size.

2. A transform fault can cut through a continent, moving one piece of crust past another. The strike-slip motion causes earthquakes that are mostly shallower than 20 to 30 km (10 to 20 mi), but some of these strike-slip earthquakes can be quite large. The San Andreas fault of California is the best-known example of a continental transform fault, but large, destructive earthquakes also occur along continental transform faults in Turkey, Pakistan, Nicaragua, and New Zealand.

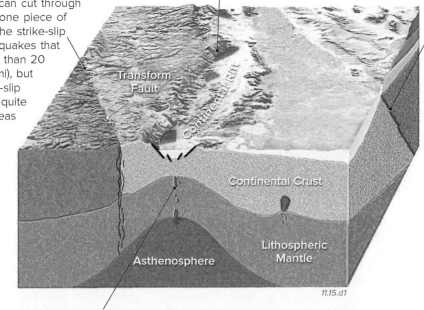

11.15.d1

3. Intrusion of magma (shown here in red) within a plate can cause small earthquakes as the magma moves and creates openings in the rock. Moving magma can produce distinctive earthquakes, which are unlike those produced by movement along faults. Heat from the magma can substantially weaken the crust, causing even more rifting and seismic activity.

4. Preexisting faults in the crust can readjust and move as the continental plate becomes older and is subjected to new stresses, such as from distant plate boundaries. Reactivation of these structures can occur in the interior of a plate and produce large earthquakes, including several large earthquakes that struck New Madrid, Missouri—in the interior of North America—in 1811.

Before You Leave This Page

☑ Sketch and explain earthquakes along mid-ocean ridges, including oceanic transform faults.

☑ Sketch and explain earthquakes associated with subduction zones, including earthquakes in the overriding plate.

☑ Summarize how continental collisions cause earthquakes.

☑ Describe some settings in which earthquakes can occur within a continental plate.

11.15

11.16 How Do Earthquakes Cause Damage?

MANY GEOSCIENTISTS SAY that "earthquakes don't kill people, buildings do." This is because most deaths from earthquakes are caused by the collapse of buildings or other structures, or by the large earthquake-generated waves called *tsunami*. Destruction and collapse may result from ground shaking during an earthquake, or it can occur later due to fires, floods, or other indirect impacts caused by the earthquake.

What Destruction Can Arise from Shaking Due to Seismic Waves?

Direct damage from an earthquake results when the ground shakes because of seismic waves, especially surface waves near the epicenter of the earthquake. Damage can also be due to *secondary effects* (also called *indirect impacts*), such as fires and flooding, that are triggered by the earthquake. In the example below, most of the damage is direct damage.

1. Mountainous regions that undergo ground shaking may experience landslides, rock falls, and other slope failures.

2. The ground can rupture along parts of the fault that slip during an earthquake, or from shaking of unconsolidated materials. The fault scarp and other cracks can destroy buildings and roads.

3. Damage to structures from shaking depends on the type of construction and type of ground motion. Concrete and masonry structures are rigid and do not flex easily. Thus, they are more susceptible to damage than wood or steel structures, which are more flexible. In this area, a flexible metal bridge in the center of the city survived the earthquake, but a concrete one downstream did not. Horizontal motions tend to be more damaging to buildings than vertical ones, because buildings are naturally designed to withstand vertical stresses.

4. *A tsunami* is a giant wave that can rapidly travel across the ocean. An earthquake that occurs undersea or along coastal areas can generate a tsunami, which can cause damage along shorelines thousands of kilometers away.

6. The amount of destruction to buildings and other structures also depends on the type of underlying soils and other materials. Some materials amplify ground shaking, and others dampen it. Shaking of unconsolidated, water-saturated soil and sediment causes grains to lose grain-to-grain contact, and the material loses most of its strength and begins to flow, a process called *liquefaction*. This can destroy anything built on top.

5. *Aftershocks* are smaller earthquakes that occur after the main earthquake, but in the same area. They are due to movement on smaller faults stressed by the larger earthquake. Aftershocks are very dangerous because they can collapse structures already damaged by the main shock. Aftershocks after a tsunami can cause widespread panic.

11.16.a1

Magnitude and Intensity of Earthquakes

In discussing how strong an earthquake is, we talk about its *magnitude*. The *Richter magnitude* is a measure of an earthquake's size, and is calculated by observing how much the ground shaking displaces a *seismograph* (an instrument that records ground shaking). The Richter scale is logarithmic, with ground motion increasing ten-fold from a Richter magnitude 4 to a 5, from a 5 to a 6, and so on. The amount of energy released increases more than 30 times for each increase in magnitude, so a magnitude 8 releases more than 30 times more energy than a magnitude 7. Another common measure of earthquake energy is *moment magnitude* (signified as Mw), which is expressed as numbers (e.g., magnitude 7.4) and calculated from the amount of slip on the fault and the size of fault area that slipped.

An alternative way to characterize the strength of an earthquake, one that is very useful in studying earthquakes that occurred before we had seismometers, is the *Modified Mercalli Intensity Scale*, abbreviated as MMI, which describes the effects of shaking in everyday terms on a scale from I to XII. A value of I represents a barely felt earthquake. An MMI of III indicates that the quake was felt strongly by people indoors, and a V is felt by nearly everyone, breaks some dishes and windows, and overturns unstable objects, like bookcases. Even stronger ground shaking is indicated by MMI X and XI, where some well-built wooden structures and most masonry structures are destroyed, bridges collapse, and railroad rails are bent. A value of XII indicates complete destruction of buildings, with visible surface waves throwing objects into the air. The Richter scale, moment magnitude, and MMI are all useful in characterizing earthquakes.

B   What Destruction Can Happen Following an Earthquake?

Some earthquake damage occurs from secondary effects that are triggered by the earthquake.

11.16.b1 Northridge, CA

11.16.b2 Sumatra

11.16.b3 San Fernando, CA

Fire is one of the main causes of destruction after an earthquake. Natural gas lines may rupture, causing explosions and fires. The problem is compounded if water lines also break during the earthquake, limiting the amount of water available to extinguish fires.

Earthquakes may cause both uplift and subsidence of the land surface by more than 10 m (30 ft). Subsidence, such as occurred during the 2004 Sumatra earthquake, can cause areas that had been dry land before the earthquake to become inundated by seawater, flooding buildings and trees.

Flooding may occur due to failure of dams as a result of ground rupturing, subsidence, or liquefaction. Near Los Angeles, 80,000 people were evacuated because of damage to nearby dams during the 1971 San Fernando earthquake (Mw 6.7).

C   How Are Tsunami Generated?

A tsunami is a wave generated by a disturbance in the sea or a lake. It is generated by abrupt changes in water level in one area relative to another, which occurs when seafloor is uplifted or downdropped during an earthquake. A large tsunami requires a large disturbance of the seafloor, which requires a major rupture of a fault across the seafloor. Typical ocean waves only involve the uppermost sea, but a tsunami involves a massive thickness of water. Tsunami can also form from volcanic eruptions and large submarine landslides.

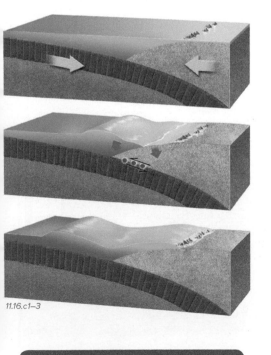
11.16.c1–3

1. Subduction-zone megathrusts can lock for long periods of time, causing the seafloor above the overriding plate to bulge, strain, and flex up or down as it accommodates the forces of convergence (◄). This upward and downward flexing is typically most prominent near the trench.

2. When the megathrust ruptures in an earthquake (along the red asterisks), the bulging plate changes shape catastrophically. The water above the plate is lifted up, forming a ridge of higher water (◄). A tsunami radiates away from the disturbance, traveling at speeds between 600 and 800 km/hr (370–500 mi/hr). In deep water, the wave energy is distributed over the entire water depth, forming a wave only a meter or so high but more than 700 km (435 mi) across (in wavelength). If you were in the open ocean, you probably would not notice its passing.

3. As the wave approaches the shore and encounters shallower water, the velocity of the front of the wave decreases, causing the following water to pile up in a higher wave (◄). Near shore, the tsunami becomes a massive, thick wave, like the front wall of a plateau of water. Most tsunami consist of a series of wave crests with intervening troughs, and the approach of a trough can cause the sea to retreat from the shoreline just prior to a tsunami, or between two crests. A sudden retreat of the ocean along a coastline is a warning sign that a tsunami may be coming.

4. The 1883 eruption of Krakatau in Indonesia, and the collapse of its immense caldera, generated a series of huge tsunami that killed 36,000 people along coastlines. A single catastrophic volcanic explosion produced the loudest sound ever heard, and most of Krakatau Island was demolished. The tsunami was as high as 40 m (more than 130 ft), and some effects of the tsunami were recorded 7,000 km away! This painting (▶) is a representation of the rising wall of water as it approached the shore.

11.16.c4

Before You Leave This Page

☑ Describe how earthquakes can cause destruction, both during and after the main earthquake.

☑ Describe the different mechanisms by which tsunami are generated.

11.16

11.17 What Were Some Recent Large Earthquakes?

THE WORLD HAS ENDURED a number of large and tragic earthquakes. These earthquakes have struck a collection of geographically and culturally diverse places, causing many deaths and extensive damage. Most large earthquakes have occurred along or near plate boundaries, especially along subduction zones. Some happened on faults that are close to, but not actually on, a plate boundary. Major earthquakes occurred recently in Japan, Haiti, and New Zealand. Each was destructive, but in different ways.

A What Happened During the Catastrophic Tōhoku Earthquake of 2011?

A huge and catastrophic earthquake struck off the coast of northeastern Japan on March 11, 2011. It had an extremely large magnitude (9.0), making it one of the five largest earthquakes ever measured. Ground shaking during the earthquake caused extensive damage, especially nearest to the epicenter. The earthquake and resulting tsunami destroyed or damaged over a million buildings, left 24,000 dead or missing, and caused more than $230 billion in economic damages, making it the most expensive natural disaster in history.

1. The epicenter of the earthquake, shown here as a red dot, was on the seafloor east of Japan, along an oceanic trench that marks the plate boundary (marked with the red lines with teeth) where the Pacific plate subducts beneath Japan. The earthquake occurred at depth along the subduction zone (plate boundary) that dips beneath Japan. The main hypocenter was 32 km deep, so this is classified as a shallow earthquake. In many regards, it was a typical, but very large, megathrust earthquake. It was followed by many aftershocks (shown as small yellow dots), which show how large an area of the plate boundary slipped during and after the main earthquake. Some of the aftershocks were large (magnitude 6), causing additional damage to already weakened buildings and raising fears of a new tsunami.

250 km

11.17.a1

2. The earthquake represented a sudden upward movement of rocks at depth. Based on the distribution of aftershocks and other seismic data, the fault slipped over an area of 300 km parallel to the trench and by 150 km perpendicular to the trench, nearly spanning the entire distance between the trench and the coastline. The fault rupture grew upward from the hypocenter toward the seafloor, uplifting a large swath of seafloor by up to 3 m. The displaced water formed into an extremely damaging tsunami that locally was higher than 10 m (33 feet), or perhaps several times that height along small stretches of the coast. A tsunami-warning system saved many residents, but thousands of people were trapped by the fast-moving, rising waters that spread far inland.

11.17.a2 Honshu, Japan

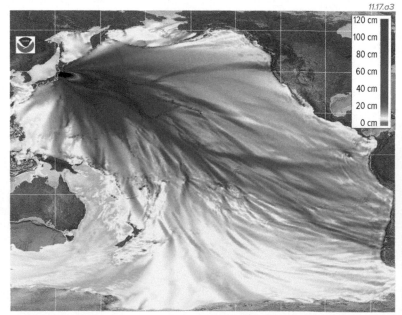

11.17.a3

120 cm
100 cm
80 cm
60 cm
40 cm
20 cm
0 cm

3. News reports showed dramatic footage of seawater spilling onto the land (▲) and then gradually rising higher and higher, flooding low areas and destroying most buildings in its path. As the tsunami moved farther inland, it became a slower moving wall of sludge containing automobiles, boats, parts of buildings, and other debris. When it was over, the tsunami had inundated nearly 500 km² of Japan. It destroyed entire towns and heavily damaged the Fukushima Nuclear Power Plant, which had a reactor-core meltdown because the tsunami destroyed the facility's cooling system and backup power generation.

4. This map (▲) shows the forecasted maximum wave amplitudes modeled for the 2011 earthquake by tsunami forecasters at NOAA. As predicted, a tsunami of measurable height reached most Pacific shorelines, including Hawaii and the west coasts of North and South America.

B How Were Recent Earthquakes in Haiti and New Zealand Similar and Different?

Large earthquakes struck Haiti in 2010 and New Zealand in 2010 and 2011. The Haiti earthquake was a magnitude 7.0, the 2010 Canterbury (New Zealand) earthquake was magnitude 7.1, and the 2011 Christchurch (New Zealand) earthquake was magnitude 6.3. All of these earthquakes were near, but not on, a plate boundary that mostly consists of strike-slip (transform) movement. However, the three earthquakes varied greatly in the amount of damage and death they inflicted. Why is this so? Let's examine these three earthquakes.

Haiti, 2010

1. The Haiti earthquake occurred on land, 25 km west of the capital, Port-au-Prince. The epicenter (shown in red) is near, but not on, an active strike-slip fault that cuts east-west across the country. On this map, aftershocks are yellow dots, green lines are strike-slip faults, and red lines have thrust movement. The area is within a zone of complex faulting near the fault boundary between the Caribbean plate to the south and the North American plate to the north. The earthquake flattened more than 300,000 buildings in and around Port-au-Prince and killed perhaps 200,000 people. The extreme poverty of the country, combined with a devastated infrastructure and an inefficient response by relief agencies, led to hunger, suffering, and epidemics after the quake.

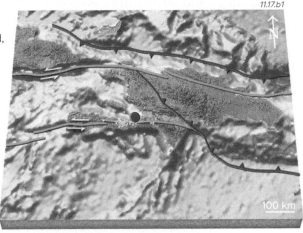

2. Most deaths during the Haiti earthquake were caused by collapsing buildings. Haiti is the most impoverished country in the Western Hemisphere, and most houses and buildings were constructed poorly. Most fared worse than the National Palace (▲), which was only partly collapsed.

New Zealand 2010, 2011

3. The two main earthquakes in New Zealand had epicenters (shown as red dots) on a broad coastal plain that lies east of the rugged Southern Alps and the Alpine fault, a mostly strike-slip plate boundary that runs the length of the island. Both earthquakes were very shallow (less than 15 km), were not on the plate boundary, and occurred on two different, but probably related, faults. Both were followed by abundant aftershocks (shown in yellow and orange), but only the 2010 quake ruptured the surface and was on a known fault (green line on map).

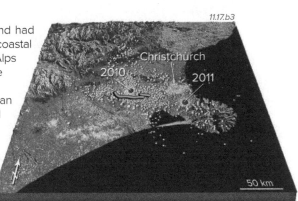

4. The magnitude-7.1 Canterbury quake caused moderate damage and only injured two people. The 2011 Christchurch quake was smaller but killed nearly 200 people. It destroyed or damaged 100,000 buildings, some by liquefaction of the soil and associated expulsion of water from the ground (▲). The main difference in the amount of destruction was that the 2011 epicenter was very near Christchurch, New Zealand's second largest city, and the quake had more vertical motion, destroying already weakened buildings.

Deadly Earthquakes

Earthquakes kill about 10,000 people per year on average. Most earthquake-related deaths are due to collapse of poorly built structures in cities and villages. Earthquake-generated tsunami account for a large part of the destruction. The table to the right shows some deadly earthquake events. The highest death tolls are due to a deadly combination of high population densities, substandard construction practices, and being situated along subduction zones or other high-risk areas. Earthquakes discussed in this chapter are not included on this table. Mw is moment magnitude. Deadly earthquakes strike each year, such as a 2015 one in Nepal (8,000 deaths).

Fatalities	Mw	Year	Location
830,000	8	1556	Shaanxi, China
11,000	6.9	1857	Naples, Italy
70,000	7.2	1908	Messina, Italy
200,000	7.8	1920	Ningxia, China
143,000	7.9	1923	Kanto, Japan
200,000	7.6	1927	Tsinghai, China
32,700	7.8	1939	Erzincan, Turkey
66,000	7.9	1970	Colombia
23,000	7.5	1976	Guatemala
242,000	7.5	1976	Tangshan, China
31,000	6.6	2003	Bam, Iran
88,000	7.9	2008	Sichuan, China

Before You Leave This Page

☑ Briefly summarize the four earthquakes presented here, including their tectonic settings and how each caused destruction.

☑ Discuss why the amount of damage and death varied among the quakes.

11.17

11.18 What Happened During the Great Alaskan Earthquake of 1964?

THE SOUTHERN COAST OF ALASKA experienced one of the world's largest earthquakes in 1964. The moment magnitude 9.2 earthquake, which is the strongest to have ever struck North America, destroyed buildings, triggered massive landslides, and unleashed a tsunami that caused damage and deaths from Alaska to California. The nearby region also has active volcanoes, further evidence that this is a tectonically active area.

A What Types of Damage Did the Earthquake Cause?

1. The earthquake occurred along the southern coast but was felt throughout Alaska, except for the far north coast. Ground shaking destroyed buildings and generated huge landslides of rock and soil. This earthquake-triggered, dark, rocky landslide (▼) covered parts of the white Sherman Glacier.

2. The blue line on the map marks the limit of shaking-caused cracking of the ground and ice during the earthquake. The red line closer to the epicenter outlines the region where property damage occurred.

11.18.a2 Southern Alaska

11.18.a1

Limit of Ground and Ice Fissuring

UNITED STATES

CANADA

ALASKA

Barrow

Fairbanks

Limit of Property Damage

Anchorage

Homer

200 km

Kodiak

11.18.a3 Anchorage, AK

3. Parts of downtown Anchorage were demolished (◄) when shaking caused the underlying land to slip and collapse. Some buildings sank so much that their second stories were level with the ground. Severe damage occurred in the Turnagain Heights area of Anchorage, where a layer of weak clay liquefied, carrying away shattered houses (►).

4. The epicenter of the earthquake was along the southern coast of Alaska, between the cities of Anchorage and Valdez. The large earthquake began at depths of 20 to 30 km (12 to 19 mi). Based on the wide distribution of about 600 aftershocks, seismologists estimate that the earthquake ruptured a fault surface that was over 900 km (560 mi) long and 250 km (160 mi) wide. The earthquake occurred on a thrust fault that dips from the Aleutian trench gently north and northwestward beneath Alaska.

11.18.a4 Anchorage, AK

B   What Happened in the Sea During the Earthquake?

Because it occurred along the coast, the earthquake also caused (1) faulting and uplift of the seafloor, (2) huge waves from landslides, and (3) a tsunami that struck the coasts of Alaska, British Columbia, Washington, Oregon, California, Hawaii, and even Japan.

The main fault that caused the earthquake did not break the land surface, but two subsidiary faults did. One fault (▶) cut a notch into a mountain and uplifted the seafloor 4 to 5 m (15 ft). The white material on the uplifted (left) side of the fault consists of calcareous marine organisms that were below sea level before the earthquake. The maximum observed uplift was 11.5 m (38 ft). Other areas subsided as much as 6 m (20 ft) during the earthquake, flooding docks, oil tanks, and buildings along the coast.

11.18.b1 Southern Alaska

11.18.b2 Kodiak Island, AK

Faulting uplifted a large area of seafloor off the southern coast of Alaska, sending a large tsunami out across the sea and up the many bays and inlets along the coast. The highest tsunami recorded was 67 m (220 ft), in a bay near Valdez. The photograph above (▲) shows damage done to Kodiak Island by a wave 6 m (20 ft) high. The tsunami killed 106 people in Alaska and 17 more in Oregon and California.

C   What Are Some Other Manifestations of Tectonics?

1. Immediately after the earthquake, a team of scientists and surveyors investigated the coastline, measuring uplift and subsidence at hundreds of sites. They plotted and contoured the elevation change measurements on a detailed map, which they then generalized into a summary map (▶) that identifies broad zones of uplift and subsidence. As a result of the earthquake, a region of more than 250,000 km² (100,000 mi²) changed elevation, either up or down. Repetitions of the earthquake, hundreds of times over millions of years, could uplift a new mountain range where the sea is now and form a basin where there is currently a ridge. Earth's elevations and landscapes are dynamic, always changing.

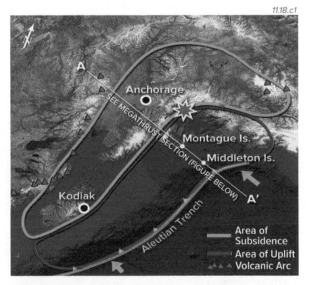

11.18.c1

Area of Subsidence
Area of Uplift
▲▲▲ Volcanic Arc

11.18.c3 Mount Redoubt, AK

3. Southern Alaska also has a number of active volcanoes, like Mount Redoubt (▲), and these continue out across the ocean to form the Aleutian Islands, an active island arc. The volcanoes display large, somewhat symmetrical shapes with steep slopes—they are dangerous composite volcanoes. They are related to subduction of an oceanic plate beneath Alaska, the same process that caused the Great Alaskan Earthquake of 1964.

2. The surveying and other studies along the coast led to the recognition that the earthquake occurred along a thrust fault beneath the southern shoreline of Alaska. This study interpreted the earthquake as resulting from oceanic material being pushed beneath Alaska (▶), along what today we call a megathrust. These conclusions, in the 1960s, were drawn before development of the theory of plate tectonics. In fact, the studies of this earthquake, using a variety of geographic tools, like surveying, preparation of contour maps, and surveys of changes to the landscape, was an important step in the development of plate tectonics. Today, geographers, geologists, and other scientists would also use GPS devices, GIS, and satellite data.

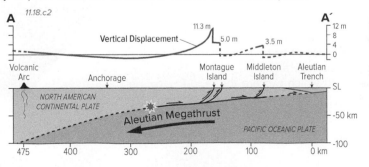

11.18.c2

Vertical Displacement   11.3 m   5.0 m   3.5 m

Volcanic Arc   Anchorage   Montague Island   Middleton Island   Aleutian Trench

NORTH AMERICAN CONTINENTAL PLATE

Aleutian Megathrust

PACIFIC OCEANIC PLATE

475   400   300   200   100   0 km

Before You Leave This Page

☑ Summarize events associated with the Alaskan earthquake, including effects on land and sea, and how studies of this area helped lead to the theory of plate tectonics.

11.18

Where Is the Safest Place to Live in This Area?

THIS COASTAL REGION CONTAINS TWO FAULTS, an active volcano, and several steep-sided mountains prone to landslides. Any of these features can represent a hazard. The area has experienced some recent ground shaking due to some moderately strong earthquakes that occurred in the area. From observations about different features in the landscape, you will decide what hazards this area poses to each of the small towns nearby.

Goals of This Exercise:

- Examine the large illustration and read the text boxes describing the types of features that are present.
- Identify what potential hazards are in this area, such as which features could produce earthquakes. For each hazard, evaluate the potential risk that hazard poses to each town.
- Decide which town you think is the safest from earthquakes, volcanoes, and other hazards caused by tectonic activity, and justify your decision with supporting evidence. There is not a single correct answer.

Procedures

The area has several small towns and diverse landscape features. The area recently experienced a series of moderately strong (moment magnitude 5 and 6) earthquakes. Use the available information to complete the following steps and enter your answers in the appropriate places on the worksheet or online.

1. Observe the features shown on the three-dimensional perspective. Read the text associated with each location, and think about what each statement implies about hazards from an earthquake, volcanic eruption, or any other hazard triggered by a tectonic event.

2. From the characteristics of the volcano, interpret what type of volcano it is and what types of eruptions occur from this type of volcano. Next, propose the hazard each type of eruption poses for each town, and identify which towns are most at risk for damage from a volcanic eruption and other volcano-related activity. Not all areas will have the same degree of hazard; for example, some towns are closer to the volcano than others.

3. Examine the various landscape features and interpret which ones are capable of generating earthquakes. Next, decide which of these potential sources of earthquakes could potentially produce the largest earthquake.

4. Use the information about the topographic features and other aspects of the landscape to interpret how a feature or area might respond if a large earthquake occurred. What types of hazards do earthquakes pose for each town? From these considerations, decide which three towns are the least safe and which two are the safest for this type of earthquake. There is not necessarily one right answer, so explain and justify your logic on the worksheet, if asked to do so by your instructor.

1. Along one part of the coastline, there is a thin, steep beach, called *Roundstone Beach*, that rises upward to some nearby small hills. The seafloor offshore is also fairly steep as it drops off toward the trench.

2. The town of *Sandpoint* is built upon land that was reclaimed from the sea by piling up loose rocks and beach sand until the area was above sea level. It received the most damage from the recent earthquake, with reports of moderate shaking that knocked over some bookcases and broke some windows and dishes. Some damage occurred to boats, but the earthquake occurred at night, so no one knows exactly how this happened.

12. There is a deep ocean trench along the edge of the continent. Ocean drilling encountered fault-bounded slices of oceanic sediment.

11. Offshore is a coral reef that blocks larger waves, creating a quiet lagoon between the reef and the shore.

Roundstone Beach

Oceanic Trench

Sandpoint

Oceanic Crust

Coral Reef

11.19.a1

Optional Procedures

After you have evaluated the first four steps in the procedure, your instructor may ask you to complete the following steps. If so, enter your answers in the appropriate places on the worksheet, online, or in a final report.

1. Considering only the natural hazards, rank the towns for their overall risk, for the eventual purpose of assigning different rates of homeowner's insurance, with higher rates for the more risky towns and lower rates for the less risky towns. For each town, propose what hazards an insurance company should consider excluding—or charge higher premiums for—to limit the potential amount of damages. Write a short report justifying the rankings of hazard and coverages that should be excluded or should cost extra. You do not have to propose specific insurance rates (how much a policy will cost).

2. To properly assess risk, you would need more data. Identify what types of data and data-analysis capabilities you would need to properly assign risk and rates. Write a report summarizing the kind of data you need and how you would use it.

---

3. A picturesque town, called *Hillside*, lies inland of some small mountains. The town is built on a flat, open area flanked by hills with fairly gentle slopes. It is a little higher in elevation than the nearby towns of *Cascade Village* and *Riverton*.

4. In the northern part of the area, there is a flat-topped mountain, known as *Red Mesa*, surrounded by steep cliffs. A new landslide lies along the southern flank of the mountain.

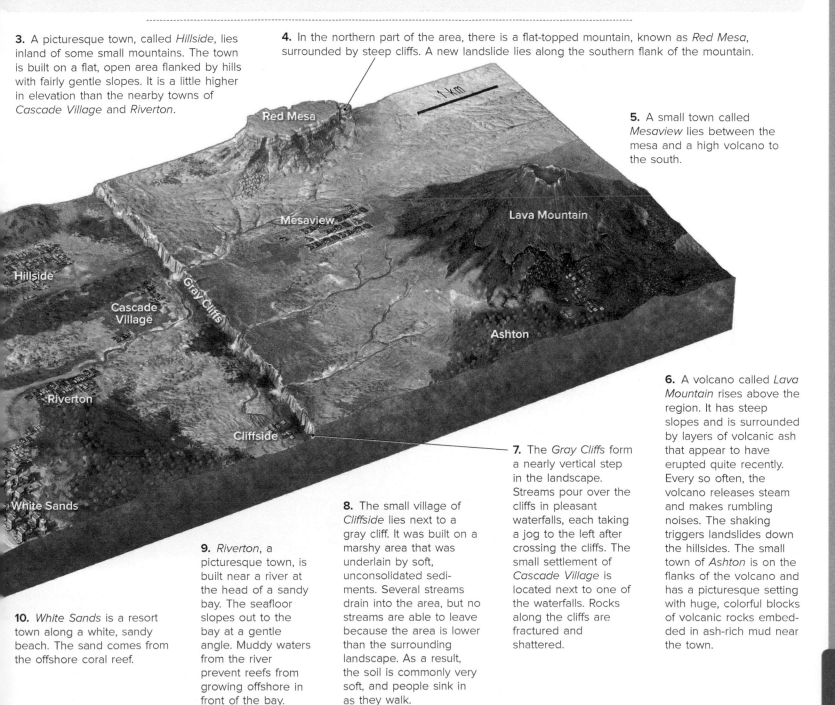

5. A small town called *Mesaview* lies between the mesa and a high volcano to the south.

6. A volcano called *Lava Mountain* rises above the region. It has steep slopes and is surrounded by layers of volcanic ash that appear to have erupted quite recently. Every so often, the volcano releases steam and makes rumbling noises. The shaking triggers landslides down the hillsides. The small town of *Ashton* is on the flanks of the volcano and has a picturesque setting with huge, colorful blocks of volcanic rocks embedded in ash-rich mud near the town.

7. The *Gray Cliffs* form a nearly vertical step in the landscape. Streams pour over the cliffs in pleasant waterfalls, each taking a jog to the left after crossing the cliffs. The small settlement of *Cascade Village* is located next to one of the waterfalls. Rocks along the cliffs are fractured and shattered.

8. The small village of *Cliffside* lies next to a gray cliff. It was built on a marshy area that was underlain by soft, unconsolidated sediments. Several streams drain into the area, but no streams are able to leave because the area is lower than the surrounding landscape. As a result, the soil is commonly very soft, and people sink in as they walk.

9. *Riverton*, a picturesque town, is built near a river at the head of a sandy bay. The seafloor slopes out to the bay at a gentle angle. Muddy waters from the river prevent reefs from growing offshore in front of the bay.

10. *White Sands* is a resort town along a white, sandy beach. The sand comes from the offshore coral reef.

CHAPTER
12 Weathering and Mass Wasting

THE BREAKDOWN OF SURFACE MATERIALS—weathering—produces soils and can lead to unstable slopes. Such slope instability is called *mass wasting*, which is the movement of material downslope in response to gravity. Mass wasting can be slow and barely perceptible, or it can be catastrophic, involving thick, dangerous slurries of mud and debris. It is a type of erosion that strips material off a landscape and transports that material away. What physical and chemical weathering processes loosen material from solid rocks and lead to mass wasting? What factors determine if a slope is stable, and how do slopes fail? In this chapter, we explore weathering and mass wasting, which help sculpt natural landscapes.

The Cordillera de la Costa is a steep 2 km-high mountain range that runs along the coast of Venezuela, separating the capital city of Caracas from the sea. This image, looking south, has topography overlain with a satellite image taken in 2000. The white areas are clouds and the purple areas are cities. The Caribbean Sea is in the foreground. The map below shows the location of Venezuela on the northern coast of South America.

In December 1999, torrential rains in the mountains caused landslides and mobilized soil and other loose material as debris flows and flash floods that buried parts of the coastal cities. Some light-colored landslide scars are visible on the hillsides in this image.

How does soil and other loose material form on hillslopes? What factors determine whether a slope is stable or is prone to landslides and other types of downhill movement?

12.00.a1

12.00.a2

The mountain slopes are too steep for buildings, so people built the coastal cities on the less steep fan-shaped areas at the foot of each valley. These flatter areas are alluvial fans composed of mountain-derived sediment that has been transported down the canyons and deposited along the mountain front.

What are some potential hazards of living next to steep mountain slopes, especially in a city built on an active alluvial fan?

The city of Caraballeda, built on one such alluvial fan, was especially hard hit in 1999 by debris flows and flash floods that tore a swath of destruction through the town. Landslides, debris flows, and flooding killed more than 19,000 people and caused up to $30 billion in damage in the region. The damage is visible as the light-colored strip through the center of town.

How can loss of life and destruction of property by debris flows and landslides be avoided or at least minimized?

TOPICS IN THIS CHAPTER

Huge boulders smashed through the lower two floors of this building in Caraballeda and ripped away part of the right side (▾). The mud and water that transported these boulders are no longer present, but the boulders remain as a testament to the strength of the event.

12.00.a3 Caraballeda, Venezuela

1999 Venezuelan Disaster

A *debris flow* is a slurry of water and debris, including mud, sand, gravel, pebbles, boulders, vegetation, and even cars and small structures. Debris flows can move at speeds up to 80 km/hr (50 mph), but most are slower. In December 1999, two storms dumped as much as 1.1 m (42 in.) of rain on the coastal mountains of Venezuela. The rain loosened soil on the steep hillsides, causing many landslides and debris flows that coalesced in the steep canyons and raced downhill toward the cities built on the alluvial fans.

In Caraballeda, the debris flows carried boulders up to 10 m (33 ft) in diameter and weighing 300 to 400 tons each. The debris flows and flash floods raced across the city, flattening cars and smashing houses, buildings, and bridges. They left behind a jumble of boulders and other debris along the path of destruction through the city.

After the event, USGS geoscientists went into the area to investigate what had happened and why. They documented the types of material that were carried by the debris flows, mapped the extent of the flows, and measured boulders (▾) to investigate processes that occurred during the event. When the scientists examined what lay beneath the foundations of destroyed houses, they discovered that much of the city had been built on older debris flows. These deposits should have provided a warning of what was to come.

This aerial photograph (◄) of Caraballeda, looking south up the canyon, shows the damage in the center of the city caused by the debris flows and flash floods. Many houses were completely demolished by the fast-moving, boulder-rich mud.

12.00.a4 Caraballeda, Venezuela

12.00.a5 Caraballeda, Venezuela

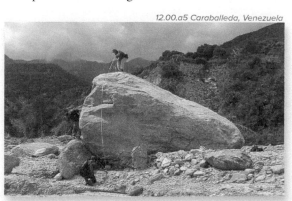

12.0

12.1 How Does Physical Weathering Affect Earth's Surface?

ROCKS AT OR NEAR EARTH'S SURFACE are subjected to changes in temperature, pressure, moisture content, and atmospheric composition that break rocks apart and alter their components. These processes, and the material's response to them, are called *weathering*, which is subdivided into *physical* weathering and *chemical* weathering. Physical weathering (also called mechanical weathering) breaks rocks into smaller fragments without causing any change in their chemical makeup. These smaller fragments can be attacked by chemical reactions, the process of chemical weathering. Weathering can change the color, texture, composition, or strength of the materials, loosening them so they can be eroded away by streams, glaciers, gravity, wind, and waves.

A What Is the Role of Joints in Weathering?

Joints are fractures, or very fine cracks, in rocks that show no significant offset. Joints help break rocks into smaller pieces and permit water and roots to penetrate the rock, thereby promoting weathering.

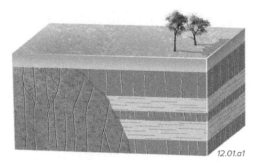

12.01.a1

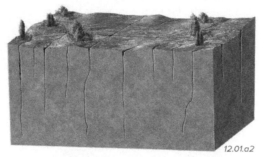

12.01.a2

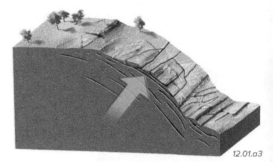

12.01.a3

Most joints form in rocks at depth and may later be uplifted to the surface. The orientation and spacing of preexisting joints and faults help determine the rates of physical and chemical weathering at the surface. More closely spaced joints promote more rapid weathering.

Some joints form as a result of contraction due to cooling of a hot rock or of expansion from a release of pressure as rocks are uplifted to the surface. These *expansion joints* can be difficult to distinguish from preexisting joints that formed by other processes.

As Earth is sculpted by erosion, the topography influences stresses that build up when the weight of overlying rocks is *unloaded*. During unloading, expansion joints can form that mimic topography, peeling off thin sheets of rock, a process called *exfoliation*.

B How Are Joints Expressed in the Landscape?

Joints greatly influence how a landscape develops. Joints affect the strength of a rock, help control its resistance to weathering and erosion, and influence whether pieces of rock are pried loose from the landscape.

12.01.b1 Connecticut Valley, MA

12.01.b2 Black Hills, SD

12.01.b3 Enchanted Rock SP, Llano, TX

Joints are the dominant features of this roadcut, but the amount of jointing is not uniform. The less-jointed areas are more resistant to weathering than the highly jointed ones.

The spacing and orientation of joints, along with rock type, determine how fast a rock will weather and which parts of the landscape will be most easily eroded. Joints play a prominent role in the weathering of the granite shown in this photograph.

Exfoliation joints can be nearly horizontal or can mimic topography, shaving off thin, curved slices of rock parallel to the surface. They can form a curved rock face called an *exfoliation dome*, such as those In Llano, Texas, or Yosemite National Park, California.

C What Other Physical Processes Loosen Rocks?

Joints, which formed by processes at depth or by expansion of rock near the surface, play a major role in weathering. Other processes may also help break rock into smaller pieces. Below are five examples of physical weathering processes that will create fragmented mineral matter, called *regolith*, lying above bedrock.

1. As rocks are heated and cooled, different minerals expand and contract by different amounts. This daily and seasonal *thermal expansion* imposes stresses on the boundaries between minerals and causes microfracturing in and along mineral grains, which physically loosens the mineral grains.

2. When water in a joint freezes, it expands by 9% and exerts a strong outward-directed force on the walls of the joint. This process of *frost wedging* can widen and lengthen the fracture and pry off loose pieces of rock.

3. Water percolating through joints and pore spaces may precipitate (remove from solution) crystals of salt, calcite, and other minerals. As the crystals grow, they exert an outward force that widens joints or weakens the rock. This process is called *mineral wedging*.

4. Burrowing organisms, including rodents, earthworms, and ants, bring material to the surface where it can be further weathered and eroded. As such, these creatures are agents of physical weathering.

12.01.c2 Baja California, Mexico

5. Another form of biotic weathering is by plant roots extending into joints and growing in length and diameter, expanding preexisting joints (▶). This biotic weathering process is *root wedging*.

12.01.c1

D How Does Fracturing a Rock Affect Weathering?

Weathering affects rock surfaces that are exposed to air and water, so rocks weather from the outside in. Physical weathering breaks rocks into pieces, providing more surface area, allowing chemical weathering processes to become more effective.

Surface Area of a Cube of Rock

1. If joints in rock form a three-dimensional network, the rock may be broken into box-shaped pieces bounded by joints, like this cube. What is the total amount of exposed surface area on the sides of the cube?

12.01.d1

2 cm

2 cm

2. To calculate the surface area of one face (side), we multiply the height by the width of that face.

$$2 \text{ cm} \times 2 \text{ cm} = 4 \text{ cm}^2$$

3. There are six faces on a cube, so we multiply the area of one face by 6 to get the total surface area.

$$4 \text{ cm}^2 \times 6 \text{ sides} = 24 \text{ cm}^2$$

Fracturing a Cube into Pieces

4. What happens to the total surface area if we fracture the same cube into eight pieces? First, we calculate the surface area for each smaller cube.

1 cm

1 cm

12.01.d2

$$1 \text{ cm} \times 1 \text{ cm} = 1 \text{ cm}^2 \text{ for each side}$$

$$1 \text{ cm}^2 \times 6 \text{ sides} = 6 \text{ cm}^2 \text{ for each cube}$$

5. But there are eight such cubes.

$$6 \text{ cm}^2 \times 8 \text{ cubes} = 48 \text{ cm}^2$$

6. Therefore, this fracturing has doubled the exposed surface area, providing more surfaces upon which weathering can operate. The rock will therefore weather faster.

12.01.d3 Bluff, UT

7. Physical weathering of steep outcrops can loosen pieces that fall, tumble, or slide downhill and accumulate on the slopes below. These piles of angular blocks are *talus*, and such slopes are *talus slopes*. The largest talus blocks here are one meter across. The talus blocks have much more surface area than the same amount of rock in the smooth cliff from which they were derived.

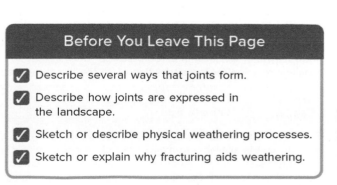

Before You Leave This Page

✓ Describe several ways that joints form.

✓ Describe how joints are expressed in the landscape.

✓ Sketch or describe physical weathering processes.

✓ Sketch or explain why fracturing aids weathering.

12.1

12.2 How Does Chemical Weathering Affect Earth's Surface?

CHEMICAL WEATHERING alters and decomposes rocks and minerals, principally through chemical reactions involving water. When physical and chemical weathering processes interact to break down and alter rocks, new minerals, called *secondary minerals*, are created that are more stable in surface conditions than the original minerals. Also, weathering liberates chemical ions, such as Na^+, K^+, and Cl^-, into groundwater. What are other manifestations of chemical weathering?

A How Does Changing a Rock's Environment Promote Weathering?

Many rocks and minerals form deep within Earth. When they are brought near the surface by uplift and erosion, these rocks encounter conditions very different from those in which they originally form, and so they may become unstable.

12.02.a1

1. Minerals that crystallize in high-temperature magmas are generally unstable when subjected to the low-temperature conditions that characterize Earth's surface. Most magma temperatures are above 700°C, whereas surface temperatures range from −40°C to 45°C (minus 40°F to plus 122°F).

2. During metamorphism, some minerals crystallize beneath the surface in high-pressure and high-temperature environments. Once such rocks reach Earth's low-pressure and low-temperature surface, they can change to different minerals that are stable at the new, wetter, low-pressure and low-temperature conditions.

3. Oxygen (O_2) is abundant in the atmosphere and as a dissolved component in rain and most surface water. This oxygen chemically reacts with rocks, causing some minerals to oxidize (rust).

4. Liquid water is more abundant on and near Earth's surface than at depth in the crust. Water, especially when it is slightly acidic, is a chemically active substance that can break the bonds in many minerals. It increases the rate of chemical weathering. The availability of water largely depends on the climate of an area, so for this and other reasons climate is a key factor in weathering.

B What Happens When Rocks Dissolve?

The main agents for chemical weathering are water and weak acids formed in water, such as carbonic acid (H_2CO_3). These agents dissolve some rocks, loosen mineral grains, form clay minerals, and widen joints.

1. Limestone (below) and other rocks rich in calcium carbonate or magnesium carbonate are soluble in water and in acids. They dissolve and form pits and cavities (▼). On a large scale, the solution of limestone forms distinctive landforms of *sinkholes* and caves called *karst*.

12.02.b1 Capitol Reef NP, UT

2. Over time, the pits deepen, widen, and may interconnect, forming furrows (small troughs).

12.02.b2

3. Fractures can widen as water flows through them and dissolves material from the walls of the joints, as in the limestone outcrop below (▼). Caves can form by dissolution of limestone and other soluble rocks at depth.

12.02.b3 Austrian Alps

One Way That Calcite Chemically Dissolves in Water

4. Limestone is a relatively soluble rock because it is composed of calcite, which is soluble in weak acids. The most common weak acid in surface water is carbonic acid, produced when rainwater reacts with carbon dioxide (CO_2) in the atmosphere, soil, and rocks. The chemical reaction for the dissolution of calcite in carbonic acid is:

$CaCO_3$	+	H_2CO_3	$\longrightarrow$	Ca^{2+}	+	$2(HCO_3)^-$
Calcite		Carbonic acid		Calcium ion		Bicarbonate ion in solution

5. Acidic water contains unbonded H^+ ions, each of which is a proton without a balancing electron and is available to make other chemical bonds. H^+ ions are small and can easily enter crystal structures, releasing other ions, such as calcium, into the water. Rainwater is a weak acid, formed by CO_2 that was acquired from the atmosphere, to make carbonic acid.

C What Happens When Rocks Oxidize Near Earth's Surface?

Oxygen (O_2) is common near Earth's surface and reacts with some minerals to change the oxidation state of an ion. This is common in iron-bearing minerals because iron (Fe) has several oxidation states.

12.02.c1 Wilson Cliffs, NV

1. Many mafic igneous rocks contain dark, iron-bearing minerals, such as pyroxene. Iron in pyroxene can become oxidized, producing iron oxide minerals, as shown in the equation below.

2. Hematite consists only of iron and oxygen and is more stable than pyroxene under oxygen-rich conditions. It commonly forms during oxidation and gives oxidized rocks a reddish color.

3. If iron-bearing rocks become oxidized, such rocks generally take on a red color from the iron oxide mineral hematite. This oxidation process is called rusting and creates less stable by-products. Reddish rocks can lose this reddish color if interacting with fluids that have less oxygen (▶).

$$4FeSiO_3 \ \underset{\text{Pyroxene}}{} + \ O_2 \ \underset{\text{Oxygen}}{} \longrightarrow \ 2Fe_2O_3 \ \underset{\text{Hematite}}{} + \ 4SiO_2 \ \underset{\text{Silica (in water)}}{}$$

D What Happens When Minerals Chemically React with Water?

When some minerals react with water they undergo a chemical reaction where the mineral combines with water to form a new mineral. This reaction, called *hydrolysis*, converts some minerals to clay, an important soil constituent.

12.02.d1 Crete

1. One kind of feldspar, containing potassium (K), is called potassium feldspar or simply K-feldspar. When this mineral reacts with acids (waters that have free H^+ ions), it can be converted into clay minerals by hydrolysis.

2. During the reaction, the H^+ ion moves into the crystalline structure of feldspar, expelling the K^+ ion and some silica, which both get carried away in the water.

3. If exposed to wet conditions, many rocks convert into clay minerals. The gray limestone shown here contained impurities that weathered into clay minerals and reddish hematite that accumulated between the blocks.

$$4KAlSi_3O_8 \ \underset{\text{K-feldspar}}{} + \ 4H^+ \ \underset{\text{Hydrogen ion}}{} + \ 2H_2O \ \underset{\text{Water}}{} \longrightarrow \ 4K^+ \ \underset{\text{Potassium (in water)}}{} + \ Al_4Si_4O_{10}(OH)_8 \ \underset{\text{Kaolinite (clay)}}{} + \ 8SiO_2 \ \underset{\text{Silica (in water)}}{}$$

E How Does Weathering Make the Ocean Salty?

Have you ever wondered why the ocean is salty or where the salt comes from? It turns out that most of the ocean's salts are derived from weathering and the dissolution of rocks on the land.

1. Rock, sediment, and soil on and near Earth's surface are exposed to water and oxygen in the atmosphere. Water that reacts with rocks and minerals can come from several sources, including rain and other forms of precipitation.

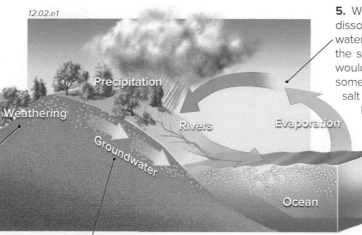
12.02.e1

5. When seawater in the oceans evaporates, the dissolved salts remain in the seawater, increasing seawater's salt content *(salinity)*. Such evaporation causes the seas to be saltier than the streams. The seas would be even saltier if salt were not removed from some parts of the sea by deposition of salt beds. If the salt beds, once formed, are uplifted and exposed on land, they can also contribute dissolved salt to streams and ultimately back to the sea.

2. Some water infiltrates into the subsurface, where it may chemically react with the materials. During weathering, hydrolysis reactions commonly produce clay minerals and also drive out positive ions (cations) from the pre-existing mineral structure. The dominant cations in feldspar, a very abundant mineral, are sodium and potassium, which can form common salts.

3. The dissolved cations, along with negative ions (anions) like chlorine, are carried by moving water, either in streams or in the subsurface by groundwater. Much of this water eventually finds its way to the oceans, where it contributes its salts and other ions to seawater.

4. Modern oceans typically contain about 3.5% dissolved salt, and are much saltier than stream water. Streams contain some dissolved salts but only a small amount, so they are considered to be fresh water, not salt water. If the oceans got their salt from the streams, why are the oceans saltier?

> **Before You Leave This Page**
>
> ✓ Describe several reasons why minerals formed at depth may not be stable at the surface.
>
> ✓ Summarize how limestone dissolves and what features are formed by dissolution.
>
> ✓ Briefly summarize the processes of oxidation and hydrolysis.
>
> ✓ Explain why oceans are salty.

12.2

12.3 How Does the Type of Earth Material Influence Weathering?

ALL MINERALS AND ROCKS can be thought of as *parent material* acted upon by physical and chemical weathering. How parent materials break down depends on a number of factors, including jointing, surface area, and the kinds of minerals that compose the rock. Some minerals weather into grains, whereas others become clay or simply dissolve away. These differences cause weathered rock outcrops to have a variety of distinctive appearances.

A What Controls How Different Minerals Weather?

Many factors determine how minerals weather, including the climate, how much time the mineral has been exposed to weathering, and the chemical composition and atomic structure of the mineral.

Chemical Bonding

12.03.a1 Miami, FL

1. The bonds in some minerals allow them to be readily dissolved in water and weak natural acids, as in this dissolved limestone. Salt and gypsum also are very soluble.

2. Other minerals have stronger bonds that make them less soluble in water. Quartz in this quartzite has very strong bonds and is not very soluble in cold water.

12.03.a2 Big Maria Mtns., CA

Reactivity

12.03.a3 Durango, CO

3. Sandstone is composed mainly of sand-sized grains of quartz. Most quartz grains weather by physical processes, rather than chemical processes.

4. Limestone is very soluble and prone to chemical weathering, especially dissolution and especially in wet climates. It also weathers by physical processes.

12.03.a4 San Juan River, UT

Relative Resistance of Minerals to Weathering

12.03.a6 Baja California Sur, Mexico

12.03.a5 Fairplay, CO

5. The stability of minerals is in a very general way related to the temperature at which the minerals formed, such as when minerals crystallize from a magma. Mafic minerals and Ca-rich plagioclase feldspars, present in this mafic igneous rock (◀), crystallize under the hottest conditions. These minerals are typically least stable when subjected to weathering at the low temperatures of Earth's surface. Minerals that crystallize at relatively lower temperatures from a magma, such as quartz and potassium feldspar, are more stable in near-surface environments. As a general rule, minerals are most stable in the conditions closest to those under which they formed. Thus, high-temperature minerals are most unstable at Earth's surface and are the quickest to weather.

6. Granitic rocks are abundant on Earth's surface and mostly composed of quartz, both kinds of feldspars, and some percentage of mica (▶). Mica and plagioclase feldspar weather relatively easily, but potassium feldspar is commonly more resistant. Quartz is much more resistant to weathering because it is essentially insoluble. It can be thought of as a "survivor" mineral, in that its superior resistance shows up as a residual product in beach sands, dunes, and soils. As different minerals weather at different rates, a rock can simply come apart into a collection of discrete mineral grains, as in the granite in the photograph above.

Before You Leave These Pages

☑ Summarize the factors that control how different rocks and minerals weather.

☑ Explain the origin of the three main weathering products (sand, clay minerals, and dissolved ions).

☑ Describe how the character of a rock influences how it weathers.

B How Do Different Rocks Respond to Weathering?

Rocks are composed of minerals. Some minerals are hard and so resist physical weathering, whereas others are weak and easily broken. Some minerals are chemically stable and resist chemical weathering, while others are less chemically stable. Some rocks are mostly a single mineral, but others contain many minerals. All these variations cause granite, one of the most common rocks, to weather in different ways.

1. Granite and related igneous rocks contain feldspar, quartz, and smaller amounts of mica (biotite and muscovite), iron oxides, or amphibole. As these rocks weather, the different minerals respond in different ways.

2. Feldspar is the most abundant mineral in granites, forming the cream-colored crystals shown here (▶). Feldspar chemically weathers by hydrolysis to form clay minerals. During this process, sodium (Na), potassium (K), and other ions leach out of the feldspar. Some of the liberated ions are released and can be carried away by water.

3. Clay minerals weathered from granite and similar rocks can accumulate in soil or be eroded and transported away by water and wind. Some clay particles are washed out to sea, and others are deposited in lakes, floodplains, deltas, and other muddy environments (▶).

12.03.b1 Baja California, Mexico

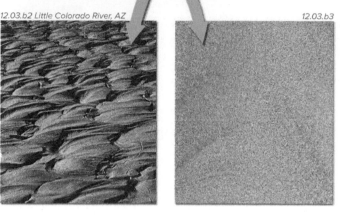

12.03.b2 Little Colorado River, AZ   *12.03.b3*

4. Granite (◀) is at least 25% quartz, which is the medium-gray, partially transparent mineral shown here. Quartz is very resistant to chemical and physical weathering. As chemically reactive minerals break down around it, quartz weathers into intact grains.

5. Quartz grains (◀) eroded from granite typically become quartz sand, which can be transported away by water and wind. Quartz sand accumulates along rivers, in dunes, and on beaches. Feldspar can also form sand grains if the granite or other source of feldspar is not too chemically weathered, as can occur in dry climates or in areas of rapid erosion.

C How Does the Character of a Rock Influence Weathering?

Differences in mineral composition, particle size, and other rock properties play an important role in how a rock responds to weathering. Equally important are joints, bedding planes, and other discontinuities.

1. *Composition*—Weathering of a rock is influenced by the types of minerals it contains. Most sandstone, such as the one in this cliff, consists largely of quartz, a mineral that is very stable on Earth's surface; it mostly weathers by physical processes. In contrast, the recesses below the cliff contain fine-grained sedimentary rocks that are more easily weathered and eroded.

2. *Variation in Composition*—Some outcrops have different parts with large contrasts in susceptibility to weathering. The more susceptible parts will weather faster than the more resistant parts. Such *differential weathering* can form alternating ledges and slopes, as shown here, or rocks with holes where less resistant material has been removed.

12.03.c1 Bluff, UT

3. *Surface Area*—Rock that is already broken into pieces provides more surface area on which chemical weathering can act. Solid, unbroken bedrock, as in the overlying cliff, provides less surface area and weathers more slowly.

4. *Discontinuities*—Fractures, bedding planes, and other discontinuities provide pathways for the entry of water into a rock body. A rock with lots of these features will weather more rapidly than a massive rock containing few such discontinuities. For example, highly jointed parts of a cliff weather faster than less jointed parts. Rocks with thin layers generally break apart and weather more readily than rocks with thick layers. A thick rock layer with few discrete bedding planes will weather more slowly than one with many bedding planes.

12.4 How Do Climate, Slope, Vegetation, and Time Influence Weathering?

SPATIAL VARIATIONS IN CLIMATE, slope, vegetation, and time also impact rock weathering, and these factors are highly interdependent. Climate, for example, will influence the effectiveness of chemical and physical weathering, which influence soil formation and the abundance of vegetation, which in turn influence wedging by roots and secretion of organic acids. Time is a key factor—more time increases the cumulative impact of weathering.

A How Does Climate Influence Weathering?

Abundant precipitation and higher temperatures cause chemical reactions to proceed faster. Thus, warm, humid areas generally have more highly weathered rock because chemical weathering operates faster than in cold or dry climates. Elevation influences temperature and precipitation patterns and is yet another influence on weathering.

1. This figure plots two important climatic factors—precipitation (blue bar) and average annual temperature (orange bar)—as a function of latitude, from the tropics (on the left) to polar regions (on the far right). The values are all relative to the polar values, so no scale is needed.

2. The horizontal line represents the weathering surface, and the brown boxes below the surface indicate the depth of weathering. Observe the various graphs and consider why there might be a relationship between the climatic factors and the depth of weathering.

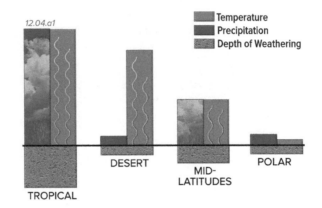

12.04.a1

■ Temperature
■ Precipitation
▨ Depth of Weathering

TROPICAL DESERT MID-LATITUDES POLAR

3. The depth of weathering (and thickness of soil) is greatest in the tropics because this climate has high temperatures, abundant precipitation, and vigorous plant growth, which contribute to weathering.

4. Weathering depths are shallowest in dry parts of the subtropics (deserts) and in polar areas, both of which have low amounts of precipitation. Although temperature is high in deserts, there is little water to facilitate chemical weathering. Instead, physical weathering may dominate, depending on the specifics of the site (see below).

B How Does Slope Influence Weathering?

How rocks weather is also controlled by geographic factors not related to the rock itself but to spatial variations in slope, specifically its aspect (the direction the slope faces) and its steepness.

1. *Windward Slopes*—Mountains are sites of *orographic lifting*, where air rises, cools, reaches its dew point, and forms clouds on the windward side of the mountain. If there is enough condensation and deposition, precipitation will occur on the windward side.

2. *Slope Aspect*—The orientation of the slope, called the *slope aspect*, is an important factor in weathering. In addition to some slopes receiving more or less rain, slopes facing the Sun receive more light and heat than those facing away. Thus, sunny surfaces tend to be warmer and drier, to have more evaporation, and to have less chemical weathering, soil, and plants than slopes facing away from the Sun. In the Northern Hemisphere, south-facing slopes receive more Sun, except in the tropics. Physical weathering is more important on Sun-facing slopes. Chemical weathering will instead likely be dominant on the slope facing away from the Sun.

12.04.b1

3. *Shaded Slopes*—Slopes of some orientations are more sheltered (shaded) from sunlight than slopes of other orientations. If a slope is sheltered from the direct rays of the Sun, it is cooler, can better retain its moisture, and may have more plants. Moisture, soil, and plants promote chemical weathering, so a slope shaded from the Sun generally has more soil than one that is not exposed to the Sun.

4. *Steepness of Slopes*—On steep slopes, rainfall runs off faster, and weathering products may quickly wash away by runoff. Soil and other loose materials can also slide down steep slopes. On a steep slope, therefore, chemical weathering is slower, so soil may be less developed and hillslopes may be more rocky and more barren.

5. On gentle slopes, weathering products can accumulate, and water may stay in contact with rock for longer periods of time, resulting in higher weathering rates. Once formed, soil and loose pieces remain in place longer.

C How Does Vegetation Influence Weathering?

Weathering can also be caused by biological activity—*biotic weathering*. Roots can pry open joints in rocks, a type of physical weathering. Animals of various sizes, from termites to mammals, burrow into the ground seeking food and shelter, and, in the process, loosen and mix sediment. Plants also contribute to chemical weathering through root secretion and through carbon dioxide (CO_2) given off during respiration.

12.04.c1 Selway, ID

1. *Lichens* (◄) consist of algae and fungi, living symbiotically. The algae provide food through photosynthesis while the fungi house the algae, providing water and protection. There are a number of varieties, resulting in different shades of gray (as shown here), green, yellow, and orange.

2. Lichens secrete *oxalic acid*, which effectively dissolves minerals, particularly the carbonates contained in marble (►). This tombstone is almost entirely covered with lichens, which help make this monument's inscription illegible.

12.04.c2 Mount Pleasant, MI

12.04.c3

3. Plant root respiration (◄) produces carbon dioxide, which diffuses into the pore spaces between particles within soil. The carbon dioxide, when combined with water, forms carbonic acid, causing a type of chemical weathering that is very destructive on chemically reactive limestone and marble.

4. The root zone for many trees (►) is far more extensive than most of us realize. The root system for this oak tree, for example, is double the extent of the crown, providing enormous opportunity for both physical and chemical biotic weathering.

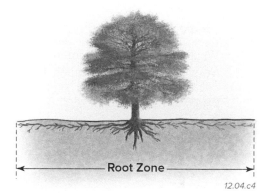

Root Zone

12.04.c4

D How Does Time Influence Weathering?

Time is a crucial factor in weathering. Physical, and especially chemical, processes take time, so the more time that is available, the more weathering will occur. The speed of weathering and the volume of material affected in a given time will depend on climate, slope aspect, vegetation, composition, and jointing of the rock or sediment. Due to great variability in these factors, and all the possible combinations, rates of weathering can range from rapid to extremely slow.

12.04.d1 Mount Pleasant, MI

12.04.d2 Mount Pleasant, MI

Both monuments featured in these photographs are composed of marble, yet the tombstone on the left, from the 1870s, is highly weathered while the tombstone on the right, from 1982, exhibits relatively little weathering. Since marble is prone to chemical weathering, it lost favor as a monument choice in the 1930s. These monuments provide a vivid example of the importance of time in weathering, for either natural or human-constructed objects.

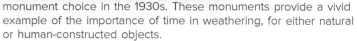

Before You Leave This Page

✓ Sketch and explain how the rate and depth of weathering are affected by the type of climate.

✓ Summarize or sketch and explain how weathering is affected by slope.

✓ Describe ways that weathering is impacted by vegetation and time.

12.4

12.5 How Is Weathering Expressed?

PHYSICAL, CHEMICAL, AND BIOTIC WEATHERING affect materials below the surface, such as during the formation of soil, but they also sculpt rocks on the surface. Several guiding principles affect both domains—the subsurface and the surface. The results of weathering are obvious, once we know what to observe.

A Why Does Weathering Produce Rounded Features?

Weathering processes usually work inward from an exposed surface. This commonly results in rounded shapes in weathered outcrops, and weathering commonly generates loose, partially rounded blocks. This process of producing somewhat spherical shapes out of solid bedrock is called *spheroidal weathering*. The three figures below illustrate what can happen to a rock that has joints but lacks other types of discontinuities.

Rock that is newly jointed generally has sharp, angular edges. Weathering attacks edges from two sides and corners from three sides. These edges and corners wear away faster than a single smooth surface.

12.05.a1

Over time, faster weathering of the edges and corners of the rock will begin to smooth away these areas and any other parts of the rock that stick out.

12.05.a2

Weathered rocks can become moderately rounded, losing their sharp edges and angular features. Rocks dislodged from the bedrock will get smaller with time as they are weathered from all sides. Much rounding continues while the rock is being transported by streams.

12.05.a

B How Is Spheroidal Weathering Expressed in the Landscapes?

Weathering exerts enormous control over the appearance of landscapes. Weathering helps define differences in appearance from one region to another, from one side of a hill to the other, or between different rock types.

12.05.b1 Julian, CA

12.05.b2 Boulder, MT

12.05.b3 Baja California Sur, Mexico

1. Weathering mostly affects rocks from the outside in, so weathered rocks have an outer weathered zone and an inner unweathered zone. The outer zone is a *weathering rind*. As weathering continues, the weathering rind thickens and can in some cases be used to infer how long the rock has been on or near the surface and exposed to weathering.

2. As weathering attacks a jointed rock, differential weathering along the fractures can cause the intact but joint-bounded blocks to become rounded. The outer weathered rind of the blocks splits away from the stronger, less weathered rock in the center, forming rounded shapes in the process of spheroidal weathering.

3. Spheroidal weathering also affects rocks on the surface, such as the granite shown above. As rocks are weathered and uncovered by erosion, they commonly display rounded shapes. Corners of rocks exposed to the elements (rain, sunlight, wind, etc.) are affected in the way described in part A above.

4. The rounding processes of weathering can also sculpt larger masses of rock, including those still attached to rocks underneath (▶). The feature in this photograph consists of several spherical shapes, separated by a recessed area along a nonresistant (easily weathered and eroded) sedimentary layer.

12.05.b4 Monticello, UT   12.05.b5 Bryce Canyon NP, UT

5. Not all weathered rocks end up as rounded shapes. The spectacular scenery of Bryce Canyon National Park in Utah contains spires, called *hoodoos*, produced by erosion of highly jointed rock layers. Two sets of intersecting joints formed columns, and weathering and erosion did the rest. Up close, the sides of the hoodoos are somewhat rounded.

C How Is Weathering Expressed in Your Town and Backyard?

You would be mistaken to believe that a trip to a distant national or state park is necessary to view features that result from weathering. Examples of physical and chemical weathering are numerous—you just have to know where to look. It is as simple as taking a short walk, starting with the shape of rocks around your home or campus. These photographs illustrate some of the everyday manifestations of weathering that can be observed in any town.

Oxidation

12.05.c1

Steel contains iron that when exposed to oxygen in the atmosphere and in water for long periods of time becomes oxidized. This forms iron oxide compounds, commonly called rust. Once formed, these iron oxides have a diagnostic orange color and are relatively stable. The steel is not so fortunate, as oxidation, given enough time, will totally dissolve away the steel. Water, air, and time are more powerful than steel.

Salt Crystallization

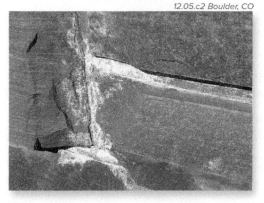

12.05.c2 Boulder, CO

Calcium carbonate is a constituent of mortar between bricks, concrete in sidewalks, stucco on walls, and some building stones. Under wet conditions, the calcium carbonate, which is quite soluble (think of limestone, composed of calcium carbonate), can go into solution. As the solution is drawn to the surface, the carbonate precipitates into crystals that weaken the mortar holding the bricks together.

Biotic Weathering

12.05.c3 Mount Pleasant, MI

The roots of lichens secrete acids and wedge into monuments, especially marble tombstones. In the monument pictured above, most of the monument is highly weathered, in part due to the corrosive effects of lichen. The white streak is less weathered and thought to be due to rainwater interacting with the metal, creating a solution poisonous to the lichen.

Differential Weathering

12.05.c4 Mount Pleasant, MI

Many materials in your hometown or on your campus are a composite of different materials, whether natural or made by humans. The different components usually weather at different rates, like rocks in a cliff do. In this boulder on a campus, light-colored *dikes* (formed by magma in a fracture) are more resistant to chemical weathering than are the dark rocks cut by the dikes. Consequently, the dikes weather out in relief, as if carved by a sculptor. In this case, though, the sculptor was natural weathering.

Root Wedging

12.05.c5 Baja California Sur, Mexico

Roots emanating from trees find their way into joints, whether they are in rocks or sidewalks. Most of us have seen a sidewalk tipped up and fractured by an underlying root. As the root gets bigger, the disturbance does, too.

Soil

12.05.c6 Arco, ID

Soil is a mixture of variably decomposed rock, sediment, and organic material, with water and air in the spaces between grains. Soil is most obvious in farms or gardens, but also underlies most neighborhoods, although covered by streets, sidewalks, and houses. All soil is the product of weathering.

12.6 How Do Caves Form?

WATER IS AN ACTIVE CHEMICAL AGENT and can dissolve rock and other materials. Weathering near the surface and groundwater at depth can work together to completely dissolve limestone and other soluble rocks, leaving openings in places where the rocks have been removed. Such dissolution of limestone forms most caves, but caves form in many other ways. Once a cave is formed, dripping and flowing water can deposit a variety of beautiful and fascinating cave formations that grow from the ceilings, walls, and floor of the cave.

A How Do Limestone Caves Form?

Water near the surface or at depth as groundwater can dissolve limestone and other carbonate rocks, to form large caves, especially if the water is acidic. Cave systems generally form in limestone because most other rock types do not easily dissolve. A few other rocks, such as gypsum or rock salt, dissolve too easily—they completely disappear and cannot maintain caves. The figure below illustrates how limestone caves form.

1. Limestone is primarily made of calcite (calcium carbonate), a relatively soluble mineral that dissolves in acidic water. Rainwater is typically slightly acidic due to dissolved carbon dioxide (CO_2), sulfur dioxide (SO_2), and organic material. Water reacts with calcite in limestone, dissolving it. This *dissolution* can be aided by acidic water coming from deeper in the Earth, by microbes, and by acids that microbes produce.

3. Most caves form below the water table, but some form from downward-flowing water above the water table. In either case, dissolution over millions of years can form a network of interconnected caves and tunnels in the limestone. If the water table falls, groundwater drains out of the tunnels and dries out part of the cave system.

2. Groundwater dissolves limestone and other carbonate rocks, often starting along fractures and boundaries between layers, and then progressively widening them over time. Open spaces become larger and more continuous, allowing more water to flow through and accelerating the dissolution and widening. If the openings become continuous, they may accommodate underground pools or underground streams.

4. If the roof of the cave collapses, the cave can be exposed to the air. This can further dry out the cave. Such a roof collapse commonly forms a pit-like depression, called a *sinkhole*, on the surface.

5. Limestone caves (▼) range in size from miniscule to huge. The Mammoth Cave System of Kentucky is the longest cave in the world, with an explored length of over 640 km (400 mi), with some parts still largely unexplored.

012.06.a1

12.06.a2 Lehman Caves, NV

B What Are Some Other Types of Caves?

Most but not all caves developed in limestone. Caves in volcanic regions are commonly *lava tubes*, which were originally subsurface channels of flowing lava within a partially solidified lava flow. When the lava drained out of the tube, it left behind a long and locally branching cave. Such caves tend to have a curved, tube-like appearance with walls that have been smoothed and grooved (▶) by the flowing lava.

12.06.b1 Hawaii Volcanoes NP, HI

12.06.b2 Hueco Tanks, TX

Almost any rock type can host a cave, as long as it is strong enough to support a roof over the open space. Granite, not known as a soluble rock, can form caves, especially where physical and chemical weathering has enlarged areas along fractures (◀). Many non-limestone caves are along a contact between a stronger rock above, which holds up the roof, and a weaker rock below, to form the opening.

C   What Features Are Associated with Caves?

Caves are beautiful and interesting places to explore. Some contain twisty, narrow passages connecting open chambers. Others are immense rooms full of cave *formations*, such as stalactites and stalagmites. Caves can be decorated with intricate features formed by dissolution and precipitation of calcite and several other minerals.

1. Most caves form by dissolution of limestone. Certain features on the land surface can indicate that there is a cave at depth. These include the presence of limestone, sinkholes, and other features. Collapse of part of the roof can open the cave to the surface, forming a *skylight* that lets light into the cave.

2. Caves contain many features formed by minerals precipitated from dripping or flowing water. Water flowing down the walls or along the floor can precipitate *travertine* (a banded form of calcium carbonate) in thin layers that build up to create formations called *flowstone* or *draperies* (▼).

12.06.c2 Carlsbad Caverns, NM

12.06.c3 Kartchner Caverns, AZ

12.06.c1

6. Dissolution of limestone along fractures and bedding planes, along with formation of sinkholes and skylights, disrupts streams and other drainages. Streams may disappear into the ground, adding more water to the cave system.

5. In humid environments, weathering at the surface commonly produces reddish, clay-rich soil. The soil, along with pieces of limestone, can be washed into crevices and sinkholes, where it forms a reddish matrix around limestone fragments.

3. Probably the most recognized features of caves are stalactites and stalagmites, which are formed when calcium-rich water dripping from the roof evaporates and leaves calcium carbonate behind. *Stalactites* hang tight from the roof. *Stalagmites* form when water drips to the floor, building mounds upward.

4. As mineral-rich water drips from the roof and flows from the walls, it leaves behind coatings, ribbons (▶), and straw-like tubes. The water can accumulate in underground pools on the floor of the cave, precipitating rims of cream-colored travertine along their edges.

12.06.c4 Kartchner Caverns, AZ

Carlsbad Caverns

About 260 million years ago (260 Ma), Carlsbad, New Mexico, was an area covered by a shallow inland sea. A huge reef, lush with sea life, thrived in this warm-water tropical environment. Eventually, the sea retreated, leaving the reef buried under other rock layers.

While buried, the limestone was dissolved by water rich in sulfuric acid generated from hydrogen sulfide that leaked upward from deeper accumulations of petroleum. Later, erosion of overlying layers uplifted the once-buried and groundwater-filled limestone cave and eventually

exposed it at the surface. Groundwater dripped and trickled into the partially dry cave, where it deposited calcium carbonate to make the cave's famous formations.

12.06.t1 Carlsbad Caverns, NM

Before You Leave This Page

✓ Summarize the character and formation of caves, sinkholes, skylights, and travertine along streams.

✓ Briefly summarize how stalactites, stalagmites, and flowstone form.

✓ Describe features on the surface that might indicate an area may contain caves at depth.

12.6

12.7 What Is Karst Topography?

LIMESTONE AND OTHER SOLUBLE ROCKS, because they commonly respond to weathering by dissolving away, produce distinctive landscapes characterized by a somewhat disorganized appearance. Instead of a typical network of drainage systems, this type of landscape—*karst topography*—features sinkholes and other depressions, streams that disappear into the ground, gray rocks that look like they are dissolving away, and in some places, exotically shaped pillars. Karst topography is common in many parts of the world, perhaps near where you live.

A What Are the Characteristics of Karst Topography?

Karst topography has diverse expressions depending on the topographic setting, climate, and other factors. It always indicates the presence of some type of soluble rock, especially limestone and dolomite, but also, in some locations, rock salt. Karst leaves an imprint on the topography as a whole and in the appearance of individual exposures of rock.

12.07.a1

1. Examine the topography represented by this three-dimensional perspective of the area near Oolitic, Indiana (▶). What are the features of this landscape?

2. One of the main characteristics of this terrain is the presence of numerous pits. These are sinkholes and shallow depressions formed by the partial collapse of the roofs of caves. The widespread distribution of the pits suggests that most of the area is underlain by caves. Since caves tend to form in limestone, we can infer that much of the area is underlain by limestone, as is indeed the case.

3. Another characteristic is the relative lack of a well-developed drainage system, such as smaller streams feeding into larger streams in a typical branching pattern. Instead, if we imagine rain falling in some area, the runoff is likely to flow into one of the pits instead of reaching a stream. This illustrates another attribute of karst regions—they commonly have a network of underground passages, including underground streams and lakes.

4. Some of the hills in this area do not have as many pits and instead have typical stream valleys. When these streams enter the karst, they cease to have any expression on the surface because their flow has been captured by the network of pits—such streams are called *disappearing streams* and are another characteristic of karst topography. They add water to the underground drainage system.

5. Sinkholes can be small and subtle topographic depressions, such as the one forming this small pond (▶), or they can be prominent, precipitous pits, as in the photograph below from Florida. In either case, they suggest collapse or settling of the roof of a cave below that spot.

12.07.a2 Radford, VA

12.07.a4 Yamaguchi, Japan

7. Areas of karst generally display evidence of rocks that have partially dissolved (◀) due to their exposure to water from precipitation or from water that was in contact with the rocks, such as water in soil, groundwater, or water in a stream, lake, or ocean. Such rocks typically have steep grooves.

6. Sinkholes are a significant natural hazard. The one shown here (▶) formed in 1981 and destroyed cars and buildings in Winter Park, Florida, where the underlying bedrock is limestone. The collapse resulted from the lowering of the water table due to pumping from water wells. Underground water helps support the roof of a cave, so when the water is removed, the roof can no longer support itself and so it collapses. Features described as "sinkholes" in news reports can also form in other ways.

12.07.a3 Winter Park, FL

12.07.a5 Philippines

8. In some settings, especially in tropical climates where there is abundant rainfall, karst terrains can feature steep pillars, knobs, and oddly shaped rocks (◀), all sculpted by dissolution of limestone. These form spectacular tourist-destination landscapes in China, Cambodia, the Philippines, and other parts of Southeast Asia and adjacent islands. Some of these pillars are on land and others are surrounded by the ocean. They form from rocks that were somewhat more resistant to dissolution.

B   What Is the Distribution of Karst?

The formation of karst relies on the presence of soluble materials at or near the surface, so a map showing the distribution of karst essentially shows the distribution of limestone, dolomite, and other soluble rocks. As the two maps below illustrate, karst is very widespread, as are its associated hazards.

1. This map shows the distribution of karst terrain in the world, excluding Antarctica, which is mostly covered with ice and is too cold for liquid water. Karst is present on all the other continents and in all types of climates, from polar climates of the Arctic to tropical climates of central America, and all climates in between. Most of these areas indicate the presence of limestone.

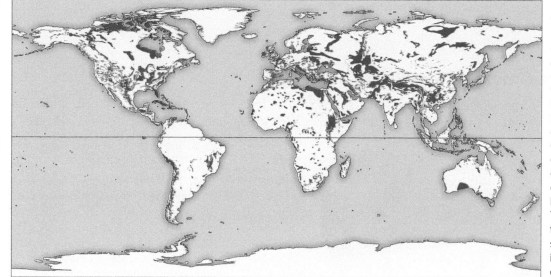

12.07.b1

2. Karst is especially widespread in Europe, such as in areas adjacent to the Alps mountain range, and in southern and central Asia. In this latter area, karst is present in the hot and humid, low elevations of Southeast Asia and in the cold and relatively dry conditions in the high elevations of Tibet. This illustrates that rock type, specifically limestone, is the primary control on karst.

3. This map depicts the distribution of karst terrain in the U.S. Purple areas are karst formed by limestone either at the surface (dark purple) or from limestone at depth (light purple). Orange areas indicate areas underlain by salt, gypsum, and other soluble rocks formed by the evaporation of water (and so are called *evaporite rocks*). Red shows areas with other types of caves, such as lava tubes, in volcanic rocks.

12.07.b2

4. In the U.S., karst occurs in belts within and adjacent to the Appalachian Mountains, following the tilted and folded layers of limestone and other carbonate rocks.

5. Karst is widely distributed in Ohio, Indiana, Kentucky, Tennessee, Missouri, and adjacent states.

6. The largest area of karst is in Florida and adjacent parts of Georgia, which are underlain by an extensive sequence of flat-lying layers of limestone. These areas are dotted with sinkholes and are most likely to be places described in sinkhole-related news stories.

CARBONATE ROCKS
Exposed
Unexposed

EVAPORITE ROCKS
Exposed
Unexposed

VOLCANIC ROCKS

12.07.b3 Miami, FL

7. In addition to sinkholes, evidence for the possible presence of karst includes exposures of limestone with smaller holes formed by dissolution (◀). Weathered limestone generally has a rough surface, with a feel of sandpaper, eliciting colorful, very descriptive names, like "tear-pants weathering." If a prospective home buyer finds such weathered limestone on real-estate property, the possibility of karst should be investigated by consulting a geologic map of the region (which will show places where limestone is exposed). Weathered limestone may indicate caves at depth and therefore the possibility of sinkholes opening up in the future.

Before You Leave This Page

☑ Describe the characteristics of karst topography and what features might indicate that karst is present.

☑ Briefly summarize the main locations of karst topography for both the world and the U.S.

12.7

12.8 What Controls the Stability of Slopes?

ONCE MATERIAL IS LOOSENED BY WEATHERING, gravity pulls it downhill. Downward movement of material on slopes under the force of gravity is called *mass wasting*, and it occurs to some degree on all slopes. Mass wasting can proceed very slowly or very quickly, sometimes with disastrous results. Mass wasting is an important part of the erosional process, moving material downslope from higher to lower elevations, and feeding sediment from hillslopes to streams, beaches, and glaciers.

A How Does Gravity Affect Slope Stability?

The main force responsible for mass wasting is *gravity*. The force of gravity acts everywhere within and on the surface of Earth, tending to pull everything toward Earth's center. Gravity operates on planets, moons, and other planetary objects.

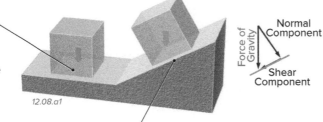

1. On a flat surface, the force of gravity acts on a block by pulling it vertically down against the base of the block. The block will not move under this force.

12.08.a1

2. On a slope, gravity acts at an angle to the base of the block. Part of the force pulls the block against the slope and another part pulls the block down the slope. These two parts of the force are referred to as *components*.

3. The part of the force pulling the block against the slope is the *normal component*.

4. The other component acts parallel to the slope, trying to shear the block down the slope. It is the *shear component*.

5. As the angle of slope becomes steeper, the shear component becomes larger while the normal component becomes smaller. As the slope angle steepens, the shear component becomes enough to overcome friction, and it causes the block to slide. The effectiveness of the shear component is influenced by the strength of the slope material to resist deformation. If the material is very weak, like mud, it will internally shear down the hill rather than slide as a rigid block.

B What Makes Some Slopes Stable?

The steepest angle at which a pile of unconsolidated grains remains stable is called the *angle of repose*. This angle is controlled by frictional contact between grains. In general, loose, dry material has an angle of repose between 30° and 37°. This angle is somewhat higher for coarser material, for more angular grains, for material that is slightly wet, and for material that is partly consolidated. It is lower for material with flakes or rounded grains and for material that contains so much water that adjacent grains lose contact.

Dry Sand — Angle of Repose

12.08.b1

1. Dry, unconsolidated sand grains form a pile (◄), and the angle of the resulting slope is at the angle of repose. If more sand is added, the pile becomes wider and higher, but the angle of repose remains the same. If part of the pile is undercut and removed, the grains slide downhill until the pile returns to a stable slope at the angle of repose. If sand is slightly wet, surface tension between the grains and a thin coating of water enable the sand to be stable on steeper slopes. Too much water allows the grains to slide downhill.

2. Loose rocks and other loose material (talus) accumulate on some slopes (►) and at the bases of cliffs. Such talus material commonly forms slopes that are at the angle of repose for the particular sediment. The smooth *talus slope* shown here became too steep in places and so locally slid downward, forming a small pile of debris at the base of the slope.

12.08.b2 Tibet

3. Most slopes of sand dunes (▼) exhibit the angle of repose for dry sand. Slopes can be more gentle than the angle of repose, but if they begin to exceed the angle of repose then the slope fails, slumping downhill. Walking up a sand dune causes the barely stable sand to slide beneath your feet.

12.08.b4 Northern AZ

12.08.b3 Morocco

4. The slopes of a cinder cone (►) reveal the angle of repose because they are typically composed of loose, volcanic cinders. The angle of repose will be steeper for coarser cinders and for material that partially fused together during the eruption.

C What Resisting Forces Control Slope Stability?

The main control on slope stability is the angle of repose for the material. Intact rock can form cliffs or steep slopes, but soil, sediment, and strongly fractured rock form slopes as steep as their angle of repose.

The addition of minor amounts of water increases the strength of soil, but oversaturation pushes grains apart and weakens the soil. Materials with high clay-mineral content can flow downhill when they become wet.

Joints, cleavage, and bedding reduce the overall mechanical strength of a rock, and may allow rocks to slip downhill. In this illustration, rock layers oriented parallel to the slope allow material to slide.

12.08.c1

12.08.c2

12.08.c3

D What Driving Forces Trigger Slope Failure?

Slope failure occurs when a slope is too steep for its material to resist the pull of gravity. Precipitation, loading of weight on a slope, oversteepening, shaking, and other factors are examples of driving forces. Some slopes slide or creep downhill continuously, but others fail because some event caused a previously stable slope to fail.

1. Precipitation can saturate sediment, increasing pore pressure and decreasing normal stress, allowing grains to slide past one another more easily. A slope that was stable under dry conditions may fail when wet. Slopes can also fail after wildfires, which destroy plants that help bind and stabilize the soil.

2. Hillslopes can fail when the load on the surface exceeds a slope's ability to resist movement. Humans sometimes build heavy structures on slopes, overloading the slope and causing it to fail. Areas with gentle slopes, such as near this town, are less prone to slope failures.

3. Modification of a slope by humans or natural causes can increase a slope's steepness so that it becomes unstable. Erosion along riverbanks, as shown here, or wave action along coasts can undercut a slope, making it unsafe.

12.08.d1

4. Volcanic eruptions can shake, fracture, and tilt the ground, unleashing landslides from oversteepened slopes. Eruptions can cover an area with hot ash and other loose material, causing melting of ice and snow. Melting can rapidly release large amounts of water and mobilize volcanic material in destructive debris flows.

5. A sudden shaking, such as tremors caused by an earthquake along this fault scarp, may trigger slope instability. Minor shocks from heavy trucks or human-caused explosions can also trigger slope failure.

6. Oversteepening of cliffs or hillslopes during road construction can cause them to fail, especially if joints or layers are inclined toward the road.

Slope Stability in Cold Climates

In cold climates, water is frozen much of the year, and ice, although solid, can flow. Freeze-thaw cycles, where ice freezes and then thaws repeatedly, cause ice to flow and can contribute to slope failure.

When water-saturated soil freezes, it expands, displacing rocks and slabs of frozen soil. When the soil thaws, the material moves down again. This process, called *frost heaving,* is a large contributor to the downslope movement of material in cold climates. In addition, when the upper layers of soil thaw during the warmer months, the water-saturated soil may move downslope more easily. Frost heaving can form polygon-shaped outlines in the soil, called *patterned ground* (▼).

12.08.t1 Kongakut, AK

12.8

12.9 How Do Slopes Fail?

THE RAPID DOWNSLOPE MOVEMENT of material, whether bedrock, soil, or a mixture of both, is commonly called a *landslide*. Movement during slope failure can occur by falling, sliding, rolling, slumping, or flowing, and movement of material (mass) downslope by any of these mechanisms is commonly called *mass wasting*, signifying that mass is moving downhill. We classify slope failures—types of mass wasting—by how the material moves and the type of material involved.

A What Are Some Ways That Slopes Fail?

Most people have seen evidence of slope failure when hiking, driving past a roadcut, or watching television news, nature shows, or movies. Slope failure can be as subtle as a small pile of rocks at the base of a hill or as spectacular as a mudflow that has destroyed a neighborhood. The photographs below show images of various types and sizes of slope failures.

12.09.a1 Yampa River, UT

1. This rocky cliff (▲) failed after being undercut by a river. Large sandstone blocks, one the size of a building, collapsed downward. The falling block detached along a prominent joint surface, which has since accumulated a brown rock varnish.

12.09.a2 Denali region, AK

2. During an earthquake, brown masses of rock and soil (◄) slid down these steep slopes in Alaska, smashed apart, and flowed as avalanches of rock, soil, and ice across a white glacier in the valley below. Parts of the *debris avalanche* flowed across the valley and partway up hillsides on the other side of the valley.

12.09.a3 Tibet

3. On this hillside (◄), millions of rock pieces slid off steep outcrops. Some pieces accumulated on a high talus slope, which is at the angle of repose for this material. Several scars indicate where the talus was remobilized and flowed downhill, constructing a debris fan at the base of the hill.

12.09.a4 El Salvador, Central America

4. This landslide in El Salvador (◄) flowed down a steep, unstable slope and cut a swath of destruction across a neighborhood. Adjacent slopes on this hill appear to be just as steep as the part that failed, and pose a hazard to the remaining homes.

5. Undercutting of hillsides by coastal erosion and highway construction (►) has made many slopes steep and unstable. Along the California coast, rocks and soil slumped downward, covering and blocking the highway along the Pacific Palisades, California. Such slope failures commonly occur during or after intense rainstorms.

12.09.a5 Pacific Palisades, CA

B How Are Slope Failures Classified?

Classification of slope failure is imprecise because the processes commonly grade into one another, and more than one mechanism of movement can occur during a single slope-failure event. All classifications consider (1) how the material moves, (2) what types of material move, and (3) the rate of movement of the material.

12.09.b1 Grand Junction, CO

Mechanism of Movement

1. Geoscientists classify slope failures primarily by how the material moved. Rocks and other material can *fall* off cliffs (◄), can *slip* along joints, cleavage, or bedding planes, can *topple* over, or can do all three.

2. Other slope failures involve the slow *creep* of the uppermost soil cover or the *flow* of material, as during turbulent flows of mud, rocks, and other debris. This brown mud (►) flowed during an earthquake.

12.09.b2 Fort Funston, CA

12.09.b3 Monument Valley, UT

Type of Material

3. Some slope failures involve slabs of *solid rock* or large pieces of broken rock derived from cliffs and rocky hillslopes (◄). Such rocks can further break apart after they begin to move.

4. Many slope failures mobilize *unconsolidated material* that is stripped from hillsides (►). Such material can include soil, sediment, pieces of wood and other plant parts, boulders, and other types of loose debris.

12.09.b4 Grand Canyon, AZ

12.09.b5 Venezuela

Rate of Movement

5. Fast rates—Another important factor is the *rate of movement* of the material. Some slope failures start in an instant and send material downhill at hundreds of kilometers per hour, or at least too fast for people to outrun (◄).

6. Slow rates—Other mass movements are more gradual and move downhill at rates that are imperceptible to an observer. This slow-moving mudflow (►) carries trees, some tilted, along for the ride.

12.09.b6 Slumgullion, CO

Submarine Slope Failures

Slope failure is not restricted to the land; it can also occur on steep or even gentle slopes on the seafloor. Such *submarine slope failures* can be caused by overloading of sediment on a slope or in a submarine canyon. They can also be triggered by shaking during an earthquake, volcanic eruption, or storm. Various types of slope failure occurred off the southwestern coast of the Big Island of Hawaii, forming the large mass of debris shown here in green.

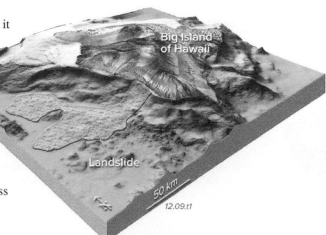
12.09.t1

Before You Leave This Page

✓ Describe slope failures and some ways they are expressed in the landscape.

✓ Summarize the classification of slope failures, and describe the different types of movement, types of material, and rates of movement.

12.9

How Does Material on Slopes Fall and Slide?

SOME SLOPE FAILURES involve materials falling off a cliff or sliding down a slope. These mechanisms commonly involve rock and pieces of rock, but they can also involve materials that are less consolidated. Rocky cliffs and slopes might appear to be immune to slope failure because they consist of hard bedrock, but they can fail spectacularly and catastrophically, posing great hazards to people and buildings.

A What Happens When Rocks Fall or Slide?

Rocks and other material can fall from cliffs or can slide on joints or other weak planes. Falling and sliding rock masses may begin as relatively intact blocks, but they commonly break apart as they begin moving or when they hit the bottom of a cliff. Some slides rotate as they move. Others simply slide down the hill. One type of rock failure can lead to another because a rock fall may remove some support, causing higher parts of a cliff to slide.

12.10.a2 Canyon Lake, AZ

Rock Falls and Debris Falls

In a *rock fall,* large blocks or smaller pieces of bedrock detach from a cliff face and fall to the ground. Rock falls can be triggered by rain, frost wedging, thawing of ice that had held rocks to the cliff, an earthquake, stream erosion, or human construction that undercuts a cliff. Less consolidated debris, including loose sand, can also fall off a cliff. Part of the cliff in this photograph (▶) has collapsed, producing a rock fall composed of large, angular blocks. As the rocks fell, and during impact with the ground, they fractured and broke apart, with most remaining as a large mass of shattered rocks. Some loose blocks fell, rolled, and slid downhill.

12.10.a1

Rock Slides

In a *rock slide,* a slab of relatively intact rock detaches from bedrock along a bedding surface, preexisting fault, joint, or other discontinuity that is inclined downslope. As it slides, parts of the slab typically shatter into angular fragments of all sizes, but large blocks can remain relatively intact. In the rock slide shown to the right (▶), road construction undercut these sedimentary rocks, which had layers and joints that were inclined downslope. At some point after construction, the rocks slid along the layers and into the road.

12.10.a4 Naxos, Greece

12.10.a3

Rotational Slides (Slumps)

In some slides, rock layers and other material rotate backward as they slide. Such *rotational slides* move along one or more curved slip surfaces. This type of slide, also called a *slump,* can occur in bedrock or less consolidated material. Individual slices can remain relatively intact or can break and spread apart. The rotational slides in this photograph (▶) offset and tilted the land surface, and houses, as a result of shaking and loss of cohesion during the 1964 Alaskan earthquake. As the slide masses slumped downhill, they rotated, tilting back, opposite to the direction of movement.

12.10.a6 Fairbanks, AK

12.10.a5

B   What Is the Geometry of a Rock Slide?

A combination of geological circumstances is required to detach a slab of rock and create a rock slide. It requires sufficiently steep slopes along with bedding planes, joints, or other flaws that are inclined downslope.

1. Many rock slides occur in bedrock with discrete layers that differ in rock type and therefore in strength. Such rock layers are most common in sedimentary rocks, but they are also present in many volcanic and metamorphic sequences. In this figure, a sequence of different sedimentary layers has beds inclined downhill, toward a small stream valley.

2. At their upper end, most rock slides detach along a series of preexisting joint or fault surfaces. In other cases, as in this example, the stresses that build up in the rock before it slides are enough to form new joints, allowing the rock slide to detach from bedrock.

3. Detachment of the base of the rock slide from underlying bedrock commonly occurs along a layer of weak rock, such as shale, mudstone, or salt. This weak layer may allow the overlying slab to slide fairly easily and remain partially intact as it moves downhill.

4. To be able to slide down the dip of the layers, the upper layer must have space downhill in which to slide. That is, the layers will probably not slide if they are supported by more rocks in a down-dip direction. In this example, a stream has eroded a low area, giving the sliding rock slab somewhere to go.

5. Although not shown here, rock slides can slip along joints, fault surfaces, cleavage, or some other discontinuity, rather than bedding surfaces. Preexisting faults of the proper orientation are especially susceptible to rock slides because they are planar, fairly continuous, and structurally weak.

7. Most slides leave a linear or curved scar, or *scarp,* on the hillslope, marking where the slide pulled away from the rest of the hill. This upper end is also called the *head* of the slide.

6. The leading edge of a slide, the *toe,* can overrun the land surface in front of the slide.

12.10.b1

The Vaiont Disaster, Italy

In 1960 a dam was built across the Vaiont Valley in northeastern Italy. As seen in the figure below, this valley runs along the bottom of a syncline, where the rocks have been folded downward and dip into the valley from both sides. The rocks are mainly limestone but with interlayered thin beds of shale and sandstone. Some of the limestone beds contain caverns formed when groundwater dissolved the rock. Joints in the rocks run both parallel and perpendicular to the bedding planes.

During August and September 1963, three years after the dam was completed, heavy rain fell in the area. One day in October, the south wall of the valley failed and slid into the reservoir behind the dam. The slide was 1.8 km high and 16 km wide with a volume of 270 million cubic meters. The slide moved along shale layers that parallel the bedding planes in the limestone.

As the slide moved into the reservoir it displaced an equal volume of water, forcing a surge of water 240 m above lake level onto the village of Casso on the northern side of the valley. Waves within the reservoir killed 1,000 people. Waves 100 m high swept over the dam. Although the dam did not fail, the water that overtopped the dam killed 2,000 people living in villages below the dam.

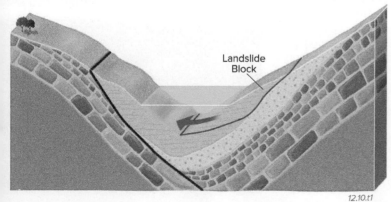

Landslide Block

12.10.t1

Before You Leave This Page

☑ Sketch and describe a rock or debris fall, a slide, and a rotational slide.

☑ Sketch how the geometry of layers, faults, and other features could allow a rock slide to begin.

☑ Describe the Vaiont landslide disaster and factors that caused it to happen.

12.10

12.11 How Does Material Flow Down Slopes?

SLOPE FAILURE CAN MOBILIZE WEAK MATERIAL, including soil, sediment, broken rock, and loose debris. Material flows downhill if it is poorly attached to a hillside, is internally too weak to resist the downward pull of gravity, and the slope is steep enough to allow flow. If the material incorporates some water, it becomes weaker and better able to flow. Such material can move rapidly, as in a debris flow, or it can creep down the hill at a nearly imperceptible rate. All these types of slope failure involve the flow of material.

Creep and Solifluction

Creep and *solifluction* are very slow, continuous movement of soil, regolith, or weathered rock down a slope. Both processes occur on almost all slopes. At the surface, evidence for creep is expressed in bent or leaning trees, warps in roads and fences, and leaning utility poles. Creep commonly causes bedding and other layers in the subsurface to bend downhill (▶). A type of creep at high latitudes or elevations is called solifluction (sometimes called *gelifluction* in polar regions). Here, poorly drained ground freezes for most of the year (permafrost) but near the surface, an upper layer of ground thaws (called the active layer) during the summer. This layer of saturated ground can move downslope in irregular, often overlapping lobes as pictured here (▶).

12.11.a1

12.11.a2 Colorado Springs, CO

12.11.a3 Norway

Debris Slides

Soil, weathered sediment, or other unconsolidated material can move downslope as a *debris slide*. Debris slides are usually less than 10 m thick and leave behind a low *scarp*. A debris slide moves downhill partly as a sliding, coherent mass and partly by internal shearing and flow. A debris slide can lose coherency as it moves, thereby evolving into an earth flow or a debris flow. The debris slide in this photograph (▶) has a clearly expressed scarp half-way up the hill. At the base of the hill, the slide has wrinkled the ground as it moved downhill, forming a slightly steeper toe.

12.11.a4

12.11.a5 Southern Montana

Earth (Mud) Flows

Earth flows are flowing masses of weak, mostly fine-grained material, especially mud and soil. The material moves like thick, wet concrete, generally slowly enough to outrun, but it contains enough water to be slightly fluid. Earth flows have more mud and other fine-grained material than rocks and also can be called *mudflows*. The earth flow in this photograph (▶) mobilized clay-rich altered volcanic and sedimentary rocks, as well as a surface veneer of angular rocks. The earth flow moved downhill until it reached and dammed a larger valley and completely flooded a small village.

12.11.a6

12.11.a7 Thunder Mtn., ID

Debris Flows

Debris flows are wet, downhill-flowing slurries of loose mud, soil, volcanic ash, rocks, and other objects picked up along the way. Some contain only a little water, whereas others are water rich and flow like a thick soup. Debris flows, especially their mud-dominated varieties, are called mudflows by the media and can move rapidly. They often result from heavy rains that saturate the soil and other loose materials. Debris flows are thick and more dense than water and so can support and carry large boulders or even houses. In this photograph (▶), debris-flow deposits from the 1999 Venezuelan disaster, discussed in the opening pages of this chapter, consist of angular blocks, some several meters across. During transport, the rocks were enclosed in a matrix of mud and other debris.

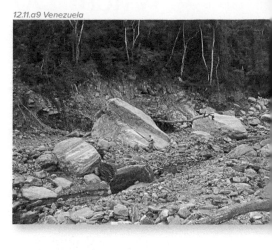
12.11.a9 Venezuela

Rock and Debris Avalanches

Debris avalanches are high-velocity flows of soil, sediment, and rock that result from the collapse of steep mountain slopes. A debris avalanche moves down the slope, in many cases traveling considerable distances down valleys and across relatively gentle slopes. A *rock avalanche* occurs when a rock mass falls off a cliff face and shatters on contact, sending a turbulent jumble of rock fragments, some bigger than cars, flowing downhill. They can flow at over a hundred kilometers per hour, fast enough to continue flowing uphill when they encounter a topographic obstacle. Both types of avalanches are often triggered by earthquakes and volcanic eruptions, and they can kill thousands of people at a time. In 1970, an earthquake shook loose a large chunk of an Andean mountain, causing a rock and debris avalanche (▲) that buried a town (not visible in this photograph) and 18,000 of its inhabitants.

12.11.a11 Huascarán, Peru

12.11.a8

12.11.a10

Classification of Slope Failures

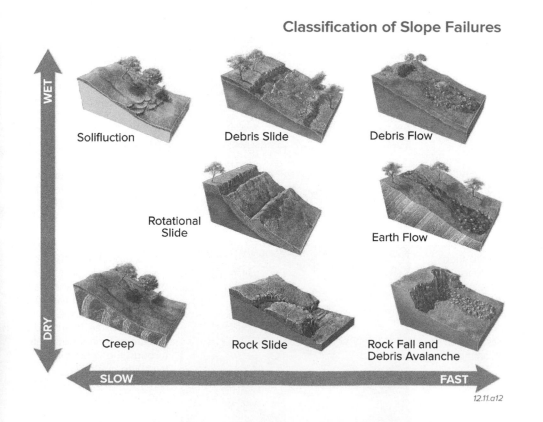

WET

Solifluction

Debris Slide

Debris Flow

Rotational Slide

Earth Flow

DRY

Creep

Rock Slide

Rock Fall and Debris Avalanche

SLOW   FAST

12.11.a12

This diagram (◀) summarizes the classification of slope failures based on the type of material involved and the mechanism of movement, but viewed in the context of the *rates* of movement and whether movement occurred under *dry* or *wet* conditions. In this table, fast means you cannot outrun it, moderate means you have a chance to escape, and slow means you might have to stay for a while to see anything happen. For the slowest rates, you would not be able to see the movement in real time, only the results of months, years, or decades of accumulated movement.

Before You Leave This Page

✓ Sketch and describe what happens during the following: creep, debris slide, earth flow, debris flow, and debris avalanche. Compare how each of these features flows.

✓ Draw a table summarizing different types of slope failures.

12.11

12.12 Where Do Slope Failures Occur in the U.S.?

LANDSLIDES AND OTHER SLOPE FAILURES have destroyed large parts of cities in some countries and have killed more than 30,000 people in a single event. Each year in the U.S., landslides cause 25 to 50 deaths and result in more than $3 billion in damage. What slope-failure events have affected the U.S.? What is the likelihood they will occur again? Which areas are most at risk in the future?

A What Are Some Large Slope Failures That Affected the U.S.?

There have been countless landslides, debris flows, and other slope failures. The landslide that accompanied the 1980 eruption of Mount St. Helens was the largest landslide in U.S. history. It generated enough debris to fill 250 million dump trucks. Other slope failures are described below.

12.12.a1 Denali Region, AK

1. A large (7.9 magnitude) earthquake struck the Denali (Mount McKinley) region of south-central Alaska in 2002. Ground shaking associated with the earthquake caused huge slope failures off the region's steep mountains. Spectacular rock falls, rock slides, and debris avalanches slid down the steep slopes and flowed across and buried the Black Rapid Glacier. The region is prone to slope failures partly because tectonics has rapidly uplifted the mountains and formed an array of joints, faults, and tilted rock layers.

12.12.a2 Pacific Palisades, CA

2. Landslides and debris flows are common in southern California (◄), where some houses have been built on risky sites that have been undercut by waves, streams, and roads, causing unstable slopes to collapse downhill.

12.12.a4 Tully Valley, NY

4. The largest landslide to affect New York in 75 years moved through the Tully Valley (◄), near Syracuse, in 1993. The earth flow mobilized over a million cubic meters of weak clays that had been deposited in a glacial lake. Similar glacial lakebeds occur in many parts of the region.

12.12.a3 Gros Ventre, WY

3. The Gros Ventre slide (◄) is one of North America's largest historic landslides. During this rock slide, a slab of limestone more than 1.5 km (1 mi) long broke loose along weak bedding planes that dipped toward, and were undercut by, the Gros Ventre River of Wyoming.

12.12.a5 Thistle, UT

5. The most costly landslide in U.S. history devastated the small Utah town of Thistle in Spanish Fork Canyon in 1983. The landslide (◄) moved slowly downhill, reaching a maximum rate of only 1 m (3 ft) per hour. It severed railroad service between Denver and Salt Lake City, flooded two major highways, and formed a new lake where the town had been. A railroad tunnel now cuts through the light-colored landslide material.

B What Is the Potential for Landslides and Debris Flows in the U.S.?

All states receive some damage from landslides and debris flows, but not all areas have the same potential hazard. The potential for landslides is highest near steep cliffs, hills, and mountains, such as in the Rocky and Appalachian mountains. Landslides are most common in areas with weak, heavily weathered materials or recent tectonic activity.

Potential for Landslides and Debris Flows in the U.S.

This USGS map portrays the landslide hazards of the lower 48 states. Red areas represent the greatest hazards, followed by yellow and then green. Other areas have less potential for landslides or are unstudied. What landslide potential exists where you live or visit?

1. Many parts in the Pacific Northwest experience landslides because of the many steep mountains, heavy rainfall, and rainfall that melts snow cover. The region also contains areas with high potential for debris flows, especially along slopes and valleys connected to the active Cascade volcanoes. An eruption on a snow-capped peak can unleash large debris flows.

2. The coastal parts of central and southern California have high landslide potential because of steep mountains, earthquakes, and coastal erosion that undercuts weak material along hillslopes overlooking the shoreline. Debris flows are also common, especially in the high mountains flanking Los Angeles, California. These recently uplifted mountains have very steep slopes and receive locally intense rainfall. Wildfires worsen the situation.

3. A high potential for landslides and debris flows occurs in the Rocky Mountains and along the Wasatch Front, the steep mountain front that flanks Salt Lake City, Utah.

4. Landslide hazards in the central U.S. are mostly along steep bluffs that flank the rivers, or in areas, such as the northern Great Plains, underlain by weak materials.

5. In the east, landslides are common in the Appalachian Mountains, where landslides mobilize soil or occur along weaknesses in folded, faulted, and weathered rock layers. Shales, because of their inherently weak character, are especially prone to landslides.

6. Florida and the coastal plain of the southeast Atlantic seaboard have some of the lowest potential for landslides because the region lacks steep slopes. Other parts of the Southeast have a similarly low potential for landslides because of the very flat nature of large areas. Sites along the Mississippi and other rivers have a potential for debris flows, such as those generated when steep riverbanks fail.

250 km

12.12.b1

Landslides and La Conchita

12.12.t1, La Conchita, CA

The coastal community of La Conchita in southern California was partially overrun by a landslide in 2005, which was a repeat of one in 1995 (right). The landslide mobilized poorly consolidated sediment along the steep bluffs overlooking the town. The material flowed down into the community, burying and destroying a number of houses. Although the 1995 landslide had previously destroyed part of the town, houses were rebuilt in the area after 1995, only to be destroyed by the 2005 landslide. The 2005 landslide remobilized parts of the 1995 landslide and also incorporated new material farther back into the cliff. Would this be a good place to rebuild again? How much should hazard insurance cost people who live here?

Before You Leave This Page

✓ Briefly describe factors involved in landslides in the U.S.

✓ Summarize some factors that make some areas of the U.S. have high risks for landslides or debris flows.

✓ Identify whether you live in an area with a high potential for landslides, and list some possible factors why your area has this potential.

✓ Describe what happened at La Conchita. Was it a good idea to build or rebuild there?

12.12

12.13 How Do We Study Slope Failures and Assess the Risk for Future Events?

HOW CAN WE LIMIT DAMAGE CAUSED BY SLOPE FAILURES? We observe and monitor active slope failures to better understand how such systems operate. We then use this information to recognize areas that either have experienced past slope failure or are susceptible to future slope failures. The most hazardous sites have dangerous combinations of geomorphic factors, such as steep slopes and weak materials.

A How Do We Recognize and Monitor Active or Recent Slope Failures?

To limit property damage and the potential for loss of life, we study and monitor active and recently active slope failures. The resulting observations and measurements provide a solid basis for investigating the processes, causes, and hazard assessment of destructive slope failures.

12.13.a1 Venezuela

After a slope-failure event, we conduct field studies to document the distribution, size, and other characteristics of the event, largely to understand what happened and to assess the potential for future events. This area (◄) was destroyed by debris flows.

12.13.a2 Mount Rainier, WA

Scientists use surveying equipment to document topography, especially changes in slope, of hazardous areas. From such measurements, they construct maps and cross sections showing where slope failures are most likely. These data help us devise plans to minimize losses from slope failure.

B How Do We Recognize Prehistoric Slope Failures?

We identify prehistoric slope failures by observing the topography, surface and subsurface distribution of rock types, condition of the rocks, and geometry of layers and other geologic structures.

12.13.b1 Hat Creek, British Columbia, Canada

1. Landslides and other slope failures can disrupt or even overrun existing topography, forming a random-looking assemblage of humps and pits (◄). We call this type of landscape *hummocky topography,* and it is often indicative of some type of slope failure.

12.13.b3 Venezuela

3. Each slope failure leaves behind characteristic deposits. We observe modern deposits and then use these characteristics to recognize deposits of past events that were similar to the modern one. These deposits (◄) were left by debris flows.

12.13.b2 Mount Hood, OR

2. This flank of Mount Hood (◄), a Cascade volcano, contains large deposits composed of broken, angular fragments of rock. These deposits formed when parts of the summit collapsed, slid, and tumbled downward, at some time in the recent past.

12.13.b4 Mount St. Helens, WA

4. Debris flows, debris avalanches, and landslides are sometimes recognized because they carried distinctive rock types into areas where such rocks do not otherwise occur. Rocks in this debris pile (◄) were derived from the distant Mount St. Helens, Washington.

C How Do We Assess an Area's Potential for Slope Failure?

We try to assess the likelihood that a location will suffer the disastrous consequences of slope failure by examining evidence of past slope failures in the area or in adjacent areas that have a similar setting. We also evaluate the steepness of slopes, including any recent changes, and any other factors leading to slope failure.

12.13.c1 Nepal

Evidence of Past Slope Failures

One of the best indications for potential slope failure is evidence of past failures. The more recent the failure, the more likely such an event will recur in the near future. This part of the Himalaya (◄), with steep hillslopes scarred with slope failures, looks risky.

12.13.c2 Durango, CO

Situated in Area with Known Problems

The slope angle of hillsides and mountains (►) is clearly a key factor in the potential for slope failure but must be evaluated in the context of the types of materials and subsurface structures present.

12.13.c3 Fort Funston, CA

Steepness of Slopes

A site may be at risk if its geologic setting is similar to other slopes that have failed. Part of this cliff (◄) collapsed during an earthquake, and nearby parts have the same steep slopes, weak rocks, and position that can be undercut by the ocean waves.

12.13.c4 Zermatt, Switzerland

Recent Changes in Slope

Natural processes, over time, act to adjust slopes to the appropriate, stable angle, but this equilibrium can be upset if the slope is steepened or undercut by natural or human activities. This rock slide (►) failed in 1991, burying 31 houses.

12.13.c5 Capitol Reef NP, UT

Conditions of Material

Another factor is the nature of material on a slope (◄)—whether material is loose sediment or solid bedrock, and is resistant to erosion or relatively weak. The presence, spacing, and orientation of fractures and other geologic structures can weaken rocks and facilitate downslope slippage of materials.

12.13.c6 Augustine volcano, AK

Potential Triggers

If other factors are equal, an area is at higher risk for slope failure if there are frequent events, such as volcanic eruptions or earthquakes, that could trigger slope failure. Steep slopes with loose material on an active, shaky volcano are trouble (►).

The Blackhawk Landslide

A huge lobe of shattered rock lies in a valley north of the San Bernardino Mountains, northeast of Los Angeles, California. This mass has hummocky topography and large, shattered pieces of rock that are different than rocks in adjacent areas. This feature, known as the *Blackhawk Landslide,* formed in prehistoric times when a large segment of the mountains collapsed, shattered, and flowed as debris avalanches more than 10 km (6 mi) out into the valley.

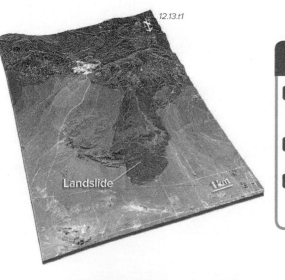

12.13.t1

Landslide   1 km

Before You Leave This Page

✓ Describe some ways that geoscientists investigate slope failures.

✓ Summarize characteristics used to identify prehistoric slope failures.

✓ Summarize some aspects that might indicate that an area has a high potential for slope failure.

12.13

12.14 What Is Happening with the Slumgullion Landslide in Colorado?

THE SCENIC SAN JUAN MOUNTAINS OF COLORADO contain the Slumgullion landslide, one of the best-studied landslides in the world. The landslide has been moving for more than 1,000 years, and part of it is still moving, allowing geomorphologists to examine up close the processes of a relatively slow slope failure.

A What Is the Setting and Morphology of the Slumgullion Landslide?

12.14.a2 Slumgullion, CO

1. The Slumgullion landslide is a conspicuous feature in the landscape of the San Juan Mountains of southwestern Colorado. It begins among high peaks of altered volcanic rocks, snakes downhill, and spreads out when it reaches the valley bottom. The landslide shows as a tan-fringed mass in this photograph (◀).

2. The flowing material in the landslide is mostly derived from the clay-rich, altered volcanic rocks and so is weak and fine grained; it could be called an earth flow. The abundant water in the landslide reduces its strength, allowing it to flow down relatively gentle slopes. Steeper slopes at the scarp, combined with weak materials and abundant water, produced a setting favorable for starting the landslide.

3. The head (top) of the landslide is a steep landslide scarp within volcanic rocks. Interactions with water altered the volcanic rocks into a weak, clay-rich material. This material is too weak to support the steep slope, and so it collapsed downward, starting the landslide.

4. When it reached the valley, the landslide blocked the Lake Fork of the Gunnison River, forming Lake San Cristobal, Colorado's second-largest natural lake. The river flows from right to left in this view.

12.14.a1

Parts of the Landslide

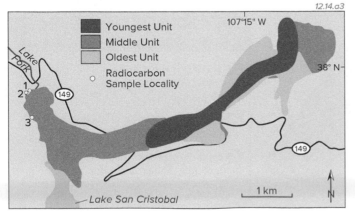

12.14.a3

Youngest Unit
Middle Unit
Oldest Unit
○ Radiocarbon Sample Locality

107°15" W
38° N

Lake Fork
149
149
1 km
N
Lake San Cristobal

12.14.a4 Slumgullion, CO

5. When USGS geomorphologists investigated the landslide, they discovered that it actually consisted of three different ages of material (◀): a young, central part that was still active, and two older parts. In the active, center part, continued movement has tipped over trees (▶) and caused damage to a road that by necessity crosses the flow.

B What Features Are Associated with the Landslide?

As the landslide moves down the hill, shearing and flow form different features along different parts. USGS scientists documented and interpreted the various features and how they formed, in terms of where they were on the landslide.

12.14.b1 Slumgullion, CO

12.14.b2 Slumgullion, CO

12.14.b3 Slumgullion, CO

The upper part of the landslide contains fissures—*scarps*—where the landslide pulled away from the adjacent rocks or where there was internal stretching of the flowing mass. This photograph shows a still-active scarp.

The sides and middle of the landslide contain linear to gently curved zones of shearing, where the fastest-flowing parts of the landslide move past slower moving or stationary material, such as this trough on the right side.

The toe of the landslide commonly overruns areas downhill, as occurred here, where a slab of soil, bushes, and other material has been thrust over land that has not slipped or not slipped as much as the material on top.

C What Other Studies Did Geoscientists Do to Understand the Landslide?

To study the landslide, the scientists combined surveying and aerial photographs to construct a very detailed topographic map. From various types of data, the geoscientists calculated that the weak materials in the landslide are moving down a slope of 7° to 10°, much less than a typical angle of repose for normal, dry materials. The landslide is moving quite fast and continues to travel downhill as scientists study it.

1. Geoscientists investigated the Slumgullion landslide not only to understand this particular landslide but to learn how landslides in general operate. They examined temperature and precipitation records to see if the landslide behaved differently during warm or wet times than during cold or dry periods. Velocity of the landslide decreased in the winter, when the formation of ice reduced the amount of water available to the landslide. The landslide also moved more slowly when conditions were drier.

2. Geoscientists used GPS and other surveying methods to measure how fast different parts of the landslide were moving. On this map (▶), longer red lines indicate faster rates of downhill flow.

3. The fastest rates of movement (the longest lines on this map) are in the middle segment of the landslide, where the landslide is moving as fast as 7.8 m (26 ft) per year.

4. The upper part of the landslide is moving more slowly, locally less than one meter per year.

5. Scientists and engineers also tested physical properties, including strength and moisture content, of the landslide material. They then used these data and computers to model how the landslide is moving.

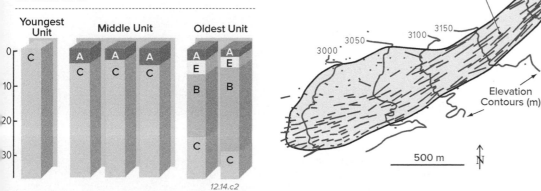

12.14.c1

Elevation Contours (m)

500 m

N

12.14.c2

6. To constrain the three ages of landslide material that were identified and mapped, scientists compared the soil development in the three parts (▲). As we discuss in another chapter, soil develops layers, called *horizons*, over time, with the A horizon on top, B horizon in the middle, and C horizon near the base. The oldest part of the landslide had well-developed A and B horizons, whereas the youngest part was too young to have developed either an A or B horizon. Carbon-14 ages obtained from samples of charcoal and wood within different parts of the landslide indicate that the landslide first formed about 1,000 years ago, yet it is still very active today.

Before You Leave This Page

☑ Describe the setting and morphology of the Slumgullion landslide.

☑ Summarize the types of studies scientists conducted on the landslide and what each type of study revealed about what was happening with the landslide.

12.14

12.15

Which Areas Have the Highest Risk of Slope Failure?

THIS DIVERSE LANDSCAPE has features that appear to be related to slope failure. Large, angular blocks occur in several different settings, and some of the hills may not be stable or safe. You will use descriptions and images of these features to determine what hillslope processes are occurring in different areas, and how they affect where people may live safely. The landscape is stylized and exaggerated to highlight potentially hazardous areas.

Goals of This Exercise:

- Observe the landscape to investigate the overall setting of different areas, and interpret the setting of each location from the figure and descriptions.
- Assess the hazards in different areas.
- Construct a map that shows areas that have a high risk for different types of slope failure.
- Identify locations that you think are most safe or moderately safe on which to construct houses and other buildings.

Procedures

Use the available information to complete the following steps, entering your answers in the appropriate places on the worksheet or online.

1. Observe the features shown on this landscape. Read the text boxes associated with each feature and decide what that statement implies about the geomorphic setting of the area and how the landscape reflects the underlying geology.

2. Think about the description of each area and consider possible types of slope failure that could occur there. Provide a reasonable interpretation of what types of slope processes are occurring and what key observations led you to that conclusion.

3. On the figure in the worksheet, draw approximate boundaries around areas that you interpret as having the highest risk for each type of slope failure. Label each area with a few words to identify the main hazard you interpret to be present.

4. Draw the letters S and M on the map for sites where you think it would be relatively safe to live. Write an (S) for one or more relatively safe places, an (M) for a moderately safe place to live. There is not a single best choice for any of these sites, so be prepared to describe your reasoning and to discuss your choice.

1. A series of small hills, referred to by local people as the *Bent Fence Hills,* contains trees that are tipped over at odd angles. Local farmers complain that they have to keep straightening their crooked fences on these hillslopes. For some reason, no one has ever built a house here.

2. A flat-topped hill, called *Flattop Hill,* is surrounded by a steep cliff formed by a resistant layer of basalt. The basalt is jointed and underlain by a weak layer of clay. Below the cliff are a series of large, angular blocks of basalt. A large, spoon-shaped scar scoops into part of the cliff.

9. The *Hazel River,* named for the greenish-brown, volcanic-derived mud along its banks, cuts through the landscape, flowing from right to left. Paralleling the river on both sides are low terraces that are only a few meters higher than river level. On these low terraces are large volcanic blocks of andesite, some as big as a house. They are not present on higher areas away from the river. No one has ever seen the river with enough water to move such large blocks, but they only occur next to the river.

12.15.a1

Flattop Hill

Bent Fence Hills

Terrace

Hazel River

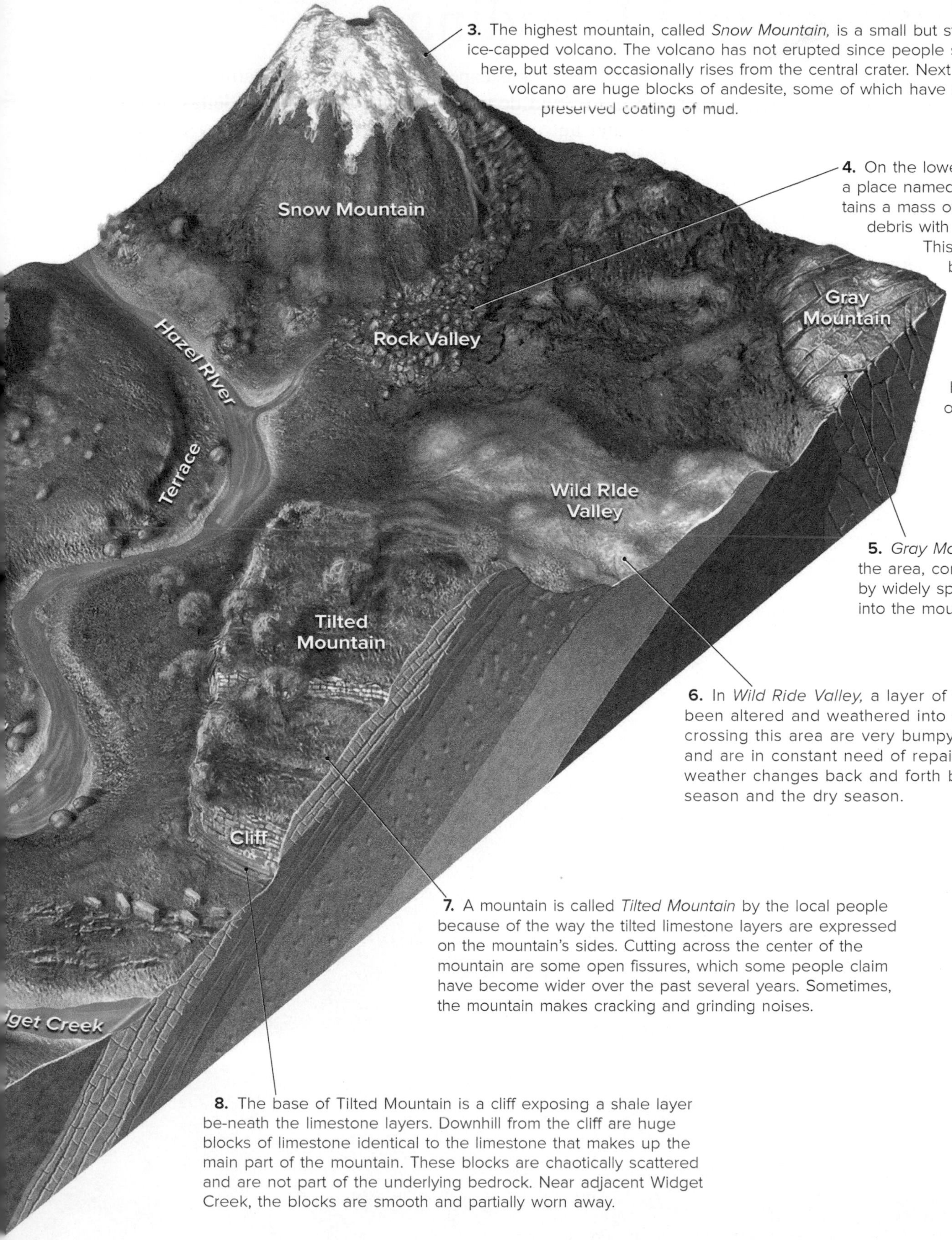

3. The highest mountain, called *Snow Mountain,* is a small but steep ice-capped volcano. The volcano has not erupted since people settled here, but steam occasionally rises from the central crater. Next to the volcano are huge blocks of andesite, some of which have a partially preserved coating of mud.

Snow Mountain

4. On the lower flanks of the volcano is a place named *Rock Valley,* which contains a mass of large rocks and other debris with hummocky topography. This mass can be followed back upslope to a huge, bare scar on the side of the volcano. This debris cuts across the paths of smaller streams that originated in adjacent hills. The area has no soil or trees.

Gray Mountain

Hazel River

Rock Valley

Terrace

Wild Ride Valley

5. *Gray Mountain,* in the corner of the area, contains a gray granite cut by widely spaced joints that dip back into the mountain.

6. In *Wild Ride Valley,* a layer of volcanic ash has been altered and weathered into sticky clays. Roads crossing this area are very bumpy, have visible cracks, and are in constant need of repair, especially when the weather changes back and forth between the rainy season and the dry season.

Tilted Mountain

Cliff

7. A mountain is called *Tilted Mountain* by the local people because of the way the tilted limestone layers are expressed on the mountain's sides. Cutting across the center of the mountain are some open fissures, which some people claim have become wider over the past several years. Sometimes, the mountain makes cracking and grinding noises.

Iget Creek

8. The base of Tilted Mountain is a cliff exposing a shale layer be-neath the limestone layers. Downhill from the cliff are huge blocks of limestone identical to the limestone that makes up the main part of the mountain. These blocks are chaotically scattered and are not part of the underlying bedrock. Near adjacent Widget Creek, the blocks are smooth and partially worn away.

12.15

Streams and Flooding

EROSION AND DEPOSITION BY STREAMS sculpt Earth's landscapes into an array of features. Flowing water in streams picks up sediment, transports it to lower elevations, and deposits it along floodplains, deltas, and other sites. Flooding rivers and smaller streams deposit sediment and nutrients critical to agriculture, but they can also inundate cities and destroy structures built too close to the riverbank. How do streams operate, and can we predict how often a flood-prone area will be flooded?

The Yukon Delta, shown in this satellite image, is a huge, fan-shaped landform where the Yukon River ends its 3,185 km-long journey by emptying into the Bering Sea along the west coast of Alaska. This longest of Alaskan rivers transports vast quantities of sediment eroded from the highlands of Alaska and northwestern Canada.

How do streams form, and how do they carry sediment?

Where the river meets the sea, the flowing water spreads out, slows down, and deposits its load of sediment in a *delta*. Sediment carried and deposited offshore (lighter blue) causes the delta to grow seaward with time, adding new land to the coast. Not all rivers have deltas.

What factors determine where a river deposits sediment?

13.00.a1

446

TOPICS IN THIS CHAPTER

13.00.a2

◄ **The Yukon River** collects water from a large region of Alaska and Canada's Yukon Territory. It drains an area of 840,000 km² (324,000 mi²). Periodically, water volume in the river exceeds the confines of its channel, causing flooding. When winter ice on the river begins to melt and break up, it piles up in ice jams that cause additional flooding.

How is the size of a river related to the size of the area it drains, what causes a flood, and what information do we need to predict flooding events?

13.00.a3 Denali NP, AK

◄ **Many Alaskan rivers** are full of sediment derived from weathering and erosion of the mountains and lowlands. This river in Denali National Park is choked with coarse gravel, sand, and fine sediment.

What types of sediment do different kinds of rivers carry?

13.00.a4 Yukon Delta, AK

◄ **During the summer,** lush vegetation grows on the strips of land between the delta waterways. Wetlands on the sediment-rich delta are important breeding sites for migratory birds.

What effect does vegetation have on streams, and what effect do streams have on vegetation?

A Variety of Streams

There is no accepted definition of the difference between a stream and a river, but generally we can think of a stream as any body of water that has a current and flows downhill within a channel, driven by the influence of gravity. A river is a large stream of considerable volume and with permanent or seasonal flow. Each stream, for example the Yukon River, has its own characteristics and history, which are specific to its geographic and geologic setting. Hydrology is the scientific study of water, including stream flow. Some streams are steep and turbulent, moving large boulders, whereas others are slow and tranquil, transporting only sediments the sizes of silt and clay. Some streams meander in huge, looping turns, while others distribute their flow in a network of channels that split off and rejoin in a braided pattern. Certain factors govern the behavior of all streams, including the steepness of the channel, supply of sediment, climate, and tectonic history. Most stream systems change downstream and over time, producing a characteristic suite of landforms that dominate the landscape. Streams can flood huge tracts of land and transport enormous volumes of sediment. The Amazon River in South America (shown below) dumps over 200,000 cubic meters of sediment-laden water into the ocean every minute.

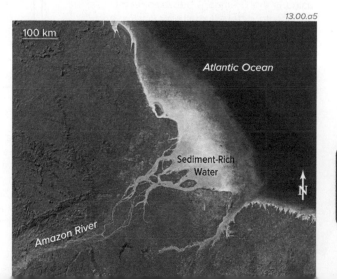

13.00.a5

100 km

Atlantic Ocean

Sediment-Rich Water

Amazon River

13.0

13.1 What Are Stream Systems?

STREAMS ARE CONDUITS OF MOVING WATER driven by gravity, flowing from higher to lower elevations. The water in streams comes from precipitation, snowmelt, springs, and lakes. A stream drains a specific area and joins other streams draining other areas, forming an interconnected *drainage network* or *stream system*.

A What Is a Stream?

A stream carries a varying amount of flowing water through a single channel or through a number of interconnected channels. Such channels vary in width from a few meters to kilometers across.

1. The Potomac River (▼) winds its way along the border between Maryland and West Virginia, on its way to Washington, D.C. A number of smaller streams join the river from both sides, forming a *stream system*. Water flowing in streams moves rock fragments and dissolved minerals from higher elevations to lower ones.

2. The amount of water that flows through a stream channel varies with time, mostly reflecting the influence of changes in the seasons (e.g., from winter to spring) and changes in weather. At some times of the year and during rainy periods, the *flow* and amount of sediment being carried increase. Flow decreases at especially dry times of the year or during times of few storms and extended periods of below-freezing temperatures. The amount of water flowing in a given amount of time is the *discharge*, which has units of cubic meters per second or m³/s.

13.01.a1

3. We calculate discharge (represented by the letter Q) by multiplying the cross-sectional area of the channel by the average velocity of the flow, assuming a rectangular cross-sectional area:

$$Q = stream\ depth \times stream\ width \times average\ stream\ velocity$$

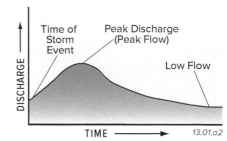

13.01.a2

4. A graph showing the change in discharge over time is a *hydrograph*. This hydrograph shows that discharge increased and then decreased over time in response to a storm. The shape of the graph displays how a stream responds to precipitation, conveying useful information about a stream and the area it drains.

B Where Does a Stream Get Its Water?

A *drainage basin* is defined by topographic high points, diverting all runoff to a single major outlet. Runoff from rainfall, snowmelt, springs, and lakes will flow downstream and out of the drainage basin at its low point. Beneath the surface, infiltration from soils and aquifers is also important in contributing to stream discharge.

1. *Drainage Basin*—In this figure (▼), each of two adjacent stream systems has a drainage basin, shaded in different colors. Runoff from the red area drains into the streams on the left; runoff from the blue area drains into the streams on the right. The ridge between the two drainage basins is the boundary between water flowing into different drainage basins, and is a *drainage divide*.

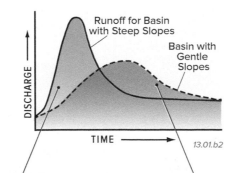

0.5 km

13.01.b1

2. *Basin Slope*—Overall slope of a drainage basin helps us determine how fast water in the basin empties after a heavy rain or after snowmelt, as shown by the graph below (▼).

13.01.b2

3. Runoff from a steep drainage basin is fast, and much water arrives downstream at about the same time, yielding high discharge values soon after a precipitation event.

4. Runoff from a more gently sloped basin is spread out over time, leading to lower peak discharges.

5. *Basin Size and Shape*—A drainage basin's size and shape influence its flow response to rainfall. The plot below (▼) shows a hydrograph for a single storm event, along with a simplified map of the basin's shape.

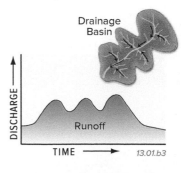

13.01.b3

6. Following a storm event, a hydrograph for a simple basin shows a single-peak increase in discharge with a gradual decrease, such as the graph in part A (above).

In contrast, the complex three-part drainage basin shown here may lead to a three-peak response to a single event. Total discharge in a larger or more complex basin will be more spread out because some water travels short distances and other water travels long distances.

C What Are Tributaries and Drainage Networks?

Streams have a main channel fed by smaller subsidiary channels called *tributaries*. Each tributary drains part of the larger drainage basin, and the combination of tributaries and the main stream forms a drainage network. The response of a stream to precipitation is influenced by the number and size of its tributaries, as well as soil conditions, amount of vegetation, and other factors.

13.01.c1

1. In this drainage network, smaller tributaries join to form larger drainages (◄), which join to form even larger drainage basins. The drainage network has a branched appearance, similar to a tree, with numbers indicating the *stream order*. A *first-order stream* (1 on the figure) has no tributaries, but two such streams can join to produce a *second-order stream* (2), and so on.

Drainage Order

TIME

2. Runoff in this type of drainage network, with many branches (stream orders), is carried by many channels. As a result, the system responds more slowly and with a smaller discharge peak to a precipitation event than a drainage system with fewer stream orders.

13.01.c2

DISCHARGE

TIME

3. A drainage network with fewer tributaries (stream orders) responds more quickly to an identical storm event (◄). The resulting increased discharge causes more erosion of sediment along the stream and more transport of sediment out of the area.

D How Does Geology Influence Drainage Patterns?

Patterns that stream systems carve across the landscape are strongly influenced by the topography and types of earth materials. Channels form preferentially in weaker material, and their paths reflect differences in materials and the geometry of fractures, layers, and other structural features. There are a number of drainage types, including the ones described below.

13.01.d1

13.01.d2

13.01.d3

Dendritic Drainage Pattern—Where rocks have about the same resistance to erosion, or if a drainage network has operated for a long time, streams can form a treelike, or dendritic, drainage pattern. Dendritic patterns are very common, and the drainages in section C (above this section) are also dendritic.

Radial Drainage Pattern—On a fairly symmetrical mountain, such as a volcano or weather-resistant rock, drainages flow downhill and outward in all directions (i.e., radially) away from the highest area. If drainages flow inward, toward a low area from several directions, the pattern is called a *centripetal drainage pattern*.

Structurally Controlled Pattern—Erosion along fractures, tilts, or folds can produce a drainage that follows a layer or structure, and then cuts across a ridge at a near 90° angle to follow a different feature, forming a *trellis drainage pattern*. If drainages follow curved layers around a central area, they form an *annular pattern*.

North American Drainages

Colors on this map show areas of the land that drain into different parts of the ocean. Boundaries between colors are drainage divides. The best known drainage divide is the Continental Divide, separating drainages that flow westward into the Pacific Ocean from those that flow east and south into the Gulf of Mexico. Other drainages flow into the Arctic Ocean, and some drainages in the western U.S. have interior drainage (they flow into low continental areas and do not reach the ocean).

Arctic Ocean

Pacific Ocean

Atlantic Ocean

Continental Divide

Interior Drainage

Eastern Continental Divide

Gulf of Mexico

13.01.d4

500 km

Before You Leave This Page

✓ Sketch and describe the variables plotted on a hydrograph and what this type of graph indicates.

✓ Describe how the shape and slope of a drainage basin affect a hydrograph.

✓ Sketch and describe how the distribution of tributaries influences a stream's response to precipitation.

✓ Sketch three kinds of drainage patterns, and discuss what factors control each type.

13.1

13.2 How Do Streams Transport Sediment and Erode Their Channels?

STREAMS ERODE BEDROCK and loose material, transporting the liberated material as sediment and as chemical components dissolved in the water. The sediment is deposited when the stream can no longer carry the load, such as when the current slows or sediment supply exceeds the stream's capacity.

A How Is Material Transported and Deposited in Streambeds?

Moving water applies force to a channel's bottom and sides and can pick up and transport particles of various sizes: clay, silt, sand, cobbles, and boulders. The amount of sediment carried, including material chemically dissolved in solution, is the *stream load*. The maximum possible stream load under given conditions is the stream's *capacity*.

1. Fine particles (silt and clay, collectively referred to as mud) can be *suspended* in the moving water, even in a relatively slow current. This material is the *suspended load*.

2. Sand grains can roll along the bottom or be picked up and carried down-current by bouncing along the streambed—the process of *saltation*.

3. Larger cobbles and boulders generally move by rolling and sliding, a process called *traction*, but they only move during times of high flow. The largest of these clasts (pieces of rock broken off from larger rocks) can be briefly picked up, but only by extremely high flows. The size of the largest particles that can be transported is called the stream's *competence*.

4. Material that is pushed, bounced, rolled, and slid along the streambed is the *bed load*. If the amount of sediment exceeds the stream's capacity to carry it, as when velocity drops, the sediment is deposited. The balance between transport and deposition shifts as conditions change, and grains are repeatedly picked up and deposited.

5. Some chemically soluble ions, such as calcium, sodium, and chlorine, are *dissolved* in and transported by the moving water. They constitute the *dissolved load*.

13.02.a1

B What Processes Erode Material in Streams?

Moving water and the sediment it carries can erode bedrock, sediment, or other material that it flows past. Erosion occurs along the base and sides of the channel and can break apart and remove sediment within the channel. The silt, sand, and larger clasts carried by the water enhance its ability to erode.

1. Sand and larger clasts are lifted by low pressure created by water flowing over the clast tops. They can also be pushed up by turbulence. Once picked up, the grains move downstream and collide with obstacles, where they chip, scrape, and sandblast pieces off the streambed—the process of *abrasion*. Abrasion is concentrated on the upstream side of obstructions, such as larger clasts or protruding bedrock.

13.02.b2 Tortilla Flats, AZ

2. Concentrated erosion can also occur when water and sediment swirl in small depressions, carving bowl-shaped pits called *potholes* (◄).

3. Turbulent flow releases and lifts material from the streambed, particularly pieces loosened by fracturing, boundaries between rock layers, and other discontinuities.

4. Soluble material in the streambed, such as salt, can be removed by *dissolution*. Most dissolved material in streams, however, comes from subsurface water that has flowed into the stream. Other dissolved material mostly comes from soluble rock layers along the stream, but some also comes from farms and other sources.

13.02.b1

C How Does Turbulence in Flowing Water Affect Erosion and Deposition?

Water, like all fluids, has *viscosity*—resistance to flow. Viscosity and surface tension are responsible for the smooth-looking surface of some streams. As the water's velocity increases, the flow becomes more chaotic or *turbulent*, often appearing white and foaming due to entrapped bubbles of air. The water can more easily pick up and move material within the channel.

1. All streams have parts that are more turbulent and parts that are less turbulent. Water moving smoothly in parallel layers is called *laminar flow*. In such smooth-appearing water, viscosity limits chaotic flow (turbulence). Fast flow and rough, obstruction-filled stream bottoms promote *turbulent flow*.

2. Moving water has momentum, so it tends to keep moving in the same direction unless its motion is perturbed, such as where the water encounters a bend in the stream. In many cases, water can flow smoothly over somewhat uneven surfaces.

3. As water velocity increases, viscosity is less able to dampen chaotic flow, and the water flow becomes more complex, or turbulent. As turbulence increases, swirls in the current, called *eddies,* form in both horizontal and vertical directions.

4. Fast-moving water has numerous eddies where flow strays from the downstream direction.

5. Near the bottom of the stream, up-current eddies can overwhelm gravitational force and lift grains from the channel. Turbulence, in general, increases the chance for grains to be picked up and carried in the flow.

Slower, Less Turbulent

Faster, More Turbulent

13.02.c1

D How Do Erosion and Deposition Occur in Streams Confined Within Bedrock?

Many streams, especially those in mountainous areas, are carved into bedrock, and are referred to as *bedrock streams*. If the bedrock is relatively hard, the shape of the stream channel is controlled by the geology. If bedrock consists of softer material, such as easily eroded shale, then the earth materials will have less control over the shape and character of the stream channel.

Erosion

Deposition

13.02.d1 Nepal

13.02.d2 Grand Canyon, AZ

13.02.d3 Black Canyon of the Gunnison, CO

1. The steep gradients and higher velocities typical of mountain streams (▶) erode down into the channel faster than the stream can erode across into its sides. The bed load of sand, cobbles, and boulders helps break up and erode the bedrock channel. Rapid changes in gradient, as occur along waterfalls and rapids, increase water velocity, turbulence, and the ability of the moving water to erode the channel.

2. As a result, steep bedrock streams commonly incise (cut) deep channels, a process called *downcutting* (▶). They can have relatively straight sections, initially controlled by the location of softer rock types, faults, or other zones that are eroded more easily than surrounding rocks. Once formed, such hard-walled canyons, such as the aptly named Black Canyon of the Gunnison, make it difficult for the stream to shift its position.

3. Deposition in bedrock channels occurs where the water velocity decreases, for example along stream banks during flooding or in pools behind rocks or other obstacles. Rocks and sediment constrict this stream (◀), forming a pool of less turbulent water behind the obstacle. Sediment is deposited in slow-moving eddies on the flanks of this pool, but such sediment is vulnerable to later erosion, and its stay is therefore very transient.

Before You Leave This Page

✓ Sketch and describe how a stream transports solid and dissolved material.

✓ Sketch and explain the processes by which a stream erodes into its channel and which sites are most susceptible to erosion.

✓ Sketch and describe laminar flow versus turbulent flow.

✓ Describe some aspects of erosion and deposition in bedrock channels.

13.2

13.3 How Do Streams Change Downstream or Over Short Time Frames?

STREAMS BECOME LARGER as more tributaries join the drainage network. As a stream flows downstream, it generally increases in size, discharge, and the amount of sediment it carries. A stream changes over short time spans, for example after a storm, from winter to summer, and from year to year.

A How Do River Systems Change Downstream?

The character of a stream changes as the stream flows downhill from its *headwaters* (where it starts), to its *mouth* (where it ends). The direction of flow from high elevations to lower ones is referred to as *downstream*.

Gradient

1. The profile of most stream systems is steep in the headwaters, gradually becoming less steep downstream toward the mouth. The steepness is also called the *gradient*, which is defined as the change in elevation for a given horizontal distance. We can calculate gradient by dividing the vertical change along some segment by the horizontal distance of that segment.

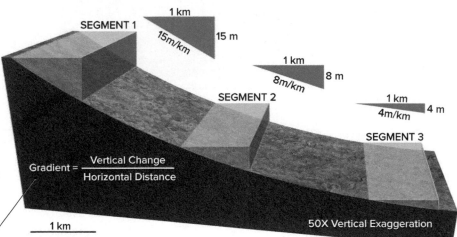

$$\text{Gradient} = \frac{\text{Vertical Change}}{\text{Horizontal Distance}}$$

50X Vertical Exaggeration

13.03.a1

2. This change in gradient downstream is represented by the blue triangles, which show how much the stream drops for a given length of stream. A steeper gradient means the stream drops more over the same horizontal distance. We express gradient as meters per kilometer, feet per mile, degrees, or as a percentage (e.g., 4%). Here, gradient is calculated for three segments. It varies from 15 m/km to 4 m/km and decreases downstream. The vertical scale of the triangles is greatly exaggerated compared to the horizontal scale.

Channel Size, Water Velocity, Discharge, and Sediment Load

3. Streams erode bedrock and other materials and then transport the sediment downstream. Sediment can be deposited anywhere along the way or can be carried all the way to the mouth of the stream. The system shown here has a main stream fed by three main tributaries, labeled T1, T2, and T3. Small graphs around the map plot how parameters change down the stream, from the headwaters (H), past each tributary (T1, T2, and T3), to the start of a delta (D), and the river's mouth (M).

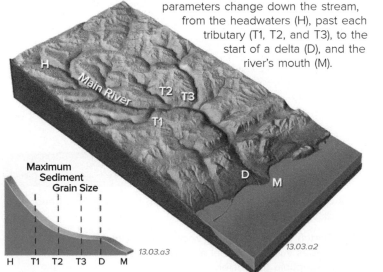

13.03.a2

13.03.a3

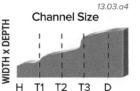

13.03.a4

13.03.a5

13.03.a6

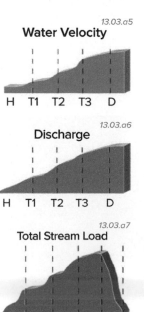

13.03.a7

4. As the gradient of the river decreases from the headwaters to the mouth (▲), the *maximum size of sediment* that the river carries decreases. Also, abrasion during transport reduces the size of clasts. For both reasons, coarse material is more common in the headwaters than it is near the mouth.

5. Downstream, the channel increases in overall *size* (◄), where size indicates a combination of the width and depth across the channel. Specifically, size means the *cross-sectional area* of the channel, obtained by multiplying the channel's width times its average depth.

6. The *velocity of water flow* increases downstream (◄), as a higher volume of water allows the water to flow more easily and faster through the channel. Velocity looks higher in steep streams, but *turbulence* and *friction* in such streams reduce the velocity.

7. Since the cross-sectional area of the channel and the velocity of the water both increase, so does the total *volume of water* flowing through the river (◄). The volume of water flowing through any part of the stream per unit of time is the *discharge* and is calculated by multiplying the water velocity times the cross-sectional area.

8. The total amount of sediment that the stream is carrying, the stream load, increases downstream (◄), until large amounts of sediment begin to be deposited within the delta and near the mouth.

B What Is the Relationship Between Water Flow and Transported Sediment?

A stream can carry sediment only up to a certain size for any particular velocity. Also, at a given flow rate, a stream is capable of transporting only a certain amount of sediment, which is called its *capacity*. Normally, a stream is carrying far less sediment than its capacity. As velocity decreases, so does capacity—a river is able to carry less sediment as it slows down.

1. This graph (▶) shows the relationship between stream velocity and the size of the particles that can be carried by different modes of transport. The vertical bands of color indicate different grain sizes, and the inclined lines indicate whether sediment of that size is being carried in suspension, is being transported on the bottom of the streambed, or is being deposited.

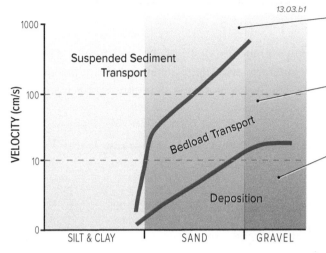

2. At high velocities (above 100 cm/s), clay, silt, and sand can be carried *suspended* (floating and drifting) in the water. That is, on this plot, those grain sizes extend above the red line.

3. At moderate velocities (10–100 cm/s), silt, clay, and perhaps some fine sand remain suspended, but most sand and gravel slide, roll, or bounce along the streambed, as part of the bed load.

4. At low velocities (below 10 cm/s), gravel and most sand remain at rest on the streambed or are deposited if a sediment-carrying stream slows down to these velocities. These grain sizes plot below the blue line. Only silt, clay, and fine sand are transported, with silt and clay in suspension and fine sand in the bed load.

C How Does Stream Behavior Vary Over Time?

The amount of precipitation, snowmelt, and influx from groundwater varies, both during a single year and over longer timescales of decades to centuries. Here, we examine the flow of a stream during a single year. This flow is highly idealized, and not every stream will have peaks at the time of year shown here.

1. Examine the hydrograph to the right (▶), which plots discharge for different seasons. What are some possible explanations for this pattern? For the year shown, why might the highest discharge occur during springtime? Other streams have other patterns than shown here, but this one is typical for streams in some parts of the U.S.

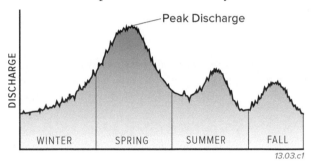

3. After the most intense period of snowmelt, discharge in the stream decreases throughout the rest of the spring and into the early summer. An increase in rainfall due to a summer-thunderstorm season causes a short-lived increase in discharge in the middle of the summer. For this stream's drainage area, the fall season typically has less precipitation than other times of the year. For the year shown, however, moisture from the tropics moved far enough north to cause a period of on-and-off rain showers in the middle of the fall.

2. The lowest discharge is in the winter, when some streams and lakes are frozen and precipitation occurs as snow and as gentle winter rains. Discharge is highest during spring, when the snow and ice melt, adding meltwater to the stream. The highest value on the plot (during the spring) is the *peak discharge*.

4. This scenario is one possible explanation for a hydrograph with this shape. Try to find a hydrograph for a river or smaller stream in your area, observe the pattern, and try to explain the increases and decreases in discharge over a year.

13.03.c2 Cottonwood Wash, Westwater Canyon, UT

5. A stream or river that flows all year, such as the Mississippi River or the stream represented by the graph above, is a *perennial stream*. Because no place has rain all of the time to keep a stream flowing, some water in a perennial stream must be supplied by subsurface flows, by a melting snowpack, by a lake, or by some combination of hydrologic stores. Some streams flow every year but not the entire year, typically flowing only during rainstorms and spring snowmelt; such a stream is an *intermittent stream*. *Ephemeral streams*, including the normally dry desert stream here (◀), only flow after a storm event but not on an annual basis.

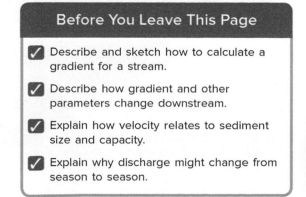

Before You Leave This Page

✓ Describe and sketch how to calculate a gradient for a stream.

✓ Describe how gradient and other parameters change downstream.

✓ Explain how velocity relates to sediment size and capacity.

✓ Explain why discharge might change from season to season.

13.3

13.4 What Factors Influence Profiles of Streams?

STREAM SYSTEMS HAVE DIFFERENT GEOMETRIES, both in map view and when viewed in profile (from the side). Streams have diverse settings, origins, and ages, and they respond to perturbations in their environment by eroding their channels and banks, by depositing sediment, and by changing their gradient.

A What Is the Shape of a Stream's Profile?

Streams are *dynamic systems* driven mostly by precipitation and gravitational forces. They respond to many factors that influence how the stream operates and how it interacts with its channel and the adjacent landscape. Over time, most streams attain a *longitudinal profile* that is steeper near the headwaters and is progressively less steep downstream.

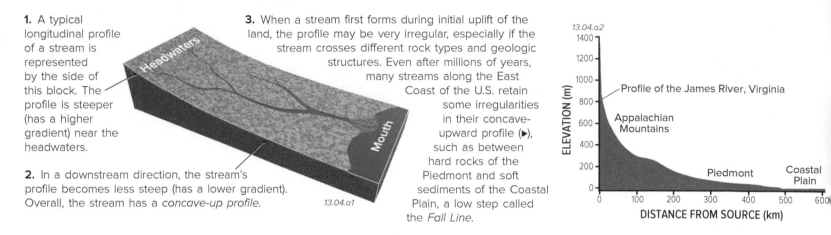

1. A typical longitudinal profile of a stream is represented by the side of this block. The profile is steeper (has a higher gradient) near the headwaters.

2. In a downstream direction, the stream's profile becomes less steep (has a lower gradient). Overall, the stream has a *concave-up profile*.   13.04.a1

3. When a stream first forms during initial uplift of the land, the profile may be very irregular, especially if the stream crosses different rock types and geologic structures. Even after millions of years, many streams along the East Coast of the U.S. retain some irregularities in their concave-upward profile (▶), such as between hard rocks of the Piedmont and soft sediments of the Coastal Plain, a low step called the *Fall Line*.

13.04.a2

B What Controls the Profiles of Streams?

Streams and other agents of transportation erode mountains and carry the sediment downhill, eventually depositing it in a basin or along the sea. The lowest level to which a stream can erode is its *base level*. The base level controls the topography along a stream, how a stream develops over time, and how it responds to change.

1. This terrain shows a typical drainage system consisting of mountainous headwaters, mid-elevation foothills, and a broad, low-elevation plain, ending at a shallow inland sea.

2. High above base level, steep gradients in the mountains cause streams to erode sharply into the bedrock. The terrain appears rough and may include deep canyons cut into bedrock.

3. Foothills in front of the mountains also experience erosion but have intermediate gradients and generally appear less rough.

4. Closer to base level, streams on the broad plain have a much lower gradient, and the surrounding landscape has less relief and appears relatively smooth. This plain has low relief because either it has been eroded down or its low parts have filled with sediment. In this case, it is some of both.

13.04.b1

6. As shown by the side of the block, variations in roughness of the landscape indicate the decrease in gradient from the mountains to the broad plains. A profile down the channel of any stream in this area is less irregular than the rough topography defined by the ridges and canyons. Most rivers and large streams have a fairly smooth, concave-up profile.

5. A stream cannot erode below its base level. In this terrain, the inland sea represents the base level. While *local base level* for a stream system may be a lake or the bottom of a closed land basin (with internal drainage), for most stream systems, the ultimate base level is sea level (the ocean).

C What Factors Influence or Change Stream Profiles?

Some rocks are more difficult to erode than others, and Earth is a dynamic planet, with frequent changes in tectonics, sea level, and climate. Thus, stream profiles are always changing in response. Below is a more detailed look at how each of these factors influences the profile of the stream, as well as changes in erosion versus deposition.

Rock Type

1. As a stream flows over different kinds of earth materials, its ability to erode a deeper channel is influenced by the type of material over which it flows. Soft rocks erode more easily than hard rocks.

13.04.c1

2. In unconsolidated sediment and easily eroded rocks, such as shale (◄), the stream can create a smooth profile that more or less represents the most efficient mechanism of draining the basin (called *equilibrium*), because there are no major obstructions. The profile attains a smooth, concave-upward shape.

3. Rocks that are more resistant to erosion will tend to form steeper slopes, with cliffs, waterfalls, steep rapids, and narrow canyons (►). Alternating strong and weak rocks yield a stair-stepped topography, but through time a stream can smooth out its profile.

13.04.c2

Tectonics

4. Tectonic forces can cause *uplift* or *subsidence* of an entire region or can occur differentially, affecting one part of the region more than other parts.

Gentler Gradient

Increasing Subsidence

Deposition of Sediment

13.04.c3

5. Differential subsidence can flatten or steepen gradients (◄), depending on where it occurs. In this example, subsidence occurred beneath the mountains, flattening the gradient and causing widespread deposition as stream velocity decreased and the stream lost capacity.

6. Tectonic uplift generally causes streams to erode down into the landscape (►), cutting canyons and steepening the topographic relief. Here, tectonic uplift of the mountains steepened the gradient, causing incision of the landscape by erosion, forming a narrow canyon.

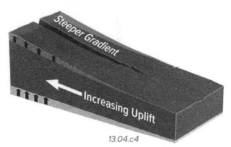

Steeper Gradient

Increasing Uplift

13.04.c4

Sea Level

7. Sea level is the ultimate base level for streams that empty into the ocean. Changes in sea level will change the location of the shoreline and the elevation of base level.

Seas

13.04.c5

8. If base level is lowered, such as by a drop in sea level, the stream will downcut to try to match the new base level. In this example (◄), erosional incision begins at the coast and works its way upstream. A drop in sea level causes most streams to downcut.

9. If the base level rises, such as during a rise in sea level, the stream will erode inland but will deposit sediment along the coastline's new position (►). In this manner, the stream tries to achieve a new equilibrium profile.

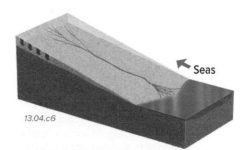

Seas

13.04.c6

Climate

10. Streams respond to changes in climate, especially an increase or decrease in rainfall. Under wet conditions, slopes will have more vegetation so they can hold soil in place, but increased discharge could allow more sediment to be carried away, beveling the hills more than during dry periods.

13.04.c7

Stability of Conditions

11. If conditions such as climate remain stable, a stream may approach an *equilibrium profile*. When a stream is in steady state, there is a balance between the supply of sediment and the amount the stream can carry. The channel becomes stable, or nearly attains a type of *dynamic equilibrium*, neither eroding nor depositing material. We call such a stream a *graded stream*.

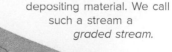

13.04.c8

Before You Leave This Page

☑ Sketch and describe the typical profile of a stream.

☑ Describe the concept of base level and how it is expressed in a typical mountain-to-sea landscape.

☑ Summarize factors that influence a stream's profile and behavior.

13.4

13.5 Why Do Streams Have Curves?

ALL STREAMS HAVE CURVES OR BENDS, ranging from gentle deflections to tightly curved but graceful meanders. Why are streams curved? What is inherent in the operation of a stream that causes it to curve? Curves and bends are a natural consequence of processes that accompany the movement of flowing water in a stream.

A What Is the Shape of Stream Channels in Map View?

Streams have bends, but not all bends are the same. Some are gentle, open arcs, whereas others are tight loops. The shape of a stream in map view can be thought of as having two main variables: whether there are single or multiple channels and how curved the channels are. The amount that a channel curves for a given length is its *sinuosity* and can be calculated by dividing channel length (including all the bends) by the down-valley length (that is, the straight-line distance down the valley). If a stream has a perfectly straight path down the valley, its *sinuosity index* is 1.0.

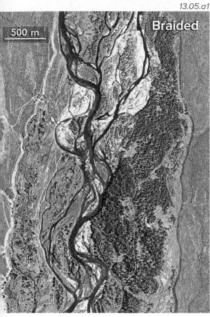
13.05.a1
Braided
500 m

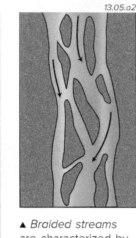
13.05.a2   13.05.a3   13.05.a4

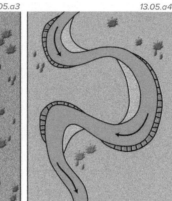
13.05.a5
Meandering
1 km

▲ *Braided streams* are characterized by a network of interweaving, sinuous channels, but the overall channel can be fairly straight.

▲ Many streams consist of a single channel that is gently curved. This type of stream is referred to as a *low sinuosity* stream.

▲ *Meandering streams* have channels that are very curved, commonly forming tight loops. Such streams have *high sinuosity,* and this type of bend is a *meander.*

B What Processes Operate When a Stream Meanders?

Channels in soft materials, especially streams with low gradients, generally do not have long, straight segments. Instead they flow along sinuous paths. Curves or meanders cause differences in water velocity in the channel and suggest a balance between deposition and erosion, as illustrated below for a meandering stream.

1. Small graphs show profiles across the channel in different locations. Arrows in the channel show relative velocity. In fairly straight segments, the channel is nearly symmetric (not deeper on one side than the other). The current in such segments is fastest in the center of the channel and slowest along the banks. In such straight segments, sediment can be deposited along the channel margins where velocity is lowest, and erosion can occur in the middle of the channel where velocity is highest.

2. Where the stream is curved, the channel becomes asymmetric (is shallower on one side than the other). The channel is shallower and the water velocity is lower on the inside of a bend. This causes sediment to be deposited on the inside of the bend in what is called a *point bar.*

13.05.b1

Cutbank (Erosion)

Inside of Bend (Deposition)

Point Bar (Deposition)

Outside of Bend (Erosion)

3. The channel is deeper and water flows faster on the outside of the bend. The deepest part of the stream would be located beneath the longest arrows on the main figure. Momentum causes the force of the water to be directed toward the outside of a bend. These factors cause the outside bend to be eroded into a steep river bank called a *cutbank.* Erosion of the cutbank can balance deposition on the point bar, keeping the channel width, cross sectional area, and volume fairly constant.

C How Do Meanders Form and Migrate?

Meanders are landforms produced by migrating channels of rivers and streams and are common in streams that have low gradients. Meanders have been studied in the field and by comparing aerial photographs taken at different times. They can be simulated in the laboratory using large, sand-filled tanks, through which water is initially directed down a straight channel in fine sand. Almost immediately, the water begins to transform the straight channel into a sinuous one, similar to the sequence shown below.

1. A curve starts to form when a slight difference in roughness on the channel bottom causes water to flow faster on one side of the channel than on the other.

2. The side of the channel that receives faster flow erodes faster, creating a slight curve. The faster moving current slightly excavates the channel bottom, deepening the outside of the bend, forming deeper areas called *pools*.

7. Once formed, a curve continues to affect the flow by causing faster flow and increased erosion on the outside of the bend. Some secondary currents develop in the bend area and further excavate the pools, speeding flow and enhancing the cutbank. This type of system, where a feature or process, once started, affects the system in such a way that it results in even more of the same (in this case erosion), is called a *positive feedback*.

10. Meanders sometimes join as they migrate toward each other, in the direction of the yellow arrows. This cuts off the meander.

11. The narrow neck of a looping meander can also get cut off during a flood event, when the stream rises above the channel and across the floodplain, connecting two segments of the stream. In either case, the part of the meander that is abandoned is a *cutoff meander*.

3. The overall discharge in the stream is constant, so the deeper channel on the outside of a bend takes more water, leaving less water for the other side. The water on the inside of the bend becomes shallower and slower.

4. The sediment carried by the slower water on the inside of the bend is dropped and deposited on a point bar, depicted here as sand-colored material.

5. Erosion scours the opposite (outside) band of the channel, forming a cutbank. Such sideways erosion is termed *lateral erosion*.

6. Through this process, each meander begins to preferentially erode its banks toward the outside. This causes the stream to migrate toward the sides and downstream, as shown by the small orange arrows.

8. As meanders migrate back and forth across the lowlands, they continuously erode and deposit the loosely bound sediment in the channel and in the zone adjacent to the stream where meanders and floods occur (*floodplain*). This is the main way in which a floodplain forms, and the old abandoned meanders remain as *scars* on the floodplain.

9. Meanders migrate until they encounter a resistant riverbank, until the volume and velocity of flow drop too low for erosion to continue, or until two parts of adjacent meanders intersect.

12. Cutoff meanders formed in either way (10 or 11) are initially filled with water, forming isolated, curved lakes, called *oxbow lakes*.

13.05.c1–6

Interfering with Sinuosity

Streams attain, through natural processes, their characteristic sinuosity, which represents the interplay between variations in channel depth, water velocity, erosion, deposition, and transport of sediment. Humans can upset this balance by straightening rivers and eliminating their natural variability. These engineering solutions (▶) often cause trouble downstream because they upset the dynamics and equilibrium of the system. Streams that have been channelized may exit the channelized segment with a higher velocity, lower sinuosity, and less sediment than is natural. Areas downstream of the channelized segment, therefore, can experience extreme erosion and destruction of riverbank property. In the photograph below, find a point bar and a cutbank, and note which part of the bend (outside or inside) is being "protected." The view is downstream, where more erosion will occur on the cutbank.

13.05.t1 Animas River, Hermosa, CO

Before You Leave This Page

☑ Sketch and describe the difference between braided, low-sinuosity, and high-sinuosity (meandering) streams.

☑ Sketch or describe how velocity and channel profile vary in a meandering stream, and what features form along different parts of bends.

☑ Sketch or explain the evolution of a meander, including how a cutoff meander forms and how it can lead to an oxbow lake.

13.5

13.6 What Happens in the Headwaters of Streams?

STREAMS CAN BEGIN in almost any setting, from a small hill in a pasture to a large, snowy mountain. In either case, the place where a stream starts is its *headwaters*. In its headwaters, a stream is fed by rain, snowmelt, groundwater, outflow from a lake, or some combination of these. The headwaters of many large rivers are in mountains, where streams are steep and actively erode the land with turbulent, energetic water, producing distinctive landforms.

A How Do Streams Start?

A stream does not start with a fully formed channel full of water, but instead it grows incrementally as surface runoff becomes concentrated into channels—is *channelized*. Smaller channels join others until a stream forms.

13.06.a1

Channel Formation

1. Rainwater causes *splash erosion* (◄) as it hits the ground, and water flowing over the surface as *overland flow* causes subtle *sheet erosion*. Water tends to accumulate in natural cracks and low spots, such as these small channels, rather than spreading uniformly across the land. Concentrated flow erodes or dissolves materials, especially those that are weak or loose, eventually carving a small channel or *gully*.

2. Once formed, a channel accommodates runoff within its small drainage basin. The increased flow causes further gully erosion and channel deepening. *Headward erosion* is the cutting toward the gully's headwater and can lead to *stream capture*, the natural diversion of water from one stream into another.

3. Channels occur at all scales. Microscopic channels feed into small channels that feed into larger ones, ultimately forming a stream. The spacing and geometry of the channels are influenced by the steepness of the slope, type of material in the slope, the type and density of vegetation, and other factors.

13.06.a2

B What Landforms Characterize the Headwaters of Mountain Streams?

Mountain streams begin in bedrock-dominated areas with relatively high relief and, in many cases, high elevation. The energetically moving water wears rock down and sculpts the bedrock into steep landforms with moderate to high relief.

1. The headwaters of many streams are in high mountainous areas, where they derive their water from rainfall, melting ice and snow, or mountain springs. Other streams originate in lower, flatter areas and are supplied by precipitation, lakes, springs, or the joining of small, local channels.

2. The area below (▼) is the headwaters of several streams. It is near the top of a mountain, receives abundant snow and rain, and has several lakes left behind when glaciers melted away fairly recently.

13.06.b1

Lake

1 km

3. Streams in mountains typically have steep gradients that result in energetic, turbulent flow (▼). Many have eroded down, or *incised*, into bedrock.

4. Steep and energetic streams can carry large materials. At times, this stream (▼) has an ability to carry large boulders.

13.06.b6 Ice Lake, Silverton, CO

6. Lakes are common in mountains where water is impounded by some obstruction, such as a landslide, or where water fills a natural low spot (▲). If a lake is created by a human-constructed dam, it is a *reservoir*.

5. With time, mountain streams begin to widen their valleys and deposit a thin, winding ribbon of stream gravels along which they flow (▼).

13.06.b2 Beartooth Plateau, WY and MT

13.06.b3 Boulder Canyon, MT

13.06.b4 Central CO

13.06.b5 ANWR, AK

C How Do Rapids and Waterfalls Form?

Streams that flow in an area of mountains and hills commonly have *rapids*, a segment of rough, turbulent water along a stream. Some segments can be so steep that water cascades through the air, forming a *waterfall*.

Most rapids (▲) develop when the gradient of a stream steepens or the channel is constricted by narrow bedrock walls, large rocks, or other debris that partially blocks the channel. Many rapids form where tributaries have deposited fans of debris that crowd or clog the main channel. These obstructions cause water to flow chaotically over and around obstacles, creating extreme turbulence.

The obstruction forming a rapid generally causes water upstream to slow down and "pile up," producing a relatively smooth and slow-moving segment called a *pool*. The photograph above, looking downstream, shows a pool upstream of a rapid. Many streams have this characteristic alternation of rapids and pools, and so are called *pool-and-riffle streams*.

Most waterfalls are awe-inspiring, as water cascades over a ledge or cliff (▲). Where a stream has an abrupt change in gradient, such as at a waterfall, it is called a *nickpoint*. A nickpoint can be distinct or more subtle, and generally indicates some difference in erosion above and below the nickpoint. With continued erosion, the nickpoint, and in this case a waterfall, migrates upstream over time.

D What Features Develop When a Stream Exits from the Mountains?

As mountain streams flow toward lower elevations, they interact with tributaries and commonly decrease in gradient as they pass through foothills or mountain fronts. In response, they form other types of landforms.

Where a steep, narrow drainage enters a broader, more gentle valley, coarse sediment carried by running water or by muddy debris flows piles up just below the edge of the slope at the decrease in gradient, forming a fan-shaped feature called an *alluvial fan* (▶).

When a mountain stream reaches less confined spaces, it commonly spreads out in a network of sediment-filled braided channels (◀). These channels are not strongly incised, so the stream spreads out and deposits sediment along its channel and over a broad floodplain.

13.06.d1 Death Valley NP, CA
13.06.d2 Waiapu River, New Zealand

How Do Mountain Streams Get Sediment?

Mountain streams are energetic primarily because their channels have steep gradients. Erosion dominates over deposition, forming deep *V-shaped valleys* with waterfalls and rapids. Steep valley walls promote landslides and other types of slope failure that widen the canyon and deliver abundant material to the stream for removal (▶). Erosion of materials to the side of a stream is *lateral erosion*. Soil and other loose materials on hillslopes can slide downhill toward the drainage. Tributaries carry debris flows and floods that scour their channels, providing even more sediment into the main stream. Sediment in mountain streams ranges from bus-sized or car-sized boulders down to silt and clay. Larger clasts start out angular but begin to round as they are transported, or as they are struck by stones and other sediment within the turbulent waters.

13.06.t1 Tibet-Nepal Border Region

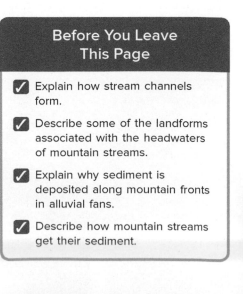

Before You Leave This Page

☑ Explain how stream channels form.

☑ Describe some of the landforms associated with the headwaters of mountain streams.

☑ Explain why sediment is deposited along mountain fronts in alluvial fans.

☑ Describe how mountain streams get their sediment.

13.6

13.7 What Features Characterize Braided Streams?

MANY STREAMS ARE BRAIDED SYSTEMS, with a network of channels that split and rejoin, giving an intertwined appearance. Braided streams have many branches within a channel, generally have a plentiful supply of sediment, a steep to moderate gradient, and they typically carry and deposit coarse sediment (sand and coarser). Braided streams can migrate across broad plains, coating them with a veneer of sediment.

A What Conditions Lead to Braided Streams?

Braided streams are most common in flat-bottomed valleys nestled within mountains and on broad, sloping plains that flank such mountain ranges. They can also form farther from the mountains, in areas where an abundant sediment supply nearly overwhelms the stream's capacity to carry it (competence).

1. Many braided streams drain from high mountains, such as these, modeled after the South Island of New Zealand.

3. The Southern Alps of New Zealand are an actively uplifting and steep range. Glaciers, steep slopes, and locally heavy precipitation in the headwaters of the streams contribute to an abundance of sediment (▶).

5 km

13.07.a1

2. A braided stream deposits sediment within and beside its shallow channel and can leave its channel when the deposited material dams it, especially during floods. Sediment in the riverbank is not cemented or otherwise tightly held together, so the material is easy to erode and redistribute, and the stream can change position relatively easily. As the channel migrates back and forth across the broad plain, it covers the low-relief area with a layer of coarse, stream-deposited sediment (▼).

10 km

4. Braided streams form where there are steep gradients, a plentiful supply of coarse sediment, and conditions that produce variable flows. In this close-up view (▼), individual channels are braided at various scales, but the overall path of the main river channel is fairly straight (has low sinuosity).

13.07.a4

13.07.a3 Denali NP, AK

1 km

B What Size of Sediment Does a Braided Stream Deposit?

Braided streams are characterized by a wider range of sediment sizes than is deposited in meandering streams. They are energetic and can carry and deposit coarse gravels and sands in addition to finer materials.

13.07.b1 New Zealand

13.07.b2 Savage River, Alaska

13.07.b3 Icefields Parkway, Alberta, Canada

Braided streams have numerous channels. They are clogged with sediment, which is constantly picked up in one place and deposited in another. As sediment is picked up, transported, and deposited, the braided channels continuously change position, width, and overall shape, slowly during low flows and more rapidly during floods and other high flows. Deposition actually increases the stream's gradient, accelerating its flow in a process called *aggradation*.

Braided streams form when the stream has a relatively high sediment load dominated by sand and larger sediment. Sands and gravels are the dominant clasts in this braided stream, but some braided streams also carry finer materials, such as mud and silt derived from glaciers and other sources. Overall, braided streams are relatively mud-poor, especially when compared with streams that have meanders. Exceptions are braided streams in regions with active silt-forming glaciers.

These braided-stream deposits contain partially rounded cobbles and pebbles in a sand-dominated matrix. The stream can transport these large clast sizes because it has a steep gradient and carries large amounts of turbulent water during the spring snowmelt from peaks of the Canadian Rockies. These deposits also contain silt-size particles formed from the grinding of rocks beneath nearby glaciers. Most silt is carried farther downstream.

Making and Investigating Braided Streams in the Laboratory

Geoscientists and engineers can study streams by making small-scale versions or models in large water tanks in a laboratory. These tanks, called *stream tables* or *flumes*, can be meters wide and tens of meters long, and they are sloped so that the water flows downhill. The tanks are loaded with sediment, usually sand, silt, and mud. Valves are opened to allow water to enter the elevated end of the tank and flow toward the low end. The scientists then observe the small-scale stream that develops, investigating the processes that occur and the features that form. Different variables, including slope, sediment supply, and consistency of flow, can be specifically varied or controlled to isolate how each factor affects the dynamics of the stream system.

The sequence of images here shows successive stages during an experiment in a 2 × 15 m tank at the National Center for Earth-Surface Dynamics in Minneapolis. In this experiment, a braided stream developed early (top), but became progressively less braided

13.07.t1-6

as alfalfa seeds embedded in the sediment sprouted and grew more dense. These experiments indicate that riverside vegetation plays a key role in stabilizing riverbanks and can actually influence whether a stream remains braided or becomes meandering. This relationship was first recognized more than 60 years ago by geomorphologists working along streams of the U.S. Recent research indicates that deposits characteristic of meandering rivers appeared in the geologic record at the same time as vascular plants (in the Paleozoic Era), implying that such streams require vegetation to stabilize their banks, the same conclusion reached from using stream tables. Prior to that time, more streams would have been braided.

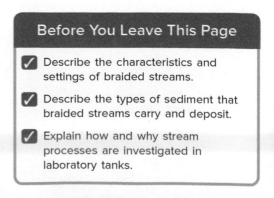

Before You Leave This Page

- ☑ Describe the characteristics and settings of braided streams.

- ☑ Describe the types of sediment that braided streams carry and deposit.

- ☑ Explain how and why stream processes are investigated in laboratory tanks.

13.7

13.8 What Features Characterize Low-Gradient Streams?

IF A STREAM CROSSES AREAS OF LOW RELIEF, the gradient of its channel decreases and the stream may spread out once it is no longer confined by a narrow canyon or valley. Sediment transported and deposited on low-relief plains is mostly clay to sand size, because insufficient energy exists to transport larger clasts, but it can include fine gravels. The stream reworks (picks up and transports) this previously deposited sediment. The resulting landforms indicate the interaction of stream velocity and sediment size, all in the context of a more gentle landscape.

A What Landforms Are Found in Streams with Low Gradients?

Many streams flow across plains that have gentle overall slopes. Such streams adapt to their environs, being dominated by the erosion, transport, and deposition of relatively fine-grained sediment. The features characteristic of these mostly single-channel streams occur at all scales, from those along small creeks to those along the Mississippi River. Features include meanders, floodplains, and low stream terraces. On these two pages, we explore two meandering rivers that have similar features but at very different scales: the Animas River of Colorado and the Mississippi River of the central U.S.

One Main Channel

1. Streams on gentle plains usually occupy a single channel rather than being braided. This single-channel characteristic is linked to the gentle downstream gradient of the stream and its floodplain. Notice that the low-gradient Animas River here occurs on a gentle plain within a mountainous region, so it is important to focus on the characteristics of a stream rather than its surrounding environment. Farther upstream, this river is steep and confined to a narrow and deep bedrock canyon.

Meanders

2. Rivers on gentle plains typically flow in noticeably curved paths, or meanders (▼). Meanders can be gentle curves or can sharply loop back on themselves, with the two meander segments separated only by a narrow strip of land.

Floodplain

4. All streams on gentle plains have floodplains (▶) beside the channel. Floodplains are areas adjacent to a stream channel regularly inundated by floodwaters, which deposit sand, silt, and clay particles. The channel is not necessarily in the middle of the floodplain.

13.08.a4 Pe

Animas River

Floodplain

Meanders

Oxbow Lakes

1 km

13.08.a1 Animas River, Hermosa, CO

Oxbow Lakes

5. Meandering rivers leave behind curved (C-shaped) depressions, called *meander scars*, which were once part of the main channel but were abandoned when that meander segment became cut off from the main channel. When such depressions contain water, they are oxbow lakes. The photograph above shows an oxbow lake on the floodplain. The main channel is on the left side of the photograph. On the large figure to the left, can you spot places where an oxbow could form?

13.08.a2 Ikpikpuk River, AK

Point Bars

3. Most meandering rivers have bow-shaped deposits of sand and gravel that parallel the inside bend of a meander (▶). Such point bars are typically visible as patches of bare, recently deposited sediment. How many point bars can you identify in the more complicated stream pattern to the left?

13.08.a3 Brazos River, Bryant, TX

Point Bar

Meander Scars

13.08.a5 Pecatonica, IL

13.08.a6

N→

5 km

6. Meander scars are exposed as low, curved ridges, lines of vegetation, or curved dry or water-filled depressions (▲).

7. On this computer-generated perspective (▲), the very broad floodplain of the Mississippi River has countless crescent-shaped scars of ancient meanders, abandoned by the shifting of the river. Note also the sand-colored point bars next to the active channel.

8. The dark, C-shaped areas are cutoff meanders (meander scars) that are filled with water—oxbow lakes. Many of these lakes are kilometers long.

Stream Terraces

13.08.a7 Thompson River, British Columbia, Canada

9. Many streams have older, stranded floodplains called *terraces* perched above and outside the current floodplain. It is common to find matching terrace levels on either side of the existing floodplain. The steep slope separating adjacent terraces or a terrace from the floodplain is called a *bluff* or *riser*.

Anastomosing Channels

11. In some places, low-gradient streams have islands between several strands of the channel that split and rejoin, or anastomose (▼). The islands are semipermanent features, with well-established vegetation, unlike the transient, mostly nonvegetated channel deposits between strands of a braided stream. To convey this distinction, such a channel is called an *anastomosing channel*, which can be considered another type of channel, along with meandering, braided, and straight channels.

13.08.a10 Lhasa River, Tibet

Scale of Floodplain, Channel, Meanders, and Other Features

10. Stream channels, meanders, floodplains, and other features can occur at very different scales. Compare the two images to the right. The first is an aerial image of the same Animas River segment shown on the left-hand page. The second is a few meander loops on the Mississippi River. The images are at the same scale. The much smaller-scale Animas River has 15 times more meanders than the Mississippi for the same downstream distance. The scale of the meanders is so different because of the huge differences in size and discharge of the rivers.

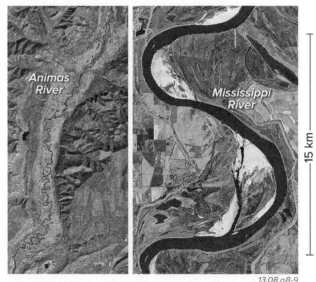

Animas River

Mississippi River

15 km

13.08.a8-9

Yazoo Streams

12. Along some low-gradient rivers, a separate channel runs parallel to the main channel but cannot join it for many kilometers because the sediment dumped on the floodplain by the main channel is high enough to create a natural dam. Instead, the channel stays on the floodplain (▼). This type of stream is called a *yazoo stream* after the Yazoo River, which runs along the floodplain of the Mississippi River, paralleling the river for 280 km (170 mi) before finally joining it.

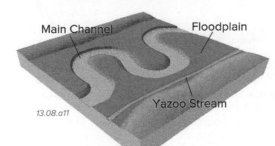

Main Channel   Floodplain

Yazoo Stream

13.08.a11

Before You Leave This Page

☑ Sketch or describe the features that accompany low-gradient rivers, explaining how each forms.

☑ Describe the character of meander scars and oxbow lakes on the floodplains of meandering rivers.

☑ Sketch or describe anastomosing channels and yazoo streams.

13.8

13.9 What Happens When a Stream Reaches Its Base Level?

FOR ALMOST ALL STREAMS, base level is ultimately the ocean. When a stream reaches the ocean or some other base level, it slows down and drops its bed load and suspended load. Temporary base levels are established when a stream is dammed by a landslide or other natural causes, or by human engineering. The new base level causes changes in the stream system both above and below the obstruction, but such changes are temporary.

A What Happens as a Stream Approaches Base Level?

Several landscape-building processes occur when a river or stream enters the ocean, lake, or a temporary base level. Large rivers, such as the Mississippi River, pump fresh water far into the ocean and carry fine sediment out to sea. They deposit coarser sediment as soon as the current slows upon reaching the mouth, forming a *delta* along the shoreline.

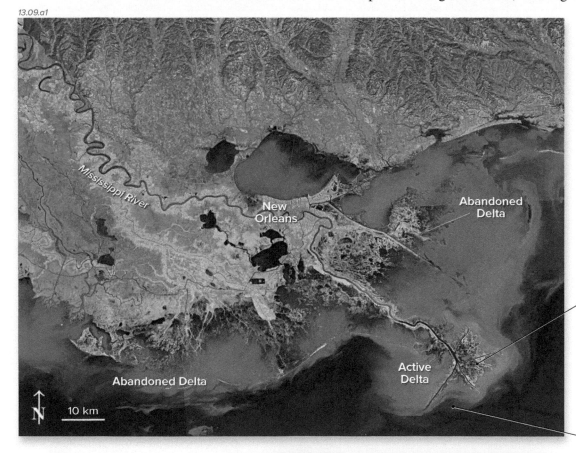

13.09.a1

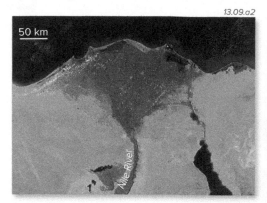

13.09.a2

1. Deltas are river-derived coastal accumulations of sediment both above and below the surface. This satellite view (▲) shows the green, triangular-shaped delta formed where sediment from the Nile River is deposited out into the Mediterranean.

2. A delta also forms where the Mississippi River meets the Gulf of Mexico. In this satellite image, the river changes from a meandering river to a series of smaller channels that branch and spread out in various directions. This branching drainage pattern is a *distributary system*. This type of delta, with separate toes spreading out, is called a *birdfoot delta*.

3. Dark blue colors on this image indicate clear, deeper waters of the Gulf, whereas lighter blue areas contain suspended sediment and mostly have shallower water. Sediment from the river accumulates and builds up the delta, which is eroded by waves and by underwater slumps of the steep, unstable delta front.

5. As a delta builds out into water, it forms new land and a characteristic sequence of sedimentary beds (▶). As the river's current slows, sand and larger particles become too heavy to be carried and are deposited in three types of beds. A set of horizontal beds forms on top of the delta.

Sedimentary Beds in a Delta

13.09.a3

6. A set of dipping beds forms when sediment is deposited over the front edge of the delta, moving the delta seaward.

7. Silt and clay are carried farther out into the ocean (or lake) and are deposited as nearly flat beds in front of the delta. Note the sequence of layers produced as a delta builds out into the ocean: marine clays overlain by cross-bedded sediment and then by horizontal beds deposited partly on land.

4. Over the last 7,000 years, the Mississippi has created and then abandoned at least six mounds of sediment in the form of deltas, each of which marks a former location of the river's mouth and its associated delta; some of these abandoned deltas are labeled on the figure. A new delta, the active delta, is forming where the Mississippi River currently enters the Gulf of Mexico. *Levees* and other constructs will delay future shifts of the river.

B What Controls the Deposition of Sediment in a Delta?

Deposition in a delta occurs where a stream slows, losing capacity and dropping its sediment load. The delta's shape and the type of sediment deposited are influenced by the sediment load and discharge of the stream, and other factors, like wave activity and the amount of vegetation or ice. Some large rivers, such as the Amazon, do not have well-developed deltas.

1. The Lena Delta of Siberia provides one of the most beautiful satellite images of Earth. This image, taken in the summer, shows a thawed East Siberian Sea and abundant vegetation on the delta. Vegetation appears as a variety of colors as this is a false-color composite using various infrared wavelengths. This delta nicely displays the factors that control deposition of sediment in a delta, including the distributary pattern of drainages.

2. *Vegetation*—The amount of vegetation and seasonal changes in vegetation affect the number and location of delta channels. Generally, deltas that have dense vegetation have fewer channels, whereas deltas with sparse vegetation have more channels. Part of the explanation is that vegetation binds the soil and stabilizes channel positions, as has been observed in sand and experimental water tanks.

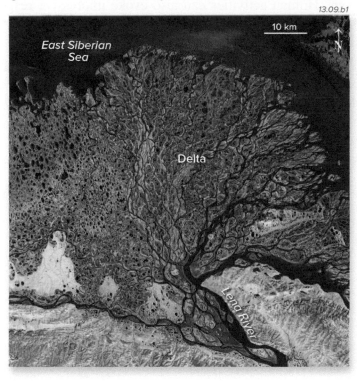

13.09.b1

4. *Discharge*—High-discharge flows tend to extend farther out into the ocean. The deposited sediment can then be affected by waves and by ocean currents.

5. *Wave Erosion*—Waves in an ocean or a lake greatly affect the shoreline of the delta. Waves that are strong or continuous can erode and redistribute sediment along the coast or move it out to sea. Wave action is strongest and most effective during storms.

6. *River and Ocean Ice*—Seasonal changes in the amount of ice in the river and along the coast affect discharge and deposition patterns. River ice makes flow more sluggish, and sea ice tends to trap sediment closer to shore. Since Siberia is a very cold place, the sea around the Lena Delta freezes during the winter. The sea ice stops the action of ocean waves and causes a slowing of the river current near the mouth. As a result, the river deposits sediment along the river-ice interface during the winter.

3. *Sediment Load*—Coarser sediment, such as sand, is carried in the bed load and deposited first as the velocity drops. Finer material, carried in suspension, can be carried farther. If the river carries more sediment and is closer to its capacity, it will deposit more sediment and drop it sooner.

7. *Water Chemistry*—As fresh water mixes with salt water, clay particles clump together, making it easier for them to be deposited as mud.

C What Are the Depositional and Erosional Consequences of Dams?

Human-constructed dams provide hydroelectric power generation, water storage, or flood control, but they stop a stream's normal flow and transport of sediment. The reservoir behind the dam represents a *temporary base level* and affects the stream above and below the dam.

13.09.c2 Hoover Dam, AZ-CA

1. When built, a dam forms a temporary base level. The stream tries to achieve a new equilibrium, both upstream and downstream of the dam.

3. Most dams (▶) release relatively clear water that contains little sediment and that has a renewed capacity to erode. As a result, erosion occurs below many dams, whose clear-water releases contrast with typically muddy or sandy flows of the stream before construction of the dam. For some dams, scientists and engineers are investigating ways to have the sediment in the reservoir bypass the dam, restoring the river to a more natural state. The clear water also starves any downstream delta from its pre-dam supply of sediment and can starve freshwater ecosystems of critical nutrients. The clear water also cannot replenish floodplains with a new influx of mud.

2. The change in base level causes the stream to deposit sediment behind the dam in an attempt to retain its equilibrium profile. The pile of sediment builds out into the reservoir in the same way that a natural delta builds out into the sea. This sediment can eventually fill up the reservoir, shortening its life span, and making it no longer able to fulfill the original function(s) for which it was built (flood control, water storage, etc.).

Post-Dam Equilibrium Gradient

Pre-Dam Gradient

13.09.c1

Before You Leave This Page

☑ Describe what happens when a river or smaller stream enters an ocean or lake, and what factors control deposition of sediment in a delta.

☑ Sketch and describe the types of sediments deposited by a delta and the setting in which each type of sediment formed.

☑ Describe how construction of a dam affects a stream.

13.9

13.10 How Do Streams Change Over Time?

ON GEOLOGIC TIMESCALES, streams, even large rivers, come and go. Some streams are old and others are surprisingly young. The age and history of a stream are important considerations when evaluating how the stream might respond to tectonic, climatic, and sea-level changes. Human activities can also evoke impressive responses in streams.

A How Old Are Streams?

Streams flow from their headwaters to base level as long as enough water and slope are available to maintain downstream flow. A stream's life can begin or end due to changes in (1) water and sediment supply at the source, (2) slopes across which the stream flows, or (3) elevation of its base level. Streams can exist for millions of years, although their characteristics may change due to climatic, glacial, and tectonic events.

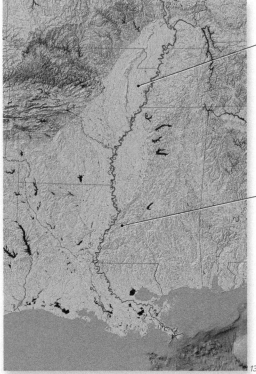
13.10.a1

Lower Mississippi River

1. This shaded-relief map shows the Mississippi River meandering across a low-relief plain, some of which is the modern floodplain. The plain is underlain by sediments that exhibit the recent history of the river. The oldest river sediments, preserved at depth, indicate that the lower Mississippi began draining the continent more than 65 million years ago at the end of the Mesozoic Era.

2. The river and its tributaries eroded across a series of older layers at a time when the river incised a valley when sea level (base level) was low. Subsequent sea-level rise decreased the river's gradient, and the river deposited sediment, filling the excavated valley to its present level. Some of these adjustments occurred during the Ice Age from 2 million to 12,000 years ago, giving new life to an old river.

Upper Mississippi River

13.10.a2

13.10.a3

3. The upper Mississippi River is young. It formed since the retreat of the last ice sheets, some 10,000 years ago. During the last glacial advance (◄), ice sheets and glaciers covered the northern half of North America, so northern rivers like the upper Mississippi did not exist. The weight of the ice sheets depressed the crust, causing large regions to slope northward (opposite of today).

4. Melting of the ice released huge discharges of water and sediment and allowed the land to be uplifted (*isostatic rebound*) near the Great Lakes region (◄). A completely new river channel was carved—the upper part of the Mississippi.

The Fall Line

5. A major boundary, called the Fall Line, traces its way between the Appalachian Mountains and the East Coast of the U.S. The Fall Line, shown as a red dashed line on the map to the right, is marked by waterfalls, which attracted early settlers because the cascading water could power early mechanical facilities, such as grain mills. Many of these early settlements grew into important East Coast cities, for example, Washington, D.C.; Richmond, Virginia; Raleigh, North Carolina; Columbia, South Carolina; Montgomery, Alabama; and Augusta, Georgia. Each of these cities owes its location to the presence of the Fall Line.

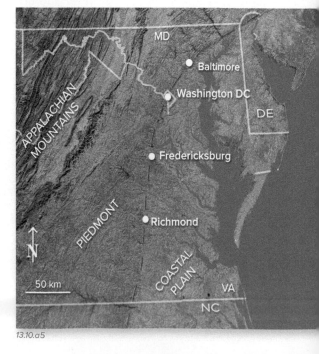
13.10.a5

13.10.a4 Great Falls of the Potomac, VA and MD

6. These waterfalls formed along the contact between soft sediment of the Coastal Plain and harder bedrock in the foothills of the mountains. The Great Falls of the Potomac River, upstream from Washington, D.C., illustrate how the Fall Line developed. Before the last glacial advance, the Potomac River occupied a broad valley. A drop in sea level during the last glacial advance caused the river to incise deeper. Erosion proceeded upstream—as a nickpoint—stripping away the soft sediment until it encountered the harder rocks at the Great Falls.

B How Do Stream Systems Respond to Changing Conditions?

Streams are sensitive to many aspects of their environment, including rainfall, climate, geology, vegetation, and the influence of humans. Streams respond to changes in these factors by eroding deeper, depositing sediment, changing the type of sediment they carry, and changing position. Here, we briefly explore a few of these issues.

Runoff

13.10.b1 East Dubuque, IL

1. The amount of flow is the most important factor in how a stream develops, and this depends mostly on the amount and timing of precipitation. Direct runoff during rainfall and delayed runoff from snowmelt supply most of the water to streams. The amount of runoff varies widely, at times being too much for the channel to hold, causing waters to spill out onto the floodplain (▲), resulting in a flood.

Glacial and Sea-Level Effects

2. Global cooling and the growth of ice sheets and glaciers lower sea level. It can load and depress the crust, causing drainages to flow toward the ice sheets, where they fill large lakes.

3. Melting of the ice releases huge amounts of meltwater and sediment, creating new or larger channels. Isostatic rebound due to ice removal can tilt the land and reverse regional drainage patterns.

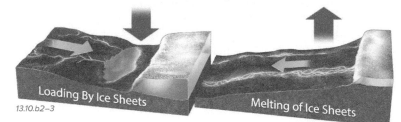

Loading By Ice Sheets

Melting of Ice Sheets

13.10.b2–3

Tectonism

4. Tectonism can uplift mountains, increasing slope, precipitation, and the supply of coarse sediment. The slope of a stream, the supply of sediment, and the abundance of vegetation largely determine whether a stream is braided or meandering.

5. Conversely, mountain uplift can create a rain shadow that decreases precipitation on the opposite (leeward) slope, reducing the amount of runoff.

13.10.b4

13.10.b5 Niagara Falls, NY

Geology

6. Streams can erode unconsolidated sediment and soft rock types more easily than hard ones. Streams that are eroding downward may encounter rocks that have different characteristics, causing a change in the geometry of the stream. The impressive Niagara Falls (◄) along the Canadian–U.S. border formed when the post-glacial Niagara River encountered a more resistant rock layer underlain by less resistant shale.

Human Engineering

7. Dams and other flood-control structures change base level, the amount of discharge, and the supply of sediment, all of which affect the stream system both upstream and downstream. For example, dams trap sediment in the reservoir and release clear, cold water.

13.10.b6 Flaming Gorge Dam, UT

Climate and Vegetation

8. When pioneers settled in the American Southwest in the mid-1800s, many streams were flowing on broad valleys. These settlers built farms on the moisture-rich floodplains.

9. Increases in precipitation around 1880 caused streams to incise (erode down) several meters into their floodplains. Paradoxically, this incision dried up the previous floodplain and many of the farms. Around 1940, represented by the lower diagram, climate and other effects caused the channels to deposit sediment and begin to build up again. Changes in vegetation, whether due to climate, humans, or some other factor, also impact the stream system in many ways, such as stabilizing the banks of the stream from erosion (►).

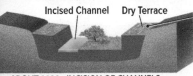

Floodplain

BEFORE 1880 - STREAMS ON FLOODPLAIN

Incised Channel   Dry Terrace

ABOUT 1880 - INCISION OF CHANNELS

Infilled Channel

ABOUT 1940 - CHANNEL FILLING BEGINS

13.10.b10 Chinle Wash, UT

Before You Leave This Page

✓ Describe how rivers can be old or young, using the Mississippi River as an example.

✓ Describe the Fall Line and how it influenced locations of cities.

✓ Sketch and describe some effects that glaciers have on river systems.

✓ Describe how river systems respond to changes imposed by climate, tectonism, geology, and human engineering.

13.10

13.11 What Happens During Stream Incision?

STREAMS CAN INCISE INTO LANDSCAPES, as when the land is uplifted, base level drops, or the climate changes (especially an increase in precipitation). Incision by streams forms a variety of features, including multiple levels of terraces. Rivers and smaller streams also carve some unusual canyons, such as those that take odd routes across the landscape, cutting right across mountains that would otherwise seem to be obstacles. What sequence of events leads to the development of these interesting features?

A How Are Stream Terraces Formed?

Stream terraces are relatively flat benches that are perched above a river or stream and that stair-step up and outward from the active channel. Most terraces are composed of river-derived sediment and are essentially former floodplains and alluvial plains. Other terraces are cut directly into bedrock and form by erosion. Terraces record different stages in the stream's history and indicate that the stream has incised into the land.

13.11.a1 Snake River, Moose, WY

Highest Terrace

Lowest Terrace (Floodplain)

1. Terraces form a series of flat to gently sloping benches or steps, flanked by steeper slopes. Terraces successively step up and away from the channel. The highest terrace may be tens of meters or more above the active channel. The lowest surface is commonly only a meter or so above the channel and is often flooded, perhaps nearly every year—it is a floodplain.

2. This series of terraces (▼) flank the Snake River in Jackson Hole, Wyoming. The terraces are numbered from highest (1) to lowest (3). The modern floodplain also is labeled (F). Which of these terraces formed long ago and which one formed more recently?

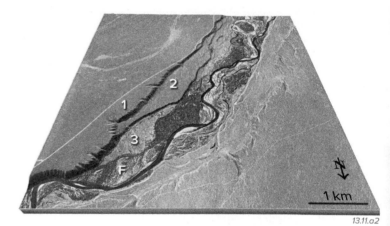

1 km

13.11.a2

First Stage (oldest) ➔ Second Stage ➔ Last Stage (youngest)

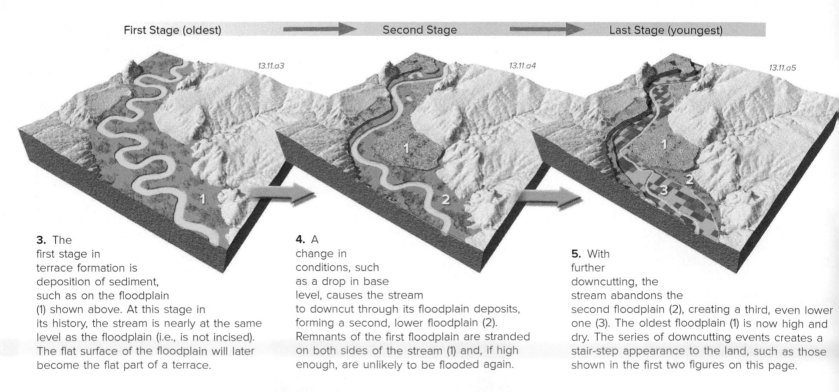

13.11.a3　　13.11.a4　　13.11.a5

3. The first stage in terrace formation is deposition of sediment, such as on the floodplain (1) shown above. At this stage in its history, the stream is nearly at the same level as the floodplain (i.e., is not incised). The flat surface of the floodplain will later become the flat part of a terrace.

4. A change in conditions, such as a drop in base level, causes the stream to downcut through its floodplain deposits, forming a second, lower floodplain (2). Remnants of the first floodplain are stranded on both sides of the stream (1) and, if high enough, are unlikely to be flooded again.

5. With further downcutting, the stream abandons the second floodplain (2), creating a third, even lower one (3). The oldest floodplain (1) is now high and dry. The series of downcutting events creates a stair-step appearance to the land, such as those shown in the first two figures on this page.

B How Are Entrenched Meanders Formed?

The landforms we know as meanders form only in loose sediments, such as those on floodplains. However, in the Four Corners region of the American Southwest (where Utah, Arizona, New Mexico, and Colorado meet) and areas west of the Appalachian Mountains, meanders with typical sweeping bends are deeply incised in hard bedrock, forming some puzzling canyons. What do these winding canyons, called *entrenched meanders*, tell us about the history of rivers and streams in these areas?

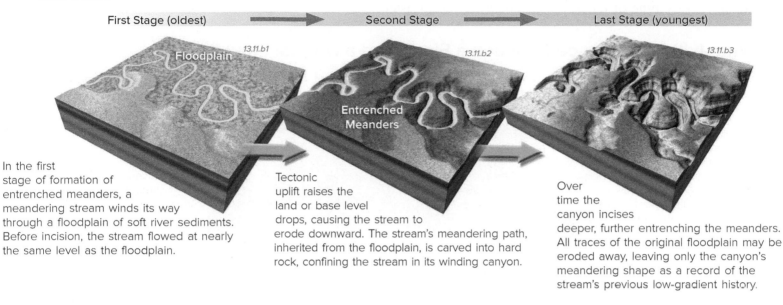

First Stage (oldest)   Second Stage   Last Stage (youngest)

13.11.b1 Floodplain   *13.11.b2* Entrenched Meanders   *13.11.b3*

In the first stage of formation of entrenched meanders, a meandering stream winds its way through a floodplain of soft river sediments. Before incision, the stream flowed at nearly the same level as the floodplain.

Tectonic uplift raises the land or base level drops, causing the stream to erode downward. The stream's meandering path, inherited from the floodplain, is carved into hard rock, confining the stream in its winding canyon.

Over time the canyon incises deeper, further entrenching the meanders. All traces of the original floodplain may be eroded away, leaving only the canyon's meandering shape as a record of the stream's previous low-gradient history.

Rivers That Cross Geologic Structures

Sometimes rivers appear to perform impossible tasks—cutting a deep canyon directly across a mountain. The Green River (below) flows across a mountain, appropriately called *Split Mountain,* as shown in the photograph to the right. This mountain ridge is an anticline of hardened sandstone in Dinosaur National Monument of northern Utah.

These odd canyons can be interpreted in at least two ways. A river may have been flowing over a region that was being actively uplifted and deformed, but the river was able to erode through the structures as fast as they were formed. Such a river is called *antecedent,* meaning it predated formation of the structure.

In some cases, a river may establish its route when it is flowing on soft, easily eroded rocks, uninfluenced by what rocks lie at depth. As the river begins to incise, it becomes trapped in its own canyon, unable to avoid any geologic structures it encounters as it erodes down, and its position becomes "locked in," through the hard rocks. Such rivers are *superposed,* meaning they were superimposed on already existing features. The Green River is best interpreted as a superposed river that established a meandering course on soft rocks and then downcut into harder ones.

13.11.t2 Green River at Split Mountain, UT

13.11.t1

5 km

Before You Leave This Page

✓ Sketch and explain a series of steps showing how river terraces form.

✓ Describe one way in which entrenched meanders form.

✓ Explain how antecedent and superposed rivers are different.

13.11

13.12 What Is and What Is Not a Flood?

THROUGHOUT HISTORY PEOPLE HAVE LIVED along rivers and smaller streams. Streams are sources of water for consumption, agriculture, and industry, and provide transportation routes and energy. Stream valleys offer a relatively flat area for construction and farming, but people who live along streams are subject to an ever-changing flow of water. High amounts of water flowing in rivers and smaller streams often lead to flooding. In most parts of the world, flooding is a common and costly type of natural and human-caused disaster.

A What Is the Difference Between a Flood and a Normal Flow Event?

Streams are dynamic systems, and they respond to changes in the amount of water entering the system. When more water enters the system than can be held within the natural confines of the channel, the result is a *flood*.

1. Flow in a channel, even when there is not a flood, may cause riverbank erosion. Such erosion can destroy structures built too close to the stream or a bridge built across the stream. Erosion can also make the stream change position over time, turning what was floodplain into channel, and what was channel into floodplain.

2. A flood occurs when there is too much water for the channel to hold. As a result, water spills out onto the adjacent land, usually inundating parts or all of the floodplain.

3. Human-constructed barriers, such as levees, can sometimes protect property from flooding during large flood events but trap water on the floodplain after the peak flooding ends.

4. Large floods can expand the width of the floodplain by burying preexisting rocks and material with sediment deposited by the stream. Sediment beneath the floodplain includes old channel deposits in addition to floodplain silt and clay.

13.12.a1 Hermann, MO

Normal, Bank-Full Flows

5. Normal (i.e., non-flooding) flows in streams can range from nearly dry to *bank-full*. Although there may be abundant water flowing down the channel, it is generally not considered a flood unless the water overflows the banks. A stream's natural floodplain is an excellent place to contain excess floodwaters—as long as it remains undeveloped by humans. Low-intensity development, such as a park, is often an appropriate use of a floodplain.

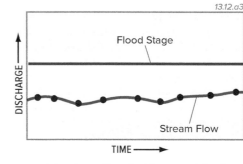

13.12.a2

Riverbank   Channel   Riverbank

13.12.a3

Flood Stage

Stream Flow

DISCHARGE

TIME →

6. This hydrograph shows a typical non-flood flow. The line labeled *Flood Stage* shows the amount of discharge required for the stream to overtop its banks and spill out onto the floodplain (i.e., a flood). During extended times of dry conditions, or at least weather that is normal for the region, hydrographs may show little change in stream flow over time, as shown here.

Flows During a Flood

7. When the amount of water in a stream exceeds the channel capacity, a flood occurs, inundating the floodplain. This hydrograph shows prolonged precipitation or snowmelt upstream that causes a flood event downstream, as represented by discharge greater than flood stage.

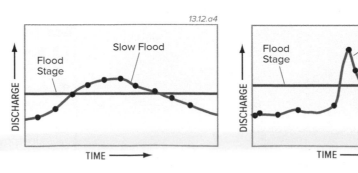

13.12.a4

Slow Flood

Flood Stage

DISCHARGE

TIME →

13.12.a5

Rapid Flood

Flood Stage

DISCHARGE

TIME →

8. Intense rainfall can unleash a brief *flash flood*, with a rapid rise in water levels and an increase in discharge that lasts only a short time. Similarly, rapid onsets of flooding result from failure of a natural or constructed dam, but the resulting flows last longer.

B What Are the Causes of Flooding?

What causes discharge to exceed the channel's capacity? A simple answer is that there is more water in the channel than can be accommodated. This can be the result of natural processes or human-caused events.

Snowmelt

1. Flooding occurs when warming temperatures or rainfall melt snow and ice somewhere in the drainage basin.

13.12.b1 Norway

2. In the Northern Hemisphere, flooding from melting ice and snow occurs in the spring, from March to May. Heavy rain that coincides with melting can cause even worse flooding.

Local Heavy Precipitation

3. Some floods are caused by heavy rainfall over a short period of time, causing a brief, but dangerous, flash flood.

13.12.b2 White Canyon, UT

4. A thunderstorm upstream of this site sent a fast-rising, muddy flash flood down this desert drainage. Vehicles attempting to cross such floods are often washed downstream.

Regional Precipitation

5. Regional floods occur when abnormally high precipitation falls over a large area over days, weeks, or months.

13.12.b3 Tucson, AZ

6. Heavy regional rains caused by moisture from a former hurricane caused this normally dry river to destroy offices built in a risky place—on loose sediment of the floodplain and on an outside bend.

Effects of an Ice Dam

7. During the winter, streams in cold climates can freeze, forming large volumes of ice. When warmer temperatures cause melting, the ice can block stream flow, causing a flood.

13.12.b4 Grand Forks, ND

8. Spring snowmelt and effects of ice dams often cause flooding along the Red River of the North, here still frozen before a flood.

Volcanic Eruption

9. If volcanic peaks are covered with snow when the volcano erupts, the snow will melt and cause flooding or catastrophic mudflows.

13.12.b5 Muddy River, WA

10. A volcanic eruption on snowy Mount St. Helens caused flooding and mudflows downstream, destroying this bridge.

Dam Failure

11. Dams occur as both natural and human-constructed features. Poorly engineered dams have failed, releasing floodwaters into downstream channels.

13.12.b6 Eastern ID

12. Catastrophic release of water during failure of the earthen Teton Dam, Idaho, in 1976 flooded towns downstream.

Urbanization

13.12.b7 Dalian, China

13. When urban growth replaces natural lands, the area responds differently to precipitation and snowmelt. Urbanization increases runoff by increasing the amount of impermeable surfaces.

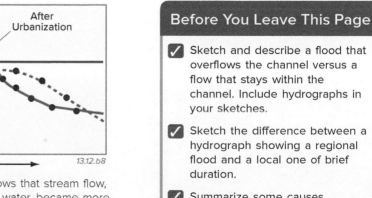

13.12.b8

14. This hydrograph shows that stream flow, for the same amount of water, became more abrupt and extreme after urbanization, causing a stream to rise above flood stage.

> ### Before You Leave This Page
>
> ☑ Sketch and describe a flood that overflows the channel versus a flow that stays within the channel. Include hydrographs in your sketches.
>
> ☑ Sketch the difference between a hydrograph showing a regional flood and a local one of brief duration.
>
> ☑ Summarize some causes of flooding.

13.12

13.13 What Were Some Devastating Floods?

FLOODS CAN BE DISASTERS that affect millions of people and cause millions or billions of dollars in property damage. Floods occur for different reasons and over very different scales of time and area. Here we explore two kinds of floods: a regional flood on a large river and a local flash flood.

A What Happened During the 1993 Upper Mississippi River Flood?

The 1993 flood on the upper Mississippi River and other Midwestern rivers arose from heavy precipitation over several weeks. It killed 47 people and resulted in extensive property damage and economic loss. Floodwaters inundated large areas of the floodplain, including areas that were considered "safe" behind levees.

1. During June and July, the polar front jet stream dipped south (▼), creating a convergence zone between warm, moist air coming from the Gulf of Mexico and colder air coming from the northwest. This resulted in persistent thunderstorms in the Upper Mississippi region.

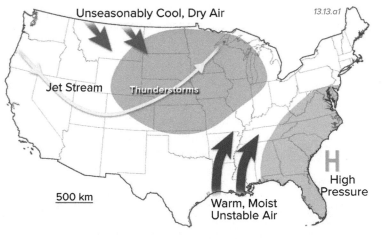

2. Contours on this map of the region (▶) show total rainfall (in inches) from June 1 to August 31, 1993. Some areas of Iowa received 36 inches of rain (nearly a meter).

3. High rainfall over such a large area resulted in flooding along major rivers and their tributaries. Heavy spring rains had already saturated the ground, preventing infiltration of additional rainfall during the summer storms. A similar setup caused severe flooding in Memphis, Tennessee, and other parts of the Mississippi in May 2011.

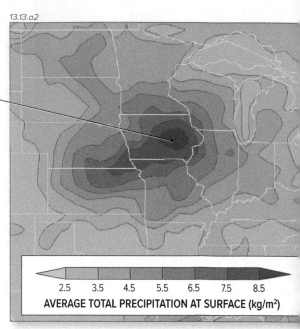

13.13.a2

AVERAGE TOTAL PRECIPITATION AT SURFACE (kg/m²)
2.5 3.5 4.5 5.5 6.5 7.5 8.5

Satellite Images of Three Rivers

4. The satellite images below both show an area at the confluence of the Mississippi, Missouri, and Illinois rivers near St. Louis, Missouri. Many homes and businesses on the modern floodplain were flooded and took months to dry out because levees, built to keep water out, trapped some water in after peak discharge.

5. Before the flood, rivers are within their channels, and floodplains next to the channel are dry (▼).

6. During the flooding, the rivers inundated the broad floodplains, flooding places far from the river channel (▼).

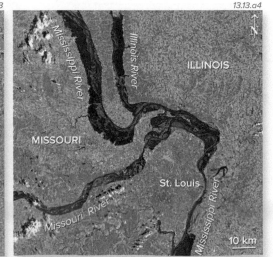

Discharge over Time

7. As shown by this hydrograph (▼), the Mississippi River at St. Louis, Missouri, reached flood stage (30 ft or 9 m) on June 26 and had peak discharge (50 ft or 15 m) 20 ft above flood stage a month later.

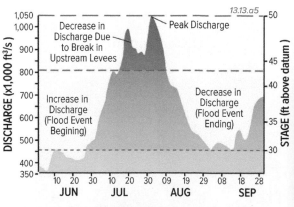

13.13.a5

8. Sudden drops in discharge in mid-July correspond to breaks in the levee system upstream from St. Louis. These breaks let floodwater escape from the channel, lowering the discharge for areas downstream, such as St. Louis.

B What Conditions Caused the Big Thompson River Flood in 1976?

The Big Thompson River near Estes Park, Colorado, flash flooded when as much as 12 inches of rain fell in a few hours in a small drainage basin. The flood killed 139 people and caused an estimated $13.5 million in damage.

1. This flood resulted from an unusual weather pattern. Cold polar winds converged with moist winds and pushed moist air upslope along the Colorado Front Range, forming a stationary thunderstorm. This image is an artist's depiction of the storm.

Thunderstorm

13.13.b1

5 km

2. The image below shows the extent and intensity of rainfall in the Big Thompson Canyon area. Darkest blue shows the highest concentration of rainfall, locally exceeding 12 inches (30 cm). The Big Thompson River, in the middle of the dark blue area, drains Rocky Mountain National Park and runs through an area of narrow canyons and steep slopes. Thin soil on steep, rocky slopes limited infiltration, allowing the storm runoff to accumulate quickly in the tributaries and main canyon of the Big Thompson River.

Rainfall Amounts

13.13.b2

5 km

13.13.b3 Big Thompson River, CO

3. The banks of the river were heavily developed with businesses, motels, campgrounds, and houses. The flood occurred in the evening, resulting in a number of deaths in a campground next to the river.

4. Houses and other structures along the Big Thompson River were totally destroyed by water and debris, which included mud, sand, and large boulders. Some houses, like this one, were ripped from their foundations and carried downstream.

13.13.b4 Big Thompson River, CO

C What Were Some Other Notable Floods?

Flood/River	Cause of Flooding and Effect on Society
Johnstown, Pennsylvania (1889)	Failure of an earth-filled dam during heavy rains; deadliest flood in early American history; destroyed the town and caused over 2,200 deaths.
Mississippi River (1927)	Heavy rains followed by flooding and destruction of levees; the most destructive flood in U.S. history; $400 million in damages and 246 killed in 7 states; resulted in enactment of Flood Control Act of 1928.
Yangtze River, China (1931)	Prolonged drought followed by intense rainfall; 3.7 million people died from drowning, disease, and famine. Overall, China has experienced the most deaths from floods.
Fargo, North Dakota (1997)	Floods occurred along the Red River of the North in the spring of 1997; rainfall coincided with snowmelt and a subsequent ice dam in the river, causing the river to overflow its banks and flood a large area.
Central America (1998)	Hurricane Mitch stalled over Central America; 75 inches of orographically enhanced rain over several days; death toll estimated to be 11,000.
Bangladesh (multiple years)	Flooding occurs regularly in Bangladesh, a low country along the Ganges and Brahmaputra rivers, which drain the Himalaya. Extensive flooding occurs due to springtime melting of ice and snow in the Himalaya, exceptionally high precipitation during the wet phase of the monsoon, or some combination.

Before You Leave This Page

☑ Describe the cause of flooding along the Mississippi River in 1993, and how this event affected the floodplains.

☑ Discuss the cause and consequences of the Big Thompson Flood of 1976.

☑ Briefly describe other circumstances that caused notable floods.

13.13

13.14 How Do We Measure Floods?

MOST FLOODS ARE A NATURAL CONSEQUENCE of fluctuating stream flow. Streams receive most of their water from precipitation and snowmelt, and the amount of precipitation falling in any given drainage basin varies from day to day and from year to year. Stream flow, in response to rainfall and snowmelt, can vary from a trickle to a raging flood. How do we determine how big a flood was, or will be? Knowing this will help us decide what places are safe and which places are not safe.

A How Is Stream Flow Measured?

Stream flow is measured by calculating discharge—the volume of water flowing through some stretch of a stream during a specified period of time. Discharge calculations help us quantify how big a flood was, determine how much water a stream channel can hold, and predict the size of future floods.

Measuring and Calculating Flow

1. The first step in calculating discharge is collecting measurements from the stream at a particular site, called a *gauging station*. To collect the data, hydrologists measure cross sections of the stream at the site. The stream shown here (▶) has an overly simple stream bottom compared to natural streams. Hydrologists then measure the water's depth and the average velocity of the water as it flows past the gauging station. Many of the measurements are automated, with data being relayed by radio and computer.

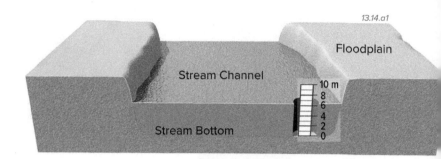

13.14.a1

2. Remember, to calculate discharge (represented in equations by the letter Q), we need the cross-sectional area of a stream (average width × average depth) and average velocity of the current:

$$Q = stream\ depth \times stream\ width \times stream\ velocity\ (or\ Q = DWV)$$

3. If the velocity of the stream shown on the left is 1.1 meters per second, the discharge would be:

$$Q = 1\ m\ deep \times 10\ m\ wide \times 1.1\ m/s = 11\ m^3/s$$

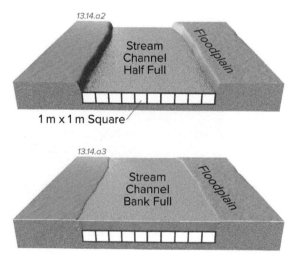

13.14.a2

1 m x 1 m Square

4. Calculate how much discharge would be needed to fill the channel to a bank-full condition. When the stream is this high, it normally flows 1.5 m/s, or nearly three times the flow in the half-full example. The width at bank-full levels is 10.5 m wide because the bank widens a little upward. These values lead to the calculation below:

$$Q = 2\ m\ deep \times 10.5\ m\ wide \times 1.5\ m/s = 31.5\ m^3/s$$

13.14.a3

Plotting Discharge During a Flood Using a Hydrograph

5. The graph below shows a hydrograph for a stream before and after a storm. Examine the figure, then read the text.

6. Before the actual flooding, this stream flows at a bank-full condition. The flowing water is contained within the channel, so there is no flood. The hydrograph for a station on such a stream shows a fairly constant discharge, represented by the horizontal part of the plot.

7. At some time, more water is added upstream to the stream by a precipitation event (thunderstorms).

8. As the additional water reaches the station, the hydrograph shows an increase. The channel can no longer contain the water, and the stream floods out over its floodplain.

9. After the pulse of higher flow moves past the station, the hydrograph shows a return to the bank-full condition. The flood has passed. Most streams will further decrease in flow, dropping well below a bank-full condition. If the stream dries up completely, the discharge is zero.

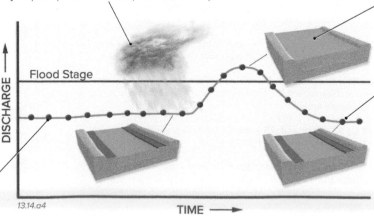

13.14.a4

B   What Is the Probability That a Flood Will Occur?

Probability is the statistical description of the likelihood that an event will happen. Suppose you and 99 other students enter a contest to win a car and each of you buys one ticket. Your probability of winning is 1 in 100 or 1%. Compare that with the probability that the Sun will rise tomorrow, which is essentially 100%. For many years, hydrologists, geomorphologists, and engineers have been collecting stream-flow data, which allows them to calculate the probability that a certain stream flow will occur in any year.

Frequency of Flows

1. Data used to estimate how often a stream may flood come from observations of discharge. The graph below plots peak daily average flows for the Yellowstone River over a 75-year period.

2. We plot discharge on the vertical axis and time on the horizontal axis. From this plot, it is common for this river to have peak flows around 4,000 to 7,000 m³/s.

3. High flows above 7,000 to 8,000 m³/s occur every so often.

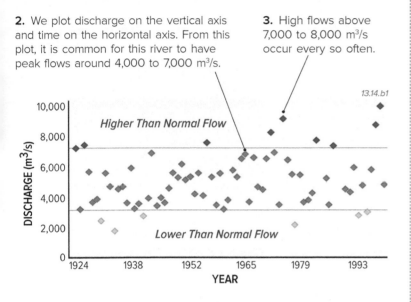

13.14.b1

4. Note that the highest flow event (flood) occurred only recently. To understand the river's behavior over time, and what to expect, we need data collected over a sufficiently long time period. A shorter data record means more uncertainty.

Flow Probability

5. Raw flow data are used to estimate probability. Hydrologists draw a rating line or curve for a river or smaller stream, giving the probability that a particular flow will be exceeded in any given year. This curve is for a smaller river than the Yellowstone River, with flow less than 2,000 m³/s. To use this curve, start on either axis and follow any value to the line, and then read off the corresponding value on the other axis.

6. For example, the probability that a 120 m³/s flow will occur or be exceeded in any year is about 99%.

7. The probability that a flow event will exceed about 2,000 m³/s in any given year is low, at about 0.5%.

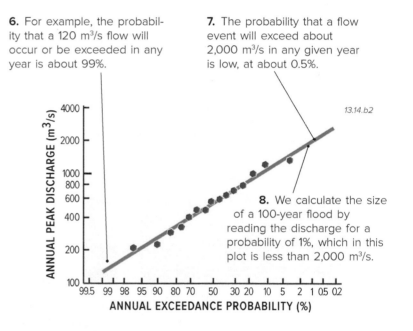

13.14.b2

8. We calculate the size of a 100-year flood by reading the discharge for a probability of 1%, which in this plot is less than 2,000 m³/s.

What Probability Does and Does Not Tell Us

The probability that any particular flow from 0 m³/s to more than 10,000 m³/s could happen along a stream in any year is estimated using graphs such as the ones above. The probability for any stream is based on a short record relative to the history of the river. The reliability of the mathematical estimations improves with more data. On the Yellowstone River, there is a very slight but real chance that floods exceeding 10,000 m³/s could happen three years in a row or twice in one year. The probability estimate doesn't guarantee future performance, but rather shows what the collected data tell us about the river's past behavior. Planning for a certain size flood involves assessing how much data we have, in addition to what the existing data predict about whether such a flood is likely to occur or not.

A term commonly used in public discussions, but less commonly by geoscientists who actually study streams, is the concept of a *100-year flood*. This term signifies the size of a flood that is predicted—from the existing data—to have a 1 in 100 probability (1%) of occurring in any given year. The term does not imply that such a flood will only happen every hundred years because "100-year floods" can, and have, happened two or three years in a row along some streams. In fact, such floods are more likely to occur in bunches, being caused by multi-year periods of abnormal amounts of precipitation and snowmelt. Such considerations influence who needs, and perhaps gets, flood insurance and how much it costs.

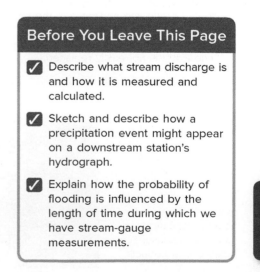

Before You Leave This Page

✓ Describe what stream discharge is and how it is measured and calculated.

✓ Sketch and describe how a precipitation event might appear on a downstream station's hydrograph.

✓ Explain how the probability of flooding is influenced by the length of time during which we have stream-gauge measurements.

13.14

13.15 How Do Streams Affect People?

STREAMS INFLUENCE MANY PEOPLE'S LIVES, in part because nearly all large cities are located along rivers. In addition to the impact of floods, stream processes and deposits control where many people live, where they do not live, and where they perhaps should not live. Rivers provide water for cities, farms, and industry, a fresh supply of nutrients for agriculture on floodplains, and routes for inexpensively transporting materials. Dams provide energy that does not produce greenhouse gases but has other consequences for the environment.

Flooding

13.15.a1 East Dubuque, IL

13.15.a2 Rillito River, Tucson, AZ 1983

13.15.a3 Rillito River, Tucson, AZ 1983

1. The main type of flooding we envision is brown, muddy water covering houses, streets, and fields in a low-lying area, such as these houses built on the floodplain. The floodwater may remain for days or weeks and cause severe damage and an unpleasant cleanup.

2. Another type of flood is shown in the photograph above, where the water mostly stays within the channel but piles up along the outside of a bend, causing severe erosion next to the river and undercutting and destroying the condominiums.

3. This photograph shows the results of the flood in the previous photograph. The condominiums had three strikes: built too close to the river, situated on an outside bend, and constructed on loose, easily eroded materials. Most floodplains are made of such materials.

Agriculture

13.15.a4 Austin, TX

13.15.a5 Potomac River, VA

13.15.a6

4. Smaller streams can develop on agricultural fields during intense rainfall and snowmelt. These can erode into and remove the soil from the bare, unprotected ground, a form of soil erosion. Soil takes a long time to develop, so once lost it is not easily replaced.

5. Streams also provide sites, such as floodplains, that are excellent for agriculture. The areas are relatively flat, are underlain by silt-rich stream deposits, soils that are productive, and there is a nearby source of water. Flooding during some years replenishes the soil nutrients.

6. Agriculture is also common on stream terraces high above the river. They have the advantage of being relatively flat, having a nearby source of water, and not being flooded. The land generally does not get replenished by flooding.

Water Supply

7. Streams are an essential supply of water for many people. Water can be pulled directly from a river and chemically treated or can be collected in reservoirs behind dams (▶) and distributed via canals and subsurface pipes and aqueducts.

13.15.a7 Bonneville Dam, Columbia River, WA and OR

13.15.a8 Thousand Springs, ID

8. Streams also interact with groundwater, adding water to the subsurface in some places or receiving water from the subsurface via springs (◀) and other types of flow. The water quality of surface and subsurface waters is also linked.

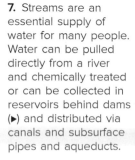

Features Formed Along Streams

13.15.a9 Rio Grande, El Paso, TX

9. Streams dictate where we live, or at least where we should not live. This river channel is active, but it was not carrying much water when this photograph was taken. At times, however, water will cover the entire ground in the image, as this is the Rio Grande!

13.15.a10 Elkhorn River, NE

10. Many people live on floodplains, for example, to be close to their agricultural fields (▲). Parts of many cities are also on floodplains, although this setting is disguised by concrete and buildings—until a big flood comes along.

13.15.a11 Thompson River, British Columbia, Canada

11. Stream terraces are a safer place to live. They are relatively flat and higher above the river than a floodplain. Most high terraces will not get flooded again, although some low ones could.

Levees

12. Along the edges of many channels is a raised embankment, or *levee*. Natural levees are created by the river, and humans construct artificial levees to try to prevent floodwaters from spilling onto the floodplain. This strategy commonly works, at least for a while.

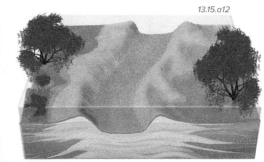

13.15.a12

13. During flooding, sediment-carrying floodwater rises above the channel and begins to spread out. As it does, the current slows and so deposits sediment in long mounds next to and paralleling the channel.

13.15.a13

14. When the flood recedes, sediment that was piled up next to the channel remains as an elevated rise or levee. Levees are barriers to water flow from the channel to the floodplain, and from the floodplain back into the channel after a flood.

13.15.a14 Near Badlands NP, SD

15. Levees can also be constructed by humans. They can be huge embankments of dirt and concrete or can be as simple as piling rocks along the stream (▲). The rocks piled here are along the cutbank, in an effort to limit the amount of lateral (sideways) erosion into the field to the right.

Levees—Boon or Bust?

The word *levee* likely leads to thoughts of flooding along the Mississippi, but other areas also have thousands of kilometers of human-constructed levees that keep seasonal rainfall from inundating cities and farmlands. Without levees, much of this land would be flooded too often to allow farming or settlements. One problem with levee systems is that they invariably fail. It is nearly impossible to engineer an *affordable* levee system that can handle the *largest* flood events. This image shows a breached levee and inundated neighborhoods near New Orleans, the result of flooding during Hurricane Katrina in 2005. Failure of a single levee can put lives in jeopardy, cause hundreds of millions of dollars in damage, and cause towns and neighborhoods to be abandoned as unsafe, as occurred during Katrina.

Some towns along the river are built upon natural levees, and so are slightly higher than surrounding lands. These largely escaped the flooding associated with Hurricane Katrina. They are also better drained, limiting how long heavy rainfall remained on the surface.

13.15.t1 New Orleans, LA

Before You Leave This Page

✓ Summarize some ways that streams influence people.

✓ Explain some advantages and disadvantages of living near a stream.

13.15

13.16 How Does the Colorado River Change as It Flows Across the Landscape?

THE COLORADO RIVER SYSTEM drains a large region of the American West. The river cuts across a geologic terrain that varies from high bedrock headwaters to low, sandy valleys, to a delta where it reaches the Gulf of California. It has a rich set of features, many of which are typical of most rivers, but some of which are unique to this river.

The large map spreading across both pages shows the drainage basin of the Colorado River. Surrounding the map are vignettes about different features, each of which is keyed to a number on the large map. Start with number 1 in the headwaters (top of right page) and proceed down the river. The smaller map below covers the same area as the large map and shows the Colorado River's largest tributaries. The edge of the map is a drainage divide between the Colorado and other river systems.

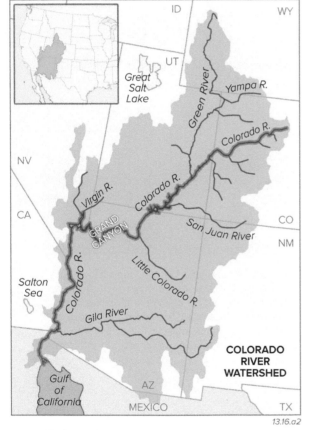

COLORADO RIVER WATERSHED

13.16.a2

Reservoirs

6. Dams constructed across the Colorado River (◄), mostly within or bordering Arizona, form large reservoirs, including Lake Powell and Lake Mead. The dams provide hydroelectric power, flood control, recreation, and water, but block sediment transport.

13.16.a7

Cutting Across Structures

7. The Colorado River cuts across some geologic structures, such as the Kaibab Uplift (▶) in the Grand Canyon (a superposed river). The river may have started to cut through the uplift when a large, natural lake overtopped its rim, flooding westward across a low divide in the uplift.

13.16.a8

Lakes

8. Older lakes were formed by geologic events, such as lava flows (▶) into the Grand Canyon, temporarily damming the river. On geologic timescales, such dams were rapidly eroded away.

13.16.a9

Salton Sea

10. This large lake (▼) is located west of this area and is shown on the above map. It filled in 1905 when a flood of the Colorado River overwhelmed canals and other structures built to divert water for irrigation in California. For two years, the river flowed into the basin, flooding 350,000 acres of land and filling a lake that had formed naturally many times in the past. These earlier lakes formed when high water volumes and high sediment load forced the river to leave its channel and flood westward into the lowlands of the Imperial Valley and ancestral Salton Sea.

13.16.a11

Colorado Delta

9. As the Colorado River nears its mouth in the Gulf of California, much of its water has been withdrawn for drinking and irrigation and its sediment load has been blocked by dams. The delta, which has been building for hundreds of thousands of years, continues to grow but at a much slower rate because of the decrease in the volume of water and sediment needed to nurture the delta's growth. The loss of water and sediment has harmed the delta's fragile ecology (▶).

13.16.a10

Headwaters

13.15.a1

1. The Green River is a tributary of the Colorado. Its headwaters are in the snow-capped mountains of Wyoming, where high-energy waters cascade down steep canyons. The Green River, like most tributaries of the Colorado River, starts in steep mountainous areas (▶).

13.16.a3

Green River

WYOMING

2. The headwaters of the Colorado River (not shown by a detailed view) are in the high Rocky Mountains. Here steep mountain streams and braided rivers erode the mountains, transporting the debris to lower elevations.

UTAH   COLORADO

ARIZONA   NEW MEXICO

Changing Conditions

3. Where the Colorado River leaves its steep bedrock canyon (▶), it changes from a steep, bedrock-confined channel into a meandering river that flows through a broad valley at Grand Junction, Colorado. Adjacent to the river is a well-developed floodplain covered with fertile farms that benefit from the Colorado's silt.

13.16.a4

Grand Junction, CO

Colorado River

Floodplain

Entrenched Meanders

4. Winding bedrock channels at the confluence of the Green and Colorado Rivers (▶) inherited their classic meander shapes when the river system was much younger and was flowing through softer materials. They are classic examples of entrenched meanders. The faults and fissures are the result of the walls of the canyon slowly sliding toward the river.

13.16.a5

Green River

Colorado River

Confluence of Green and Colorado Rivers

Entrenched Meanders

Colorado River

Faults and Fractures

Records of Flooding on the Colorado River

5. The Colorado River drains a large area and has experienced large floods. The graph below shows stream-flow data from Lee's Ferry, a historic river crossing upstream of the Grand Canyon.

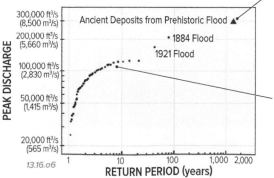

Ancient Deposits from Prehistoric Flood ▲

- 1884 Flood

1921 Flood

PEAK DISCHARGE: 300,000 ft³/s (8,500 m³/s); 200,000 ft³/s (5,660 m³/s); 100,000 ft³/s (2,830 m³/s); 50,000 ft³/s (1,415 m³/s); 20,000 ft³/s (565 m³/s)

RETURN PERIOD (years): 1   10   100   1,000 2,000

13.16.a6

N

MEXICO   100 km

Before You Leave This Page

✓ Describe where the Colorado River is located, from its headwaters to its mouth.

✓ Describe the features that occur along the river and how each formed.

✓ Describe the record of flooding for the Colorado River at Lee's Ferry.

✓ Explain why stream-flow data collected over the last 100 years may not accurately indicate the maximum flood possible on a river.

Geoscientists investigating ancient river-flood deposits infer that a very large flood with a discharge estimated at 8,500 m³/s occurred before humans were in the area. For comparison, modern, dam-controlled flows through the canyon rarely exceed 570 m³/s.

Even the pre-dam measured flows (represented by the dots) were generally less than 3,700 m³/s. During the largest flood recorded at Lee's Ferry, in 1884, the river's discharge was 6,200 m³/s. Regional drought and other changes in climate greatly affect the flow of the river, and how much water is available for the rapidly growing cities of Nevada, Arizona, and California, all of which count on a steady supply of Colorado River water. What happens if flows decrease? Continue reading on the left-hand page.

13.16

13.17 How Would Flooding Affect This Place?

STREAMS PRESENT BENEFITS AND RISKS to people living along their banks. Meandering streams provide floodplains with fertile soil and a relatively flat place to farm and perhaps build. Living on a floodplain is a hazardous proposition because it has flooded in the past, may be flooded in the near future, and owes its very existence to flooding. In this exercise, you will calculate the likelihood of flooding on two levels of the landscape and decide if potential economic and societal benefits are worth the risk of living there.

Goals of This Exercise:

- Observe and interpret features associated with a short stretch of a meandering river.
- Evaluate different locations for building a house and siting a farm, comparing and summarizing the advantages and disadvantages of each site.
- Calculate the risk of flooding for each location and discuss the risk versus the benefit.

Procedures

Use the available information to complete the following steps, entering your answers in appropriate places on the worksheet or answering questions online.

1. Observe the terrain below, in order to interpret the various parts of the landscape. Assign each landform feature or topographic level of the landscape its appropriate river term (for example, *channel*).

2. Apply your knowledge of the processes, features, and sediment associated with meandering rivers to predict what processes characterize each landform and how the landform might be affected by flow along the river.

3. Use relative elevations and other attributes to infer the order in which the features formed and the steps involved in the formation of each feature.

4. Determine which sites would be the best places to put *croplands*, considering all relevant factors, such as the flatness of the area, proximity to water, nature of the soil, what is growing there now, and possible added costs of growing crops in a specific site. You should also consider each site's vulnerability to bank erosion.

5. Evaluate the benefits of building a new house at each of the different levels of the landscape and at various locations on each level, for *both sides of the river.* Identify five homesites that are favorable, considering each site's proximity to croplands, to drinking water from the river, and any aesthetic considerations (e.g., just a nice place to live). Rank the five sites on the basis of your evaluation of their suitability.

6. Use the supplied dimensions on the profile on the next page and stream-flow data to calculate the river discharge required to flood two levels of the landscape.

7. Use the discharges you calculated and an *exceedance probability plot* for this river (provided) to estimate the probability of flooding for two different levels of the landscape.

8. Evaluate the flood-risk probabilities against the other considerations (in steps 4 and 5), and describe how including the risk of flooding has changed or not changed your rankings.

Step 1: Consider the Following Observations About Different Levels Near the River

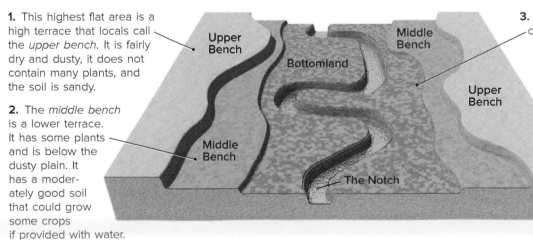

1. This highest flat area is a high terrace that locals call the *upper bench*. It is fairly dry and dusty, it does not contain many plants, and the soil is sandy.

2. The *middle bench* is a lower terrace. It has some plants and is below the dusty plain. It has a moderately good soil that could grow some crops if provided with water.

3. A green, plant-covered, lower flat area, locally called the *bottomland*, flanks the river channel. It has some soil composed of silt and decayed plants, but in many areas the soil is overlain by several layers of loose silt. Close to the channel, many bushes and trees on the bottomland lean over a little in a downstream direction but were not uprooted by whatever made them lean over.

4. The lowest part of the valley, called *The Notch*, contains the river, whose water flows toward you in this view. When exposed during the dry season, sediment on the river bottom within the notch is loose and displays no soil development.

13.17.a1

Step 2: Calculate Discharge for a Profile Across the River

The diagram below on the left is a profile across the river, showing the widths of The Notch and the Bottomland. You will calculate discharges along this main profile, which crosses the river near the front of the model on the right. Your instructor may provide you with a second profile (farther back in the model), because the river has different dimensions at different places. This means that the same amount of discharge may reach different heights along different segments of the river. For your profile(s), complete the following steps:

1. To calculate the discharge needed to fill the notch, first calculate the cross-sectional area of the notch in the profile. In all these calculations, we are using averages for width, depth, and velocity.

2. Next, calculate how much discharge is needed to fill the notch and begin to spill water out onto the bottomland. To calculate discharge, multiply the cross-sectional area of the notch by the average velocity of the river, which is 0.7 m/s when the notch is filled.

cross-sectional area = width × depth

discharge (Q) = cross-sectional area × stream velocity

3. Repeat the calculations, but this time determine the additional discharge needed to flood the bottomland to a height where floodwater would begin to spill onto the middle bench. The river flows faster when there is more water, so use an average water velocity of 2.0 m/s. Enter your calculated discharges in the table on the worksheet or on a sheet of paper. You should have two discharge calculations, one to fill and overtop the notch, and another that fills up the notch and bottomland and then begins to spill out onto the middle bench.

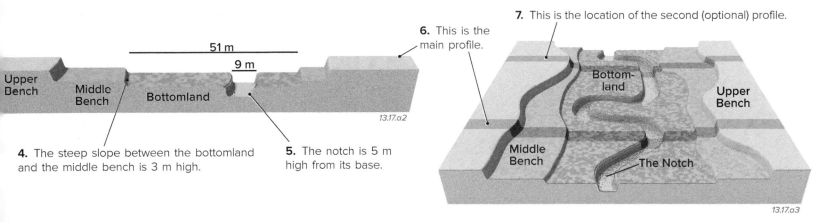

7. This is the location of the second (optional) profile.

6. This is the main profile.

51 m

9 m

Upper Bench

Middle Bench

Bottomland

Bottom-land

Upper Bench

Middle Bench

The Notch

13.17.a2

13.17.a3

4. The steep slope between the bottomland and the middle bench is 3 m high.

5. The notch is 5 m high from its base.

Step 3: Evaluate Flooding Risk Using Exceedance Probability

To determine the probability that each area will be flooded, compare both of your calculated discharges against the following plot, which is an *exceedance probability plot*. Follow the steps below and list in the worksheet or answer online the estimated probabilities for overfilling the notch and for overfilling the bottomland on the profile.

1. For each discharge calculation, find the position of that discharge value on the vertical axis of the plot.

2. Draw a horizontal line from that value to the right until you intersect the probability line (which slopes from lower left to upper right).

3. From the point of intersection, draw a vertical line down to the horizontal axis of the plot and read off the corresponding *probability of exceedance* (probability of flooding of a certain magnitude) on the horizontal axis. The *probability of exceedance* indicates the probability of the calculated amount of discharge being exceeded in any given year.

4. Repeat this procedure for both of your discharge calculations.

5. Consider the implications of each of these probabilities for your choice of site for cropland and a homesite. Use this information to choose final sites for cropland and a house. Explain your reasons on the worksheet or in the version online.

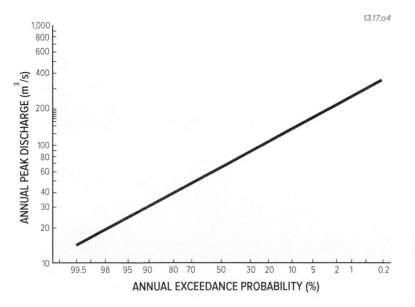

13.17.a4

13.17

GLACIERS ARE A POWERFUL FORCE in high-latitude and high-elevation areas, but in the past they also covered most of Canada, the northern parts of the U.S., and large areas of Europe and Asia. Erosion and deposition associated with glaciers reshape the underlying landscape and form many distinctive landforms, some of which indicate that ice once covered an area, even if none remains today. What features do glaciers form, and how can we use these features to reconstruct the distribution and flow directions of ice thousands or millions of years ago?

The northeastern U.S. and adjacent Canada, shown in this shaded-relief map, have a striking collection of features on land and along the coast. The map extends from North and South Dakota (in the northwestern corner of the map) across the Great Lakes as far east as Maine and Virginia. The northern part of the map includes southeastern Canada.

Huge, smooth troughs (each labeled with a **T** on the map) cut across the landscape. Examples are near the northwestern corner of the map, in western Minnesota and the Dakotas, and southwest of Lake Erie, in Ohio and Indiana.

What caused these smooth areas of the landscape, and is the process still occurring?

Curiously curved ridges (labeled with an **R** on the map) cross some of the smooth areas and are especially noticeable southwest of Lake Erie and Lake Michigan.

What are these ridges, how did they form, and what do they tell us about the geologic history of this region?

14.00.a1

TOPICS IN THIS CHAPTER

North of the Great Lakes, the landscape of Canada is rough on a small scale, containing many lakes, more than in most parts of the U.S.

Why does this region have so many lakes, and when and how did this landscape form?

The Finger Lakes of Upstate New York (▼) are elongated, like fingers, in a north-south direction and are among the deepest lakes in the U.S. They were carved by glaciers that no longer remain in the region.

How do glaciers form lakes, and how do we recognize a glacially formed landscape?

14.00.a2

Long Island, New York, largely consists of loose sedimentary materials that form two main ridges that define the island. From their character, these materials were deposited along the front of a glacier and from streams that emerged from the glaciers.

Did glaciers really get this far south, why did they grow so large, and what caused them to go away?

Ice Ages, Glaciers, and Lakes

Landscapes in the Great Lakes region contain evidence that huge ice sheets once flowed across this part of the continent—in the recent geologic past. This conclusion arises from comparing the distinctive landscape features and their associated sedimentary deposits with those observed today near currently active glaciers. With glaciers, as with most geologic features, the present is the key to interpreting the past.

During the last two million years, Earth has been experiencing an *Ice Age,* during which large regions of the Northern Hemisphere, as well as Antarctica, were covered year-round with ice and snow. Where the ice was thick enough, or rested on a steep enough slope, it moved downhill as a mass of flowing ice called a *glacier.* Some glaciers were small and restricted to mountain areas, whereas others covered large parts of the continents, forming *continental ice sheets.* Continental ice sheets flowed southward from Canada and smoothed off and carved grooves into the underlying landscape by grinding ice, rocks, and sand against the bedrock. The smooth troughs (T) on this map were carved by these continental glaciers. Southward-flowing parts of the glaciers carved the valleys that hold the Finger Lakes of New York and some other lakes in the region. These glaciers advanced down the Hudson River, between the Finger Lakes and Long Island, and covered wide areas on both sides of the valley.

As the climate warmed over the past 20,000 years, the ice sheets and glaciers melted and covered less area. Rocks and other sediment carried in the ice were dropped along the front of the melting glaciers, forming a series of curved ridges south of the Great Lakes (R on the map). The ice also left piles of glacially derived sediment on Long Island and Cape Cod. As the glaciers melted back, they uncovered a region with numerous topographic lows scoured by the ice. Many of these areas also lacked an established drainage network, because they had recently been covered with ice. For these reasons, the low spots became the countless lakes of Canada, Minnesota, Wisconsin, Michigan, and nearby states.

Water released from melting ice filled the glacially carved troughs and other low areas, resulting in the Great Lakes. The waters carved new river valleys or modified preexisting valleys and flowed into the sea, causing a rise in sea level.

14.1 What Are Glaciers?

GLACIERS ARE MASSES OF ICE that persist from year to year and tend to flow downhill under their own weight. They range in size from ice sheets that cover a large region—*continental glaciers*—to much smaller glaciers that are restricted to a single mountain or valley—*alpine glaciers*. Most glaciers are primarily ice and snow, but glaciers typically contain significant amounts of angular rock and finer sediment that are incorporated into the glacier as it flows from higher to lower elevations.

A What Are the Characteristics of Glaciers?

1. Glaciers form in *snowfields* where snow and ice accumulate faster than they melt, so many glaciers begin in higher elevations or at higher latitudes (closer to the North or South Pole). Glaciers only form if an area is cold and receives enough snowfall to allow ice and snow to accumulate faster than the ice and snow can melt.

2. As the snow gets buried, it compresses into ice, often turning blue in the process—ice absorbs longer wavelengths, such as red, while transmitting and scattering shorter blue wavelengths. Most glaciers also have lines (grooves, ridges, and sediment-rich streaks) formed by flow within the glacier. These are fairly straight or gently curved if the glacier has a simple pattern of flow, but they are contorted and folded if the glacier experienced more complex patterns of flow.

3. Ice can cover broad areas, but glaciers can become confined within valleys as they flow from higher elevations to lower ones. As adjacent ice-filled valleys merge, so do the glaciers, producing a wider and commonly thicker mass of flowing ice.

4. Whether a glacier forms in mountainous regions depends partly on the slope and shade of the area. Gentle slopes allow snow and ice to pile up rather than slide downhill. Areas with lots of shade, typically the north and east sides of slopes in the Northern Hemisphere, will allow snow and ice to accumulate faster.

14.01.a1

14.01.a2 Patagonia Andes, South America

5. This glacier of blue ice flows down a steep valley and ends in a lake of meltwater from the glacier. As a glacier moves, internal stresses cause its upper surface to break, forming vertical fractures, each of which is called a *crevasse*. Crevasses are especially abundant and well developed where a glacier flows along valley walls, around a curve, or where the land beneath the glacier changes slope, either from a steep slope to a more gentle one or from a gentle slope to a steeper drop. The glacier shown here (◄) breaks apart, opening up crevasses as it flows over a steep drop-off.

14.01.a3 Swiss Alps

6. As a glacier moves past bedrock (▲), it plucks away pieces of rock, and its surface may be partially or totally covered by rock pieces derived from nearby steep slopes. It grinds up some rock into a fine rock powder. The glacier carries away all this material, depositing it where the glacier melts. The glacier in the photograph above has dark fringes of rocky material on both sides and at its end.

B   What Are the Types of Glaciers?

1. The largest accumulations of ice are in *ice sheets*, continental to subcontinental masses of ice like those covering nearly all of Antarctica and Greenland. Slightly smaller ones would be dome-shaped ice caps or ice fields, similar to those covering Iceland or Patagonia, in the southern parts of Chile and Argentina.

14.01.b1 Antarctica

2. This panoramic photograph from Antarctica shows a *nunatak*, a small area of bedrock high enough to protrude through the ice sheet.

3. An ice sheet can spill over a cliff, as a steeply flowing mass of ice called an *icefall*.

14.01.b2 British Columbia, Canada

4. Glaciers that form in mountainous regions without being part of a regional ice sheet are *Alpine glaciers*. The smallest Alpine glaciers are *cirque glaciers*, which form in glacial depressions that are sheltered from the Sun; they are usually more wide than long. Glaciers that flow down valleys from cirque glaciers are called *valley glaciers*. Valley glaciers tend to be fairly narrow (several kilometers wide) but can flow down valleys for tens of kilometers.

5 km

14.01.b3 Dry Valleys, Antarctica

5. As some valley glaciers or ice sheets flow out of the mountains into broader, less confined topography, the glaciers can spread out, forming *piedmont glaciers*.

6. Cirque glaciers (▼) can grow large enough to spill out down an adjacent valley, forming a valley glacier, and valley glaciers can flow downhill and join even larger masses of ice—continental ice sheets.

14.01.b5

Cirque Glaciers

Ice Sheet

Valley Glacier

km

14.01.b4 La Perouse Glacier, AK

Glaciers, Rock Glaciers, Snowfields, and Sea Ice

In addition to typical ice-covered glaciers, some glaciers consist of angular rocks on the outside and slowly flowing ice on the inside. These features are *rock glaciers*, as shown in the photograph below. Not every large mass of ice on Earth's surface is a glacier. Some accumulations of snow and ice never move, and these are simply called *snowfields*. Snowfields are present in most of the images shown on this page. Snowfields can contain ice in addition to snow.

Large masses of ice also form when the upper surface of a lake or the sea freezes. In the ocean, such ice is *sea ice*. In all but the coldest places, sea ice freezes in the winter and thaws in the spring or summer. Freezing excludes most salt from the crystalline structure of ice, so sea ice melts to form water that is largely fresh (not salty). In the photograph below, broken sheets of sea ice surround small, rocky islands near Antarctica.

14.01.t1 Mount Sneffles, CO

14.01.t2 Antarctica

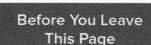

Before You Leave This Page

☑ Describe the characteristics of glaciers.

☑ Summarize the four main types (ice sheets, cirque glaciers, valley glaciers, and piedmont glaciers).

☑ Summarize which places have ice sheets and glaciers.

14.1

14.2

What Is the Distribution of Present and Past Glaciers?

TODAY, GLACIERS COVER around 10% of the Earth's land surface, but over the past two million years ice sheets and glaciers intermittently covered up to 30% of the land in the Northern and Southern hemispheres. The ice sheets and glaciers advanced during time periods within an ice age called *glacial periods*, and they retreated during other times still in the ice age, called *interglacial periods*. Overall, the past two million years have had a striking increase in glacial periods compared to most other times in Earth's geologic history. We call the most recent period of ice advance the Pleistocene Epoch. The Pleistocene glacial advance ended about 12,000 years ago, to correspond with the beginning of the present Holocene Epoch.

A Where Are Most Ice Sheets and Glaciers Today?

Glaciers form where snow and ice accumulate faster than they melt. Most glaciers are therefore in high latitudes (closer to the North or South Pole), at high elevations, or some combination of latitude and elevation. Glaciers exist along the equator, but at very high elevations (>6000 m) as in the Peruvian Andes.

1. In the Southern Hemisphere, glaciers occupy high peaks of the Andes, especially the Patagonian ice cap, which occupies the most southerly part of the Andes mountain range.

2. The largest continental ice mass on Earth is on Antarctica, which sits squarely over the South Pole. Ice and snow cover about 98% of the continent. Valley and piedmont glaciers form where the ice sheet is channeled to lower elevations.

3. Glaciers cover many of the highest parts of the Tibetan Plateau and the Himalaya, the highest mountain range on Earth, even though this region has a fairly low latitude. Glaciers and ice sheets are present elsewhere in Asia, especially on islands and peninsulas along the Arctic Ocean.

4. About 80% of Greenland is occupied by continental ice. Large areas of the neighboring islands also have glaciers, including the Iceland ice cap.

5. In North America, alpine glaciers are present in Alaska, northern Canada, the Rocky Mountains of the U.S. and Canada, the Coast Range of British Columbia, and on the larger volcanoes and other high peaks of the Cascade Range.

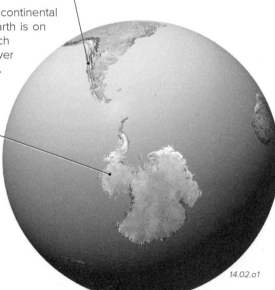

14.02.a1

14.02.a2

14.02.a3

6. The combination of cooler temperatures and abundant snowfall contributes to large areas of alpine glaciers in Alaska and the mountains of western Canada. Glaciers cover over one hundred times more area in Alaska than in all the western U.S. combined.

7. In the conterminous U.S., alpine glaciers are restricted to high elevation zones. Over two-thirds of alpine glaciers in the lower 48 states are in Washington state.

14.02.a4

8. In mainland Europe, the largest areas of glaciers are in the high-latitude country of Norway, with smaller areas in adjacent countries.

9. Farther south, glaciers are restricted to high mountains, including the Alps of central Europe and a smaller area farther west, in the Pyrenees mountain range, which straddles the border between France and Spain.

B What Areas of the Northern Hemisphere Were Covered by Ice During the Last Glacial Advance?

The two maps below show estimated ice cover for North America. The left map shows a time when ice coverage was close to a maximum, approximately 28,000 years ago, whereas the map in the center shows the position of ice at about 11,000 years ago, after the ice sheets had retreated and as the Pleistocene glacial advance was ending.

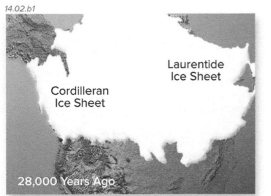

14.02.b1

Cordilleran Ice Sheet

Laurentide Ice Sheet

28,000 Years Ago

14.02.b2

11,000 Years Ago

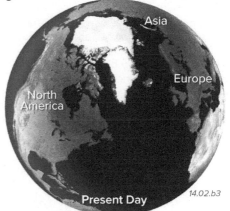

Asia

Europe

North America

Present Day   14.02.b3

1. The Laurentide ice sheet was centered over Hudson Bay in northern Canada. The Cordilleran ice sheet occupied highland regions of western Canada. The Laurentide ice sheets extended into the northern parts of the U.S., covering New England, the Great Lakes region, and the upper Midwest, terminating near the present-day Ohio and Missouri rivers. Most of Alaska was not covered by ice. This allowed people and animals from Asia to migrate into North America.

2. Around 11,000 years ago, early in the Holocene Epoch, the ice sheets melted back to north-central Canada. When we say that the ice sheets *retreated,* we mean that ice within the sheets was still flowing forward, but the front of the ice sheet melted faster than it could be replenished. As the ice retreated, the northern U.S. emerged from beneath the ice, the Great Lakes formed, and river systems like the upper Mississippi and the St. Lawrence began to develop.

3. This globe, centered on Greenland, shows the present-day distribution of ice in the Northern Hemisphere. It also shows sea ice—which is not a glacier but rather a mass of floating ice—covering much of the Arctic Ocean between North America and Siberia (northeastern Asia). Note that the only large land areas covered by ice are Greenland, adjacent islands in northern Canada, parts of Iceland, and mountainous parts of Alaska and westernmost Canada.

4. Today, glaciers in Asia (▶) are mostly restricted to high elevations or high latitudes. During the Pleistocene glacial advance, continental ice sheets covered the northern part of Asia and Europe, including northern Russia and all of Scandinavia, Great Britain, and much of Eastern Europe. Elsewhere in Asia and Europe, glaciers were widespread in mountain ranges and flowed out across adjacent lowlands. The alpine glaciers were concentrated in the same areas where glaciers occur today, but were much more extensive.

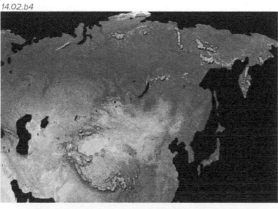

14.02.b4

5. Like glaciers throughout the Northern Hemisphere, the once-extensive glaciers of Asia and the Alps have been receding for much of the last 100 years and so are mere remnants of their former selves. As a reminder of their past positions, they left behind piles of glacial sediment that are still obvious in the landscape.

14.02.b5 West Antarctica

6. We don't know what the continental glaciers that covered portions of North America looked like, but they must have been similar to the West Antarctica sheet pictured above. These continental ice sheets were several kilometers deep and buried all vestiges of the pre-glacial landscape. The higher mountains, for example the Rockies and Alps, produced alpine glaciers that spread out of the mountains, forming piedmont glaciers that covered nearby lowlands.

Before You Leave This Page ✓ Describe what parts of the Northern Hemisphere and North America are covered with ice today and what parts were covered with ice during the last glacial advances.

14.2

14.3 How Do Glaciers Form, Move, and Vanish?

GLACIERS FORM, MOVE DOWNHILL, and eventually melt away. How does a glacier form? Once formed, how does a glacier move across the landscape, and what happens to it as it flows downhill and toward warmer areas with more melting and generally less snowfall?

A How Do Snow and Ice Accumulate in Glaciers?

1. Glaciers, including the ones below, form by the right combination of cold temperatures and adequate precipitation in the form of accumulating snow and ice. In mountainous regions, snow and ice can also be added by avalanches from the higher ground surrounding the glacier.

14.03.a1 La Perouse Glacier, AK

2. Snow falls as individual flakes. Once on the ground, flakes get pressed together by the weight of other snowflakes. Loose snow can contain 90% air between the flakes.

3. As more snow accumulates on top, snowflakes farther down are compressed, forcing out more than 50% of the air. The snowflakes become compressed into small, irregular spheres of more dense snow called *firn*.

4. With increasing depth and pressure, the snow begins to recrystallize into small interlocking crystals, forming solid ice. Ice is a crystalline material and is considered to be a type of mineral and a rock. As the ice is compressed further, the planes of weakness in the crystals become parallel, permitting the ice to shear and flow. Crystalline ice contains less air and commonly has a bluish color.

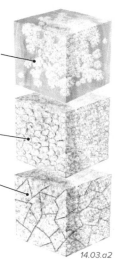

14.03.a2

B How Does a Glacier Form and Change as It Moves Downhill?

Glaciers have inputs of snow that are balanced by outputs of ice, glacial meltwater, evaporation, and sublimation (the loss of ice molecules directly to the atmosphere). When inputs of snow exceed outputs, there is a positive mass balance and the glacier advances. When it is warmer or drier, outputs exceed inputs, and the glacier retreats.

1. The upper part of the glacier or ice sheet, where snow and ice are added faster than they melt, is the *zone of accumulation*. When the glacier reaches around 50 m in thickness (less on steeper slopes), gravity, working on the weight of accumulating snow, causes the glacier to flow downhill.

2. As the glacier moves downhill, it loses more and more ice and snow by melting, by wind erosion, and by *sublimation*. At some point along the glacier, the losses of ice and snow exactly balance the amount of accumulation; this boundary is called the *equilibrium* or *firn line*. The equilibrium line is sometimes, but not always, marked by a gradational boundary between snow-covered ice upslope on the glacier and exposed bluish ice downslope. The bluish ice formed at depth and became exposed at the surface as upper levels of ice and snow were removed. In some cases, the entire length of a glacier may be covered with snow, but blue ice can be observed at depth in fractures (crevasses) that cut the glacier's upper surface.

Zone of Accumulation

Snow

Equilibrium Line

Blue Ice

Zone of Ablation

Land

1 km

Sea

14.03.b2 Morteratsch, Swiss Alps

4. At lower elevations, ice melts away faster than it can be replenished by downward movement of ice within the glacier and by snowfall. This is the *zone of ablation* (where the glacier is losing mass). Glaciers terminate either on land or in the sea.

14.03.b1

3. The valley glacier in the photograph to the right has an upper, snow-covered part (zone of accumulation) and a lower area of blue ice below the equilibrium line.

C How Do Glaciers Move?

The upper part of a glacier is brittle, but at depth the overlying weight of the ice causes the glacier to behave plastically. Glaciers can then move downhill because the ice is not strong enough to support its own weight against the relentless downward pull of gravity. As a glacier spreads downward, as in the main figure below, it moves plastically by internal shearing and flow of the solid ice, by simply sliding across the bedrock, or by some combination of these two mechanisms. Whether or how much it moves downhill is strongly influenced by the slope—steeper slopes favor downhill movement.

1. As gravity pulls the ice downhill, friction along the base of the glacier causes the bottom of the glacier to lag behind the upper, less constrained parts. The upper part of a glacier (▾) therefore flows faster than the lower part, causing internal shearing within the glacier.

14.03.c2

2. If the interface between the glacier and the underlying bedrock is very irregular and temperatures are close to freezing, basal ice might melt by compression at one point and then refreeze through a process called *regelation*. If basal ice is particularly cold and dry, it might become locked to the bedrock and not move at all.

14.03.c3

3. If the bedrock-glacier interface is less irregular (i.e., smoother) or contains water from pressure melting, the glacier may be able to slide over the bedrock in a process called *basal slip* (◄). Such glaciers can move relatively rapidly. In general, glaciers can move more easily during the warmer temperatures of summer.

4. The rates at which glaciers move are extremely variable. Many glaciers move about a meter per day, but some move centimeters per day. Some glaciers have been known to lurch forward up to 30 m in a day in what's called a *glacial surge*.

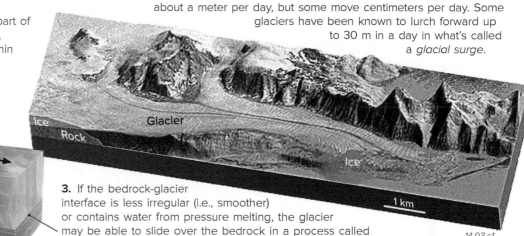

14.03.c1

D What Happens When a Glacier Encounters the Sea or a Lake?

1. When a glacier reaches the ocean or a lake (▾), it may float on the water if the sea or lake is deep enough. Ice, even in the densest bottom portion of the glacier, will float because it is less dense than either fresh water or salt water.

2. As ice along the leading edge of a glacier floats, it tends to spread or be pulled apart, forming large crevasses within the ice. These allow large blocks of ice (▸) to collapse off the front of the glacier, a natural process called *calving*.

3. As the blocks of ice fall into the water, they float, forming *icebergs*. As much as 90% of an iceberg is beneath the water. As icebergs melt, rocks and other sediment within them drop into the water. Some ice sheets and glaciers flow into the sea with such large quantities of ice that they form a large *ice shelf* that floats on seawater (▸). These can be hundreds of kilometers wide. The lighter spots in the sea, in front of the ice shelf shown in the photograph to the right, are sea ice, which form when seawater freezes—they are not ice sheets.

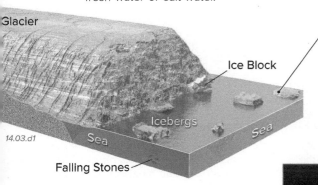

14.03.d1

4. In most years, icebergs calving from the Greenland ice sheet enter the North Atlantic shipping lanes (red lines) during the spring and early summer. This is the only area where iceberg fields intersect major transoceanic shipping lanes (▸). In 1912 the RMS *Titanic* struck an iceberg in the North Atlantic Ocean, resulting in over 1,500 deaths. Only half as many people survived.

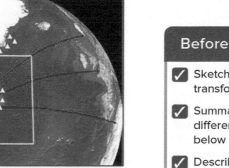

14.03.d4

5. Numerous and massive icebergs from Antarctic ice shelves populate the Southern Ocean, which surrounds the Antarctic continent. Sometimes these icebergs drift into the South Pacific Ocean near New Zealand and into the South Atlantic near the coast of South America.

14.03.d2 Glacier Bay, AK

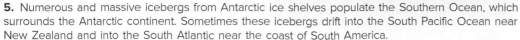

14.03.d3 Antarctica

Ice Shelf

Sea Ice

Before You Leave This Page

✓ Sketch and describe how snow is transformed by pressure into ice.

✓ Summarize or sketch the differences in a glacier above and below the equilibrium line.

✓ Describe how glaciers move and what happens when they encounter a lake or the sea.

14.3

14.4 How Do Glaciers Erode, Transport, and Deposit?

GLACIERS ARE CAPABLE OF INCREDIBLE amounts of erosion, often gouging into landscapes hundreds of meters deep. Some of us picture glaciers as uniformly white and free of debris, but most glaciers are engorged with debris and act as conveyor belts transporting debris hundreds of kilometers. Once deposited, glacially carried and deposited debris forms distinct landforms that are used for agriculture, recreation, and urbanization. What processes are involved in glacial erosion, transport, and deposition? How can we identify these processes and glacial deposits in the landscape?

A How Do Glaciers Erode?

Ice is not a hard material, but the base and sides of a glacier contain rocks and other material that can gouge (pluck) and scrape (abrade) the underlying land surface, smoothing off rough edges and removing rocks and other sediment. Once plucked and abraded from the bedrock, the debris can become incorporated into the ice, transported some distance, and eventually deposited. Meltwater at the glacier's base and front (terminus) can also cause erosion.

1. Glaciers cause erosion in three main ways: plucking, abrasion, and from glacial meltwater. In the example here, a glacier is moving over a small hill that predated the advance of the glacier.

2. At a glacier's base, pressure is great enough that ice melts, forming a thin film of water through a phenomenon known as *pressure melting*. This water can refreeze inside joints that were either preexisting or where reduced pressure (*unloading*) at a bedrock step creates jointing by expansion. As the glacier moves, rock is torn away (plucked) from the joint and incorporated into the glacier.

14.04.a1

3. On the upflow side of the hill, the motion of the glacier is pressing material against the bedrock, which is in the way. As a result, this side of the hill is heavily abraded and smoothed off, resulting in a more streamlined, almost aerodynamic, shape that is less resistant to the flowing ice. With abrasion concentrated on the upflow side and plucking concentrated on the downflow side, the glacially eroded hill takes on an asymmetric shape.

Abrasion

4. Rock and smaller sized sediment at the glacier's base scrapes at underlying bedrock through a process called *abrasion*.

Glacial Grooves · 14.04.a2 Kelley's Island, OH

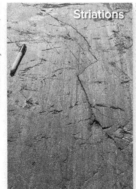
Striations · 14.04.a3 Marquette, MI

5. When ice sheets flow across the surface, they smooth and polish rocks over broad areas. They typically carve the top of bedrock into a relatively smooth, polished surface, which is gouged by scratch marks, called *glacial striations*. Such polished surfaces and scratch marks are evidence that a glacier once moved across the area (note, however that fault movement can form similar features). If the gouge marks are large and deep, they are *glacial grooves*, which provide evidence of subglacial erosion by dragged sediment and boulders at the base of a glacier. Glacially derived water can help carve or accentuate some glacial grooves.

Plucking

6. *Plucking* is concentrated on the downflow side of irregularities in the underlying bedrock, but it can occur anywhere beneath a glacier.

Roche Moutonnée · Direction of Ice Flow · 14.04.a4 Lembert Dome, Yosemite, CA

7. Plucking can occur where rocks become loosened by and incorporated into ice at the base of a glacier. The combination of plucking and abrasion can sculpt an asymmetric feature—a *roche moutonnée* (▲). Abrasion predominated to the right, plucking to the left. Plucking also liberates large rocks that are left behind when the glacier melts, perhaps hundreds of kilometers from their origin. These rocks typically are compositionally different than the local bedrock and so seem out of place; such a boulder is a *glacial erratic*.

Erratic · 14.04.a6 Pinedale, WY

Glacial Meltwater

8. Part of the glacier can melt along its base from pressure melting or on the surface from insolation, in either case forming meltwater.

Subglacial Channel · 14.04.a5 Mendenhall Glacier, AK

9. The meltwater can flow along the base of the glacier, in a *subglacial channel*. The pressurized, flowing meltwater is heavily laden with rocks and other glacial sediment and so exerts strong erosive power on the bedrock underlying glacial sediment. The water typically leaves the glacier at its front (*terminus*).

Before You Leave These Pages

✓ Describe how glaciers erode, transport, and deposit material.

✓ Summarize or sketch the differences in a glacier above and below the equilibrium line.

B How Do Glaciers Transport Material?

Once material is eroded by glacial abrasion and plucking, glaciers can entrain (incorporate) and transport vast amounts of debris, ranging from house-sized boulders to microscopic clay particles. Debris is carried at the surface (*supraglacially*), within the ice (*englacially*), and at the base of the ice (*subglacially*). Together these processes make glaciers effective conveyor belts of debris transport.

Englacial Transport

1. Material encased within the ice is called *englacial debris*. The glacier can retain this debris somewhere within the glacier, or internal shearing along the inclined shear planes shown here can bring englacial debris to the surface of the glacier, where it becomes *supraglacial debris*. Such inclined shear planes develop when moving ice segments encounter immobile ice and "ride" over and past this obstruction.

Supraglacial Transport

2. *Supraglacial debris* is transported at the surface by ice, sometimes falling into crevasses (▶) or whisked away through meltwater tunnels. Many glaciers are laced with a plumbing system of tunnels (photograph to the far right) funneling meltwater and sediment to englacial and subglacial locations.

14.04.b3 Denali NP, AK

14.04.b1

Subglacial Transport

14.04.b4 Mendenhall Glacier, AK

3. *Subglacial debris* transport occurs at the ice-bedrock interface (▲), as exposed in the meltwater tunnel in the photograph above. In some glaciers, ice moves along a "soft" bed of deformable sediment. Here, finely ground sediment, called *rock flour*, turns meltwater gray and is carried out to the glacier's margin, or even from subglacial lakes.

C How Do Glaciers Deposit Material?

Any sediment carried by ice, icebergs, or meltwater is called *glacial drift*. In any of these three cases, the debris eventually comes to rest. If deposited directly by ice, the material is generally unsorted (particles are not segregated by size) and unstratified (lacking layers)—it is called *till*. If deposited by meltwater, the glacial deposits are sorted and stratified (layered) in appearance.

1. This glacier is directly depositing dark till at its terminus. There is debris on top and along the sides of the glacier (▶), and there is more debris inside and at the base of the glacier. If the terminus of a glacier remains at about the same place, the till piles up into an irregular mass (the lumpy hills at the downhill end of the glacier).

2. This material (▶) was directly deposited by ice and is a glacial till. It is characteristically unsorted and unstratified. Most fragments (clasts) are generally angular, but others become somewhat rounded as their corners get knocked off and ground (abraded) away as they are carried by the ice. Note the people for scale in the lower right.

14.04.c1 Denali NP, AK
Terminus

14.04.c3
Glaciofluvial Outwash

14.04.c2 Athabaska Glacier, Alberta, Canada
Till

14.04.c4 Milwaukee, WI
Glaciofluvial Deposits

3. In this photograph (◀), meltwater in the foreground is issuing from the glacier in the background and carries abundant sediment. Deposition of sediment by glacial streams is called *glaciofluvial deposition*, where the term refers to the involvement of glaciers and streams (fluvial).

4. Glaciofluvial deposition involves running water and so creates sorted and stratified deposits, as in the sediments shown here (◀). Running water can sort larger, heavier clasts, such as pebbles, from those that are smaller and lighter, such as sand, producing a layered deposit composed mostly of glacial debris.

14.4

14.5 What Are the Landforms of Alpine Glaciation?

AS GLACIERS MOVE, the ice scours underlying rock and unconsolidated materials, picking up the pieces and carrying debris toward lower elevations. In mountainous areas, glaciers pluck rocks from peaks and ridges, producing some distinctive landforms that we can use to recognize landscapes that are glacially carved. Glaciers and ice sheets grind into the underlying land surface, wearing down hills and other topographic high points, and locally polishing smooth surfaces onto bedrock.

How Does Glacial Erosion Modify Landscapes?

Glaciers occupy and modify landscape features that existed before glaciation and imprint into the landscape clues that glaciers were once there. Glacial erosion reshapes a valley, typically changing it from a pre-glacial, river-carved, V-shaped valley to a glacially carved U-shaped valley. Glaciers deposit the eroded material locally or farther away.

14.05.a1   14.05.a2   14.05.a3 Denali NP, AK

U-Shaped Valleys

During Glaciation   After Glaciation   U-Shaped Valley

In these two computer-generated perspectives of the San Juan Mountains of Colorado (▲), glaciation in the left image (an artist's interpretation) results in the present-day landscape of the second image (a satellite image combined with topography).

One result of glaciation (▲) is formation of a *U-shaped valley* (i.e., a "U" shape in profile), which contrasts with V-shaped stream valleys.

What Landforms Do Valley Glaciers Form When They Deposit Sediment?

Valley glaciers erode and transport debris along their sides and bases, but flow within the glacier and the merging of adjacent glaciers distributes debris throughout much of the glacier. The sides contain especially abundant sediment because the ice receives loose materials from the mountainous slopes and streams flanking the glacier. The base of a glacier is also relatively rich in sediment because it plucks away pieces of bedrock and any loose materials over which the glacier moves. An accumulation of sediment that was carried and deposited by a glacier is a *moraine*. That is, moraine is an accumulation of till that can form ridges or be a somewhat flat sheet of till covering the land surface. We classify moraines into different types, according to where they form.

14.05.b1 Denali NP

14.05.b2 Sawtooth Mountains, ID

1. A *lateral moraine* forms along the sides of the glacier and is expressed as a dark fringe of rocks and other debris (▶). When the glacier melts, lateral moraines commonly form low ridges along what were the edges of the glacier. The computer-generated image to the far right displays lateral moraines from three now-gone glaciers.

Moraine Carried by Active Glacier

Medial Moraine   Lateral Moraine

Lateral Moraine

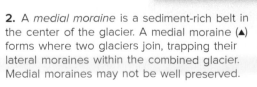

Moraine Left Behind After Glacial Melt

Terminal Moraine

2. A *medial moraine* is a sediment-rich belt in the center of the glacier. A medial moraine (▲) forms where two glaciers join, trapping their lateral moraines within the combined glacier. Medial moraines may not be well preserved.

3. A *terminal moraine* forms at the termination of a glacier and marks the glacier's farthest downhill extent (▶).

C What Alpine Landforms Form in Bedrock?

In mountains, glacial erosion produces distinctive landforms chiseled out of bedrock, including bowl-shaped basins flanked by steep ridges, and U-shaped valleys carved by the moving ice. We can use these features to recognize landscapes carved by glaciers, even after the ice melts away. The three-dimensional terrain below shows many of these features from the aptly named Glacier National Park in Montana, and photographs of examples are below the main figure.

1. Near the uppermost end of an alpine glacier, the ice plucks pieces from the bedrock, excavating a bowl-shaped depression called a *cirque*. When the ice melts, it exposes the cirque.

2. A lake within a glacially scoured depression in a cirque is referred to as a *tarn*. If a series of these lakes is connected by a stream, we refer to these as *paternoster lakes*.

3. Hard bedrock ridges that flank cirques are commonly narrow, sharp, and jagged, like the ridges shown in the figure below. Such a ridge is an *arête,* and it is jagged because it has been glacially eroded from both sides (▶). A low point in this ridge can afford a pathway from one side of the ridge to the other and is referred to as a *col*.

14.05.c2 Absaroka Mountains, WY

14.05.c3

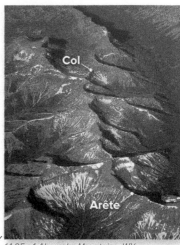
14.05.c4 Absaroka Mountains, WY

14.05.c1

14.05.c6 Silverton, CO

14.05.c5 Zermatt, Switzerland

4. Glaciers from smaller valleys can merge with a larger, thicker glacier flowing down a main valley. The larger glacier scours deeper into the bedrock, so the main valley is deepened more than the side valleys, forming a U-shaped glacial trough. When the glaciers melt away, the side valleys (▶) are higher than the main valley, and we refer to one of these as a *hanging valley*. A U-shaped valley eroded below sea level and subsequently flooded by a rise in sea level is a *fjord* (▼), famous examples of which are in Norway, Alaska, and the Arctic.

5. Where three or more cirques merge by headward erosion, they form a pyramid-like feature called a *horn*. The most famous of these is the Matterhorn (▲) near Zermatt, Switzerland.

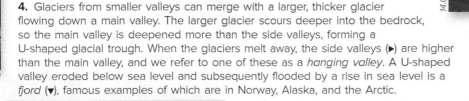
14.05.c7 Baffin Island, Canada

Before You Leave This Page

☑ Summarize what happens at the base and sides of a glacier.

☑ Describe the origins of landforms formed by glacial erosion.

14.5

14.6 What Are the Landforms of Continental Glaciation?

CONTINENTAL ICE SHEETS can cover large areas of continents, completely reshaping the landscape by scouring down the surface and depositing sheets and lumps of sediment beneath, within, and in front of the ice sheet. When the ice melts away, the entire surface tends to have low relief because any protruding topography was planed off and many low points are filled with sediment. Accumulations of glacially affected sediment form distinctive features.

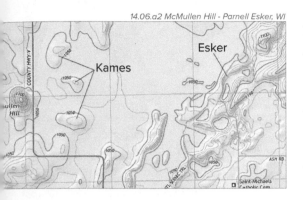
14.06.a2 McMullen Hill - Parnell Esker, WI

1. Meltwater carves tunnels through and along the bases of many glaciers, depositing sediment within the tunnels and out in front of the glacier. When the glacier melts back, sediment sorted and deposited along these meltwater channels forms long, sinuous ridges called *eskers*. The sinuous ridges on the map to the left and the photograph to the right are eskers, left behind by a retreating glacier. Eskers are an "endangered species" of glacial landforms in that many have been removed because they are highly prized as a source of gravel for roadbeds, fill, and other uses.

14.06.a3 West Dundee, WI

14.06.a4 McMullen Hill, WI

14.06.a5 Great Plains, ND

2. *Kames*, such as this one in Wisconsin (▲), are believed to have formed where meltwater in stagnant ice deposited sediment in ice crevasses or in the space between the glacier and the land surface. In the map above, a kame forms a fairly round hill. Like eskers, because their deposits are well sorted, kames are often excavated for gravel.

3. These flat to gently rolling plains (◄) are composed of sediment deposited from the base of the ice as *ground moraine*, which occurs in many parts of the Great Lakes region and upper Midwest of the U.S. The ground moraine has enriched the soils for farming, and the smoothened topography makes the area just well-drained enough, while not too susceptible to erosion.

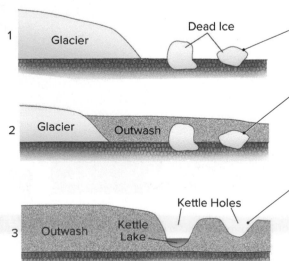
14.06.a6

4. As a glacier retreats, it leaves behind blocks of ice encased in the glacial and glaciofluvial sediment.

5. As the glacier remains nearby, or after it has retreated farther, the ice block can become partially or totally buried by glacial outwash deposited by glacial streams.

14.06.a7 Northeastern SD

6. When an ice block melts, as in the bottom diagram, it creates a small depression, called a *kettle*, within the sediment. This depression, if intersecting the water table, will fill with water, becoming a *kettle lake*, as in the photograph to the right (▶). If the water table is below the kettle, the depression will be dry except during wet periods. Kettles can also form in till, not just outwash. Kames and kettles commonly reside together, forming a rugged terrain of steep hills, ponds, and poor drainage. Such terrain is termed *kame and kettle topography*.

7. On the 3D perspective and photograph below and on the map to the right (▶) are some curiously shaped hills that resemble teardrops, each with its blunt, steep end pointing to the direction from which the ice flowed (from the north in both cases). Each of these streamlined hills is composed of till and glaciofluvial deposits and is called a *drumlin*. A drumlin forms as a moving glacier sculpts these soft materials into a shape designed to minimize drag, similar to the shape of a shark or submarine. Drumlins form in groups or fields, called *drumlin fields*, as shown in the 3D perspective below. The greatest concentrations of Drumlins are in eastern Wisconsin and in central New York. Drumlins are the only glacial feature on this page formed by advancing ice. All the rest are formed when the ice is stagnant or retreating.

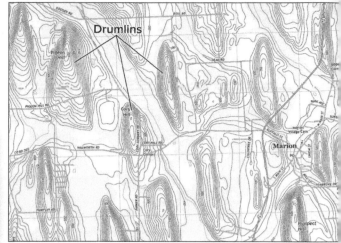

Drumlin

14.06.a9 *Central Wisconsin*

14.06.a10 *Marion, NY*

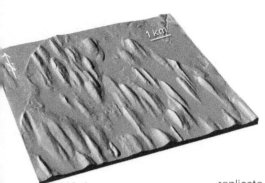

14.06.a8

8. *Recessional moraines* form as the front of the glacier melts back and stagnates for a while in one location, depositing a pile of sediment along the front of the glacier. The shape and distribution of a recessional moraine replicate the shape of the front of the glacier when it stagnated. Recessional moraines form curvilinear patterns around the Great Lakes (▶). Between moraines are flatter terrain consisting of ground moraine and outwash plains. Most of the curved ridges displayed in the large figure in the opening pages of this chapter are recessional moraines.

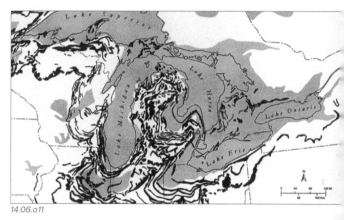

14.06.a11

9. A *terminal moraine* represents the maximum forward extent of the front of the glacier, whether it is a continental ice sheet or an alpine glacier. It has the same shape and character as a recessional moraine, but it is farthest in front. Areas between a terminal moraine and the present-day glacier, if it still exists, were once covered with ice.

Moraines, such as this one pictured to the right (▶), can have considerable relief, up to 100 m above the surrounding landscape. The steep terrain is usually forested, providing opportunities for hiking, skiing, and wildlife habitat. Flatter outwash plains and ground moraine are used for agriculture and urban settlement.

Recessional Moraines

Terminal Moraine

Glacial Outwash Plain

Roche Moutonnée

round oraine

5.a1

Moraine (Ridge)

14.06.a12 *Kettle Moraine SF, WI*

10. Melting ice sheets produce large braided streams that carry glaciofluvial sediment away from recessional or terminal moraines and deposit it either nearby or some distance away. This landform is a glacial outwash plain, which can be pitted by kettles that form depressions or ponds. If glacial outwash is confined to a valley, the term given is *valley train*.

14.06.a13

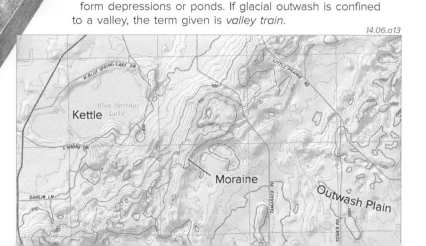

Kettle

Moraine

Outwash Plain

Before You Leave This Page

✓ Summarize where continental glaciers carry and deposit sediment, explaining the three main types of moraine.

✓ Sketch and describe the features associated with continental ice sheets, and explain how each type of feature formed.

14.6

14.7 What Features Are Peripheral to Glaciers?

REGIONS WITH COLD CLIMATES exhibit other features that are either related to glacial processes or are simply related to freezing of the ground. Some other features form in the cool, wet climates that accompany continental glaciation but in regions too warm for glaciers.

A What Types of Deposits Are Related to Glacial Episodes?

Glaciers produce an abundance of sediment and water, so glacially derived sediment can accumulate over wide regions and can be transported far from the actual glaciers by streams, wind, and waves.

14.07.a1 Tibet

14.07.a2 Shaanxi Province, China

Glacially produced sediment of all sizes mixes with abundant glacial meltwater to form large rivers with waters loaded with sediment. The rivers and streams deposit sediment on broad *outwash plains* in front of the glaciers. The gravel deposits shown here were deposited during the Pleistocene and are being eroded into by a modern stream.

Glaciers pulverize entrained rocks, producing abundant silt-sized material that can be blown away by the wind. Accumulations of windblown silt are called *loess*, and many loess deposits are glacially derived, such as these soft, tan deposits that drape over topography. This particular deposit from China occurred during the recent glacial episode.

B What Is Permafrost and Where Does It Occur?

In cold regions, below some depth, water in and below the soil remains frozen year after year, a condition called *permafrost*. The uppermost parts, called the active layer, thaw during summer.

1. The ground below the surface in this photograph (◄) is permanently frozen and so it is *permafrost*. During warm months, the top few meters of the ground (the *active layer*) thaw, allowing some vegetation to grow. Permafrost commonly does not allow trees to grow. Trees on the edge of permafrost are typically short and stunted.

2. In North America (►), large areas of continuous permafrost are restricted to northern Canada and Alaska. In other areas, permafrost is either discontinuous or occurs in high, cold mountains. When frozen, permafrost is a very hard material, but it weakens considerably if it thaws.

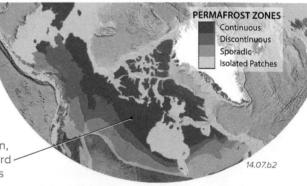

PERMAFROST ZONES
- Continuous
- Discontinuous
- Sporadic
- Isolated Patches

14.07.b2

14.07.b1 Denali NP, AK

14.07.b3 Kolyma River, Siberia, Russia

3. The smooth, lower part of this ledge (◄) is permafrost, and it has a hard, icy appearance. On top is a thin, brownish layer that is not always frozen (the active layer). Trees on permafrost commonly lean in various directions because permafrost keeps the roots shallow, and thawing of the active layer limits how well the roots support the trees.

14.07.b4 Aerial photograph, North Slope, AK

4. *Patterned ground* consists of geometric patterns, such as polygons and circles, above permafrost. It has several different expressions, but it forms when expansion and contraction from frost action concentrates gravels, stones, or boulders at the surface. These geometric patterns repeat over wide swaths of continuous and discontinuous permafrost.

C What Types of Lakes Were Associated with Glacial Times?

The cool, wet climates that favor an increase in precipitation also favor the formation and maintenance of lakes. In the western U.S., huge lakes existed at times during the recent glacial advances, but they largely dried up when the climate changed about 15,000 years ago.

1. A Pleistocene lake filled low, interconnected basins in the Rocky Mountains of western Montana. This former lake, named *Lake Missoula,* caused catastrophic flooding, forming the channeled scablands of Washington. Shorelines from this lake were etched as horizontal lines into the hills surrounding Missoula, Montana (▼).

14.07.c2 Missoula, MT

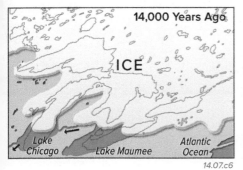

14.07.c3 Dry Falls, WA

Western United States

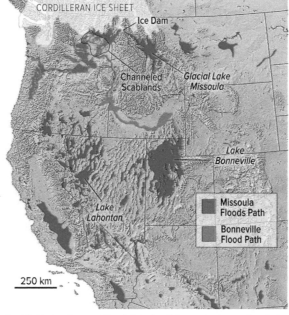

14.07.c1

CORDILLERAN ICE SHEET
Ice Dam
Channeled Scablands
Glacial Lake Missoula
Lake Bonneville
Lake Lahontan

Missoula Floods Path
Bonneville Flood Path

250 km

2. Multiple times, Glacial Lake Missoula breached the glacial dam holding back its waters, and catastrophic torrents of water raced across the landscape to the west, carving the scablands. The huge floods carried gigantic boulders, carved smooth potholes into the bedrock (◀), and formed enormous ripples (▶). The shapes of the ripples record immense currents moving downstream, from right to left in this view.

14.07.c4 Bonneville Salt Flats, UT

3. The Great Salt Lake of Utah is a remnant of a much larger Pleistocene lake named Lake Bonneville, which caused a huge flood onto the Snake River Plain. As the large lake dried up, it left the Bonneville Salt Flats (▲), home to rocket testing, land-speed records, and many miles of salt.

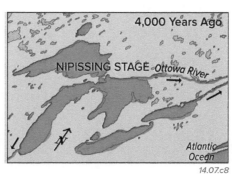

14.07.c5 West Bar, Columbia River, WA

Evolution of the Great Lakes

14,000 Years Ago
ICE
Lake Chicago   Lake Maumee   Atlantic Ocean
14.07.c6

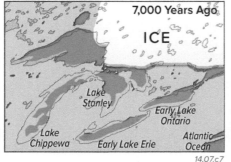

7,000 Years Ago
ICE
Lake Stanley
Early Lake Ontario
Lake Chippewa   Early Lake Erie   Atlantic Ocean
14.07.c7

4,000 Years Ago
NIPISSING STAGE Ottawa River
Atlantic Ocean
14.07.c8

6. By around 4,000 years ago the Great Lakes had nearly achieved their present configuration except for an eastern outlet using the Ottawa River channel. Eventually, this outlet was abandoned as the crust readjusted to the removal of ice.

4. As the Laurentide Ice Sheet retreated, meltwater was dammed between the ice front and moraines to the south, forming lakes. Early Lake Chicago and Early Lake Maumee were the first ancestral lakes to form. The glacier prevented drainage to the northeast, so drainage was directed southward into the Mississippi drainage basin. These lakes built beach ridges that were the sites for early trails through the wetlands and later railroads and highways.

5. As the ice continued to retreat, lower outlets were revealed at around 7,000 years ago. These spillways were low enough to effectively empty most of the ancestral lakes. The ancestor of Lake Michigan was Glacial Lake Chippewa and the ancestor to Lake Huron was Glacial Lake Stanley.

Before You Leave This Page

☑ Describe the characteristics of different deposits related to glacial episodes and how each type forms.

☑ Explain permafrost and where it occurs.

☑ Describe several large ice-age lakes and features they formed, either while full or while emptying.

14.7

14.8 What Is the Evidence for Past Glaciations?

SEVERAL LINES OF EVIDENCE confirm that huge ice sheets once covered the land. This evidence is directly expressed as features and deposits within the landscape and is recorded indirectly in ice cores, in isotopic compositions of marine fossils, and in the pollen record.

A — What Features Indicate the Former Presence of Glaciers?

To be able to recognize past glaciations, geoscientists visit active glaciers and ice sheets to observe what types of landforms and deposits suggest glaciation. These modern-day examples can help explain the origin, form, and distribution of Pleistocene landforms and deposits, an example of how "the present is the key to the past."

14.08.a1 Absaroka Mtns., WY

1. The most obvious evidence of glaciers includes landscapes that may have active ice and contain features diagnostic of glaciation, including cirques, arêtes, tarns, and U-shaped valleys. Each of those features is visible in this scene.

14.08.a2 Fort Atkinson, WI

2. Other landscape features are diagnostic of glaciers, including eskers, low ridges of moraine, and drumlins. Some of these features may be subtle, such as these two hills (◄), each of which is a drumlin. The concentric strips on these hills are the result of farming, but they help define the oval to teardrop shape of each hill. The presence of these features across much of New England and the Great Lakes region demonstrates that glaciers once covered these areas, leaving behind clues to their former presence after the glaciers had totally disappeared.

Landforms

14.08.a3 Marquette, MI

3. When glaciers and ice sheets flow across the land, the ice smooths and polishes underlying rocks and deposits glacial sediments (till) on the polished surface (◄). The site shown in the photograph to the left has a shiny, polished surface that displays striations and chattermarks cut into the bedrock by stones carried along the base of the glacier. The striations and chattermarks remain as clues that glaciers once covered this region.

4. As glaciers advanced and melted back across the landscape, the ice deposited layers or patches of till. Ancient examples, such as this one from Kimberly, South Africa (►), have lithified into rock called *tillite*. The tillite provides evidence for ancient glaciations, in this case near the end of the Permian Period, about 250 million years ago.

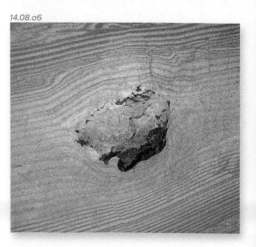

14.08.a4 Kimberly, South Africa

Deposits

14.08.a5 Pinedale, WY

5. Glaciers can carry huge rocks (◄), some as big as a house, and leave these boulders scattered about the landscape. Glaciers may transport large blocks hundreds of kilometers, taking boulders to places where such rock types are not present in the bedrock. Such an out-of-place block, like the ones shown here, is a glacial erratic.

6. An unusual feature of some marine and lake sediment is the presence of scattered stones (►) in an otherwise fine-grained sediment. We call these *dropstones* because they have been carried within floating icebergs and then dropped into fine sediment on the seafloor or lake bottom.

14.08.a6

B   How Do We Know Where and When the Most Recent Glacial Advance Occurred?

From diverse lines of evidence, geoscientists determine which areas were once covered by ice and which ones were not. They then use fossils, isotopic dating methods, and pollen to determine when glaciers were most widespread. Because it affects sea level and the influx of fresh water into the ocean, glaciation can also be investigated by examining the nature and chemistry of marine fossils and sediment.

1. During evaporation of seawater or fresh water, heavier isotopes of an element preferentially remain in the water, while lighter isotopes escape more easily into the water vapor. In the case of oxygen isotopes, evaporation causes the water to become enriched in the heavier isotope oxygen-18 (^{18}O) while enriching the water vapor in the lighter isotope oxygen-16 (^{16}O).

2. As the water vapor condenses into clouds and precipitation (rain, snow, or hail), the water, snow, or ice contains the higher proportion of lighter isotopes that was in the water vapor. If snow and ice accumulate on land, they tend to keep the light isotopes from returning to the sea. As a result, an increase in the amount of snow and ice on land, as during a glacial event, will cause seawater to be more enriched in heavy isotopes.

3. As glaciers and ice sheets melt, they release their water, which is relatively enriched in lighter isotopes. Streams and melting icebergs return these light-isotope-enriched waters back to the sea. A decrease in glaciation, therefore, causes seawater to shift toward lighter isotopic compositions, just the opposite of an increase in glaciation. As a result, isotopic compositions of ice on land and on water in the sea are indications of increases and decreases in the amount of snow and ice on land.

Precipitation

Evaporation

14.08.b1

5. Some marine organisms build shells or skeletons of calcium carbonate by extracting the necessary chemicals from seawater (▶). As the chemistry and temperature of the water change, so does the chemical composition of the shells or skeletons formed in that water. Geoscientists analyze oxygen and carbon isotopes in fossils to infer changes in seawater temperature and chemistry over time. We can then use such changes to infer the times of glaciation or times when melting of glaciers released fresh water into the ocean.

14.08.b2 Cayman Islands

4. The graph below shows temperatures inferred from oxygen-isotope compositions of ice from part of a 3 km-deep hole that scientists drilled into the ice sheet of central Greenland. The data show how scientists interpret paleoclimatic conditions in central Greenland to have varied over the last 100,000 years. Points to the right indicate that temperatures were warmer, as they are today, and glaciers were less widespread. From these and other data, we infer that glaciers decreased and increased in extent many times during the last 100,000 years.

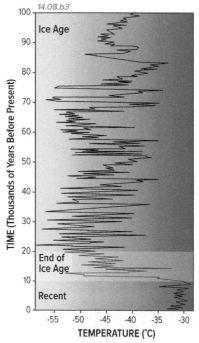

14.08.b3

6. Glaciers left behind other evidence in the Northeast and Great Lakes areas. As the glaciers retreated, meltwater collected in low areas, forming numerous lakes. Sediment accumulating in lakes and in other settings contains a record, in the form of pollen (▶), of the types of plants that grew at different times. These pollen records document a shift from spruce and birch trees to oak trees as glaciers retreated and the climate warmed.

14.08.b5

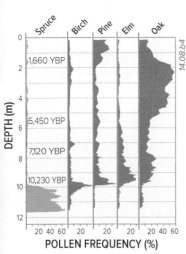

Spruce   Birch   Pine   Elm   Oak

DEPTH (m)

0

1,660 YBP

2

4

5,450 YBP

6

7,120 YBP

8

10,230 YBP

10

12

20 40 60   20   20   20   20 40 60

POLLEN FREQUENCY (%)

14.08.b4

Before You Leave This Page

✓ Describe evidence used to infer that glaciers once covered a landscape.

✓ Discuss how glaciations can be expressed in ice and the ocean and how we can use this record to interpret when glaciation occurred.

14.8

14.9 What Starts and Stops Glaciations?

TO UNDERSTAND THE REASONS for glacial and interglacial times, we need to further explore what causes global changes in climate. Because human history is short compared to the timescales on which global climate change occurs, we do not completely understand all the causes.

A What Variations in Earth's Tilt and Orbit Influence Global Climate?

Short-term (up to 100,000 years) variations in climate are likely controlled by the amount of solar radiation reaching Earth. Milutin Milankovitch, a Serbian astronomer and geophysicist, recognized that Earth's climate could be influenced by changes in the amount and direction of Earth's tilt and Earth's orbit shape. We call such changes *Milankovitch cycles*.

Changes in Earth's Tilt (Obliquity)

Over time, Earth's axis of rotation changes its tilt relative to its plane of orbit around the Sun. About every 40,000 years, the tilt changes from 22.5° to 24.5° and back again. The amount of tilt toward the Sun affects Earth's climate because tilt affects how much summer sunlight strikes higher latitudes. The diagrams below are for winter in the Northern Hemisphere.

1. The *maximum tilt angle* of Earth's rotation axis is 24.5°. This amount of tilt increases the effects of the seasons. When combined with other climatic factors, it can lead to a decrease in glacial activity because warmer summer temperatures melt more polar ice.

14.09.a1

2. The present-day position of Earth's tilt is 23.5°. Earth's tilt is currently adjusting back from the maximum 24.5° tilt, which last occurred near the end of the Pleistocene glacial advance.

14.09.a2

3. When Earth's tilt is at its *minimum tilt angle* (22.5°), high latitudes receive less direct sunlight during the summer, causing cooler summers and an increase in glaciers.

14.09.a3

Wobble of Rotation Axis (Precession)

4. *Precession* is similar to what happens when a spinning top slows down and wobbles. As Earth wobbles, its spin axis changes from pointing at the North Star (Polaris) to pointing at the star Vega and back again.

5. Precession causes December to be winter in the Northern Hemisphere during some times in Earth history, but June to be winter at other times in Earth history. If the effects of precession are added to other astronomical factors, these phenomena could affect global climate. The precession cycle lasts about 23,000 years.

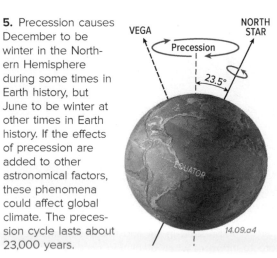

14.09.a4

Shape of Orbit (Eccentricity)

6. *Eccentricity* is the term for the noncircular shape of Earth's orbit around the Sun. Earth's orbit changes over long timescales, sometimes being more circular and sometimes slightly more elliptical. These changes cause variations in the amount of insolation reaching Earth.

7. This change from a more circular to a less circular orbit is thought to only slightly affect climate, but when added to other astronomical cycles, its effect might be significant. The eccentricity cycle lasts about 100,000 years.

8. Scientists have computed the effects of each of these factors (tilt, precession, and eccentricity) and then combined the effects to investigate how they interact, both in the past and in the future. When used to try to reconstruct past events, the calculated effects can be compared with records of past climate, like the isotopic composition of ice cores. Such comparisons support the hypothesis that Milankovitch cycles can explain many past climatic variations, including the waxing and waning of glaciations. Computer models can also be used to predict future Milankovitch-related climate changes, but these results are currently being debated by the scientific community.

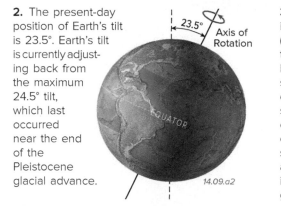

14.09.a5

B Can Variations in Solar Heating and Atmospheric Composition Influence the Intensity of a Glacial Episode?

Several factors may not, by themselves, cause the onset or demise of a glacial episode, but they can influence how severe the resulting glaciation is. For example, the amount of energy given off by the Sun is not constant, and the composition of Earth's atmosphere is not constant. As a result, the amount of energy that can pass through the atmosphere and be retained by Earth is variable and can lead to a warmer or cooler climate.

1. Every 11 to 14 years, the level of *sunspot activity* on the Sun increases, producing very small changes in solar energy output. This energy fluctuation can influence Earth's climate system, affecting temperature, precipitation, and other weather phenomena. When sunspot activity declines, evidence shows that Earth's climate cools by a small amount.

2. During glacial episodes, snow and ice cover more of Earth's surface, and cloud cover also increases. Both of these increase the reflectivity, also called *albedo,* of Earth so that more insolation gets reflected off Earth's surface and is lost to space. This loss of heat makes the climate cooler. A cooling climate can result in more snow, ice, and clouds, leading to more cooling. In this way, the system reinforces itself—a positive feedback.

3. Volcanoes release millions of tons of carbon dioxide (CO_2) into the atmosphere every year, and plants and marine life extract some of this CO_2. An increase in the amount of CO_2 and methane (CH_4), both greenhouse gases, tends to warm the planet. The amounts of CO_2 and CH_4 were relatively lower during glacial episodes and higher during interglacial periods.

4. Large, explosive volcanic eruptions can add significant quantities of volcanic ash, dust, and sulfur dioxide (SO_2) (which is converted to sulfuric acid aerosols) into the atmosphere. These aerosols reflect insolation back into space, allowing less sunlight to reach Earth's surface. A major volcanic eruption will increase the amount of ash and dust in the atmosphere, perhaps resulting in global cooling. It is thought that volcanic eruptions in the tropics during the 13th and 15th centuries, along with decreased solar activity, may have triggered an unusually cold period in Europe.

14.09.b1

C What Is the Role of Ocean Currents and Continental Positions in Glaciations?

Ocean currents transport water of different temperatures from one part of the ocean to another. Such currents can warm or cool land areas, helping to increase or decrease glaciation on land.

1. Upwelling currents can bring deep cold water to the surface. Cold currents help cool the land, perhaps allowing glaciers to form, if there is sufficient precipitation.

2. Cold currents, however, can also inhibit the growth of glaciers because the ice puts less moisture into the atmosphere, leading to less snowfall.

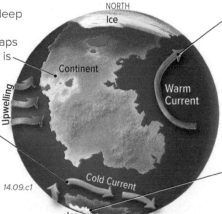

14.09.c1

3. Warm currents bring tropical waters northward, waters that have been heated by the Sun. These currents can warm adjacent parts of a continent, inhibiting glaciations. Warm currents also bring warm, moist air, increasing precipitation, which can allow more ice to accumulate if the temperature is cold enough.

4. The position of continents affects the geometry of ocean currents, deflecting currents in certain directions or blocking the connection between different oceans. In this way, continental positions influence ocean currents, which impact regional and local climates.

5. A continent located at or near the North or South Pole provides a large landmass on which continental ice sheets can form.

Causes of Ice Ages

The ice age lasting from approximately 2.6 million years ago to the present has included many *glacial* and *interglacial* periods. This time span lies mostly in what has traditionally been called the *Pleistocene* epoch, which ended with the end of the last major glaciation about 12,000 years ago. Major glacial periods commonly lasted about 40,000 to 100,000 years, with interglacial periods lasting about 10,000 years on average. Changes between the two conditions apparently could occur rapidly in geologic terms, in some cases over less than several thousand years. Some scientists regard the present time as an interglacial.

The cause of ice ages remains controversial, but the factors described on these two pages are among the culprits that can help instigate an ice age or influence how pronounced one is. Also involved are other factors, such as changes in solar activity and the amount of atmospheric greenhouse gases.

Before You Leave This Page

☑ Describe how variations in Earth's tilt and orbit influence global climate.

☑ Explain how global climate can be affected by atmospheric gases, volcanic ash, and the amount of snow, ice, and cloud cover.

☑ Describe the role of ocean currents and continental positions on glaciations.

14.9

What Would Happen to Sea Level if the Ice in West Antarctica Melted?

WEST ANTARCTICA HAS THE POTENTIAL TO CAUSE a dramatic rise in sea level if its ice sheet melts. It contains a huge volume of ice that is especially vulnerable because it is in direct contact with the sea. If the area's ice sheets melted, rising sea level would pose a great hazard for the world's coastlines.

 ## A What Is the Setting of Glaciers, Ice Sheets, and Ice Shelves in West Antarctica?

The continent of Antarctica, sitting over the South Pole, is a frozen world mostly covered by snow and thick sheets of ice. It contains 90% of the world's ice and 75% of the world's fresh water.

14.10.a1 West Antarctica

1. West Antarctica, like the rest of the continent, is mostly covered by snow and ice, with bedrock in the mountains and along the coast. There are many glaciers and ice shelves, both of which contribute large icebergs, such as the tilted one in the foreground (◄), which is larger than a medium-sized building.

2. The Transantarctic Mountains (►), a major mountain range more than 4,500 m tall, divide West Antarctica from East Antarctica. West Antarctica is much smaller than East Antarctica and consists of a central landmass that leads to a peninsula extending toward South America. West Antarctica contains 11% of the ice in Antarctica.

3. The land is flanked by three large ice shelves (►), where glacial ice from the land has pushed out into, and is now floating on, the ocean. These three shelves are the Ross, Ronne-Filchner, and Larsen ice shelves.

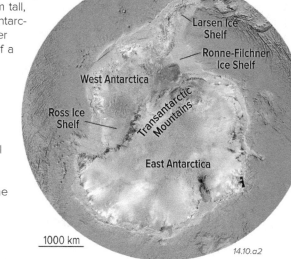

1000 km

14.10.a2

4. The central part of West Antarctica contains an ice sheet as thick as 3,500 m (about 11,500 ft). The ice sheet accumulates from snowfall on land and feeds rapidly moving glaciers that carry massive amounts of ice toward the sea and the ice shelves (►).

5. The base of the ice sheet is below sea level. The central part of the ice sheet is resting on solid bedrock, but outer parts are floating on the sea.

14.10.a3

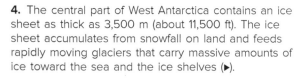

Ice Sheet

6. Each ice shelf loses large volumes of ice every year by calving of icebergs and by melting of the underside, which is in contact with seawater. One part of the coast that has been studied in detail loses an average of 250 km³ of ice each year.

B What Could Happen to West Antarctica if Global Sea Levels Rise?

Ice shelves can be reshaped suddenly. In early 2002, much of the Larsen Ice Shelf collapsed, breaking into millions of icebergs that floated out to sea and melted. It is possible, but unlikely, that larger, more disastrous melting events could occur, including loss of large parts of the West Antarctic ice sheet, but this is not occurring now.

These satellite images, taken a month apart, show the collapse of part of the Larsen Ice Shelf. The left image shows the ice shelf before the ice-loss event.

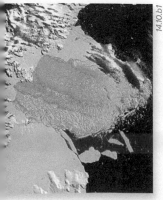

14.10.b1

14.10.b2

An area of 3,250 km² was lost, which is much larger than the entire state of Rhode Island. Other events have sprung single icebergs that were 70 km by 25 km (1,750 km²).

One possible scenario, much debated at this time, is that rising global sea level could float more of the West Antarctica ice sheet, detaching it from the underlying bedrock. If this occurs, the collapsed parts would melt, raising global sea level by some amount. But by how much? We might want to know this.

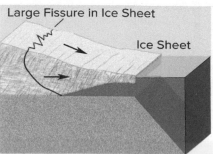

Large Fissure in Ice Sheet

Ice Sheet

14.10.b3

C How Do We Calculate the Rise in Sea Level If West Antarctica's Ice Melts?

To evaluate how melting of ice sheets would affect our shorelines, we can make some simple calculations to determine how much sea level would rise in an unlikely scenario—melting of all the ice from West Antarctica.

1. Examine the situation below. A rectangular tub of water has one block of ice floating in it and two blocks on land that will add water to the tub if the blocks melt. The ice blocks and the grids on the side of the tub are 10 cm on a side for easy measuring.

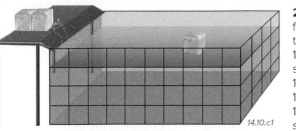

14.10.c1

2. The block floating in the water is 10 cm on all sides, or 10 cm by 10 cm by 10 cm. We simply multiply these three dimensions to get the volume of the block, which is 1,000 cm³ (10 cm × 10 cm × 10 cm = 1,000 cm³). The two blocks on the table total 20 cm (two blocks wide) by 10 cm (one block deep) by 10 cm (one block) high. If we multiply 20 cm × 10 cm × 10 cm, we get 2,000 cm³.

3. The Floating Block—Most of the floating block is below the surface. As ice melts, it yields a volume of water that is slightly less than the volume of ice, because water is more dense than ice. In other words, the floating ice displaces a mass that is very similar to the amount of volume that it would add if melted. As a result, melting ice that is floating in fresh water does not appreciably raise the level of the water. It does cause a slight rise in a body of salt water because melting ice yields fresh water, which is less dense, but we'll ignore this factor to simplify things.

4. Blocks on the Table—If the blocks on the table melt, all of the water helps raise the level in the tub. To see how much, we need only worry about the surface area of the water, not how much water is already there at depth within the tub. Also, a volume of ice produces about nine-tenths that volume of water, or a ratio of 0.9 (volume water produced/volume ice melted).

5. To get the surface area of a rectangle of water, we multiply the dimensions of its two sides. The tub is 100 cm long by 40 cm wide, yielding a surface area of 4,000 cm². To calculate how much the melting blocks will raise water level in the tub, we spread our volume of water over this surface area. The calculation is as follows:

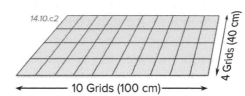

14.10.c2

10 Grids (100 cm)

4 Grids (40 cm)

| 2,000 cm³ (volume of the ice blocks on table) | × | 0.90 (to convert the ice to water) | ÷ | 4,000 cm² (surface area of the water) | = | 0.45 cm (rise in level of water) |

6. So melting an ice block floating in the water (representing sea ice) does not appreciably change sea level, but melting ice on land does. The larger the amount of ice on land that is melted, the larger the rise. But the larger the surface area of the tub, the smaller the rise. For West Antarctica and our modern seas the calculation is:

| 3,000,000 km³ (volume of all the ice) | × | 0.90 (ice to water) | ÷ | 361,000,000 km² (surface area of the world's oceans) | = | 0.0075 km (rise in sea level) |

To get meters, we multiply 0.0075 km × 1,000 m/km = 7.5 m (25 ft)

7. This calculation does not take into account that as we add water and raise sea level, the ocean spreads out over the land and so the surface area increases. The number calculated when considering this factor is more like 6 m (20 ft). Recall that this is a worst-case scenario that would occur only under a huge change in climate.

D What Impact Would Raised Sea Levels Have on the East Coast?

Think about some photographs of coasts you have seen, or visits to the coast you may have taken, and imagine those areas if sea level were 6 m (about 20 ft) higher. To plan for such contingencies, the USGS conducted a detailed assessment of the relative risk of sea-level rise for each part of the East Coast of the U.S. For each segment of coast on the map shown here, coastal scientists investigated various factors, including elevation, slope of the land, etc. From this analysis, each area was assigned a risk, from low to very high. The most vulnerable settings include the eastern coast of Florida, the barrier islands of Virginia and North Carolina, especially Cape Hatteras, and coastlines around Maryland, Delaware, and New Jersey. How vulnerable is your favorite part of the East Coast?

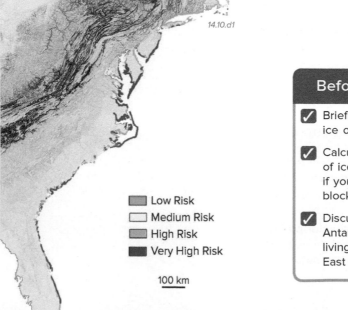

14.10.d1

◻ Low Risk
◻ Medium Risk
◻ High Risk
◼ Very High Risk

100 km

Before You Leave This Page

✓ Briefly summarize the settings where ice occurs in West Antarctica.

✓ Calculate how much melting a block of ice will raise water levels in a tub, if you know the dimensions of the block and tub.

✓ Discuss why calculations about West Antarctica are important to people living along coastlines, including the East Coast of the U.S.

14.10

14.11

How Could Global Warming or a Glacial Period Affect Sea Level in North America?

THIS SHADED-RELIEF MAP OF NORTH AMERICA colors the land surface and seafloor according to elevation above and below the present sea level. These elevations represent possible levels to which the sea could rise if the climate warms or fall if the climate cools substantially, in either case causing a change in the extent of land ice. You will use estimates of the amount of ice that could be lost or gained to calculate how much sea level could rise or fall, and then evaluate the implications for the economy, transportation, and hazards for some major cities of North America. The scenarios presented in this investigation are extreme end members, not situations that will happen, except perhaps after thousands, millions, or even hundreds of millions of years.

Goals of This Exercise:

- Observe a shaded-relief map of North America to identify areas that are close to sea level (above it and below it).
- Use estimates of current amounts of ice on Earth to calculate how high sea level would rise if all the ice melted.
- Use estimates of the amount of ice that was present during the last glacial maximum (20,000 years ago) to calculate how much sea level would drop if these conditions returned.
- Use your results to identify how such rises and falls in sea level would affect some major cities of North America.

Data

Listed below are data about the present surface area of the oceans and estimates for the amount of ice that (1) is present today and (2) was present when glaciers were at a maximum 20,000 years ago. Use these data to complete the calculations on the next page.

1. The present surface area of the oceans is 361,000,000 km².

2. The total amount of continental ice (ice sheets, ice caps, glaciers) currently present on the planet is estimated to be 32,000,000 km³.

3. During the last glacial maximum, 20,000 years ago, the amount of ice is estimated to have been 52,000,000 km³ more than is present today. Your calculations will determine how much water is used to make this additional amount of ice, and how much this would lower sea level.

4. When ice melts, the volume of water produced is about 0.9 times the volume of the ice. That is, the volume of water produced is only 90% of the volume of ice.

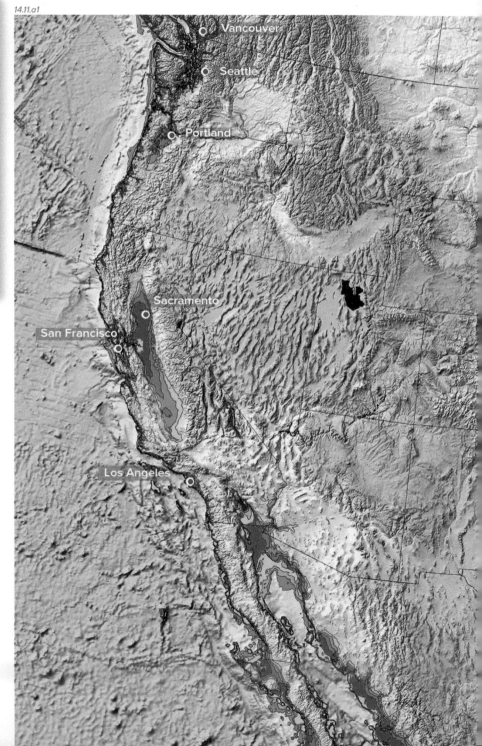

14.11.a1

Procedures

Follow the steps below, entering your answers for each step in the appropriate place on the worksheet.

1. Calculate how much water would be released if all the ice on the planet melted (an extremely unlikely scenario). For this calculation the equation is: the volume of water gained = the volume of ice × 0.9 (volume water / volume ice).

2. Calculate the volume of water that would be tied up in ice if glaciers returned to the same volume as 20,000 years ago. The equation is: the volume of water lost = the volume of *additional* ice 20,000 years ago × 0.9 (volume water / volume ice).

3. Calculate how much sea level would rise for the water volume gained in step 1 or how much sea level would fall for the water volume lost in step 2. Ignoring many important complications, the much simplified equation is: the change in sea level = change in water volume / surface area of the oceans.

4. Examine how each city shown on this map would be affected by the two extremes. Would it be flooded, not flooded but much closer to the shoreline, or much farther from the shoreline? Discuss how such changes would affect a typical city's transportation, vulnerability to coastal flooding, economic livelihood, and any other factors you can think of.

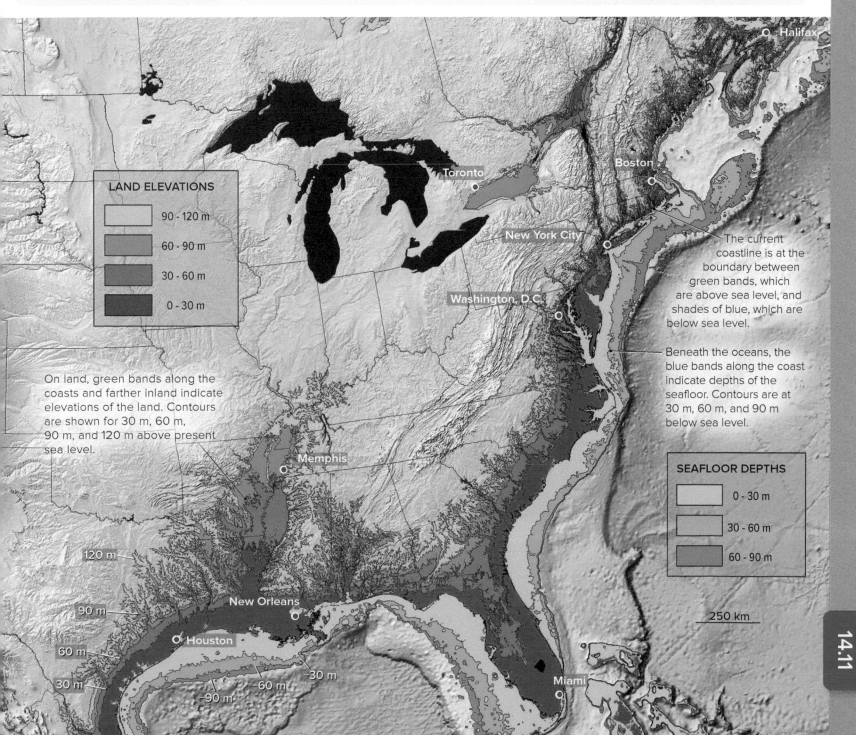

LAND ELEVATIONS

90 - 120 m

60 - 90 m

30 - 60 m

0 - 30 m

On land, green bands along the coasts and farther inland indicate elevations of the land. Contours are shown for 30 m, 60 m, 90 m, and 120 m above present sea level.

The current coastline is at the boundary between green bands, which are above sea level, and shades of blue, which are below sea level.

Beneath the oceans, the blue bands along the coast indicate depths of the seafloor. Contours are at 30 m, 60 m, and 90 m below sea level.

SEAFLOOR DEPTHS

0 - 30 m

30 - 60 m

60 - 90 m

250 km

14.11

Coasts and Changing Sea Levels

PEOPLE LIVE ALONG COASTS for a variety of good reasons. Oceans moderate climate, supply fish and other sea resources, provide transportation for people and goods, and make available a range of recreation options. Living along a coast also presents challenges. Coasts are not static, and they can change significantly as the result of a single storm. Coasts also experience waves, tides, and shifting beach materials. Past sea levels have been more than 200 m (660 ft) higher than today and about 120 m (390 ft) lower, flooding the continents or exposing continental shelves. The most rapid sea level changes are related to changes in glaciers and continental ice sheets.

15.00.a2 Cape Fear, NC

The East Coast of the U.S. (◄) has mostly gentle topography along the length of the coastline with a variety of different features found at different segments. In the figure below (▼), observe the features along each part of the coast, from Georgia on the left to Maine on the right.

What types of features are common along coasts, and why does a coastline vary in the features?

The Pacific Coast has a much different appearance (►). Instead of gentle topography along the coast, the Pacific Coast commonly has cliffs that drop off into the ocean.

What causes some coastlines to have subdued topography and others to have cliffs and steep slopes along the beach?

15.00.a3 Big Sur, CA

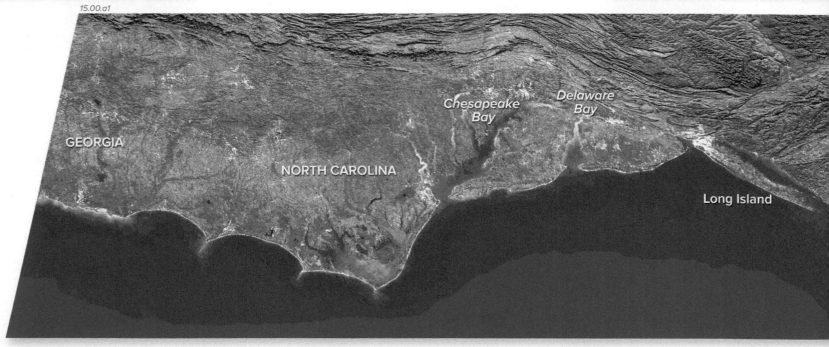
15.00.a1

The Outer Banks of North Carolina, including Cape Hatteras, consists of a chain of linear islands offshore of the main coast. These islands help shield the main coastline from storms but are themselves very vulnerable to hurricanes and other storms coming from the ocean.

What types of features are common along coasts, and why does a coastline produce these smooth areas of the landscape? Is the process still occurring?

Chesapeake Bay and Delaware Bay are large *embayments*, places where the coastline curves inward toward the land. In their outlines, they resemble stream valleys with tributaries joining from the sides. Between them, the land juts out toward the sea, forming a peninsula.

How do features such as Chesapeake Bay and Delaware Bay form, and why do they have this branching shape?

Long Island, the lengthy island extending eastward from New York City, and the adjacent shorelines of New Jersey, were recently damaged by waves associated with Hurricane Sandy. Houses and other buildings along the coast were completely destroyed by waves and storm waters that surged over the beach.

What processes cause damage along coastlines, and what factors make one coast more vulnerable than another?

TOPICS IN THIS CHAPTER

15.00.a4 Acadia NP, ME

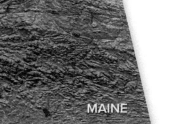

The beautiful Maine coastline is very irregular, with bays that curve into the coast and rocky headlands that protrude out into the sea (◄).

How did this irregular coastline form, and what does such a coastline indicate about its history?

MAINE

Cape Cod

Cape Cod, Massachusetts protrudes into the ocean as a flexed arm or a boot with curled toes (▼). Offshore are a variety of islands, for which the region is well known.

How did Cape Cod form, and what coastal processes caused this shape?

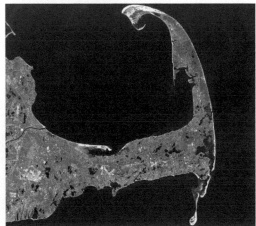

15.00.a5 Cape Cod, MA

Sea levels have been rising for at least the past century, and there are concerns that climate change may cause this rise to accelerate. Farther back in the past, sea levels have been more than 100 meters higher and lower than at present.

What causes sea level to rise and fall, and what happens to a coast when sea level changes by a significant amount?

Changing Sea Levels and Coasts

Features along the East Coast of the U.S. and Canada respond to the rising and falling of sea level relative to the land. As the climate warmed over the past 20,000 years, it caused melting of ice sheets and glaciers that had covered much of the northern U.S. and nearly all of Canada. Water released from melting ice carved new stream valleys as it flowed toward the sea. The shapes of Chesapeake Bay and Delaware Bay originated as stream valleys, with tributaries joining from the sides. As the meltwaters reached the sea, global sea level rose. The rising seas flooded coastlines and stream valleys, turning Chesapeake Bay from a stream valley to an embayment filled by the sea. The rising sea levels flooded similar features up and down the coastline, forming the many inlets and bays in Maine, Canada, and elsewhere along the Atlantic coast.

Other features along the East Coast likewise indicate rising and falling sea level over the last several million years. The Outer Banks region of North Carolina consists of long, sandy islands that form a barrier offshore of the main coastline. For this reason, this type of island is called a *barrier island*. The Outer Banks is very low, barely above sea level, juts eastward into the Atlantic, and lies within frequent tracks for hurricanes and other tropical storms. For these reasons, the Outer Banks is one of the most vulnerable stretches of coast in the world. The area is further threatened as sea level continues to rise.

Long Island and Cape Cod both contain materials deposited along the front of glaciers during the last glacial advance. As sea level rose from melting of the glaciers, glacial deposits were reworked and transported by waves and ocean currents. In Cape Cod, the prevailing currents move sand and other materials northward along the coast, curling around the north end to form Cape Cod's distinctive curved shape.

15.0

15.1 What Processes Occur Along Coasts?

COASTS ARE THE INTERFACE BETWEEN LAND AND WATER and so respond to processes that arise from both sides and from changes in sea level. Waves and tides affect a coast from the water side, while streams, wind, and sometimes ice contribute sediment from the land. Together these processes sculpt the coast, redistribute sediment, and present challenges for people who live in the *coastal zone*—which includes the shoreline and a strip of adjacent land and water. The width of the strip of land and water within the coastal zone varies with the type of coast.

A What Types of Processes Affect Coasts?

15.01.a2

1. From the water side, most coasts are strongly affected by waves (◄), which are near-surface features generated by wind blowing across the water. Waves typically form far from shore but can approach and "break" upon reaching shallower water, where they erode rock and loose material, deposit sand and other sediment, or simply move sediment around. The shoreline is also called a *beach*, especially if it is sandy.

2. From the land side, streams can be important contributors of sediment into the coastal zone. Silt, sand, and coarser sediment carried by streams accumulate close to where a stream meets the sea or lake, commonly forming a *delta*. Fine-grained sediment suspended in stream water can be carried farther away from shore.

5. Many beaches have similar sets of landforms, as shown in this idealized beach profile (▼). Each beach is unique, so these features don't always occur in these proportions.

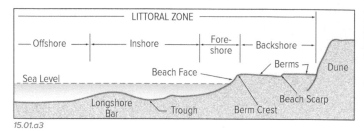
15.01.a3

3. Wind can move sand and finer sediment away from, toward, or along a beach, depending on the direction the wind blows. Many beach areas are backed by coastal sand dunes, many of which are held partially in place by vegetation (▼).

6. The littoral zone (▲) contains several components. The *backshore* is the highest part of the beach, only seldom covered by water. The backshore consists of nearly flat platforms called *berms* that are deposited by waves during the strongest storms. The *foreshore* is submerged during high tides but exposed during low tides. Farther out is the *inshore* or *swash zone*, which is permanently submerged by shallow water where waves break. A rocky coast may have stretches that have some of these features and some stretches without any beach (▼).

15.01.a4 Clearwater, FL

15.01.a1

7. Faulting and other tectonic activity can raise parts of the coastal zone above sea level, or drop parts of the land, submerging areas along the coast.

15.01.a5 Central California

4. Currents form when ocean or lake water flows in a certain direction. A single current can greatly modify water depth by pushing shallow water in one direction and deeper water in another.

15.01.a6 Mont-Saint-Michel, France

Low Tide

8. Changes in sea level greatly affect shorelines. In most places, tides raise and lower sea level relative to the land twice a day. Longer term changes in sea level are primarily due to changes in climate and tectonics. Tidal flats, such as the one surrounding Mont-Saint-Michel in France, shown in both of these photographs, are uncovered by low tides (◄) and flooded by high tides (►) and storms. Such low areas could be submerged by an overall rise in sea level.

15.01.a7 Mont-Saint-Michel, France

High Tide

B   What Factors Affect the Appearance of a Coast?

Coasts around the world have varied appearances, from sandy white beaches to dark, craggy cliffs that plunge vertically into the sea with no beach at all. A number of factors control these differences, including orientation of the coast, slope of the seafloor, hardness of the rocks, and contributions of sediment from the land.

Factors on the Water Side

1. The appearance of a coast is greatly influenced by the strength of the waves and tides that impact the shore. Stronger waves will typically cause greater erosion and move larger clasts of sediment along the coast.

2. The size and intensity of storms influence the appearance of a coast because storms bring with them large waves, strong winds, and intense rainfall. Some coasts are ravaged by hurricanes, whereas others rarely experience the erosive effects of powerful storms.

3. The slope of the seafloor is also a factor. Steep slopes allow large waves to break directly against rocks along the shore, whereas more gentle slopes cause waves to break a short distance offshore.

4. The orientation of a coastline is also important, because waves typically approach from specific directions in response to prevailing winds. The dominant wave direction may change with the season (summer versus winter or dry versus rainy). Also, some parts of the coast will receive less wave action because they are sheltered in a bay or are protected by an island, barrier reef, or other offshore feature. The coast below is rocky and most of it is affected by strong waves, especially during powerful storms.

Factors on the Land Side

5. On the land side, the appearance of the coast is related to the hardness of the bedrock along the coast. Hard rock that resists erosion tends to form rocky cliffs (▶), whereas erosion sculpts softer sediment and rock into more gentle slopes and rounded hills.

15.01.b2 Scotland

6. Coastal landforms reflect the amount and size of available sediment. A coast cannot be rocky if the only materials present are soft and fine grained. A supply of sand from streams provides a fresh influx of sediment into the coastal environment.

7. Coastlines undergoing uplift have a different appearance from those where the land has dropped relative to water level. A rise in sea level flooded river valleys along the North Carolina coast (▼), producing a coastal outline marked by long, narrow inlets and bays.

8. Climate is a major factor influencing coastal landscapes. Wet climates provide abundant precipitation for erosion, formation of soil, and growth of vegetation. Vegetation stabilizes soil and limits the amount of material that can be picked up by wind and water or moved downslope by gravity. Dry climates result in less vegetation, less soil, and less stable slopes.

Labels on central block diagram: Storm, Waves, Rocky Coast, Coast Composed of Soft Sediment, Stream, Delta, Sheltered Bay, Waves, Flooded River Valley

15.01.b1

15.01.b3

15.01.b4 Acadia NP, ME

Before You Leave This Page

☑ Summarize or sketch the types of processes that affect coasts and an idealized beach profile.

☑ Summarize or sketch how different factors, from the water side and from the land side, affect the appearance of a coast.

15.1

15.2 What Causes High Tides and Low Tides?

THE SEA SURFACE MOVES UP AND DOWN across the shoreline, generally twice each day. These changes, called *tides*, are observed in the oceans and in bodies of water, such as bays and estuaries, that are connected to the ocean. What causes tides to rise and fall, and why are some tides higher than others?

A What Are High and Low Tides?

Tides are cyclic changes in the height of the sea surface, generally measured at locations along the coast. The difference between high and low tide is typically 1 to 3 m, but it can be more than 12 m or almost zero.

High Tide

15.02.a1

During *high tide*, the height of water in the ocean has risen to its highest level relative to the land. At this point, the water floods into the foreshore. In most places, high tide occurs every 12 hours and 25 minutes.

Average Sea Level

15.02.a2

Following high tide, the water level begins to fall relative to the land—the tide is going out. At some time, water level will reach the average sea level for that location, but it keeps falling on its way to low tide.

Low Tide

15.02.a3

When the water level reaches its lowest level, it is at *low tide*, and the foreshore is exposed. Low tide in most places also occurs every 12 hours and 25 minutes. Water level begins to rise again after low tide, and the tide is coming in. Rising tide spreads water across the land.

B What Causes High and Low Tides?

Tides rise and fall largely because water in the ocean is pulled by the *gravity* of the Moon and to a lesser extent the Sun. As Earth rotates on its axis, most coasts experience two high and two low tides in each 25-hour period.

1. This figure depicts Earth and the Moon as if looking directly down on Earth's North Pole. It shows the Moon much closer to Earth than it would be for the size of the two bodies. Earth rotates (spins) counterclockwise in this view and, relative to the Sun, completes a full rotation once every 24 hours.

2. The Moon makes one complete orbit around Earth each 29 days, also counterclockwise in this view. Due to this motion, it takes 24 hours and 50 minutes for a point on Earth facing the Moon to rotate all the way around to catch up with and again face the Moon. The extra 50 minutes is because the Moon will have orbited 1/29th of the way around Earth by 24 hours later.

Moon's Gravity

Earth's Rotation

15.02.b1

4. On parts of Earth that are facing neither toward nor away from the Moon, sea level is lower as water is pulled away from these regions toward areas of high tide. Coastal areas here experience low tide. At this place, for the next six hours and 12 minutes tides will be rising—a situation called *flood tide*.

5. On the side of Earth opposite the Moon, the water is relatively far from the Moon and so feels less of the Moon's gravitational pull (recall that the force of gravity decreases with distance). The water bulges out, called the *tidal bulge*, and the side of Earth facing directly away from the Moon therefore experiences high tide. At this place, for the next six hours and 12 minutes tides will be falling—a situation called *ebb tide*.

6. Earth rotates much faster than the Moon orbits Earth, so it is best to think of the mounds of water—but not the water itself—as remaining fixed in position relative to the Moon as Earth spins. During a complete rotation of Earth, a coastal area will pass through both mounds of water, causing most coasts to have two high tides (and two low tides) in each 24-hour and 50-minute period. Odd coastal configurations can cause local differences in the number of daily tides.

3. The Moon exerts a gravitational pull on Earth and its water. This pulls the water in the ocean toward the Moon, causing it to mound up on the side of Earth nearest to (i.e., facing) the Moon. Coastal areas beneath the mound of water experience high tide. On this figure, the thickness and mounding of the (blue) water are greatly exaggerated.

C Why Are Some High Tides Higher Than Others?

From week to week, not all high tides at any location reach the same level—some are higher and others are lower than average. Similarly, some low tides are very low and others are not very low. Such variations are related to the added influence of the Sun's gravity and follow a predictable pattern that repeats about every 15 days.

Spring Tides

1. Like the Moon, the Sun exerts a gravitational pull on Earth and its water. The Sun is larger and more massive than the Moon, but it is so much farther away from Earth that the Sun's gravitational effect on Earth is about half that of the Moon.

2. The Sun's gravity attracts a mound of water on the side of Earth facing the Sun and causes another mound on the side facing away from the Sun. These thin mounds, shown in dark blue, are always in the same position relative to the Sun. Locations on Earth rotate through each position once every 24 hours.

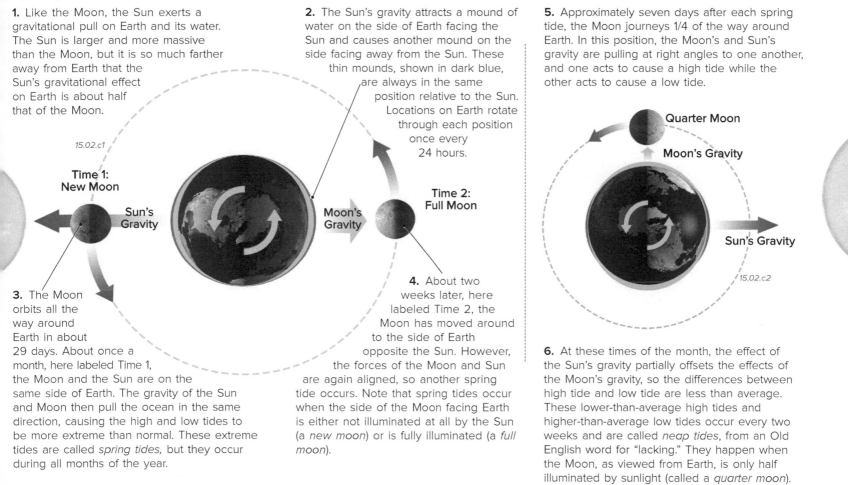

15.02.c1

Time 1: New Moon

Sun's Gravity

Moon's Gravity

Time 2: Full Moon

3. The Moon orbits all the way around Earth in about 29 days. About once a month, here labeled Time 1, the Moon and the Sun are on the same side of Earth. The gravity of the Sun and Moon then pull the ocean in the same direction, causing the high and low tides to be more extreme than normal. These extreme tides are called *spring tides,* but they occur during all months of the year.

4. About two weeks later, here labeled Time 2, the Moon has moved around to the side of Earth opposite the Sun. However, the forces of the Moon and Sun are again aligned, so another spring tide occurs. Note that spring tides occur when the side of the Moon facing Earth is either not illuminated at all by the Sun (a *new moon*) or is fully illuminated (a *full moon*).

Neap Tides

5. Approximately seven days after each spring tide, the Moon journeys 1/4 of the way around Earth. In this position, the Moon's and Sun's gravity are pulling at right angles to one another, and one acts to cause a high tide while the other acts to cause a low tide.

Quarter Moon

Moon's Gravity

Sun's Gravity

15.02.c2

6. At these times of the month, the effect of the Sun's gravity partially offsets the effects of the Moon's gravity, so the differences between high tide and low tide are less than average. These lower-than-average high tides and higher-than-average low tides occur every two weeks and are called *neap tides,* from an Old English word for "lacking." They happen when the Moon, as viewed from Earth, is only half illuminated by sunlight (called a *quarter moon*).

The Most Extreme Tides in the World

S ome places have higher tides than others. The difference between high and low tide, or the *tidal range,* can be so small as to be nearly undetectable, or it can be so extreme as to be dangerous. The Mediterranean and Gulf of Mexico have very little tide, and much of the Caribbean has only one tide each day. Southampton in the United Kingdom has four high tides a day, but the world's highest tides are in the Bay of Fundy along the Atlantic coast of Canada. In this place, the geometry of the coast and sea bottom funnel water in and out of the bay at just the right rate to cause a tidal range of as much as 16 m (52 ft). Canadians use the large tides to generate electricity for Nova Scotia. These two photographs illustrate the extreme tidal range within the Bay of Fundy.

15.02.t1

15.02.t2

High Tide

Low Tide

Before You Leave This Page

✓ Describe or sketch what tides are.

✓ Sketch and describe how tides relate to the position of the Moon and why.

✓ Sketch or summarize how the gravity of the Moon and Sun cause spring tides and neap tides.

15.2

15.3 How Do Waves Form and Propagate?

WAVES ARE THE MAIN cause of coastal erosion. Most ocean waves are generated by wind and only affect the uppermost levels of the water. Waves transport and deposit eroded material along the coast. How do waves form, how do they propagate across the water, and what causes waves to break?

A What Is an Ocean Wave and How Is the Size of a Wave Described?

1. *Waves* are irregularities on the surface of a body of water usually generated by wind. They vary from a series of curved ridges and troughs to more irregular bumps and depressions to breaking curls (▼).

2. In deep, open water, many waves occur in sets, where adjacent waves are similar in size and shape to one another and follow one behind another.

3. The lowest part of a wave is the *trough*, the highest part is the *crest*, and the vertical distance between the two is the *wave height*. The distance between two adjacent crests is the *wavelength*. All of these features can vary greatly with location and time. A typical ocean wave is several meters high, with a wavelength of several tens of meters.

15.03.a1 Two Points SP, ME

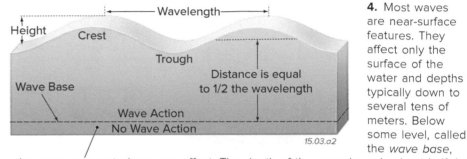

Wavelength
Height
Crest
Trough
Distance is equal
to 1/2 the wavelength
Wave Base
Wave Action
No Wave Action
15.03.a2

4. Most waves are near-surface features. They affect only the surface of the water and depths typically down to several tens of meters. Below some level, called the *wave base*, the wave ceases to have any effect. The depth of the wave base is about half the wavelength, so if two wave crests are 20 m apart, then the wave base is about 10 m deep. The equation that expresses this relationship is:

$$\text{depth of wave base} = \text{wavelength}/2$$

B How Do Waves Propagate Across the Water?

A set of waves can travel a long distance across an ocean or other body, but the water through which the wave passes moves only a short distance. To visualize this, imagine shaking a rope into a series of waves. The waves move along the rope, but any part of the rope stays about the same distance from your hand.

1. Water waves propagate in a manner similar to waves in a rope or to seismic surface waves. Water molecules move up and down and from side to side, but they mostly stay in about the same place. Compressive forces cause water within the wave to push against the water in front of it. The three figures shown here (▶) are snapshots of a set of waves propagating to the right. Examine the motion of water at points A, B, and C within the wave. At the start, when the points are under the wave crest, they are farthest apart.

2. Here (▶), the waves have propagated through the water to the right, but the lettered reference points have moved only a short distance. Point A, which is close to the surface, moves more than point B, which is deeper. Point C is below the wave base and does not move at all.

3. Later, points A and B are beneath the wave trough (▶) and have nearly returned to their positions on the reference line. As the next wave crest approaches, they will move left and upward, and then right, returning to near their starting points.

PROPAGATION OF WAVES ⟶

2nd Wave   1st Wave
• A
• B
Wave Base
• C

2nd Wave   1st Wave
• A
• B
Wave Base
• C

3rd Wave   2nd Wave   1st Wave
• A
• B
Wave Base
• C
15.03.b1

Reference Line

4. During the passage of an entire wave (crest to crest), points A and B will each have followed a small, circular path (▼). The figure below shows the circular paths of 20 different points at different depths and positions along the wave.

5. Water particles on the surface of the water travel the most, going up and forward and then down and back, in a circular motion, as the wave passes.

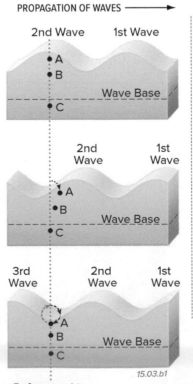

Wave He
Wave Base
No Wave Action
15.03.b2

6. Deeper within the wave, water particles travel smaller circular paths, and the paths become smaller with depth. Water that is right above the wave base barely moves, and water below the wave base does not move at all as the wave passes.

C How Do Waves Form and What Happens When They Reach Shallow Water?

As a wave moves from deep water into shallower water, it starts to interact with the bottom and changes in size and shape. A wave can also bend (refract) as one part of the wave encounters the bottom before other parts do.

How Waves Form

1. Most ocean waves are caused by wind blowing across the surface of the water.

2. When a gentle breeze is blowing, gas molecules in the air collide with the surface of the water, transferring some of their momentum to the water. This forms waves that are small in height and wavelength. Once a wave forms, it catches the wind even more and so can increase in height and wavelength.

Wind

15.03.c1

3. With greater wind speed, waves get larger, both in height and wavelength. The stronger the wind, the larger the wave generated. Long-wavelength waves move faster than shorter wavelength ones and can travel farther before dying out.

4. Waves continue to move, even if the wind dies or the waves move away from the windy area. Such waves existing independently of winds or storms are called *swell*.

5. If the wind gets too strong, a wave becomes too steep to be stable and its top collapses (breaks), even if it is still out in open water. This collapse traps air and forms a white, foamy wave, a *whitecap*.

6. Most of the time, there are multiple sets of waves, propagating in different directions, resulting in *wave interference*. Where wave crests from two sets join, the waves get higher. Interference results in a choppy sea surface with bumps and depressions.

How Waves Break Upon the Shore

7. As waves approach the shore, they encounter shallower water and change in a fairly systematic way. In this figure, waves are moving to the right, toward the shore.

8. In deep water, waves propagate unimpeded as long as the sea bottom is deeper than the wave base. The deep-water waves shown here have a longer wavelength and lower height than waves nearer shore.

Wave Direction ⟶

Increase in Height, Decrease in Wavelength

Breaking Wave

Wavelength

Wave Base
Wave Base = 1/2 Wavelength

15.03.c2

9. When a wave reaches water that is shallower than the wave base, its lower parts begin shearing against the bottom. The waves slow, crowd together, and get higher as their wavelengths get shorter.

10. Near shore, the wave becomes too high and steep to support itself. The top of the wave, which is moving faster than lower parts, topples over (breaks), forming a plunging *breaker* or a jumble of spilling water. In contrast to waves in the swell, these waves have horizontally moving water molecules. The area where waves break on the shore is the swash zone, or simply the *surf zone*.

How Waves Bend

11. Due to varying wind directions and underwater topography, waves typically approach the shore at an angle, rather than straight on. In this figure, waves propagate toward the shore obliquely, from left to right. In deep water, the wave crests are straight or only gently curved.

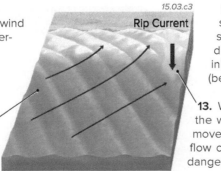

15.03.c3

Rip Current

12. As a wave begins to encounter the bottom, its side closest to shore is slowed more than the segment in deeper water. This difference in velocity causes the initially straight waves to refract (bend)—*wave refraction*.

13. Water flowing back out to sea after the wave comes ashore (*backwash*) moves perpendicular to the shore. The flow can take the form of fast and dangerous *rip currents,* which cause many drownings. Rip currents tend to disrupt the normal wave pattern. Always swim parallel to the shore if caught in a rip current.

Before You Leave This Page

☑ Sketch and label the parts of a wave, including the wave height, amplitude, wavelength, and wave base.

☑ Explain how the propagation of a wave differs from the motion of the water through which the wave travels.

☑ Sketch and explain why a wave rises and breaks as it reaches shallow water.

☑ Explain why waves bend if they approach the shore at an angle.

15.4 How Is Material Eroded, Transported, and Deposited Along Coasts?

SOME COASTS ARE ENERGETIC ENVIRONMENTS where solid rock is eroded into loose sand, pebbles, and other kinds of sediment. Sediment can also be brought in by streams, wind, waves, and ocean currents. Once on the beach, sediment moves in and out from the beach, and often laterally along the beach, as waves alternately transport and deposit sediment.

A How Do Waves Erode Materials Along the Coast?

Most erosion along coasts is done by waves. Waves crash against rocky shores and onto beaches, breaking off pieces of bedrock that can then be reworked by more wave action and ultimately transported away.

1. When waves break onto a rocky coast, they cause erosion by swirling away loose pieces of bedrock and by picking up and crashing these loose pieces back against the bedrock. They also grind away rocks by scraping sand back and forth against the bedrock. Salt crystals from the sea and ice (which expands as it forms) force pores open through *salt wedging* and *frost wedging*, respectively.

2. Crashing and grinding water and sediment wear away at the bedrock, especially at the level where wave action is strongest. Over time, this repeated erosion in the same place may carve a *wave-cut notch* into a rocky coast. The notch undercuts overlying rocks, leaving them unsupported and prone to collapsing into the sea.

3. As waves wash sand and stones back and forth across the sea bottom, they smooth off the underlying bedrock, carving a *wave-cut platform*. Knobs of resistant bedrock locally rise up above the platform, but they may eventually be worn away by the erosive action of waves.

4. Once pieces of rock are loose and within the surf zone, waves smash them together, rounding off angular corners and fracturing larger pieces into smaller ones. In this way, large angular rocks, derived from bedrock, over time become the rounded and flattened stones that dominate many beaches. With further action, stones wear down into sand.

5. Through this mechanism, waves liberate and rework pieces of bedrock. In the process they create stones and sand that help the waves erode and rework even more bedrock, and break other stones into sand.

15.04.a1

B How Does the Shape of a Coast Influence Wave-Related Erosion?

Most coasts are not straight but have curves, bays, and other irregularities. As a result, some parts of the coast are somewhat protected, while other parts bear the brunt of oncoming storms and waves.

1. Waves approach a coast from a specific direction, usually at an angle to the shore. Curves in the coastline form inward-curving bays, whose quiet waters are protected from the largest waves. From season to season, the prevailing wind may change direction, causing what was once protected to be subjected to strong waves. Parts of the shore that are struck head-on by the waves will experience more wave action and erosion than those that are at an angle to the full force of the oncoming waves.

2. A *promontory*, which is a ridge of land that juts out into the water, is in a vulnerable position. The steep sides of many promontories allow large, powerful waves to focus all their energy on the rock, instead of losing energy through interaction with a gently sloping bottom.

3. The seafloor flanking a promontory can cause waves to bend (refract) around the promontory and strike it from all sides. All other things being equal, more waves means more erosion, so promontories tend to be preferentially worn away, eventually resulting in a straighter coastline.

Promontory

Bay

15.04.b1

Before You Leave These Pages

☑ Describe how waves erode material from the coast and how the shape of a coastline influences wave erosion.

☑ Sketch and describe how waves move sand and other sediment on the beach.

☑ Summarize the factors controlling whether a coast gains or loses sand over time.

C   How Do Sand and Other Sediment Get Moved on a Beach?

1. During normal (non-stormy) conditions, waves wash sand and other sediment back and forth laterally near the beach, in a process called *beach drift*. Sediment on the sea bottom is churned up and carried toward the beach by incoming waves.

2. After a wave breaks on the beach, most water flows directly downslope off the beach, carrying sediment back toward the sea. Sediment gets reworked back and forth by the waves, but it may not be transported very far. The area on a beach where broken waves run up onto the beach is the swash zone. The return flow, back toward the sea, is called backwash.

3. During storms, large, vigorous waves can carry sediment farther up the beach than normal, depositing it out of the reach of smaller waves that characterize more typical, less stormy conditions. Storms can also erode material from the beach, carrying it farther out to sea. Some beaches lose sand and become rocky during the winter because of the increased energy of larger winter waves, but they regain sand and become sandy and less rocky in the summer.

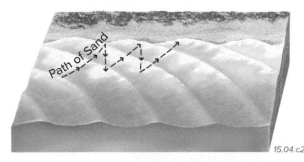

4. Sediment along beaches and farther offshore can *slump* downhill if the sea bottom is too steep to hold the sediment. Sediment will also slump if it is physically disturbed, perhaps by deep wave action during a storm, by shaking during an earthquake, or if sediment piles up too fast.

5. Wind is common along coastlines and can carry sand and finer materials long distances. Low- to moderate-strength wind cannot dislodge sand that is wet because surface tension from the water between sand grains tends to hold the grains together. Wind is more effective above the shoreline where the sand is dry and loose.

6. If waves approach the coastline at an angle, the sand and other sediment can be moved laterally along the coast (▲). Incoming waves move the sand at an angle relative to the coastline, and then the sand washes directly downslope when the water washes back into the sea. By this process, the sand moves laterally along the coast (beach drift). Sand and sediment farther offshore can also move laterally due to ocean currents paralleling the shore, a process called *longshore drift*.

D   What Determines Whether a Coast Gains or Loses Sand with Time?

A coastline can gain or lose sand, depending on the rate at which sand enters the system and the rate at which it leaves. Many coastlines retain approximately the same amount of sand over time. The amount of sediment available to the system is described as the *sediment budget*, and it controls many factors of the coast.

1. On most coasts, sand and other kinds of sediment largely derive from erosion taking place inland. Larger volumes of sediment are produced and carried to the ocean if the land receives sufficient precipitation to generate runoff and is not overly protected by vegetation that limits the effectiveness of erosion.

3. Streams provide an influx of sediment into the coastal system. Deltas formed at the interface between a stream and the sea can be reworked, contributing sediment that is transported offshore or along the coast.

4. Coastal sand dunes commonly consist of sand blown inland from the beach, beyond the backshore, representing a net loss of sand from the beach. Other dunes may derive their sand from the uplands and add sand to the coastal system. Many dunes, however, simply swap sand with the beach as wind alternately blows landward and seaward, or from season to season.

5. Most areas have a preferred atmospheric circulation that steers a current flowing along the coast. This *longshore current* transports sediment parallel to the coast. Such currents can add sand to the beach system, remove sand, or add as much sand as they remove (so the amount of sand is more or less in equilibrium).

2. Sediment is generated by wave erosion and associated slumping of rocks along the coast, which adds to the sand budget. Waves can bring sand in from offshore, pick up and take sand out to deeper water, or simply swash it back and forth.

6. Waves erode reefs and offshore islands, especially during storms, and carry loose sediment toward the coast. Many white-sand beaches consist of calcium carbonate sand eroded from coral reefs and shells. These creatures build their shells from chemicals dissolved in the water, so they increase the amount of sand in the system when the shells are smashed and broken by the waves.

15.4

What Landforms Occur Along Coasts?

COASTLINES CAN DISPLAY SPECTACULAR LANDFORMS. Erosion carves some of these landforms, while sediment deposition forms others. Erosion is dominant in high-energy coastal environments, whereas deposition is more common in low-energy environments. Another controlling factor is whether the coastal zone has been uplifted relative to sea level or has been submerged under encroaching seas.

A What Coastal Features Are Carved by Erosion?

In a high-energy coastal environment, the relentless pounding of waves wears away the coastline, eroding it back toward the land. Such erosional retreat is not uniform but is often concentrated at specific locations and certain elevations. This results in some distinctive landforms along the coast.

15.05.a1 Southern Australia

1. *Sea Cliffs*—Coasts composed of hard bedrock can be eroded into cliffs (◄) that plunge directly into the surf or that are fronted by a narrow beach. Sea cliffs are more common in regions with active tectonism, especially where the land has been uplifted.

15.05.a3 Southern Australia

3. *Sea Caves* and *Sea Arches*—Erosion concentrates in the tidal zone where waves can undercut cliffs, forming caves (◄). Erosion can cut through small promontories jutting out into the sea, forming arches or windows through weaker spots in the bedrock. Caves and sea arches are most common along uplifted coasts.

2. *Wave-Cut Platforms*—Continued erosion at sea level can bevel off bedrock, forming a flat, wave-cut platform (◄). It may be covered by water at high tide but fully exposed at low tide. A platform can be uplifted above sea level, at which point we call it a *marine terrace*.

15.05.a2 Crete

4. *Sea Stacks*—Erosion along a coastal zone is not uniform, and some areas of rock are left behind as erosion cuts back the shoreline (◄). Such remnants can form isolated, steep-sided knobs called *sea stacks*, such as these along Australia's famous Great Ocean Road.

15.05.a4 Southern Australia

Formation of a Sea Cave and Sea Stack

5. A sea stack forms where a promontory extends out into the sea. As waves approach the shore, they refract, focusing erosion on the front and sides of the promontory.

6. If parts of the rock are weaker than others, because of differences in rock type or the relative concentration of fractures, rock behind the tip of the promontory may erode faster than the tip, forming a sea cave.

7. Continued erosion can collapse the roof of a cave and carve a passage behind the former tip of the promontory. The more resistant knob of rock becomes surrounded by the sea and is a sea stack.

15.05.a5–7

B   What Coastal Features Result from Deposition?

As sediment moves along a coastline, it can preferentially accumulate in places where water velocity is slower, forming a variety of low, mostly sandy features. These include sand bars, barrier islands, and sand spits.

15.05.b1 Goodland Bay, FL

15.05.b2 Dauphin Island, AL

15.05.b3 Tigertail Beach, FL

1. *Sandbar*—Offshore of many coasts is a low, sandy area, called a sandbar or gravel bar. Bars are typically submerged much of the time and can shift position as waves and longshore currents pick up, move, and deposit the sand.

2. *Barrier Island*—Offshore of many coasts are low islands that act as barriers, partially protecting the coast from large waves and rough seas. Many barrier islands are barely above sea level and consist of loose sand, including sand dunes, and saltwater marshes.

3. *Spit*—Along some coasts, a low ridge of sand and other sediment extends like a prong off a corner of the coast. Such a feature is a sand spit or a spit and is easily eroded, especially by storm waves. A spit can change length over time, reflecting gains and losses in the sediment budget.

Formation of a Spit, Baymouth Bar, and Barrier Island

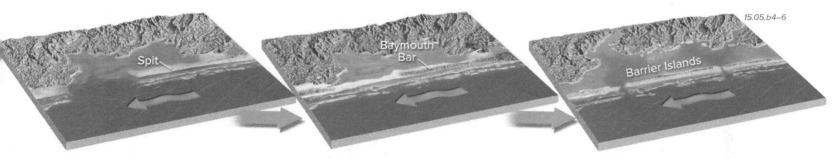
15.05.b4–6

4. A spit forms when waves and longshore currents transport sand and other beach sediment along the coast, building a long but low mound of sediment that lengthens in the direction of the prevailing longshore current.

5. If a spit grows long enough, it may cut off a bay, becoming a *baymouth bar*. This bar shelters the bay from waves, creating a *lagoon*, and may allow it to fill in with sediment, forming a new area of low-lying land, perhaps creating a *marsh*.

6. If sea level rises enough to submerge low-lying parts of the spit or baymouth bar, former spits and bars may become long, sandy barrier islands. Barrier islands may also form if mounds of sediment, deposited by streams when sea level is lower, become islands when sea level rises.

Cape Cod

Cape Cod sticks out into the Atlantic Ocean from the rest of Massachusetts like a huge, flexed arm. The "curled fist" is mostly a large spit. Other features are bars and barrier islands. Much of the sediment was originally deposited here by glaciers, which retreated from the area 18,000 years ago. As the glaciers melted, global sea level rose, flooding the piles of sediment and causing them to be reworked by waves and longshore currents.

15.05.t1

20 km

Before You Leave This Page

✓ Describe the different types of coastal features.

✓ Sketch and summarize one way that a sea stack, spit, baymouth bar, and barrier island can each form.

✓ List the types of features that are present on Cape Cod, and discuss how these types of features typically form.

15.5

How Do Reefs and Coral Atolls Form?

REEFS ARE SHALLOW, MOSTLY SUBMARINE FEATURES, built primarily by colonies of living marine organisms, including coral, sponges, and shellfish. Reefs can also be constructed by accumulations of shells and other debris. Coral reefs thrive in many settings as long as the seawater is warm, clear, and shallow.

A In What Settings Do Coral Reefs Form?

Corals are a group of marine invertebrate animals that form structures made of calcium carbonate. To thrive, corals require nutrients, warmth, sunlight for photosynthesis, and water that is relatively free of suspended sediment. Too much sediment partially blocks the Sun, can bury the coral, or can clog openings in the tiny organisms. Thus, coral reefs form in shallow tropical seas with relatively clear water. Large waves batter many reefs, producing white sand and often carbonate sediment.

1. Some reefs (▼) occur parallel to the continents, forming *barrier reefs* offshore. Reefs and islands protect a continent from large waves. They enclose a lagoon on the landward side but have a side that faces the open ocean and is exposed to large waves and storms. Erosion of the reefs can form low, sandy islands with beaches covered by white sand produced by erosion and reworking of pieces of reef, shells, and other carbonate materials.

2. Reefs and other carbonate accumulations can form broad, shallow *platforms,* such as the Bahamas east of Florida. In some cases, older reef deposits and dunes rise slightly above sea level. Between most islands, the water is shallow and the seabed is composed of white, carbonate-rich sand derived from wave erosion of reefs and the land.

15.06.a2 Raja Ampat, Indonesia

15.06.a1

Carbonate Platform

Lagoon

Barrier Reef

Lagoon

Barrier Reef

Lagoon

Lagoon

Atoll

Lagoon

Barrier Reef

Fringing Reef

15.06.a3 South Pacific Ocean

3. *Fringing reefs* are attached to a coast or are just offshore, surrounding an island. The seaward edge of the reef slopes down toward deeper water. Most reefs begin as fringing reefs (◄), such as this one (light green-brown color) in the South Pacific Ocean.

5. *Atolls* are curved reefs that enclose a shallow, inner lagoon (▼). Some atolls form when an island flanked by coral sinks, but upward coral growth keeps pace with the sinking. These reefs are fairly unique to extinct volcanoes because they require subsidence, as occurs when magmatism ends and the rocks cool and contract.

15.06.a4 Great Barrier Reef, Australia

4. The Great Barrier Reef (◄) is along the eastern coast of Australia and has a unique history. Its base was formed along the edge of a shallow platform when glaciers elsewhere tied up enough water to lower sea level around 17,000 years ago. As sea levels returned to normal and began to drown the platform, the corals grew upward, keeping themselves in shallow water. Over time, the reef formed the largest organic buildup on Earth, one that is easily visible from space.

15.06.a5 Nukuoro Atoll, Federated States of Micronesia

B How Do Atolls Form?

Charles Darwin proposed a hypothesis for the origin of atolls after observing a link between certain islands and atolls during his research aboard the ship HMS *Beagle* from 1831 to 1836. According to his model, shown below, atolls form around a sinking landmass, such as a cooling or extinct volcano. Another model (not shown) interprets some atolls as being the result of preferential erosion of the less dense center of a carbonate platform.

15.06.b1–3

Stage 1: A volcanic island forms through a series of eruptions in a tropical ocean, establishing a shoreline along which corals can later grow and construct a fringing reef.

Stage 2: After volcanic activity ceases, the new crust begins to cool and sink. Coral reefs continue building upward as the island subsides, forming a barrier reef some distance out from the coast.

Stage 3: The volcano eventually sinks below the ocean surface, but upward growth of the reef continues, forming a ring of coral and other carbonate material. This forms an atoll, with a central, shallow lagoon.

C Where Do Reefs Occur in the World?

Most of the world's reefs are in tropical waters, located near the equator, between latitudes of 30° north and 30° south. Coral reefs require certain conditions, such as warm, clear water, and are sensitive to water chemistry. Increased CO_2 in the atmosphere can combine with water to make it slightly more acidic, a process called *ocean acidification*, a potential result of climate change. Reef corals are more diverse in the Pacific, probably because many species went extinct in the Atlantic during the last glacial advance. The map below shows coral reefs as red dots.

15.06.c1

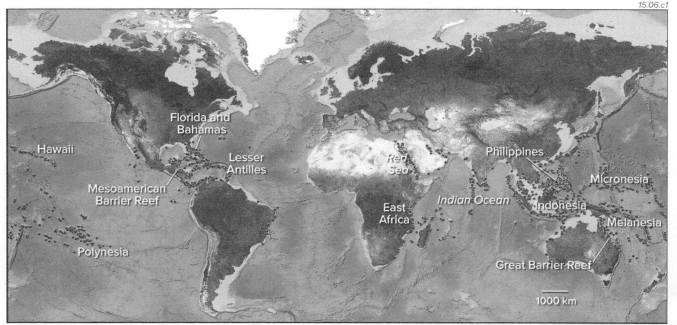

5. Reefs in the Philippines cover an estimated 25,000 square kilometers and consist of fringing reefs with several large atolls. Reefs also flank Indonesia and nearby Malaysia.

4. The Great Barrier Reef, along the northeastern flank of Australia, is the largest reef complex in the world. The world's second largest reef is in New Caledonia, a series of islands east of Australia and in Melanesia.

1. The central and southwestern Pacific, including Polynesia, Micronesia, and Melanesia, has many atolls and reefs, including a wide variety of barrier and fringing reefs. Farther north, Hawaii is also warm enough for reefs.

2. Well-known reefs are present throughout much of the Caribbean region, including Florida, the Bahamas, and the Lesser Antilles. The longest barrier reef in the Caribbean extends some 250 km (150 mi) along the Yucatán Peninsula, from Mexico southward to Honduras.

3. Reefs occur along the continental shelf of East Africa, such as in Kenya and Tanzania. Other reefs encircle islands in the Indian Ocean and the coast of the Red Sea.

Before You Leave This Page

☑ Describe the different kinds of reefs and where they form.

☑ Describe the stages of atoll formation.

☑ Name some locations with large reefs.

15.6

15.7 What Are Some Challenges of Living in a Coastal Zone?

COASTAL ZONES CAN BE RISKY PLACES TO LIVE because of their dynamic nature. Destruction of property and loss of life result from waves, storm surges, and other events that are integral to the coastal environment. Beaches and other coastal lands can be totally eroded away, along with poorly situated buildings. How can home buyers and investors identify and avoid such unsuitable and potentially risky sites?

A What Hazards Exist Along Coasts?

Most coastal hazards involve interactions between water and land, but strong winds also pose risks. Significant hazards accompany storms, which can produce large waves, strong winds, and surges of water onto the land. Many coastlines are also vulnerable to *tsunami*, earthquake-generated waves that pour vast amounts of water onto the land. Tsunami are described elsewhere in this book.

15.07.a1 Pemaquid Point, ME

15.07.a2 Long Beach Island, NJ

15.07.a3

1. Waves are constantly present but are not always a threat to land, buildings, and people along coasts. Most damage from waves occurs during extreme events, such as hurricanes and other storms. Waves can erode land and undermine hillsides, causing slopes and buildings to collapse into the water.

2. A *storm surge* is a local rise in the level of the sea or large lake during a hurricane or other storm. A storm surge results from strong winds that pile up water in front of an approaching storm, inundating low-lying areas along the coast. Surges are accompanied by severe erosion, transport, deposition, and destruction.

3. Strong winds and rain accompany storms that strike the coast. Communities right on the coast are especially susceptible to these hazards because they often lack a windbreak between them and open water. Also, many coastal lands are flat, so structures built in low-lying areas are prone to rainfall-related flooding.

Before
15.07.a4 Topsail Island, NC

After
15.07.a5 Topsail Island, NC

4. These images document damage caused by Hurricane Fran in 1996 along the beach on Topsail Island, North Carolina. This first photograph shows the area before the hurricane. White numbers mark two houses in both photographs. Compare these two photographs to observe what happened to the two houses, and to houses nearby, during the hurricane.

5. This photograph, taken after the hurricane, shows the loss of beach and destruction of houses caused by waves, storm surge, and erosion. The hurricane came ashore with sustained winds measured at 185 km/hr (115 mph) and a 4-meter-high (12-foot) storm surge. It caused more than $3 billion in damage.

B What Approaches Have Been Attempted to Address Coastal Problems?

Various strategies are used to minimize the impacts of natural coastal processes, including erecting barriers, trying to reconstitute the natural system, or simply avoiding the most hazardous sites.

15.07.b1 Dauphin Island, AL

1. One way to limit the amount of erosion is to construct a *seawall* along the shore. Such walls consist of concrete, steel, large rocks (◄), or some other strong material that can absorb the impacts of waves, especially during storms. As shown in this photograph, large rocks and other debris can be placed along the coast to armor it, in an attempt to protect it from erosion. Material used in this way is called *rip rap*. Building a seawall commonly results in loss of beach in front of the wall because the supply of sand from inland is choked off.

15.07.b2 Clearwater, FL

2. Another type of wall, called a *jetty* (◄), juts out into the water, generally to protect a bay, harbor, or nearby beach. Jetties are usually built in pairs to protect the sides of a shipping channel. In an attempt to protect one area of the coast, jetties, like many other engineering approaches to coastal problems, can have unintended and problematic consequences. Jetties and other walls can focus waves and currents on adjacent stretches of the coast. These directed waves and currents, deprived of their normal load of sediment by the wall, erode the adjacent areas as they try to regain an equilibrium amount of sediment.

3. Low walls, called *groins*, are built out into the water to influence the lateral transport of sand by longshore currents and by waves that strike at an angle to the coast. A groin is intended to trap sand on its up-current side, but it has the sometimes unintended consequence of causing the beach immediately down-current of the groin to receive less sand and to become eroded.

15.07.b3 Presque Isle SP, PA

4. A wall, called a *breakwater*, can be built out in the water to bear the brunt of the waves and currents. Breakwaters are built parallel to the coast to protect the beach from severe erosion and to cause sand to accumulate on the beach behind the structures. Some communities bring in sand to replenish what is lost to storms and currents. This procedure of beach nourishment is expensive and may last only until the next storm.

Avoiding Hazards and Restoring the Coastal System to Its Natural State

The preferred way to avoid hazards along coasts is to simply not build in those places that have the highest likelihood of erosion, coastal flooding, coastal landslides, and other coastal hazards. Coastal geoscientists can map a coastal zone and conduct studies to identify the most vulnerable stretches of coastline. With such information in hand, the most inexpensive approach—in the long run—is to forbid the building of houses or other structures in those areas identified as high risk. After the destruction to New Orleans and nearby coastal communities by Hurricane Katrina, there was a debate about whether to rebuild those neighborhoods that are at highest risk, such as those that are below sea level.

In many cases, such concerns are either ignored or are overruled by financial and aesthetic interests of developers, communities, and people who own the land. Beachfront property is desirable from an aesthetic standpoint and so can be expensive real estate, which some people think is too precious to leave undeveloped.

Another approach is to try to return the system to its original situation, or at least a stable and natural one, rather than trying to "engineer" the coastline. Engineering solutions can be expensive, may not last long, or may have detrimental consequences to adjacent beaches. Returning the system to a natural state may involve restoring wetlands and barrier islands that buffer areas farther inland from waves and wind. Examining the balance of sediment moving in and out of the system can help identify non-natural factors, such as dammed rivers, which if restored to original conditions would bring more sediment into the system and stabilize beaches, dunes, and marshes. As is typical of issues associated with development versus conservation, there are many trade-offs and multiple viewpoints of the situation.

Before You Leave This Page

☑ Summarize some of the hazards that affect beaches and other coastlines.

☑ Describe approaches to address coastal erosion and loss of sand, including not building and trying to restore systems to a natural state.

15.7

15.8 How Do We Assess the Relative Risks of Different Stretches of Coastline?

UNDERSTANDING THE LANDFORMS AND DYNAMICS of a coastal zone is the first step in assessing the potential risks posed by waves, currents, coastal flooding, and other coastal processes. Coastal scientists study coastlines using traditional field methods and high-technology methods that involve lasers and satellites.

A What Field Studies Do We Conduct in the Coastal Zone?

To investigate potential coastal hazards, coastal scientists map and characterize the topographic and geologic features of the land, coast, and nearshore sea bottom. They combine this information with an understanding of the important coastal processes to identify those areas with the greatest hazard.

1. To assess coastal hazards, a first step is to document the land surface elevations. High areas clearly have less risk of being flooded by the sea. Precise measurements of elevations of the land are taken with various surveying tools, some using satellites (Global Positioning System, or GPS) or lasers that scan the ground surface from an airplane. These surveys identify areas, such as this high marine terrace, that are too high to be flooded, even during a hurricane.

2. Areas that are close to sea level may be subject to flooding by storm surges and storm-related intense rainfall that cause flooding along coastal rivers. Vulnerable low-elevation areas may extend far inland, in this case along a low river valley.

3. Mapping the bedrock geology, as well as the loose sediments along the beach, helps us assess how different areas will erode. Coasts backed by resistant bedrock, as along a cliff, will be less likely to be eroded by strong waves and currents. Parts of a cliff may fail over time, however, as they are undercut by constant wave action.

4. Scientists map the distribution and height of coastal dunes. Dunes, especially those that are large or are stabilized by vegetation, decrease the risk inland for storm surge and associated erosion. Marshlands, such as those on a delta, also help buffer areas farther inland from waves, storm surges, and strong coastal winds.

5. We can also map the location and height of sandbars, islands, reefs, and other offshore barriers. These barriers can protect the coast from wave action. In the photograph below (▼), a thin strip of sandy land produces a shallow, quiet-water lagoon to the right. The sand and shallow waters of the lagoon generally protect the shoreline from large, potentially damaging waves, except during strong storms, when a storm surge can raise water levels and deepen water close to shore.

6. To help assess potential hazards, we can also document the slope of the land adjacent to the shore. A steeper slope limits how far storm surges can encroach on the land, whereas a more gentle slope allows the sea to wash farther into the land.

Sand Dunes

Offshore Island

Longshore Current

15.08.a1

7. Coastal geoscientists commonly document the width of beaches. An area with a wide beach between the shoreline and houses is generally less risky than an area where houses sit right behind a narrow beach. Seawalls, groins, and other constructed features can greatly affect beach width and therefore potential risk. A seawall can limit the amount of landward erosion and may protect buildings from storm surges, but sand in front of the seawall may be lost, allowing waves to break directly against the seawall, weakening it. A groin affects the width of the beach differently on either side. It may decrease the risk of storm erosion to the beach on the up-current side, which gains sand and becomes wider, but increase the risk of storm erosion to the beach on the down-current side, which loses sand and becomes narrower. Beach width, wave size, and potential hazard are also affected by barrier islands and reefs, which can protect the coast.

15.08.a2 Tigertail Beach, FL

B | What Can New High-Resolution Elevation Data Tell Us About Coastlines?

Satellite data (GPS) and other new methods of mapping elevation now allow geoscientists to characterize coastal regions more accurately and track in detail how a shoreline changes during storms. One relatively new method is *lidar*, which is an acronym for LIght Detecting And Ranging.

1. In the lidar method, a laser beam is bounced off the ground, detected back at the airplane, and timed as to how long it takes to travel to the surface and back to the instrument. The shorter the time the beam takes to reach the ground and return, the shorter the distance between the plane and the ground—and therefore the higher the elevation of the land. Lidar elevations are accurate to within about 15 cm (6 in.) or less and can be quickly collected over larger areas than is practical to cover with conventional surveying.

2. As the plane flies forward, mirrors direct the laser toward different areas beneath the lidar sensor. Thousands of data measurements are recorded each second in a narrow belt, called a swath, across the land. The plane flies back and forth over the area, overlapping adjacent swaths to ensure that there are no gaps in the measurements. A GPS unit mounted on the plane allows technicians to register the lidar data with geographic map coordinates and to match the data to features on the ground.

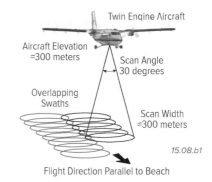

15.08.b1

Mapping Hurricane-Related Changes in Coastal Alabama

15.08.t1–3

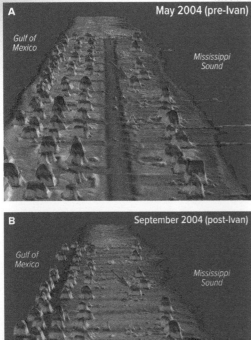

A   May 2004 (pre-Ivan)
Gulf of Mexico
Mississippi Sound

B   September 2004 (post-Ivan)
Gulf of Mexico
Mississippi Sound

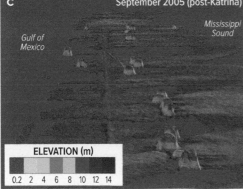

C   September 2005 (post-Katrina)
Gulf of Mexico
Mississippi Sound

ELEVATION (m)
0.2 2 4 6 8 10 12 14

Coastal Alabama has been hit by a series of powerful hurricanes, including Hurricane Ivan in 2004 and Hurricane Katrina in 2005. The U.S. Geological Survey (USGS) and other government agencies have used lidar to investigate the changes that such large hurricanes inflict on the coastline. One detailed study was of Dauphin Island, an inhabited barrier island along the Gulf Coast of Alabama.

The first three images (A–C) show perspective views of detailed lidar elevations taken at three different times (before Hurricane Ivan, after Ivan, and after Hurricane Katrina). Red arrows point to the same house in all five images. The first image (A) shows a central road with houses (the colored "peaks") on both sides. In the second image (B), the storm surge from Hurricane Ivan has washed over the low island from left to right, eroding the left beach, covering the road with sand, and redepositing some of the sand on the right.

The aftermath of Hurricane Katrina, bottom left, is even more dramatic (image C). All but a few houses were totally washed away. Both the width and height of the island decreased, leaving the remaining houses even more vulnerable to the next storm surge (see photograph below).

The last two images (D–E) show the calculated changes caused by each storm. Each image was produced by comparing the *before* and *after* data sets and computing the difference. Features shown in reds and pinks represent losses due to erosion, green areas show where deposition occurred, and whitish features were unchanged by that storm.

15.08.t4–5

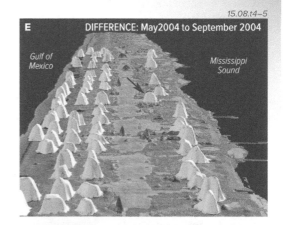

E   DIFFERENCE: May2004 to September 2004
Gulf of Mexico
Mississippi Sound

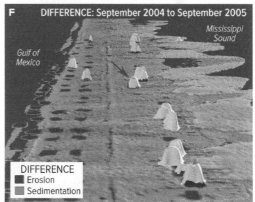

F   DIFFERENCE: September 2004 to September 2005
Mississippi Sound
Gulf of Mexico

DIFFERENCE
■ Erosion
□ Sedimentation

15.08.t6 Dauphin Island, AL

Before You Leave This Page

☑ Describe how studying the geologic features along a coast can help identify areas of highest hazard.

☑ Summarize how lidar data are collected, and provide an example of how they can be used to document changes in a coastline.

15.8

15.9 What Happens When Sea Level Changes?

SEA LEVEL HAS RISEN AND FALLEN many times in Earth's history. A rise in sea level causes low-lying parts of continents to be inundated by shallow seas, whereas a fall in sea level can expose previously submerged parts of the continental shelf. Changes in sea level produce certain landscape features along the coast and farther inland, and they can deposit marine sediments on what is normally land. Changing sea level produces two kinds of coasts: *submergent coasts* and *emergent coasts*.

A What Features Form if Sea Level Rises Relative to the Land?

Coastlines adjust their appearance, sometimes substantially, if sea level rises or falls relative to land. A relative change in sea level can be caused by a global change in sea level or by tectonics that causes the local land to subside or be uplifted relative to the sea. Distinctive features form along a coast when sea level rises relative to the land.

1. *Submergent coasts* form where the land has been inundated by the sea because of a rise in sea level or subsidence of the land.

2. The shape of the land exerts a strong control on how the coastline will look after it is flooded by rising sea level. Examine this figure (▼) and predict what will happen to different features if sea level rises.

3. After the land is inundated, flooded river valleys give the coast an irregular outline, featuring branching estuaries and other embayments.

4. Hills and ridges in the original landscape are surrounded by rising seas, forming islands along the shore.

5. Preexisting deltas and coastal dunes, when flooded, may become offshore bars or sandy barrier islands. Barrier islands may become totally submerged by rising seas.

15.09.a1

15.09.a2

6. An *estuary* is a coastal body of water that is influenced by the sea and by fresh water from the land. A common site for an estuary is a delta or a river valley that has been flooded by the sea, either of which allows fresh water from the land to interact with salt water from the sea. Water levels in the estuary and the balance between fresh and salt water are affected by tides and by changes in the amount of water coming from the land. The satellite image below shows the Chesapeake Bay estuary. The bay was a valley originally carved by rivers, but it was flooded when sea level rose at the end of the last glacial advance (▼).

15.09.a4 Norway

7. The coasts of Norway (◄), Greenland, Alaska, and New Zealand all feature narrow, deep embayments called *fjords*. Fjords are steep-sided valleys that were carved by glaciers and later invaded by the sea as the ice melted and sea level rose.

15.09.a5 North Carolina

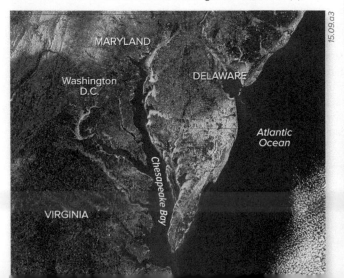

15.09.a3

MARYLAND

Washington D.C.

DELAWARE

Atlantic Ocean

Chesapeake Bay

VIRGINIA

8. Many barrier islands (◄) are interpreted to have been formed by rising sea level. Some barrier islands began as coastal dunes or piles of sediment deposited by rivers. As sea level rose, the rising water surrounded the piles of sediment, resulting in new islands.

B   What Features Form if Sea Level Falls Relative to the Land?

Some features suggest a fall in sea level, or uplift of the land by tectonics or by isostatic processes. Tectonic processes can result in a gentle uplift of the shore, resulting in emergent features. Another mechanism for creating emergent features is through isostatic rebound where the shore rises after glacial retreat. A fall in sea level can expose features that were submerged and can greatly affect what happens on the adjacent land.

1. *Emergent coasts* form where the sea has retreated from the land due to falling sea level or due to uplift of the land relative to the sea.

4. After sea level drops, erosion incises (cuts) valleys into the land. If sea level drops in a series of stages, emergent wave-cut notches form topographic steps on the land, and wave-cut platforms form a series of relatively flat benches, known as *marine terraces*.

15.09.b1

15.09.b2

2. Submerged features that may be exposed by falling sea levels include reefs, offshore sandbars, and the underwater parts of deltas.

3. As sea level falls or the land rises, coral reefs can become exposed on land, a sure sign that a coast has emerged.

5. Sandbars that originally formed offshore can become coastal dunes, or the sand can be blown onshore or eroded and returned to the sea.

15.09.b3 California

15.09.b4 Galápagos

6. Wave-cut platforms form within the surf zone along many rocky shorelines and, when exposed above sea level, form relatively flat terraces on the land. The surface of such marine terraces may contain marine fossils and wave-rounded stones.

7. Coral reefs and other features that originally formed at or below sea level can be exposed when seas drop relative to the land. These coral reefs, now well above sea level, provide evidence of relative uplift of the land.

8. A wave-cut notch is an originally horizontal recess eroded into rock by persistent wave erosion at sea level along a coast. In this photograph, taken at low tide, a horizontal wave-cut platform is associated with a slightly higher wave-cut notch. In a few hours, the rising tide will inundate that wave-cut platform, and waves will once again crash into and erode the rocky wall. Uplift of the land, or a drop in sea level, can leave a wave-cut notch high and dry, a hint of what occurred.

15.09.b5 Palau

Before You Leave This Page

☑ Summarize what a submergent coast is and what types of features can indicate that sea level has risen relative to the land.

☑ Summarize what an emergent coast is and what types of features indicate that sea level has fallen relative to the land.

15.9

15.10 What Causes Changes in Sea Level?

SEA LEVEL HAS VARIED GREATLY IN THE PAST. Global sea level has been more than 200 m higher and more than 120 m lower than today. What processes caused these changes in sea level? Large variations in past sea level resulted from a number of competing factors, including the extent of glaciation, rates of seafloor spreading, and global warming and cooling. These processes operate at different timescales, from thousands to millions of years.

A How Does Continental Glaciation Affect Sea Level?

The height of sea level is greatly affected by the existence and extent of glaciers and regional ice sheets. At times in Earth's past, ice sheets were more extensive than today, and at other times they were less extensive or absent.

1. The ice in glaciers and continental ice sheets accumulates from snowfall on land. When glaciers and ice sheets are extensive, they tie up large volumes of Earth's water, causing sea level to drop (▶).

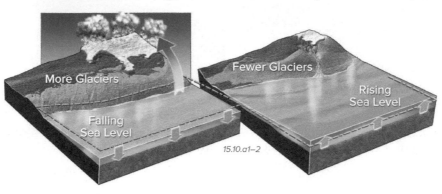

2. When glaciers and ice sheets melt, they release large volumes of water that flow back into the ocean, causing sea level to rise (◀).

3. The growth and shrinkage of ice sheets and glaciers is the main cause of sea level change on relatively short timescales (thousands of years).

B How Do Changes in the Rate of Seafloor Spreading Affect Sea Level?

At times in Earth's history, the rate of seafloor spreading was faster than it is today, and at other times it was probably slower. Such changes cause the rise and fall of sea level.

1. The shape and elevation of a mid-ocean ridge and adjacent seafloor reflect the rate of spreading. As the plate moves away from the spreading center, it cools and contracts, causing the seafloor to subside, creating space for seawater.

2. If seafloor spreading along a ridge is slow, the ridge is narrower because the slow-moving plate has time to cool and contract before getting very far from the ridge. Slow seafloor spreading and narrow ridges leave more room in the ocean basin for seawater (▶). So over time, a decrease in the spreading rate causes sea level to fall.

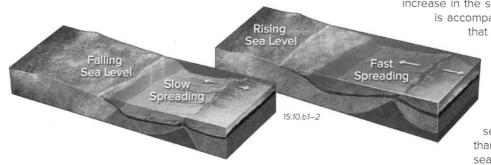

3. If seafloor spreading along a ridge is relatively fast (10 cm/year or faster), the ridge is broad because still-warm parts of the plate move farther outward before cooling and subsiding. So an increase in the seafloor spreading rate is accompanied by broader ridges that displace water out of the ocean basins, causing sea level to rise (◀). In other words, faster spreading yields more young seafloor, and young seafloor is less deep than older seafloor, raising sea level.

C How Do Changes in Ocean Temperatures Cause Sea Level to Rise and Fall?

Sea level is affected by changes in ocean temperatures, which cause water in the oceans to slightly expand or contract. Such effects accompany global warming or global cooling and result in relatively moderate changes in sea level.

1. Water, like most materials, contracts slightly as it cools, taking up less volume. The amount of contraction is greatly exaggerated in this small block of water.

3. Water expands slightly when heated, taking up more volume. Again, the amount of expansion is exaggerated in this small block of water.

2. When ocean temperatures fall, but still remain above 4°C, water in the ocean contracts, causing sea level to fall (▶).

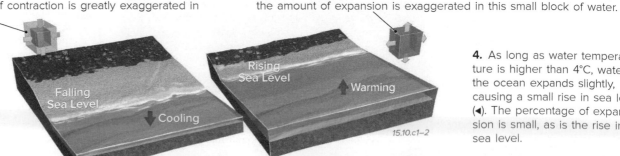

4. As long as water temperature is higher than 4°C, water in the ocean expands slightly, causing a small rise in sea level (◀). The percentage of expansion is small, as is the rise in sea level.

D How Does the Position of the Continents Influence Global Sea Level?

As a result of plate tectonics, continents move across the face of the planet, sometimes being near the North or South Poles and at other times being closer to the equator. These positions influence sea level in several ways.

1. Glaciers and continental ice sheets form on land, and so they require a landmass to be cold enough to allow ice to persist year round. This occurs most easily if a landmass is at high latitudes (near the poles) or is high in elevation.

2. At most times in Earth's past, ice ages occurred when one or more of the continents were near the poles (▶), like Antarctica is now. Continents at high latitudes usually include glaciers, so such a configuration of the continents can cause global sea level to drop because water is stored in glaciers.

Lower Sea Level

Higher Sea Level

15.10.d1   15.10.d2

3. At other times in Earth's past, the larger continents were not so close to the poles (◀). This lower latitude position of continents minimized or eliminated widespread glaciation (the non-ice-age periods in Earth's history). This low-latitude configuration of the continents tends to keep sea levels high.

E How Do Loading and Unloading Affect Land Elevations Relative to Sea Level?

Weight can be added to a landmass, a process called *loading*. A weight can also be removed, a process called *unloading*. Loading and unloading can change the elevation of a region relative to sea level.

1. Weight loaded on top of a region imposes a downward force that, if large enough, can downwarp the land surface beneath the load and in adjacent areas (▶). Loading, such as by continental ice sheets, lowers the loaded region relative to sea level. This can allow seawater to inundate regions near the ice sheets. The ice in this figure and the amount of subsidence are very stylized and vertically exaggerated.

100 km

15.10.e1

2. If the weight is unloaded from the land, the region flexes back upward, a process known as *isostatic rebound*. The uplifted, rebounding region rises relative to sea level (▶).

3. Unloading and isostatic rebound can occur when continental ice sheets melt. Rebound begins as soon as significant amounts of ice are removed, but it can still be occurring thousands of years after all the ice is gone.

100 km

15.10.e2

Ongoing Isostatic Rebound of Northeastern Canada

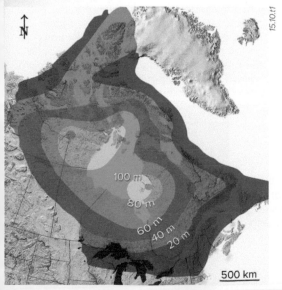

15.10.t1

100 m

80 m

60 m   40 m   20 m

500 km

The northern part of North America has been covered by glaciers off and on for the last two million years. The weight of these continental ice sheets loaded and depressed this part of the North American plate. When the ice sheets began melting from the area, approximately 15,000 years ago, unloading caused the land, especially in Canada, to begin to isostatically rebound upward.

The amount of rebound has been measured both directly and indirectly. We can measure uplift directly by making repeated elevation surveys across the land and then calculating the amount of uplift (rebound) between surveys. Rates of rebound are typically millimeters per year, which is enough to detect with surveying methods. Satellite measurements (GPS) are also sensitive enough to measure such changes. The amount and rate of rebound can also be inferred more indirectly by documenting how shorelines and other features have been warped and uplifted. Contours on this map to the left indicate the amount of rebound interpreted to have

occurred in northeastern North America over the last 6,000 years. Some areas have experienced more than 100 m of uplift. Isostatic rebound can also occur around large lakes, as they fill or empty over time, loading or unloading the land.

Before You Leave This Page

✓ Summarize how continental glaciation, rates of seafloor spreading, ocean temperatures, and position of the continents affect sea level.

✓ Explain how loading and unloading affect land elevations using the example of Canada.

15.10

15.11 What Coastal Damage Was Caused by These Recent Atlantic Hurricanes?

THE EAST COAST AND GULF COAST of North America routinely suffer damage from hurricanes and other storms. In 2011 and 2012, several large but not especially powerful hurricanes hit different coastlines of the U.S. and caused extensive damage. They included Hurricanes Irene (2011), Isaac (2012), and Sandy (2012). The damage caused by these three storms illustrates some issues of building along coasts.

A What Hazards Do Hurricanes and Other Storms Pose to Coastlines?

Hurricanes bring strong winds, heavy rains, and the risk of tornadoes, but the greatest damage is commonly due to storm surge, where the hurricane pushes a mound of water against the coastline. This temporarily raises sea level in affected areas and allows large waves to crash farther up the beach than is normal. As a result of the surge and waves, extreme amounts of erosion occur, damaging or destroying structures too close to the beach or too low in elevation. The U.S. Geological Survey (USGS) has an extensive program to assess the hazards of such storms, as presented here.

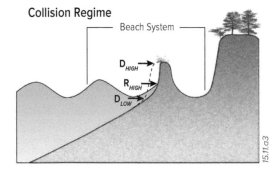

15.11.a1 Myrtle Beach, SC

1. Many people live along the East Coast (◄) and Gulf Coast of North America. Living near the coast has many advantages, but it also poses a risk to structures built in places where coastal erosion could occur, removing sand, roads, and buildings that are in the way. Areas farther back could be flooded if water rises high enough.

2. To assess how risky living along a particular stretch of coast is, the USGS coastal scientists developed a conceptual model summarized in this figure (►). It compares the maximum height of the swash zone (R_{high}) with the heights of the base (D_{low}) and top (D_{high}) of coastal dunes. If the swash is below the base of the dune, little permanent erosion occurs.

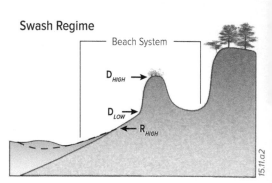

15.11.a6 Nags Head, NC

3. If the local sea level is higher, such as from a storm surge, the swash and waves can directly strike the lower parts of coastal dunes in what the USGS calls the *collision regime*. This causes erosion of the base of the dune, and the loss of sand can be permanent. Erosion is, however, somewhat limited to areas close to the shoreline.

4. In the next stage, the *overwash regime*, the maximum height of swash and waves reaches the top of the dunes. This causes erosion of the dunes, with the transport of sand toward the land. Through this process, the beach or entire barrier island can migrate toward the land. Houses that were on the beach may now be in the water.

5. The next and most damaging stage is when the water level is so high that the entire dune is submerged, the *inundation regime*. Large quantities of sand are typically transported toward the land, permanently moving the beach and burying areas farther inland in sand. If the site is a barrier island, the entire island may become submerged.

6. This photograph (◄) illustrates erosion of the front (seaward side) of a dune. Although the house was high enough to avoid being flooded, erosion is devouring the dune on which the house sits. The dune acted as a buffer between the house and the waves, but as the dune became narrower the house became more threatened.

7. This photograph of a barrier island (►) documents the process of *overwash*, where material is moved from the side facing the sea (i.e., the waves in the foreground) to the side away from the sea. The seaward side lost beach material while the backside of the island grew.

15.11.a7 Cone Banks, NC

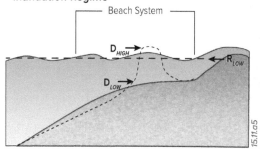

B What Damage Was Caused by Hurricanes Irene, Isaac, and Sandy?

Hurricane Irene (2011) struck the Outer Banks of North Carolina and adjacent areas, and Hurricane Isaac (2012) struck the Gulf Coast, including many of the same areas previously damaged by Hurricane Katrina (2005). Hurricane Sandy (2012) caused widespread damage along the East Coast, especially in New Jersey and New York.

15.11.b1

August 30, 2011

Hurricane Irene

15.11.b2–3

May 6, 2008

August 31, 2011

Hurricane Isaac

15.11.b4–5

August 8, 2012 Pre-Isaac

September 2, 2012 Post-Isaac

PRE-ISAAC JANUARY 2010

Mississippi Sound

Gulf of Mexico

200 m

POST-ISAAC SEPTEMBER 2012

Mississippi Sound

Gulf of Mexico

15.11.b6–7

0.23 — 4.00
ELEVATION (m)

1. These three photographs (▲) are of the same part of the Outer Banks, North Carolina. The photograph on the top right shows the areas before Irene, whereas the other two show the conditions after Irene. Yellow arrows point to the same house. Observe changes that occurred from this single storm. Severe erosion of the beach obliterated any physical evidence that the house closest to the beach ever existed. Erosion cut a new channel through the landscape.

2. This photograph pair (▲) illustrates the damage done to barrier islands in Louisiana by Hurricane Isaac. The small island in the background was eroded to a much smaller size, but the low islands and sand bars in the foreground became totally submerged.

3. These figures (▲) show before and after lidar-measured topographic maps of another island. Note how this storm breached the island and redistributed material from one area to another. Barrier islands clearly are transient features. The locations are indicated with Universal Transverse Mercator (UTM) coordinates.

Hurricane Sandy

4. When Hurricane Sandy struck the Northeastern U.S., one of the hardest-hit areas was in New Jersey, where the center of the storm came ashore. Compare the before and after scenes (▼).

5. These two lidar images (▼) document the changes in elevation in the same area. Note that houses and the bridge were destroyed by erosion during the storm surge. Elevation can be almost as important as proximity to the beach.

Before May 21, 2009

After November 5, 2012

15.11.b8–9

PRE-SANDY

POST-SANDY

15.11.b10–11

0 — >8
ELEVATION (m)

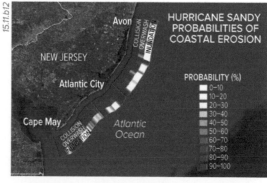

15.11.b12

HURRICANE SANDY PROBABILITIES OF COASTAL EROSION

Avon

NEW JERSEY

Atlantic City

Cape May

Atlantic Ocean

PROBABILITY (%)
0–10
10–20
20–30
30–40
40–50
50–60
60–70
70–80
80–90
90–100

6. As the storm neared, the USGS assigned extent-of-damage probabilities, which were reviewed when assigning evacuation routes.

Before You Leave This Page

✓ Sketch and describe the four regimes of beach erosion.

✓ Summarize the kinds of erosion and deposition that occurred during three Atlantic hurricanes, identifying areas you consider to be most vulnerable.

15.11

What Is Happening Along the Coast of This Island?

THIS PREVIOUSLY UNKNOWN ISLAND is being considered as a possible site for a small settlement and seaport to resupply ocean travelers with water and other supplies as they pass by. To avoid any missteps in this process, you are asked to observe the coastline, identify the various coastal features, infer what processes are occurring, and identify possible sites for the port and nearby settlement. Your considerations should include any issues that could affect the operation of the port, such as avoiding natural hazards and keeping shipping channels from becoming blocked with sand. The area does experience hurricanes.

Goals of This Exercise:

- Observe the three-dimensional perspective of the newly discovered island in order to identify the coastal features and processes.

- Fully consider any factors that would impact the location of a port and small settlement for people operating the port.

- Propose an acceptable location for the port and settlement and be prepared to defend your choices.

Geographic Setting

Listed below is some geographic information about the island.

A. The island, surrounded by the ocean, is in the Northern Hemisphere, located at approximately 30° N latitude, at about the latitude of Florida.

B. The island is somewhat elongated and nearly 10 km in its longest direction. It reaches an elevation of 1,000 m above sea level in two mountains.

C. The climate of the island is warm during the summer, when it sometimes rains. It is wetter during the winter, when it receives enough rainfall to provide ships with water, as long as the water resources are managed intelligently. The ocean moderates the climate, keeping it a little cooler during the day and warmer at night relative to a place that is not next to the sea. The air is fairly humid, so the mountains sometimes cause and receive short summer rains.

D. There is a moderate amount of vegetation on the island, except on the rocky mountains. The most dense vegetation is near a delta on the north side of the island and on a raised flat area on the western (left) side. Both sites are suitable for growing crops.

15.12.a2

15.12.a3

1. Along the western side of the island are cliffs and steep slopes that rise out of the sea, with only a narrow beach along the shoreline (◄). The beach is composed of sand, but it also contains many rounded pebbles and larger stones plus some large angular rocks that match the types of rocks in the cliff. There are also shells and fossils of other marine creatures.

2. Above the cliffs is a relatively flat area, like a shelf in the landscape (▲). This type of feature is only observed on this western side of the island. Scattered about on this flat surface are rounded pebbles and larger stones, along with abundant loose sand. On some parts of the surface there is a layer that contains fossil shells (▼) and fossils of other marine organisms identical to those that occur on the beach.

15.12.a4

15.12.a1

Procedures

Follow the steps below, entering your answers for each step in the appropriate place on the worksheet or online.

1. Observe the overall character of the island, noting what is along each part of the coast and the locations of streams (a source of fresh water). Read the information about the geographic setting of the island (on the previous page).

2. Observe the photographs and read the accompanying text describing different parts of the island. Use what you know about coasts to interpret what type of feature is present, what processes are typical for this type of feature, and what the significance of each feature is for a port and small settlement.

3. Synthesize all the information and decide on the best locations for the settlement and port, which should be fairly close to one another (no more than 1 km apart). Be able to logically support your choices. There is not a single correct answer or best site.

4. OPTIONAL: Your instructor may have you write a short report or prepare a presentation describing all the factors you considered, why you chose the sites you did, or interpreting of aspects of the natural history of the island.

6. Along the north shore of the island is a delta where the largest streams in the area reach the sea. This area has abundant trees (▶) and other plants, and is easily accessed from the east by walking along the beach.

15.12.a6

15.12.a5

5. The north shore of the island (▲) has gentle relief along the shore and has a well-developed beach that is mostly sand. The beach slopes gently into the water, allowing people to wade quite a distance out into the sea. The sand is constantly in motion, moving up and back on the beach and also moving laterally along the coast. Lying on some parts of the beach are pieces of wood, similar in size to the trees observed near the delta to the west.

4. Cut into the eastern end of the island are a series of embayments, or bays, that branch farther inland. In the largest bay, the land slopes into the bay at a moderate angle, gentle enough to walk up. The bay projects toward two mountains that have some well-developed vegetation on their flanks. The mountains are not too steep to walk up. Small streams flow on every side of the mountains but have more water in the summer. The mountains are not volcanoes.

15.12.a7

3. Extending off the land from the southern side of the island is a low ridge of relatively loose sand (▶). Some of the high parts of the ridge are clearly sand dunes. Some lack any vegetation, but others are anchored in place by grasses and other plants. To the east is a similar long island, again composed of sand, including dunes.

16 Soils

SOILS FORM A VERY THIN LAYER but constitute a critical resource supporting all terrestrial plant life, which in turn supports entire ecosystems. If the Earth were the size of an apple, the atmosphere would be comparable to the apple's skin, and the soil is 200,000 times thinner than the atmosphere. Humans depend on soil for the crops that feed and clothe us, as a foundation for buildings, and as a medium for recycling natural and man-made waste products. Soil is threatened by many natural and human activities, especially those that cause soil erosion. What is soil, how does it form, and are all soils the same? This chapter explores these key questions.

16.00.a2 Lake County, IL

16.00.a1 Arco, ID

The world's food production depends on fertile soils, a critical but fragile part of any ecosystem. We need good soil stewardship to support the increasing demands placed on soil. An example of sound soil conservation techniques is no-till farming (◄), a technique that increases the amount of water and organic matter in the soil by leaving the remnants of the previous crop in place rather than removing them. Other conservation techniques include contour plowing, in which the land is tilled following the same elevation rather than up and down the slope, and filter strips, which use dense vegetation next to a stream to capture sediment and chemical pollutants.

What is soil and what are some commonly used methods to conserve soil?

16.00.a3

Within a single handful of soil, there can be billions of different organisms, ranging from single-celled protozoa and bacteria to earthworms (▲), insects, and vertebrates. These organisms form a complex food web that make soil more fertile, aiding plant growth. Almost all terrestrial plants use soil as a medium for growth.

How do living organisms contribute to soil fertility?

16.00.a4 Phoenix, AZ

When disturbed, soil can become eroded by wind and form choking dust storms (◄) that transport organic and inorganic pollutants, and even living microbes, for thousands of kilometers. Fertilizer-enriched soils, carried away by surface runoff, can also pollute waters thousands of miles away from where the soil was originally established.

What processes are responsible for loosening and transporting soil?

TOPICS IN THIS CHAPTER

Soil profiles (▼), some tens of thousands of years old, can unlock secrets into past climatic and geologic conditions and plant-animal communities.

How do layers in soil form, and what do they tell us about climate, geology, and plants?

16.00.a5 Badlands NP, SD

Soils, and the vegetation they nurture, serve as a sink for atmospheric carbon dioxide, removing CO_2 from the atmosphere. Soil also serves as an important host to the nitrogen cycle, a vital process for supplying nitrogen to plants. Molds (▶) found in soil are used for a variety of medicines. Not all soils are equally well suited for these roles.

How do plants and microbes in soil interact with gases and water in the atmosphere?

16.00.a7 Hermosa, CO

16.00.a8 Borah Peak, ID

16.00.a6

Soil is the "grand recycler," transforming organic and inorganic waste products into more benign forms. Soil (◀) can also immobilize contaminants, keeping them from quickly spreading to groundwater or from running off the surface into local streams. Soil also improves water quality by remediating waste products from plants and animals.

How do the texture and chemical properties of a soil promote or inhibit the dispersal of harmful contamination?

Soils have uses that may surprise you. Soil is a product of the types of starting materials and the environmental conditions, such as amount of rainfall, under which it formed. As these factors vary spatially, the physical, chemical, and biological properties of soils will also vary, providing clues useful for solving crimes committed outdoors. Soils are also used to determine how often an earthquake-causing fault moves, such as along a fault scarp (◀).

What controls the characteristics of a soil, and how could these be used to determine where or how often some event took place?

As illustrated by the large central photograph, soil, along with water, is one of the most important resources on the planet. Nearly all living things on land either require soil for their habitat or depend on plants in soil as their source of energy and nutrition. Soil, however, is a fragile resource, easily degraded by erosion and contamination.

What are the main threats to soil, and what can we do to better preserve this precious resource?

16.0

16.1 What Is Soil?

EARTH'S LAND SURFACE is covered by a wide range of materials, some that are classified as soil and others that are not soil. How soil is defined varies depending on the perspectives of the definer. To some people, soil is any loose material on the surface, but to most geographers such materials must also show evidence of having been affected by in-place weathering and other processes that produce layers and support the growth of plants.

A What Is a Soil and What Does It Contain?

Soil is a thin layer on Earth's surface that is capable of supporting life and consists of four components—minerals, organic material (living and dead), water, and air. Soil is truly at the interface between the atmosphere, lithosphere, hydrosphere, and biosphere, each of which is represented by one of the soil components. The four soil components interact with their environment to determine the overall properties of a soil.

Most soil consists of approximately half solid material and half *pore space*, the openings between and within grains. Most of the solid material consists of pieces of rocks and minerals, such as those pictured on the far upper right, with lesser but highly variable amounts of organic material. Pore space is filled with water and air.

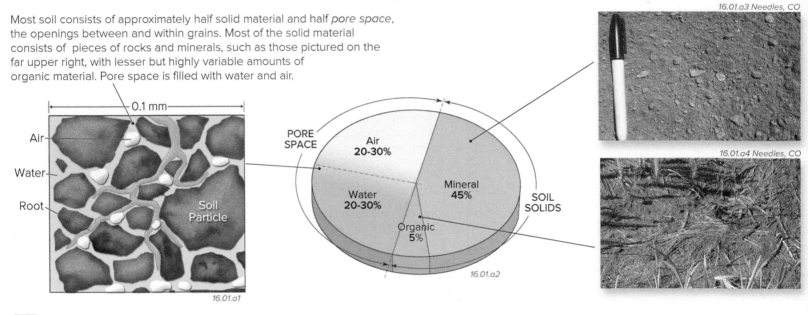

16.01.a3 Needles, CO

16.01.a4 Needles, CO

16.01.a1

16.01.a2

B What Is the Mineral Fraction of Soils?

The solid component of soil is comprised mostly of inorganic solids, especially several types of minerals. We subdivide minerals in soil into *primary minerals* and *secondary minerals*. Primary minerals, like quartz, are derived directly from the weathering and erosion of various types of rocks. Primary minerals undergo no chemical transformation during the soil-forming process, but they can be reduced in size or shape. Secondary minerals, including most silicate clays, are minerals that were not originally present but instead have been produced by fairly intense chemical weathering.

16.01.b1

Sand (2 to 0.05 mm)

Silt (0.05 to 0.002 mm)
Clay (Less Than 0.002 mm)

1. We subdivide the sizes of particles in soils into three main categories (◄): sand, silt, and clay, from largest to smallest.

2. This graph (►) plots the abundance of different types of minerals that typically occur in each size of particle (smaller on the left, larger on the right).

3. If particles are very small, we refer to them as *clay* sized, or simply clay. Such small particles in a soil consist mostly of secondary minerals. Most of these will be silicate minerals, like clay minerals. Remember that the term "clay" refers to a family of sheetlike silicate minerals and to a very small grain size. Most clay minerals are also very small (clay sized). Clay particles consist of stacked flattened particles similar to a deck of cards, giving clays a slippery feel.

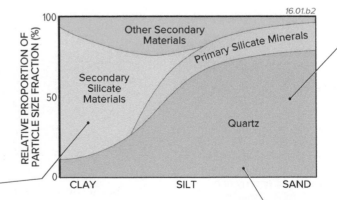

16.01.b2

5. If particles are sand sized, they are mostly quartz and other primary silicate minerals (e.g., feldspar and mica), with minor amounts of non-clay secondary minerals. Most sand also contains oxide minerals, such as magnetite, a magnetic iron oxide mineral that generally forms the dark grains observed in most sand.

4. The next larger particle is *silt*. Silt-sized particles mostly consist of quartz and other primary silicate minerals. They typically contain some secondary minerals, but only a minor amount of clay.

C   What Is the Organic Fraction of Soils?

The organic fraction of soils is rich in carbon and contains plants and animals, some living, some dead, and some in various stages of decay. There can be a balance between organic matter that is accumulating and that which is lost due to erosion and microbial decay, or there can be a net gain or loss of organic material. The organic content of most soils is rather low, typically about 5%, and includes *humus*—a durable, partially decomposed form of organic matter. Humus and other organic matter are vital for soil health by improving infiltration, retaining water, decreasing evaporation, retaining and cycling nutrients, reducing erosion by binding soil together, and providing an energy source for soil organisms.

16.01.c4 Black Hills, SD

1. Organic matter in soils contains a wide variety of material, mostly variably decomposed parts of plants. It nearly always includes decomposed leaves, twigs, branches, and roots, and can also include, depending on the setting, bark from trees, pine needles and pine cones (in pine forests), mosses, mushrooms, and the woody interiors of cacti. If enough plant material accumulates and is not destroyed, like in marshy bogs, it can form *peat* (◀), a somewhat consolidated mass of partly decayed moss, wood, and other plant parts.

16.01.c1 Milwaukee, WI

3. Plants and animals form complex interrelationships (◀) based on nutrient and energy cycling. One gram of soil may contain millions of plants and animals, representing a food chain involving producers, consumers, and decomposers. The nutrients pass from the soil to grass and other plants and to small organisms, and then to higher level consumers, like deer.

2. Soil that is high in organic matter sticks together, as shown by the jar on the left (▶). The soil on the right, placed in the bottle just before this photograph was taken, is already separating rapidly because it is low in organic matter.

16.01.c2   16.01.c3

16.01.c5

Pie chart:
- Stabilized Organic Matter (Humus) 35-50%
- Decomposing Organic Matter (Active Fraction) 35-50%
- Living Organisms ≤ 5%
- Fresh Residue ≤ 10%

4. Organic material occurs in soil in several forms. After land has been cultivated for a couple of decades, much of the decomposing organic matter—the *active fraction*—is consumed by animals, plants, and microbes; in the process, nutrients become available for plants. A more stable form of organic matter is humus, which can make up half the soil's organic matter. Humus acts like a sponge, absorbing up to six times its weight in water.

D   What About the Pore Space of Soil?

Soil is approximately half pore space filled with air and water. The amount of pore space in a material is its *porosity* and is expressed as a percentage (10% means one-tenth of the material is pore space). Soils with clay-sized particles have the greatest amount of pore space because clay particles, unless they are perfectly aligned, do not fit together very well. As a result, they prop up one another, producing countless tiny air pockets—abundant pore space. Sandy soils, with their rounded sand grains, also have a moderate amount of porosity, but not as much as unaligned clays.

1. In sandy soils, like the one shown here, there is commonly 35 to 40% porosity, whereas clay-rich soils can have well over 50% porosity. In either type of soil, pore space is reduced when soils are compacted, especially for clayey soils.

2. *Soil air* content is inversely proportional to water content—if there is more water in the pore space, there is less pore space remaining for air. Compared to atmospheric air, soil air typically has a higher CO_2 content, higher relative humidity, and lower oxygen content.

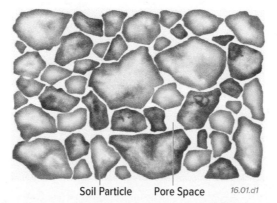
Soil Particle   Pore Space   16.01.d1

3. *Soil water* is not the free-flowing variety found in streams. Instead, it adheres to the surrounding grains. If the pore space is too small or the grains are too close, the soil water may move very slowly, if at all. Soil water contains nutrients for roots and determines rates of biological and chemical reactions, including how quickly a soil can form from the starting material.

Before You Leave This Page

✓ Explain what soil is.

✓ Describe the relationship between soil particle size and the predominance of primary and secondary minerals.

✓ List and describe the benefits of humus to soil health.

✓ Describe the relationship between air and water in soil pore spaces.

16.1

16.2 What Are the Physical Properties of Soil?

THE MINERAL COMPONENTS OF A SOIL can be characterized by texture, structure, and color. In a soil, *texture* is a measure of the sizes of particles, whereas *structure* refers to how the soil clumps together or breaks apart. Texture and structure control many important properties of soils, including the amount of pore space, fertility, how easily the soil can be plowed, and how soil is classified. Soil *color* tells us further information about the soil.

A What Is the Influence of Soil Texture on Soil Properties?

Soil texture is one of the most fundamental physical properties of soil; it describes the average distribution of particle sizes. While several soil properties, such as structure or pH, can be modified by humans, soil texture is an inherent property that changes little with land use and management. We classify soil based on the percentages of different particle sizes. The size distribution in turn exerts a major control on the soil's fertility, stability, and other properties. Each particle size can behave in distinct chemical and physical fashions.

Size Limits of Soil Particles		
	Sand	2.00-0.05 mm
	Silt	0.05-0.002 mm
	Clay	< 0.002 mm

(left side label: <-Mud->, <<-Mud->)

1. A key aspect of a soil is the size of particles it contains: the sand, silt, and clay. This table (◄) lists the sizes of these three particles in millimeters. The term *mud* is a general term that refers to both silt and clay.

16.02.a1

2. These three photographs show greatly enlarged views of three different sizes of particles. The left one is sand, the middle one is silt, and the right one is clay. Even at this scale, clay and silt look similar because both have grains too small to see without a magnifying glass. Different combinations can exist in soil. A soil could be mostly sand with some silt and clay. Another could be mostly silt.

3. The relative proportion of the three sizes (sand, silt, and clay) is the *soil texture*. We can describe soil texture by plotting the proportions of each size on a triangular diagram, like the one shown below. In this example, percent sand is along the bottom, percent clay is on the left side, and percent silt is on the right side. A soil that is 100% clay plots at the top of the triangle. One that is 100% sand plots in the lower left corner, and one that is 100% silt plots in the lower right corner. Most soils have a mix of the three particle sizes, and so they plot somewhere in the triangle. Based on percentages, we can describe the texture with the words listed where the soil percentages plot. The names are not as important as understanding how texture is described and the implications to humans.

6. The soil block below depicts a soil that is nearly equal amounts sand, silt, and clay, and so it plots in the very center of the triangular diagram. This type of relatively even distribution of sand, silt, and clay is generally healthy, and is called a *loam*, several varieties of which occupy the middle region of the diagram. A clay loam is near the very center. Loamy soils allow good drainage of water through the soil, nutrient retention, and aeration, all of which are optimal for plant growth.

4. A soil with the percentages represented by this block of soil (►) contains 60% clay, 10% silt, and 30% sand. Note where it plots on the triangular diagram. We call such a soil texture a *clay* (yet a third usage for this term). A soil, like this, that contains too much clay often drains poorly, is hard to *aerate* (get air into it), and gets too compacted for most agriculture.

7. This soil block is mostly silt and clay (◄). To a soil scientist, mud means a combination of silt and clay, like this type of soil. What other people refer to as mud could be this or could be all clay.

5. This soil has more sand than clay and silt (►). It has a mix of sizes but is quite sandy. Soils that are too sandy easily lose their water and nutrients, although sandy soils are easily plowed and are good for some construction and engineering purposes.

8. This soil is mostly silt, with some clay and sand (◄), a silty loam. A silty textured soil, like this one, has properties that are intermediate between sand and clay. If the word *loam* is in a soil's name it indicates that it is overall a good soil for agriculture.

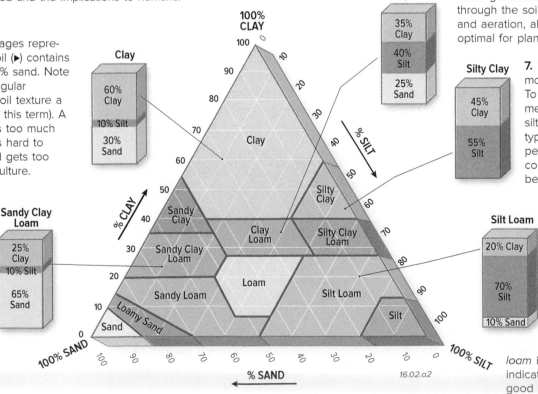
16.02.a2

B What Is the Influence of Soil Structure on Soil Properties?

Structure refers to how the particles are clumped together in aggregates called *peds*. The aggregation of soil particles into peds is determined by the amount and arrangement of organic matter, iron oxides, clays, carbonate minerals, and cultivation method. Soils rich in sodium, as occurs in salt, tend to have poor soil structure as too much sodium breaks up any soil structure. The figure below illustrates the main types of soil structure.

1. There are two main types of soils: those that have some type of structure and those without any structure. Let's look at some of the major types of structure.

2. A *granular* soil structure has peds shaped like small spherical grains less than 0.5 cm in diameter. A granular structure is usually in the root zone and is ideal for root penetration.

3. A *blocky* soil structure has peds that are shaped like irregular chunks. Problems associated with blocky soils include limited root penetration, permeability, and poor *tilth* (ability to be plowed).

7. *Platy* soils have peds that take the form of horizontal plates. This soil structure usually indicates that the soils have been compacted by the weight of overlying materials.

STRUCTURED | NO STRUCTURE

Granular   Platy   Single Grained

Blocky   Columnar   Massive

16.02.b1

6. Some soils have no structure (are *structureless*) because they consist of grains that do not stick together. One type of structureless soils are *single-grain soils*. These loose soils are common in sandy settings and are usually infertile because there is no organic matter to bond soil into peds or to provide nutrients.

5. Another type of soil with no obvious structure are soils that are massive. *Massive soils* generally do not allow water to pass through easily, and so they can present a challenge to plants.

4. *Columnar* soils come apart into fragments that are taller than they are wide, that is, in columns. The columns can be irregular in outline or can have a more angular shape, like six-sided hexagons. They mostly form in arid climates where the soils can dry out and crack.

C What Does Soil Color Indicate?

Soil color is a very useful property for gauging the fertility of a soil. Dark, well drained, humus-rich soils are fertile, while red and whitish soils are generally infertile. Bright soils usually indicate well-aerated conditions, while dull, gray soils indicate oxygen-deficient (anaerobic) conditions. On aerial photographs, a dry part of a soil appears to be lighter than a wet area of the same soil type. As a generalization, red soils are confined to the tropics and subtropics, whereas brown soils are located in the mid-latitudes.

16.02.c1

16.02.c2

16.02.c3

16.02.c4

1. Iron and/or aluminum oxides cause reddish tints in soils. These may either be present from the parent material or may form after a long period of weathering, especially in hot, humid climates where chemical weathering is faster.

2. Dark soils often indicate that a large amount of organic matter is present, as in the example above, or that the soil has a large amount of basalt (a dark igneous rock) as its parent material.

3. Soil can have other tints, including tan, gray, green, and purple. A greenish or grayish tint usually indicates that a soil has been waterlogged at some point. A purplish tint can indicate the presence of manganese oxides.

4. Whitish colors generally indicate that there is salt (NaCl), calcium carbonate ($CaCO_3$), as shown above, or other minerals deposited by evaporation of water. Some soils are light colored because of their parent material.

Before You Leave This Page

☑ Summarize how we describe the texture of a soil, including the main particle sizes and some examples of how texture affects plants.

☑ Sketch and describe the types of soil structures, and give some examples of how soil color indicates the materials in soil.

16.3 What Is the Role of Water in Soil?

WATER IS A CRITICAL INGREDIENT in soils as it is the main determinant of whether plants will grow or perish. Water enters and leaves soil through various pathways and remains in the soil in several ways as *soil water*. Too much water or too little can both be harmful for plants. The character of a soil, such as its texture and structure, determines how well the soil accommodates water.

A How Does Water Enter and Exit Soil?

1. This figure illustrates some aspects of the movement of water into and out of soil. Observe the various pathways shown here and then read the associated text in a counterclockwise direction.

2. *Precipitation* is the ultimate source of fresh water on land, and so it is also the source of soil water. If precipitation falls on the land, three things can happen to the water: it can flow downhill on the surface as *runoff*, sink into the soil through the process of *infiltration*, or *evaporate* back into the air.

3. Once in the soil, the water can remain there or continue flowing into the subsurface, driven by the force of gravity. Water percolating downward can leave the soil zone and become *groundwater*.

6. Once water reaches the above-ground parts of plants, it is used in the process of photosynthesis to make food for the plant. Some of this water is lost as water vapor, mostly through the pores of leaves, through the process of *transpiration*. The combination of evaporation and transpiration is *evapotranspiration*. The water vapor released to the atmosphere via evapotranspiration is then available for additional precipitation. The flow of water from one part of the environment to another is part of the *hydrologic cycle* (also called the water cycle).

4. Soil water can instead be drawn back toward the surface by *capillary action*, which is made possible by the high surface tension of liquid water. Once near the surface, the water can evaporate and be lost from the soil.

5. Plant roots can extract water from the soil and carry it up through the roots and into stems and higher parts of the plant. This rise of water in plants is also due to capillary action.

B How Does Water Reside in Soil and How Much Is Available to Plants?

Water occurs in several settings in soils, but only some of this is available to plants. Some water exists in the crystalline structure of minerals and other materials, so this water is not mobile or accessible to plants. Water also resides in pore spaces in the soil, but not all of this water is available, either.

Settings of Soil Water

1. There are three main settings for soil water. Soil water can be a thin coating that adheres to soil particles, called *hygroscopic water,* which is typically so tightly held that it is unavailable to plants.

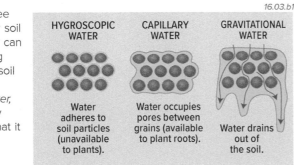

2. If extra soil water is available, the water occupies more of the pore spaces and is not attached to the soil particles. This water is available to capillary action, either through the soil or through the roots, and is therefore called *capillary water*. This is the setting of water that is most important to the long-term survival of plants with roots (some plants acquire moisture in other ways, such as directly from precipitation or from water vapor).

3. If incoming water is abundant, it can pass through the soil, pulled by gravity deeper into the subsurface. Water that follows this route is termed *gravitational water*. It can become groundwater.

Amount of Soil Water

4. The three different settings of soil water greatly influence whether the soil environment is hospitable or hostile to plant growth. If all the pore spaces are filled with water, the soil is considered to be *saturated*. Any additional water will accumulate on the surface (*ponding*), flow over the surface as overland flow, or percolate through the soil and into underlying rocks to become groundwater.

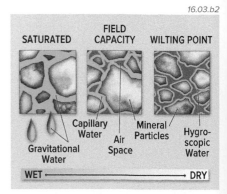

5. If the amount of water decreases, at some point the water stops flowing downward under the influence of gravity. Instead, it remains in the pore spaces as capillary water. This amount of water is called the soil's *field capacity*—it is the maximum amount of water the soil can hold without draining and is measured in cm.

6. If the soil has even less water, it can have limited or no capillary water, only hygroscopic water. This condition is the *wilting point* because at this water content plants begin to wilt and perish.

C How Do Texture and Structure Affect Water in Soil?

Soil texture and structure control for differences in the amount of water and water-borne nutrients that can be stored between the grains and how easily water can pass between the grains. If too much water remains between grains for too long, or if it cannot pass easily, nutrients will not be dispersed throughout the root zone. These are termed "heavy" or *poorly drained soils*. On the other hand, if water infiltrates too quickly, nutrients will be carried downward (*leached*) too quickly to be captured by plants. Such soils are termed *droughty*.

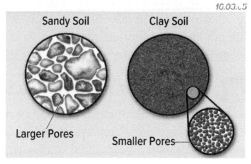
10.03.c5

Sandy Soil   Clay Soil

Larger Pores   Smaller Pores

Porosity

1. Recall that porosity is the proportion of the volume of rock that is open space (pore space). It ranges from less than 1% to more than 50% and determines how much water a material can hold.

2. Some soils consist of particles that are mostly the same size (◄), for which we would say the soil particles are well sorted. A sandy soil having well-rounded and well-sorted grains would have moderate to high porosity. The jar of marbles to the left is analogous to a well-rounded, well-sorted sandy soil. The soil to the right (►) has a wide range in particle sizes (it is less well sorted). The smaller particles, such as clay, can nestle between the larger particles, such as sand, decreasing the porosity.

16.03.c1

3. Clay-rich soils, when first formed, consist of small particles shaped like plates or sheets that do not fit tightly together (◄). There is abundant open space (porosity), but such pores, like the clay particles, are very small, making movement of water difficult. If the soil becomes compacted (►), the clay particles rotate and fit together more tightly, resulting in less porosity.

16.03.c3

16.03.c2

4. High porosity means that there is abundant space in the soil for water and air, but this also means that there is higher potential for the soil to be compacted. Compare the two soils above, which consist of different-sized particles. Clay soils have small pore sizes but high total porosity. Sands have larger pores but moderate porosity (less than uncompacted clays). The most productive soils have a variety of different-sized pore spaces (they are not too well sorted).

16.03.c4

Permeability

5. *Permeability* is a measure of the ability of a material to transmit a fluid. It is related to the size and interconnectedness of the pore spaces. Materials with low porosity usually have low permeability. Permeability is also called *hydraulic conductivity*.

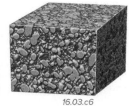

6. Loosely cemented sandy soils (◄) contain interconnected pore spaces that allow soil water to move relatively easily. Such soils have *high permeability* and would drain easily. A massive, clay-rich soil (►) would not allow water to pass easily, and so it has low permeability.

16.03.c6

16.03.c7

7. A clay-rich soil can become compacted naturally, from the weight of overlying materials, or from human construction projects, the weight of cattle herds, or other human-caused disturbances. As the clay particles are compressed together, they lose porosity and have a very low permeability (►). Such materials are described as *impermeable*.

16.03.c8

16.03.c9

Gravel   Sand   Silt   Clay

1 m

8. *Permeability* affects soils in many ways, but especially in how quickly the soil allows water to infiltrate and drain (▲). Observe the relationship between permeability and four particle sizes above. Coarser particle sizes, as in gravel and sand, and certain soil structures (e.g., granular structure) allow water to move more quickly downward. Smaller-sized particles, such as clays, and platy or massive structure force the water to be retained near the surface for extended periods. This retention, as you might expect, is due largely to the *surface tension* of water.

Before You Leave This Page

✓ Sketch and explain the hydrologic cycle as it relates to soil.

✓ Sketch and explain the settings in which water resides in soil, whether each setting of water is available to plants, and how differing amounts of water can affect plant growth.

✓ Describe porosity and permeability, providing some examples of materials with high and low values of each and explaining how they affect the character of a soil.

16.3

16.4 What Are the Chemical and Biological Properties of Soil?

EVEN IF WEATHERING BREAKS DOWN ROCK to form a texture and a structure favorable for plant growth, soil productivity depends on the right combination of chemical and biological properties. Soil chemistry and biology are intimately intertwined and help to determine the type of crop that can be grown, nutrient availability, rates of chemical weathering, the ease with which soil pollutants can be mobilized, and soil structure.

A How Does pH Impact Soils?

The pH scale measures whether a material is acidic or the opposite (basic). Soils that have a high concentration of hydrogen ions are considered *acidic* on the pH scale. These soils are unproductive because acidic soils tend to leach away vital positively charged ions—*cations*—that are nutrients. Examples of essential cations are magnesium (Mg^{2+}), potassium (K^+), and calcium (Ca^{2+}). Generally, humid environments tend to have acidic soils (low pH). Soils that have relatively few ions of hydrogen are *basic* or *alkaline* and generally form in arid regions. Alkaline soils can be unproductive because the positive charge on the soil surface repels essential cations. Productive soils have the right pH and carry an adequate number of nutrients, which will then easily move into nearby roots to help plants grow.

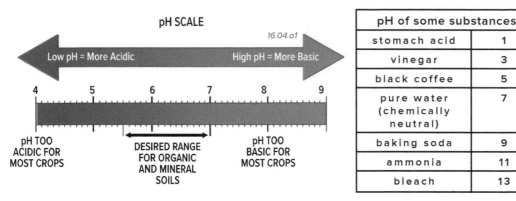

pH of some substances	
stomach acid	1
vinegar	3
black coffee	5
pure water (chemically neutral)	7
baking soda	9
ammonia	11
bleach	13

The chemical measure of pH expresses the concentration of hydrogen ions (H^+) in a substance and is measured on a logarithmic scale, where an increase or decrease of one pH unit (such as from 7 to 8) represents a ten-fold change in the number of H^+ ions. pH ranges from 0 to 14, with 7 being a neutral pH. A pH less than 7 is acidic and one greater than 7 is basic. Most agriculturally productive soils have values between 5.5 and 7, neutral to slightly acidic. Soils that are too acidic (low pH) can be treated with *agricultural lime* (pulverized limestone) to improve fertility. Alkaline soils with too high a pH can be acidified by adding ammonium sulfate.

Blueberries are an example of a crop that thrives in acidic (low pH) conditions, which are common in relatively wet climates.

Date palms are a crop that tolerates alkaline soil (high pH) conditions, which are common in relatively dry climates.

B What Is the Role of Cation Exchange Capacity and Clays in Soil Fertility?

Cation exchange capacity (CEC) is the ease with which a cation, such as Mg^{2+} or Ca^{2+} in soil solution is exchanged with the surface of organic matter or clay particles. Maintaining a high organic-matter content is important for agriculturally productive soils because the CEC of organic matter far exceeds that of clay. CEC is directly related to pH—as pH decreases, CEC also decreases (i.e., acidic soils have lower CEC), and vice versa.

1. Clay particles with their elongated surface area tend to have negatively charged surfaces, which attract positively charged ions (cations). Common cations include sodium (Na^+), magnesium (Mg^{2+}), calcium (Ca^{2+}), and potassium (K^+). There are also free cations in the water, including hydrogen (H^+).

2. In cation exchange, a cation attached to the surface of a clay or other soil particle exchanges place with a cation in the water. In the example shown here, H^+ is trading places with K^+, and Na^+ is doing so with Ca^{2+}. Once captured by the clay particle, the nutrient can be exchanged with roots.

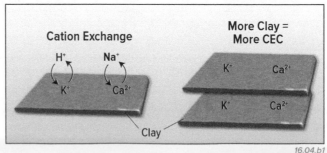

3. Clay-rich soils have a higher CEC than sandy soils and will require lime to correct a low pH. Sand-rich soils have little or no CEC, and so they represent a challenge for plant growth.

C What Nutrients Are Vital for Soil Productivity?

The vast majority of nutrients is in the top 10 cm of soil, right below the surface. Additional nutrients can be added to soils as *fertilizers*, adding nitrogen (N), phosphorus (P), and potassium (K). These nutrients are often lacking in soils, which can limit productivity. Therefore, a label for fertilizer purchased at a store will list the contents of these chemical elements. Calcium (Ca), magnesium (Mg), iron (Fe), zinc (Zn), and sulfur (S) are other important soil nutrients.

1. Plants obtain their carbon from the atmosphere, but they need water and various nutrients from the soil (▼). Nitrogen is vital for protein and enzymes that are at the heart of all biological processes. A sign of N-deficient plants is a stunted growth and a yellowish green color.

16.04.c2

16.04.c1

2. Potassium is necessary for water uptake and conservation within leaves. A sign of K-deficient plants is a burnt appearance of leaf edges.

3. Phosphorus is a difficult nutrient to maintain in the soil. Plants lacking in P are typically stunted, particularly in their root systems.

16.04.c3

4. Spreading N, P, and K on soils (▲) is a way to make up for nutrient deficiencies, but soils with a low CEC will leach N, P, and K into local groundwater systems. These nutrients can then encourage algal blooms in local water supplies.

D What Role Does Organic Matter Play in Soil?

The organic fraction of soils contains living plants and animals and dead plants and animals in various stages of decay. Though the organic content is rather low in soils, typically 5%, it plays a vital role in soil health. Organic matter binds soil together, bonds nutrients, cycles nutrients vital for health, and serves as an energy source for organisms.

1. This figure (▶) illustrates the flow of material between the atmosphere, hydrosphere, lithosphere, and biosphere. Examine the figure and think of all the types of matter that might move from one sphere to another.

2. Plants lose leaves and branches, or they can die completely, providing material to the ground. This dead plant material on the ground, called *litter*, adds organic material to the soil. Litter can also change the soil chemistry, typically making it more acidic (lower pH).

DEAD LEAVES, OTHER PLANT AND ANIMAL MATTER

Plants grow

Water and air penetrate the soil

Decomposers break down organic matter

Rocks are broken down

ROCKY SUBSOIL

Minerals and nutrients are released into the soil

16.04.d1

3. Decomposition of dead organisms, carried out by microbes, returns nutrients from the dead plant materials to the soil. There can be a balance between organic matter accumulating and organic matter lost to microbial decay and erosion, but the addition and loss of organics can be out of balance, resulting in a change in the amount of organics over time.

4. Roots deliver water and nutrients from the soil to the plant. As they grow downward, roots move organic material deeper into the soil. They also break up the soil, creating pathways for downward moving air, water, nutrients, particles, and creatures. Over time, root growth causes a very slow churning effect in soils, generally improving soil productivity.

16.04.d2

5. Earthworms (▲) and other small organisms are an especially important part of this cycling of organic material. They improve soil by fertilizing it with their waste, mixing soil layers by their movement, and aerating the soil with their burrows.

Before You Leave This Page

✓ Explain the limitation imposed on plants by a low or high soil pH.

✓ Describe the role of CEC in soil fertility and what types of soils promote CEC.

✓ Explain the importance of the chemical elements N, P, and K in soils.

✓ Discuss the vital role organic matter plays in soil productivity.

16.4

16.5 How Do Climate and Vegetation Affect Soil?

SOIL FORMATION is influenced by many factors, including climate, organic materials, topographic relief, parent material from which the soil was derived, and time. We refer to these factors with the acronym ClORPT (read the above sentence again, but with ClORPT in mind). We address the role of climate and vegetation (in the form of organic material) here and explore the "RPT" part of ClORPT—relief, parent material, and time—later.

A How Does Climate Affect Soils in Hot, Humid Environments?

Temperature and precipitation impact the rates of chemical, physical, and biological reactions in soil. For every 10 C° increase in temperature, there is a doubling of the reaction rate. Water is necessary for weathering and soil development, so soil-formation rates are slow where water is lacking, as in arid environments, or where it is locked up in the form of ice, as in polar environments. In general, warm, wet climates allow for fast formation of soils primarily because chemical weathering occurs easily. At the same time, however, nutrients are rapidly lost due to decomposition and leaching.

16.05.a1 La Sal, Mtns., UT

This photograph illustrates factors that influence the formation of soil, including climate, vegetation, topography, parent material, and time. High rainfall, such as in mountains and other rainy areas, leads to intense leaching or *eluviation* of clays from surface horizons and *illuviation* or deposition of clays into lower levels of the soil. Low rainfall, as in the desert in the foreground, leads to less of these.

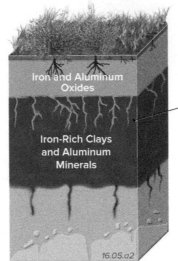

Iron and Aluminum Oxides

Iron-Rich Clays and Aluminum Minerals

16.05.a2

When soils rich in iron are exposed to the atmosphere, oxidation converts the iron minerals into reddish and orange iron-oxide minerals, the common component of rust. These minerals, along with aluminum-rich materials, accumulate in the soil, producing a reddish or orangish color (▶). In wet climates, continued leaching of soil closest to the surface can remove some of the iron oxides and dampen the reddish color in the uppermost part of the soil. Reddish or orangish soils, like the one shown here, are most common in tropical climates, locally forming an important type of soil called *laterites*, which are agriculturally unproductive.

16.05.a3

B How Does Climate Impact Soils in Arid Environments?

In deserts and other arid climates, a lack of rainfall and weathering limits nutrient availability. There is limited vegetation, so there is little or no organic matter accumulation. Clay, iron oxide, and salts, all partly derived from windblown material, can accumulate at various levels in the soil. Overall, Ca^{2+} and CO_3^{2+} ions are dissolved from upper soil layers during intermittent rains and chemically precipitated lower as calcium carbonate ($CaCO_3$).

16.05.b1

When the amount of water passing through the soil is not enough to completely remove salts, the concentration of salt and other dissolved materials increases with time. As the water evaporates, the dissolved materials are precipitated (removed from solution), first coating pebbles and rocks and eventually forming a discrete layer, called *caliche*, through the process of *calcification*.

16.05.b2 South Mtns, AZ

16.05.b3 South Mtns, AZ

Soils in arid lands can contain high amounts of various salts, like NaCl, or calcium carbonate, as in the soil pictured above. The calcium carbonate accumulates below the surface and can form a hard layer of caliche that can be exposed if the upper, easily eroded soil layers are removed (left photograph), revealing the hard caliche (right photograph).

C How Does Climate Impact Soils in Temperate Environments?

Temperate climates are cooler and generally have less rainfall than tropical climates. Some of the most fertile soils in the world are present here because vital nutrients are retained within the crop root zone. Informal names for such soils are *grassland soil* or *forest soil*, depending on the type of associated vegetation.

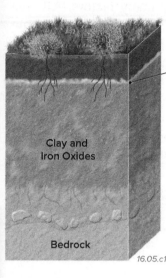

Clay and Iron Oxides

Bedrock

16.05.c1

1. Soils in temperate zones have nutrients available in the root zone, rather than an excessive leaching of nutrients as in the wet tropics or mostly immobile nutrients as in arid environments. Some leaching of iron oxides and clays occurs to lower levels in the soil, but nutrients are available to crops in the upper part of the soil. A vertical section through a soil, as shown here, is a *soil profile*.

16.05.c2   16.05.c3

2. Observe this upper part of a soil profile from a humid location in the mid-latitudes. Compare it to those for tropical and arid environments (on the facing page). What similarities and differences do you observe?

3. Within temperate climates, a transition occurs from forest soils to grassland soils as precipitation decreases. In grasslands, leaching is not as intense, and organic matter is retained, giving rise to some of the most fertile soils in the world.

D How Do Vegetation and Organisms Impact Soil Formation?

Climate helps to determine the distribution of vegetation and organisms. Living organisms, in turn, contribute to the accumulation of organic matter. Organic matter helps in the formation of soil structure, is used to differentiate soil layers, is vital in nutrient cycling, and reduces soil erosion by leaf litter, which minimizes the effect of rain-splash erosion and overland flow. In addition to the abundance of microorganisms involved in nutrient cycling, gophers, moles, and earthworms act to transport and mix soil.

This illustration (▼) shows the interplay of climate and vegetation in soil formation. Weathering breaks down bedrock, providing an opportunity for vegetation and organisms to further wear down bedrock through root action and animal burrowing. Organic matter begins to accumulate as humus, giving rise to the development of soil and of the layers within the soil called *soil horizons*, here shown with letters A, B, and C. We discuss these letters in a bit.

16.05.d2 Black Hills, SD

These trees on this granite knob are supported by soil within the fractures in the rock (▶). Soil and plants have a mutually beneficial relationship—soil provides nutrients and a physical medium for plants to grow. Plants in turn contribute organic matter that helps the soil develop and protects it from erosion. Most of the carbon-rich wood and leaves, however, comes from CO_2 in the air.

16.05.d1

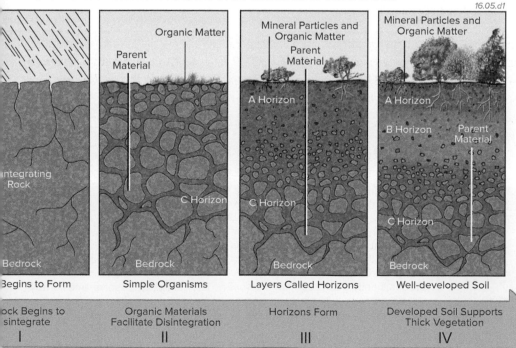

Organic Matter

Parent Material

Mineral Particles and Organic Matter

Parent Material

A Horizon

C Horizon

Mineral Particles and Organic Matter

A Horizon

B Horizon   Parent Material

C Horizon

Disintegrating Rock

C Horizon

Bedrock

Bedrock

Bedrock

Bedrock

Begins to Form   Simple Organisms   Layers Called Horizons   Well-developed Soil

Rock Begins to Disintegrate   Organic Materials Facilitate Disintegration   Horizons Form   Developed Soil Supports Thick Vegetation

I   II   III   IV

Before You Leave This Page

☑ Explain why soils develop faster in hot, tropical environments than in colder climates.

☑ Describe why temperate soils are often more fertile than tropical or arid soils.

☑ Describe examples of how vegetation and climate interact to create a fertile soil.

16.5

16.6 How Do Terrain, Parent Material, and Time Affect Soil?

SOIL CHARACTERISTICS, such as texture, structure, permeability, and fertility, are impacted by many factors. All five factors of ClORPT are interdependent, such as climate influencing the type of vegetation. ClORPT reminds us that how soil forms in a particular region is a complex question. Slow weathering of resistant parent material, for example, will limit clay formation that then limits the availability of nutrients and water-holding capacity. In turn, this limits the amount and type of vegetation that can inhabit the soil.

A What Is the Relationship Between Slope and Soil Development?

Topography—variations in slope and elevation—impacts soil properties by influencing erosion rates. Soil formation is favored when the rate of weathering exceeds the rate at which soil can be dislodged and carried away. Low-lying areas accumulate more soil, gather soil moisture, and generally have more organic matter available to vegetative communities.

1. Slopes oriented to the north in the Northern Hemisphere are shadier, cooler, wetter, and accumulate more organic matter than do slopes that face south. If there is abundant organic matter, as in humid climates, it will move downslope, rendering a darker color to soils in the lower parts and at the bottom of a slope.

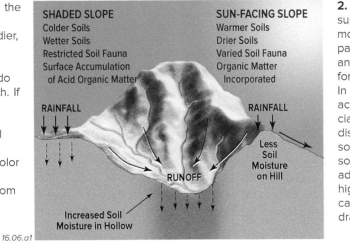

SHADED SLOPE
Colder Soils
Wetter Soils
Restricted Soil Fauna
Surface Accumulation of Acid Organic Matter

RAINFALL

SUN-FACING SLOPE
Warmer Soils
Drier Soils
Varied Soil Fauna
Organic Matter Incorporated

RAINFALL

RUNOFF

Less Soil Moisture on Hill

Increased Soil Moisture in Hollow

16.06.a1

2. Steeply sloping surfaces promote the rapid movement of water, soil particles, organic matter, and nutrients, so the in-situ formation of soils is limited. In contrast, bottomlands accumulate these beneficial materials and are less disturbed by movement, so they develop thicker soils. Bottomlands can be adversely affected by a high water table, but they can be productive if well drained.

16.06.a2

3. Note the difference in color and organic matter accumulation between a soil sample taken on a slope (left) and one taken at the bottom of the same slope (right).

B What Is the Relationship Between Parent Material and Soil Development?

Parent material is the rock from which a soil is derived. It can be derived in place or be transported from another location. Parent material influences the rate at which weathering will occur, which in turn influences permeability, pH, and the types of nutrients available to plants. Soils that develop in place are called *residual soils*.

1. A well-jointed rock in a cold, moist climate is prone to freeze-thaw action. Here (▶), a rocky peak is being broken down by a series of fractures, some of which are produced by the expansion of freezing water. The resulting pieces serve as parent material for soil.

16.06.b1 Sawtooth Mtns., ID

2. Shale, a sedimentary rock composed mostly of clay-size particles (▶), is easily decomposed and weathers into a clay-rich soil that has abundant soluble cations, such as Ca⁺, Mg⁺, and K⁺. When combined with humus, these soils are quite fertile, but the fine texture of clay soils hinders drainage.

16.06.b2 Durango, CO

3. Soil derived from sandstone (▲) is coarse textured, forming sandy loam and loamy sand. Such soils will exhibit good drainage and are easy to plow (good tilth), but because water drains so easily, these soils tend to be drought prone. If weathered from predominantly quartz sandstone, soils will be relatively infertile as quartz (SiO_2) is inert, providing little in the way of soil nutrients.

16.06.b3 Gallup, NM

4. Soils formed from limestone, a chemically soluble rock, are often highly productive. Calcium carbonate weathered from limestone (▶) can act as a buffer, preventing soil pH from becoming too low. Limestone is often ground up and used as a soil additive (agricultural lime) to raise pH to the ideal 6.5–7.0 range. In dry climates, where there is less rainfall to dissolve limestone, a limestone parent rock can result in dense accumulations of soil carbonates and a very alkaline soil.

16.06.b4

C How Can Parent Material Be Moved?

Parent material formed in place from the bedrock beneath the soil is termed *residuum*, but other material is brought into a site by streams, gravity, glaciers, or wind. This material is *transported* parent material. Much of the soil in the U.S. is created from parent material transported from elsewhere.

16.06.c2 Salmon, ID

1. One type of transported parent material is *colluvium*, which refers to loose material moving down a slope (◄) via mass wasting. In most cases, the sediments eventually stabilize and are weathered sufficiently for a soil to develop. Others remain too mobile to accumulate soil.

16.06.c1

16.06.c5 Central Colorado

4. A fourth type of transported parent material is soil developed from stream-deposited sediment (▲), referred to as *alluvium*. *Alluvial soils* form from gravels, sand, silt, and clay deposited by streams. Materials on floodplains would also be considered alluvium. The fertility of alluvium is highly variable, varying from droughty, inert, coarse sands to productive, organic-rich, silty soils.

16.06.c3 Alberta, Canada

2. Any place where glaciers have covered land—past or present—can reveal parent material formed from glacial deposits. The debris in this photograph (◄) was deposited when an active Canadian glacier receded, leaving behind a blanket of sediment—new parent material. The deposited material has a wide range of sizes of particles, so soil-forming processes begin with this assortment of material. Streams emanating from glaciers can rework this material, removing the smaller particles.

16.06.c4 Illinois

3. Wind can deposit an *aeolian* parent material—sand and dust—from which soils can develop. Such parent material can derive from sand dunes or glacial outwash plains. One special type of aeolian parent material is *loess*—a fertile deposit (▲) composed of silt-sized, glacially formed dust.

D How Does Time Affect Soil Formation?

The longer a soil has been exposed to weathering processes and organic-matter accumulation, the better the soil development with distinguishable *horizons* (layers). Soils form within decades in tropical, moist environments, but over thousands of years in arid environments. In any climate or setting, soil progressively becomes better developed with more time, as long as it is not disturbed or eroded away. Time is the "T" in ClORPT.

1. This figure shows soil development as a function of time, from left to right. Bedrock is weathered by physical and chemical processes to form weathered parent material called *regolith*. Transported parent material will already be at this stage.

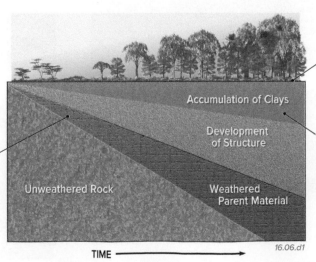

Accumulation of Clays

Development of Structure

Unweathered Rock

Weathered Parent Material

TIME ⟶

16.06.d1

2. When vegetation takes hold, organic matter begins to accumulate in the surface layers of the soil. Binding action by organic matter helps in the creation of soil structure.

3. Continued weathering by inorganic and organic acids contributes to the formation of clays and further development of soil structure—better developed soil layers.

Before You Leave This Page

☑ Explain the fertility variations of soils formed on steep versus gentle slopes and for soils derived from sandstone versus shale.

☑ Sketch and describe four environments from which transported soil parent material can originate.

☑ Outline how the passage of time contributes to the development of soil and soil layers.

16.7 What Are the Major Layers of Soil?

SOILS HAVE LAYERS roughly parallel to the ground surface that have distinct physical, chemical, and biological characteristics—they are soil horizons. A vertical section of the soil from the ground surface to parent material, usually around one to two meters, is called a *soil profile*. Layers become differentiated through weathering, groundwater movement of minerals, additions of organic matter, and erosion. When most people talk about soil from an agricultural standpoint, they are referring to topsoil, the uppermost layer of soil where seeds are normally planted. The large figure straddling the page describes the main soil horizons, each of which is assigned a letter to help us remember them and to communicate more efficiently.

16.07.a2

16.07.a1

Material Above the Topsoil

1. On top of soils in most climates is a layer of loose leaves, twigs, branches, pine needles, and other plant debris (◄). This is called the *litter layer*. While not really a part of the soil, the litter layer is an important source of organic material for soil development. The leaf-litter layer is critical in reducing the erosive impact of raindrops and slowing soil from being carried away by overland flow and surface runoff.

2. The first layer of soil is called the *O horizon*. The "O" signifies a high content of organic matter. It contains decomposed or partially decomposed plant and animal biomass—humus. This material helps retain water and nutrients that can be used later by plant roots. The O horizon is usually a few centimeters thick, although it can be over a meter thick in wetland soils.

Material Within the Topsoil

3. The *topsoil* layer contains the O horizon and the next layer down—the *A horizon*. The A-horizon contains a mixture of humus, living and dead organisms, and some inorganic mineral particles. Mature development of this layer is critical for plant growth as the A horizon is the primary source of nutrients for plants.

4. The O and A horizons (►) are teeming with bacteria, fungi, molds, earthworms, and small insects (▼). Depending on the climate and region, topsoil can also have larger burrowing animals, such as moles and gophers.

16.07.a3

16.07.a4

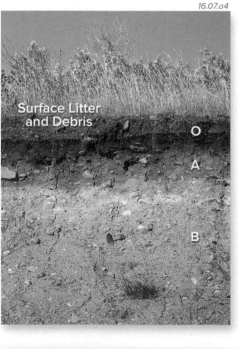

Surface Litter and Debris

O

A

B

Layers Below the Topsoil

5. The text below describes the main layers in soil, from the O and A horizons on top (the topsoil) to other horizons deeper in the soil. The O, A, E, B, C, and R horizons are termed "master horizons." They appear in most soil profiles, but not every soil profile will have all of these horizons. Examine the soil shown here (▶). Try to match some of the soil horizons in this photograph with those in the large, conceptual diagram below.

SOIL HORIZONS

O Surface Litter

A Topsoil
Organic Matter (Humus),
Living Organisms,
Inorganic Minerals

E Zone of Leaching
Dissolved or suspended
materials move
downward

B Subsoil
Accumulation of iron,
aluminum, humic
compounds, and clay
leached down from
the A and E horizons

C Regolith
Partially broken down
inorganic minerals

R Bedrock

16.07.a5 Badlands NP, SD

6. Although not shown here, each of the horizons can be subcategorized (such as A1, A2 or Bt1, Bt2 horizons) based on subtle differences in composition, horizon boundary characteristics, and development.

7. The *zone of eluviation*, or *E horizon* (◀), is below the A horizon and results from vigorous chemical weathering that leaches iron, clays, and organic matter, resulting in a bleached, "eluviated" horizon. This process of removal is termed *eluviation*. Note the lighter colored middle of the soil in the photograph above. This is an example of eluviation and formation of an E horizon.

8. The next layer down is called the *subsoil layer*, or *B horizon* (◀). It is also referred to as the *zone of accumulation* or *zone of illuviation* because this part of the soil profile can accumulate material from above. This layer is where material leached from the A and E horizons accumulates, including clays, compounds of iron and aluminum, and some organic matter. The depth of the B horizon is highly variable, but it is usually found around half a meter below the surface.

16.07.a6 Mackay, ID

9. The *C horizon* or *regolith* (◀) is composed of either weathered bedrock or unconsolidated sediment, and grades downward into unweathered bedrock or sediment (▶). No organic matter is present. This layer is little affected by the processes that impact the O, A, E, and B horizons.

10. The *R horizon* (◀) is bedrock. As such, it is not really a soil at all.

Before You Leave This Page

✓ Describe the material that is important in topsoils.

✓ Sketch and describe the sequence and characteristics of the O, A, E, B, C, and R horizons, and explain how each horizon forms.

16.7

16.8 What Are the Major Types of Soil?

ALTHOUGH MANY DIFFERENT categorization systems for soils exist, the most common one in the U.S.—the *Soil Taxonomy*—divides soils into 12 different types, or soil orders. We assign a soil to a soil order mainly on the basis of the soil-forming factors, which impart distinct properties to the different soil horizons, as viewed in the soil profile. In many cases, however, soil properties are highly localized, depending on vegetation, slope, and human influences.

A How Are Soils Classified?

Soils are classified on the basis of distinct physical, chemical, and biotic variations within soil horizons, usually to a depth of about 2 m. Examples of distinctive soil physical properties would be texture, structure, color, and horizon depth and distinctiveness. Chemical properties would include pH, carbonate content, and cation exchange capacity (CEC). The amount, depth, and degree of decay of organic matter would help differentiate soils from a biological standpoint. The resulting soils can be grouped into three categories: *site-dominated soils, developing soils,* and *soils strongly influenced by climate*. These three groups are differentiated in the table on the right-hand page.

Material Introduced from Above

1. Soil receives several types of material from the land surface. Leaves, pine needles, twigs, and other plant parts accumulate on the surface and are worked downward into the soil. Roots emit CO_2 gas, other gases come from the atmosphere, and moisture mostly arrives as rainfall and snowmelt. Soil material moves both down and up as it is carried by water, plants, animals, and gravity.

Material Derived from the Land

2. Soil material is mostly derived from weathering of underlying rock and sediment, but some material is introduced by water and wind. Sediment washes onto the surface from adjacent hillslopes or arrives as windblown sand, dust, and salts.

3. Weathering weakens and loosens underlying bedrock, providing starting material to make soil. This material can be worked upward into the developing soil, or the soil can gradually affect deeper and deeper levels of the bedrock. Some residual material remains in place at depth.

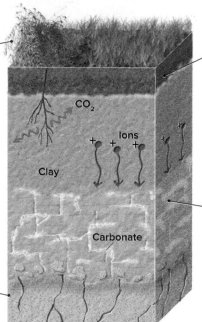

16.08.a1

Movement of Material

4. *Zone of Eluviation*—The upper part of soil loses easily dissolved material downward. Water soaking into the soil leaches (dissolves and removes) soluble material liberated by chemical weathering, carrying it deeper. Clay minerals and other fine particles are carried downward by infiltrating water. Plant parts and other organic material are also worked downward into the soil.

5. *Zone of Accumulation*—Chemical ions leached from above may accumulate in an underlying zone, if the water does not carry them all the way to the water table, where they enter the groundwater system. Clay minerals, iron and aluminum oxides, salt minerals, and calcium carbonate commonly accumulate in layers, depending on how much water and oxygen pass through the soil.

Spatial Characterization of a Soil

6. The smallest unit with which soils are described is the soil *pedon*, which is a three-dimensional body, about 2 m deep and large enough (1 m² to 10 m²) to distinguish soil horizons.

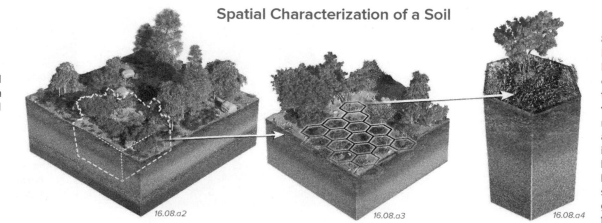

16.08.a2

16.08.a3

16.08.a4

8. An individual landscape can have several different soil types based on vegetation, parent material, slope, and human influence. Boundaries between different soil types can be gradational or fairly abrupt.

7. Up to 10 pedons, if similar in their properties, can be considered a *polypedon*.

Soil Taxonomy

Type of Soil	Soil Order	Derivation of Name	Characteristic of Soil	Geographic Setting	% Land Coverage
Site-Dominated Soils	Andisol	"andesite" a common volcanic rock of Andes of South America.	Weathers into fertile soil with high CEC and organic matter content.	Volcanic zones	0.8
	Histosol	Greek *histos* "tissue" from organic residue accumulates in wetland soil.	Thick accumulation of organic matter, poor drainage, little evidence of horizons.	Wetlands	1.2
Developing Soils	Entisol	"recENT" recently formed soil.	Recent or underdeveloped soils, little or no horizon development.	Highly variable	19.0
	Inceptisol	Latin *inceptum* "beginning" for soil beginning to show layers.	Soils beginning to show evidence of forming distinct layers.	Highly variable	16.1
Soils with Strong Climatic Influence	Alfisol	"Al" from aluminum, "F" from iron, important constituents of this order.	Clay accumulation in B horizon, moderate weathering, bases leached so liming needed.	Humid, mid-latitude forests	10.7
	Aridisol	Latin *aridus* "dry" for desert soil.	Desert soil lacking in organic matter accumulation, high pH, accumulation of carbonates at or beneath surface.	Hot deserts	12.5
	Gelisol	Latin *gelatio* "freezing" for areas with permafrost (permanently frozen ground).	Permafrost close to soil surface, much frost action, high organic matter accumulation.	High latitudes or high elevations	9.6
	Mollisol	Latin *mollus* "soft" for easy-to-till (plow) layer.	Rich accumulation of humus, proper balance of bases, nutrients, well-developed horizons.	Mid-latitude grasslands	7.4
	Oxisol	"oxide" for large amounts of iron and aluminum oxides.	Extensive weathering, red color, low pH, laterite formation possible.	Wet, tropical, hot locations	7.9
	Spodosol	Greek *spodos* "wood ash" from the grayish appearance of bleached E horizon.	Bleached E horizon from extensive leaching, low pH and low CEC, accumulation of iron in B horizon.	Coniferous forests of Northern Hemisphere	3.7
	Ultisol	Latin *ultimus* "ultimate" amount of leaching removes most bases.	More strongly weathered and redder than Alfisols, clay accumulation in B horizon.	Subtropical forests and savanna	8.5
	Vertisol	Latin *verto* "to turn" from extensive clay-particle movement due to repeated clay expansion and contraction.	Rich in swelling clays, forms cracks upon drying, low in humus.	Subtropics and tropics	2.6

Note: Percentages of world coverage exclude ice-covered areas from calculation.

Pedogenic Regimes

Although there are a variety of soils, we can group the formation of soils into five different environmental settings—called *pedogenic regimes*—where certain physical, chemical, and biological conditions dominate. These include the following regimes, in approximate order from warmer environments to colder ones: laterization, salinization, calcification, podzolization, and gleization.

Laterization occurs in hot, wet settings (in the tropics) where there is rapid chemical weathering and extreme dissolution. It produces red, leached soils, such as laterites.

Salinization produces salts on the surface because warm, dry conditions (arid and semiarid lands) draw moisture toward the surface, where it evaporates. Soils have a deficit in soil moisture.

Calcification also is most common in arid and semiarid regions, including the dry prairies of the Great Plains. Soil moisture moves upward toward the dry air, depositing calcium carbonate (caliche) in the B horizon.

Gleization occurs in places where the soil is often saturated and waterlogged, slowing the rate of decay of organic material and resulting in a dark, organic-rich A horizon.

Podzolization occurs where winters are long, cold, and subject to frost action, and chemical weathering is relatively slow. This results in shallow acidic soils with a well-developed soil profile.

Before You Leave This Page

✓ Sketch and explain the processes that occur within a soil profile and the properties that result in each part of the soil profile.

✓ Describe how geographers refer to the different spatial scales with which they study soils.

✓ Describe the three general categories of soils, as listed in the Soil Taxonomy.

✓ Describe the five pedogenic regimes.

16.8

16.9 What Types of Soils Largely Reflect the Local Setting or Initial Stages in Soil Formation?

MANY SOILS ARE DOMINATED by the specifics of the local site, such as certain rock types or a certain local geographic setting, such as being situated at the base of a cliff. Other soils are distinctive in that they are not well developed, representing the early stages in the progressive formation of soil and differentiation of the soil profile into discrete layers. The process of forming soil horizons is sometimes called *horizonation*.

A What Are the Properties of Site-Dominated Soils?

Histosols and Andisols have properties determined more by site location than climatic influences. Together, these soils comprise only 2.0% of all global soils. Entisols and Inceptisols are generally young soils that have not been impacted significantly by soil-forming processes.

16.09.a1

Andisols

1. *Andisols* are formed from volcanic ash and other volcanic material. Minerals derived from weathering have high nutrient and water-holding capacity, making these soils extremely fertile. The productive nature of these soils, however, is tempered by the threat of hazards imposed by volcanic eruptions, landslides, and volcano-related earthquakes.

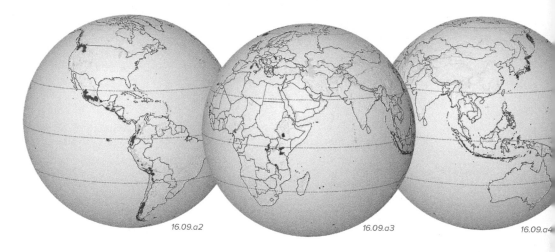
16.09.a2   16.09.a3   16.09.a4

2. Since Andisols form from volcanic materials, these soils naturally occur in volcanic areas, mostly around the Pacific Ring of Fire, a belt of active volcanoes that runs up the west coast of the Americas and down the Pacific coast of Asia and islands east of Asia, such as Japan.

16.09.a5

Histosols

3. *Histosols* are wetland soils without permafrost. Due to poor drainage, anaerobic conditions are created, and organic matter accumulates in a thick O horizon. When drained, Histosols can be extremely productive agriculturally, but they are prone to compaction due to the low density of highly porous organic matter. They contain carbon derived from the atmosphere.

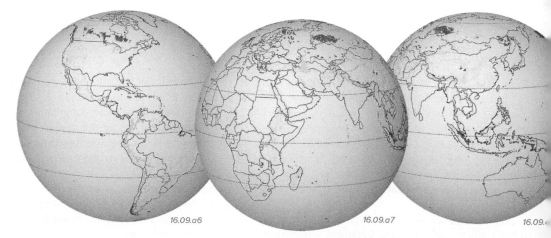
16.09.a6   16.09.a7   16.09.

4. Histosols mostly occur in high latitudes of the Northern Hemisphere, areas that were recently covered by continental ice sheets, and so have low relief and poorly developed drainage networks. Not enough time has passed since the last glacial advance for much soil development in such cold environments.

B What Are the Properties of Developing Soils?

Entisols and Inceptisols are generally young soils that have not been impacted significantly by soil-forming processes. Together these two soil types comprise about 35% of soils on the globe. With time, the influence of climate becomes increasingly important, influencing weathering rates and soil-profile development.

Entisols

16.09.b1

1. *Entisols* are recently created soils with no B horizon and little or no profile development. Entisols can also form where parent material is resistant to weathering. Often the parent material has been transported recently by streams, wind, glaciers, or mass wasting into "new" landscapes. Given the variety of environments where these processes take place, the geographic settings of Entisols are quite variable.

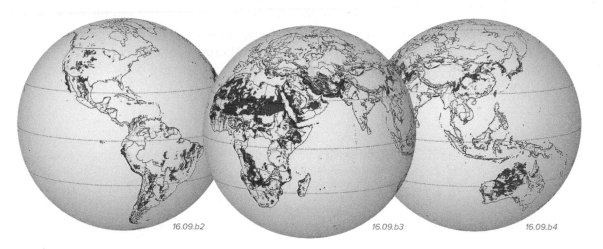

16.09.b2   *16.09.b3*   *16.09.b4*

2. Examine these globes showing the distribution of Entisols. Entisols are widely distributed around the world. They are widespread in the western mountains and sandy areas of North and South America, including nearly all of Baja California. Entisols are also the dominant soil in sand dominated deserts of North Africa (Sahara) and the Arabian Peninsula. They cover much of western Australia.

Inceptisols

16.09.b5

3. *Inceptisols* display mild weathering and the beginning of a weak B horizon. Inceptisols share similar characteristics with Entisols but are generally older and show better horizonation. Like Entisols, Inceptisols occur in extremely variable geographic settings, but they are most common in locations where soil-forming processes such as weathering, leaching, and the accumulation of organic matter are more active.

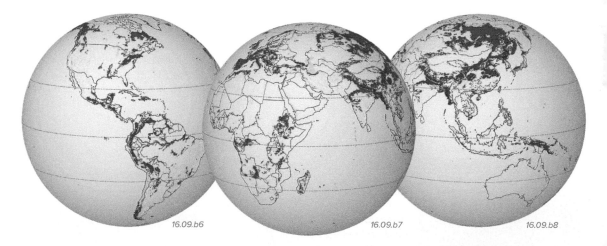

16.09.b6   *16.09.b7*   *16.09.b8*

4. Inceptisols are widely distributed, but they are not as extensive and do not cover as large an area as Entisols. In the Americas, they form obvious bands along the Amazon and lower Mississippi rivers, as well as other locations. Inceptisols are common in southern Europe, eastern Africa, India, and the eastern part of Asia. They are notably lacking in most of Australia, which has landscapes that have existed almost unchanged for millions of years, providing abundant time for better developed soils.

Before You Leave This Page

✓ Describe the main characteristics of Andisols and Histosols, and what dominates their properties.

✓ Describe the main characteristics of Entisols and Inceptisols, what they have in common, and where they generally occur.

16.9

What Types of Soils Form Under Relatively Warm Conditions?

CLIMATE IS AN EXTREMELY important factor in the formation of soils. Unlike the four soils discussed on the previous two pages, the remaining eight soil orders all bear the strong imprint of climatic influences. Of these eight, four soil orders form under relatively warm conditions—near the equator, in the tropics, or in the adjacent subtropics or warm parts of the temperate zone. Although these areas all have relatively warm temperatures in common, tropical areas are very wet, whereas those in the subtropics can be very dry. The combination of these conditions results in four soil orders: *Oxisols, Ultisols, Vertisols,* and *Aridisols.*

16.10.a1

Oxisols

1. *Oxisols* are deeply weathered, leached, and oxidized tropical soils, usually high in iron and aluminum and low in nutrients. The natural vegetation is efficient in cycling the limited nutrient supply, but if deforestation occurs, nutrients become leached rapidly, and soil hardens in a process called *laterization.*

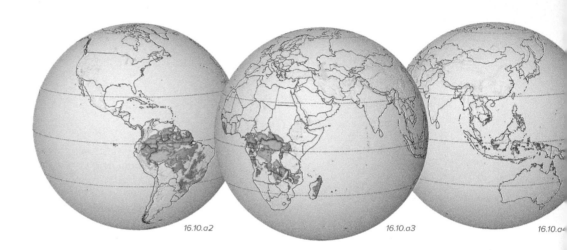

16.10.a2   16.10.a3   16.10.a4

2. Examine these globes showing the distribution of Oxisols. This soil's distribution clearly occurs relatively close to the equator, in the hot and humid conditions of the tropics. The largest areas are in the Amazon Basin of South America and the Congo Basin of central Africa. Rising air in the tropics, including along the Intertropical Convergence Zone (ITCZ), produces abundant precipitation throughout most of the year. The combination of warm temperatures and heavy rainfall causes rapid weathering and leaching of the soil, producing Oxisols.

16.10.a5

Ultisols

3. *Ultisols* form in humid parts of the subtropics and warmer mid-latitudes and are less highly weathered than Oxisols. Fairly high rainfall totals encourage the eluviation of clay minerals into the B horizon. With the lack of cations, such as Ca^{++} and Mg^{++} in the topsoil, these soils tend to be acidic, which can limit their agricultural productivity.

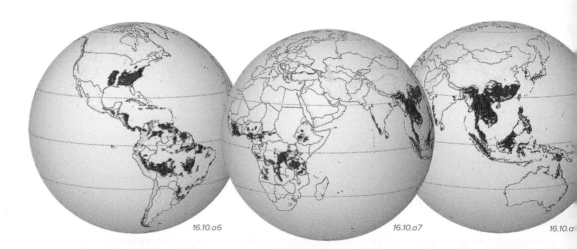

16.10.a6   16.10.a7   16.10.a8

4. Some Ultisols occur in the tropics, but these soils extend farther away from the equator than do Oxisols. Ultisols are abundant in warm, humid parts of continents, like the southeastern U.S., tropical Central America, parts of South America and Africa, and a large area in southeast Asia (e.g., Cambodia and Indonesia) and southeast China. These latter areas are under the influence of the East Asian monsoon.

Vertisols

16.10.a9

5. *Vertisols* are clay-rich soils located in temperate and subtropical environments. Due to the high clay content, Vertisols swell when wet and shrink when dry, creating pronounced changes in volume and disrupting a normal layered structure. The high clay content and poor structure of Vertisols create problems with agricultural management. Their tendency to shrink and swell also causes engineering problems, such as cracked foundations.

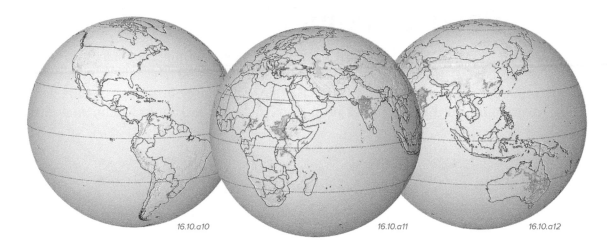

16.10.a10   16.10.a11   16.10.a12

6. Vertisols are rare in the Western Hemisphere, occurring mostly along the western side of the Gulf of Mexico, forming a north-south belt in Texas and northeastern Mexico. These soils are more widespread in eastern Africa, especially in southern Sudan, and in west-central India, an area subject to the changing wind directions and seasonal precipitation of the South Asian Monsoon. Vertisols are also present in the eastern, non-arid half of Australia.

Aridisols

16.10.a13

7. Arid areas give rise to *Aridisols*. The lack of vegetation and water causes soils to form with not much organic matter and leaching of nutrients. The lack of water restricts weathering, so horizonation is limited. Calcium carbonate, salt, and gypsum, which are easily leached in humid environments, tend to accumulate at the surface in the process of salinization, and within a subsurface whitish, caliche layer.

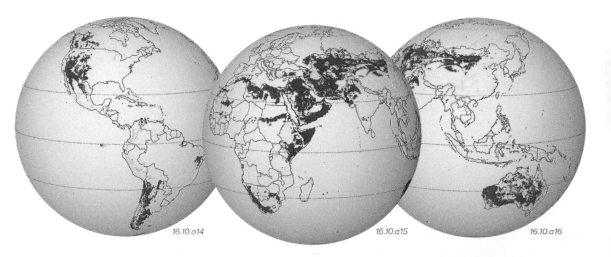

16.10.a14   16.10.a15   16.10.a16

8. Aridisols occur in dry conditions, so we would expect the distribution to be in the dry part of the subtropics, which are kept dry by descending air and high pressure. Aridisols indeed are abundant in desert regions, such as the Desert Southwest of North America and on the fringes of the Sahara and Arabian deserts, extending south of the subtropics in South America, where the rain shadow of the southern Andes is an influence. Note, however, that these soils also occur much farther north than we might expect, in Canada, south-central Asia, and Mongolia. Aridisols are also the dominant soil for large parts of Australia.

Before You Leave This Page

☑ Describe the main characteristics of Oxisols, Ultisols, and Vertisols, the settings in which they form, and where they occur.

☑ Explain why Aridisols have characteristics that are quite different from other soils, where they form, and why these areas are so dry.

16.10

16.11 What Types of Soils Form Under Temperate and Polar Conditions?

FOUR OTHER SOIL ORDERS form in climatic settings ranging from the relatively warm, semi-arid lands of Africa to the coldest polar regions. These four soil orders include much of the agricultural heartland of the central U.S. and south-central Canada, and most of Europe. They also form large swaths across central and northern Asia, tracking the distribution of different types of climate, such as the Humid Continental climate.

16.11.a1

Alfisols

1. *Alfisols* are usually fertile soils with ample accumulation of organic matter in the A horizon. A yearly moisture surplus assists in clay eluviation into the B horizon, where the clays retain moisture and nutrients for plant growth. Alfisols have a natural tendency to become acidic, but this can be counteracted by regular lime applications to raise soil pH. Only Mollisols (see below) are more productive agriculturally.

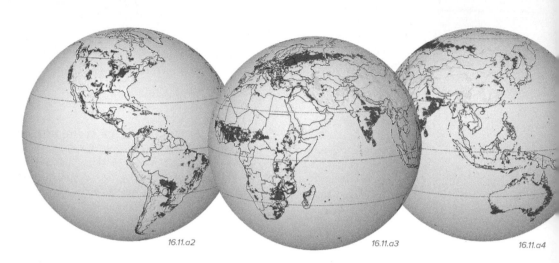

16.11.a2　　　16.11.a3　　　16.11.a4

2. Alfisols show one of the widest distributions of any of the soil orders, which indicates that they form under a variety of climatic conditions. Patches of Alfisols are scattered across North and South America, from the Great Lakes and Canada into southernmost Mexico. A large east-west belt of Alfisols crosses the southern part of western Africa, largely coinciding with a subarid region called the Sahel, which lies south of the Sahara. A dense concentration of Alfisols occupies western Russia and eastern India.

16.11.a5

Mollisols

3. *Mollisols* are the most agriculturally productive soils, rich in calcium carbonates, which form in the process of calcification, and organic material, with little leaching, and a humus-rich A horizon. These soils of the mid-latitudes experience a period of moderate moisture deficit that prevents nutrients from being leached out of the root zone. Mollisols are generally found beneath tall and short prairie grass.

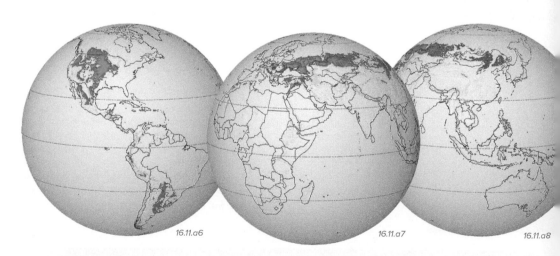

16.11.a6　　　16.11.a7　　　16.11.a8

4. Mollisols cover much of the agriculturally fertile Great Plains region of central North America and also cover relatively large parts of Mexico and southern South America (e.g., Argentina). The distribution of Mollisols forms an east-west belt across southern Russia and central Asia, mostly to the south of the belt of Alfisols. As with Alfisols and the remaining two soil orders on the next page, Mollisols are relatively sparse in the region of Australia, Indonesia, and southeast Asia.

Spodosols

16.11.a9

5. *Spodosols* are formed under acidic forest litter in cooler portions of the mid-latitudes and in high latitudes. Acids deriving from the A horizon leach the soil of nutrients and organic matter, leaving a bleached, grayish subsurface E horizon. Because of the high acidity and short growing season, Spodosols tend to be fairly unproductive.

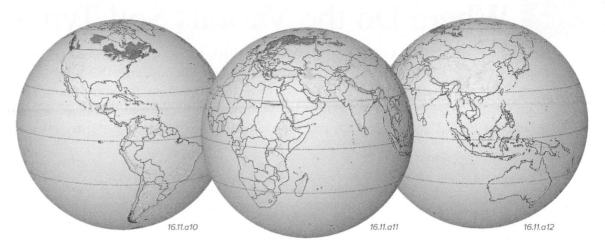

16.11.a10   16.11.a11   16.11.a12

6. Spodosols are largely restricted to the northern parts of North America, Europe, and Asia, and this distribution illustrates their development in colder climates. These soils are particularly prominent in eastern Canada, and they also occur along the western coast of Canada. Spodosols also cover much of Scandinavia, including most of Norway and Sweden. This soil order is mostly absent in the Southern Hemisphere due to the sparseness of land at high southern latitudes. A few small patches are barely visible in southernmost South America.

Gelisols

16.11.a13

7. *Gelisols* are soils above permafrost, having frozen soil for at least a portion of the year. Often waterlogged, these soils are rich in organic matter and sensitive to disturbance brought about through buildings, construction, and roads. Heat radiated from building foundations or sun-heated, dark-albedo roads can melt the top of these permafrost soils, causing the foundations of the building or road to sag and eventually collapse.

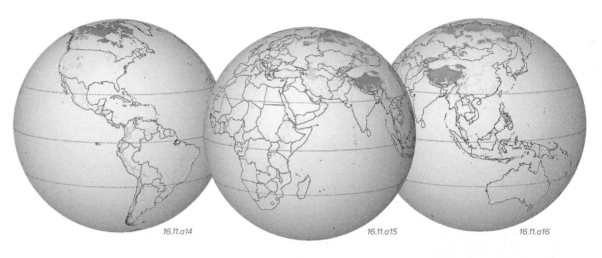

16.11.a14   16.11.a15   16.11.a16

8. If someone showed you the three globes above and asked you what types of soils form there, you would probably reply "very cold-climate ones." Since Gelisols form on ground that is frozen for part or all of the year, they are restricted to the polar regions or to very high elevations. They cover nearly all the polar regions of northernmost Canada and Alaska and northernmost Asia (Siberia). A few small patches occur in the western U.S., marking the highest and coldest peaks of the Rocky Mountains. A thin belt of Gelisols down the mountainous spine of South America likewise marks the highest and coldest part of the Andes. The large area of Gelisols in southern Asia is in the very high elevations of Tibet and the Himalaya.

Before You Leave This Page

☑ Describe the main characteristics of Alfisols and Mollisols, where they occur, and why they are so important.

☑ Describe the character of Spodosols and Gelisols, and why each occurs mostly in the northernmost parts of the Northern Hemisphere.

16.11

16.12 Where Do the Various Soil Types Exist?

THE DISTRIBUTION OF THE 12 SOIL ORDERS across the planet can be viewed with globes, as done in the previous pages, or with a more traditional map, as done here. Globes have the advantage that they show the latitudinal setting of each soil order more properly and do not have the distortions inherent in any projection of a sphere to a flat map. Their main disadvantage is that only half of the world is visible at once. Examine this global map of soils to compare the distributions of the 12 soil orders, but pay most attention to the broad patterns, not the local ones.

16.12.a2 Yoho NP, British Columbia, Canada

16.12.a3, Badlands NP, SD

16.12.a4 Big Maria Mtns., CA

16.12.a5 Amazon Basin, South America

1. Entisols and Inceptisols are characterized as being relatively poorly developed soils. Prospects for bringing this land into agricultural production can be limited by climate, terrain, inert parent material, and steep slopes, such as these mountainous slopes in the northern Rockies (◀).

2. In addition to topography, parent material, and climate, the type of overlying vegetation can greatly influence the type of soil. For example, mollisols occur beneath grasslands (◀).

3. Aridisols are the main soil in many arid regions, like in this photograph from the deserts of the American Southwest. The relative lack of rainfall and warm temperatures result in somewhat sparse vegetation, and so these soils have poorly developed O horizons. Aridisols are relatively infertile because of a lack of organic matter. Desert soils also have distinctive soil-related features like caliche and desert pavement (◀).

4. Slash-and-burn agriculture has been a traditional method of clearing land for cultivation in the tropics, especially in the world's rain forests, as shown here in a tropical rain forest (◀). This approach releases nutrients from trees to the relatively infertile Oxisols beneath the crops, but is incredibly destructive of the local ecosystem and can lead to soil erosion and eventually abandonment of the cleared land.

16.12.a1

Mollisols		Verti
Inceptisols		Spod
Alfisols		Oxise
Entisols		Ultis

5. Gelisols, limited to high latitudes or very high elevations, are permafrost soils and can melt, becoming unstable, disrupting the surface (▶) and damaging roads and buildings.

16.12.a6 Siberia, Russia

6. Mollisols support extensive wheat (▶) and corn production. These soils feature an ideal balance between leaching and retention of nutrients. This soil order's high organic matter also contributes to fertility. With proper stewardship and sufficient rainfall, Mollisols are very productive soils.

16.12.a7

16.12.a8 Southeast Asia

7. Ultisols are generally infertile and need a heavy input of fertilizer to make them productive. This flooded rice paddy in southeast Asia (▶) has been made fertile through regular manure applications.

16.12.a9 SW Australia

8. Many soils, such as these of southwest Australia (▶), build up salt over time if they are not properly drained or do not receive enough rainfall to flush the salts farther down and out of the system. If these soils build up enough salt deposits, further agriculture is inhibited.

Aridisols
Andisols
Histosols
Gelisols
Ice
Rock
Shifting Sand
Unclassified

Before You Leave This Page

✓ Describe some of the main patterns of distribution of some common soil types, and provide examples of how the climate or other conditions result in this type of soil.

16.12

16.13 What Are the Causes and Impacts of Soil Erosion?

HUMANS OBTAIN A GREAT majority of their calories from the land, so good soil stewardship is vital. Yet much land throughout the world is suffering degradation from one of the biggest threats to soil—*soil erosion*. Human activities have caused increased soil erosion in many places, but we can limit much of this by employing conservation practices that are known to be successful. What causes soil erosion, and what can we do about it?

A What Are the Causes of Soil Erosion?

Two processes are involved in soil erosion—*detaching* a soil particle from underlying material and then *moving* (transporting) the particle. The main causes of detachment are rainsplash and freeze-thaw cycles. The main agents of transportation are surface runoff and wind. Erosion of most landscapes is natural, but humans cause or accelerate soil erosion through deforestation, farming, overgrazing, and construction projects.

16.13.a1

1. With each falling raindrop (◄), pore spaces in soils are increasingly sealed, which reduces percolation of surface waters down into groundwater. As a result, more water remains on the surface to detach particles as the water starts to move.

2. As water accumulates on the surface of this field (►) and begins to flow downhill as *runoff*, it has the potential to detach and transport soil particles that are not anchored by grass, roots, or surface litter. The top layer of soil can be eroded off one part of a field and deposited in another part, or it can be transported away. Erosion can also carve into the soil, removing some of the deeper layers as well.

16.13.a2 Iowa

3. Wind erosion (►) also occurs through particle disaggregation and transport. Heavier soil particles are rolled along the surface through *creep*, lighter particles bounce through *saltation*, and the lightest become airborne, *suspended* in wind currents and sometimes transported for thousands of kilometers.

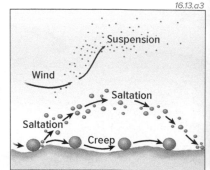

16.13.a3

4. The most erosion and transport of soils by wind occurs during times of the strongest winds during storms. This dust storm (►) was from the central U.S. in the 1930s, a time of very warm temperatures and great drought known as the *Dust Bowl* years.

16.13.a4 Colorado

B What Are the Forms of Soil Erosion?

Raindrop impact tends to promote surface sealing, so rainfall runs off at the surface, carrying soil in a thin layer of unchanneled water called *sheet erosion*. If sheet erosion becomes channelized, the water begins to concentrate in small and shallow (2–4 cm deep) channels, called *rills*. Rills can coalesce to form deeper channels called *gullies*. If enough gullies coalesce, they can form a *streambed*.

16.13.b1

16.13.b2 Green River, UT

16.13.b3

This sequence of three photographs illustrates how the flow of water can become more concentrated. In the photograph on the left, unconfined *sheetwash* draining an agricultural field begins to run together, forming rills, similar to those pictured in the center photograph. As rills merge, the water has more power and can cut deeper, forming a *gully*, as in the right image.

C What Is the Impact of Soil Erosion in the United States?

Impacts of soil erosion are a loss of agricultural productivity, loss of wildlife habitat, filling in of reservoirs, impaired health due to the spread of bacteria and viruses, more fertilizer use, flooding, and mass wasting.

16.13.c1

1. Drought weakens plant roots so that the network of roots can no longer bind soil against wind erosion. Such erosion (◄) removes the most nutritionally rich A soil horizon. Overgrazing in dryland areas can be especially devastating, causing soil erosion and loss of productivity.

16.13.c2

2. Water-borne sediment derived from soil erosion carries fertilizer residue, (◄) which can lead to *eutrophication* (overfertilization of water bodies due to excess runoff of fertilizers). The muddy or cloudy water column also decreases photosynthesis, resulting in lower dissolved oxygen in the water column.

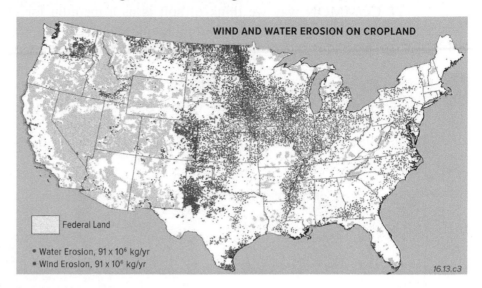
WIND AND WATER EROSION ON CROPLAND

Federal Land

• Water Erosion, 91 x 10⁶ kg/yr
• Wind Erosion, 91 x 10⁶ kg/yr

16.13.c3

3. In the U.S., wind erosion (marked in red) is more of a problem in High Plains states, whereas water erosion (marked in blue) is more common in the Midwest and the eastern and southern states (▲). While substantial reductions in soil erosion have occurred in the last several decades, it is still estimated that more than 80% of the nation's soils are being lost at a rate greater than the rate of formation.

D What Are Some Solutions to Soil Erosion?

Techniques for reducing erosion can be divided into two related categories: preventive and control measures. Preventive techniques include minimizing the impact of rainsplash erosion and runoff. Control techniques are designed to reduce the velocity of runoff after soil has already been dislodged, capturing sediment and nutrients before leaving the field.

Preventive Techniques

16.13.d1

1. Managing the number of grazing animals (▲) prevents erosion due to overgrazing. Too many grazing animals per acre often results in loss of vegetative cover, which in turn leads to soil erosion.

16.13.d2 Iowa-Minnesota Border

2. *Strip cropping*, pictured here (◄), protects the vulnerable land between the row crop (corn), capturing water and slowing surface runoff. Note that the land is also contour plowed, preventing runoff from traveling quickly down the slope.

16.13.d3

3. *No-till agriculture* retains last year's crop residue (◄) to reduce rainsplash erosion, control sheetwash, retain moisture, and protect against unseasonable early frosts.

Control Techniques

4. Trees and other plants are placed along waterways and are designed to capture sediment before these sediments and the contaminants on sedimentary particles reach the water (▶). Also, if trees are planted in rows perpendicular to the predominant wind direction as *windbreaks*, they help slow wind and reduce wind's erosive force.

16.13.d4

Before You Leave This Page

✓ Compare erosion control techniques. Explain which would be most appropriate around your hometown. Why?

✓ Explain why some land uses result in faster soil erosion than others.

16.14 How Does Soil Impact the Way We Use Land?

THE MOST COMMON way we think of soil is as a medium for plant growth, but soil is also a key component for construction, remediating pollution, and regulating and purifying our drinking water supply. The physical properties of soil can cause problems, but soil also has some uses that might surprise you.

How Do We Engineer Problem Soils?

The suitability of a site for a building or road depends on soil. In addition to being a valuable resource, soil can cause problems because of its low strength and how it behaves when shaken, wetted, dried, or compacted. Problematic soil can be recognized by builders and home buyers so that building on, or buying, such risky sites can be avoided. In some cases, the soil can be compacted properly before building on it.

16.14.a1

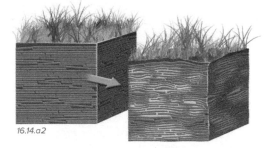

16.14.a2

16.14.a3

1. *Liquefaction* occurs when loose sediment becomes saturated with water and individual grains lose grain-to-grain contact as water squeezes between them (▲). *Quicksand* is an example of liquefaction. Liquefaction is especially common when loose, water-saturated sediment is shaken during an earthquake. The houses below, destroyed during the 1989 Loma Prieta earthquake, sank into artificial fill that liquefied during shaking.

2. Vertisols are a type of soil that contains a high proportion of certain clay minerals, called *swelling clays*, which increase in volume when wetted, expanding upward or sideways. As these clays dry out, they decrease in volume, causing the soil to shrink or compact (▲). Repeated *expansion* and *compaction* during wet-dry cycles can crack foundations, make buildings unsafe, and ruin roads (▼).

3. In some soil, clay minerals (▲) start out arranged randomly, with much pore space between individual grains. When wheel traffic runs over clay-rich soils, or a building is constructed on top, pore space is reduced as clay minerals begin to lie flat, diminishing open spaces and therefore compacting the soil. Such *soil compaction* typically does not occur uniformly, because some parts of the soil have more clay than others. Differential compaction can crack walls (▼), foundations, and roads.

16.14.a4 California

16.14.a5 Chandler, AZ

16.14.a6

What Can Soils Tell Us About Watersheds?

Soils play a key role in the hydrologic cycle. The type of soil determines the fate of precipitation, whether it will infiltrate into the soil and percolate to recharge groundwater or run off at the surface, producing soil erosion and possibly flooding.

16.14.b1 Baraboo, WI

◄ The health of soils is a critical factor in protecting watersheds. Soils beneath a forested watershed will slowly release water into the watershed by reducing surface runoff. Here, soils act similar to a sponge, reducing the chance of floods in response to heavy rain.

16.14.b2 Johnstown, PA

◄ Suburbanization brings an increase in impermeable surfaces, as open space is converted into houses, streets, and sidewalks. The new impermeable surfaces promote more surface runoff, which contributes to urban flooding.

C   How Do Soils Interact with Certain Types of Pollution?

Soil is a resource that allows many pollutants to become less harmful—for this reason, it is called the *grand recycler*. Without soil acting as a decomposer and recycler, the depth of all the dead plants and other organisms that ever lived would be overwhelming. Soil also purifies groundwater, where billions of bacteria contained in soil can either detoxify or immobilize many harmful contaminants, including those contained in fertilizers, pesticides, and gasoline.

16.14.c1

1. Soil is best at a pH level of around 6.5 to 7.0 (close to neutral). When soils become increasingly acidic, lead, arsenic, and other heavy metals become mobile and more likely to be transported through and off the soil.

16.14.c2 Maryland

3. The ideal soil keeps surface pollutants from draining too quickly and contaminating ground-water. Conversely, the soil should have a coarse enough texture that it drains fast enough to prevent ponding.

16.14.c3 Eastern Washington

2. Soil can become polluted near farms that use pesti-cides, herbicides, fungicides, and fuel oils. Soil can also become contaminated by salt from irrigation water that has acquired a high salinity due to high rates of evaporation.

16.14.c4 Arizona

4. Silt- and clay-sized soils have a density low enough that these tiny sediments can become entrained into wind currents, depositing nutrients and even living bacteria thousands of kilometers from their source region.

D   What Are Some Unusual Uses of Soil?

Soil and water are the two most important natural resources on the planet. In addition to the importance of soil in agriculture and overall health of an ecosystem, soil is used in other ways.

1. Scientists have discovered that some clays, like the one in this soil pit (▼), have healing properties for certain skin infections. Flanking the central photograph below are microscopic images of dangerous bacteria. On the left are bacteria cells before being exposed to the clays, and on the right are bacteria cells that have been successfully damaged by exposure to the clays.

16.14.d2

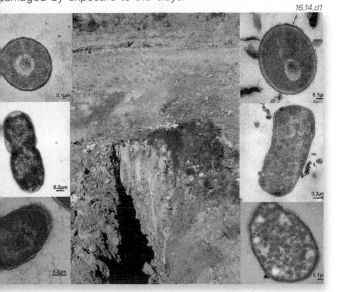
16.14.d1

2. Baseball diamonds use soil (▲) that is the proper mixture of sand, silt, and clay, giving the field good traction and prevent-ing excessive muddiness that might slow down the game. Likewise, we construct running tracks and clay-court tennis facilities using various mixtures of clay and other materials. The clay is derived from soils. The color of the track or court might tell us what type of soil was used.

16.14.d3

3. Soil is used in *forensic science* to investigate crimes. Footprints, soil on shoes, and the soil surrounding the crime scene can help identify the guilty party. Many criminals have been brought to justice as a result of clues left in soil (▶), as applied by soil scientists.

Before You Leave This Page

☑ List the factors responsible for soil liquefaction, shrink-swell, and compaction.

☑ Discuss some practices that affect runoff versus groundwater recharge.

☑ Discuss the best texture, pH, and structure to contain pollution.

☑ Describe some unusual uses of soil.

Where Do These Soils Originate?

SOIL FORMATION INVOLVES a complex interplay of climate, vegetation, topography, parent material, and time. Another way to think of these variables is with the acronym ClORPT—Climate (Cl), Organic Matter/Vegetation (O), Relief/Topography (R), Parent Material (P), and Time (T). In this investigation you will use clues provided by ClORPT to locate these soil profiles correctly and provide a justification for your choice.

Goals of This Exercise:

1. Apply information from soil horizons to identify the soil order, and from the order infer the location.

2. Identify relevant ClORPT variables acting on each soil profile.

3. Synthesize ClORPT variables, specifically climate and vegetation patterns, to explain observed soil-horizon characteristics.

16.15.a2

16.15.a3

16.15.a4

Soil Profile 1—These agriculturally productive soils have a rich A horizon, a distinct E horizon where eluviation occurs, and a B horizon where a darker iron-rich layer is present.

Soil Profile 2—A barely visible A horizon is present. This soil has a colluvium parent material that is derived from a bouldery glacial till. Soil processes have begun to act on this profile, but there is still no discernible B horizon.

Soil Profile 3—Little organic matter can accumulate here and there is a weak A horizon. The dominance of evaporation has caused soluble calcium carbonate to accumulate in the B horizon.

16.15.a1

Procedures

This shaded-relief map shows elevations for North America and beyond. Light purple and brown colors depict areas of high elevation, whereas shades of green represent areas of lower elevation. Consider the general climatic setting and elevation of an area and its impact on soil development when completing the steps in this investigation. For each step below, use the worksheet provided by your instructor or answer the questions in this investigation online.

1. For each soil profile, list the approximate depth and color of each horizon.

2. Based on clues provided for each pedon and a review of previous sections in the chapter, identify the soil order for each soil profile. Using the soil order as a clue, locate the position of the soil profile on the map. There are six possible locations on the map, one for each soil profile.

3. For each soil profile, create a table and write a short statement discussing how each variable of ClORPT likely has influenced the soil-profile patterns.

4. List the opportunities and challenges offered by each soil.

5. OPTIONAL PROCEDURE: Your instructor may ask you to consult the climate chapter and biome chapter to list the climate and biome type associated with each of these soil profiles. Write a short paragraph discussing how climate and vegetation patterns would influence the origin of each soil order you identified.

16.15.a6

16.15.a7

Soil Profile 4—No true humus exists in the A horizon because most of the time the entire soil profile is frozen. There is a brief summer period, however, when the A horizon will form a spongy substrate unsuitable for engineering purposes.

Soil Profile 5—A weak A horizon displays rapid decomposition of organic matter. An E horizon is discernible, where clays and iron-bearing minerals have been leached and moved downward to a B horizon.

Soil Profile 6—A rich A horizon extends far into the soil profile. The A horizon, with its soft granular structure, supports many earthworms. The wind blown loess has formed a blocky clay structure in the B horizon.

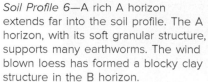

Ecosystems and Biogeochemical Cycles

THE BIOSPHERE IS THE REALM OF LIFE. *Biogeography* explores the biosphere and the Earth systems that support organisms. Adequate water, energy, and nutrients are necessary to sustain life, and their availability is controlled by climate and other aspects of physical geography. This chapter examines how organisms interact with each other and with their physical environment—*ecology*. The provisions and constraints of the physical environment along with the entire community of interacting organisms (plants, animals, and others) in an area is an *ecosystem*. *Biogeographers* study ecosystems from a broad perspective that considers spatial and temporal variations of physical and chemical processes in the *biosphere* and how the biosphere impacts Earth's systems, from local to global scales.

Costa Rica provides a wonderful example of ecology, ecosystems, and biogeography. Earlier in this textbook, we explored the region's weather and climatic patterns, always an excellent place to start when considering the biogeography of a region. The large perspective view here (▶) is a satellite image draped over the topography of the land.

NICARAGUA

Volcanoes

COSTA RICA

Pacific Ocean

17.00.a1

Land, Water, and Atmosphere

17.00.a2

The backbone of the country is a line of volcanoes, such as Volcán Arenal (◀), which has nearly continuous low-level eruptions of molten lava and ash. The volcanoes are formed by plate-tectonic activity (specifically subduction) along the west coast. The eruptions of Arenal are sufficiently continuous that the volcano has become a major site for *ecotourism*. The volcano interacts with the moist air in this tropical place, so it is often partially obscured by clouds and volcanic steam.

How do interactions between the land and atmosphere affect life, including humans?

17.00.a3

17.00.a4

Eruptions of lava and ash flow down the mountain and bury anything on the surface (◀). The initially barren lava flows are colonized by certain types of plants and animals that can establish themselves on such "new land," and these are known as a *pioneer community*. Once the pioneer community is established, an entire succession of other types of life can colonize the area. Over time, weathering begins to form soils (◀), which in this volcanic terrain are mostly Andisols, and the new land begins to blend into the landscape.

How do life and land interact chemically through the interface of soil?

17.00.a5

Abundant tropical moisture sustains a dense growth of tropical rain forest, and the rain forest releases some of this moisture during transpiration. In addition, plants remove carbon dioxide (CO_2) from the air to produce carbon-rich wood, leaves, roots, and other material.

How does carbon move from one part of the environment to another?

17.00.a6

Tropical rains, combined with locally impermeable tropical soils, result in plentiful runoff in streams. These corridors of running water attract many animals, forming a small streambed ecosystem within the larger tropical ecosystem.

What are the main components of an ecosystem, and what role does water play in a healthy ecosystem?

TOPICS IN THIS CHAPTER

Caribbean Sea

anoes

COSTA RICA

PANAMA

Life

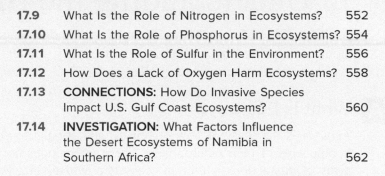

Various forms of life thrive in the tropical and mountain habitats, including coatimundi (◄), a mammal in the raccoon family, and colorful and interesting tropical birds, such as a toucan (▶). The lush vegetation hosts other classes of animals, for example reptiles, including this eyelash snake, a venomous pit viper (▶).

How do different types of animals interact, and what happens to their populations if they compete with one another or depend on one another?

The rain forest is home to other small creatures, like this "poison-dart frog" (▶), an amphibian, and countless types of insects, including ants that recycle vegetation and other materials that have fallen on the ground (photograph on the bottom right). The number of species in an ecosystem is one measure of biodiversity.

What do we mean by biodiversity, and what can decrease biodiversity?

The rain forest earns its name from the thick vegetation (the four photographs below) that forms a canopy and shades lower levels of the ecosystem. Vines and other plants climb or drape on the trees, trying to stay off the ground and gain better access to the light. Plants grow brightly colored flowers to better attract insects that, in turn, pollinate the plants, a *symbiotic* relationship that helps the plants and the insects.

What does it mean for a relationship to be called symbiotic, and what other types of relationships exist between living organisms?

17.0

17.1 How Is the Ecosystem Approach Useful in Understanding the Biosphere?

IN AN ECOSYSTEM, organisms interact with individuals of the same species and with other types of organisms. The ecosystem is supported by various types of energy, matter, and processes, some that involve living organisms and others that involve nonliving components, like rocks, soil, and water.

A What Is Ecology?

Ecology is the study of how organisms and populations of organisms interact with one another and the nonliving components of their environment. These interactions occur from the local to regional scale—the ecosystem. The intricate array of interactions within an ecosystem evolves in a way that promotes increasing efficiency in energy exchanges and nutrient cycling, because efficiency promotes survival. *Ecologists* study ecosystems, and *biogeographers* use ecological principles to explain the distribution of life—the realm of *biogeography*. Photographs below explore the Galápagos Islands (west of South America), one of the world's truly unique and fragile ecosystems.

17.01.a1

17.01.a2

17.01.a3

Individual—Most living organisms exist as individuals that can function somewhat independently, at least for a while, such as this marine iguana, a type of lizard (reptile) that started on land but forages in shallow waters.

Population—Although many individuals spend significant time as solitary creatures, they are always part of a population of individuals of the same species. The population can go up or down as the ecosystem changes.

Community—Not only do organisms interact with others of their kind, but individuals from different species interact, as part of a community. Here, an ocean-dwelling seal sniffs a marine iguana.

17.01.a4

17.01.a5

17.01.a6

Biotic Components—Living plants and animals, along with materials such as digestive wastes, discarded plant parts, and decaying remains of creatures, are the *biotic components* of an ecosystem.

Abiotic Components—Components of an ecosystem not directly produced by living organisms are *abiotic components*. These include the air, rocks, soils, and water that provide a home for the plants and other organisms.

Energy—All ecosystems require a source of energy, in most cases ultimately derived from the Sun. Insolation warms water and air, drives ocean currents, wind, and weather, and provides energy for plant photosynthesis.

Habitat—Some species live on land, such as this different species of iguana, whereas others live in the water or air. The physical environment in which a species lives, such as a rocky coast where it finds its particular mix of food sources, is its *habitat*.

17.01.a7

17.01.a8

Ecological Niches—The unique position of a species in the ecosystem, especially its role relative to the other species, is called its *ecological niche*. How would you characterize the ecological niche of these Galápagos fur seals?

B What Are the Structures and Functions of Ecosystems?

All ecosystems consist of the storage and flow of matter and energy, and various interactions between an organism and other organisms. The components of an ecosystem can be categorized into *structures* and *functions*. Ecosystems also have a variety of spaces for a wide array of organisms to exist, both supporting and being supported by the ecosystem.

Ecosystem Structures

1. *Ecosystem structures* consist of the biotic, abiotic, and energy components of the ecosystem. In most cases, these are the tangible, observable aspects of an environment. Observe the ecosystem structures shown in this diagram.

2. In the site depicted here, the most obvious *abiotic* components are the land, water, and air. For the land, there is the surface as well as the materials, like soil and rocks, that are beneath the surface. Less obvious abiotic structures would include gases and small particles in the air, the energy contained in sunlight, nutrients in the soil, and chemicals dissolved in the water.

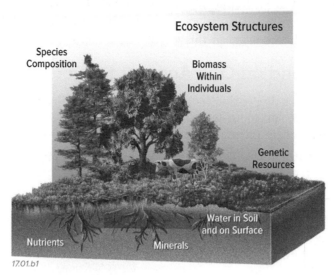

17.01.b1

3. The *biotic* components include the cow, the trees, the low bushes and grass, and the roots of the plants. Other biotic components that are less visible, but certain to be present, include insects, birds, burrowing animals, bacteria in the soil, fish, algae, and other organisms in the water.

4. To understand this ecosystem, one of the first things we would want to do is inventory the ecosystem structures. How many trees and cows are there, how thick is the soil, and what is the pH of the water? Biogeographers and ecologists do similar inventories and quantitative measurements of structures in an ecosystem.

Ecosystem Functions

5. *Ecosystem functions* are the dynamic processes that occur in the environment to support the ecosystem. This figure shows a few of the processes that are occurring in this scene.

6. Some ecosystem functions occur on or near the land surface. These include the growth and other changes in the plants, including the sprouting of new plants, shedding of leaves, cycling (movement) of nutrients, the uptake and release of water, and the eventual demise of organisms and return of their components to the soil or other sites.

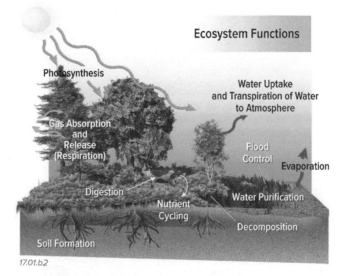

17.01.b2

7. Other ecosystem processes involve the atmosphere, such as changes in the amount of sunlight from day to night and season to season, the constant exchange of gases between the atmosphere and soil, water, and organisms (plants, animals, and microbes). Other processes include the wind and evaporation, and at times precipitation, which provides fresh water to the system and causes the processes of runoff and infiltration.

8. A number of ecosystem functions are also important but not so obvious. These include the capturing and converting of sunlight for photosynthesis (called *fixation*), decomposition of organisms, and chemical and physical weathering that slowly forms soil.

The Aral Sea of Central Asia—A Dying Ecosystem

In the 1960s, the USSR diverted flow in the Amu Darya and Syr Darya Rivers to provide irrigation for agriculture, especially cotton. Today, Uzbekistan is one of the leading exporters of cotton, but at what price? When the streams were diverted, the Aral Sea, which at the time was the world's fourth largest lake (left image), lost most of its inflow and became about one-tenth of its former surface area (right image). The lack of inflow sharply increased salinity in

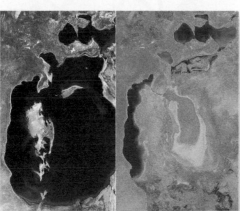

17.01.t1

the sea, and the entire aquatic ecosystem died. Toxic dust from the fertilizer and pesticides left behind as the Aral Sea evaporated has caused increasing instances of respiratory diseases and other ailments. This created impacts on the land surface that contributed to the decline of terrestrial ecosystems. The local temperature has become more severe in winter and summer, exacerbating risks for drought and causing other severe human impacts.

Before You Leave This Page

✓ Explain what ecology and an ecosystem are.

✓ Describe the difference between habitat and niche.

✓ Sketch and explain an example that contrasts ecosystem structures and ecosystem functions.

✓ Discuss recent impacts on the Aral Sea ecosystem.

17.1

17.2 What Types of Organisms Inhabit Ecosystems?

THE WORD "ECOSYSTEM" implies interdependencies and other interactions between different types of organisms and various other components in the environment. Together, the components and processes represent a complex and changing system. Most organisms are dependent on a particular setting, such as a fish needing water, but they are also dependent on other organisms, perhaps for nutrition, defense, or a place to live. There are some definite hierarchies that influence what happens in an ecosystem, and whether the ecosystem thrives or becomes threatened.

A How Do Organisms Depend on One Another for Food?

Some organisms, most notably plants, extract what they need directly from the soil, water, air, and energy in the environment, but most other organisms are more interdependent, deriving their food from plants or from other organisms. In most ecosystems, there is a complex chain of interdependencies, called a *food web*.

Trophic Levels

1. Organisms that acquire what they need from the soil, water, air, and energy of the environment, without consuming other organisms, are called *autotrophs*. A plant, like the corn shown here (▶), is a typical example of an autotroph. Most plants absorb nutrients and water from the soil and build their carbon-rich stems, leaves, and roots from CO_2 extracted from the atmosphere. Other organisms may be involved in moving and processing nutrients in the soil, but they are not consumed by the plants, except in a few unusual cases.

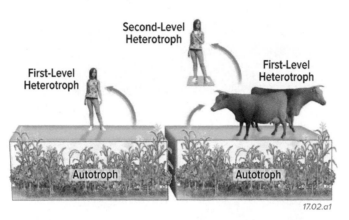
17.02.a1

2. Organisms that acquire their food from other organisms, such as from a plant, are *heterotrophs*. In the figure above, humans and cows are both heterotrophs.

3. If we consume the corn directly, we are a *first-level heterotroph*, but if we feed corn to cattle and then drink the cow's milk, we are a *second-level heterotroph*. It takes large amounts of energy to grow corn—solar energy from the Sun, energy to power the tractors, harvesters, and trucks that get the corn from the field to the supermarket, and electrical energy to keep the corn refrigerated. If we eat as second- or higher-level heterotrophs, we require and waste much more energy. At each stage in this process, much energy is wasted, generally as waste heat. A vegetarian diet would feed more people on the same amount of farmland because this eliminates the intermediate trophic level(s) and associated waste of energy.

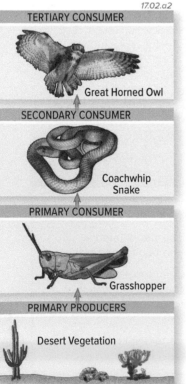
17.02.a2

Food Chains and Webs

4. One way to portray the dependence of one type of organism on another is to arrange them in the order in which the consumption occurs, from bottom to top (◀). The autotrophs, represented by the three varieties of cactus, are at the base. They derive their nutrients from the soil, water, and air, and their energy from sunlight-driven photosynthesis. For this reason, autotrophs are also called *primary producers*.

5. Grasshoppers eat parts of the cacti, and so are first-level heterotrophs. Snakes that eat the grasshoppers are second-level heterotrophs, and owls that feed on the snakes are third-level heterotrophs. The sequencing of organisms from autotrophs to higher level heterotrophs is called a *food chain*. One reason for this metaphor is that the entire system (the chain) can only be strong and survive if every link (organism) is present.

6. Relationships between different types of organisms within an ecosystem are typically more complex than a single, linear food chain. To represent such complexities, we envision a network of links, forming a *food web* (▼). The food web below incorporates the four types of organisms from the food chain to the left, but adds other organisms that are in the ecosystem.

7. In this food web, an organism can have several food sources and be food for several organisms. A mouse can eat cactus or grasshoppers, and be preyed upon by owls or by snakes, which are in turn preyed upon by owls.

17.02.a3

Carnivores, Herbivores, Omnivores, and Detritivores

8. Different types of creatures prefer different kinds of foods. Some eat only plants, some eat only meat, some have diverse diets, and some find food in soils. Such food preferences control the position of that type of creature on a food chain or web.

17.02.a4 Africa

17.02.a5 Namibia

17.02.a6 Costa Rica

17.02.a7 Michigan

9. *Carnivores* exclusively or mainly feed on animals. They include lions (▲) and similar predator cats, raptors, snakes and other reptiles, and many insects. Carnivores that eat insects are also called *insectivores*.

10. *Herbivores*, like this elephant (▲), eat plants. Grazing and browsing animals, like horses, goats, deer, and giraffes, are herbivores, as are many birds (seed or fruit eater), insects, and small mammals. Some herbivores live in water.

11. *Omnivores* are creatures with diverse diets that include plants, other creatures, and most any edible things they find. These include most bears, raccoons, coatimundi (▲), some birds, small mammals, insects, fish, and humans.

12. *Detritivores* eat detritus, the remains of decomposing (nonliving) plants and animals in the soil. Earthworms are the best known of these, but others include millipedes, flies, and aquatic bottom feeders, such as sea cucumbers.

Reproductive Strategies

13. We can group organisms based on their reproductive strategies. Some species, like humans, have few and infrequent offspring, in part to allow a longer period of protection and nurturing to increase their odds of survival. These are called *K-selected species*.

14. Other species have many offspring, resulting in a high reproduction rate, with the consequence that many of these offspring will not survive to adulthood. These are *r-selected species*. The letters K and r are derived from letters in an equation theorized to describe populations. The r is lowercase, as it is in the equation. K- vs. r-selection is a continuum rather than two discrete categories.

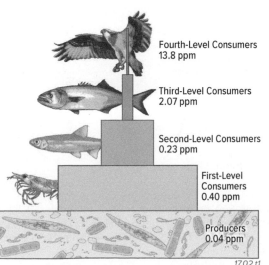
17.02.a8 Kgalagadi Transfrontier Park, South Africa

17.02.a9 Sweden

15. *K-selected species*, such as these wildebeest (▲), have slow rates of reproduction and maturation, but have strong and prolonged nurturing, with a higher likelihood of survival into adulthood.

16. *r-selected species* produce relatively frequent batches of numerous offspring, like lemmings (▲) and cockroaches. Offspring mature faster but are less supervised, and many do not survive.

Biomagnification and Food Chains

As a chemical element or compound works its way up the food chain, it can become more concentrated, a process called *biomagnification*. Many toxins are biomagnified with each step up the food chain. Higher-level heterotrophs must consume many lower-level heterotrophs, which themselves consume many autotrophs. Traces of heavy metals or other toxins taken up by the autotrophs may result in concentrations high enough to cause illness or death in their predators or their predators' predators.

DDT is a type of chlorinated hydrocarbon used as a pesticide that has been banned in the U.S. since 1970. DDT is a substance that is biomagnified as it moves up the food chain, from autotroph to higher level heterotrophs (▶). The banning of DDT has been linked to the comeback of the American bald eagle. There were trade-offs, however, because some of the pesticides that replaced DDT are even more dangerous, and the banning of DDT partially resulted in a resurgence of malaria, which causes millions of deaths worldwide, mostly in tropical climates and in impoverished populations.

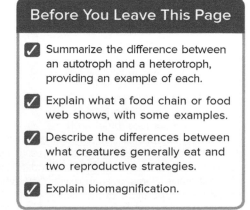

Fourth-Level Consumers 13.8 ppm

Third-Level Consumers 2.07 ppm

Second-Level Consumers 0.23 ppm

First-Level Consumers 0.40 ppm

Producers 0.04 ppm

17.02.t1

Before You Leave This Page

✔ Summarize the difference between an autotroph and a heterotroph, providing an example of each.

✔ Explain what a food chain or food web shows, with some examples.

✔ Describe the differences between what creatures generally eat and two reproductive strategies.

✔ Explain biomagnification.

17.2

17.3 What Interactions Occur in Ecosystems?

ORGANISMS IN AN ECOSYSTEM interact with each other and with their environment. These interactions occur both within species and between species and involve *reproduction*, *competition*, *predation*, *decomposition*, and *symbiosis*. Some of these types of interactions result in benefits to all species involved, while others benefit the individual of one species while lowering the other's chances for survival.

Types of Interactions

17.03.a1 Vulture Mtns., AZ

17.03.a2 Namibia

17.03.a3

1. *Reproduction*—Some method of reproduction is required to sustain a population, and different plants and animals reproduce in different ways. These wildflowers produce seeds. In contrast, the cacti grow new versions from small parts of the original plant.

2. *Competition*—If two or more species, such as hyena and wild dogs, compete for the same sparse resources, there are likely to be winners and losers. Competition between members of a single species can also be a problem if resources are in short supply.

3. *Predators and Prey*—Being a food source for some other creature is never a good survival strategy. The opposite is also true—if an animal relies on only one kind of food, survival becomes problematic if that food source is scarce or disappears. These lions prey on a variety of first-order heterotrophs like antelope.

Reproduction

4. We can examine reproduction of a population by plotting the number of individuals versus time. The graph here (▶) depicts the impact of reproduction in an environment of unlimited resources. The population continues to increase with time, but at faster and faster rates—that is, the rate of population growth accelerates as the offspring of offspring have more offspring. This type of accelerating growth rate is an *exponential growth curve*.

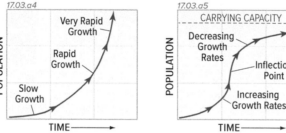
17.03.a4
Slow Growth / Rapid Growth / Very Rapid Growth

17.03.a5
CARRYING CAPACITY / Decreasing Growth Rates / Inflection Point / Increasing Growth Rates

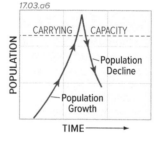
17.03.a6
CARRYING CAPACITY / Population Decline / Population Growth

6. Many populations of plants and animals experience a "boom or bust" cycle of population growth (◀). Many spring wildflowers exhibit this kind of curve, mostly sprouting, blooming, and dying in a relatively short time period, resulting in a population pattern as in the curve above. This kind of curve can also represent a case when the carrying capacity is exceeded, such as when grazing animals eat all their food, and the population decreases rapidly, or "crashes."

5. In the real world, limitations dictate that an ecosystem can only support a certain number of individuals of a given species. This theoretical maximum population is called the *carrying capacity*. A population of a species will increase in its ecosystem rapidly at first and then level off, hovering near its carrying capacity. In this graph (▲) the population does not decrease, but the rate of growth decreases.

Competition

7. Individuals within a species will compete for limited resources, as will individuals from different species, if the two species are in the same ecological niche (and are consuming the same resource). In the population graph here, a cold-adapted species (blue) and warm-adapted species (red) occupy the same niche, but the two species do best in different temperature ranges. They therefore normally do not compete directly because they are in different areas or in the same area at different times. Some overlap exists, however, so there is some competition. If the temperature of the region cools, the cold-adapted species will out-compete (thrive at the expense of) the warm-adapted one.

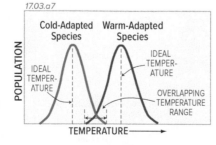
17.03.a7
Cold-Adapted Species / Warm-Adapted Species / IDEAL TEMPERATURE / OVERLAPPING TEMPERATURE RANGE

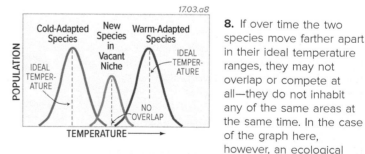
17.03.a8
Cold-Adapted Species / New Species in Vacant Niche / Warm-Adapted Species / IDEAL TEMPERATURE / NO OVERLAP

8. If over time the two species move farther apart in their ideal temperature ranges, they may not overlap or compete at all—they do not inhabit any of the same areas at the same time. In the case of the graph here, however, an ecological niche has opened up in the temperatures (areas) between the two species. A third species, shown in green, may arise or move in to occupy the newly vacant ecological niche. If the new species is able to broaden the range of its habitat, it may out-compete one or both of the preexisting species. All species will compete for position across space and a range of environmental conditions.

17.03.a9 Secretary Bird, Namibia

Predation

9. As heterotrophs (◄) consume other species, they keep their prey's population from exceeding its carrying capacity, thereby contributing to the ecosystem's stability. If the prey's population falls too low, the predator population will also fall. This negative feedback keeps the ecosystem relatively stable.

10. Most predators have certain prey they prefer, but other prey may suffice in times of shortage. Higher-level predators will always be less numerous in the ecosystem than lower-level heterotrophs (their prey). We can study predator/prey ratios (which are generally very low) to assess the health and food chain in an ecosystem. *Keystone predators*, such as otters and jaguars, generally are at the top of the food chain and are of great interest because they have a strong effect on the ecosystem despite their small numbers. Their removal causes other populations to expand. A harmful human impact on ecosystems is often the removal of keystone predators, like wolves.

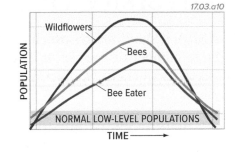
17.03.a10

11. Some predator-prey (and heterotroph-autotroph) relationships (▲) display very short lag times between an increase in the population of the consumed (i.e., prey) and a resulting increase in the population of the consumer (i.e., predator). In the graph above, an increase in wildflowers is almost immediately accompanied by an increase in the number of bees and by an increase in the population of "bee-eaters." Such relationships with minimal lag times usually occur when the predator relies almost exclusively on a single species for predation.

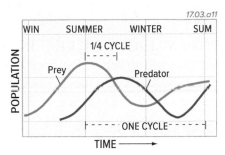
17.03.a11

12. The increase of a predator population usually lags behind an increase in prey (▲). The lag is commonly 1/4 of the time required for the prey to increase, decrease, and begin to increase again. In the example above, the prey species (in green) is relatively sparse in the winter, but rapidly increases with the onset of summer. The predator population increases to take advantage of the increased abundance of prey, but it may take months for the predator to bear new young, resulting in the observed lag. The lag in this case is about 1/4 of a year.

17.03.a12 Costa Rica

Decomposition

13. The final dissipation of usable energy in an ecosystem occurs through consumption of dead organisms by *detritivores* and by *decomposition*, the breakdown of organic materials into inorganic nutrients. Bacteria and fungi are the main agents for decomposition. The rate of decomposition is faster when temperature and moisture content are higher, as in a tropical setting (◄). Reactions break down glucose first, followed by more complex material. Decomposition also occurs in marine environments, for instance kelp along the shoreline (▶).

17.03.a13 Point Lobos, CA

Symbiosis

14. Unlike competition (in which both species have detrimental impacts on each other) or predation (in which only one of the species involved has a detrimental impact on the other), some species have a close, life-long interaction called *symbiosis*. There are three types, depending on the degree of benefit to the species involved.

17.03.a14 Indonesia

15. *Mutualism*—Both species benefit from the interaction. The clownfish above finds refuge in the anemone's stinging tentacles that ward off predators. The fish, in turn, drives away other fish that would eat the anemone.

17.03.a15 Namibia

16. *Commensalism*—One species benefits and the other is largely unaffected by the interaction. This chameleon gains protection from the plant but the plant gains little in return from the chameleon's presence.

17.03.a16

17. *Parasitism*—One species may benefit while the other is harmed from the interaction. In the photograph above, a tomato plant is being eaten by a tomato hornworm, which can decimate the plant's leaves and tomatoes.

Before You Leave This Page

✓ Sketch and explain the different patterns of population growth in an ecosystem.

✓ Explain how populations of two species in an ecosystem can change over time, under the effects of competition, consumption, and predation.

✓ Explain the processes of decomposition and the three different types of symbiosis in ecosystems.

17.3

17.4 How Can Biodiversity Be Assessed?

THE VARIETY OF LIFE IN AN AREA is its *diversity*, which can be expressed in various ways. An ecosystem can be considered diverse if it includes a large number of species or if there is an unusual variety in the gene pool within the species. Most commonly, however, diversity refers to *biodiversity* — the number of species and the evenness of the populations of those species in an ecosystem.

A What Are Some Simple Indicators of Biodiversity?

Ecologists and biogeographers complete inventories of the number of individuals in an ecosystem, commonly by working in a field site. In a few cases, it is possible to count each individual, but generally a population inventory involves surveying a representative area or subpopulation and then extrapolating to a population as a whole. The results of these population surveys are examined for two measures of biodiversity: *richness* and *equitability*.

1. These two graphs show the populations of different species that exist in two simple, hypothetical ecosystems. Each color bar represents a different species, and the height of the bar indicates the number of individuals in that species (i.e., the population). The number of species present is called the *richness*.

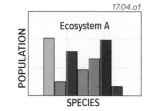

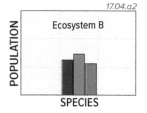

2. The richness is greater in Ecosystem A than in Ecosystem B (right graph), by a score of seven species to three.

3. Ecosystem B has fewer species, but the species present have a more even distribution — they all have about the same populations.

4. The evenness of the populations of different species in an ecosystem is known as the *equitability* of the ecosystem. The equitability of ecosystem B is greater than that in ecosystem A, which has a more uneven distribution. Richness and equitability are two components of biodiversity. Notice that neither of these variables takes into account the total population of all species present, which may or may not be related to biodiversity. Biogeographers and ecologists inventory the populations and then calculate a measure of ecosystem diversity called the *Shannon Index* (S), which incorporates both richness and equitability. A high value of the Shannon Index indicates a diverse ecosystem.

B How Does Biodiversity Vary Geographically?

As with most other attributes of our world, biodiversity of ecosystems varies geographically. The map below shows the richness of amphibian species around the world. Similar maps exist for mammals, birds, reptiles, and other creatures.

1. This map shows that the richest areas for amphibians is in the tropics, specifically in the tropical rain forests of South America, Africa, and Southeast Asia.

2. With regard to biodiversity, climate is the most limiting abiotic factor. In tropical areas, day length and atmospheric conditions are similar throughout the year, so many species can exist without major climatic disruptions. In polar or alpine environments, harshness and temperature variability of the environment limit both populations and richness. Tropical ecosystems are far more diverse than polar ecosystems, but intense competition in the tropics reduces the number of individuals of any single species.

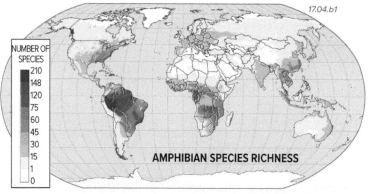

AMPHIBIAN SPECIES RICHNESS

4. In some types of ecosystems, equitability (not shown here) increases with increasing latitude, but because of the overwhelming richness closer to the equator, biodiversity tends to decrease poleward and upslope in mountains.

3. All other factors being equal, a more complicated vertical structure and variety of the physical environment, such as multi-layered tree canopies, also contribute to the higher richness and biodiversity in tropical areas than in single-layered ecosystems, such as grasslands.

5. The types of organisms also differ spatially. For example, species that are long-lived and have low reproduction rates (K-selected species) tend to be more dominant in tropical regions (▶). They subsist near the carrying capacity of their environment and so they thrive best in an environment, like the tropics, that is relatively consistent from day to day and month to month.

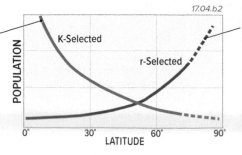

6. Species that are short-lived and have high reproduction rates (r-selected species) tend to live in polar and alpine regions, because the ever-changing environment, in such measures as temperature, number of hours of daylight, and precipitation, necessitates rapid reproduction for species survival. Most weeds display this type of reproduction behavior, rapidly growing and blooming in the spring. r-selected heterotrophs are generally more opportunistic feeders, partly in response to the more limited resources at high latitudes.

C Why Is Biodiversity Important and How Is It Threatened?

Lack of, or loss of, biodiversity is a major concern in many parts of the world. Degradation of ecosystems results in loss of species and a loss of individuals of some species but not others—loss of diversity. It is difficult to gauge what such losses represent, in part because we do not yet understand the role of many species in their ecosystems or the potential uses of these species, such as for medicine.

Importance of Biodiversity

17.04.c1

17.04.c2 Forks, WA

1. *Human Health and Medicinal Uses*—Some species have properties that are used to treat ailments and disease. Aloe vera (◄) is a succulent used to treat burns and other skin discomforts. Aloe vera is thought to be extinct in the wild, but it persists because it is widely cultivated. Maintaining biodiversity in plant, bird, and mosquito populations protects us against emerging diseases, such as West Nile virus.

2. *Aesthetic and Spiritual*—Many places, such as this temperate rain forest (►), owe their special aesthetics to a delicate balance between the environment and the living organisms that inhabit these special places—that is, the ecology. Disturbing one component could have positive or negative feedbacks, inducing unanticipated consequences.

3. *Food Security*—Genetic diversity in agricultural lands (◄), like different kinds of corn, ensures variety in food supply and flexibility for future breeding stocks—an "insurance policy" against the impacts of a new plant disease, or climate variability or change.

4. *Environmental Ethics*—Animals and plants belong in their natural habitats, and humans should try to keep these ecosystems as undisturbed as possible. This goal conflicts to some degree with the goal of *ecotourism*, especially in fragile places like the penguin colonies and biologically rich offshore waters of Antarctica (►).

5. *Buffers Against Extreme Events*—The "checks and balances" in diverse ecosystems protect the ecosystem, for example, against the effects of storms (◄), and allows the ecosystem to continue to provide structures and functions, many of which help people.

6. *Ecotourism*—Loss of wild animals threatens the ecotourism industry of ecologically threatened places, such as tropical rain forests and coral reefs around the world, the habitat of the mountain gorillas of central Africa, the lemurs and other unusual wildlife of Madagascar, and the fragile island ecosystem of the Galápagos (►).

17.04.c3 Eastern Colorado

17.04.c4 Antarctica

17.04.c5 Pemaquid Point, ME

17.04.c6 Galápagos

Threats to Biodiversity

7. Biodiversity is threatened by many human activities, including clear-cutting of forests for farms and timber (◄), especially when this occurs in species-rich tropical rain forests. Densely planted "energy crops" are grown in tropical ecosystems to provide biofuels as a way out of poverty, but at a cost to biodiversity. Equally harmful is the growth of communities and the building of roads into previously pristine ecosystems.

8. Another threat to biodiversity is contamination (►). Government contractors released millions of gallons of toxic mine waters down the Animas River of Colorado, threatening ecosystems downstream in this river and in the San Juan River and Colorado River, into which the Animas River flows.

17.04.c7 Eastern Mississippi

17.04.c8 Animas River, CO

Protecting Biodiversity

There are no easy answers to protecting global biodiversity, but several solutions, in combination with one another, may offer the best hope. Wildlife and nature preserves have been successful in protecting delicate ecosystems from too much human intrusion. Some of these preserves are the result of governments and legislation, but others are created with private funds, completely outside of any government. Both approaches work if done thoughtfully.

In some regions, species have been reintroduced to areas that were once their habitats, but from which they have disappeared. This usually involves harvesting a select number of plants and animals from an ecosystem where they are thriving and safely transporting them to their new locale. This acts as an insurance policy if some calamity strikes the place from where the organisms were harvested. Gene and seed banks provide "insurance" for the long-term survival of species.

Before You Leave This Page

✓ Explain the components of biodiversity.

✓ Sketch or describe how and why biodiversity varies across latitudes and altitudes, including the difference in spatial distribution between K-selected and r-selected species.

✓ Summarize the importance of biodiversity, how it is threatened, and ways to protect it.

17.4

17.5 How Does Energy Flow Through Ecosystems?

ENERGY IS AS FUNDAMENTAL to all ecosystems as it is to the rest of the universe. Unlike nutrients and water, energy is not recycled; instead it flows through or is stored by systems. Energy can be converted from one form to another in the way that electromagnetic energy from the Sun causes moisture to evaporate and rise, gaining latent heat and potential energy, but energy is neither created nor destroyed, only converted from one form to another.

A How Do Energy and Chemical Substances Flow Through Ecosystems?

Inputs and Outputs

1. Any system, including an ecosystem, has inputs and outputs of energy and matter, accompanied by storage or processing of energy and matter within the system. *Inputs* to an ecosystem consist of energy and materials such as nutrients, soils, and water. Some of these materials are stored in the living tissue of organisms.

2. If inputs of energy and matter come from outside the system, we refer to this as an *open system*. A plant is an open system, receiving energy from the Sun, water and nutrients from the soil, and exchanging gases with the air.

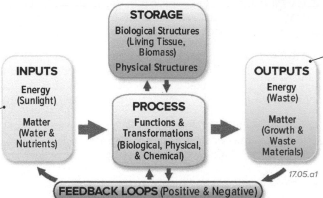

5. As processes occur in ecosystems, heat, chemicals, nutrients, and water are the output. In other words, these can be released from the living matter through waste heat, respiration, transpiration of water from leaves, decomposition, or waste products. Strong ecosystems are generally more efficient than weaker ecosystems, but no ecosystem can be 100% efficient. Some energy and matter are always wasted.

3. Energy and matter can be *stored* in the ecosystem, as *resource reservoirs,* for later use. Energy and matter also flow through the ecosystem as producers, or autotrophs, use the energy and matter for sustaining life. Heterotrophs consume the autotrophs or materials produced by autotrophs, acquiring energy and matter in a less direct manner.

4. Processes in the ecosystem that magnify the flow of energy or matter through the ecosystem are *positive feedbacks. Negative feedbacks* reduce the flow, maintaining them within an acceptable range.

Flow of Energy Through an Ecosystem

6. The figure below illustrates the efficiency of energy transfers, expressed as percentages, at each step from Sun to producers to consumers, up the food chain. The waste heat is represented by purple arrows escaping from each step in the energy transfer up the food chain or food web. Examine the figure and then read on.

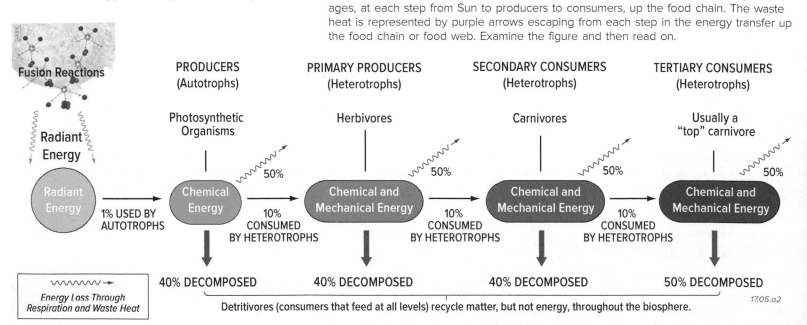

7. The *First Law of Thermodynamics* states that energy can be neither created nor destroyed, but merely transferred from one form to another. In ecosystems, organisms cannot generate their own energy but instead they must acquire it from other sources. The fusion reactions on the Sun's surface release solar radiation (insolation) that provides energy for the vast majority of autotrophs on Earth. Likewise, heterotrophs rely on the energy acquired by autotrophs or by other heterotrophs.

8. The *Second Law of Thermodynamics* states that when energy transfer occurs, some will be changed into less useful energy (waste heat)—energy can never be transferred with 100% efficiency. Within ecosystems, energy is transferred from one organism to another as food energy moves through the system. As shown in this figure, only about 10% of the energy is transferred from one level to the next, with the rest being waste, either physical waste that is decomposed or waste heat that escapes into the environment. With so much waste along the way, only 1% of the energy, or less, used by autotrophs reaches the higher-level heterotrophs at the top of the food chain or food web.

B   How Does Photosynthesis Work?

The amount of energy flowing through an ecosystem depends on the amount captured, or "fixed," by autotrophs. This primary production is accomplished mostly by photosynthesis within plants and algae. The resultant energy is transferred to higher trophic levels through consumption of autotrophs by heterotrophs.

1. Plant cells and some algae have specialized organelles called *chloroplasts*, which capture insolation and produce sugar during photosynthesis, as represented by a simplified chemical reaction (▶):

$$\text{Insolation} + 6CO_2 + 6H_2O \longrightarrow C_6H_{12}O_6 + 6O_2$$

2. There are two components of photosynthesis. The first, the *light reaction*, occurs when chlorophyll in the chloroplasts "fixes" insolation, exciting electrons. The energy released by the electrons is used to make molecules of *ATP* (adenosine triphosphate) and a phosphate compound called *NADPH*. The ATP transfers energy among the cells and is recycled over and over, in slightly differing forms, releasing O_2 in the process.

3. The second component, the "Calvin cycle" phase of photosynthesis, occurs even in the absence of sunlight, as long as ATP and NADPH remain available. ATP and NADPH fix CO_2 from the air to create organic molecules (sugars) that are energy to the plant.

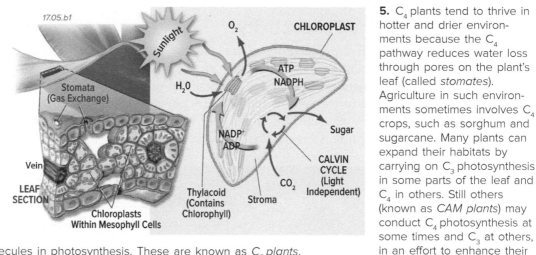

17.05.b1

4. Most green plants create 3-carbon-based molecules in photosynthesis. These are known as C_3 *plants*. However, in the 1960s, it was discovered that some plants can form a 4-carbon molecule instead of 3-carbon molecules. These are called C_4 *plants*.

5. C_4 plants tend to thrive in hotter and drier environments because the C_4 pathway reduces water loss through pores on the plant's leaf (called *stomates*). Agriculture in such environments sometimes involves C_4 crops, such as sorghum and sugarcane. Many plants can expand their habitats by carrying on C_3 photosynthesis in some parts of the leaf and C_4 in others. Still others (known as *CAM plants*) may conduct C_4 photosynthesis at some times and C_3 at others, in an effort to enhance their chance of survival.

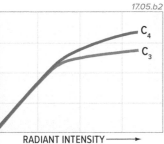

17.05.b2

6. This graph (◄) plots the rate of photosynthesis as a function of the amount of incoming radiation (light) for C_3 and C_4 plants. At low levels of radiation (not much light), the rate of photosynthesis increases at about the same rate for both types of plants as light increases. However, at higher light intensities, increased energy is available to power the C_4 plants relative to the C_3 plants.

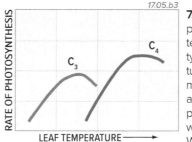

17.05.b3

7. This graph (◄) compares the rate of photosynthesis as a function of leaf temperature for C_3 and C_4 plants. Both types of plants have a peak temperature for production (where they are most efficient), but that rate is higher at higher leaf temperatures in C_4 plants than in C_3 plants. What habitat would a C_4 plant most likely occupy? Warm conditions favor C_4 plants.

Other Energy Sources in Ecosystems

Although photosynthesis is by far the most important mechanism for fixing the energy that supports ecosystems, it is not the only mechanism. In places where sunlight is unavailable, such as the deep ocean floor and within Earth's subsurface, organisms must rely on an alternative energy source. *Chemosynthesis* is a process by which inorganic compounds, such as hydrogen sulfide or diatomic hydrogen, make energy available for the synthesis of organic molecules when combined with oxygen (in the process of *oxidation*). In chemosynthesis, carbon is extracted (usually from carbon dioxide or methane) and converted into organic materials. Energy released in chemosynthesis

then becomes the fundamental power source for the autotrophs in the ecosystem. Amazingly productive and diverse ecosystems have been discovered at hydrothermal vents on the deep ocean floor. These rely on energy from the oxidation of hydrogen sulfides in chemosynthesis and support eerie life forms, such as tube worms. It has been proposed that any life discovered on other planets may rely on chemosynthesis to convert energy to usable forms. Organisms that live in these extreme environments, for example scalding hydrothermal vents at the very high pressures of the deep seafloor, are called *extremophiles*. The suffix "phile" means an affinity for

something, in this case for extreme conditions. It is an antonym to the suffix "phobe," which refers to an aversion to something.

17.05.t1

Before You Leave This Page

☑ Sketch and describe how energy and chemical substances flow through ecosystems in the context of the First and Second Laws of Thermodynamics.

☑ Describe how photosynthesis works, explaining the differences between C_3 and C_4 species.

☑ Briefly describe chemosynthesis.

17.6 How Do We Describe Ecosystem Productivity?

ENERGY DRIVES ALL BIOTIC PROCESSES and is therefore vital for the health of an ecosystem. Energy input, in the form of insolation, is measured in watts per square meter or in calories. Autotrophs convert this energy into organic molecules, but efficiency cannot be 100%. In examining the flow of energy and matter up through the food chain, we begin by quantifying the energy initially fixed by the autotrophs.

A What Is Primary Productivity?

To compare the relative productivity of ecosystems, we calculate a variable known as the *net primary productivity* (NPP). This is the total amount of energy fixed for photosynthesis, which is the *gross primary productivity*, minus the energy wasted or used for plant respiration and maintaining existing tissue. Net primary productivity represents the plant growth rate and is expressed in grams of carbon in new growth (or *biomass*) per m² of surface area per year (g/m²/yr). The map below shows the average annual distribution of NPP on land, as estimated from satellite data of vegetation greenness and solar radiation absorption. Observe this map and try to explain the larger patterns.

1. On land, differences in NPP can be estimated using a factor called the *Normalized Difference Vegetation Index* (NDVI). The index and associated NPP show the amount of trees, shrubbery, grasses, and other plants. On this map, darker greens indicate higher NPP, whereas brown and white indicate much lower NPP. Gray indicates essentially no NPP, as in the ice-covered land and some sand-covered deserts.

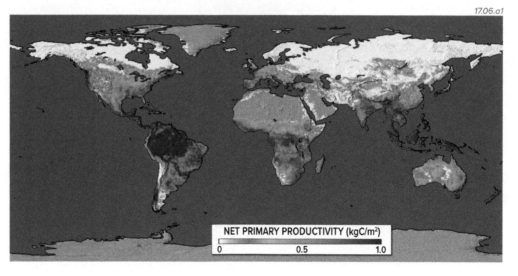

17.06.a1

NET PRIMARY PRODUCTIVITY (kgC/m²)
0   0.5   1.0

2. The lowest NPP is at high latitudes. This is mostly due to the limited and inconsistent amounts of insolation and the low humidity of cold air. Most deserts also have low NPP.

3. The highest NPP is in tropical rain forests, such as those in maritime Southeast Asia, central Africa, and the Amazon basin of South America.

4. Land and sea have such different characteristics that we use different measures of NPP for each. In the oceans, we can use satellites to measure the concentration of chlorophyll, which is a measure of NPP. Chlorophyll is produced mostly by *phytoplankton*, which are microscopic photosynthesizing organisms, whose name is derived from the Greek words for "wandering plant." For the oceans, red and orange indicate the highest chlorophyll (NPP), purple and blue indicate the lowest NPP, and yellow and green represent intermediate values.

6. The highest chlorophyll concentrations, which indicate a high abundance of phytoplankton and high amounts of productivity, are along the edges of the continents and in shallow seas, such as the Baltic Sea in northern Europe.

7. In the open ocean, some of the highest chlorophyll concentrations are in the mid- and high-latitude oceans, such as between Europe and North America, adjacent to the Arctic Ocean, and east of Japan.

5. Observe the patterns on this map and identify which settings in the ocean generally have the highest and lowest NPP. Try to explain these patterns, using such factors as latitude, proximity to land, ocean currents, and areas of upwelling (all presented elsewhere in this book).

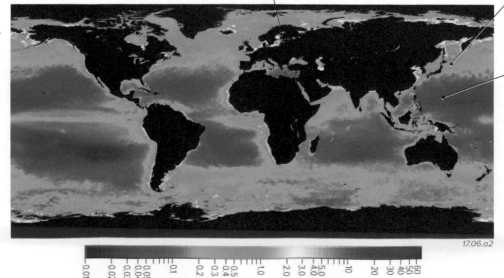

17.06.a2

CHLOROPHYLL A CONCENTRATION (mg/m³)

0.01 0.02 0.03 0.04 0.05 0.1 0.2 0.3 0.4 0.5 1.0 2.0 3.0 5.0 10 20 30 40 50 60

8. Much of the ocean is relatively unproductive (blue and purple), particularly in areas beneath the subtropical highs. Values are only slightly higher along the equator.

9. Locally high productivity occurs in areas where upwelling brings nutrients into shallow-enough levels for photosynthesis. One obvious area of upwelling is along the west coast of Africa.

B What Factors Control Primary Productivity?

The primary productivity of an area depends on many factors, some that operate at global or regional scales and others that result from local conditions. We present a few factors below, but there are many more than are presented here.

Global and Regional Influences

1. These globes present three important factors in productivity (from left to right): (1) a measure of the length of the growing season, called *growing degree days*, which conveys the number of days that were warm enough for plant growth; (2) the average amount of precipitation, and (3) the soil conditions, here expressed as soil pH, where red is acidic and purple is alkaline. Compare these three maps with the map that shows NPP on land (on the left page). What patterns do you observe?

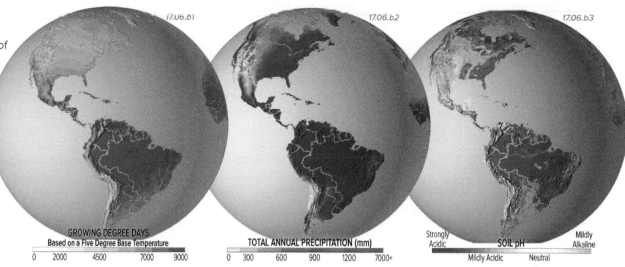

GROWING DEGREE DAYS
Based on a Five Degree Base Temperature
0   2000   4500   7000   9000

TOTAL ANNUAL PRECIPITATION (mm)
0   300   600   900   1200   7000+

Strongly Acidic   SOIL pH   Mildly Alkaline
Mildly Acidic   Neutral

2. Using just these three examples, you can gauge how productivity includes complex interactions between many factors that vary spatially, requiring a geographic approach.

Local Influences

The figure below illustrates some of the local factors that affect NPP.

4. Intensity of insolation depends on Sun angle and cloud cover. Low Sun angles and more clouds mean less insolation and therefore less NPP.

5. Air and surface temperatures can trigger germination, cause plants to lose too much water, and wilt. They also influence leaf size.

7. Shading by higher trees limits the amount of sunlight for plants closer to the ground, affecting their growth.

8. Other factors include wind, the steepness of slopes, which influences water retention and soil formation, and the abundance of plant eaters.

6. Soil is required by most plants, and plants grow better if there is a good supply of soil nutrients and moisture.

3. Light-use efficiency is the energy fixed during photosynthesis as a percentage of total available energy in insolation. It is measured as the ratio of calories of harvested vegetation to total solar energy intercepted. Even in the most productive ecosystems, less than 10% of the available insolation is converted successfully (▼). Note in the table below how inefficient is the use of sunlight, particularly in the central and northern parts of North America.

June – September Light-Use Efficiency (%)	
Cornfield near Champaign, Illinois	2.2
Tall-grass prairie near Manhattan, Kansas	1.7
Mid-latitude forest near Athol, Massachusetts	1.8
Boreal forest in north-central Manitoba, Canada	1.0

9. We can calculate the ratio of energy demand by the human population (Human Appropriation of NPP or HANPP) to total NPP. When the ratio, expressed as a percentage, exceeds 100%, the local ecosystem cannot support its human population. Such areas are shown here in orange and purple. How are such populations sustained?

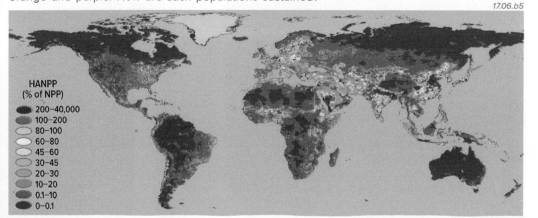

HANPP
(% of NPP)
- 200–40,000
- 100–200
- 80–100
- 60–80
- 45–60
- 30–45
- 20–30
- 10–20
- 0.1–10
- 0–0.1

Before You Leave This Page

✓ Explain the concept of net primary productivity.

✓ Describe the spatial variations of NPP, both globally and locally, and explain why those variations exist.

✓ Identify the relationship between human population and capacity of the local natural environment to support those populations.

17.6

17.7 How Do Ecosystems React to Disturbance?

DISTURBANCES SUCH AS FIRE, storms, or human interference are common to most ecosystems. Following the disturbance, an ecosystem may remain largely unaffected, try to rebound from the disturbance, or start over. As an ecosystem recovers from a disturbance or tries to colonize new land, a succession of organisms may populate the area. While it is tempting to think of disturbance as endangering or destabilizing an ecosystem, disturbances can promote the stability of some ecosystems.

A How Does Disturbance Affect Ecosystem Stability?

Ecosystem stability refers to the degree of fluctuation in population of species over time. *Buffering mechanisms* help species avoid stress and thrive. For example, animals may hibernate during times of the year when food is scarce, or plants may go dormant in the winter or evolve to tolerate a wider range of conditions. When these buffering mechanisms allow populations of all species within the ecosystem to stabilize, the ecosystem is said to be in balance, or *stable*.

1. This graph shows how the populations of three species, in three different ecosystems, change over time. The top curve represents the population of an abundant species. The number of individuals varies only slightly over time, so the ecosystem, as measured by this species, is *stable*. This species and its ecosystem can withstand environmental stresses without experiencing drastic changes in populations.

2. The bottom curve depicts a species with fewer individuals. The population of this species is also stable.

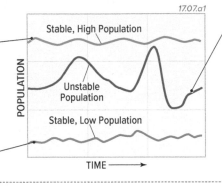
17.07.a1

3. The middle curve shows a species whose population varies widely with time, including some abrupt increases and decreases. This ecosystem is *unstable* and vulnerable to a population crash due to some event or change in conditions. This generally occurs in ecosystems that are considered to be less mature.

4. Some ecologists argue that energy constraints prohibit the stabilization of ecosystems, such as when there is not enough energy to withstand the addition of a new predator species.

5. Organisms generally have a limited range of conditions (▶), such as temperature and moisture, in which they can survive. The levels at which these conditions become too high or low for that organism to survive are known as the *lethal limits*.

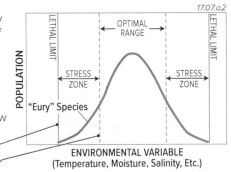
17.07.a2

ENVIRONMENTAL VARIABLE
(Temperature, Moisture, Salinity, Etc.)

6. Within the lethal limits exists an *optimal range* of conditions under which the organism tends to thrive, resulting in a relatively large population. If conditions fall outside the optimal range but inside the lethal limits (the "stress zone"), the organism can survive but does not prosper, so the population will be low under these conditions.

7. Some species have wider optimal ranges and stress zones than others (▶). An organism that can tolerate a wide range of conditions for a given abiotic factor (e.g., temperature) is described with the prefix of "eury-" before a root word that refers to the factor, as in *euryhaline* for a creature that can tolerate a wide range of salinities. On the curve to the right, a "eury" species would be expressed by a broad curve of tolerances, represented by the green curve.

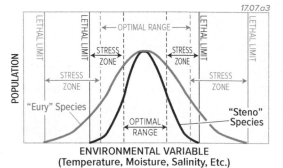
17.07.a3

ENVIRONMENTAL VARIABLE
(Temperature, Moisture, Salinity, Etc.)

8. In contrast, some species can only tolerate a narrow range of conditions for a given environmental factor, as represented by the black curve (▲). They are given a prefix of "steno," as in a *stenohaline* species that cannot tolerate changes in salinity or *stenothermal* for a species with a narrow range of tolerable temperatures.

Inertia and Resilience

9. There are two other important components of ecosystem stability. One is the ability of an ecosystem to resist change after a stress is applied—*inertia*. If an ecosystem is relatively stable, it may not respond or may respond slowly to an imposed change, like a decrease in rainfall at the start of a drought. It has inertia.

10. The other is *resilience*, the rate at which an ecosystem recovers following stress. For example, the area around Mount St. Helens was devastated by a volcanic eruption in 1980. Since that time, the vegetation has recovered quite rapidly (▶).

17.07.a4 Mount St. Helens, WA

11. Over the last 200 years, many scientists have viewed the natural world in the context of *uniformitarianism*—the concept that the natural world has evolved in the past by processes and rates similar to those we observe today. At the opposite end from uniformitarianism is *catastrophism*, which emphasizes the importance of sudden events in creating change. The eruption of Mount St. Helens was a perfect example of an abrupt catastrophe to the ecosystem, but the recovery of the ecosystem is an example of uniformitarianism.

B What Is the Process of Succession?

Disturbances tend to cause great fluctuations in populations, as stresses affect species in different ways. The ecosystem typically responds with a predictable change in species composition, called *succession*. If the successional process begins with the formation of new land or with so much disturbance that little life is able to survive the disturbance (as in glaciations), the recovery process is a *primary succession* and begins with a *pioneer community*. *Secondary succession* occurs when soil is present in the landscape from the outset, as occurs in an area affected by a forest fire—the vegetation is destroyed but the soil is largely unaffected. Succession can lead to a stable community.

1. This figure shows the stages of recovery in a primary succession with early stages on the left and later ones on the right. The underlying graph shows a typical sequence of species replacement as the ecosystem evolves.

3. With time, K-selected species, such as tree species, become possible and begin to dominate the ecosystem.

4. The final stable community, called a *climax community*, develops and includes an intricate food web and efficient nutrient cycling, with the diversity and productivity determined by the abundance or lack of necessary abiotic factors, specifically sunlight, water, and nutrients.

2. A pioneer community contains mostly r-selected species, including mosses and lichens.

5. The concept of *patch dynamics* refers to the idea that subareas within an ecosystem may be influenced by different processes and are at different successional stages at the same time. Over time, a climax community may be disturbed and then reestablish a new climax community.

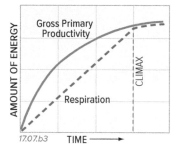

Lichen   Mosses   Grasses   Shrubs   Trees

17.07.b1

6. Biodiversity increases in the early successional stages, but eventually it decreases as nutrient cycling becomes more efficient and competition weeds out less robust species. Different species have differing abilities to deal with disturbance, even in the same ecosystem.

8. Before the establishment of a climax community, more energy is fixed by autotrophs than is consumed by respiration in the ecosystem.

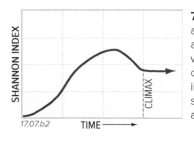

SHANNON INDEX

CLIMAX

17.07.b2   TIME ⟶

7. The resulting distribution of species (◀) after this steady-state "endpoint" is achieved (some time after the ecosystem was disturbed) represents the climax community. K-selected species become increasingly favored, and r-selected species are at an increasing disadvantage at greater times after the disturbance.

AMOUNT OF ENERGY

Gross Primary Productivity

CLIMAX

Respiration

17.07.b3   TIME ⟶

9. At the time when the climax community is established, the total energy fixed (gross primary productivity) will nearly balance energy expended in respiration.

10. After the climax community is established, nearly all of the energy available is used in respiration to break down food.

C How Does Disturbance Affect Biogeochemical Cycles?

An important aspect of how ecosystems become established, evolve, and die is the concept of a *biogeochemical cycle*. A biogeochemical cycle is a way to conceptualize chemical connections between various components of an ecosystem—land, groundwater and surface water, oceans, ice, air, and organisms.

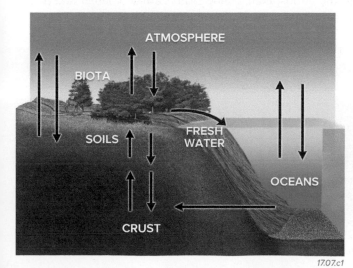

ATMOSPHERE

BIOTA

SOILS

FRESH WATER

OCEANS

CRUST

17.07.c1

This figure illustrates the main aspects of a biogeochemical cycle. Arrows indicate how some chemical substance, for example carbon, flows between various components of an ecosystem (a flux), including between the soils, biota, atmosphere, and oceans. Typically, the amounts of that chemical substance are listed next to each arrow to convey the magnitude of the flux and how some change in the ecosystem could affect the ecological balance. Ecosystems store and cycle nutrients, so any changes in that ecosystem will be felt in the biogeochemical cycles involving them. This, in turn, can affect other ecosystems. The following pages have biogeochemical cycles for carbon, nitrogen, phosphorus, and sulfur, but others exist for many other chemical elements and compounds.

Before You Leave This Page

✓ Explain the concept of ecosystem stability.

✓ Give examples of primary and secondary succession resulting from a disturbance of the land.

✓ Describe a biogeochemical cycle.

17.7

17.8 What Is the Role of Carbon in Ecosystems?

OF THE 92 STABLE ELEMENTS, only 17 are essential for all plants, and a few more are vital for animals. The nutrients formed by these elements must be recycled continuously through all of the "spheres" of the Earth to sustain life—forming biogeochemical cycles. Despite its very minute abundance on Earth, carbon (C) is the element that forms the building block for life. Organic molecules contain the carbon-hydrogen bond and include carbohydrates, lipids, proteins, and nucleic acids, all found in the many life-forms in ecosystems.

A What Processes Are Involved in Carbon Storage and Cycling?

1. This figure depicts the global carbon cycle as a series of arrows that show the direction in which carbon moves between components of the environment. The number with each arrow indicates how much carbon moves along that pathway (the flux), and numbers next to the main storage sites (stores) of carbon (e.g., oceans) indicate how much carbon is in that reservoir. In both cases, the amounts are in gigatons (10^{15} gm or a quadrillion grams). Some of these values are not known precisely, so they are educated guesses.

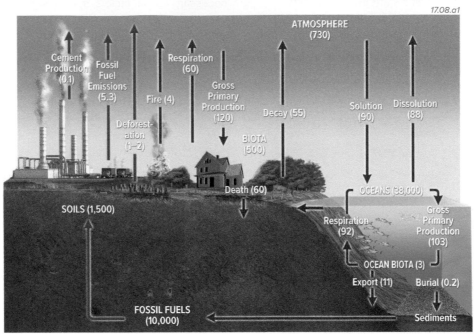

17.08.a1

5. The C emitted by human activities (red arrows) to the atmosphere, while small, causes imbalance in the system. Of this anthropogenic CO_2, some is cycled back to the surface by enhanced photosynthesis by plants, some remains in the atmosphere, and some goes into the ocean.

4. The C going into and out of the oceans would balance naturally, but two units of the extra units produced by people are absorbed by the ocean, causing acidification. Over long timescales, this extra carbon is cycled to the ocean floor as sediments.

2. Over time, Earth's atmospheric CO_2 was moved to the lithosphere as carbonate rocks formed over geologic time in a cycle so slow that it is considered to be in "storage" rather than in a cycle. So the amount stored in the lithosphere is much greater than that stored in the atmosphere. When considering the flux of carbon, we mostly focus on the amount in fossil fuels, not the entire lithosphere.

3. Notice that the exchange of carbon between two reservoirs is essentially balanced. For example, 120 gigatons goes from the atmosphere to land in the form of gross primary production (photosynthesis) and almost the same amount goes from the land to the atmosphere.

B How Do Humans Impact and Become Impacted by the Carbon Cycle?

The production of greenhouse gases through fossil fuel combustion is the most obvious way that humans impact the C cycle, but deforestation and other land-use changes also add carbon to the atmosphere. Aside from climate change and its associated impacts, elevated atmospheric CO_2 also increases weathering and decay rates by forming carbonic acid.

17.08.b1

17.08.b2 Central British Columbia, Canada

17.08.b3 Washington

17.08.b4

1. Fossil fuel combustion for industrial, transportation, and domestic uses amounts to about 95% of the C that people emit to the atmosphere.

2. Most of the other 5% comes from cement production, which grinds up and processes carbonate rocks (limestone and marble), releasing C as a by-product.

3. Deforestation reduces the amount of plants, which in turn reduces the amount of carbon that is extracted from the atmosphere. Changes in the rate of deforestation can affect the carbon balance.

4. Crops remove C from the atmosphere, but the problem is that they are usually only seasonal. They often replace forests that would have removed much more atmospheric carbon.

C How Does the Carbon Cycle Regulate Earth's Long-Term Temperature?

The massive quantities of stored C may change on timescales of millions of years through feedbacks tied to the global water and rock cycles. As atmospheric CO_2 concentration changes, so does the global climate.

1. Falling rain or snow moves CO_2 from the atmosphere to the lithosphere after it reacts with the precipitation to form a weak acid called carbonic acid (H_2CO_3). This loss of CO_2, a greenhouse gas, can cool the atmosphere.

7. By this process, some of the stored carbon gradually moves back to the atmosphere, which then slowly intensifies the greenhouse effect and warms the Earth again.

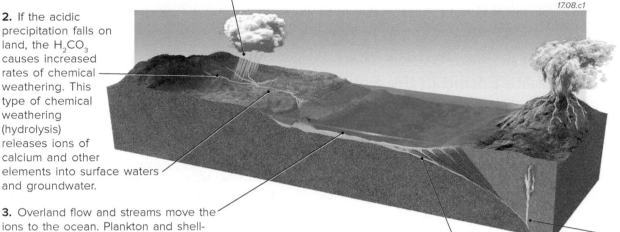

17.08.c1

2. If the acidic precipitation falls on land, the H_2CO_3 causes increased rates of chemical weathering. This type of chemical weathering (hydrolysis) releases ions of calcium and other elements into surface waters and groundwater.

6. CO_2 dissolved in magma rises with the magma toward the surface. As the magma nears the surface, the reduced pressures allow the CO_2 to escape, either as a slow outgassing or in a quick release during an explosive volcanic eruption. In either case, the CO_2 goes back into the atmosphere.

3. Overland flow and streams move the ions to the ocean. Plankton and shell-building organisms, as they grow, convert the calcium ions to calcium carbonate ($CaCO_3$). When they die, the $CaCO_3$ falls to the ocean floor to become limestone, chalk, or other types of seafloor sediments.

4. Plate tectonics moves limestone and its carbonates very slowly over millions of years to subduction zones. Most of the sediment is plastered against the overriding plate, effectively taking the enclosed carbon out of the cycle for a while.

5. Some of the oceanic sediments are taken to depth and heated, liberating CO_2 from the hot rock. Some CO_2 rises toward the surface, mostly in dissolved magma.

---

8. How effective the system is—and whether it leads to warming or cooling overall—is affected by several factors. If most limestone and other carbonate-bearing rocks end up being scraped off the subducting plate, then this carbon is removed from the system by the combined action of the hydrosphere (rivers, groundwater, and oceans), biosphere (microorganism and shell builders), and lithosphere.

9. Periods in geologic history with abundant mountain building may offer more opportunity for chemical weathering, which stores extra carbon in the lithosphere. This would limit atmospheric CO_2 and keep Earth cool, but these natural changes are extremely slow. In the last 50 million years, the buildup of the Himalaya, due to plate tectonics, is linked to a global temperature drop that has occurred since very warm time periods between 50 and 120 million years ago. During that time, global temperature and sea level were much higher than today.

D How May the Carbon Cycle Be Impacted at Shorter Timescales?

There are various schemes considered to affect the carbon cycle, especially because of concerns about the climatic effects of CO_2. Three are most discussed.

Process 2: Ocean Fertilization—Scattering iron on the ocean surface may promote phytoplankton growth, which would increase the efficiency of Step 3 above. Would this be a logical and safe means of removing atmospheric CO_2 to combat global warming? Why or why not?

17.08.d1 Bushveld, South Africa

Process 1: Mineral Carbonation—Ultramafic igneous rocks (those with very high magnesium and iron, and low silica content), like the ones shown here, absorb atmospheric CO_2 efficiently during mineral formation. Humans may be able to accelerate this slow natural process either by heating the rock or by injecting it with water rich in CO_2. In the U.S., ultramafic rocks are most common along the West Coast and near the Appalachian Mountains, near major population centers. Might this spatial distribution, which resulted from a complex sequence of geologic events unrelated to humans, be helpful?

Process 3: Invigorated Vegetation—Higher temperatures may increase the rate of carbon uptake by plants under longer growing seasons with more favorable conditions for photosynthesis. The increased rates of decay in a warmer world may liberate nutrients from dead organisms quickly to allow for increased carbon fixation by plants. Would this be a possible strategy for decreasing the amount of CO_2 in the atmosphere?

17.08.d2 Angkor Wat, Cambodia

Before You Leave This Page

☑ Sketch, label, and explain the carbon cycle, identifying the main carbon reservoirs and fluxes, mentioning some possible human impacts.

☑ Sketch and explain how CO_2 regulates global temperatures.

☑ Describe some possible ways to decrease atmospheric CO_2.

17.8

What Is the Role of Nitrogen in Ecosystems?

NITROGEN (N) IS ANOTHER ESSENTIAL NUTRIENT. The largest available stores for diatomic nitrogen (N_2) are the atmosphere and hydrosphere, rather than the biosphere or lithosphere. Thus, the nitrogen cycle, like the carbon cycle, is primarily a gaseous cycle. The nitrogen cycle differs from the carbon cycle in that it is involved in neither photosynthesis nor respiration. For humans, we inhale and exhale nitrogen without really using it in this form. Yet nitrogen is essential for life because it is fundamental to cell metabolism and is in amino acids and nucleic acids (DNA and RNA). Nitrogen is also the limiting nutrient in many ecosystems. A limiting nutrient prevents ecosystem health even if all other biotic and abiotic components are adequate.

A What Forms of Nitrogen Are Found in the Environment?

Nitrogen occurs in many forms in the biosphere; these are summarized in the table below. Atmospheric N_2 is inert and is not useful to plants and animals. Nitrogen is the essential nutrient in the shortest supply in many ecosystems, so it is the resource that limits growth at cellular to global scales. Most fertilizers therefore include nitrogen.

	Form of Nitrogen		Properties and Uses
	N	Nitrogen	chemical element, common component of Earth and other planetary atmospheres
	N_2	Nitrogen gas	78% of atmospheric mass; necessary for biological processes, but unusable by plants
	NO	Nitric oxide	by-product of combustion; oxidizes readily to form NO_2
NO_x	NO_2	Nitrogen dioxide	brown, pungent odor, toxic gas; prevents growth of some bacteria in large concentrations
	NO_3	Nitrate	natural rocks; explosives; fertilizers when combined with positive ions such as potassium, sodium, and calcium
	N_2O	Nitrous oxide	greenhouse gas; anesthetic (laughing gas)
	NH_3	Ammonia	pungent odor; cleaning products; pharmaceuticals; fertilizers; neutralizes acids including in the kidneys
	HNO_3	Nitric acid	highly corrosive; fertilizers; rocket fuel, acid rain forms from NO_2 reacting with hydroxide (OH^-)

B How Does the Global Nitrogen Cycle Operate?

1. This figure depicts the nitrogen cycle as a series of arrows showing the fluxes of nitrogen in all of its forms between various reservoirs (stores). As with the carbon cycle, the number with each arrow indicates how much nitrogen moves along that pathway (the flux), and numbers next to the main storage sites (stores) of nitrogen (e.g., atmosphere) indicate how much nitrogen is in that reservoir. The amounts are in megatons (10^{12} gm), three orders of magnitude (1,000 times) less than in the equivalent diagram for carbon.

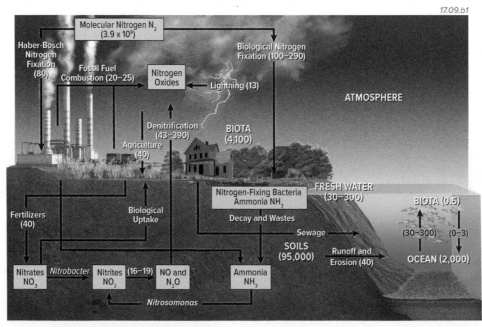

5. The Haber-Bosch industrial process extracts nitrogen from the atmosphere and combines it with hydrogen gas to produce ammonia and other nitrogen compounds that each of us uses, in one way or another, every day (e.g. fertilized crops).

4. Note that there is little direct exchange of nitrogen between the atmosphere and oceans, in part because the two reservoirs are essentially in equilibrium. Nitrogen does enter the oceans, but does so mostly through the flow of streams, including those that are draining agricultural areas where nitrogen-based fertilizers are used.

2. The N cycle consists of natural and anthropogenic components, and some stores (reservoirs) between which nitrogen is exchanged. The largest reservoir is the atmosphere, with almost 4 billion megatons. Recall that nitrogen gas (N_2) makes up 78% of the atmosphere. The next largest reservoir is in soils, followed by the ocean.

3. Nitrogen moves from the atmosphere to the land mostly through plants and nitrogen-fixing bacteria. Nitrogen going the other direction is released into the atmosphere by the *denitrification* process, where microbes convert other forms of nitrogen into nitrogen gas, and by combustion of fossil fuels.

C How Does Nitrogen Act as a Water Pollutant?

Since about 1990, humans have been creating more NO_x than have natural processes. The accumulation of NO_x in the environment has been an increasing problem. However, NO_x is needed to supply the world's population with food. This creates the dilemma of how to produce more food while minimizing the addition of NO_x to the lithosphere and hydrosphere. Most of the "extra" nitrogen ends up as water pollution.

1. Humans add nitrogen to the environment in various ways, both directly and indirectly. Production of fertilizer converts atmospheric N_2 to other forms (almost as much as natural biological processes) through the Haber-Bosch process. Runoff from fertilized agricultural areas adds nitrogen to aquatic systems (where it is not needed as much as on land). This causes too much aquatic plant growth, which uses up oxygen and nutrients and encourages population explosions among second-degree heterotrophs, stressing and destabilizing ecosystems with low inertia. We also add nitrogen to the environment through fossil-fuel combustion, animal feedlots, and sewage.

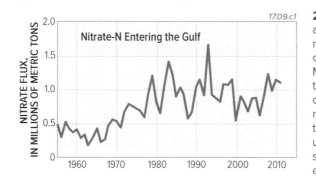

2. This graph shows the annual flow, or flux, of nitrogen entering the Gulf of Mexico through the Mississippi River and its tributaries. Observe the overall increase in nitrogen as a function of time, as well as the yearly ups and downs related to short-term climatic and economic factors.

D How Does Nitrogen Act as an Air Pollutant?

Humans also release nitrogen to the atmosphere. Atmospheric NO_x (nitric oxide [NO], NO_2, and NO_3) is such a widespread concern that it appears on the U.S. Environmental Protection Agency's list of criteria pollutants (pollutants of most concern).

GLOBAL NO_2 CONCENTRATION AVERAGED OVER 2014

NO₂ CONCENTRATION
0 1 2 3 4 5
× 10¹⁵ molecules/cm²

2005

3. NO_x reacts with ammonia (NH_3) in the formation and intensification of particulate matter, which is linked to haze, decreasing visibility, and respiratory ailments. NO_x also combines with water to form HNO_3, a very strong acid and a component, along with sulfuric acid, of acid rain. The chemical reaction that results in nitrogen-related acid rain is summarized in the chemical reaction below:

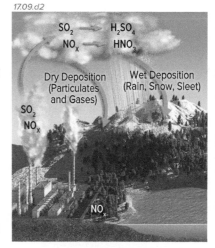

1. Atmospheric NO_x is monitored by satellite and other methods in the U.S. and other parts of the world (▲). As you can observe from these maps, high concentrations of NO_x are mostly associated with industrial areas. NO_x is released primarily by power plants and automobiles, and it contributes to the danger posed by other pollutants, including *tropospheric ozone*, tiny aerosols known as particulate matter, and acid rain.

2. NO_x is a major pollutant that is monitored by the EPA in the U.S. Greater effort to limit NO_x production at the state level has been undertaken, with wide disparity in the degree of success at controlling it.

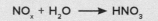

$$NO_x + H_2O \longrightarrow HNO_3$$

4. Nitrous oxide (N_2O) is a greenhouse gas. An increase in N_2O in the atmosphere has potential environmental and human impacts, including melting of glaciers, rising sea levels, and stressing of ecosystems.

5. Tropospheric ozone (O_3) is a product of NO_2 (▶). Unlike stratospheric ozone that protects us from harmful ultraviolet radiation, tropospheric O_3 has undesirable impacts, including causing respiratory issues. One way that it is produced is when sunlight triggers the breakdown of NO_2 (*photodissociation*) into NO and O.

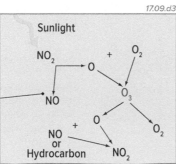

6. The unstable O bonds with O_2 to form O_3, which itself is a pollutant near the ground. NO_x reacts with sunlight and volatile organic compounds (VOC) to produce ground-level ozone, an air pollutant.

Before You Leave This Page

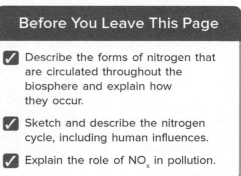

✓ Describe the forms of nitrogen that are circulated throughout the biosphere and explain how they occur.

✓ Sketch and describe the nitrogen cycle, including human influences.

✓ Explain the role of NO_x in pollution.

17.9

17.10 What Is the Role of Phosphorus in Ecosystems?

PHOSPHORUS (P) IS AN ESSENTIAL NUTRIENT for every living plant and animal cell, but in nature it occurs as inorganic phosphate (PO_4^{-2}). Unlike carbon and nitrogen, the phosphorus cycle occurs largely in the lithosphere, so it is termed a sedimentary biogeochemical cycle. As little as 1% of the phosphorus in the crust is available to plants. To be economically feasible to mine, a rock must contain at least 8% phosphate.

A How Does the Phosphorus Cycle Operate?

1. This schematic depicts the global phosphorus cycle. Storages are shown in metric tons of phosphorus, and fluxes are in metric tons per year, with red arrows depicting the fluxes primarily caused by human activity. Note how much smaller the fluxes are for phosphorus (metric tons) compared to those of nitrogen and carbon (megatons or gigatons), but the reservoir represented by ocean-floor sediments is substantial.

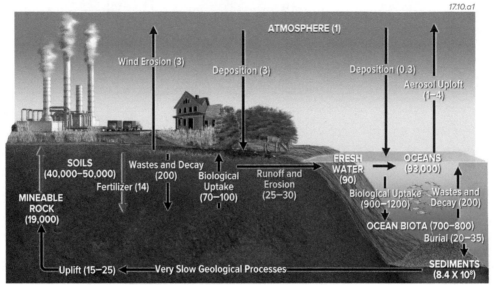

17.10.a1

5. Notice the small storages of phosphorus in the atmosphere and the very small fluxes in and out of the atmosphere, unlike the C and N cycles.

4. Locally significant phosphate deposits occur at sites with large populations of birds and bats. These deposits represent the accumulation of excrement (called *guano*) from these airborne creatures. If large enough, these deposits can be mined as a source of phosphate, as occurs along the western coast of South America.

2. Various parts of the phosphorus (P) cycle occur on different timescales, as occurred with the carbon cycle. The part that occurs near Earth's surface is the relatively fast part of the cycle. Some plants, like dandelions, have roots that are exceptionally adept at attracting nutrients (like P) that have leached downward below the root zone. When these "hyperaccumulator" plants die or are consumed, the P is available for other organisms.

3. The deep replenishment of phosphorus via the rock cycle requires millions of years, but it contains by far the most phosphorus, with an estimated one billion metric tons. Finding an economically feasible way to retrieve such phosphorus would solve the problem of availability.

B Where Are Major Phosphate Deposits Located?

17.10.b1 Florida

1. Phosphate is a general term for rocks and chemical substances that have phosphorus. Plants generally use the phosphate ions in building ATP, DNA, and proteins. Phosphate occurs in recent marine deposits in Florida (◄) and elsewhere. Teeth of various creatures contain a phosphate mineral, so accumulations of these materials can occur when these organisms die, and they represent an important resource of phosphate.

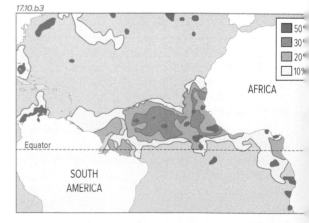

17.10.b3

2. Phosphate occurs in many parts of the world (►). In addition to guano and recent marine layers, phosphate occurs in older sedimentary units, including some important ones in Idaho and adjacent parts of Wyoming and Utah.

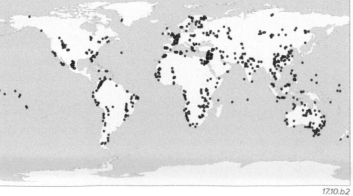

17.10.b2

3. Atmospheric transport of phosphorus can enrich water thousands of miles from the point of origin. The map above shows the percentage of insolation that is scattered and absorbed by dust particles, some of which contain phosphorus. In this region, the trade winds carry phosphorus all the way from Africa to the coast of South America.

C   What Role Does Phosphorus Play in Biological Processes?

Phosphorus is one of the most important elements for life. It is essential to the cell structure of many creatures and plants, and it occurs in teeth. Here, we highlight a few important and interesting aspects of the role of phosphorus in biology.

1. Part of the value of phosphorus (P) is its highly reactive nature, which allows it to bond with many other atoms and molecules in forming essential compounds (▶). For example, phosphorus forms part of adenosine triphosphate (ATP), which stores the energy acquired in the "light reaction" phase of photosynthesis. This same substance is involved in food consumption in humans and other creatures.

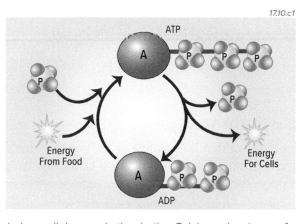

17.10.c1

3. When phosphorus is added to or removed from proteins, various types of cellular functions are regulated to optimal levels for the health of the organism.

4. *Deoxyribonucleic acid* (DNA) contains the genetic code of every cell (◀). Without phosphorus, DNA, the building block of most life, would not have its characteristic double helix structure, or even exist, for that matter.

Sugar-Phosphate Backbone

17.10.c2

2. ATP also releases that energy during cellular respiration in the Calvin cycle phase of photosynthesis, liberating it as fuel for other biological processes. At that point, ATP becomes adenosine diphosphate (ADP) or the even lower-energy adenosine monophosphate (AMP). When additional phosphorus is present, AMP can become ADP or ATP, increasing the amount of energy available to the organism.

D   What Happens When Too Much Phosphorus Is Added Locally?

Local overabundance of phosphorus can be a problem in some aquatic ecosystems. If phosphates have been added to a water body, the water can undergo the process of *eutrophication*, which is the response of an aquatic ecosystem to the addition of a foreign substance or the addition of too much of a substance that is normally present but not in such high concentrations. Water bodies with low nutrient concentrations and low net primary productivity (NPP) are termed *oligotrophic*, while those with high nutrient concentrations and higher NPP are *eutrophic*.

The figure below illustrates the processes that occur in a water body in which phosphorus has been added, causing eutrophication. Typically, succession in aquatic ecosystems causes eutrophication as a natural process over time. Humans, however, have accelerated the eutrophication process by adding P and N primarily through runoff from fertilized fields and golf courses, leaky septic systems, and municipal sewage systems. This can have serious consequences, producing algal blooms that then reduce the amount of oxygen upon which other organisms depend.

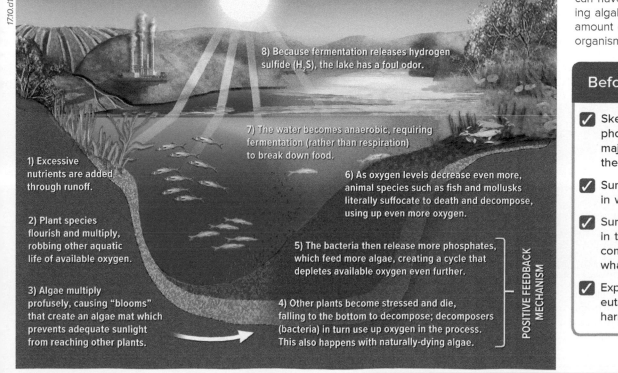

17.10.d1

8) Because fermentation releases hydrogen sulfide (H_2S), the lake has a foul odor.

7) The water becomes anaerobic, requiring fermentation (rather than respiration) to break down food.

1) Excessive nutrients are added through runoff.

2) Plant species flourish and multiply, robbing other aquatic life of available oxygen.

3) Algae multiply profusely, causing "blooms" that create an algae mat which prevents adequate sunlight from reaching other plants.

6) As oxygen levels decrease even more, animal species such as fish and mollusks literally suffocate to death and decompose, using up even more oxygen.

5) The bacteria then release more phosphates, which feed more algae, creating a cycle that depletes available oxygen even further.

4) Other plants become stressed and die, falling to the bottom to decompose; decomposers (bacteria) in turn use up oxygen in the process. This also happens with naturally-dying algae.

POSITIVE FEEDBACK MECHANISM

Before You Leave This Page

☑ Sketch and explain the phosphorus cycle, identifying the major reservoirs and explaining the major fluxes.

☑ Summarize the types of deposits in which phosphate occurs.

☑ Summarize the role of phosphorus in the biosphere, identifying biotic components where it occurs and what its role is.

☑ Explain the process of eutrophication and why it can be harmful to aquatic ecosystems.

17.10

17.11 What Is the Role of Sulfur in the Environment?

SULFUR IS ANOTHER ESSENTIAL NUTRIENT for life. Sulfur (S) is necessary for plant growth, especially for legumes like peas and beans, which return N to the soil as they grow. S is present in the molecules that protect plants and animals from bacterial infections. S is also used as the energy source in *chemosynthesis* to make food for deep-ocean organisms and for bacteria in oxygen-poor environments. The S cycle is not very well understood, largely because of the complicated chemical changes involving different forms of S, such as sulfur dioxide (SO_2), sulfate ions (SO_4^{2-}), and hydrogen sulfide (H_2S).

A How Does the Global Sulfur Cycle Operate?

The sulfur cycle looks similar to the phosphorus cycle in that it is almost entirely sedimentary, involving only minor interaction with the atmosphere. Again, the red arrow indicates the flow linked to the human-produced (anthropogenic) part of the cycle.

1. This figure depicts the global sulfur cycle, identifying the reservoirs of sulfur and the fluxes of sulfur between those reservoirs. The numbers indicate the size of the reservoir or the size of the flux in 10^{12} moles. The largest reservoirs are in the lithosphere, including sediments in the ocean. Large amounts of sulfur also occur in sedimentary rocks that were deposited in the ocean but are now on land and in rocks that have been affected by hot (hydrothermal) waters.

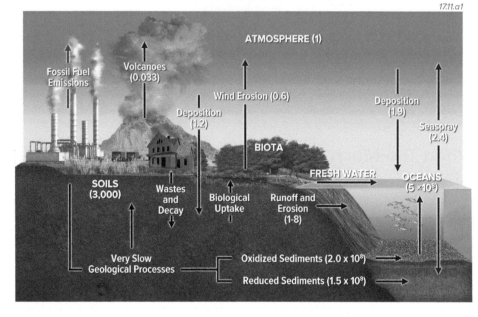

17.11.a1

5. Although the effects of acid precipitation are most obvious outside the oceans, such as damage to land plants, most acid rain actually falls into the oceans, causing damage to aquatic animals.

4. Some sulfur goes from the ocean to atmosphere, and sulfur also goes from the atmosphere to the ocean and to the land, in the form of precipitation. In the atmosphere, sulfur oxides (SO_x) react with water to form sulfuric acid (H_2SO_4), which becomes incorporated into the precipitation, causing *acid rain* (or more generally acid precipitation).

2. As sulfur shifts between reservoirs, it moves from the land to the atmosphere from burning coal and petroleum, both of which contain variable amounts of sulfur. Sulfur is also released from volcanoes, and an increased release of sulfur may indicate that a volcanic eruption is imminent. Sulfur can be picked up by wind.

3. Sulfur is relatively mobile in water, so it enters streams and groundwater when sulfur-bearing rocks are weathered. Streams then carry this sulfur to the ocean. When sediment accumulates on the seafloor, it incorporates significant amounts of sulfur, essentially removing it from circulation through the system for a while.

Sulfur in the Land and Water

17.11.a2 Kidd Creek Mine, Ontario, Canada

17.11.a3 Silverton, CO

17.11.a4 Silverton, CO

6. Sulfur occurs naturally in rocks, including these (▲) that are almost all sulfide minerals. They are exposed underground in a copper mine with a rock hammer for scale.

7. Sulfide minerals are not stable when exposed to atmospheric oxygen in the process of weathering. As a result, they convert to red and orange oxide and sulfate minerals.

8. The reddish rocks in the center photograph (to the left) drain into local streams, causing the water and sediments to be acidic and stained orange by the sulfur and iron.

B | What Impacts Do Humans Have on the Sulfur Cycle?

The human impact on the S cycle is miniscule on a global scale, but it can be significant at the regional and local scales. About half of the atmospheric emissions come from humans, but much more than half is released in industrial areas. SO_x emissions, especially SO_2, are dangerous because they destroy plant hormones, stunt plant growth, reduce plant reproduction rates, and cause respiratory illnesses in humans.

17.11.b1 Fruitland, NM

1. The burning of coal (◀), especially coal that is high in sulfur, is the prime human source of atmospheric S. The sulfur falls back toward the surface in *acid precipitation*, like acid rain. The term *acid deposition* is also used because dust can be coated with acid and then fall to the surface. Later, when moisture comes in contact with this acidified dust, anything in contact with it, including raindrops, will become acidified.

2. Only a few counties in the U.S. (▶) are in noncompliance with the clean-air standard for SO_2, such as areas that have smelters (furnaces in which to heat rocks) that process sulfur-rich rocks. Areas in yellow on this map exceeded 1971 EPA standards, whereas those in red exceeded tougher standards established by the EPA in 2010.

AREAS IN NON-ATTAINMENT FOR SO_2 EMISSIONS

Lewis and Clark, MT
Yellowstone (MT)
Oneida (WI)
Tooele (UT)   Salt Lake (UT)
Warren (PA)
Pinal (AZ)
St. Bernard (LA)

17.11.b2

(June 13, 2016)

■ By 1971 EPA Standards
■ By 2010 EPA Standards

Spatial and Temporal Variations in Human-Generated Sulfur

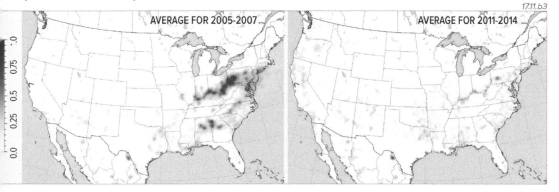

17.11.b3

AVERAGE FOR 2005-2007   AVERAGE FOR 2011-2014

3. These two maps (▲) depict the amount of sulfur emitted in the continental U.S., averaged over two different three-year periods. Examine the left map and consider what might be causing the observed patterns. The largest emissions are in the eastern half of the country, around the industrial centers, including at coal-fired power plants.

4. The decrease in emissions between the first and second maps is due largely to natural gas replacing coal-fired power plants in the generation of electricity.

5. The impacts of acid precipitation and acid deposition are most pronounced where bedrock compositions cannot buffer the acid deposition and where rainfall is orographically concentrated, for example in the Adirondack Mountains of upstate New York and the Appalachian Mountains. The good news is that there has been an improvement in the U.S. over the last 30 years (▼).

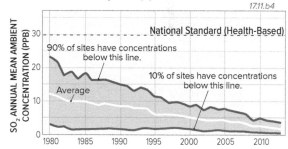

17.11.b4

National Standard (Health-Based)

90% of sites have concentrations below this line.

10% of sites have concentrations below this line.

Average

SO_2 ANNUAL MEAN AMBIENT CONCENTRATION (PPB)

1980 1985 1990 1995 2000 2005 2010

6. This unusual-looking map, called a *cartogram*, shows the size of each country in proportion to the amount of sulfur it releases into the atmosphere. Note that the industrialized countries, which look like big balloons, are responsible for most of the SO_2 emissions.

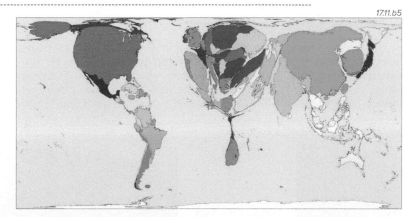

17.11.b5

7. Emissions elsewhere in the world are extremely variable. They are decreasing in many locations, such as in European Union countries, where emissions have fallen by 76% from 1990 to 2009. Emissions are increasing in the developing world, such as in China, which has been building coal-fired power plants at the rate of several per week! Scientists have tracked the increased emissions from China all the way across the Pacific and into the Pacific Northwest.

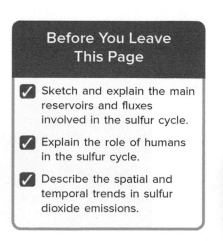

Before You Leave This Page

☑ Sketch and explain the main reservoirs and fluxes involved in the sulfur cycle.

☑ Explain the role of humans in the sulfur cycle.

☑ Describe the spatial and temporal trends in sulfur dioxide emissions.

17.11

17.12 How Does a Lack of Oxygen Harm Ecosystems?

ORGANISMS IN TERRESTRIAL AND AQUATIC ECOSYSTEMS depend on available oxygen (O_2). Oxygen is the second most abundant atmospheric constituent, so it is in constant contact with the upper surfaces of the land and water bodies. As a result, it can become incorporated into soils and surface waters. Aquatic ecosystems are particularly susceptible to having too little O_2 to accommodate their resident organisms. What determines dissolved O_2 content, what causes O_2 shortages, and what are the impacts of having too little O_2 available in an ecosystem?

A Where Is Dissolved Oxygen Most Available in a Lake Ecosystem?

Cold water can absorb more dissolved gas than warm water, because molecules are more energetic at high temperatures and so gas can more easily escape into the atmosphere. All other factors being equal, more O_2 is available to organisms in colder water, including bacteria that decompose organic matter. On a hot afternoon, the amount of O_2 needed by the bacteria—the *biological oxygen demand* (BOD)—is highest. This occurs precisely when dissolved O_2 concentrations are lowest, as described below for a typical lake.

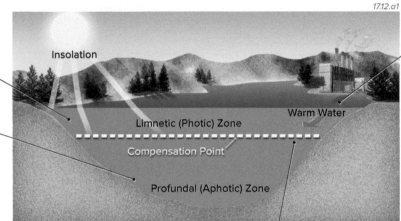

17.12.a1

1. Near the top of the lake where sunlight is available, more O_2 is generated from photosynthesis than is used by respiration or fermentation. The upper zone in which this occurs is called the *limnetic* (or *photic*) *zone*.

2. Near the bottom of the lake, little or no photosynthesis is occurring, so more O_2 is used by respiration or fermentation than is released by photosynthesis. This is the *profundal* (or *aphotic*) *zone*.

3. The level where the production of O_2 through photosynthesis is balanced by the use of O_2 by respiration or fermentation is the *compensation point*. The compensation point (or level) moves downward during the daytime, when photosynthesis is most active, and upward all through the night. It also varies from season to season.

5. Dumping waste hot water from factories into rivers flowing into the lake decreases its ability to hold dissolved oxygen. Also, as temperature increases, the metabolic rate of organisms increases, accelerating the need for more O_2. If O_2 concentrations fall low enough, fish and other aquatic organisms can suffocate.

4. As temperature increases, the metabolic rate for organisms increases, accelerating the need for more O_2. So, a lake ecosystem is most sensitive to organic pollution near the end of summer, after months of intense photosynthesis, and just before dawn, when the limnetic zone is the thinnest.

B How Does Oxygen Availability Affect the Health of a Lake?

If O_2 is lacking in aquatic ecosystems—a condition known as *hypoxia*—the amount and type of circulation within the lake determine the severity of the problem. This circulation is affected by the depth of the thermocline, the boundary between the sun-warmed upper waters and the cold bottom waters. Little circulation occurs across the thermocline.

In extratropical climates, the compensation point is usually above the thermocline in summer. This allows lake circulation to move to the surface any hypoxic water below the compensation point but above the thermocline. This setting restricts circulation to the upper part of the lake, making this scenario the more serious case for the input of organic pollutants.

But at some point in winter, the lake "turns over" suddenly, as the upper layer becomes chilled by cold air masses. The increased density of this chilled surface layer, which is rich in O_2, causes it to flow downward, below the thermocline, and to trade places with the now-warmer, less dense waters below it. Some water with a net O_2 surplus below the thermocline will be diffused to the area with a deficit, minimizing the problem. Above the thermocline, no water has an O_2 deficit. As a result, this is the less serious case for the input of organic pollutants.

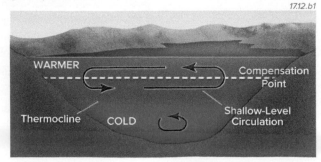

17.12.b1

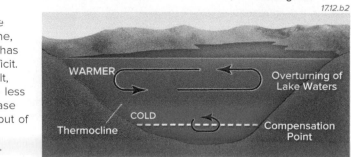

17.12.b2

C How Can Nutrient Inputs Affect Oxygen Availability?

One of the world's most severe hypoxic zones is in the northern Gulf of Mexico, near where the Mississippi River, the largest river in North America, empties into the Gulf. The Mississippi River drains huge areas of the country, from the eastern side of the Rocky Mountains to the Appalachian Mountains on the east. The region between the two mountain ranges includes the agricultural areas of the Great Plains and Ohio River and Mississippi River valleys. Water draining off agricultural fields in these areas can end up in the Mississippi, and ultimately in the Gulf.

1. This map of atmospheric nitrogen deposition reflects the heavy agricultural practices in the Mississippi River Basin. Fertilizers and manure (in addition to fossil-fuel combustion) greatly increase nitrogen emissions, which are redeposited to the ground by wind and rain, adding even more nitrogen to the soil.

2. Runoff from the Mississippi River drainage system carries nutrient-rich agricultural fertilizers and manure into the Mississippi River and out to the Gulf of Mexico.

3. From there, the prevailing longshore current brings the Mississippi's waters westward along the Louisiana coast. The nitrates and other chemicals in the water promote growth that uses up oxygen causing a linear hypoxic zone offshore of Louisiana and southeastern Texas. The hypoxic zone acts as a barrier—called a "dead zone"—to the migration of many marine species that use the adjacent coastal swamps and marshes as nurseries, preventing individuals from reaching the open Gulf after maturation.

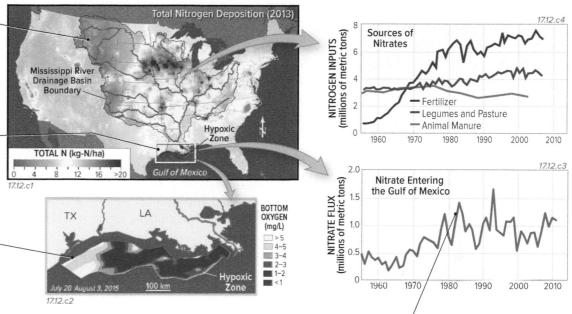

4. The nitrate loading into the Gulf is increasing over time, with wetter years over the Mississippi River basin producing larger hypoxic zones than drier years. The Chesapeake Bay area experiences similar problems with hypoxia. The graph on the top right shows that fertilizer usage in the drainage basin of the Mississippi has increased, exacerbating the hypoxia problem.

D How Do Storms Cause Fish Kills in Coastal Ecosystems?

1. Hurricanes or other severe coastal storms can force salt water inland, killing freshwater fish, particularly those that are more stenohaline (unable to survive changes in salinity). The decay of these organisms uses more dissolved O_2, further stressing the remaining fish in the area. As stressed fish and other aquatic organisms die, a positive feedback occurs, as detritivores (decomposers) use even more O_2 to break down dead organisms, causing more organisms to die (▶), and so on. This process continues until the inflow of other O_2-rich water can restore O_2 levels.

3. After the storm passes, water is pulled back out to sea and the hydrogen sulfide and tannic acid accumulate in locations where organic matter is left behind. If the debris is floating, it can cut off surface water from receiving atmospheric O_2. Thus, the ecosystem cannot break down the additional organic matter into nutrients because too little O_2 is available. The positive feedback continues as more fish die, using more of the scarce O_2.

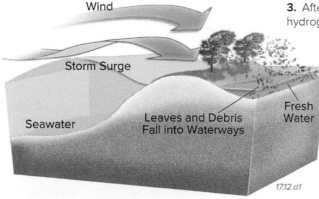

2. When a storm blows leaves, branches, and other organic debris into aquatic ecosystems, hydrogen sulfide (H_2S) and tannic acid are produced as by-products of the decomposition process. The heavy concentration of these by-products prohibits the growth of phytoplankton that would be able to help restore depleted O_2 levels.

Before You Leave This Page

✓ Explain the trend in dissolved O_2 and BOD by season and time of day.

✓ Sketch the seasonal position of the thermocline and the compensation point in a lake, explaining how they influence the oxygen content and the response to organic nutrient loadings.

✓ Explain the origin of the hypoxia problem in the Gulf of Mexico.

✓ Explain how storms can compound problems related to hypoxia.

17.12

17.13 How Do Invasive Species Impact U.S. Gulf Coast Ecosystems?

WHEN A NEW OR "EXOTIC" SPECIES is introduced in an ecosystem, either intentionally or accidentally, it often dies off quickly in its new surroundings because it does not have an ecological niche that will permit its survival. Species that survive and become a part of the ecosystem are called *non-native species*. A non-native species is referred to as an *invasive species* if it disrupts the ecosystem by developing a broad geographical range and occupying a biological niche. As an invasive species invades habitats, it can negatively impact or even eliminate one or more *native species*. In some cases, invasive species remain uncontrolled by predators, because they are not a familiar prey to the predator. Invasive species are most common in ecosystems where the biotic and abiotic factors are not particularly limiting and where species from other regions enter and leave an area easily. The U.S. Gulf of Mexico Coast is one location where invasive species have been particularly troublesome.

A What Are Some Terrestrial Invasive Species and Their Impacts on the Gulf Coast?

Invasive species have been introduced on land and in local waters of the region. Three invasive land species—the coypu (nutria), Burmese python, and kudzu—represent mammals, reptiles, and plants, respectively.

17.13.a1

1. Beaver-like mammals called coypu, locally known as nutria, were introduced to the area from Brazil accidentally through the Port of New Orleans in the 1930s. Burrowing by the nutria has caused tremendous structural problems for levees in southern Louisiana. Hunting them is encouraged by governments. After nutria eat marsh vegetation, they leave a barren landscape.

17.13.a2

2. Burmese pythons were likely introduced into the Florida Everglades by pet owners, and also as pet stores were destroyed in 1992 by Hurricane Andrew. They ingest small alligators, deer, and nearly any other native species. In 2012, researchers caught one that was over 5 m (17.5 ft) in length. Fortunately, as the map below shows, the Burmese python has not been able to migrate beyond Florida, so far.

17.13.a4 Lynchburg, VA

3. Kudzu is a vine in the pea family that is native to east Asia. It has many uses, including animal fodder, erosion control, and medicines. Since its introduction to the U.S. in 1876, however, its very rapid rate of growth in the Gulf states has caused it to out-compete native vegetation for sunlight. Expensive and potentially harmful herbicides are used to control kudzu's spread. Kudzu has been reported in the counties shown below. It has also been introduced to other tropical and subtropical areas, such as Pacific islands and northeastern Australia.

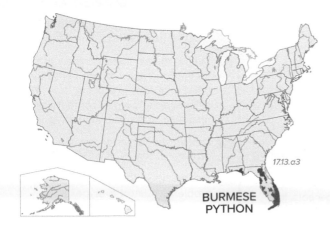

17.13.a3

BURMESE PYTHON

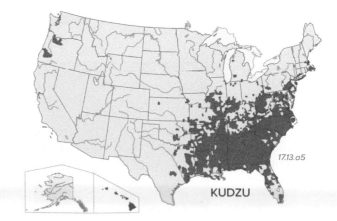

17.13.a5

KUDZU

B What Are Some Aquatic Invasive Species and Their Impacts on the Gulf Coast?

Other invasive species were introduced into waters of the region, mostly in lakes and streams. These include non-native plants, such as water hyacinth, and various species of fish, including the silver carp and blue tilapia. All three of these species, and others not described here, have caused extensive damage to the ecosystems by displacing native plants and animals.

17.13.b1

17.13.b3

17.13.b5

1. The water hyacinth, a floating plant with lavender flowers, has spread from its native South America to the southern U.S. and beyond, probably since it was introduced at the World's Industrial and Cotton Centennial Exposition in New Orleans in 1884–1885. It has since spread to nearly all tropical and subtropical areas of the world. It is also present in waterways of California.

3. Species of carp were brought to catfish ponds in the Gulf states to clean up algae buildup. Flooding moved them into larger rivers, including the Mississippi, where they can be as large as 100 pounds. Silver carp have disrupted the food chain significantly. They are considered a delicacy in some Asian cuisines, so people fish for these Asian carp in the middle of America.

5. Tilapia, a type of fish originating in Africa, was introduced into Florida waters in the 1960s purposely, as a sport fish. In a way similar to the silver carp, their effects on existing native fish were very harmful. A map showing the extent of their range is shown below. They have spread across many parts of the Gulf Coast. Separate introductions occurred elsewhere.

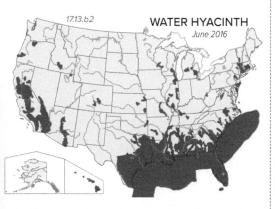
17.13.b2   **WATER HYACINTH**   *June 2016*

17.13.b4   **SILVER CARP**   *June 2016*

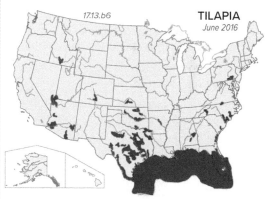
17.13.b6   **TILAPIA**   *June 2016*

2. The water hyacinth is notorious for clogging navigation waterways and damaging boats. It also prevents photosynthesis in phytoplankton, resulting in the depletion of aquatic oxygen and interfering with the food chain. Its rapid reproduction results in abrupt fish kills.

4. The map above shows that carp have spread far beyond the Gulf states. This invasive species worked its way up the Mississippi River and into its tributaries, including the Ohio, Missouri, and Platte rivers. As a result, it has reached Colorado and the Great Lakes.

6. The examples shown on these pages are just a few of the numerous invasive species that plague just one region. As globalization "shrinks our world," careful control of the spread of exotics will become even more important. Are any of these invasive species a problem where you live, or does your area have others?

The Law of Unintended Consequences

Although it is not a law in the way the Second Law of Thermodynamics is, the "Law of Unintended Consequences" is something that we should always consider, especially in matters of ecosystems. When well-meaning people introduced these non-native species into the Gulf Coast region, they thought it was for the good (to clear algae, etc.). But they did not consider all the possible side effects—the unintended consequences. It is nearly impossible to know how the introduction of an exotic species will interact with ecosystems, which as we've seen from this chapter are very complex, with a tremendous number of physical, chemical, and biological connections.

Before You Leave This Page

✓ Give several examples of terrestrial and aquatic invasive species in the U.S. Gulf Coast, describing the harmful effects of these species.

17.13

What Factors Influence the Desert Ecosystems of Namibia in Southern Africa?

IN EXAMINING AN ECOSYSTEM, we would like some measure of its diversity, productivity, and what the inter-relationships are between different species. In this exercise, you will examine some ecosystems of Namibia, a country along the west coast of southern Africa. Namibia features a suite of interesting landscapes, animals, and plants. You will construct a food web, identify what factors are most likely to limit or disrupt the productivity of the ecosystems, and contrast your findings with the ecosystems of Costa Rica, as presented in the opening pages of this chapter.

Goals of This Exercise:

- Make observations and inferences about ecosystems in Namibia based on maps and on photographs and descriptions of landscapes and organisms.
- Construct a simple food web of selected plants and animals, identifying animals that are probably competing for the same limited resources.
- Identify factors that are likely to be limiting the productivity of the ecosystems, and what changes to the environment could cause a disruption.
- Contrast the overall ecosystems, limiting resource factors, and potential threats to biodiversity in Namibia versus those in Costa Rica.

Procedures

Follow the steps below, entering your answers for each step in the appropriate place on the worksheet or online.

1. Examine the regional map of Namibia (below left) to note its general location. Next, observe the satellite image (below right), noting the different regions. Read the descriptions that accompany the three photographs of three main regions.

2. Observe the 12 photographs of different organisms in Namibia, and read the associated descriptions. For additional insight, examine other photographs of Namibia earlier in this chapter and elsewhere in this book (consult the index in the back of the book).

3. Construct a plausible food web of the 12 organisms shown on the next page, and identify those that appear to compete for the same resource. Also indicate any predator-prey relationships.

4. Make a list of what environmental factors are likely to be limiting the productivity of ecosystems in Namibia, explain your reasons, and propose how each organism would respond if a drastic change caused a disruption in one resource.

5. Observe the photographs and read the text in the opening two pages of this chapter, which describe some aspects of the ecosystems of Costa Rica. Compare the overall ecosystems, their limits, and possible threats to ecosystems in each country.

Regional Setting of Namibia

These three photographs show three regions of Namibia. The one below shows huge sand dunes of the coastal desert. The one to the upper right is of Etosha, a world-famous ecosystem centered around water holes next to a large, often dry lake. The photograph on the lower right shows the interior, which has grasslands, savanna, and nearly barren, rocky mountains.

Satellite Image of Most of Namibia

17.14.a2

Etosha

17.14.a4

17.14.a5

Interior

17.14.a3

Coastal Desert

17.14.a1

17.14.a6

1. *Oryx:* Herbivore that grazes on grass and browses on shoots of trees; can live in drier conditions than similar large herbivores.

17.14.a7

2. *Agama Lizard:* Reptile and omnivore that forages mostly on insects, with some grass and berries; can withstand high temperatures.

17.14.a8

3. *Namibian Giraffe:* Herbivore; uses its long neck to browse on trees and tall shrubs that shorter grazers cannot reach.

17.14.a9

4. *Southern Yellow-billed Hornbill:* Omnivore that grazes on small mammals, snakes, eggs, insects, and berries.

17.14.a10

5. *Rhinoceros:* Herbivore that mostly grazes on grass and can go without water for several days; no natural predators, except humans.

17.14.a11

6. *Lion:* Carnivore that attacks and feeds upon mostly large grazing mammals, except fully grown rhinos, elephants, and giraffes.

17.14.a12

7. *Stone Grasshopper:* Well-camouflaged insect that lives in rocky ground; herbivore grazing mostly on grasses and leaves.

17.14.a13

8. *Snake:* Reptile and carnivore that feeds on lizards, small mammals, insects, eggs, and other small creatures.

17.14.a14

9. *Mountain Zebra:* Herbivore preferring to graze on grass, but will consume small bushes if necessary; a threatened species.

17.14.a15

10. *African Elephant:* Herbivore that grazes on trees, shrubs, and herbs; world's largest living land animal; adults have no natural predators.

17.14.a16

11. *Meerkat:* Small mammal of the mongoose family and primarily an insectivore, but will eat lizards, snakes, spiders, eggs, and plants.

17.14.a17

12. *Goats:* Human-introduced omnivore that browses on various plants, especially weeds and shrubbery, but will sample various items.

18 Biomes

ADJACENT ECOSYSTEMS commonly share certain attributes, such as regional climate and the kinds of available nutrients. As a result, they share similar types of flora and fauna. A number of adjacent ecosystems that together have a distinctive collection of flora and fauna, all adapted to regional environmental conditions, is a *biome*. A biome has natural vegetation with similar types of plants, spacing of plants, and fauna across large distances.

18.00.a2 Blue Ridge Mtns., VA

Temperate forests are present across the world, mostly in temperate climates of the mid-latitudes. Temperate forests vary in size, growth rates, and biodiversity. Some trees shed leaves; others don't.

What are some of the factors that determine the distribution of these forests?

18.00.a3 Alaska

Tundra occurs in polar and subpolar regions, and near the tops of high mountains, mostly in the middle and high latitudes. It is characterized by treeless areas with ground that is frozen for much or all of the year.

How do living things in the polar tundra cope with up to six months of no Sun and winter temperatures well below zero?

18.00.a4 Namibia

Grasslands occupy large parts of the Great Plains. Grassy African savannas have some of the most awe-inspiring scenery and wildlife on the planet, as exemplified by the exceptional animals of Africa (◄).

Under what climatic conditions do grasslands and savannas form?

Lush, diverse vegetation characterizes tropical rain forests, with abundant moisture and little variation in temperature or day length throughout the year.

◄ Can you tell where this photograph was taken? Central Africa? South America? Indonesia?

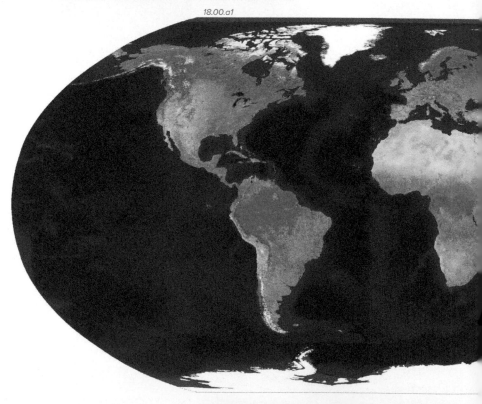
18.00.a1

18.00.a5 Costa Rica

Oceans have diverse ecosystems, which vary as a function of water depth, temperature, salinity, and other factors. Shallow waters have coral reefs—oases of productivity that are currently being threatened by pollution, fishing, and natural changes. Reefs only form in shallow, warm water (◄). Different ecosystems occur at different latitudes and depths, populated by many free-swimming creatures, such as this jumping mobula (►), related to the manta ray.

How many biomes does the ocean have, and what types of marine creatures inhabit each?

18.00.a6 Honduras

18.00.a7 Gulf of California, Mexico

TOPICS IN THIS CHAPTER

The Arctic has distinctive biomes, including tundra, and is the habitat of the polar bear (▶).

What is the distinction among different types of polar biomes?

18.00.a8 Churchill, Canada

18.00.a9 Alaska Range, AK

Boreal forest, a high-latitude or high-elevation biome (▲), has the most extensive geographic distribution of any biome, in spite of relatively infertile soils and a short growing season.

How did this and other biomes develop?

The desert biome conjures up images of plants with spines or vast areas of sand and no plants, but not all deserts look that way (▶). Deserts are so much more, with seasonal bursts of wildflowers and an entire community of interesting flora and fauna that are adapted to the hot, dry conditions and the relative scarcity of water.

What are some of the surprising adaptations made by desert flora and fauna?

18.00.a10 Central Australia

Biomes

A biome is defined by a broad-scale collection of flora and fauna that—although different in detail from ecosystem to ecosystem—share some commonalities. Ecosystems within a biome are often similar in nutrients and energy available to plants and animals. This leads to similar types of flora and fauna across the biome, even though individual ecosystems within the biome differ in scale, structure, and function.

The traditional definition of biomes is rather static, implying no evolution of flora and fauna with time, but this is clearly not the case. As the climate has changed in the past and continues to change in the future, some regions become wetter, some become drier, and some increase or decrease in temperature. Warming on a global scale will cause the boundaries of the high-latitude, boreal forest, and tundra biomes to shift poleward and upward toward higher elevations. Global cooling would do the opposite.

The contraction or expansion of biome pattern and distribution is not solely a function of changing temperatures; the distribution also reflects changes in atmospheric pressure, humidity, amount of precipitation, wind directions, and other atmospheric factors. Biomes are also strongly controlled by the type of soil and other aspects related to the lithosphere, hydrosphere, and cryosphere. Soil takes a long time to develop, so it is unable to shift its distribution over short time periods, compared to plants and animals. If a plant or animal is dependent on a particular soil type, then it may not shift its position under conditions of a changing climate, even though temperature changes would encourage such a shift. Finally, invasive species, fire, deforestation, and urbanization are also likely to impact future biomes. How ecologically sensitive is your home biome?

18.0

18.1 What Biomes Exist on Earth?

BIOMES ARE USUALLY IDENTIFIED in terms of characteristic vegetation, but some are differentiated by geographical or climatic considerations. For this reason, biogeographers disagree on the exact number of biomes, as some consider the form of dominant vegetation, such as a particular type of forest, to be what defines a biome, while others distinguish biomes by the vegetation's survival strategies relating to climate. For example, is a "forest" the same biome whether it is a tropical rain forest or a subarctic forest? This text takes a hybrid approach.

A How Many Biomes Exist in the World?

We recognize nine main biomes, of which seven are terrestrial (on land) and two are water biomes. Each biome could be subdivided into subtypes based on flora (plants), fauna (animals), and climatic factors.

18.01.a1 Costa Rica

18.01.a2 Tucson Mtns., AZ

18.01.a3 Near Badlands NP, SD

The *rain forest biome* consists of evergreen trees that require abundant water balance surpluses year-round. Rain forests grade into other biomes if a pronounced dry season occurs.

The *desert biome* consists of vegetation that is especially adapted to survive in places where moisture deficits are typical year-round. Some deserts are mostly sand, but most have a large number of plants, including cacti.

The *grassland biome* exists both in tropical and extra-tropical environments, but all grasslands have seasonal moisture deficits. This biome has tall-grass and short-grass subtypes, related to different climatic conditions.

18.01.a4 San Gabriel Mtns., CA

18.01.a5 Quebec, Canada

18.01.a6 McCall, ID

The *subtropical scrub and woodland biome* includes areas with hardy bushes, low trees, and areas with larger trees but no closed canopy. In either case, the trees, scrub, and grasses must tolerate warm, dry summers.

The *temperate forest biome* has deciduous and evergreen trees, often intermixed, that thrive in humid areas with warm summers and cool to cold winters. Mid-latitude rain forests are temperate forests with abundant precipitation.

The *boreal forest biome* consists of a few hardy cone-bearing (*coniferous*) tree species that have needles instead of normal flat leaves. It also contains local stands of deciduous species, similar to some in temperate forests.

18.01.a7 Denali NP, AK

18.01.a8 Everglades, FL

18.01.a9 Point Lobos, CA

The *tundra biome* is treeless but includes low bushes, shrubs, mosses, grasses, and lichens (a symbiotic combination of algae and fungi) that can withstand cold.

The *freshwater biome* includes lakes, streams, marshes, and other surface waters. They have at least 50% non-frozen surface water cover for at least six months per year.

The *saltwater biome* is the world's largest biome and covers more than 70% of Earth's surface. This biome includes the coastal regime as well as deeper water.

B What Climatic and Other Conditions Determine the Type of Biome?

Mean annual temperature and precipitation largely control the type of biome at a location, and so they can be used to predict what biome is likely in a specific regime of temperature and precipitation. Local features and other environmental conditions, however, are also important and make the actual life-environment relationship much more complex.

1. This diagram plots average annual temperature versus the annual average amount of precipitation for terrestrial biomes. Cold and dry conditions plot in the lower left part of the diagram, hot and dry plot in the lower right, and hot and wet plot in the upper right. Note that there are no biomes—or regions on Earth—that are very cold and also have high precipitation. Very cold air simply has a water-vapor capacity too low for such a regime to occur.

2. The cold and dry extremes are represented by the *tundra biome.* Somewhat warmer and overall wetter conditions are in the *boreal forest biome.*

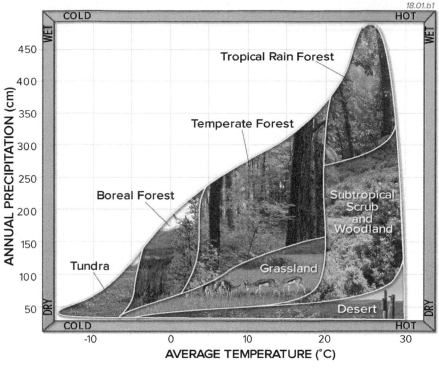

3. Hot and wet conditions (high average amounts of precipitation) of the tropics result in the *rain forest biome,* with its dense vegetation and high biodiversity.

4. If conditions are equally warm, but rainfall is less overall and commonly more seasonal, then it results in the *subtropical scrub and woodland biome.*

5. Moderate temperatures and moderate amounts of precipitation favor the *temperate forest biome.* Under even drier conditions are the *desert biome* (hot and dry) and *grassland biome* (not quite as hot and not quite as dry as a desert).

C Which Biomes Are Most Productive?

Net primary productivity (NPP) is the total amount of energy available for consumers. NPP varies widely by biome, with a few biomes having most of the planet's biological productivity. The graph below plots the percentage that each biome contributes to Earth's total NPP.

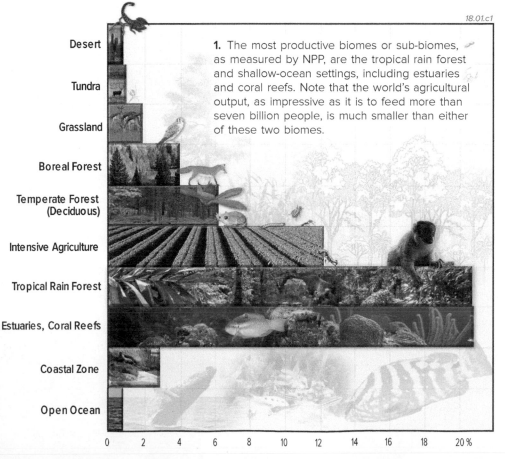

1. The most productive biomes or sub-biomes, as measured by NPP, are the tropical rain forest and shallow-ocean settings, including estuaries and coral reefs. Note that the world's agricultural output, as impressive as it is to feed more than seven billion people, is much smaller than either of these two biomes.

2. For terrestrial biomes, the length of the growing season, average yearly temperature, precipitation, and nutrient availability determine the NPP. Soil type is important for terrestrial biomes because it provides the physical support, water, and nutrients necessary for survival.

3. For aquatic biomes, light, water temperature, and access to nutrients determine the net primary productivity. Notice how unproductive the open ocean is compared to the incredible productivity of coral reefs, estuaries, and other near-coastal waters.

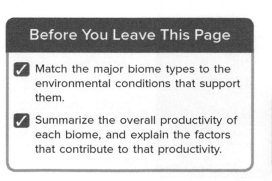

Before You Leave This Page

✓ Match the major biome types to the environmental conditions that support them.

✓ Summarize the overall productivity of each biome, and explain the factors that contribute to that productivity.

18.1

Where Are Each of the Biomes Distributed?

TWO FACTORS MOST AFFECT the global distribution of terrestrial biomes—temperature and moisture, including the seasonality of both. Nutrient availability and other local features are also important, but these vary locally, and therefore they have little affect on biome distribution at the global scale.

1. This map shows the distribution of the main biomes on Earth. Examine the patterns on the map, noting which biomes occur in what climatic zones. Factor in what you know about the distribution of temperature, prevailing wind directions, and semipermanent atmospheric highs and lows. Adjacent to the map are photographs and text describing some interesting or puzzling areas. The map legend is on the bottom of the right-hand page.

18.02.a2 Sawtooth Mtns., ID

2. Western North and South America have complicated patterns of biomes because of the widely varying (localized) topography, climate, and soils. In this photograph (◀), note how the type of vegetation changes as a function of elevation, from grasslands and sage down low, through evergreen forests, and to high peaks with no obvious vegetation.

18.02.a3 Buffalo Gap National Grassland, SD

3. The grassland biome can contain large expanses of tall and short grasses, with few trees, except along streams. Biogeographers disagree on why the grassland (◀) biome extends eastward into Iowa, Illinois, and Indiana (the "Prairie Peninsula"), even though conditions in these places can support mid-latitude forest.

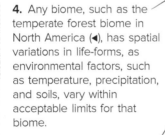

18.02.a4 Tennessee

4. Any biome, such as the temperate forest biome in North America (◀), has spatial variations in life-forms, as environmental factors, such as temperature, precipitation, and soils, vary within acceptable limits for that biome.

18.02.a5

5. The open oceans are unproductive in regard to net primary productivity (NPP), especially the deeper parts of the ocean, but there are local "oases" of productivity, such as the "black smokers" along mid-ocean ridges (◀), where animal communities live without the presence of light.

6. As you examine this map, keep in mind that environmental boundaries are not "lines." Instead, one biome gradually gives way to another across a zone of transition—an *ecotone*—rather than everywhere being a sharp boundary. So if you were traveling across one of the color boundaries on the map above, you would gradually, rather than suddenly, start to notice the difference in biotic features. Also, remember that the biome boundaries are not static across time. Biomes, like any other bioclimatic zones, shift over time in response to changes in climate, hydrology, or human influence.

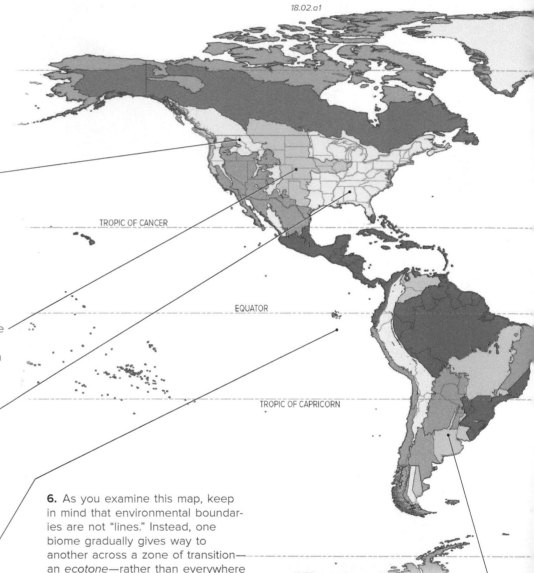

18.02.a1

TROPIC OF CANCER

EQUATOR

TROPIC OF CAPRICORN

7. The *Pampas* region (▼) of Uruguay and Argentina is noted for supporting productive ranches and farms. What is the biome that makes this area so productive? It is part of the grassland.

18.02.a6 Argentina

18.02.a7 Lake District, England

18.02.a8 Denali NP, Alaska

8. The forests of eastern North America and western Europe (▲) have undergone much more deforestation than in any other part of the world, including the Amazon rain forest. Today, some land that was formerly cleared for agriculture is being returned to forest.

9. Boreal forest (▲) and tundra are likely to be the biomes most impacted by climate change. This biome is poorly represented in the Southern Hemisphere because of a lack of landmasses at middle to high latitudes.

TROPIC OF CANCER

EQUATOR

TROPIC OF CAPRICORN

10. Deserts in Africa are expanding at astonishing rates, partly for natural reasons, but primarily because of overgrazing and other land-use decisions (▼). The amount of rain forest is also decreasing in Africa and elsewhere, but for different reasons.

11. Madagascar and Australia (▼) are home to many unique animals and plants that have evolved in isolation from their mainlands. Such islands are also especially sensitive to the introduction of non-native species.

18.02.a9 The Sahel, Africa

18.02.a10 Anglesea, Australia

WORLD BIOMES

■ Tropical Rain Forest	■ Boreal Forest
■ Desert	■ Tundra
■ Grassland	■ Freshwater
■ Subtropical Scrub and Woodland	□ Saltwater
□ Temperate Forest	■ Ice Caps or Highlands

18.2

18.3 What Factors Influence Biome Distribution?

ABIOTIC AND BIOTIC FACTORS influence the distribution of flora and fauna among and within biomes. These variables, in turn, interact with each other in complicated ways, determining not only the spatial pattern of living organisms but the complex relationships between living organisms and their environment. Some factors operate on broad scales while others exert their influence at more localized scales.

A How Do Abiotic Components Influence Terrestrial Biomes?

Abiotic factors involve nonliving entities, such as climate and soils, and greatly affect *terrestrial biomes*. Biome location is most related to climate, which is a function of latitude, altitude, atmospheric and ocean currents, distance to large water bodies, and local nuances. Latitude controls sunlight duration and intensity, which influence the rates of photosynthesis. Temperature and precipitation patterns also control soil formation, impacting the effectiveness and depth of weathering, the balance between physical and chemical weathering, and what organisms live in an area.

Temperature

Most life resides in environments where liquid water is stable (0 to 100°C). Life can survive in a remarkably wide range of temperatures, from freezing conditions to this 80°C hot pool in Yellowstone National Park in Wyoming (▼), where bacteria thrive. Life also exists in temperatures up to 350°C in deep-sea vents.

18.03.a1 Yellowstone NP, WY

Precipitation

Plants and animals, including humans, are composed primarily of water and cannot survive without it. Water is required to carry out metabolic processes at the cellular level. Where there is little water, there is generally little visible life, as in the deserts in Death Valley, California (▼).

18.03.a2 Death Valley, CA

Sunlight and Nutrients

Through photosynthesis, plants convert sunlight and CO_2 into sugars for energy. Plants also need soil nutrients, which over time are unlocked through weathering processes, under the influence of climate, particularly rainfall. Some tropical locations, such as the western Pacific nation of Palau (▼), have dense rain forest due to ample sunlight for photosynthesis and abundant rainfall. These warm, moist conditions favor the weathering of nutrients from parent material that resides in decomposed plant biomass rather than in the soil.

18.03.a3 Palau

Disturbances

Abiotic disturbances such as fire, wind, flooding, and earthquakes can leave the ground covered with dead and decaying matter and an open canopy allowing sunlight to reach the ground. Abundant sunlight and newly enriched soil promote new growth. Forest fires, like this frightening image captured from the Bitterroot Range of Montana (▼), quickly return nutrients locked up in biomass back to the soil.

18.03.a4 Bitterroot Range, MT

Local Site Considerations

Elevation influences temperature and precipitation through the orographic effect. The direction that a site faces—its *slope aspect*—affects weathering, soil development, soil moisture, and therefore vegetation. Orographic effects are striking along many mountain ranges (▼), where clouds and precipitation occur on the windward side and a rain shadow exists on the leeward side.

18.03.a5 Coast Ranges, British Columbia, Canada

Before You Leave These Pages

- ☑ Explain how abiotic factors help explain the distribution and type of natural vegetation.

- ☑ Describe some abiotic factors in aquatic environments.

- ☑ Give several examples of how organisms influence biomes.

B How Do Abiotic Factors Influence Aquatic Biomes?

Aquatic biomes, including the *freshwater biome* and *saltwater biome*, are also strongly influenced by abiotic factors. Sunlight is important in determining rates of aquatic productivity through photosynthesis. Climate influences the productivity of aquatic biomes through water temperature and the amount of fresh water entering the biomes. Temperature and saltwater concentration influence the layering of water, which in turn affects the horizontal and vertical transfer of nutrients. Factors such as soil, slope aspect, and slope steepness do not have much relevance in aquatic biomes, except where the land can influence adjacent water bodies.

Nutrients

The lack of nutrients, such as nitrogen, phosphorus, and iron, limits the productivity of aquatic systems. The map below of net primary productivity, as measured by chlorophyll concentration, shows the highest productivity where turnover, upwelling, or river-delta environments bring these nutrients to the surface to be used by aquatic plants. The open ocean, away from these zones, shows low levels of productivity.

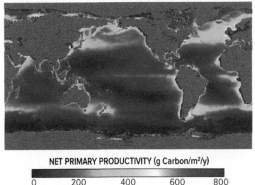

NET PRIMARY PRODUCTIVITY (g Carbon/m²/y)

0   200   400   600   800

18.03.b1

Dissolved Oxygen

Dissolved oxygen content is determined by water temperature. Higher dissolved oxygen occurs at low water temperatures and near the surface, where oxygen-producing phytoplankton have access to sunlight. Fish and other marine creatures are highly sensitive to dissolved oxygen content, typically becoming stressed at concentrations below 6 mg/l. Stratification of lakes and the ocean causes different oxygen levels at different depths.

18.03.b2 Big Cyprus NP, FL

Salinity

Salt content determines whether an aquatic biome is classified as freshwater or saltwater. The ocean's average salinity is 32 parts per thousand (ppt), while freshwater salinities rarely exceed 0.6 ppt. Most freshwater fauna are intolerant of saltwater conditions, and most saltwater fauna cannot tolerate fresh water for long periods of time. Some species, however, such as salmon (▼), smelt, and sturgeon, spawn in fresh water but live most of their lives in salt water; such fish are called *anadromous*. Fish, such as American eels, that live in fresh water and spawn in salt water, are *catadromous*.

18.03.b3

C What Are Some Biotic Controls on Biomes?

Biomes are also influenced by *biotic components*—those related to living things. The distribution and functioning of biomes are impacted by species succession, interaction, and migration. These influences are usually expressed at a more localized ecosystem level. Continued land use change and exotic species introduction, however, are creating the lasting impacts of reducing biodiversity at the biome level.

Succession

Disturbance, either human-induced as through land-use change, or natural, as through fire, can shift a biome to an earlier successional stage. This changes the types of species present. Introduced animals and plants can also modify succession. For example, fireweed, with pink or reddish flowers (▼), is one of the first plants to become established after a forest fire.

18.03.c1 Bergdorf, ID

Species Interaction

Removal of key species impacts predation, competition, and species composition. For example, the poaching of elephants may cause woodlands to encroach on the park-like savanna. Keeping elephant herds healthy is vital to rhinoceros, hippopotamus, gazelles, termites, and dung beetles that would struggle if elephants did not limit the number of trees.

18.03.c2 Africa

Migration of Exotic Species

The ability of organisms to move into a new biome is influenced by climate change and by mountain and water barriers. Introduction of non-native trees, such as the tamarisk into desert biomes, uses large amounts of groundwater, so native species suffer. In the American Southwest, specialized beetles have been introduced to help eradicate (▼) this invasive plant.

18.03.c3 Beetle-damaged Tamarisk, Westwater Canyon, UT

18.3

18.4 How Does Topography Influence Biomes?

DIFFERENCES IN TOPOGRAPHY, from valleys to mountain peaks, and differences in the directions slopes face, result in large local variations in biomes. The wide range of environmental conditions on mountain slopes supports many biomes in close proximity to one another that are normally thousands of kilometers apart. Topography also controls the flow of water, from peaks toward valleys.

A. What Are Riparian Zones?

18.04.a1 San Juan River, UT

18.04.a2 Las Cruces, Baja California, Mexico

18.04.a3 Jackson Hole, WY

Riparian zones consist of trees, shrubs, and grasses that form along streams (▲), creating a local environment that is more lush with vegetation than the surrounding land. Due to the presence of flowing water for all or part of the year, riparian zones can host a very different biome than the surrounding region.

Riparian zones occur in many settings, such as grasslands and various forests. They are best expressed, however, in deserts and other arid lands, where the contrast between dry uplands and lush valleys can be sharp. They can form even where water is not flowing but is just below the surface (▲).

Riparian zones allow habitats for birds, shade-loving vegetation, and other organisms not otherwise common away from the stream valley. The water and vegetation in turn attract diverse types of animals, including frogs and other amphibians, snakes and other reptiles, and mammals, like moose (▲).

B. What Factors Cause Conditions to Vary on Different Slopes?

Differences in elevation, slope orientation, precipitation, humidity, insolation, and wind intensity and direction cause mountain environments to differ greatly from one side to the other.

Clouds, Precipitation, and Humidity

1. One of the most significant factors causing local to regional variations in conditions is a mountain, which can cause orographic effects. The side of the slope facing upwind at any given time is the *windward* slope. Air blowing toward the mountain is forced upslope on this side, where it cools, condenses into clouds, and precipitates. The increased rain on this side of the mountain results in an increase in vegetation and other effects.

18.04.b1

2. The opposite side, facing downwind, is the *leeward slope*. This side of a mountain tends to be drier because the air flowing down the leeward side has lost water to precipitation on the other side. It also compresses, heats up, and becomes less humid as it descends. The leeward side is considered to be in the rain shadow of the mountain and will have less vegetation, less soil, and drier ecosystems.

Insolation and Slope

3. Except near the equator, south-facing and north-facing slopes receive different amounts (▼) of insolation, when averaged over the year. A slope that on average receives more direct Sun is warmer and drier and is called an *adret slope*. The opposite side is cooler, wetter, and more vegetated and is called a *ubac slope*.

Elevation

4. Elevation affects temperature and the amount and type of precipitation. If the mountain is high enough, there is an upper limit to trees—the *tree line* (▼). The tree line is a transition, not an abrupt line. The tree line typically is farther downslope on the ubac side of the mountain and higher on the adret side. The *snow line* occurs at the elevation where snow and ice cover the ground year-round.

Wind Variation

5. Deformed and stunted trees called *krummholz* occur near the tree line, where they must be able to survive the intense cold, dry winds. Note the branches taking advantage of the shelter offered by the trunk itself, growing only on the leeward side. Such a krummholz would be termed a *flag tree* or *banner tree* and indicates the direction of prevailing winds.

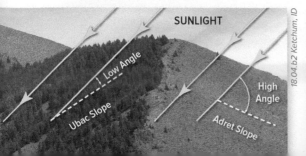

18.04.b4 Patagonia, Chile

C How Do Biomes Change with Elevation on a Mountain?

On mountain slopes, environmental conditions change abruptly with changes in elevation. As a result, a number of biomes can occur over very short distances upslope and downslope.

18.04.c1 Mount Kilimanjaro from Kenya

18.04.c2 Santa Catalina Mountains, AZ

18.04.c6 Mt. Lemon, AZ

1. As conditions vary with elevation up a mountain, so do the biomes, changing from warmer and commonly drier conditions near the base to colder and wetter settings upward. Mount Kilimanjaro (▲), the highest mountain in Africa at 5,895 meters (19,341 ft) above sea level, is located near the equator. From the base of the mountain, the vegetation progresses upward from savanna through several types of forests to alpine conditions with a glacier near the top. Remember that the global average *environmental lapse rate* is about 6.5 C°/km, so during the 3 km climb to the top, temperatures would drop more than 20 C° (36 F°). Nearly 80% of the glacier has melted since the early 1900s.

2. Another example of the effects of elevation is the Santa Catalina Mountains (▶) near Tucson, Arizona. The top of the mountain is at an elevation of 2,791 m (9,157 ft) and the base is about 2,000 m (about 6,600 ft) lower. Observe the different types of vegetation as one progresses from the base to the peak, beginning with the photograph below.

6. The highest peaks are in forests of large ponderosa pine and other high-altitude trees, and there is an alpine ski area.

18.04.c3 Tucson, AZ

18.04.c4 Molino Basin, AZ

18.04.c5 Windy Pt., AZ

3. The Tucson basin below the mountain consists of cacti and other heat-adapted plants of the Sonoran Desert.

4. One-third of the way up the mountain, the cacti give way to scrub and scattered trees, including piñon and juniper.

5. Two-thirds of the way up, the piñon and juniper become larger and closer, and are joined by some ponderosa pine trees.

D How Are Highland Environments Changing?

These vertical progressions of closely spaced ecosystems are particularly sensitive to climate change, in part because each is adapted to a particular range of elevation. Warming associated with climate change is expected to reduce snowpack, change the timing and magnitude of snowmelt, and alter stream discharge. When superimposed on the effects of land use changes from resource and recreational exploitation, highland environments may change abruptly.

1. Humans and animals, like these guanacos (▶) in the Patagonian Andes of South America, have adapted to various mountain biomes. *Transhumance* is a type of seasonal nomadism in which people move to the optimal elevation to experience the best temperatures and animal fodder at a particular time of year—generally upslope in summer and downslope in winter. If the climate changes, species that cannot find their new niche quickly enough will not survive.

18.04.d1 Patagonia, Chile

2. Global or regional warming or cooling can shift mountain biomes to higher elevations, impacting some species. In many cases it is difficult to predict how the range of a species will change, because of local variations in soil, hydrology, microclimate, and topography. The net impact of migrating mountain biomes is likely to be increased habitat fragmentation and more endangered species, though there are likely to be new opportunities for some now-threatened populations. The details are not fully known.

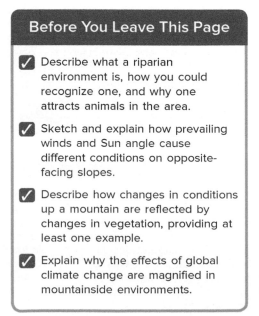
18.04.d2 Coal Bank Pass, CO

Before You Leave This Page

✓ Describe what a riparian environment is, how you could recognize one, and why one attracts animals in the area.

✓ Sketch and explain how prevailing winds and Sun angle cause different conditions on opposite-facing slopes.

✓ Describe how changes in conditions up a mountain are reflected by changes in vegetation, providing at least one example.

✓ Explain why the effects of global climate change are magnified in mountainside environments.

18.5 What Characterizes the Rain Forest Biome?

THE RAIN FOREST BIOME is among the most fascinating on Earth. Rain forest location is determined by abundant precipitation (several meters per year). Rain forests have an immense diversity of plants, animals, insects, and microbes. They also have a vertically layered structure where different animals and plants thrive in specific amounts of shade. Rain forests require rapid rates of nutrient recycling where organisms decompose rapidly after death and return nutrients back into biomass rather than into the soil. The tropical rain forest biome is restricted to the tropics, but rain forests also occur in the temperate forest biome and have enough similarities with tropical rain forests that they are described here.

A What Air-Circulation Features Produce Rain Forests?

Most of the world's rain forests lie in the tropics and are associated with the Intertropical Convergence Zone (ITCZ), the sinuous zone along the equator where the northeast and southeast trade winds converge. The ITCZ swings north and south seasonally, shifting low pressure and abundant rainfall to different regions. Tropical rain forests largely correspond to the Tropical Rain Forest (*Af*) or Tropical Monsoon (*Am*) climate type. Most temperate rain forests form a small subset of the Marine West Coast (*Cfb* and *Cfc*) climate.

1. This map depicts the distribution of rain forests in green. The typical June position of the ITCZ is in orange and the December position is in light blue. Observe this map and note any large-scale patterns. Can you pose a possible explanation for each large green area on the map, and even some of the smaller ones?

2. Tropical rain forests are located along tropical latitudes, either near coasts or well inland. Their greatest extent is along the equator through South America, Africa, and the Indonesian archipelago. Temperate rain forests occur at higher latitudes but are confined to coastal locations where the water helps moderate temperature extremes.

3. Pacific northwestern North America contains almost two-thirds of the world's temperate rain forests. In this area, onshore winds bring winter rain and summer fog.

4. India's rain forest locations are related to the ITCZ and to an inland flow of moist air during the South Asian monsoon. Rain forests farther east, for example, in the Philippines, are influenced by the East Asian monsoon.

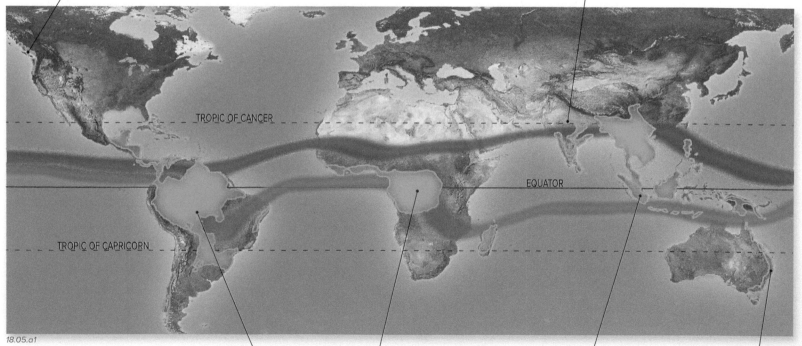

18.05.a1

TROPIC OF CANCER

EQUATOR

TROPIC OF CAPRICORN

5. Rain forests near the equator primarily contain broadleaf evergreens. Those farther from the equator have a higher percentage of deciduous trees that shed leaves to minimize their vulnerability to moisture loss during drier seasons.

6. The Amazon in South America is the world's largest and most diverse rain forest. The ITCZ is located over the Amazon for most of the year.

7. The central African rain forest, centered on the drainage basin of the Congo River, is the second largest rain forest. It receives heavy rainfall from the ITCZ for much of the year, but especially in December and January.

8. Tropical rain forests, including those in Southeast Asia, thrive along the equator, where every day has 12 hours of sunlight. Rising water-rich air within low-pressure zones brings frequent rain to promote lush plant growth.

9. Although mostly a dry continent, Australia has both tropical and temperate rain forests. Rain is provided by prevailing winds that blow over warm ocean currents.

B | How Do Tropical and Temperate Rain Forests Differ?

Tropical rain forests and temperate rain forests share high rainfall, but otherwise they can be quite different. Average annual precipitation is well above 200 cm for temperate rain forests and much higher for most tropical rain forests. Tropical rain forests never experience temperatures below freezing, but some temperate rain forests do, including those along the panhandle of Alaska and in the Pacific Northwest.

Tropical Rain Forests

18.05.b1 Gabon, Africa

Tropical rain forests are in warm and very humid tropical regions, factors that favor rapid growth of trees, vines, and other types of plants. Tropical rain forests typically have higher tree density, more diverse vegetation and animals, and more biomass than temperate rain forests. A recent NASA study found that the world's tropical rain forests store almost 250 billion tons of carbon. By comparison, humans release about 10 billion tons per year through fossil fuel burning and land-use changes.

Temperate Rain Forests

18.05.b2 Queensland, Australia

Temperate rain forests are much less extensive, mostly restricted to relatively thin strips (in map view) along some coasts. Temperate rain forests occur along the ocean side of the Olympic Peninsula in Washington and the ocean side of the Great Divide Range in eastern Australia (◄). In both places, precipitation is the result of a strong orographic effect on moist ocean air flowing onto the mountains. Rain is abundant along the sea-facing (windward) slopes. In most temperate rain forests, temperatures rarely fall below freezing, and summer temperatures remain relatively mild.

C | What Is the Structure of a Rain Forest?

Rain forests are generally composed of five vertical layers: emergent layer, canopy, understory, shrub layer (which some botanists combine with the understory), and forest floor.

18.05.c2 Costa Rica

18.05.c3 Costa Rica

18.05.c4 Costa Rica

1. The *emergent layer* consists of a few trees (◄) that can tower up to 70 m above the forest floor. This layer has free access to sunlight but must cope with high temperatures and desiccating sunlight. Animals inhabiting this zone include eagles, butterflies, and small gliding mammals.

2. The *understory* and *shrub layers* are shaded, more humid, and have little wind. In this dim world, vines climb tree trunks (◄) and plants have broad leaves, both strategies to capture sunlight. Showy flowers are needed to attract pollinators. Jaguars, numerous amphibians, snakes, and bats are common.

3. The *forest floor* (◄) is a dank, humid environment where organic debris decomposes rapidly. Decomposing leaves are food for countless invertebrates such as millipedes, beetles, and termites. The soil is highly leached by rainfall and so it contains few nutrients.

18.05.c1

Emergent Layer

Canopy

Understory

Shrub Layer

Forest Floor

4. The *canopy* is the rain forest's roof and forms an umbrella over the forest floor (▼). Vines called *lianas* wind their way around this maze of branches and leaves, as do *epiphytes*—plants that live suspended on other plants and derive their nutrients and water without contacting soil. Most of the rain forest's species live in the canopy, including monkeys, tree frogs, toucans, snakes, and an abundance of insects.

18.05.c5 Costa Rica

Before You Leave This Page

✓ Locate on a map where temperate and tropical forests occur.

✓ Explain the difference between tropical and temperate rain forests.

✓ Sketch and describe the different vertical habitats within a rain forest.

18.5

18.6 How Are Tropical Rain Forests Threatened?

YOU MAY HAVE HEARD that rain forests are disappearing. Deforestation of tropical rain forests is occurring at an estimated 70,000 to 180,000 km²/yr. The midpoint of this range suggests that an area the size of Pennsylvania is lost each year. The "where," "why," and "so what" of this issue are questions that geographers, ecologists, environmentalists, and politicians want to answer.

A Where Are Tropical Rain Forests Disappearing?

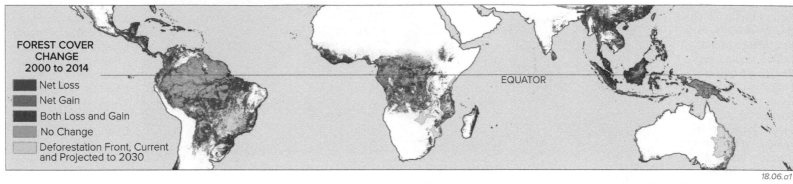

FOREST COVER
CHANGE
2000 to 2014

- Net Loss
- Net Gain
- Both Loss and Gain
- No Change
- Deforestation Front, Current and Projected to 2030

EQUATOR

18.06.a1

The rate of tropical forest deforestation has declined during the last decade but is still unacceptably high, with more than 100,000 km² converted to agriculture or lost through drought or fire each year. Brazil, central Africa, and Indonesia had the highest rates of deforestation in the 1990s, but they have experienced a slowing of deforestation in recent years. On average in 2011, Brazil lost 6,000 km² of rain forest, equivalent to nearly 1.5 million acres, or over 3,000 football fields, each day.

100 yards (91.4 m)

160 feet (49 m)

18.06.a2

x 3000 = Daily Brazilian rain forest lost

B Why Are Tropical Rain Forests Disappearing?

Some biogeographers estimate that most tropical rain forests will be destroyed by 2040. The main cause is economic pressure, but what kind of economic pressure?

18.06.b2 Ipoh, Malaysia

1. Commercial logging for tropical hardwoods like mahogany consumes large areas of rain forest (◄). Logging commonly involves *clear-cutting*, where all the trees are removed.

6. Exploitation of copper, iron, gold, oil, and other natural resources has destroyed tracts of rain forest, especially in Africa, South America, and Indonesia. Surface mines create large pits and generate large volumes of waste rock that is removed from the pit and piled nearby in a dump. The dump and pit can collect rainfall, causing contamination.

18.06.b4

2. Cattle ranchers clear rain forest and then plant pasture grasses. Clearing the land exposes the soil to heavy rain and then soil erosion. This precludes the reestablishment of indigenous forest species. Cattle ranching is now the biggest threat in some places like Brazil.

18.06.b3 Solomon Islands

3. Major highway construction and road building in support of logging and other development destroy swaths of forest (◄). New roads provide arteries for other types of development, leading to more loss of pristine rain forest.

4. Dam construction, such as for hydroelectric power, destroys forests in areas flooded by the reservoirs and may deny water to rivers downstream.

18.06.b1

5. Subsistence farming, where forests are cleared by "slash-and-burn" practices (▲), has historically been a major threat.

C Why Is Tropical Rain Forest Deforestation Such a Concern?

Tropical rain forests are important as reservoirs for carbon, as a source of oxygen expelled by plants during transpiration, as a regulator of climate, as a tremendous storehouse of animal and plant diversity, and as a source of medicines, including many potential cancer-fighting drugs.

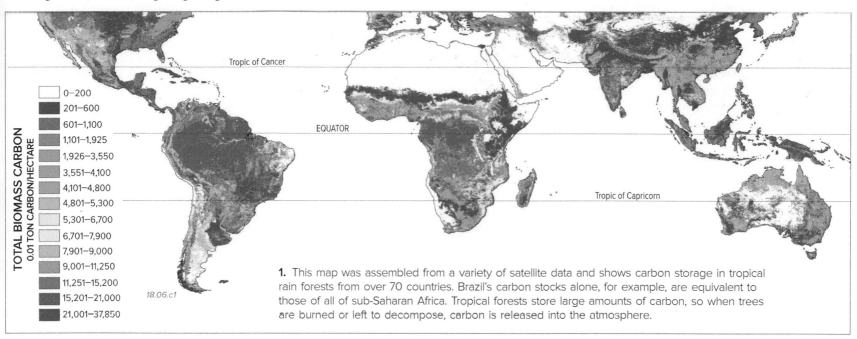

TOTAL BIOMASS CARBON
0.01 TON CARBON/HECTARE

	0–200
	201–600
	601–1,100
	1,101–1,925
	1,926–3,550
	3,551–4,100
	4,101–4,800
	4,801–5,300
	5,301–6,700
	6,701–7,900
	7,901–9,000
	9,001–11,250
	11,251–15,200
	15,201–21,000
	21,001–37,850

18.06.c1

1. This map was assembled from a variety of satellite data and shows carbon storage in tropical rain forests from over 70 countries. Brazil's carbon stocks alone, for example, are equivalent to those of all of sub-Saharan Africa. Tropical forests store large amounts of carbon, so when trees are burned or left to decompose, carbon is released into the atmosphere.

The "Lungs" of the Planet

2. Photosynthesis releases billions of tons of atmospheric oxygen (O_2) each year. Of the O_2 that is needed on our planet, around one-third is derived from trees, with the rest (two-thirds) coming from phytoplankton. Because tropical rain forest trees are evergreens and day length is consistent throughout the year, the planet does not suffer from a shortage of O_2—in part because of the tropical rain forests.

18.06.c2 Raja Ampat, Indonesia

Diversity Storehouse

3. Although tropical rain forests only cover around 6% of Earth's land surface, this biome may contain about half of all plant and animal species. As such, tropical rain forests serve as Earth's "pharmacy" by providing useful medicines to fight a variety of illnesses, including cancer and diabetes. The genes of rain forest plants, fungi, and bacteria may also be key to increasing crop yields for our growing population. Tropical rain forests used to cover much more of Earth's land surface than they do now.

18.06.c3 Costa Rica

Impacts on Climate

4. Rain forests intercept and use insolation that would otherwise strike the ground. As a result, rain forest trees and plants provide a cooler daytime environment. In contrast, cleared land, like the area on the left side of the satellite image below, is warmer, with less evaporational cooling and greater convectional lifting, but usually less precipitation, as indicated by the presence of low-level cumulus clouds over the deforested area (▼). Replacing rain forests with cleared fields creates a drier, hotter climate.

18.06.c4 Congo Basin, Africa

Before You Leave This Page

✓ Describe where deforestation of tropical rain forests has occurred.

✓ Summarize the major causes of the deforestation.

✓ Explain why the loss of rain forests is a major concern globally, not just locally.

18.6

18.7 What Are Deserts and How Do They Form?

DESERTS ARE DRY LANDS, often with little vegetation, that cover around one-third of Earth's land surface. It is a misconception that most deserts are barren sand dunes. Instead, deserts have plants and animals adapted to life in a dry environment. What conditions create deserts, what controls desert locations, and are deserts expanding?

A What Is a Desert?

An *arid region* receives less water as precipitation than it could lose to evaporation and other processes. In general, arid regions that have less than 25 cm (10 in.) of rainfall per year tend to be deserts. Vegetation is sparse in deserts, commonly covering less than 15% of the ground. Many deserts lack permanent streams, although some deserts contain rivers that begin flowing from areas where rainfall is more abundant. Some examples of deserts are described below.

Sahara Desert and Sahel

1. This satellite image shows the northern half of Africa. Most of the region is tan colored because it consists of sand and rock, with very sparse vegetation. This tan region is the Sahara Desert, stretching from Morocco to Egypt.

18.07.a1

2. South of the Sahara is the *Sahel,* a region that is relatively dry but not quite a desert. A region intermediate between a true desert and a more humid climate is called *semiarid* or a *steppe.* The Sahel is currently threatened by the encroachment of deserts from the north. The region has recently experienced a number of devastating droughts.

Mojave and Sonoran Deserts

18.07.a2 Cima, CA

3. The Mojave Desert of Southern California has rocky mountain slopes and broad valleys, which can be sandy or rocky. Joshua trees (▶) are a distinctively Mojave plant, although vast areas have almost no large plants.

4. The Sonoran Desert of Arizona receives more rain than the Mojave Desert, much of it during a summer monsoon. The Sonoran has more cacti and other heat-adapted plants. It can be green (▶) after spring rains.

18.07.a3 Bighorn Mtns., AZ

B Where Do Deserts and Other Arid Lands Occur?

Examine the three globes below, which show deserts and other arid lands in tan. Where are most of the world's arid lands? Do they occur in certain settings? Compare the distribution of these lands with the (1) locations of atmospheric cells of rising and descending air, and directions of prevailing wind, (2) locations and directions of ocean surface currents, and (3) locations of mountains between the deserts and oceans. The latitude lines on these and other globes in this chapter mark the equator and 30° and 60° latitude.

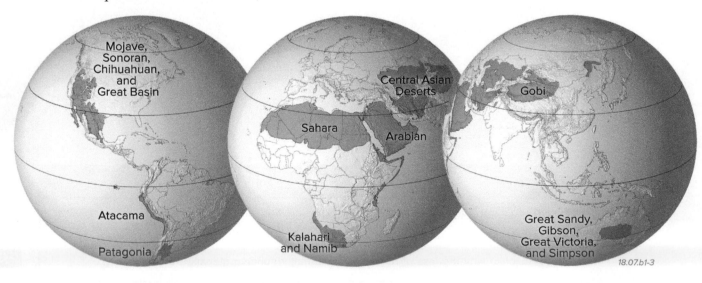

These three globes do not show Antarctica because they are slightly tilted to better display the Northern Hemisphere, which contains most of the world's land and therefore terrestrial biomes. Antarctica, like Greenland, is partly classified as a desert because many areas, such as the dry valleys of Antarctica, receive less than 25 cm of precipitation (mostly snow) per year, but they are not part of the desert biome.

18.07.b1-3

C In What Settings Do Deserts and Other Dry Lands Form?

Large deserts occur beneath descending air in subtropical belts, in rain shadows associated with mountain ranges, far inland from sources of moisture, near cold ocean currents, and in cold, dry polar regions.

1. The world's largest deserts, including the Sahara, Sonoran, and Australian deserts, are *subtropical deserts*. These deserts form where the general atmospheric circulation brings dry air into the subtropics (near the Tropic of Cancer and the Tropic of Capricorn). These areas are situated beneath descending flow cells and are associated with high pressure (indicated as "H" in the globe below).

18.07.c1

4. *Continental deserts* form in the interiors of some continents, far from sources of moisture, or where prevailing winds blow toward the sea. In many such settings, summers are hot and winters are very cold, like in the Gobi Desert of China and Mongolia. The flow of relatively dry air from a continental interior can form deserts in adjacent lands.

3. *Coastal deserts* form where cold, upwelling ocean currents cool the air and stabilize the atmosphere. The Kalahari Desert of southwestern Africa is coastal desert, in land adjacent to the cold Benguela current. This also applies to the Atacama Desert, one of the driest places on Earth, which is in western South America, adjacent to the cold Humboldt current. A cold-ocean-current effect can be enhanced by dry air blowing in from a continent.

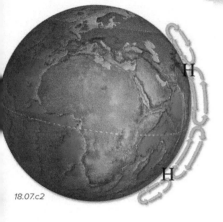

18.07.c2

2. When moist air rises up over mountains, rain may form. As the air descends on the downwind side of the mountain, it dries, forming a *rain-shadow desert*. The Andes mountains extract moisture from mid-latitude winds, forming the Patagonia Desert to the east. The setting of a typical rain shadow is shown below (▼).

Wind   Rain Shadow

18.07.c3

Wind

18.07.c4

D What Is Desertification?

Extended periods of drought, overgrazing by livestock, poor farming techniques, and diversion of surface water can cause soil loss and change grasslands to desert. Converting other lands to desert is called *desertification*. The cycle often begins with livestock overgrazing and destroying plant root nets. With the soil no longer anchored by roots, wind and water erode the soil. Soil compaction from livestock and the collection of firewood further reduce plant cover.

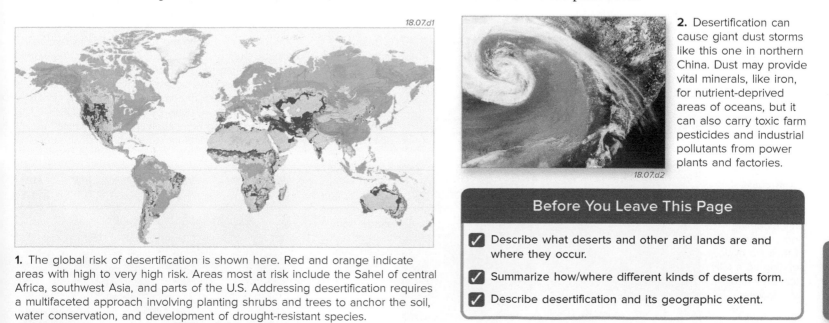

18.07.d1

18.07.d2

2. Desertification can cause giant dust storms like this one in northern China. Dust may provide vital minerals, like iron, for nutrient-deprived areas of oceans, but it can also carry toxic farm pesticides and industrial pollutants from power plants and factories.

1. The global risk of desertification is shown here. Red and orange indicate areas with high to very high risk. Areas most at risk include the Sahel of central Africa, southwest Asia, and parts of the U.S. Addressing desertification requires a multifaceted approach involving planting shrubs and trees to anchor the soil, water conservation, and development of drought-resistant species.

Before You Leave This Page

✓ Describe what deserts and other arid lands are and where they occur.

✓ Summarize how/where different kinds of deserts form.

✓ Describe desertification and its geographic extent.

18.7

18.8 How Do Desert Organisms Survive?

LIFE EXISTS EVEN IN THE HARSHEST conditions on Earth. In such environments, ecosystems and biomes are fragile, as growth is usually slow. Species that have built-in mechanisms that make them tolerant to arid environments are called *xerophytic*. What characteristics do xerophytic vegetation and animals possess?

A What Strategies Involve the Efficient Use of Water in Plants?

Plants have evolved numerous strategies to deal with scarce water, high temperatures, little shade, and drought usually associated with deserts. What are some of the active strategies employed by plants to survive harsh desert conditions?

Gathering Water	Minimizing Water Loss	Strategic Positioning

18.08.a1 Chandler, AZ   18.08.a2 Phoenix, AZ   18.08.a3 Namibia

1. Tamarisk, an invasive species, dominates some riverbanks and have roots more than 20 m deep. They can out-compete native flora for limited water. Such a plant, with a deep taproot system, is termed a *phreatophyte*. Mesquite trees of the Desert Southwest have a similarly deep-rooted strategy.

2. Small, hard, waxy leaves on the creosote bush lose less water through transpiration than wider, softer leaves. In dry times, creosote leaves fold in half (▲) to conserve water, or the plant becomes deciduous by dropping its leaves. Some plants take advantage of cooler night temperatures to flower.

3. Plants may be concentrated near oases or streams, or in places where groundwater is fairly close to the surface. Desert plants can have leaves that point downwind from the prevailing direction to minimize evaporation. In some cases, buds only develop on the leeward side of a branch.

Water Strategies of Cacti

4. Cacti and other desert plants have various strategies (▼) to maximize how much rain they collect, how much water they store, how much water they lose, and how much sunlight they receive. For example, some cacti, like the tall giant saguaro and shorter barrel cactus, have accordion-like ribs that expand to store more water and contract during drier times to reduce their surface area and reduce the amount of water loss.

5. Other cacti, like agave, have sturdy, trough-shaped leaves that open upward, catching precipitation and funneling it toward the center of the plant. Some agave conserve water and energy by only blooming once, at the end of their life cycle, sending up a tall stem crowned with flowers.

6. Many types of cacti, such as cholla and prickly pear, have spines, which discourage browsing by animals, help collect some moisture, and provide some shade from the hot desert Sun.

18.08.a4 Tucson, AZ

CAM Photosynthesis

7. Some desert plants, like cacti and agave (▼), use a special type of photosynthesis to minimize water loss. In crassulacean acid metabolism (CAM), the *CAM plant* collects CO_2, converts it to an acid, and stores it at night rather than being active during daytime. This allows the leaf pores—*stomates*—through which CO_2 (and water) passes to remain closed during daytime hours, thus minimizing water losses during transpiration. A CAM plant can also leave its stomates closed during both night and day in extremely dry times. Photosynthesis continues over such intervals by recycling the CO_2 released during respiration. This "CAM-idling" allows desert plants to withstand dry periods and recover quickly when water becomes available.

18.08.a5 Tucson, AZ

B Which Survival Strategies Do Not Directly Involve Water Use?

Some methods for surviving searing heat and lack of rainfall do not directly involve water conservation. Rather, these adaptations involve rapid reproduction, spreading progeny through hitchhiking, and keeping plant neighbors at a distance. What is the nature of these adaptations?

Rapid Germination/Flowering

18.08.b1 Tucson, AZ

Despite active methods to conserve water, such as waxy leaves or deep taproots, a major survival strategy involves just waiting for the right moment to reproduce. Such a moment involves rapid germination and flowering following a significant rainfall event, either after winter rain or snow or following a summer monsoon. Deserts can be lush with spring flowers after a winter where rainfall was sufficient and somewhat evenly spread over several months.

Animal-Assisted Dispersal

18.08.b2 Copper Creek, AZ

The cholla has spine-covered stem segments that detach easily (and painfully) when brushed by a passing hiker or animal. This serves not only as a defense mechanism against hungry herbivores but also as a means of opportunistic reproduction. Tiny barbs on the cholla spines assure dispersal. A new cholla easily grows where a stem segment was discarded. Note the many new cholla growing beneath the larger, older cholla.

Allelopathy

18.08.b3 Phoenix, AZ

Some desert plants employ a survival strategy of poisoning its neighbors—*allelopathy*. Examples include the creosote bush (▲), which keeps away large plants by poisoning the adjacent soil with its roots. Spring grass avoids the poison. Another example is African rue, an aggressive and invasive plant species that out-competes native grasses by germinating earlier in the spring and by putting toxins into the soil via its root systems to prevent other plants from growing nearby.

C What Survival Strategies Do Desert Fauna Employ?

Mammals' and birds' body temperatures remain constant as long as they aren't overexposed to the hot Sun. Accordingly, these creatures seek shade during the hottest times of the day. Reptiles, amphibians, and insects cannot regulate their body temperatures, so their body temperatures mirror desert heat and cold.

1. Light colors give animals like this scorpion (▼) a high albedo. Sunlight reflects from their bodies and keeps them cooler.

18.08.c1

2. Camouflage can be important to survival. One of the best disguises ever is this grasshopper in Namibia (▼) that looks exactly like the rough-textured stones around it.

18.08.c2 Namibia

3. Desert fauna are typically small, so body heat can escape easily. The shape of body parts, such as the long ears of a desert hare (▶), radiate body heat to the atmosphere. Note the ear's blood vessels that facilitate this process. Arctic hares, by contrast, have short ears in order to conserve heat and avoid frostbite.

4. Ephemeral (seasonal) behavior like *estivation* (summer dormancy) is used by species like the North American desert tortoise (▼). By day, many desert creatures seek the comfort of underground burrows, choosing instead to carry out most of their activity at night.

18.08.c3

18.08.c4 Arizona

Before You Leave This Page

- ✓ Explain some strategies that desert plants use to collect water.

- ✓ Give other examples of survival strategies that do not involve water directly.

- ✓ Provide several examples of evolutionary adaptive strategies for desert fauna.

18.8

18.9 What Are the Features of the Grassland Biome?

THE GRASSLAND BIOME occurs in moderately dry environments, where forests cannot survive. Instead, the biome is dominated by perennial grasses of variable height, scattered trees, and migratory herbivores, like antelope. What determines the distribution of this biome, and what are the characteristics of its plants and animals?

A Where Are Grasslands Found and What Controls Grassland Distribution?

The grassland biome occurs in the transition zone between deserts and humid regions. Rainfall is a limiting factor. Grasslands in the tropics occur as *savannas*, whereas those outside the tropics are *prairies* if they have tall grass or are *steppes* if they have short grass. The annual migration of the ITCZ is important in savannas, because the zone of rainy convection associated with the ITCZ is nearby during the rainy season. In the dry season, the ITCZ has moved away and savannas are beneath subtropical high pressures, causing a dry winter with no freezes. Prairie and steppe organisms must be able to survive in a wide range of temperatures.

1. Prairies and steppes occur poleward of savannas, in interior continental locations that are more isolated from humid marine winds. The largest continuous parcels of prairies and steppes are located in the Great Plains of central North America and along the southern flank of Russia, respectively.

2. Tropical rain forests in Brazil are straddled by a savanna called the *llanos* (Spanish for plain) in Venezuela and Colombia and the *campos* (Spanish for countryside) in Brazil and Argentina.

3. South America also has prairies and steppes, including the *Pampas* of Argentina, which has mild winters.

18.09.a1-3

4. Savannas cover large regions of central Africa, including the famous wildlife areas of Kenya and Tanzania. They also form an east-west belt, the *Sahel*, south of the Sahara.

5. The central desert of Australia is fringed by savanna (mostly to the north) and steppe (mostly to the south).

B What Are the Characteristics of Prairies and Steppes?

Some grasslands exist in the transitional environments between forest and desert biomes. They must be able to withstand harsh winters and long, hot summers. The taller grasses (prairies) are in more humid areas, while shorter grasses (steppes) are in less humid areas. Trees are mainly confined to *riparian zones*, ribbons of more lush vegetation along streams. Grasslands grade into the temperate forest biome on their wetter margins and into the desert biome on drier margins.

18.09.b1 Argentina

18.09.b2 Custer SP, SD

18.09.b3

Rodents, including prairie dogs (◄), stock food for the winter in underground tunnels and will often hibernate. Bigger

Prairie and steppe soils are extremely fertile black soils and are used extensively for agriculture. Interlaced roots form a dense, rich sod layer that won't dry out or blow away, unless disturbed. This biome is typically highly disturbed due mostly to agriculture.

An extensive root system prevents grasses from being pulled out by herbivores. Common mammals of the North American grasslands are bison (▲), pronghorn antelope, prairie dogs, ground squirrels, and mice. In Australia, kangaroos and wallabies roam the Outback.

animals, such as bison or antelope, survive by growing thicker coats and foraging for food. Some seasonal migration of larger animals, like bison, occurred in the past, when these animals were more numerous and were not restricted by fences.

C   What Are the Characteristics of Savannas?

Savannas are grasslands that occur in tropical settings and that have a park-like appearance, with clumps of fire-resistant trees and shrubs mixed with wide open expanses of grasses. In savannas, temperatures remain warm year-round, but flora and fauna must adapt to a pronounced winter dry season. In Africa, great migrating herds of herbivores, like zebra and wildebeest, are closely followed by lions and other predators.

18.09.c1 Kenya

18.09.c2 Kakadu World Heritage Site, Australia

1. Trees are more abundant in savannas (▲) than in steppes or prairies, but there is no closed canopy, except in riparian areas. Note that trees tend to grow outward rather than upward. They exist where local hydrology accumulates water, such as slight depressions in the land. The photograph above shows an African savanna partly in the rainy season, when there begins to be abundant green grass.

2. During the rainy season, the landscape becomes greener. Grasses sprout and grow, and trees add new leaves, some of which will be lost during the winter dry season. During the rainy season, low areas on the land become watering holes that attract animals. Many of the watering holes will dry up completely during the dry season, which concentrates animals near the ones that still contain water.

18.09.c3

18.09.c4 Namibia

3. Giraffes and elephants take advantage of their long necks and trunks, respectively, and their height to browse on trees. They must often navigate around thorns grown by the trees to discourage browsing. Other animals, like monkeys and leopards, use the trees as sanctuaries.

4. Grazers such as gazelles, wildebeests (▶), and zebras migrate with the rainy season. Reproduction occurs at this time to ensure that more young survive. The abundance of herbivores leads to high rates of nutrient turnover, as large amounts of grass and other plants get eaten and digestively processed. This turnover is aided in part by earthworms, dung beetles and other insects, and burrowing rodents and other animals.

18.09.c5 Kenya

18.09.c6 Etosha NP, Namibia

5. Predators such as lions (▲), hyenas, and wild dogs don't generally range far from their homes. Instead they wait for herbivores to pass through their territory. Lions form small family units called *prides*, which include one dominant male, several females (who do most of the hunting), and any young. Lions spend much of the day sleeping, especially after they have eaten.

6. Watering holes become the locus of activity, especially when few exist during the dry season. They attract birds, small mammals, reptiles, amphibians, various grazers such as gazelles (▲), and browsers such as giraffes. This abundance of prey in turn attracts predators. In other words, lots of activity occurs around the water holes, both during the day and at night.

Before You Leave This Page

☑ Distinguish between conditions in savannas, prairies, and steppes.

☑ Give some examples of animals that are well-suited to savannas, prairies, and steppes.

18.9

18.10 What Characterizes the Subtropical Scrub and Woodland Biome?

SCRUB CONSISTS OF SMALL TREES, bushes, and grasses. Woodlands are areas where trees are not dense enough to close the canopy. Both vegetation types exist in locations with seasonal moisture deficits that are too great to support forests, but not severe enough to restrict flora to mostly grasses. Summers are hot and dry, and winters are moist and cool. Small patches of broadleaf evergreen forests may be interspersed with the woodlands where microclimates, such as an orographic effect, allow more moisture to accumulate on the surface or in the subsurface.

A Where Is the Subtropical Scrub and Woodland Biome?

This biome exists in isolated, scattered areas, but flora and fauna are similar from region to region. The *theory of convergent evolution* holds that, despite wide separation and unrelated lineage, organisms develop similar physical and behavioral characteristics if they develop in similar environments.

1. In the Western Hemisphere, this biome is present in coastal southern California, like the hills above Hollywood, and in parts of Baja California and South America.

2. This biome is loosely associated with the Mediterranean climate and includes parts of the Mediterranean basin, the southern tip of Africa, and southwestern Australia. It covers most of India, a region affected by the dry and wet seasons associated with the South Asian monsoon. Human development and disturbances are intense in all these locations because of the desirable climate.

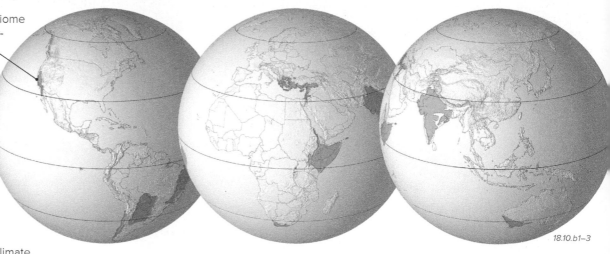

18.10.b1–3

B What Flora Is Common in the Subtropical Scrub and Woodland Biome?

Vegetation in this biome has developed a tolerance to drought and fire, an attribute termed *schlerophyllic*. It includes broadleaf evergreen with tough, leathery leaves adapted to avoid moisture loss. Leaves commonly contain flammable resins, however, that burn fiercely during wildfires. Large fauna are rare, given the summer heat and human presence.

18.10.b1

18.10.b2 Santa Ynez, Mtns., CA

18.10.b3 Napa Valley, CA

Chaparral (also known as *maquis*) is the distinguishing type of vegetation here. This biome has scrubby trees with thick bark to protect against fire, and short, gnarly shrubs, thickets, and grasses with deep root systems. All plants are adapted to a long summer period of drought associated with the dominance of the summer subtropical high.

This is a zone of origin for *therophytes*— annual plants that complete their life cycle quickly by having their seeds lie dormant during the dry season. Therophytes, like these California poppies, often sprout up, bloom quickly, and flourish only for a brief time during or after the wet season. In dry years, very few poppies will sprout.

This biome is commonly disturbed because it is suitable for productive but specialized agriculture, including olives and grapes, like those pictured above in Napa Valley, California. These crops tolerate or even thrive amid severe seasonal moisture deficits. The biome is even more productive when artificially watered during the summer dry season.

C What Role Does Fire Play in Biomes?

There is a tendency to view fire as a destructive, sporadic event that is to be avoided at all costs. While fires can be destructive, life in the subtropical scrub and woodland biome and grassland biome has adapted particularly well to fire and is even dependent upon it. The negative aspects of fire outweigh the potential benefits to most organisms.

Negative Aspects of Fire

1. Destructive fires cause loss of human life and structures and are expensive to fight.

2. If not killed outright, wildlife are impacted by a loss of food sources; if occurring in the late fall, fire can cause a severe short-term food shortage.

3. Smoke and ash pollute the air and release greenhouse gases.

4. The intense heat can cause soil particles to become water-repellent, causing rainwater to run off rather than infiltrate into the soil.

5. Soil erosion and flooding result from bare and "baked" soil surfaces.

6. Fires in woodlands destroy wood that could be important for the local economy.

18.10.c1

7. Humans have altered natural fire regimes. One technique to control fire is the fuel break, which involves eliminating vegetation over a strip of land in the hope of "starving" a fire of fuel. Also pictured here is a *prescribed burn* (an intentionally started fire) used to reduce fuel buildup and release soil nutrients. Prescribed burns are conducted during times when the fire will be least likely to rage out of control.

Benefits of Fire

8. Fire plays an important ecological role by restoring nutrients to the soil. Plants benefit from the ash produced by fire.

9. Fire allows for a new cycle of successional stages to begin. Some species or plants and animals thrive only in disturbed or burned-over areas.

10. Some species of chaparral are *pyrophilic*, meaning that they actually require intense heat to crack the seed and to allow for germination. Fire can also cause the roots of some plants to sprout.

11. Fire kills disease and harmful insects.

12. Sporadic controlled fires prevent hotter, more destructive fires.

13. Periodic fires improve wildlife access.

D What Factors Impact Fire Severity in the Subtropical Scrub and Woodland Biome?

While drought and high temperatures are major factors determining fire severity, other factors like amount and type of available fuel, humidity, wind, and topography also play a role. To follow the progress of a fire that starts on the ground, read counterclockwise from the lower left.

6. As is the case with many other disasters, in the end, the net influence of human activity is probably to suppress smaller fire events but to make the larger events worse than they would have been otherwise.

5. Once formed, a number of factors influence the subsequent behavior of the fire and the effectiveness of any fire-fighting measures. High temperatures, low humidity, and lack of rainfall make plants more combustible. Strong winds spread fires and can further dry out brush, making it even more combustible.

18.10.d1

4. Once fires reach the treetops, the burn is extremely intense and difficult to control. Fires that are particularly hot leap from treetop to treetop—a *crown fire*. A fire ignited on the crest of a slope is likely to spread slowly, while a fire at the bottom of a slope will spread rapidly uphill because flames, cinders, and warm air rise and preheat uphill fuels.

1. In the subtropical scrub and woodland biome where fire is suppressed intentionally or by chance, grass and woody debris, some with flammable resins, build up, providing a fuel source.

2. Fire spreads quickly among dry grass and woody debris that has accumulated on the ground.

3. Convection and ladder-like branches drive the fire upward toward the tops of the trees.

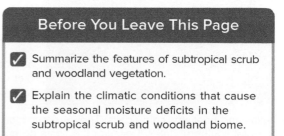

Before You Leave This Page

✓ Summarize the features of subtropical scrub and woodland vegetation.

✓ Explain the climatic conditions that cause the seasonal moisture deficits in the subtropical scrub and woodland biome.

✓ Explain the role of fire in biomes and the impact of humans in fire suppression.

18.11 What Characterizes the Temperate Forest Biome?

TEMPERATE FORESTS roughly correspond to the Humid Subtropical, Marine West Coast, and Humid Continental climate types. Winters are cool to cold and summers are warm to hot. Precipitation is well distributed throughout the year. The two major subtypes within this biome are *evergreen* and *deciduous*. Evergreen temperate forests are further categorized into *broadleaf evergreen* and *coniferous evergreen*, both of which retain their leaves or needles (respectively) year-round. Deciduous temperate forests drop leaves as shorter days herald the approach of winter.

A Where Are the Temperate Forests Located?

18.11.a4 Big Thicket NR, TX

Temperate forests (▸) are mainly a Northern Hemisphere mid-latitude biome. Biodiversity decreases and conifers become more dominant poleward. The biome transitions into variable vegetation assemblages to the south.

1. One type of *evergreen temperate forest* is the *coniferous evergreen forest*, primarily a North American Pacific coast phenomenon, represented by sequoia, Douglas fir, and redwoods (▾).

18.11.a5 Sequoia NP, CA

18.11.a1–3

2. *Deciduous temperate forests* occupy large expanses of eastern North America, the Rocky Mountains, Eurasia, and East Asia. Small sections of Australia and New Zealand also have this forest type.

3. *Broadleaf evergreen temperate forests* occupy sizable sections of eastern China and Japan. These forests are restricted to the lowland U.S. Atlantic coast and Mississippi Valley.

B What Are Some Examples of Evergreen Temperate Forests?

The term *evergreen* denotes the year-round presence of leaves or needles, although individual needles and leaves die, fall, and are replaced by new growth. Leaf litter is a major mechanism for nutrient recycling, particularly for broadleaf evergreen trees.

Broadleaf Evergreen

18.11.b1 Ouchita Mtns., AR

In the U.S., broadleaf evergreens are generally bottomland hardwood species, such as the majestic live oak, bald cypress, and other trees that can tolerate a high water table but also can withstand brief dry, cold spells. They are common in the coastal southern states. Because the roots do not need to penetrate deeply into the soil, these top-heavy trees can be uprooted during storms. Along coastal areas, they are also susceptible to damage from *saltwater intrusion*.

Coniferous Evergreen

Coniferous evergreens are represented by redwoods and sequoias—the largest living trees, with diameters in excess of 10 m and life spans exceeding 3,000 years. Redwoods are taller than sequoias, reaching heights of nearly 100 m. The seed of a sequoia is about the size of an oat flake; the seed of a redwood is the size of a tomato seed. Seeds flake off from cones.

Coastal redwoods are restricted to the fog-bound coasts of California and Oregon. Sequoias grow only on the west side of the Sierra Nevada. In many places, forests of sequoias and redwoods have been removed due to logging and severe fires. Only about 4% of old-growth (original) redwoods remain.

18.11.b2 Calaveras Big Trees SP, CA

C What Characterizes Mid-Latitude Deciduous Forests?

Mid-latitude deciduous forests are relatively lush. Humus forms easily from the fallen leaves, and the understory is well-developed with mosses, shrubs, wildflowers, and sometimes grasses. This understory layer plus the canopy promote various habitats, so biodiversity is fairly high. With four distinct seasons, organisms must adjust to the progression of the seasons, from an explosion of sprouting and breeding in the spring to periods of *senescence* (aging) or *hibernation* in the late fall and winter.

18.11.c1

1. The canopy here is less dense than the tropical rain forest, but sunlight becomes progressively filtered out downward. Understory plants must therefore be shade tolerant. Early spring sprouting can allow access to the Sun before the canopy develops.

2. Pioneer species include big- and small-tooth aspen, cherry, and some conifers. Succession to a climax forest favors beech, maple, oak, hickory, basswood, hemlock, elm, and ash.

4. Alfisols are the dominant soil type here. They can be made productive for agriculture by liming the soil to maintain a good pH balance of 6.5–7 and by maintaining the organic matter content.

5. Humans have cleared large areas of mid-latitude deciduous forest, mostly for agriculture and urban development, in these hospitable climates.

18.11.c2

3. Food webs are complicated in these environments, as are the abundant insects and vegetation that thrive in the various layers of the forest. This environment attracts birds, mice, deer (▶), squirrels, rabbits, and larger predators, including owls, foxes, and wolves.

D How Is Climate Change Impacting the Temperate Forest Biome?

Climate change models suggest that the temperate forest biome will warm, with more precipitation extremes over the next hundred years, resulting in biogeochemical, hydrologic, and biological changes.

Biogeochemical Changes

18.11.d1 Cascade, CO

1. Warming can result in accelerated carbon loss from soils, but also increased carbon sequestration from additional vegetation growth. It is unclear which factor will dominate.

2. Increased precipitation causes larger losses of soils and nitrogen as these are leached away.

3. Warm, wet conditions cause faster decay of leaf litter.

Hydrologic Changes

18.11.d2 Molas Pass, CO

4. Warmer winters result in more intense storms, and more flooding and erosion. Drier summers cause a reduced snowpack, earlier spring thawing of frozen surfaces, less infiltration, and increased evaporation and transpiration during times when water is available. All of these result in drought stress and lowering of the water table, a threatening situation for trees with shallow roots.

Biological Changes

18.11.d3 Stanley, ID

5. Drought stresses trees, causing increased rates of death and increased susceptibility to pest and disease infestations. This results in increased openings within the forest from tree death and storm damage. As a result, warming is predicted to lead to an expansion of subtropical species at the expense of existing cold-hardy ones. Other uncertain impacts are hidden within these predictions.

Before You Leave This Page

☑ Describe the types and geographic locations of temperate forests.

☑ Describe the environments that support deciduous temperate forests.

☑ Explain how temperate forests may be impacted by global climate change.

18.12 What Characterizes the Boreal Forest Biome?

THE BOREAL FOREST BIOME corresponds well with the Subarctic climate, with long, cold winters, and short, cool summers. Biodiversity is minimal in this two-layered forest consisting of an *overstory* of trees and a *ground layer* of lichens, mosses, and herbs. Spruce, fir, larch, and pine dominate, with some stands of deciduous poplar, aspen, and birch. Fauna include caribou, deer, moose, beaver, and muskrat; bear and wolf are the dominant carnivores. In colder locations, permanently frozen soil (*permafrost*) dictates shallow root systems and waterlogged, acidic soils. Because organic matter decays slowly, boreal forests store carbon effectively.

A Where Is the Boreal Forest Biome Located?

Boreal forest is extensive in the Northern Hemisphere, where there are vast land areas at relatively high latitudes. It is virtually absent in the Southern Hemisphere because of the relative lack of land at the appropriate latitudes and elevations. In both hemispheres, it occurs at high elevations (mountains), but mostly in patches too small to show on the globes below.

This biome has the largest continuous extent for any forest. It includes parts of Canada (▼), Alaska, Scandinavia, and Russia. In Russia, the biome is known as the *taiga.* Population densities of humans are low in this biome due to the harsh climate and relative isolation, but disturbance is occurring as new natural resources are exploited.

18.12.a4 Western Canada

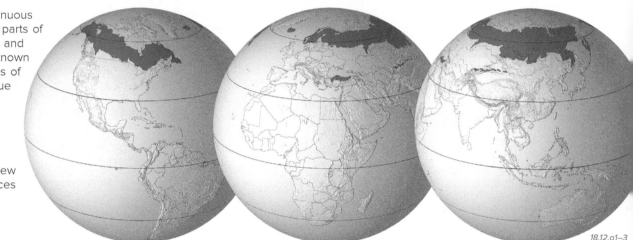

18.12.a1–3

B How Does Life Respond to the Climatic Limits to Growth?

Flora and fauna must adapt to low Sun angles, frigid temperatures, and strong winds. Life proliferates when and where precipitation increases, the frost-free period lengthens, soil drainage is enhanced, or a warmer, south-facing slope exists.

18.12.b1 Katmai, AK

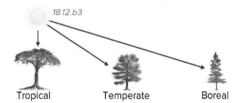

18.12.b3

Tropical   Temperate   Boreal

2. Boreal forest organisms must adapt to very low Sun angles. The dark color and rough texture of the forest promote a low surface albedo, helping to warm the trees. Needles (spruce, fir, and pine trees) and disposable leaves (e.g., aspen trees) help limit heat loss in the winter. Low evapotranspiration rates also minimize cooling.

18.12.b5 Ontario, Canada

1. Animals adapt to cold, long winters by hibernating, migrating, growing insulating fur (▲), camouflage (▼), and having specialized footing. Staying warm, blending in, and laying low in the winter are critical in cold biomes.

18.12.b2

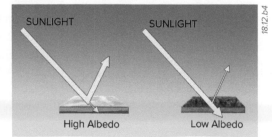

SUNLIGHT   SUNLIGHT

High Albedo   Low Albedo

18.12.b4

3. Many trees in boreal forests have needles, rather than leaves, that allow snow to mostly fall to the ground, rather than being captured and causing snow-laden branches to break off. Trees with broad leaves in boreal forests generally shed these "snow-catchers" before the heaviest snows of winter. In areas of permafrost, trees often lean over because permafrost limits the depth of stabilizing roots. Shorter trees often die off because of a lack of sunlight, while taller trees can die from exposure to the frigid winter wind, resulting in slow, but steady, uniform growth.

C   What Other Factors Impact Growth in Boreal Forests?

Like grasslands and subtropical scrub and woodlands, fire plays an important ecological role in the boreal forest, not only in the succession of plants and animals after a disturbance, but also in global carbon storage. Fires free stored carbon, and fuel for fires is increased by the damage caused by a pestilent insect called the *mountain bark beetle*.

Fire and Succession

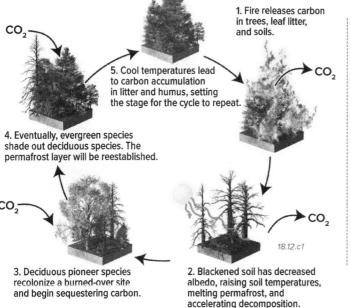

CO₂

1. Fire releases carbon in trees, leaf litter, and soils.

5. Cool temperatures lead to carbon accumulation in litter and humus, setting the stage for the cycle to repeat.

CO₂

4. Eventually, evergreen species shade out deciduous species. The permafrost layer will be reestablished.

CO₂

CO₂

3. Deciduous pioneer species recolonize a burned-over site and begin sequestering carbon.

2. Blackened soil has decreased albedo, raising soil temperatures, melting permafrost, and accelerating decomposition.

18.12.c1

Pest Invasions

6. Warmer temperatures and drought have increased the range and population of the mountain bark beetle, a voracious pest (▶). The beetle destroys conifers, mainly pines, causing huge parts of pine and other types of forests to turn brown, as shown in the photograph below. Although only about the size of a grain of rice, the beetle has impacted thousands of square kilometers of forest like here in the high country of central Colorado (▶). Note the brown, decimated appearance of the trees in the foreground and hillsides in the background. Fallen trees—or dead but standing trees—provide abundant fuel for fire. Beetles greatly increase the likelihood of a forest-decimating crown fire. Ironically, fire is one of the best ways to keep the number of these beetles in check.

18.12.c2

18.12.c3 Breckenridge, CO

D   How Might the Loss of Boreal Forests Accelerate Global Warming?

We are becoming increasingly aware of the importance of boreal forests in buffering the effects of climate change. Global warming is very likely to result in the poleward displacement of boreal and temperate forest biomes.

1. Boreal forests store large amounts of carbon in the tree trunks, branches, leaves, needles, roots, and their dead equivalents on the forest floor. If the trees are removed, this carbon can get back into the atmosphere as additional greenhouse gases in the form of carbon dioxide (CO_2).

18.12.d1   18.12.d2 Alaska

3. Human expansion into the Alaskan, Russian, and Canadian boreal forests will likely accelerate as the demand for timber, oil, natural gas, and minerals increases and transportation and logistical difficulties are overcome. This is a photograph of the Alaska pipeline.

2. This table of carbon stores in tons per acre shows that the boreal forest biome stores even more carbon in soil than the tropical rain forest biome. With the loss of tree biomass through fire, deforestation, or insect invasion, a positive feedback loop is created in which the warming releases more CO_2 and methane (CH_4), which both had been trapped in peat and permafrost beneath the forest.

Biome	Plants	Soil	Total
Tropical forests	54	55	109
Temperate forests	25	43	68
Boreal forests	29	153	182
Tundra	3	57	60
Croplands	1	36	37
Tropical savannas	13	52	65
Temperate grasslands	3	105	108
Desert/Semidesert	1	19	20
Wetlands	19	287	306

Before You Leave This Page

- ☑ Describe the environmental and geographic features of the boreal forest biome.

- ☑ Describe the characteristics of vegetation that exist in these harsh conditions.

- ☑ Explain why the boreal forest biome is more vulnerable to the effects of climatic changes than most other biomes.

18.12

18.13 What Characterizes the Tundra Biome?

TUNDRA IS A GENERIC TERM for treeless areas with low Arctic vegetation, including dwarf bushes, grass perennials, and flowering herbs. It is derived from a term in Russian that means "treeless mountain area." The tundra biome closely follows the location of the Tundra climate, which is characterized by very cold conditions with little winter daylight. Tundra is inhabited by a unique assemblage of plants and animals.

A Where Is Tundra Present?

Like boreal forests, tundra is mainly limited to Alaska, Canada, and Russia in the Northern Hemisphere and mountains and other highland regions in both hemispheres. Life must adapt to the cold with little winter daylight but long periods of low-angle summer daylight. Water is locked up as ice for most of the year. Roots must remain shallower than the depth of the permanently frozen ground (permafrost).

Boreal forests grade into tundra with trees becoming stunted and more widely spaced at the treeline. Farther poleward or higher in altitude, temperatures are too severe for grasses, sedges, and mosses. There, only *lichens* clinging to rocks can survive (◄). Lichen is a combination of algae and fungi, living symbiotically.

18.13.a4 ANWR, AK

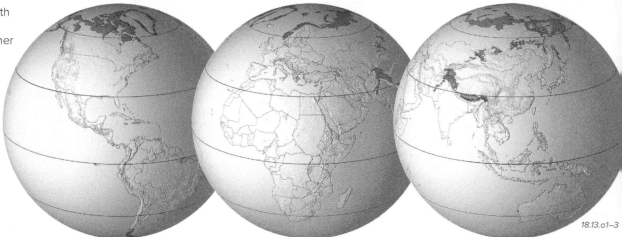

18.13.a1–3

B What Are Some Characteristics of the Tundra Environment?

The tundra is a cold, windy, treeless region dominated by low-lying perennials that need to store energy for the brief summer flowering period. Soils are underlain by a permafrost layer and are usually waterlogged and infertile.

1. Along the boundary between boreal forests and tundra, low vegetation surrounds sparse trees, and none are very tall. The lack of trees is a result of the freezing of the active soil layer each winter, followed by thawing each summer.

3. As the climate becomes more severe poleward, the active layer thins and the water table and permafrost approach the surface. Organic matter accumulates in this waterlogged environment. Drainage is poor and weathering is limited, resulting in poor soil.

4. Along coastal plains and lakes, the water table may reach the surface. The warmer seawater heats the land surface, melting permafrost and producing thaw basins and patterned ground, as shown in the photograph below.

18.13.b1 Central Alaska

18.13.b2 ANWR, AK

18.13.b3 ANWR, AK

2. This figure (▶) illustrates how the depth of the permafrost layer results in four different types of tundra. Between the permafrost and the surface is the *active layer*, which undergoes a freeze-thaw cycle each year (frozen in the winter, thawed in the summer).

18.13.b4

C What Are the Characteristics of Permafrost?

Permafrost or permanently frozen ground is one of the defining characteristics of the tundra biome. Above this frozen layer, the slushy active layer thaws seasonally, limiting not only the depth of the root layer and microbe activity but also the types of structures that can be built.

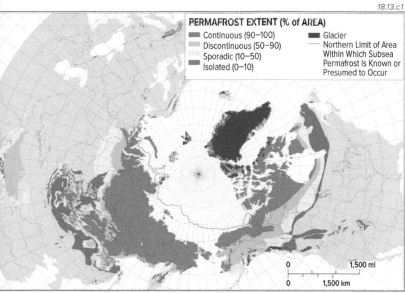

18.13.c1

PERMAFROST EXTENT (% of AREA)
- Continuous (90–100)
- Discontinuous (50–90)
- Sporadic (10–50)
- Isolated (0–10)
- Glacier
- Northern Limit of Area Within Which Subsea Permafrost Is Known or Presumed to Occur

0   1,500 mi
0   1,500 km

18.13.c2 Alaska

2. This photograph shows a section of icy permafrost in a block loosened by coastal erosion (◄). Although the soil and sedimentary material appear to be relatively soft, the network of ice in the spaces between the grains strengthens the material considerably. If the permafrost is thawed, however, the strength melts away with the ice. Melting of permafrost has varied impacts, including more fresh water delivered to rivers, impeded drainage, altered water tables, and the release of carbon that was sequestered in the frozen soil.

1. Permafrost is most widespread in polar regions (▲). It also occurs farther from the poles in areas of high elevation. On this map, these include the highest peaks of the Rocky Mountains and Alps and large areas in central Asia, including Mongolia, Tibet, and the Himalaya. Although not shown, permafrost occurs in similar settings in Antarctica and some parts of the Andes of South America.

3. Fire, climatic warming, or the heat radiated from human development can thaw permafrost, removing the support beneath buildings, airports, utility lines, and roads (►). Thawing of permafrost can also lead to slumping and other types of slope failure, damaging any structures on top or in the way.

18.13.c3 Alaska

D What Animals Live in the Tundra?

To survive in the harsh environment of the tundra, animals migrate, hibernate, burrow, and develop thick insulating layers of fat and warm, furry coats.

18.13.d1 Churchill, Canada

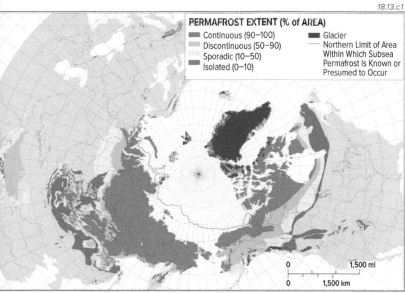

18.13.d2

18.13.d4 Brooks Range, AK

2. Caribou are the animal most commonly associated with tundra (►). They migrate to different feeding grounds from season to season. The most critical areas to preserve are places where they mate and give birth.

1. Food webs in the tundra biome are much less complicated than those in more temperate locations. Thus, ecosystems and the biome itself are more vulnerable to disturbance. Because few species can adapt to such conditions, biodiversity is low. Some of the mammals here include musk oxen, caribou, lemmings, and polar bear (in the photograph above). These populations can experience large oscillations. Insects and migratory birds, like trumpeter swan (►), are abundant during the brief summer. There are few or no reptiles or amphibians. Wetlands are common because of poorly drained ground and provide ideal summer breeding grounds for insects and migratory birds.

18.13.d3 Talkeetna, AK

Before You Leave This Page

☑ Describe the environmental characteristics of the tundra biome and the two main settings in which it occurs.

☑ Describe the types of plants and animals that live in tundra and how they survive these harsh conditions.

☑ Sketch and describe permafrost, including its relation to types of tundra and the damage that can result from thawing.

18.13

18.14 What Characterizes the Freshwater Biome?

THE FRESHWATER BIOME includes streams, lakes, and wetlands, and it usually has water with less than 1% salinity. This biome typically occurs in much smaller, more isolated pockets than the others, and it covers only about 1% of the globe's surface. Most of the world's people, however, reside near fresh water and depend on it to survive. This biome is under assault by development pressures and pollution. The floral and faunal characteristics indicate temperature, sunlight, murkiness (*turbidity*) of the water, oxygen and food availability, and water velocity.

A How Is the Freshwater Biome Subclassified?

There are three major subtypes of the freshwater biome. Streams comprise *lotic systems*, lakes and ponds are *lentic systems*, and areas of standing water with emergent vegetation are *wetlands*.

Lotic Systems

1. The streams of *lotic systems* may originate from snowmelt, springs, or lakes; they flow from higher elevation headwaters toward lower elevation lakes or the ocean. Life-forms respond to the increasing turbidity and discharge downstream. Likewise, organisms favoring a rocky or sandy bottom substrate give way downstream to those preferring a muddier bottom. Temperature depends on season, shade, depth, and number of impoundments (which tend to increase water temperature). Flow varies considerably, both across the channel and from upstream to downstream.

2. Curves, or *meanders*, cause variations in depth and velocity of the water, resulting in different microenvironments within the stream. The outside bend is usually where depth, velocity, turbulence, and erosion are greatest, resulting in a *cutbank*, a steep drop-off from land to water. Organisms must adjust to this dynamic environment.

18.14.a1 Salmon River, Riggins, ID

4. On the inside bends of meanders, water moves more slowly and deposits some of the sediment being carried by the moving water, eventually forming a new habitat for plant and animal colonization.

3. *Riffles* are fast-moving, shallow zones where water usually passes over a rocky bottom. The turbulent flow recharges the water with atmospheric oxygen (O_2), so it enables life to proliferate.

Lentic Systems

5. Lakes and ponds form in many ways—glacial ice blocks (*kettles*) may melt, shifting stream channels can form oxbow lakes, mass wasting can dam a stream, solutional weathering may dissolve bedrock joints or bedding planes and leave a lake, or growing shoreline sandbars can close off the mouth of a bay.

6. The *lentic system* refers to watery environments of lakes and ponds and is subdivided into zones. The *littoral zone* has access to sunlight— the *photic zone*—through its entire depth, allowing for plant growth.

18.14.a2 Alturas Lake, ID

7. The open water or *limnetic zone* is the part that is too deep to allow light penetration to the bottom, with the *aphotic zone* beneath it.

8. Lakes and ponds may be seasonal or may last hundreds of thousands to millions of years. Biodiversity depends on nutrient input from the surrounding watershed. Nutrient-rich lakes, such as Lake Erie in the Great Lakes, are called *eutrophic* and have more biodiversity and varied food webs than nutrient-poor (*oligotrophic*) lakes, such as Lake Superior.

Wetlands

18.14.a3 10,000 Islands NWR, FL

9. *Wetlands* occur in freshwater environments along lotic or lentic systems, and where the water table is at or near the surface for most of the year. They consist of standing water (◄) with emergent vegetation. Fens, bogs, swamps, and marshes are types of wetlands, with the distinction based on soils, vegetation, and water-table level. Some meadows also have standing water.

10. *Marshes* commonly form along the borders of streams or lakes. They are dominated by grasses and reeds, but they can have some trees as well. ►

18.14.a4 Hassayampa River, AZ

11. *Bogs* are muddy and fed by precipitation. Nutrients are sparse, and dead plant material (*peat*) acidifies the water. *Fens* are similar but are not acidic and have grasses and sedges.

18.14.a5 Needles, CO

12. *Swamps* are forested wetlands (▼), such as those within deltas and along stream floodplains. Trees, like cypress, and shrubs are dominant rather than grasses.

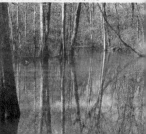
18.14.a6 Noxubee NWR, MS

B What Types of Organisms Exist in the Freshwater Biome?

The abundance and distribution of freshwater biome organisms are shaped by many factors, including nutrients, depth of photic zone, oxygen concentration, temperature, type of substrate, and flow speed. Amphibians, such as frogs and toads, are perhaps the animals that thrive most in the freshwater biome, whether living in running water or stagnant marshland. Other organisms of this biome include reptiles, like turtles and snakes; waterbirds like ducks, geese, and swans; and mammals, including otters and beavers.

Life in Lotic Systems

1. *Lotic systems*, since they occur along streams, are very dynamic. Organisms must adjust to current and nutrient levels that might vary hourly and from meter to meter. Organisms adapt to the characteristics of a stream as it progresses from its source to the mouth.

18.14.b4 Arizona

18.14.b1 Bozeman, MT

18.14.b2

2. Cooler, generally clearer water with higher dissolved O$_2$ is present in the headwaters (◄), providing habitats for species such as trout (▲).

18.14.b3

3. In the middle part of the stream, biodiversity generally increases, with aquatic algae and green plants at the bottom of the food chain. Relative to the headwaters, cross-sectional area and discharge increase. At meanders, organisms adjust to whether their habitat is located in an erosional zone on the outside of a meander (cutbank) or the inside of a meander (point bar). Here, a caddisfly larva (▲) builds its pebble home in the faster-flowing cutbank environment.

4. Downstream, turbidity increases from sediments eroded from upstream, and dissolved O$_2$ concentration decreases. Catfish (▲) are ideally suited for low light, low-O$_2$ environments.

Life in Lentic Systems

5. The littoral zone absorbs more insolation and is thus warmest. Algae are abundant in this zone, as are submergent and emergent rooted plants (◄), clams, snails, insect larvae, and amphibians. Such flora and fauna are food for organisms higher on the food chain such as snakes, ducks, and alligators (▲). Diversity of vertebrates decreases when descending into the aphotic zone, but bacteria and other decomposers flourish.

18.14.b5 Noxubee NWR, MS

18.14.b6 Everglades, FL

6. Light in the photic portion of the limnetic zone allows the growth of phytoplankton (►), zooplankton, and the fish that feed on them. Organisms migrate daily and seasonally to find the optimal light and temperature.

18.14.b7

Value of Wetlands and Their Organisms

7. Wetlands are highly productive ecosystems and are important wildlife habitats, pollution abators, and flood controllers.

8. Wetlands are home to many species of insects, amphibians, reptiles, otters, beavers, and birds, including snowy egrets (▼), herons, and ducks.

9. Plant species adapted to the constantly wet conditions of wetlands are called *hydrophytes*. These include pond lilies, cattails (►), sedges, and specialized kinds of trees including cypress, tamarack, willows, and black spruce.

18.14.b8 Ketchener, Ontario, Canada

Stream Energy Is Dissipated

Contaminants and Sediment Are Filtered

Provides Critical Wildlife Habitat

Cleaner Water Outflow

Groundwater Flow

Stream

Slow Release of Water

Bacteria Break Down Contaminants

Saturated Peat Stores Water

18.14.b9

18.14.b10 Texas

Before You Leave This Page

✓ Describe the distinguishing features of lotic systems, lentic systems, and wetlands, and how their conditions affect life within them.

18.14

18.15 What Ecosystems Exist in the Saltwater Biome?

THE SALTWATER BIOME covers over 70% of the Earth's surface and so is Earth's largest biome. It is one of the most productive parts of the Earth, but the productivity varies greatly from region to region. Most parts have an overall lack of nutrients, resulting in relatively sparse life and productivity in those places.

A What Are the Characteristics of the Saltwater Biome?

A way to subclassify creatures in the saltwater biome is by whether an organism resides mainly suspended in water—is *pelagic*—or on the ocean floor—is *benthic*. In either case, marine organisms have adapted to different depths, temperatures, salinities, horizontal and vertical currents, substrate characteristics, distances to land, and nutrient sources.

18.15.a1 Point Lobos, CA

1. In the *coastal* or *intertidal zone* (◀), both terrestrial and aquatic organisms are challenged by water availability, wave action, temperature variations, and changes in substrate. Biodiversity is greater in more secure, rocky substrate than in less secure, shifting sand. Sea grass, crabs, starfish, and plankton are abundant in this zone.

2. The *pelagic zone* is located seaward of the coastal zone's low-tide mark and contains the vast open waters of the ocean. Life here can be classified according to location within the water column, mobility, and size. In the *epipelagic* (photic) *zone,* light penetrates and becomes differentiated by depth into colors. The *mesopelagic* and *hadal zones* are too deep for light to penetrate—that is, they are in the *aphotic* zone.

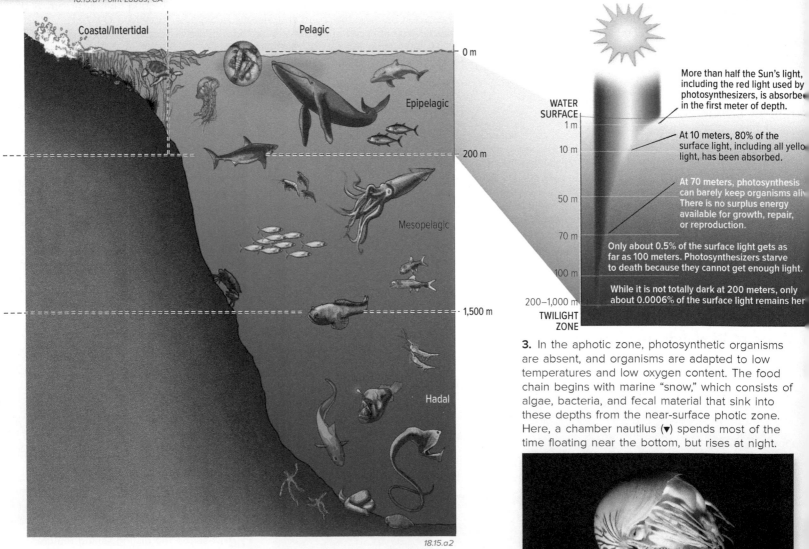

18.15.a2

More than half the Sun's light, including the red light used by photosynthesizers, is absorbed in the first meter of depth.

At 10 meters, 80% of the surface light, including all yellow light, has been absorbed.

At 70 meters, photosynthesis can barely keep organisms alive. There is no surplus energy available for growth, repair, or reproduction.

Only about 0.5% of the surface light gets as far as 100 meters. Photosynthesizers starve to death because they cannot get enough light.

While it is not totally dark at 200 meters, only about 0.0006% of the surface light remains here.

3. In the aphotic zone, photosynthetic organisms are absent, and organisms are adapted to low temperatures and low oxygen content. The food chain begins with marine "snow," which consists of algae, bacteria, and fecal material that sink into these depths from the near-surface photic zone. Here, a chamber nautilus (▼) spends most of the time floating near the bottom, but rises at night.

18.15.a3

4. The *benthic zone* contains all the habitats of the sea bottom, whether in coastal, continental shelf, or deep-sea environments. Organisms may live within the bottom material or on its surface. This zone is among the most poorly explored and understood on Earth.

B What Are the Characteristics of Coral Reefs?

Coral reefs are formed by stony coral polyps living in symbiosis with green and red algae. Coral reefs occur in shallow water where sea surface temperatures range from 20°C to 36°C (68°F to 97°F). They are considered the "tropical rain forest" of the ocean, providing a habitat for some 25% of marine species. Unfortunately, reefs and their community of creatures are sensitive to overfishing, pollution, climate change, and other disturbances.

18.15.b1 (Montastrea cavernosa)

1. Coral are small, but not microscopic, marine animals that build colonies of calcium carbonate (of the mineral calcite and the rock limestone). They gain most of their food from photosynthetic algae that share their habitat, but they also have tentacles for obtaining food that floats or swims by. This close-up photograph (▲) shows individual coral polyps with tentacles and a greenish coating of algae.

18.15.b2

2. Over three-quarters of coral reefs are located in Asia and the Pacific Ocean. Coral thrive in these areas when water temperature is 20–25°C (68–77°F) and the water is clear. The waters, however, are fairly deficient in nutrients. Like terrestrial tropical rain forests, food webs are complex, and nutrients are recycled efficiently.

3. Coral have symbiotic algae (*zooxanthellae*) that provide additional oxygen and nutrients for coral. In return, the coral provide CO_2 and a place to live. Reefs provide shelter and food for an entire community of creatures, including sponges, colorful fish (◄), crabs, and worm-like creatures.

4. Coral bleaching occurs when waters become too warm. Corals expel the algae (zooxanthellae) from their tissues, resulting in coral turning completely white (►). Corals can survive a bleaching event but are in a weakened state and will suffer higher rates of mortality.

18.15.b3 Raja Ampat, Indonesia

18.15.b4 Cayman Islands

C Where Are Other Highly Productive Marine Areas?

Deep-sea thermal vents, estuaries, and upwelling zones are also highly productive marine ecosystems. Chemosynthesis, tidal movement, and Ekman transport help explain much of the productivity in each setting.

18.15.c1

1. In places where there is volcanic activity on the ocean floor—at both convergent and divergent plate boundaries—there exist bizarre ecosystems with productivity that rivals any marine ecosystem—*deep-sea thermal vents*. At these depths, typically deeper than several kilometers, there is no light, so the local energy source is not photosynthesis. Instead, at the bottom of the food chain are microorganisms that derive chemical energy from hydrogen sulfide compounds (in the hot waters of the thermal vents) without requiring light. Tube worms, the eel-like zoarchid fish, vent crabs, squat lobsters, shrimp, and snails exist in symbiotic relationships with vent bacteria. Organisms not only survive without light but also endure crushing pressures and extremely high temperatures. Similar vent environments may exist elsewhere in the Solar System.

2. Saltwater *estuaries,* which can occur in sounds, bays, inlets, or lagoons, are located where fresh water from rivers (or groundwater) mixes with saline ocean water, producing water that is not quite as salty as seawater. Daily changes in water movement and salinity levels are controlling factors in estuaries, as are varying degrees of light penetration; mixing of nutrients (and pollution) from sea and land; and varying sediment deposition patterns. Such diversity in conditions creates varied habitats and high biodiversity, including manatees (◄). Estuaries have high productivity and are the site of most U.S. commercial fishing. They also buffer against storm surges.

18.15.c2 Florida

18.15.c3

Warm Water Low Nutrients

SURFACE WATER

WIND

UPWELLING

CONTINENTAL SHELF

Cold Water High Nutrients

CONTINENTAL SLOPE

SPRING-SUMMER UPWELLING

3. In the figure above, water being pushed parallel to the shore by wind has an apparent deflection to the right of the flow in the Northern Hemisphere (leftward in the Southern Hemisphere) due to *Ekman transport*—the ocean's Coriolis effect. Cold, deep water *upwells*, replacing near-surface water that moved offshore, bringing nutrients up into the photic zone. Such zones of *upwelling* fuel a rich planktonic community, which in turn serves as the base of a food chain for fish, birds, and mammals.

Before You Leave This Page

☑ Describe the zones of the ocean and their ability to support life.

☑ Explain the settings and creatures of coral reefs.

☑ Explain the settings of thermal vents, estuaries, and upwelling zones, and how each supports life.

18.15

18.16 How Do Animal Realms Overlap with Biomes?

PLANTS CANNOT RAPIDLY move in response to environmental stress, so biomes provide a good indication of environmental conditions in a given place. The geographic range and migration of major animal types is influenced by environmental changes across space, but with distributions of different animal realms separated more directly by physical, rather than climatic, barriers. These barriers include bodies of water, deserts, and mountain ranges that are difficult for plants and land-bound animals to cross. *Zoogeographers* study the factors that cause animal distributions and migrations.

A What and Where Are the Major Zoogeographic Realms?

1. Earth's surface can be subdivided based on differences in how the assemblage of plants and animals evolved over time instead of simply using modern-day distributions. Each of the resulting large, continental-sized areas, shown here on the central map, is called a *zoogeographic realm* or *ecozone;* each encompasses many biomes.

18.16.a3

18.16.a2

2. The North American continent, called the *Nearctic realm,* has been mostly isolated from other continents for the past 70 million years, except for intermittent connections, as with Asia during the last glacial maximum. The harsh, seasonal environment has left this realm with limited richness and biodiversity, other than fishes and reptiles. However, canine-type animals, raccoons (▲), and horses are *endemic* (native) to this region.

4. The largest zone, the *Palearctic,* encompasses all of Europe and most of Asia and North Africa. It has many similarities to the Nearctic region, because the two zones were once connected in a supercontinent called *Laurasia.* More recently a land bridge connected the two continents during the last glacial maximum, further allowing an exchange of fauna, including humans. The red panda (▲) is one of the few unique, endemic animals in the Palearctic.

18.16.a4 Namibia

5. The *Paleotropic realm,* covering most of Africa, has great variety, with more mammalian families than any other region. It includes giraffe (▲) and many endemic bird and freshwater fish species.

6. The *Indomalayan realm* includes the Indian subcontinent, which was once joined with southern Africa as part of a large southern supercontinent called *Gondwana.* As a result, the two regions share some faunal characteristics. India was not connected to Asia until much later, 45 million years ago. The realm also includes Southeast Asia and part of Indonesia.

3. The Pacific Islands are too small and isolated to have developed a rich natural fauna, but the *Pacific realm* is home to whales, dolphins, walrus, and seals.

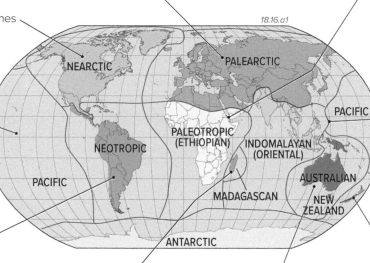
18.16.a1

18.16.a5 Ancient Mammal

18.16.a6

7. Until relatively recently, North and South America were not connected and so evolved very different plants and creatures. South America constitutes the *Neotropic realm,* which contains unique mammals and colorful birds. Once the Panama isthmus connected the Americas (in the last 3 million years), fauna from two realms intermixed somewhat, with *Neotropic* fauna, like opossum, migrating into North America. Some Nearctic fauna migrated into South America, including large cats.

8. The unforgiving environment of Antarctica restricts fauna to birds, including penguins, but there is abundant evidence of extinct animals there.

18.16.a7 Antarctica

9. Known for its primitive primates, such as the lemur (▼) and other fauna that differ widely from other realms, the *Madagascan realm* is the smallest and most distinctive region.

18.16.a8 Madagascar

10. Lengthy isolation has made *Australian* fauna unique, particularly its pouched mammals—*marsupials*—like the kangaroo (▼) and wallaby. With little fresh water in these dry environments, amphibians and fish are rare. Other animals were able to migrate in from Asia.

18.16.a9 Anglesea, Australia

11. Birds, especially flightless ones like the kiwi (pictured in the New Zealand highway sign above), comprise an unusually large share of the fauna in the *New Zealand realm.* Otherwise, large animal life is sparse, with no native mammals.

18.16.a10 New Zealand

B What Changes in Zoogeographic Realms Have Occurred in Geologic Time?

Different zoogeographic realms support different faunal assemblages, suggesting that animals evolved in response to the physical conditions (climate, geology, soils, vegetation) imposed by the biogeographic realm at the time. Mountains, and especially oceans, were barriers to dispersal. Such barriers, however, were mobile and temporary, at least in geologic time.

Before 250 Mya

1. Prior to 250 million years ago (Mya), all of the Earth's major landmasses were joined in a supercontinent known as *Pangaea*. North America was connected on three sides to Europe, Africa, and South America.

18.16.b1

2. Antarctica joined Africa, India, and Australia. At that time, animals could migrate between all of these areas, which today are separate continents. Sedimentary rocks deposited at that time contain some of the same animal and plant fossils, documenting this former linkage. East of the supercontinent was a large, wedge-shaped area of ocean.

After 250 Mya

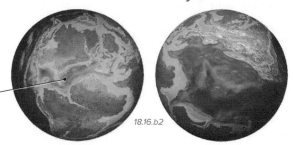

3. After 250 Mya, diversification of mammals and other animals began to increase sharply, just as Pangaea began to rift apart, dispersing the continents in various directions. The sea between the northern and southern continents was a barrier to faunal migration. For example, flightless birds of South America and Australia only had access to continental regions that were part of *Gondwana* (South America, Africa, India, Australia, and Antarctica).

18.16.b2

4. Eventually, India became a separate, small continent that would carry its "southern" flora and fauna northward, until it collided with Asia 45 Mya. New land produced by volcanoes linked South America with Central America 3 Mya.

A Land Bridge Between Zoogeographic Realms

5. Humans have also migrated in response to changing environmental conditions. For hundreds of thousands of years they were restricted to parts of the world with dry-land connections. Around 20,000 years ago, during the *Pleistocene Epoch*, a broad land bridge called *Beringia* formed between Asia and Alaska. With sea level lower, due to water being locked up in glaciers and continental ice sheets, humans could walk to Alaska and colonize the New World.

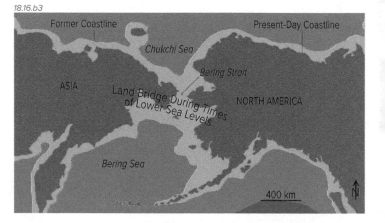

18.16.b3

18.16.b4

6. Animals like the woolly mammoth (◄), caribou, and saber-toothed cat also made the trip across the Beringia land bridge to join the fauna already in North America. Horses and camels from the Americas ventured westward to Asia. Later, humans also ventured across the land bridge, with their arrival largely coinciding with the extinction of many of the large animals (called *megafauna*), but whether this arrival was a cause or was coincidental with the megafauna extinction is being debated.

The Panama Connection

18.16.b6

7. Before 5 Mya, the Atlantic flowed into the Pacific through the Central American Seaway between the Americas, but this connection was closed with the Panama Land Bridge. The connection allowed the migration of armadillos, opossums, and spine-bearing porcupines (▼) northward and deer, horses, and large cats southward. In addition, the Gulf Stream strengthened, and this may have triggered the Pleistocene glaciation. Off the coast of South America the Humboldt Current created a rich upwelling zone for nutrients. *ENSO* events may have also begun with the formation of the land bridge.

18.16.b5

Before You Leave This Page

☑ Explain the concept of zoogeographic realms, using several examples.

☑ Describe how the changing connections and positions of the continents affected the distribution of fauna and flora.

☑ Explain the importance of land bridges in affecting animal distributions and migrations.

18.16

18.17 What Are the Issues with Sustainability?

SUSTAINABILITY, in the context of ecosystems and biomes, refers to whether the environment and its various components operate in a manner that allows them to continue operating without degrading the environment. What is required for sustainability, what threatens sustainability, and what are some opportunities for us to support sustainability? Like exploring other complicated issues, one way to examine sustainability is with a *SWOT analysis*—analyzing Strengths, Weaknesses, Opportunities, and Threats.

A What Is Sustainability?

Sustainability can be defined as effective management that will allow long-term endurance and success, and it can refer to many different aspects of our world. The notion of sustainability implies the existence of responsibility, wisdom, foresight, cooperation, persistence, and balance. Here, we examine sustainability in the context of a biome and its ecosystems.

1. Sustainability requires the consideration of three partially overlapping aspects—*environmental*, *economic*, and *social*, as illustrated in the diagram below. The first step in achieving sustainability is to gather data and ideas about how to maximize the benefit to each of these components, without harming the others.

2. Consider the management of a national park. The images on this page are all from Grand Canyon National Park in Arizona, mostly taken during hikes up the many side canyons that branch off from the Colorado River, which flows through and carved the canyon. Examine each photograph and think about the balance that should be achieved to ensure that the park is healthy environmentally, economically, and socially. What if nobody uses the park? What if it is overused? What if it isn't accessible to some segments of the population? What is the optimum approach that balances all of these considerations? Will such a plan for sustainability need to be revisited over time?

18.17.a1

Venn diagram: ENVIRONMENTAL, ECONOMIC, SOCIAL; Conservation Practices, Environmental Justice, Business Ethics, SUSTAINABILITY

3. Can you think of any drawbacks to trying to achieve sustainability in the national park example? What about in our management of the biosphere?

B   What Are Some Threats and Opportunities for the Sustainability of Biomes?

The often-delicate interactions within an ecosystem—or within a regional biome—can be easily disturbed. Threats are things that might prevent us from achieving sustainability. Awareness of threats is a prerequisite for conquering them.

1. This map shows the amount of land that has been taken over by humans (domesticated) and the biomes of those regions that have not been domesticated. Observe this map and consider all the factors—climate, biome, soil, water availability, suitability for agriculture, and livability, to name a few—that probably influenced whether land was domesticated or not. There are many other factors, including those brought up earlier in this chapter or in other chapters.

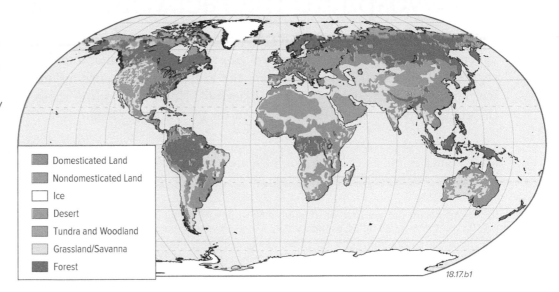

18.17.b1

Legend:
- Domesticated Land
- Nondomesticated Land
- Ice
- Desert
- Tundra and Woodland
- Grassland/Savanna
- Forest

2. As this map reveals, most of the land area has not yet been domesticated. Is this what you expected? Consider all the factors that determine whether the nondomesticated land in this map is used wisely, and remember that even nondomesticated lands are influenced by humans, sometimes heavily. Should only the local inhabitants decide what happens in their home? Should government officials make these decisions instead? The answer may be different for a national park versus a small plot of tundra.

18.17.b2

18.17.b3 Cayman Islands

18.17.b4 Namibia

3. Rain forests are among the most threatened biomes, mostly due to incursions and development by humans. Threatening activities include clear-cutting rain forests for commercial logging and commercial or subsistence agriculture. Sometimes, the most apparent and quick path out of poverty involves exploiting such natural resources.

4. Marine and freshwater ecosystems are likewise vulnerable. Marine ecosystems are damaged by overfishing of species such as the Nassau grouper (▲); pollution dumped into the ocean; and potentially by ocean acidification, where the pH of the ocean becomes more acidic due to increased amounts of CO_2 that move from the atmosphere to the water.

5. Certain large land animals have had their populations decimated by human activities. The most damaging activity is usually encroachment of human developments into the habitat of such creatures. Poaching, such as that which occurs for rhinoceros (▲), has nearly caused some large mammals to become extinct. Some animals, such as mountain gorillas, remain highly threatened, but the situation has recently improved for others.

18.17.b5 Kilimanjaro, Kenya

18.17.b6 Monterey Bay Aquarium, CA

6. There are opportunities to preserve threatened ecosystems. Some species have been greatly aided by conservation efforts, but far too many remain on the endangered list. Conservation efforts include a mix of education, enforcement, and empowerment.

7. Education, like at aquariums (▲), is a key in many situations, enabling people to take pathways to prosperity that minimize inadvertent damage. Science is showing that sustainability makes sense, and it has a history of driving human behavior, eventually.

Before You Leave This Page

✓ Describe the concept of sustainability and its components as it pertains to ecosystems and biomes.

✓ List some threats to sustainability and some opportunities to preserve the diversity in biomes.

18.17

18.18 How Do the Atmosphere, Hydrosphere, and Cryosphere Interact with Land to Form Biomes?

BIOMES ARE THE CULMINATION of the interactions between the atmosphere, hydrosphere, cryosphere, and lithosphere as they intersect to form the biosphere. A particular biome arises because of regional and local atmospheric regimes, its distance from the ocean, the kinds of ocean currents nearby, whether it is cold enough to have snow and ice, and what Earth materials are underfoot. Here, we explore these connections using a longitudinal strip from near the equator to the North Pole, centered on the Atlantic Ocean.

1. Observe this large diagram, the main part of which shows large-scale atmospheric circulation, semipermanent high-pressure and low-pressure areas, warm and cold ocean currents, and the lands flanking the Atlantic Ocean. Examine how the patterns of vegetation on land correspond to the large-scale patterns in the atmosphere and ocean. This figure, and its associated text, is somewhat of a synthesis of this entire book.

2. *Tropics*—Near the equator, warm and moist equatorial (E) air masses rise because this part of the Earth receives more intense insolation than latitudes farther north and south. The rising air causes the region to have low atmospheric pressure. As the air rises, it cools and results in heavy and regular precipitation over the tropics. The abundant water allows luxuriant and diverse plant growth, with many K-selected species, but it also causes severe leaching of soluble components of the soil, resulting in reddish, leached *Oxisols*. The combination of perpetually high temperatures and abundant rainfall is characteristic of a *Tropical Rain Forest climate* (*Köppen Group A*) and produces a *tropical rain forest biome* with its characteristic flora and fauna.

3. *Subtropics*—In the subtropics, near 30° N latitude, descending air from the northern limb of the Hadley cell causes the underlying region to have high atmospheric pressure (subtropical highs, in this case the Bermuda-Azores High) and to be exceptionally dry. The dry, descending air stifles the formation of clouds and precipitation. Evaporation exceeds precipitation in these "biological desert" oceanic areas, increasing the salinity of the water. The lack of high-discharge rivers flowing into the North Atlantic reinforces the high salinity, and this region becomes part of the global thermohaline circulation. The clear skies allow unrestricted insolation to the land and ocean, which heats the overlying air. The land warms and cools faster than the water because it has a lower heat capacity (less thermal inertia). In the ocean offshore of Africa, the south-flowing Canary current, part of the North Atlantic gyre, brings cold water southward along the coast, further stabilizing the atmosphere by reducing the environmental lapse rate and in some cases causing a temperature inversion, which in turn limits precipitation. The dry conditions from continental tropical (cT) air masses result in *Aridisols*, with their organic material and calcium carbonate–rich B-horizon. The combination of descending dry air, sunny skies, and limited humidity and precipitation results in an *Arid climate* (*Köppen Group B*) and a *desert biome*. Plants and animals in this biome are adapted to the high temperatures and limited water.

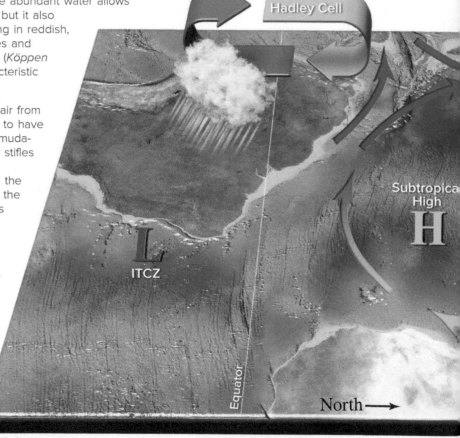

18.18.a1

Tropical Rain Forest

18.18.a2 Cambodia   18.18.a3 Costa Rica

Desert

18.18.a4 Namibia   18.18.a5 Tucson, AZ

4. *Mid-latitudes*—The mid-latitudes, near 45 to 60° N latitude, coincide with Great Britain east of the Atlantic Ocean (east is toward the bottom of the page) and eastern Canada west of the Atlantic. These areas are beneath rising air at the polar front, which also forms part of the polar air-circulation cell. The rising air causes semipermanent low pressure, in this case the Icelandic Low. As the moist air rises, its water vapor condenses into clouds and precipitation. The air is cold enough during the winter months to cause snow. Cold air flowing along the surface from the north into the Icelandic Low has an apparent deflection to the right, becoming polar easterlies. The west and southwest-moving air collides with air flowing from the southwest along the polar front, generating mid-latitude cyclones and trailing cold and warm fronts. This often involves the clash between maritime tropical (mT) air masses and continental polar (cP) air masses. The region is also under the influence of the polar front jet stream, which is embedded in the Rossby waves with its ridges and troughs that modulate the surface weather patterns. The combination of rising and converging air, low pressure, and the overlying jet stream results in an overall wet climate. Under the warming influence of the Gulf Stream, Great Britain maintains a *Marine West Coast climate* (*Köppen Group* C). Seasonal shifts in the strength and position of the Icelandic Low and Bermuda-Azores High (the North Atlantic Oscillation) cause climate variability. Great Britain has mostly *Alfisols* and *Inceptisols* and a *temperate forest biome*.

> ### Before You Leave This Page
>
> ☑ Describe each of the four regions and explain the combinations of factors that produce the characteristic biomes in each.

5. *Polar Regions*—Near the North Pole, the land receives insolation at a very low angle of incidence in summer and is dark for months at a time during the winter. As a result, the land and water do not warm much and the air stays cold much of the year. Added to the weak insolation is cold air descending over the pole, the downward limb of the polar cell—the Polar High. The cold temperatures, low insolation, and semipermanent high pressure of these Arctic (A) air masses limit the amount of water vapor in the air and precipitation, which in these cold latitudes is generally snow. The high albedo of snow reflects much of the paltry amount of insolation on it, in a positive feedback mechanism. The climate is *Polar* (*Köppen Group* E) or *Subarctic* (*Köppen Group* D). Permanently frozen ground (permafrost) at shallow depths—overlain by *Gelisols*—limits tree growth but allows ground-hugging cold-tolerant plants, the signature of the *tundra biome*. The harsh conditions limit the biodiversity, but allow polar bears, caribou, and other polar creatures to survive.

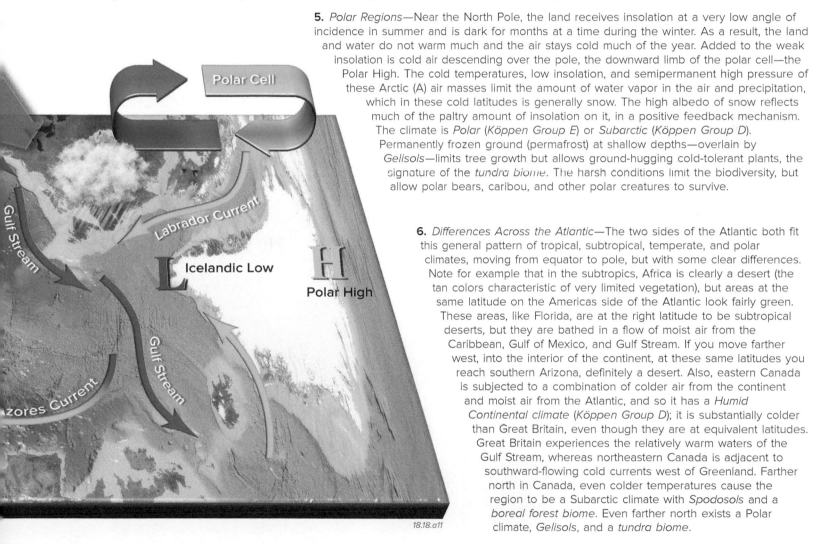

Polar Cell

Gulf Stream

Labrador Current

L Icelandic Low

H Polar High

Azores Current

Gulf Stream

18.18.a11

6. *Differences Across the Atlantic*—The two sides of the Atlantic both fit this general pattern of tropical, subtropical, temperate, and polar climates, moving from equator to pole, but with some clear differences. Note for example that in the subtropics, Africa is clearly a desert (the tan colors characteristic of very limited vegetation), but areas at the same latitude on the Americas side of the Atlantic look fairly green. These areas, like Florida, are at the right latitude to be subtropical deserts, but they are bathed in a flow of moist air from the Caribbean, Gulf of Mexico, and Gulf Stream. If you move farther west, into the interior of the continent, at these same latitudes you reach southern Arizona, definitely a desert. Also, eastern Canada is subjected to a combination of colder air from the continent and moist air from the Atlantic, and so it has a *Humid Continental climate* (*Köppen Group* D); it is substantially colder than Great Britain, even though they are at equivalent latitudes. Great Britain experiences the relatively warm waters of the Gulf Stream, whereas northeastern Canada is adjacent to southward-flowing cold currents west of Greenland. Farther north in Canada, even colder temperatures cause the region to be a Subarctic climate with *Spodosols* and a *boreal forest biome*. Even farther north exists a Polar climate, *Gelisols*, and a *tundra biome*.

Temperate Forest

a6 Quebec, Canada

18.18.a7

Tundra

18.18.a8 ANWR, AK

18.18.a9 Churchill, Canada

18.18

18.19 What Factors Might Control the Biomes in This Location?

THE INVESTIGATION at the end of Chapter 1 asked about possible impacts of logging of a mountainside, the same area shown here. Now that you have acquired a more heightened sense of the Earth-ocean-atmosphere-cryosphere-biosphere system, let us revisit the problem—and the entire environment—from a wide array of perspectives.

Goals of This Exercise:

- Review the list that you created in Chapter 1 of the many possible impacts of deforestation in this area and elaborate on, refine, or dismiss the relevancy of each point.

- Examine the area and identify all the aspects that helped shape the observed landscape.

- Identify five areas or processes that might be a concern from the standpoint of sustainability. For each, conduct a SWOT analysis, factoring in all the relevant aspects presented in this book.

- Your instructor may have you develop a land-use plan to make this place as sustainable as possible.

This three-dimensional perspective shows an area with a variety of terrains and climates. Make as many observations as you can about the landscape and record these on a sheet of paper or on the worksheet. Complete each step in the "Procedures" on the facing page. Listed around the figure are some of the issues you should consider, but there are others not listed. Think broadly and be creative, using the knowledge and observational skills you have gained through the geographic approach.

Patterns of local atmospheric circulation, including prevailing wind direction and precipitation patterns:

Rock types:

Soil types:

Slope stability:

Weathering and erosion:

Local climate:

Types of vegetation:

Natural space versus developed space:

Procedures

1. Examine the scene and think about all the factors that are important with regard to the issue about deforestation and environmental threats to sustainability. Consider all the factors indicated by labels arranged around the figure, and think of other factors that are not listed. To help you organize your thoughts, list these factors. Compare your ideas with those you produced when you did the Investigation in chapter 1. Elaborate, refine, or replace any item based on what you can now surmise about this area (for example, what is the likely climate, soil, ecosystems, and biome).

2. Think broadly about the entire scene and each of its components. For each of the main components, sketch and explain the main processes involved in each system (e.g., the glacier and snow), including what conditions were required to have that system develop, what processes occur within that system (e.g., precipitation as snow), and what factors could influence how that system operates in the future (e.g., weather and climatic patterns that cause more or less snow). Complete one annotated sketch for at least five systems within this area. There are more systems than this, so choose the ones you think are most important or most interesting. Be sure to consider all aspects by reviewing the title of each chapter in the book to remind yourself of all the different avenues that might be important.

Flora and fauna on the land and in the freshwater biome, including birds:

Large-scale patterns of atmospheric circulation:

Possible seasonality:

Water supplies from the streams and lake:

Microclimates associated with the topography, shoreline, and city:

Impact on solar incidence and storage of solar-derived energy:

Coastal environment, which features a current flowing parallel to the coast:

Vertical currents in the ocean and their effect on the marine ecosystems:

Plants and creatures in the upper part of the ocean (pelagic) and near the seafloor (benthic):

Oscillations in sea temperature and air pressure:

Optional Activities

Your instructor may have you complete the additional activities listed below. For these activities, type your answers into a word-processing document and submit this to your instructor in the manner in which you are directed (paper copy versus electronic copy).

3. For the system you sketched and explained in step 2 above, conduct a SWOT analysis to consider any factors that would affect the sustainability of that system.

4. Produce a land-use planning document that considers each system sketched in step 2, to develop a plan about how each part of the area would best be used. Your instructor will give you guidance about the length and format of the document.

18.19.a1

18.19

GLOSSARY

A

aa A blocky type of lava with a rough, jagged surface, or a lava flow characterized by jumbled, angular blocks of hardened lava. (11.3)

abiotic component The part of an ecosystem not directly produced by living organisms, including the air, rocks, soils, and water that provide a home for the plants and other organisms. (17.1)

abrasion The mechanical wearing away of rock surfaces by contact with sand and other solid rock particles transported by running water, waves, wind, ice, or gravity. (14.4)

absolute humidity The mass of water vapor present per cubic meter of air. (4.2)

absolute zero The temperature at which all molecular motion theoretically ceases (i.e., there is no internal kinetic energy); occurs at 0 Kelvin, or –273°C. (2.3)

absorption The concentration of electromagnetic energy within the molecular structure of matter. (2.1)

abyssal plain A relatively flat, smooth, and old region of the deep ocean floor. (10.2)

abyssal waters Any water near the ocean floor, generally circulating toward the equator. (6.7)

accretionary prism A mass of material (terrane) formed from sediments (mostly marine) that are scraped onto the non-subducting tectonic plate at a convergent plate boundary. (10.13)

acidic A material, such as water or soil, having a high concentration of hydrogen ions. (16.4)

acid rain An acidic secondary pollutant in the atmosphere formed when sulfur dioxide or nitrogen oxides interact with water in precipitation, either before or after the precipitation falls to the surface. (7.10)

acre-foot The volume of water required to cover an acre of land to a height of one foot; equivalent to about 326,000 gallons or 1.23 million liters. (8.5)

active dune A dune of any type that is actively moving and being reshaped by the wind. (9.11)

active fraction The component of a soil consisting of decomposing organic matter. (16.1)

active layer In tundra, the zone between the surface and the permafrost that undergoes a freeze-thaw cycle each year. (14.7)

active remote sensing system A type of remote sensing in which the environment is monitored as the sensor emits energy and then detects properties of that energy after it interacts with matter in the atmosphere or near the surface and returns to the platform. (1.11)

adenosine triphosphate (ATP) A molecule produced as energy is released by "excited" electrons during the light reaction of photosynthesis. (17.5)

adiabatic A condition of a system in which energy does not enter or leave the system. (4.5)

adret slope The side of a slope that receives more direct sunlight, and is therefore warmer and drier. (18.4)

advection The horizontal transfer of energy through a gas, liquid, or weak solid. (2.2)

advection fog A type of fog that occurs when warmer air flows laterally over a colder surface, causing it to chill from beneath to the dew-point temperature or frost-point temperature. (4.9)

aeolian Any process relating to the wind's ability to shape the landscape or a geomorphic feature, primarily caused by wind deposition. (9.11, 16.6)

aerosol Any tiny solid or liquid particle suspended in the air. (2.1)

African easterly jet A west-flowing, fast stream of air in the northeast trade winds over or near northwestern Africa. (5.13)

aftershock A smaller earthquake that occurs after the main earthquake, in the same area. (11.16)

agent of transport Any phenomenon, including gravity, streams, glaciers, waves, and wind, that moves material, which can later be deposited to create landforms. (9.10)

aggradation The deposition of material by a stream or current, building up the land surface in that area. (13.7)

agricultural lime Pulverized limestone used to improve fertility in acidic soils. (16.4)

A horizon Topsoil composed of dark gray, brown, or black organic material mixed with mineral grains. (16.7)

air mass A body of air that is relatively uniform in its meteorological characteristics over areas of thousands of square kilometers. (5.1)

air pollution A gaseous, liquid, or solid particle suspended above the surface; usually introduced into the atmosphere by human activities and having effects that are harmful to life or property. (7.9)

Alaska Current A south-to-north-flowing, warm surface ocean current in the eastern North Pacific Ocean. (6.2)

albedo The percentage of insolation that is reflected by an object. (2.13)

Alberta clipper A type of mid-latitude cyclone that forms on the leeward side of the Canadian Rockies and tracks generally east-northeastward. (5.3)

Aleutian Low A prominent area of semipermanent low atmospheric pressure (cyclone), particularly strong in winter, in the subpolar North Pacific Ocean. (3.6)

Alfisol A fertile type of soil, typically with ample accumulation of organic matter in the A horizon and clay eluviation into the B horizon. (16.11)

alkaline A material, such as water or soil, having a low concentration of hydrogen ions; also known as basic. (16.4)

allelopathy A survival strategy of desert vegetation in which a poison is injected from the roots into the soil to prevent other plants from growing nearby. (18.8)

alluvial fan A low, gently sloping mass of sediment, shaped like an open fan or a segment of a cone; typically deposited by a stream where it exits a mountain range or joins a main stream. (9.10)

alluvial soil Any soil formed from gravels, sand, silt, and clay that was deposited by streams. (16.6)

alluvium Sedimentary material that was transported and deposited by a stream. (16.6)

alpine glacier A glacier that flows down a valley and tends to be narrow; also known as a valley glacier. (14.1)

altocumulus cloud A type of mid-level, vertically oriented cloud that contains both liquid water and ice. (4.8)

altostratus cloud A type of mid-level cloud that is wide and layered, and composed of both liquid water and ice. (4.8)

amphibole A group of silicate minerals characterized by double chains of silicon-oxygen bonds; most common in igneous and metamorphic rocks. (9.1)

amplitude The height between the trough and the crest of a wave, including electromagnetic, ocean, and seismic waves. (2.5)

anabatic wind A local-to-regional-scale wind that blows from a valley uphill to the peak of a hill or mountain; also known as a valley breeze. (3.4)

anadromous Fish that spawn in fresh water but live most of their lives in salt water. (18.3)

analysis products Maps based on spatial-interpolation techniques that represent the observed and inferred state of the atmosphere at a given instant or interval of time at the Earth's surface or at a specified vertical constant pressure level. (5.15)

anastomosing channel A low-gradient stream that has semipermanent islands between several strands of the channel that split and rejoin. (13.8)

andesite An intermediate-composition igneous rock that is the fine-grained equivalent of diorite; typically gray or greenish gray, commonly with visible crystals of cream-colored feldspar or dark amphibole. (9.5)

Andisol A site-dominated soil type formed from volcanic ash and other volcanic material. (16.9)

anemometer A device used to measure wind speed. (3.3)

angle of incidence The angle between the direction of incoming energy and a surface. (2.8)

angle of repose The steepest angle at which a pile of unconsolidated grains remains stable. (12.8)

anion A negatively charged ion. (6.4)

annular drainage pattern A drainage network in which the pattern of streams generally encircles a central area; characteristic of terraced areas around a central high or low point. (13.1)

Antarctic Bottom Waters The moderately saline, very cold, and very dense ocean waters in the middle to high latitudes of the Southern Hemisphere. (6.5)

Antarctic Circle The 66.5° parallel of latitude south of the equator; the most northerly latitude in the Southern Hemisphere that experiences 24 continuous hours of daylight (on the December Solstice) and 24 continuous hours of darkness (on the June Solstice). (1.7)

Antarctic Circumpolar Current A west-to-east-flowing, surface ocean current in the high latitudes of the Southern Hemisphere; also known as the West Wind Drift. (6.2)

Antarctic Intermediate Waters A tongue of less saline ocean waters from near Antarctica that forms a wedge below saline waters near the surface in the Southern Hemisphere, centered around 40° S. (6.5)

antecedent stream A stream that predates formation of a geologic structure through which it flows. (13.11)

anticline A fold, generally concave downward (in the shape of an *A*), with the oldest rocks in the center. (11.11)

anticyclone An enclosed area of high atmospheric pressure. (3.9)

aphelion The date of maximum distance between Earth and the Sun during the course of Earth's orbit, at present corresponding to a day on or near July 4. (2.7)

aphotic zone The part of a lentic system that does not have access to sunlight. (17.12)

aquiclude An impermeable surface that completely blocks flow of a liquid through it. (8.6)

aquifer A saturated body of permeable material through which groundwater can flow and that yields significant volumes of water to wells and springs. (8.6)

Arctic (or Antarctic - A) air mass A body of air that is relatively uniform in its very cold and very dry characteristics for thousands of square kilometers, originating in high-latitude source regions that are either inland or over sea ice. (5.1)

Arctic Circle The 66.5° parallel of latitude north of the equator; represents the most southerly latitude in the Northern Hemisphere that experiences 24 continuous hours of daylight (on the June Solstice) and 24 continuous hours of darkness (on the December Solstice). (1.7)

arête A hard, jagged bedrock ridge flanked by two or more cirques. (14.5)

arid (B) climate In the Köppen climate classification, a broad group of climatic types characterized by inadequate moisture availability, in which potential evapotranspiration typically exceeds precipitation. (7.5)

Aridisol A type of soil found in arid regions, characterized by a lack of organic matter, little leaching of nutrients, and limited horizonation; has a propensity for salinization and formation of a caliche layer. (16.10)

artesian Groundwater that is in a confined aquifer and is under enough water pressure to rise above the level of the aquifer. (8.6)

ash flow A fast-moving cloud of hot volcanic gases, ash, pumice, and rock fragments that generally travel down the flanks of a volcano; also known as a pyroclastic flow. (11.2)

asthenosphere The area of mantle beneath the lithosphere that is solid, but hotter than the rock above it, and can flow under pressure; functions as a soft, weak zone over which the lithosphere may move. (10.1)

Atlantic hurricane season The time of year when Atlantic tropical cyclones are most likely to occur—from June 1 through November 30. (5.13)

Atlantic Multidecadal Oscillation (AMO) An oscillation in sea-surface temperatures in the Atlantic Ocean, in which temperatures tend to be warmer than normal in much of the basin for many years and then colder than normal for many years. (7.12)

atmosphere The gaseous envelope that surrounds the solid Earth. (1.3)

atmospheric attenuation The process by which electromagnetic energy is absorbed or scattered in the atmosphere. (2.8)

atmospheric window The part of the electromagnetic spectrum between 8 and 13 micrometers, in which atmospheric absorption of longwave radiation is minimal, except when water vapor is abundant. (2.14)

atoll A curved to roughly circular reef that encloses a shallow, inner lagoon or a circular arrangement of low, coral islands. (15.6)

aurora A light that originates from interactions of solar energy and energetic gas molecules, primarily in the thermosphere. (2.1)

autotroph An organism that acquires what it needs from the soil, water, air, and energy of the environment, without consuming other organisms; also known as a primary producer. (17.2)

B

backshore The highest part of the beach, only seldom covered by water; situated between the foreshore and the dune. (15.1)

backwash The flow of water back out to sea after a wave comes ashore. (15.3)

badlands A distinctive type of topography with a soft, rounded appearance, usually composed of shale and other easily eroded rocks. (9.7)

bank-full The maximum stream flow without causing a flood. (13.12)

banner tree A wind-shaped tree, commonly near the tree line, that has branches growing only on the downwind (leeward) side of a trunk; also known as a flag tree. (18.4)

bar Abbreviation for barometric pressure, another term for atmospheric pressure. (3.2)

barchan dune A crescent-shaped dune of windblown sand with its tails pointing in the direction of the prevailing wind; also known as a crescent dune. (9.11)

barometer An instrument used to measure atmospheric pressure. (3.2)

barrier island A long, narrow island adjacent to a coastline and commonly containing sandy areas, like dunes, marshes, and sandy beaches. (15.5)

barrier reef A long, commonly curved, coral reef that parallels the coast of an island or continent from which it is separated by a lagoon. (15.6)

bar scale A visual representation of the relationship between actual distance and a distance on a map. (1.10)

basal slip The relatively fast sliding of a glacier over bedrock, either in a smooth bedrock-glacier interface or, if the glacier contains water, from pressure melting. (14.3)

basalt A fine-grained, dark-colored mafic igneous rock, with or without vesicles and visible crystals of pyroxene, olivine, or feldspar. (9.5)

basaltic lava flow A lava flow composed of dark, basaltic material. (9.7)

base level The lowest level to which a stream can erode, commonly represented by the level of the sea, lake, or main river valley. (13.4)

base map A map or image upon which data are plotted. (1.10)

basic A material, such as water or soil, having a low concentration of hydrogen ions; also known as alkaline. (16.4)

basin A low-lying area or an area in which sediments or volcanic materials accumulate with no or a limited surface outlet; also can refer to a structural feature toward which rock strata are inclined in all directions. (10.6)

Basin and Range Province The physiographic region in the interior Southwest of the U.S., consisting of almost all of the state of Nevada and parts of adjacent states, and characterized by alternating valleys and mountain ranges, commonly the expression of horsts and graben. (11.10)

batholith One or more contiguous intrusions that cover more than 100 square kilometers. (11.8)

baymouth bar A ridge of sand or gravel deposited entirely or partly across the mouth of a bay. (15.5)

beach A stretch of coastline along which sediment, especially sand and stones, has accumulated. (9.2)

beach drift The process by which sand and other sediment move in a zig-zag fashion, but in a general lateral direction along a coast. (15.4)

bed A distinct layer formed during deposition, generally in sediment or sedimentary rock; can be present in volcanic tuff and some metamorphic rocks. (9.8)

bed load Material, commonly sand and larger, that is transported along the bed of a stream. (13.2)

bedrock stream A stream that is carved into bedrock, commonly in mountainous areas. (13.2)

Benguela Current A south-to-north-flowing, cold surface ocean current in the eastern South Atlantic Ocean. (6.2)

benthic Pertaining to an organism in the saltwater biome that resides mainly on the ocean floor. (18.15)

benthic zone The part of the saltwater biome that contains habitats of the sea bottom, whether in coastal, continental shelf, or deep-sea environments. (18.15)

Bergeron process A process of precipitation formation that occurs by evaporation of liquid water, which then allows the additional water vapor to deposit onto solid ice particles, enlarging the snowflake, which can then melt, forming a raindrop, or remaining solid on descent. (4.11)

Bering Current A small north-to-south-flowing, cold surface ocean current into the North Pacific Ocean through the Bering Strait. (6.2)

Beringia The broad land bridge that connected Asia to Alaska during the Pleistocene Epoch. (18.16)

berm A nearly flat platform in the backshore, caused by wave deposition during the strongest storms. (15.1)

Bermuda-Azores High A prominent area of semipermanent, high atmospheric pressure (anticyclone) in the subtropical North Atlantic Ocean; is particularly strong in summer. (3.6)

B horizon A zone in the soil characterized by the accumulation of material, including iron oxide, clay, and calcium carbonate, depending on the climate and parent materials; also known as the subsoil layer, zone of accumulation, or the zone of illuviation. (16.7)

biodiversity The number of species and the evenness of the populations of those species in an ecosystem. (17.4)

biogeochemical cycle The movement of matter throughout the atmosphere, hydrosphere, lithosphere, and biosphere by interaction with living and dead organisms and involvement in chemical and physical processes. (1.4)

biogeography The branch of physical geography that explores the biosphere and the Earth systems that support organisms. (17.0)

biological oxygen demand (BOD) The amount of oxygen needed by the bacteria in an aquatic ecosystem. (17.12)

biomagnification The concentration of toxins, such as mercury, in an organism from ingesting other plants or animals in which the toxins are more widely disbursed. (17.2)

biomass The mass of carbon contained in a living or dead organism or in a collection of organisms. (17.6)

biome A number of adjacent ecosystems defined by the predominant types of vegetation and characterized by the presence of organisms that are adapted to that environment. (7.1, 18.0)

biosphere The spherical zone that includes life and all of the places life can exist on, below, and above Earth's surface. (1.3, 17.0)

biotic component The part of an ecosystem consisting of living plants and animals, along with materials such as digestive wastes, discarded plant parts, and decaying remains of creatures. (17.1)

biotic weathering Any physical or chemical weathering process caused by organisms. (12.4)

biotite A common, dark phyllosilicate mineral within the mica group. (12.3)

birdfoot delta A nearly flat tract of land formed by deposition of sediment at the mouth of a stream, formed by distributaries and having separate "toes" spreading out at the mouth. (13.9)

blocky soil A soil structure in which the peds are shaped like irregular chunks, which may hinder root penetration. (16.2)

blue jet A dim, blue, smoke-like streak of light that bursts out above a thunderhead, arcs upward high into the atmosphere, and then fades away. (5.9)

bluff The steep slope separating adjacent terraces or a terrace from the floodplain; also known as a riser. (13.8)

bog A wetland that is fed mostly by precipitation rather than groundwater, surface runoff, or streams; usually nutrient-poor. (18.14)

boiling point The temperature at which a liquid boils; that is, the temperature at which the vapor pressure in a liquid increases to a value equal to the pressure of the gas adjacent to the liquid, thereby forcing the molecules in the liquid to leave the liquid spontaneously. (2.3)

bomb (volcanic) A large rock fragment representing either a large mass of magma or a solid angular block ejected during an explosive volcanic eruption. (11.1, 11.6)

boreal forest biome A collection of adjacent ecosystems that consists of a few hardy cone-bearing (coniferous) tree species that have needles instead of normal flat leaves, and local stands of deciduous species, similar to some species in temperate forests. (18.1)

boulder A clast or rock fragment with a diameter exceeding 0.25 m. (13.2)

bow echo The arc-shaped pattern on Doppler radar that is characteristic of a derecho. (5.12)

Boyle's Law The inversely proportional relationship between pressure and volume of a gas, under constant temperature. (3.1)

brackish Water that is intermediate in salinity between normal fresh water and seawater. (8.1)

braided stream A stream with an interconnecting network of branching and reuniting shallow channels. (13.5)

Brazil Current A north-to-south-flowing, warm surface ocean current in the western South Atlantic Ocean. (6.2)

breaker A large sea wave that breaks into white foam when reaching the shore. (15.3)

breakwater An offshore structure (such as a wall), typically parallel to the shore, that breaks the force of the waves to protect a shoreline or harbor. (15.7)

breccia A rock composed of large, angular fragments; typically formed in sedimentary environments, but can be formed by volcanic, hydrothermal, or tectonic processes. (9.4)

brine Highly saline water. (6.6)

brine exclusion The nonparticipation of salt in the freezing of seawater, which makes the surrounding water saltier. (6.6)

brittle deformation A type of structural deformation that is characterized by breaking of material, such as by jointing, faulting, and the formation of tectonic breccia; typical of shallow crustal conditions. (11.9)

broadleaf Any tree with leaves rather than needles. (18.11)

buffering mechanism A feature of an ecosystem that helps species avoid stress and thrive. (17.7)

C

C₃ plants Any plant that creates sets of two 3-carbon-based molecules in photosynthesis; these include most green plants. (17.5)

C₄ plants Any plant that creates sets of 4-carbon-based molecules in photosynthesis; such plants tend to thrive in hotter and drier environments because water loss through the plant's stomates is reduced in this process. (17.5)

calcification A pedogenic regime characterized by the upward movement of soil moisture toward dry air in arid and semiarid regions, depositing calcium carbonate (caliche) in the B horizon. (16.6, 16.8)

calcite A common rock-forming calcium carbonate ($CaCO_3$) mineral occurring in limestone, travertine, and a variety of water-related deposits. (9.1)

caldera A large volcanic depression that is typically circular to elongated in shape and formed by collapse of a magma chamber. (11.1)

caliche A soil-related accumulation of calcareous material that cements sand, gravel, and other materials, commonly forming a hard layer in arid environments. (16.5)

California Current A north-to-south-flowing, cold surface ocean current in the eastern North Pacific Ocean. (6.2)

calorie A unit used to measure energy input. (17.6)

Calvin cycle A series of reactions during photosynthesis whereby carbon dioxide is converted to glucose. (17.5)

calving The breaking off of a mass of ice from a glacier, iceberg, or ice shelf. (14.3)

Cambrian explosion The widespread, relatively rapid appearance of diverse types of hard-shelled organisms near the beginning of the Cambrian Period. (9.13)

CAM plant Any plant that conducts C_4 photosynthesis at some times and C_3 at others, in an effort to enhance its chance of survival. (17.5)

campos A grassy plain with an occasional stunted tree, as can be found in some parts of South America, especially Brazil. (18.9)

Canadian Shield The region of North America, mostly in Canada, that consists of relatively old crystalline metamorphic and igneous rocks. (10.13)

Canary Current A north-to-south-flowing, cold surface ocean current in the eastern North Atlantic Ocean. (6.2)

canopy The second-highest layer of life in a rain forest, typically forming an umbrella over the forest floor. (18.5)

capacity (stream) The amount of sediment, at a given flow rate, that a stream is capable of transporting. (13.2)

capillary action The ability of water, aided by surface tension, to travel upward against the force of gravity. (4.1)

capillary water Subsurface water that is within pore spaces between the grains but not adhering to the soil particles, and therefore is available to move upward through soil toward the surface. (16.3)

carbonate minerals A group of minerals containing a significant amount of the carbon-oxygen combination called carbonate. (9.1)

carbon cycle The movement of carbon between the biosphere, atmosphere, hydrosphere, and lithosphere. (1.4)

carbonic acid A weak acid (H_2CO_3) formed in water and serving as a major agent of chemical weathering. (12.2)

carnivore Any heterotroph that exclusively or mainly feeds on animals. (17.2)

carrying capacity The theoretical maximum population of a given species that an eco-system can support. (17.3)

cartogram A type of map that distorts areal features (e.g., states, countries, or other political unit) so that their size conforms to the magnitude of the variable associated with that state, country, or other political unit being mapped. (17.11)

cartographer A person who makes maps. (1.9)

catadromous Fish that live most of their lives in fresh water and spawn in salt water. (18.3)

catastrophism The concept that emphasizes the importance of sudden events in creating change to a landscape, ecology, or other system. (17.7)

cation A positively charged ion. (6.4)

cation exchange capacity The ease with which a cation in soil solutions is exchanged with the surface of organic matter or clay particles; high cation exchange capacity is generally associated with agriculturally productive soils. (16.4)

cave A natural underground chamber that is generally large enough for a person to enter. (12.2)

Celsius scale A system for indexing temperature calibrated such that water at sea level freezes at 0°C and water at sea level boils at 100°C. (2.3)

cementation The precipitation of a binding material (the cement) around grains in sedimentary rocks. (9.4)

Cenozoic Era A major subdivision of geologic time from 66 Ma to the present; characterized by an abundance of mammals. (9.13)

centripetal drainage pattern A drainage network in which the pattern of streams is generally inward toward a central point. (13.1)

chalk A type of soft, very fine-grained limestone that forms from the accumulation of calcium carbonate remains of microscopic organisms that float in the sea. (9.4)

channeled scablands Flat, elevated land deeply scarred by channels of glacial or fluvioglacial activity, especially in the Columbia Plateau of Washington State. (14.7)

channelized Flowing water that is confined to a channel, either naturally or through artificial means. (13.6)

chaparral The distinguishing type of vegetation in the subtropical scrub and woodland biome, characterized by scrubby trees with thick bark and short, gnarly shrubs, thickets, and grasses with deep root systems; also known as maquis. (18.10)

Charles's Law The proportional relationship between temperature and volume in a gas at a constant pressure. (3.1)

chemical weathering Chemical reactions that affect a rock or other material by breaking down minerals and removing soluble material from the rock. (9.9, 12.1)

chemosynthesis A process by which inorganic compounds, such as hydrogen sulfide or diatomic hydrogen, make energy available for synthesis of organic molecules when combined with oxygen. (17.5)

chert A silica-rich, nonclastic sedimentary rock that typically forms from the accumulation and compaction of tiny, silica-rich plankton shells that fall to the ocean bottom, or also as layers and irregular masses within limestone that form from the mixing of different types of water. (9.4)

Chile Current A local name for the Humboldt Current. (6.2)

Chinook wind A warm, dry wind that descends the leeward side of mountain ranges in western North America. (3.5)

chlorofluorocarbon A type of chemical released from aerosol cans, air conditioning units, refrigerators, and polystyrene, which contains halogens that can destroy ozone molecules by attracting one of the oxygen atoms away from them. (2.12)

chloroplast A specialized organelle, on a plant cell or in some algae, that captures insolation during photosynthesis. (17.5)

C horizon A zone in the soil composed of weathered bedrock or sediment, grading downward into unweathered bedrock or sediment. (16.7)

cinder A loose black or red, pebble-sized material, often a type of basalt, that composes a cinder cone. (11.1)

cinder cone A relatively small type of volcano that is cone shaped and mostly composed of volcanic cinders (scoria); also known as a scoria cone. (11.1)

circle of illumination The circle that separates the half of the Earth that is experiencing daylight from the half that is experiencing darkness at a given time. (2.10)

circulation cell A large, somewhat rectangular, vertically oriented circulation system in which air rises, moves laterally, sinks, and then moves laterally to replace the air that rose. (3.9)

cirque An open-sided, bowl-shaped depression formed in a mountain, commonly at the head of a glacial valley; produced by erosion of a glacier. (14.5)

cirque glacier The smallest type of alpine glacier; forms in a glacial depression that is somewhat sheltered from the Sun. (14.1)

cirriform cloud A family of cloud types that is "feathery" or wispy and composed of ice crystals, usually high in the troposphere. (4.8)

cirrocumulus cloud A type of high-level cloud that is partly cumuliform (lumpy) and partly cirriform (wispy) and almost entirely composed of ice. (4.8)

cirrostratus cloud A type of high cloud that is somewhat wispy but with a continuous, sheetlike appearance. (4.8)

cirrus cloud A type of high-level cloud with a wispy appearance, composed of ice particles. (4.8)

clast An individual grain or fragment of rock, produced by the physical breakdown of a larger rock mass. (9.4)

clastic rock Rock consisting of pieces (clasts) derived from preexisting rocks; usually formed on Earth's surface in low-temperature environments. (9.4)

clay Any fine-grained sedimentary particle with a diameter less than 1/256 mm; also, a family of finely crystalline, hydrous-silicate minerals with a two- or three-layered crystal structure. (9.1)

Clean Air Act of 1970 The (U.S.) federal legislation that mandated the formation of the federal Environmental Protection Agency along with other similar agencies at the state level, for the purpose of monitoring certain widespread pollutants nationwide and keeping the concentrations of each below a "threshold" level deemed hazardous. (7.9)

clear-cutting Deforestation in which all the trees in an area are removed. (18.6)

climate The long-term pattern of weather at a place, including not just the average condition but also the variability and extremes of weather. (7.0)

climatic water balance An accounting of the amount of water in each store and its rate of movement through the hydrologic cycle. (8.3)

climax community The culmination of succession, leading to an ecosystem with an intricate food web, efficient nutrient cycling, and biodiversity and productivity determined by the abundance or lack of necessary abiotic factors. (17.7)

climograph A graph that shows average monthly temperature (with a line) and precipitation (with bars) for a location. (7.4)

closed system Any interconnected set of matter, energy, and processes that does not allow material and energy to be exchanged outside itself. (1.3)

cloud-to-cloud lightning The electrical charges that are exchanged suddenly but only between two clouds; comprises the majority of lightning strikes. (5.9)

cloud-to-ground lightning Electrical charges exchanged suddenly between a cloud and the surface; comprises about 25 percent of lightning strikes. (5.9)

coal A natural, brown to black rock derived from peat and other plant materials that have been buried, compacted, and heated; used for heating and to generate electricity. (9.4)

coastal plain A plain that is barely above sea level and adjacent to the coast. (10.13)

coastal zone The area including the shoreline and a strip of adjacent land and water; specifically, the part of the saltwater biome nearest to the shore, characterized by wave action, temperature variations, and changes in substrate; also known as the intertidal zone. (15.1)

cobble A clast or rock fragment larger than a pebble and smaller than a boulder, having a diameter in the range of 64 to 256 mm. (13.2)

col A low point on an arête, which can provide a pathway from one side of the ridge to the other. (14.5)

Cold Desert (BWk) climate In the Köppen climate classification, a climate type characterized by inadequate moisture availability, in which potential evapotranspiration typically exceeds precipitation by a large amount, and a mean annual temperature below 12°C (54°F). (7.5)

cold front The sloping boundary surface between an advancing mass of relatively cold air and a warmer air mass. (5.2)

cold phase (ENSO) A periodic cooling of the eastern equatorial Pacific surface ocean waters, and reduction in precipitation adjacent to South America, as the Southern Hemisphere's summer approaches; also known as La Niña. (6.10)

cold sector The part of a mid-latitude cyclone that is behind the cold front, characterized by cold air. (5.2)

Cold Steppe (BSk) climate In the Köppen climate classification, a climate type characterized by inadequate moisture availability, in which potential evapotranspiration typically exceeds precipitation by a small or moderate amount, and a mean annual temperature below 12°C (54°F). (7.5)

collision-coalescence A process of precipitation formation by the repeated merger of tiny liquid water droplets. (4.11)

collision regime The scenario of coastal inundation during which the height of the swash zone exceeds the base of coastal dunes but does not exceed the top of coastal dunes. (15.11)

colluvium Sedimentary material that was transported to a location after being loosened and moved downhill via mass-wasting processes. (16.6)

columnar joint A distinctive fracture-bounded column of rock formed when hot but solid igneous rock contracts as it cools. (9.7)

columnar soil A soil structure in which the peds are taller than they are wide, commonly forming in arid climates where soil dries out and cracks. (16.2)

commensalism A type of symbiosis in which one species benefits from the interaction but the other is largely unaffected. (17.3)

compaction Process by which soil, sediment, and volcanic materials lose pore space in response to the weight of overlying material. (9.4)

compensation point The level within a body of water where the production of oxygen through photosynthesis is balanced by the use of oxygen by respiration or fermentation. (17.12)

competence The largest size of particle that a stream can carry. (13.2)

competition A situation in which members of two or more species, or two or more individuals of the same species, are seeking the same scarce resources. (17.3)

composite volcano A common type of volcano constructed of alternating layers of lava, pyroclastic deposits, and mass-wasting deposits, including mudflows; also known as a stratovolcano. (11.1)

compression The type of differential stress that occurs when forces push in on a rock. (11.9)

condensation A change in state from a gas to a liquid, which releases latent energy to the environment. (2.4)

condensation nuclei The solid aerosols that facilitate the conversion of water vapor to liquid water around them. (4.11)

conditional instability (atmosphere) A situation in which the atmosphere is stable if unsaturated but unstable if saturated; the environmental lapse rate is between the unsaturated adiabatic lapse rate and the saturated adiabatic lapse rate. (4.5)

conduction The transfer of thermal energy by direct contact, primarily between two solids. (2.2)

cone of depression A depression in the water table that has the shape of a downward-narrowing cone; develops around a well that is pumped, especially one that is over-pumped. (8.8)

confined aquifer An aquifer that is separated from Earth's surface by materials with low permeability. (8.6)

conformal projection Any type of mathematical algorithm for representing the three-dimensional Earth on a flat map that accurately preserves the shapes of features depicted. (1.9)

conglomerate A coarse-grained clastic sedimentary rock composed of rounded to sub-rounded clasts (pebbles, cobbles, and boulders) in a fine-grained matrix of sand and mud. (9.4)

conical projection A family of map projections based on the premise that the globe is projected onto a cone, with the tangent line having no distortion; other locations are neither conformal nor equal area, but retain some advantages of both conformal and equal area maps when depicting small areas. (1.9)

coniferous A cone-bearing tree with needles rather than broad leaves. (18.1)

coniferous evergreen forest A forest occurring in the mid-latitudes, dominated by trees that do not lose their needles seasonally. (18.11)

contact effects Evidence of baking, passage of hot fluids, or some other manifestation of the thermal and chemical effects of a magma chamber or a hot volcanic unit, as expressed in changes to adjacent wall rocks. (9.12)

contact metamorphism Metamorphism that principally involves heating of the rocks next to a magma. (9.6)

continental collision A plate-tectonic boundary where two continental masses collide; also known as a continent-continent convergent boundary. (10.7)

continental crust The type of Earth's crust that underlies the continents and the continental shelves; has an average composition of granite, but includes diverse types of material. (10.1)

continental drift The concept of the movement of continents and other landmasses across the surface of the Earth. (10.3)

continental glacier An ice sheet that covers a large region. (14.1)

Continental Monsoon (Dwa or Dwb) climate In the Köppen climate classification, a climate type characterized by adequate moisture in summer but not winter, a mean temperature of the coldest month below −3°C, and a hot or warm summer. (7.7)

continental platform A broad region that surrounds the continental shield and typically exposes horizontal to gently dipping sedimentary rocks. (10.13)

continental polar (cP) air mass A body of air that is relatively uniform in its cold and dry characteristics for thousands of square kilometers, originating in inland, subarctic source regions. (5.1)

continental rift A low trough or series of troughs bounded by normal faults, especially where two parts of a continent begin to rift apart. (10.6)

continental shelf A gently sloping, relatively shallow area of seafloor that flanks a continent, is underlain by thinned continental crust, and is covered by sedimentary layers. (10.2)

continental shield A central region of many continents, consisting of relatively old metamorphic and igneous rocks, commonly of Precambrian age. (10.13)

continental tropical (cT) air mass A body of air that is relatively uniform in its hot and dry characteristics for thousands of square kilometers, originating in inland, subtropical regions that are isolated from moisture sources. (5.1)

continent-continent convergent boundary A plate-tectonic boundary where two continental masses collide; also known as a continental collision. (10.7)

contour A line on a map or chart connecting points of equal value, generally elevation. (1.6)

contrail A condensation trail; a linear streak of clouds left behind by a jet. (4.8)

convection The vertical transfer of energy through a gas, liquid, or weak solid. (2.2)

convection cell A circulation system in a fluid that includes rising motion where the fluid is relatively hot, sinking motion where the fluid is relatively cold, and lateral movement between the areas of rising and sinking. (2.2)

convection current A flowing liquid or solid material that transfers heat from hotter regions to cooler ones, commonly involving movement of material in a loop. (10.10)

convergence (atmospheric) The horizontal piling up of air, either from two air streams colliding as they move laterally, or from an air current slowing down as it moves laterally. (3.3)

convergence (tectonic) The relative movement of two tectonic plates toward each other. (10.5)

convergent boundary A plate-tectonic boundary where two plates have relative movement toward each other. (10.5)

cool sector The part of a mid-latitude cyclone that is ahead of the warm front, characterized by air that is intermediate in temperature, as compared to that ahead of and behind the cold front. (5.2)

core The inner layer of Earth, below the mantle; interpreted to be divided into a molten outer core and a solid inner core. (10.1)

Coriolis effect An apparent force resulting from Earth's rotation that causes moving objects to appear to be deflected to the right of their intended path in the Northern Hemisphere and to the left in the Southern Hemisphere. (3.7)

counter-radiation The longwave radiation emitted from the atmosphere down to the surface. (2.14)

crater A typically bowl-shaped, steep-sided pit or depression, generally formed by a volcanic eruption or meteorite impact. (11.1)

creep The slow, continuous movement of material, such as soil and other weak materials, down a slope. (12.9)

crescent dune A crescent-shaped dune of windblown sand with its tails pointing in the direction of the prevailing wind; also known as a barchan dune. (9.11)

crest The highest point of any type of wave. (2.5)

crevasse A steep fracture in a glacier caused by internal stresses generated by the glacier's movement. (14.1)

cross bed A series of beds inclined at an angle to the main layers or beds. (9.11)

crosscutting relations The principle that a geologic unit or feature is older than a rock or feature that crosscuts it. (9.12)

crown fire A wildfire that leaps from treetop to treetop. (18.10)

crust The outermost layer of the solid Earth, consisting of continental and oceanic crust. (10.1)

crystalline A mineral that has an ordered internal structure due to its atoms being arranged in a regular, repeating way. (9.1)

crystalline basement Any rock below sedimentary rocks or sedimentary basins that are metamorphic or igneous in origin. (10.13)

crystalline rock A rock composed of interlocking minerals that grew together; usually formed in high-temperature environments by crystallization of magma, by metamorphism, or by precipitation from hot water. (10.13)

crystallization The formation of crystals during the slow cooling and solidification of rock, usually at depth. (9.3)

cuesta Any ridge with a sharp summit and one slope inclined approximately parallel to the dip of layers, resembling in outline the back of a hog; also known as a hogback. (11.11)

cumuliform cloud A family of puffy cloud types that is taller than it is wide. (4.8)

cumulonimbus cloud A cumuliform cloud type that is actively precipitating and has an extensive vertical extent, composed of liquid water near the base and ice near the top; commonly associated with severe weather. (4.8)

cumulus cloud A type of cumuliform cloud that occurs in fair weather, without extensive vertical extent, but under free-convection conditions. (4.8)

cumulus stage The first stage in a single-cell thunderstorm's life cycle, characterized by updrafts that form and grow cumuliform clouds, but without precipitation. (5.6)

cutbank A steep cut or slope formed by lateral erosion of a stream, especially on the outside bend of a channel. (13.5)

cutoff meander A meander cut off from the main channel by a new channel that formed when a stream cuts through the neck of a meander. (13.5)

cyclogenesis The formation stage in the life cycle of a cyclone, particularly a mid-latitude cyclone. (5.4)

cyclone An enclosed area of low atmospheric pressure; can also refer to the most intense type of storm of tropical origin in the Indian Ocean basin. (3.9)

cylindrical projection A family of conformal map projections based on the premise that the globe is projected onto a cylinder, with distortion increasing with distance from the tangent at which the globe touches the cylinder. (1.9)

D

dart leader A second flow of electrical energy after the first lightning stroke, which flows through approximately the same pathway as the original stroke between the cloud and the ground. (5.9)

daughter product The element produced by radioactive decay of a parent atom; also known as a daughter atom. (9.13)

Daylight Savings Time A one-hour offset of standard time used in much of the U.S. for the purpose of skewing daylight hours toward evening hours on the clock. (1.13)

debris avalanche A high-velocity flow of soil, sediment, and rock, commonly from the collapse of a steep mountainside. (12.9)

debris flow Downhill-flowing slurries of loose rock, mud, and other materials, and the resulting landform and sedimentary deposit; also known as a mudflow. (9.10)

debris slide The downslope movement of soil, weathered sediment, or other unconsolidated material, partly as a sliding, coherent mass and partly by internal shearing and flow. (12.11)

December Solstice The day, on or near December 21, when the North Pole (marking where the Earth's axis of rotation intersects the Earth's surface in the Northern Hemisphere) is pointed most directly away from the Sun. (2.9)

deciduous Any tree that loses its leaves or needles seasonally. (18.11)

deciduous temperate forest A forest occurring in the mid-latitudes, dominated by trees that lose their leaves seasonally. (18.11)

decomposition The breakdown of organic materials from dead organisms into inorganic nutrients, primarily by bacteria and fungi. (17.3)

deep-sea thermal vents A type of productive ecosystem within the saltwater biome on the ocean floor where the local energy source is chemosynthesis. (18.15)

deficit The amount of water (in centimeters or inches of precipitation equivalent) that could have evaporated if it were available; equivalent to the difference between potential evapotranspiration and actual evapotranspiration; a measure of drought severity. (8.3)

deforestation The clearing of a forested landscape. (4.6)

deformation Processes that cause a rock body to change position, orientation, size, or shape, such as by folding, faulting, and shearing. (1.4)

delta A nearly flat tract of land formed by deposition of sediment at the mouth of a stream. (13.9)

dendritic drainage pattern A drainage network in which the pattern of streams is treelike; usually an indicator that underlying rocks have similar resistance to erosion or that the streams have been flowing for a long time. (13.1)

denitrification The movement of nitrogen from the land to the atmosphere when microbes convert other compounds of nitrogen into nitrogen gas, and during combustion of fossil fuels. (17.9)

density An object's mass divided by the volume it occupies. (2.2)

denudation The wearing away of the surface after erosion and transportation remove material that may be loosened by weathering. (9.9)

deoxyribonucleic acid (DNA) A nucleic acid that carries the genetic information in cells and some viruses. (17.10)

deposition (atmospheric) A change in state from a gas to a solid, which releases latent energy to the environment. (2.4)

deposition (geomorphology) The dropping of sediment that occurs when the transporting agent can no longer carry the sediment, such as when a flowing stream slows down. (1.4)

depth The distance of an object below the surface or below sea level. (1.6)

derecho A regional windstorm characterized by nonrotating winds that can advance over huge areas quickly. (5.12)

desert biome A collection of adjacent ecosystems that consists of vegetation that is especially adapted to survive in places where moisture deficits are typical yearround. (18.1)

desertification The process by which a climate becomes increasingly arid, either through natural or anthropogenic causes. (7.13)

desert pavement A natural concentration of pebbles and other rock fragments that mantle a desert surface of low relief. (9.11)

desert varnish A thin, dark film or coating of iron and manganese oxides, silica, and other materials; formed by prolonged exposure at the surface; also known as rock varnish. (9.11)

detritivore Any organism that consumes the remains of decomposing (nonliving) plants and animals in the soil. (17.2)

dew A condensed liquid water drop on a solid surface. (4.4)

dew-point depression The difference between the air temperature and the dew-point temperature. (4.4)

dew-point temperature The temperature to which a volume of air must be cooled for it to become saturated with water vapor, thereby causing condensation. (4.4)

diabatic A condition of a system in which energy enters or leaves the system. (4.5)

differential erosion The varying rate of wearing away of soil, sediment, and rock in a landscape caused by varying composition of rock types in that landscape. (11.12)

Differential GPS An enhanced GPS that uses a correction signal to check each GPS satellite's signal independently. (1.11)

differential weathering The breakdown of rock at differing rates because of differences in mineral composition, hardness, or other resistance to weathering. (12.3)

dike A sheetlike, igneous intrusion that cuts across any layers in a host rock, commonly formed with a steep orientation. (9.12)

diorite A medium- to coarse-grained, intermediate-composition igneous rock; equivalent in composition to andesite. (9.5)

dip The angle that a layer or structural surface makes with the horizontal. (11.10)

dip slope A slope caused by the erosion of hard and soft layers parallel to the dipping layers. (11.11)

disappearing stream A stream that vanishes suddenly from the surface as it begins to flow beneath the surface, often captured by pits in karst terrain. (12.7)

discharge The volume of water flowing through some stretch of a stream per unit of time. (13.1)

displacement The movement of a body in response to stress, either from one place to another, or manifested as a change in shape of the body. (11.9)

dissipating stage The final stage in a single-cell thunderstorm's life cycle, characterized by downdrafts, an end of precipitation, and decay of the system. (5.6)

dissolution The process by which a material is dissolved. (9.9)

dissolved load Chemically soluble ions, such as sodium and chlorine, that are dissolved in and transported by moving water, as in a stream. (13.2)

distributary system The branching drainage pattern formed when a stream branches and spreads out into a series of smaller channels. (13.9)

divergence (atmospheric) The horizontal spreading apart of air, either from two air streams moving laterally in opposite directions, or from an air current accelerating as it moves laterally. (3.3)

divergence (tectonic) The relative movement of two tectonic plates away from each other. (10.5)

divergent boundary A plate-tectonic boundary in which two plates move apart (diverge) relative to one another. (10.5)

diversity (ecology) The variety of life in an area; biodiversity. (17.4)

Dixie Alley The zone of high tornado frequency across the U.S. Gulf Coast region. (5.11)

doldrums The belt of weak winds along the Intertropical Convergence Zone, between the northeast trade winds and the southeast trade winds. (6.1)

dolomite A carbonate mineral containing calcium and magnesium. (9.1)

Doppler radar A form of remote sensing that emits a microwave signal and relates changes in frequency or amplitude of that wave to the rate of movement of an object that intercepts the wave; used to detect precipitation, storm movement, and motion within a storm. (5.16)

double sea breeze A convergence of air flowing onshore from two different but nearby coastlines in a peninsula, such as in Florida. (5.7)

downcutting The process by which a stream incises (cuts) downward as it erodes into its bedrock. (13.2)

downdraft A local downward flow of air, usually on very short timescales. (3.3)

downstream The direction of flow of water from higher elevations to lower ones. (13.3)

drainage basin An area in which all drainages merge into a single stream or other body of water. (8.6)

drainage network The configuration or arrangement of streams within a drainage basin; also known as stream system. (13.1)

drapery A thin sheet of travertine formed by the evaporation of water dripping from the ceiling of a cave. (12.6)

driving force A force that causes a tectonic plate to move. (10.9)

dropstone A fragment of rock found in fine-grained, water-deposited sedimentary rock, that was dropped in rather than transported by normal water currents. (14.8)

drought The dry conditions caused by excessive water-balance deficits. (8.3)

drumlin A commonly teardrop-shaped hill formed when a glacier reshapes glacial deposits. (14.6)

drumlin field A group of drumlins in the same area. (14.6)

dry line The boundary between a warm, dry (cT) air mass and a warm, moist (mT) air mass, in the warm sector of a mid-latitude cyclone. (5.6)

ductile deformation A type of structural deformation that is characterized by the material flowing as a solid, without fracturing; typical of the crust below about 15 kilometers and of the asthenosphere. (11.9)

Dust Bowl A period of severe dust storms in the Great Plains of the U.S. and Prairies of Canada during the 1930s that greatly damaged agriculture. (16.13)

dust devil A small-scale, rotating dust storm. (5.12)

dynamic equilibrium (streams) A condition in which a stream has neither a net eroding nor depositing capability. (13.4)

dynamic equilibrium (systems) A condition of a system in approximate steady state maintained by equal quantities of matter and energy flowing into and out of a system. (1.3)

dynamic system Any interconnected set of matter, energy, and processes in which matter, energy, or both are constantly changing their position, amounts, or form. (1.3)

dynamo An electrical generator. (10.10)

E

earth flow A flowing mass of weak, mostly fine-grained material, especially mud and soil; also commonly called a mudflow. (12.11)

East Australian Current A north-to-south-flowing, warm surface ocean current in the western South Pacific Ocean. (6.2)

easterly wave A bend in the trade-wind flow, where relatively low atmospheric pressure from the Intertropical Convergence Zone protrudes farther poleward than elsewhere. (5.13)

East Pacific Rise A mid-ocean ridge (divergent boundary) in the South Pacific Ocean that crosses the eastern Pacific and approaches North America. (10.2)

ebb tide The local falling elevation of the sea surface between high tide and low tide. (15.2)

eccentricity A change in the degree to which Earth's orbit deviates from being perfectly circular. (14.9)

ecological niche The unique position of a species, especially its role relative to the other species, in an ecosystem. (17.1)

ecologist A scientist who studies how organisms interact with each other and with their physical environment. (17.1)

ecology The complex set of relations between living organisms and their environment. (17.0)

ecosystem The provisions and constraints of the physical environment, along with the entire community of interacting organisms (plants, animals, and others) in an area. (17.0)

ecosystem functions Dynamic processes that occur in the environment to support an ecosystem. (17.1)

ecosystem stability The degree of fluctuation of the population of species in an ecosystem over time. (17.7)

ecosystem structure A biotic, abiotic, or energy component of an ecosystem. (17.1)

ecotone A broad transition zone between two zones of natural vegetation (biomes), and therefore between two Köppen climate types. (7.1)

ecotourism Tourism to exotic natural environments intended to support conservation efforts and observe wildlife. (17.4)

eddy A swirl in a fluid's current that can form in both horizontal and vertical directions, characteristic of turbulent flow, as in streams or the atmosphere. (13.2)

E horizon A light-colored, leached zone of soil, lacking clay and organic matter; also known as the zone of eluviation. (16.7)

Ekman layer The layer of the atmosphere characterized by an increased turning of the winds with height; from about 100 meters to about 1–2 kilometers above the surface. (3.8)

Ekman spiral The change in the direction of flow from the surface of the ocean downward to the null point, which turns from parallel to the wind direction at the surface to an increasingly rightward (in the Northern Hemisphere) deflection with depth. (6.1)

Ekman transport The apparent deflection of currents in the saltwater biome with depth; the ocean's Coriolis effect. (18.15)

electromagnetic spectrum The full assemblage of energy of all possible wavelengths, resulting from electrical and magnetic fields. (2.6)

electron shell A discrete "level" away from a nucleus in an atom, on which a maximum number of electrons can exist, with successively higher energy content existing on more outward levels. (2.5)

elevation The vertical distance of an object above or below a reference datum (usually mean sea level); generally the height of a ground point above sea level. (1.6)

El Niño A periodic warming of eastern equatorial Pacific surface ocean waters, accompanied by excessive precipitation and flooding adjacent to South America, as the Southern Hemisphere's summer approaches; also known as a warm phase of ENSO. (6.9)

El Niño–Southern Oscillation (ENSO) The combined alterations of oceanic temperatures, atmospheric pressures, and resulting oceanic and atmospheric circulations in the equatorial Pacific Ocean. (6.10)

eluviation The separation or dissolution of soluble constituents from a rock, sediment, or soil by percolation of water; also known as leaching. (16.7)

elve A flat disk of dim reddish light that forms high in the atmosphere after lightning strikes. (5.9)

embayment A place where the coastline curves inward toward the land. (15.0)

emergent coast A coast that forms where the sea has retreated from the land due to falling sea level or due to uplift of the land relative to the sea. (15.9)

emergent layer The highest layer of life in a rain forest, consisting of a few trees that tower up to 70 meters above the forest floor but without forming a closed canopy at that level. (18.5)

endemic Native to an area. (18.16)

energy The ability to do work. (1.4)

englacial debris Material transported within the ice of a glacier. (14.4)

Enhanced Fujita scale A system for rating the strength of tornadoes based on the amount of damage. (5.10)

Entisol A type of recently developed soil, characterized by no B horizon and little or no profile development. (16.9)

entrainment The mixing of drier air with more humid air inside a cloud, characterized by the evaporation of cloud water and the decay of a cloud. (5.6)

entrenched meander A curved canyon that represents a meander carved into the land surface. (13.11)

environmental lapse rate The rate of change of temperature with height in a static atmosphere. (4.5)

ephemeral Seasonal. (18.8)

ephemeral stream A stream that only flows during some times of the year. (13.3)

epicenter The point on Earth's surface directly above where an earthquake occurs (directly above the focus or hypocenter). (11.13)

epipelagic zone The part of the saltwater biome located seaward of the coastal zone's low-tide mark and containing the vast open waters of the ocean, at shallow depth, above the mesopelagic zone, where light penetrates and separates into colors. (18.15)

epiphyte A plant that lives suspended on other plants and derives its nutrients and water without contacting soil. (18.5)

equal-area projection Any type of mathematical algorithm for representing the three-dimensional Earth on a flat map that accurately preserves areas of features depicted. (1.9)

equator The imaginary line of latitude, equidistant between the North and South Poles, that divides the Earth into halves, and is defined as the 0° parallel. (1.7)

equatorial (E) air mass A body of air that is relatively uniform in its warm and very humid characteristics for thousands of square kilometers, originating in oceanic, equatorial source regions. (5.1)

equilibrium The state of a system characterized by an equal movement of atoms and molecules between two states of matter; also used in ecosystems and other contexts. (4.1)

equilibrium line The zone across a glacier where the losses of ice and snow balance the accumulation of ice and snow; also known as firn line. (14.3)

equilibrium profile A profile reflecting a stream in an approximate steady state where deposition of sediment is balanced by erosion. (13.4)

equitability The evenness of populations of different species in an ecosystem. (17.4)

era A broad division of geologic time, including the Paleozoic, Mesozoic, and Cenozoic, with each divided into periods. (9.13)

erosion The removal of soil, sediment, and rock by running water, waves, currents, ice, wind, and gravity. (1.4)

erosional remnant A mountain or hill that remains when adjacent areas have eroded to lower levels. (11.12)

eruption column A rising column of hot gases, pyroclastic material, and rock fragments ejected high into the atmosphere during a volcanic eruption. (11.2)

esker A long, narrow, sinuous ridge composed of sediment deposited by a stream flowing within or beneath a glacier. (14.6)

estivation A long period of dormancy of an animal during a hot or dry period. (18.8)

estuary A channel where fresh water from the land interacts with salt water from the sea and commonly is affected by ocean tides. (15.9)

Ethiopian realm The zoogeographic realm consisting of most of Africa. (18.16)

euryhaline Capable of tolerating a wide range of salinity conditions. (17.7)

eutrophic An aquatic ecosystem with high nutrient concentrations and high net primary productivity. (17.10)

eutrophication The addition of so many nutrients, especially phosphates, into an aquatic ecosystem that the ecosystem is harmed; also known as hypertrophication. (16.13, 17.10)

evaporation A change in state from a liquid to a gas, requiring an input of energy from the environment and the absorption of that energy at the molecular level as latent energy. (2.4)

evaporation fog A type of fog that occurs when cold air overlies a warm water surface, destabilizing the atmosphere and encouraging evaporation into the colder air. (4.9)

evaporite Any type of rock, such as rock salt, that forms from the evaporation of water. (12.7)

evapotranspiration The combined effect of evaporation and transpiration in moving liquid water from the surface up to the atmosphere in its gaseous form. (4.0, 16.3)

evergreen Any tree that does not lose its leaves or needles seasonally, even though individual leaves or needles fall and are replaced by new growth. (18.11)

evergreen temperate forest A forest occurring in the mid-latitudes, dominated by trees that do not lose their leaves seasonally. (18.11)

exfoliation The processes by which a rock sheds concentric sheets, as occurs due to the release of pressure during exposure. (12.1)

exfoliation dome A curved rock face left behind after an exfoliation joint shaves off thin, curved slices of rock parallel to the surface. (12.1)

exfoliation joint A joint that forms when a rock expands near the surface, shedding concentric plates, mimicking the topography. (12.1)

expansion joint A joint that forms as a result of expansion due to cooling or to a release of pressure as rocks are uplifted to the surface. (12.1)

exponential growth curve A plot that represents an accelerating population growth rate over time. (17.3)

Extended Köppen Classification The categorization of climatic types over both land and ocean, by applying the algorithm for the Köppen method to climatic data collected over the ocean. (7.2)

extremophile An organism that lives in extreme environments, such as scalding hydrothermal vents at the very high pressures of the deep seafloor. (17.5)

eye The center of a tropical cyclone, characterized by sinking air and clear and calm conditions. (5.13)

eyewall The area of a tropical cyclone immediately surrounding the eye. (5.13)

F

Fahrenheit scale A system for indexing temperature calibrated such that water at sea level freezes at 32°F and water at sea level boils at 212°F. (2.3)

failed rift A narrow trough (rift) at a triple junction, which becomes less active and fails to completely break up a continent into separate plates. (10.12)

Fall Line An imaginary line connecting waterfalls on several adjacent rivers, especially along the boundary between the Piedmont and Coastal Plains in eastern North America. (13.4)

false echo A reflection of a radar image from ground clutter, such as buildings or birds; also known as noise. (5.16)

fault A fracture along which the adjacent rock surfaces are displaced parallel to the fracture. (10.8)

fault block A section of rock bounded on at least two sides by faults. (11.10)

fault-line scarp A step in topography created by the erosion of one side of a fault more than the other. (11.10)

fault scarp A step in the landscape caused when fault movement offsets Earth's surface. (9.14)

faunal succession The systematic change of fossils with age. (9.13)

feedback A reinforcement or inhibitor to a change in a system. (1.3)

feldspar A very common rock-forming silicate mineral that is abundant in most igneous and metamorphic rocks and some sedimentary rocks, with varieties containing different amounts of potassium, sodium, and calcium. (9.1)

felsic rock An igneous rock with a light color and generally abundant quartz and feldspar. (9.5)

fen A wetland that has neutral or alkaline, but not acidic, waters; usually fed by nutrient-rich groundwater and dominated by grasses and sedges. (18.14)

fermentation The breakdown of carbohydrates to carbon dioxide by yeasts and bacteria. (17.12)

Ferrel cell A hypothetical atmospheric circulation cell over the mid-latitudes in each hemisphere that appears only weakly if at all in the real world. (3.13)

fertilizer A supplement to soils containing nutrients, typically comprised of some combination of nitrogen, phosphorus, and potassium. (16.4)

field capacity The maximum amount of water (in centimeters or inches of precipitation equivalent) that a soil can hold. (8.3)

firn Snowflakes that become compressed into small, irregular spheres of more dense snow. (14.3)

firn line The zone across a glacier where the losses of ice and snow balance the accumulation of ice and snow; also known as equilibrium line. (14.3)

First Law of Thermodynamics The fundamental scientific principle that energy and matter can be neither created nor destroyed, but only transferred from one form to another. (1.4)

first-level heterotroph An organism that consumes a primary producer. (17.2)

first-order stream A stream with no tributaries. (13.1)

fissure (volcanic) A magma-filled fracture in the subsurface, which typically solidifies into a dike, or a linear volcanic vent erupting onto the land surface. (11.1)

fixation The conversion of carbon dioxide to organic compounds by living organisms, as in photosynthesis. (17.1)

fjord A long, narrow arm of the sea contained within a U-shaped valley, interpreted to be carved by a glacier and later invaded by the sea as the ice melted and sea level rose. (14.5)

flag tree A wind-shaped tree with branches growing only on the downwind (leeward) side of a trunk; also known as a banner tree. (18.4)

flash flood A local and sudden flood of short duration, such as that which may follow a brief but heavy rainfall. (13.12)

flood The result of water overfilling a channel and spilling out onto the floodplain or other adjacent land. (13.12)

flood basalt A series of basalt lavas that cover large stretches of land or the ocean floor, usually resulting in a characteristic stairstep landscape. (10.12)

floodplain An area of relatively smooth land adjacent to a stream channel that is intermittently flooded when the stream overflows its banks. (9.10)

flood stage The level at which the amount of discharge causes a stream to overtop its banks and spill out onto the floodplain. (13.12)

flood tide The local rising elevation of the sea surface between low tide and high tide. (15.2)

flowstone Any deposit of calcium carbonate or other mineral formed by flowing water on the walls or floor of a cave. (12.6)

flume A large, sand-filled tank, used to study stream behavior in a laboratory; also known as a stream table. (13.7)

focus The place where an earthquake is generated; also known as a hypocenter. (11.13)

fog A cloud at the surface of the Earth. (4.8)

folding A form of deformation in which horizontal rock layers are bent into a wavy pattern. (9.12)

foliation The planar arrangement of textural or structural features in metamorphic rocks and certain igneous rocks. (9.6)

food chain The sequencing of organisms from autotrophs to higher level heterotrophs. (17.2)

food web A network of predator-prey relationships between different types of organisms within an ecosystem; typically more complex than a single, linear food chain. (17.2)

force A push or pull that causes, or tends to cause, change in the motion of a body, commonly expressed as the amount of acceleration experienced by a mass. (11.9)

forced convection The rising motion in the atmosphere caused by horizontal convergence, an orographic barrier, or a frontal boundary, rather than by unstable conditions. (4.7)

forecast products The maps of the forecasted state of the atmosphere at a given time into the future, at the surface or at a specified vertical constant pressure level. (5.15)

forensic science The use of technologies to address scientific questions needed for criminal investigations or civil litigation. (16.14)

foreshore The part of a beach submerged during high tides but exposed during low tides, located between the inshore (or swash zone) and the backshore. (15.1)

forest floor The lowest layer of life in a rain forest, where organic debris decomposes rapidly. (18.5)

forest soil An informal name for soils that support temperate forests. (16.6)

formation A rock unit that is distinct, laterally traceable, and mappable; also used as an informal term for an eroded, perhaps unusually shaped, mass of rock. (12.6)

fossil Any remains, trace, or imprint of a plant or animal that has been preserved from some past geologic or prehistoric time. (9.13)

fracture A break or crack in a rock; subdivided into joints and faults. (9.8)

free convection The rising motion in the atmosphere caused by unstable conditions. (4.7)

freezing A change in state from a liquid to a solid, which releases latent energy to the environment. (2.4)

freezing point The temperature at which a substance changes its state from a liquid to a solid. (2.3)

freezing rain A form of precipitation in which a snowflake produced by the Bergeron process melts on its way down to the surface, but refreezes in a cold layer upon impact with the surface, causing it to stick to the surface. (4.12)

frequency The number of waves per second passing a point. (2.5)

freshwater biome A collection of adjacent ecosystems that includes lakes, streams, marshes, and other surface waters. (18.1)

friction layer The lowest one kilometer or so of the troposphere, where friction with the Earth's surface exerts an important effect on motion. (3.3)

fringing reef A reef that fringes the shoreline of an island or continent. (15.6)

front A narrow zone separating two different air masses. (5.2)

frontal lifting The rising motion in the atmosphere caused by a warmer, less dense air mass being pushed up over a colder, more dense one. (4.7)

frost heaving The uneven upward movement and distortion of soils and other materials due to subsurface freezing of soil water. (12.8)

frost-point temperature The subfreezing temperature to which a volume of air must be cooled in order to become saturated with water vapor, thereby causing frost to form. (4.4)

frost wedging The process by which jointed rock is pried and dislodged by the expansion of ice during freezing. (12.1)

full moon The position of the Moon in its orbit around Earth when it is directly on the opposite side of Earth from the Sun; results in the Moon being fully illuminated by the Sun. (15.2)

funnel cloud A tornado-like circulation that does not reach the surface. (5.10)

fusion (nuclear) The joining together of protons and neutrons to produce larger particles, releasing electromagnetic radiation in the process. (2.5)

G

gabbro A medium- to coarse-grained mafic igneous rock, the compositional equivalent of basalt. (9.5)

gaining stream (river) A stream or part of a stream that receives water from the inflow of groundwater. (8.7)

gamma ray The electromagnetic radiation of the shortest wavelengths (less than 10^{-11} micrometers). (2.6)

gas A state of matter in which the energy content is high enough that molecules break apart from each other and move freely at high speeds. (2.2)

gauging station A site where measurements from a stream are collected in order to calculate discharge of that segment of the stream. (13.14)

gelifluction The seasonal freeze-thaw cycle experienced by waterlogged topsoil, very similar to solifluction. (12.11)

Gelisol A type of soil situated above permafrost, characterized by waterlogged conditions and organic matter that is slow to decompose. (16.11)

general circulation model (GCM) A type of complicated, computer-based model that predicts global atmospheric circulation using physically based and statistically based equations incorporating processes and feedbacks in the atmosphere, hydrosphere, cryosphere, lithosphere, and biosphere. (7.14)

geographic information system (GIS) A hardware and software system that allows a variety of spatial and nonspatial information to be combined quickly and efficiently to examine relationships among the different features. (1.12)

geologic timescale A chronologic subdivision of geologic time depicting the sequence of geologic events, including those represented by fossils; ages of boundaries are assigned through numeric dating of key rock units. (9.12)

geomagnetic polarity timescale A chronology based on the pattern and numeric ages of reversals of Earth's magnetic field. (10.10)

geomorphologist A scientist that studies landforms. (9.8)

georeferencing The process of matching an image or map to standard coordinates; also known as rectification. (1.12)

geostrophic level The lowest height in the atmosphere at which airflow is parallel to isobars, neither toward nor away from lower atmospheric pressure. (3.8)

geostrophic wind A wind that blows parallel to the isobars, characterized by a balance between the pressure gradient force and the Coriolis effect, above the level at which friction exerts an important impact on winds. (3.8)

geyser A type of hot spring that intermittently erupts a fountain or spray of hot water and steam. (8.7)

glacial drift Any sediment carried by ice, icebergs, or meltwater. (14.4)

glacial erratic A rock fragment carried by moving ice and deposited some distance from where it was derived. (14.4)

glacial groove A very large, deep scratch mark amid an otherwise smooth polished surface created by glacial abrasion. (14.4)

glacial period A time interval within an ice age when glaciers were abundant. (14.2)

glacial striation A scratch mark amid an otherwise smooth polished surface created by glacial abrasion. (14.4)

glacier A moving mass of ice, snow, rock, and other sediment on land. (1.3)

glaciofluvial deposition Dropping of sediment by glacial streams. (14.4)

gleization A pedogenic regime characterized by a dark, organic-rich A horizon produced in saturated and waterlogged soils where the rate of decay of organic material is slow. (16.8)

global cooling Decreasing global atmospheric and oceanic temperatures as measured or inferred from some point in the past to the present. (7.12)

global outgoing longwave budget The apportioning of upward-directed longwave radiation to various components of the global system on an annual basis. (2.15)

Global Positioning System (GPS) A tool that provides locational information by receiving radio signals from orbiting satellites that relay the position and distance of each satellite. (1.11)

global shortwave radiation budget The apportioning of shortwave radiation to various components of the global system on an annual basis. (2.13)

global warming Increasing global atmospheric and oceanic temperatures as measured or inferred from some point in the past to the present. (7.11)

gneiss A metamorphic rock that contains a gneissic foliation defined by a preferred orientation of crystals and generally by alternating lighter and darker colored bands. (9.6)

Gondwana The inferred supercontinent consisting of many of the southern continents. (10.3)

Goode projection A popular type of sinusoidal map projection. (1.9)

graben An elongate, downdropped crustal block that is bounded by faults on one or both sides. (11.10)

graded stream A stream in equilibrium, showing a balance between its capacity to transport sediment and the amount of sediment supplied to it, thus with no overall erosion or deposition of sediment. (13.4)

gradient The change in a variable, such as elevation or atmospheric pressure, across a particular horizontal distance, expressed in degrees; also referred to as slope. (1.6)

gradient flow The airflow between higher and lower atmospheric pressure areas parallel to the curving isohypses; typically occurs above the friction layer. (3.8)

granite A coarse-grained, felsic igneous rock containing mostly feldspar and quartz. (9.5)

granular soil A soil structure in which the peds are shaped like small spherical grains less than 0.5 centimeters in diameter, a structure that is often ideal for root penetration. (16.2)

grassland biome A collection of adjacent ecosystems that has tall-grass and short-grass subtypes, related to different climatic conditions, but generally characterized by moisture deficits for part of the year. (18.1)

grassland soil An informal name for soils that support the mid-latitude grassland biome; generally corresponds to Mollisols. (16.6)

gravitational water Water within pore space in soil that is pulled by gravity deeper into the subsurface. (16.3)

gravity The force exerted between any two objects, such as that between the Sun, Earth, and Moon. (12.8)

great circle Any curved line formed by the intersection of a plane and a sphere that divides the sphere into two equal halves and therefore represents the shortest distance between any two points on the sphere. (1.7)

Great Dying The mass extinction event that occurred about 252 million years ago and separates the Paleozoic Era from the Mesozoic Era. (9.13)

Great Pacific Climate Shift The reversal of the Pacific Decadal Oscillation (PDO) from a cool phase to a warm phase around 1977. (6.0)

greenhouse effect The phenomenon by which Earth's lower atmosphere is warmed by the presence of gases that absorb longwave radiation from the Earth and emit longwave radiation downward to the Earth. (2.14)

greenhouse gas An atmospheric gas that effectively absorbs infrared (longwave) radiation and emits it downward to the Earth, contributing to the warming of the lower atmosphere. (2.14)

Greenland Current A small, north-to-south-flowing, cold surface ocean current in the North Atlantic Ocean near the Greenland coast. (6.2)

groin A low wall built out into a body of water to affect the lateral transport of sand by waves and longshore currents. (15.7)

gross primary productivity The total amount of energy fixed for photosynthesis. (17.6)

ground clutter A non-weather object that intercepts a radar signal, giving the false impression that there is precipitation where none exists. (5.16)

ground heat The insolation absorbed at Earth's surface. (2.13)

ground layer The lower of the two layers in most boreal forests. (18.12)

ground-level ozone A harmful air pollutant, produced by sunlight striking hydrocarbons, such as car exhaust, in the lower troposphere; also known as tropospheric ozone. (2.12)

ground moraine Sediment deposited from the base of the ice in a glacier and scattered around a large area. (14.6)

groundwater Water that occurs in the pores, fractures, and cavities in the subsurface. (1.4)

groundwater divide A relatively high area of the water table, separating ground-water that flows in opposite directions. (8.6)

groundwater recharge The replenishment of water below the water table by percolation. (8.7)

growing degree day A variable indexing the length of the growing season and how warm it is above a threshold temperature required for the growth of a particular type of vegetation. (17.6)

guano The excrement of seabirds and bats, used as fertilizer. (17.10)

Gulf low A type of mid-latitude cyclone that forms near the Gulf of Mexico coast of the U.S. mainly in winter and tracks generally northeastward. (5.3)

Gulf Stream A south-to-north-flowing, warm surface ocean current in the western subtropical North Atlantic Ocean. (6.2)

gully A small channel eroded into the land surface, formed from rills that coalesce. (13.6, 16.13)

gully erosion The wearing away of soil, sediment, and rock as increasing flow in a small but growing channel occurs. (13.6)

gust front A strong, cool breeze that acts as a harbinger of a thunderstorm a few minutes before the rain falls. (5.6)

gypsum A common calcium sulfate mineral, generally formed by the evaporation of water. (9.1)

gyre Any circular system of currents near the ocean's surface, caused by effects of atmospheric flow around subtropical anticyclones or subpolar lows. (6.1)

H

habitat The physical environment in which a species lives. (17.1)

haboob A severe, downdraft-related, non-rotating dust storm. (5.12)

hadal zone The part of the saltwater biome located seaward of the coastal zone's low-tide mark and containing the vast open waters of the ocean, at the greatest depth, below the mesopelagic zone (i.e., part of the aphotic zone). (18.15)

Hadley cell An atmospheric circulation cell with rising motion at the Intertropical Convergence Zone, poleward flow aloft to the subtropics in each hemisphere, sinking motion at the subtropical high-pressure belt, and surface motion back to the Intertropical Convergence Zone, with one cell in each hemisphere and the most vigorous motion occurring at longitudes experiencing noon. (3.10)

hailstone A ball of ice that forms in subfreezing temperatures within a cumulonimbus cloud and that subsequently falls toward the Earth's surface. (5.8)

half-life In radioactive decay, the time required for one-half of the parent atoms in a sample to decay into the daughter product. (9.13)

halite A salt mineral (NaCl) generally formed by the evaporation of water; cleaves into cubes and has a distinctive salty taste. (9.1)

halogen Any of a group of five elements, including fluorine and chlorine, that easily bonds with another element or molecule. (2.12)

hand specimen A hand-sized piece of rock for study, sampling, or inclusion in a collection. (9.1)

hanging valley A glacial valley whose mouth is higher than the bottom of a larger glacially carved valley it joins. (14.5)

harsh mid-latitude (D) climate In the Köppen climate classification, a broad group of climatic types characterized by adequate moisture availability, the mean temperature of the coldest month less than –3°C, and the mean temperature of the warmest month above 10°C. (7.1)

Hatteras low A type of mid-latitude cyclone that forms near the mid-Atlantic coast of the U.S., mainly in winter, and tracks generally northeastward. (5.3)

Hawaiian High A prominent area of semipermanent high atmospheric pressure (anticyclone) in the subtropical North Pacific Ocean; is particularly strong in summer. (3.6)

hazard The existence of a potentially dangerous situation or event. (11.6)

head The upslope end of a rock slide. (12.10)

headward erosion The wearing away of soil, sediment, and rock at the headwaters of a stream, into the area between channels (the divide). (13.6)

headwaters The location or general area where a stream begins. (13.3)

heat The thermal energy transferred from one object to another. (2.2)

heat capacity The amount of energy required to change the temperature of a volume by one Kelvin, expressed in Joules per cubic meter per Kelvin. (2.17)

heat flux The transfer of thermal energy from higher temperature to lower temperature objects. (2.2)

hematite An iron oxide (Fe_2O_3) metal that has a reddish streak and commonly forms under oxidizing conditions. (9.1)

herbivore Any organism that consumes only plants. (17.2)

heterotroph An organism that consumes other organisms. (17.2)

hibernation The inactivity or dormancy of an organism or species, usually in winter. (18.11)

high tide The maximum height to which water in the ocean rises relative to the land in response to the gravitational pull of the Moon; also refers to the time when such high levels occur. (15.2)

hinge The part of a fold that is most sharply curved. (11.11)

Histosol A site-dominated soil type formed in wetlands without permafrost, influenced by anaerobic conditions and the accumulation of organic matter in a thick O horizon. (16.9)

hogback Any ridge with a sharp summit and one slope inclined approximately parallel to the dip of layers, resembling in outline the back of a hog; also known as a cuesta. (11.11)

hoodoo A tall, thin rock formation produced by erosion of highly jointed rock layers, with somewhat rounded sides. (12.5)

hook echo The curved pattern on Doppler radar characteristic of a tornado. (5.10)

horizon A soil layer that is distinct from adjacent layers, including differences in color, texture, content of minerals and organic matter, or other attributes. (12.14)

horizonation The process of forming soil horizons. (16.9)

horn A glacial erosional feature formed when three or more cirques merge by headward erosion. (14.5)

horse latitudes The oceanic regions near 30° N and 30° S. (3.12)

horst An elongate, relatively uplifted crustal block that is bounded by faults on two sides. (11.10)

Hot Desert (BWh) climate In the Köppen climate classification, a climate type characterized by inadequate moisture availability, in which potential evapotranspiration typically exceeds precipitation by a large amount, and a mean annual temperature above 12°C (54°F). (7.5)

hot spot A volcanically active site interpreted to be above an unusually high-temperature region in the deep crust and upper mantle. (10.9)

Hot Steppe (BSh) climate In the Köppen climate classification, a climate type characterized by inadequate moisture availability, in which potential evapotranspiration typically exceeds precipitation by a small or moderate amount, and a mean annual temperature above 12°C (54°F). (7.5)

human geography The study of the spatial distribution of phenomena that are related to the activities of people. (7.10)

Humboldt Current A south-to-north-flowing, cold surface ocean current in the eastern South Pacific Ocean; also known locally as the Peru Current or the Chile Current. (6.2)

Humid Continental (Dfa or Dfb) climate In the Köppen climate classification, a climate type characterized by adequate moisture year-round, a mean temperature of the coldest month below –3°C, and a hot or warm summer. (7.7)

humidity The moisture in the atmosphere. (4.0)

Humid Subtropical (Cfa) climate In the Köppen climate classification, a climate type characterized by adequate moisture year-round, hot summers, and the mean temperature of the coldest month between –3°C and 18°C. (7.6)

hummocky topography A type of landscape characterized by randomly distributed humps and pits, commonly created by a landslide or less commonly by a pyroclastic eruption. (12.13)

humus A durable, partially decomposed form of organic matter. (16.1)

hundred-year flood The size of a flood that is predicted, from the existing data, to have a 1 in 100 chance (1%) of occurring in any given year. (13.14)

hurricane A regional term used in the Atlantic basin and eastern Pacific basin applied to the most intense type of storm of tropical oceanic origin. (5.13)

hydraulic conductivity A measure of the ability of a material to transmit a fluid; also known as permeability. (16.3)

hydraulic gradient The slope of energy content of subsurface water. (8.6)

hydrogen bond A weak bond in water that forms between one molecule's hydrogen atom and another molecule's oxygen atom. (4.1)

hydrograph A graph showing the change in the amount of flowing water (discharge) over time. (13.1)

hydrologic cycle The cycle representing the movement of water between the oceans, atmosphere, land, rivers and other surface water, groundwater, and organisms. (1.4)

hydrologic fracturing The fracturing of rock by injecting liquid at high pressure, especially to extract oil or gas. (11.13)

hydrolysis A form of chemical weathering involving the chemical reaction of silicate minerals with water, converting the original materials to clay minerals. (9.9)

hydrophyte A plant species adapted to the constantly wet conditions of wetlands. (18.14)

hydrosphere The part of Earth characterized by the presence of water in all its expressions, including oceans, lakes, streams, wetlands, glaciers, groundwater, moisture in soil, water vapor, and drops and ice crystals in clouds and precipitation. (1.3)

hygroscopic water Water within pore space in soil that adheres to soil particles; typically so tightly held that it is unavailable to plants. (16.3)

hypocenter Place where an earthquake is generated; also known as the focus. (11.13)

hypothesis A proposition that is tentatively assumed, and then tested for validity by comparison with observed facts and by experimentation. (1.2)

hypoxia A lack of oxygen in an aquatic system. (17.12)

hypsometric curve A graph that plots elevation versus the percentage of land or seafloor that is above that elevation (or depth). (10.13)

I

iceberg A massive piece of ice floating or grounded in the ocean or other body of water. (6.4)

Ice Cap (EF) climate In the Köppen climate classification, a climate type characterized by all 12 months having a climatological average temperature below freezing. (7.8)

icefall The portion of a glacier characterized by rapid flow and a crevassed surface, generally from flowing over steep terrain. (14.1)

Icelandic Low A prominent area of semipermanent low atmospheric pressure (cyclone) in the subpolar North Atlantic Ocean. (3.6)

ice sheet A mass of ice of considerable thickness and more than 50,000 square kilometers in area, forming a nearly continuous cover of ice and snow over a land surface. (14.1)

ice shelf An ice sheet or large glacier that flows into the ocean and floats on seawater. (14.3)

Ideal Gas Law The relationships between pressure, density, and temperature in certain gases, including the atmosphere. (3.1)

igneous rock A rock that formed by solidification of molten material (magma). (9.2)

illuviation The deposition and accumulation of materials in the lower levels of a soil. (16.7)

impermeable A property of a material or surface that completely blocks flow of a liquid through it; also known as an aquiclude. (8.6)

Inceptisol A type of recently developed soil characterized by mild weathering and the beginning of a weak B horizon, exhibiting similar characteristics to Entisols but generally older and having better horizonation. (16.9)

incision Cutting into bedrock by downward erosion. (13.11)

inclusion A fragment of older rock or material that is contained within another rock or material, as in a fragment of preexisting rock in a magma. (9.12)

index contour A dark line on a topographic map which helps emphasize the broader elevation patterns of an area and allows easier tracking of lines across the map; on most topographic maps, every fifth line is an index contour. (1.6)

Indian Ocean Dipole An atmospheric pressure oscillation between the eastern Indian Ocean and the western Indian Ocean. (6.13)

Indomalayan realm The zoogeographic realm consisting of the Indian subcontinent and adjacent parts of southern and southeastern Asia. (18.16)

inertia The ability of an ecosystem to resist change after a stress. (17.7)

infiltration Water and other fluids that seep into the ground through open pores, fractures, and cavities in soil and rocks. (8.1)

infrared energy (IR) A form of electromagnetic energy with longer wavelengths than visible light (from about 0.7 to about 1,000 micrometers); much of the Sun's radiation that reaches Earth's surface is ultimately converted into this type of energy. (2.6)

input (ecology) Any energy and materials, such as nutrients and water, that are available in a system, especially in an ecosystem. (17.5)

insectivore A carnivore that consumes insects. (17.2)

inshore The part of the coastal zone that is permanently submerged by shallow water where waves break; also known as the swash zone. (15.1)

insolation The energy transmitted from the Sun to Earth, as incoming solar radiation. (2.7)

interglacial period A time during an ice age when glaciers are melting, retreating, or diminished in extent. (14.2)

intermediate (igneous rock) An igneous rock with a medium color and generally containing both felsic and mafic minerals. (9.5)

intermittent stream A stream that flows every year but not during the entire year, typically flowing only during rainstorms and spring snowmelt. (13.3)

International Date Line The imaginary line that generally runs north-south but zigzags through the Pacific Ocean near the 180° meridian; locations west of this line are one day later in the calendar than locations east of this line. (1.7)

intertidal zone The area including the shoreline and a strip of adjacent land and water; specifically, the part of the saltwater biome nearest to the shore, characterized by wave action, temperature variations, and changes in substrate; also known as the coastal zone. (18.15)

Intertropical Convergence Zone (ITCZ) The belt of relatively low surface atmospheric pressure around the equatorial region, characterized by the collision of the northeast and southeast trade winds and rising motion. (3.9)

intrusion A subsurface magma body or the mass of igneous rock into which it solidifies; also known as a pluton. (11.8)

intrusive rock Any igneous rock that solidified from magma below Earth's surface; also known as a plutonic rock. (11.8)

inundation regime The scenario of coastal inundation during which the height of the swash zone exceeds the tops of coastal dunes. (15.11)

invasive species A non-native species that disrupts its new ecosystem by developing a broad geographical range and occupying a niche. (17.13)

invigorated vegetation The notion that higher temperatures may increase the rate of carbon uptake by plants under longer growing seasons with more favorable conditions for photosynthesis. (17.8)

ion A charged atom. (6.4)

iron formation A rock composed of millimeter- to centimeter-thick layers of iron-bearing minerals, especially iron oxide, commonly with quartz. (9.4)

island and seamount chain A series of islands and submarine peaks in a long, commonly linear belt. (10.2)

island arc A generally curved belt of volcanic islands above a subduction zone; also used as an adjective to refer to this setting. (10.2)

isobar A line connecting locations having the same atmospheric pressure; usually shown for sea-level-corrected pressures. (3.2)

isohypse A line on an upper-level weather map connecting locations having the same height at which a particular atmospheric pressure value occurs. (3.8)

isostasy The concept that crust of different thickness will ride on the mantle at different elevations. (10.1)

isostatic rebound Uplifting caused by the removal of weight on top of the crust, as when an ice sheet melts away or when erosion strips material off the top of a thick crustal root of a mountain. (10.15)

isotope One of two or more species of the same chemical element but differing from one another by having a different number of neutrons. (9.13)

isotopic dating The determination of the age of a rock usually by chemical analysis of the products of natural radioactive decay. (9.12)

J

jet stream A fast-flowing current of air, often near the tropopause. (3.9)

jetty An engineering structure built from the shore into a body of water to redirect current or tide, for example, to protect a harbor. (15.7)

joint A fracture in a rock where the rock has been pulled apart slightly, without significant displacement parallel to the fracture. (11.10)

Joule A unit of measurement of work or energy. (2.3)

June Solstice The day, on or near June 21, when the North Pole (marking where the Earth's axis of rotation intersects the Earth's surface in the Northern Hemisphere) is pointed most directly toward the Sun. (2.9)

K

kame A hill formed where meltwater in stagnant ice deposited sediment in crevasses or in the space between the glacier and the land surface. (14.6)

kame and kettle topography A terrain of hills, ponds, and poor drainage, characteristic of the effects of kames and kettles. (14.6)

karst A topography characterized by sinkholes, caves, limestone pillars, poorly organized drainage patterns, and disappearing streams; generally formed from the dissolution of limestone or other soluble rocks. (12.2)

katabatic wind A local-to-regional-scale wind that blows from the peak of a hill or mountain down to the valley; also known as a mountain breeze. (3.4)

Kelvin scale An absolute system for indexing temperature in which all molecular motion ceases at 0 K, water freezes at 273 K, and water at sea level boils at 373 K. (2.3)

Kelvin wave A subsurface equatorial wave in the Pacific Ocean that propagates from west to east along the thermocline, at the onset of an El Niño event. (6.12)

Kerguelen Plateau The second-largest oceanic plateau on Earth, in the southern Indian Ocean. (10.11)

kettle A pitlike depression in glacial deposits, commonly a lake or swamp; formed by the melting of a large block of ice that had been at least partly buried in the glacial deposits. (14.6, 18.14)

kettle lake A body of water occupying a kettle. (14.6)

keystone predator The species at the top of the food chain in an ecosystem. (17.3)

kinetic energy The energy possessed by a solid, liquid, or gas expressed by the motion of that object. (2.2, 8.2)

knot A unit of speed equal to one nautical mile (1.852 km) per hour, approximately 1.151 mph. (3.3)

Köppen climate classification A scheme for categorizing climates based on the annual regime of temperature and precipitation, with the categories of climate types corresponding to the major realms of natural vegetation. (7.1)

krummholz A deformed, stunted tree near the treeline, such as banner trees. (18.4)

K-selected species Any species with a slow rate of reproduction and maturation, but with a strong and prolonged nurturing, leading to a relatively high likelihood of survival into adulthood. (17.2)

Kuroshio Current A south-to-north-flowing, warm surface ocean current in the western North Pacific Ocean. (6.2)

L

Labrador Current A small, north-to-south-flowing cold surface ocean current in the northern Atlantic Ocean between Greenland and eastern Canada. (6.2)

laccolith A bulge-shaped igneous body that has domed and tilted overlying layers and that is observed or interpreted to have a relatively flat floor. (11.8)

lagoon A shallow part of the sea between the shoreline and a protecting feature, such as a reef or barrier island, farther out to sea. (15.5)

lahar A mudflow mostly composed of volcanic-derived materials and generally formed on the flank of a volcano. (11.4)

lake-effect snow Intense snowfall events that occur near lakes, especially the Great Lakes, resulting from enhancement of moisture contributed by the lakes, in air that has been destabilized by very cold air overlying relatively warm lakes. (4.13, 5.4)

laminar flow Flow characterized by smooth movement in parallel layers. (13.2)

land breeze A local-to-regional-scale wind that blows from land to sea, typically at night. (3.4)

land bridges Hypothesized ridges, now barely submerged in the oceans, proposed to explain the worldwide migration of animals from their geographic origin; most proposed ridges have been disproved by more recent data. (10.3)

land cover The natural or human-made features on the surface of the Earth, such as bare rock, water, soil, forest, or concrete. (4.6)

landfall The arrival of a tropical cyclone on land. (5.13)

landslide A general term for the rapid downslope movement of soil, sediment, bedrock, or a mixture of these; also the material or landform formed by this process. (11.4)

La Niña A periodic cooling of the eastern equatorial Pacific surface ocean waters, and reduction in precipitation, adjacent to South America, as the Southern Hemisphere's summer approaches; also known as a cold phase of ENSO. (6.10)

lapse rate The change in temperature of an air parcel or the atmosphere in general, per unit change in elevation. (4.5)

large-scale map A map that shows great detail within a small geographic area. (1.10)

latent heat The energy stored or released as a consequence of a change between the solid, liquid, or gaseous states of matter. (2.4)

latent heat flux The flow, or transfer, of latent heat in the atmosphere by convection and advection. (2.15)

latent heat of fusion The energy absorbed as a consequence of a change in state from a solid to liquid state of matter (melting), or released as a consequence of a change in state from a liquid to solid state of matter (freezing). (2.4)

latent heat of sublimation The energy absorbed as a consequence of a change in state from a solid to gaseous state of matter (sublimation), or released as a consequence of a change in state from a gaseous to solid state (deposition). (2.4)

latent heat of vaporization The energy absorbed as a consequence of a change in state from a liquid to gaseous state of matter (evaporation), or released as a consequence of a change in state from a gaseous to liquid state of matter (condensation). (2.4)

lateral erosion The wearing away of soil, sediment, and rock along the side of a stream. (13.5)

lateral moraine Sediment carried in and deposited along the sides of a glacier. (14.5)

laterite A type of tropical soil rich in iron (Fe) and aluminum (Al) oxides, commonly giving the soil a deep red color. (16.6)

laterization A pedogenic regime characterized by rapid chemical weathering and extreme dissolution in hot, wet settings, producing red, leached soils, such as laterites. (16.8)

latitude A measure of position north or south of the equator; all locations sharing the same latitude in a hemisphere fall along an imaginary east-west line that encircles the Earth. (1.7)

Laurasia The inferred northern supercontinent that existed in the Mesozoic and included North America, Europe, and Asia. (10.14)

lava Magma that is erupted onto the surface, or the rock mass into which it solidifies. (9.2)

lava dome A dome-shaped mountain or hill of at least partly solidified lava, generally of felsic to intermediate composition. (11.2)

lava flow A volume of magma that erupts onto the surface and flows downhill from the vent of a volcano; also the solidified body of rock formed by this magma. (11.2)

lava fountain A shooting stream of molten lava propelled into the air by pressure and escaping gases during an explosive volcanic eruption. (11.2)

lava tube A long, tubular opening under a roof of solidified lava and representing an active or partially emptied subsurface channel of lava. (11.3)

leaching The separation or dissolution of soluble constituents from a rock, sediment, or soil by percolation of water; also known as eluviation. (16.3)

leeward The downwind side of a slope. (3.5)

lentic system Any freshwater ecosystem that contains still water, such as lakes and ponds. (18.14)

lethal limit The threshold at which an environmental variable becomes too high or too low for a particular organism to survive. (17.7)

levee A long, low ridge of sediment deposited by a stream next to the channel, or built by humans to prevent floodwaters from spilling onto a floodplain. (13.15)

liana A vine in a rain forest that winds around the maze of branches and leaves in the canopy layer. (18.5)

lichens A symbiotic combination of algae and fungi. (12.4)

lidar A type of active remote sensing in which electromagnetic radiation usually at near-infrared wavelengths is emitted, and the amount and timing of returning energy are used to determine distance and other characteristics of the land. (1.11)

lifting condensation level The height at which a parcel of air becomes saturated when it cools to its dew-point temperature as it is lifted adiabatically. (4.7)

light reaction A chemical reactions that takes place as part of photosynthesis and requires the presence of light, such that the energy captured from light is converted to chemical energy. (17.5)

lightning The bright flash observed at the instant that electrical energy is transferred through the atmosphere, after the return stroke is established. (5.9)

limb The planar or less curved parts of a fold on either side of the hinge. (11.11)

limestone A sedimentary rock composed predominantly of calcium carbonate, principally in the form of calcite, and which may also include chert, dolomite, and fine-grained clastic sediment. (9.4)

limnetic zone The upper zone of a lake in which more oxygen is generated from photosynthesis than is used by respiration or fermentation. (17.12)

lineation A linear structure in rock, usually in a metamorphic rock. (9.6)

liquefaction A loss of cohesion when grains in water-saturated soil or sediment lose grain-to-grain contact, as when shaken during an earthquake. (11.16)

liquid A state of matter in which a collection of mobile atoms and molecules more or less stay together but are not held in the rigid form of a solid. (2.2)

lithification The conversion of unconsolidated sediment or volcanic ash into a coherent, solid rock, involving compaction and cementation. (9.3)

lithosphere Earth's upper, rigid layer composed of the crust and uppermost mantle. (1.3)

lithospheric mantle The part of the uppermost mantle that is in the lithosphere. (10.1)

litter Dead plant material on the ground. (16.4)

litter layer A layer of loose leaves, twigs, branches, pine needles, and other plant debris on top of soil. (16.7)

Little Ice Age A period from about 1450 to about 1850 A.D. in which global temperatures were relatively cold. (2.7)

littoral zone The part of a lentic system or coastal zone that has access to sunlight throughout its entire depth; typically near the shore where the water is not too deep. (15.1)

llanos A treeless, grassy plain, especially in northern South America. (18.9)

loading The process by which weight is added to the lithosphere. (15.10)

loam A soil texture consisting of a balance between particles of different sizes, enhancing the soil's ability to support agriculture. (16.2)

local climate The average atmospheric conditions, along with the extremes and variability of those conditions, at a local scale; also known as microclimate. (7.10)

location The setting of a place, both in terms of its absolute geographic coordinates and its position relative to other features. (1.2)

loess An essentially unconsolidated sediment consisting predominantly of silt, interpreted to be windblown dust, commonly of glacial origin. (9.11)

longitude An imaginary north-south-running line that passes through the North and South Poles, encircles the Earth, and indicates how far east or west of the Prime Meridian a location is. (1.7)

longitudinal dune A type of sand dune that is linear or only very gently curving for many kilometers, oriented parallel to prevailing winds. (9.11)

longitudinal profile A plot of the elevation of a stream versus distance from the source, typically showing a steeper gradient near the headwaters and a progressively less steep gradient downstream. (13.4)

longshore current A current, generally in an ocean or large lake, flowing more or less parallel to a coastline. (15.4)

longshore drift The process by which sand and other sediment move parallel to a coast with the prevailing longshore current. (15.4)

longwave radiation Electromagnetic energy at wavelengths greater than about 4 micrometers; corresponds to the wavelengths at which the Earth emits nearly all of its energy. (2.6)

losing stream (river) A stream or part of a stream that loses water from outflow to the unsaturated zone beneath it; also known as a losing river. (8.7)

lotic system Any freshwater ecosystem that contains moving water, such as a stream. (18.14)

low tide The minimum height to which water in the ocean falls relative to the land in response to the gravitational pull of the Moon; also refers to the time when such low levels occur. (15.2)

M

Madagascan realm The zoogeographic realm consisting of Madagascar. (18.16)

mafic A group of generally dark-colored, silicate minerals with a high magnesium (Mg) and iron (Fe) content. (9.1)

mafic rock A generally dark-colored igneous rock with a mafic composition. (9.5)

magma Molten rock, which may or may not contain some crystals, solidified rock, and gas. (9.2)

magma chamber A large reservoir in the crust or mantle that is occupied by a body of magma. (9.2)

magmatism The melting of rock at high temperatures beneath the Earth's surface, causing magma to form. (9.3)

magnetic reversal A switch between normal polarity and reversed polarity of the Earth's magnetic field. (10.10)

magnetite An iron oxide (Fe_3O_4) mineral that is typically black and is strongly magnetic. (9.1)

magnetometer An instrument used to measure the direction and strength of magnetism in rocks and other materials. (10.10)

magnitude (seismic) A measure of the amount of energy released by an earthquake; used to compare sizes of earthquakes; can refer to Richter magnitude or moment magnitude. (11.16)

mantle The most voluminous layer of Earth; located below the crust and above the core. (10.1)

mantle convection Movement of mantle material in response to variations in density, especially those caused by differences in temperature. (10.9)

mantle plume A mechanism to explain volcanic regions of the Earth that are not thought to be on a plate-tectonic boundary, for example, Hawaii. (10.11)

map projection Any mathematical algorithm or technique for representing the three-dimensional Earth on a flat map. (1.9)

maquis The distinguishing type of vegetation in the subtropical scrub and woodland biome, characterized by scrubby trees with thick bark and short, gnarly shrubs, thickets, and grasses with deep root systems; also known as chaparral. (18.10)

marble A metamorphic rock composed of recrystallized calcite or dolomite, possibly including other metamorphic minerals. (9.6)

March Equinox The day, on or near March 21, when Earth's axis of rotation is pointed exactly sideways relative to the Sun, and the durations of daylight and darkness are equal throughout the world. (2.9)

marine terrace An erosional coastal landform consisting of a platform that was cut or constructed by waves but is now elevated above sea level; commonly covered by a thin veneer of marine sediment. (15.5)

Marine West Coast (Cfb or Cfc) climate In the Köppen climate classification, a climate type characterized by adequate moisture year-round and the mean temperature of the coldest month between −3°C and 18°C, with warm or mild summers. (7.7)

maritime polar (mP) air mass A body of air that is relatively uniform in its cool/cold and humid characteristics for thousands of square kilometers, originating in oceanic, subarctic source regions. (5.1)

maritime tropical (mT) air mass A body of air that is relatively uniform in its warm and humid characteristics for thousands of square kilometers, originating in oceanic, subtropical source regions. (5.1)

marsh A wetland that is covered by grasses and reeds rather than trees; often forming along the borders of a stream or lake. (15.5)

marsupial Any pouched mammal, largely confined to the Australian realm. (18.16)

mass extinction The disappearance of many species in a geologically short period of time. (9.13)

massive soil Soil with no visible structure, typically forming large clods. (16.2)

mass wasting The downward movement of material on slopes under the influence of gravity. (12.8)

matter Any solid, liquid, or gas that occupies space and has mass. (1.4)

mature stage The second stage in a single-cell thunderstorm's life cycle, characterized by continuing updrafts along with downdrafts, precipitation, lightning, and thunder. (5.6)

Maunder Minimum A relatively cold period from about 1650 to 1750 A.D. during which few or no sunspots were observed. (2.7)

meander A sinuous curve or bend in the course of a stream, or in atmospheric Rossby wave flow. (3.13)

meandering stream A stream that has a strongly curved channel. (13.5)

meander scar A crescent-shaped feature in the landscape that indicates the former position of a stream meander. (13.8)

mechanical turbulence The vertical motion generated by friction or vertical wind shear. (5.10)

medial moraine Sediment carried in the center of a glacier, representing where two smaller glaciers joined; also refers to the deposited sediment and resulting landform. (14.5)

Medieval Warm Period A warm climatic interval about 1,000 years ago. (7.11)

Mediterranean (Csa or Csb) climate In the Köppen climate classification, a climate type characterized by adequate moisture in winter but not in summer, hot or warm summers, and the mean temperature of the coldest month between −3°C and 18°C. (7.6)

megafauna Any type of large animal. (18.16)

megathrust A major thrust fault, representing the boundary between the subducted slab and overriding plate. (11.15)

melting A change in state from a solid to a liquid, requiring an input of energy from the environment and the absorption of that energy at the molecular level as latent energy. (1.4)

Mercator projection A popular type of cylindrical map projection. (1.9)

meridian Any line of longitude. (1.7)

mesa A broad, flat-topped and steep-sided, isolated hill or mountain. (11.1)

mesocyclone A rotating vortex embedded within a supercell thunderstorm. (5.6)

mesopause The transition zone between the mesosphere and overlying thermosphere. (2.11)

mesopelagic zone The part of the saltwater biome located seaward of the coastal zone's low-tide mark and containing the vast open waters of the ocean, at medium depth, below the (epipelagic) zone, where light cannot penetrate (i.e., part of the aphotic zone). (18.15)

mesosphere The third layer of the atmosphere above the surface. (2.1)

Mesozoic Era A major subdivision of geologic time from 252 Ma to 66 Ma; characterized by dinosaurs. (9.13)

metamorphic rock A rock changed in the solid state by temperature, pressure, deformation, or chemical reactions that modified a preexisting rock. (9.2)

metamorphism The mineralogical and structural changes of solid rock in response to changes in environmental conditions, especially at depth. (9.2)

mica A family of silicate minerals that break into flakes and sheets that are typically partially transparent and shiny. (9.1)

microburst The brief but strong, downward-moving winds in a "starburst" pattern out from a central point; sometimes referred to as straight-line winds. (5.12)

microclimate The average atmospheric conditions, along with the extremes and variability of those conditions, at a local scale; sometimes known as local climate. (7.10)

microwave Any electromagnetic radiation of wavelengths between 1 millimeter and 10 meters. (2.6)

mid-latitude cyclone A large, migrating storm system composed of a central cyclone, with cold front and warm fronts which may become stationary at various times and possibly undergo occlusion, emanating from it. (5.3)

mid-ocean ridge A long mountain range on the floor of the ocean, associated with sea-floor spreading, also known as spreading center. (10.2)

Milankovitch cycle Periodic variations in Earth's orbit and angle and direction of tilt, interpreted to influence Earth's insolation and climate. (14.9)

millibar A unit of atmospheric pressure (force per unit area) often used in meteorology, equivalent to 100 Pascals. (3.2)

mineral A naturally occurring, inorganic, crystalline solid with a relatively consistent chemical composition. (9.1)

mineral carbonation The notion that ultramafic igneous rocks could be used to absorb atmospheric carbon dioxide efficiently in an attempt to regulate global temperatures. (17.8)

mineral wedging The process by which the outward force caused by growing minerals can fracture rock or loosen grains. (12.1)

Modified Mercalli Intensity Scale An earthquake intensity scale based on the amount of damage and how the earthquake was perceived by people. (11.16)

Mollisol A fertile, mid-latitude type of soil, rich in calcium carbonates and organic material, with little leaching and a humus-rich A horizon. (16.11)

moment magnitude (Mw) A measure of the amount of energy released by an earthquake. (11.16)

momentum The product of mass and velocity. (1.4)

monocline A fold defined by local steepening in gently dipping layers. (11.11)

moraine Sediment carried by and deposited by a glacier; also refers to the resulting landform. (14.5)

mountain bark beetle A pestilent insect common in the boreal forest biome. (18.12)

mountain breeze A local-to-regional-scale wind that blows from the peak of a hill or mountain down to the valley; also known as a katabatic wind. (3.4)

mouth The location where a stream or canyon ends, such as where a stream enters the sea. (13.3)

Mozambique Current A north-to south-flowing, warm surface ocean current in the western Indian Ocean. (6.2)

mudflow A downhill-flowing slurry of loose rock, mud, and other materials, and the resulting landform and sedimentary deposit; also known as a debris flow or earthflow. (11.4, 12.11)

multi-cell thunderstorm A cluster of individual thunderstorms that interact as part of an organized weather system to produce severe weather. (5.6)

multispectral Data that are remotely sensed via multiple bands of wavelengths simultaneously. (1.11)

muscovite A light-colored phyllosilicate mineral of aluminum and potassium. (12.3)

mutualism A symbiotic relationship in which both species benefit from the interaction. (17.3)

N

NADPH A phosphate compound produced as energy is released by "excited" electrons during the light reaction of photosynthesis. (17.5)

native species A species that is in its natural ecosystem. (17.13)

natural cement Any mineral, such as calcite, silica, clay minerals, and iron oxides, that precipitates as water transporting dissolved chemicals moves through pore spaces in rock or sediment. (9.4)

neap tide Lower than average high tides and higher than average low tides caused when the Sun's gravity partially offsets the effects of the Moon's gravity. (15.2)

Nearctic realm The zoogeographic realm consisting of the vast majority of North America. (18.16)

near infrared (IR) energy The part of the electromagnetic spectrum with wavelengths between about 0.7 and 3 micrometers. (1.11)

negative feedback A response to a change in a system that diminishes and dampens more of the same type of change. (1.3)

negative Indian Ocean Dipole (IOD) The phase of the pressure oscillation between the eastern and western Indian Ocean characterized by higher atmospheric pressure and cooler water than normal in the western Indian Ocean and lower than normal atmospheric pressure with warmer water than normal in the eastern Indian Ocean. (6.13)

negative magnetic anomaly A weak signal from magnetic instruments, indicative of rock formed during a time of reversed polarity. (10.10)

negative North Atlantic Oscillation index A condition characterized by simultaneous weaker than normal atmospheric pressures in the Bermuda-Azores High and Icelandic Low. (6.13)

Neotropic realm The zoogeographic realm consisting of South America and a small section of central America. (18.16)

net primary productivity The total amount of energy fixed for photosynthesis (i.e., gross primary productivity) minus the energy wasted or used for plant respiration and maintaining existing tissue. (17.6, 18.1)

neutral phase (ENSO) An atmospheric and oceanic condition in the equatorial Pacific Ocean that is near-normal, with little evidence of anomalous warming (El Niño) or cooling (La Niña) in the eastern equatorial Pacific. (6.10)

new moon The position of the Moon in its orbit around Earth when it is between the Sun and Earth; results in the side of the Moon facing Earth being not illuminated by the Sun. (15.2)

Newton A unit of force equivalent to one kilogram-meter per second squared. (3.2)

New Zealand realm The zoogeographic realm consisting of New Zealand. (18.16)

nickpoint A place where a stream has an abrupt change in gradient, such as at a waterfall. (13.6)

nimbostratus cloud A type of wide, flat cloud that is actively precipitating. (4.8)

noise A reflection of a radar image from ground clutter, such as buildings or birds; also known as a false echo. (5.16)

nonclastic rock Any sedimentary rock that is not composed of clasts (rock fragments). (9.4)

non-native species Species that survive an introduction to a new ecosystem and become part of that ecosystem. (17.13)

non-point source A widespread area of multiple sites which produce pollution. (7.9)

Nor'easter A Hatteras Low that has tracked near or over the northeastern U.S. (5.3)

normal component The part of the gravitational force that pulls a mass against the sloping surface on which it sits. (12.8)

normal fault A fault in which the block above the fault moves down relative to the block below the fault. (11.10)

Normalized Difference Vegetation Index (NDVI) A variable, typically measurable using remote sensing techniques, that can be used to estimate net primary productivity. (17.6)

normal polarity The orientation of the Earth's magnetic field such that the field flows from south to north, causing the magnetic ends of a compass needle to point toward the north. (10.10)

normal temperature gradient An atmospheric condition in which temperature decreases as height increases. (2.11)

North Atlantic Deep Waters The deep, very saline, polar waters that descend in the Atlantic Ocean near Iceland and flow at depth well south of the equator. (6.5)

North Atlantic Drift The northern extension of the Gulf Stream, which moderates the climate of northwestern Europe. (6.2)

North Atlantic Oscillation (NAO) A teleconnection in which simultaneous opposing atmospheric pressure anomalies exist between the Bermuda-Azores High and the Icelandic Low. (6.13)

North Atlantic Oscillation (NAO) index A numerical indicator of the degree to which atmospheric pressure patterns differ between the Bermuda-Azores High and the Icelandic Low. (6.13)

northeast trade winds The prevailing surface atmospheric circulation that exists in the tropical latitudes of the Northern Hemisphere, characterized by flow from northeast to southwest. (3.9)

North Equatorial Current An east-to-west-flowing, surface ocean current north of the equator in the Pacific Ocean. (6.2)

North Pacific Current A west-to-east-flowing, surface ocean current in the high latitudes of the Northern Hemisphere. (6.2)

no-till agriculture A technique to minimize soil erosion, whereby the previous year's crop residue is left intact and the field is not plowed into rows. (16.13)

nucleus A particle composed of protons and generally neutrons in the core of an atom. (2.5)

null point The depth in the ocean at which the effects of wind stress vanish; usually at a depth of around 100 meters. (6.1)

nunatak A small area of bedrock that protrudes through an ice sheet. (14.1)

O

oasis effect An isolated center of relatively high humidity in an otherwise arid area; often caused by human activities such as irrigation, reservoir construction, or lawn sprinklers. (7.10)

occluded front A narrow zone that evolves late in the life cycle of the mid-latitude cyclone, marking the location where the cold front "catches up with" the warm front and lifts the warmer air above the colder air. (5.4)

ocean acidification The process by which atmospheric carbon dioxide dissolves in seawater, reducing its pH. (7.12)

ocean-continent convergent boundary A plate-tectonic boundary where an oceanic plate converges with a continental plate, generally expressed by subduction of the oceanic plate beneath the continent. (10.7)

ocean fertilization The idea that the scattering of iron on the ocean surface could be used to promote phytoplankton growth, which would hypothetically regulate global temperatures. (17.8)

ocean oscillation A temperature oscillation between two large areas of an ocean, such that warmer than normal sea-surface temperatures occur simultaneously in one part of the ocean while colder than normal sea-surface temperatures are occurring in the other; or, a temperature anomaly over a single large part of the ocean, in which above-normal sea-surface temperatures persist for a period of many years followed by the dominance of below-normal sea-surface temperatures. (7.12)

ocean(ic) trench A narrow, steep-sided, elongate depression of the deep seafloor, formed by downward bending of a subducting oceanic plate at a convergent plate boundary; includes the deepest parts of the ocean. (10.2)

oceanic crust The type of thin, mafic crust that underlies the ocean basins. (10.1)

oceanic fracture zone A crack or step in elevation of the seafloor that formed as a transform fault along a mid-ocean ridge but is no longer a plate boundary. (10.2)

oceanic plateau A broad, elevated region in the ocean floor. (10.2)

ocean-ocean convergent boundary A plate boundary where two oceanic tectonic plates converge. (10.7)

O horizon An upper, organic-rich soil horizon composed of dead leaves and other plant and animal remains. (16.7)

oligotrophic An aquatic ecosystem with low nutrient concentrations and low net primary productivity. (17.10)

olivine A mostly olive-green or brown mineral occurring widely in basalt, peridotite, and other basic igneous rocks. (10.1)

omnivore Any organism with a diverse diet of both meat and plants. (17.2)

Ontong Java Plateau The largest oceanic plateau on Earth, in the tropical western Pacific Ocean near Papua New Guinea. (10.11)

open system Any interconnected set of matter, energy, and processes that allows material and energy to be exchanged outside itself. (1.3)

optimal range The thresholds of an environmental variable between which an organism thrives. (17.7)

orbital plane The imaginary plane in which the Earth revolves around the Sun. (2.9)

original horizontality The principle that most sediments and many volcanic units are deposited in layers that originally are more or less horizontal. (9.12)

orographic effect The rising motion and enhancement of precipitation on the windward slopes of a mountain or other topographic feature. (4.7)

outgoing longwave radiation (OLR) Electromagnetic energy at wavelengths longer than 4 micrometers that radiates into space after having been emitted by the Earth's surface and atmosphere. (2.14)

outwash plain An area in front of a glacier where streams deposit glacially produced sediment. (14.6)

overland flow Water flowing over the surface but not in a channel. (13.6, 16.3)

overstory The higher of the two layers in most boreal forests. (18.12)

overturning The mixing of waters from different depths in a large body of water; also known as turnover. (6.5)

overwash The process by which a wave or storm surge moves material from the side of a landmass nearest to the sea to an area farther from the sea. (15.11)

overwash regime The scenario of coastal inundation during which the height of the swash zone reaches the tops of coastal dunes. (15.11)

oxalic acid A colorless crystalline organic compound with the formula $H_2C_2O_4$. (12.4)

oxbow lake An isolated, curved lake formed when a cutoff meander is filled with water. (13.5)

oxidation The chemical process during which a material combines with oxygen; a form of chemical weathering. (9.9)

oxide minerals A group of nonsilicate minerals that contains oxygen bonded with a metallic element, such as iron, manganese, or titanium. (9.1)

Oxisol A type of soil formed in wet tropical conditions and characterized by deep weathering, leaching, and oxidation, usually high in iron and aluminum and low in nutrients. (16.10)

ozone A molecule consisting of three oxygen atoms, concentrated in the stratosphere where it absorbs harmful ultraviolet radiation, but also existing near the Earth's surface where it is a pollutant. (2.12)

ozone layer The lower part of the stratosphere where ozone is formed and photodissociated quickly. (2.12)

P

Pacific Decadal Oscillation (PDO) An oscillation in sea-surface temperatures in the North Pacific Ocean, in which temperatures are warmer than normal either near the Alaska coast or over the central North Pacific, while they are simultaneously below normal at the other location. (6.0)

Pacific Decadal Oscillation (PDO) index A numerical indicator of the degree to which the Pacific Decadal Oscillation is in its warm phase, with warmer than normal sea-surface temperatures in the eastern North Pacific near the Alaska coast (when the index is positive), or in its cool phase, which is characterized by cooler than normal sea-surface temperatures in the eastern North Pacific near the Alaska coast (when the index is negative). (6.0)

Pacific realm The zoogeographic realm consisting of most of the Pacific islands. (18.16)

Pacific Ring of Fire The belt of frequent volcanic activity from the southwestern Pacific, through the Philippines, Japan, and Alaska, and then down the western coasts of North and South America, resulting from subduction of oceanic plates around the Pacific Ocean. (10.7)

pahoehoe A type of lava or lava flow that has a relatively smooth upper surface or folds that form a "ropy" texture. (11.3)

Palearctic realm The zoogeographic realm consisting of Europe and most of Asia and northern Africa. (18.16)

Paleozoic Era A major subdivision of geologic time from 541 Ma to 252 Ma. (9.13)

Palmer Drought Index A numerical indicator of the severity of drought conditions or of the excessiveness of water, customized to the place of interest at the time of year of interest. (4.14)

Pampas An agriculturally fertile plains region of South America, encompassing Uruguay, part of eastern Argentina, and extreme southern Brazil. (18.2)

Pangaea An inferred supercontinent that existed from about 300 million to about 200 million years ago and included most of Earth's continental crust. (10.14)

parallel Any line of latitude. (1.7)

parasitism A type of symbiosis in which one species benefits from the interaction and the other is harmed by the interaction. (17.3)

parent atom An atom before it undergoes radioactive decay; also known as parent isotope. (9.13)

parent material Any minerals and rocks upon which physical and chemical weathering acts, and from which soils are derived. (12.3)

Pascal A unit of pressure (force per unit area) equivalent to one Newton per square meter. (3.2)

passive margin A continental edge that is not a plate boundary. (10.6)

passive remote sensing system A type of remote sensing in which the sensor detects whatever light, heat, or other energy is naturally coming from the region being monitored. (1.11)

patch dynamics The idea that subareas within an ecosystem may be influenced by different processes and are at different successional stages at the same time. (17.7)

paternoster lakes A series of tarns connected by a stream. (14.5)

patterned ground Polygon-shaped outlines in the soil of Arctic areas, caused by frost heaving. (12.8)

peak discharge The maximum volume of water per unit time, shown on a hydrograph. (13.3)

peat An unconsolidated deposit of partially decayed plant matter. (16.1)

pebble A small stone between 6 and 64 mm in diameter. (13.7)

ped A group of soil particles clumped together. (16.2)

pediment A gently sloping, low-relief plain or erosion surface carved onto bedrock, commonly with a thin, discontinuous veneer of sediment. (9.9)

pediment dome A broad, gently sloping peak or ridge formed by erosion of bedrock over a long period of geologic time. (9.9)

pedogenic regime An environmental setting where certain physical, chemical, and biological conditions dominate; includes laterization, salinization, calcification, podzolization, and gleization. (16.8)

pedon The smallest unit within which soils are described; a three-dimensional body about 2 meters deep and large enough (1 to 10 square meters) to distinguish soil horizons. (16.8)

pegmatite An igneous rock containing very large crystals, which may be centimeters to meters long. (9.5)

pelagic Pertaining to an organism in a saltwater biome that resides mainly suspended in the water. (18.15)

pelagic zone The part of the saltwater biome located seaward of the coastal zone's low-tide mark and containing the vast open waters of the ocean, consisting of the epipelagic, mesopelagic, and hadal zones at various depths. (18.15)

perched water table Groundwater that sits above the main water table and generally is underlain by an impermeable layer or lens that blocks the downward flow of groundwater. (8.6)

percolation The deep downward movement of water from the unsaturated soil layer into the saturated layer. (8.7)

perennial stream A stream that flows all year. (13.3)

perihelion The date of closest approach of Earth to the Sun during the course of its orbit, at present corresponding to a day on or near January 4. (2.7)

period The largest subdivision of geologic time within an era, usually lasting tens of millions of years. (9.13)

permafrost A condition in which water in the uppermost part of the ground remains frozen all or most of the time. (8.1, 14.7, 18.12)

permeability A measure of the ability of a material to transmit a fluid; also known as hydraulic conductivity. (8.6, 16.3)

Peru Current A local name for the Humboldt Current. (6.2)

photic zone The (topmost) layer of an aquatic system that has access to sunlight, generally to a depth of 100 meters or less. (6.7)

photodissociation The splitting apart of chemical bonds caused by exposure to radiation. (2.12)

photosynthesis The process by which plants produce carbohydrates, using water, light, and atmospheric carbon dioxide. (17.1)

phreatophyte Any type of plant with a deep taproot system, to enhance survival in a desert biome. (18.8)

phyllite A shiny, foliated, fine-grained metamorphic rock, intermediate in character between slate and schist. (9.6)

physical geography The study of spatial distributions of phenomena across the landscape, processes that created and changed those distributions, and implications of those distributions for people. (1.1)

physical weathering The physical breaking or disintegration of rocks when exposed to the environment. (9.9)

phytoplankton A broad group of mostly microscopic photosynthesizing organisms. (17.6)

piedmont glacier A broad glacier that forms when an ice sheet or valley glacier spreads out as it moves into less confined topography. (14.1)

pillow A rounded, pillow-shaped structure that forms when lava erupts into water. (11.3)

pioneer community The first group of organisms to appear in an ecosystem following primary succession. (17.7)

plagioclase feldspar A form of feldspar common in igneous rocks consisting of aluminosilicates of sodium and/or calcium. (9.1)

planar projection A family of map projections based on the premise that the globe is projected onto a plane, with no distortion at the tangent point at which the globe touches the plane, and increasing distortion of both shape and area with distance from that point. (1.9)

plateau A broad, relatively flat region of land that has a high elevation. (10.2)

platform The instrument-carrying vehicle used in remote sensing. (1.11)

platy soil A soil structure in which the peds take the form of horizontal plates, commonly because the soils have been compacted by the weight of overlying materials. (16.2)

plucking The tearing away of rock by a glacier, particularly as a result of jointing caused by expansion during freeze-thaw action; common on the downflow side of a hill. (14.4)

PM$_{10}$ Any solid aerosol (particulate matter) with a diameter of $\leq$ 10 micrometers. (7.10)

podzolization A pedogenic regime characterized by shallow acidic soils with a well-developed soil profile, created in environments where winters are long, cold, and subject to frost action, and chemical weathering is relatively slow. (16.8)

point bar A series of low, arcuate ridges of sand and gravel deposited on the inside of a stream bend or meander. (13.5)

point-pattern analysis A type of spatial analysis in which the degree of clustering of a spatial distribution can be determined. (1.12)

point source A localized site which produces pollution. (7.9)

polar (chemical property) A molecule, such as water, that has an uneven distribution of charges from one side to the other. (4.1)

polar cell An atmospheric circulation cell with sinking motion near the pole, equatorward surface flow to subpolar latitudes, rising motion at the subpolar low-pressure belt, and motion aloft back to the pole, with one cell in each hemisphere, and strong Coriolis deflection affecting these motions. (3.11)

polar (E) climate In the Köppen climate classification, a broad group of climatic types characterized by adequate moisture availability, the mean temperature of the coldest month less than –3°C, and the mean temperature of the warmest month below 10°C. (7.1)

polar easterlies The prevailing surface general circulation feature that exists between the pole and about 60° of latitude in each hemisphere, characterized by flow from east to west. (3.9)

polar front jet stream The fast-flowing current of air near the tropopause in each hemisphere at the boundary between the polar cell and the mid-latitudes. (3.13)

polar high The large permanent areas of high surface atmospheric pressure centered near the North Pole and South Pole. (3.11)

polar stereographic projection A popular type of planar map projection. (1.9)

polyconical projection A family of map projections based on the premise that the globe is projected onto a cone that intersects it in two arcs, with the two tangent lines having no distortion; other locations are neither conformal nor equal area, but retain some advantages of both conformal and equal area maps when depicting small areas. (1.9)

polypedon Up to 10 pedons, if similar in properties. (16.8)

ponding Water that accumulates in a concentrated area on the surface when the intensity of precipitation exceeds the rate of infiltration. (16.3)

pool A slow-moving segment in a stream upstream from rapids. (13.5)

pool-and-riffle stream A stream characterized by an alternation of pools with riffles and rapids. (13.6)

poorly drained soil A soil in which too much water remains between the grains for too long, or in which water cannot pass easily through it; usually results in the ineffective dispersion of nutrients throughout the root zone. (16.3)

pore space Any open space within rocks, sediment, or soil, including between grains in a sedimentary rock, within fractures, and in other cavities. (9.4)

porosity The percentage of the volume of a rock, sediment, or soil that is open space (pore space). (16.1)

porphyritic An igneous rock having relatively larger crystals in a finer grained matrix; signifies that the magma cooled somewhat at depth but then rose and solidified on, or closer to, the surface where it cooled rapidly to solidify the finer grained matrix. (9.5)

positive feedback A response to a change in a system that reinforces and amplifies more of the same type of change. (1.3)

positive Indian Ocean Dipole (IOD) The phase of the pressure oscillation between the eastern and western Indian Ocean characterized by lower atmospheric pressure and warmer water than normal in the western Indian Ocean and higher than normal atmospheric pressure with cooler water than normal in the eastern Indian Ocean. (6.13)

positive magnetic anomaly A strong signal from magnetic instruments, indicative of more magnetic rock or of a rock formed during a time of normal polarity. (10.10)

positive North Atlantic Oscillation index A condition characterized by simultaneous stronger than normal atmospheric pressures in the Bermuda-Azores High and Icelandic Low. (6.13)

potassium feldspar An alkali feldspar, such as orthoclase or microcline, that contains a high amounts of potassium relative to sodium. (9.1)

potential energy The energy stored within the molecular structure of a solid, liquid, or gas as a consequence of its position relative to where it could possibly move. (2.2)

potential evapotranspiration The maximum amount of water that could possibly be lost from the surface to the atmosphere over a given interval of time through the combined effect of evaporation and transpiration, if the water were abundantly available. (7.3)

pothole A bowl-shaped pit eroded into rock by swirling water and sediment. (13.2)

prairie A tall-grass subcategory of the grassland biome. (18.9)

Prairie Peninsula The narrow, eastward protrusion of the grassland biome into Iowa, Illinois, and Indiana. (18.2)

Precambrian Era A major subdivision of geologic time from 4.55 Ga to 541 Ma. (9.13)

precession The long-term, cyclic process involving the change in direction of the Earth's axial tilt. (14.9)

precipitation (atmospheric) The water in solid or liquid form that falls from the atmosphere to the surface, including rain, freezing rain, sleet, snow, and hail. (1.4)

precipitation (chemical) The movement of a dissolved material out of the solution and into solid form. (9.4)

precipitation fog A type of fog that occurs when falling raindrops evaporate, increasing the humidity of the colder air beneath it, or when precipitation evaporates rapidly upon reaching the surface. (4.9)

predator An animal that consumes another animal. (17.3)

prescribed burn A fire that is intentionally started for beneficial reasons. (18.10)

pressure A force exerted on a unit area, expressed in a gas by the frequency of molecular collisions. (3.1)

pressure gradient The horizontal difference in pressure across some distance. (3.3)

pressure-gradient force The force that initiates movement of air from areas of higher atmospheric pressure to areas of lower pressure at right angles to the isobars. (3.3)

pressure melting The temperature at which ice melts at a given pressure. (14.4)

prey An animal that is consumed by another animal. (17.3)

pride A small family unit of lions. (18.9)

primary data Any new information, such as that generated when a new map is made based on original observations. (1.10)

primary mineral Minerals in soil, such as quartz, that are derived directly from the weathering and erosion of various types of rocks but undergo no chemical transformation during the soil-forming process. (16.1)

primary producer An organism that acquires what it needs from the soil, water, air, and energy of the environment, without consuming other organisms; also known as an autotroph. (17.2)

primary succession The predictable change in species composition in an ecosystem following stress which causes so much disturbance that little life is able to survive. (17.7)

primary wave (P-wave) A seismic body wave that involves particle motion, consisting of alternating compression and expansion, in the direction of propagation. (11.13)

Prime Meridian The line of longitude running through Greenwich, England, which serves as the zero degree of longitude. (1.7)

principle of superposition The concept that a sedimentary or volcanic layer is younger than any rock unit on which it is deposited. (9.12)

probability The statistical likelihood that an event will occur. (13.14)

profundal zone The lower zone of a lake in which more oxygen is used by respiration or fermentation than is generated from photosynthesis. (17.12)

promontory A local ridge of land that juts out into a body of water. (15.4)

proxy evidence A clue about past climatic conditions during times when direct measurements are not available. (7.11)

Public Land Survey System (PLSS) A coordinate system used in much of the U.S. that characterizes location based on the number of 6-mile by 6-mile squares north, south, east, or west of the intersection of a baseline with a principal meridian, with subdivisions within the square also described; also known as township-range system. (1.8)

pumice Volcanic rock, especially of felsic or intermediate composition, containing many vesicles (holes) formed by expanding gases in magma. (9.5)

pycnocline The level in the ocean at which density changes rapidly with depth. (6.7)

pyrite A common, pale bronze to brass yellow, iron sulfide mineral (FeS_2), commonly called "fool's gold." (9.1)

pyroclastic eruption A volcanic eruption in which hot fragments and magma are thrown into the air. (11.2)

pyroclastic flow A fast-moving cloud of hot volcanic gases, ash, pumice, and rock fragments that generally travels down the flanks of a volcano; also known as an ash flow. (9.5)

pyrophilic Pertaining to a species that germinates after the intense heat of fire cracks open the seed. (18.10)

pyroxene One of a group of mostly dark, silicate minerals, with a single chain of silicon-oxygen bonds. (9.1)

Q

qualitative data Data or observations that include descriptive words, labels, sketches, or other images. (1.2)

quantitative data A numeric measurement that is typically visualized and analyzed using data tables, calculations, equations, and graphs. (1.2)

quarter moon The position of the Moon at one-quarter and three-quarters of the distance through its orbit around Earth; characterized by a 90-degree angle formed between the position of the Moon, Earth, and Sun; results in the Moon being half illuminated by the Sun, with half of the illuminated side (that is, a quarter of the Moon) being visible from Earth. (15.2)

quartz A very common rock-forming silicate mineral, consisting of crystalline silica, with the chemical formula of SiO_2. (9.1)

quartzite A very hard rock consisting chiefly of quartz grains joined by secondary silica that causes the rock to break across rather than around the grains; formed by metamorphism or by silica cementation of a quartz sandstone and other quartz-rich rocks. (9.6)

quicksand Loose sand saturated with water and undergoing liquefaction, rendering it unable to support weight. (16.14)

R

radar A type of active remote sensing in which a microwave or radio wave is emitted, and the time and amount of energy returning at those wavelengths are used to map the surface. (1.10)

radial drainage pattern A drainage network in which the pattern of streams is generally outward away from a central high point. (13.1)

radiation (electromagnetic) The energy emitted from charged particles and manifested as interacting electrical and magnetic fields called electromagnetic waves; can be transferred either through matter or through a vacuum. (2.2)

radiation balance The difference between incoming and outgoing radiant energy. (2.16)

radiation deficit A property of matter at a given place and time characterized by a greater magnitude of outgoing than incoming radiant energy. (2.16)

radiation fog A type of fog that occurs when nocturnal longwave radiation loss cools the surface to the dew-point or frost-point temperature. (4.9)

radiation surplus A property of matter at a given place and time characterized by a greater magnitude of incoming than outgoing radiant energy. (2.16)

radioactive decay The spontaneous disintegration and emission of particles from an unstable atom. (9.13)

radiosonde An instrument package attached to a weather balloon that measures a range of variables at various heights. (2.3)

radio wave The electromagnetic radiation of the longest wavelengths (10 meters to 100 kilometers). (2.6)

rain forest biome A collection of adjacent ecosystems that consists of evergreen trees requiring abundant water balance surpluses year-round. (18.1)

rain shadow A relatively dry region on the downwind side of a topographic obstacle, usually a mountain range; rainfall is noticeably less than on the windward side. (18.3)

rapid A segment of rough, turbulent water along a stream. (13.6)

recessional moraine A moraine that forms as the front of a glacier melts back and stagnates for some time in one location, depositing a pile of sediment. (14.6)

rectification The process of matching an image or map to standard coordinates; also known as georeferencing. (1.12)

recurrence interval The average length of time that would be expected to occur between events that equal or exceed a given magnitude; also known as a return period. (4.14)

red sprite Faint red lights that sometimes project up from the top of a thunderhead after an energetic bolt of lightning strikes the ground. (5.9)

reef A shallow, mostly submarine feature, built primarily by colonies of living marine organisms, including coral, sponges, and shellfish, or by the accumulation of shells and other debris. (15.6)

reflection The change in direction of electromagnetic energy when it interacts with matter, with the energy remaining concentrated before and after interaction. (2.1)

regelation The refreezing of basal ice after it had melted by compression earlier. (14.3)

regional metamorphism Metamorphism affecting an extensive region and related mostly to burial, heating, and deformation of rocks. (9.6)

regolith A near-surface zone of weathered rock and soil, grading downward into unweathered bedrock or sediment. (12.1, 16.6)

relative dating The determination of the age of a rock relative to another rock, by using the principles of superposition, cross-cutting relationships, and other commonsense geological principles. (9.12)

relative humidity The ratio of the atmospheric vapor pressure to the water-vapor capacity, usually expressed as a percentage. (4.2)

relief The relative difference in elevation between two features, such as from a valley to a mountain peak. (1.6)

remediation The cleanup of a contaminated area. (8.10)

remote sensing Any technique used to collect data or images from a distance, including the processing and analysis of such data and the construction of maps using these techniques. (1.11)

representative fraction The ratio of the distance on a map to the actual distance (in the same units) on Earth; also known as scale. (1.10)

reproduction The production of offspring. (17.3)

reservoir A lake created by a human-constructed dam. (13.6)

residence time The average amount of time that an atom or molecule of a certain type of matter remains in a store. (8.2)

residual soil Soils that develop in place rather than being deposited into a place from elsewhere. (16.6)

residuum Parent material formed in place from bedrock beneath the soil. (16.6)

resilience The ability of an ecosystem to recover following stress. (17.7)

resisting force A force that resists the motion of an object, such as resisting the movement of tectonic plates. (10.9)

resource reservoir The storage of energy and/or matter in an ecosystem for later use. (17.5)

return period The average length of time between events that equal or exceed a given magnitude; also known as a recurrence interval. (4.14)

return stroke The positively charged particles that flow upward from the ground as negatively charged particles approach the ground from the cloud, after an electrical connection is established in cloud-to-ground lightning. (5.9)

reversed polarity The orientation of the Earth's magnetic field during many times in the geologic past, when the field flowed from north to south, causing the magnetic ends of a compass needle to point toward the south; the opposite of the present-day polarity. (10.10)

reverse fault A fault in which the block above the fault moves up relative to the block below the fault. (11.10)

rhyolite A mostly fine-grained, felsic igneous rock, generally of volcanic origin; can contain glass, volcanic ash, pieces of pumice, and variable amounts of visible crystals (phenocrysts). (9.5)

richness The number of species present in an ecosystem. (17.4)

Richter magnitude A numeric scale of earthquake magnitude, devised by the seismologist C. F. Richter. (11.16)

ridge (atmospheric) An elongated area of high atmospheric pressure. (3.2)

ridge push A plate-driving force that results from the tendency of an oceanic plate to slide down the sloping lithosphere-asthenosphere boundary near a mid-ocean ridge. (10.9)

riffle A fast-moving, shallow zone in a stream or other freshwater ecosystem. (18.14)

rift A narrow trough that runs along the axis of most mid-ocean ridges, formed when large blocks of crust slip down as spreading occurs. (10.6)

rill A small and shallow (2–4 centimeters deep) channel that forms when sheet flow begins to become concentrated in some places more than other adjacent places. (16.13)

riparian zone A relatively narrow strip of trees, shrubs, and grasses that form along streams. (18.4)

rip current A narrow, fast flow of water from the shoreline back out to sea. (15.3)

rip rap Large rocks and other debris that are dumped along a coast or stream channel for protection from erosion. (15.7)

riser The steep slope separating adjacent terraces or a terrace from the floodplain; also known as a bluff. (13.8)

risk An assessment of whether a hazard might have some societal impact. (11.6)

roaring forties The belt of particularly strong westerly winds aloft in the Southern Hemisphere between about 40° S and 50° S, where little land exists to provide friction to slow the winds. (3.12)

Robinson projection A popular type of map projection for depicting the entire Earth that blends aspects of cylindrical maps with other types of projections. (1.9)

roche moutonnée An asymmetrical glacial erosional landform caused by abrasion on the upflow side and plucking on the downflow side. (14.4)

rock cycle A conceptual framework presenting possible paths and processes to which an Earth material can be subjected as it moves from one place to another and between different depths within Earth. (1.4)

rock fall A mass-wasting process whereby large rocks and smaller pieces of bedrock detach and fall onto the ground. (12.10)

rock flour A finely powdered rock formed by glacial or other erosion. (14.4)

rock glacier A feature consisting of angular rocks on the outside and slowly flowing ice on the inside. (14.1)

rock slide A slab of relatively intact rock that detaches from bedrock and slides downhill, shattering as it moves. (12.10)

rock varnish A thin, dark film or coating of iron and manganese oxides, silica, and other materials; formed by prolonged exposure at the surface; also known as desert varnish. (9.11)

Rodinia An inferred supercontinent, consisting of all the continents joined, that existed near the Precambrian-Paleozoic boundary. (10.14)

root wedging The process of plant roots extending into fractures and growing in length and diameter, expanding preexisting fractures. (12.1)

Rossby wave Upper-tropospheric, mid-latitude waves of motion that are primarily west-to-east in each hemisphere, but with meanders toward the poles (ridges) and toward the equator (troughs), displaying wavelengths of hundreds to thousands of kilometers. (3.13)

rotation (tectonic) The tilting or horizontal rotation of a body in response to an applied stress. (11.9)

rotational slide A slide in which shearing takes place on a well-defined, curved shear surface, concave upward, producing a backward rotation in the displaced mass; also known as a slump. (12.10)

r-selected species Any species that produces relatively frequent batches of numerous offspring, which mature faster but are less supervised and relatively unlikely to survive into adulthood. (17.2)

runoff Precipitation that flows on the surface, such as in streams. (8.1)

S

Saffir-Simpson scale A classification system for the strength of Atlantic and eastern Pacific hurricanes, from Category 1 (weakest) to Category 5 (strongest). (5.14)

Sahel A semiarid, east-west belt in Africa south of the Sahara, characterized by a Tropical Savanna climate and a grassland biome. (18.7)

salinization A pedogenic regime characterized by the upward movement and evaporation of moisture, leaving salts behind; typically occurs in arid or semiarid regions. (16.8)

saltation Transport of sediment in which particles are moved in a series of short, intermittent bounces on a bottom surface. (9.11)

salt minerals A group of soluble, nonsilicate minerals including halite (NaCl). (9.1)

saltwater biome A collection of adjacent ecosystems in saline waters, in the ocean or inland water bodies. (18.1)

saltwater incursion (intrusion) Displacement of fresh groundwater by the advance of salt water, usually in coastal areas. (8.8)

salt wedging The forced widening of pores in rock by growing salt crystals, primarily in coastal environments, which enhances the ability of wave action to erode rocks. (15.4)

sandbar A low, sandy feature, possibly submerged, offshore of a shoreline or within a sandy stream. (15.5)

sand dune An accumulation of loose sand piled up by the wind. (9.2)

sandstone A medium-grained, clastic sedimentary rock composed mostly of grains of sand, along with other material. (9.4)

Santa Ana wind A warm, dry regional wind of southern California that blows from the northeast, across desert areas, over mountains, and then warms as it descends the leeward slopes. (3.5)

saturated (atmosphere) A condition characterized by atmospheric vapor pressure that is equal to the water-vapor capacity; implies that no more water can evaporate at that temperature and that the relative humidity is 100 percent. (4.2)

saturated (soil) A condition of the soil in which all the pore spaces are filled with water. (16.3)

saturated adiabatic lapse rate The rate of change of temperature with height of an adiabatically rising air parcel with a relative humidity of 100 percent; the rate itself differs depending on the temperature but can be no more than 10 C°/km. (4.5)

saturated zone The area in the subsurface where groundwater fills the pore spaces. (8.3)

scale The ratio of the distance on a map to the actual distance (in the same units) on Earth; also known as representative fraction. (1.10)

scarp (fault-line) A step in the landscape caused when fault movement offsets Earth's surface. (9.14)

scarp (landslide) A linear or curved scar left behind by a landslide on a hillslope, marking where the landslide pulled away from the rest of the hill. (12.10)

scattering The change in direction of electromagnetic energy when it interacts with matter, characterized by the energy dispersing in different directions after interaction. (2.1)

schist A shiny, foliated, metamorphic rock generally containing abundant visible crystals of mica and other metamorphic minerals. (9.6)

schlerophyllic Pertaining to vegetation that has developed a tolerance to drought and fire. (18.10)

scientific method A logic-based framework for investigating and revisiting some types of problems, involving observation, hypothesis, prediction, testing, retesting, and reaching a conclusion. (1.2)

scoria cone A relatively small type of volcano that is cone shaped and mostly composed of volcanic cinders (scoria); also known as a cinder cone. (11.1)

screaming sixties The area between 60 and 70 degrees south, prone to strong winds and extreme waves. (6.2)

sea arch An opening through a thin promontory of land that extends out into the ocean. (15.5)

sea breeze A local- to regional-scale wind that blows onshore in a coastal area, typically by day. (3.4)

sea cave A cave at the base of a sea cliff, usually flooded by seawater. (15.5)

sea cliff A cliff or steep slope situated along the coast. (15.5)

seafloor spreading The process by which two oceanic plates move apart and new magmatic material is added between the plates. (10.3)

sea ice Ice that forms from the freezing of seawater. (6.4)

seamount A submarine mountain, in some cases flat-topped, that rises above the seafloor. (10.2)

sea stack An isolated, pillar-like, rocky island or pinnacle near a rocky coastline. (15.5)

seawall A human-constructed wall or embankment of concrete, stone, or other materials along a shoreline, intended to prevent erosion by waves. (15.7)

secondary data Any information derived from preexisting information, such as a preexisting map that is used for providing the input for answering some other question. (1.10)

secondary mineral A mineral formed, generally in soils, after physical and chemical weathering processes interact to break down and alter rocks; usually more stable than the minerals composing the original rock. (12.2)

secondary pollutant A contaminant that occurs in the atmosphere as a by-product of a chemical reaction between an emitted pollutant and sunlight, water, or other primary pollutant in the atmosphere. (7.10)

secondary succession The predictable change in species composition in an ecosystem following stress that does not remove the soil from the ecosystem. (17.7)

secondary wave (S-wave) A seismic body wave propagated by a shearing motion that involves movement of material perpendicular to the direction of propagation; an S-wave cannot travel through magma and other liquids. (11.13)

Second Law of Thermodynamics The fundamental scientific principle that energy and matter tend to become dispersed into a more uniform spatial distribution. (1.4)

second-level heterotroph An organism that consumes a first-level heterotroph or products derived from a first-level heterotroph. (17.2)

second-order stream A stream that results from the confluence, or merging, of two first-order streams. (13.1)

sediment Grains and other fragments that originate from the weathering and transport of rocks, and the unconsolidated deposits that result from the deposition of this material. (1.4)

sedimentary rock A rock resulting from the consolidation of sediment. (9.2)

sediment budget The balance between the energy in waves and currents and the sediment available for transport. (15.4)

sediment load The amount of sediment carried by a stream, comprised of suspended load and bed load, but not dissolved load. (13.3)

seep A small spring. (8.7)

seismicity Any of the Earth's movements, either on the surface or at depth, caused by earthquakes. (11.14)

seismic station The location of a scientific instrument (seismograph) that measures seismic vibrations. (11.13)

seismic wave A wave produced by earthquakes or generated artificially. (11.13)

seismograph An instrument that measures and records details of earthquakes, such as force and duration. (11.16)

senescence The old-age state of an organism or ecosystem. (18.11)

sensible heat A type of heat in a material detectable as its temperature. (2.3)

sensible heat flux The flow, or transfer, of sensible heat in the atmosphere by convection and advection. (2.15)

September Equinox The day, on or near September 21, when Earth's axis of rotation is pointed exactly sideways relative to the Sun, and the durations of daylight and darkness are equal throughout the world. (2.9)

shaded-relief map A representation of elevation of Earth's surface by using color tones or shading to mimic the slope's appearance in three dimensions. (1.6)

shale A fine-grained clastic sedimentary rock, formed by the consolidation of clay and other fine-grained material. (9.4)

Shannon index A measure of biodiversity used by biogeographers and ecologists, which incorporates both richness and equitability. (17.4)

shear The type of differential stress that occurs when stresses on the edge of a mass are applied in opposing directions. (11.9)

shear component The part of the gravitational force that pulls a mass parallel to the sloping surface on which it sits, and therefore may cause it to slide downhill. (12.8)

sheen A shiny appearance, especially on a rock. (9.6)

sheet erosion The detachment and transportation of soil, sediment, or rock in a thin layer of unchanneled water that runs off from the surface during and shortly after a precipitation event. (13.6)

sheetflow Any water that flows across the Earth's surface in a thin, continuous fashion rather than in a stream. (8.3)

sheet lightning The electrical charges that are exchanged in less discrete mechanisms than by a single stroke or series of strokes, causing a broad area of the sky to be illuminated. (5.9)

sheetwash An unconfined flow of water in a thin layer downslope. (16.13)

shelf cloud A distinctive type of cloud that is characteristic of derechos. (5.12)

shield volcano A type of volcano that has broad, gently curved slopes constructed mostly of relatively fluid basaltic lava flows. (11.1)

shortwave radiation The electromagnetic energy at wavelengths less than about 4 micrometers; corresponds to the wavelengths at which the Sun emits nearly all of its energy. (2.6)

shrub layer The shaded, humid layer of life in a rain forest below the understory, experiencing little wind. (18.5)

Siberian High An area of surface high atmospheric pressure (anticyclone) over north-central Asia in winter, caused by the sinking of very cold, dense air. (3.6)

silicate A group of minerals usually consisting of one silicon atom bonded with four oxygen atoms; the most common mineral group on Earth. (9.1)

sill A tabular igneous intrusion that parallels layers or other planar structures of the surrounding rock and which usually has a subhorizontal orientation. (11.8)

silt A fine-grained rock fragment or clast, 1/256 to 1/16 mm in diameter. (9.10)

single-cell thunderstorm The simplest category of thunderstorm, existing independently of other storm systems. (5.6)

single-grain soil Pertaining to a soil with no apparent soil structure, because the grains do not stick together into peds; also known as structureless soils. (16.2)

sinkhole A closed, circular depression, usually in a limestone karst area, resulting from the collapse of an underlying cave. (12.2)

sinuosity The amount that a stream channel curves for a given length. (13.5)

sinuosity index The ratio of a stream's channel length (including all the bends) to the down-valley length (i.e., the straight-line distance down the valley). (13.5)

sinusoidal projection A family of equal-area map projections in which map coverage is interrupted, showing only lobes where shape distortion is minimized. (1.9)

slab pull A plate-driving force generated by the sinking action of a relatively dense, subducted slab. (10.9)

slash-and-burn A type of mostly subsistence agriculture typical of the tropics in which an area is deforested, cultivated, and then abandoned after the soils harden into laterites and become unproductive. (18.6)

slate A compact, fine-grained, weakly metamorphosed rock that possesses slaty cleavage. (9.6)

sleet A form of precipitation in which a Bergeron process–produced snowflake melts on its way down to the surface, but refreezes in a cold layer before hitting the surface, resulting in a small ball of ice that bounces from the surface. (4.12)

slip face The steep side of an asymmetrical mound, such as a sand dune, which is also the side that is downwind (leeward) of the prevailing atmospheric circulation. (9.11)

slope The change in elevation across a particular horizontal distance, expressed in degrees; also referred to as gradient. (1.6)

slope aspect The direction (orientation) that a slope faces. (12.4, 18.3)

slope failure The sudden or gradual collapse of a slope that is too steep for its material to resist the pull of gravity. (12.8)

slump A slide in which shearing takes place on a well-defined, curved shear surface, concave upward, producing a backward rotation in the displaced mass; also known as rotational slide. (12.10, 15.4)

small circle Any curved line formed by the intersection of a plane and a sphere that does not divide the sphere into two equal halves. (1.7)

small-scale map A map that shows little detail but covers a large area. (1.10)

smog A type of hazy air pollution caused by interactions between sunlight, hydrocarbons, and nitrogen oxides. (2.12)

Snowball Earth A hypothesis that proposes Earth's surface became nearly entirely frozen at least once, sometime earlier than 650 million years ago. (10.14)

snowfield A large area covered with snow and ice that, unlike a glacier, does not move. (14.1)

snow line The elevation on a slope at which snow and ice cover the ground year-round. (18.4)

soil Unconsolidated material at and near the surface, produced by weathering; includes mineral matter, organic matter, air, and water, and is generally capable of supporting plant growth. (16.1)

soil color A property of soil that can reveal information about soils, such as the degree of aeration, moisture conditions, and mineral composition. (16.2)

soil compaction The reduction of pore space as clay particles begin to lie flat, often because of land uses such as overgrazing or passage of traffic. (16.14)

soil erosion The detachment of a soil particle from underlying material and subsequent transportation of the particle. (16.13)

soil profile A vertical section of the soil from the ground surface to parent material, usually around one to two meters in depth. (16.7)

soil structure The manner in which soil particles are clumped together, such as granular, blocky, columnar, or platy. (16.2)

Soil Taxonomy The most common soil classification system used in the United States. (16.8)

soil texture The average distribution of particle sizes (sand, silt, or clay) in a soil. (16.2)

soil water Water trapped within the pore spaces in a soil, as hygroscopic water, capillary water, or gravitational water. (16.3)

soil-water recharge Any water that soaks into the soil. (8.3)

soil-water use The evapotranspiration of water stored in soil during times when potential evapotranspiration exceeds precipitation. (8.3)

solar constant The amount of energy emitted by the Sun each second and received at the "top" of the Earth's atmosphere, equivalent to 1,366 Joules per square meter per second, or 1,366 watts per square meter. (2.7)

solar cycle An approximately 11-year cycle of sunspot activity. (2.7)

solar flare A burst of intense energy and matter directed from the Sun into space, which can affect wireless communication systems on Earth. (2.7)

solid A state of matter in which the constituent atoms and molecules are bound together tightly enough with low enough energy levels that the mass can withstand the vibrations and other motions without coming apart. (2.2)

solidification The process in which magma cools and hardens into solid rock, with or without the formation of crystals. (9.3)

solifluction The very slow, continuous movement of wet soil, regolith, or weathered rock down a slope. (12.11)

sonar A type of active remote sensing in which a sound wave is emitted, and the amount of energy at those wavelengths and the time required for it to return to the sensor are used to map the surface. (1.11)

source region An area of thousands of square kilometers with similar meteorological characteristics where an air mass forms and acquires its characteristics. (5.1)

South Atlantic High A semipermanent subtropical high atmospheric pressure system (anticyclone) in the Atlantic Ocean. (7.6)

southeast trade winds The prevailing surface atmospheric circulation feature that exists in the tropical latitudes of the Southern Hemisphere, characterized by flow from southeast to northwest. (3.9)

South Equatorial Current An east-to-west-flowing, surface ocean current south of the equator in the Pacific Ocean. (6.2)

Southern Oscillation (SO) The oscillation of surface atmospheric pressure between the area near Darwin, Australia, and the area near Tahiti; when pressure is higher at one of these places, it tends to be lower at the other. (6.9)

Southern Oscillation index (SOI) A numerical indicator of the degree to which atmospheric pressure patterns differ between the western equatorial Pacific Ocean, near Darwin, Australia, and the eastern equatorial Pacific Ocean, near Tahiti; when the index is positive, pressures are higher near Tahiti and lower near Darwin; when it is negative, pressures are higher near Darwin and lower near Tahiti. (6.9)

spatial interpolation A family of geostatistical methods of estimating a data value at any point based on values for nearby points. (1.12)

specific heat (capacity) The amount of energy required to change the temperature of a mass of one kilogram by one Kelvin, expressed in Joules per kilogram per Kelvin. (2.17)

specific humidity The ratio of the mass of water vapor in a body of air to the total mass of that air, often expressed in grams per kilogram. (4.2)

spectral model A weather or climate forecasting model that represents atmospheric motion as a series of sine waves rather than relying on computations at discrete grid points. (5.15)

spheroidal weathering A form of mostly chemical weathering in which concentric or spherical shells of decayed rock are successively separated from a block of rock. (9.7)

spit A small point or low ridge of sand or gravel projecting from the shore into a body of water. (15.5)

splash erosion The breakdown and transportation of soil, sediment, and rock as raindrops hit the ground. (13.6)

Spodosol An agriculturally unproductive type of soil formed under acidic forest litter in cooler portions of the mid-latitudes and in high latitudes. (16.11)

spreading center A divergent boundary where two oceanic plates move apart (diverge), also known as mid-ocean ridge. (10.6)

spring A place where groundwater flows out of the ground onto the surface. (8.7)

spring tide Higher-than-average high tides and lower-than-average low tides caused when the Sun's gravity adds to the effects of the Moon's gravity. (15.2)

squall-line thunderstorm A series of organized thunderstorms oriented along the dryline in the warm sector of a mid-latitude cyclone. (5.6)

stabilized dune A dune of any type that is stationary or barely moving, commonly with vegetation holding the sand in place. (9.11)

stable (ecology) An ecosystem that is in balance, such that populations of all species within it remain fairly constant. (17.7)

stable condition (atmosphere) A situation in which sinking atmospheric motion is encouraged because an adiabatically moving parcel of air is colder than its surrounding environmental air. (4.5)

stalactite A conical or cylindrical cave formation that hangs from the ceiling of a cave and is composed mostly of calcium carbonate. (12.6)

stalagmite A conical, cylindrical, or mound-like cave formation that developed upward from the floor of a cave and is composed mostly of calcium carbonate. (12.6)

standard line The tangent line at which the globe touches a cylinder in a cylindrical projection, where no distortion exists. (1.9)

standing water Any water at the surface of the Earth that is not flowing. (8.3)

star dune A type of sand dune with an irregular, complex shape, and variably trending sand ridges radiating out from a central peak. (9.11)

State Plane Coordinate System (SPCS) A coordinate system used in the U.S. that divides a state into elongated zones and characterizes location based on distance from a central meridian or parallel that runs along the middle of the zone; map distortion is zero at the central meridian or parallel and expands away from it. (1.8)

states of matter The three major types of arrangements of atoms and molecules: solid, liquid, or gas. (2.2)

stationary front A narrow zone separating two air mass types in which neither air mass is displacing the other. (5.2)

Stefan-Boltzmann Law A relationship linking the Kelvin temperature of an object to the amount of electromagnetic radiation it emits. (2.5)

stenohaline Capable of tolerating only a narrow range of salinity conditions. (17.7)

stenothermal Capable of tolerating only a narrow range of temperature conditions. (17.7)

steppe A short-grass subcategory of the grassland biome. (18.7)

stepped leader An invisible channel of charged particles that starts and stops and zig-zags between a cloud and the ground, immediately before a lightning strike. (5.9)

stereo pair A pair of aerial photographs that have overlapping coverage and slightly different perspectives. (1.10)

stereoscope An instrument that allows stereo pairs to be seen in three-dimensional form. (1.10)

stomate A pore on a plant's leaf through which water is lost to the atmosphere during transpiration. (17.5)

store A part of a system that retains matter for a period of time rather than allowing that matter to flow through the system. (8.2)

storm surge The pileup of water from waves onshore as a severe storm, particularly a tropical cyclone, makes landfall. (5.13)

storm track The path that a storm takes across the surface. (5.3)

strain A change in size or shape of a body as a result of stress. (11.9)

stratiform cloud A family of cloud types that is layered, and wider than tall. (4.8)

stratocumulus cloud A type of low, layered cloud that is composed solely of tiny drops of liquid water. (4.8)

stratopause The transition zone between the stratosphere and overlying mesosphere. (2.11)

stratosphere The second layer of the atmosphere above the surface, which contains most atmospheric ozone. (2.1)

stratovolcano A common type of volcano constructed of alternating layers of lava, pyroclastic deposits, and mass-wasting deposits, including mudflows; also known as a composite volcano. (11.4)

stratus cloud A type of low-level stratiform cloud that is not precipitating. (4.8)

streambed The zone of contact between the water in a stream and the earth material over which it is passing; formed by the coalescence of gullies. (16.13)

stream capture The natural diversion of water from one stream to another, often occurring as headward erosion removes the divide between two adjacent drainage basins. (13.6)

streamer The positively charged streams of energy that move upward from the surface to meet the stepped leader immediately prior to a lightning strike, in cloud-to-ground lightning. (5.9)

stream load The material carried by a stream, including the material chemically dissolved in solution; comprised of suspended load, dissolved load, and bed load. (13.2)

stream system The configuration or arrangement of streams within a drainage basin; also known as drainage network. (13.1)

stream table A large, sand-filled tank, used to study stream behavior in a laboratory; also known as a flume. (13.7)

stream terrace A relatively level or gently inclined surface, or bench, bounded on one edge by a steeper descending slope, above the floodplain that flanks a stream. (9.14)

stress The amount of force divided by the area where the force is applied; similar to pressure. (11.9)

stress zone The range of an environmental variable outside the optimal range but within the lethal limit, within which an organism can survive but not thrive. (17.7)

striation A linear furrow generated from fault movement, often used to determine direction of movement along the fault. (10.3)

striations Tiny, parallel lines found on some crystal faces. (10.3)

strike-slip fault A fault in which the relative movement between the material on either side is essentially horizontal. (11.10)

strip cropping The spatial alternating of different crops, each planted in a narrow strip often along the contours of the land, as a technique to minimize soil erosion. (16.13)

structureless Pertaining to a soil with no apparent soil structure, because the grains do not stick together into peds; also known as single-grain soils. (16.2)

Subarctic (Dfc or Dfd) climate In the Köppen climate classification, a climate type characterized by adequate moisture year-round, a mean temperature of the coldest month below –3°C, and a mild or cool summer. (7.8)

Subarctic Monsoon (Dwc or Dwd) climate In the Köppen climate classification, a climate type characterized by adequate moisture in summer but not winter, an average temperature of the coldest month below –3°C, and a mild or cool summer. (7.8)

subduction zone An area where subduction takes place, either referring to the actual downgoing slab and its surroundings, or to the region, including Earth's surface, above the subducting slab. (10.7)

subglacial debris Material transported at the base of the ice of a glacier. (14.4)

sublimation A change in state from a solid to a gas, requiring an input of energy from the environment and the absorption of that energy at the molecular level as latent energy. (2.4, 14.3)

submarine slope failure Slope failure that occurs on the seafloor. (12.9)

submergent coast A coast that forms where land has been inundated by the sea because of a rise in sea level or subsidence of the land. (15.9)

subpolar low A large, permanent area of low surface atmospheric pressure centered over the oceans near 60° N and 60° S, including the Icelandic Low and Aleutian Low in the Northern Hemisphere. (3.11, 3.12)

subsoil layer A zone in the soil characterized by the accumulation of material, including iron oxide, clay, and calcium carbonate, depending on the climate and starting materials; also known as the B horizon, the zone of accumulation, or the zone of illuviation. (16.7)

subtropical jet stream The fast-flowing current of air near the tropopause in each hemisphere at the boundary between the Hadley cell and the mid-latitudes. (3.13)

subtropical scrub and woodland biome A collection of adjacent ecosystems that has hardy bushes, grasses, and low trees to areas with larger trees but no closed canopy, which must tolerate a hot, dry summer. (18.1)

succession The predictable change in species composition in an ecosystem following stress. (17.7)

sulfide minerals A group of nonsilicate minerals containing sulfur (S) bonded with a metal. (9.1)

Sun angle The angle formed by the Sun's position, a point on Earth's surface, and the horizon. (2.8)

sunspot A dark spot on the Sun that is slightly cooler than the rest of the Sun but associated with increased overall total solar irradiance. (2.7)

sunspot activity The extent to which dark spots on the Sun are occurring at any given time. (14.9)

sunspot number The average number of sunspots present at a given time. (2.7)

supercell thunderstorm A very large, single thunderstorm characterized by an organized set of features that allow the system to sustain itself for many hours and produce very severe weather. (5.6)

superposed stream A stream that was superimposed on already existing geologic features. (13.11)

supersaturated A theoretical condition of the atmosphere characterized by vapor pressure that slightly exceeds the water-vapor capacity; implies that active condensation or deposition must be occurring. (4.2)

supraglacial debris Material transported on the surface of a glacier. (14.4)

surface boundary layer The lowest part of the friction layer in the atmosphere; the part of the atmosphere adjacent to the surface. (3.8)

surface tension The tendency for molecules, such as water, to coalesce with a discrete outer surface. (4.1)

surface water Water that occurs in streams, rivers, lakes, oceans, and other settings on Earth's surface. (8.0)

surface wave A seismic wave that travels on Earth's surface. (11.13)

surf zone The zone where ocean waves break, forming a foamy surface called surf. (15.3)

surplus Any "extra" water (in centimeters or inches of precipitation equivalent) on the surface after the soils are saturated. (8.3)

suspended load Fine particles, generally clay and silt, that are carried suspended in moving water. (13.2)

suspension A mode of sediment transport in which water or wind picks up and carries the sediment as floating particles. (9.11)

sustainability Effective management that will allow long-term endurance and success, in environmental, economic, and social aspects. (18.17)

swamp A forested wetland along a stream or floodplain. (18.14)

swash regime The scenario of coastal inundation during which the height of the swash zone remains below the bases of coastal dunes. (15.11)

swash zone The part of the coastal zone that is permanently submerged by shallow water where waves break; also known as the inshore. (15.1)

swell The gradual formation of rolling ocean waves that do not break. (15.3)

swelling clay Soils, especially Vertisols, composed of a high percentage of certain clay minerals; they increase in volume when wet. (16.14)

SWOT analysis An acronym for Strengths, Weaknesses, Opportunities and Threats; in environmental analysis, includes aspects that can and cannot be controlled. (18.17)

symbiosis Any type of intricate, life-long interaction between individuals of different species, including through mutualism, commensalism, or parasitism. (17.3)

syncline A fold, generally concave upward (in the shape of a *U*), with the youngest rocks in the center. (11.11)

system An interconnected set of matter, energy, and processes. (1.3)

T

talus A pile of angular rock fragments on a steep slope loosened by weathering processes, or referring to an accumulation of such fragments. (12.1)

talus slope A steep slope composed of rock fragments loosened by weathering processes (talus). (12.1)

tarn A small lake within a cirque, a glacially scoured depression. (14.5)

tectonic plate Any of the dozen or so fairly rigid blocks into which Earth's lithosphere is broken. (10.5)

tectonics Earthquakes, volcanoes, and other processes that deform Earth's crust and mantle. (9.3)

tectonic terrane A fault-bounded body of rock that has a different geologic history than adjacent regions. (10.13)

teleconnection An association in atmospheric features, usually referring to opposing pressure conditions, between two distant places. (6.11)

temperate forest biome A collection of adjacent ecosystems that have deciduous and evergreen trees, often intermixed, that thrive in humid areas with warm summers and cool to cold winters. (18.1)

temperate mid-latitude (C) climate In the Köppen climate classification, a broad group of climatic types characterized by adequate moisture availability, and the mean temperature of the coldest month between −3°C and 18°C. (7.1)

Temperate Monsoon (Cwa, Cwb, or Cwc) climate In the Köppen climate classification, a climate type characterized by adequate moisture in summer but dry winters, hot, warm, or mild summers, and the mean temperature of the coldest month between −3°C and 18°C. (7.6)

temperature A measure of the object's average internal kinetic energy—the energy contained within atoms or molecules that are moving. (2.3)

temperature inversion An atmospheric condition in which temperature increases as height increases. (2.11)

temporary base level Any base level, other than sea level, that limits the downward extent of erosion. (13.9)

tension The type of differential stress where stress is directed outward, pulling the material. (11.9)

terminal moraine Glacially carried sediment that accumulates at the terminus (end) of a glacier and a landform composed of such material; generally marks the glacier's farthest downhill or most equatorward extent. (14.5)

terminus The farthest extent, especially the downslope end, of a glacier. (14.4)

terrace A relatively flat bench that is perched above a stream and that was formed by past deposition or erosion by the stream. (13.8)

terrestrial biome Any biome that exists on land. (18.3)

Tethys Sea An inferred wedge-shaped ocean that separated Asia from landmasses farther to the south, in the late Paleozoic Era and later. (10.14)

theory of convergent evolution The notion that, despite wide separation and unrelated lineage, organisms develop similar physical and behavioral characteristics if they develop in similar environments. (18.10)

theory of plate tectonics The theory, currently accepted by nearly all earth scientists, that Earth's lithosphere is broken into a number of fairly rigid plates that move relative to one another. (10.0)

thermal inertia The degree to which a particular type of matter remains at the same temperature when a change in energy is imposed on it. (4.6)

thermal infrared (IR) energy The part of the electromagnetic spectrum with wavelengths between about 3 and 20 micrometers. (1.11)

thermocline The level in a body of water at which temperature changes rapidly with depth. (6.7)

thermohaline conveyor The globally interconnected, three-dimensional oceanic circulation system. (6.7)

thermopycnocline The level in a body of water where both the temperature and density change rapidly with depth; represents a correspondence between the thermocline and the pycnocline. (6.7)

thermosphere The upper most layer of the atmosphere, characterized by a temperature inversion. (2.1)

therophyte Any annual plant that completes its life cycle quickly by having its seeds lie dormant during the dry season; common in the subtropical scrub and woodland biome. (18.10)

thrust fault A reverse fault that has a gentle dip. (11.10)

thrust sheet The sheet of rock that has been displaced above a thrust fault. (11.10)

thunder The sound made when air rushes back into the channel where a lightning stroke passed. (5.9)

thunderhead A cumulonimbus cloud that is producing hail and lightning. (5.8)

Tibetan Low A prominent area of low atmospheric pressure in summer, centered over Tibet. (3.14)

tidal bulge The raised level of water associated with high tide, on the side of the Earth closest to the Moon and on the opposite side of the Earth. (15.2)

tidal range The difference in local sea elevation between high and low tide. (15.2)

tide A cyclic change in the height of the sea surface, generally measured at locations along the coast; caused by the pull of the Moon's gravity and to a lesser extent the Sun's gravity. (15.2)

till Unsorted, generally unlayered sediment, deposited directly by or underneath a glacier. (14.4)

tillite A sedimentary rock composed of lithified glacial till. (14.8)

tilth A property of soil indicating its amenability to being plowed. (16.2)

time step The determined period of time into the future (usually minutes for a weather forecasting model) for which an initial forecast is generated, with subsequent forecasts being made out farther into the future using this same time increment, repeatedly. (5.15)

toe The leading edge of a rock slide. (12.10)

topographic map A representation of the elevation of Earth's surface by using contours. (1.6)

topographic profile A collection of connected points used to depict the change in elevation across a feature. (1.6)

topsoil The soil found in the A horizon, composed of dark gray, brown, or black organic material mixed with mineral grains. (16.7)

tornado A violent, rapidly rotating, funnel-shaped column of air that extends to the ground. (5.10)

Tornado Alley The swath from south-central Canada to central Texas where tornado frequency is among the highest on Earth. (5.11)

total solar irradiance (TSI) A measure of the Sun's total output of energy. (2.7)

township-range survey system A coordinate system used in much of the U.S. that characterizes location based on the number of 6-mile by 6-mile squares north, south, east, or west of the intersection of a baseline with a principal meridian, with subdivisions within the square also described; also known as the Public Land Survey System. (1.8)

traction The process by which particles roll, slide, or otherwise move on the surface, by such transport agents as streams, wind, or waves. (13.2)

transform boundary A plate boundary in which two tectonic plates move horizontally past one another. (10.5)

transform fault A strike-slip fault that accommodates the horizontal movement of one tectonic plate past another. (10.8)

transhumance Seasonal nomadism in which people move to the optimal elevation to experience the best temperatures and animal fodder at a particular time of year—generally upslope in summer and downslope in winter. (18.4)

transmission The passing of energy through matter, such as air. (2.1)

transpiration The transfer of water from the soil to the atmosphere via plant roots, stems, and leaves. (4.0)

transverse dune A type of sand dune that is oriented linearly or only very gently curving for many kilometers, perpendicular to prevailing winds. (9.11)

travertine A variety of limestone that is commonly concentrically banded and porous. (12.6)

trellis drainage pattern A drainage network in which the pattern of streams follows a layer or structure, such as along a fracture or folded area, and then cuts across a ridge to follow a different feature; results in a rectangular pattern of streams. (13.1)

triple junction The place where three tectonic plates and three plate-tectonic boundaries meet. (10.8)

Tropic of Cancer The 23.5° parallel of latitude north of the equator; represents the most northerly latitude that experiences the directly overhead rays of the Sun (on or near June 21). (1.7)

Tropic of Capricorn The 23.5° parallel of latitude south of the equator; represents the most southerly latitude that experiences the directly overhead rays of the Sun (on or near December 21). (1.7)

tropical (A) climate In the Köppen climate classification, a broad group of climatic types characterized by adequate moisture availability and the mean temperature of the coldest month above 18°C. (7.4)

tropical cyclone A well-organized, rotating storm that originates in the tropical part of the Earth, ranging in strength from weak storms to hurricanes. (5.10)

tropical depression An enclosed area of low atmospheric pressure of tropical origin, with at least one closed isobar. (5.14)

Tropical Monsoon (Am) climate In the Köppen climate classification, a climate type characterized by a mean temperature of the coldest month above 18°C, and by precipitation totals that exceed potential evapotranspiration in most months, but with enough water stored in the soils during the brief dry season to support the growth of forests. (7.4)

Tropical Rain Forest (Af) climate In the Köppen climate classification, a climate type characterized by a mean temperature of the coldest month above 18°C and precipitation totals that typically exceed potential evapotranspiration year-round. (7.4)

Tropical Savanna (Aw) climate In the Köppen climate classification, a climate type characterized by a mean temperature of the coldest month above 18°C, and by precipitation totals that exceed potential evapotranspiration for about half of the year, providing enough moisture for tropical grasslands and a few scattered trees, but not forests; also known as Tropical Wet-Dry climate. (7.4)

tropical storm A tropical cyclone with well-developed circulation features and wind speeds of 64 km/hr (39 mi/hr) to 118 km/hr (73 mi/hr). (5.14)

Tropical Wet-Dry (Aw) climate *See* Tropical Savanna (Aw) climate.

tropics The region between the Tropic of Cancer (23.5° N) and the Tropic of Capricorn (23.5° S). (2.9)

tropopause The transition zone between the troposphere and overlying stratosphere. (2.11)

troposphere The lowest layer of the atmosphere, which contains nearly all of the atmosphere's water. (2.1)

tropospheric ozone A harmful air pollutant, produced by sunlight striking hydrocarbons, such as car exhaust, in the lower troposphere; also known as ground-level ozone. (17.9)

trough (atmosphere) An elongated area of low atmospheric pressure. (3.2)

trough (waves) The lowest point of any type of wave. (2.5, 15.3)

tsunami A large sea wave produced by uplift, subsidence, or some other disturbance of the seafloor, especially by a shallow submarine earthquake. (11.16, 15.7)

tuff Volcanic rock composed of consolidated volcanic ash and other material, commonly including pumice, crystals, and rock fragments. (9.5)

tundra biome A collection of adjacent ecosystems that is treeless, but includes low-lying bushes, shrubs, mosses, grasses, and lichens (algae and moss) that can withstand bitter cold. (18.1)

Tundra (ET) climate In the Köppen climate classification, a climate type characterized by adequate moisture, the warmest month's temperature averaging below 10°C (50°F), and frosts possible in any month. (7.8)

turbidity The degree of murkiness of water. (18.14)

turbulent flow Flow characterized by chaotic motion. (13.2)

turnover The mixing of waters from different depths in a large body of water; also known as overturning. (6.5)

typhoon A regional term used in the western Pacific basin applied to the most intense type of storm of tropical oceanic origin. (5.13)

U

ubac slope The side of a slope that receives less direct Sun, and is therefore cooler and more damp. (18.4)

Ultisol A type of soil formed in the subtropics and warmer mid-latitudes that is less highly weathered than Oxisols, with the eluviation of clay minerals into the B horizon, and generally acidic. (16.10)

ultramafic A composition of magma and igneous rocks, generally expressed as a dark or greenish igneous rock composed chiefly of mafic minerals rich in magnesium and iron. (9.5)

ultraviolet light The electromagnetic radiation of wavelengths between about 10^{-8} and 0.4 micrometers. (2.6)

unconfined aquifer An aquifer where the water-bearing unit is open (not restricted by impermeable rocks) to Earth's surface and atmosphere. (8.6)

unconformity A boundary between underlying and overlying rock strata, representing a significant break or gap in the geologic record; an unconformity represents an interval of nondeposition or erosion, commonly accompanied by uplift. (9.15)

unconsolidated material Material sitting on top of solid rock, including soil, loose sediment, pieces of wood and other plant parts, boulders, and other types of loose debris. (12.9)

understory The shaded, humid layer of life in a rain forest above the shrub layer, with little wind. (18.5)

uniformitarianism The concept that the natural world has evolved in the past by processes and rates similar to those observed today. (17.7)

Universal Transverse Mercator (UTM) System A coordinate system that divides Earth into zones 6° of longitude wide and 20° of latitude long, with eastings identifying the number of meters east or west of the central meridian within a zone and northings representing the number of meters north or south of the equator. (1.8)

unloading The removal of mass from a landmass, such as by the melting of a glacier. (12.1)

unloading joint A joint formed from stresses that arise during uplift of buried rocks and that cause rocks to fracture due to reduced pressure. (11.10)

unsaturated (atmosphere) A condition characterized by atmospheric vapor pressure that is less than the water-vapor capacity; implies that more water could evaporate at that temperature under the right conditions and that the relative humidity is less than 100 percent. (4.2)

unsaturated adiabatic lapse rate The rate of change of temperature with height of an adiabatically rising air parcel with a relative humidity less than 100 percent; always amounts to 10 C°/km. (4.5)

unsaturated zone The area below the surface but above the water table, in which much of the pore space is filled by air rather than water. (8.3)

unstable (ecology) An ecosystem that is not in balance, such that populations of all species within it fluctuate considerably. (17.7)

unstable condition (atmosphere) A situation in which rising atmospheric motion is encouraged because an adiabatically moving parcel of air is warmer than its surrounding environmental air. (4.5)

updraft A local upward flow of air, usually on very short timescales. (3.3)

uplift The raising of rock up to the surface by tectonic forces. (9.3)

upslope fog A type of fog that occurs when air cools to the dew-point temperature or frost-point temperature as it moves upslope. (4.9)

upwelling The flow of colder, deeper waters up toward the surface of the ocean, replacing surface ocean water that circulated elsewhere. (6.3)

urban canyon The area between buildings in an urban area, containing different atmospheric properties than the surrounding rural environment. (7.10)

urban heat island (UHI) The phenomenon of increased air temperatures over cities as compared to their surrounding rural areas. (3.4)

urbanization The replacement of natural land cover with built-up land cover associated with the growth of cities and towns. (4.6)

U-shaped valley A steep-sided, rounded valley carved by a glacier. (14.5)

V

Valley and Ridge Province A physiographic region in the Appalachian Mountains of the eastern U.S., characterized by linear or curved ridges and valleys. (11.11)

valley breeze A local-to-regional-scale wind that blows from a valley uphill to the peak of a hill or mountain; also known as an anabatic wind. (3.4)

valley fog A type of fog that occurs when cold nocturnal downslope winds trap atmospheric moisture in the valleys, causing the air to reach the dew-point temperature through the addition of humid air; or when cold winds move downslope into the valley, chilling the valley air to the dew-point or frost-point temperature. (4.9)

valley glacier A glacier that flows down a valley and tends to be narrow; also known as alpine glacier. (14.1)

valley train A valley in which glacial outwash is concentrated. (14.6)

vapor pressure The portion of the total atmospheric pressure exerted by water vapor molecules, usually measured in millibars. (4.2)

vein A generally tabular accumulation of minerals that filled a fracture or other discontinuity in a rock; formed by precipitation of material from fluids, especially hydrothermal fluids. (9.7)

vertical wind shear A change in horizontal wind speed or direction with height. (5.6)

Vertisol A type of clay-rich soil located in temperate and subtropical environments, characterized by swelling when wet and shrinking when dry. (16.10)

vesicle A small hole found in a volcanic rock, representing a gas bubble in a magma that was trapped when the lava solidified. (9.5)

virga Any precipitation that evaporates before reaching the ground. (5.12)

viscosity A measure of a material's resistance to flow. (6.4)

visible light The part of the electromagnetic spectrum consisting of wavelengths between about 0.4 and 0.7 micrometers. (1.11)

volatile organic compound A large family of carbon-based chemicals that evaporate easily at room temperatures, including acetone, benzene, formaldehyde, and many others. (7.10)

volcanic ash Sand-sized or smaller particles originating from volcanic pyroclastic eruptions, and accumulations of such material. (9.2)

volcanic breccia A volcanic rock containing angular fragments in a matrix of finer material. (9.5)

volcanic dome A dome-shaped volcanic feature, largely composed of solidified lava of felsic to intermediate composition. (11.1)

volcanic glass A natural glass produced by the cooling and solidification of molten lava at a rate too rapid to permit crystallization. (9.5)

volcanic neck A steep, typically butte-shaped topographic feature composed of volcanic materials that formed in the conduit within or beneath a volcanic vent and that were more resistant to erosion than surrounding materials. (11.8)

volcanic rock An igneous rock that formed from a volcanic eruption, including lava, pumice, volcanic ash, and other volcanic materials. (11.1)

volcanic tremor A small earthquake caused by the movement of magma. (11.13)

volcano A vent in the surface of Earth through which magma and associated gases and ash erupt; also the form or structure constructed from magma erupted from the vent. (11.1)

volume The amount of three-dimensional space, measured in cubic units of length, that a mass occupies. (3.1)

vorticity Any clockwise or counterclockwise horizontal rotation (spin) of air. (5.3)

V-shaped valley A very narrow, steep valley typically found in environments in which erosion dominates over deposition, such as the headwaters of a mountain stream, where the slope is steep. (13.6)

W

Walker cell circulation Any east-west-oriented atmospheric circulation system in the equatorial part of the Earth characterized by relatively high pressure and subsiding air on one side, relatively low surface pressure and rising motion on the opposite side, east-to-west trade winds at the surface, and west-to-east flow of air aloft; often refers to the cell of this type over the equatorial Pacific Ocean. (6.8)

wall cloud A sharp, low-forming protrusion on a cumulonimbus cloud often associated with tornadoes and other severe weather. (5.10)

warm front The sloping boundary between an advancing relatively warm air mass and a colder one. (5.2)

warm phase (ENSO) A periodic warming of eastern equatorial Pacific surface ocean waters, accompanied by excessive precipitation and flooding adjacent to South America as the Southern Hemisphere's summer approaches; also known as El Niño. (6.10)

warm pool A large region of ocean water warmer than the water surrounding it. (6.6)

warm sector The part of a mid-latitude cyclone that is between the warm front and the cold front, characterized by warm air. (5.2)

warning An alert by the U.S. National Weather Service that a specific type of life-threatening severe weather, including severe thunderstorms, tornadoes, winter storms, strong winds, or tropical cyclones, has been observed and is approaching a specific area at a designated time. (5.16)

watch An alert by the United States National Weather Service that conditions are favorable in a specified area for some specific type of life-threatening severe weather, including thunderstorms, tornadoes, winter storms, strong winds, or tropical cyclones. (5.16)

waterfall A steep descent of water within a stream, such as the place where it crosses a cliff or steep ledge. (13.6)

waterspout A rotating, columnar vortex of air and small water droplets that is over a body of water, usually an ocean or a large lake. (5.12)

water table The surface between the unsaturated zone and the saturated zone, as in the top of groundwater in an unconfined aquifer. (8.3)

water-vapor capacity The maximum atmospheric vapor pressure that can exist at a given temperature. (4.2)

wave (ocean) An irregularity on the surface of a body of water usually generated by wind. (15.3)

wave base The depth, usually about half the wavelength, at which the action of a wave in an ocean or lake no longer has an effect. (15.3)

wave-cut notch An indentation produced in rocks or sediment by continued wave action at a specific level along a coast. (15.4)

wave-cut platform A gently sloping surface or bench produced by wave erosion in the coastal zone. (15.4)

wave height The vertical distance between the height of the crest and the height of the trough of a wave. (15.3)

wave interference The presence of multiple sets of waves propagating from different directions onto the same point. (15.3)

wavelength The distance between two adjacent wave crests or troughs. (2.5)

wave refraction The bending of any type of wave, such as when one side of an ocean wave encounters the sea bottom before the other side of the wave. (15.3)

weather The conditions in the atmosphere at some specific time and place. (5.0)

weathering The breaking apart of rock. (1.4)

weathering rind A weathered, outer crust on a rock fragment or bedrock mass exposed to weathering. (12.5)

westerlies The prevailing surface and upper-tropospheric general circulation feature that exists in the middle latitudes in each hemisphere, characterized by flow from west to east. (3.9)

Western Atlantic Warm Pool A large region of warm ocean temperatures centered on the Caribbean Sea. (6.6)

Western Australian Current A south-to-north-flowing, cold surface ocean current in the eastern Indian Ocean. (6.2)

Western Pacific Warm Pool The broad region of the warmest ocean temperatures on Earth, from China to Australia, spanning the western Pacific and eastern Indian Oceans. (6.6)

West Wind Drift A west-to-east-flowing, surface ocean current in the high latitudes of the Southern Hemisphere from southern Australia to South America and south of Africa; also known as the Antarctic Circumpolar Current. (6.2)

wetland Any area of standing water with emergent vegetation. (18.14)

whitecap Any small wave with a foamy, white crest. (15.3)

Wien's Law A relationship between the temperature of a body and the wavelength at which it emits the most electromagnetic energy. (2.6)

wilting point The minimum amount of water (in centimeters or inches of precipitation equivalent) that is necessary in the rooting zone to permit extraction by plants; any less water in the soil will adhere to rock particles and resist evapotranspiration. (8.3)

wind barb A particular type of symbol on a weather map that shows the direction and speed that wind is blowing at a point. (3.8)

windbreak A technique to minimize soil erosion, whereby a row of trees is planted perpendicular to the prevailing wind direction with the intent of slowing down the wind and therefore minimizing transportation of soil particles. (16.13)

wind streak A thin, small ridge of sand formed when winds in arid areas are strong enough that sand is blown through, without accumulating in sizeable dunes, often on the downwind (leeward) side of small plants and stones. (9.11)

windward The upwind side of a slope. (3.5)

X

xerophytic Pertaining to a species that has built-in mechanisms that make it tolerant to arid environments. (18.8)

x-ray The electromagnetic radiation of wavelengths between about 10^{-11} and 10^{-8} micrometers. (2.6)

Y

yazoo stream A channel that runs parallel to the main channel of a low-gradient stream but cannot join it for many kilometers because the sediment deposited on the floodplain by the main channel is high enough to create a natural separating barrier. (13.8)

Z

zonal flow A wind that is primarily blowing from west to east or east to west. (4.15)

zone of ablation The lower or equatorward part of a glacier or ice sheet, where snow and ice are removed by melting and other processes faster than they can be added. (14.3)

zone of accumulation (glaciers) The upper or poleward part of a glacier or ice sheet, where snow and ice are added faster than they are removed by melting and other processes. (14.3)

zone of accumulation (soil) A zone in the soil characterized by the accumulation of material, including iron oxide, clay, and calcium carbonate, depending on the climate and starting materials; also known as B horizon, zone of illuviation, or subsoil layer. (16.7)

zone of eluviation A light-colored, leached zone of soil, lacking clay and organic matter; also known as E horizon. (16.7)

zone of illuviation *See* zone of accumulation (soil).

zoogeographer A physical geographer/biogeographer who studies the factors that cause animal distributions and migrations. (18.16)

zooxanthellae A yellowish-brown type of algae that provide oxygen and nutrients for coral in a symbiotic relationship. (18.15)

CREDITS

Select Abbreviations Used in Photograph and Text/Line Art Credits

CPC: Climate Prediction Center
CSC: Coastal Services Center
EPA: Environmental Protection Agency
ERSL: Environmental Remote Sensing Laboratory (University of Nebraska, Lincoln)
FAO: Food and Agriculture Organization
GSFC: Goddard Space Flight Center
HPC: Hydrometerological Prediction Center
NASA: National Aeronautic and Space Administration
NCAR: National Center for Atmospheric Research
NCEP: National Centers for Environmental Prediction
NCDC: National Climatic Data Center
NESDIS: NOAA Satellite and Information Service
NGA: National Geospatial-Intelligence Agency
NHC: National Hurricane Center
NOAA: National Oceanic and Atmospheric Administration
NWS: National Weather Service
SPC: Storm Prediction Center
UN: United Nations
USGS: United States Geological Survey

PHOTOGRAPHS

Unless otherwise credited: © Stephen Reynolds

FRONT MATTER

Page xxix; © Robert Rohli; xxix: © Julia K. Johnson; xxx: © Peter R. Waylen; xxx: © Mark A. Francek.

DESIGN ELEMENTS

Chapter 2 Banner: NASA.

CHAPTER 1

Figure 0100a1: NASA GSFC image by Robert Simmon and Reto Stöckli; 0100a3, 0100a5, 0101a3: Photo by Susanne Gillatt; 0102b1: Cyrus Read/USGS; 0102b2: Kate Bull/USGS; 0102b3: T. A. Plucinski/USGS; 0103b1: Photo by Susanne Gillatt; 0103b2: NASA; 0105a1: Photo by Susanne Gillatt; 0105a2: M. E. Yount, USGS; 0105a3: Photo by Susanne Gillatt; 0105a5: Photo by Cynthia Shaw; 0106a2: © Wendell Duffield/USGS; 0109b1: Photo Cynthia Shaw; 0109b4: NASA; 0110b1: NASA/JPL/University of Arizona; 0111a1: © National Air and Space Museum, Smithsonian Institution; 0111a2: NASA; 0111b2: Photo by Ramon Arrowsmith; 0111c1: NASA; 0111d1: © USGS; 0111d2-d4: NASA; 0111d5: USGS; 0113a1: © McGraw-Hill Education; 0113c2-c3: USGS; 0114a1: NASA.

CHAPTER 2

Figure 0200a3: Photo by Susanne Gillatt; 0200a4: © Glow Images RF; 0200a5: © Konstantin Kalishko/Alamy; 0200a7: © Grant Faint/Getty RF; 0200a8: © McGraw-Hill Education. Bear Dancer Studios/Mark Dierker, photographer; 0200a9: NASA; 0201a1: © InterNetwork Media/Getty Images; 0201a2: © McGraw-Hill Education/Bear Dancer Studios/Mark Dierker, photographer; 0202c2: © Don Farrall/Photodisc/Getty Images; 0202c1: © McGraw-Hill Education/Kevin Cavanagh, photographer; 0203b1: © Comstock/Getty RF; 0203b2: Game McGimsey/Alaska Volcano Observatory/USGS; 0203b3: Courtesy of Hukseflux USA; 0203b4, 0204d1: NOAA; 0204d2: T. O'Keefe/PhotoLink/Getty Images RF; 0204d3: © Marc Gutierrez/Flickr/Getty RF; 0205a1: © Getty Images RF; 0205a2: © Corbis RF; 0207c1-c4: NASA; 0207c5: NASA/SDO; 0208b2: © Fusc/Getty Images; 0208c1, 0208c3: Photo by Susanne Gillatt; 0210c1: © Photodisc/Getty Images; 0211a3: © PhotoLink/Getty Images; 0212a2: © Patrick Clark/Getty Images; 0212a3: © Ingram Publishing/Getty Images; 0213a3: © Mark Morgan/Alamy; 0213b1: NASA; 0213c2: © Dave Harlow/USGS; 0218a1, 0218a10: Photo by Susanne Gillatt; 0218a14: © Sexto Sol/Getty Images; 0218a17: Photo by Susanne Gillatt.

CHAPTER 3

Figure 0301a1: Photo by Susanne Gillatt; 0302b2: NOAA; 0304a2: © Mark Francek; 0304a4: Photo by Susanne Gillatt; 0304c1: Courtesy of Chuck Carter; 0305b1, 0305b2: NASA; 0305c3: NPS Photo by R. G. Johnsson/NPS; 0305c4: NASA/MODIS Rapid Response; 0307d2: Source: NASA; 0309a7: © Jacques Desclotres, MODIS Rapid Response Team, NASA/GSFC; 0310a2: Photo by Susanne Gillatt; 0314a5, 0314a6: NASA; 0315a3: Library of Congress Prints and Photographs Division [LC-DIG-pga-00675]; 0315a4: Library of Congress Prints and Photographs Division [LC-D41-13]; 0315a5: Library of Congress Prints and Photographs Division [LC-USZ62-38479]; 0315a6: Photo by Susanne Gillatt; 0315a7: © Transportation Collection, Division of Work & Industry, National Museum of American History, Smithsonian Institution; 0315a8: Library of Congress Prints and Photographs Division [LC-DIG-pga-00809].

CHAPTER 4

Figure 0400a2: Photo by Susanne Gillatt; 0400a3: Laura J. Hartley, USGS; 0400a4, 0401a2, 0404a1: Photo by Susanne Gillatt; 0406a2: NASA/JPL; 0408a2: NOAA/NASA; 0408a4: NASA; 0408a5: © Robert Rohli; 0408a8: Ralph F. Kresge/NOAA; 0409b1: © Ingram Publishing/SuperStock RF; 0409b2: Jeff Schmaltz/NASA; 0409c5: © Digital Vision/Getty Images RF; 0412t1: NOAA/Warren Hornsby; 0413t1: Henry Reges/CoCoRaHS HQ.

CHAPTER 5

Figure 0500a1: © Getty Images RF; 0500a2-a4, 0501a1: NOAA; 0506b4: © 2010 Ryan McGinnis/Getty Images RF; 0506b6: NOAA; 0509a1: © CWellsPhotography/Getty Images; 0509a2: © McGraw-Hill Education. Bear Dancer Studios/Mark Dierker, photographer; 0510a1: © Warren Faidley/Corbis RF; 0510b2: NASA; 0510b3: National Severe Storms Lab/NOAA; 0510c1: © Willoughby Owen/Getty Images RF; 0510c2: © Alan Sealls/Weather-VideoHD.TV; 0511b5: © McGraw-Hill Education/Thomas Augustine, photographer; 0512a1: Tim Osborn, NOAA Coast Survey; 0512a2: Melody Ovard, OMAO/NOAA Ship Nancy Foster, NOAA/Department of Commerce; 0512a6: Sean Waugh, NOAA/NSSL; 0512a7: © Patrick Byrd/Alamy RF; 0512a10: NOAA; 0512a11: © Photo courtesy of Bill Cartwright; 0512a12: NASA/U. of Michigan; 0513a2: NASA; 0514b1: Barbara Ambrose NOAA/NODC/NCDDC/Department of Commerce; 0514b2: Andrea Booher/FEMA; 0514b3: FEMA/Tim Burkitt; 0514b5: FEMA; 0515a1: NOAA/ARM.gov; 0515a2: Eric Kurth, NOAA/NWS/ER/WFO/Sacramento, NOAA/Department of Commerce; 0515a3-a4: NOAA; 0515b1: NASA/NSF; 0516a2: NOAA/NASA; 0516b3: USFWS/Steven R. Emmons; 0517t1: USGS; 0517t2: Andrea Booher/FEMA; 0517t3: DoD/U.S. Air Force photo by Master Sgt. Mark C. Olsen.

CHAPTER 6

Figure 0600a4: John Scurlock/USGS; 0604b6, 0606a4-a6: Photo by Susanne Gillatt; 0606b2: © Brand X Pictures/Getty Images RF; 0609c4-c6: Photo by Susanne Gillatt.

CHAPTER 7

Figure 0701c2-c3, 0701c6, 0704c1-c3: Photo by Susanne Gillatt; 0708c1: Photo by Cynthia Shaw; 0709a1: © Getty RF; 0709a2: © Kelly Redinger/Design Pics RF; 0709a3: © Comstock/Jupiterimages RF; 0709a4: © Thinkstock/Jupiterimages RF; 0709a5: © Ingram Publishing RF; 0709a6: © Digital Vision/Punchstock RF; 0709c1: © Getty RF; 0709c3: NOAA; 0710c2: NASA; 0713a2-a3: NASA; 0716a3: © Bruce Heinemann/Getty Images; 0716a6: © D. Normark/PhotoLink/Getty Images; 0716a7: © Rab Harling/Alamy; 0716a12: Photo by Susanne Gillatt; 0716a13: NASA.

TEXT AND LINE ART

a19: http://www.worldclimate.com/; 02.18.a22: http://www.worldclimate.com/; 02.19.a3: Adapted from Neiburger, M., J. G. Edinger and W. D. Bonner, 1982: Understanding our Atmospheric Environment, W. H. Freeman and Company; 02.19.a4: NCEP/NCAR Reanalysis dataset.

CHAPTER 3

Figure 03.02.c1: Cynthia Shaw; 03.13.d1: NCEP/NCAR Reanalysis dataset; 03.14.a3: http://www.worldclimate.com/; 03.14.a4: http://www.worldclimate.com/; 03.14.b1: NCEP/NCAR Reanalysis dataset; 03.14.b2: NCEP/NCAR Reanalysis dataset; 03.14.b3: http://www.worldclimate.com/; 03.14.b4: NCEP/NCAR Reanalysis dataset; 03.14.b5: NCEP/NCAR Reanalysis dataset; 03.14.b6: http://www.worldclimate.com/; 03.14.b7: NCEP/NCAR Reanalysis dataset; 03.14.b8: NCEP/NCAR Reanalysis dataset; 03.14.b9: http://www.worldclimate.com/; 03.16.a2: NCEP/NCAR Reanalysis dataset; 03.16.a3: NCEP/NCAR Reanalysis dataset; 03.16.a4: NCEP/NCAR Reanalysis dataset; 03.16.a5: NCEP/NCAR Reanalysis dataset; 03.16.a6: NCEP/NCAR Reanalysis dataset; 03.16.a7: NCEP/NCAR Reanalysis dataset; 03.02.d1: NOAA/NWS; 03.06.b1: NCEP/NCAR Reanalysis dataset; 03.06.b2: NCEP/NCAR Reanalysis dataset; 03.06.b3: NCEP/NCAR Reanalysis dataset; 03.09.b2: NOAA/NHC; 03.12.a10: NCEP/NCAR Reanalysis dataset; 03.12.a11: NCEP/NCAR Reanalysis dataset; 03.12.a8: NCEP/NCAR Reanalysis dataset; 03.12.a9: NCEP/NCAR Reanalysis dataset.

CHAPTER 4

Figure 04.03 b1: NCEP/NCAR Reanalysis dataset; 04.03.b2: NCEP/NCAR Reanalysis dataset; 04.04.c1: NCEP/NCAR Reanalysis dataset; 04.04.c2: NCEP/NCAR Reanalysis dataset; 04.04.c3: NCEP/NCAR Reanalysis dataset; 04.04.c4: NCEP/NCAR Reanalysis dataset; 04.07.a2: Data from NASA; 04.07.c2: NOAA/NCEP; 04.10.c1: National Climatic Data Center Climate Atlas; 04.10.c2: National Climatic Data Center Climate Atlas; 04.10.c3: National Climatic Data Center Climate Atlas; 04.10.c4: National Climatic Data Center Climate Atlas; 04.10.c5: National Climatic Data Center Climate Atlas; 04.12.c1: Data from Changnon, S. A. and T. R. Karl, 2003: Journal of Applied Meteorology 42; 04.13.a1: NCEP/NCAR Reanalysis dataset; 04.13.a2: NCEP/NCAR Reanalysis dataset; 04.13.b1: NCEP/NCAR Reanalysis dataset; 04.13.b1: NCEP/NCAR Reanalysis dataset; 04.13.b1: NCEP/NCAR Reanalysis dataset; 04.14.a2: NCEP/NCAR Reanalysis dataset; 04.14.b1: NASA/USGS; 04.14.b2: NOAA/CPC; 04.14.c1: USDA/Natural Resources Conservation Services; 04.14.c2: USDA/Natural Resources Conservation Services; 04.14.c3: USDA/

Natural Resources Conservation Services; 04.15.a1: NCEP/NCAR Reanalysis dataset; 04.15.a2: NCEP/NCAR Reanalysis dataset; 04.15.a3: NCEP/NCAR Reanalysis dataset; 04.15.a4: NCEP/NCAR Reanalysis dataset; 04.15.b1: NOAA/CPC; 04.15.b2: NOAA/CPC.

CHAPTER 5

Figure 05.00.a6: NCEP/NCAR Reanalysis dataset; 05.00.a7: NCEP/NCAR Reanalysis dataset; 05.00.a8: NCEP/NCAR Reanalysis dataset; 05.00.a9: NCEP/NCAR Reanalysis dataset; 05.01.b1: NCEP/NCAR Reanalysis dataset; 05.02.a1: NOAA/NWS/HPC; 05.02.b1: NOAA/NWS/HPC; 05.02.c1: NOAA/NWS/HPC; 05.02.c3: NOAA/NWS/HPC; 05.03.a4: NCEP/NCAR Reanalysis dataset; 05.03.b2: NCEP/NCAR Reanalysis dataset; 05.03 b3: NOAA/NWS/HPC; 05.03.b4: NCEP/NCAR Reanalysis dataset; 05.03.b5: NOAA/NWS/HPC; 05.04.a1: NOAA/NWS/HPC; 05.04.a2: NOAA/NWS; 05.04.a3: NOAA/NWS/HPC; 05.04.a4: NOAA/NWS; 05.04.a5: NOAA/NWS/HPC; 05.04.a6: NOAA/NWS; 05.05.a1: NCEP/NCAR Reanalysis dataset; 05.05.c2: NCEP/NCAR Reanalysis dataset; 05.05.c3: NCEP/NCAR Reanalysis dataset; 05.05.c4: NCEP/NCAR Reanalysis dataset; 05.06.b3: NOAA/NWS/HPC; 05.07.a1: Data from NASA; 05.07.a2: Data from NASA; 05.07.a3: Data from NASA; 05.07.a4: Modified from NASA; 05.07.b1: NOAA/NCDC/SPC/National Severe Weather Database; 05.07.b2: NOAA/NCDC/SPC/National Severe Weather Database; 05.08.b1: NOAA/NCDC/SPC/National Severe Weather Database; 05.08.b2: NOAA/NCDC/SPC/National Severe Weather Database; 05.08.b3: NOAA/NCDC/SPC/National Severe Weather Database; 05.09.a3-a7: Cynthia Shaw; 05.09.t1: NOAA/NWS; 05.10.b1: NOAA/NWS/HPC; 05.11.a1: NOAA/NCDC/SPC/National Severe Weather Database; 05.11.a2: NOAA/NCDC/SPC/National Severe Weather Database; 05.11.a3: NOAA/NCDC/SPC/National Severe Weather Database; 05.11.a4: NOAA/NCDC/SPC/National Severe Weather Database; 05.11.a5: NOAA/NCDC/SPC/National Severe Weather Database; 05.11.b2: NOAA/NCDC/SPC/National Severe Weather Database; 05.12.a3: NOAA/NWS; 05.12.a4: NOAA; modified from Coniglio, M. C. and D. J. Stensrud, 2004: Weather and Forecasting 19: 595-605; 05.12.a5: NOAA/NWS; 05.13.b1: NASA; 05.13.c4: NOAA; 05.16.b2: NOAA; 05.16.c1: NOAA; 05.17.a3: NOAA/NWS; 05.17.a4: NOAA; 05.18.a1: NOAA/NWS; 05.18.a2: NOAA/NWS; 05.18.a3: NOAA/NWS/HPC; 05.18.a4: NOAA/NWS/HPC; 05.18.a5: NCEP/NCAR Reanalysis dataset; 05.18.a6: NCEP/NCAR Reanalysis dataset; 05.18.a7: NCEP/NCAR Reanalysis dataset; 05.18.a8: NCEP/NCAR Reanalysis dataset; 05.18.a8: NCEP/NCAR

Reanalysis dataset; 05.18.a8: NCEP/NCAR Reanalysis dataset.

CHAPTER 6

Figure 06.00.a3: NOAA/Northwest Fisheries Science Center; 06.00.a5: D'Aleo, J. D. and D. Easterbrook. Multidecadal tendencies in ENSO and global temperatures related to multidecadal oscillations. Energy and Environment 21(5), 437–460; 06.00 a6: NCEP/NCAR Reanalysis dataset; 06.00.a7: NCEP/NCAR Reanalysis dataset; 06.01.c2: NASA/GSFC/Scientific Visualization Studio; 06.03.a1: NCEP/NCAR Reanalysis dataset; 06.03.a2: NCEP/NCAR Reanalysis dataset; 06.03.a3: NCEP/NCAR Reanalysis dataset; 06.03.a4: NCEP/NCAR Reanalysis dataset; 06.05.a1: NOAA/National Ocean Data Center; 06.05.a2: NOAA/National Ocean Data Center; 06.05.b1: http://www.soes.soton.ac.uk/teaching/courses/oa631/hydro.html#alt;Center; 06.05.b2: http://www.soes.soton.ac.uk/teaching/courses/oa631/hydro.html#alt;Center; 06.05.c1: Modified from Webster, P. J., 1994: The role of hydrological processes in ocean atmosphere interactions. Reviews of Geophysics, 32(4), 427–476; 06.06.a2: NCEP/NCAR Reanalysis dataset; 06.06.a3: NCEP/NCAR Reanalysis dataset; 06.06.b5: S. Adapted from Barker et al., 2011: 800,000 years of abrupt climate change. Science 334, 347–351; 06.07.b1: Argonne National Laboratory; 06.07.b2: Modified from Durack P. J., S. E. Wij els, and R. J. Matear, 2012: Ocean salinities reveal strong global water cycle intensification during 1950 to 2000. Science 336, 455–458; 06.07.b3: Modified from Durack P. J., S. E. Wij els, and R. J. Matear, 2012: Ocean salinities reveal strong global water cycle intensification during 1950 to 2000. Science 336, 455–458; 06.08.t1: Adapted from International Research Institute for Climate and Society, Lamont-Doherty Earth Observatory; 06.09a1: NASA; 06.09.a2: International Research Institute for Climate and Society, Lamont-Doherty Earth Observatory; 06.09.b3: NCEP/NCAR Reanalysis dataset; 06.08.t1: Adapted from International Research Institute for Climate and Society, Lamont-Doherty Earth Observatory; 06.09.c1: Data from Walker, G. T., and E. W. Bliss, 1932: World Weather V., Memoir Royal Meteorological Society, 4, No. 36, 53–84; 06.09.c2: NCEP/NCAR Reanalysis dataset; 06.10.b1-2: NCEP/NCAR Reanalysis dataset; 06.10.b3-4: NCEP/NCAR Reanalysis dataset; 06.10.b5-6: NCEP/NCAR Reanalysis dataset; 06.11.a2-3 NCEP/NCAR Reanalysis dataset; 06.11.b1: NOAA/CPC; 06.11.b2: NOAA/CPC; 06.12.a4: NASA/TAO/Triton database; 06.12.a5: NASA/TAO/Triton database; 06.12.a6: NASA/TAO/Triton database; 06.12.a7: NASA/TAO/Triton database; 06.12.b1-b4: NCEP/NCAR Reanalysis dataset; 06.12.t1: NCEP/NCAR Reanalysis dataset; 06.12.t2: NOAA/Climate

Prediction Center; 06.13.a2: NCEP/NCAR Reanalysis dataset; 06.14.a6: NCEP/NCAR Reanalysis dataset; 06.14.a7: NCEP/NCAR Reanalysis dataset; 06.14.a8–a9: P. Waylen, unpublished; 06.14.a8–a9: P. Waylen, unpublished.

CHAPTER 7

Figure 07.00.a1: NCEP/NCAR Reanalysis dataset; 07.00.a2: NCEP/NCAR Reanalysis dataset; 07.00.a3: NCEP/NCAR Reanalysis dataset; 07.01.b1: NCEP/NCAR Reanalysis dataset; 07.01.c1: Rohli and Vega, 2012: Climatology, Second Edition, Jones and Bartlett Learning; 07.02.00: NCEP/NCAR Reanalysis dataset; 07.03.a1: NCEP/NCAR Reanalysis dataset; 07.03.a2: NCEP/NCAR Reanalysis dataset; 07.03.a3: NCEP/NCAR Reanalysis dataset; 07.03.a4: NCEP/NCAR Reanalysis dataset; 07.04.a1: NCEP/NCAR Reanalysis dataset; 07.04.a2: NCEP/NCAR Reanalysis dataset; 07.04.a3: NCEP/NCAR Reanalysis dataset; 07.04.c5: http://www.worldclimate.com/; 07.04.c6: http://www.worldclimate.com/; 07.04.c7: http://www.worldclimate.com/; 07.05.a1: NCEP/NCAR Reanalysis dataset; 07.05.a2: NCEP/NCAR Reanalysis dataset; 07.05.a3: NCEP/NCAR Reanalysis dataset; 07.05.c1: NCEP/NCAR Reanalysis dataset; 07.05.c2: NCEP/NCAR Reanalysis dataset; 07.05.c3: NCEP/NCAR Reanalysis dataset; 07.05.c4: NCEP/NCAR Reanalysis dataset; 07.05.d1: http://www.worldclimate.com/; 07.05.d3: http://www.worldclimate.com/; 07.05.d4: http://www.worldclimate.com/; 07.05.d2: http://www.worldclimate.com/; 07.06.a1: NCEP/NCAR Reanalysis dataset; 07.06.a2: NCEP/NCAR Reanalysis dataset; 07.06.a3: NCEP/NCAR Reanalysis dataset; 07.06.c2: NCEP/NCAR Reanalysis dataset; 07.06.c3: http://www.worldclimate.com/; 07.06.c4: http://www.worldclimate.com/; 07.06.c5: http://www.worldclimate.com/; 07.07.a1: NCEP/NCAR Reanalysis dataset; 07.07.a2: NCEP/NCAR Reanalysis dataset; 07.07.a3: NCEP/NCAR Reanalysis dataset; 07.07.a3: NCEP/NCAR Reanalysis dataset; 07.07.a3: NCEP/NCAR Reanalysis dataset; 07.07.b3: http://www.worldclimate.com/; 07.07.b4: http://www.worldclimate.com/; 07.07.b5: http://www.worldclimate.com/; 07.08.a1: NCEP/NCAR Reanalysis dataset; 07.08.a2: NCEP/NCAR Reanalysis dataset; 07.08.a3: NCEP/NCAR Reanalysis dataset; 07.08.b1: http://reynolds.asu.edu/biosphere3d/bio3d_home.htm; 07.08.c4: National Snow and Ice Data Center, University of Colorado, Boulder; 07.08.c5: NCEP/NCAR Reanalysis dataset; 07.08.c6: http://www.worldclimate.com/; 07.08.c7: http://www.worldclimate.com/; 07.08.c8: http://www.worldclimate.com/; 07.09.b1: U.S. EPA; 07.09.b2: U.S. EPA; 07.10.a1: Kaufmann, R. K. and C. J. Cleveland, 2008: Environmental Science, McGraw-Hill Higher Education; 07.10.a2: Kaufmann, R. K. and C. J. Cleveland, 2008: Environmental Science, McGraw-Hill Higher Education; 07.10.c2: NASA; 07.11.a1: NOAA/NCDC/NESDIS; 07.11.a2: NOAA/

NCDC/NESDIS; 07.11.a3: National Research Council, 2006: Surface Temperature Reconstructions for the Last 2,000 Years, The National Academies Press; 07.11.a4: National Research Council, 2006: Surface Temperature Reconstructions for the Last 2,000 Years, The National Academies Press; 07.11.a5: National Research Council, 2006: Surface Temperature Reconstructions for the Last 2,000 Years, The National Academies Press; 07.11.a6: National Research Council, 2006: Surface Temperature Reconstructions for the Last 2,000 Years, The National Academies Press; 07.11.a7: Dr. Roy Spencer, University of Alabama, Huntsville; 07.12.a3: NASA/Marshall Space Flight Center; 07.13.a1: Cynthia Shaw; 07.13.b2: NOAA; 07.13.b3: NOAA.

CHAPTER 8

Figure 08.01.a2: After USGS; 08.03.a4: NWS; 08.3.c1: NOAA/CPC; 08.04.a1: Adapted from: Willmott, C. J. and J. J. Feddema, 1992: A more rational climatic moisture index, Professional Geographer, 44, 84–87; 08.04.a2–a5: WebWIMP, University of Delaware; 08.04.a7: Adapted from: Willmott, C. J. and J. J. Feddema, 1992: A more rational climatic moisture index, Professional Geographer, 44, 84–87; 08.04.a8: Adapted from: Willmott, C. J. and J. J. Feddema, 1992: A more rational climatic moisture index, Professional Geographer, 44, 84–87; 08.04.a9: WebWIMP, University of Delaware; 08.04.a10: WebWIMP, University of Delaware; 08.04.a11: WebWIMP, University of Delaware; 08.04.a12: WebWIMP, University of Delaware; 08.04.a14: WebWIMP, University of Delaware; 08.04.a15: WebWIMP, University of Delaware; 08.04.a17: WebWIMP, University of Delaware; 08.05.a4: USGS; 08.11.a1: USGS/High Plains Regional Ground Water Study; 08.11.a2: Texas Water Development Board; 08.11.b1: USGS; 08.11.b2: Texas Water Development Board.

CHAPTER 10

Figure 10.04.a1: USGS; 10.04.b1: Centers for the Disease Control and Prevention/USGS; 10.05.a1: Data from Michelle K. Hall-Wallace; 10.05.c1: Data from Michelle K. Hall-Wallace; 10.09.b1: Data from Michelle K. Hall-Wallace; 10.10.b2: After J. Kious/USGS 1996; 10.10.c1: USGS; 10.10.c2: USGS; 10.11.c1: Don Anderson/Seismological Laboratory, California Institute of Technology; 10.13.c2: P. King, 1977: Princeton University Press; 10.13.t2: After E. Moores and others, 1999: Geological Society of America Special Paper 338.

CHAPTER 11

Figure 11.05.d1: USGS Fact Sheet100-03; 11.05.d2: USGS Fact Sheet 2005-3024; 11.14.a1: NOAA/Paula Dunbar; 11.14.b1: NOAA/Paula

Dunbar; 11.17.a3: NOAA; 11.18.a1: E. Eckel 1970: USGS Professional Paper 546; 11.18.c1: After G. Plafker 1969. USGS Professional Paper 543; 11.18.c2: After G. Plafker 1969: USGS Professional Paper 543.

CHAPTER 12

Figure 12.07.b1: http://web.env.auckland.ac.nz/our_research/karst/; 12.07 b2: USGS; 12.10.t1: Kiersch, G. A., 1964. 'Vaiont reservoir disaster'. Civil Engineering, 34, 32-39; 12.12.b1: USGS Open-File report 97-0289; 12.14.a3: Varnes, D. and W. Savage, 1996: USGS Bulletin 2130; 12.14.c1: Varnes, D. and W. Savage, 1996: USGS Bulletin 2130; 12.14.c2: Varnes, D. and W. Savage, 1996: USGS Bulletin 2130.

CHAPTER 13

Figure 13.03.b1: Adapted from Mount, J., 1995: California Rivers and Streams: The Conflict between Fluvial Process and Land Use, University of California Press; 13.04.a2: Bailey, C., 1998: William and Mary College; 13.10.a2: Martin Jakobsson/Stockholm GeoVisualisation Laboratory; 13.10.a3: Martin Jakobsson/Stockholm GeoVisualisation Laboratory; 13.13.a1: K. Wahl, USGS Circular 1120-B; 13.13.a2: C. Parett, USGS Circular 1120-A; 13.13.b2: Based on data from Maddox, R. A., L. R. Hoxit, C. F. Chappell, and F. Caracena, 1978: Comparison of meteorological aspects of the Big Thompson and Rapid City flash floods, 1977: Monthly Weather Review 106, 375–389; 13.14.b1: Data from USGS/National Water Information System; 13.14.b2: Data from USGS/National Water Information System; 13.16.a6: After D. Topping, 2003: USGS Professional Paper 1677.

CHAPTER 14

Figure 14.06.a2: USGS; 14.06.a10: USGS; 14.06.a13: USGS; 14.07.b2: After European Science Foundation/PACE21 Network; 14.07.c1: After J. Fet, USGS Professional Paper 424B, 1961; 14.07.c6: NOAA/Great Lakes Environmental Research Laboratory; 14.07.c7: NOAA/Great Lakes Environmental Research Laboratory; 14.07.c8: NOAA/Great Lakes Environmental Research Laboratory; 14.08.b3: University of British Columbia; 14.08.b4: Emi Ito, University of Minnesota; 14.10.d1: USGS Fact Sheet 076-00.

CHAPTER 15

Figure 15.01.a3: After U.S. Army Corps of Engineers, 1992: Engineer Manual EM1110-2-1502: "Coastal Littoral Transport"; 15.06.c1: Chris

Jenkins/Institute of Arctic and Alpine Research, University of Colorado at Boulder; 15.08.t1–5: USGS; 15.10.t1: After S. Dutch/University of Wisconsin, Green Bay; 15.11.a2: USGS; 15.11.a3: USGS; 15.11.a4: USGS; 15.11.a5: USGS; 15.11.b6–b7: St. Petersburg Coastal and Marine Science Center/USGS; 15.11.b10–b11: St. Petersburg Coastal and Marine Science Center/USGS; 15.11.b12: USGS.

CHAPTER 16

Figure 16.01.b2: http://www.terragis.bees.unsw.edu.au/terraGIS_soil/sp_particle_size_fractions.html; 16.09.a2-a4,a6-a8, b2-b8,16.10.a2-a4, a6-a8, a10-12, a14-16, 16.11.a2-a4, a6-a8, a10-12, a14-a16, 16.12.a1: © FAO 2001, Lecture Notes on the Major Soils of the World, http://www.fao.org/docrep/003/Y1899E/Y1899E00.HTM. This is an adaptation of an original work by FAO. Views and opinions expressed in the adaptation are the sole responsibility of the author or authors of the adaptation and are not endorsed by FAO; 16.13.c3: USDA/NRCS.

CHAPTER 17

Figure 17.02.a2: Kaufmann, R. K. and C. J. Cleveland, 2008: Environmental Science, McGraw-Hill Higher Education; 17.02.a3: Kaufmann, R. K. and C. J. Cleveland, 2008: Environmental Science, McGraw-Hill Higher Education; 17.04.b1: OpenStax College. (2013, June 20). e Biodiversity Crisis. Retrieved from the Connexions Web site: http://cnx.org/content/m44892/1.5/; 17.06.a1: NASA; 17.06.a2: Adapted from Cunningham, W. P. and M. A. Cunningham, 2012: Environmental Science: A Global Concern. McGraw-Hill Higher Education; 17.06.b1: NCEP/NCAR Reanalysis dataset; 17.06.b2: NCEP/NCAR Reanalysis dataset; 17.06.b3:

NCEP/NCAR Reanalysis dataset; 17.06.b5: NASA; 17.07.c1: Kaufmann, R. K. and C. J. Cleveland, 2008: Environmental Science, McGraw-Hill Higher Education; 17.08.a1: Kaufmann, R. K. and C. J. Cleveland, 2008: Environmental Science, McGraw-Hill Higher Education; 17.09.b1: Kaufmann, R. K. and C. J. Cleveland, 2008: Environmental Science, McGraw-Hill Higher Education; 17.09.d1 US.EPA; 17.09.b1: Kaufmann, R. K. and C. J. Cleveland, 2008: Environmental Science, McGraw-Hill Higher Education; 17.10.b2: Adapted from USGS; 17.10.b3: Swap, R. M. et al., 1992: Saharan Dust in the Amazon Basin, Tellus 44B, 133–149; 17.06.a2: Adapted from Cunningham, W. P. and M. A. Cunningham, 2012: Environmental Science: A Global Concern. McGraw-Hill Higher Education; 17.11.a1: Kaufmann, R. K. and C. J. Cleveland, 2008: Environmental Science, McGraw-Hill Higher Education; 17.11.b2: US.EPA; 17.11.b3: US.EPA; 17.11.b4: US.EPA; 17.11.b5: http://www.worldmapper.org/; 17.12.c1: Adapted from Kaufmann, R. K. and C. J. Cleveland, 2008: Environmental Science, McGraw-Hill Higher Education; 17.13.a3: Adapted from USGS; 17.13.a5: University of Georgia, Center for Invasive Species and Ecosystem Health; 17.13.b2: Adapted from USGS; 17.13.b4: Adapted from USGS; 17.13.b6: Adapted from USGS.

CHAPTER 18

Figure 18.01.b1: Adapted from Cunningham, W. P. and M. A. Cunningham, 2012: Environmental Science: A Global Concern. McGraw-Hill Higher Education; 18.02.a1: Adapted from https://www.sciencebase.gov/catalog/item/508fece8e4b0a1b43c29ca22; 18.03.b1: NASA; 18.05.a1: Modified from NCAR; 18.06.a1: http://www.grida.no/graphicslib/detail/changing-global-forest-cover_b110; Philippe Rekacewicz assisted by Cecile Marin, Agnes Stienne, Guilio

Frigieri, Riccardo Pravettoni, Laura Margueritte and Marion Lecoquierre); 18.06.c1: Modified from JPL/NASA; 18.07.b1-3: Adapted from https://www.sciencebase.gov/catalog/i tem/508fece8e4b0a1b43c29ca22; 18.07.c1:Adapted from https://www.sciencebase.gov/catalog/item/ 508fece8e4b0a1b43c29ca22; 18.07.d1: Modified from USDA/Reich, P: World Soil Resources; 18.09.a1-a3: Adapted from https://www.sciencebase.gov/catalog/item/508fece8e4b0a1b43c29ca22; 1810.b1-b3: Adapted from https://www.sciencebase.gov/catalog/item/508fece8e4b0a1b43c29ca22; 18.11.a1-a3: Adapted from https://www.sciencebase.gov/catalog/item/508fece8e4b0a1b43c29ca22; 18.11.c1: Adapted from Cunningham, W. P. and M. A. Cunningham, 2012: Environmental Science: A Global Concern. McGraw-Hill Higher Education; 18.12.a1-3: Adapted from https://www.sciencebase.gov/catalog/item/508fece8e4b0a1b43c29ca22; 18.13.b4: http://dec.alaska.gov/; 18.13.c1: NOAA; 18.13.d2: Adapted from Cunningham, W. P. and M. A. Cunningham, 2012: Environmental Science: A Global Concern. McGraw-Hill Higher Education; 18.14.b9: Adapted from National Resources of Canada; 18.15.a2: Adapted from Kaufmann, R. K. and C. J. Cleveland, 2008: Environmental Science, McGraw-Hill Higher Education; 18.15.b2: Adapted from Kaufmann, R. K. and C. J. Cleveland, 2008: Environmental Science, McGraw-Hill Higher Education; 18.16.b3: National Park Service; 18.16.b6: Woods Hole Oceanographic Institution; 18.17.a1: Adapted from: Adams, W. M., 2006: The future of sustainability: Rethinking environment and development in the twenty-first century. In Report of the IUCN Renowned Thinkers Meeting the Future of Sustainability Re-thinking Environment and Development in the Twenty-first Century, World Conservation Union; 18.17.b1: Adapted from Cunningham, W. P. and M. A. Cunningham, 2012: Environmental Science: A Global Concern. McGraw-Hill Higher Education.

INDEX

Shaded-Relief Map of the Conterminous
United States and Adjacent Areas

This map simulates shading on the land surface as if lighted from the upper left.
It includes the southern part of Canada and the northern part of Mexico.

Source: Paul Morin